Food Additives

FOOD SCIENCE AND TECHNOLOGY

A Series of Monographs, Textbooks, and Reference Books

1. Flavor Research: Principles and Techniques, *R. Teranishi, I. Hornstein, P. Issenberg, and E. L. Wick*
2. Principles of Enzymology for the Food Sciences, *John R. Whitaker*
3. Low-Temperature Preservation of Foods and Living Matter, *Owen R. Fennema, William D. Powrie, and Elmer H. Marth*
4. Principles of Food Science
   Part I: Food Chemistry, *edited by Owen R. Fennema*
   Part II: Physical Methods of Food Preservation, *Marcus Karel, Owen R. Fennema, and Daryl B. Lund*
5. Food Emulsions, *edited by Stig E. Friberg*
6. Nutritional and Safety Aspects of Food Processing, *edited by Steven R. Tannenbaum*
7. Flavor Research: Recent Advances, *edited by R. Teranishi, Robert A. Flath, and Hiroshi Sugisawa*
8. Computer-Aided Techniques in Food Technology, *edited by Israel Saguy*
9. Handbook of Tropical Foods, *edited by Harvey T. Chan*
10. Antimicrobials in Foods, *edited by Alfred Larry Branen and P. Michael Davidson*
11. Food Constituents and Food Residues: Their Chromatographic Determination, *edited by James F. Lawrence*
12. Aspartame: Physiology and Biochemistry, *edited by Lewis D. Stegink and L. J. Filer, Jr.*

13. Handbook of Vitamins: Nutritional, Biochemical, and Clinical Aspects, *edited by Lawrence J. Machlin*
14. Starch Conversion Technology, *edited by G. M. A. van Beynum and J. A. Roels*
15. Food Chemistry: Second Edition, Revised and Expanded, *edited by Owen R. Fennema*
16. Sensory Evaluation of Food: Statistical Methods and Procedures, *Michael O'Mahony*
17. Alternative Sweeteners, *edited by Lyn O'Brien Nabors and Robert C. Gelardi*
18. Citrus Fruits and Their Products: Analysis and Technology, *S. V. Ting and Russell L. Rouseff*
19. Engineering Properties of Foods, *edited by M. A. Rao and S. S. H. Rizvi*
20. Umami: A Basic Taste, *edited by Yojiro Kawamura and Morley R. Kare*
21. Food Biotechnology, *edited by Dietrich Knorr*
22. Food Texture: Instrumental and Sensory Measurement, *edited by Howard R. Moskowitz*
23. Seafoods and Fish Oils in Human Health and Disease, *John E. Kinsella*
24. Postharvest Physiology of Vegetables, *edited by J. Weichmann*
25. Handbook of Dietary Fiber: An Applied Approach, *Mark L. Dreher*
26. Food Toxicology, Parts A and B, *Jose M. Concon*
27. Modern Carbohydrate Chemistry, *Roger W. Binkley*
28. Trace Minerals in Foods, *edited by Kenneth T. Smith*
29. Protein Quality and the Effects of Processing, *edited by R. Dixon Phillips and John W. Finley*
30. Adulteration of Fruit Juice Beverages, *edited by Steven Nagy, John A. Attaway, and Martha E. Rhodes*
31. Foodborne Bacterial Pathogens, *edited by Michael P. Doyle*
32. Legumes: Chemistry, Technology, and Human Nutrition, *edited by Ruth H. Matthews*
33. Industrialization of Indigenous Fermented Foods, *edited by Keith H. Steinkraus*
34. International Food Regulation Handbook: Policy • Science • Law, *edited by Roger D. Middlekauff and Philippe Shubik*
35. Food Additives, *edited by A. Larry Branen, P. Michael Davidson, and Seppo Salminen*
36. Safety of Irradiated Foods, *J. F. Diehl*
37. Omega-3 Fatty Acids in Health and Disease, *edited by Robert S. Lees and Marcus Karel*
38. Food Emulsions: Second Edition, Revised and Expanded, *edited by Kåre Larsson and Stig E. Friberg*
39. Seafood: Effects of Technology on Nutrition, *George M. Pigott and Barbee W. Tucker*
40. Handbook of Vitamins: Second Edition, Revised and Expanded, *edited by Lawrence J. Machlin*
41. Handbook of Cereal Science and Technology, *Klaus J. Lorenz and Karel Kulp*
42. Food Processing Operations and Scale-Up, *Kenneth J. Valentas, Leon Levine, and J. Peter Clark*
43. Fish Quality Control by Computer Vision, *edited by L. F. Pau and R. Olafsson*
44. Volatile Compounds in Foods and Beverages, *edited by Henk Maarse*
45. Instrumental Methods for Quality Assurance in Foods, *edited by Daniel Y. C. Fung and Richard F. Matthews*
46. *Listeria*, Listeriosis, and Food Safety, *Elliot T. Ryser and Elmer H. Marth*
47. Acesulfame-K, *edited by D. G. Mayer and F. H. Kemper*
48. Alternative Sweeteners: Second Edition, Revised and Expanded, *edited by Lyn O'Brien Nabors and Robert C. Gelardi*
49. Food Extrusion Science and Technology, *edited by Jozef L. Kokini, Chi-Tang Ho, and Mukund V. Karwe*

50. Surimi Technology, *edited by Tyre C. Lanier and Chong M. Lee*
51. Handbook of Food Engineering, *edited by Dennis R. Heldman and Daryl B. Lund*
52. Food Analysis by HPLC, *edited by Leo M. L. Nollet*
53. Fatty Acids in Foods and Their Health Implications, *edited by Ching Kuang Chow*
54. *Clostridium botulinum*: Ecology and Control in Foods, *edited by Andreas H. W. Hauschild and Karen L. Dodds*
55. Cereals in Breadmaking: A Molecular Colloidal Approach, *Ann-Charlotte Eliasson and Kåre Larsson*
56. Low-Calorie Foods Handbook, *edited by Aaron M. Altschul*
57. Antimicrobials in Foods: Second Edition, Revised and Expanded, *edited by P. Michael Davidson and Alfred Larry Branen*
58. Lactic Acid Bacteria, *edited by Seppo Salminen and Atte von Wright*
59. Rice Science and Technology, *edited by Wayne E. Marshall and James I. Wadsworth*
60. Food Biosensor Analysis, *edited by Gabriele Wagner and George G. Guilbault*
61. Principles of Enzymology for the Food Sciences: Second Edition, *John R. Whitaker*
62. Carbohydrate Polyesters as Fat Substitutes, *edited by Casimir C. Akoh and Barry G. Swanson*
63. Engineering Properties of Foods: Second Edition, Revised and Expanded, *edited by M. A. Rao and S. S. H. Rizvi*
64. Handbook of Brewing, *edited by William A. Hardwick*
65. Analyzing Food for Nutrition Labeling and Hazardous Contaminants, *edited by Ike J. Jeon and William G. Ikins*
66. Ingredient Interactions: Effects on Food Quality, *edited by Anilkumar G. Gaonkar*
67. Food Polysaccharides and Their Applications, *edited by Alistair M. Stephen*
68. Safety of Irradiated Foods: Second Edition, Revised and Expanded, *J. F. Diehl*
69. Nutrition Labeling Handbook, *edited by Ralph Shapiro*
70. Handbook of Fruit Science and Technology: Production, Composition, Storage, and Processing, *edited by D. K. Salunkhe and S. S. Kadam*
71. Food Antioxidants: Technological, Toxicological, and Health Perspectives, *edited by D. L. Madhavi, S. S. Deshpande, and D. K. Salunkhe*
72. Freezing Effects on Food Quality, *edited by Lester E. Jeremiah*
73. Handbook of Indigenous Fermented Foods: Second Edition, Revised and Expanded, *edited by Keith H. Steinkraus*
74. Carbohydrates in Food, *edited by Ann-Charlotte Eliasson*
75. Baked Goods Freshness: Technology, Evaluation, and Inhibition of Staling, *edited by Ronald E. Hebeda and Henry F. Zobel*
76. Food Chemistry: Third Edition, *edited by Owen R. Fennema*
77. Handbook of Food Analysis: Volumes 1 and 2, *edited by Leo M. L. Nollet*
78. Computerized Control Systems in the Food Industry, *edited by Gauri S. Mittal*
79. Techniques for Analyzing Food Aroma, *edited by Ray Marsili*
80. Food Proteins and Their Applications, *edited by Srinivasan Damodaran and Alain Paraf*
81. Food Emulsions: Third Edition, Revised and Expanded, *edited by Stig E. Friberg and Kåre Larsson*
82. Nonthermal Preservation of Foods, *Gustavo V. Barbosa-Cánovas, Usha R. Pothakamury, Enrique Palou, and Barry G. Swanson*
83. Milk and Dairy Product Technology, *Edgar Spreer*
84. Applied Dairy Microbiology, *edited by Elmer H. Marth and James L. Steele*
85. Lactic Acid Bacteria: Microbiology and Functional Aspects, Second Edition, Revised and Expanded, *edited by Seppo Salminen and Atte von Wright*

86. Handbook of Vegetable Science and Technology: Production, Composition, Storage, and Processing, *edited by D. K. Salunkhe and S. S. Kadam*
87. Polysaccharide Association Structures in Food, *edited by Reginald H. Walter*
88. Food Lipids: Chemistry, Nutrition, and Biotechnology, *edited by Casimir C. Akoh and David B. Min*
89. Spice Science and Technology, *Kenji Hirasa and Mitsuo Takemasa*
90. Dairy Technology: Principles of Milk Properties and Processes, *P. Walstra, T. J. Geurts, A. Noomen, A. Jellema, and M. A. J. S. van Boekel*
91. Coloring of Food, Drugs, and Cosmetics, *Gisbert Otterstätter*
92. *Listeria*, Listeriosis, and Food Safety: Second Edition, Revised and Expanded, *edited by Elliot T. Ryser and Elmer H. Marth*
93. Complex Carbohydrates in Foods, *edited by Susan Sungsoo Cho, Leon Prosky, and Mark Dreher*
94. Handbook of Food Preservation, *edited by M. Shafiur Rahman*
95. International Food Safety Handbook: Science, International Regulation, and Control, *edited by Kees van der Heijden, Maged Younes, Lawrence Fishbein, and Sanford Miller*
96. Fatty Acids in Foods and Their Health Implications: Second Edition, Revised and Expanded, *edited by Ching Kuang Chow*
97. Seafood Enzymes: Utilization and Influence on Postharvest Seafood Quality, *edited by Norman F. Haard and Benjamin K. Simpson*
98. Safe Handling of Foods, *edited by Jeffrey M. Farber and Ewen C. D. Todd*
99. Handbook of Cereal Science and Technology: Second Edition, Revised and Expanded, *edited by Karel Kulp and Joseph G. Ponte, Jr.*
100. Food Analysis by HPLC: Second Edition, Revised and Expanded, *edited by Leo M. L. Nollet*
101. Surimi and Surimi Seafood, *edited by Jae W. Park*
102. Drug Residues in Foods: Pharmacology, Food Safety, and Analysis, *Nickos A. Botsoglou and Dimitrios J. Fletouris*
103. Seafood and Freshwater Toxins: Pharmacology, Physiology, and Detection, *edited by Luis M. Botana*
104. Handbook of Nutrition and Diet, *Babasaheb B. Desai*
105. Nondestructive Food Evaluation: Techniques to Analyze Properties and Quality, *edited by Sundaram Gunasekaran*
106. Green Tea: Health Benefits and Applications, *Yukihiko Hara*
107. Food Processing Operations Modeling: Design and Analysis, *edited by Joseph Irudayaraj*
108. Wine Microbiology: Science and Technology, *Claudio Delfini and Joseph V. Formica*
109. Handbook of Microwave Technology for Food Applications, *edited by Ashim K. Datta and Ramaswamy C. Anantheswaran*
110. Applied Dairy Microbiology: Second Edition, Revised and Expanded, *edited by Elmer H. Marth and James L. Steele*
111. Transport Properties of Foods, *George D. Saravacos and Zacharias B. Maroulis*
112. Alternative Sweeteners: Third Edition, Revised and Expanded, *edited by Lyn O'Brien Nabors*
113. Handbook of Dietary Fiber, *edited by Susan Sungsoo Cho and Mark L. Dreher*
114. Control of Foodborne Microorganisms, *edited by Vijay K. Juneja and John N. Sofos*
115. Flavor, Fragrance, and Odor Analysis, *edited by Ray Marsili*
116. Food Additives: Second Edition, Revised and Expanded, *edited by Alfred Larry Branen, P. Michael Davidson, Seppo Salminen, and John H. Thorngate III*

Additional Volumes in Preparation

Characterization of Cereals and Flours: Properties, Analysis, and Applications, *edited by Gönül Kaletunç and Kenneth J. Breslauer*

Postharvest Physiology and Pathology of Vegetables: Second Edition, Revised and Expanded, *edited by Jerry A. Bartz and Jeffrey K. Brecht*

Food Additives

Second Edition
Revised and Expanded

edited by

A. Larry Branen
University of Idaho
Moscow, Idaho

P. Michael Davidson
University of Tennessee
Knoxville, Tennessee

Seppo Salminen
University of Turku
Turku, Finland, and
Royal Melbourne Institute of Technology
Melbourne, Australia

John H. Thorngate III
University of California
Davis, California

MARCEL DEKKER, INC. NEW YORK • BASEL

The first edition was edited by A. Larry Branen, P. Michael Davidson, and Seppo Salminen.

ISBN: 0-8247-9343-9

This book is printed on acid-free paper.

Headquarters
Marcel Dekker, Inc.
270 Madison Avenue, New York, NY 10016
tel: 212-696-9000; fax: 212-685-4540

Eastern Hemisphere Distribution
Marcel Dekker AG
Hutgasse 4, Postfach 812, CH-4001 Basel, Switzerland
tel: 41-61-261-8482; fax: 41-61-261-8896

World Wide Web
http://www.dekker.com

The publisher offers discounts on this book when ordered in bulk quantities. For more information, write to Special Sales/Professional Marketing at the headquarters address above.

Current printing (last digit):
10 9 8 7 6 5 4 3 2 1

PRINTED IN THE UNITED STATES OF AMERICA

Preface to the Second Edition

In the Preface to the First Edition of *Food Additives*, we stated that food additives would ''continue to play an important and essential role in food production'' because of consumer demand and desire for ''convenient, tasty, and nutritious foods'' in an ever increasing populace. This prediction has been proven correct over the period since the first edition was published, and much has changed in food science, food safety, and food technology. Despite continued research on the toxicological effects of food additives on humans, few were restricted in the 1990s and several were actually approved for use.

In the 1990s, the food processing industry became much more global. As more processed foods were produced for export, the requirement for a working knowledge of world food additive regulations became paramount. This led to increased pressure for countries to adopt similar food additive regulations. *Codex Alimentarius* has become an important guide on food additives and a stepping-stone to improving world food marketing. Because foods are now shipped worldwide, there is a greater demand for maintenance of product quality attributes for extended periods. Food processors have had to utilize various food additives (in addition to other processing techniques, e.g., packaging) to achieve this extended shelf life. In 1990, rapid communication and information retrieval were still in their infancy. Today, the Internet as a means of communication and information retrieval concerning food additive regulations has become an important vehicle for exporters.

While there is continued pressure from consumers and consumer interest groups for the food industry to produce foods that are ''preservative-free'' or ''additive-free,'' food additives continue to play a major role in nutritional and sensory quality and safety of foods produced worldwide. In recent years, even food processors have been desirous of marketing foods with ''green'' labels (i.e., containing fewer additives). This has driven food processors to examine natural compounds as food additives.

When the first edition of *Food Additives* was published, terms such as probiotics, functional foods, nutraceuticals, and enteral nutrition products were essentially unknown. They are now commonplace, and food additives are playing a role in the production, maintenance, and safety of these products. Home meal replacement and refrigerated processed foods of extended durability (REPFED) have required the use of additives to maintain nutritional quality, sensory characteristics, and microbiological safety. Genetic engineering has played a role in food additives by making a few biologically derived compounds more available or less expensive. This area has great potential for growth.

In the 21st century, food additives continue to play an important role in the global food supply.

However, much still needs to be learned about existing food additives as well as potential novel food additives that exist in nature.

The original objective of *Food Additives* was to provide organized information on the various classes of intentional (i.e., added purposely for specific functions) food additives. The book included a general introduction and discussion of the chemistry and chemical analysis of the food additive class, the function and mechanism action, the application of the additive, and toxicological research and concerns. We accomplished that objective by soliciting chapters on all major groups of food additives, including antimicrobial agents, antioxidants, coloring agents, emulsifiers, enzymes, flavoring agents, flavor enhancers, miscellaneous additives, nutritional additives, pH control agents, polysaccharides, and sweeteners. Also presented in the first edition was ancillary information on food additives, including an overview along with chapters on food additive intake, hypersensitivity to additives, risks and benefits, and safety testing. The authors and editors of the first edition were selected from throughout the world to provide an international perspective. The editors wanted the book to be more than just a handbook for food additive applications. Rather, it was designed to be a treatise on these compounds that are so useful in our food supply.

In the second edition of *Food Additives*, we have striven to maintain that comprehensive approach to the subject. We again have solicited chapters on individual groups of food additives but have added several new chapters. The expanded offering includes the division of coloring agents into two chapters: synthetic food colorants (Chapter 16) and natural color additives (Chapter 17), and other new chapters on antibrowning agents (Chapter 19), commercial starches (Chapter 24), essential fatty acids (Chapter 10), fat substitutes and replacers (Chapter 11) and phosphates (Chapter 25). Another chapter (Chapter 12) discusses food additives for special dietary purposes such as in functional foods, engineering of clinical nutrition products, and probiotics.

Ancillary information has been significantly increased to reflect the interests of processors, consumer interest groups, and researchers in the many issues regarding food additives. In addition to updating the original discussions of intake assessment, risks and benefits, and hypersensitivity, several new chapters on these topics have been added. Consumer perception of food additives is discussed in Chapter 6. The role of food additives in behavior, learning, activity, and sleep of children is covered in Chapter 5. Two extremely useful chapters review food additive regulations in the European Union (Chapter 7) and the United States (Chapter 8). As in the first edition, the work is truly an international collaboration. Authors come from Europe, North America, Africa, Asia, and Australia, giving the book a global viewpoint.

In summary, we feel that the reader will see the second edition of *Food Additives* not just as an update but as an even more useful tool on the application and scientific study of food additives. It remains useful to those in universities, regulatory organizations, consumer interest groups, and food processing companies and their allied organizations. As was stated in the first edition preface, this book should provide a ''basis for practical selection'' of the appropriate additive for a particular food use, improve understanding of the risks and benefits of the many food additives used worldwide, aid in identification of additives requiring further research, and allow for understanding of the role of government in regulation of food additives.

Finally, we wish to express our sincerest gratitude to the chapter authors. Without their conscientious and diligent efforts, this work would not have been possible.

A. Larry Branen
P. Michael Davidson
Seppo Salminen
John H. Thorngate III

Preface to the First Edition

Food additives have been used for centuries to enhance the quality of food products. Smoke, alcohol, vinegar, oils, and spice were used more than 10,000 years ago to preserve foods. Until the time of the Industrial Revolution, these and a limited number of other chemicals were the major food additives used. The Industrial Revolution brought many changes to the food supply, including a better understanding of foods and a demand for both an increased quantity and quality. This resulted in the development of a variety of chemicals which were used to preserve, as well as enhance, the color, flavor, and texture of foods. The development of the land-grant university system in the late 1800s, and the subsequent interest in food chemistry and preservation, significantly increased the knowledge and use of food additives. With this increased technology, as well as the increased standards of living demanded by the population, additive usage in foods significantly increased in the 1950s. By the early 1960s, over 2500 different chemicals were being used in foods and in the United States the per capita consumption of these chemicals was estimated to exceed three pounds, excluding salt and sugar. The demand for new, tasty, convenient, and nutritious foods continued to increase from that time until today, and now in the United States alone, over 2500 different additives are being used to help produce over 15,000 different food items. The estimated consumption of such additives has increased from three to nine pounds per capita per year.

There has always been some concern regarding the safety of consuming the additives used in foods. Even in the early 1200s, it is reported that kings hired garglers to test the foods and several books were written in the early 1800s regarding food safety. The first real concern regarding additives, however, was expressed by Dr. Wiley when he was hired in the early 1900s as the Head of the Bureau of Chemistry of the U.S. Department of Agriculture. The work of Dr. Wiley and his so-called ''poison squad'' led to the first regulations to control the use of food additives. The 1906 Pure Food and Drug Act, which led to the 1938 development of the Federal Food, Drug, and Cosmetic Act, had a direct impact on additive use. Although this early legislation provided some control, implementation of these laws by the FDA was limited by the ambiguous legislative standard of ''harmfulness,'' the fact that the burden of proof was on the FDA, and the lack of scientific knowledge of safety evaluation. Thus, the greatest legal impact on food additive use did not occur until the passage of the 1958 Food Additives Amendment. This amendment, which was followed by the 1960 Color Additives Amendment, remedied limitations of the earlier legislation by shifting the burden of proving safety to the manufacturer of the food and requiring premarketing safety data on all additives. The 1958 amendment also included the controversial ''GRAS'' list and Delaney Amendment, and utilized a more restrictive and less ambiguous terminology by shifting from ''harmfulness'' to ''safety.''

Passage of the 1958 legislation and increasing knowledge of toxicology in the 1960s and 1970s brought greater attention to the possible risks of additives. These concerns were enhanced by books such as Rachel Carson's *Silent Spring* and *The Chemical Feast* by James Turner. The increase of consumerism and the anti-establishment views of the 1970s led to a continuing controversy over additives among scientists, lawyers, consumer advocates, and policy makers. The legislation and controversy surrounding additives resulted in the banning of some chemicals as well as a lessened submission of new chemicals for approval. During the period from 1960 to 1980, very few chemicals were approved for use while several were banned.

A significant event in food additive safety review was the call for total review of the GRAS list by President Nixon in 1972. This review led to new regulations regarding several additives, but also led to an affirmation of the safety of several others. These studies and others provided a clearer understanding of both the risks and benefits of additives.

An event that called the greatest attention to the balance of risks and benefits of food additives was the U.S. Senate moratorium on the proposed ban of saccharin by the FDA. The moratorium was essentially the first political recognition of the importance of balancing the potential risks of an additive against its perceived benefits and allowing the customer the choice of balancing the benefits and the risks. The moratorium has continued for several years and has undoubtedly had a significant impact on the continued and proposed uses of additives.

Concern regarding the safety of additives has declined in the United States since enactment of the saccharin moratorium. This, plus a changing political trend away from consumerism and more responsible use of additives by manufacturers, has lessened the controversy regarding additives. The potential risks of additives are well recognized, but the benefits these additives play in food production, processing, and utilization are also felt to be essential to the maintenance of current food systems. With the convenient, tasty, and nutritious foods demanded, or at least desired, by consumers, and the increasing overall demand for foods as populations increase, food additives will continue to play an important and essential role in food production. There will, however, also be concern regarding the potential long-term risks associated with long-term consumption of small amounts of these chemicals and possible interactive toxicological effects. As methods improve for evaluating these toxicological effects, some additives may be banned. At the same time, this information may be used to develop safer new additives or techniques for using existing additives in a way that will lessen risk.

As we enter an age with the referendum to balance risks versus benefits, it is mandatory that a book detailing the most recent information regarding additives be made available. It is the intent of this book to provide the most up-to-date information available on a worldwide basis. Authors and the editors have been selected from throughout the world so as to bring an international perspective to this work. This is especially important since current regulations, additive use, and approaches to toxicological evaluation differ worldwide. We hope this work can serve to overcome the barriers that sometimes cause misunderstanding in the additive issue.

The book includes an overview of the use and consumption of food additives: a specific discussion of each of the major food additive categories; and information on the safety evaluation of food additives. The first two chapters provide an introduction to the food additives and current information on food additive intakes. In Chapters 3 through 14, specific information is provided on the major food additive categories. Each chapter contains a general introduction and discussion of the chemistry and chemical analysis, the function and mechanism of action, the uses and regulations governing use, and toxicological concerns of each additive category. The final three chapters provide a current view on the safety of food additive use. The first chapter in this section (Chapter 15) describes methods for safety evaluation. Hypersensitivity to additives is discussed in Chapter 16 and the final chapter provides a discussion on balancing risks and benefits of foods and food additives.

The book should be useful to those in universities, governmental organizations, consumer groups, and food companies. It should provide a basis for practical selection of additives for use in foods, for understanding the benefits and risks of the more than 2500 additives in use worldwide, for selection of additives needing further study and research, and for understanding the role of governments in determining use of food additives. The reader is also referred to other major works on the risks and benefits of additives, including the *Handbook of Food Additives*, published several years ago by the Chemical Rubber Company.

A. Larry Branen
P. Michael Davidson
Seppo Salminen

Contents

Contributors

William E. Artz *University of Illinois, Urbana, Illinois*

A. Larry Branen *University of Idaho, Moscow, Idaho*

Jill K. Branen *University of Illinois, Urbana, Illinois*

Joan Breakey *Consulting Dietitian, Beachmere, Australia*

Christine M. Bruhn *Center for Consumer Research, Davis, California*

Helen Connell *Queensland Health, Spring Hill, Australia*

P. Michael Davidson *University of Tennessee, Knoxville, Tennessee*

Manik Desai *The Pennsylvania State University, University Park, Pennsylvania*

Stephanie Doores *The Pennsylvania State University, University Park, Pennsylvania*

Joseph D. Eifert *Nestle USA, Inc., Dublin, Ohio*

Pam Evenson *South Dakota State University, Brookings, South Dakota*

J. Bruce German *University of California, Davis, California*

John P. Godber *Albright & Wilson Canada, Ltd., Rhodia, Aubervilliers, France*

Robert J. Gordon *Givaudan Roure Inc., Dramptom, Ontario, Canada*

John M. V. Grigor *University of Lincolnshire and Humberside, Grimsby, England*

Tari Haahtela *Skin and Allergy Hospital and Helsinki University Hospital, Helsinki, Finland*

R. J. Haggerty *University of Idaho, Moscow, Idaho*

Anja Hallikainen *National Food Administration, Helsinki, Finland*

Matti Hannuksela *South Karelia Central Hospital, Lappeenranta, Finland*

Steven L. Hansen *Cargill Analytical Services, Minnetonka, Minnesota*

Kevin B. Hicks *Agricultural Research Service, U.S. Department of Agriculture, Wyndmoor, Pennsylvania*

Peter Barton Hutt *Covington & Burling, Washington, D.C.*

Wendy S. Johnson *SHS International Ltd., Liverpool, England*

Vijay K. Juneja *U.S. Department of Agriculture, Wyndmoor, Pennsylvania*

Hwee-Peng Khng *PSB Corporation Pte. Ltd., Singapore*

Arun Kilara *Arun Kilara Worldwide, Northbrook, Illinois*

David J. Kyle *Martek Biosciences Corporation, Columbia, Maryland*

Lucinda E. Lampila *Albright & Wilson Americas, Inc., Rhodia, Cranbury, New Jersey*

Yuan-Kun Lee *National University of Singapore, Singapore*

Thomas E. Luallen *Cargill, Cedar Rapids, Iowa*

Symon M. Mahungu *Egerton University, Njoro, Kenya*

Robert L. Miller *Agricultural Research Service, U.S. Department of Agriculture, Wyndmoor, Pennsylvania*

Conor Reilly *Oxford Brookes University, Oxford, England*

Seppo Salminen *University of Turku, Turku, Finland, and Royal Melbourne Institute of Technology, Melbourne, Australia*

Gerald M. Sapers *Agricultural Research Service, U.S. Department of Agriculture, Wyndmoor, Pennsylvania*

Gabriel S. Sinki *Sinki Flavor Technology, Little Falls, New Jersey*

Yoshi-hisa Sugita *Ajinomoto Co., Inc., Tokyo, Japan*

Susan S. Sumner *Virginia Tech, Blacksburg, Virginia*

Marilyn A. Swanson *National Food Service Management Institute, University, Mississippi*

Raija Tahvonen *University of Turku, Turku, Finland*

John H. Thorngate III *University of California, Davis, California*

Ronald Verbruggen *Bioresco Ltd., Brussels, Belgium*

1

Introduction to Food Additives

A. LARRY BRANEN and R. J. HAGGERTY

University of Idaho, Moscow, Idaho

I. INTRODUCTION

According to the Food Protection Committee of the Food and Nutrition Board, food additives may be defined as follows:

> a substance or mixture of substances, other than a basic foodstuff, which is present in a food as a result of any aspect of production, processing, storage, or packaging. The term does not include chance contaminants.

Since prehistoric times, chemicals have been added to foods to perform special functions. Although basic foods contain no additives, as foods are processed for conversion into a variety of products, an increasing number of additives are generally used. Technological advances in food processing have increased the variety and use of these additives. Today, more than 2500 different additives are intentionally added to foods to produce a desired effect. The use of these additives is a well-accepted practice but is not without controversy. In this chapter, we explore some of the major benefits and risks of using additives. In subsequent chapters, each category of additives and some of the benefits and potential risks are explored in more detail.

II. TYPES OF ADDITIVES

Additives can be divided into six major categories: preservatives, nutritional additives, flavoring agents, coloring agents, texturizing agents, and miscellaneous additives. Several lists of these additives are available and as will be noted throughout this book, several additives commonly serve more than one function in foods. In Europe and other parts of the world, the E system, developed by the European Union (formally the European Economic Community), provides a listing of several commonly used additives (Hanssen, 1984; Jukes, 2001). Chapter 7 in this book describes the process and the additives currently on the E system list. The list, which is updated on a regular basis, includes those additives that are generally recognized as safe within the member states and allows foods to move from country to country within the European Union. The European Union adopted directives which set the criteria by which additives are assessed. The European Scientific Committee for Food (SCF) oversees additive safety against the established criteria. Specific directives have been established for sweeteners, colors, and other food additives (Jukes, 2001). Nutrients are not included in the E system. The Codex Alimentarius Commission Committee on Food Additives and Contaminants has developed an international numbering system (INS) for food additives based on the E system (Codex Alimentarius Commission, 2001). The INS system is broader than the E system and is intended as an identification system for food additives approved for use in one or more countries. It does not imply toxicological approval by Codex, and the list extends beyond the additives currently cleared by the Joint FAO/WHO Expert Committee on Food Additives (Codex Alimentarius Commission, 2001). The INS numbers are largely the same numbers used in the E system without the E. The INS system also includes a listing of the technical function for each additive based on 23 functional classes. The E numbers for a number of additives are included in Chapter 7 and the INS numbers are in the appendix.

A. Preservatives

There are basically three types of preservatives used in foods: antimicrobials, antioxidants, and antibrowning agents. These additives are grouped under the category of preservatives in the INS system. The antimicrobials, with E and INS numbers ranging from 200 to 290, are used to check or prevent the growth of microorganisms. Antimicrobials are discussed in further detail in Chapter 20. In addition, the book *Antimicrobials in Foods* (Davidson and Branen, 1993) gives a complete treatment of these additives. Antimicrobials play a major role in extending the shelf-life of numerous snack and convenience foods and have come into even greater use in recent years as microbial food safety concerns have increased.

The antioxidants (INS 300–326 and E300–E326), discussed in further detail in Chapter 18, are used to prevent lipid and/or vitamin oxidation in food products. They are used primarily to prevent autoxidation and subsequent development of rancidity and off-flavor. They vary from natural substances such as vitamins C and E to synthetic chemicals such as butylated hydroxyanisole (BHA) and butylated hydroxytoluene (BHT). The antioxidants are especially useful in preserving dry and frozen foods for an extended period of time.

Antibrowning agents are chemicals used to prevent both enzymatic and nonenzymatic browning in food products, especially dried fruits or vegetables and are discussed in Chapter 19 of this book. Vitamin C (E300), citric acid (E330), and sodium sulfite (E221) are the most commonly used additives in this category. These additives are classified as either antioxidants or preservatives in the INS system, but retain the same numbers as in the E system without the E.

B. Nutritional Additives

Nutritional additives have increased in use in recent years as consumers have become more concerned about and interested in nutrition. Because of this increased interest, this edition includes four chapters devoted to nutritional additives as well as additives used for special dietary purposes. Chapter 9 provides detailed coverage of the primary nutritional additives, vitamins, and minerals, while Chapters 10, 11, and 12 highlight the food additives used for special dietary purposes. The nutritional additives are not included as a functional class within the INS or E numbering system, although several of the additives are included under other functional classes and as expected serve several functions in these products.

Vitamins, which as indicated above are also used in some cases as preservatives, are commonly added to cereals and cereal products to restore nutrients lost in processing or to enhance the overall nutritive value of the food. The addition of vitamin D to milk and of B vitamins to bread has been associated with the prevention of major nutritional deficiencies in the United States. Minerals such as iron and iodine have also been of extreme value in preventing nutritional deficiencies. Like vitamins, the primary use of minerals is in cereal products.

Amino acids and other proteinaceous materials are not commonly used in foods. However, lysine is sometimes added to cereals to enhance protein quality. Proteins or proteinaceous materials such as soya protein are also sometimes used as nutritional additives, although they are most commonly used as texturizing agents.

Fiber additives have seen increased popularity in recent years with the increase in consumer interest in dietary fiber. Various cellulose, pectin, and starch derivatives have been used for this purpose. Recently, naturally derived fiber from apples and other fruits as well as sugarbeets has been introduced as a fiber additive. Fiber additives are not well defined and in reality have little or no direct nutritional value, although they do have indirect nutritional benefits. In some cases, fiber additives also provide improved texture to food products and are categorized in the INS and E system as bulking agents, thickeners, or stabilizers.

The number of food additives used for special dietary purposes has increased significantly in recent years with an emphasis on the replacement of fat to reduce calories. United States food processors use more than 16 billion pounds of fat each year (Anonymous, 1995), and the fat replacement industry grew rapidly in the 1990s, although it appears to have peaked in recent years, and the growth of this sector has subsided (Sloan, 1997). Fat replacers include many texturizing agents and, as indicated in Chapter 11, include carbohydrate-, protein-, and fat-based systems.

The increased interest in nutrition has also led to the rapid growth of the functional food or nutraceutical industry with the development of several additives for the purpose of enhancing overall health. Chapter 10 describes some of the current interest in essential fatty acids as additives, while Chapter 12 describes some of the major additives used in engineering clinical nutritional products.

C. Coloring Agents

Most coloring agents are used to improve the overall attractiveness of the food. A number of natural and synthetic additives are used to color foods. In addition, sodium nitrite is used not only as an antimicrobial, but also to fix the color of meat by interaction with meat pigments. The colors are included in the E system as E100–E180 and in the INS as 100–182. As indicated in the review of coloring agents in Chapters 16 and 17, there has been much controversy regarding their use. Although synthetic coloring agents continue to be used extensively, there has been significant increased interest in natural colorants as described in Chapter 17.

D. Flavoring Agents

Flavoring agents comprise the greatest number of additives used in foods. There are three major types of flavoring additives: sweeteners, natural and synthetic flavors, and flavor enhancers.

The most commonly used sweeteners are sucrose, glucose, fructose, and lactose, with sucrose being the most popular. These substances, however, are commonly classified as foods rather than as additives. The most common additives used as sweeteners are low-calorie or noncaloric sweeteners such as saccharin and aspartame. These sweeteners, as discussed in Chapter 15, have had a major impact on the development of new foods.

In addition to sweeteners, there are more than 1700 natural and synthetic substances used to flavor foods. These additives are, in most cases, mixtures of several chemicals and are used to substitute for natural flavors. In most cases, flavoring agents are the same chemical mixtures that would naturally provide the flavor. These flavoring substances are discussed in detail in Chapter 13. The acidulants (see Chapter 21), which add a sour taste, often serve other purposes, including preservation.

Flavor enhancers (INS 620–642 and E620–E640) magnify or modify the flavor of foods and do not contribute any flavor of their own. Flavor enhancers, which include chemicals such as monosodium glutamate (E621) and various nucleotides (E626–E635), are often used in Asian foods or in soups to enhance the perception of other tastes. These chemicals are covered in detail in Chapter 14.

E. Texturizing Agents

Although flavoring agents comprise the greatest number of chemicals, texturizing agents are used in the greatest total quantity. These agents are used to add to or modify the overall texture or mouthfeel of food products. Emulsifiers and stabilizers are the primary additives in this category and are discussed in more detail in Chapters 23 and 24. Phosphates and dough conditioners are other chemicals that play a major role in modifying food texture. Phosphates are some of the most widely used and serve a number of functions in foods, as discussed in Chapter 25.

Emulsifiers (INS 429–496 and, primarily, E431 and E495) include natural substances such as lecithin (INS 322 and E322) and mono- and diglycerides as well as several synthetic derivatives. The primary role of these agents is to allow flavors and oils to be dispersed throughout a food product.

Stabilizers include several natural gums such as carrageenan as well as natural and modified starches. These additives have been used for several years to provide the desired texture in products such as ice cream and are now also finding use in both dry and liquid products. They also are used to prevent evaporation and deterioration of volatile flavor oils.

Phosphates (E338–E343) are often used to modify the texture of foods containing protein or starch. These chemicals are especially useful in stabilizing various dairy and meat products. The phosphates apparently react with protein and/or starch and modify the water-holding capacity of these natural food components.

Dough conditioners such as steroyl-2-lactylate and various humectants such as sodium silicoaluminate are also used as texturizing agents under very specific conditions.

F. Miscellaneous Additives

There are numerous other chemicals used in food products for specific yet limited purposes. Included are various processing aids such as chelating agents, enzymes, and antifoaming agents; surface finishing agents; catalysts; and various solvents, lubricants, and propellants. More information on enzymes is provided in Chapter 22.

III. BENEFITS OF ADDITIVES

There are obviously many recognized benefits to be derived from additives. Some of the major benefits are a safer and more nutritious food supply, a greater choice of food products, and a lower-priced food supply.

A. Safer and More Nutritious Foods

There is no question that the preservative and nutritional additives used in foods increase the safety and overall value of many food products. The use of several antimicrobials is known to prevent food poisoning from various bacteria and molds. Antioxidants, used to prevent the development of off-flavors, also prevent the formation of potentially toxic autoxidation products and maintain the nutritional value of vitamins and lipids. As indicated in Chapter 9, the use of various nutritional additives such as vitamins is also of proven value in preventing nutritional deficiencies.

B. Greater Choice of Foods

Most major supermarkets today carry more than 20,000 food items, providing the consumer a wide choice of food products. The availability of additives has allowed the production of numerous out-of-season foods and a variety of new food products. Additives have increased the development of convenience foods, snack foods, low-calorie and health promoting (functional) foods, exotic foods, and a variety of food substitutes. Convenience has been built into TV dinners and breakfast cereals as well as several microwave products. Additives allow these foods to be pre-prepared and still maintain acceptable flavor, texture, and nutritional value. Although many of these foods can have added convenience through the use of new packaging approaches or other processing methods, most depend on preservatives and texturizing agents. It is estimated that the shelf life of cereal products can be increased over 200% by the use of antioxidants (Branen, 1975).

The snack food industry has continued to be successful because the use of coloring and flavoring additives make available a wide array of snack items. These items, which are commonly subjected to high-temperature processing and are expected to have an extended shelf-life, also contain preservatives.

The increased interest by consumers in dieting has resulted in a proliferation of low-calorie food items. The use of saccharin and cyclamates opened the market for various food products with reduced calories, and by 1996 these and other calorie reduction agents became the highest selling category of food additives at $1.3 billion annually (Anonymous, 1995). These sweeteners are now being phased out, being at least partially replaced with aspartame, yet remain the primary additives used in low-calorie foods. However, many emulsifiers and stabilizers have allowed a reduction in the lipid content of foods, thus also lowering calories. As is noted in Chapter 11, fatty acid esters of sucrose have allowed an even greater reduction in the lipid content of foods. Coloring and flavoring agents have enhanced the appeal of these foods to consumers.

Stabilizers, emulsifiers, and coloring and flavoring additives have also allowed development of a number of food substitutes, especially dairy and meat substitutes. Margarine and soya meat products would simply not exist without the use of additives. The same is true of many soft drinks, which are primarily a mixture of food additives. Although the market for beverages has leveled out in recent years, the beverage industry continues to be the greatest user of food additives with a market exceeding $1.4 billion annually (Anonymous, 1995).

The greatest increase in food additive use in the next several years is likely to be in the functional food and nutraceutical industry. Several recent publications have noted the rapid growth of this industry both in the United States and in Europe (Hollingsworth, 1999; Sloan, 1997, 1998, 1999). Although definitions vary, Sloan (2000) defines a functional food as a food or beverage that imparts a physiological benefit that enhances overall health, helps prevent or treat a disease/condition, or improves physical or mental performance via an added functional ingredient, processing modification, or biotechnology. When low-calorie and fatfree foods are included, the total market for functional foods exceeds $92 billion in the U.S. and is expected to grow annually 6–10% (Sloan, 1999, 2000). Although lower fat content continues to be a major factor that motivates food purchase decisions, overall good health is now the major motivating factor (Sloan, 2000).

C. Lower-Priced Foods

Although there have been few recent studies to indicate that additives reduce the overall price of foods, a study reported in 1973 (Angeline and Leonardos, 1973) indicated that, at least for some processed foods, total removal of additives would result in higher prices. This study was based on the premise that the consumer would still desire the same type of foods in the absence of additives. The researchers reported that if, for example, additives were removed from margarine, consumers would have no alternative but to purchase a higher-priced spread such as butter, which usually contains few or no additives. They also reported that if additives were removed from bread, franks, wieners, and processed cheese, new processing procedures, increased refrigeration, and improved packaging would be required, at a higher cost, to keep the same type of products available. In 1973 prices, it was estimated that a consumer of sandwich fixings, including bread, margarine, franks or wieners, and processed cheese, would pay an additional $9.65 per year if additives were not available.

Although packaging or processing procedures could be developed to replace the need for additives, in most cases processing or packaging alternatives are not as cost-effective as the use of additives. However, it must be recognized that some of the additive-containing foods could be replaced in the diet with foods free of additives. It is also important to realize that the assumption that food additives lower the price of foods is based on maintaining the same type and quality of foods that we currently have available. Without additives, we could still have an excellent food supply at a reasonable cost. However, to provide consumers with the variety of foods along with the other benefits mentioned, would cost more without additives.

IV. RISKS OF ADDITIVES

Despite the benefits attributed to food additives, for several years there have also been a number of concerns regarding the potential short- and long-term risks of consuming these substances. Critics of additives are concerned with both indirect and direct impacts of using additives. As for many of the benefits mentioned, there is not always adequate scientific proof of whether or not a particular additive is safe. Little or no data are available concerning the health risks or joint effects of the additive cocktail each of us consumes daily.

The indirect risks that have been described for additives are the converse of some of the benefits attributed to their use. While it is accepted that through additives a greater choice and variety of foods have been made available, there is no question that additives have also resulted in the increased availability of food products with a low density of nutrients. These so-called junk foods, which include many snack items, can in fact be used as substitutes in the diet for more nutritious foods. Recently the food industry has attempted to address this criticism by adding nutritional additives to snack items so that these foods are a source of selected vitamins and minerals. The long-term effectiveness of this is questionable. Obviously, educational programs are needed to ensure that consumers select nutritious foods. Some scientists, however, feel that there is a place in the diet for foods that provide pleasure even if no direct nutritional benefit can be ascribed to their consumption.

Of greater concern than the indirect risks are the potential direct toxicological effects of additives. Short-term acute effects from additives are unlikely. Few additives are used at levels that will cause a direct toxicological impact, although there have been incidents where this has happened. Of particular concern are the hypersensitivity reactions to some additives that can have a direct and severe impact on sensitive individuals even when the chemicals are used at legally acceptable levels. The reactions to sulfites and other additives, as described in Chapters 3–5 of this book, are examples of such a problem. With proper labeling, however, sensitive individuals should be able to avoid potential allergens.

Toxicological problems resulting from the long-term consumption of additives are not well documented. Cancer and reproductive problems are of primary concern, although there is no direct evidence linking additive consumption with their occurrence in humans. There are, however, animal studies that have indicated potential problems with some additives. Although most of these additives have been banned, some continue to be used, the most notable being saccharin.

Most existing additives and all new ones must undergo extensive toxicological evaluation to be approved for use. Although questions continue to be asked regarding the validity of animal studies, there is a consensus among scientists that animal testing does provide the information needed to make safety decisions. The procedures used for this evaluation and the current philosophy regarding safety testing are outlined in Chapters 2 and 3.

V. BALANCING RISKS AND BENEFITS

Due to the difficulties in precisely defining the risks and benefits of individual additives, a legal rather than a scientific decision is commonly made regarding the safety of a food additive. In such a decision, the potential risks must be weighed against the potential benefits. A common example of this balance is saccharin. Although there is no direct evidence that saccharin, in the low amounts consumed in foods, causes cancer in humans, risk evaluation in rats indicates a potential for cancer in humans. On the benefit side, saccharin is an excellent noncaloric sweetener that is useful for diabetics and those interested in reducing consumption of calories. Many consumers feel that the benefits of having saccharin available as a sweetening agent outweigh the risks. On the basis of available risk information, however, the FDA initially issued a ban on saccharin in the early 1970s. The U.S. Senate, recognizing the consumer demand for low-calorie foods, subsequently placed a moratorium on the ban, thus allowing saccharin's continued use. The moratorium was essentially the first political recognition of the importance of balancing the potential risks of an additive against its perceived benefits and allowing the consumer the choice of consuming or not consuming the food. The moratorium has continued for several years and has undoubtedly had a significant impact on the continued and proposed use of additives.

Concern regarding the safety of additives has declined in the United States since the enactment of the saccharin moratorium. As noted in Chapter 6, a 1997 study indicates that only 21% of supermarket shoppers were concerned about additives and preservatives, a significant decline from a 1987 study. Reduced consumer concerns plus a changing political environment away from consumerism and a move toward more responsible use of additives by manufacturers have lessened the controversy over additive use. The potential risks of additives are well recognized, but the beneficial role these additives play in food production, processing, and utilization are also felt to be essential to the maintenance of our current food systems. With the convenient, tasty, and nutritious foods demanded, or at least desired, by consumers and the increasing overall demand for foods as populations increase, food additives will continue to play an important and essential role in food production. There will, however, continue to be concern regarding the potential risks associated with long-term consumption of small amounts of these chemicals and the possible interactive toxicological effects. As methods improve for evaluating these toxicological effects, some additives may be banned. At the same time, the same information may be used to develop safer new additives or techniques for using existing additives in a way that will lessen risk.

New technology is likely to have a profound impact on the use of food additives in the future. Of these, recombinant DNA biotechnology may have the greatest effect on the future development and use of food additives. Recombinant DNA biotechnology is already routinely used for production of additives through bioprocessing, including organic acids, bacteriocin preservatives, enzymes, microorganisms, vitamins, and minerals (Institute of Food Technologists, 2000). Biotechnology may also decrease the need for food additives. Plants have been produced through recombinant DNA with increased shelf-life and nutritional value, thus decreasing the need for a variety of additives. Although it is expected these recombinant DNA methods will be accepted in the future, there are currently several questions being raised regarding the risks and benefits of these products as well.

VI. LEGAL QUESTIONS

The final decision regarding additive use will most likely fall on governmental agencies that will evaluate the available information of potential risks and benefits to reach informed decisions. There is a need to harmonize these legal decisions on a worldwide basis, especially with the continued increase in movement of processed foods between countries. Most likely these decisions will be specific to each country and depend on the perceived benefits, which may vary from country to country. It is hoped that research can continue to better define safety evaluation procedures and their interpretation. Research will also be needed to better define the use and benefits of additives and identify possible alternatives. Existing laws in most countries appear to reflect consumer concerns and provide adequate protection. Chapters 7 and 8 provide excellent overviews of major laws governing use of additives in the United States and Europe. As research results become available, however, these laws will undoubtedly change to reflect changing information.

Informed scientists, food producers and consumers, and legal authorities need to continue to meet on a worldwide basis to develop strategies for addressing concerns regarding additive use. It is doubtful that the interest in the wide variety of foods made available with additives will decline in the future. With the expected continued increase in per capita income on a worldwide basis, the demand for a variety of convenient foods will also continue to increase. At the same time, as consumers become better educated, they may also want less risk. In rare cases, decisions will be left to consumers, but most will be made by legal authorities. We undoubtedly will continue to live in a society in which additives are a way of life.

REFERENCES

Angeline, J. F., Leonardos, G. P. 1973. Food additives—some economic considerations. *Food Technol.* April:40–50.

Anonymous, 1995. *Food Additives: U.S. Products, Applications, Markets.* Technomic Publishing, Lancaster, PA.

Branen, A. L. 1975. Toxicology and biochemistry of butylated hydroxyanisole and butylated hydroxytoluene. *J. Am. Oil Chem. Soc.* 52:59.

Codex Alimentarius Commission. 2001. Class names and the international numbering system for food additives. Codex Alimentarius: Vol. 1A—General Requirements. www.fao.org/es/esn/codex/standard/volume 1a/vol la 1a_e.htm.

Davidson, P. M., Branen, A. L. 1993. *Antimicrobials in Foods.* Marcel Dekker, New York.

Hanssen, M. 1984. *E for Additives: The Complete ''E'' Number Guide.* Thorsons Publishers, Wellingborough, England.

Hollingsworth, P. 1999. Keys to Euro-U.S. food product marketing. *Food Technol.* 53(1):24.

Institute of Food Technologists. 2000. IFT expert report on biotechnology and foods: benefits and concerns associated with recombinant DNA biotechnology-derived Foods. *Food Technol.* 54(10):61.

Jukes, D. 2001. Food Additives in the European Union. 2000. *Food Law.* www.fst.rdg.ac.uk/foodlaw/additive.htm.

Sloan, A. E. 1997. Fats and oils slip and slide. *Food Technol.* 51(1):30.

Sloan, A. E. 1998. Food industry forecast: consumer trends to 2020 and beyond. *Food Technol.* 52(1):37.

Sloan, A. E. 1999. Top ten trends to watch and work on for the millennium. *Food Technol.* 53(8):40.

Sloan, A. E. 2000. The top ten functional food trends. *Food Technol.* 54(4):33.

2

Food Additive Intake Assessment

SEPPO SALMINEN

University of Turku, Turku, Finland, and Royal Melbourne Institute of Technology, Melbourne, Australia

RAIJA TAHVONEN

University of Turku, Turku, Finland

I. INTRODUCTION

Food additives are special chemicals that are added into our food supply on purpose, and they are meant to be in the food at the time of consumption. Both international organizations and local governments generally evaluate the safety of food additives. The goal of local assessment is to take into account local food supply and cultural differences in dietary habits that may influence the intake of food additives.

Food additive intake assessment has three major goals:

1. Monitoring the intake of chemicals and relating it to the acceptable daily intake (ADI) values
2. Identifying consumer groups that may be at risk for food additive intake close to or higher than the ADI values
3. Provide information for the regulatory bodies for reassessing the food additive regulations in case of high intake in all or some consumer groups

The major aim of the intake assessment is to protect consumer health and to assist in developing food additive regulations.

II. SCOPE AND PURPOSE OF FOOD ADDITIVE INTAKE ASSESSMENT

The safety evaluation of food additives is based on the assessment of toxicity of the chemicals added to food. The rationale of intake assessment is to determine the likelihood and

Table 1 Toxicological Classification of Food Additives Based on Available Safety Data

Group A	Substances with an established ADI value
Group B	Substances generally regarded as safe
Group C	Substances with inadequate data
Group D	Flavoring components
Group E	Natural components used as additives without any scientific safety data or with very limited data

extent to which ADI values may be exceeded. The ADI value is determined by the Joint FAO/WHO Expert Committee on Food Additives (JECFA). The ADI value defines an estimate of the amount of food additives, expressed on a body weight basis, that can be ingested over a lifetime without appreciable health risk (WHO, 1987). This sets the limits on food additive intake assessment and also the goals for the assessment procedures. Several approaches can be taken, and food additives can be classified based on the safety data (Table 1). However, it is important to assess the use of specific components, whether additive or ingredient, and to define the safety data needed for each purpose. This then forms the basis for intake assessment and risk evaluation.

III. REGULATION OF MAXIMUM LEVELS OF FOOD ADDITIVES

In the United States, the principle of certain food components being generally recognized as safe (GRAS) was established in early legislation and later rigorously defined to include scientific evidence (Wodicka, 1980). Other regulations are included in the Code of Federal Regulations on specific food additives. In the European Union, three major directives regulate the use of food additives in member countries. In Australia and New Zealand, the regulation is similar to that in Europe. In Japan, differences exist in the regulation of all additives from natural sources, while strict regulation concerns chemical additives.

Usually the regulation is based on the acceptable daily intake values determined by the JECFA. This committee started establishing the ADI values as early as 1956. The ADI value is not an exact figure or a mathematical value, but it gives an estimate of how much of a chemical can be relatively safely ingested daily by normal consumers. The ADI value is not related to the intake over a few days or even a few months, but rather the daily intake over a lifetime (Fondu, 1992). It has also been pointed out that the ADI value, which depends on a series of factors, is not a constant regulatory number. Rather it is a guide serving to calculate the acceptable limits of different chemical agents incorporated into our food supply (Truhaut, 1992; Fondu, 1992). As the use of food additives is generally regulated on the basis of the ADI values, the applicability of the ADI value has been discussed especially in terms of infants and children. These studies were brought together by ILSI Europe in a consensus meeting (Clayton et al., 1998). It was recommended that special ADIs should not be created for infants and children, but rather that special sensitivity should be taken into account when assessing and defining individual ADI values (Clayton et al., 1998; Larsen and Pascal, 1998).

IV. EUROPEAN REGULATORY SITUATION

The assessment of food additive intakes has become increasingly important in Europe due to legal developments. Directive 89/107/EEC on food additives sets out the general framework for additive legislation in the European Union. Similar legal developments have also taken place in the United States. In this European directive, particularly the Annex II and Part 4, a general statement is made on intake estimation. According to the directive ''all food additives must be kept under continuous observation and must be re-evaluated whenever necessary in the light of changing conditions of use and new scientific information.'' Following the general outline directive, three specific directives in Europe have been passed: on sweeteners (94/35/EC), colors (94/36/EC), and other additives (95/2/EC). All these directives include obligations for the European Union member states to monitor the usage and consumption of food additives and to report their findings to the European Commission. The Commission is also required to submit reports to the European Parliament on changes in the additive market and the levels of use and to propose amendments to food additive use within five years (Wagstaff, 1996).

The European approach was extensively discussed in a workshop organized by ILSI Europe concerning the scientific assessment of the regulatory requirements in Europe (Howlett, 1996; Wagstaffe, 1996). This meeting also summarized the different approaches taken by many European countries to fulfill the commission requirements on monitoring food additive intakes (Penttilä, 1996; Cadby, 1996; Lawrie and Rees, 1996; Verger, 1996). Also, the differences between various additive groups were discussed, and special attention was given to flavoring substances. As there are only a few hundred additives, there are over 2000 flavoring substances and this poses special problems for intake assessment (Cadby, 1996). Methods for assessing the intake of flavor componds have been compared by Hall and Ford (1999).

V. METHODS OF ESTIMATING DIETARY INTAKE OF ADDITIVES

Methods of estimating dietary food additive intake can be classified as either one-phase or two-phase. A one-phase method uses information from one data source, usually concerning food additive production and usage. A two-phase method combines information from two data sources; these usually concern additive concentrations in foods and food consumption. In the latter case the investigator is required to decide how to combine the two different types of data to estimate the food additive intake. The sources of information and methods available to the investigator of dietary food additive intakes are summarized.

Figure 1 shows a diagram of information sources concerning food additives and food consumption. Various methods of additive intake estimation are based on the data sources written in capital letters. One-phase methods require information about food additive production and usage. Two-phase methods require information about food additive concentrations in foods and food consumption. In duplicate meal studies, both these types of information are collected from one source, which increases the accuracy of this method of intake estimation over the others.

There is an inherent inaccuracy in all the methods of estimating food additive intakes, and this varies in degree between methods of calculation. Each method is based on different assumptions. For example, in one method all food items are assumed to contain the maximum permitted food additive concentrations, which results in overestimated intake values. Several methods employ two different data files, one on food consumption levels and the other on food additive concentrations in various food items. Usually investigators do not collect all the information they require, but use other food surveys to supply the missing information. In combining data from different sources, they incorporate the assumptions of the various investigators. Thus the validity of each approach has to be

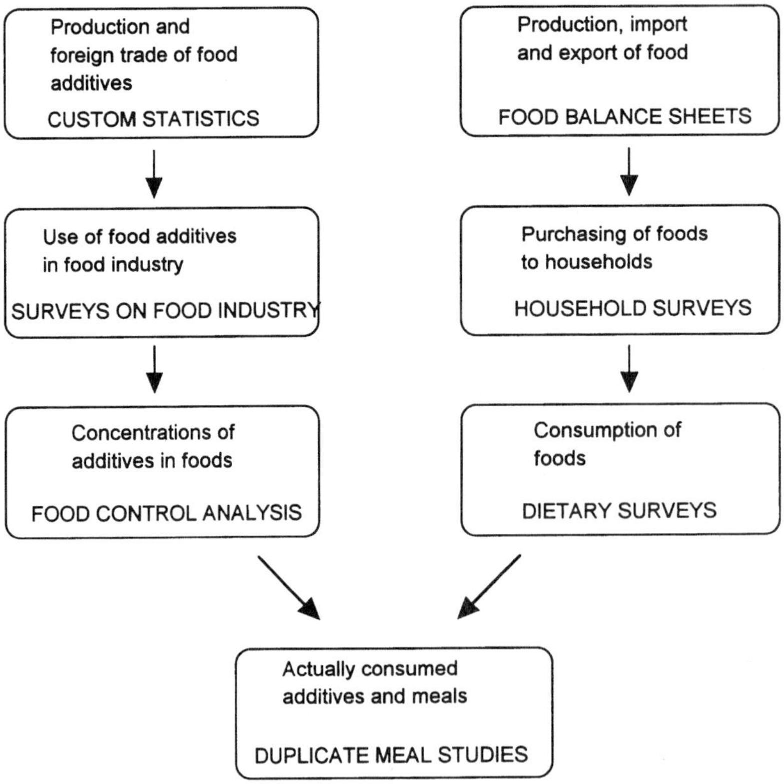

Figure 1 The methods used for assessing the intakes of food components and food additives. The combination of the methods from top down indicate increasing cost and accuracy in intake assessment.

carefully assessed. A critique of dietary survey methodology has shown that large random and systematic errors can be introduced into inadequately planned experimental protocols (Bingham, 1985). For example, when participating subjects estimate the weight of foods consumed by methods other than by weighing, the errors involved can vary from 20 to 50. Also the responses of subjects asked to recall their food intake during the previous 24 hours have been shown to vary by between 4 and 400% of observed intakes. This means that experimental protocols need to be carefully designed to provide the most accurate information from the data to be collected. The methods applied to assess the intake of contaminants, additives, and nutrients have also been reviewed by Petersen and Barraj (1996). Dietary assessment methodologies have also been critically reviewed and recommendations made by a Nordic compendium (Lennernäs, 1998; Berglund, 1998; Goldberg and Black, 1998; Hambraeus, 1998).

A. One-Phase Methods

1. Estimation Based on Production and Foreign Trade

One estimation of food additive usage in a given country can be calculated from the amounts of food additives produced, exported, and imported by that country. This value of additive usage by the food industry can then be divided by the total population to estimate an average food additive intake. It may also be necessary to take into account any alternative uses of the additives; for example, carrageenan is widely used in the manufacture of toothpaste. It is evident that this method of calculating the intake of a food additive is not very accurate and therefore should not be employed as the sole method if it is possible to use other methods.

Examples of this type of calculation have recently been published (Conning, 1986). Estimates of the annual utilization of additives in the United Kingdom were combined with the assumptions that 32% of the UK production is exported and that 20% of food purchased is wasted. The calculation resulted in an estimated intake of 0.53 g of synthetic additives per person per day, or 440 ppm in a daily diet of 1200 g. It was suggested that this food additive intake would comprise preservatives, antioxidants, colors, emulsifiers, and phosphates. A similar calculation was used to estimate the dietary concentration of flavors to be of the order of 0.003 ppm.

2. Surveys of Food Industry Usage

Records of the purchase or use of food additives by the food industry can be collected to estimate the total usage of additives. Dividing this amount by the number of consumers yields an estimate of the average intake of that food additive. The U.S. Food and Drug Administration (FDA) has sponsored a number of surveys on the use of food additives; the third was completed in 1977. The National Research Council undertook this survey to elicit information concerning the concentrations of additives while determining which additives were used in the products available to consumers. The potential food additive intakes were then calculated using this data file in combination with estimates of food consumption (National Research Council, 1979). Furthermore, the survey committee divided each major category into approximately 200 subcategories, which were in turn subdivided into food classes. This classification system enabled the surveyors to ensure that a sufficient number of responses were collected for each category. About 2500 companies received the survey questionnaires, and about 500 returned them. It was clear from the information obtained that the 500 respondents included 71 of the "Top 100" food companies in the United States. The investigators assumed that companies generally use similar concentrations of food additives in the same types of food. Thus the results of the survey were considered to represent the national diet adequately and therefore to provide reasonably accurate and useful estimates of food additive levels entering the U.S. food supply. The calculated intake estimates for each additive were derived using the following data: the average concentrations (weighted mean use levels) of additives in food, based on the responses from the manufacturers, the frequency with which foods containing each additive were eaten each day, and the average amount of foods (weight) consumed on each occasion.

The individual average daily intake of each additive was estimated by multiplying the total substance concentration used in foods of each category by the average consumption of such foods on each eating occasion. This figure was in turn multiplied by the frequency with which the foods were consumed over the 14-day survey period, and the result was subsequently divided by 14 to obtain the average daily intake. The investigators commented that it was likely that almost all the additive intakes calculated in this way would be overestimated. Therefore the results should be regarded as the upper limits of a potential range of intakes. In Finland, questionnaires were circulated to the members of the Association of Food Industry, who were asked to record both their use of food additives and which food products contained the majority of these additives (Heiskanen, 1983).

Table 2 Food Additives Intakes: Comparison of Two Analysis Methods in Finland

Additive	Use of additives by the food industry (mg/person/day)	Food control analysis (mg/person/day)
Sodium benzoate	46	40
Sorbic acid	40	37
Nitrates	0.9	6.4
Nitrites	6.3	3.5
Phosphates	204	97
Glutamate	40	186
Saccharin	2.3	6.6

Source: Penttilä, 1996.

Government authorities conducted another study (Penttilä 1996) during the same period. Although the two studies reported variations in the estimated intakes (Table 2), explanations for these differences could be suggested without invalidating the experimental method of either report. However, in most countries, no valid information on food additive intake is available.

In a method described by Buchet and Lauwerys (1983), the housewife was the reference person who stored duplicate meals of all the foods she ate. The sample meals were stored in the domestic refrigerator until they were transported to the laboratory for analysis. The use of duplicate meal samples from large catering establishments, such as hospitals, may also be considered because of the ease with which the preparation and collection of meal samples can be organized. The number of analyses required for each food additive is equal to the number of duplicate meals collected and increases when the experimental design includes representative sampling of population subgroups (socioeconomic groups, ethnic minorities, etc.) across seasons and different geographical areas. Although difficulties are encountered when adopting the analysis of duplicate meals to estimate the additive intakes, there are several advantages to using this experimental protocol. In the first instance, it provides accurate estimates of food additive intakes of individuals and realistic variations of these intakes. Also, food consumption data and analytical data on single food items are unnecessary for such a study. This experimental procedure could therefore be used to assess the accuracy of the other methods of estimating food additive intakes and thus be used to validate their results. Furthermore, a duplicate meal study can be used to ascertain whether any significant losses or gains in additive content of foods occur during the preparation of those foods in the home.

B. Two-Phase Methods

1. Assumption of Maximum Permitted Levels

This method is based on the assumption that the concentration of the food additive in each food item is the permitted maximum. The estimated daily intake is therefore calculated by multiplying the maximum permitted level (mg/kg) of additive in the food item by the average consumption of various food items (g/day). This calculation is carried out for all the food items that may contain the additive, and the sum of the results gives a total intake (mg/day) for that food additive. Information on food consumption data of population subgroups of special concern, such as children or an ethnic minority group can be added. This enables the estimation of potential additive intakes of special population subgroups. This survey method was used by WHO to investigate the potential intakes of 54 additives; only three were found to exceed their corresponding ADIs. The use of a different list of maximum permitted additive concentrations resulted in nine additives exceeding their ADIs (Lu, 1973). Subsequent studies (Toyoda et al., 1983; van Dokkum et al., 1982; Niemi et al., 1982; Gergely, 1980) using analytical methods revealed that actual consumption of various food additives rarely exceeds their ADIs. This leads to the conclusion that the use of maximum permitted additive concentrations in food overestimates the intakes because additives are seldom used in all the foods for which they have been approved. Even when they are used, their concentrations are usually below the maximum permitted levels. However, improvements in analytical techniques and analytical capacity now enable the determination of food additive concentrations in duplicate meals or food samples. Therefore the maximum permitted additive concentrations method should be used for the estimation of additive intakes only when analyzed additive concentrations are not available. Recently, in Italy the maximum theoretical intake estimates of antioxidants showed a potential excess intake of BHT, but not other antioxidants (Leclerq et al., 2000).

2. Market Basket Method

The market basket method, also referred to as a total diet study, uses selection of food items representing the typical pattern of food consumption. The foods are purchased from retail shops and are then prepared and cooked (with herbs, spices, dressings, etc.) to incorporate the usual gains and losses of food additives. The food items are then sorted into food groups (cereals, fats, fish, fruits, meats, oils, vegetables, etc.), and the items of each group are combined according to the proportions in which they occur in the diet as indicated by food consumption statistics. The foods in each group are then homogenized, and the concentrations of food additives in the homogenate are determined. The daily food additive intake for the food group is estimated by multiplying the measured concentrations of additives by the average consumption of foods in that food group. An estimate of the total intake is the subsequent sum of the intakes calculated for each food group. It is not appropriate to use the market basket study for estimating the intake of food additives with very limited use because the additive concentrations in food groups are diluted and may fall below the analytical detection limits. The use of this method could therefore underestimate the food additive intakes. The accuracy of the information obtained in this type of study is dependent on the initial construction of the typical diet. In addition, the accuracy of these estimates of additive intakes for the population of a country can be increased by taking into account any regional differences in food consumption (Toyoda et al., 1985), by using the different types of foods available in different shops and by incorporating seasonal variations of food consumption into the study (van Dokkum et al., 1982). These studies are conducted in several places around a country and several food items and brands are included. A good case example of these studies has been conducted in the United States for the Food and Drug Administration. The so-called FDA total diet studies have been reported from 1989 to 1991 assessing dietary intakes of pesticides, selected elements, and other industrial chemicals (Gunderson, 1995).

The market basket method is a broad concept that can be applied to many different approaches to the study of food additive intakes. The intake estimates may concern the whole population of a country, regional populations, or population subgroups such as ethnic minorities and particular age groups. For example, van Dokkum and coworkers (1982) studied the dietary pattern of 16- to 18-year-old males, who have the highest rate of food consumption measured in grams per day. Other studies may be designed to identify the factors responsible for observed variations in food additive intakes. In such studies, dietary habits, the effect of age distribution in the population, the degree of urbanization, and the type of shops available for the purchase of food samples need to be considered. For example, van Dokkum et al. did not establish any relationship between the sampling period and the additive intakes. Another approach may be to use the market basket method to investigate the contribution of a particular food group to the additive intakes. A more detailed description of the different methodologies and the various factors to be considered when applying the market basket study is provided by the guidelines compiled by WHO (1985). Most of the considerations discussed in the guidelines are appropriate and practical for estimating food additive intakes although the guidelines were designed for the study of food contaminant intakes.

VI. FOOD CONSUMPTION DATA

A. Food Balance Sheets

National food balance sheet studies coordinated by the United Nations Food and Agriculture Organization (FAO) include figures concerning the national production, import, export, and net consumption of food and the contribution of each food to the intakes of energy, protein, and fat (FAO, 1984). The per capita consumption is calculated by simply dividing the net consumption of food by the total population. The use of food balance sheets in studies estimating food additive intakes is inadequate for a number of reasons. The balance sheet studies do not provide information about the variations in food consumption caused by different dietary patterns. The list of food commodities (50–100 items) is limited (for instance, household surveys contain 250–300 items) and does not cover all the items containing food additives; the list basically contains raw food materials. For example, the consumption of cereals such as wheat and barley is presented, whereas the consumption of bread, cakes, and biscuits is not. Processed foods are included only when their foreign trade is of particular interest; for example, processed fruits, cheese, and butter are included in the balance sheets for Finland (FAO, 1984).

The use of food balance sheets in the estimation of food additive intakes should therefore be restricted to studies of food items that are included in the national balance sheets. If crude estimates of the intakes of food contaminants (such as heavy metals and pesticides) are required, then food balance sheets could be used.

B. Household Surveys

Household surveys are conducted in most western countries primarily for estimating expenditures on food and certain other commodities to enable the calculation of economic indices. This survey requires the householder to keep a diary on food purchases, but these records do not provide information about the foods actually consumed. It is assumed that the members of the household subsequently consume all the foods that are purchased. The number of households participating in the surveys varies; in Finland the number is approximately 10,000 for a population of 5 million (Central Statistical Office of Finland), whereas in Great Britain the sample is approximately 7500 (MAFF, 1986) for a population of 55 million. In the United States the sample of the 1977–1978 National Food Consumption Survey was 15,000 households for a population of 220 million (National Research Council, 1984).

The average amount of food purchased by each household provides information that can be used to estimate additive intakes. Household studies have certain deficiencies:

1. The consumption unit is the household and not the individual, except for households that contain one person (predominantly young or old people).
2. Meals eaten outside of the household, in canteens, lunchrooms, and restaurants, are not recorded in detail.

Thus, household surveys are not the best source of food consumption data, but they are much more useful than the balance sheet studies because they include a wider spectrum of food items. The British and Finnish household surveys can be used to estimate variation in food consumption because it is possible to study the purchase diaries of the households participating in the survey. The statistics published in Great Britain are not as extensive as those in Finland, but there are summaries of the total population and of various subgroups in the population defined by income, family size, status of home ownership, etc. The Finnish surveys also include socioeconomic details such as education, employment, and the size of households, which can be used to correlate variables other than food consumption with food additive intakes. Other information is available in Great Britain from census surveys, which are carried out periodically, and could be combined with the household survey information to estimate the different variables with food additive intakes.

C. Nutrition Surveys

Many countries carry out nutrition surveys to monitor the nutrient intake of the population and thus determine the adequacy of the diet. Three survey methods are normally used: food diary records, records of the weight of foods consumed, and dietary recall. Nutrient intakes can also be estimated using food balance sheets and household surveys, but these methods provide such crude estimates that they are seldom used for this purpose. A review of the errors associated with the use of diary records, food weighing, and dietary recall methods indicates that there is a need to improve the accuracy of these methods (Bingham, 1985). Random errors, systematic errors, or both may be introduced by the use of nutrient tables for foods (estimated error range 2–20%), by coding errors of the investigator, by the use of estimated food weights as opposed to weighing each food sample (20–50%), and by the reporting errors of the 24-h recall method (4–400%), to name but a few. Indeed an investigation of the accuracy of retrospective dietary assessments indicated that although dietary habits alter, the retrospective dietary recall of earlier dietary habits was strongly influenced by the current dietary regime (Jain et al., 1980; Jensen et al., 1984). Therefore the experimental design of such studies requires careful planning to obtain accurate and useful data.

D. The Use of Food Consumption Data in Food Additive Intake Studies

There are a variety of criteria that can justify the use of food consumption data files from nutrition surveys for estimating food additive intakes despite the risk of errors just mentioned. When a study objective is the estimation of national *average food additive intakes*, it is necessary to use food consumption data in which no subgroup is overrepresented. However, nationwide nutrition surveys are not conducted in every country. Therefore it is not always possible to provide food consumption data that would satisfy this criterion. On the other hand, it may be possible to use food consumption data collected as part of epidemiology surveys, which are more abundant than national surveys. Information collected in small surveys can be employed only if the population subgroup concerned is of interest for food additive intake studies. For example, the population subgroup may be children, teenagers, or old people. Another criterion for the use of food consumption data files may be the availability of *individual* food consumption records from the study. The essential characteristics of food additive intake estimations, such as the range of estimated intakes; standard deviation; and the proportions for certain population subgroups defined by age range, socioeconomic status, and so on, cannot be calculated if the data file does not include individual records. If the food consumption data are published with meaningful classifications concerning the population subgroups, while including the relevant statistical information, then individual records are not absolutely necessary. However, if new classifications of subgroups among the cohort are required, it may be useful to have access to the original data file containing the individual records. In general, the food consumption data of nutrition surveys should be preferred in studies of food additive intakes, especially when there is an inherent balance of the population subgroups in the sample. If data from a nutrition survey are not available, then data from household surveys should be used to estimate additive intakes.

VII. EXCESSIVE FOOD ADDITIVE INTAKES AND POPULATION RISK GROUPS

Currently there is a public debate concerning food additive usage, including food safety, the health risks associated with food additives, and the wholesomeness of food in general (Hanssen and Marsden, 1984). The food industries assessments suggest that the food currently available is safe if placed in a historical context (Conning, 1986; Turner, 1985a,b). The scientific debate regarding additive-associated risks is concerned with the difficulties of epidemiological studies performed in this area, comparisons of the toxicities of processed and unprocessed foods, appraisals of the resource priorities for scientific investigations (Goldberg, 1985), and critical evaluations of the methods of analysis (Bingham, 1985; Conning, 1985). Despite the factors employed to ensure a minimum human sensitivity to the additives in food, particular sections of the population are susceptible to additive-associated clinical disorders or toxic responses, some of which are usually thought to have an allergic origin (see ''Food Additive Hypersensitivities'' in this book). Excessive intakes have often been suggested for infants and children. The applicability of safety studies has been specifically assessed for infants and children (Clayton et al., 1998) and the different dietary patterns in relation to age and the consequences for intake of food chemical reported (Lawrie, 1998). In this report it was indicated that in some cases, such as saccharin, the intake was highest in children and potentially exceeding the acceptable daily intake. Similar reports for other additives have been reported by Penttilä (1996). Thus, specific measures should be taken when planning risk management procedures for different target groups.

However, it is likely that the exposure to food additives and the individual's susceptibility expressed as an undesirable response are not separate phenomena, and that the latter could be mediated by allergic reactions. The following considerations should be taken into account when estimating the *potential or maximum additive intakes* or identifying *population subgroups exposed to unacceptable risk*:

1. When a food additive is known to be present in a food but cannot be detected in analysis, the maximum additive concentration is assumed to be equal to the lower limit of detection (van Dokkum et al., 1982). This method enables estimation of the potential average intake.
2. Another intake estimation can be achieved by multiplying the highest additive concentration determined by analysis with the amount of the food item or food group under consideration. This value may underestimate the additive intake because consumption of food varies. When the two extreme figures—maximum food additive concentration and maximum food consumption—are combined to give an estimate of intake, the calculated value is probably an overestimate.
3. In order to increase the accuracy of estimating the maximum intake value it is necessary to combine the additive concentration of analyzed meals with the true food consumption. Thus, once the population subgroup at risk of excessive food additive intakes is identified, the maximum intakes can be accurately estimated by conducting a duplicate meal study.

Because young men have the greatest demand for food, it may by thought that they also have the highest food additive intakes. However, a comparison of food consumption on a weight-to-body-weight basis indicates that children have the greatest intake of food. Moreover, children consume large quantities of certain food items such as sweets and soft drinks that contain high concentrations of additives. Therefore an average value of food additive intake does not adequately describe the additive intakes of children. This can be studied separately or as part of a comprehensive nutrition survey (National Research Council, 1979; Verger et al., 1999). Generally, when food items are identified as being consumed in large quantities by children, the respective additive regulations should be adjusted to permit lower maximum additive levels than those allowed in other foods.

VIII. VARIATION IN FOOD ADDITIVE INTAKES

The variation between the food additive intakes of extreme and average consumers is difficult to determine. One estimate proposes that it is unlikely that the consumption of a single foodstuff by extreme consumers would be more than three times higher than the average (WHO, 1985). This is probably true for long-term food consumption. However, in a short-term study of food additive intakes among English school children (14–15 years of age), a wide range of intakes was observed (Disselduff et al., 1979). The maximum intakes of benzoic acid, saccharin, and sorbic acid were calculated to be approximately tenfold greater than average intakes. The children's food and drink consumption was measured by weight over seven consecutive days, and additive concentration was assumed to be the maximum permitted. The differences in additive intakes varied with the food consumption over the seven days, but over a longer period of time it is likely that the differences in food consumption would be smaller. An evaluation of the risks associated with food additive and contaminant intakes is dependent on an estimation of the *long-term variation in food consumption* by individuals. It is therefore necessary to perform relevant studies. When the use of a food additive is confined to a limited number of foods that are not basic to the ordinary diet, the intake of that additive varies considerably. For example, in the United States the mean intake of saccharin is 0.12 mg/kg body weight, whereas the intake of the 99th percentile is 2.2 mg/kg body weight. That is approximately 20 times greater than the mean (National Research Council, 1979). However, intakes of additives used in several common foods vary much less. Examples of the latter are sodium benzoate and potassium sorbate, which have mean intakes of 3.3 and 1.5 mg/kg body weight, while the 99th percentiles are 16 and 7.3 mg/kg body weight, respectively. The latter figures are fivefold greater than their respective means.

For particular cases of interest such as saccharin, if those who consume foods containing this sugar substitute are considered, the mean intake and 99th percentile are 0.41 and 4.3 mg/kg body weight, respectively. This demonstrates that mean intake is not always a good indicator when evaluating risk. Certain groups of the population, such as diabetics, may consume much more saccharin than the average person, sometimes as much as 40-fold in excess of the mean. Therefore, when the safety of additives is being assessed and when maximum levels of additives are being established, it is important to consider the variation of food additive intake and the variation in consumption of relevant items and categories of food.

IX. DISSEMINATING INFORMATION TO CONSUMERS

Consumers are increasingly aware of additives in food and often express a concern about food safety or a desire for a greater availability of information. This is also the interest of regulatory authorities in Europe and the United States. In the European Union, the European Parliament is also keen to receive information on food additive intakes (Wagstaffe, 1996; Renwick, 1996; Penttilä, 1996). Much of the information concerning food is obtained and edited by government agencies, but the dissemination of this information to consumers is not always carried out in the best way. One way to improve the current situation could be the use of health care organizations and audiovisual media to give objective reports concerning current scientific information about the composition and safety of food items. Authorities and organizations actively disseminating information about food should be able to convince consumers of the safety of major sections of the diet and indicate concern only when it is warranted.

If disseminating information to the public could achieve the same end as regulations, why are regulations required? Regulation of food additive use is preferable because information concerning individual components of the food supply may cause unnecessary concern among consumers, who may find it difficult to put this information into perspective. However, when information is disseminated through unofficial agencies—for example, when a certain food additive is suspected of being a carcinogen—then an alternative approach is required. The relevant authority, usually a governmental agency, should publish the information currently available, give its evaluation of the possible risks involved with the additive, and outline the action it plans to undertake. It is not easy to draw up guidelines for the actions governments should undertake to form the basis of reassuring information. It is equally difficult to decide what information should be made available to consumers. However, the steps taken by government agencies may include the following: All food additives should have a safety margin incorporated into their ADIs or their equivalent, and these should be published. Any toxic effects detected in experimental animals are extremely unlikely to occur in humans due to the safety factors. The intake of food additives should be surveyed regularly, and the maximum permitted concentrations in foods should be re-evaluated and reduced if intakes approach current ADIs. As methods of analysis and scientific knowledge improve, new information should be used to reassess the value and safety of food additives. The structure of government committees should be established in such a way that new information can be assessed and appropriate actions can be put into effect without delay.

X. CONCLUSIONS AND FUTURE DIRECTIONS

The estimation of food additive intakes can be used not only to ensure that the food supply is safe, but also to review the changing use of additives in food to maintain relevant regulations for both consumer safety and food processors. Although ADI values are only as good as the toxicological studies in animals and the methods used for their calculation, they are useful for food safety evaluation. The JFCMP proposal for extreme food consumption suggests three categories of additive intake:

1. Those below 30% of the ADI are certainly safe for whole population.
2. Those between 30 and 100% of the ADI indicate concern for the safety of extreme consumers, often children in particular.
3. Those greater than 100% of the ADI are by definition unsafe for the whole population.

Current evidence suggests that additive intakes do not often exceed their ADI, but that in the cases where they do, additive usage and the regulations should be altered. In some cases, this may be difficult because of the lack of suitable alternatives. For example, in meat products nitrite can be only partly replaced by ascorbate (Puolanne, 1985). The methods used for additive intake estimation are not only numerous, they vary widely in their accuracy and each has its limitations. Those methods that are easy to perform are usually based on broad assumptions and produce average intake values without any estimation of the variation within the population, whereas the more accurate methods tend to be more difficult to perform and require a large investment of resources. When considering the introduction of new additives or new uses of additives, the potential maximum additive intake can be estimated using the methods based on the physiological maximum food and drink consumption or the ratio of energy intakes of different foods. These methods may also be used with data on additive manufacture, export, and import; food industry additive usage; the assumption of maximum permitted levels in foods; food balance sheet surveys; and household survey methods to give a general view of additive intakes. If the experimental protocols of the market basket survey and nutritional survey methods are adequately designed, more accurate additive intake values can be estimated. However, all these methods suffer from the problem of diluting individual variation within the broad average intake value that is calculated. Where information concerning individual variation of additive intakes is vital, such methods as selected food studies and duplicate meal studies may be more appropriate. In addition, although the duplicate meal studies require a large investment of resources, this method may be employed to estimate the accuracy and thus calibrate any of the other methods of additive intake calculation.

Finally, it appears that despite the factors employed to ensure that food additive usage in the food supply is safe, certain population subgroups are susceptible to allergic or toxic manifestations of food components, including food additives. Children and diabetics are population subgroups that are particularly at risk of exposure to these undesirable manifestations. Furthermore, both nutrient and additive intakes of these groups may differ considerably from the average. These are sufficient reasons, if for no other, that the consumption of food and its components should be regularly reviewed in the light of currently available scientific data to ensure the safety of the food supply.

As the direct measurement of food additive intake remains a complex and time-consuming process of a great expense, it is important to assess the intakes in a stepwise manner (Penttilä, 1995, 1996; Leclerq et al., 2000). Priority should be given to the additives of most concern and to the additives with the lowest ADI values but significant industry use (Waggstaffe, 1996). Intakes below the ADI values may cause concern for specific target groups, especially children. Renwick (1996) has proposed also a scheme for assessing food additive intakes with a targeted manner incorporating both exposure estimation and toxicity data in determining which additives should be allocated a high priority in intake studies. The risk could be related also to the data on intakes, and ADI values could be interpreted in the light of toxicity data, as has been reported for cyclamate (Renwick and Walker, 1993). The application of such a scheme would harmonize the efforts in different countries and result in focusing the limited resources in additives of most concern.

REFERENCES

Berglund, L. 1998. Dietary investigations—what are the effects of invalid selection procedures and measurement errors. *Scand. J. Nutr.* 42:60–62.

Bingham S. 1987. The dietary assessment of individuals: methods, accuracy, new techniques and recommendations. *Nutr. Abstr. Rev. Series A.* 57:705–742.

Buchet, J. P., Lauwerys, R. 1983. Oral daily intake of cadmium, lead, manganese, copper, chromium, mercury, calcium, zinc and arsenic in Belgium: a duplicate meal study. *Food Chem. Toxicol.* 21:19–24.

Cadby, P. 1996. Estimating intakes of flavoring substances. *Food Add. Contam.* 453–460.

Clayton, B., Kroes, R., Larsen, J. C, Pascal, G. 1998. Applicability of the ADI to infants and children. *Food Add. Contam.* 15(Suppl.):S1–S89.

Conning, D. M. 1985. Artificial sweeteners—a long running saga. In: *Food Toxicology*, Gibson, G. G., Walker, R. (Eds.). Taylor and Francis, Philadelphia, pp. 169–180.

Disselduff, M. M., Try, G. P., Berry, T. C. 1979. Possible use of dietary surveys to assess intake of food additives. *Food Cosmet. Toxicol.* 17:391–396.

Fondu, M. 1992. Food additives: dietary surveys and calculation of consumption. *Medecine et Nutrition* 20:163–175.

Gibney, M. J. and Lambe J. 1996. Estimation of food additive intake: methodology overview. *Food Add. Contam.* 13:405–410.

Goldberg, G. R., Black, A. E. 1998. Assessment of the validity of reported energy intakes—review and recent developments. *Scand. J. Nutr.* 42:6–9.

Gunderson E. L. 1995. FDA total diet study, July 1986–April 1991, dietary intakes of pesticides, selected elements, and other chemicals. *J. AOAC Intern.* 78:1353–1363.

Hall, R., Ford, R. (1999). Comparison of two methods to assess the intake of flavouring substances. *Food Addit. Contam.* 16:481–495.

Hambraeus, L. 1998. Dietary assessments: how to validate primary data before conclusions. *Scand. J. Nutr.* 42:66–68.

Howlett, J. 1996. ILSI Europe workshop on food additive intake: scientific assessment of the regulatory requirements in Europe. *Food Add. Contam.* 13:385–396.

Larsen J. C., Pascal, G. 1998. Workshop on the applicability of the ADI to infants and children: consensus summary. *Food Add. Contam.* 15(Suppl.):S1–S10.

Lawrie, C. A. 1998. Different dietary patterns in relation to age and the consequences for the intake of food chemicals. *Food Add. Contam.* 13(Suppl.):S75–S82.

Lawrie C. A., Rees, N. M. A. 1996. The approach adopted in the UK for the estimation of the intake of food additives. *Food Add. Contam.* 13:411–416.

Leclerq, C., Arcella, D., Turrini, A. (2000). Estimates of theoretical maximum daily intake of erythorbic acid, gallates, butylated hydroxyanisole (BHA) and butylated hydroxytoluene in Italy: a stepwise approach. *Food Chem. Toxicol.* 38:1075–1084.

Lennernäs, M. 1998. Dietary assessment and validity: to measure what is meant to measure. *Scand. J. Nutr.* 42:63–65.

Louekari, K., Scott, A. O., Salminen S. Estimation of food additive intakes. In: *Food Additives*, Branen, A. L., Davidson, M., Salminen S. (Eds.). Marcel Dekker, New York, 1989, pp. 9–32.

Löwik, H. 1996. Possible uses of food consumption surveys to estimate exposure to additives. *Food Add. Contam.* 427–442.

MAFF. 1986. Household food consumption and expenditure 1984. Annual Report of the National Food Survey Committee. HMSO, London.

National Research Council. 1984. *National Survey Data on Food Consumption: Uses and Recommendations*. National Academy Press, Washington, D.C.

Penttilä P.-L. 1995. Estimation of food additive and pesticide intake by means of a stepwise method. Ph.D. Thesis. University of Turku, Finland.

Penttilä, P.-L. 1996. Estimation of food additive intakes. Nordic approach. *Food Add. Contam.* 13: 421–426.

Petersen B., Barraj, L. 1996. Assessing the intake of contaminants and nutrients. An overview of methods. *J. Food Compos. Anal.* 9:243–254.

Renwick, A. G. 1996. Prioritization of estimates of food additive intakes. *Food Add. Contam.* 13: 467–475.

Renwick, A. G., Walker, R. 1993. An analysis of the risk exceeding the acceptable or tolerable daily intake. *Regulatory Toxicology and Pharmacology* 18:463–480.

Toyoda, M., Ito, Y., Isshiki, K., et al. 1983. Estimation of daily intake of many kinds of food additives according to the market basket studies in Japan. *J. Jpn. Soc. Nutr. Food Sci.* 36(6): 489–497.

Toyoda, M., Yomota, C., Ito, Y., et al. 1985. Estimation of daily intake of methylcellulose, CMC, polyphosphates and erytorbate according to the market basket studies in Japan. *J. Jpn. Soc. Nutr. Food Sci.* 38(1):33–38.

Truhaut, R. 1992. The concept of the acceptable daily intake: a historical review. *Food Add. Contam.* 8:151–162.

Turner, A. 1986a. A technologist looks at additives. *Food Manuf.* 61(7):37, 38, 44.

Turner, A. 1986b. A technologist looks at additives. *Food Manuf.* 61(8):40, 41, 45.

van Dokkum, W., de Vos, R. H., Gloughley, F., Hulshof, K., Dukel, R., Wijsman, J. 1982. Food additives and food components in total diets in the Netherlands. *Br. J. Nutr.* 48:223–231.

Verger, P. 1996. One example of utilisation of the 'French approach.' *Food Add. Contam.* 417–420.

Verger, P., Garnier-Sayre, I., Leblanc, J-C. (1999). Identification of risk groups for intake of food chemicals. *Regulat. Tox. Pharmacol.* 30:S103–S108.

Wagstaffe, P. J. 1996. The assessment of food additive usage and consumption: the Commission perspective. *Food Add. Contam.* 13:397–403.

WHO 1987. Principles for the safety assessment of food additives and contaminants in food. Environmental Health Criteria 70. World Health Organization, Geneva.

Wodicka, V. O. 1980. Legal considerations on food additives. In: *CRC Handbook on Food Additives*, Vol. II, 2nd ed., Furia, T. E. (Ed.). CRC Press, Boca Raton, Florida, pp. 1–12.

3

Risks and Benefits of Food Additives

SUSAN S. SUMNER

Virginia Tech, Blacksburg, Virginia

JOSEPH D. EIFERT

Nestle USA, Inc., Dublin, Ohio

I. INTRODUCTION

A treatise on food and color additives would be incomplete without a discussion of the risks and benefits associated with their use. Any risks associated with the use of food and color additives should be placed in perspective with other food-associated risks and with the risks associated with the use of alternative additives or specific additive-free formulations. This chapter will focus on the risks and benefits associated with food and color additives as well as generally recognized as safe (GRAS) and prior sanctioned food ingredients. These latter classes of food ingredients were established by the 1958 Food Additive Amendments to the Federal Food, Drug and Cosmetic (FD&C) Act and are not legally classified as food additives in the United States.

The FD&C Act requires that food additives be functional but does not take into account any benefits that may accrue from their use. However, the functions of food additives are directly related to their benefits. The FD&C Act does stipulate that food additives must possess a low or insignificant level of risk to human health. Thus, in most cases the benefits of food and color additives, GRAS ingredients, and prior sanctioned ingredients far outweigh the risks by any means of comparison. Yet comparisons are not required by U.S. food laws. And decisions on the acceptability of food ingredients are based solely on assessments of risk. Thus, a beneficial additive with even a moderate degree of risk would not be allowed. The Delaney clause in the Federal Food, Drug and Cosmetic Act dictates that any food additive having a quantifiable carcinogenic risk be banned. This clause has rarely been invoked, but the zero tolerance for carcinogens has focused considerable attention on the health risks of weakly carcinogenic food and color additives.

The level of risk associated with the ingestion of food in the United States is quite low. The same is true for most developed countries of the world. That does not mean that the risk level could not be lowered even further. Certainly, major advances could be made toward improving the wholesomeness of the food supply on a worldwide basis, although most of the improvement is needed in the developing countries. The statement that the food supply has never been safer is true, especially in the United States and other developed countries.

The present-day perception that major risks may be associated with the food supply seems to be the result of several factors. First, current analytical capabilities allow the detection of mere traces of potentially hazardous substances in foods. Even a few decades ago such levels of detection were not possible. Second, the food laws in the United States focus considerable attention on food additives at the expense of naturally occurring chemicals in foods (although naturally occurring substances can be food additives in some instances). While food additives should be thoroughly tested for safety, the lack of testing of naturally occurring chemicals has resulted in a loss of perspective on the comparative hazards associated with food additives.

II. FUNCTIONS AND USAGE OF FOOD ADDITIVES

A. Functions

Hundreds of chemical additives are incorporated into foods directly or migrate into foods from the environment or packaging materials. A food additive can be sometimes defined as a substance whose intended use will lead to its incorporation into the food or will affect the characteristics of the food. These additives generally provide some type of benefit to the food producer, processor, or consumer. For the consumer, additives can improve organoleptic qualities of foods, improve the nutritive value, or ease the preparation of ingredients and meals. Typical additive benefits to the food producer or processor include improving product quality, safety, and variety.

Additives may be found in varying quantities in foods, perform different functions in foods and ingredients, and function synergistically with other additives. Their functions can usually be classified as one of the following: (1) to maintain or improve nutritional quality, (2) to maintain or improve product safety or quality, (3) to aid in processing or preparation, and (4) to enhance sensory characteristics (FDA, 1979, 1992). Additives that affect nutritional quality are primarily vitamins and minerals. In some foods, these may be added to enrich the food or replace nutrients that may have been lost during processing. In other foods, vitamins and minerals may be added for fortification in order to supplement nutrients that may often be lacking in human diets. Preservatives or antimicrobial substances are used to prevent bacterial and fungal growth in foods. These additives can delay spoilage or extend the shelf life of the finished product. Antioxidants are additives that also can extend the shelf life of foods by delaying rancidity or lipid oxidation. Additives that maintain product quality may also ensure food product safety for the consumer. For example, acids that may be added to prevent the growth of microorganisms that cause spoilage may also prevent the growth of microorganisms that can cause foodborne illness. Additives that are used as processing or preparation aids usually affect the texture of ingredients and finished foods. Some of these are classified as emulsifiers, stabilizers, thickeners, leavening agents, humectants, and anticaking agents. Chemicals in this group of food additives are also used to adjust the homogeneity, stability, and volume of foods. The fourth major function of food additives is to enhance the flavor or color of foods to make them more appealing to the consumer. Flavoring chemicals may be used to magnify the original taste or aroma of food ingredients or to restore flavors lost during processing. Natural and artificial coloring substances are added to increase the visual appeal of foods, to distinguish flavors of foods, to increase the intensity of naturally occurring color, or to restore color lost during processing.

B. Usage

Chemicals that are added to foods may be manmade or derived from natural plant or animal sources. Also, additives may be synthesized for foods that are chemically or functionally identical to those that may be derived from natural sources. In the case of color additives, synthetic colors are also identified as certifiable colors. Each batch of these coloring chemicals is tested by the manufacturer and the U.S. Food and Drug Administration to ensure the safety, quality, consistency, and strength of the color additive prior to its use in foods. Currently, nine synthetic color additives may be added to foods. Color additives that are exempt from this certification process include pigments derived from natural sources such as vegetables, minerals or animals, or the manmade equivalents of naturally derived colors. Examples include annatto extract, paprika, turmeric, and beet powder.

Substances that are considered food additives may be naturally occurring in some foods and ingredients or artificially added to other foods. If an additive from a natural source is added to a food, then its package label may state that the additive is "artificially added" since it normally is not found in the food. Additives are further classified by how they are added to foods. Some additives are directly added to foods and ingredients, while others may be added indirectly through contact with packaging materials as are, for example, the preservatives BHA and BHT on the inside of breakfast cereal bags. Water is usually considered to be a food ingredient when added during the manufacturing process, but it can be considered a direct additive, too. For example, in poultry processing in the United States, trimmed and washed carcasses are immersed in a cold water "chiller tank" for approximately one to two hours to reduce carcass temperature. During this time, the carcasses may gain weight when they absorb water from the tank. At this time, regulators are debating whether to mandate that processed poultry be labeled to state that water (and weight) has been added to the product.

Some undesirable food additives may be unintentionally added to foods. For example, pesticides and fumigants may come in contact with produce and grains during growing, harvesting, and storage. While these chemicals may have permitted uses for some crops, they may be illegal for use on others. Sometimes pesticides are mistakenly or unintentionally applied at the wrong concentration or too soon before harvest. Pesticides have been found, infrequently, on certain crops and at certain concentrations that are not permitted by the Environmental Protection Agency. Other examples of unintentional addition of food chemicals are those that migrate into foods from packaging materials, especially plastics, and those that occur from poor storage practices. Numerous chemicals may find their way into foods through contamination from the environment. Frequently cited examples include dioxin, polychlorinated biphenyl compounds (PCBs), and heavy metals such as lead and mercury.

Food additives are generally intended to provide important benefits to the producer or consumer of foods. But sometimes these additives cause unwanted or unhealthful effects. Often these undesirable effects of food additives are due to their excessive or accidental use; usage at an inappropriate stage of production, processing, or storage; or from a lack of purity or quality. Extensive research has been conducted on many additives to show they are safe for consumption and are effective for their prescribed function. For other additives, widespread and long-term use by food processors has demonstrated safety and efficacy. For the majority of additives used in the United States, the federal government has identified the source, prescribed use, or quantities permissible in foods. The next section discusses the regulation of food additive use and how the benefits and risks of additive consumption influence these regulations.

III. REGULATION OF FOOD ADDITIVE USAGE

A. Federal Authority

Numerous research studies have confirmed that appropriate use of many additives is safe for human health and provides a benefit to the processor, preparer, or consumer. Other additives are considered safe and efficacious due to successful widespread use over many years. Multiple federal agencies, laws, and regulations work together to ensure the safety and efficacy of thousands of food additives, but there are great differences in how many of these additives are regulated. The Food, Drug and Cosmetic Act of 1938 gives the U.S. Food and Drug Administration authority to regulate foods, ingredients, and their labeling. The 1958 Food Additives Amendment to the FD&C Act requires FDA approval for the use of a new additives prior to its inclusion in food. Also this Amendment requires the additive manufacturer to prove an additive's safety for its recommended use.

Food additives are defined as substances which may, by their intended uses, become components of food, either directly or indirectly, or which may otherwise affect the characteristics of the food. The term specifically includes any substance intended for use in producing, manufacturing, packing, processing, preparing, treating, packaging, transporting, or holding the food, and any source of radiation intended for any such use (FDA, 1998a). This definition does not include some classes of additives such as pesticide chemicals for raw agricultural products, new animal drugs, certified colors, or colors exempt from certification. These additives are similarly regulated, but under other laws or acts. Additionally, food additives that have a history of safe usage are exempted from the regulation process described in the Food Additives Amendment. One group of these additives is known as ''generally recognized as safe'' (e.g., salt, vitamins). They have been generally recognized, by experts, as safe based on their extensive history of use in foods before 1958, or based on published scientific evidence. Another group of exempted additives is designated as prior sanctioned additives. These are substances that the FDA or U.S. Department of Agriculture (USDA) had determined were safe for use in specific foods prior to 1958 (e.g., sodium nitrite to preserve meat).

Other important federal regulation of food and color additives is described in the 1960 Color Additive Amendment to the FD&C Act, the 1990 Nutrition Labeling and Education Act, and the 1996 Food Quality Protection Act. Current good manufacturing practice (CGMP) regulations can limit the quantity of food and color additives used in production. Manufacturers may only use the amount of an additive necessary to achieve a desired effect.

In the case of pesticides for foods and animal feeds, the U.S. Environmental Protection Agency and the FDA have regulatory responsibilities for the approval and monitoring of pesticide use. Three major legislative acts describe the authority of the EPA and FDA to regulate pesticide chemicals (EPA, 1998). The Federal Insecticide, Fungicide and Rodenticide Act prescribes that the EPA is responsible for registering or licensing pesticide products for use in the United States. The Federal Food, Drug and Cosmetic Act governs the maximum level of pesticide residues allowed in or on specific human foods and animal feeds. In 1996, the Food Quality Protection Act (FQPA) amended these two acts in several respects. For example, to assess the risks of pesticide residues in foods or feeds, the FQPA requires that the combined exposure from dietary and other nonoccupational sources be considered. Also, when setting new or adjusted tolerances for pesticides, the EPA must consider any special risks to infants and children.

B. Monitoring

Federal regulations are in place to determine if an additive is safe and effective under its prescribed uses. Also, regulations may stipulate the permitted food uses or usage levels for hundreds of additives. Nevertheless, the usage and labeling of food additives may be monitored through analytical testing and label review of finished products or ingredients. Improved chemical analytical methodology has resulted in faster extraction and identification of additives from foods. Many testing protocols can now detect chemical concentrations of very low levels (e.g., parts per billion) and with increasing discrimination.

Products imported into the United States are more frequently examined than domestic products by federal authorities for compliance to food additive regulations. Processed foods from other countries may contain additives not permitted for use in the United States. Or the labeling of additives may be lacking or use descriptions that are unfamiliar to American consumers.

C. Labeling

The risks or benefits of food additives and ingredients must be clearly displayed for consumers. The FD&C Act requires, in virtually all cases, a complete listing of all the ingredients of a food. The Nutrition Labeling and Education Act, which amended the FD&C Act, requires most foods to bear nutrition labeling and requires food labels that bear nutrient content claims and certain health messages to comply with specific requirements. Two of the exemptions from ingredient labeling requirements have resulted in special product labeling efforts to protect the health of consumers. First, the act provides that spices, flavorings, and colorings may be declared collectively without naming each one. One exception is the artificial color additive FD&C Yellow #5. This chemical must be specifically identified in the ingredients statement of finished foods because a small percentage of the population may be allergic or sensitive to the additive. Second, FDA regulations exempt from ingredient declaration incidental additives, such as processing aids, that are present in a food at insignificant levels and that do not have a technical or functional effect in the finished food. One important example of an incidental additive is peanuts. An increasing number of products are identified that they "may contain peanuts." While peanuts or peanut-derived ingredients were not intentionally added to these products, residues from peanut use in processing on nearby equipment or previous production runs may have contaminated these products with peanut residues. Since peanuts are one of the leading causes of allergenic responses to foods, many companies have chosen to label some products with "may contain peanuts." Some processors have elected to process products with peanuts in separate locations from their product lines that are not to contain peanuts.

Some foods may be identified or labeled to contain additives that can improve public or individual health. Many ready-to-eat breakfast cereal products are fortified with several vitamins and minerals. A quantity of these substances is added to the cereal so that a consumer may expect to consume 25 to 100% of the recommended daily intake of that nutrient from a defined size serving. Additives that can improve human health are sometimes advertised elsewhere on finished product packaging besides the ingredients list and nutritional label. Advertising claims that imply better health through consumption of a food product additive are permitted for specific additives, ingredients, or inherent compounds. As one example, regular consumption of folic acid can be advertised as beneficial for preventing neural tube defects in newborn babies.

IV. REGULATORY ASSESSMENT OF THE RISKS OF FOOD ADDITIVES

A. Determination of Safe and Effective Quantities for Foods

For some additives, especially vitamins and minerals, the quantity added to a finished food product may not be sufficient to achieve an intended health benefit. Some foods that are commonly consumed by the majority of the population may be supplemented with various vitamins or minerals. A single serving of a fortified food may not provide the same health benefit as regular consumption of a fortified food. For example, iodine is added to table salt to provide a regular source of this important mineral. Table salt is regularly consumed by most Americans, and iodine is often lacking in typical diets. The determination of a beneficial, yet safe, quantity of iodine to add to table salt must consider the typical consumption of salt over the lifetime of the consumer.

While a food additive may provide a benefit to a processor or a consumer when used as intended, the use of an inappropriate quantity may be deleterious to the food or to the consumer. If an additive is used in excess to process a product, then the desired effect may not be achieved, or else there may be an undesirable quality defect attributed to the product. For example, if an artificial sweetener such as aspartame is added to a beverage in a high concentration, then the product may be too sweet and rejected by consumers. Or an excessively high concentration of a chemical additive may lead to an acute illness or injury to the consumer, such as can occur with sulfiting agents. People who are sensitive to additives containing sulfur, especially those with asthma, are at a greater risk to suffer an allergic reaction to high levels of sulfites. Safety determinations of additive use should consider the effects of accidental or intentional consumption of a high quantity of an additive that is beyond its prescribed use level. Toxicological principles for assessing the safety of food additives have been developed (FDA, 1993).

Also an additive may be injurious to health when consumed in moderate doses over extended time periods. When the risks or benefits of food additive use are considered, estimation must be made of the long-term or lifetime consumption of the additive. These substances may have cumulative effects on health, may interact with other biological or chemical compounds in the body, or may elicit different responses in consumers of different ages or health status.

B. New Additive Approval

The U.S. Food and Drug Administration determines if new additives will be permitted for food use, or if existing additives can be used in additional food products or for new functions (FDA, 1998b). Usually, a commercial company will ask the FDA to approve a new additive or additive use through the submission of a new additive petition. For example, in 1996 the new additive olestra was approved for food use after the Proctor & Gamble Company filed a petition with the FDA. Olestra consists of sucrose esterified with medium and long chain fatty acids, and is approved as a replacement for fats and oils in some foods (FDA, 1998c). The FDA evaluates new additive petitions to determine if substances are considered safe for addition to the food supply and maintains a database for evaluation and monitoring of these substances. This petition must detail all pertinent information concerning the additive including the chemical identity and composition; its physical, chemical, and biological properties; and information regarding possible byproducts or impurities. Other information that must be included relates to method of preparation, identify of the manufacturer, determination of stability, proposed uses or concentrations of use, and methods of analysis or recovery. Most importantly, data must be presented that establish that the additive will achieve the intended effects and that those investigations have demonstrated the safety of the additive for human consumption.

C. Risk Assessment

The determination of the appropriate uses and concentrations of additives to allow in foods can be a complex process. Directly or indirectly added additives may have a demonstrated usefulness to a food processor or consumer, but they may also have harmful toxicological effects when consumed in excess quantities or by sensitive population groups. To determine if a food additive can become a health hazard to the consumer, the inherent toxicity of the additive and the typical consumer consumption or exposure must be estimated. Both the short-term and the lifetime cumulative exposures should be considered since some additive uses may result in acute or chronic effects detrimental to health. To protect individuals from the possible adverse effects of these substances, studies to assess the risk of exposure to chemical residues should be performed. The basic components of a risk assessment include hazard identification, dose-response assessment, exposure assessment, and risk characterization (NRC, 1980, 1983; Winter and Francis, 1997).

Hazard identification is the process where specific chemicals are causally linked to the exhibition of particular health effects. These may include illnesses, birth and developmental defects, and reproductive abnormalities. Also it must be determined if consumption of a chemical could lead to the development of cancer. Cancer usually develops after long-term exposure to a carcinogenic substance. Other adverse health effects may be observed after a short-term or high-level exposure to an additive. In other words, the use of some additives may only be hazardous when consumed in specific quantities.

A dose-response assessment is used to predict the relationship between human exposure to the chemical and the probability of adverse effects. For carcinogens, it is assumed that no threshold level of exposure may exist, which implies that carcinogens may be hazardous when consumed in any quantity. For noncarcinogenic hazards, toxic effects will not generally be observed until a minimum, or threshold, dose is reached. Toxicology studies may be designed to identify the dose just above the threshold where effects are seen (lowest observed effect level [LOEL]) and the dose just below threshold at which no effects are seen (no observed effect level [NOEL] or no observed adverse effect level [NOAEL]). Often, an uncertainty factor has been applied to the NOEL to give a value known as the acceptable daily intake (ADI). This term may be expressed as the acceptable chemical exposure per amount of body weight per day. The ADI is usually calculated as either the NOAEL divided by 100, when the the NOAEL is derived from animal studies, or as the NOAEL divided by 10 when the NOAEL relates to human data (Renwick, 1996;

WHO, 1987). The decision to incorporate a specific quantity of an additive should consider not only the ADI level, but also the minimum amount that is deemed necessary to achieve a desired technical effect.

Exposure assessment is necessary to predict the likely amount of human exposure to an additive. For many foods and food additives it can be difficult to determine how much may be consumed by a particular population and how consumption varies among individuals. Some population subgroups may be exposed to greatly different quantities of food additives. Food consumption patterns vary greatly due to consumer age, gender, ethnicity, socioeconomic status, health status, and so on. Additionally, the quantity of additives present in some foods may be a known or unknown quantity, or could vary greatly. While the concentration of many food additives may be known or relatively constant, the quantity of indirect additives or pesticide residues that remain on food at the time of consumption may vary considerably and be difficult to predict.

Risk characterization describes the origin, magnitude, and uncertainties of estimates of the health risk. Considerations for evaluating the overall risk of using a food additive must include whether it has specific hazardous properties, a prediction of the likelihood of adverse effects based on exposure, and an estimation of the amount of exposure. For noncarcinogens, risk characterization typically relates the estimated exposure to the acceptable daily intake. Exposures at the level of the ADI represent a very low risk. Increasing chemical exposures above the ADI would result in an increasing risk or increased probability of an adverse health consequence. For carcinogens, estimated cancer risks are obtained by multiplying exposure estimates by cancer potency factors. This practice often results in numerical cancer risks that may describe the frequency of people (e.g., 1 in 1 million) who would be expected to develop new cancers after long-term exposure to the food additive. Due to the considerable uncertainties and wide ranges of data used to estimate risks, a risk characterization should include qualitative evaluations of risk (Winters and Francis, 1997).

V. CATEGORIZATION OF RISKS AND BENEFITS

A. Categories of Risk

What are the risks associated with our food supply? This can be a very confusing question because of our inability to adequately measure the risks associated with different components. Most foodborne risks are not associated with food additives. Roberts (1981) and Wodicka (1977) have categorized the major hazards associated with foods, including additives, into five groups, ranked in order of importance: (1) foodborne hazards of microbial origin, (2) nutritional hazards, (3) environmental contaminant hazards, (4) foodborne hazards of natural origin, and (5) food and color additive hazards.

The public perception of the risks associated with foods is often in the reverse order of the list above (Oser, 1978). Therefore, it is important to examine each of these areas to gain a better understanding of the total food safety problem. While the major focus of this book is on food additives, an examination of all of the risks associated with the food supply is necessary to provide a perspective on the comparative risks associated with food additives.

The most prevalent hazard associated with food is foodborne disease of microbial origin. Microbial contamination can result from poor sanitary control during preparation and/or storage in the home, food service facility, or food processing plant. The four primary factors that contribute to outbreaks of foodborne illness are holding food at the wrong temperature (includes inadequate cooling), inadequate cooking, use of contaminated equipment in handling food, and poor personal hygiene by the food handler (Marth, 1981).

Foodborne diseases of microbial origin are important in food safety because of their wide diversity. These microbial illnesses can range in severity from the very severe, like botulism, to milder illnesses such as staphylococcal food poisoning. Foodborne hazards of microbial origin pose the greatest risk to infants, to the elderly, and to debilitated persons. Listeriosis, associated with *Listeria monocytogenes* infection, is an excellent example of a foodborne disease of opportunistic origin that primarily affects compromised individuals. The foodborne diseases of microbial origin are readily recognizable and are often easily diagnosed provided that a sample of food remains to confirm its contamination with the suspect organism. The establishment of clear cause-and-effect relationships for this category of foodborne disease has been an important factor in its assumption of the top ranking in our classification.

The second major risk associated with food is nutritional hazards. Nutritional hazards have earned this high ranking because their adverse effects can come from either deficiencies or excesses in nutrient intake (Stults, 1981). The majority of nutritional hazards come from an improper balancing of the food intake in the diet. Diseases caused by nutritional deficiencies such as scurvy (vitamin C), pellagra (niacin), rickets (vitamin D), beriberi (thiamin), and goiter (iodine) are probably the most widely known hazards associated with insufficient nutrient intakes. These diseases were prevalent in the United States in the early twentieth century, but with nutrient fortification of certain foods such as milk and table salt, improved dietary intake, and improved distribution and storage of perishable foods, they have been virtually eliminated.

At the other end of the spectrum are the hazards associated with the consumption of excessive amounts of the fat-soluble vitamins and some of the trace elements. It is important when discussing the toxicity of vitamins to differentiate between the fat-soluble vitamins (A, D, E, K) and the water-soluble vitamins (C and the B vitamins). Since the fat-soluble vitamins are stored in body fat, excessive intake of these vitamins, especially vitamins A and D, might result in accumulation with toxic side effects. On the other hand, excess amounts of the water-soluble vitamins are usually excreted in urine and sweat (Stults, 1981), although mild cases of toxicity are occasionally reported. The degree of toxicity of the trace elements is greatly affected by their interactions with one another. For instance, it is known that toxic amounts of iron can interfere with the absorption and utilization of copper, zinc, and manganese, and that excessive amounts of manganese can interfere with the absorption of vitamin B-12 (Davies, 1978).

For most healthy individuals, consumption of an adequate and varied diet presents no significant risk for these nutritional excesses and deficiencies. Consequently, one might argue that nutritional hazards should not be ranked second on the list of concerns for food safety. However, because of the common occurrence of heart disease, cancer, and stroke and the possible involvement of dietary factors in these diseases, an argument could be made for elevating nutritional hazards to the top ranking and demoting the microbial hazards to the second ranking. Furthermore, in many parts of the world, poor food distribution systems, inefficient manufacturing and inadequate storage facilities prevent many people from receiving an adequate and varied diet.

Dietary intakes of cholesterol and saturated fats may contribute to the development of coronary heart disease, but there is not universal agreement on this point. Other risk factors such as smoking, genetics or family health status, obesity, low economic status, lack of exercise, and stress appear to be more important risk factors than dietary cholesterol and saturated fats. High blood cholesterol levels undeniably increase the risk of heart disease, but there is wide variability between dietary intakes of cholesterol and saturated fats and blood cholesterol levels. Dietary intake of fat, calories, and fiber may have some role in the likelihood of development of cancer. However, there is not universal agreement on this point either. Other risk factors, such as genetic predisposition, exposure to environmental carcinogens, and smoking, appear to be much more important than dietary intakes of fat, calories, or fiber in the development of cancer. Sodium intake may be an important factor in the development of hypertension, which plays a role in both stroke and heart disease. However, sodium is not the only critical factor in the development of hypertension, and control of sodium intake alone is likely to have a minimal benefit on most hypertensive individuals. Although heart disease, cancer, and stroke are the leading causes of death in the United States, the role of foods in the onset of these illnesses is difficult to quantify. There is no compelling evidence to suggest that food additives or nutrients play a major role in these illnesses, although they may be a contributing factor. Because of the uncertainty and controversy regarding the role of diet in these chronic diseases, the nutritional hazards associated with foods seem to merit, at best, a ranking secondary to the microbial hazards (Stults, 1981). Depending on the final judgment on the role of dietary factors in chronic diseases, this ranking could change in either direction.

In third place on the list of hazards associated with foods are environmental contaminants. Environmental contaminants can find their way into the food supply by the release of industrial chemicals or from natural sources. Although this category contains chemical substances of a quite diverse nature, there are some common characteristics. For example, these contaminants often persist in the environment and resist degradation. These chemicals tend to have a slow rate of metabolism and elimination, which could result in their accumulation in certain body tissues. Also, certain environmental contaminants can accumulate in the food supply, such as mercury in swordfish and shark or neurotoxins in shellfish.

Some of the environmental contaminants that pose a hazard to the food supply are polychlorinated biphenyls (PCBs), dioxins, mercury, and lead. PCBs and mercury have been associated with disease in humans due to the consumption of contaminated fish (Munro and Charbonneau, 1981). Contaminants from natural sources usually come from the erosion of rock formations or from soils with naturally high levels of certain substances. The major contaminants of natural origin are mercury, arsenic, selenium, cadmium, and tin. Pesticides and drug residues in food-producing animals are also included in this category.

When considering guidelines for the control of environmental contaminants, one must remember that toxicity is a function of dose. Therefore, one must know the level of the contaminant in the food and the amount of that food that is normally eaten. This can become quite complicated, but regulatory action levels for some chemical residues in foods have been established. These permitted levels appear to help minimize human exposure to particular contaminants that might be in foods. However, with certain environmental contaminants, such as mercury, the action levels have been set at the level of natural occurrence, minimizing exposure and allowing no room for industrially derived contamination or impact of dose. In other cases (PCBs), action levels have been set on the basis of knowledge of toxicity and calculation of risk, which permits more flexibility for determination of hazardous levels.

Naturally occurring toxicants rank fourth on the list of major foodborne hazards (Rodricks and Pohland, 1981). Since these contaminants seem to cause problems only under certain extreme conditions, they are ranked relatively low. They also rank low because public opinion seems to view ''natural'' risks with much less alarm than manmade contaminants (Rogers, 1983). Some of the more common naturally occurring toxicants found in foods are oxalates in spinach, glycoalkaloids in potatoes, mercury in swordfish, mushroom toxins, mycotoxins, and marine toxins. Certain other compounds like biologically active amines and nitrosamines that can be produced during food storage, processing, or preparation can also pose a food hazard.

Whether a chemical is synthetic or ''natural'' has no bearing on its toxicity. Therefore, this area of food safety concerning naturally occurring contaminants needs to be explored further to determine if the risk from natural contaminants is really low. Naturally occurring seafood toxins and mushroom toxins are relatively common causes of acute foodborne disease (Hughes et al., 1977). The effects of human exposure to natural toxins are difficult to study because consumption of naturally occurring toxins is variable and often cannot be determined. Also, excessive natural toxin consumption generally results in long-term or chronic illness whose source can be difficult to trace.

Ranked below the other four food hazards is the risk obtained from food additives. The GRAS ingredients would be included in this classification. Although GRAS ingredients are not legally food additives, the public perceives no distinction. This class includes thousands of substances. Any potential hazard to humans from a certain food additive depends on the toxicity of the food additive and the level at which the additive is ingested. The four most widely used direct food additives, which account for 93% by weight of all the direct food additives, are sucrose, salt, corn syrup, and dextrose (Clydesdale, 1982; Roberts, 1981). Human exposure to indirect additives is difficult to measure, but this exposure is minimal.

The majority of direct food ingredients are used on the basis of a determination that they are GRAS or prior sanctioned. Review of some items on the GRAS list has indicated that the majority present no significant hazard with normal use, although only a small percentage have been thoroughly evaluated (Roberts, 1981). The other direct food additives used in foods have been approved, and their uses are regulated by the FDA.

Why, then, does the public view food additives and certain GRAS ingredients with such concern? Both Roberts (1981) and Oser (1978) speculated that the problem with the public perception of food ingredients is that these foodborne substances must be proven ''safe.'' It is impossible to ensure the complete safety of any substance for all human beings under all conditions of use. Therefore, any uncertainty about the safety of a food additive can result in the public suspicion of a much greater risk. The recently approved ingredient olestra (brand name Olean), a noncaloric fat replacer, is just one example where debate continues about the safety of its use in foods (ACSH, 1998). Olestra is derived from sugar and vegetable oil, but its molecules are too large to be digested or absorbed in the body. Like fats and oils, it can add taste and texture to savory snack foods, but no calories are provided. Many consumers have reported gastrointestinal distress from eating products with olestra, and others are concerned that this product will prevent the absorption of fat-soluble vitamins and carotenoids into the body. In one study, participants who believed that they were eating olestra snack chips reported gastrointestinal symptoms approximately 50% more than participants who believed that they were eating regular chips, regardless of the type of chip they were actually eating. The reporting of symptoms may have been influenced by what the participants thought they were eating and their knowledge of the product labels that declare that olestra may cause abdominal cramping and loose stools (Sandler et al., 1999). Several scientific studies have shown that consumption of typical diets containing the fat replacer olestra does not lead to gastrointestinal discomfort and that any loss of vitamin uptake by the body would only occur with the meal that contained olestra formulated products. FDA regulators have concluded that olestra is intrinsically safe when consumed as part of a well-balanced and varied diet. Nevertheless, we can expect continued controversy over the safety and uses of olestra.

Equally as important, the public views food additives and certain GRAS ingredients as unnecessary, involuntary sources of risk. The benefits of these food ingredients are not widely appreciated. Also, these food ingredients are not viewed as natural or normal food components, which heightens suspicions in some consumers. Since additives can increase both the quantity and the quality of foods, they will always be used. Therefore, the FDA must continue to review the use of food additives in order to assure the public of the safety of food additives.

B. Categories of Benefits

The benefits derived from the food supply generally fall into four categories: (1) health benefits that reduce some health risk or provide some health benefits such as improved nutrition, (2) supply benefits relating to abundance, diversity, and economic availability, (3) hedonic benefits that provide sensory satisfaction, and (4) benefits that lead to increased convenience (Darby, 1980; Food Safety Council, 1980). Food additives can play an important role in each of these categories of benefits by improving health, increasing supply, enhancing appeal, or improving convenience. Of these benefits, health benefits should be given the greatest consideration, while supply benefits are second in importance, and increased convenience and improved appeal are the least important.

Health benefits of two types may be provided by food additives and other food components: those that prevent or reduce the incidence of specific diseases and those that provide enhanced nutrition. Nitrites have antibotulinal effects and may thus reduce the risk of botulism in cured meats. Nutritional benefits accrue primarily from the presence or addition of nutrients. Nutritional wholesomeness is increased by the enrichment and fortification of certain staple foods, such as bread, milk, and salt, with vitamins and minerals. Fortification with vitamins and minerals could be viewed as preventing deficiency diseases such as scurvy, beriberi, or goiter. Of course, excessive fortification of foods with certain nutrients can increase risks as noted earlier. Supply benefits are also enhanced by the use of food ingredients that prevent the spoilage of foods, increase the yield of processing techniques, or provide new sources for desired functions. Preservatives prevent food spoilage and thus increase supply and lower costs. Preservatives also have indirect health benefits by protecting nutrients, preventing the growth of hazardous microbes, and helping to ensure the availability of an abundant and nutritious food supply. In 1999 the U.S. Food and Drug Administration announced that it would provide an expedited review of a new food additive petition if the additive is intended to significantly decrease foodborne human pathogenic organisms or their toxins.

Hedonic benefits include improved color, flavor, and texture to enhance consumer appeal. Convenience benefits accrue from those components of foods that result in time savings during preparation. These benefits usually assume greater importance in affluent societies.

C. Striking a Balance

As noted above, the quantitation of the degree of risk or benefit associated with a food additive is hardly an exact science. With our current regulatory statutes, the entire emphasis is placed on risk assessment with no consideration given to benefits. While risk–benefit approaches have never been used, sometimes it may be possible to simply evaluate the net risk by comparing the risk of using an additive with the risk of not using it. This risk–risk approach is more acceptable because risk is much more amenable to quantitation than benefit; it is much easier to balance one risk against another than to balance risks against benefits, and risk assessment is a well-accepted regulatory concept. Furthermore the risks of additive use are generally compared by their detrimental effects to human health. While a benefit of some food additives is to enhance health status or prevent disease, most benefits reflect economic considerations for food processors and sensory attributes and convenience for consumers. Thus, an adequate comparison of risk and benefit for every food additive can be difficult to perform and difficult to quantitate (IFT, 1988).

An important concept in either the risk–benefit or risk–risk approach to decisionmaking is the concept of a defined, socially acceptable risk level (Food Safety Council, 1980). Some level of risk is inherent with any chemical. But often there is disagreement about the degree of concern about health risks due to chemical consumption. Which risks are less tolerable—ones that could cause acute illness versus chronic illness? or risks that may lead to many cases of temporary illness versus a few cases of mortality? or risks that affect children more than adults? The definition of a socially acceptable risk level is particularly needed when reviewing carcinogenicity data from animal experiments and the extrapolation of these data to human experiences. The Delaney clause of the FD&C Act defines the acceptable risk level as zero, but historical experience has shown that consumers will accept higher levels of risk if they perceive an important benefit. In 1996 the Food Quality and Protection Act repealed the Delaney Clause with respect to pesticide residues in foods. This legislation instituted a ''reasonable certainty of no harm'' standard that considers risks from different exposures, risks to different population subgroups, and multiple toxicological effects of pesticides on human health (EPA, 1998; Winter and Francis, 1997).

The FDA has operationally defined acceptable risk from chemical consumption as up to one additional case of cancer per million cases, or 10^{-6}, when that chemical is consumed at typical levels during a lifetime. However, the Food Safety Council (1980) took a more detailed look at the acceptability of risk on a theoretical basis. The Food Safety Council defined four situations that could arise in risk–benefit considerations: (1) where the chemical has no identifiable risk, (2) where the substance has a clearly unacceptable level of risk, (3) where the chemical has a measured risk level that is less than threshold for acceptability, and (4) where the substance has measured risk level that is greater than the threshold for acceptability. The benefits that might accrue from the use of an additive are considered differently in each of these situations.

In the situation where the substance has no identifiable risk, any measurable benefit should allow use of the substance in the food supply. The lack of identifiable risk should not be overly comforting, however. All chemicals have some inherent toxicity. The failure to demonstrate any toxicity may simply mean that appropriate tests have not been conducted. Additional tests or tests conducted at higher dose levels or in different species might identify some risk. In such cases, a review of the risk–benefit situation might be necessary at a later time. Such reviews could be necessary for any acceptable substance in light of new toxicological information.

Substances having clearly unacceptable risks should not be allowed in foods under any circumstances. In this situation, the risks far outweigh the benefits. The decision tree approach to toxicological assessment makes ample use of this concept. A substance demonstrating lethal effects in an acute toxicity test at relatively low dose levels is dropped from consideration without further testing. Usually substances in this category would be identified and rejected before any food additive petition is formulated. For most types of toxic effects, no clearly defined level of acceptability has been established, but clearly unacceptable risks should be evident to any toxicologist.

The situation could theoretically arise where a substance will have some well-documented risk but the degree of risk is below the socially acceptable risk level. In such cases, use of the substance would be allowed according to the logic of the Food Safety Council only if a substantial net benefit was evident (Food Safety Council, 1980). This situation assumes that a socially acceptable risk level has been defined for this particular type of toxic response. Such definitions have not been established for most toxic responses. The Food Safety Council (1980) suggested that substances in this category must be carefully evaluated to ensure that benefits outweighed risks. Health and nutritional benefits would have the greatest impact on such a theoretical comparison. If the benefits were largely those of supply, appeal, or convenience, then much larger benefits would be needed to offset the known risks.

The final situation involves substances where the level of measured risk exceeds the socially acceptable level. Again, the problem arises that the level of acceptability has not been established for any toxicological responses except cancer. However, assuming that such definitions existed, a substance in this category would have to possess a sufficient level of benefits to cause reconsideration of the socially acceptable risk level for that particular case. Presumably, very large health benefits would be one possible offsetting factor. Special regulatory restrictions might be necessary to inform consumers of the possible risks. Saccharin may be a good example. The carcinogenicity of saccharin exceeds the acceptable risk level. The perceived benefits of saccharin accrue to those consumers wishing to control their weight or diabetic conditions. Warning labels have been imposed to alert consumers to the possible carcinogenic hazard. However, further safety evaluation of saccharin led to the repeal of the label requirement in 2000.

Risk–benefit decisions require careful attention to the type and degree of both the risks and the benefits. Such decisions are complicated by the lack of knowledge about how to properly extrapolate risks from animal experiments to human situations, the lack of suitable methods for quantifying benefits, and the problems inherent in comparing health risks with non–health risks.

The use of the risk–risk approach to these decisions can alleviate two of the three complications detailed above. In risk–risk approaches, health risks are compared to health benefits, and other types of benefits are not given much consideration. In other words, the risk of using an additive is compared to the risk of not using the additive. If an additive has a net positive effect on health, then it would be allowed for use in foods. While this approach is commendable, it cannot be applied easily to most types of food additives since some additives have no health benefits or there are limited risks associated with not using them. However, it can be used in some cases and could provide an interesting perspective. Again, it must be emphasized that such an approach is not mandated under current regulations.

In some cases, the comparative risks are obvious. For example, the risk of using nitrites and acquiring cancer from exposure to nitrosamines must be balanced against the risk of not using nitrites and acquiring *Clostridium botulinum* toxin from cured meat (IFT, 1988). The risk of acquiring botulism is very small, but the illness is often fatal. In other cases, the comparative risks are more obscure or difficult to quantitate. For example, the small risk of using saccharin and acquiring bladder cancer must be weighed against the alternative risks. Theoretically, the alternative risks would be increased consumption of sucrose and a higher risk of all the diseases associated with obesity. However, other non-nutritive sweeteners exist, and the alternative to these sweeteners is not necessarily equivalent sweetness intake with highly caloric sweeteners. The use of saccharin does not prevent obesity.

In some cases, the risk–risk approach is even more difficult to apply. How could it be applied to tartrazine or some other food colorant, for example? Some evidence of risk might be available for use of the food colorant. But it would be very difficult to identify any risks attributed to the lack of availability of the substance. Perhaps consumers would switch to less safe food choices, or perhaps the food industry would select a more hazardous food colorant as an alternative. However, these alternative risks would be difficult to foresee or quantitate. Consequently, the risk–risk approach is likely to be useful only with certain food ingredients.

REFERENCES

ACSH. 1998. *What's the story? Olestra.* http://www.acsh.org/publications/story/olestra/index.html.

Clydesdale, F. M. 1982. Nutritional consequences of technology. *J. Food Prot.* 45:859–864.

Darby, W. J. 1980. The nature of benefits. *Nutr. Rev.* 38:37–44.

Davies, G. K. 1978. Manganese interactions with other elements. In: *National Nutrition Consortium, Vitamin-Mineral Safety, Toxicity & Misuse.* American Dietetic Assoc., Chicago, p. 33.

Dubois, G. E. 1992. Sweeteners, non-nutritive. In: *Encyclopedia of Food Science and Technology,* Hui, Y. H. (Ed.). Wiley, New York, pp. 2470–2487.

EPA. 1998. *Laws affecting EPA's pesticide programs.* http://www.epa.gov/pesticides/citizens/legisfac.htm.

FDA. 1979. *More Than You Ever Thought You Would Know About Food Additives.* FDA Consumer HHS Publication No. (FDA) 79–2115.

FDA. 1993. *Toxicological Principles for the Safety Assessment of Direct Food Additives and Color Additives used in Food* (Redbook II). National Technical Information Services, Springfield, Virginia.

FDA. 1992. Food additives. FDA/IFIC Brochure, Jan. 1992.

FDA. 1993. Food color facts. FDA/IFIC Brochure, Jan. 1993.

FDA. 1998a. Food Additives Permitted for Direct Addition to Food for Human Consumption. Code of Federal Regulations. Title 21, Part 172, Government Printing Office, Washington, D.C.

FDA. 1998b. *Food additives and premarket approval.* http://vm.cfsan.fda.gov/~lrd/foodadd.html.

FDA. 1998c. Olestra. Code of Federal Regulations. Title 21, Part 172.867, Government Printing Office, Washington, D.C.

Food Safety Council, Social and Economic Committee. 1980. Principles and processes for making food safety decisions. *Food Technol.* 34(3):89–125.

Hughes, J. M., Horwitz, M. A., Merson, M. H., Barker, W. H., Jr., and Gangarosa, E. J. 1977. Foodborne disease outbreaks of chemical etiology in the United States, 1970–1974. *Am. J. Epidemiol.* 105:233–244.

IFT. 1988. The risk/benefit concept as applied to food. *Food Tech.* 42(3):119–126.

Marth, E. H. 1981. Foodborne hazards of microbial origin. In: *Food Safety*, Roberts, H. R. (Ed.). Wiley, New York, pp. 15–65.

Munro, I. C., Charbonneau, S. M. 1981. Environmental contaminants. In: *Food Safety*, Roberts, H. R. (Ed.). Wiley, New York, pp. 141–180.

NRC (National Research Council). 1980. Risk assessment/Safety Evaluation of Food Chemicals. Report of the Subcommittee on Food Toxicology. Committee on Food Protection, Food and Nutrition Board. National Academy Press, Washington, D.C.

NRC (National Research Council). 1983. Risk Assessment in the Federal Government: Managing the Process. Report of the Committee on the Institutional Means for Assessment of Risks to Public Health. Commission on Life Sciences. National Academy Press, Washington, D.C.

Oser, B. L. 1978. Benefit/risk: whose? what? how much? *Food Technol.* 32(8):55–58.

Renwick, A. G. 1996. Needs and methods for priority setting for estimating the intake of food additives. *Food Add. Contam.* 13(4):467–475.

Roberts, H. R. 1981. Food safety in perspective. In: *Food Safety*, Roberts, H. R. (Ed.). Wiley, New York, pp. 1–13.

Rodricks, J. R., Pohland, A. E. 1981. Food hazards of a natural origin. In: *Food Safety*, Roberts, H. R. (Ed.). Wiley, New York, pp. 181–237.

Rogers, E. M. 1983. *Life Is in the Balance: Weighing the Questions of Risk and Benefit in Today's World.* Dow Chemical Report No. 233-10-83, Dow Chemical Company, Midland, MI.

Stults, V. J. 1981. Nutritional hazards. In: *Food Safety*, Roberts, H. R. (Ed.). Wiley, New York, pp. 67–139.

Sandler, R. S., Zorich, N. L., Filloon, T. G., Wiseman, H. B., Lietz, D. J., Brock, M. H., Royer, M. G., Miday, R. K. 1999. Gastrointestinal symptoms in 3181 volunteers ingesting snack foods containing olestra or triglycerides. *Ann. Intern. Med.* 130:253–261.

WHO. 1987. Principles for the safety assessment of food additives and contaminants in food. Environmental Health Criteria #70. World Health Organization, Geneva.

Winter, C. K., Francis, F. J. 1997. Assessing, managing, and communicating chemical food risks. *Food Technol* 51(5):85–92.

Wodicka, V. O. 1977. Food safety—rationalizing the ground rules for safety evaluation. *Food Technol.* 31(9):75–79.

4

Food Additives and Hypersensitivity

MATTI HANNUKSELA

South Karelia Central Hospital, Lappeenranta, Finland

TARI HAAHTELA

Skin and Allergy Hospital and Helsinki University Hospital, Helsinki, Finland

I. DEFINITIONS AND PREVALENCE OF HYPERSENSITIVITY REACTIONS

Many people have uncomfortable reactions to various foods, and this is often considered to be due to food additives and other artificial chemicals, such as pesticide residues, rather than to the food itself. This discomfort is usually called allergy by nonmedical persons, but there are also nurses and doctors who use the term *allergy* when they mean various kinds of untoward reactions to foods without knowing the mechanisms of such reactions.

Strictly, an *allergy* is a harmful physiological reaction caused by an *immunologic mechanism*. If the mechanism is not an allergic one but the reaction resembles allergic reaction, the term *intolerance* is often used. The symptoms mimic those seen in allergic reactions, but the amount of agent producing the reaction is small enough not to cause a *toxic reaction*. When the mechanism is not known, it is better to talk about *hypersensitivity*, which can mean both allergic reaction and intolerance (Tables 1–3).

There are several foods and food additives that can cause both immunologic and nonimmunologic reactions indistinguishable from each other. For example, fish can act both as nonspecific histamine liberators and as true allergens. Tuna, mackerel, and certain cheeses contain histidine and tyrosine to such an extent that histamine and tyramine produced by decarboxylation from them can cause allergic-type symptoms in atopic people (Royal College of Physicians and the British Nutrition Foundation, 1984). Among food additives, cinnamon and nitrogen mustard are the most well-known causes of both allergic and nonallergic reactions.

Table 1 Terminology of Hypersensitivity Reactions

Hypersensitivity	A small amount of a substance produces symptoms that can be objectively verified and repeated.
Allergy	Immunologic mechanisms are involved in the pathogenesis of symptoms.
Atopic allergy	The reaction is mediated by immunoglobulin E.
Intolerance	A small amount of a substance produces a reaction similar to or closely resembling a true allergic reaction, but immunologic mechanisms are not involved.

Table 2 Allergic Reactions

Allergic reaction (mechanism)	Clinical and immunological features
1. Anaphylactic (atopic)	Symptoms (urticaria, allergic rhinitis, conjunctivitis, and asthma; abdominal pain and aches, diarrhea; severe itch and acute worsening of atopic dermatitis) appear usually within minutes and subside within hours.
2. Cytotoxic	Very rarely caused by food additives. Purpura due to destruction of platelets is the most common clinical symptom.
3. Immunocomplex disease	Immunoglobulin G and M form complexes with the allergen (antigen) resulting in purpura, urticaria, arthritis, and other symptoms of immunocomplex disease (serum sickness).
3. Delayed allergy	Expressed usually as delayed-type contact dermatitis. A rash resembling virus exanthema may be produced by ingested allergens. Granulomatosis.

Table 3 Intolerance and Other Nonallergic Reactions

Nonspecific histamine liberation—Cocoa, citrus fruits, strawberry, etc.
Intestinal diseases—Glutein intolerance produces abdominal and skin symptoms.
Primary or acquired enzyme deficiency—Diarrhea and colic in lactase deficiency.
Microbes and their toxins—Bacteria, viruses, yeasts, molds, and fungi cause a diversity of gastrointestinal symptoms.
Psychological causes may underlie various kinds of skin, gastrointestinal, and other reactions.

The prevalence of hypersensitivity reactions to food additives has been investigated in certain diseases, such as chronic urticaria and asthma, suspected to be caused by these chemicals. Juhlin (1981) estimated that 30–50% of patients with chronic urticaria react to one or more food additives. In his studies, however, Juhlin used single-blind challenges. This fact makes his results more or less unreliable.

Farr et al. (1979) reached the conclusion that 4–6% of acetylsalicylic acid (ASA, aspirin) sensitive asthmatics, or about 5000 persons in the United States, react to tartrazine. Stevenson et al. (1986) reviewed the studies on the provocative effect of tartrazine in urticaria; they consider tartrazine to be responsible for worsening of urticaria in occasional patients only and not the cause of chronic urticaria.

Weber et al. (1979) estimated that bronchoconstriction provoked by food dyes and preservatives occurs in about 2% of all asthmatics. Sulfite-induced asthma is thought to affect 5–10% of all asthmatics (450,000 people in the United States) (Stevenson and Simon, 1981). Bush et al. (1986) stated that the figure (5–10%) concerns only severe asthmatics (90,000 people). In their double-blind study on the significance of sulfites in steroid-dependent and non-steroid-dependent asthmatic patients, Bush et al. (1986) showed that the prevalence of sulfite sensitivity in the asthmatic population as a whole would be less than 3.9%, and that these patients are more common among steroid-dependent asthmatic patients.

Juhlin (1981) calculated that 0.5% of all people will show intolerance to aspirin, 0.06% to tartrazine, and 0.05% to benzoates. In Denmark, Poulson (1980) estimated that 0.01–0.1% of the population are sensitive to both tartrazine and benzoates. In France, Moneret-Vautrin et al. (1980b) concluded that 0.03–0.15% of the population experience sensitivity to tartrazine. The Commission of the European Communities (1981) estimated the prevalence of food additive intolerance to be 0.03–0.15% in the whole population. Bronchoconstriction to salicylate, sulfite, and tartrazine has been shown to be dose dependent. The median dose eliciting 15% fall in FEV_1 varied between 0.1 and 0.2 mM, and the most sensitive (5%) asthmatics responded to 4.6 mg metabisulfite, 3.4 mg tartrazine, and 2.6 mg salicylate (Corder and Buckley, 1995).

Of adult patients with hay fever and asthma, 1–2% have experienced harmful symptoms from spices (Eriksson, 1978). Immediate skin test reactions to spices are seen in 20% of atopic patients, especially in those with birch pollen allergy (Niinimäki and Hannuksela, 1981). In oral challenge tests, positive reactions were seen in only 14 of 35 patients with positive skin tests with spices. Most skin prick test reactions to cinnamon and mustard are, in fact, nonallergic, but the proportion of allergic and nonallergic skin test reactions to other spices is much higher than those for cinnamon and mustard.

Madsen (1994) reviewed three papers estimating the frequency of food additive intolerance. In an EEC report (Commission of the European Communities, 1982), based on the results of clinical patient studies of chronic urticaria and asthma, the frequency of additive intolerance was estimated to be 0.13%. In a Danish study (National Food Agency of Denmark, 1980) the estimate for additive intolerance in children was 1%. The third study was a large epidemiological study in England (Young et al., 1987) in which 7.4% of 18,582 responders reported problems with food additives. The results of subsequent interviews and a clinical study including double blind peroral challenges suggested the total prevalence of intolerance to any food additive to be 0.23% in the whole population.

Thus, adverse reactions to ingested food additives seem to be rare and true allergy to them very rare. Nevertheless, there are a small minority of the population who suffer from these reactions. These individuals are often atopic.

II. MECHANISMS

Adverse reactions to food additives are caused by several mechanisms. Food additives are ingested irregularly and in small doses. Additives are usually low molecular weight chemicals, unlike many high molecular weight proteins which are potent allergens. There is very little evidence of an immunological basis in reactions caused by food additives. Adverse effects due to various pharmacological or other mechanisms are much more common.

Many authors avoid the use of the terms *allergy* and *atopy* in connection with food additive reactions and prefer *hypersensitivity* or *intolerance* instead. Terms such as *false allergies* (Moneret-Vautrin, 1983), *allergomimetic reactions*, or *pseudoallergy* (Pearson and Rix, 1985) have been suggested. All these titles emphasize the nonspecific, nonimmunological nature of these reactions.

It has been observed in several studies that there is a dose-response relationship with these reactions. A small amount of the agent is harmless, but a larger amount causes symptoms. This better suits the concept of intolerance than immunologic sensitization.

Many patients suffering from food additive reactions have an atopic constitution and symptoms such as eczema, rhinitis, and asthma. Atopic subjects are more sensitive to numerous irritating environmental factors, including food additives, than the nonatopic population. Atopic subjects have a particular tendency to release histamine after certain foods and additives that possess a histamine-releasing action.

There appear to be more reactions caused by additives in adults than in children. The reason is obscure, but cumulative or slow development of intolerance may occur. On the other hand, adults often have psychosomatic symptoms and are influenced by television programs and articles in journals and newspapers dealing with the adverse effects of food additives.

A. Basis of Sensitization in Atopic Allergy

During the first year of life, allergic sensitization occurs readily via the intestinal tract because the protective immunologic mechanisms are not fully developed. Breast milk offers passive protection during the vulnerable age but is also a possible route of sensitization.

Overproduction of IgE is characteristic of atopic allergy. Many healthy children develop IgE antibodies, against egg white for instance, during the first year of life. These antibodies usually disappear in a few months and are probably a reflection of the normal humoral immunoresponse. However, in those children who manifest a clinical allergy to egg white, antibodies disappear more slowly. It has been postulated that the function of the cells, especially lymphocytes, controlling IgE synthesis is not in balance. Atopy may be related to deficient or retarded maturation of cells suppressing IgE synthesis.

The protective mechanisms of the intestinal mucosa, although their exact nature is not known, prevent sensitization effectively (see also Section VIII). From animal studies it is known that mice develop antibodies when given cow's milk protein parenterally. However, if the animal is fed cow's milk before the injection, it develops immunotolerance and no antibodies can be detected. The role of mucosal IgA is vitally important. Quantitative or qualitative deficiencies of IgA enhance the manifestation of circulating antibodies to foreign proteins as well as the formation of immunocomplexes. If the permeability of the gastrointestinal or respiratory tract is increased for some reason, allergens penetrate the mucous membranes more easily and induce an immunologic response. The function of the intestinal mucosal barrier is disturbed by infections, dietary habits, ingestion of alcohol and drugs, and so forth.

B. Immunologic Reactions to Food Additives

Some low molecular weight additives can act as haptens and convert to allergens only after they bind to some carrier protein (Chafee and Settipane, 1967). In particular, it has been suggested that azo dyes act in this way. The same phenomenon is observed with drugs or frequently with their metabolites. Experimentally, the existence of IgE antibodies against tartrazine has been demonstrated by Moneret-Vautrin et al. (1979) but not by other investigators.

There is no evidence that untoward reactions to food colors are mediated through specific IgE antibodies normally implicated in type I allergic reactions (Miller, 1985), with the exception of carmine, a natural red dye (Quirce et al., 1994; Baldwin et al., 1997).

Weliky et al. (1979) found IgD antibodies specific to tartrazine in six allergic subjects. The role of these antibodies is not known. Tartrazine has also been shown to induce specific IgG antibody production in rabbits when bound to a protein carrier (Johnson et al., 1971). Miller (1985) states, however, that demonstration that a tartrazine conjugate is antigenic when injected and gives rise to a normal humoral response in an animal model does not imply that it can provoke an allergic reaction.

In some sulfite-sensitive patients, skin tests, peroral challenge tests, and passive cutaneous transfer tests with sulfite have been reported to be positive, suggesting that some sulfite-induced immediate reactions might well be mediated by IgE (Yang et al., 1986; Przybilla and Ring, 1987; Gay et al., 1994). Gelatine and modified gelatine have been shown to produce IgE-mediated allergic reaction in a 32-year-old patient after eating fruit gums (gummy bears) (Wahl and Kleinhans 1989).

A delayed type of hypersensitivity (type IV) is possible after ingestion of additives [azo dyes, butylated hydroxytoluene (BHT), butylated hydroxy-anisole (BHA), parabens, etc.], manifesting itself as eczema (Moneret-Vautrin, 1983). The sensitization may originally develop after skin contact with the additive, for example, occupational contact with foods, ointments, or cosmetics.

Delayed contact hypersensitivity has been reported by a number of investigators. A total of 2894 consecutive eczematous patients were patch tested with sodium and potassium metabisulfites by Vena et al. (1994). Positive reactions to 1% metabisulfites were seen in 50 (1.7%) patients. Only two of them were reacting to sodium sulfite 1% pet. Prick and intradermal tests with sodium metabisulfite, 10 mg/mL, were negative. Twelve out of the 50 positive cases were considered clinically relevant. Oral challenges with 30 mg and 50 mg of sodium metabisulfite remained negative.

Immunologic contact urticaria (ICU) reaction is mediated by specific IgE, thus being a type I allergic reaction. The mechanisms underlying nonimmunologic contact urticaria (NICU) are mostly unknown. Histamine is obviously not involved in NICU. The NICU reaction can be abolished with peroral aspirin, suggesting that prostaglandins are responsible at least in part for the wheal and flare reaction.

Delayed contact allergy (type IV immunologic reaction) is the most common mechanism in contact dermatitis due to food additives. Dermatitis may also be due to irritant properties of food additives and be a consequence of ICU or NICU reaction.

1. Histamine and Tyramine Excess

Spoiled foods may contain large amounts of histamine (tinned foods, e.g., fish) and cause urticaria, erythema, headache, and intestinal symptoms that mimic allergy.

Harington (1967) drew attention to tyramine as a cause of migraine and urticaria. Tyramine may cause symptoms by several mechanisms (Bonnet and Nepveux, 1971): excessive consumption of foods that are rich in it (e.g., cheeses), excessive endogenous synthesis by the decarboxylation by the intestinal flora, or incomplete degradation of tyramine.

2. Phenylethylamine

Ingestion of phenylethylamine may lead to headaches. It is present in small amounts in many cheeses and wines and in very low concentration also in chocolate. Three milligrams of phenylethylamine provoked headaches in 12 h in 16 of 38 patients who had suffered from headaches after ingestion of chocolate (Moneret-Vautrin, 1983). However, it seems unlikely that one could eat enough chocolate to reach 3 mg of phenylethylamine.

3. Sodium Nitrite

Sodium nitrite is used as an antioxidant and antimicrobial agent in various foodstuffs, such as cooked pork meats. The authorized and widely accepted daily dose is 0.2 mg/kg. Daily consumption commonly exceeds this figure, particularly in children. Moneret-Vautrin et al. (1980a) have shown that an oral challenge test with 30 mg of sodium nitrite may cause urticaria, intestinal disorders, or headache. This has been confirmed by others [e.g., Henderson and Raskin (1972)]. The mechanisms by which sodium nitrite acts are unclear. It may cause cellular anoxia and inhibit the protective enzymatic activities of the intestinal mucosa. This may lead to increased permeability of the mucosa to other antigens. In addition, sodium nitrite may in some way or other enhance the effect of histamine present in many foods.

4. Alcohol

Alcohol causes vasodilatation and facilitates the rapid passage of foods and additives across the intestinal mucosa. A metabolite of alcohol, acetaldehyde, is a potent histamine releaser. Intolerance to wines is fairly common. Usually only small quantities of a certain wine are enough to induce adverse effects: flush, tachycardia, palpitations, rhinitis or asthma, and skin symptoms (Breslin et al., 1973). The reactions are not immunologically mediated. Moneret-Vautrin (1983) suggested several mechanisms as an explanation:

- The histamine-releasing action of acetaldehyde
- Richness of histamine in certain red or white wines
- Intolerance to benzoates, which are present in great quantities in certain grapes
- Intolerance to quinine, present in many aperitifs
- Intolerance to some preservatives, such as sulfur dioxide

Harada et al. (1981) showed that vasomotor and cardiac problems arise particularly in subjects who have hepatic aldehyde dehydrogenase deficiencies in association with a raised level of blood acetaldehyde.

Acute urticaria attacks provoked by alcohol itself have been occasionally observed (Karvonen and Hannuksela, 1976; Ormerod and Holt, 1983). The mechanism is not known, but a true anaphylactic-type allergy to ethanol or reactions mediated by endogenous opiates and prostaglandins have been suggested (see also Section III.A).

5. Sulfur Dioxide and Sulfites

Sulfur dioxide, sodium sulfite, and sodium and potassium bisulfite and metasulfites are widely used for the preparation, processing, and storage of foods (Settipane, 1984; Przybilla and Ring, 1987). Sometimes they are sprayed on fresh fruits, vegetables, and shellfish, especially in restaurants. Wines may contain sulfur dioxide.

The mechanisms by which sulfur dioxide produces symptoms that mimic allergy are open to speculation. Freedman (1977a) suggested that the occurrence of asthmatic symptoms 1–2 min after ingestion of soft drinks indicates that the response is due to inhalation of gas or rapid absorption across the buccal mucosa. He provoked patients with 25 mL of an aqueous mixture of sodium metasulfite and citric acid that yielded 100 ppm of sulfur dioxide. This concentration is often found in many orange drinks in Europe but not, for example, in the United States. The air above such a mixture has a greater than 1 ppm concentration of sulfur dioxide. It has been shown that inhalation of sulfur dioxide at a concentration of 1 ppm may cause bronchoconstriction in asthmatics (Koenig et al., 1980). Stevenson and Simon (1981) suggested, however, that other mechanisms are also involved. Ingestion of encapsulated potassium metabisulfite produced severe systemic symptoms in some sensitive asthmatic patients. The patients showed no sign of immunologic sensitivity against metasulfites. The presence of a reactive sulfite oxidase deficiency in sulfite-sensitive asthmatics has been suggested (Jacobsen et al., 1984). This may be present in some asthmatics who react to ingested sulfite. Delohery et al. (1984) suggested that ingested metabisulfite may initiate an orobronchial reflex or variable inhalation of SO_2 during swallowing. Koepke et al. (1985) demonstrated that asthmatics sensitive to ingested sulfite developed bronchoconstriction with inhaled sulfite. Furthermore, sensitivity to inhaled sulfite was more common than sensitivity to ingested sulfite.

C. Pharmacological Mechanisms

According to the results of several studies, in 8–25% of patients with asthma or chronic perennial rhinitis the symptoms are provoked by aspirin [acetylsalicylic acid (ASA)]. In patients with chronic or recurrent urticaria the figure is about 20% (Hannuksela, 1983). Aspirin interferes with the synthesis of prostaglandins and leukotrienes. Adverse reactions to aspirin may be caused by the inhibition of cyclooxygenase, which provokes inhibition of prostaglandin synthesis, particularly PGE. It normally acts as a bronchodilator.

It is possible that some food additives have the same kind of action as aspirin on the cyclooxygenase pathway. This is supported by the observation that often the same patients who are intolerant to aspirin also experience adverse reactions to tartrazine, other azo dyes, and benzoates (Samter and Beers, 1968; Champion et al., 1969; Juhlin et al., 1972; Michaelsson and Juhlin, 1973). This opinion has been objected to by Stevenson et al. (1986), who did not find any correlation between aspirin and tartrazine sensitivity. Moreover, tartrazine has not been observed to inhibit the prostaglandin pathways (Gerber et al., 1979). Borgeat (1981) presented some evidence that tartrazine induces asthma via the leukotriene effect.

Williams et al. (1989) studied food additives in tests measuring platelet activation by noradrenaline. All the investigated food additives (ascorbic acid, trisodium citrate, monosodium glutamate, sodium nitrite, tartrazine, sodium benzoate and sodium metabisulfite) inhibited platelet aggregation, and this was associated with inhibition of the cyclooxygenase–thromboxane pathway. The effect of acetyl salicylic acid was increased by suboptimal inhibitory concentrations of the food additives studied.

Other mediators have also been studied, but the evidence is sparse. Neumann et al. (1978) proposed that the activation of bradykinin may be involved. There is some evidence that in aspirin-sensitive chronic urticaria, benzoates, tartrazine, and aspirin cause release of leukocyte inhibition factor (LIF) from mononuclear cells in vitro (Warrington et al., 1986).

Augustine and Levitan (1980) have observed that some coloring agents, xanthines and erythrosin, alter the membrane permeability of neurons in animal studies. They suggested that this action may explain the behavioral disorders that have been attributed to colorants.

Caramel Color III [2-acetyl-4(5)-(1,2,3,4-tetrahydroxybutyl)-imidazole] has been found to cause a reduction in total white blood cell counts in rats and several other effects on the immune system of rodents (Hoube and Penninks, 1994). The use of the color is, however, considered safe for humans.

Monosodium glutamate (MSG), butylated hydroxytoluene (BHT), butylated hydroxyanisole (BHA), and other food additives have been suspected of provoking aggressiveness and mental disorders in children, but the results of various studies are controversial (*Lancet*, 1982). A fall in serotonin and competition with some neurotransmitters have been suggested as possible mechanisms leading to behavioral disturbance (Augustine and Levitan, 1980).

III. REACTIONS IN THE SKIN

The most important skin symptoms are listed in Table 4. In addition to these, exanthematous rashes mimicking virus exanthemas are often suspected to be caused by foods and food additives. Such reactions will not be discussed in detail because they can hardly be distinguished from exanthemas due to several other causes.

A. Urticaria and Angioneurotic Edema

Chronic and recurrent urticaria from ingested food additives has gained more attention than other possible side effects from these chemicals (Table 5).

The provocative effect of aspirin in urticaria was realized in the late 1950s (Calnan, 1957; Warin, 1960). It was soon found that this phenomenon was not restricted to aspirin but that other salicylates were also capable of producing exacerbation of urticaria (Moore-Robinson and Warin, 1967). In various investigations on chronic and recurrent urticaria,

Table 4 Symptoms and Signs in the Skin Caused by Food Additives

Symptoms and signs	Mechanism(s) underlying the reaction
Acute urticaria	Type I allergy (IgE mediated)
Exacerbation of chronic and recurrent urticaria	Type III allergy (immunocomplex disease)
Angioedema	Intolerance
Contact urticaria	
Immunologic (ICU)	Type I allergy (IgE mediated)
Nonimmunologic (NICU)	Mechanism(s) unknown
Purpura	Thormbocytopenia (type II reaction); allergic vasculities (type III reaction); unknown (nonthrombocytopenic purpura)
Fixed and lichenoid eruptions	Type IV allergy?
Primary irritant dermatitis	Irritancy
Allergic contact dermatitis	Contact allergy (type IV)
Photodermatitis	Photoallergy or phototoxicity
Orofacial granulomatosis	Unknown

Table 5 Food Additives Causing or Aggravating Urticaria and Angioneurotic Edema

Benzoic acid, benzoates
Sorbic acid, sorbates
Azo dyes (tartrazine and others)
Canthaxanthine, β-carotene
Annatto
Quinoline yellow
Butylated hydroxytoluene (BHT), butylated hydroxyanisole (BHA)
Nitrites and nitrates
Spices (e.g., cinnamon, clove, white pepper)
Ethanol

the number of aspirin-sensitive patients has been on average 20% (Juhlin, 1986). Natural salicylates are unlikely to cause any trouble even in aspirin-sensitive people.

Swain et al. (1985) measured the salicylate content of our daily foodstuffs, the highest concentrations being 218 mg/100 g in curry powder, 203 mg/100 g in paprika hot powder, and 183 mg/100 g in dry thyme leaves. The authors estimated our daily intake of natural salicylates varies between 10 mg and 200 mg in western diets. The amount may be high enough to produce symptoms in some highly sensitive people.

Azo dyes, especially tartrazine, new coccine, and Sunset Yellow, have been known to be precipitating factors in chronic urticaria since 1959 (Lockey, 1959; Juhlin et al., 1972; Desmond and Trautlein, 1981; Freedman, 1977b). Because of frequent harmful reactions, both urticaria and asthma, azo dyes have been removed from most foodstuffs and medicines in many western countries, and U.S. regulations require specific labeling of FD&C Yellow No. 5 and Yellow No. 6.

In alcoholic beverages there are several agents that can cause urticaria and angioneurotic edema. Salicylates and flavoring substances (Feingold, 1968; Lockey, 1971) are responsible for such reactions more often than ethanol itself. Clayton and Busse (1980) reported a case of anaphylaxis from red and white wine, but the causative chemical remained unknown. A peroral challenge with 20 g of pure ethanol produced urticaria in a female patient who also developed symptoms from acetic acid but not from acetaldehyde or from flavored soft drinks (Karvonen and Hannuksela, 1976). Reports of two other cases of ethanol urticaria were verified with challenge tests (Hicks, 1968; Ormerod and Holt, 1983), and two cases of probable ethanol urticaria have been published (Strean, 1937; Biro and Pecache, 1969). The mechanism of ethanol urticaria is unclear. In the case of Ormerod and Holt (1983), neither H1 nor H1 + H2 antihistamines prevented whealing, but indomethacin, 25 mg/day; Naloxone infusion, 0.4 mg/min; and sodium chromoglycate, 100 mg/day, did. It thus seems that the mechanism might well not be type I allergy.

In recent years, butylated hydroxytoluene (BHT), butylated hydroxyanisole, sorbic acid, canthaxanthine, quinoline yellow, nitrites and nitrates, cloves, and cinnamon have been added to the list of inducers of urticaria (Juhlin, 1981; Warin and Smith, 1982; Hannuksela, 1983) (Table 5).

Angioneurotic edema is a common accompanying symptom in chronic and recurrent urticaria, occurring in two-thirds of the patients, most commonly on the face (lips, eyelids, cheeks) and tongue and on the hands and feet (Champion et al., 1969; Juhlin, 1981). Other symptoms frequently reported by patients with chronic urticaria are vasomotor or intrinsic rhinitis (Settipane and Pudupakkam, 1975) and abdominal troubles, such as aches and pain in the upper abdomen, diarrhea, and abdominal swelling (Juhlin, 1981). Anaphylaxis after an oral challenge test with 500 mg of sodium benzoate was described by Pevny et al. (1981). Orofacial granulomatosis responding well to exclusion diet was found in a young girl who was sensitive to carmosine, Sunset Yellow, and monosodium glutamate (Sweatman et al., 1986).

B. Contact Urticaria

Contact urticaria appears as a wheal and flare reaction on normal or eczematous skin within some minutes to an hour after agents capable of producing this type of contact reaction have been in touch with the skin. In mild reactions there are redness and itching but no wheals. The difference between an immediate irritant (toxic) reaction and contact urticaria is far from clear, and therefore there is much confusion in using terms such as contact urticaria, immediate contact reaction, atopic contact dermatitis, and protein contact dermatitis (Table 6). Immediate contact reaction includes both urticarial and other reactions, whereas protein contact dermatitis includes allergic and nonallergic eczematous dermatitis caused by proteins or proteinaceous materials. Dermatitis is usually seen in food handlers, veterinarians, and slaughterers (Hjorth and Roed-Petersen, 1976). Atopic contact dermatitis is an immediate type of allergic contact reaction in atopic people (Hannuksela, 1980). It may be regarded as a special form of immunologic or allergic contact urticaria.

Contact urticaria is also known as *contact urticaria syndrome* (Maibach and Johnson, 1975), which, in addition to localized urticaria and eczematous dermatitis, also includes generalized urticaria or maintenance of status asthmaticus.

According to the mechanism underlying the urticarial reaction, contact urticaria is divided into two main types, namely immunologic (ICU) and nonimmunologic contact urticaria (NICU). There are still some substances that cannot be classified as agents causing either ICU or NICU.

1. Symptoms and Signs

Subjective symptoms include itching, tingling, and burning sensations on the contact sites. Objectively, wheals and redness can often be demonstrated with an open application test or rub test, or with a use test when examining a patient in the office. In ICU the symptoms and signs appear within a few minutes and in NICU usually in 30–45 min (Hannuksela

Table 6 Terminology of Immediate Contact Reactions

Term	Remarks
Immediate contact reaction	Includes urticarial, eczematous, and other immediate reactions
Contact urticaria	Allergic and nonallergic contact urticaria reactions
Protein contact dermatitis	Allergic and nonallergic eczematous immediate reactions caused by proteins or proteinaceous material
Atopic contact dermatitis	Immediate urticarial or eczematous IgE-mediated immediate contact reaction
Contact urticaria syndrome	Includes both local and systemic immediate reactions precipitated by contact with contact urticaria agents

and Lahti, 1980), but there may be a delay of up to 6 h in rare cases (von Krogh and Maibach, 1982).

Eczematous dermatitis with or without tiny vesicles may result from continuously recurring contact urticaria, but it may also appear without any signs of urticaria. This type of dermatitis cannot be distinguished from primary irritant or delayed allergic contact dermatitis, either clinically or histologically. The patient history is usually suggestive of contact urticaria, and the mechanism can be classified in most cases by using the various skin tests mentioned above.

Generalized urticaria is a rare complication of ICU, and only very strong allergens are capable of producing it.

In addition to the skin, the lips, tongue, oral mucosa, and throat often react with contact urticaria, but the patient usually notices only a vague itching (Hannuksela and Lahti, 1977; Niinimäki and Hannuksela, 1981).

2. Immunologic Contact Urticaria

Of the many food additives, spices are the most common causes of immunologic contact urticaria (ICU) (Table 7). Mustard, cinnamon, and cayenne produce both ICU and NICU reactions, but many other spices, such as cardamom, caraway, and coriander, produce mostly ICU. Persons working in grill bars and pizzerias are exposed to many powdered spices and may get occupational ICU dermatitis on the hands and face. The most common nonoccupational ICU reactions from spices appear as cheilitis, perleche, and eczema around the lips, as well as mucosal edema in the mouth (''lump'' in the larynx), expecially in children (Niinimäki and Hannuksela, 1981). Systemic reactions from ingested spices are to be expected occasionally. Peppers have caused nasobronchial and gastrointestinal symptoms and shock (Rowe and Rowe, 1972), and cinnamon and cloves have caused chronic urticaria (Warin and Smith, 1982; Hannuksela, 1983).

Immediate allergy to various spices is usually connected with birch pollen allergy (Niinimäki and Hannuksela, 1981). There are obviously common antigenic determinants in birch pollen, various fruits such as apples and pears, carrots, celery tuber (Halmepuro and Lowenstein, 1985), and spices.

Cremodan SE40, a food consolidator containing mono-, di-, and triglycerides of fatty acids, sodium alginate, polysorbate 80, guar gum, carrageenan (Camarasa, 1981), and BHT (Osmundsen, 1980) have been reported to be causes of contact urticaria in occasional cases.

3. Nonimmunologic Contact Urticaria

Of the spices, cinnamon, mustard, and cayenne are the most well-known producers of nonimmunologic contact urticaria (Table 8). Balsam of Peru, benzoic acid and benzoates,

Table 7 Food Additives Causing Immunologic Contact Urticaria

Various spices: cinnamon, mustard, cayenne, cardamom, caraway, coriander, etc.
Cremodan SE40 (di- and triglycerides)
Polysorbate 80
Sodium alginate
Guar gum
Carrageen
α-Amylase

Table 8 Food Additives Producing Nonimmunologic Contact Urticaria

Benzoic acid and benzoates
Sorbic acid and sorbates
Cinnamic acid and cinnamates
Cinnamaldehyde
Various spices: cinnamon, mustard, etc.

sorbic acid and sorbates, cinnamic acid and cinnamates, and cinnamaldehyde readily produce contact urticaria when applied to the skin. The concentration capable of causing NICU depends on the vehicle, the skin region exposed, and the mode of exposition (Lahti, 1980; Lahti and Maibach, 1985). Sorbic acid, for example, causes wheal and flare reactions when tested at 5% in petrolatum, but only a few people react to a 0.2% mixture.

The back skin, the face, and the extensor sides of the upper extremities react to NICU agents more readily than other parts of the body, the soles and palms being the least sensitive areas (Lahti, 1980; Gollhausen and Kligman; 1985). Scratching does not enhance the reactivity; neither does the use of occlusion. On the other hand, stripping the skin weakens the contact urticarial reaction. Topical or systemic antihistamines do not influence the skin response to NICU substances, but local anesthesia and the use of strong topical steroids diminish the reactivity of the skin (Lahti, 1980). Systemic aspirin inhibits NICU reactions to most NICU agents (Lahti et al., 1986).

Nonimmunologic contact urticaria reactions are to be expected on the hands of employees handling food additives in the food industry. Such reactions from benzoic and sorbic acids, benzoates, and sorbates are obviously rare, but they are somewhat more commonly seen from spices.

In a kindergarten, salad dressing containing benzoic acid and sorbic acid raised perioral NICU reactions when the children smeared the dressing around their mouths for fun (Clemmensen and Hjorth, 1982). Flavored food was reported to cause generalized urticaria and gastrointestinal symptoms in six patients with a marked contact urticarial reaction to 25% balsam of Peru in petrolatum (Temesvari et al., 1978).

C. Purpura

Allergic purpura appeared 3 h after peroral provocation with 10 mg of tartrazine in a 51-year-old man suffering from cold urticaria and recurrent episodes of purpura (Parodi et al., 1985). The provocation was repeated four times, resulting in purpura each time. Purpura with fever, malaise, and pains in the abdomen and joints in five patients (Michaelsson et al., 1974) and allergic purpura in one patient (Criep. 1971) were apparently caused by tartrazine. A placebo-controlled oral challenge with 50 mg of ponceau red produced cutaneous leucocytoclastic vasculitis in a 24-year-old woman (Veien and Krogdahl, 1991). The vasculitis faded in two months during an additive-free diet.

Purpuric eruptions from quinine are well documented. The mechanism in thrombocytopenic purpura is cytotoxic allergy (type II) (Belkin, 1967). There are also quinine-induced nonthrombocytopenic purpuras (Levantine and Almeyda, 1974).

Table 9 Causes of Primary Irritant Contact Dermatitis (Especially in Bakers)

Emulsifiers
Acetic, ascorbic, and lactic acids
Potassium bicarbonate
Potassium bromate and iodide
Calcium acetate and sulfate
Bleaching agents

D. Fixed, Bullous, and Lichenoid Eruptions

Quinine frequently produces lichenoid eruptions (Almeyda and Levantine, 1971), and less frequently bullous and fixed eruptions (fixed erythema) (Derbes, 1964). Fixed drug eruption from tartrazine was recently published by Orchard and Varigos (1997).

E. Primary Irritant Contact Dermatitis

The usual causes of irritant dermatitis in food handlers are wet work and detergents. However, some food additives are possible irritants and may be partly responsible for hand dermatitis. Table 9 lists some irritants that bakers may encounter (Fisher, 1982).

F. Allergic Contact Dermatitis

Esters of *p*-hydroxybenzoic acid (parabens), sorbic acid and sorbates, BHA, BHT, gallate esters, vitamin E, propylene glycol, spices, essential oils, and other flavoring agents cause delayed-type contact allergy, which may be of significance to food handlers and consumers (Table 10).

Parabens are important contact allergens in cosmetics and other dermatological preparations. They are also used as preservatives in foods and drugs. Although parabens may

Table 10 Food Additives Inducing Contact Allergies

Parabens
Sorbic acid
Butylated hydroxytoluene and butylated hydroxyanisole
Gallates (octyl, lauryl)
Vitamin E (tocopherols)
Propylene glycol
Spices, essential oils, and balsams
Azo dyes
Karaya gum
Benzoyl peroxide
Ethanol
Metabisulfites and sulfites
α-Amylase
Foodmuls E3137

rarely cause allergic contact dermatitis in the food industry, they seem not to cause any trouble in consumers, even in those sensitized (Schorr, 1972).

Sorbic acid has become a more and more popular preservative in bakery products in recent years. Contact allergy is fairly infrequent and is most commonly due to the use of creams containing sorbic acid. Ingested sorbic acid has not been reported to produce or worsen eczematous dermatitis.

Contact sensitivity to BHT and BHA is rare. Roed-Petersen and Hjorth (1976) reported two female patients with hand dermatitis. One patient was allergic to BHT but not BHA; the other reacted to both BHA and BHT. The two patients got rid of their hand dermatitis by using topical corticosteroids and a diet free of antioxidants. Further, these two patients had itching and vesicular eruption when taking 5–10 mg of BHA daily for 4 days, suggesting that antioxidants in foods were the causative factor or at least an exacerbating factor in their hand dermatitis.

Butylated hydroxyanisole was the apparent cause of occupational hand dermatitis, circumoral dermatitis, and cheilitis in a cook who had skin trouble after handling and eating mayonnaise containing BHA (Fisher, 1975).

Lauryl gallate in a margarine was the cause of occupational hand dermatitis in five bakers and pastry cooks investigated by Brun (1964, 1970). Burckhardt and Fierz (1964) also reported five cases of occupational contact allergy to gallates in margarine, and van Ketel (1978) reported one case of occupational hand and face dermatitis from peanut butter containing octyl gallate. Ingested gallates appear to be harmless even in hypersensitive individuals.

Vitamin E (tocopherol) is a potent contact sensitizer. The antioxidant in "E-deodorants" produced so many cases of allergic contact dermatitis in the early 1970s that these deodorants were withdrawn from the market (Fisher, 1976). Ingested vitamin E, however, has not been reported to produce allergic reactions.

Propylene glycol (PG) is a good solvent and vehicle for many substances, in addition to which it is antimicrobic and hygroscopic (isotonic concentration about 2% in water) (Reynolds and Prasad, 1982). According to the FAO/WHO, its estimated acceptable daily intake is up to 25 mg/kg body weight, or 1.75 g/70 kg. The average daily intake of PG is estimated to be about 0.5 g (Andersen, 1983).

Propylene glycol is widely used in dermatological preparations, up to 70% w/w, mostly because of its hygroscopic and dissolving properties. Under occlusion it irritates the skin easily (Warshaw and Herrmann, 1952; Hannuksela et al., 1975). It is difficult to assess whether patch test reactions are primary irritant or allergic. All patch test reactions caused by PG at 2% in water and most of those elicited by 10% PG seem to be allergic and clinically relevant (Hannuksela and Salo, 1986).

Peroral ingestion of 2–15 mL of PG produced an exanthematous reaction in 8 of 10 persons with a positive patch test result with 2% PG in water. The corresponding number among those reacting to 10–100% PG in patch testing was 7 of 28 cases and none of 20 control subjects (Hannuksela and Förström, 1978). In one of the patients, PG in food was the apparent cause of recurrent rashes. In another study, 15 mL of PG produced an exanthematous rash in 2 of 5 patients with a positive patch test reaction to 2–100% PG (Andersen and Storrs, 1982). It thus seems that propylene glycol is a rare but possible cause of exanthematous rash even in amounts ingested in daily food.

Spices and other flavoring agents are numerous and mostly poorly standardized. Many of them are potent contact allergens, but the exact composition of the chemicals responsible for contact sensitization is not usually known. Eugenol, cinnamic aldehyde, cinnamic acid, and cinnamyl alcohol are well-known sensitizers present in various balsams, spices, and fragrances (Hjorth, 1961; Mitchell et al., 1976).

Niinimäki (1984) patch tested 2258 eczema patients, 150 of whom (6.6%) were allergic to balsam of Peru. Further patch tests with nine powdered spices were positive in 3–46% of patients allergic to balsam of Peru but only sporadically in patients not allergic to the balsam. Cloves elicited allergic patch test reactions most often (46%), then came Jamaica pepper (21%) and cinnamon (15%). The other spices tested were ginger, curry, cardamon, white pepper, vanillin, and paprika, being positive in 3–6% of patients allergic to balsam of Peru.

Niinimäki (1984) also made challenge tests with various spices (300 mg in gelatin capsules) in 71 patients with Peru balsam allergy and in 50 dermatological patients with negative patch test results. Seven of the balsam of Peru allergic patients (10%) had objective symptoms regarded as positive challenge test results. Two patients had urticarial lesions, one on the back skin at the patch test site of balsam of Peru and the other on the waist. The latter patient also had facial dermatitis and conjunctival irritation. four patients experienced vesiculation (pompholyx) on the palms and fingers, and in one patient the patch test reaction to Jamaica pepper was reactivated, although the reaction had already faded away prior to the challenge test. None of the 50 control patients had symptoms from ingested spices. Flavoring agents in food caused a flare-up of contact dermatitis in an elderly patient allergic to balsam of Peru reported by Bedello et al. (1982).

Rare causes of occupational allergic hand dermatitis in bakers are benzoyl peroxide, Karaya gum, and azo dyes (Fisher, 1982). Ethanol is also capable of producing contact allergy, and allergic patients may have an erythematous rash after drinking alcoholic beverages (Drevets and Seebohm, 1961; Cronin, 1980).

Metabisulfites and more rarely also sulfites cause both occupational and nonoccupational allergic contact dermatitis (Fisher, 1989; Vena et al., 1994). They are also capable of producing irritant contact dermatitis. An emulsifying agent, Foodmuls E3137, caused hand dermatitis in a 37-year-old male baker, in whom both patch test and scratch-chamber test were positive, the former after 48 and 72 h and the latter after 24 h (Vincenzi et al., 1995). α-Amylase, a flour additive, causes both delayed and immediate skin contact allergies in bakers (Morren et al., 1993).

G. Photodermatitis

Cyclamate is the only food additive that has been reported to cause photodermatitis via an apparently allergic mechanism (Kobori and Araki, 1966; Lamberg, 1967).

H. Orofacial Granulomatosis

Orofacial granulomatosis is a descriptive name for a variety of conditions presenting as chronic or fluctuating swelling of lips and other peroral regions and the oral mucosa. Often there are noncaseating granulomas or at least epithelioid cells in the dermis. These conditions include, for example, Crohn's disease, Melkerson–Rosenthal syndrome, and sarcoidosis.

The cause of orofacial granulomatosis is unknown. In individual cases, food additives have been found to be at least provocative. Food or food additive intolerance was reported as a causative factor in 14 of 80 patients by Patton et al. (1985). Carmoisine, Sunset Yellow, and monosodium glutamate were attributed to orofacial granulomatosis in a 8-year-old girl by Sweatman et al. (1986). The patient responded to an additive-free diet, as did a 15-year-old girl whose symptoms were shown to be related to monosodium glutamate (Oliver et al., 1991). A 26-year-old man with Melkerson–Rosenthal syndrome showed allergic patch test reactions to cinnamaldehyde, cinnamyl alcohol, eugenol, benzoic acid, and sorbic acid, and responded favorably to avoidance of cinnamaldehyde-containing foods (McKenna and Walsh, 1994).

In another study on Melkersson–Rosenthal syndrome, food additives did not show any influence on the symptoms or signs when they were given in double-blind manner to the patients (Morales et al., 1995).

I. Exacerbation of Atopic Dermatitis

Van Bever et al. (1989) investigated the role of food and food additives in severe atopic dermatitis. They made double-blind placebo-controlled challenge tests with four food additives (tartrazine, sodium benzoate, sodium glutamate, and sodium metabisulfite), acetylsalicylic acid, and tyramine through a nasogastric tube in six children. All six children reacted to at least one additive while placebo challenges remained negative. The authors considered the aggravation of atopic dermatitis by food additives to be due to erythemogenic and urticariogenic properties of the additives rather than to direct effect on atopic dermatitis. In a series of 91 adult atopic dermatitis patients, Hannuksela and Lahti (1986) found one positive reaction in double-blind peroral challenge with 100 mg of β-apocarotenal plus 100 mg of β-carotene.

IV. REACTIONS IN THE AIRWAYS

Asthma, rhinitis, and nasal polyposis are the main hypersensitivity symptoms of the respiratory tract.

A. Azo Dyes and Benzoates

In the late 1950s it was discovered that food colorants may provoke asthma (Speer, 1958; Lockey, 1959). Of them, tartrazine has been most studied, and its role in asthmatic reactions was established in the 1960s (Chafee and Settipane, 1967). It is still added to many pharmaceutical products as well as to foods and soft drinks in several countries. Many adverse reactions reported have been associated with medications, pills, capsules, and elixirs. Sodium benzoate and other benzoic acid derivatives can also produce respiratory symptoms. The tartrazine molecule possesses similarities to other azo compounds, benzoates, pyrazole compounds, and the hydroxyaromatic acids, which include salicylate and aspirin (Figure 1) (Miller, 1985).

Usually the subjects examined for tartrazine and benzoate reactions have been highly selected, often asthmatics who are known to be intolerant to aspirin. Aspirin was reported to cause serious reactions in some patients with asthma as early as 1919 (Cooke, 1919). The reported figures of intolerance to aspirin and other nonsteroidal anti-inflammatory drugs (NSAIDs) in patients with asthma differ widely, from 1 to 44% (Vedanthan et al., 1977; Weber et al., 1979). In an unselected population an incidence of 0.5% has been reported (Schlumberger, 1980).

Often patients have rhinitis associated with nasal polyps for years before asthmatic intolerance to aspirin becomes manifest. It is possible that when intolerance to aspirin develops, it also extends to some food additives, which then cause adverse reactions via more or less the same mechanism. However, reactions to food additives also occur without

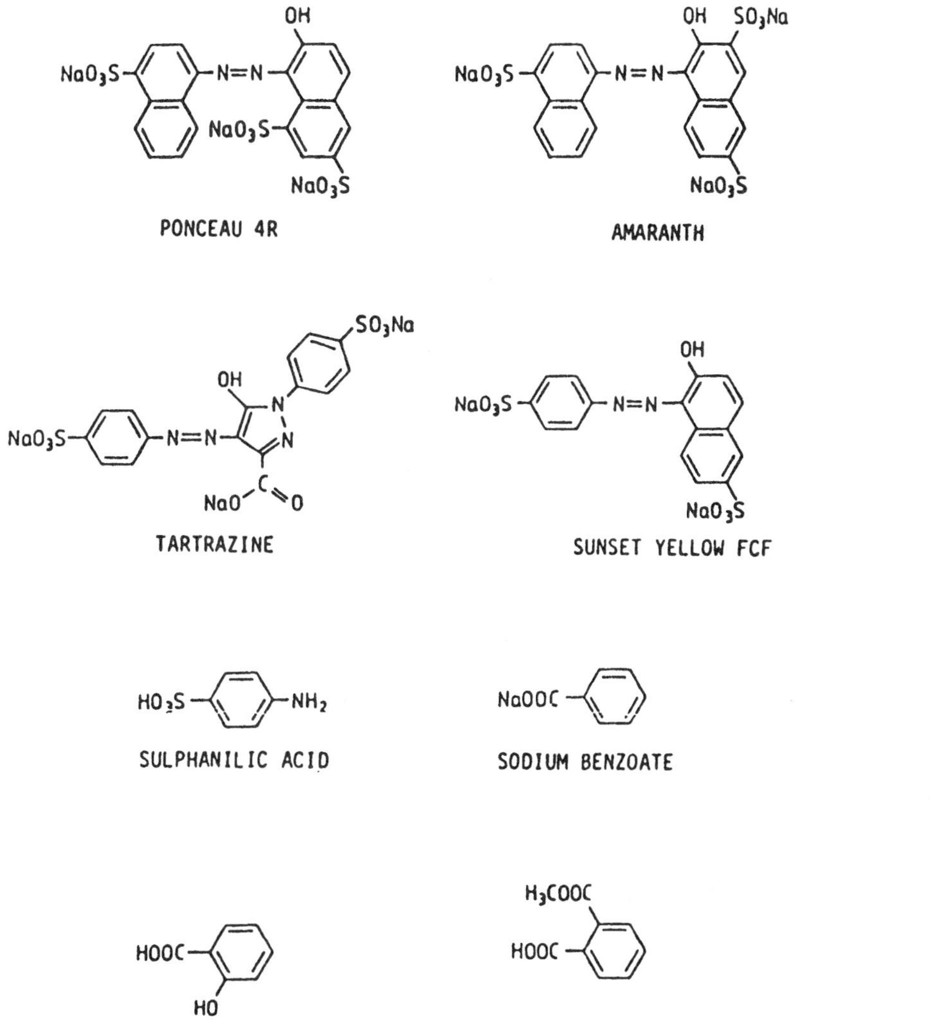

Figure 1 Molecular similarities between azo dyes, benzoates, pyrrazole compounds, and hydroxyaromatic acids including acetylsalicylic acid.

aspirin sensitivity (Speer et al., 1981; Rosenhall, 1977). Concomitant reactivity to tartrazine has been reported in 8–50% of aspirin-sensitive asthmatic patients (e.g., Samter and Beers, 1968; Settipane and Pudupakkam, 1975; Stenius and Lemola, 1976; Spector et al., 1979). No correlation between aspirin and tartrazine was found by Stevenson et al. (1986).

Rosenhall (1977) tested 504 patients with asthma and rhinitis. Reaction to at least one of the substances (food colorants, preservatives, analgesics) occurred in 106 (21%) patients. In 33 patients who reacted to tartrazine, 42% were also intolerant to aspirin and 39% to sodium benzoate. Stenius and Lemola (1976) tested 140 asthmatics, of whom 17 reacted to tartrazine. Six of the 17 subjects did not react to aspirin. One-third of the patients with a positive reaction to tartrazine also had a history of reactions following ingestion of drinks or foods containing colorants. Wütrich and Fabro (1981) reported that the frequency of intolerance to tartrazine, according to an open oral challenge test, was 6% in urticaria patients, 7% in asthma patients, and 14.5 in urticaria and asthma patients. Intolerance to benzoate varied from 2.5% in rhinitis to 11.5% in asthma.

Some of the asthmatic patients experience only rhinitis when challenged with aspirin, tartrazine, or benzoates. There were 104 patients with chronic rhinitis in the Rosenhall study; 4% of them had a positive reaction with nasal symptoms and 16% had a doubtful reaction when challenged.

It has been assumed that children have adverse reactions of the respiratory tract to food additives less frequently than adults. Vedanthan et al. (1977), Osterballe et al. (1979), and Weber et al. (1979) found few asthmatic children who reacted to aspirin, food colorants, and benzoates.

Many of the studies on food colorants have been open and poorly controlled. Recently, Morales et al. (1985) made a placebo-controlled study and challenged 47 asthmatic patients with intolerance to aspirin. Only one patient had a respiratory reaction to tartrazine on two successive occasions. The authors stated:

> The inconvenience associated with a colour-free diet and the small incidence of proven reactions to tartrazine, tend to invalidate the practice of recommending such diets unless evidence is available of a positive challenge test on at least two occasions. Even so, the risks induced are minimal.

Genton et al. (1985) also made a placebo-controlled study and challenged 17 asthmatics whose case histories suggested an intolerance to aspirin or additives with tartrazine and sodium benzoate. Only two patients had a mild positive reaction. Additionally, two patients reacted with urticaria. Stevenson et al. (1986) made single-blind challenges with 25–50 mg of tartrazine in 150 aspirin-sensitive asthmatics and found six positive responses. Double-blind rechallenge with tartrazine was negative in all five patients in whom the test was performed.

At present it seems that respiratory adverse reactions to food colorants and benzoates are infrequent. Asthmatic patients with aspirin intolerance may have an increased risk of experiencing these reactions. If the reaction occurs, it is usually mild. Life-threatening reactions are extremely rare.

B. Sulfites

Sulfiting agents include sulfur dioxide (SO_2), sodium sulfite, and the potassium and sodium salts of bisulfite and metabisulfite. Sodium and potassium bisulfite and metabisulfites are converted into sulfurous acid in solutions and sulfur dioxide itself. Sulfites are antioxidants that are used as preservatives in foods and drugs, especially in injected and inhaled medications, and as antimicrobial and sanitizing agents in the production of wine (Settipane, 1984). They are used on dehydrated vegetables and dried fruits, in fruit drinks, and also in many other products (Przybilla and Ring, 1987).

Prenner and Stevens (1976) were the first to report a patient who experienced an anaphylactic reaction after ingestion of sulfites in foods. After that, several reports suggesting anaphylactoid reactions to sulfite agents were published (e.g., Stevenson and Simon, 1981). Freedman (1977a) demonstrated that asthmatics could react to the sulfur dioxide contained in drinks. In some situations, for example, during respiratory infection, a glass of wine could have an additive effect on viral-induced bronchospasm and became clinically important (Gershwin et al., 1985). Twarog and Leung (1982) documented a severe reaction to sulfites in medication. The presence of sulfite in a bronchodilator aerosol has caused bronchoconstriction (Koepke et al., 1985). Sulfite-containing aerosols produce sulfur dioxide, and the sensitivity of the asthmatic airway to sulfur dioxide is much increased (Boushey, 1982). Inhaled sulfur dioxide may produce bronchoconstriction at 1 ppm in asthmatics (Koenig et al., 1980). Timberlake et al. (1992) made a challenge test with tartrazine and sodium metabisulfite in seven asthmatic Melanesian patients. Asthma was precipitated by sodium metabisulfite in three patients but none reacted to tartrazine.

Bronchoconstriction after oral administration of sulfur dioxide was seen in 4 of 17 adult asthmatics by Genton et al. (1985). Towns and Mellis (1984) reported that as many as 19 children of 29 with chronic asthma reacted with bronchoconstriction when challenged with metabisulfite in a solution of 0.5% citric acid. None of the children reacted to metabisulfite in capsule form. Wolf and Nicklas (1985) described a 7-year-old child who experienced cough and chest tightness immediately after ingestion of a restaurant salad. The reaction was confirmed in an oral challenge test. Schwartz and Sher reported a patient sensitive to bisulfite in local dental anesthetics and another patient who got bronchospasm from bisulfite in eyedrops (Schwartz and Sher, 1983, 1985).

It seems that sulfites cause adverse respiratory reactions more often than food colorants and benzoates. These reactions are obviously nonspecific in nature, similar to the response to metacholine and histamine (Bush et al., 1986). Reactions to sulfites may be severe and not readily recognized if not suspected.

C. Other Agents

''Chinese restaurant asthma'' has been suggested to be caused by monosodium *l*-glutamate (Schaumburg et al., 1969; Allen and Baker, 1981; Allen et al., 1983). There are, however, studies in which double-blind provocation tests have produced no asthmatic symptoms at all (Ghezzi et al., 1980; Grattini, 1982; Morselli and Grattini, 1970). Numbness of the neck, headache, facial pressure, and chest pain have also been claimed to be symptoms of Chinese restaurant syndrome in some individuals. Tarasoff and Kelly (1993) made a critical review of the glutamate literature and performed a very strict double-blind peroral challenge test with 1.5, 3.0, and 3.15 g per person. They found no difference between the responses to placebo and glutamate, and they stated the ''Chinese Restaurant Syndrome'' to be anecdotal. Just recently, Yang et al. (1997) recorded headache, muscle tightness, numbness and tingling, general weakness, and flushing from glutamate more often than from placebo, 2.5 g of monosodium glutamate being a threshold dose for positive responses. The dilemma on the effects of glutamate thus seems to continue.

Occupational asthma induced by inhalation of dust from spices such as powdered coriander, curry, and paprika has been described (Toorenenberger and Dieges, 1985).

Enzymes are frequent causes of IgE-mediated nasal, bronchial, and skin symptoms. α-Amylase is a usual flour additive which has been found to cause both immediate and delayed contact allergies in bakers. Morren et al. (1993) tested 32 bakers, 7 of whom reacted to α-amylase in scratch-chamber test at 20 min. RAST was positive in five of them. Two patients showed also a delayed reaction in the scratch-chamber test. Larese et al. (1993) made skin prick tests with α-amylase in 226 bakers and pastrymakers. Seventeen (7.5%) reacted to it. Rhinitis, conjunctivitis, and asthma were the most common allergic symptoms.

Severe systemic allergic reaction to papain in a meat tenderizer was experienced by a male patient within 20 min after eating beefsteak (Mansfield and Bowers, 1983).

V. OTHER REACTIONS

Hypersensitivity reactions to food additives in organs other than the skin or respiratory tract are rare and poorly documented. The consideration that adverse reactions may easily affect the site of entrance of the food or additive suggests that hypersensitivity reactions in the gastrointestinal tract would be common. Nausea, diarrhea, pain, abdominal distension, and vomiting have been observed in some patients who have experienced respiratory or skin symptoms provoked by azo dyes or sulfites. However, virtually nothing is known about the incidence of gastrointestinal manifestation associated with additives when other symptoms are absent. Farah et al. (1985) showed convincingly that only a small number of patients with gastrointestinal symptoms have verifiable specific food intolerance and the greater number have symptoms attributable to psychogenic causes. It is more than probable that the same conclusions also apply to food additives.

A. Hyperactivity in Children

Overreactivity, with concentration and learning difficulties, may be present in 1–5% of young children (Lambert et al., 1978). Feingold (1973) claimed that hyperactivity of children is associated with the ingestion of salicylates and crossreacting food additives. He reported that 30–50% of hyperactive children in his practice had complete remission of symptoms on a salicylate- and additive-free diet. However, these observations have not been confirmed in double-blind studies. The U.S. National Institutes of Health (NIH, 1983) evaluated the evidence and concluded that it does not support the Feingold hypothesis, although it does not exclude the possibility that hyperactive behavior could be attributable to food dyes in a minority of cases. Weiss et al. (1980) observed deterioration of behavior in 2 of 22 hyperactive children challenged with a mixture of food dyes after 3 months on an additive-free diet. A doubtful positive response to diet was achieved in 15 hyperkinetic children in the study of Conners et al. (1976), but no diet effect was found by Harley et al. (1978) in 36 school-age hyperactive boys.

Rowe and Rowe (1994) made a 21-day double-blind placebo-controlled repeated-measures study in 34 children, 23 of whom were suspected to react to artificial food colors along with 11 uncertain reactors and 20 control children. Tatrazine (1, 2, 5, 10, 20, and 50 mg) or placebo were given randomly each morning. Altogether 24 out of the 24 children reacted to tartrazine, and a dose-response effect was noticed in 19 of 23 suspected reactors, in 3 of 11 uncertain reactors, and in 2 of 20 control children. Irritability, restlessness, and sleep disturbances were the most common symptoms.

Pollock and Warner (1990) recruited 39 children whose behavior was observed by their parents to improve on a food additive–free diet and to deteriorate with dietary lapses. Placebo-controlled double-blind peroral challenges were performed with the following colorants: 50 mg of tartrazine, 25 mg of Sunset Yellow, 25 mg of carmoisine, and 25 mg of amaranth. Only 19 children completed the study. In only two instances, the parents were able to notice changes in their childrens' behavior. The mean daily Conners behavior scores were significantly ($p < 0.01$) higher when the children were taking the food colors compared with placebo.

B. Other Symptoms Attributed to Allergies

Headaches and migraines have often been claimed to be provoked by food allergies, but the objective evidence is sparse. The role of food additives is not known. Tartrazine has been shown to cause pain and swelling in the knee joints in an asthmatic patient (Wraith, 1980). Joint and muscular pains, arthralgias, and even arthritis provoked by food and food additives have been reported (Rowe and Rowe, 1972) but have not been confirmed by controlled studies.

C. Psychological Factors

According to several studies it is obvious that many, if not most, of the adverse reactions to food and food additives are in fact psychosomatic. Pearson et al. (1983) found that objectively confirmable hypersensitivity could be found in only 4 of 23 adults who claimed to have food allergy. All four subjects were atopic. Lum (1985) reviewed hyperventilation and pseudoallergic reactions. He wrote:

> Following an address to a meeting of food intolerance sufferers, more than 50% of the audience recognized that many of their presumed ''intolerance'' symptoms tallied with those of overbreathing. A patient who experiences faintness, dizziness, nausea or migraine in a warm crowded and stuffy restaurant or at dinner party may consider that an allergy to food is involved. Repetition of the reaction in similar circumstances is likely to set up a conditioned reflex response.

Pearson and Rix (1985) published an extensive review on allergomimetic reactions and pseudo food allergy. Lessof (1983), in his review of food intolerance and allergy, stated that the most common cause of an aversion to food is psychological.

The importance of double-blind challenging was studied by Jewett et al. (1990). Eighteen patients were tested with various foods by double-blind test. Positive responses to active substances were seen in 16 of 60 (27%) instances, while the corresponding number for placebo was 44 of 180 (24%). The difference was statistically not significant.

The methodological aspects for clinical studies of adverse reactions to foods and food additives have been discussed by many groups. One can get a good view on the problems in such investigations in the paper of Metcalfe and Sampson (1990).

VI. TESTS FOR HYPERSENSITIVITY REACTIONS

Because of the large variety of hypersensitivity and other untoward reactions to foods and food additives there is no simple all-embracing in vivo or in vitro test method to detect the causes of such reactions. In immediate (type I) allergic reactions, the allergen is usually a protein inducing IgE production, which can be shown with both skin and challenge tests as well as with in vitro tests in a laboratory. Patch tests and other skin tests are the tests of choice for delayed allergies. In experimental work, several in vitro tests are also available.

The grade of intolerance to food additives may vary from time to time. The symptoms, especially those caused by nonprotein food additives, are so ambiguous that our comprehension of the exact number and severity of hypersensitivity reactions to nonprotein food additives is incomplete.

A. Skin Tests

Prick testing is the simplest way to test immediate allergies in vivo. Drops of allergen solutions for prick testing are applied to the forearm, upper arm, or back skin of the patient at a distance 3–5 cm apart and pierced with special test lancets, each drop with a new lancet (Figures 2–4). After 15–20 min, the wheals are measured in two directions: the longest diameter (D) and the diameter perpendicular to it (d). The result is usually expressed as the mean diameter of the two; $(D + d)/2$. The result is compared with that produced by histamine hydrochloride, 10 mg/mL. A reaction larger than 3 mm and at least half the size of that produced by histamine is considered to be positive. However, reactions smaller than that of histamine are not always clinically significant (Haahtela et al., 1980).

Scratch testing is a less standardized method than prick testing for immediate allergy. It is, however, still in use because many nonstandardized test substances are often used and the assessment of a scratch test result may be easier than that of a prick test result. Scratches approximately 5 mm long are made with a lancet or with a puncture needle on the forearm, upper arm, or back skin. Bleeding is avoided. Allergen solutions for prick/scratch tests are applied to the scratches, and the results are read 15–20 min later. Freeze-dried and other powdered allergens can also be used. They are moistened with 0.1 N aqueous sodium hydroxide solution. Histamine hydrochloride, 10 mg/mL, is the positive control, and only the longest diameter perpendicular to the scratch is measured. Reactions of at least the size elicited by histamine are usually clinically significant.

Intracutaneous (intradermal) tests are needed only for special purposes. A disposable tuberculin syringe with a small (26-gauge) needle is filled with approximately 0.1 mL of allergen solution for intradermal testing. Usually 0.02–0.05 mL of test solution is injected intradermally, preferably in the forearm skin, but the back skin can also be used. A bleb 3–5 mm in diameter is formed. Histamine hydrochloride, 0.1 mg/mL, is used as the positive reference, and the test vehicle (usually saline or phosphate-buffered saline with phenol or parabens as preservative) as the negative control. The results are read at

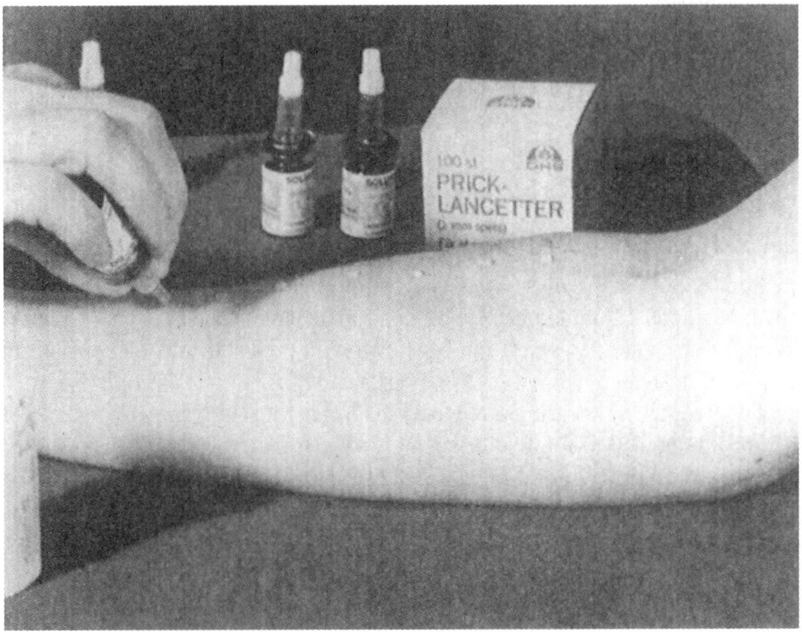

Figure 2 Skin-prick testing. Drops of allergen solutions are applied to the forearm skin.

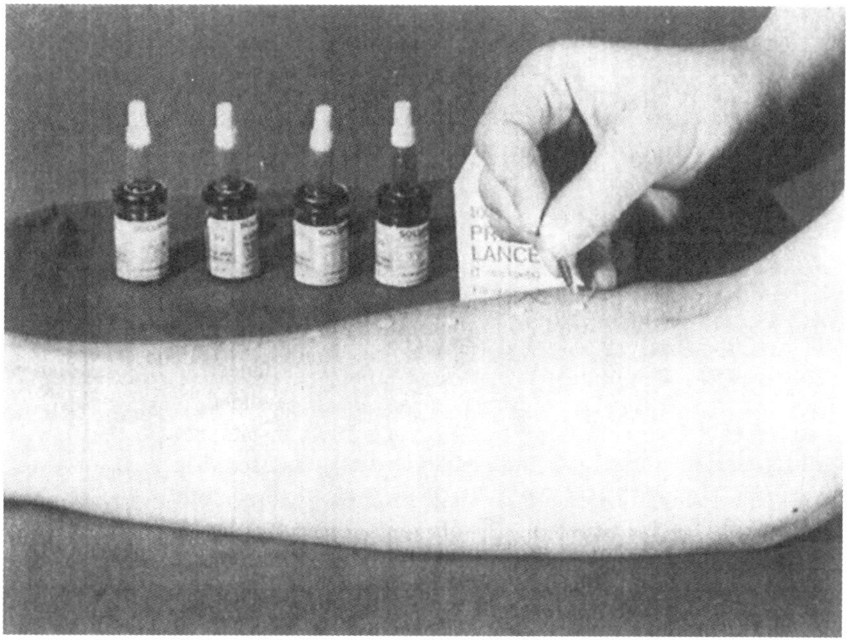

Figure 3 Skin-prick testing. Allergen drops are pierced with prick test lancets, each drop with a new one.

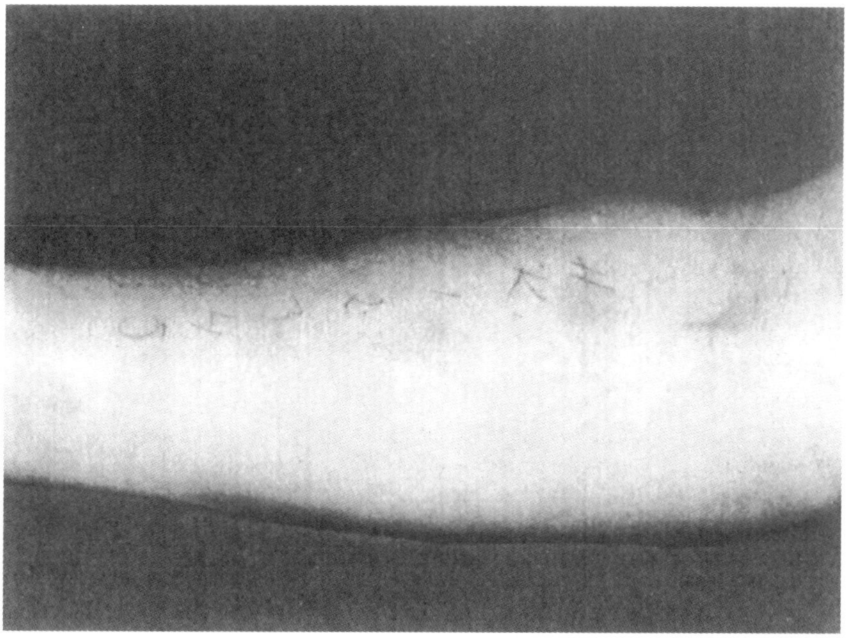

Figure 4 Skin-prick testing. Results are read 15 min later, and the reactions are compared with that produced by histamine (H in the picture) and with that produced by the diluent for the allergens (K).

15 min, and reactions larger than that induced by histamine are regarded as positive. Higher or lower concentrations of allergens are used if the threshold of positivity is needed.

Allergens for prick, scratch, and intradermal tests. Among food additives, spices are the only group in which there are protein allergens capable of producing type I allergic reactions. There are only a few commercially available allergen solutions, and that is why many test solutions must be prepared by the investigator (Niinimäki and Hannuksela, 1981).

Scratch-chamber tests were introduced for testing foodstuffs as such when there are no commercially available allergens and the allergenicity of the test substance is easily lost (Hannuksela and Lahti, 1977). The test is performed like an ordinary scratch test, but the test material is covered with a small aluminum chamber (Finn Chamber, Epitest Ltd, Hyrylä, Finland) for 15 min. The result is read according to the criteria for scratch tests. Scratch chamber tests are used especially for testing fresh material such as potato, apple, and carrot, but they can also be used for testing spices.

Patch tests are needed for detecting contact allergies. In the modern test technique, test substances are usually incorporated in white soft paraffin (WSP) and applied to the skin in small aluminum chanbers (Finn Chamber). Fluids, usually alcoholic and aqueous solutions of allergens, are adsorbed in filter paper disks placed on the bottom of the test chamber. The test chambers, filled to half volume, are fixed on the back skin of the patient for 48 h (Figures 5 and 6). The results are read 20–30 min after the removal of the chambers and once again at 72–96 h (D2 and D3–D4). Both irritant and allergic reactions may develop. Primary irritant reactions are at their maximum on D2 and allergic reactions on D2–D4 (Bandmann and Fregert, 1982) (Figures 7 and 8).

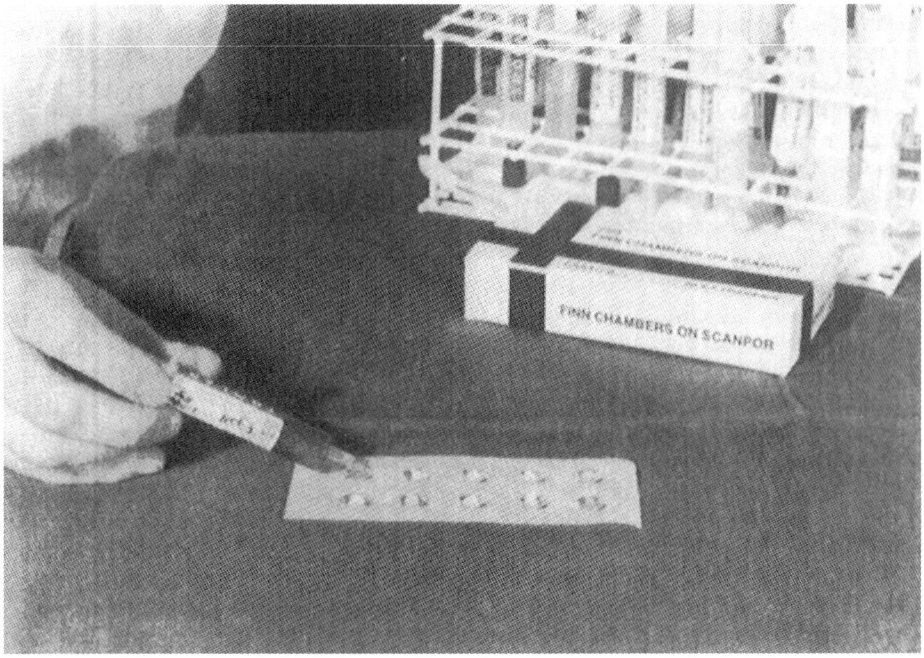

Figure 5 Patch testing. Allergens, most of which are incorporated in petrolatum, are applied to small aluminum chambers (Finn Chambers).

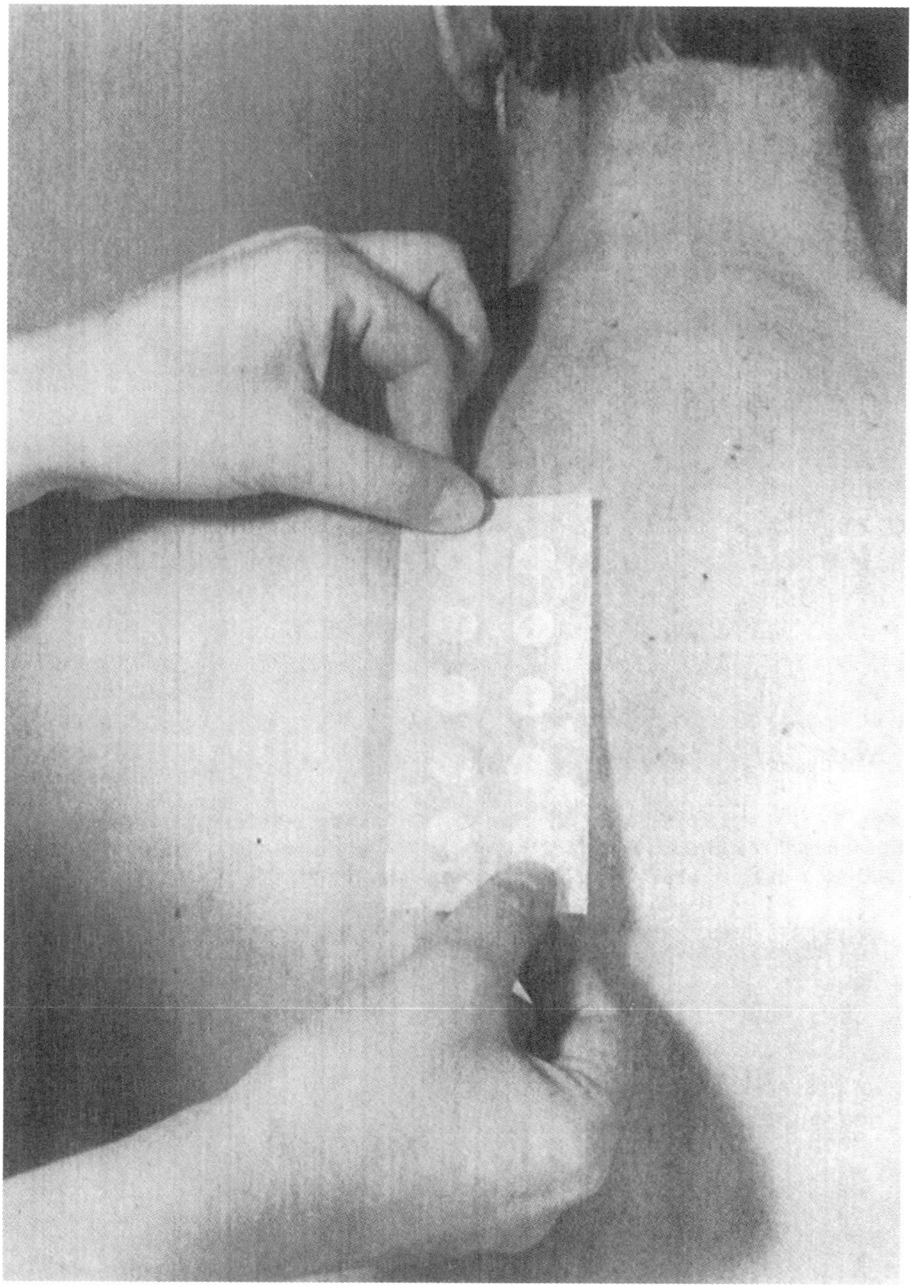

Figure 6 Patch testing. Test chambers are fixed on the back skin with a porous tape for 48 h.

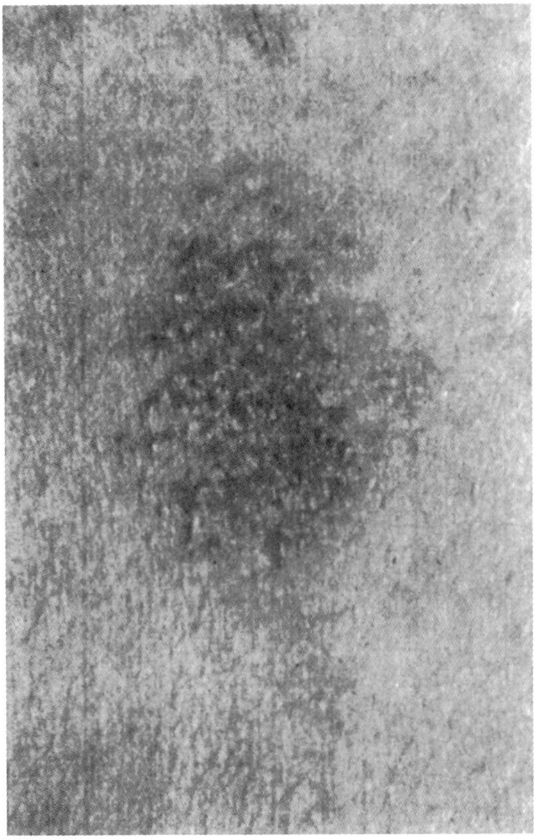

Figure 7 Patch test result at 48 h. An allergic reaction comprises ''mini-eczema'' (redness and infiltration with tiny vesicles), which tends to spread to a larger area than the chamber size.

Parabens, sorbic acid, propylene glycol, and spices are the most common contact allergens among food additives. Parabens are tested as a 15% mixture of methyl-, ethyl-, and propyl-*p*-hydroxybenzoates, sorbic acid at 2.5% in WSP, propylene glycol at 2–30% in water, and spices as ground powders.

Open application tests and *rub tests* are used in searching for the causes of contact urticarial and other immediate reactions and sometimes also in detecting delayed contact allergies. In open application tests a small amount (approximately 0.1 mL) of test substance is applied to apparently healthy skin on a 5 × 5 cm area of the extensor side of the upper extremity or on the back skin (Hannuksela, 1986a, b). Allergic (ICU) reactions are seen in 15–20 min, nonimmunologic reactions in an hour (Figure 9). It is also possible that in spite of a patient history suggestive of contact urticaria, normal-looking skin is nonreactive but a positive reaction can be seen in diseased skin. Therefore open application tests are often done on the back of the hand.

The rub test is a modification of the open application test. The test substance is gently rubbed into a skin area (usually the hands) where the reaction previously occurred.

A positive open application test result comprises a wheal and flare reaction, but in weaker reactions only redness is seen (Lahti, 1980). In the hands, especially in the rub

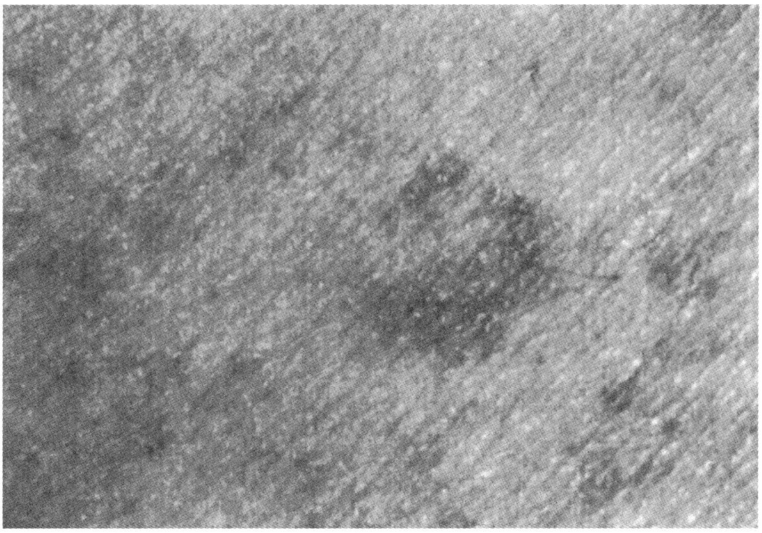

Figure 8 Patch test result at 48 h. An irritant reaction appears as redness and possible infiltration without vesicles. There is no spreading of the reaction.

test, small vesicles may be the only response, appearing within 15–30 min and disappearing within a couple of hours (Hjorth and Roed-Petersen, 1976; Köpman and Hannuksela, 1983).

The *repeated open application test* (ROAT) is done in questionable cases of delayed contact allergies in whom the significance of the reaction seen is not clear (Hannuksela and Salo, 1986). Approximately 0.1 mL of test substance is applied twice daily to a 5 ×

Figure 9 Nonimmunologic contact urticaria reaction from 5% benzoic acid in petrolatum 40 min after the application.

5 cm area on the volar aspect of the forearm or on the back skin. In positive cases, eczematous dermatitis appears on the test area, usually on D2–D5. If no sign of dermatitis is seen in a week, the test is discontinued and the result is considered to be negative.

In the *use test*, the product suspected to be responsible for the reaction is used as usual and the skin response is followed for up to 1 month. Both weak allergens and primary irritant substances may produce reactions in the use test.

The *Prausnitz–Küstner test* (the passive transfer test) provides a method for demonstration of IgE-mediated reactions in nonsensitized individuals. Aliquots of 0.1 mL of serum derived from a patient are injected intradermally into a healthy person or a monkey. Prick, scratch, or intradermal tests are performed on the infection sites 24–72 h later. If the patient's serum contains specific IgE, the test result in the healthy person or monkey is positive. Previously, the Prausnitz–Küstner test was often used, but now radioimmunoassay tests have replaced it in most cases. It is noteworthy that in the Prausnitz–Küstner test, infective viruses may be transfered from the patient to another person. Hence, tests for the presence of HBAg and HTLV III antibodies must be carried out before the patient's serum is injected in the test subject.

B. Oral Challenge Tests

Allergic reactions to spices appear most often as local reactions in the mouth and adjacent areas (see Section III). The simplest way to assess the role of spices and other substances suspected of producing such reactions is to perform an *oral mucosal test*, spreading the substances on the oral mucosa (Niinimäki and Hannuksela, 1981). Within 15–20 min, swelling of the oral mucosa, lips, or throat; an attack of allergic rhinitis; and/or conjunctivitis may appear. In rare cases, a systemic reaction (urticaria) may take place.

In the *peroral challenge test* the substance to be tested is given to the patient in nontransparent gelatin capsules, preferably in a double-blind manner. The dose should be about the average daily dose used by the population (Table 11). In some cases, individual doses are needed.

Table 11 An Example of Test Battery for Peroral Challenge Tests with Food Additives

Substance	Amount
Placebo (wheat starch)	100 + 200 mg
Benzoic acid	200 + 400 mg
BHT + BHA, 1:1	50 + 100 mg
Annatto	5 + 10 mg
Sorbic acid	200 + 400 mg
β-Apocarotenal + β-carotene	200 + 400 mg
Sodium nitrite	50 + 100 mg
Sodium metabisulfite	10 + 20 mg
Sodium glutamate	100 + 200 mg
Azo dyes[a]	5 + 10 mg

Note: Dose given to patients in nontransparent gelatin capsules at 1-h intervals.

[a] Tartrazine, new coccine, and Sunset Yellow (or paraorange). Only in countries in which azo dyes are allowed to be added to foods.

In cases of urticaria, possible reactions are followed up to 24 hr, although the reaction is usually seen within 1–4 h (Hannuksela and Lahti, 1986). In patients with purpura and in those with suspected internal exacerbation of contact dermatitis, a longer exposition and follow-up time (up to several days) may be needed.

Subjective sensations, such as nausea, vertigo, and itching, are frequent after peroral challenge tests. However, they are always regarded as negative challenge test results. Only clearcut reactions are to be regarded as positive results. In chronic and recurrent urticaria, a marked worsening of urticaria with or without angioneurotic edema is suggestive of real hypersensitivity reaction. In cases of purpura, a widespread attack of purpural patches, and in cases of internal exacerbation of contact dermatitis, eczema and/or pompholyx reactions, are significant. The assessment of respiratory symptoms must also be based on objective criteria. In rhinitis, the number of sneezes may be counted, the mucus flow from the nose measured, and the number of handkerchiefs used counted. The nasal obstruction is measured usually by a peak flow meter (nasal peak expiratory flow) or a rhinomanometer. The asthmatic response is expressed as PEF (peak expiratory flow) or FEV_1 (forced expiratory volume in 1 s). Usually, a drop of 20% is regarded as significant. The respiratory reaction may be delayed up to 8 h. All equivocal and positive challenge tests should be repeated.

The importance of the double-blind challenge test technique was clearly demonstrated in a study by Hannuksela and Lahti (1986). Peroral challenge tests with common food additives were made in patients with chronic urticaria, in patients with atopic dermatitis, and in patients with eczematous dermatitis as a control group. Positive and equivocal reactions were seen in all groups, and there were no differences between the various additives and the placebo in their capacity to produce these reactions (Table 12). This result indicates that the double-blind technique ought to be the only way to make peroral challenge tests with these substances, reactions to which are poorly established and difficult to assess.

Whether or not to avoid antihistamines, peroral steroids, and antiasthmatic drugs during peroral challenges is a very important question. When urticaria or asthma is in a steady state and the patient is nearly or totally free of symptoms without medication, the drugs mentioned before should be avoided during challenge tests. Patients with severe asthma need their medication, even if the drugs may have some influence on the results of the challenge tests.

A *restriction diet* free from important food additives is recommended for some days to some weeks preceding the peroral challenge test (Juhlin, 1981). The need for the diet is, however, a matter of controversy. We now know that aspirin produces a refractory period of a couple of days during which further intake of aspirin does not produce symptoms. Whether or not this is also true with food additives remains to be investigated.

C. In Vitro Tests

1. Determination of Specific IgE

Demonstration of specific circulating antibodies of the IgE class provides evidence of an immunologic sensitization. However, it has been shown that low levels of IgE antibodies to inhalant or food allergens are often present without clinical manifestation (Haahtela and Jaakonmäki, 1981).

The most used technique to measure IgE antibodies is the Radio Allergo Sorbent Test (RAST) (Pharmacia, Uppsala, Sweden). As indicated in Section II, there is no firm

Table 12 Results of Double-Blind, Placebo-Controlled Peroral Challenge Tests with Food Additives

		Disease group		
Substance(s) result	Amount per capsule	Urticaria	Atopy	Contact dermatitis
Wheat starch (placebo)	100 mg			
equivocal		1[a]	—	—
positive		1[b]	—	1[c]
Sodium metabisulfite	9 mg			
equivocal		1[a]	1[a]	—
positive		—	—	—
Benzoic acid	200 mg			
equivocal		1[a]	—	—
positive		1[b]	—	—
BHT + BHA[d]	50 mg			
equivocal		2[a]	2[a]	—
positive		—	—	—
β-Apocarotenal + β-Carotene 1:1	200 mg			
equivocal		—	1[a]	—
positive		—	1[c]	—

Note: Carried out in 44 patients with chronic urticaria, in 91 patients with atopic dermatitis, and in 123 patients with contact dermatitis as a comparison group. The dose was 1 + 2 capsules given at 1-h intervals. The reactions were followed up for 24 h.

[a] Retest 4 days later negative.

[b] Retest 4 days later positive.

[c] Retest not done due to severe symptoms in the first challenge.

[d] BHT = butylhydroxytoluene, BHA = butylhydroxyanisole.

Source: Hannuksela and Lahti, 1986.

evidence that, with the exception of spices (Toorenenberger and Dieges, 1985), adverse reactions to food additives are mediated by IgE. Therefore there are as yet no commercial RAST tests for food additives available.

2. Basophil Degranulation Test and Histamine Release

The basophil degranulation test using either whole blood or washed leukocytes has been advocated to diagnose immediate hypersensitivity to various allergens (Benveniste, 1981). The results correlate well with histamine release from leukocytes, particularly from basophils, and with the RAST. However, virtually nothing is known about the possibilities of using these tests in the diagnosis of hypersensitivity reactions to food additives.

3. Lymphocyte Transformation Test

The lymphocyte transformation test (LTT) is widely used as an in vitro test for delayed allergies. It is also suitable for examination of several contact allergies such as those toward cobalt (Veien and Svejgard, 1978), chromium (Jung, 1969), gold (Denman and Denman, 1968), and nickel (MacLeod et al., 1970; Kim and Schöpf, 1976; Silvennoinen-Kassinen, 1980). In cases of drug reactions, several antibiotics, chemotherapeutics, and many other drugs have been found to induce blast formation from lymphocytes in LTT (Saurat et al., 1973).

It is possible that LTTs could be used for the investigation of food additives suspected of producing delayed-type reactions. For complex mixtures of many chemicals, such as essential oils and spices, the LTT might not be successful, but it is worth trying in selected cases for simpler chemicals such as sorbic acid and parabens.

4. Migration Inhibition Test

The leukocyte migration inhibition test has been used, for example, in nickel allergy (Mirza et al., 1975). This test could obviously also be used for studies on contact allergies from food additives.

5. Cytotoxic Test

It has been claimed that the white blood cells of food allergy patients die and disintegrate in the presence of the food to which the patient is sensitive (Black, 1956). In the cytotoxic test, circulating white blood cells are exposed in vitro to food extracts, and the number of dead cells is counted under a microscope. The test has been found to be difficult to reproduce and unreliable (Sethi et al., 1987), and the interindividual results may vary grossly from time to time (Ruokonen, 1982).

Summarizing, at the moment a controlled peroral challenge is the only reliable way to test systemic reactions to food additives. Skin tests are irreplaceable in the diagnosis of many skin reactions.

VII. PREDICTION OF ALLERGY RISK

There are several tests for identification of contact allergens in animals [for reviews see Klecak (1977) and Maurer et al. (1985)]. The guinea pig is used most often, and the guinea pig maximization test (GPMT), described by Magnusson and Kligman (1969), has become the most popular of the test procedures. In brief: In the first stage of induction, double intradermal injections of (1) 0.1 mL of Freund's complete adjuvant (FCA), (2) 0.1 mL of test substance, and (3) 0.1 mL of test substance emulsified in FCA are made in the shoulder area. In the second stage of induction, 7 days later, the test substance is applied to a 2×4 cm filter paper and placed under an occlusive bandage for 48 h on the shoulder region of the animal. The challenge is made 2 weeks after the second stage of induction by applying an occlusive patch to the back skin for 24 h. The challenge site is evaluated 24 and 48 h after removal of the patch. The allergenic potency of the test substance is graded from I (weak sensitizer) to V (extreme sensitizer).

The results of GPMTs correspond fairly well to the sensitizing capacity of various substances in humans. There are some exceptions, including parabens and propylene glycol. Parabens are extreme sensitizers in guinea pig but rarely sensitize humans, and propylene glycol does not sensitize guinea pigs at all but sensitizes humans more often than parabens (Kero and Hannuksela, 1980).

Among other animal tests for identification of contact allergens, Freund's complete adjuvant test (FCAT) (Klecak et al., 1977), the optimization test (Maurer, 1983), and the split adjuvant test (Maguire and Chase, 1972) are more sensitive than nonadjuvant methods, and the results are roughly comparable to those achieved in GPMT (Maurer et al., 1985).

Human models for contact allergy risk involve epicutaneous application(s) of test substances on untreated or sodium lauryl sulfate–treated skin. The duration of the induction phase is 2–5 days, and the challenge is usually made 1–3 weeks later [for review, see Marzulli and Maibach (1977)]. The drawbacks of human assays are the great number of test subjects required (25–100 people) and the harmful allergic reactions to be expected later in life.

An animal model for nonimmunologic contact urticaria (NICU) was described by Lahti and Maibach (1984). They found that the guinea pig ear responds readily to agents producing NICU in humans. Their test method seems to be suitable for screening food additives for their capacity to produce NICU.

VIII. DEVELOPMENT OF TOLERANCE AND TREATMENT OF HYPERSENSITIVITY

A. Development of Tolerance

Immunologic tolerance means immunologic nonreactivity to specific antigens. Normal immune responses fail to develop when antigenic substances come into contact with immunocompetent tissues in early fetal life and perhaps also in early childhood. This type of tolerance cannot be transfered to another individual by either serum or cells. The mechanism of such ''instant'' tolerance is, so far, unknown (Miller and Claman, 1976). The development of ''instant'' tolerance for contact allergens may partly be due to lack of interleukin 2 at the time of challenge (Malkovsky and Medavar, 1984). In contact allergy, tolerance is also mediated by suppressor T cells, which can inhibit either the afferent limb of sensitization or the efferent (elicitation) phase (Claman and Miller, 1980).

Acquired tolerance can be achieved either by giving very large doses of antigen, resulting in *high zone tolerance*, or by giving small doses of antigen below the immunization dose, leading to *low zone tolerance*. Acquired tolerance is induced by intravenous or peroral administration of allergens. The induction of high-dose tolerance can be used in experimental work with laboratory animals, but in clinical work with humans the low-dose method is recommended. Obviously the disappearance of many IgE-mediated food allergies during childhood express low zone tolerance rather than peroral hyposensitization. Another example of low zone tolerance is the disappearance of nickel allergy in patients with total hip prosthesis.

Pseudotolerance is seen in patients with continuous high allergen administration and, consequently, neutralization of antibodies. Patients allergic to penicillin can be treated with high doses of penicillin without clinical signs of allergy.

Certain factors, called *adjuvants*, increase the host response to allergens. Many bacteria may be adjuvants in the development of immediate allergies. *Bordetella pertussis* vaccine, for example, increases the IgE response and also the sensitivity of the organism to chemical mediators such as serotonin and histamine in laboratory animals (Reed et al., 1972; Pauwels et al., 1983). Freund's complete adjuvant, containing killed mycobacteria, is used in the induction of contact allergy in guinea pigs (Magnusson and Kligman, 1969).

Other factors also influence the development and grade of allergy. In atopic dermatitis, ultraviolet (UV) light decreases inflammation, erythema, and itch in over 90% of cases (Hannuksela et al., 1985). Ultraviolet B (280–320 nm) seems to be more efficient than longer wavelengths. It is also beneficial in chronic urticaria, especially in dermographism, and cholinergic and cold urticarias (Hannuksela and Kokkonen, 1985).

Ultraviolet radiation is also capable of modifying delayed-type hypersensitivity reactions. There are both local and distant alterations, the former resulting as a direct effect of UV light at the site of irradiation (Kripke and Morison, 1985). Kripke and Morison

(1985) postulated that UV light causes at least two separate alterations that influence distant immune responses. The first is a reversible "inflammatory" alteration that affects delayed hypersensitivity, and the second is an afferent suppression of contact allergy and the production of antigen-specific suppressor T cells. The most active wavelengths are those below 280 nm, indicating that there must be some soluble substance that acts directly on Langerhans cells or other cells involved in the presentation of haptens. Psoralens and long-wave (320–400 nm) UV light (PUVA) have similar effects on contact allergy (Morison et al., 1981).

Cyclophosphamide and other drugs inhibiting cell proliferation may influence the induction of contact allergy and the development of allergic contact dermatitis (Andersen, 1985).

B. Treatment of Hypersensitivity

1. Avoidance Diets

Dietary restrictions have been widely used for real food allergy. They have also been used as a diagnostic and therapeutic tool in chronic and recurrent urticaria. Good or excellent results have been reported in 31–81% of the patients in these experiments (Michaelsson and Juhlin, 1973; Ros et al., 1976; Warin and Smith, 1976; Lindemayr and Schmidt, 1979; Gibson and Clancy, 1980; Rudzki et al., 1980; Kirchhof et al., 1982; Verschave et al., 1983). August (1979) successfully treated 22 urticaria patients with disodium cromoglycate (DSCG) and an exclusion diet. The studies mentioned above were all uncontrolled, and therefore the results are open to doubt.

In chronic asthma, the avoidance of tartrazine and benzoates was found to be of no use (Tarlo and Broder, 1982). Genton et al. (1985) reported that in an open trial excluding NSAIDs and food additives from the diet resulted in a favorable response in 6 of 9 asthmatics and in 14 of 15 patients with urticaria. However, only four patients with urticaria noticed a complete disappearance of the symptoms on the diet alone.

The duration of sensitivity to food additives, whether permanent or temporary, has never been evaluated. For aspirin intolerance it is known that once it has developed it lasts for years, and caution for an indefinite period is advocated.

Wolf and Nicklas (1985) recommended that a sulfite-sensitive patient document the sensitivity on a Medic-Alert tag.

Ten years ago, a temporary refractory state was found to follow intake of aspirin in aspirin-induced asthma (Zeiss and Lockey, 1976). Since then, this type of induction of tolerance and treatment of asthma with low-dose aspirin regimens has been used with some success (Chiu, 1983; Stevenson et al., 1982), but long-term aspirin treatment seems to be less effective in severe corticosteroid-dependent aspirin-sensitive asthmatics (Dor et al., 1985).

Five of six patients with aspirin urticaria became tolerant to aspirin after a single dose of 30–515 mg of aspirin (Asad et al., 1983). These five patients remained free of symptoms during the following weeks while taking 650 mg of aspirin daily. We have succeeded in desensitizing some asthma patients with aspirin sensitivity, but our experience with urticaria has been disappointing.

Our experience shows that restricted intake of food colorants and benzoates seldom has any effect on symptoms. Many diets are not only unnecessary but may also be harmful. The patients may become neurotic in rejecting foods that they suspect to contain additives. Weight loss due to total caloric deficit may occur as well as vitamin and mineral deficiencies. The assistance of a properly trained dietician is needed if a long-term diet is planned. The dietary protocols are usually difficult and time consuming to follow in everyday life. Nevertheless, a diet that restricts the excessive intake of additives may be worth trying. This is especially important in sulfite-sensitive persons. A diet that totally eliminates food additives is usually not reasonable.

2. Treatment with Drugs

Ortolani et al. (1984) showed that premedication with 800 mg of oral DSCG is capable of protecting against the effects of challenge with aspirin and tartrazine in patients with urticaria. They also reported that 8 weeks treatment with DSCG ameliorates symptoms caused by foods and additives, compared to placebo. Conflicting results have also been published (Harries et al., 1978; Thormann et al., 1980). Oral DSCG has been found to prevent asthmatic reaction from metabisulfite (Schwartz, 1986). Dahl (1981) reported that oral DSCG protected against an asthmatic reaction to foods but neither oral nor inhaled DSCG protected the patient against bronchospasm caused by aspirin. Oral DSCG is poorly absorbed, and it is suggested that the activity occurs within the gastrointestinal mucosa and that DSCG modifies the way in which food allergens are transfered and metabolized in the gut.

The reactivity to additives is linked to the overall balance of the hypersensitivity disease, for example, asthma or urticaria. If the patient's condition is poor, symptoms are easily provoked by numerous nonspecific irritants. Food additives are in part also nonspecific irritants that may cause symptoms at one time and be without effect at another. This emphasizes the primary importance of the treatment as a whole.

Immunologic contact urticaria reactions may sometimes be widespread, and infrequently systemic reactions are encountered (von Krogh and Maibach, 1982). Peroral antihistamines are effective, and in very rare cases systemic corticosteroids may be needed.

Oral aspirin and other prostaglandin inhibitors have been found to prevent diarrhea due to food intolerance (Buisseret et al., 1978).

Nonimmunologic contact urticaria reactions are mild, and no treatment is needed. Avoidance of causative agent(s) is sufficient.

Irritant and allergic contact dermatitis are treated with corticosteroid creams and ointments.

IX. SUMMARY

Only seldom have food additives been shown to cause true allergic (immunologic) reactions. Adverse effects due to various pharmacological or other mechanisms are much more common. The individual tolerance may be decreased for one reason or another, and it may fluctuate from time to time. Therefore, the terms "allergy" and "atopy" have been avoided in conjunction with food additive reactions, and "intolerance" or "hypersensitivity" are preferred. However, many patients suffering from food additive reactions have atopic constitution and clinical symptoms such as flexural dermatitis, rhinitis, and asthma. Atopic subjects are more sensitive to numerous irritant environmental factors, including food additives, than the nonatopic population.

The most important skin symptoms caused by food additives are urticaria, angioneurotic edema, and contact urticaria. Azo dyes, benzoic acid, and several other common food additives that are ingested may aggravate or, more rarely, even cause urticaria. The reactions have been found in aspirin-sensitive patients more often than in patients nonreactive to aspirin. Contact urticaria refers to a wheal and flare response after external exposure of certain food additives. It is divided into two main types, immunologic and nonimmunologic. Spices are the most common causes of immunologic contact urticaria. Nonimmunologic contact urticaria is produced by numerous spices, benzoic acid, sorbic acid, cinnamic acid, and many essential oils.

Asthma and rhinitis are the main hypersensitivity symptoms in the respiratory tract. Azo dyes, benzoic acid, and sulfiting agents are the most common causative food additives. Respiratory reactions to food colorants and benzoates seem to be more common in patients with aspirin sensitivity. Untoward reactions to sulfites is a problem to which more attention should be paid.

Hypersensitivity reactions in organs other than the skin and respiratory tract are rare or poorly documented. It is obvious that most of the claimed adverse reactions to additives are, in fact, psychosomatic. Psychological factors play an essential role in both food and food additive reactions, especially in adults.

There is no simple and all-embracing diagnostic method to study adverse reactions to ingested additives. A controlled peroral challenge test is the only reliable diagnostic tool available. Better knowledge of the mechanisms underlying these reactions is the only basis for more dependable tests. This should not be too difficult a task because most additives are pure chemicals.

Prick, scratch, intracutaneous, scratch-chamber, rub, and open application tests are the tests for immediate contact reactions in the skin. Patch and repeated open application tests are used in cases of contact dermatitis.

Life-threatening reactions to food additives are extremely rare. Reactions to sulfites may be severe, but they are not readily recognized if not suspected. In the treatment, a diet and/or oral disodium cromuglycate may be worth trying in some cases. On the other hand, it is possible that daily low intake of natural salicylates, benzoates, and other food additives, but not of sulfites, might be beneficial similarly to low-dose aspirin in aspirin-sensitive patients.

REFERENCES

Allen, D. H., Baker, G. J. 1981. Chinese-restaurant asthma. *New Engl. J. Med.* 305:1154–1155.

Allen, D. H., Delohery, J., Baker, G. J., Wood, R. 1983. Monosodium glutamate-induced asthma. *J. Allergy Clin. Immunol.* 71:98.

Almeyda, J., Levantine, A. 1971. Drug reactions. XVI. Lichenoid drug eruptions. *Br. J. Dermatol.* 85:604–607.

Andersen, K. E. 1983. Vissa bieffekter av konserveringsmedel och emulgatorer. *Svensk Farm. Tidskr.* 87:32–34.

Andersen, K. 1985. Sensitivity and subsequent "down regulation" of sensitivity induced by chlorocresol in guinea pigs. *Arch. Dermatol. Res.* 277:84–87.

Andersen, K. E., Storrs, F. J. 1982. Hautreizungen durch Propylenglykol. *Hautartzt* 33:12–14.

Asad, S. I., Youlten, L. J. F., Lessof, M. H. 1983. Specific desensitization in "aspirin-sensitive" urticaria; plasma prostaglandin levels and clinical manifestations. *Clin. Allergy* 13:459–466.

August, P. J. 1979. Successful treatment of urticaria due to food additives with sodium cromoglycate and an exclusion diet. In: *The Mast Cell: Its Role in Health and Disease*, Pepys, J., Edwards, A. M. (Eds.). Pitman Medical, Bath, p. 584.

Augustine, G. J., Jr., Levitan, H. 1980. Neurotransmitter release from a vertebrate neuromuscular synapse affected by a food dye. *Science* 207:1489–1490.

Baldwin, J. L., Chou, A. H., Solomon, W. R. 1997. Popsicle-induced anaphylaxis due to carmine dye allergy. *Ann. Allergy Asthma Immunol.* 79:415–419.

Bandmann, H.-J., Fregert, S. 1982. *Epicutantestung*, 2nd ed. Springer-Verlag, Berlin.
Bedello, P. G., Goitre, M., Cane, D. 1982. Contact dermatitis and flare from food flavouring agents. *Contact Dermatitis* 8:143–144.
Belkin, G. A. 1967. Cocktail purpura. An unusual case of quinine sensitivity. *Ann. Intern. Med.* 66: 583–586.
Benveniste, J. 1981. The human basophil degranulation test as an *in vitro* method for the diagnosis of allergies. *Clin. Allergy* 11:1–11.
Van Bever, H. P., Docx, M., Stevens, W. J. 1989. Food and food additives in severe atopic dermatitis. *Allergy* 44:588–594.
Biro, L., Pecache, J. C. 1969. Alcoholic urticaria (?cholinergic). *Arch. Dermatol.* 100:644–655.
Black, A. P. 1956. A new diagnostic method in allergic disease. *Pediatrics* 17:716–723.
Bonnet, G. F., Nepveux, P. 1971. Les migraines tyraminiques. *Sem. Hop. Paris* 47:2441–2445.
Borgeat, P. 1981. Leukotrienes: a major step in the understanding of immediate hypersensitivity reactions. *J. Med. Chem.* 24:121–126.
Boushey, H. A. 1982. Bronchial hyperreactivity to sulfur dioxide: physiological and political implications. *J. Allergy Clin. Immunol.* 69:335–338.
Breslin, A. B. X., Hendrick, D. J., Pepys, J. 1973. Effect of disodium cromoglycate on asthmatic reactions to alcoholic beverages. *Clin. Allergy* 3:71–82.
Brun, R. 1964. Kontaktekzem auf Laurylgallat und *p*-Hydroxybenzoe-säureester. *Berufsdermatosen* 12:281–284.
Brun, R. 1970. Eczema de contact a un antioxidant de la margarine (gallate) et changement de metier. *Dermatologica* 140:390–394.
Buisseret, P. D., Youlten, L. F., Heinzelmann, D. I., Lessof, M. H. 1978. Prostaglandin-synthesis inhibitors in prophylaxis of food intolerance. *Lancet* 1(8070):906–908.
Burckhardt, W., Fierz, U. 1964. Antioxydantien in der Margarine als Ursache von Gewerbeekzemen. *Dermatologica* 129:431–432.
Bush, R. K., Taylor, S. L., Busse, W. W. 1986. A critical evaluation of clinical trial in reaction to sulfites. *J. Allergy Clin. Immunol.* 78:191–202.
Calnan, C. D. 1957. Release of histamine in urticaria pigmentosa. *Lancet* i:996.
Camarasa, J. M. G. 1982. Acute contact urticaria. *Contact Dermatitis* 8:347–348.
Chafee, F. H., Settipane, G. A. 1967. Asthma caused by FD&C approved dyes. *J. Allergy* 40:65–72.
Champion, R. H., Roberts, S. O. B., Carpenter, R. G., Roger, J. H. 1969. Urticaria and angio-oedema. A review of 554 patients. *Br. J. Dermatol.* 81:588–597.
Chiu, J. T. 1983. Improvement of aspirin sensitive asthmatics after rapid desensitization and aspirin maintenance treatment. *J. Allergy Clin. Immunol.* 71:560–567.
Claman, H. N., Miller, S. D. 1980. Immunoregulation of contact sensitivity. *J. Invest. Dermatol.* 74:263–266.
Clayton, D. E., Busse, W. 1980. Anaphylaxis to wine. *Clin. Allergy* 10:341–343.
Clemmensen, O., Hjorth, N. 1982. Perioral contact urticaria from sorbic acid and benzoic acid in a salad dressing. *Contact Dermatitis* 8:1–6.
Commission of the European Communities. 1981. Report of a working group on adverse reactions to ingested additives.
Conners, C. K., Goyette, C. H., Southwick, D. A., Lees, J. M., Andrulonis, P. A. 1976. Food additives and hyperkinesis: a controlled double-blind experiment. *Pediatrics* 58:154–166.
Cooke, R. A. 1919. Allergy in drug idiosyncrasy. *J. Am. Med. Assoc.* 73:759–760.
Corder, E. H., Buckley, C. E., III. 1995. Aspirin, salicylate, sulfite and tartrazine induced bronchoconstriction. Safe doses and case definition in epidemiological studies. *J. Clin. Epidemiol.* 48: 1269–1275.
Criep, L. H. 1971. Allergic vascular purpura. *J. Allergy Clin. Immunol.* 48:7–12.
Cronin, E. 1980. *Contact Dermatitis*. Churchill Livingstone, Edinburgh, pp. 805–806.
Dahl, R. 1981. Oral inhaled sodium chromoglycate in challenge test with food allergens or acetylsalicylic acid. *Allergy* 36:161–165.

Delohery, J., Simmal, R., Castle, W. D., Allen, D. H. 1984. The relationship of inhaled sulfur dioxide reactivity to ingested metasulfite sensitivity in patients with asthma. *Am. Rev. Resp. Dis.* 130: 1027–1032.

Denman, E. J., Denman, A. M. 1968. The lymphocyte transformation test and gold hypersensitivity. *Ann. Rheum. Dis.* 27:582–589.

Derbes, V. J. 1964. The fixed eruption. *J. Am. Med. Assoc.* 190:765–766.

Desmond, R. E., Trautlein, J. J. 1981. Tartrazine anaphylaxis: a case report. *Ann. Allergy* 46:81–82.

Dor, P. J., Vervloet, D., Baldocchi, G., Charpin, J. 1985. Aspirin intolerance and asthmal induction of a tolerance and long-term monitoring. *Clin. Allergy* 15:37–42.

Drevets, C. C., Seebohm, P. M. 1961. Dermatitis from alcohol. *J. Allergy* 32:277–282.

Eriksson, N. E. 1978. Food sensitivity reported by patients with asthma and hay fever. *Allergy* 33: 189–196.

Farah, D. A., Calder, I., Banson, L., Mackenzie, J. F. 1985. Specific food intolerance: its place as a cause of gastrointestinal symptoms. *Gut* 26:164–168.

Farr, R. S., Spector, S. L., Wangaard, C. H. 1979. Evaluation of aspirin and tartrazine idiosyncrasy. *J. Allergy Clin. Immunol.* 64:667–675.

Feingold, B. F. 1968. Recognition of food additives as a cause of symptoms of allergy. *Ann. Allergy* 26:309–313.

Feingold, B. F. 1973. Food additives and child development. *Hosp. Pract.* 8:11–12.

Fisher, A. A. 1975. Contact dermatitis due to food additives. *Cutis* 16:961–966.

Fisher, A. A. 1976. Reactions to antioxidants in cosmetics and foods. *Cutis* 17:21–28.

Fisher, A. A. 1982. Hand dermatitis—a "baker's dozen." *Cutis* 29:214–221.

Fisher, A. A. 1989. Reactions to sulfites in foods: delayed eczematous and immediate urticarial, anaphylactoid, and asthmatic reactions. Part III. *Cutis* 44:187–190.

Freedman, B. J. 1977a. Asthma induced by sulphur dioxide, benzoate and tartrazine contained in orange drinks. *Clin. Allergy* 7:407–415.

Freedman, B. J. 1977b. A dietary free from additives in the management of allergic disease. *Clin. Allergy* 7:417–421.

Gay, G., Sabbah, A., Drouet, M. 1994. Valeur diagnostique de l'epidermotest aux sulfites. *Allerg. Immunol. (Paris)* 26:139–140.

Genton, C., Frei, P. C., Pecond, A. 1985. Value of oral provocation tests to aspirin and food additives in the routine investigation of asthma and chronic urticaria. *J. Allergy Clin. Immunol.* 63:289–294.

Gerber, J. C., Payne, N. A., Oelz, O., Nies, A. S., Oales, J. A. 1979. Tartrazine and the prostaglandin system. *J. Allergy Clin. Immunol.* 63:289–294.

Gershwin, M. E., Ough, C., Bock, A., Fletcher, M. P., Nagy, S. M., Tuft, D. S. 1985. Adverse reactions to wine. *J. Allergy Clin. Immunol.* 75:411–420.

Ghezzi, P., Salmona, M., Recchia, M., Dagnino, G., Grattini, S. 1980. Monosodium glutamate kinetic studies in human volunteers. *Toxicol. Lett.* 5:417–421.

Gibson, A., Clancy, R. 1980. Management of chronic idiopathic urticaria by the identification and exclusion of dietary factors. *Clin. Allergy* 10:699–704.

Gollhausen, R., Kligman, A. M. 1985. Human assay for identifying substances which induce non-allergic contact urticaria: the NICU-test. *Contact Dermatitis* 13:98–106.

Grattini, S. 1982. Chinese-restaurant asthma. *New Engl. J. Med.* 306:1181 (Letter).

Haahtela, T., Jaakonmäki, I. 1981. The relationship of allergen-specific IgE antibodies, skin prick tests and allergic disorders in unselected adolescents. *Allergy* 36:251–256.

Haahtela, T., Heiskala, M., Suoniemi, I. 1980. Allergic disorders and immediate skin test reactivity in Finnish adolescents. *Allergy* 35:433–441.

Halmepuro, L., Lowenstein, H. 1985. Immunological investigation of possible structural similarities between pollen antigens and antigens in apple, carrot and celery tuber. *Allergy* 40:264–272.

Hannuksela, M. 1980. Atopic contact dermatitis. *Contact Dermatitis* 6:30–32.

Hannuksela, M. 1983. Food allergy and skin diseases. *Ann. Allergy* 51:269–272.
Hannuksela, M. 1986a. Contact urticaria from foods. In: *Contemporary Issues in Clinical Nutrition. Nutrition and the Skin*, Roe, D. A. (Ed.). Alan R. Liss, New York, pp. 153–162.
Hannuksela, M. 1986b. Tests for immediate hypersensitivity. In: *Occupational and Industrial Dermatology*, 2nd ed., Maibach, H. I. (Ed.). Year Book Medical Publishers, Chicago, pp. 168–178.
Hannuksela, M., Förström, L. 1978. Reactions to peroral propylene glycol. *Contact Dermatitis* 4: 41–45.
Hannuksela, M., Kokkonen, E.-L. 1985. Ultraviolet light therapy in chronic urticaria. *Acta Dermato-venereol. (Stockholm)* 65:449–450.
Hannuksela, M., Lahti, A. 1977. Immediate reactions to fruits and vegetables. *Contact Dermatitis* 3:79–84.
Hannuksela, M., Lahti, A. 1986. Peroral challenge tests with food additives in urticaria and atopic dermatitis. *Int. J. Dermatol.* 25:178–180.
Hannuksela, M., Salo, H. 1986. The repeated open application test (ROAT). *Contact Dermatitis* 14:221–227.
Hannuksela, M., Pirilä, V., Salo, O. P. 1975. Skin reactions to propylene glycol. *Contact Dermatitis* 1:112–116.
Hannuksela, M., Karvonen, J., Husa, M., Jokela, R., Katajamäki, L., Leppisaari, M. 1985. Ultraviolet light therapy in atopic dermatitis. *Acta Derm. (Stockholm)* 114 (Suppl.):137–139.
Harada, S., Agarwal, D. P., Goedde, H. W. 1981. Aldehyde dehydrogenase deficiency as cause of facial flushing reaction to alcohol. *Lancet ü*: 982.
Harington, E. 1967. Preliminary report on tyramine headache. *Br. Med. J.* 1:550–551.
Harley, J. P., Ray, R. S., Tomasi, L., Eichman, P. L., Matthews, C. G. Chun, R., Cleeland, S., Traisman, E. 1978. Hyperkinesis and food additives: testing the Feingold hypothesis. *Pediatrics* 6:818–828.
Harries, M. G., Oprien, I. M., Burge, P. S., Pepys, I. 1978. Effects of orally administered sodium chromoglycate in asthma and urticaria due to foods. *Clin. Allergy* 8:423–427.
Henderson, W. R., Raskin, N. H. 1972. ''Hot-dog'' headache: individual susceptibility to nitrite. *Lancet* 2:1162–1163.
Hicks, R. 1968. Ethanol, a possible allergen. *Ann. Allergy* 26:641–643.
Hjorth, N. 1961. *Eczematous Allergy to Balsams, Allied Perfumes and Flavouring Agents*. Munksgaard, Copenhagen.
Hjorth, N., Roed-Petersen, J. 1976. Occupational protein contact dermatitis in food handlers. *Contact Dermatitis* 2:28–42.
Hoube, G. F., Penninks, A. H. 1994. Immunotoxicity of the colour additive Caramel Colour III: a review on complicated issues in the safety evaluation of a food additive. *Toxicology* 91:289–302.
Jacobsen, D. W., Simon, R. A., Singh, M. 1984. Sulfite oxidase deficiency and cobalamin protection in sulfite-sensitive asthmatics (SSA). *J. Allergy Clin. Immunol.* 73:135.
Jewett, D. L., Fein, G., Greenberg, M. H. 1990. A double-blind study of symptom provocation to determine food sensitivity. *New Engl. J. Med.* 323:429–433.
Johnson, H. M., Peeler, J. T., Smith, B. D. 1971. Tartrazine—quantitative passive hemagglutination studies on a food-borne allergen of small molecular weight. *Immunochemistry* 8:281–287.
Juhlin, L. 1981. Recurrent urticaria: clinical investigation of 330 patients. *Br. J. Dermatol.* 104: 369–381.
Juhlin, L. 1986. Adverse reactions to food additives in chronic urticaria. In: *Nutrition and the Skin*, Roe, D. A. (Ed.). Alan R. Liss, New York, pp. 163–177.
Juhlin, L., Michaelsson, G., Zetterström, O. 1972. Urticaria and asthma induced by food- and-drug additives in patients with aspirin hypersensitivity. *J. Allergy Clin. Immunol.* 50:92–98.
Jung, E. G. von. 1969. Bedeutung der Lymphoblastentransformation beim allergischen Kontaktekzem. *Therap. Umschau* 26:94–96.

Karvonen, J., Hannuksela, M. 1976. Urticaria from alcoholic beverages. *Acta Allergol. (Copenhagen)* 31:167–170.

Kero, M., Hannuksela, M. 1980. Guinea pig maximization test, open epicutaneous test and chamber test in induction of delayed contact hypersensitivity. *Contact Dermatitis* 6:341–344.

Ketel, W. G. van. 1978. Dermatitis from octyl gallate in peanut butter. *Contact Dermatitis* 4:60–61.

Kim, C. W., Schöpf, E. 1976. A comparative study of nickel hypersensitivity by the lymphocyte transformation test in atopic and non-atopic dermatitis. *Arch. Dermatol. Res.* 257:57–65.

Kirchhof, B., Haustein, U.-F., Rytter, M. 1982. Azetylsalizylsäure-Additava-Intoleranzphänomene bei chronisch rezidivierender Urtikaria. *Dermatol. Monatsschr.* 168:513–519.

Klecak, G. 1977. Identification of contact allergens: predictive tests in animals. In: *Advances in Modern Toxicology*, Vol. 4, *Dermatoxicology and Pharmacology*, Marzulli, F. N., Maibach, H. I. (Eds.). Hemisphere, Washington, D.C., pp. 305–339.

Klecak, G., Geleick, H., Frey, J. R. 1977. Screening of fragrance materials for allergenicity in the guinea pig. I. Comparison of four testing methods. *J. Soc. Cosmetic Chem.* 28:53–64.

Kobori, T., Araki, H. 1966. Photoallergy in dermatology. *J. Asthma Res.* 3:213–215.

Koenig, J. Q., Pierson, W. E., Frank, R. 1980. Acute effects of inhaled SO_2 plus NaCl droplet aerosol on pulmonary function in asthmatic adolescents. *Environ. Res.* 22:145–153.

Koepke, J. W., Standenmayer, H., Selner, J. C. 1985. Inhaled metabisulfite sensitivity. *Ann. Allergy* 54:213–215.

Köpman, A., Hannuksela, M. 1983. Contact urticaria to rubber. *Duodecim* 99:221–224.

Kripke, M. L., Morison, W. L. 1985. Modulation of immune function by UV radiation. *J. Invest. Dermatol.* 85 (Suppl.):62s–66s.

Krogh, G. von, Maibach, H. I. 1982. The contact urticaria syndrome—1982. *Semin. Dermatol.* 1: 59–66.

Lahti, A. 1980. Non-immunologic contact urticaria. *Acta Dermatovenereol. (Stockholm)*. 60(Suppl.): 91.

Lahti, A., Maibach, H. I. 1984. An animal model for nonimmunologic contact urticaria. *Toxicol. Appl. Pharmacol.* 76:219–224.

Lahti, A., Maibach, H. I. 1985. Long refractory period after one application of nonimmunologic contact urticaria agents to guinea pig ear. *J. Am. Acad. Dermatol.* 13:585–589.

Lahti, A., Väänänen, A., Hannuksela, M. 1986. Acetylsalicylic acid inhibits nonimmunologic contact urticaria. Paper read at the VIII International Symposium on Contact Dermatitis, Cambridge, 20–22 March.

Lamberg, S. I. 1967. A new photosensitizer. The artificial sweetener cyclamate. *J. Am. Med. Assoc.* 201:747–750.

Lambert, N. M., Sandoval, J. H., Sassone, D. M. 1978. Prevalence estimates of hypersensitivity in school children. *Pediatr. Ann.* 7:330–338.

Lancet. 1982. Food additives and hyperactivity (Editorial). *Lancet* i:662–663.

Larese, F., De Zotti, R., Molinari, S., Negro, C., Baur, X. 1993. Occupational allergens in the baking industry. *Allergy* 48(Suppl.):16:20.

Lessof, M. H. 1983. Food intolerance and allergy—a review. *Quart. J. Med.* 52:111–119.

Levantine, A. L., Almeyda, J. 1974. Cutaneous reactions to food and drug additives. *Br. J. Dermatol.* 91:359–362.

Lindmayr, H., Schmidt, J. 1979. Additivaintoleranz bei chronischer Urtikaria. *Wiener Klin. Wochenschr.* 91:817–822.

Lockey, S. D. 1959. Allergic reactions due to FD&C Yellow No. 5 tartrazine, an aniline dye used as a coloring and identifying agent in various steroids. *Ann. Allergy* 17:719–721.

Lockey, S. D. 1971. Reactions to hidden agents in foods, beverages and drugs. *Ann. Allergy* 29: 461–466.

Lum, L. C. 1985. Hyperventilation and pseudo-allergic reactions. In: PAR. *Pseudo-allergic Reactions. Involvement of Drugs and Chemicals*, Vol. 4. Karger, Basel, pp. 106–119.

MacLeod, T. M., Hutchinson, F., Raffle, E. J. 1970. The uptake of labelled thymidine by leucocytes of nickel sensitive patients. *Br. J. Dermatol.* 82:487–492.

Madsen, C. 1994. Prevalence of food additive intolerance. *Human Exp. Toxicol.* 13:393–399.

Magnusson, B., Kligman, A. M. 1969. The identification of contact allergens by animal assay. The guinea pig maximization test. *J. Invest. Dermatol.* 52:268–276.

Maguire, H. C., Chase, M. W. 1972. Studies on the sensitization of animals with simple chemical compounds. *J. Exper. Med.* 135:357–375.

Maibach, H. I., Johnson, H. L. 1975. Contact urticaria syndrome. *Arch. Dermatol.* 111:726–730.

Malkovsky, M., Medavar, P. B. 1984. Is immunological tolerance (nonresponsiveness) a consequence of interleukin 2 deficit during the recognition of antigen? *Immunol. Today* 5:340–343.

Mansfield, L. E., Bowers, C. H. 1983. Systemic reaction to papain in a nonoccupational setting. *J. Allergy Clin. Immunol.* 71:371–374.

Marzulli, F. N., Maibach, H. I. 1977. Contact allergy: predictive testing in humans. In: *Advances in Modern Toxicology*, Vol. 4, *Dermatotoxicology* and Pharmacology, Marzulli, F. N., Maibach, H. I. (Eds.). Hemisphere, Washington, D.C., pp. 353–372.

Mathison, D. A., Stevenson, D. D., Simon, R. A. 1985. Precipitating factors in asthma. Aspirin, sulfites and other drugs and chemicals. *Chest.* 87/1 (Suppl.):50–54.

Maurer, T. 1983. *Contact and Photocontact Allergens. A Manual of Predictive Test Methods.* Marcel Dekker, New York.

Maurer, T., Hess, R., Weirich, E. G. 1985. Prädiktive Tierexperimentelle Kontaktallergenitätsprüfung. Relevanz der Methoden der OECD- und EG-Richtlinien. *Dermatosen* 33:6–11.

McKenna, K. E., Walsh, M. Y. 1994. The Melkersson–Rosenthal syndrome and food additive hypersensitivity. *Br. J. Dermatol.* 31:921–922.

Metcalfe, D. D., Sampson, H. A. 1990. Workshop on experimental methodology for clinical studies of adverse reactions to foods and food additives. *J. Allergy Clin. Immunol.* 86:421–442.

Michaelsson, G., Juhlin, L. 1973. Urticaria induced by preservatives and dye additives in food and drugs. *Br. J. Dermatol.* 88:525–532.

Michaelsson, G., Petterson, L., Juhlin, L. 1974. Purpura caused by food and drug additives. *Arch. Dermatol.* 109:49–52.

Miller, K. 1985. Allergies and idiosyncratic responses. In: *Food Toxicology. Real and Imaginary Problem.* Walker, G. (Ed.). Taylor S. Francis, London.

Miller, S. D., Claman, H. N. 1976. The induction of hapten-specific T cell tolerance using hapten modified lymphoid cells. I. Characteristics of tolerance induction. *J. Immunol.* 117:1519–1526.

Mirza, A. M., Perera, M. G., Maccia, C. A., Dziubynskyj, O. G., Bernstein, I. L. 1975. Leucocyte migration in nickel dermatitis. *Int. Arch. Allergy Appl. Immunol.* 49:782–788.

Mitchell, J. C., Calnan, C. D., Clendenning, W. E., Cronin, E., Hjorth, N., Magnusson, B., Maibach, H. I., Meneghini, C. L., Wilkinson, D. S. 1976. Patch testing with some components of balsam of Peru. *Contact Dermatitis* 2:57–58.

Moneret-Vautrin, D. A. 1983. False food allergies: non-specific reactions to foodstuffs. In: *Clinical Reactions to Food*, Lessof, M. H. (Ed.). Wiley, Chichester, pp. 135–153.

Moneret-Vautrin, D. A., Demange, G., Selve, C., Grilliat, J. P., Savinet, H. 1979. Induction d'une hypersensibilite reaginique a la tartrazine chez le lapin immunisation par voie digestive par le conjugue covalent tartrazine-seralbumine humaine. *Ann. Immunol. Inst. Pasteur* 130c:419–430.

Moneret-Vautrin, D. A., Einhorn, C., Tisserand, J. 1980a. Le role du nitrite de sodium dans les urticaires histaminiques d'origine alimentaire. *Ann. Nutr. Aliment.* 34:1125–1132.

Moneret-Vautrin, D. A., Grilliat, J. P., Demange, G. 1980b. Allergie et intolerance a la tartrazine, colorant alimentaire et medicamenteux a propos de 2 observations. *Med. Nutr.* 16:171–174.

Moore-Robinson, M., Warin, R. P. 1967. Effect of salicylates in urticaria. *Br. Med. J.* 4:262–264.

Morales, C., Penarrocha, M., Bagan, J. V., Burches, E., Pelaez, A. 1995. Immunological study of Melkersson–Rosenthal syndrome. Lack of response to food additive challenge. *Clin. Exp. Allergy* 25:260–264.

Morales, M. C., Basomba, A., Delaez, A., Villamanzo, I. G., Campos, A. 1985. Challenge test with tartrazine in patients with asthma associated with intolerance to analgesics (ASA-triad). *Clin. Allergy* 15:55–59.

Morren, M.-A., Janssens, V., Dooms. Goossens, A., Van Hoeyveld, E., De Wolf-Peeters, C., Heremans, A. 1993. α-Amylase, a flour additive: an important cause of protein contact dermatitis in bakers. *J. Am. Acad. Dermatol.* 29:723–728.

Morison, W. L., Parrish, J. A., Woehler, M. E., Krugler, J. I., Bloch, K. J. 1981. Influence of PUVA and UVB radiation on delayed hypersensitivity in the guinea pig. *J. Invest. Dermatol.* 76:484–488.

Morselli, P. L., Grattini, S. 1970. Monosodium glutamate and the Chinese restaurant syndrome. *Nature* 227:611–612.

National Food Agency of Denmark. 1980. The significance of food and food additives in allergy and intolerance. Report from a working group. (In Danish).

Neuman, I., Elian, R., Nahum, H., Shaked, P., Creter, D. 1978. The danger of yellow dyes (tartrazine) to allergic subjects. *Clin. Allergy* 8:65–68.

NIH. 1983. National Institutes of Health Consensus Development Conference Statement: definite diets and childhood hyperactivity. *Am. J. Clin. Nutr.* 37:161–172.

Niinimäki, A. 1984. Delayed-type allergy to spices. *Contact Dermatitis* 11:34–40.

Niinimäki, A., Hannuksela, M. 1981. Immediate skin test reactions to spices. *Allergy* 36:487–493.

Oliver, A. J., Reade, P. C., Varigos, G. A., Radden, B. G. 1991. Monosodium glutamate-related orofacial granulomatosis. Review and case report. *Oral Surg. Oral Med. Oral Pathol.* 71:560–564.

Orchard, D. C., Varigos, G. A. 1997. Fixed drug eruption to tartrazine. *Australas. J. Dermatol.* 38: 212–214.

Ormerod, A. D., Holt, P. J. A. 1983. Acute urticaria due to alcohol. *Br. J. Dermatol.* 108:723–724.

Ortolani, C., Pastorello, E., Luraghi, M. T., Della Torre, F., Bellani, M., Zanussi, C. 1984. Diagnosis and intolerance to food additives. *Ann. Allergy* 53:587–591.

Osmundsen, P. E. 1980. Contact urticaria from nickel and plastic additives (butylhydroxytoluene, oleylamide). *Contact Dermatitis* 6:452–454.

Osterballe, O., Taudorf, E., Haahr, J. 1979. Asthma bronchiale af konserveringsmidler, farvestoffer og acetylsalicylsyre hos born. *Ugesk. Laeger* 28:1908–1910.

Parodi, G., Parodi, A., Rebora, A. 1985. Purpuric vasculitis due to tartrazine. *Dermatologica* 171: 62–63.

Patton, D. W., Fergusson, M. M., Forsyth, A., James, J. 1985. Oro-facial granulomatosis: a possible allergic basis. *Br. J. Oral Maxillofac. Surg.* 23:235–242.

Pauwels, R., Van Der Straeten, M., Platteau, B., Bazin, H. 1983. The nonspecific enhancement of allergy. *Allergy* 38:239–246.

Pearson, D. J., Rix, K. J. B. 1985. Allergomimetic reactions to food and pseudo-food allergy. In: *PAR. Pseudo-Allergic Reactions. Involvement of Drugs and Chemicals*, Vol. 4. Karger, Basel, pp. 59–105.

Pearson, D. J., Rix, K. J. B., Bentley, S. J. 1983. Food allergy: how much in the mind? A clinical and psychiatric study of suspected food hypersensitivity. *Lancet* i:1259–1261.

Pevny, I., Rauscher, E., Lechner, W., Metz, J. 1981. Excessive Allergie gegen Benzoesäure mit anaphylaktischem Schock nach Expositionstest. *Dermatosen* 29:123–130.

Pollock, I., Warner, O., 1990. Effect of artificial food colours on childhood behaviour. *Arch. Dis. Child.* 65:74–77.

Poulsen, E. 1980. Danish report on allergy and intolerance to food ingredients and food additives. Toxicology Forum, Aspen, Colorado.

Prenner, B. M., Stevens, J. J. 1976. Anaphylaxis after ingestion of sodium bisulfite. *Ann. Allergy* 36:180–182.

Przybilla, B., Ring, J. 1987. Sulfit-uberempfindlickeit. *Hautartzt* 38:445–448.

Quirce, S., Cuevas, M., Olaguibel, J. M. Tabar, A. I. 1994. Occupational asthma and immunologic

responses induced by inhaled carmine among employees at a factory making natural dyes. *J. Allergy Clin. Immunol.* 93:44–52.
Reed, C. E., Benner, M., Lockey, S. D., Enta, T., Makino, S., Carr, R. H. 1972. On the mechanism of the adjuvant effect of *Bordetella pertussis* vaccine. *J. Allergy Clin. Immunol.* 49:174–182.
Reynolds, J. E. F., Prasad, A. B. (Eds.). 1982. *Martindale: The Extra Pharmacopoeia*, 28th ed. The Pharmaceutical Press, London, pp. 708–709.
Roed-Petersen, J., Hjorth, N. 1976. Contact dermatitis from antioxidants. *Br. J. Dermatol.* 94:233–241.
Ros, A.-M., Juhlin, L., Michaelsson, G. 1976. A follow-up study of patients with recurrent urticaria and hypersensitivity to aspirin, benzoates and axo dyes. *Br. J. Dermatol.* 95:19–24.
Rosenhall, L. 1977. Hypersensitivity to analgesics, preservatives and food colourants in patients with asthma or rhinitis. *Acta Univ. Upsaliensis* 269:1–117.
Rowe, A. H., Rowe, A., Jr. 1972. *Food Allergy. Its Manifestations and Control and Elimination Diets.* Charles C. Thomas, Springfield, Illinois.
Rowe, K., S., Rowe, K. J. 1994. Synthetic food coloring and behavior: a dose response effect in a double-blind, placebo-controlled, repeated-measures study. *J. Pediatr.* 125:691–698.
Royal College of Physicians and the British Nutrition Foundation. 1984. Food intolerance and food aversion. A joint report. *J. Roy. Coll. Physicians London* 18:3–41.
Rudzki, E., Czubalski, K., Grzywa, Z. 1980. Detection of urticaria with food additives intolerance by means of diet. *Dermatologica* 161:57–62.
Ruokonen, J. 1982. Secretory otitis media and allergy with special reference to food allergy and cytotoxic leukocyte test. Thesis. Helsinki University.
Samter, M., Beers, R. F. 1968. Intolerance to aspirin. Clinical studies and consideration of its pathogenesis. *Ann. Intern. Med.* 68:975–983.
Saurat, J. H., Burtin, C. J., Soubrane, C. B., Paupe, J. R. 1973. Cell mediated hypersensitivity in skin reactions to drugs (except contact dermatitis). *Clin. Allergy* 3:427–437.
Schaumburg, H., Byck, R., Gerstl, R., Mashman, J. H. 1969. Monosodium l-glutamate: its pharmacology and role in the Chinese-restaurant syndrome. *Science* 163:826.
Schlumberger, H. D. 1980. Drug induced pseudo-allergic syndrome as exemplified by acetylsalicylic acid intolerance. In: *Pseudo-Anaphylactoid Reactions*, Karger, Basel, pp. 125–203.
Schorr, W. F. 1972. The skin and chemical additives to foods. *Arch. Dermatol.* 105:131.
Schwartz, H. I. 1986. Observations on the use of oral cromoglycate in a sulfite-sensitive asthmatic patient. *Ann. Allergy* 57:36–37.
Schwartz, H. I., Scher, T. H. 1983. Bisulfite intolerance manifested as bronchospasm following topical dipivetrin hydrochloride therapy for glaucoma. *Arch. Opthamol.* 103:14–16.
Schwartz, H. I., Sher, T. H. 1985. Bisulfite sensitivity manifesting as allergy to dental anesthesia. *J. Allergy Clin. Immunol.* 75:525–527.
Sethi, T. J., Lessof, M. H., Kemeny, D. M., Lambourn, E., Tobin, S., Bradley, A. 1987. How reliable are commercial allergy tests? *Lancet* i:92–94.
Settipane, G. A. 1984. Adverse reactions to sulfites in drugs and foods. *J. Am. Acad. Dermatol.* 10: 1077–1080.
Settipane, G. A., Pudupakkam, R. K. 1975. Aspirin intolerance. III. Subtypes, familial occurrence, and cross-reactivity with tartrazine. *J. Allergy Clin. Immunol.* 56:215–221.
Silvennoinen-Kassinen, S. 1980. Lymphocyte transformation in nickel allergy: amplification of T-lymphocyte responses to nickel sulphate by macrophages in vitro. *Scand. J. Immunol.* 12:61–65.
Spector, S. L., Wangaard, C. H., Farr, R. S. 1979. Aspirin and concomitant idiosyncrasies in adult asthmatic patients. *J. Allergy Clin. Immunol.* 64:500–509.
Speer, F. 1958. *The Management of Childhood Asthma.* Charles C. Thomas, Springfield, Illinois.
Speer, F., Denison, T. R., Baptist, J. E. 1981. Aspirin allergy. *Ann. Allergy* 46:123–126.
Stenius, B. S. M., Lemola, M. 1976. Hypersensitivity to acetylsalicylic acid (ASA) and tartrazine in patients with asthma. *Clin. Allergy* 6:119–129.

Stevenson, D. D., Simon, R. A. 1981. Sensitivity to ingested metabisulfates in asthma subjects. *J. Allergy Clin. Immunol.* 68:26–32.

Stevenson, D. D., Pleskow, W. W., Gurd, J. G., Simon, R. A., Mathison, D. A. 1982. Desensitization to acetylsalicylic acid (ASA) in ASA-sensitive patients with asthma/rhinosinusitis. In: *PAR. Pseudo-Allergic Reactions. Involvement of Drugs and Chemicals*, Vol. 3. Karger, Basel, pp. 133–156.

Stevenson, D. D., Simon, R. A., Lumry, W. R., Mathison, D. A. 1986. Adverse reactions to tartrazine. *J. Allergy Clin. Immunol.* 78:182–191.

Strean, L. P. 1937. A case of angioneurotic oedema from alcohol. *Can. Med. Assoc. J.* 36:180–181.

Swain, A. R., Dutton, S. P., Truswell, A. S. 1985. Salicylates in foods. *J. Amer. Diet. Assoc.* 85: 950–960.

Sweatman, M. C., Tasher, R., Warner, J. O., Ferguson, M. M., Mitchell, D. N. 1986. Oro-facial granulomatosis. Response to elemental diet and provocation by food additives. *Clin. Allergy* 16:331–338.

Tarasoff, L., Kelly, M. F. 1993. Monosodium *l*-glutamate: a double blind study and review. *Food Chem. Toxicol.* 31:1019–1035.

Tarlo, S. M., Broder, I. 1982. Tartrazine and benzoate challenge and dietary avoidance in chronic asthma. *Clin. Allergy* 12:303–312.

Temesvari, E., Soos, G., Podanyi, B., Kovacs, I., Nemeth, I. 1978. Contact urticaria provoked by balsam of Peru. *Contact Dermatitis* 4:65–68.

Thormann, J., Laurberg, G., Zachariae, H. 1980. Oral sodium cromoglycate in chronic urticaria. *Allergy* 35:139–141.

Timberlake, C. M., Toun, A. K., Hudson, B. J. 1992. Precipitation of asthma attacks in Melanesian adults by sodium metabisulphite. *P. N. G. Med. J.* 35:186–190.

Toorenenberger, A. W. van, Dieges, P. H. 1985. Immunoglobulin E antibodies against coriander and other spices. *J. Allergy Clin. Immunol.* 76:477–481.

Towns, S. J., Mellis, C. M. 1984. The role of acetylsalicylic acid and sodium metasulfite in chronic childhood asthma. *Pediatrics* 73:631–637.

Twarog, F. J., Leung, D. Y. M. 1982. Anaphylaxis to a component of isoetharine (sodium bisulfite). *J. Am. Med. Assoc.* 248:2030–2031.

Vedanthan, P. I., Menon, M. M., Bell, T. D., Bergin, D. 1977. Aspirin and tartrazine oral challenge: incidence of adverse response in chronic childhood asthma. *J. Allergy Clin. Immunol.* 60:8–13.

Veien, N. K., Krogdahl, A. 1991. Cutaneous vasculitis induced by food additives. *Acta Derm. Venereol. (Stockholm)*, 71:73–74.

Veien, N. K., Svejgaard, E. 1978. Lymphycyte transformation test in patients with cobalt dermatitis. *Br. J. Dermatol.* 99:191–196.

Vena, G. A., Foti, C., Angelini, G. 1994. Sulfite contact allergy. *Contact Dermatitis* 31:172–175.

Verschave, A., Stevens, E., Degreef, H. 1983. Pseudo-allergen-free diet in chronic urticaria. *Dermatologica* 167:256–259.

Vincenzi, C., Stinchi, C., Ricci, C., Tosti, A. 1995. Contact dermatitis due to an emulsifying agent in a baker. *Contact Dermatitis* 32:57.

Wahl, R., Kleinhans, D. 1989. IgE-mediated allergic reactions to fruit gums and investigation of cross-reactivity between gelatine and modified gelatine-containing products. *Clin. Exp. Allergy* 19:77–80.

Warin, R. P. 1960. The effect of aspirin in chronic urticaria. *Br. J. Dermatol.* 72:350–351.

Warin, R. P., Smith, R. J. 1976. Challenge test battery in chronic urticaria. *Br. J. Dermatol.* 94: 401–406.

Warin, R. P., Smith, R. J. 1982. Chronic urticaria. Investigations with patch and challenge tests. *Contact Dermatitis* 8:117–121.

Warrington, R. J., Sauder, P. J., McPhillips, S. 1986. Cell-mediated immune response to artificial food additives in chronic urticaria. *Clin. Allergy* 16:527–533.

Warshaw, T. G., Herrmann, F. 1952. Studies of skin reactions to propylene glycol. *J. Invest. Dermatol.* 19:423–430.

Weber, R. W., Hoffman, M., Raine, D. A., Nelson, H. S. 1979. Incidence of bronchoconstriction due to aspirin, azo dyes, non-azo dyes, and preservatives in a population of perennial asthmatics. *J. Allergy Clin. Immunol.* 64:32–37.

Weiss, B., Williams, J. H., Margen, S., et al. 1980. Behavioral responses to artificial food colours. *Science* 207:1487–1489.

Weliky, N., Heiner, D. C., Tamura, H., Anderson, S. 1979. Correlation of tartrazine hypersensitivity with specific serum IgD levels. *Immunol. Commun.* 8:65–71.

Williams, W. R., Pawlowicz, A., Davies, B. H. 1989. Aspirin-like effects of selected food additives and industrial sensitizing agents. *Clin. Exp. Allergy* 19:533–537.

Wolf, S. I., Nicklas, R. A. 1985. Sulfite sensitivity in a seven-year-old child. *Ann. Allergy* 54:420–422.

Wraith, D. G. 1980. Cited in: Commission of the European Communities: Report of a working group on adverse reactions to ingested additives, 1981.

Wütrich, B., Fabro, L. 1981. Acetylsalicylic acid and food additive intolerance in urticaria, bronchial asthma and rhinopathy. *Schweizer Med. Wochenschr.* 39:1445–1450.

Yang, H. W., Purchase, E. C. R., Rivington, R. N. 1986. Positive skin tests and Prausnitz-Küstner reactions in metabisulfite-sensitive subjects. *J. Allergy Clin. Immunol.* 78:443–449.

Yang, W. H., Drouin, M. A., Herbert, M., Mao, Y., Karsh, J. 1997. The monosodium glutamate symptom complex: assessment in a double-blind, placebo-controlled, randomized study. *J. Allergy Clin. Immunol.* 99:757–762.

Young, E., Patel, S., Stoneham, M., Rona, R., Wilkinson, J. D. 1987. The prevalence of reactions to food additives in a survey population. *J. Royal Coll. Phys.* London 21:241–247.

Zeiss, C. R., Lockey, R. F. 1976. Refractory period to aspirin in a patient with aspirin induced asthma. *J. Allergy Clin. Immunol.* 57:440–448.

5

The Role of Food Additives and Chemicals in Behavioral, Learning, Activity, and Sleep Problems in Children

JOAN BREAKEY

Consulting Dietitian, Beachmere, Australia

CONOR REILLY

Oxford Brookes University, Oxford, England

HELEN CONNELL

Queensland Health, Spring Hill, Australia

I. FEINGOLD'S HYPOTHESIS AND ITS HISTORY

In 1974 Dr. Ben Feingold, an allergist in the United States, published a book titled *Why Your Child is Hyperactive* (1). The same ideas were contained in his scientific publications (2–4). The symptoms he investigated were hyperactivity, aggression, excitability, impulsiveness, low frustration tolerance, poor attention, clumsiness, and sleep problems. The outcome measure was parent reports of the usefulness or otherwise of the diet.

Feingold was not the first to publish a relationship between diet and behavior. In the 1920s behavioral or ''neuropathic'' changes due to food ''allergy'' had been described (5–7). In 1945 Schneider (8) claimed that allergy was important in causing the ''syndrome of childhood hyperkinesis''; Speer (9) included behavior changes in children in an ''allergic tension-fatigue syndrome'' in 1954; and others also implicated diet in behavior changes (10,11). An improvement in children with ''minimal brain dysfunction'' was reported in 1970 after allergens such as milk and chocolate were excluded (12). There were widespread misconceptions about ''food allergy'' that confused the orthodox position that reactions which definitely had an immunological basis were allergy, with a view that all types of idiosyncratic reactions to foods or chemicals could come under the umbrella of ''allergy'' (13–15). The latter view was espoused by the less orthodox clinical ecologists who saw hyperactivity and fatigue as part of the ''stimulatory and withdrawal levels of allergic manifestations'' (16).

Historically there were also developments in terminology describing the syndrome involving hyperactivity (17). By 1968 new classifications, ICD-9 and DSM-111 (18,19), introduced the term ''hyperkinetic syndrome of childhood.'' In 1983 attention was emphasised as the core deficit in the new name ''attention deficit disorder,'' with or without hyperactivity (ADD or ADHD).

The Feingold diet excluded artificial colors, artificial flavors, BHA, BHT, and salicylates (medicinal and natural), including apples, stone fruit, berries, almonds, cloves, tea, grapes, oranges, tomato, chilli, cucumber, mint, and perfumes. It was derived from the salicylate-free diet of the Kaiser–Permanente Centre used for urticaria and pruritus (20). It used earlier work on intolerance to aspirin (21–23), with the data on salicylate in food from much earlier studies (24). Additive exclusion had been considered since a connection to allergic symptoms had been published from the 1940s (25–28) and in Feingold's own earlier publication (29).

II. EARLY STUDIES INVESTIGATING FEINGOLD'S HYPOTHESIS

Feingold's ideas became very controversial for several reasons. He said, the ''Pattern was one of 'turn-on' and 'turn-off' '' (1, p. 34); and ''These children are normal. Their environment is abnormal'' (1, p. 74); and ''The time is now overdue to look at these chemicals, not only in relation to [hyperactivity-learning disorders] H-LD's but in regard to the human species as a whole'' (1, p. 162). The discussion became polarized, with Feingold assertively stating the problems and the food industry becoming increasingly defensive.

Three notable double-blind studies were designed specifically to test Feingold's hypothesis. The first, led by Conners, concluded that the results ''strongly suggest that the K-P Diet reduces the perceived hyperactivity of some children'' but was not supportive because of the inconsistency in the results (30). The second study also concluded that Feingold's assertions had not been supported (31). Here parents of the preschooler group rated their children as better on the test diet, but more formal tests did not did not show any difference. Both groups of parents rated their school-age children as better on the test diet but only when the diet order was the control diet followed by the Feingold Diet. The third study utilized specially produced chocolate cookies with or without a range of artificial colors (32). These had been produced by the Nutrition Foundation in the United States after it had suggested strategies for researching Feingold's hypothesis in 1975 (33). Two cookies contained 26 mg dye, the estimated daily intake. Teachers rated children as worse on the colored cookies while on placebo medication, and parents rated children as worse on the colored cookies while on active psychostimulant medication.

The most quoted evaluation of Feingold's hypothesis was by the National Advisory Committee on Hyperkinesis, chaired by Lipton, reporting to the Nutrition Foundation in 1977 (33). Its assessment was based on the three studies cited. It reported that the Feingold hypothesis was not supported and concluded ''that there are presently no data to suggest that initiation of major changes in food manufacture or labelling is urgently needed.'' (33, p. 9). This report was widely publicized and it solved the public health issue in relation to effects on the whole population. Unfortunately this resolution was assumed to *also* apply to clinical use of the diet with individual children, despite the identification of statistically significant effects in some children in the studies.

III. FURTHER DOUBLE-BLIND AND CLINICAL RESEARCH

The early studies were followed by many others, in both research and clinical populations. The gradually accumulating information altered the value of the early studies showing the issue to be much more complex than had been anticipated. During diet implementation, a withdrawal effect with worsening symptoms in the first week was reported (34), while initial improvement took two to six weeks (35) and reactions took up to a week to clear after infractions (36). An effect lasting only three hours was reported after 5 mg tartrazine was used as a challenge (37), and after 26 mg dye, particularly in the young children (38). This seems to contradict reports that exposure to challenge over days increased the reaction (39,40). However, there were also reports that the reaction to different chemicals excluded were different (39,41), and that where an effect reached significance it was in younger children, as had been reported in the early studies.

Feingold had reported the reaction as a nonallergic all-or-nothing phenomenon. Issues related to dose were not discussed. Similarly the various dyes were presumed to be equi-potent. The earlier estimated daily dose was reevaluated, and the dose of 26 mg was estimated as only 40% of the daily intake of dyes (42). Another researcher (43) reported estimated intakes of 76 to 318 mg dyes/day (44). In addition it is reported that artificial flavors are used in *10 times* the dose of color in food, and in *15 times* the dose in pediatric syrups (Hulscher, 1991 personal communication). A study in 1980 (45) used 100 mg and 150 mg dye challenge and showed significant impairment with either dose.

In terms of outcome, studies reported partial responses in ADHD symptoms (40,34), with medication still needed in some patients. Initial studies had focused on ADHD features, but changes in other symptoms were now being reported. Physical allergic symptoms, ''neurotic'' symptoms, and bedwetting were reported as decreasing (35,46), and it was noted that after diet violations in some, the behavior problems were more severe than prior to the diet. As well, improvements in various symptoms were reported in parents and siblings also using the diet (34).

In one study where parent assessment of the presence of dyes in cookies reached significance the authors report that assessment of irritability and restlessness were not the relevant hyperactive symptoms in school (42). Where parent and teacher assessment differed, each was belittled by its lack of correlation with the other, yet researchers in other areas of psychiatry have noted that parents and teachers assess on different criteria even on the same questionnaire (47).

IV. DEVELOPMENT OF THE DIET

Research, particularly that involving dietitians, also showed the need for changes to the diet itself. Further food exclusions of all citrus, currants, capsicum, melon, essences, essential oils, herbs, spices, and chocolate were made (Woodhill's diets of 1975 and 1976), partly based on research in dermatology (48–54). A nature-identical mango flavor in unpreserved uncolored drink was not tolerated (Breakey unpublished data 1977). Limitation of whole foods commonly implicated in allergic reactions, such as milk, wheat, and chocolate, were found necessary in some (34,40,55). Note that chocolate had been used to mask dyes in research challenges. Environmental factors also producing reactions included smells (petrol fumes, felt pens, fly sprays) (39), molds, aromatic plants (36), stress, and infections (56). These findings were also reported by the less orthodox clinical ecologists (57,58). It was becoming apparent that there were several ''aggravating factors,'' both food related and environmental affecting susceptible people with individual variation in chemicals not tolerated.

V. USEFUL DOUBLE-BLIND STUDIES OF THE 1980S

With a background of this information the research during the 1980s was even more clarifying (59). In a double-blind study an initial strict oligoantigenic diet was used. It excluded additives, most fruits and vegetables, and whole foods often implicated in allergy (60). As well as tartrazine [150 mg/day] and benzoate, many whole foods were found to provoke symptoms that included hyperactivity, abdominal pain, headaches, fits, and lethargy.

Another double-blind study used an additive-free diet not excluding salicylates and a challenge of 50 mg dye. Two of eight subjects were ''more irritable and restless'' with more sleep disturbance on the dye (61). The author argues that the inclusion of only those who meet the strict ADHD criteria may exclude some with behavior problems which could be helped by diet.

The much needed data on salicylate content in foods (62,63) paved the way for a strict initial diet emphasising exclusion of salicylates, suspect additives, amines, and monosodium glutamate (63). Additional exclusions were as follows: salicylates, excluding pineapple, dates, rockmelon, guava, gherkin, endive, olive, champignon, radish, chicory, and zucchini; amines, excluding all cocoa and chocolates, matured cheese, all aged foods, vinegar, offal, yeast spreads, pork, salted fish, and bananas; natural and added monosodium glutamate; and the preservatives sulphites, benzoates, nisin, nitrates, propionates, sorbates, and gallates. This diet was used in double-blind research which implicated chemical intolerance in asthma, eczema, migraine, mouth ulcers, urticaria, irritable bowel syndrome, as well as hyperactivity (64,65). These researchers described this as ''food intolerance'' as no immunological mechanism was involved. Wheat and milk were investigated at small dosage but rarely implicated.

VI. REEVALUATION OF THE EARLY STUDIES

The subsequent discoveries have shown the many limitations to the early studies. There was no definition of the on-diet time before assessment began or washout time between diet trials. The order effect could be explained by neglect of withdrawal in the first week of control or test diets. The use of two diets similar enough to satisfy double-blind conditions meant both were lower in additives than the normal diet. As well, the additive dose difference between test and control diets was not great enough in the absence of accurate salicylate data. One study was limited by its lack of exclusion of salicylates, which would interfere with the effect of the added colors.

VII. CLINICAL TRIALS IN THE 1980S

As well as double-blind studies, this author believes it has been important that clinical trials of dietary treatment of ADHD children occur. From 1984–1989, 516 children attending a community child psychiatry service trialled a low additive and amine, low salicylate

(LAALS) diet as part of management of behavior, learning, and hyperactivity problems. As well as perfumes, smells such as paint were investigated. A positive response was obtained in 79.5% of children, with a normal range of behavior achieved in 54.5%. The proportion of responders in the under 9 years old group was significantly higher ($p < 0.005$). As well almost 50% benefited by limiting or excluding other foods such as milk, wheat, and chocolate. Change occurred in behavioral, social, learning, activity, sleep, and allergic problems. It was concluded that suspect diet substances and smells are better thought of as aggravating the underlying predisposition in susceptible children (66).

A second group of 112 children with ADHD symptoms was followed over 18 months. More data particularly on additive and chemical tolerance were collected, and a questionnaire designed by Rowe (RBRI) was used to provide more detail on outcome. A ''diet detective'' approach was used. That is, patients were told that it was not possible to predict outcome, which presenting problems may change nor the degree of change. Other studies had not investigated interacting psychosocial issues, so clinical issues that arose were recorded with progress notes. The sample presented with sleep problems occurring in 74%, developmental problems in 65%, and allergic symptoms, such as asthma or eczema, in 53%. Parent assessment of the value of the diet was substantiated (Fig. 1) with 69.7% reporting benefit. The most significant finding provided by analysis of the questionnaire was that diet was reported to change many problem areas, with the greatest change being in irritability (see Fig. 2). This was followed by poor concentration, impulsivity, unreasonableness, and restlessness, with hyperactivity itself a less significant change. Two-thirds of all responders under 6 years old improved into the normal range. Parents consistently reported that children under 5 years had clearer reactions and that these resolved faster. There was a positive correlation between a history of reports of reactions to food at the first visit and a positive outcome ($p < 0.005$). Thirty-two percent reported other family members benefiting from the diet.

An important part of this research involved trials of individual challenge foods reintroduced for 7 days to test tolerance. Only trials which were uncomplicated by other factors were recorded (see Fig. 3). While this allowed expansion of included foods, the significant number who did not tolerate tomato sauce and chocolate is noteworthy. Additional benefit from limiting dairy foods occurred in 25%, but only 3.6% needed to exclude it. Sugar, in the absence of suspect chemicals, caused reactions in less than 2%. While tolerance of

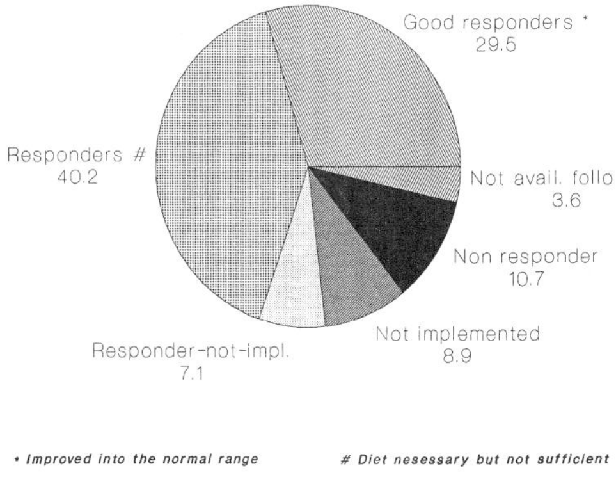

Figure 1 Reported outcome of trial of diet.

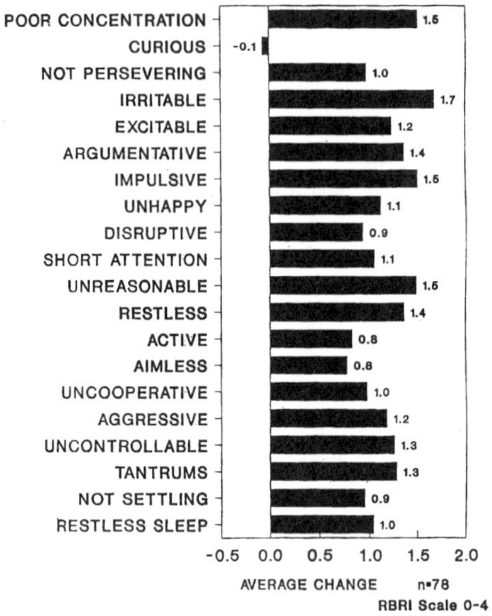

Figure 2 Diet effects for all responders.

fruit containing salicylate generally followed predictions from data on salicylate content, this did not always occur. There was variation between varieties with the least acidic better tolerated. Ripeness was important with salicylates decreasing with ripening (e.g., in bland apples), and amines increasing with ripening (e.g., in bananas). As well, smells (perfumes, petrol, etc.), infections, inhalants (especially pollen), contact dye (face and finger paints), and stress were identified as equally potent sources of reactions. Their ac-

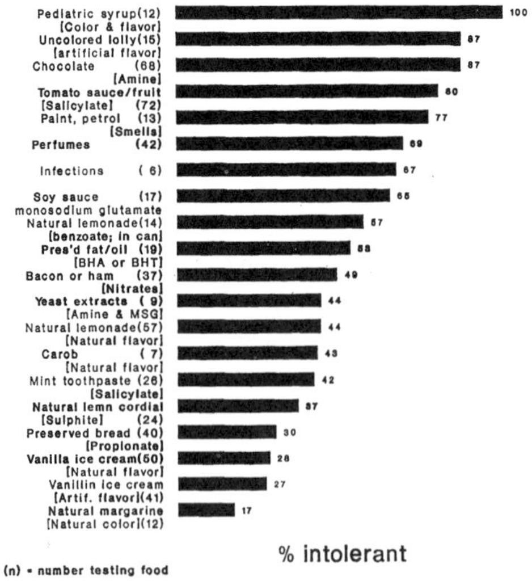

Figure 3 Individual food trials and reported intolerances.

knowledgment does not negate the role of dietary factors. Most patients are prepared to adhere to a well-prepared diet if improvement is of a sufficient degree.

Clinical research shows that problems in child psychiatry are not neat and tidy. Investigation of diet effect revealed diet interacting with many clinical issues such as individual variation in presenting problems; family coping skills and motivation; management of fussy, often underweight children; as well as psychodynamic factors. From this study it is suggested that food-sensitive ADHD children are better described as *hyper-reactive* than hyperactive; reacting to many aspects of the environment, of which the food components form one part, depending on susceptibility.

VIII. A NEW VIEW OF THE DIET

The research during the 1980s allowed updating of the diet. Most fruit and vegetables high in natural amines or glutamates are also high in salicylates (67). Overall, chemicals implicated were added natural and artificial colors, artificial flavors, and most preservatives; added and natural MSG; and natural and medicinal salicylates and amines. A diet with these chemical exclusions formed the initial exclusion diet.

In the 1990s a second part to the initial diet became clearer. This is the role of whole foods. Their limitation or exclusion can be based on a history of a family member having problems with a whole food or from skin prick tests. Research which does not significantly restrict *all the chemical groups and family sensitivities* will not show maximum possible effect.

After the initial trial and challenge (or off-diet week), and washout period if deterioration occurs, individual foods containing various additives, suspect chemicals, or whole foods can be reintroduced one at a time to determine which are tolerated. Each individual

Table 1 Some Factors Contributing to Adverse Reactions: The Total Body Load

Food
Natural food chemicals: salicylates, amines, monosodium glutamate, flavors
Additives: added natural and artificial colors and flavors, benzoates, BHA and BHT, sulphites, nitrates, propionates
Whole foods: consider if a problem to any family member
Environment
Stress, if significant for the child or in the family
Infections: viral, bacterial, parasitic, and fungal
Inhalants: pollen, house dust mite, dander, grasses
Perfumes and smells: paint, air fresheners, glue, bubble bath
Contacts: dyes in finger paint, preservatives in creams
Excess sensory stimuli: shopping centers, crowds
Biology
Genetic susceptibility to ADD or ADHD
Genetic susceptibility to food sensitivity (allergy or food intolerance)
Age and developmental level
Temperament
Other medical problems

Note: Individual variation in susceptibility to different factors occurs. When the cumulative effect of several factors (the total body load) exceeds the individual's threshold, reactions are noted.

can learn doses they tolerate and so expand food intake over time. This improves compliance and ensures better nutrition. With dietetic supervision of dose and form, dairy and wheat foods can be reintroduced or increased. No screening tests have been found useful for predicting food intolerance. A schema for diet investigation is included as ''The Diet Detective Process'' in Appendix A, a ''Diet for Suspected Food Intolerance'' comprises Appendix B, and an overview of factors contributing to adverse reactions (''the total body load'') is provided in Table 1.

IX. THE ROLE OF DIET IN ADVERSE SYMPTOMS IS COMPLEX

The suspect diet substances being investigated have broadened, as have the symptoms affected by diet. Both the dose of suspect substances and the age of child are relevant. There is individual variation in problem profiles presented, in symptoms that change, in amount of change, and in substances not tolerated. A review of the research from 1985 to 1995 shows that a role of diet in some ADHD children has been demonstrated (59). However, there is much work still to be done. Various mechanisms have been proposed, notably allergy and idiosyncrasy coexisting (60), neuropharmacological abnormality (65), and many others (68–76). A valuable pharmacological view is provided by Neims (77). The health of the gut may be important. Tolerance is often decreased as grain fiber is increased. New work on probiotics in the form of degradation of some protein components and adsorption of suspect chemicals from intestinal contents should be considered (78).

Note that most foods not tolerated are highly flavored: spices, teas, tangy fruit, tomato, chocolate, aged food, foods high in monosodium glutamate, and commercial sweets and drinks. This author suggests a change in emphasis from artificial colors and preservatives toward flavor to the extent that all strong aromatic natural and additive flavors be equally emphasised suspect chemicals. Higher tolerance is related to mild, good quality flavor. High quality foods (e.g., fine wines) and preferred smells (e.g., expensive perfumes) are better tolerated. At the other end of the spectrum is decreased tolerance as the food is perceived as smelling stale or musty. Perhaps this is susceptibility to very small doses of spoilage organisms or their products. Clinically, food-sensitive people are more aware than others of both smell and taste. They are people whose threshold for adverse reactions seems lower than the general population, with more noticeable interaction with ''the total body load'' of factors in the environment. On a positive note, it is of interest that few food-sensitive people develop coronary heart disease, diabetes, or obesity.

A second suggestion is to no longer presume that salicylate is the main suspect natural chemical. Tolerance to salicylate-containing foods varies with individuals and does not correlate closely with analysis data. Researchers using similar methodology to Swain report finding little salicylate in their analysis of similar foods (79,80). This does not mean sensitivity to natural chemicals is not significant. Indeed, most of the later research implicates natural foods as well as additives. At this time it appears wisest to clarify the foods that are consistently causing reactions (outlined in Appendix B) and to admit that it is not known exactly what the suspect chemicals are or why they cause reactions.

In conclusion, the research, particularly from the 1980s and 1990s, has shown diet has a role. There are three groups of suspect substances: additives, natural chemicals, and whole foods, with those implicated in the family given special emphasis (81). This discussion has considered suspect natural substances at some length as their role is only gradually being clarified. Artificial additive color and flavor are still the most often reported suspect substances. Rather than susceptability being individual, as occurs in allergic reactions, it is present in a population subgroup as yet not clearly defined. A positive diet response is more likely where there is a family history of allergic conditions, migraine, or irritable bowel syndrome.

Further research should incorporate all of the issues raised here and should be multidisciplinary, preferably including psychiatry, pediatrics, allergy, food technology, neurology, education, immunology, psychology, dietetics, molecular biology, toxicology, and pharmacology. The latter inclusions should have a higher profile than in early work.

The picture of what could be called the side effects of foods is complex. Food sensitivity exists and provides a challenge for all those working in related fields.

APPENDIX A: THE DIET DETECTIVE PROCESS

Use for investigation of suspected food sensitivity.

Stage 1: Screening Diet

The family detective diet. Use ''The diet for suspected food sensitivity'' plus family suspect whole foods, inhalants, and chemicals. Also exclude foods causing positive skin-prick tests. Trial diet for 4 weeks. Withdrawal effects may occur in the first 3–10 days.

Stage 2: Challenges

1. If little or no change occurred after diet trial, reintroduce *all exclusions in maximum dose.*
2. If noticeable change occurred, reintroduce *groups of food chemicals* one group at a time.

All challenges must continue for 7 days or until a reaction is obvious if this is earlier.

Stage 3: Individualizing Diet

The aim is to finally clarify necessary exclusions and to provide foods which may be used occasionally.

Single Food Challenges

Food challenges can be selected to provide trials of particular suspect additives or food chemicals. Trial each food for 7 days to allow for a possible cumulative effect.

From these trials the long-term diet can show necessary exclusions, frequency of occasional inclusions, and freely allowed items.

APPENDIX B: THE DIET FOR SUSPECTED FOOD INTOLERANCE

The diet excludes artificial and natural additive colors, flavors, some preservatives and antioxidants, added and natural monosodium glutamate, and also natural amines and natural and medicinal salicylates. Environmental factors to also minimize are strong smells, skin contact with children's colored paint, and significant stress. Ensure any infection present is treated before beginning trial.

You may use:	Do not use:
Bread and cereals	
All plain breads without preservative, all flours, meals, cereals, tapioca, sago, rice, commercial breakfast cereals without coloring or flavoring, homemade cakes and biscuits	Fruit or fancy loaves, most commercial bakery products, pasta with coloring, spaghetti in tomato sauce, cake mixes, custard powder, instant puddings, bread with preservatives
Fruit	
Peeled ripe pear, pawpaw, mango* (including tinned in syrup and sundried), just-ripe banana, pomegranate, peeled nashi fruit (outer part)	All other fruit, fruit juices, and dried fruit
Vegetables	
Potatoes (peeled), lettuce, celery, cabbage, bamboo shoots, swede, green beans, brussels sprouts, peas, leeks, shallots, chives, choko, mild carrot* and pumpkin*, parsnip*, broccoli*, beetroot*, marrow*, spinach*, onion*, sweet potato*, cauliflower*, turnip*, asparagus*, sweet corn*, garlic^{+}	All other vegetables, especially tomato and tomato products, capsicum, peppers, gherkins, olives, minted peas
Protein foods	
Plain beef, including mince, veal, lamb, chicken, pork, plain fish, lobster, oysters, prawns, eggs, dried peas, beans and lentils, most nuts, peanut butter*, sunflower seeds, bacon^{+}, ham^{+}	Sausages, sausage mince, baked beans in tomato sauce, luncheon sausages, corned beef, meat and fish pastes, commercial fish in batter, smoked fish or meats, almonds, water chestnuts
Milk products	
Whole, trim & skim milk, plain yoghurt, evaporated and sweetened condensed milks, home-made ice cream, fresh cream, cottage, processed and fresh mild cheeses, plain whey milk, plain soy milk	Commercial ice cream; fruit and flavored yogurt; milk ices; matured, tasty, camembert, blue vein, or flavored cheeses; flavored milk drinks
Fats and oils	
Butter, cream, copha, oils, lard, dripping or ghee without preservative, butter–oil blends, pale colored milk-free margarine	Margarines, any fat or oil containing preservatives, olive oil, almond oil, sesame and anchovy oils
Drinks	
Water; milk; malted milk; soda and plain mineral waters; decaffeinated coffee; diluted drinks from tinned pear, pawpaw, or mango; diluted homemade golden passionfruit[#] cordial	All soft drinks, cordials, fruit juice drinks and juices, tea, flavored milks, chocolate drinks, cider, wine, beer, liqueurs

You may use:	Do not use:
Snacks and sweets	
Sugar, golden syrup, pure maple syrup, treacle, very bland honey*, malt syrup, plain junket tablets, plain gelatine (boil), glycerine, plain potato chips, yeast spreads* (scrape only)	Twisties, cheezels, flavored french fries or potato chips, jelly, flavored junket tablets, low joule drinks, jellies, etc., all colored and flavored lollipops, ices, and ice cream
Miscellaneous	
Salt, pepper+, malt vinegar+, parmesan cheese+, garlic and onion salts+, home-made mayonnaise	Flavor enhancer MSG, meat and chicken cubes, tomato and other commercial sauces, pickles, cider or wine vinegar
Toothpaste, soap, and cosmetics	
Unflavored white toothpaste, plain unperfumed or minimally perfumed shampoo, soaps, deodorant, detergents, powders	All mint and herbal toothpastes, room deodorizers, perfumed items, bubble bath, paint, glues, varnish, etc.

* Foods allowed during diet trial if only one serving per day (several may be included, one serving of each).

+ Foods allowed if used only to flavor meals.

\# Foods allowed only as dilute flavor in drinks, not whole.

Note: To ensure sufficient vitamin C is present include four servings of fruit or vegetable daily. Good sources include banana, potato, bean sprouts, broccoli, and mango nectar.

REFERENCES

1. Feingold, B. F. 1974. *Why your child is hyperactive*. Random House, New York.
2. Feingold, B. F. 1973. Psychological factors in allergic disease. In: *Introduction to Clinical Allergy*. Charles C. Thomas, Springfield, Illinois, p. 189.
3. Feingold, B. F. 1974. Hyperkinesis and learning difficulties (H-LD) linked to the ingestion of artificial colours and flavours. *Presentation to the AMA Annual Meeting*.
4. Feingold, B. F. 1976. Hyperkinesis and learning disabilities (H-LD) linked to the ingestion of artificial food colours and flavours. *J. Learning Disabilities* 9:551.
5. Shannon, R. 1922. Neuropathic manifestations in infants and children as a result of anaphylactic reactions to foods contained in their dietary. *Am. J. Dis. Child.* 24.
6. Duke, W. 1923. Food allergy as a cause of illness, *JAMA* 81:886.
7. Rowe, A. 1928. Food allergy: its manifestations, diagnosis and treatment. *JAMA* 91:1623.
8. Schneider, W. 1945. Psychiatric evaluation of the hyperkinetic child. *J. Pediat.* 26:559.
9. Speer, F. 1954. The allergic-tension-fatigue syndrome, *Paed. Clin. N. Am.* 1:1029.
10. Rinkel, H., Randolph, T., Zeller, M. 1951. *Food Allergy*. Charles C. Thomas, Springfield, Illinois.
11. Rowe, A. 1969. Allergic toxaemia and fatigue. *Annals. All.* 17:70.
12. Kittler, F. 1970. The effect of allergy on children with minimal brain damage. In: *Allergy of the Nervous System*, Speer, F. (Ed.). Charles C. Thomas, Springfield, Illinois.
13. Loblay, R. 1989. Unorthodox approaches to allergy. *Aust. Prescriber.* 12:78.
14. Randolph, T. G. 1947. Allergy as a cause of fatigue, irritability and behaviour problems in children. *J. Pediatr.* 31:560.

15. Crook, W. 1963. The allergic tension-fatigue syndrome. In: *The Allergic Child*, Speer, F. (Ed.). Har-Raw, New York, p. 329.
16. Randolph, T. G. 1976. Adaptation to specific environmental exposures enhanced by individual susceptibility. In: *Clinical Ecology*, Dickey, L. (Ed.). Charles C. Thomas, Springfield, Illinois.
17. Weiss, G. 1991. Attention deficit hyperactivity disorder. In: *Child and Adolescent Psychiatry*, Lewis, M. (Ed.). Williams & Wilkins, Baltimore.
18. World Health Organisation. 1965. *International Classification of Disease.*
19. American Psychiatric Association. 1968. *Diagnostic and Statistical Manual of Mental Disorders II.* American Psychiatric Association, Washington, D.C.
20. Friedman, A. 1973. Management with the elimination diet. In: *Introduction to Clinical Allergy*, Feingold, B. (Ed.). Charles C. Thomas, Springfield, Illinois, p. 162.
21. Shelly, W. 1964. Birch pollen and aspirin psoriasis, *JAMA* 189:985.
22. Moore-Robinson, M., Warin, R. 1967. Effects of salicylates in urticaria. *Br. Med. J.* 4:262.
23. Samter, M., Beers, R. 1968. Intolerance to aspirin. *Ann. Intern. Med.* 68:975.
24. Wein, G. Klein. 1932. *Handbuch der Pflanzenanalyse*, Vol. 1. p. 538.
25. Lockey, S. Allergic reactions due to dyes in foods. Pennsylvania Allergy Society, Fall Meeting, 1948.
26. Lockey, S. 1959. Allergic reactions due to FD&C Yellow No. 5 tartrazine, an aniline dye used as a colouring and identifying agent in various steroids. *Ann. Allerg.* 17:719.
27. Randolph, T. 1954. Allergic type reactions to chemical additives of foods and drugs. *J. Lab. Clin. Med.* 913.
28. Chaffe, F., Settipane, G. 1967. Asthma caused by FD&C approved dyes. *J. Allergy* 40:65.
29. Feingold, B. 1968. Recognition of food additives as a cause of symptoms of allergy. *Ann. Allerg.* 26:309.
30. Conners, C. K., Goyette, C. H., Lees, J. M., Andralonis, P. 1976. Food additives and hyperkinesis: a controlled double-blind experiment. *Paediatrics* 58:154.
31. Harley, J. P., Ray, R. S., Tomasi, L., Eichman, P. L., Matthews, C. G., Chun, R., Cleeland, S., Traisman, E. 1978. Hyperkinesis and food additives: testing the Feingold hypothesis. *Paediatrics* 61:818.
32. Williams, J. I., Cram, D. M., Tausig, F. T., Webster, E. 1978. Relative effects of drugs and diet on hyperactive behaviours: a experimental study. *Paediatrics* 61:811.
33. Nutrition Foundation. 1975. *Report to the Nutrition Foundation.* National Advisory Committee on Hyperkinesis and Food Additives, Nutrition Foundation, New York.
34. Breakey, J. 1978. Dietary management of hyperkinesis and behavioural problems. *Aust. Family Physician* 7:720.
35. Hindle, R. C., Priest, J. 1978. The management of hyperkinetic children: a trial of dietary therapy. *NZ Med. J.* 88:43–45.
36. Breakey, J. 1977. *A Manual for the Additive Free Low Salicylate Diet.* J. Breakey, Brisbane, Australia.
37. Levy, F., Drumbrell, S., Hobbes, G., Ryan, M., Wilton, N., Woodhill, J. M. 1978. Hyperkinesis and diet. A double-blind crossover trial with tartrazine challenge. *Med. J. Aust.* 1:61.
38. Goyette, C. H., Conners, C. K., Petti, A. P., Curtis, L. E., 1978. Effects of artificial colours on hyperactive children: a double-blind challenge study. *Psychopharmacology Bulletin* 14:39.
39. Cook, P. S., Woodhill, J. M. 1976. The Feingold dietary treatment of the hyperkinetic syndrome. *Med. J. Aust.* 2:85.
40. Brenner, A. 1977. A study of the efficacy of the Feingold diet on hyperkinetic children. *Clinical Paediatrics* 16:652.
41. Priest, J. 1979. Dietary treatment of hyperkinesis. Home Science Quadrennial Conference Proceedings, Hamilton, New Zealand.
42. Mattes, J. A., Gittelman-Klein, R. 1978. A crossover study of artificial food colourings in a hyperkinetic child. *Am J Psychiatry* 135:987.

43. Prinz, R. J., Roberts, W. A., Hantman, E. 1980. Dietary correlates of hyperactive behaviour in children. *J. Consulting Clin. Psychol.* 48:760.
44. Beloran, A. 1976. Effects of average, 90th percentile and maximum daily intakes of FD&C artificial food colours in a day's diet among 2 age groups of children. Department of Health Education and Welfare, FDA Biochemical Toxicology Branch, Memo dated July 30, Washington, D.C.
45. Swanson, J. M., Kinsbourne, M. 1980. Food dyes impair performance of hyperactive children on a laboratory learning test. *Science* 207:1485.
46. Salzman, L. K. 1976. Allergy testing, psychological assessment and dietary treatment of the hyperactive child syndrome. *Med. J. Aust.* 2:248.
47. Rutter, M., Tizard, J., Whitmore, K. 1970. *Education, Health and Behaviour.* Longman, London, p. 164.
48. Ros, A. M., Juhlin, L., Michaelsson, G. 1976. A follow-up study of patients with recurrent urticaria and hypersensitivity to aspirin, benzoates, and azo dyes. *Br. J. Dermatology* 19–24.
49. Freedman, B. 1977. A dietary free from additives in the management of allergic disease. *Clin. Allergy* 7: 417.
50. Amos, H. E., Drake, J. J. 1977. Food additives and hypersensitivity. *International Flavours and Food Additives* Jan.:19.
51. Michaelssohn, G., Juhlin, L. 1973. Urticaria induced by preservatives and dye additives in food and drugs. *Br. J. Dermatology* 88:525.
52. Juhlin, L., Michaelsson, G., Zetterstrom, O. 1972. Urticaria and asthma induced by food- and-drug additives in patients with aspirin hypersensitivity. *J. Allergy Clin. Immunol.* 50:92.
53. Lockey, S. D. 1971. Reaction to hidden agents in foods, beverages and drugs. *Ann. Allergy* 29:461.
54. Noid, H. E., Schulze, T. W., Winkelmann, R. K. 1974. Diet plan for patients with salicylate induced urticaria. *Arch. Dermatol.* 109:866.
55. Mackarness, R. 1976. *Not All in the Mind.* Pan, London.
56. Breneman, J. 1978. *Basics of Food Allergy.* Charles C. Thomas, Springfield, Illinois.
57. Dickey, L. D. (Ed.). 1976. *Clinical Ecology.* Charles C. Thomas, Springfield, Illinois.
58. Randolph, T. G. 1962. *Human Ecology and Susceptibility to the Chemical Environment,* Charles C. Thomas, Springfield, Illinois.
59. Breakey, J. 1977. The role of diet and behaviour in childhood. *J. Paediatr. Child Health* 33: 190–194.
60. Egger, J., Graham, P. J., Carter, C. M., Gumley, D., Soothill, J. F. 1985. Controlled trial of oligoantigenic treatment in the hyperkinetic syndrome. *Lancet* 540.
61. Rowe, K. S. 1988. Synthetic food colourings and hyperactivity: a double-blind crossover study, *Aust. Paediatr. J.* 24:143.
62. Swain, A., Dutton, S., Truswell, A. S. 1982. Salicylates in Australian foods. Proceedings of the Nutrition Society of Australia, p. 163.
63. Swain, A. R., Dutton, S. P., Truswell, A. S. 1985. Salicylates in foods. *J. Am. Dietetic Assn.* 85:950.
64. Swain, A. R. 1988. *The Role of Natural Salicylates in Food Intolerance.* University of Sydney, Sydney.
65. Loblay, R. H., Swain, A. R. 1986. Food intolerance. *Recent Adv. Clin. Nutrition* 2:169.
66. Breakey, J., Hill, M., Reilly, C., Connell, H. 1991. A report on a trial of the low additive, low salicylate diet in the treatment of behaviour and learning problems in children. *Am. J. Nutr. Diet.* 48:3.
67. Swain, A. R., Loblay, R., Soutter, V. 1991. *Friendly Food: Avoiding Allergies, Additives and Problem Chemicals.* Murdoch, Sydney.
68. Wurtman, R. D. 1983. Nutrition: the changing scene. *Lancet* 1:1145.
69. Lifshitz, F. Disaccharide intolerance. In: *Food Intolerance.* Chandra, R. K. (Ed.). Elsevier Science, New York.

70. Grady, G. F., Keusch, G. T. A. 1971. Pathogenesis of bacterial diarrhoeas. *New Eng. J. Med.* 825:831.
71. Hunter, J. O. 1991. Food allergy—or enterometabolic disorder. *Lancet* 338:495.
72. Crayton, J. W. 1986. Immunologically mediated behavioural reactions to food. *Food Technol.* 40(1):153.
73. Weiss, B. 1980. Behaviour as a common focus of toxicology and nutrition. In: *Nutrition and Behaviour.* Miller, S. A. (Ed.). The Franklin Institute Press, p. 95.
74. Henker, B., Wahlen, C. K. 1989. Hyperactivity and attention deficits. *Am. Psychologist* 44(21): 216.
75. Zametkin, A. J., Nordahl, T. E., Gross, M., King, A. C., Semple, W. E., Rumsey, J., Hamburger, S. 1990. Cerebral glucose metabolism in adults with hyperactivity of childhood onset. *New Eng. J. Med.* 323(20):1361.
76. Brostoff, J., Challacombe, S. (eds.). 1987. *Food Allergy and Intolerance.* Bailliere Tindall, London.
77. Neims, A. H. 1986. Individuality in the response to dietary constituents: some lesson from drugs. *Nutrition Rev.* 44(Suppl.):S237.
78. Isolauri, E. 1996. Studies on Lactobacillus GG in food hypersensitivity disorders. *Nutrition Today Supplement* 31(Nov./Dec.):6.
79. Venema, D. P., Holman, P. C. H., Janssen, K. P., Katan, M. B. 1996. Determination of acetylsalicyclic acid and salicylic acid in foods, using HPLC with fluorescence detection. *J. Agric. Food Chem.* 44:1762–1767.
80. Janssen, K. P., Holman, P. C., Venema, D. P., van Staveran, W. A., Katan, M. B. 1996. Salicylates in foods. *Nutrition Rev.* 54(11):357–359.
81. Breakey, J. 1998. *Are You Food Sensitive*? CE Breakey Medical Pty. Ltd. Brisbane, Australia.

6

Consumer Attitudes Toward Food Additives

CHRISTINE M. BRUHN

Center for Consumer Research, Davis, California

I. INTRODUCTION

Concern for food additives is expressed by a relatively small percentage of U.S. consumers. When a variety of potential food hazards are specified, those who classified food additives as a serious health hazard decreased from 36% in 1987 to 21% in 1997 (Abt Associates Inc., 1997; Food Marketing Institute, 1990). Similarly, those viewing artificial coloring as a serious hazard decreased from 28% in 1985 to 21% in 1990 (Food Marketing Institute, 1990). This compares to 82% classifying harmful bacterial and 66% classifying pesticide residues as a serious hazard in 1997 (Abt Associates Inc., 1997). When consumers are asked to volunteer food safety concerns, the most frequently cited area is bacteria related, mentioned by 69% of consumers in 1997. Additives, preservatives, and artificial colors are volunteered by 1% or fewer consumers.

Among European consumers, bacteria contamination leads the list of health hazards, with 85% considering this very dangerous (Torjdman, 1995). Artificial coloring is considered very dangerous by 39% of consumers, additives and preservatives by 31%. It is not realistic to expect a question to lead to a ''no concern'' response by all respondents. Asking about risk results in some consumers considering each topic addressed a hazard. In this survey, the item considered very dangerous by the fewest consumers was sugar, classified as very dangerous by 12% of the respondents.

Food safety topics were not specified in a survey targeting Australian consumers; rather people were asked to volunteer areas of concern. Additives and preservatives are considered hazardous by relatively few Australians. Only 5% mentioned additives and 4% preservatives as a perceived threat to food safety in 1996 (Australian Supermarket

Institute, 1997). This is a decrease from 12% and 10% who mentioned additives and preservatives in 1992.

II. BASIS FOR CONCERN

Concern is directed toward synthetic rather than naturally occurring additives. In-home interviews found concern about additives was related to a general fear of chemicals and diseases, specially cancer (Sloan et al., 1986). Consumers believe chemicals present hidden and unknown dangers which people are powerless to predict (McNutt et al., 1986). Consumer association with chemicals may relate to environmental concerns such as acid rain and toxic waste dumps. Over half of consumers indicated they think there is a connection between chemical spills and their feelings about chemicals in food. Almost 20% of consumers believed that chemicals are never good for people and over 80% believed that chemicals cause cancer. Consumers appear to believe that chemicals cause cancer, artificial ingredients are chemicals, and therefore artificial ingredients cause cancer.

Consumers indicated a lack of trust in regulators and regulatory procedures (Kajanne and Pirttila-Backman, 1996). Both men and women believed that expert opinion was influenced by who paid their salary. Concern was believed justified since some approved additives were later withdrawn.

Food allergies may also trigger concern about additives. In the early 1980s 26% of U.S. consumers said they feared an allergic reaction to artificial ingredients, and 22% avoided particular foods fearing the food may contain an allergen (*Good Housekeeping*, 1984). Similarly in Australia, the main reason given for looking for food additives was allergies (Crowe et al., 1992).

Consumer response to additives may also relate to perceived quality. Over 75% of consumers indicated that artificial flavors give food a poorer taste than natural flavors. Their descriptions of artificial flavoring range from ''too salty'' to ''too sweet'' and from ''strong tasting'' to ''flat tasting'' (Zibrik et al., 1981). Almost 70% of consumers attributed better quality to natural flavors (McNutt et al., 1986).

Consumers appear unaware of potential benefits from additives. More than 80% of Canadian consumers surveyed in 1978 thought food colors were unnecessary and believed additives made food less safe (Zibrik et al., 1981). Sloan and co-authors believe consumers will accept a product if benefits outweigh disadvantages and possible risks are small (Sloan et al., 1986).

Terminology can influence consumer acceptability. Terms that connote approval by an official body, such as ''U.S. Certified'' or ''FD&C'' were moderately acceptable, while ''artificial color'' was least acceptable (McNutt et al., 1986). Concerns about additives among Finish consumers also related to perceptions of nature or natural processes (Kajanne and Pirttila-Backman, 1996). When Finish men spoke of additives, they discussed industrial poisons, radiation, and pollution. U.S. consumers defined ''natural'' as ''food without additives,'' ''no artificial or synthetic ingredients,'' ''free of chemicals/not chemically treated'' (*Good Housekeeping*, 1983). Additionally 71% agreed that ''natural foods'' are always better for you than processed foods.''

Industry advertising can increase consumer concern. In the 1980s and 1990s consumers are sometimes urged to select a product for the ingredients it lacks (Sloan et al., 1986). A 1997–1998 advertisement for a national brand of ice cream shows a child having difficulty pronouncing a list of ingredients. The child then reads ingredients from another label, ''cream, sugar, vanilla,'' This brand is then described as providing ''taste not technology.'' By suggesting there is a reason to reject a product with less familiar ingredients, concern about safety or quality is validated.

III. EFFECT OF LABELING ON ATTITUDES

The Australian food additive labeling requirements are based on the European system of numerical codes. In 1987, approved food additives were given specific numbers, and lists relating numbers to specific ingredients became available to the public. Interviews with over 500 consumers found only 39% were able to correctly identify a food additive. Almost half of the respondents in the sample who used food-additive labeling found it confusing or difficult to understand. When asked what improvements they would make, the primary suggestions were that labels should carry more information. Consumers thought that manufacturers were trying to disguise the additives by giving them numbers rather than names. Consumers also requested information on the function of individual additives and the side effects related to human consumption.

IV. SPECIAL CASE: FOOD IRRADIATION

Irradiation is the exposure of food to a source of radiation energy: gamma rays, electrons, and X-rays. This energy inactivates parasites and destroys pathogens in food, including *Escherichia coli* O157:H7, *Salmonella*, *Campylobacter*, and *Trichinella spiralis*. Irradiation destroys insects and extends the shelf life of fresh fruit and vegetables. In 1958, the United States Congress defined the process of food irradiation as a food additive. The U.S. Food and Drug Administration must approve any application of irradiation, and irradiated foods sold to the consumer must be labeled with the international symbol for radiation and the words, ''treated by radiation'' or ''treated with irradiation.'' The food industry has hesitated to use this technology, in part due to concern about consumer reaction.

Although consumer understanding of irradiation is limited, interest in purchasing irradiated foods has gradually increased due to increased presentations of the safety and benefits of irradiation in the media (Bruhn, 1998; Bruhn, 1995). Following the FDA approval of irradiation of red meat, television, newspapers, and magazines carried features which included endorsements of irradiation by health professionals. A nationwide study conducted in March 1998 found almost 80% of consumers said they would buy products labeled, ''irradiated to destroy harmful bacteria'' (American Meat Institute, 1998). This compares to the 1996 response rate of 69% among those who had heard of irradiation (Abt Associates Inc., 1996).

Consumers appear to be responsive to the potential advantages of irradiating poultry. Sixty-seven percent of consumers said it was ''appropriate'' to irradiate poultry, with pork and ground beef seen as ''appropriate'' by slightly fewer consumers (American Meat Institute, 1998). Over 60% felt irradiation was appropriate at a fast food restaurant with almost 50% considering it appropriate at the grocery store deli or sit-down restaurant.

Consumers see irradiation's main advantage as the destruction of harmful bacteria, with almost 80% indicating that as a reason to buy irradiated products. No one expects irradiation to replace safe food handling. In the 1998 survey, 91% of consumers responded that safe food handling is still important (American Meat Institute, 1998).

When asked what information about irradiation would be useful, people indicated information about product safety, nutritional quality, potential harm to employees, and potential danger from living near an irradiation facility. Consumers have indicated in focus group discussions that the most important information is: irradiated foods established record of safety and wholesomeness, the process destroys bacteria and protects against foodborne illness, and irradiation is endorsed for safety by health authorities (Gallup Organization, 1993).

Consumers have purchased labeled irradiated products since 1986, although sustained marketing did not occur until 1992 when a record amount of irradiated strawberries were sold in Florida in 1992 (Marcotte, 1992). Thereafter numerous irradiated produce items have been marketed in the Chicago area, and irradiated products typically outsell nonirradiated by a twenty-to-one ratio (Corrigan, 1996; Pszczola, 1992).

Since the fall of 1995 tropical fruit from Hawaii has been in several Midwest and West Coast markets in collaboration with a study to determine quarantine treatment. Over 320 thousand pounds of irradiated fruit has been sold, including papaya, atemoya, rambutan, lychee, starfruit, banana, and Chinese taro. The fruit was shipped to an Isomedix plant near Chicago for irradiation between 0.25 kGy and 1.0 kGy (Dietz, 1998).

The market response to irradiated poultry was tested in Kansas. In 1996 market-irradiated poultry captured 63% of the market share when priced 10% less than the store brand, 47% when priced equally, and 18% and 17% when priced 10% or 20% higher (Fox, 1996). In 1997, when consumers read background information about irradiation before making a selection, 80% selected irradiated poultry (Fox and Olson, 1998).

V. SPECIAL CASE: BIOTECHNOLOGY

Consumer attitudes toward biotechnology and labeling policy of genetically modified products differ. The U.S. Food and Drug Administration (FDA) policy of labeling has focused on the safety of the product, not the method of production. This policy calls for a review of any food with nutritional changes, potential allergens, or the addition of pesticides or herbicides. In contrast, the European Union has chosen to focus on method of production. All food products containing ''genetically modified organisms'' must be labeled (Lewis, 1997).

A. Attitudes

Consumer response toward biotechnology in the United States is quite positive. In a 1997 study, almost 80% of Americans said they were aware of biotechnology, more than half (54%) said biotechnology has already provided benefits to them, and 78% predicted they would benefit from biotechnology in the next five years (IFIC, 1997a,b,c). Nearly half of survey respondents realized foods produced through biotechnology were already in supermarkets. Almost two-thirds (62%) indicated they were very or somewhat likely to buy a product modified to taste better or fresher, with 17% of these very likely (Abt Associates Inc., 1996). Additionally 74% were very or somewhat likely to buy a product modified to resist insect damage and require fewer pesticide applications. When specifically asked, 15% of consumers viewed products modified by biotechnology as presenting a serious health hazard.

Canadians are positive toward biotechnology, however they believe the technology entails risks (Walter, 1994). Most Australians believe genetic engineering is a ''good idea,'' with as many as 90% supporting medical and environmental applications and 80% interested in tastier foods. About two-thirds indicated they would eat products modified by biotechnology (Kelley, 1995). Almost all (93%) Japanese consumers believe biotechnology will provide benefits to them or their family in the next five years, with the greatest interest in environmental applications (Hoban, 1996a).

When asked about the severity of potential food risks, 44% of Europeans considered genetic engineering a serious risk (Tordjman, 1995). This is about in the middle of potential food risks, with bacterial contamination at the top with 85% of consumers and sugar at the bottom with 12%. Response varies by individual countries, with more consumers in Scandinavian countries, Germany, and Austria perceiving biotechnology as risky compared to other countries (Hoban, 1997). With the exception of Austria and Germany, half or more European consumers indicate they would purchase a product modified by genetic engineering.

B. Concerns

Questions of personal or environmental safety have led to opposition among some consumer and activist groups. Some believe changing nature is dangerous and each product modified by biotechnology should be tested for human and environmental safety. Some philosophically oppose modifying animals for human purposes, either by biotechnology or through traditional breeding. Others contend that specific applications, such as use of rBST to increase milk production, will lead to animal disease, increased use of antibiotics, and changes in milk which could cause cancer.

Environmental concerns are also expressed. People fear that widespread use of Bt toxin will increase the rate by which insects become resistant, thereby leading to the loss of an environmentally benign pesticide. Opponents of biotechnology are skeptical that creating herbicide resistance will really lead to less herbicide use. They see this application as expanding the market for both seeds and herbicide. Some also fear that herbicide resistance will spread to weeds, thereby lessening the effectiveness of more benign herbicides.

Opponents also express philosophical concerns apart from product characteristics. They believe techniques of biotechnology encourage a myopic approach to agricultural production, whereas a holistic view is needed. Some are opposed to the practice of seeking a patent on scientific modification, saying it is inappropriate to ''patent life.'' Some believe biotechnology change is motivated by the pursuit of excess profits. Opponents also believe developing countries will be exploited in the quest for valuable genetic codes.

C. Labeling

It is unclear if mandatory labeling required in Europe will have a negative or positive impact on consumers, although it does present a significant tracking problem for manufacturers. Labeled tomato sauce in England is well received. If consumers perceive use of biotechnology as a safe technique that increases quality or improves the environment, identification on the label could be positive. If they see biotechnology as an unnatural innovation in which they are exposed to increased risks while others get benefits, biotechnology labeled products will be avoided. Acceptance depends on providing information to the public and consumer philosophical orientation.

VI. EFFECT OF INFORMATION ON ATTITUDES

Trust in information source and message content influences attitudes. Trusted sources are described as knowledgeable, concerned with public welfare, truthful, and with a ''good track record'' (Frewer et al., 1996). Exaggeration, distortion, and vested interest characterize less credible sources. Health professionals are the most trusted sources in the United States and Japan (Hoban, 1994, 1996b). In the United Kingdom, however, the consumers express greatest trust for current affairs television programs (Frewer and Shepherd, 1994).

VII. RECOMMENDATION

Food processors and scientists should convey science-based information to consumers describing the potential benefits and risks of food additives and new technologies such as food irradiation and genetically modified food. Information can be presented on labels or through other forms of product information. Similar information presented through multiple channels by trusted information sources increases consumer perception of information reliability. Consumers can then evaluate a product based upon a more accurate understanding of product attributes and personal values. Newer technologies can help enhance environmental quality and human health. These benefits can only be achieved if the public is comfortable with the choices available.

REFERENCES

Abt Associates Inc. 1996. Trends in the United States, consumer attitudes and the supermarket. In: *Trends*. Food Marketing Institute, Washington, D.C., p. 88.

Abt Associates Inc. 1997. Trends in the United States, consumer attitudes and the supermarket. In: *Trends*. Food Marketing Institute, Washington, D.C., p. 88.

American Meat Institute. 1998. New consumer research good news for irradiated foods. Document #16719.

Bruhn, C. 1998. Consumer acceptance of irradiated food: theory and reality. *Rad. Phys. Chem.* 52(1–6):129–133.

Bruhn, C. M. 1995. Consumer attitudes and market response to irradiated food. *J. Food Protection* 58(2):175–181.

Corrigan, J. 1996. Personal communication. Carrot Top, Illinois.

Crowe, M., Harris, S., Maggiore, P., Binns, C. 1992. Consumer understanding of food-additive labels. *Australian J. Nutrition and Dietetics* 49:19–22.

Dietz, G. 1998. Personal communication. Isometics, New Jersey.

Food Marketing Institute. 1990. Trends 1990: consumer attitudes and the supermarket. In: *Trends*. Food Marketing Institute, Washington, D.C.

Fox, P. 1996. Personal communication. Department of Agricultural Economics, Kansas State University, Manhattan, Kansas.

Fox, J., Olson, D. 1998. Market trials of irradiated chicken. *Rad. Phys. Chem.* 52:63–66.

Frewer, L., Shepherd, R. 1994. Attributing information to different sources: effects on the perceived qualities of information, on the perceived relevance of information, and on attitude formation *Public Understand. Sci.* 3:385–401.

Frewer, L., Howard, C., Hedderley, D., Shepherd, R. 1996. What determines trust in information about food-related risks? Underlying psychological constructs. *Risk Analysis* 16(4):473–485.

Gallup Organization. 1993. Irradiation: consumers attitudes. American Meat Institute.

Gallup Organization. 1994. How are Americans making food choices. American Dietetic Association (ADA) and International Food Information Council (IFIC).

Good Housekeeping. 1983. Food labeling study. Consumer Research Department, Good Housekeeping Institute, New York.

Good Housekeeping. 1984. Women's opinions of food allergens study. Consumer Research Department, Good Housekeeping Institute, New York.

Hoban, T. 1994. Consumer awareness and acceptance of bovine somatotropin (BST). Grocery Manufactures of America Survey.
Hoban, T. 1996a. Anticipating public reaction to the use of genetic engineering in infant nutrition. *Am. J. Clin. Nutrition* 63:657S–662S.
Hoban, T. 1996b. How Japanese consumers view biotechnology. *Food Technol.* 50:85–88.
Hoban, T. 1997. Consumer acceptance of biotechnology: an international perspective. *Nature Biotechnol.* 15:232–234.
IFIC (International Food Information Council). 1997a. Americans say yes to food biotechnology. In: *IFIC Review.*
IFIC (International Food Information Council). 1997b. How to understand and interpret food and health-related scientific studies. In: *IFIC Review.*
IFIC (International Food Information Council). 1997c. News release on consumer survey conducted by the Wirthlin Group. In: *IFIC Review.*
Kajanne, A., Pirttila-Backman, A. 1996. Toward an understanding of lay people's notions about additives in food: clear-cut viewpoints about additives decrease with education. *Appetite* 27: 207–222.
Kelley, J. 1995. Public perceptions of genetic engineering: Australia, 1994. International Social Science Survey/Australia.
Australian Supermarket Institute. 1997. Australian supermarket shopper—attitudes and behavior 1996–1997. Leading Edge Market Research Consultants Pty Ltd. Published by the Food Marketing Institute, USA in cooperation with the Australian Supermarket Institute and Coca Cola Company, pp. 1–72.
Lewis, S. 1997. Europe to require labeling of genetically modified soybeans and maize beginning November 1. *Food Chemical News* 11:5–7.
Marcotte, M. 1992. Irradiated strawberries enter the US market. *Food Technology* 46(5):80.
McNutt, K., Powers, M., Sloan, A. 1986. Food colors, flavors, and safety: a consumer viewpoint. *Food Technol.* 1:72–78.
Pszczola, D. 1992. Irradiated produce reaches Midwest market. *Food Technol.* 46(5):89.
Sloan, A., Powers, M., Hom, B. 1986. Consumer attitudes toward additives. *Cereal Foods World* 31(8):523–532.
Tordjman, A. 1995. Trends in Europe. *Trends.* Food Marketing Institute, p. 96.
Walter, R. 1994. Baseline study of public attitudes to biotechnology. Canadian Institute of Biotechnology, Ottawa, Canada.
Zibrik, D., Peters, H., Kuhnlein, H. 1981. Knowledge and attitudes of Vancouver residents towards food additives. *Can. J. Public Health* 72:49–54.

7

Food Additives in the European Union

RONALD VERBRUGGEN

Bioresco Ltd., Brussels, Belgium

I. INTRODUCTION

Every member state of the European Union (EU) has had for many years its own legislation on food additives. Foodstuffs constitute a very important segment of the economy. And food additive legislation has proved an excellent means for protectionist measures and for obstructing trade across borders. The idea is now generally accepted that open markets promote economic development.

The EU wants to realize the dual objectives of providing a high level of protection to its consumers and of being a completely open internal market. Hence, it was essential to undertake the harmonization of the additives legislation in the Union. This chapter tries to give a general orientation to the harmonization achieved and to draw the main outlines of the structure established.

After a note on history (Section II) some basic concepts will be outlined in Section III. Section IV is devoted to a discussion on the categories of, and descriptors for, foodstuffs as well as languages. In Section V the basic tenets and principles applied in the new European additives legislation are presented. Labeling rules for additives are detailed in Section VI. The authorizations for the different additives have been listed in three main Directives, completed with some updating Directives. These three main sets of authorizations are presented in Section VII. At the end of the chapter, in the Appendix, Tables 1 and 2 list all EU authorized additives.

II. HISTORY

A. Past Status

For many years attempts have been made in the EU to achieve some harmonization in the field of additives. This has been tried in two ways: by horizontal measures (on certain classes of additives) and by vertical measures (food standards for some specific products, including rules on authorized additives). See also Section III.A.

Up to the end of the 1980s, the horizontal measures had remained incomplete and were limited to, for example, labeling, and only on a few classes of additives such as colors (1), emulsifiers, and similar substances (2). And although some vertical measures had been adopted,* the obligation to adopt new legislation by *unanimity voting* made it virtually impossible to enact Community rules on important categories of foodstuffs (such as soft drinks and beer). For those food categories where an agreement was reached (i.e., a harmonizing measure was enacted) a number of derogations, maintaining some national rules, were sometimes part of the agreement. An example was the earlier Directive on cocoa and chocolate products (3). That Directive did not, for instance, harmonize the use of certain emulsifiers in chocolate. However, a new version of that Directive, pushing harmonization further, is now on the way to being adopted (see Section VII.A.4). In this way, at a certain point the process of harmonization came to a virtual standstill.

B. The New Elan (1985–1987)

The European Commission's *White Paper* (1985) and the resulting Single European Act (1986–1987), permitting the adoption of the measures necessary to create the single open market by qualified *majority voting*, gave a new impetus and constituted a real breakthrough toward harmonization. Since that date, DG III, the European Commission's directorate general for industry, has addressed its efforts on additives and other essential aspects (such as flavors, contaminants, hygiene, etc.) mainly toward horizontal measures as the best means for harmonization. The concepts of horizontal versus vertical legislation are explained in Section III.A.

The *White Paper* stated that harmonization of national legislation should only be imposed when that was necessary for proper functioning of the open internal market. Harmonization of vertical standards was generally considered unnecessary in view of the principle of mutual recognition (confirmed by the ruling of the court on the ''cassis de Dijon'' and subsequent similar cases). Even when member states maintain their national vertical standards as far as their own production is concerned, they cannot impose them any more on noncomplying imports from other member states if these imports comply with the legislation of the exporting country. For many foods there is no vertical EU standard, yet there are vertical standards for a variety of foodstuffs in many member states. Examples are the vertical legislation on soft drinks in Benelux and on beer in Belgium and Germany. In such cases, the horizontal Union rules on additives clearly prevail over the national food standards. It follows that merely *national vertical standards* are implicitly modified by the very fact of the prevalence of the newer horizontal rules.

Although the plan to do so was explicitly formulated by the Commission already at the Edinburgh summit of December, 1992, the older *Community vertical food standards*

---

* Vertical measures in existence include, e.g., the Directives on: caseins and caseinates; cocoa and chocolate products; coffee and chicory extracts; quick-frozen foodstuffs; honey; fruit jams, jellies and marmalades and chestnut puree; fruit juices and similar products; preserved milk; natural mineral waters; erucic acid in oils and fats; dietary foodstuffs; and sugars. There is also an older Regulation on spirit drinks. More recently a new Regulation on olive oil was issued, as well as one on spreadable fats (butter and margarine type products). These older measures have—in general—been updated and amended several times. (See Section III.D.2.a and b for the difference between Directives and Regulations.)

are only now being brought in line with newer horizontal texts. (It should be noted, however, that the "spreadable fats" Regulation of December 5, 1994 is a recent Community measure. And it cannot be excluded that still others may be created later on.) In the cases where EU vertical standards are in existence next to horizontal rules, one can only take it that the more recent texts prevail over the older ones.

C. Role of the European Parliament

With the entry into force of the Single European Act on July 1, 1987, *qualified majority* voting replaced *unanimity voting* in the Council and in some committees for measures required to realize the single market. But at the same time the European Parliament (EP) gained the right to propose amendments to legislative proposals.

The Treaty of Maastricht (signed on February 7, 1992 and entered into force January 1, 1993) has further increased the EP's role in the legislative process in creating a *conciliation procedure*. The EP can, in certain cases, reject and effectively block proposals. This is the case when its amendments are not accepted by the Commission, unless the Council adopts them unanimously anyhow. So far, however, Parliament can still not take its own legislative initiatives. With time however, the EP's role in the process is sure to increase further in importance.

III. BASIC CONCEPTS

A. Vertical Versus Horizontal Legislation

Food legislation is termed *vertical* when it pertains to a specific group of foods, that is, a specific food category (for example, chocolate). Vertical legislation on a food category is sometimes called a *food standard*. It normally deals, in one text, with several or all relevant aspects of that food category, such as additives authorizations, specifications of additives, sales denominations, labeling, claims, contaminants, methods of analysis, sampling plans, official control, and so on.

A vertical law may be convenient for an operator in a specific sector, with little specialized knowledge of overall food law. Indeed such legislation may answer all questions at once with one text. Still, vertical standards may easily constitute considerable impediments to progress and innovation. Updating adaptations of vertical measures has generally proven very hard to achieve. This specific aspect may have led the European Commission to change tack and opt for horizontal measures in order to further harmonization.

Such horizontal legislation is legislation dealing with only one aspect, but is applicable to all foodstuffs (except when specifically indicated). Such legislation is politically far more difficult to elaborate. It requires, indeed, for its development a very thorough knowledge of the entirety of the food chain, and of all the existing legislation, so as to foresee all consequences and to avoid conflicting implications of its general application.

B. European Versus Member States' Authority

Driven by the will to open up the European market, or to create political unity, the actions of the European Community reflect the changing political mood of the time. Periods in which centralizing tendencies draw power to the central authority alternate with centrifugal forces which emphasize the member states' own powers. Since the summit of European leaders in Edinburgh (December, 1992), a return of power over certain areas to the individual member states is enshrined in the *principle of subsidiarity*. According to this principle, a matter is to be dealt with at the level which can most efficaciously achieve the required results.

But even in cases where Union authority is to take charge, the member states by their presence and voting rights in committees, in the Council of Ministers and also in the EP, wield a lot of power. It is inconceivable for Union legislation to see the light if it does not enjoy broad majority support from the member states.

C. Qualified Majority: Member States' Power

In several committees, as well as in the European Council of Ministers, many important decisions have to be made by qualified majority. In such cases the member states are allotted a number of votes, roughly proportional to their relative weight in the Union: 10 votes each for the larger countries (France, Germany, Italy, and the United Kingdom), 8 votes for Spain, and a lesser number for each of the smaller countries. The allotment of votes tends to give a disproportionately greater weight to the smaller countries, compared to the size of their populations. It is expected that these numbers of votes allotted to member states will be correspondingly adapted as more countries join the Union.

D. The European Legislative Environment

1. The Parties Involved

New European legislation is always prepared by the European Commission (or the ''Commission,'' the ''Parliament,'' the ''Council'') before further proceedings leading to adoption. For important measures with political implications—such as the authorizations of additives—the joint EP and the Council of Ministers adopt the measures according to well-established procedures (see Section D.2). In its preparations, the Commission consults with interested parties, such as member states' government experts, industry, consumers, traders, agriculture, etc. In these consultations, associations—organized by sector and/or by member state—are important partners. But individual businesses and advisors can also have input. The directly elected representatives of Parliament are gaining ever more importance in the process (see Section II.C). Throughout all stages of the proceedings to develop the legislation, both the Council and the Commission cooperate intensively, through committees, with experts from the member states' administrations.

One of the important committees assisting the Commission in matters of food is the *Standing Committee for Foodstuffs*. It is essentially a voting college of experts appointed by the member states and presided over by the Commission. This committee takes its decisions by *qualified majority* (the number of votes weighted to the different countries). Other committees and institutions, such as the *Advisory Committee on Foodstuffs* and the *Economic and Social Committee* have only an advisory role.

Without being a legislating institution, the European Court can be asked—on a case by case basis—to interpret the legislation.

Certain technical matters do not need to be regulated by law, but can be more flexibly agreed by standardization. In this way, the European Standardization Committee (CEN) became one more partner in the process of harmonization of European food legislation as far as, for example, methods of analysis and voluntary codes of practice per sector are concerned.

See Section X.B for sources of information regarding Commission bodies and procedures.

2. The Legislative Instruments

a. The Main Four Instruments. In the European Union regulatory landscape, four types of measures are prominent: Recommendations, Decisions, Directives, and Regulations. Both the Council and the Commission can take each of these. But recently, as already stated, important measures with political implications are taken jointly by Parliament and Council.

Recommendations are not legislation, and therefore have no binding effect.
Decisions are only binding for those to whom they are addressed.
Directives and Regulations are generally binding in the whole Union.
A Regulation is directly applicable as such. Hence, for issues for which strict reliability and conformity of legislation are required, Regulations are the preferred legal tool. This is, for example, the case for *novel foods*. (See Section III.E.6.1.)

b. Directives. The main legislative instruments in the Union are Directives. Directives are binding only as far as the result aimed at is concerned. For a Directive to be applicable, however, every member state has first to implement it—transpose it into its own legislation according to its own internal legal system. This process does, of course, entail clear risks for the equivalence of the resulting legislation.

The concept of Directives was developed in order to cope with the fact that the legal systems in the individual member states are quite different. Directives are the key European Union legislative instruments, aimed at permitting the implementation of the measures in each member state, fitting into, and respecting, the national legislative structures and traditions of each state. Directives have been extremely useful tools that have strongly advanced the European harmonization process. However, they are frequently not implemented in a timely way, triggering the Commission to undertake corrective action. Most of the recent food legislation is drawn up in the form of Directives.

Before undertaking to regulate an extensive domain of importance (such as, for example, additives or flavorings), the rules of procedure, the general criteria applicable, as well as the underlying rationale are laid down in a so-called *framework Directive*.

c. Numbering of Directives. The framework Directive on additives was published as the 107th Directive of the year 1989. It therefore received the number 89/107/EEC,* even

---

* The extension EEC stands for European Economic Community. From sometime in 1993, and consequent to the Treaty on European Union (Maastricht Treaty), measures refer to European Community (EC).

A similar but independent numbering system applies to the other types of measures. Hence the mere number does not unambiguously identify a text of European law. In principle one also has to know whether, for example, a Directive or a Decision is meant. For example, Council decision 95/2/EC carries the same number code as the Directive on sweeteners, 95/2/EC. The Council decision 95/2/EC, however, establishes the order in which the presidency of the Council shall be held and is not related to sweeteners.

On the other hand, Regulations are indicated with: No/year. Examples are the vertical regulations of 1989: Regulation (EEC) No 1576/89 on spirit drinks, and of 1994: Regulation (EC) No 2991/94 on spreadable fats.

Lately the year indicator carries the full four digits 1999/ . . . for Decisions and for Directives, and N° . . . /1999 for Regulations.

though it was adopted on December 21, 1988. In addition to their number, the legislative measures are characterized with dates, as mentioned in Section III.D.2.f.

d. Format of the Legislative Texts. The overall format of the European legislative texts on additives comprises a *legal part* and *annexes*. Tabulated data and fine print, as well as technical detail, are generally laid down in the annexes (such as the tables of additives and their conditions of use). Unfortunately these important tables can neither be found in the official legislative database of the EU (''Celex''; see Section X.B) nor in a number of derived information systems.

The legal part consists of articles and is generally preceded by the underlying motivation in a number of citations (''whereases'').

e. Translation and Publication. The services of the Council take care to translate all measures into each of the official languages of the Union, and publish the text—issued simultaneously in all member states—in the different language versions of the *Official Journal of the European Communities* (*OJ*), recent issues of which are available on the Internet. For ease of consultation, the publishers of the *OJ* print the documents so that the same items can be found in identical locations (page numbers) in the different language versions of the journal.

f. Characterizing Dates, Implementation, and Enactment. Normally the last articles of a Directive or Regulation indicate (1) the limiting dates at which the member states have to adapt their legislation, so as to align them with the Directive or Regulation requirements, (2) the date from which products in conformity with the measure should gain free circulation, and (3) the date from which products not in compliance with the requirements of the measure must be prohibited.

In addition to these dates, a Directive or Regulation can be characterized by its date of adoption which is mentioned in the title, the date at which it enters into force, and the date of its publication (i.e., the *OJ* reference).

3. Development: Two Main Procedures

One of two main development procedures have to be followed, depending on the nature of the measure. Measures of a technical or detailed implementation nature can normally be dealt with by the Commission, in discussion with the member states' administration experts, so far without involvement of the Council or the Parliament's politicians. The mandate given to the Commission for these measures is generally laid down in another Directive, adopted according to the second procedure (see below). Examples of such measures are the specifications for additives. The precise procedure which is to be followed for the adoption of such a measure can normally be found in one of the articles of the Directive that mandates the Commission to take the measures.

Important measures with political implications, however, must be adopted jointly by the EP and the Council (i.e., by political appointees) by the so called *co-decision procedure*. In this procedure, Parliament—in a *first reading*—examines a Commission proposal and proposes amendments. Similarly, the Economic and Social Committee (ECOSOC) is required to give an opinion on the proposal at this stage of the procedure. Normally the ECOSOC opinion is the first one to be issued, and this tends to set the trend.

At the same time, the Council prepares a political *common position* based on the Commission proposal, on the Parliamentary amendments, and on any other relevant elements. This common position is re-examined by Parliament in a *second reading*.

Council and Parliament then have to come to terms, which sometimes requires an intricate *conciliation* procedure.

The EP adoption of the concluding joint EP–Council conciliation texts is also called the *third reading*.

It is beyond the bounds of this chapter to develop the precise intricacies of the procedure which may lead to adoption or rejection of a proposal. It will be clear though that developing a measure is a very lengthy undertaking.

Any approval to use an additive in a specific foodstuff is considered such a political decision and is subject to this procedure. Hence, more and more, the food and additive manufacturers will have to watch out for political aspects of their innovations and take care to properly sell the concepts of their new products to consumers and politicians (see Section VIII).

4. Arbitration on the Meaning of the Additive Texts

The Standing Committee for Foodstuffs is the official body for *arbitrating* whether a substance is a food additive (rather than a novel food, for example) and whether a foodstuff comes under one of the food descriptors or food categories occurring in the Directives on additives.

This specific authority is neither controlled by the Council nor by the EP. It should be clear that the Standing Committee for Foodstuffs cannot give new authorizations. That can only be done through the lengthy authorization procedure, involving European Commission, Council, and Parliament.

Given the important mission of the Standing Committee for Foodstuffs, it is most worrisome that—so far—this body has not published reports on its deliberations nor did it issue official summaries of its decisions. A whole series of decisions on, for example, the exact meaning of the food descriptors in the additives Directives has already been made. Democracy and true transparency would be served by very fast publication of, and easy access to, any decisions made by that Committee.

E. Definitions and Scope of Additives Legislation

With respect to the length of the approval procedure for additives in the EU, the question as to what is and what is not an additive is a very important one.

1. Definition of Additive

The definition of additive is laid down as follows in Art. 1 of Directive 89/107/EEC (4):

> For the purposes of this Directive 'food additive' means any substance not normally consumed as a food in itself and not normally used as a characteristic ingredient of food whether or not it has nutritive value, the intentional addition of which to food for a technological purpose in the manufacture, processing, preparation, treatment, packaging, transport or storage of such food results, or may be reasonably expected to result, in it or its by-products becoming directly or indirectly a component of such foods.

2. Processing Aids Versus Additives

Processing aids are defined as follows in Directive 89/107/EEC (4):

> For the purpose of this Directive, 'processing aid' means any substance not consumed as a food ingredient by itself, intentionally used in the processing of raw materials, foods or their ingredients, to fulfil a certain technological purpose during treatment or processing and which

may result in the unintentional but technically unavoidable presence of residues of the substance or its derivatives in the final product, provided that these residues do not present any health risk and do not have any technological effect on the finished product.

The same substance can sometimes be used either as an additive or as a processing aid. It depends—in principle—on the intention of its application, whether the substance exerts its effect in the final food or merely during the manufacturing process.

Sulfites are clear examples of such a double option; they can be used as additives (e.g., as preservatives), but also as processing aids (e.g., as bleaching agents for candy products or for "*marinated*" nuts).*

The definitions for additives and for processing aids are built on earlier ones from Codex Alimentarius. Yet they differ from these in some aspects. They do not always allow one to establish unambiguously whether a substance added to food is an additive or a processing aid. Dimethyldicarbonate (E242) is a case in point. After years as a processing aid in Germany, the European Union has now made E242 into an additive. Another example is hexylresorcinol. It is a very effective inhibitor of the polyphenoloxidase in the shell of crustaceans. It could be applied to prevent "black spot" discoloration in shrimps, for example. Hence it can act as a replacement for sulfite in that function. Hexylresorcinol could be considered the perfect example of a processing aid. However, several public authorities insist now on considering it an additive anyhow.

It is most remarkable that the EU definition of additive in English (and related languages) differs from the one in French (and related languages) in a point that heavily bears on the classification of a substance either as an additive or as a nonadditive (generally, a processing aid). The English-type version states that—in case of an additive—the substance or its byproducts can become part of the final food. The French-type version speaks about the substance or its derivatives (instead of byproducts) to become part of the food.

Byproducts, in contrast to derivatives, are admixtures present in the original preparation. Derivatives are produced essentially after addition of the preparation to the food. It should be noted that this difference follows, in all likelihood, from the same difference between the English and French definitions already in place in Codex Alimentarius.

In the cases referred to above, the derivatives produced (methanol and carbon dioxide from dimethyldicarbonate and the bound inhibitor moiety from hexylresorcinol) led to classification of the preparations added as non–processing aids, although their outspoken reactivity in the manufacturing or premanufacturing processes only might have perfectly allowed their classification as processing aids.

Certain substances can act as flour treatment agents. They exert their effect during the production of baked goods. In the finished product, they do not play a functional role. Hence it could be concluded logically that they should be termed processing aids. However, flour treatment agents are listed among the additive functions in Annex I of the framework Directive on Additives (4) (see Table 1). Therefore, L-cysteine has been listed as an additive (5).

---

* In the case of sulfites—and that is the only one in the combined additives Directives up to now—a footnote states that the sulfite quantities in the Directive refer to the totality of all sulfite present (i.e., any natural sulfite; sulfite produced by the manufacturing process, such as in brewing; sulfite added as an additive; and sulfite added as a processing aid). See p. 19 of Ref. 10.

Table 1 The Function Categories of Additives Established for Authorization Purposes

Categories of food additives
Color
Preservative
Antioxidant
Emulsifier
Emulsifying salt
Thickener
Gelling agent
Stabilizer
Flavor enhancer
Acid
Acidity regulator
Anticaking agent
Modified starch
Sweetener
Raising agent
Antifoaming agent
Glazing agent
Flour treatment agent
Firming agent
Humectant
Sequestrant
Enzyme[a]
Bulking agent
Propellant gas and packaging gas

[a] Only those used as additives.
Source: Ref. 4.

It should be stressed that in Community law processing aids are not ingredients. Up to now some countries, such as Spain, Germany, and Denmark, have considered processing aids as a subgroup of additives. It is interesting to watch whether these countries will change their legislation in that respect. In any case, in a communication of the Danish Ministry of Public Health of December 18, 1995, concerning the new Community legislation on additives, the classification of processing aids as a subclass of additives has been maintained.

The formal consequences of classifying a substance as an additive (in contrast to classification as a processing aid) are that (1) a heavy safety dossier has to be submitted to the Scientific Committee for Food, (2) the substance has to be indicated in the ingredient list on the label, (3) a Community authorization for use has to be obtained, and (4) the use levels have to be agreed upon by the authorities.

Although a manufacturer always has to ensure the complete safety of the final foodstuffs they are marketing, there is no EU Community obligation to formalize authorization, safety studies, and risk evaluation for processing aids. Except for solvents, there is no Community legislation on processing aids yet. Individual member states may, of course, have their own requirements (e.g., France still requires a specific authorization for each application of a processing aid). Such national requirements risk becoming impediments to the operation of the open single market.

Solvents have already been regulated by specific Directives (6–8). But, so far, for processing aids as a group, Community legislation has not been planned.

3. Enzymes Versus Additives

There have been rumors that (after solvents) enzymes would be the second class of processing aids for which Community legislation might be put forward. The creation of legal certainty in this field is of benefit to the food industry, on the strict condition that it does not lead to multiplication of requirements in the legislation. Although enzymes derived from genetically modified organisms (GMOs) (mainly from microorganisms) constitute the majority of food enzymes for the future, other, non-GMO enzymes are also of importance to the food industry.

Rules on enzymes of GMO origin have not been laid down in the Regulation on novel foods (9). It would indeed be quite illogical to regulate enzymes in the Regulation on novel foods and novel food ingredients, since processing aids are neither foods nor food ingredients in EU law.

Although *enzyme* is an established function of additives as well (see Table 3), so far only two enzymes have been listed as authorized additives in the *miscellaneous additives* Directives: lysozyme (10) and invertase (5). Other enzymes are, so far, considered processing aids.

4. Nutrients Versus Additives

Substances added to food for their nutritional effect (e.g., certain vitamins, certain minerals, and certain amino acids) are not additives. They can sometimes, however, also be used for a technological purpose, that is, as additives (e.g., ascorbic acid as an antioxidant). In that case, an authorization (case of need, conditions of use, etc.) is required.

5. Foods Versus Additives

In EU food law, there is (so far, again) no firm definition of what is a *food*. Hence, the question whether something is a food or an additive is bound to arise regularly. A recognized nutrient function can be an argument in order to declare something a food. Still, chewing gum can be without nutrient function, yet it is generally still food (it can also be a cosmetic or a medicinal product). Clearly, organoleptic pleasure is also a criterion to declare something a food. Additives must be used for predominantly technological functions (see Section V.B). However, many, if not all, undisputed foods have clear technological functions as well: sugar is a sweetener;* a humectant, a thickener, and it is the most important of all; cocoa powder gives color, etc. The same is true for the polyols. Still, suggestions have been made to reclassify the polyols as foodstuffs since they also have a nutritional function. The authorities will generally prefer to classify substances as additives when they wish to maintain firm control over their conditions of use. A regular foodstuff is indeed not subject to conditions of use.

---

* It is most remarkable in this context that the framework Directive's definition of additives (4) states that for a substance to be an additive, it is irrelevant whether the substance has nutritive value. Hence the distinction between foods and additives is far more disputable than is tacitly accepted.

Additives can have nutritive value (see the definition of additives in Section III.E.1). There has been a long debate as to whether the modified starches are additives or foodstuffs. The EU legislation now considers organochemically modified starches as additives. On the other hand, products obtained from starch by mere hydrolysis and by amylolytic enzymes are considered nonadditives, that is, foods.

6. Novel Foods Versus Additives

Nowhere are such questions more pressing than for novel substances. Since January, 1997, the EU has a Regulation on *novel foods* (9). Up to now, and according to the principle of ''one door, one key,'' the European Commission has refused to have additives regulated outside the additives Regulations proper. Hence, it has resisted requests, from several sides, to bring additives derived from genetically modified organisms (GMOs) within the scope of the novel foods Directive. Giving in would mean additional hurdles to the introduction of GMO-derived additives. (Additives derived from GMOs would have to go through both the procedure for authorization of additives and the procedure for authorization of novel foods.)

In order to neutralize criticism about this refusal, the European Commission has declared that approved additives, whose method of preparation is switched from classical production techniques to different ones (e.g., employing GMOs as their source) would have to be resubmitted for SCF scrutiny. In preparation for this, the European Commission made certain specifications (see also Section III.H) of certain additives with detail on the mode of preparation.

New substances can be considered as either novel foods or additives, depending on whether they have essentially a nutrient function or essentially a technological function. Whether the function of a substance is nutritional or technological can be far more controversial than would immediately appear. In the field of fat replacers for example, it is up for debate whether a lipid with lower or zero energy effect is still a nutrient or whether it is to be assigned a mere bulking agent (additive) function.

EU authorization of a new additive is an expensive and tedious process, requiring two (or even three) readings in the EP. Such an authorization may take several years. In principle, a novel food authorization should be faster, as it does not have to go through the heavy Parliament procedures, and as it does not have to be implemented in the national legislations of the member states. Time will show whether that proves to be the case.

7. Flavorings and Solvents Versus Additives

Flavorings, and processing aids (including, e.g., solvents and enzymes), as well as nutrients, are not additives in the forthcoming legislation. For solvents, Community legislation is already in place (6–8).

On flavorings and flavoring substances, other specific Community legislation is already enacted (11–15) or is in the process of being completed (16,17). Certain substances, classified earlier as additives in certain member states have now been reclassified as flavoring substances for the EU legislation. Examples are the maltols, ammonium chloride, and formic acid.

It should be noted in this context that there is, so far, no really good definition in EU legislation for flavorings nor for flavoring substances.

The European Commission has been asked to start harmonizing the field of additives for flavorings. Up to now—this area not being regulated at the European level—industry has had to comply with the legislation of the different member states.

A list of additives for flavorings is in preparation. It is not yet certain that the list will finally hold only additives authorized already as the regular additives Directives or whether entirely different substances might also be considered for inclusion. For example, a substance used by flavorists, and which is so far neither listed among authorized solvents nor among authorized additives is benzyl alcohol.

It must also be decided which conditions of use or labeling provisions will be added. Art. 6 § 4(c)ii of the general labeling Directive 79/112/EEC (18) states: ". . . shall not be regarded as ingredients . . . substances used in the quantities strictly necessary as solvents or media for additives or flavoring." Hence, so far, they need not appear in the ingredient list on food labels.

F. Additive Authorization in the EU

In order to give consumers confidence in the safety of food additives authorized in the EU, authorities provided long ago acceptable food additives with an unambiguous identification number, preceded by an E.* This initiative was torpedoed by the infamous "*Tract de Villejuif*" pamphlet, which spread the message that E numbered additives constituted a special health hazard. This malevolent disinformation, which is still circulating widely, continues unfortunately to cause considerable confusion and uncertainty in the minds of the European consumer.

The new Directives authorize 365 additives. This seemingly high number has provoked criticism from consumers and from members of the EP. However, that number is extremely misleading because its magnitude is a consequence of the rather arbitrary definition of what an additive is in the EU and of the way E numbers are assigned. The public at large has come to distrust substances *merely because they are categorized as additives and carry an E number*. It should be realized, though, that the mere administrative categorization of a substance as an additive does not affect the properties—physical or biological—of that substance so as to entail a risk by its use.

In many cases slightly different forms of the same substance carry their own different E numbers (e.g., sodium, potassium, calcium salts of an acid; and for the phosphates alone 24 different numbers are listed so far). Furthermore, ordinary ambient gases (oxygen, carbon dioxide, and nitrogen) were also given additive status with an E number. The same is true for the noble gases argon and helium. Although these two are so far rarely if ever used, they are normal minor constituents of air as well.

Moreover, the mere number of authorized additives is meaningless. Indeed, additives with different E numbers may fulfil similar functions; that is, they are mutually exclusive in use for a certain application or alternatively they are used in combinations with a lower concentration for each component of the mixture applied (e.g., for sweeteners).

A considerable number of natural food components are also termed additives in the legislation when added as separate substances. Some of these, such as acetic acid, propionic acid, ascorbic acid, citric acid, lecithin, etc., could also be properly considered as

---

* Note in this context that the E numbering system in the EU came into existence before the INS numbering system of *Codex Alimentarius* and served as a model for the latter one. In this chapter, indicating E numbers along with the verbal additive names with every additive reference in the text has been avoided in order to improve readability. The reader may consult Table 2 in order to find the E numbers corresponding to an additive name where needed.

foods in their own right. And as has been mentioned (Section III.E.5), many other compounds, such as the modified starches and the polyols, might very well be classified as foods. Furthermore, a number of the other additives are extremely close to, or identical with, regular food components, such as, for example, E471: mono- and diglycerides of fatty acids.

The example of a simple tomato illustrates the point. It contains naturally nine E number substances.* Furthermore, many salt ions can all be found in the tomato (potassium, sodium, calcium, chloride, etc.). Combined in different ways, they would lead to a considerable number of E numbers for the same natural foodstuff.

A list of all additives authorized by the EU Directives can be found in the Appendix Table 1, (sorted by E number) and Appendix Table 2, (sorted alphabetically) at the end of this chapter. In both these tables the corresponding Directives in which they are authorized are indicated, along with all the page numbers on which they appear.

Listing of additives, along with their conditions of use in the tables of the Directives, constitutes the corresponding authorization. For colors there is an additional table with a mere enumeration of all authorized substances, as a separate annex to the legislative text (see Section VII.B.1.a).

G. Compilation of Needs: CIAA Database

In 1990, CIAA (the Confederation of the Food and Drink Industries of the EU) established in a computer database for internal usage its ''wish list'' of food additive needs. This database was established in preparation for the consultations offered by the Commission during the drafting of the additives Directives. This exercise was probably the first attempt to draw a pan-European picture of food additive usage in Europe. Although it has been replaced by the European legislation discussed here, this original database is at the origin of the actual system of authorizations now in place in the EU.

H. Specifications and Methods of Analysis

Specifications updated after the new additive authorizations came in place have been issued for sweeteners (19) and for colors (20). For the miscellaneous additives, the updated specifications are not yet complete (21,22).

Methods of analysis are not described in the new European additive specifications. Lately, the European authorities prefer avoiding issuing Community methods of analysis by legislation. Instead, they refer elaboration and validation of such methods to standardization organizations (CEN, etc.). The main general rules applicable to sampling and methods of analysis for the Community are given in an earlier Directive (23). See Sections VII.A, B, and C for detail on Community specifications, as well as on Community methods of analysis laid down in earlier legislation.

In each of the recent EU additive specifications Directives (19–22), a citation can be found which is quite important to developers of additives; it reads

* E621 (monosodium glutamate), E622 (monopotassium glutamate), E623 (calcium diglutamate), E625 (magnesium diglutamate), E300 (ascorbic acid), E301 (sodium ascorbate), E160a (mixed carotenes), E160d (lycopene), and E101 (riboflavine).

> Whereas food additives, if prepared by production methods or starting materials significantly different from those included in the evaluation of the Scientific Committee for Food, or if different from those mentioned in this Directive, should be submitted for evaluation by of the Scientific Committee for Food for the purposes of a full evaluation with emphasis on the purity criteria . . .

This means, for example, that before the replacement of the raw commodity for the manufacturing of an authorized additive by a GMO-derived raw material, the SCF should be consulted before the change. Indeed, GMO-derived materials are considered to be different (see also Section III.G.6).

IV. FOOD DESCRIPTORS AND FOOD CATEGORIES

A. Foods, Food Categories and Food Descriptors

From the very start of the harmonization efforts of the additives legislation in the European Commission, far-reaching decisions were made on food categories and on food descriptors. In this chapter, any designation of a foodstuff will be called a *food descriptor*. The term *food category* is to be understood as *a set of more or less similar foodstuffs or groups of foodstuffs themselves.*

It was decided not to define, according to a clear system, food categories for which authorizations for use would be laid down in the Directives. Consequently, food descriptors without precise prior definition were introduced into the lists so as to best fit the requests. It was argued that doing so would allow a lot more flexibility in the application of the new rules. An example of such fuzzy descriptors used is *''water based flavored drinks, energy reduced or with no added sugar,''* rather than *''soft drinks.''*

There is another important reason for this choice for *''fuzzy''* descriptors. The process undertaken by the Commission was a harmonization. At the time of the start of that process, the different member states had very different food descriptors in use in their national legislation for many hundreds of different foodstuffs. It would have been a nearly impossible task to harmonize those before tackling additive usage harmonization itself. Furthermore, to do so would be to reembark on discussions which would be similar to establishing vertical food standards once again, a concept intentionally dropped since the *''Single European Act''* (cf. Section II.B).

In this way, a food category can be designated by one or more food descriptors. The same foods can be included by one or by several wider or narrower descriptors.

As an example: In the additive Directives there is a broad food category defined by the overall food descriptor ''snacks'' (including potato crisps). Within that category, we can find the subcategories ''nuts,'' under the descriptor ''nuts,'' and—quite confusingly—''snacks'' proper again, under the mere descriptor ''snacks'' (excluding the potato crisps, but referring, e.g., to expanded and extruded products). ''Snacks'' proper holds two subcategories: ''cereal-based snack products'' and ''cereal- or potato-based snacks.'' Obviously, the latter category includes the former. Among the ''nuts,'' we find the descriptors ''savoury coated nuts'' and ''coated nuts.'' Here again, the latter category includes the entire former one as well.

Still, in an ideal setting, logic in meaning is the rule. An industrial operator would then be in a position to deduce the answers to his questions from the Directives. In the opposite setting, the operator would have to return to the authorities for clarifications on the meaning of the legislation for operational decisions in the matter of additives. Not only industrial operators were concerned about this; the different peripheral public authority bodies also need logic, whether they work in different sub-units of the official control organizations of the same members states or in different member states with different food cultures.

This sets the first requirement for the system: the system must be logical. The minimum for such logic is that (1) the same descriptor means the same foodstuff in the whole EU and throughout all texts, both legislative and interpretative, and (2) the same foodstuff is always designated by the same descriptor.

Furthermore, in the face of the great many different foodstuffs in existence in the EU, group descriptors are necessary in order to deal with the multitude of items to be identified in the legislation. Indeed, in the EU there are thousands of specific foods on the market. And the new additive Directives authorize 365 different additives for use. This requires many thousands of authorization cells to be filled.

This sets the second requirement: for the system to be manageable, it must unavoidably be hierarchical (treelike). This means that any food category of higher hierarchic rank includes all its subordinate members of lower hierarchic rank; these are either individual foodstuffs and/or other subordinate food categories. In such a system, an additive authorized in a higher ranking category is also authorized in all its lower order subcategories, except where specifically excluded.

It is evident that any hierarchical system itself—in order to be workable—must also satisfy the requirements for logic mentioned, that is, the same group descriptor ought always to have the same meaning.

B. Food Categorization System

Industry generally opts for operational security rather than for flexibility and uncertainty. In order to improve understanding of the new food Directives, CIAA has developed a food category system which satisfies the imperatives for logic and hierarchy. It was modeled after a system already in use in Denmark. The principles according to which this system is built are essentially supported by the authorities of the member states. The system CIAA has developed aims at being general; it is to comprise all foodstuffs, including foods for which no additives are authorized. The food categories and descriptors of the system are to be used solely as *a tool for the assignment of authorizations for the use of additives, and for nothing else*. Hence it should not serve as a basis for discussions on labeling, contaminants, customs tariffs, etc.

CIAA has put its food categorization system forward to Codex Alimentarius, where it has been welcomed as a worthwhile contribution. The new Codex Alimentarius General Standard for Food Additives (GSFA) will employ a similar, though slightly different, categorization system. Within Codex the contents of the system is wider in order to accommodate non-European foods as well. However, the principles on which it is based will remain essentially the same.

C. Computer Database

In an attempt to facilitate easy and flexible retrieval of data from the Directives, an effort has been made by the EU food industry association to create a hierarchy among the food descriptors occurring in the EU additives Directives according to the principles behind the CIAA food categorization system. The aim was also to lay this hierarchy down in a database for personal computers, together with precise referrals to the published Directive sources.

It should, however, be stressed that it is not easy to organize the quite complicated structure of the Directives' authorizations with computer software. In addition, any system should be so designed as to readily accommodate any future changes to the legal texts. Furthermore, no computer database will ever be able to *interpret* the Directives. Though fast sorting and query answering capabilities of PC databases provide useful facilities for retrieval from the original texts. The resulting database (in Access 97) has been made public and can be purchased on a diskette from CIAA.

D. Different Language Versions of the Additives Directives

The Council of Ministers is responsible for the translation and publication of the Directives into the different languages of the Community. But *traduire c'est trahir.* That is, no translation can ever be completely correct. It should first of all be noted that, at the time of the adoption of the main *miscellaneous* additives Directive (10), of all amending additive Directive updates (5,24,25), as well as of all the recent specifications Directives (19–22), Sweden and Finland had already joined the EU. Hence Finnish and Swedish versions of that Directive have been prepared as well. This is not the case for the *framework* Directive (4) and its updating Directive (26), for the sweeteners Directive (27), or for the colors Directive (28).

As mentioned previously, for ease of consultation the publishers of the *OJ* print the documents so that the same items can be found in identical locations in the different language versions of the journal.

With respect to the intricate and culture-related nature of the food descriptors in the Directives, some of the translations published are, according to industry experts, imprecise or plainly incorrect. Hence the translations risk perturbing the operations of industry and of international trade. An example of imprecision can be found on p. 25 of the *miscellaneous* additives Directive for the erythorbates. The English version indicates ''*Semi-preserved and preserved meat products*,'' whereas the French translation refers to ''*Produits de viande en conserve et en semi-conserve*.'' This translation is opposed by the French meat processors association and seemingly also by the French administration. Their preference is for ''*Produits a base de viande*.'' An example of plain error in translation is found on p. 32 of the same Directive for ''*Confectionery (including chocolate)*.'' The German version reads ''*Süsswaren (ausser Schokolade)*'' (i.e., confectionery *ex*cluding chocolate). A similar grave error in translation, with consequences, was highlighted already in Sections III.E.1 and 2 (in the definition of additives). Although the principle of hierarchy in the food categories is not inscribed in the Directives, it clearly exists.

The European Commission realized that its decision not to define the food descriptors which it uses in the Directives would lead to controversies. Hence it organized an arbitration system to resolve these controversies.

The task of setting disputes on the exact scope of the food descriptors has been entrusted to the *Standing Committee for Foodstuffs*. This body will have to arbitrate in cases where such conflicts cannot be resolved more readily (see Section III.D.1). The individual member states had to make their own versions of the new legislative texts, each one in its own language. The resulting ''implementations'' might not be identical with what the Council interpreters have produced. Still, it can only be hoped that some of the translation problems will have been ironed out at the member state level. But in some cases implementation will, no doubt, have aggravated the confusion. A further discussion of differences between original versions and the member states' texts in each of the languages is clearly beyond the scope of this paper.

V. BASIC ASPECTS AND PRINCIPLES

The additives legislation described in this paper pertains to all food additives put on the market in the EU, either additives preparations as such, or additives incorporated in foodstuffs, or also intermediates for the manufacture of foods for direct consumption. In principle, it also applies to all imports. However, for certain imported products, specific additional authorizations were agreed by special agreements. An example is the authorization to import wine containing erythorbate from Australia (29).

In the harmonization process undertaken in the Union, one should distinguish between the principles: the legal part of the Directives and the systematic part (the Directive Annexes with tables). The lists with additives and conditions of use—the systematic part—have been split into three separate Directives: on sweeteners (27), on colors (28), and on all the other additives, or ''*miscellaneous*'' additives (10). Each of these three Directives also carries a legal part, with principles, which may be different in some respects for sweeteners, colors, and *miscellaneous* additives.

A. Framework Directive (89/107/EEC)

The main principles on additives legislation are laid down essentially in the *framework* Directive on additives (4). This *framework* Directive on additives establishes that additives shall be ruled by positive list, which also lays down the conditions of use, that is, only additives which are listed may be used, to the exclusion of all others.

Furthermore, a Community procedure is foreseen for trial introduction of new additives, starting in one member state, but then either to be extended to the whole Community or revoked (Art. 5).* It should be stressed that this trial procedure is not meant for broadening of authorizations of additives which are already approved in the EU. Although support from member state administrations is most helpful, any broadening of existing authorizations (use level or extension to new food categories) is best requested through the Commission.

Directive 89/107/EEC sets criteria for an additive to be approved (Annex II). These criteria are presented in Section V.C. It also establishes 24 functions of food additives (called ''categories of additives,'' Annex I of the Directive) (see Table 3). The precise functions of additives, which are open to endless debate, do not have an important role in the Directives. They are merely used in the titles of the Directives and in their Annex

---

* It should be noted that a somewhat similar but still different authorization for a trial introduction of an additive existed earlier, e.g., in German law. This is the so-called *Ausnahmegenehmigung nach* § 37 LMBG. This earlier German trial authorization procedure is not the transcription of the European Article 5 of 89/107/EEC. The German text differs in that it limits the authorization to *one foodstuff*, to a *limited tonnage* of that foodstuff, and to *one company*. Further requirements are set in that German leglation. The German text, but not the European Directive, furthermore allows a possible extension of the trial period. The trial authorization, according to Directive 89/107/EEC, as well as according to German law, is temporarily bypassing democratic control by Parliament. The idea is to escape from sluggish administrative procedures stifling innovation. However, the limitation to two years for the whole process, including the final parliamentary confirmation of the authorization, is very tight. Some government experts simply rejected the whole procedure for this very reason.

headings. Still, the precise meanings of additive functions (or categories) are defined in the original sweeteners Directive (27), in Art. 1 § 2 of the colors Directive (28), and in Art. 1 ∍ 3 of the main *miscellaneous* Directive (10).

The framework Directive sets furthermore the conditions for selling additives preparations as such (Arts. 7 through 10): (1) additives not for sale to the ultimate consumer (e.g., to food manufacturers or to dealers) and (2) additives for sale to the ultimate consumer (such as, e.g., intense sweeteners for tabletop use, or baking powder for the kitchen. The core of these conditions—for both (1) and (2)—are the precise labeling requirements for such preparations.

B. Additive Function Definitions

These definitions may be helpful in deciding whether a substance should be classified as an additive. The reverse question remains unanswered, whether a substance with a clearly technological function, a function which does not occur in Annex I of the *framework* Directive (Table 3), can therefore be certified not to be an additive. A typical example of such a case could be an enzyme inhibitor.

It should be pointed out that, remarkably enough, the categories listed in the Annex to this Directive are not completely identical with the ones already adopted earlier for labeling purposes (see Table 2)* (18). The definition for sweeteners is presented below in Section VII.A.1 and for colors in Section VII.B.2.

C. Criteria for Additive Authorization

The essential criteria for approval are (1) safety, (2) not misleading the consumer, and also (3) the presence of a need, a technological and/or economic justification. Normally a case of need is established by demonstration of a technological problem to which the additive brings an economic solution. Such a dossier is best underscored by letters of support from interested user industries. Merely economic justifications (price) are more difficult to defend, but it would be hypocritical to maintain that they cannot in themselves sufficiently justify the use of certain safe additives. It should be obvious that advantages to both the consumer and the manufacturer are the driving forces behind all efforts to manufacture foods economically.

The exact formulation of each of the criteria can be criticized. The philosophical question can be asked, for example, whether added colors do not always—in one way or another—mislead the consumer. But the consumer may in fact like to believe that things have more color than they do in reality, or even insist on certain colors for specific foods which are naturally colorless or weak in color (e.g., green mint). Unmitigated absence of risk to health is not warranted either. It is sometimes a tradeoff. Few cases are clearer than the nitrites, which have a recognized toxicity. Still they are properly recommended as the preservatives of choice to counter the far greater risk of botulism in meat products.

---

* Table 1 [Annex I of the additives *framework* Directive 89/107/EEC (4)] lists ''*sequestrant*'' and ''*enzyme*,'' as well as ''*propellant and packaging gas*.'' These do not occur in Table 2 [Annex II of amendment 93/102/EC to the Labeling Directive (18)]. Instead of ''*propellant and packaging gas*,'' Table 2 mentions only ''*propellant gas*.''

Table 2 The Function Categories of Additives Established for Labeling Purposes

Categories of ingredients which must be designated by the name of their category followed by their specific name or EC number
Color
Preservative
Antioxidant
Emulsifier
Thickener
Gelling agent
Stabilizer
Flavor enhancer
Acid
Acidity regulator
Anticaking agent
Modified starch
Sweetener
Raising agent
Antifoaming agent
Glazing agent
Emulsifying salts
Flour treatment agent
Firming agent
Humectant
Bulking agent
Propellant gas

Source: Ref. 18.

But the same is true even for many regular foods. Eggs contain cholesterol; frequent sugar consumption constitutes a risk of caries, etc. That is no reason to ban their reasonable use as nutritious and tasty foodstuffs.

The requirement for a case of need is a typical European matter, originating in the legislation of several member states. It does not exist in the same way, for example, in the United States. The reason for using this approach is to avoid overly long lists of authorizations which are not really being used. Yet it is certainly stifling innovation. Although it does not entail an obligation to start real use afterwards, manufacturers are sometimes reluctant to lend their support to a request for authorization. Indeed, trials with a substance which may in the end not get an authorization may still consume scarce development money, spent before they are willing to endorse a request for authorization.

That the case-of-need approach is stifling innovation is substantiated furthermore by the fact that several additives have turned out to be very useful for applications quite different from the one(s) for which they were conceived in the first place. For example, polyols designed to act as sweeteners are now also finding widespread application as humectants. Carboxymethyl cellulose, a thickener, can be used as a barrier substance on French fries, preventing the penetration of excess fat into the potato.

Furthermore, the very fact that the lengthy and expensive process of authorization ends with a political decision, is economic nonsense. Industry needs certainty to operate. If all the criteria for the authorization of an ingredient are fulfilled, and its safety is proved, it is unacceptable that an arbitrary political decision at the end could waste the enormous effort spent.

D. Additive Safety and the Scientific Committee for Food

As to safety, the Commission's Scientific Committee for Food (SCF) plays a cardinal role. It sets acceptable daily intakes (ADIs) which ought to guide the administrators in establishing safe conditions of use. (The SCF does not have a mandate to grant an authorization for use; that is a political decision, taken, as a rule, in line with opinions delivered in the SCF; see Section X.B). However, the vastness of the European area, with its extremely varied food consumption patterns, complicates this establishing of acceptable conditions of use for the whole Community. Although an official effort for surveying is underway, the problem of unavailability of reliable data for the whole EU on the intake of specific foodstuffs, and hence of additives, remains to be solved.

Although the primary aim of the undertaking was merely to open the market by harmonizing existing legislation, the proceedings have allowed two important new developments. First, some new additives were taken up directly into the Community legislation, without prior authorization in any individual member state (e.g., the case of neohesperidine dihydrochalcone, or NHDC). Second, in order to underscore the safety basis of the new rules, the SCF has now set ADIs for additives which had been in use for a long time without earlier adequate safety evaluations (such as for the glycerol esters of wood rosin).

Guidance on how to submit a new additive to the SCF is given in a brochure published by the European Commission (30). No rigid program of studies has been established. In general the following are required for the dossier: the results of acute toxicity, genotoxicity studies, metabolic and pharmacokinetics studies, subchronic toxicity studies, reproduction and teratogenicity studies, and chronic toxicity and carcinogenicity studies. Omission of certain investigations, or inclusion of additional ones, should be justified and explained. The studies carried out should be reported on in their entirety, with evaluation of the results and conclusions.

In general, the SCF establishes a no Effect level or no Adverse effect level (NOEL; NOAEL) and derives an ADI from this. This ADI is the guide figure for assigning maximum use levels. A list of the ADIs of the EU authorized additives can be found in Tables 1 and 2. Note, however, that ADI values are subject to change. The tables reflect the status on April 17, 1999. The ADIs set by the SCF sometimes differ from the ADIs set by the Joint FAO/WHO Expert Committee on Food Additives (JECFA).

As a standard, the ADI is expressed numerically, in milligrams per kilogram body weight. For very high ADIs the indication can be given in grams. For unproblematic additives, A (acceptable) or NS (not specified) can be indicated as non-numerical indications. An ADI on its own, does not mean very much. For risk assessment, the ADI of an additive must be seen together with the dose required to achieve its technological effect. Recently fears have been voiced that an ADI per se is not a good enough tool to assess the risks of an additive. It should be stressed that, as a rule, the ADI is established for a whole lifetime.

E. Food Additives as Agents Causing Other Adverse Reactions and as Allergens

Contrary to widespread and persistent belief, approved food additives do not frequently cause real allergic reactions. For example, regular food ingredients (fish, egg, etc.) are by far the most frequently proven food allergens. *Allergy* refers to a specific type of immune reaction. The real mechanism of a reaction being many times unclear, *intolerance*, as a more general term, is to be preferred in many discussions. It should be stressed that intolerance reactions are individual, and are not part of a general toxicity syndrome. A food component causing very serious problems to one individual may be completely innocuous to the majority of the population at large.

Indeed, people suffering from a specific food intolerance are often relative minorities of the population. Elimination of each of these problems by prohibition of all such specific causative agents in foodstuffs would be absolutely impossible and largely disruptive for major segments of the food chain. Hence the proper legislative approach to this issue is to impose correct indication of intolerance agents on the food so that sufferers can avoid foods they do not tolerate well. As most real intolerance accidents are not due to prepackaged foods but to foods in restaurants, private occasions, and catering, additional measures should be considered for these. The responsibility for information on nonprepackaged foods remains primarily the responsibility of the individual member states. Hence the rules, if any are in existence, can be different from one member state to another.

A legislative measure of the EU is in preparation to impose warning labeling of prepackaged foods for agents of severe intolerance. Most important in this respect is the list of specific agents, the indication of which will be rendered mandatory. The development of that list is paralleled by the Codex Alimentarius standard on labeling.

In the latest EU Commission proposal on warning labeling for intolerance-causing agents—based on SCF advice (31)—only one set of additives was listed for warnings on the label: the sulfites, at concentrations of at least 10 mg/kg (32). The fact that the sulfites are as yet the only additives in the list proposed (along with nine natural foods) underscores quite strikingly the statements in this section. As far as additives are concerned, the Codex Alimentarius list is also limited to sulfites ($\geq$10 mg/kg) (33). (See also Sections V.J and V.N.).

F. Combination of Additive Usage

An important principle in the EU additives Directives, though not explicitly stated in the legal part of the texts, is the option to combine in a single foodstuff the use of all additives corresponding to all authorizations applicable to that foodstuff. Of course, only very seldom will it be necessary to use all authorized additives together and or use them up to their maximum use level.

One of the important consequences of this principle in the context of the sweeteners Directives (27) is that all sweeteners authorized for a foodstuff can be combined, each one up to its maximum use level, intense sweeteners and polyols, independently or mixed. In several places in the annexes of the *miscellaneous* additives Directive (10), specific limitations to that principle are given. See, for example, the benzoates (p. 17) and the synthetic antioxidants (p. 25) (or for colors, for an exception, see Section VII.B.1.b). These cases normally pertain to additives to which a *"group ADI"* was assigned. A *group ADI* sets the maximum intake only for the sum of the interchangeable additives of that group, rather than for an individual substance. Normally this concerns very similar substances with the same technological effects. In such cases, as a rule, the authorizations in the Directives are given for a group of additives (e.g., for the sulfites or for the phosphates), the sum of which is not to exceed the use level.

G. Unprocessed Foodstuffs and Foodstuffs Without Additives

A definition of *''unprocessed''* foodstuffs cannot be found in the framework Directive but is given in the *miscellaneous* additives Directive (10), as well as in the colors Directive* (28). As a rule, additives are prohibited in all *unprocessed* foodstuffs. The definition of *unprocessed* foodstuffs in the context of the EU additives Directives differs slightly from the one of Codex Alimentarius in that deep-freezing is not considered as processing in the EU. Consequently, deep-frozen foods should be devoid of additives.

Furthermore, the application of colors is prohibited in the foods listed in Art. 2 § 3a) of the colors Directive (28) whilst a list of prohibitions against *''miscellaneous''* additives can be found in Art. 2 of the corresponding Directive (10).

The sweeteners Directives have been written with nearly identical descriptors of food categories repeated for each sweetener and with a limited number of authorizations. This alleviates the need for a clause on *unprocessed* foodstuffs or for a table with sweetener-free foodstuffs.

H. Organic Produce

Certain additives are authorized for use in foodstuffs from organic production. None of the sweeteners are authorized. Colors are neither authorized, provided that the calcium carbonates are considered noncolors [i.e., as *miscellaneous* additives (10)]. With the exception of karaya gum, all the authorized additives also occur in Annex I of the *miscellaneous* additives Directive (10). This means that they are all carrying a non-numerical ADI, and can be used *''quantum satis''* (QS; see Section V.K). The only additive which carries a numerical ADI in the list of additives authorized for organic produce are the tartrates. The complete list of these additives is given in Table 3 in this chapter.

I. Traditional Foods

The most striking example of political compromise in the elaboration of the European additives legislation is without doubt the amendment to the additives framework Directive. This amendment authorizes ''. . . the Member States to maintain the prohibition of certain additives in the production of certain foodstuffs considered traditional . . .'' (34). This exception—a fundamental regression to national protectionism, away from the ideal of full European harmonization—was originally requested by Germany as an amendment to the sweeteners Directive (27). The intention was to prevent the use of intense sweeteners in beers brewed according to the *Reinheitsgebot* (German law on *''purity''* of beer).

---

* Art. 2 § 11 of the colors Directive (27) and Art. 2 § 3a of the *miscellaneous* Directive hold a definition of unprocessed foods, as well as a nonlimiting list of treatments which do *not* make a food into a processed one: '''unprocessed' means not having undergone any treatment resulting in a substantial change in the original state of the foodstuffs; however, the foodstuffs may have been, for example, divided, parted, severed, boned, minced, skinned, pared, peeled, ground, cut, cleaned, trimmed, deep-frozen or frozen, chilled, milled or husked, packed or unpacked'' (10).

Table 3 Ingredients of Nonagricultural Origin: Food Additives, Including Carriers

E number	Name	Specific conditions
E170	Calcium carbonates	—
E270	Lactic acid	—
E290	Carbon dioxide	—
E296	Malic acid	—
E300	Ascorbic acid	—
E322	Lecithins	—
E330	Citric acid	—
E334	Tartaric acid (L(+)—)	—
E335	Sodium tartrate	—
E336	Potassium tartrate	—
E400	Alginic acid	—
E401	Sodium alginate	—
E402	Potassium alginate	—
E406	Agar	—
E410	Locust beam gum	—
E412	Guar gum	—
E413	Tragacanth gum	—
E414	Arabic gum	—
E415	Xanthan gum	—
E416	Karaya gum	—
E440	Pectin	—
E500	Sodium carbonates	—
E501	Potassium carbonates	—
E503	Ammonium carbonates	—
E504	Magnesium carbonates	—
E516	Calcium sulphate	CR
E938	Argon	—
E941	Nitrogen	—
E948	Oxygen	—

Source: Ref. 54. CR = carrier.

This damaging change was vigorously opposed by industry and by the EP as well. After months of stalemate in the proceedings on the sweeteners Directive (27), a compromise was reached. Instead of limiting the exception merely to Germany's beers, it was decided then that all member states would equally be authorized to make certain exemptions for "*traditional*" foodstuffs from the horizontal legislation on additives in general, not just on sweeteners. This agreement has now been laid down in a Directive amending the *framework* Directive on additives (26).

This additional and complicated rule prevents the use of the newly authorized additives in the manufacture of its declared *traditional* products, sold as such, in the territory of the requesting member state. This amendment is in line with the openness of the internal market in the Union, and with the principle of freedom of establishment. Indeed it does not prohibit, for example, a foreign brewer *selling*, or even *brewing*, a beer in Germany, with an intense sweetener, or with any other additive authorized for beer in the Union. The only condition is that it is not sold as a "*Reinheitsgebot*" beer.

One of the consequences of the amendment was that a list of *traditional* foodstuffs had to be drawn up, as it has proved impossible to define precisely what ''*traditional*'' really is. This compromise showed that many member states are reluctant to give up their protectionist habits. Indeed, a considerable number of products were proposed to the Commission for exemption from the horizontal rules on additives. Most of these were turned down on objective grounds by the Commission. The list of foodstuffs, with the additives which are banned, has been published as a Decision of the EP and the Council (34), and is presented in Table 4.

A basic remark to be made in this context is that such prohibiting legislation is not at all necessary. Indeed, the authorization of an additive does not make its use mandatory. And the ingredient list allows consumers to choose what they prefer. If there is commercially that much mileage in ''*brewed according to the Reinheitsgebot*'' and similar claims, market forces will provide the customer with such foods. A proper claim on the label would do better to achieve this than prohibitive legislation.

Table 4 Products for Which the Member States Concerned May Maintain the Prohibition of Certain Categories of Additives

Member state	Foodstuffs	Categories of additives which may continue to be banned
Germany	Traditional German beer (''*Bier nach deutschem Reinheitsgebot gebraut*'')	All except propellant gases
Greece	Feta	All
France	Traditional French bread	All
France	Traditional French preserved truffles	All
France	Traditional French preserved snails	All
France	Traditional French goose and duck preserves (''*confit*'')	All
Austria	Traditional Austrian ''*Bergkäse*''	All except preservatives
Finland	Traditional Finnish ''*Mämmi*''	All except preservatives
Sweden Finland	Traditional Swedish and Finnish fruit syrups	Colors
Denmark	Traditional Danish ''*Kødboller*''	Preservatives and colors
Denmark	Traditional Danish ''*Leverpostej*''	Preservatives (other than sorbic acid) and colors
Spain	Traditional Spanish ''*Lomo embuchado*''	All except preservatives and antioxidants
Italy	Traditional Italian ''*Salame cacciatore*''	All except preservatives, antioxidants, flavor enhancers, and packaging gas
Italy	Traditional Italian ''*Mortadella*''	All except preservatives, antioxidants, pH-adjusting agents, flavor enhancers, stabilizers, and packaging gas
Italy	Traditional Italian ''*Cotechino e zampone*''	All except preservatives, antioxidants, pH-adjusting agents, flavor enhancers, stabilizers, and packaging gas

Source: Ref. 34.

J. Use Levels

Caution is to be applied in the interpretation of the maximum use levels given in the Directives. The general rule in the Directives is that the use levels authorized apply to the foodstuffs *after preparation* according to the manufacturers instructions.

In both the sweeteners Directive (27), Art 2 § 4, and the colors Directive (28), Art. 2 § 6, that is the case. In the *miscellaneous* Directive the same is also true for the additives listed in Annex IV. In the same Directive, Annex III might be subject to the same rule, if the notes at the bottom of page 16 in the *Official Journal* refer to the whole of Annex III. If that note applies only to Part A of Annex III (sorbates and benzoates), it would follow that Parts B (sulfites), C (other preservatives), and D (other antioxidants) fall under the rule for *miscellaneous* additives, which is expressed in Art. 2 § 7 of that Directive (10). That rule is that the quantities apply to the foodstuff *as marketed*. Which of the two interpretations is to prevail remains open until an official clarification is given.

For nitrites and nitrates, residual amounts are mentioned, rather than ingoing amounts, although some indicative ingoing amounts for nitrites are mentioned as well.

Although the Directive on *miscellaneous* additives does not mention it, similar considerations might apply (e.g., for antioxidants), for which, at least in principle, the entirely of the amount added might not be found in the finished product.

The official control will normally run its checks on analysis figures (residual amounts), rather than on recipe figures (ingoing amounts). The two may sometimes differ.

The case of sulfites merits special citation. Sulfites are controversial as additives. The *miscellaneous* additives Directive indicates that the levels authorized ". . . relate to the total quantity, available from all sources," (i.e., the sum of natural presence, added as an additive, added as a processing aid, from "*carry-over*," spontaneously created in the manufacturing process, etc.). Still the same text stipulates that a quantity of 10 mg/kg or mg/L is considered as not present. This arbitrary analytical borderline is laid down in a Directive authorizing additives (p. 19 in Ref. 10). Hence the statement cannot bear on labeling. For example, one can imagine a case where 15 mg/kg of sulfite arises by *carry-over* in a complex foodstuff without technological effect in the final foodstuff. Although it is analytically present, it would not need to be labeled since it is present merely by *carry-over* (see also Sections V.E and V.N). However, in all likelihood legislation on the labeling for intolerance-causing agents in preparation in the EU may make the labeling of 10 mg/kg (or mg/L) of sulfite, or more, compulsory anyhow.

For colors the maximum use levels indicated in the Directive refer to the "*coloring principle*" (as examined by the SCF), rather than to the formulated preparation as a whole.

K. Quantum Satis

For a number of authorizations, the regulators have accepted a concept from Belgian law: *Quantum Satis* (QS). This means that ". . . no maximum level is specified. However additives shall be used in accordance with good manufacturing practice, at a level not higher than is necessary to achieve the intended purpose and provided that they do not mislead the consumer." The notion is defined in Art. 2 § 8 of the *miscellaneous* Directive (10) and identically in Art. 2 § 7 of the colors Directive (28). Remarkably, QS is also used, but not defined, in the sweeteners Directive. There is no doubt that this same interpretation is to be applied there.

Before the concept of QS was introduced, the same was expressed regularly with the term *good manufacturing practice (GMP)* (as in Codex).

Quantum Satis is, as a rule, used for additives for which the SCF has set a non-numerical ADI. But this is not true across the board. In certain self-limiting applications, additives with numerical ADIs are authorized under QS. An example is karaya gum (which has an ADI of only 12.5 mg/kg body weight, but the authorization at QS is given only for the low volume dietary supplements in the *miscellaneous* Directive (10, p. 28). Tartaric acid (E334) is authorized at QS (Annex I of Ref. 10), yet the ADI is 30 mg/kg body weight.

Conversely, there are additives for which the ADI is NS (i.e., the non-numerical ADI not specified), yet the authorizations given are not QS [an example is propionic acid in bread, where German opposition may have been the main political reason to limit the quantity in the *miscellaneous* Directive (10, p. 24)]. For a similarly non-numerical ADI A (acceptable), there are cases where QS has not been allowed. [An example is thaumatin in confectionery. See the sweeteners Directive (27) p. 11].

L. Additives Authorized and/or Restricted for Foods for Particular Nutritional Uses

The additives for foods for particular nutritional uses are to be found in the tables of the regular additives Directives.

Also additives for infant foods, weaning foods (follow-on formulae), and foods for babies are regulated by the the regular additives Directives.* However, it should be noted that neither sweeteners (see in Section VII.A.1) nor colors (see in Section VII.B.1.a) are permitted in foods for infants and in weaning foods. So, authorizations for additives in these foodstuffs can only be found in the *miscellaneous* additives Directive (see in Section VII.C.3).

M. Intake Surveys

In each of the three additives Directives—on sweeteners (Art. 8 of Ref. 27), on colors (Art. 15 of Ref. 28), and on *miscellaneous* additives (Art. 7 of Ref. 10)—the EP obtained the insertion of an amendment imposing surveys on additive intake. The basic motivation behind this request is expressed only in the sweeteners Directive (27). It somehow translates distrust and fear of the safety of additive usage as authorized. Yet such surveys may bring real benefits.

Although the request to run the surveys on additive intake is addressed to the member states, it is the European Commission which has to report to the Parliament within five years of adoption of the sweeteners Directive (i.e., before June 30, 1999), or of the enactment of the colors and *miscellaneous* additives Directives (December 31, 2000, and September 25, 2001, respectively).

It is likely that updating Directives on food additive authorizations will be issued after 1999, when the results of the surveys have been evaluated.

---

* In Directive 91/321/EEC it was announced that a Directive would be issued, authorizing additives for infant foods and for weaning foods (Art. 5 of Ref. 35).

1. Industry Perceptions on Additive Intake

The vast majority of data available suggest that for most additives the real intake represents a far less important part of the ADI than is generally thought. Moreover, there is no harm in going beyond the ADI on certain days. The ADI is a guideline limit, only not to be exceeded every day in a lifetime. Furthermore, many additives which are authorized in certain foods are not always all present, or present up to the maximum level permitted, or present in all the brands of foodstuffs for which they are permissible. Hence, it should be evident that, across the board, additive usage is far lower than would be believed upon straight extrapolation from the legislation.

In many cases the ADI was set at a rather low level (e.g., the case of annatto or nisin). This is not because adverse effects were observed in the toxicological tests, but rather because the tests have not been extended to include sufficiently high intake levels.

Industry is caught in a double bind on the matter of intake surveys. If it contributes, even only in offering information, its contribution may well be used to justify cutting additive authorizations. If it does not contribute, the unreliability of the data that might be put forward by the authorities might lead to wild overestimates of consumption, and hence to authorization cuts anyhow.

In the first case, industry's contribution might, for example, establish that far fewer, or lesser amounts of, additives are being applied than are now authorized. This might be an excellent ground for legislators to contest many a case of need, and consequently to trim authorizations to less kinds of additives, or to ''*more realistic*'' levels. Of course, acceptance of such a cutback scenario would make a mockery of past concentrations and would be disastrous for innovation.

In the exercise, it is evident that industry cannot be only an interested bystander. CIAA, for example, has expressed its views. These are, amongst others, that (1) it is willing to contribute to data collection, (2) it encourages participation at national level and in the context of the system of scientific cooperation between the member states, and (3) it wants to know the specific objectives and the nature, as well as the format, of the data sought.

2. Scope of the Intake Surveys

Crossing 365 additives with thousands of foodstuffs creates a staggering number of cells. The enormity of the undertaking, to fill possibly millions of cells, requires gross simplification to make the task manageable. The problem of cost is not a minor one either.

The SCF, as well as outsiders, are being consulted on the best methods of approaching the matter. The European Commission has looked for support (information and funds) from industry.

Even if data are collected nationally—exploiting to the maximum the systems already in use for that purpose in the member states—it is necessary to work at the European level to establish uniform operating principles and coordination, and to consolidate the results in a database.

3. First Steps in Additive Surveys

Not all additives require evaluation in detail. The European Commission, on the basis of a *preliminary* study, adopted a decision tree approach as a risk assessment tool. Using essentially the Danish budget method (36), a list has been established of additives for which further evaluation of the consumption might be required. It includes, amongst others, curcumin, annatto, adipate, the sulfites, the borates, etc.

The occurrence of an additive in that list does not necessarily mean that the intake of that additive is too high. It may very well mean that the Danish budget method is not perfect for risk assessment in that case. Further studies and deliberations are ongoing.

4. Value of Surveying Additives Now

The new additives legislation did, no doubt, change additive usage in the EU. Hence, it can be foreseen that intake studies run during the first years after adoption of the new additives legislation, as requested by the EP, may not yield results relevant for the past as well as for the coming years. Running such studies now is like trying to pinpoint a moving target.

N. Carry-Over: Two Kinds

The principle of *carry-over* also authorizes in compound foodstuffs those additives, in the quantities which are authorized, in their constituent ingredients.* On the other hand, a principle termed *reverse carry-over* authorizes the use of additives which are permitted in a final compound foodstuff, also in the (intermediate) ingredients to be used in such a compound foodstuff. The condition for the application of the principle of this *reverse carry-over* is that the (intermediate) ingredients shall not be for direct consumption. The quantities permitted in the (intermediate) ingredient have to be adjusted to produce a lawful level in the final compound foodstuff.†

It should be clearly understood that *carry-over* relates to the *authorization* for use of an additive, not to its *labeling*. An additive introduced according to *reverse carry-over* is of course to be labeled, since it will necessarily be functional in the final product. On the other hand, an additive which is present in a compound foodstuff can be either functional or not. The *functionality* of such an additive decides whether labeling is triggered; additives which are functional in the end product are to be labeled. In general, if the additive in the end product is no longer functional, it need not be labeled (see also Section V.J). Note however that additives when used in a compound ingredient (e.g., in the chocolate coating of a cereal bar) must be declared in the ingredient list (see also Section VI.A)

Both the *carry-over* principles are expressed only in the Directives on colors (28) and on *miscellaneous* additives (10). Industry has asked that the principle should be generalized to include sweeteners as well. It was hoped that this would be granted in the context of an amendment to the sweeteners Directive. Some of that has materialized; however, the amending Directive (25) has severely restricted *reverse carry-over* for sweeteners (see Section VII.A.7).

In each of the Directives with clauses on *carry-over* (10,25,28) exceptions on authorization to use clauses on *carry-over* are indicated. Each of these are discussed in sections on the specific Directives (see Section VII).

---

* An example: fruit-based pie fillings may contain up to 100 mg of sulfites per kilogram of filling. Although there is no specific authorization for sulfite in pies, the pies prepared with such fillings may contain all the sulfite introduced with such fillings. Assuming that only one-third of the pie is filling, the total pie may not contain more than 33.3 mg of sulfite per kilogram of pie.

† An example: annatto is authorized in extruded or expanded snacks, up to 20 mg per kilogram of snack. Although annatto is not an authorized color for seasonings sold to the final consumer, seasonings destined for flavoring these snacks may contain up to 200 mg of annatto per kilogram of seasoning, if the seasoning constitutes 10% of the final weight of the extruded or expanded snack product.

O. Enriched Preparations: Coloring Foods

Only the colors Directive (28) insists that the isolation of a constituent of regular food, resulting in an enriched preparation with a distinct and strengthened technological effect, changes the resulting preparation into an additive, to be authorized, labeled, etc. It might be that the same principle could, in the long run, come to prevail for all other additive types as well: for sweetening preparations (such as liquorice or Stevia*) flavor enhancing properties (such as yeast extract), or antioxidant preparations (such as rosemary extract).

The consequences for business of these stipulations are far-reaching. There has been, in the past, a widespread practice of using extracts from flowers and foodstuffs as coloring agents, for example, in Germany and in France. The Directive on colors prohibits this practice, unless these preparations are defined, receive an official safety clearance from the SCF. According to the letter of the Directive, only when the coloring effect of preparations made from foodstuffs is secondary to a dominant property of taste, sapidity, or nutritional effect will an exception to this rule be accepted (such as for paprika, turmeric, and saffron).

A particularly important example of a coloring foodstuff is malt extract. Industry has argued forcefully in favor of considering malt extract as a food in most applications.

The Standing Committee for Foodstuffs has, based on past history of use as a food, accepted *Hibiscus* (rosebud) as a food which can be used for coloring purposes. Several other of such preparations remain problematic however.

It is obvious that this is a gray area, inviting dispute. The clarifications and decisions made, on a case by case basis, by the official control in individual member states will remain very important in these matters.

P. Natural Occurrence

Only for sulfites is there a statement in the *miscellaneous* Directive (10) stating that the maximum values include sulfite from all sources. For none of the other additives is such a clarification given. Still the same Directive acknowledges that nisin, benzoates, and propionates may be produced in fermentation processes, and thus be present without willful addition.

But substances classified as additives can also be present in foodstuffs without being added, and without being generated by fermentation. Phosphates are naturally present in important quantities in milk, for example. (Other mineral salts come in the same category in various foodstuffs.)

To date there is no general statement in the additives Directives stating that natural compounds present are not included in the maximum use levels for the additives, and that these levels refer merely to the quantities added. Such a statement is desirable. Indeed, it is disturbing that in dairy products, for example, the phosphates present naturally can exceed the maximum levels authorized for addition as additives.

---

* Stevioside, a sweetening substance derived from *Stevia*, has not been authorized as a sweetener in the EU, as no acceptable toxicological data were yet submitted to the SCF. With the high sweetening power of stevioside preparations, it is hard to imagine how such preparations could be considered foods (nutrient properties negligible compared to the sweetening power). Glycerrhizin, a sweetening substance present in liquorice, is not authorized as a sweetener in the EU, for reason of toxicity.

VI. LABELING OF ADDITIVES

A. Requirements

Additives are ingredients. Article 6 of the general labeling Directive (18) makes labeling of their use compulsory in an ingredient list. There are however some exemptions; the most notable one being drinks with an alcohol content of more than 1.2%. So far, these need not carry ingredient labeling (Art. 6 of Ref. 18). Consequently alcohol-free beers routinely carry a list of ingredients. A proposal of the Commission is now being developed in order to generalize the requirement for labeling of ingredients to all alcoholic drinks (37). It is the author's perception that enactment of such a measure in the EU is several years off due to the procedural problems.

It seems evident that substances which are sometimes being used as additives, but which may also be produced as byproducts in the manufacturing process, need not be labeled as ingredients either, since they have not been *added intentionally* (e.g., in fermentation: benzoates, propionates, sulfites; see Section V.P).

In the same way, additives whose presence in a foodstuff is merely due to *carry-over* (see Sections V.J and V.N) and which have no technological function in the final foodstuff need not be indicated in the ingredient list [Art. 6 § 4.(c).ii of Ref. 18].

Furthermore, additives used as processing aids need not be indicated in the ingredients list [Art. 6 § 4.(c).ii of Ref. 18] since processing aids are not to be considered as ingredients. Similarly, ''*solvents*'' or ''*media*'' for additives or for flavoring need not be labeled either [Art. 6 § 4.(c).ii of Ref. 18].

Additives in the ingredient list must be preceded by the name of their additive labeling category (i.e., their function), as laid down in Table 4 (from Annex II of 93/102/EC) (38). If a substance exerts more than one function, it is the principal function that should be indicated.

Additional special labeling requirements for certain categories of additives are presented in Section VI.E. See also Sections VII.A.4 (sweeteners) and VII.B.4 (colors).

There is a rule on labeling—contested nowadays—which exempts from indicating the individual components of compound ingredients in the ingredient list when they total less than 25% of the finished food. (Art. 6 § 7 of Ref. 18). However, all additives must be separately labeled; the 25% rule on ingredients does not apply to additives.

An EU Commission Directive is in the pipeline to make mandatory the indication on the label of prepackaged foods of agents causing severe adverse reactions (32). As stated, the only additives considered in this respect are the sulfites (≥10 mg/kg).

Irrespective whether it is for the purpose of intolerance warnings (32) or for ingredient listing (10), sulfites are considered not present when their total quantity from all sources does not reach 10 mg/kg.

B. Additives Names

The additives must be designated by their specific name or by their E number. The E number is a number officially assigned by the EU (see Section III.F). The numbering is established by the publication of the Directives authorizing their use in foodstuffs. (It should be noted in this respect that E numbers have been created longer ago than the INS numbers of Codex Alimentarius).

Both in the E numbering system and in the INS, complementary lower case specifiers sometimes have to be added to distinguish very similar substances with specifications somewhat different from the ones of the main parent compounds. These specifiers can be of two kinds. Specifiers "*i, ii, iii*, etc." indicate that the ADI is not different and therefore different conditions of use do not have to be established. And, in ingredient labeling, E-i numbers can be put together with the parent substance (i.e., neither a separate item nor a "*i, ii, iii* . . ." specifier is required in the ingredient list). Specifiers "*a, b, c* . . . etc.," on the other hand, indicate that a separate risk evaluation was carried out, which resulted in an independent ADI. Consequently such substances are assigned independent conditions of use, and should therefore be labeled separately—with their specifier—as different ingredients.

Examples of the first kind are the diphosphates: E450 i through vii (10). Examples of the second kind are E407 (carrageenan) (10) and E407a (alternatively refined carrageenan) (24).

As to the specific name to be used, it seems reasonable to assume that, in addition to the names used in the Directives, generally accepted synonyms can also be used. No doubt synonyms listed in the official specifications of the EU and of JECFA are to be considered "*generally accepted*" and understood. They could hardly be considered misleading. Still, it may remain controversial for some time to come as to whether all these synonyms—rather than only the exact Directive designations for the additives—are to be used. It is clear however that commercial trade names are not acceptable for use in ingredient lists.

At the time of this writing, industry is active in proposing to allow the use of simplified names for the additives. The E numbers have carried an unpleasant stigma since the infamous "*Tract de Villejuif*" (see Section III.F), and the Directive designations sometimes sound awfully long and chemical. It is to be hoped that the authorities will authorize simplifications in order to enhance the transparency of the labels. Deliberations among the member states on the matters do not seem promising in this respect.

C. Order of Indication

The ingredients must be listed in the ingredient list in descending order of their weight at the mixing bowl stage [Art. 6. § 5(a) of Ref. 18]. Additives may therefore often occur in several positions, interspersed amongst other additive and nonadditive ingredients. In cases where several different additives with the same function/category occur in the same foodstuff, the ingredient list will have to mention several times (and at different positions in the ingredient list) that same function/category indication, followed by the name of the different additives. All this may make the ingredient list quite unwieldy.

The list of the function/categories to be use in labeling are given in Table 2. As already indicated, it differs only slightly from the one in Table 1 (which refers only to authorization). In labeling, the functions "*sequestrant*," "*enzyme*," and "*packaging gas*" (occurring in Table 1, but not in Table 2) do not have to be indicated in the ingredient list with the names of the substances added to a food. See Section VI.E.1, however, for detail on the obligation to indicate on the label when packaging gases are being used.

D. Quantitative Ingredient Declaration

The quantity of any ingredient in a foodstuff which is emphasized on the label (by name, picture, or claim)—including additives—must be indicated, either in the ingredient list or next to the sales descriptor on the label (39). The percentage indication must show (with some exceptions) the quantities at the mixing bowl stage during manufacturing. In some specific cases, this obligation is waived (40).

As far as additives are concerned, this is not frequently the case. An example where emphasis on an additive triggers, quantitative ingredient declaration (QUID) is "*with xyli-*

tol'' on a special banner on a chewing gum. However, cases where sweetener(s), or sweetener(s) and sugar(s), are to be ''*double labeled*'' merely because the law requires so (see Section VI.E.3) do not trigger QUID (40).

E. Labeling Warnings for Specific Additives

For certain additive categories, the legislators have considered that the mere indication in the ingredient list does not sufficiently inform the consumer, specifically, for packaging gases and sweeteners.

1. Packaging Gases

In an amendment (41) to the labeling Directive, the application of the special statement ''*packaged in a protective atmosphere*'' to all ''. . . *foodstuffs whose durability has been extended by means of packaging gases* . . .'' was imposed. No detail is given on how (font size or color, e.g.) and where exactly this indication is to be put on the label.

2. Tabletop Sweeteners

Tabletop sweeteners are not foods, but additives sold directly to the ultimate consumer (see Section VII.A.1.c). The sweeteners Directive (27) requires that tabletop sweeteners (just like foodstuffs; see Section VI.E.3) containing more than 10% of polyol bear a warning for laxative effect: ''*excessive consumption may induce laxative effects*.'' The dose at which such effects may occur need not be referred to. For tabletop sweeteners which contain aspartame, also the indication ''*contains a source of phenylalanine*'' is required (see also in Section VI.E.4 for aspartame).

For none of these additional labeling particulars is the size or presentation of the letters explicitly laid down by the text.

The general warning for laxative effect of a tabletop sweetener, without reference to the absolute amount of polyol ingested, seems strange. Indeed, such products are essentially made of intense sweetener, whilst the polyol ingested constitutes a very small amount of excipient. Only a large amount of polyol can trigger laxation.

3. Foodstuffs Containing Sweeteners

The sweeteners Directive (27) requires also for foodstuffs containing more than 10% of polyol, as for similar tabletop sweeteners, that they bear the same warning for laxative effect: ''*excessive consumption may induce laxative effects*.''

Similarly, for foodstuffs which contain aspartame, the indication ''*contains a source of phenylalanine*'' is required as well. Certain associations of people suffering from phenylketonuria (PKU) quite logically declared that mentioning aspartame in the ingredient list was sufficient for them.

Furthermore, an additional special measure imposes a ''*double*'' indication that sweeteners are being employed in a foodstuff (42).* The same text also imposes such an

---

* The sweeteners Directive announced the adoption, before the end of 1995 and according to a precise procedure laid down in the sweeteners Directive (27), of provisions which would spell out ''*the details which must appear on the labeling of foodstuffs containing sweeteners in order to make their presence clear*'' and ''*warnings concerning the presence of certain sweeteners in foodstuffs*'' (meaning again aspartame for its phenylalanine content and the polyols for their laxative effect in case of abuse).

That procedure would normally have led to a *Commission Directive*. Such a *Commission Directive* can be adopted by a faster procedure, in which the EP, e.g., does not really intervene. However, due to the indication of sugar, if sugar is used together with sweetener in foods. The indications ''*with sweetener*'' or ''*with sugar(s) and sweetener(s)*'' are to ''*accompany the name under which the product is sold.*'' (The name meant here is the trade description, or the lawful sales denomination; it can never be the brand name!) The requirement that the statement ''*accompany the name under which the product is sold*'' is not expressed for the warnings against the laxative effect and against phenylketonuria. Neither the size nor the presentation of the letters of these warnings is explicitly laid down by the legislation.

This ''*double labeling*'' legislation has been strongly opposed by industry, since it means that certain additives would have to be indicated twice on the label. People at special risk can consult the ingredient list, which is provided exactly for that purpose. Furthermore, the principle of *double* labeling of only certain ingredients seems grossly unfair; it suggests risks inherent in sweeteners which are nonexistent or disproportionate. It also sets a precedent. Why would colors or preservatives not have to be similarly double labeled in the future? Hence, it should be realized that the measures imposed neither address real risks nor sufficiently address real but concealed risks inherent to other kinds of ingredients, such as, for example, sugar in regular foods without sweeteners with reference to caries or to diabetes. Some further comments on this issue as related to additive function are given in Section VII.C.1.

4. Aspartame-Sweetened Foodstuffs

A far more serious objection to a generalized special double labeling for aspartame concerns its misleading effect on the real people concerned. Normal advice to people suffering from PKU, who ought to avoid aspartame and for whom the special warning is meant, is to shy away from foods that contain copious amounts of phenylalanine—eggs, milk, meats, etc. Although the tolerance is personal, a rule of thumb for a PKU patient is to stay below 20 mg of total phenylalanine intake per kilogram body weight per day. But even PKU sufferers need an essential minimum of 300 mg of total phenylalanine per day.

What is the contribution of aspartame to the burden of, for example, a snack product? Assume a child of 20 kg suffering from PKU. The child should avoid phenylalanine in excess of 400 mg per day. The sweeteners Directive allows snack products to contain up to 25 mg of aspartame per 50 g of product. This is equivalent to 14 mg of phenylalanine. This is between 9 and 0.9% of the phenylalanine burden of the same product without aspartame. This means that the relative importance of the aspartame addition is not very significant. It is indeed not the added aspartame which will be the important burden; the snack food itself is the vastly more important burden which should be avoided by the rare PKU sufferer.

PKU patients clearly know how to avoid basic foods containing a lot of protein, such as steak, eggs, cheese, etc. For a variety of transformed, processed, and packaged foodstuffs, however, such knowledge may be more difficult to get without information from the label, from the manufacturer, or from their PKU associations.

---

rejection of the proposal by the Standing Committee for Foodstuffs (SCF) the proposal was finally adopted by the Council, leading to a *Council Directive.*

Most of the more recent Directives, adopted to complete the internal market, have been adopted as ''*EP and Council Directives.*''

In principle, the deadline of the end of 1995 for these special labeling measures for sweeteners has not been respected since the proposal was adopted only on March 29, 1996.

Hence, if for example a sweet-sour seafood in a jar whose phenylalanine level is slightly increased due to the presence of aspartame is labeled as ''*contains a source of phenylalanine*,'' PKU patients may be lulled into believing that such seafood products that do not contain aspartame are not sources of phenylalanine. And this is plainly a counterproductive kind of disinformation. It coaches PKU patients into making their own decisions in such matters and away from the more reliable sources of information (such as for example their associations' databases). Thus, the label, as the European Commission document intends to complete it, is not at all the appropriate vehicle to provide the rare PKU sufferers with the sufficiently detailed information they need.

5. Adverse Reaction and Allergen Warnings

This subject is discussed above in Section IV.E.

F. GMO-Derived Additives

As stated in Section II.E.6 GMO-derived additives are not regulated by the novel foods Regulation (9). A Regulation on the labeling of GMO-derived maize and soya products was issued insofar as these are foods only (43).

Although there is strong pressure on the legislators to also impose the indication of the GMO origin of additives on the label, this raises a number of difficult problems. A proposal for such legislation is circulating anyhow at the time this manuscript goes to press. It might be finalized and issued by the end of 1999 or in the year 2000.

VII. THE SPECIFIC EUROPEAN DIRECTIVES ON ADDITIVES

The additives legislation for the EU has been split essentially into three parts: sweeteners, colors, and all the others (or *miscellaneous* additives). The full texts of the Directives, as published in the *OJ,* is available on diskette from the author of this chapter (see Section X.1).

The legislators' aim is to consolidate, at a later stage, all these Directives into one global food additives Directive. Neither planning date nor precise format for this future consolidation are known at present. As all Directives are to be transposed in national law anyhow, a consolidated version is no high priority.

In Appendix Tables 1 and 2 all EU authorized additives have been listed with their ADIs, as well as with the page numbers where they occur in the Directives. The corresponding Directives have been indicated by SWET (sweeteners) (27), COLR (colors) (28) and MISC (*miscellaneous* additives) (5,10,24). In Appendix Table 1 the additives have been sorted by E number; in Appendix Table 2 they are put in alphabetical order. Authorizations introduced in updating Directives have been indicated with UPM (*miscellaneous* additives) or UPS (sweeteners) in Appendix Tables 1 and 2.

The Directives were originally elaborated starting with sweeteners and colors. These two categories of additives were chosen to start the proceedings because public opinion is highly sensitive to them. Furthermore, as some member states were gearing up to issue their own legislation on sweeteners, it was logical for the European Commission to start there.

A. Sweeteners Directives

1. Structure, Authorizations, and Annexes

After the first Directive on sweeteners (94/35/EC) (27), an updating Directive was issued (96/21/EC) (25). The Directives on sweeteners each consist of a ''*legal part*'' and one Annex with tabulated authorizations.

A definition for sweeteners can be found in Art. 1 § 2 of 94/35/EC. Two kinds of sweeteners are authorized: bulk sweeteners, or polyols, and intense sweeteners. The bulk sweeteners have a relative sweetness per unit of mass similar to the sweetness of sugar. Intense sweeteners are many times sweeter. All the sweeteners authorized carry E numbers of the 900 series, except E420 (sorbitol) and E421 (mannitol).

In drafting the sweeteners Directive, the European Commission wanted to limit authorizations essentially to foodstuffs for which a clear consumer benefit and/or a technological need was perceived: reduction of energy intake, manufacturing of noncariogenic and of dietetic foodstuffs, foodstuffs *with no added sugar*, or foodstuffs with increased shelf-life (prevention of microbial spoilage).

However, certain flavors of nonsweet snack foods, containing a quantity of intense sweeteners (for which sugars were not appropriate), but which did not satisfy the stated authorization criteria, were already on the market in at least two member states. This existing practice, properly highlighted to those responsible, has forced the hand of the legislators into accepting this additional criterion for authorization.

Sweeteners are prohibited in foods for infants and for young children (27; Art. 2 § 3).

A *carry-over* clause (see Section IV.N), missing in Directive 94/35/EC, is introduced for sweeteners in the update Directive (see Section VII.A.6).

a. The Polyols. The polyol authorizations in the two sweeteners Directives bear exclusively on their prime function: to sweeten foods. In addition to this function, polyols can also fulfill other technological functions, such as humectants, stabilizers, etc. For these last functions, the authorizations are regulated by two *miscellaneous* additives Directives (see Section VII.C).

For polyols, authorizations in drinks were not granted out of concern for their laxative effect in youngsters. Sweetener authorizations for polyols have not been granted for all solid foods. Instead, a positive list was imposed. This will, of course, limit potential new applications, unless additional authorizations are obtained. And that is a long and difficult process.

In the sweeteners (27) and in the *miscellaneous* additives Directives (5,10) the authorizations for polyols are given up to a level of *Quantum Satis*. The polyols for which authorizations have been given are sorbitol, mannitol, isomalt, maltitol, lactitol, and xylitol. For uses authorized under *miscellaneous* additives, the labeling should not indicate ''*sweetener*'' as the function. The real function—such as, e.g., ''*humectant*''—should precede the name of the polyol.

b. Intense Sweeteners. Authorizations for six intense sweeteners are given in two Directives on sweeteners (25,27): acesulfame K, aspartame, the cyclamates, saccharin and its salts, thaumatin, and neohesperidine dihydrochalcone (or NHDC). See Table 5 (see also footnote in Section V.O.). So far, sucralose and alitame are not authorized in the EU.

Considerable opposition has been voiced against cyclamate by France and the UK on the basis of toxicology, and more recently by Sweden on the basis of intake versus

ADI. The EU Commission has steadfastly defended the position taken by the SCF in that respect (44). In 1995, This body first reconfirmed a temporary ADI of 0–11 mg/kg body weight for the cyclamates, pending new data. The cyclamates were up for revision by the SCF before June 30, 1999 (45). In March, 2000, a new ADI of 7 mg/kg body weight was set.

For some intense sweeteners some other technological applications have also been authorized as flavor enhancers in the *miscellaneous* additives Directives (5,10). For use in certain desserts, soft drinks, and chewing gums these are acesulfame K, aspartame, thaumatin, and NHDC. NHDC is furthermore also authorized—below its sweetness threshold level—in margarine, spreadable fats, meat products, fruit jellies, and in vegetable proteins.

Again, in these cases the intense sweeteners should be labeled with their real function, that is, as ''*flavor enhancers*'' and can, in principle, do without warning labeling (see Section VII.C.1).

c. Tabletop Sweeteners. In the emerging EU additives legislation tabletop sweeteners are not considered foodstuffs. They are additives sold as such to the final consumer. In tabletop sweeteners, and other sweeteners sold directly to the consumer, only the same sweeteners are authorized as those authorized in regular foods (i.e., those occurring in the Annexes of Ref. 27).

d. Food Descriptors and Format. Most of the food descriptors used in the sweeteners Directives are uniform, that is, terms used for a food descriptor are the same each time, irrespective of the specific sweetener. This fact makes possible to list the authorizations of intense sweeteners in an easy orientation table (see Table 5).

In fact, this similarity in terms used for the descriptors is somewhat misleading, as it might suggest that all sweeteners received all the same authorizations. This is not the case; there are small differences. Table 5 should be consulted with reference to the published Directives in order to know precisely which authorizations are applicable for each sweetener.

In the published form, the subtitles and the layout in the Annexes of the Directives are somewhat unfortunate as, for example, beers and sauces have come under ''*confectionery.*''

2. "No Added Sugar"

Directive 94/35/EC (27) defines ''*with no added sugar*'' as ''*without any added mono- or disaccharides or any other foodstuff used for its sweetening properties.*'' Indeed, foodstuffs with sweetening properties, such as honey, as a rule hold plenty of mono- and disaccharides. By and large these compounds have properties very similar to these of regular sugar (caloric effect, cariogenicity, general physiological properties, glycaemic index, etc.). Hence someone wishing to avoid regular sugar is generally well advised to avoid such sweetening foods and mono- and disaccharides as well.

3. "Energy Reduced"

Similarly, ''*energy reduced*'' is defined in the same text as ''*with an energy value reduced by at least 30% compared with a similar foodstuff or a similar product.*'' The ''*energy reduced*'' claim is very important. The definition of the term in the sweeteners Directive is relevant only for the *addition* of sweeteners to foods. It does not pertain to *labeling*. As to labeling, there is no Community legislation on claims. Consequently, the rules on

Table 5 Maximum Use Levels of Intense Sweeteners in Foods

[Directives 94/36 (in regular font) and **96/83 (in bold type)**]

BIORESCO Ltd.

		E number		E950	E951	E952	E954	E957	E959
FOOD CATEGORY*	**Conditions**	**ER**	**NAS**	**ACSF**	**ASPM**	**CYCL**	**SACR**	**THAU**	**NHDC**
Non-alçoholic drinks									
Water-flavour		ER	NAS	350	600	400	80		30
Dairy based		ER	NAS	350	600	400	80		50
Fruit-juice based		ER	NAS	350	600	400	80		30
"Gaseosa"				350	600	400	100		30
Mixed drinks: non-alcoh.+alcoh.	beer,cider,perry spirit,wine			**350**	**600**	**250**	**80**		**30**
Alcoholic Drinks									
Cider and perry				350	600		80		20
Certain beers (+ low alcohol)				350	600		80		10
Beer		ER		**25**	**25**				**10**
Spirit drinks / alcohol restrctd, >>	< 15 % alcohol			**350**	**600**		**80**		**30**
Desserts		ER	NAS	350	1000	250	100		50
Edible Ices		ER	NAS	800	800	250	100	50	50
Fruit									
Cans / bottles		ER	NAS	350	1000	1000	200		50
Jams		ER		1000	1000	1000	200		50
Fruit preparations		ER		350	1000	250	200		50
Sweet-sour preserves				200	300		160		100
Sandwich spreads		ER	NAS	1000	1000	500	200		50
"Eßoblaten"				**2000**			**800**		
Confectionery									
Confectionery			NAS	500	1000	500	500	50	100
Cocoa + fruit confectionery		ER	NAS	500	2000	500	500	50	100
Starch confectionery		ER	NAS	1000	2000	500	300		150
Chewing gum			NAS	2000	5500	1500	1200	50	400
Breath freshening microsweets			NAS	**2500**	**6000**	**2500**	**3000**		**400**
Throat pastilles	strongly flavd, freshng.		NAS		**2000**				
Tablet-form confectionery		ER		**500**					
Breakfast cereals	>= 15 % fibr. >= 20 % bran		NAS	**1200**	**1000**		**100**		**50**
Bakery products									
*Fine bakery/special nutrit. Uses				1000	1700	1600	170		150
Cornets & wafers > ice cream			NAS	**2000**			**800**		**50**
Snacks				350	500		100		50
Fish, sea food									
Sweet-sour marinades				200	300		160		30
Vegetables									
Sweet-sour preserves				200	300		160		100
Vegetable preparations		ER		350	1000	250	200		50
Feinkostsalat				**350**	**350**		**160**		**50**
Soups		ER		**110**	**110**		**110**		**50**
Sauces									
Emulsified + non-emulsified				350	350		160		50
Mustard				350	350		320		50
Special Foods									
Food supplemnts / diet integrators	vitam/miner; syrup-chew	rena-med		2000	5500	1250	1200	400	400
Weight control formulae				450	800	400	240		100
Compl formulae / Nutritnl supplm.	use: medical supervision			450	1000	400	200		100
Solid supplements				500	2000	500	500		100
Liquid supplements				350	600	400	80		50

* Always refer to the original publication for the precise deignation of the food category

		E number		E950	E951	E952	E954	E957	E959
KEY DATA ON SWEETENERS				**ACSF**	**ASPM**	**CYCL**	**SACR**	**THAU**	**NHDC**
ADI in mg/kg bodyweigh	SCF			9	40	7	5	A	5.
	JECFA			15	40	11	5	NS	-
Approximate relative sweetness**				300	200	30	300	400	300

** Strongly dependent on the food matrix

ER = ENERGY REDUCED

Italics are printed as such also in the EU Official Journal

NAS = NO ADDED SUGAR

the *energy reduced* claims in labeling may be different from member state to member state. The stakes on the definition of this claim are high. Dietary significance should prevail.

Although claims should have been regulated in the Directive on labeling claims, it was decided to issue a ruling on energy reduction in the sweeteners Directive. Although the motive to do so is obviously the desire to give the sweeteners Directive a clear meaning, the decision is quite unfortunate as will be explained. The text of the Directive defines *energy reduced* foods as having at least 30% reduction in calories in comparison to the original product or a similar product.

With some exceptions, the sweeteners Directives restrict the use of sweeteners to foods that do not contain added sugar or that are energy reduced. In the opinion of the legislators, these conditions should help limit sweetener intake. The main motive in imposing a 30% minimum energy reduction is the wish to respect national practice and existing food standards. But it may be thought, at first, that a more stringent definition of *energy reduced* (i.e., 30% instead of 25%) would further restrict consumer exposure to sweeteners. Close examination of the case reveals that this is not so. At least 90% of the sweetener intake stems from the consumption of tabletop sweeteners, essentially calorie-free soft drinks and yogurt with 0% fat. Since in these products a calorie reduction of 30% or more is easily achieved, the main sources of sweetener intake are not affected by this legislation. As for the remaining 10% of products, which earlier had been reduced in calories by less than 30%, such as marzipan in Germany, the new legislation will require the substitution of a bigger share of sugars by intense sweeteners and thereby *increase* sweetener intake. Some other products, such as certain fine bakery wares, may not be manufactured because a 30% reduction of calories cannot be achieved. On balance, sweetener intake will therefore not be diminished. The application of the 30% limit therefore does not fulfil its intended purpose.

Furthermore, if it is accepted that the incorporation of a 30% value in the sweeteners Directive will lead sooner or later to the adoption of an identical value in national or European regulations on food labeling or claims; the nutritional implications for the consumer must already be considered at this stage.

In this regard it must be noted that dietary guidelines issued by many European countries advise consumers to reduce their fat intake and to control their calorie intake so as to avoid obesity and being overweight. It is clear that calorie-reduced foods play an increasingly important role in the consumer's attempt to reach these dietary goals.

Although the food industry would like to meet the consumer's demand for calorie-reduced foods, it is often difficult (for food technological reasons) to reduce the calorie content by more than 20–25% without affecting the taste and texture of the product in an unacceptable way. If the consumer is to be offered a sufficiently wide choice of calorie-reduced good tasting foods, the definition of "*calorie reduced*" should not be made too stringent. Although it might at first appear that the consumer benefits more from a 30% limit than from a 25% limit for calorie-reduced foods, the reverse is true. In fact, a moderate 25% reduction of calories in a wide range of good tasting foodstuffs will facilitate compliance with dietary objectives, while a more ambitious 30% reduction of calories will inevitably result in a smaller number of less attractive taste-wise calorie-reduced foods.

This opinion was also shared by the Codex Committee on Nutrition and Foods for Special Dietary Uses, which at their 1989 meeting noted that calorie reductions of as high as 30% were not always feasible, and which therefore recommended a minimum reduction of 25% in calories (46). It is significant to note that this Codex recommendation was adopted, although earlier step 3 transactions were advocating reductions of one-third (47).

Although seemingly moderate, a 25% reduction in calories is nutritionally significant if it is maintained. Consequently, the established Codex standard for nutrition claims requires an energy reduction of at least 25% for allowing a reduced energy claim.

Thus any future decisions on this matter of reduced energy clearly constitute a potential threat not only to industry's, but also to consumers' best interests.

A number of foodstuffs are rich in energy even if sugar is not added (for example, by the presence of fats and starch). It follows that the legislators have not authorized sweeteners for such foods. In this way, no usage of sweeteners was granted for regular fine bakery ware. However, for such bakery ware destined for particular nutritional uses (dietetic, etc.), certain sweeteners were authorized.

In authorizing sweeteners for ''*cocoa-(based) confectionery, energy reduced or with no added sugar*'' in a horizontal rule (27), the legislators had to solve a conflict with the vertical one (3), which requires sugar as the sweetener for chocolate. A solution was agreed to the extent that ''*energy reduced chocolate*'' or ''*chocolate with no added sugar*'' would be considered outside the scope of the chocolate Directive (as far as authorization of additives is concerned; this is not necessarily the case for labeling). Certain member states, such as Denmark, have recently contested this approach. Other member states, however, such as Belgium, are endorsing it. Note in this context that an updated vertical EU Directive on cocoa and chocolate products is in the pipeline, but has not yet been adopted at the time when this paper goes to press.

4. Labeling for Sweeteners

This issue has been discussed in depth in Sections VI.E.2, 3, and 4. Also as highlighted in Sections VII.A.1.a and .b, substances typically known as sweeteners can have nonsweetener functions as well. Such applications have been granted in the *miscellaneous* Directive (cf. Section VII.C.1).

5. Mixtures of Sweeteners

In establishing the conditions of use, and as recommended by the SCF, the European Commission has trimmed maximum levels of the individual sweeteners on purpose so as to push manufacturers to apply mixtures of different sweeteners instead of using more single sweeteners. This way, there is a lesser chance of exceeding the ADI for individual sweeteners. Of course this entails additional complications on the label.

All sweeteners (intense and bulk sweeteners) may be used in combination, up to their maximum use levels, also in mixtures with sugars.

6. Update of the Sweeteners Directive

As Directive 94/35/EC was being finalized and adopted, it became clear that an amending update was desirable. Such a Directive—96/83/EC—was issued mainly in order to encourage technical developments (25). The new Directive renames the food descriptor ''*Vitamins and dietary preparations*'' to ''*Food supplements/diet integrators (e.g., vitamins, trace elements), syrup-type or chewable.*'' In fact this is not a mere renaming (that is, another descriptor for the same foods), but rather a redefinition of the meaning of the food category.

The new text introduces the QS principle. It also introduces the two *carry-over* clauses for sweeteners as well, except for *food supplements/diet integrators*. Amazingly however, the Council has limited the application of the first *carry-over* rule (for compound foodstuffs) to "*energy reduced*" or "*with no added sugar*" foodstuffs. This limitation complicates the legislation and again discriminates against sweeteners in an irrational way.

In addition, in the Annex some 40-odd new authorizations are given, all for the same intense sweeteners as were found in the original sweeteners Directive (27). These newly proposed authorizations have been indicated with "UPS" in Appendix Tables 1 and 2.

The Annex of the update Directive draft also introduces, for saccharin and for the cyclamates, notes which state that the quantities are to be expressed as free imide or acid, respectively. This is a worthwhile clarification since it removes ambiguity from the tables. It should be realized that these notes increase the levels of the sweetener anions saccharin and cyclamates which can be used by 12% (sodium) to 22% (calcium). Note that saccharin potassium salt can be used as well in a similar way. As acesulfame K is only being commercialized in the potassium form, the same rule has not been made applicable to that substance.

7. Implementation

All member states should have already adapted their national legislation so as to bring it in compliance with the two Directives on sweeteners. They are already obliged to grant free circulation to products complying with the two Directives, and noncomplying products are already banned from all of the EU market as well.

8. Specifications

For sweeteners, the European Commission has published the specifications for all authorized sweeteners in a Directive (19).

B. Colors Directive

1. Structure, Authorizations, and Annexes

The Directive on colors deals with additives with a clear function. Only two closely related additives are common to both the colors Directive (28) and the *miscellaneous* additives Directive (10). These are calcium carbonate and calcium hydrogen carbonate (E170 i and ii, respectively). When the calcium carbonates are used in noncoloring functions, they should obviously be labeled in their real function; such as, for example, "*acidity regulator.*"

The Directive on colors (28) comprises a legal part as well as five Annexes. It has a quite complex structure. It gives authorizations for 43 colors. These carry E numbers in the 100 series (between 100 and 180).

The legal part defines the term color, based on the function, either to add or restore color to a food (Art. 1 § 2). Colors used for staining inedible external parts of foodstuffs are excluded.

It is extremely important to note that the maximum use levels authorized refer to the quantities of the "*coloring principle*" (i.e., the pure coloring substance) in the preparation. As a rule, the authorization corresponds to the material, as it has been examined by the SCF.

Carry-over is regulated by Art. 3 (see also Sections V.J and V.N as well as Section VII.B.1.a).

a. The Annexes. The colors which are approved are listed in Annex I of the Directive (28). All these may also be sold directly to the consumer, except E123, E127, E128, E154, E160b, E161g, E173, and E180. These last ones are the ones with the lower ADI values.

Annex II lists foodstuffs in which no colors may be used, except if provided for specifically in Annexes III, IV, or V. In addition to unprocessed foodstuffs, there are 32 other descriptors for such foods. An important one is foods for infants and young children (Annex II § 29 of Ref. 28). This prohibition also excludes ''*reverse carry-over*'' as a way to add color to such foods for the young.

Annex III lists foodstuffs in which, each time, only a limited number of specific colors may be used. These are put into 27 groups and indicate maximum use levels. The colors indicated in Annex III may also occur in Annexes IV and V.

Annex IV is a list of colors which are authorized only for specific foodstuffs. It also indicates maximum use levels.

Annex V is split into two parts. Part 1 is a list of 15 colors (which are all different from the ones in Annex IV and Annex V, Part 2) that are permitted at QS level in all foodstuffs, except those listed in Annexes II and III. Hence they may also be used at QS in all the foods listed in Annexes IV and V, part 2. Part 2 of Annex V gives authorizations for 18 colors (which are all different from those in Annex IV and Annex V, Part 1). It is organized in two tables: a table with 18 colors and a table with some 30 food descriptors. Every color from the list may be used up to the maximum use level indicated for the food in the table—independently, alone, or in combinations. There is however an exception.

b. Exception Made for Children in Annex V, Part 2. For five of the foodstuffs in the table of Part 2, an exception is made. The foodstuffs for which an exception is made are nonalcoholic drinks, edible ices, desserts, fine bakery ware, and confectionary. All these foodstuffs are frequently consumed in copious amounts by children. Four items of the colors list are restricted in use level to no more than 50 mg per kilogram or per liter. The colors for which this exception is made are Sunset Yellow FCF, azorubine, cochineal red A, and brown HT. The concern for protecting children is clear in this decision, since each of these colors has an ADI of less than 5 mg/kg body weight. For the other colors in the list of Part 2, there are no restrictions except for the maximum use levels indicated.

2. Implementation

All member states must have adapted their national legislation by now so as to bring it in compliance with the colors Directive. They are obliged already now to grant free circulation to complying products. Noncomplying products are banned from all of the EU market already.

3. Update

An update Directive, as were issued for sweeteners (see Section VII.A.7) and for *miscellaneous* additives (see Section VII.C.4.c), is not planned. No doubt this is because of the extreme sensitivity of the issue of colors in foods.

4. Azo-Colors

Sweden joined the EU on January 1, 1995. Before that date it had already on the books a prohibition against the use in foods of the azo-colors: tartrazine, Sunset Yellow FCF, azorubine, amaranth, ponceau 4R, red 2G, allura red AC, brilliant black BN, brown FK, brown HT, litholrubine BK. The main health protection reason Sweden presented as a justification for this prohibition is the allergenicity of the azo dyes.

The EU Commission maintains that the SCF approved of the use of these colors in full knowledge of their occasional allergenicity (44). It considers the Swedish measure disproportionate, as other additives not prohibited in Sweden are sometimes as allergenic. And allergenicity cannot be a valid reason to prohibit an additive. Labeling of additives used in the ingredient list is the best protection for some rare individuals suffering from allergies against additives (see also Section IV.E).

5. Labeling

Amazingly, notwithstanding the sensitivity of the public toward colors, and unlike for the case of sweeteners, special labeling requirements for colors outside the ingredient list have never been proposed during the development of this Directive.

6. Specifications and Methods of Analysis

The specifications for all authorized colors permitted in foodstuffs have been published in the *Official Journal* (20).

Two older methods of analysis for colors (pH and ether extractable compounds) have been laid down in Directive 81/712/EEC (48).

C. Miscellaneous Additives Directives

1. Structure and Annexes

Miscellaneous additives Directives 95/2/EC (10) holds again articles and annexes.

In Art. 1 § 5 of the legal part of Directive 95/2/EC nine groups of products are classified as nonadditives.

Article 2 § 2 of the legal part of the same Directive refers to Annex I to indicate the additives which may be used at QS in all foodstuffs except those mentioned in Annex II and in Art. 2 § 3.

Article 2 § 6 in the legal part states that the additives authorized for foods for particular and nutritional uses are also to be found in this Directive.

A standard *carry-over* clause can be found in Art. 3. Annex I lists additives authorized at *Quantum Satis* in all foods except the foodstuffs listed in Annex II.

Annex II was drawn up to limit the applications of Annex I additives for a series of foods for which the legislators disliked overliberalizing the use of additives. Annex II lists, for example, foods for which vertical standards are in existence as well as some foodstuffs in which specific member states put a special pride (e.g., pasta, *pain courant français*, certain wines, and *mozzarella*).

Annexes III and IV list the different additives with the descriptors for the foodstuffs in which they can be used, under the conditions mentioned therein.

Annex V is the list of additives and one solvent which may be used as carriers, together with their conditions of use.

Annex VI lays down, if four parts, which additives may be used in infant formulae, in follow-on formulae, and in weaning foods, and in foods for infants and young children for special medical purposes. In each of the four parts, the conditions of use are indicated as well.

See Section V.H for the discussion of use levels on *miscellaneous* additives.

One substance can be used sometimes in several functions. Only in the case of sweeteners and colors have the specific functions been laid down in separate Directives. Consequently, one color and some of the sweeteners appear in more than one Directive.

The calcium carbonates (E170i and E170ii) have been authorized as colors (cf. Section VII.B.1), but are also authorized in the *miscellaneous* additives Directive.

The polyols and some intense sweeteners as well were authorized in some functions in the *miscellaneous* additives Directive (cf. Sections VII.C.A.1.a and b.)

The legislator's intention to impose complementary labeling for sweeteners in foods (cf. Sections VI.E.3 and 4) has its legal basis in the sweeteners Directive (27). Hence, in legal terms the compulsory indication of additional warning labels on foods containing sweeteners cannot apply to foods which contain sweeteners on the basis of authorizations given by the *miscellaneous* additives Directive (10). From this viewpoint, the use of polyols in nonsweetening functions, or of some intense sweeteners as flavor enhancers, would not trigger the obligation to apply the sweetener warnings.*

2. Additive Function

As stated, listing of an additive in the sweeteners or in the colors Directive is clearly defining the function meant for the authorization. As mentioned (Section V.B), in the *miscellaneous* additives Directive the precise technological function for which an additive is used is not always laid down in the legislative text. Indeed, the Directive's Annex IV just refers to ''*Other permitted additives*.'' The definitions of *miscellaneous* additives' functions are laid down in Art. 1 § 3.

For antioxidants (see Annex III, Part B) and preservatives (see Annex III, Part A), for example, the function is clearly stated in Directive 95/2/EC (10). Still EDTA, though listed in Annex IV, is frequently referred to as an antioxidant too. Hence there are more additives with antioxidant function than the ones indicated in Annex III, Part B.

It remains a question whether nitrates/nitrites—indicated as preservatives in the Directive—can be used for their coloring effect in certain meat products, at levels below the maximum authorized, and below the level required for assuring an effective preservative function.

3. Flour Treatment Agents

Although ''*flour treatment agent*'' is defined as an additive function (Art. 1.§ 4 of Ref. 10), flour treatment agents would logically fit better among the processing aids (see Section III.E.2).

Some substances that were considered for that purpose have various functions in the *miscellaneous* additives Directive. The classical flour treatment agents historically include a series of emulsifiers (such as E471 through E475 and E481, E482, and E483) as well as the reactive agents, ammonium persulfate, ascorbate, azodicarbonamide, benzoyl peroxide, chlorine, chlorine dioxide, cysteine, potassium bromate, and sulfites. The emulsifiers have been assigned authorizations in the *miscellaneous* additives Directive (10). As

---

* This seems logically sound. Yet, if that were really so, the presence of aspartame in sugar-containing chewing gums, specifically authorized in the *miscellaneous* additives Directive (10) would not be highlighted as a warning outside the ingredient list to phenylketonuria patients. Still for ''*no added sugar*'' chewing gums, the same sweetener will trigger warning labeling, since for such chewing gums the authorization to use aspartame is given in the sweeteners Directive (though at a higher level) (27).

to the others, three of them are listed in the *miscellaneous* additives Directives: ascorbate, sulfite (10), and L-cysteine (5).

The EU Commission's SCF has rejected ammonium persulfate and potassium bromate. Further studies were requested on azodicarbonamide, chlorine, and chlorine dioxide before a final decision could be made. On benzoyl peroxide no data were found in the reports of the SCF, or in the minutes of its meetings.* Ascorbate is authorized by the main *miscellaneous* additives Directive (10). In the same Directive, there are a number of authorizations also for the sulfites, though not explicitly as flour treatment agents. (Yet the authorizations for use in starches and in dry biscuits may cover some of the needs.) A community authorization has been given for L-cysteine in the second update of the *miscellaneous* additives Directive (10). Hence, except for the emulsifiers, ascorbate, L-cysteine, and sulfite, the other reactive candidate flour treatment agents are prohibited. The additives *framework* Directive—in its Art. 2 § 1 (4)—states that the additives lists of the EU are exclusive, in so far as such lists have been drawn up. The last but one citation of the main *miscellaneous* additives Directive refers flour treatment agents to a, then future, Directive to be drawn up after the SCF has evaluated them (10). Since the SCF has made its evaluations approving only L-cysteine, the case of the other flour treatment agents seems closed, until new data will be cleared by the SCF.

4. Corrigendum

A corrigendum to Directive 95/2/EC has been published in the *Official Journal* to correct "5 mg/kg" to read "5 g/kg" for phosphates in potato products (49). However, since a corrected table for the phosphates was introduced afterwards in Directive 98/72/EC (5), this corrigendum has completely lost its meaning.

5. Update Directives

a. First Minor Amendment. Directive 96/85/EC (24) authorizes alternatively refined carrageenan, E407a, by a special first updating *Directive*.

b. Second Amendment. As an important second update (98/72/EC) (5), in parallel with and for the same purpose as the amending update for sweeteners, the Commission issued an important update Directive on *miscellaneous* additives. In addition to authorizations requested to permit technological progress, some additional requests emanated from the new member states (Austria, Finland, and Sweden), which joined the Union on January 1, 1995.

A second enzyme is introduced as a new additive by Directive 98/72/EC (5): invertase. Other new additives appearing for the first time are mono- and dimagnesium phosphate, konjac gum and konjac glucomannane, β-cyclodextrin, cross-linked sodium carboxymethyl cellulose, enzymatically hydrolyzed carboxymethyl cellulose, microcristalline wax, L-cysteine, and acetylated oxidized starch.

The entries on phosphates of Directive 95/2/EC (10) have been replaced by this draft proposal with a new improved table. A number of additional individual modifications and new foodstuff entries are also put forward in this second update Directive.

All newly proposed entries have been indicated with "UPM" (*miscellaneous* additives) in Appendix Tables 1 and 2.

c. Toward Further Amendments to 95/2/EC. Article 5 of the framework Directive (4) permits member states to authorize new additives on trial within their own territories. As

---

* JECFA though, assigned an ADI of 40 mg/kg of body weight to benzoyl peroxide.

these authorizations are temporary, they ought to be extended to all of the EU before expiring or be withdrawn altogether. This is one of the reasons why the Commission considers an additional update to 95/2/EC within a relatively short time (1999). This would lead to the fourth amendment of the *miscellaneous* additives Directive.

It is likely that a fifth amending Directive will be issued some time later, after the additive intake surveys (see Section V.M) have been finalized in order to apply corrections and allow another adaptation to new developments.

6. Implementation of the Directives on Miscellaneous Additives

All member states should bring their legislation in compliance with Directive 95/2/EC (10) not later than September 25, 1996. Complying products will have to be granted free circulation as from the same date. Nonconforming products should already have been banned from all of the European Union.

7. Specifications for Miscellaneous Additives and Community Methods of Analysis

Updated specifications for additives other than colors and sweeteners have not yet all been finalized. Their elaboration is being continued.

After the new additives Directives were issued, *miscellaneous* additives have been issued: for E200 through E385, as well as for E1105, in Directive 96/77/EC (21); for E400 through E585 they can be found in Directive 98/86/EC (22). Additional specifications (E170 and E296 through E1520) are still being developed at the time this paper goes to press.

As the Commission has taken JECFA specifications as an important starting point for its work in this field, JECFA specifications could be used in absence of final EU specifications. However, earlier official EU specifications or the member states' national specifications should take priority. In older Directives criteria for purity for certain *miscellaneous* additives were already laid down for certain antioxidants (50), emulsifiers, stabilizers, thickeners, and gelling agents (51), and for certain preservatives (52,53).

A couple of generalities on methods of analysis, as well as some Community methods of analysis for preservatives and antioxidants were also issued in an older Directive (48).

VIII. GETTING AN AUTHORIZATION FOR USING A FOOD ADDITIVE IN THE EU

It should be realized that an authorization for a new food additive in the EU should be projected in the wider international framework. Before embarking on the quest for an authorization in the EU, a global strategy should be considered for the material. In order to get an authorization for using an additive in the EU, the following data, methods, and/ or steps are essential:

1. Define for what purpose and how much of the material will be used. Provide a reasonable and extreme estimate of intake ensuing from such use for the average consumer (mean, median) and for the the so-called heavy consumer (90th percentile).
2. Define the precise food category descriptors, as well as the use levels required.
3. Make sure to have the necessary toxicological data, demonstrating the safety of the material under its intended conditions of use.
4. Define the method of analysis of the material in food.

5. Provide specifications, including methods of analysis, for the control of the additive's purity.
6. Build the case of need: identify interested parties and obtain their written support for the authorization.

IX. CONCLUSIONS

With the adoption of the last Directives on specifications of *miscellaneous* additives, the first stage of the harmonization of the European additives legislation will approach completion. Legislation on this dynamic sector of the economy will never be final. Flexible adaptation to new developments is absolutely essential. Forthcoming Directives, and a series of decisions by the *Standing Committee for Foodstuffs*, will continue the monumental task.

Hundreds of people have contributed thousands of hours of very intense work. The European Commission has, in a very open style, consulted with all parties involved and come up with measures that have passed scientific as well as political scrutiny. In this way the new legislation is an enormous leap forward. It harmonizes the additives authorizations in fifteen countries for thousands of foodstuffs, and it unquestionably constitutes a major liberalization in many cases. More solid evidence of safety has been supplied with additional studies and was formalized by refined ADI settings.

In the very way this legislation had to be elaborated—involving technology, business, and politics—the result could never be perfect. Yet the result is far better than anybody had dared hoped for. What has been achieved will, to a great extent, allow both consumers and industry to reap the benefits of a single open market of the European Union in the field of food.

The legislators had an unenviable task, which would be criticized whatever the outcome was. This paper does not seek to blindly join in the chorus of criticism of the authorities. It is quite remarkable to observe that most industries which took care to prepare and present their needs well received due attention and had most of their needs met. The real aim is, rather, to discuss the perceived strengths and weaknesses of the emerging legislation, as the best way of analyzing what it can mean to the practical operator.

(text continues on p. 193)

APPENDIX

A. List of Abbreviations

ADI	Acceptable daily intake
EP	European Parliament
EU	European Union
CEN	European Standardisation Committee
CIAA	Confederation of the Food and Drink Industries of the European Union
ECOSOC	Economic and Social Committee of the European Communities
GMO	Genetically modified organism
JECFA	Joint FAO/WHO Expert Committee on Food Additives
NHDC	Neohesperidine dehydrochalcone
NOEL/NOAEL	No effect level/no adverse effect level
OJ	*Official Journal of the European Communities*
PKU	Phenylketonuria
QS	Quantum Satis
SCF	Scientific Committee for Food (of the European Commission)

Appendix Table 1

E #			ADI SCF mg/kg	DIR	Directives PAGES	NAME / E numerically sorted - TABLE 1
UPS1(18)	= on page 18 of the 1st. adopted update of the sweeteners Directive					
UPM2(36)	= on page 36 of the 2nd. adopted update of the "*miscellaneous*" Directive					
E	100		A	COLR	22,23,24,28	Curcumin
E	101	i	NS	COLR	22,23,24,27	Riboflavine
E	101	ii	NS	COLR	22,23,24,27	Riboflavine-5'-phosphate
E	102		7.5	COLR	22,24,28	Tartrazine
E	104		10	COLR	22,24,28	Quinoline yellow
E	110		2.5	COLR	22,24,28	Sunset yellow FCF, Orange yellow S
E	120		5	COLR	21,22,23,24,28	Cochineal, carmines, carminic acid
E	122		4	COLR	22,28	Azorubine, carmoisine
E	123		0.8	COLR	22,25	Amaranth
E	124		4	COLR	22,24,28	Ponceau 4R, cochineal red A
E	127		0.1	COLR	25	Erythrosine
E	128		0.1	COLR	25	Red 2G
E	129		7	COLR	22,24,28	Allura red AC
E	131		15	COLR	28	Patent blue V
E	132		5	COLR	28	Indigotine, indigo carmine
E	133		10	COLR	24,28	Brilliant blue FCF
E	140	i	NS	COLR	22,23,27	Chlorophylls
E	140	ii	NS	COLR	22,23,27	Chlorophyllins
E	141	i	15	COLR	22,23,27	Ccpper complexes of chlorophylls
E	141	ii	15	COLR	22,23,27	Copper complexes of chlorophyllins
E	142		5	COLR	24,28	Green S
E	150	a	A	COLR	21,23,24,27	Plain caramel
E	150	b	200	COLR	21,23,24,27	Caustic sulphite caramel
E	150	c	200	COLR	21,23,24,27	Ammonia caramel
E	150	d	200	COLR	21,23,24,27	Sulphite ammonia caramel
E	151		5	COLR	28	Brilliant black BN, brilliant black PN
E	153		NS	COLR	21,27	Vegetable carbon
E	154		0.15	COLR	25	Brown FK
E	155		3	COLR	28	Brown HT
E	160	a,i	NS	COLR	21,23,24,27	Mixed carotenes
E	160	a,ii	5	COLR	23,24,27	Beta-carotene
E	160	b	0.065	COLR	21,25,28	Annatto, bixin, norbixin
E	160	c	A	COLR	21,23,24,27	Paprika extract, capsanthin, capsorubin
E	160	d	A	COLR	24,28	Lycopene
E	160	e	5	COLR	28	Beta-apo-8'-carotenal (C30)
E	160	f	5	COLR	28	Ethyl ester of beta-apo-8'-carotenal
E	161	b	NS	COLR	24	Lutein

Appendix Table 1 (Continued)

E #			ADI SCF mg/kg	DIR	Directives PAGES	NAME / E numerically sorted - TABLE 1
UPS1(18)		= on page 18 of the 1st. adopted update of the sweeteners Directive				
UPM2(36)		= on page 36 of the 2nd. adopted update of the "*miscellaneous*" Directive				
E	161	g	0.03	COLR	25	Canthaxanthin
E	162		NS	COLR	23,24,27	Beetroot red, betanin
E	163		A	COLR	21,22,23,27	Anthocyanins
E	170	i	NS	COLR	6,27	Calcium carbonate
E	170	i	NS	MISC	6,10,13,35,39,UPM2(21)	Calcium carbonate
E	170	ii	NS	COLR	6,27	Calcium hydrogen carbonate
E	170	ii	NS	MISC	6,10,13,35,39,UPM2(21)	Calcium hydrogen carbonate
E	171		NS	COLR	27	Titanium dioxide
E	172		NS	COLR	27	Iron oxides and hydroxides
E	173		A	COLR	25	Aluminium
E	174		A	COLR	25	Silver
E	175		A	COLR	25	Gold
E	180		1.5	COLR	25	Lithorubine BK
E	200		25	MISC	16,17,18,19,UPM2(21,22)	Sorbic acid
E	202		25	MISC	16,17,18,19,UPM2(21,22)	Potassium sorbate
E	203		25	MISC	16,17,18,19,UPM2(21,22)	Calcium sorbate
E	210		5	MISC	16,17,18,19,UPM2(21,22)	Benzoic acid
E	211		5	MISC	16,17,18,19,UPM2(21,22)	Sodium benzoate
E	212		5	MISC	16,17,18,19,UPM2(21,22)	Potassium benzoate
E	213		5	MISC	16,17,18,19,UPM2(21,22)	Calcium benzoate
E	214		10	MISC	16,17,18,19	Ethyl p-hydroxybenzoate
E	215		10	MISC	16,17,18,19	Sodium ethyl p-hydroxybenzoate
E	216		10	MISC	16,17,18,19	Propyl p-hydroxybenzoate
E	217		10	MISC	16,17,18,19	Sodium propyl p-hydroxybenzoate
E	218		10	MISC	16,17,18,19	Methyl p-hydroxybenzoate
E	219		10	MISC	16,17,18,19	Sodium methyl p-hydroxybenzoate
E	220		0.7	MISC	19,20,21,22,UPM2(22,23)	Sulphur dioxide
E	221		0.7	MISC	19,20,21,22,UPM2(22,23)	Sodium sulphite
E	222		0.7	MISC	19,20,21,22,UPM2(22,23)	Sodium hydrogen sulphite
E	223		0.7	MISC	19,20,21,22,UPM2(22,23)	Sodium metabisulphite
E	224		0.7	MISC	19,20,21,22,UPM2(22,23)	Potassium metabisulphite
E	226		0.7	MISC	19,20,21,22,UPM2(22,23)	Calcium sulphite
E	227		0.7	MISC	19,20,21,22,UPM2(22,23)	Calcium hydrogen sulphite
E	228		0.7	MISC	19,20,21,22,UPM2(22,23)	Potassium hydrogen sulphite
E	230		NE	MISC	23	Biphenyl, diphenyl
E	231		NE	MISC	23	Orthophenyl phenol
E	232		NE	MISC	23	Sodium orthophenyl phenol

Appendix Table 1 (Continued)

E #			ADI SCF mg/kg	DIR	Directives PAGES	NAME / E numerically sorted - TABLE 1
UPS1(18)	= on page 18 of the 1st. adopted update of the sweeteners Directive					
UPM2(36)	= on page 36 of the 2nd. adopted update of the "*miscellaneous*" Directive					
E	233		0.3	MISC	23,UPM2(23)	Thiabendazole
E	234		0.13	MISC	23,UPM2(23)	Nisin
E	235		A	MISC	23	Natamycin
E	239		A	MISC	23	Hexamethylene tetramine
E	242		A	MISC	23	Dimethyl dicarbonate
E	249		0.1	MISC	24	Potassium nitrite
E	250		0.1	MISC	24	Sodium nitrite
E	251		5	MISC	24,UPM2(23)	Sodium nitrate
E	252		5	MISC	24,UPM2(23)	Potassium nitrate
E	260		NS	MISC	6,13,14,15,39,UPM2(21)	Acetic acid
E	261		NS	MISC	6,13,14,15,39	Potassium acetate
E	262	i	NS	MISC	6,13,14,15,39	Sodium acetate
E	262	ii	NS	MISC	6,13,14,15,39	Sodium hydrogen acetate (Sodium diacetate)
E	262		NS	MISC	6,13,14,15,39	Sodium acetates
E	263		NS	MISC	6,13,14,15,35,39	Calcium acetate
E	270		NS	MISC	6,10,11,12,13,14,15,37,38,39,UPM2(20)	Lactic acid
E	280		NS	MISC	24,UPM2(23)	Propionic acid
E	281		NS	MISC	24,UPM2(23)	Sodium propionate
E	282		NS	MISC	24,UPM2(23)	Calcium propionate
E	283		NS	MISC	24,UPM2(23)	Potassium propionate
E	284		A	MISC	23	Boric acid
E	285		A	MISC	23	Sodium tetraborate (borax)
E	290		A	MISC	6	Carbon dioxide
E	296		NS	MISC	6,10,11,39,UPM2(21)	Malic acid
E	297		6	MISC	26,UPM2(23)	Fumaric acid
E	300		A	MISC	6,10,11,12,13,14,15,39,UPM2(20,21)	Acorbic acid
E	301		A	MISC	6,11,12,13,14,15,39,UPM2(21,28)	Sodium ascorbate
E	302		A	MISC	6,12,13,14,15,39,UPM2(21)	Calcium ascorbate
E	304	i	A	MISC	6,11,13,14,39,UPM2(20,29)	Ascorbyl palmitate
E	304	ii	A	MISC	6,11,13,14,39,UPM2(20, ?29?)	Ascorbyl stearate
E	306		A	MISC	6,13,37,38,39,UPM2(20)	Tocopherol-rich extract
E	307		A	MISC	6,13,37,38,39,UPM2(20)	Alpha-tocopherol
E	308		A	MISC	6,13,37,38,39,UPM2(20)	Gamma-tocopherol
E	309		A	MISC	6,13,37,38,39,UPM2(20)	Delta-tocopherol
E	310		0.5	MISC	25,UPM2(23)	Propyl gallate
E	311		0.5	MISC	25,UPM2(23)	Octyl gallate
E	312		0.5	MISC	25,UPM2(23)	Dodecyl gallate

Appendix Table 1 (Continued)

E #			ADI SCF mg/kg	DIR	Directives PAGES	NAME / E numerically sorted - TABLE 1
UPS1(18)	= on page 18 of the 1st. adopted update of the sweeteners Directive					
UPM2(36)	= on page 36 of the 2nd. adopted update of the *"miscellaneous"* Directive					
E	315		6	MISC	25,UPM2(23)	Erythorbic acid
E	316		6	MISC	25,UPM2(23)	Sodium erythorbate
E	320		0.5	MISC	25,UPM2(23)	Butylated hydroxyanisole
E	321		0.05	MISC	25	Butylated hydroxytoluene
E	322		NS	MISC	6,10,11,12,13,14,35,37,38,39,40,UPM2(20,28,29)	Lecithins
E	325		NS	MISC	6,12,13,14,15,39	Sodium lactate
E	326		NS	MISC	6,12,13,14,15,39	Potassium lactate
E	327		NS	MISC	6,10,11,12,13,14,15,39	Calcium lactate
E	330		NS	MISC	7,10,11,12,13,14,15,37,38,39,UPM2(20)	Citric acid
E	331	i	NS	MISC	7,10,11,12,13,14,15,37,38,39,UPM2(20,29)	Monosodium citrate
E	331	ii	NS	MISC	7,10,11,12,13,14,15,37,38,39,UPM2(20,29)	Disodium citrate
E	331	iii	NS	MISC	7,10,11,12,13,14,15,37,38,39,UPM2(20,29)	Trisodium citrate
E	332	i	NS	MISC	7,10,11,12,13,14,15,37,38,39,UPM2(20,29)	Monopotassium citrate
E	332	ii	NS	MISC	7,10,11,12,13,14,15,37,38,39,UPM2(20,29)	Tripotassium citrate
E	333	i	NS	MISC	7,11,12,13,14,35,39,UPM2(20,30)	Monocalcium citrate
E	333	ii	NS	MISC	7,11,12,13,14,35,39,UPM2(20)	Dicalcium citrate
E	333	iii	NS	MISC	7,11,12,13,14,35,39,UPM2(20)	Tricalcium citrate
E	334		30	MISC	7,11,13,14,15,40	Tartaric acid [L(+)-]
E	335	i	30	MISC	7,11,13,14,40	Monosodium tartrate
E	335	ii	30	MISC	7,11,13,14,40	Disodium tartrate
E	336	i	30	MISC	7,11,13,14,40	Monopotassium tartrate
E	336	ii	30	MISC	7,11,13,14,40	Dipotassium tartrate
E	337		30	MISC	7,11,13,14	Sodium potassium tartrate
E	338		70	MISC	26,37,38,39,UPM2(24)	Phosphoric acid
E	339	i	70	MISC	26,39,UPM2(24,29)	Monosodium phosphate
E	339	ii	70	MISC	26,39,UPM2(24,29)	Disodium phosphate
E	339	iii	70	MISC	26,39,UPM2(24,29)	Trisodium phosphate
E	340	i	70	MISC	26,39,UPM2(24,29)	Monopotassium phosphate
E	340	ii	70	MISC	26,39,UPM2(24,29)	Dipotassium phosphate
E	340	iii	70	MISC	26,39,UPM2(24,29)	Tripotassium phosphate
E	341	i	70	MISC	26,35,39,UPM2(24,30)	Monocalcium phosphate
E	341	ii	70	MISC	26,35,39,UPM2(24)	Dicalcium phosphate
E	341	iii	70	MISC	26,35,39,UPM2(24)	Tricalcium phosphate
E	343	i	70	MISC	UPM2(24)	Monomagnesium phosphate
E	343	ii	70	MISC	UPM2(24)	Dimagnesium phosphate
E	350	i	NS	MISC	7,11	Sodium malate
E	350	ii	NS	MISC	7,11	Sodium hydrogen malate

Appendix Table 1 (Continued)

E #			ADI SCF mg/kg	DIR	Directives PAGES	NAME / E numerically sorted - TABLE 1
UPS1(18)	= on page 18 of the 1st. adopted update of the sweeteners Directive					
UPM2(36)	= on page 36 of the 2nd. adopted update of the *"miscellaneous"* Directive					
E	351		NS	MISC	7	Potassium malate
E	352	i	NS	MISC	7	Calcium malate
E	352	ii	NS	MISC	7	Calcium hydrogen malate
E	353		A	MISC	27	Metatartaric acid
E	354		30	MISC	7,40	Calcium tartrate
E	355		5	MISC	27	Adipic acid
E	356		5	MISC	27	Sodium adipate
E	357		5	MISC	27	Potassium adipate
E	363		NS	MISC	28	Succinic acid
E	380		NS	MISC	7	Triammonium citrate
E	385		2.5	MISC	28,UPM2(25)	Calcium disodium ethylene diamine tetra-acetate (CaNa2EDTA)
E	400		NS	MISC	7,11,12,34,40	Alginic acid
E	401		NS	MISC	7,11,12,34,40,UPM2(20,30)	Sodium alginate
E	402		NS	MISC	7,11,12,34,40,UPM2(20)	Potassium alginate
E	403		NS	MISC	7,11,12,34	Ammonium alginate
E	404		NS	MISC	7,11,12,34,40	Calcium alginate
E	405		25	MISC	28,34,UPM2(25,30)	Propane-1,2-diol alginate
E	406		NS	MISC	7,11,12,34	Agar
E	407	a	75	MISC	UPM1(4)	Alternatively refined carrageenan
E	407		75	MISC	7,11,12,34,38,UPM2(20)	Carrageenan
E	410		NS	MISC	7,11,12,34,38,40,UPM2(30)	Locust bean gum
E	412		NS	MISC	7,11,34,38,40,UPM2(29,30)	Guar gum
E	413		NS	MISC	7,34	Tragacanth
E	414		NS	MISC	7,10,15,34,40,UPM2(28)	Acacia gum (gum arabic)
E	415		NS	MISC	7,11,12,34,40,UPM2(30)	Xanthan gum
E	416		12.5	MISC	28	Karaya gum
E	417		NS	MISC	7	Tara gum
E	418		NS	MISC	7,11	Gellan gum
E	420	i	A	MISC	28,34	Sorbitol
E	420	ii	A	MISC	28,34	Sorbitol syrup
E	421		A	MISC	28,34,UPM2(28)	Mannitol
E	422		NS	MISC	7,10,34	Glycerol
E	425	i	NS	MISC	UPM2(27,28)	Konjac gum
E	425	ii	NS	MISC	UPM2(27,28)	Konjac glucomannane
E	431		NA	MISC	27	Polyoxyethylene (40) stearate
E	432		10	MISC	29,34,35,UPM2(28)	Polyoxyethylene sorbitan monolaurate (Polysorbate 20)
E	433		10	MISC	29,34,35,UPM2(28)	Polyoxyethylene sorbitan monooleate (Polysorbate 80)

Appendix Table 1 (Continued)

E #			ADI SCF mg/kg	DIR	Directives PAGES	NAME / E numerically sorted - TABLE 1
UPS1(18)	= on page 18 of the 1st. adopted update of the sweeteners Directive					
UPM2(36)	= on page 36 of the 2nd. adopted update of the "*miscellaneous*" Directive					
E	434		10	MISC	29,34,35,UPM2(28)	Polyoxyethylene sorbitan palmitateate (Polysorbate 40)
E	435		10	MISC	29,34,35,UPM2(28)	Polyoxyethylene sorbitan monostearate (Polysorbate 60)
E	436		10	MISC	29,34,35,UPM2(28)	Polyoxyethylene sorbitan tristearate (Polysorbate 65)
E	440	i	NS	MISC	7,10,11,12,34,38,40,UPM2(21,30)	Pectins
E	440	ii	NS	MISC	7,10,11,12,34,38,40,UPM2(21,30)	Amidated pectins
E	442		30	MISC	29,34,UPM2(26)	Ammonium phosphatides
E	444		10	MISC	29	Sucrose acetate isobutyrate
E	445		12.5	MISC	29,UPM2(26)	Glycerol esters of wood rosin
E	450	a	70	MISC	40	Disodium diphosphate
E	450	i	70	MISC	27,UPM2(24)	Disodium diphosphate
E	450	ii	70	MISC	27,UPM2(24)	Trisodium diphosphate
E	450	iii	70	MISC	27,UPM2(24)	Tetrasodium diphosphate
E	450	iv	70	MISC	27	Dipotassium diphosphate
E	450	v	70	MISC	27,UPM2(24)	Tetrapotassium diphosphate
E	450	vi	70	MISC	27,UPM2(24)	Dicalcium diphosphate
E	450	vii	70	MISC	27,UPM2(24)	Calcium dihydrogen diphosphate
E	451	i	70	MISC	27,UPM2(24)	Pentasodium triphosphate
E	451	ii	70	MISC	27,UPM2(24)	Pentapotassium triphosphate
E	452	i	70	MISC	27,UPM2(24)	Sodium polyphosphate
E	452	ii	70	MISC	27,UPM2(24)	Potassium polyphosphate
E	452	iii	70	MISC	27,UPM2(24)	Sodium calcium polyphosphate
E	452	iv	70	MISC	27,UPM2(24)	Calcium polyphosphate
E	459		5	MISC	UPM2(27,28)	Beta-cyclodextrin
E	460	i	A	MISC	8,12,34,UPM2(21)	Microcrystalline cellulose
E	460	ii	NS	MISC	8,12,34,UPM2(21)	Powdered cellulose
E	461		NS	MISC	8,12,34	Methyl cellulose
E	463		NS	MISC	8,12,34	Hydroxypropyl cellulose
E	464		NS	MISC	8,12,35	Hydroxypropyl methyl cellulose
E	465		NS	MISC	8,12,35	Ethyl methyl cellulose
E	466		NS	MISC	8,12,35,UPM2(20,30)	Carboxy methyl cellulose + sodium carboxyy methyl cellulose
E	468		NS	MISC	UPM2(25,28)	Cross linked sodium carboxy methyl cellulose
E	469		NS	MISC	UPM2(20,28)	Enzymatically hydrolyzed carboxy methyl cellulose
E	470	a	NS	MISC	8,35,UPM2(28)	Sodium, potassium and calcium salts of fatty acids
E	470	b	NS	MISC	8,35	Magnesium salts of fatty acids
E	471		NS	MISC	8,11,12,13,14,15,35,37,38,40,UPM2(20,28,29,30)	Mono- and diglycerides of fatty acids
E	472	a	NS	MISC	8,13,14,35,40	Acetic acid esters of mono- and diglycerides of fatty acids
E	472	b	NS	MISC	8,40	Lactic acid esters of mono- and diglycerides of fatty acids

Appendix Table 1 (Continued)

E #			ADI SCF mg/kg	DIR	Directives PAGES	NAME / E numerically sorted - TABLE 1
UPS1(18)	= on page 18 of the 1st. adopted update of the sweeteners Directive					
UPM2(36)	= on page 36 of the 2nd. adopted update of the *"miscellaneous"* Directive					
E	472	c	NS	MISC	8,35,40,UPM2(29)	Citric acid esters of mono- and diglycerides of fatty acids
E	472	d	NS	MISC	8,14	Tartaric acid esters of mono- and diglycerides of fatty acids
E	472	e	25	MISC	8,14,35	Mono- and diacetyl tartaric acid esters of mono- and diglycerides of fatty acids
E	472	f	NS	MISC	8,14	Mixed acetic and tartaric acid esters of mono- and diglycerides of fatty acids
E	473		20	MISC	29,35,UPM2(26,28,29)	Sucrose esters of fatty acids
E	474		20	MISC	29,UPM2(26)	Sucroglycerides
E	475		25	MISC	29,35	Polyglycerol esters of fatty acids
E	476		7.5	MISC	30,UPM2(26)	Polyglycerol polyricinoleate
E	477		25	MISC	30	Propane 1,2-diol esters of fatty acids
E	479	b	25	MISC	30	Thermally oxidized soya bean oil interacted with mono- and diglycerides of fatty acids
E	481		20	MISC	30	Sodium stearoyl-2-lactate
E	482		20	MISC	30	Calcium stearoyl-2-lactate
E	483		20	MISC	30	Stearyl tartrate
E	491		25	MISC	31,35,UPM2(28)	Sorbitan monostearate
E	492		25	MISC	31,35,UPM2(28)	Sorbitan tristearate
E	493		5	MISC	31,35,UPM2(28)	Sorbitan monolaurate
E	494		5	MISC	31,35,UPM2(28)	Sorbitan monooleate
E	495		25	MISC	31,35,UPM2(28)	Sorbitan monopalmitate
E	500	i	NS	MISC	8,10,39,UPM2(21)	Sodium carbonate
E	500	ii	NS	MISC	8,10,11,39,UPM2(21)	Sodium hydrogen carbonate (Sodium bicarbonate)
E	500	iii	NS	MISC	8,39,UPM2(21)	Sodium sesquicarbonate
E	501	i	NS	MISC	8,10,35,39	Potassium carbonate
E	501	ii	NS	MISC	8,10,11,35,39	Potassium hydrogen carbonate (Potassium bicarbonate)
E	503	i	NS	MISC	8,10,39	Ammonium carbonate
E	503	ii	NS	MISC	8,10,39	Ammonium hydrogen carbonate
E	504	i	NS	MISC	8,13,35,UPM2(21)	Magnesium carbonate
E	504	ii	NS	MISC	8,13,35,UPM2(21)	Magnesium hydroxide carbonate
E	507		NS	MISC	8,39	Hydrochloric acid
E	508		NS	MISC	8,12,35	Potassium chloride
E	509		NS	MISC	8,11,12,13,14,35,UPM2(21)	Calcium chloride
E	511		NS	MISC	8,35	Magnesium chloride
E	512		A	MISC	31	Stannous chloride
E	513		NS	MISC	8	Sulphuric acid
E	514	i	NS	MISC	8,35	Sodium sulphate
E	514	ii	NS	MISC	8,35	Sodium hydrogen sulphate
E	515	i	NS	MISC	8,35	Potassium sulphate
E	515	ii	NS	MISC	8,35	Potassium hydrogen sulphate

Appendix Table 1 (Continued)

E #			ADI SCF mg/kg	DIR	Directives PAGES	NAME / E numerically sorted - TABLE 1
UPS1(18)	= on page 18 of the 1st. adopted update of the sweeteners Directive					
UPM2(36)	= on page 36 of the 2nd. adopted update of the *"miscellaneous"* Directive					
E	516		NS	MISC	8,35	Calcium sulphate
E	517		NS	MISC	35	Ammonium sulphate
E	520		1	MISC	31	Aluminium sulphate
E	521		1	MISC	31	Aluminium sodium sulphate
E	522		1	MISC	31	Aluminium potassium sulphate
E	523		1	MISC	31	Aluminium ammonium sulphate
E	524		NS	MISC	8,10,11,39	Sodium hydroxide
E	525		NS	MISC	8,10,39	Potassium hydroxide
E	526		NS	MISC	9,10,39	Calcium hydroxide
E	527		NS	MISC	9,10	Ammonium hydroxide
E	528		NS	MISC	9,10	Magnesium hydroxide
E	529		NS	MISC	9	Calcium oxide
E	530		NS	MISC	9,10	Magnesium oxide
E	535		0.025	MISC	31	Sodium ferrocyanide
E	536		0.025	MISC	31	Potassium ferrocyanide
E	538		0.025	MISC	31	Calcium ferrocyanide
E	541		1	MISC	31,40	Sodium aluminium phosphate, acidic
E	551		NS	MISC	31,36,UPM2(26,28)	Silicon dioxide
E	552		NS	MISC	31,36	Calcium silicate
E	553	a i	NS	MISC	31	Magnesium silicate
E	553	a ii	NS	MISC	31	Magnesium trisilicate
E	553	b	NS	MISC	31,36,	Talc
E	554		1	MISC	31	Sodium aluminium silicate
E	555		1	MISC	31	Potassium aluminium silicate
E	556		1	MISC	31	Calcium aluminium silicate
E	558		1	MISC	31,36	Bentonite
E	559		1	MISC	31,36,UPM2(26)	Aluminium silicate (kaolin)
E	570		NS	MISC	9,UPM2(28)	Fatty acids
E	574		NS	MISC	9	Gluconic acid
E	575		NS	MISC	9,13,14,15,40,UPM2(21)	Glucono-delta-lactone
E	576		NS	MISC	9	Sodium gluconate
E	577		NS	MISC	9,36	Potassium gluconate
E	578		NS	MISC	9	Calcium gluconate
E	579		A	MISC	32	Ferrous gluconate
E	585		A	MISC	32	Ferrous lactate
E	620		NS	MISC	32	Glutamic acid
E	621		NS	MISC	32	Monosodium glutamate

Appendix Table 1 (Continued)

E #			ADI SCF mg/kg	DIR	Directives PAGES	NAME / E numerically sorted - TABLE 1
UPS1(18)	= on page 18 of the 1st. adopted update of the sweeteners Directive					
UPM2(36)	= on page 36 of the 2nd. adopted update of the *"miscellaneous"* Directive					
E	622		NS	MISC	32	Monopotassium glutamate
E	623		NS	MISC	32	Calcium diglutamate
E	624		NS	MISC	32	Monoammonium glutamate
E	625		NS	MISC	32	Magnesium diglutamate
E	626		NS	MISC	32	Guanilic acid
E	627		NS	MISC	32	Disodium guanilate
E	628		NS	MISC	32	Dipotassium guanilate
E	629		NS	MISC	32	Calcium guanilate
E	630		NS	MISC	32	Inosinic acid
E	631		NS	MISC	32	Disodium inosinate
E	632		NS	MISC	32	Dipotassium inosinate
E	633		NS	MISC	32	Calcium inosinate
E	634		NS	MISC	32	Calcium 5'-ribonucleotides
E	635		NS	MISC	32	Disodium 5'-ribonucleotides
E	640		NS	MISC	9,36	Glycine and its sodium salt
E	900		1.5	MISC	32,UPM2(26,28)	Dimethyl polysiloxane
E	901		A	MISC	32,36,UPM2(26)	Beeswax, white and yellow
E	902		A	MISC	32,UPM2(26)	Candellila wax
E	903		A	MISC	32,UPM2(26)	Carnauba wax
E	904		A	MISC	32,UPM2(26)	Shellac
E	905		20	MISC	UPM2(27)	Microcristalline wax
E	912		A	MISC	33,UPM2(27)	Montan acid esters
E	914		A	MISC	33,UPM2(27)	Oxidized polyethylene wax
E	920		NS	MISC	UPM2(20)	Cysteine [L-cyst...]
E	927	b	A	MISC	33	Carbamide
E	938		A	MISC	9	Argon
E	939		A	MISC	9	Helium
E	941		A	MISC	9	Nitrogen
E	942		A	MISC	9	Nitrous oxide
E	948		A	MISC	9	Oxygen
E	950		9	MISC	33	Acesulfame K
E	950		9	SWET	6,7,8,UPS1(18)	Acesulfame K
E	951		40	MISC	33	Aspartame
E	951		40	SWET	8,9,UPS1(18)	Aspartame
E	952		7	SWET	9,10,UPS1(18,19)	Cyclamic acid and its sodium and calcium salts
E	953		A	MISC	28,34	Isomalt
E	953		A	SWET	6	Isomalt

Appendix Table 1 (Continued)

E #		ADI SCF mg/kg	DIR	Directives PAGES	NAME / E numerically sorted - TABLE 1
UPS1(18)	= on page 18 of the 1st. adopted update of the sweeteners Directive				
UPM2(36)	= on page 36 of the 2nd. adopted update of the "*miscellaneous*" Directive				
E	954	5	SWET	10,11,UPS1(19)	Saccharin and sodium potassium and calcium salts
E	957	A	MISC	33,UPM2(27)	Thaumatin
E	957	A	SWET	11,UPS1(19)	Thaumatin
E	959	5	MISC	33,UPM2(27)	Neohesperidine dihydrochalcone (NHDC)
E	959	5	SWET	11,12,UPS1(9)	Neohesperidine dihydrochalcone (NHDC)
E	965	A	MISC	28,34	Maltitol
E	965	A	MISC	28	Maltitol syrup
E	965	A	SWET	6	Maltitol syrup
E	965	A	SWET	6	Maltitol
E	966	A	MISC	28,34	Lactitol
E	966	A	SWET	6	Lactitol
E	967	A	MISC	28,34	Xylitol
E	967	A	SWET	6	Xylitol
E	999	5	MISC	33,UPM2(27)	Quillaia extract
E	1103	A	MISC	UPM2(20)	Invertase
E	1105	A	MISC	24	Lysozyme
E	1200	NS	MISC	9,36	Polydextrose
E	1201	A	MISC	33,36	Polyvinylpyrrolidone
E	1202	A	MISC	33,36	Polyvinylpolypyrrolidone
E	1404	A	MISC	9,12,35,40	Oxidized starch
E	1410	A	MISC	9,12,35,40	Monostarch phosphate
E	1412	A	MISC	9,12,35,40	Distarch phosphate
E	1413	A	MISC	9,12,35,40	Phosphated distarch phosphate
E	1414	A	MISC	9,12,35,40	Acetylated distarch phosphate
E	1420	A	MISC	9,12,35,40	Acetylated starch
E	1422	A	MISC	9,12,35,40	Acetylated distarch adipate
E	1440	A	MISC	9,12,35	Hydroxy propyl starch
E	1442	A	MISC	9,12,35	Hydroxy propyl distarch phosphate
E	1450	A	MISC	9,12,35,40,UPM2(30)	Starch sodium octenyl succinate
E	1451	A	MISC	UPM2(20,28,30)	Acetylated oxidized starch
E	1505	20	MISC	33,36	Triethyl citrate
E	1518	A	MISC	36,UPM2(27)	Glyceryl triacetate (triacetin)
Solvent		25	MISC	34	Propan-1,2-diol (propylene glycol)

Appendix Table 2

E #			ADI SCF mg/kg	DIR	Directives PAGES	NAME / Alphabetically sorted - TABLE 2 -
UPS1(18)	= on page 18 of the 1st. adopted update of the sweeteners Directive					
UPM2(36)	= on page 36 of the 2nd. adopted update of the *"miscellaneous"* Directive					
E	414		NS	MISC	7,10,15,34,40,UPM2(28)	Acacia gum (gum arabic)
E	950		9	MISC	33	Acesulfame K
E	950		9	SWET	6,7,8,UPS1(18)	Acesulfame K
E	263		NS	MISC	6,13,14,15,35,39	Acetate, calcium -
E	385		2.5	MISC	28,UPM2(25)	Acetate, calcium disodium ethylene diamine tetra- - (CaNa2EDTA)
E	261		NS	MISC	6,13,14,15,39	Acetate, potassium -
E	262	i	NS	MISC	6,13,14,15,39	Acetate, sodium -
E	262	ii	NS	MISC	6,13,14,15,39	Acetate, sodium hydrogen - (Sodium diacetate)
E	444		10	MISC	29	Acetate, sucrose - isobutyrate
E	262		NS	MISC	6,13,14,15,39	Acetates, sodium -
E	260		NS	MISC	6,13,14,15,39,UPM2(21)	Acetic acid
E	472	a	NS	MISC	8,13,14,35,40	Acetic acid esters of mono- and diglycerides of fatty acids
E	472	f	NS	MISC	8,14	Acetic, mixed - and tartaric acid esters of mono- and diglycerides of fatty acids
E	472	e	25	MISC	8,14,35	Acetyl, mono- and di- tartaric acid esters of mono- and diglycerides of fatty acids
E	1422		A	MISC	9,12,35,40	Acetylated distarch adipate
E	1414		A	MISC	9,12,35,40	Acetylated distarch phosphate
E	1451		A	MISC	UPM2(20,28,30)	Acetylated oxidized starch
E	1420		A	MISC	9,12,35,40	Acetylated starch
E	260		NS	MISC	6,13,14,15,39,UPM2(21)	Acid, acetic -
E	472	a	NS	MISC	8,13,14,35,40	Acid, acetic acid esters of mono- and diglycerides of fatty acids
E	355		5	MISC	27	Acid, adipic -
E	400		NS	MISC	7,11,12,34,40	Acid, alginic -
E	300		A	MISC	6,10,11,12,13,14,15,39,UPM2(20,21)	Acid, ascorbic -
E	210		5	MISC	16,17,18,19,UPM2(21,22)	Acid, benzoic -
E	284		A	MISC	23	Acid, boric -
E	120		5	COLR	21,22,23,24,28	Acid, carminic - ,cochineal, carmines
E	330		NS	MISC	7,10,11,12,13,14,15,37,38,39,UPM2(20)	Acid, citric -
E	472	c	NS	MISC	8,35,40,UPM2(29)	Acid, citric - esters of mono- and diglycerides of fatty acids
E	952		7	SWET	9,10,UPS1(18,19)	Acid, cyclamic - and its sodium and calcium salts
E	315		6	MISC	25,UPM2(23)	Acid, erythorbic -
E	297		6	MISC	26,UPM2(23)	Acid, fumaric -
E	574		NS	MISC	9	Acid, gluconic -
E	620		NS	MISC	32	Acid, glutamic -
E	626		NS	MISC	32	Acid, guanilic -
E	507		NS	MISC	8,39	Acid, hydrochloric -
E	630		NS	MISC	32	Acid, inosinic -
E	270		NS	MISC	6,10,11,12,13,14,15,37,38,39,UPM2(20)	Acid, lactic -
E	472	b	NS	MISC	8,40	Acid, lactic - esters of mono- and diglycerides of fatty acids
E	296		NS	MISC	6,10,11,39,UPM2(21)	Acid, malic -

Appendix Table 2 (Continued)

E #			ADI SCF mg/kg	DIR	Directives PAGES	NAME / Alphabetically sorted - TABLE 2 -
UPS1(18)	= on page 18 of the 1st. adopted update of the sweeteners Directive					
UPM2(36)	= on page 36 of the 2nd. adopted update of the "*miscellaneous*" Directive					
E	353		A	MISC	27	Acid, metatartaric -
E	472	f	NS	MISC	8,14	Acid, mixed acetic and tartaric - esters of mono- and diglycerides of fatty acids
E	472	e	25	MISC	8,14,35	Acid, mono- and diacetyl tartaric - esters of mono- and diglycerides of fatty acids
E	912		A	MISC	33,UPM2(27)	Acid, montan - esters
E	338		70	MISC	26,37,38,39,UPM2(24)	Acid, phosphoric -
E	280		NS	MISC	24,UPM2(23)	Acid, propionic -
E	200		25	MISC	16,17,18,19,UPM2(21,22)	Acid, sorbic -
E	363		NS	MISC	28	Acid, succinic -
E	513		NS	MISC	8	Acid, sulphuric -
E	334		30	MISC	7,11,13,14,15,40	Acid, tartaric - (L(+)-)
E	472	d	NS	MISC	8,14	Acid, tartaric - esters of mono- and diglycerides of fatty acids
E	541		1	MISC	31,40	Acidic, sodium aluminium phosphate, -
E	472	a	NS	MISC	8,13,14,35,40	Acids, acetic acid esters of mono- and diglycerides of fatty -
E	472	c	NS	MISC	8,35,40,UPM2(29)	Acids, citric acid esters of mono- and diglycerides of fatty -
E	570		NS	MISC	9,UPM2(28)	Acids, fatty
E	472	b	NS	MISC	8,40	Acids, lactic acid esters of mono- and diglycerides of fatty -
E	470	b	NS	MISC	8,35	Acids, magnesium salts of fatty -
E	472	f	NS	MISC	8,14	Acids, mixed acetic and tartaric acid esters of mono- and diglycerides of fatty -
E	472	e	25	MISC	8,14,35	Acids, mono- and diacetyl tartaric acid esters of mono- and diglycerides of fatty -
E	471		NS	MISC	8,11,12,13,14,15,35,37,38,40,UPM2(20,28,29,30)	Acids, mono- and diglycerides of fatty -
E	475		25	MISC	29,35	Acids, polyglycerol esters of fatty -
E	477		25	MISC	30	Acids, propane 1,2-diol esters of fatty -
E	470	a	NS	MISC	8,35,UPM2(28)	Acids, sodium potassium and calcium salts of fatty -
E	473		20	MISC	29,35,UPM2(26,28,29)	Acids, sucrose esters of fatty -
E	472	d	NS	MISC	8,14	Acids, tartaric - esters of mono- and diglycerides of fatty -
E	479	b	25	MISC	30	Acids, thermally oxidized soya bean oil interacted with mono- and diglycerides of fatty -
E	300		A	MISC	6,10,11,12,13,14,15,39,UPM2(20,21)	Acorbic acid
E	1422		A	MISC	9,12,35,40	Adipate, acetylated distarch -
E	357		5	MISC	27	Adipate, potassium -
E	356		5	MISC	27	Adipate, sodium -
E	355		5	MISC	27	Adipic acid
E	406		NS	MISC	7,11,12,34	Agar
E	403		NS	MISC	7,11,12,34	Alginate, ammonia -
E	404		NS	MISC	7,11,12,34,40	Alginate, calcium -
E	402		NS	MISC	7,11,12,34,40,UPM2(20)	Alginate, potassium -
E	405		25	MISC	28,34,UPM2(25,30)	Alginate, propane-1,2-diol -
E	401		NS	MISC	7,11,12,34,40,UPM2(20,30)	Alginate, sodium -
E	400		NS	MISC	7,11,12,34,40	Alginic acid
E	129		7	COLR	22,24,28	Allura red AC

Appendix Table 2 (Continued)

E #			ADI SCF mg/kg	DIR	Directives PAGES	NAME / Alphabetically sorted - TABLE 2 -
UPS1(18)	= on page 18 of the 1st. adopted update of the sweeteners Directive					
UPM2(36)	= on page 36 of the 2nd. adopted update of the "*miscellaneous*" Directive					
E	307		A	MISC	6,13,37,38,39,UPM2(20)	Alpha-tocopherol
E	407	a	75	MISC	UPM1(4)	Alternatively refined carrageenan
E	173		A	COLR	25	Aluminium
E	523		1	MISC	31	Aluminium, ammonium sulphate
E	555		1	MISC	31	Aluminium, potassium - silicate
E	522		1	MISC	31	Aluminium, potassium sulphate
E	559		1	MISC	31,36,UPM2(26)	Aluminium, silicate (kaolin)
E	556		1	MISC	31	Aluminium, silicate, calcium
E	541		1	MISC	31,40	Aluminium, sodium - phosphate, acidic
E	554		1	MISC	31	Aluminium, sodium - silicate
E	521		1	MISC	31	Aluminium, sodium sulphate
E	520		0.3	MISC	31	Aluminium, sulphate
E	123		0.8	COLR	22,25	Amaranth
E	440	ii	NS	MISC	7,10,11,12,34,38,40,UPM2(21,30)	Amidated pectins
E	385		2.5	MISC	28,UPM2(25)	Amine, calcium disodium ethylene di- tetra-acetate (CaNa2EDTA)
E	239		A	MISC	23	Amine, hexamethylene tetr-
E	150	c	200	COLR	21,23,24,27	Ammonia caramel
E	150	d	200	COLR	21,23,24,27	Ammonia, sulphite - caramel
E	403		NS	MISC	7,11,12,34	Ammonium, alginate
E	523		1	MISC	31	Ammonium, aluminium - sulphate
E	503	i	NS	MISC	8,10,39	Ammonium, carbonate
E	380		NS	MISC	7	Ammonium, citrate, tri- citrate
E	503	ii	NS	MISC	8,10,39	Ammonium, hydrogen carbonate
E	527		NS	MISC	9,10	Ammonium, hydroxide
E	624		NS	MISC	32	Ammonium, mono- glutamate
E	442		30	MISC	29,34,UPM2(26)	Ammonium, phosphatides
E	517		?	MISC	35	Ammonium, sulphate
E	380		NS	MISC	7	Ammonium, tri- citrate
E	320		0.5	MISC	25,UPM2(23)	Anisole, butylated hydroxy- (Bha BHA)
E	160	b	0.065	COLR	21,25,28	Annatto, bixin, norbixin
E	163		A	COLR	21,22,23,27	Anthocyanins
E	160	f	5	COLR	28	Apo, ethyl ester of beta- - -8'-carotenal
E	938		A	MISC	9	Argon
E	302		A	MISC	6,12,13,14,15,39,UPM2(21)	Ascorbate, calcium -
E	301		A	MISC	6,11,12,13,14,15,39,UPM2(21,28)	Ascorbate, sodium -
E	304	i	A	MISC	6,11,13,14,39,UPM2(20,29)	Ascorbyl palmitate
E	304	ii	A	MISC	6,11,13,14,39,UPM2(20, ?29?)	Ascorbyl stearate
E	951		40	MISC	33	Aspartame
E	951		40	SWET	8,9,UPS1(18)	Aspartame

Appendix Table 2 (Continued)

E #			ADI SCF mg/kg	DIR	Directives PAGES	NAME / Alphabetically sorted - TABLE 2 -
UPS1(18)	= on page 18 of the 1st. adopted update of the sweeteners Directive					
UPM2(36)	= on page 36 of the 2nd. adopted update of the *"miscellaneous"* Directive					
E	122		4	COLR	22,28	Azorubine, carmoisine
E	410		NS	MISC	7,11,12,34,38,40,UPM2(30)	Bean, locust - gum
E	479	b	25	MISC	30	Bean, thermally oxidized soya - oil interacted with mono- and diglycerides of fatty acids
E	901		A	MISC	32,36,UPM2(26)	Beeswax, white and yellow
E	162		NS	COLR	23,24,27	Beetroot red, betanin
E	558		1	MISC	31,36	Bentonite
E	213		5	MISC	16,17,18,19,UPM2(21,22)	Benzoate, calcium -
E	214		10	MISC	16,17,18,19	Benzoate, ethyl p-hydroxy-
E	218		10	MISC	16,17,18,19	Benzoate, methyl p-hydroxy-
E	212		5	MISC	16,17,18,19,UPM2(21,22)	Benzoate, potassium
E	216		10	MISC	16,17,18,19	Benzoate, propyl p-hydroxy-
E	211		5	MISC	16,17,18,19,UPM2(21,22)	Benzoate, sodium
E	215		10	MISC	16,17,18,19	Benzoate, sodium ethyl p-hydroxy-
E	219		10	MISC	16,17,18,19	Benzoate, sodium methyl p-hydroxy-
E	217		10	MISC	16,17,18,19	Benzoate, sodium propyl p-hydroxy-
E	210		5	MISC	16,17,18,19,UPM2(21,22)	Benzoic acid
E	160	f	5	COLR	28	Beta, ethyl ester of - -apo-8'-carotenal
E	160	e	5	COLR	28	Beta-apo-8'-carotenal (C30)
E	160	a,ii	5	COLR	23,24,27	Beta-carotene
E	459		5	MISC	UPM2(27,28)	Beta-cyclodextrin
E	162		NS	COLR	23,24,27	Betanin, beetroot red
E	320		0.5	MISC	25,UPM2(23)	Bha BHA (Butylated hydroxyanisole)
E	321		0.05	MISC	25	Bht BHT (Butylated hydroxytoluene)
E	501	ii	NS	MISC	8,10,11,35,39	Bicarbonate, potassium hydrogen carbonate (potassium -)
E	500	ii	NS	MISC	8,10,11,39,UPM2(21)	Bicarbonate, sodium hydrogen carbonate (sodium -)
E	230		NE	MISC	23	Biphenyl, diphenyl
E	224		0.7	MISC	19,20,21,22,UPM2(22,23)	Bisulphite, potassium meta-
E	223		0.7	MISC	19,20,21,22,UPM2(22,23)	Bisulphite, sodium meta-
E	160	b	0.065	COLR	21,25,28	Bixin, annatto, - norbixin
E	151		5	COLR	28	Black, brilliant black BN, brilliant black PN
E	133		10	COLR	24,28	Blue, brilliant FCF
E	131		15	COLR	28	Blue, patent - V
E	285		A	MISC	23	Borax, sodium tetraborate -
E	284		A	MISC	23	Boric acid
E	151		5	COLR	28	Brilliant black BN, brilliant black PN
E	133		10	COLR	24,28	Brilliant blue FCF
E	154		0.15	COLR	25	Brown FK
E	155		3	COLR	28	Brown HT
E	320		0.5	MISC	25,UPM2(23)	Butylated hydroxyanisole (Bha BHA)

Appendix Table 2 (Continued)

E #			ADI SCF mg/kg	DIR	Directives PAGES	NAME / Alphabetically sorted - TABLE 2 -
UPS1(18)	= on page 18 of the 1st. adopted update of the sweeteners Directive					
UPM2(36)	= on page 36 of the 2nd. adopted update of the "*miscellaneous*" Directive					
E	321		0.05	MISC	25	Butylated hydroxytoluene (Bht BHT)
E	444		10	MISC	29	Butyrate, sucrose acetate iso-
E	634		NS	MISC	32	Calcium, 5'-ribonucleotides
E	263		NS	MISC	6,13,14,15,35,39	Calcium, acetate
E	404		NS	MISC	7,11,12,34,40	Calcium, alginate
E	556		1	MISC	31	Calcium, aluminium silicate
E	302		A	MISC	6,12,13,14,15,39,UPM2(21)	Calcium, ascorbate
E	213		5	MISC	16,17,18,19,UPM2(21,22)	Calcium, benzoate
E	170	i	NS	COLR	6,27	Calcium, carbonate
E	170	i	NS	MISC	6,10,13,35,39,UPM2(21)	Calcium, carbonate
E	509		NS	MISC	8,11,12,13,14,35,UPM2(21)	Calcium, chloride
E	333	ii	NS	MISC	7,11,12,13,14,39	Calcium, citrate, di- citrate
E	333	i	NS	MISC	7,11,12,13,14,35,39,UPM2(20,30)	Calcium, citrate, mono- citrate
E	333	i	NS	MISC	7,11,12,13,14,35,39,UPM2(20,30)	Calcium, citrate, mono- citrate
E	333	iii	NS	MISC	7,11,12,13,14,35,39,UPM2(20)	Calcium, citrate, tri- citrate
E	952		7	SWET	9,10	Calcium, cyclamic acid and its sodium and - salts
E	333	ii	NS	MISC	7,11,12,13,14,35,39,UPM2(20)	Calcium, di- citrate
E	450	vi	70	MISC	27,UPM2(24)	Calcium, di- diphosphate
E	341	ii	70	MISC	26,35,39,UPM2(24)	Calcium, di- phosphate
E	623		NS	MISC	32	Calcium, diglutamate
E	450	vii	70	MISC	27,UPM2(24)	Calcium, dihydrogen diphosphate
E	450	vii	70	MISC	27,UPM2(24)	Calcium, diphosphate, - dihydrogen diphosphate
E	385		2.5	MISC	28,UPM2(25)	Calcium, disodium ethylene diamine tetra-acetate (CaNa2EDTA)
E	538		0.025	MISC	31	Calcium, ferrocyanide
E	578		NS	MISC	9	Calcium, gluconate
E	623		NS	MISC	32	Calcium, glutamate, - diglutamate
E	629		NS	MISC	32	Calcium, guanilate
E	170	ii	NS	COLR	6,27	Calcium, hydrogen carbonate
E	170	ii	NS	MISC	6,10,13,35,39,UPM2(21)	Calcium, hydrogen carbonate
E	352	ii	NS	MISC	7	Calcium, hydrogen malate
E	227		0.7	MISC	19,20,21,22,UPM2(22,23)	Calcium, hydrogen sulphite
E	526		NS	MISC	9,10,39	Calcium, hydroxide
E	633		NS	MISC	32	Calcium, inosinate
E	327		NS	MISC	6,10,11,12,13,14,15,39	Calcium, lactate
E	352	i	NS	MISC	7	Calcium, malate
E	333	i	NS	MISC	7,11,12,13,14,35,39,UPM2(20,30)	Calcium, mono- citrate
E	341	i	70	MISC	26,35,39,UPM2(24,30)	Calcium, mono- phosphate
E	529		NS	MISC	9	Calcium, oxide
E	450	vii	70	MISC	27,UPM2(24)	Calcium, phosphate, - dihydrogen diphosphate

Appendix Table 2 (Continued)

E #			ADI SCF mg/kg	DIR	Directives PAGES	NAME / Alphabetically sorted - TABLE 2 -
UPS1(18)	= on page 18 of the 1st. adopted update of the sweeteners Directive					
UPM2(36)	= on page 36 of the 2nd. adopted update of the "*miscellaneous*" Directive					
E	450	vi	70	MISC	27,UPM2(24)	Calcium, phosphate, di- diphosphate
E	341	ii	70	MISC	26,35,39,UPM2(24)	Calcium, phosphate, di- phosphate
E	341	i	70	MISC	26,35,39,UPM2(24,30)	Calcium, phosphate, mono- phosphate
E	341	iii	70	MISC	26,35,39,UPM2(24)	Calcium, phosphate, tri- phosphate
E	452	iv	70	MISC	27,UPM2(24)	Calcium, polyphosphate
E	282		NS	MISC	24,UPM2(23)	Calcium, propionate
E	954		5	SWET	10,11,UPS1(19)	Calcium, saccharin and sodium potassium and - salts
E	552		NS	MISC	31,36	Calcium, silicate
E	452	iii	70	MISC	27,UPM2(24)	Calcium, sodium - polyphosphate
E	470	a	NS	MISC	8,35,UPM2(28)	Calcium, sodium potassium and - salts of fatty acids
E	203		25	MISC	16,17,18,19,UPM2(21,22)	Calcium, sorbate
E	482		20	MISC	30	Calcium, stearoyl-2-lactylate
E	516		NS	MISC	8,35	Calcium, sulphate
E	226		0.7	MISC	19,20,21,22,UPM2(22,23)	Calcium, sulphite
E	354		30	MISC	7,40	Calcium, tartrate
E	333	iii	NS	MISC	7,11,12,13,14,35,39,UPM2(20)	Calcium, tri- citrate
E	341	iii	70	MISC	26,35,39,UPM2(24)	Calcium, tri- phosphate
E	902		A	MISC	32,UPM2(26)	Candellila wax
E	161	g	0.03	COLR	25	Canthaxanthin
E	160	c	A	COLR	21,23,24,27	Capsanthin, paprika extract, - , capsorubin
E	160	c	A	COLR	21,23,24,27	Capsorubin, paprika extract, capsanthin, -
E	150	c	200	COLR	21,23,24,27	Caramel, ammonia -
E	150	b	200	COLR	21,23,24,27	Caramel, caustic sulphite -
E	150	a	A	COLR	21,23,24,27	Caramel, plain -
E	150	d	200	COLR	21,23,24,27	Caramel, sulphite ammonia -
E	927	b	A	MISC	33	Carbamide
E	290		A	MISC	6	Carbon dioxide
E	153		NS	COLR	21,27	Carbon, vegetable -
E	503	i	NS	MISC	8,10,39	Carbonate, ammonium -
E	503	ii	NS	MISC	8,10,39	Carbonate, ammonium hydrogen -
E	170	i	NS	COLR	6,27	Carbonate, calcium -
E	170	i	NS	MISC	6,10,13,35,39,UPM2(21)	Carbonate, calcium -
E	170	ii	NS	COLR	6,27	Carbonate, calcium, hydrogen -
E	170	ii	NS	MISC	6,10,13,35,39,UPM2(21)	Carbonate, calcium, hydrogen -
E	242		A	MISC	23	Carbonate, dimethyl - di-
E	504	i	NS	MISC	8,13,35,UPM2(21)	Carbonate, magnesium -
E	504	ii	NS	MISC	8,13,35,UPM2(21)	Carbonate, magnesium hydroxide -
E	501	i	NS	MISC	8,10,35,39	Carbonate, potassium -
E	501	ii	NS	MISC	8,10,11,35,39	Carbonate, potassium hydrogen - (potassium bicarbonate)

Appendix Table 2 (Continued)

E #			ADI SCF mg/kg	DIR	Directives PAGES	NAME / Alphabetically sorted - TABLE 2 -
UPS1(18)	= on page 18 of the 1st. adopted update of the sweeteners Directive					
UPM2(36)	= on page 36 of the 2nd. adopted update of the "*miscellaneous*" Directive					
E	500	i	NS	MISC	8,10,39,UPM2(21)	Carbonate, sodium -
E	500	ii	NS	MISC	8,10,11,39,UPM2(21)	Carbonate, sodium hydrogen - (sodium bicarbonate)
E	500	iii	NS	MISC	8,39,UPM2(21)	Carbonate, sodium sesqui-
E	466		NS	MISC	8,12,35,UPM2(20,30)	Carboxy methyl cellulose + sodium carboxyy methyl cellulose
E	468		NS	MISC	UPM2(25,28)	Carboxy methyl cellulose, cross linked sodium
E	468		NS	MISC	UPM2(25,28)	Carboxy methyl cellulose, cross linked sodium - - -
E	469		NS	MISC	UPM2(25,28)	Carboxy methyl cellulose, enzymatically hydrolysed -
E	469		NS	MISC	UPM2(25,28)	Carboxy methyl cellulose, enzymatically hydrolysed - - -
E	132		5	COLR	28	Carmine, indigotine, indigo -
E	120		5	COLR	21,22,23,24,28	Carmines, cochineal, carminic acid
E	120		5	COLR	21,22,23,24,28	Carminic acid, cochineal, carmines
E	122		4	COLR	22,28	Carmoisine, azorubine
E	903		A	MISC	32,UPM2(26)	Carnauba wax
E	160	e	5	COLR	28	Carotenal, beta-apo-8'- - (C30)
E	160	f	5	COLR	28	Carotenal, ethyl ester of beta-apo-8'- -
E	160	a,ii	5	COLR	23,24,27	Carotene, beta- -
E	160	a,i	NS	COLR	21,23,24,27	Carotenes, mixed
E	407		75	MISC	7,11,12,34,38,UPM2(20)	Carrageenan
E	407	a	75	MISC	UPM1(4)	Carrageenan, alternatively refined -
E	150	b	200	COLR	21,23,24,27	Caustic sulphite caramel
E	466		NS	MISC	8,12,35,UPM2(20,30)	Cellulose, carboxy methyl - + sodium carboxyy methyl cellulose
E	468		NS	MISC	UPM2(25,28)	Cellulose, cross linked sodium carboxy methyl -
E	468		NS	MISC	UPM2(25,28)	Cellulose, cross linked sodium carboxy methyl -
E	469		NS	MISC	UPM2(25,28)	Cellulose, enzymatically hydrolysed carboxy methyl -
E	469		NS	MISC	UPM2(25,28)	Cellulose, enzymatically hydrolysed carboxymethyl -
E	465		NS	MISC	8,12,35	Cellulose, ethyl methyl -
E	463		NS	MISC	8,12,34	Cellulose, hydroxypropyl -
E	464		NS	MISC	8,12,35	Cellulose, hydroxypropyl methyl -
E	461		NS	MISC	8,12,34	Cellulose, methyl -
E	460	i	A	MISC	8,12,34,UPM2(21)	Cellulose, microcristalline -
E	460	ii	NS	MISC	8,12,34,UPM2(21)	Cellulose, powdered -
E	959		5	MISC	33,UPM2(27)	Chalcone, neohesperidine dihydrocha... (NHDC)
E	959		5	SWET	11,12,UPS1(19)	Chalcone, neohesperidine dihydrocha... (NHDC)
E	509		NS	MISC	8,11,12,13,14,35,UPM2(21)	Chloride, calcium -
E	511		NS	MISC	8,35	Chloride, magnesium -
E	508		NS	MISC	8,12,35	Chloride, potassium -
E	512		A	MISC	31	Chloride, stannous -
E	140	ii	NS	COLR	22,23,27	Chlorophyllins
E	141	ii	15	COLR	22,23,27	Chlorophyllins, copper complexes of -

Appendix Table 2 (Continued)

E #			ADI scf mg/kg	DIR	Directives PAGES	NAME / Alphabetically sorted - TABLE 2 -
UPS1(18)	= on page 18 of the 1st. adopted update of the sweeteners Directive					
UPM2(36)	= on page 36 of the 2nd. adopted update of the "*miscellaneous*" Directive					
E	140	i	NS	COLR	22,23,27	Chlorophylls
E	141	i	15	COLR	22,23,27	Chlorophylls, copper complexes of -
E	333	ii	NS	MISC	7,11,12,13,14,35,39,UPM2(20)	Citrate, calcium, dialcium -
E	333	i	NS	MISC	7,11,12,13,14,35,39,UPM2(20,30)	Citrate, calcium, monocalcium -
E	333	iii	NS	MISC	7,11,12,13,14,35,39,UPM2(20)	Citrate, calcium, tricalcium -
E	333	ii	NS	MISC	7,11,12,13,14,35,39,UPM2(20)	Citrate, dicalcium -
E	331	ii	NS	MISC	7,10,11,12,13,14,15,37,38,39,UPM2(20,29)	Citrate, disodium -
E	1505		20	MISC	33,36	Citrate, ethyl, triethyl -
E	333	i	NS	MISC	7,11,12,13,14,35,39,UPM2(20,30)	Citrate, monocalcium -
E	332	i	NS	MISC	7,10,11,12,13,14,15,37,38,39,UPM2(20,29)	Citrate, monopotassium -
E	331	i	NS	MISC	7,10,11,12,13,14,15,37,38,39,UPM2(20,29)	Citrate, monosodium
E	332	i	NS	MISC	7,10,11,12,13,14,15,37,38,39,UPM2(20,29)	Citrate, potassium, monopotassium -
E	332	ii	NS	MISC	7,10,11,12,13,14,15,37,38,39,UPM2(20,29)	Citrate, potassium, tripotassium -
E	331	ii	NS	MISC	7,10,11,12,13,14,15,37,38,39,UPM2(20,29)	Citrate, sodium, disodium -
E	331	i	NS	MISC	7,10,11,12,13,14,15,37,38,39,UPM2(20,29)	Citrate, sodium, monosodium -
E	331	iii	NS	MISC	7,10,11,12,13,14,15,37,38,39,UPM2(20,29)	Citrate, sodium, trisodium -
E	380		NS	MISC	7	Citrate, triammonium -
E	380		NS	MISC	7	Citrate, triammonium -
E	333	iii	NS	MISC	7,11,12,13,14,35,39,UPM2(20)	Citrate, tricalcium -
E	1505		20	MISC	33,36	Citrate, triethyl -
E	332	ii	NS	MISC	7,10,11,12,13,14,15,37,38,39,UPM2(20,29)	Citrate, tripotassium -
E	331	iii	NS	MISC	7,10,11,12,13,14,15,37,38,39,UPM2(20,29)	Citrate, trisodium -
E	330		NS	MISC	7,10,11,12,13,14,15,37,38,39,UPM2(20)	Citric acid
E	472	c	NS	MISC	8,35,40,UPM2(29)	Citric acid esters of mono- and diglycerides of fatty acids
E	124		4	COLR	22,24,28	Cochineal red A, ponceau 4R
E	120		5	COLR	21,22,23,24,28	Cochineal, carmines, carminic acid
E	141	ii	15	COLR	22,23,27	Complexes, copper - of chlorophyllins
E	141	i	15	COLR	22,23,27	Complexes, copper - of chlorophylls
E	141	ii	15	COLR	22,23,27	Copper complexes of chlorophyllins
E	141	i	15	COLR	22,23,27	Copper complexes of chlorophylls
E	460	i	A	MISC	8,12,34,UPM2(21)	Cristalline, micro- cellulose
E	468		NS	MISC	UPM2(25,28)	Cross linked sodium carboxy methyl cellulose
E	100		A	COLR	22,23,24,28	Curcumin
E	538		0.025	MISC	31	Cyanide, calcium ferro-
E	538		0.025	MISC	31	Cyanide, potassium ferro
E	535		MISC	MISC	31	Cyanide, sodium ferro-
E	952		7	SWET	9,10	Cyclamic acid and its sodium and calcium salts
E	459		5	MISC	UPM2(27,28)	Cyclodextrin, beta-
E	920		NS	MISC	UPM2(20)	Cysteine [L-cyst...]

Appendix Table 2 (Continued)

E #			ADI SCF mg/kg	DIR	Directives PAGES	NAME / Alphabetically sorted - TABLE 2 -
UPS1(18)			= on page 18 of the 1st. adopted update of the sweeteners Directive			
UPM2(36)			= on page 36 of the 2nd. adopted update of the *"miscellaneous"* Directive			
E	575		NS	MISC	9,13,14,15,40,UPM2(21)	Delta-lactone, glucono- -
E	309		A	MISC	6,13,37,38,39,UPM2(20)	Delta-tocopherol
E	459		5	MISC	UPM2(27,28)	Dextrin, beta cyclo-
E	1200		NS	MISC	9,36	Dextrose, poly-
E	262	ii	NS	MISC	6,13,14,15,39	Diacetate, sodium hydrogen acetate (sodium -)
E	472	e	25	MISC	8,14,35	Diacetyl, mono- and - tartaric acid esters of mono- and diglycerides of fatty acids
E	385		2.5	MISC	28,UPM2(25)	Diamine tetra-acetate, calcium disodium ethylene - (CaNa2EDTA)
E	333	ii	NS	MISC	7,11,12,13,14,35,39,UPM2(20)	Dicalcium, citrate
E	450	vi	70	MISC	27,UPM2(24)	Dicalcium, diphosphate
E	341	ii	70	MISC	26,35,39,UPM2(24)	Dicalcium, phosphate
E	242		A	MISC	23	Dicarbonate, dimethyl -
E	623		NS	MISC	32	Diglutamate, calcium -
E	625		NS	MISC	32	Diglutamate, magnesium -
E	472	a	NS	MISC	8,13,14,35,40	Diglycerides, acetic acid esters of mono- and - of fatty acids
E	472	c	NS	MISC	8,35,40,UPM2(29)	Diglycerides, citric acid esters of mono- and - of fatty acids
E	472	b	NS	MISC	8,40	Diglycerides, lactic acid esters of mono- and - of fatty acids
E	472	f	NS	MISC	8,14	Diglycerides, mixed acetic and tartaric acid esters of mono- and - of fatty acids
E	471		NS	MISC	8,11,12,13,14,15,35,37,38,40,UPM2(20,28,29,30)	Diglycerides, mono- and - of fatty acids
E	472	e	25	MISC	8,14,35	Diglycerides, mono-, mono and diacetyl tartaric acid esters of mono- and - of fatty acids
E	472	d	NS	MISC	8,14	Diglycerides, tartaric acid esters of mono- and - of fatty acids
E	479	b	25	MISC	30	Diglycerides, thermally oxidized soya bean oil interacted with mono- and - of fatty acids
E	959		5	MISC	33,UPM2(27)	Dihydrochalcone, neohesperidine di... (NHDC)
E	959		5	SWET	11,12,UPS1(19)	Dihydrochalcone, neohesperidine di... (NHDC)
E	450	vii	70	MISC	27,UPM2(24)	Dihydrogen, calcium - diphosphate
E	343	ii	70	MISC	UPM2(24)	Dimagnesium, phosphate
E	242		A	MISC	23	Dimethyl, dicarbonate
E	900		1.5	MISC	32,UPM2(26,28)	Dimethyln polysiloxane
Solvent			25	MISC	34	Diol, propan-1,2- - (propylene glycol)
E	477		25	MISC	30	Diol, propane 1,2- - esters of fatty acids
E	405		25	MISC	28,34,UPM2(25,30)	Diol, propane-1,2- - alginate
E	290		A	MISC	6	Dioxide, carbon -
E	551		NS	MISC	31,36,UPM2(26,28)	Dioxide, silicon -
E	220		0.7	MISC	19,20,21,22,UPM2(22,23)	Dioxide, sulphur -
E	171		NS	COLR	27	Dioxide, titanium -
E	230		NE	MISC	23	Diphenyl, biphenyl
E	450	vii	70	MISC	27,UPM2(24)	Diphosphate, calcium dihydrogen -
E	450	vi	70	MISC	27,UPM2(24)	Diphosphate, dicalcium -
E	450	vii	70	MISC	27,UPM2(24)	Diphosphate, dihydrogen, calcium dihydrogen -
E	450	iv	70	MISC	27,UPM2(24)	Diphosphate, dipotassium -

Appendix Table 2 (Continued)

E #			ADI SCF mg/kg	DIR	Directives PAGES	NAME / Alphabetically sorted - TABLE 2 -
UPS1(18) = on page 18 of the 1st. adopted update of the sweeteners Directive						
UPM2(36) = on page 36 of the 2nd. adopted update of the "*miscellaneous*" Directive						
E	450	a	70	MISC	27	Diphosphate, disodium -
E	450	i	70	MISC	27,UPM2(24)	Diphosphate, disodium -
E	450	v	70	MISC	27,UPM2(24)	Diphosphate, tetrapotassium -
E	450	iii	70	MISC	27,UPM2(24)	Diphosphate, tetrasodium -
E	450	ii	70	MISC	27,UPM2(24)	Diphosphate, trisodium -
E	450	iv	70	MISC	27,UPM2(24)	Dipotassium, diphosphate
E	628		NS	MISC	32	Dipotassium, guanilate
E	632		NS	MISC	32	Dipotassium, inosinate
E	340	ii	70	MISC	26,39,UPM2(24,29)	Dipotassium, phosphate
E	336	ii	30	MISC	7,11,13,14,40	Dipotassium, tartrate
E	635		NS	MISC	32	Disodium, 5'-ribonucleotides
E	385		2.5	MISC	28,UPM2(25)	Disodium, calcium - ethylene diamine tetra-acetate (CaNa2EDTA)
E	331	ii	NS	MISC	7,10,11,12,13,14,15,37,38,39,UPM2(20,29)	Disodium, citrate
E	450	a	70	MISC	27	Disodium, diphosphate
E	450	i	70	MISC	27,UPM2(24)	Disodium, diphosphate
E	627		NS	MISC	32	Disodium, guanylate
E	631		NS	MISC	32	Disodium, inosinate
E	339	ii	70	MISC	26,39,UPM2(24,29)	Disodium, phosphate
E	335	ii	30	MISC	7,11,13,14,40	Disodium, tartrate
E	1422		A	MISC	9,12,35,40	Distarch, acetylated distarch adipate
E	1414		A	MISC	9,12,35,40	Distarch, acetylated distarch phosphate
E	1442		A	MISC	9,12,35	Distarch, hydroxy propyl - phosphate
E	1412		A	MISC	9,12,35,40	Distarch, phosphate
E	1413		A	MISC	9,12,35,40	Distarch, phosphated - phosphate
E	312		0.5	MISC	25,UPM2(23)	Dodecyl, gallate
E	385		2.5	MISC	28,UPM2(25)	Edta EDTA, ethylene diamine tetra-acetate, calcium disodium (CaNa2EDTA)
E	469		NS	MISC	UPM2(25,28)	Enzymatically hydrolysed carboxy methyl cellulose
E	316		6	MISC	25,UPM2(23)	Erythorbate, sodium -
E	315		6	MISC	25,UPM2(23)	Erythorbic acid
E	127		0.1	COLR	25	Erythrosine
E	160	f	5	COLR	28	Ester, ethyl - of beta-apo-8'-carotenal
E	472	a	NS	MISC	8,13,14,35,40	Esters, Acetic acid - of mono- and diglycerides of fatty acids
E	472	c	NS	MISC	8,35,40,UPM2(29)	Esters, citric acid - of mono- and diglycerides of fatty acids
E	445		12.5	MISC	29,UPM2(26)	Esters, glycerol - of wood rosin
E	472	b	NS	MISC	8,40	Esters, lactic acid - of mono- and diglycerides of fatty acids
E	472	f	NS	MISC	8,14	Esters, mixed acetic and tartaric acid - of mono- and diglycerides of fatty acids
E	472	e	25	MISC	8,14,35	Esters, mono- and diacetyl tartaric acid - of mono- and diglycerides of fatty acids
E	912		A	MISC	33,UPM2(27)	Esters, montan acid -
E	475		25	MISC	29,35	Esters, polyglycerol - of fatty acids

Appendix Table 2 (Continued)

E #			ADI SCF mg/kg	DIR	Directives PAGES	NAME / Alphabetically sorted - TABLE 2 -
UPS1(18)			= on page 18 of the 1st. adopted update of the sweeteners Directive			
UPM2(36)			= on page 36 of the 2nd. adopted update of the "*miscellaneous*" Directive			
E	477		25	MISC	30	Esters, propane 1,2-diol - of fatty acids
E	473		20	MISC	29,35,UPM2(26,28,29)	Esters, sucrose - of fatty acids
E	472	d	NS	MISC	8,14	Esters, tartaric acid - of mono- and diglycerides of fatty acids
E	160	f	5	COLR	28	Ethyl ester of beta-apo-8'-carotenal
E	465		NS	MISC	8,12,35	Ethyl methyl cellulose
E	214		10	MISC	16,17,18,19	Ethyl p-hydroxybenzoate
E	215		10	MISC	16,17,18,19	Ethyl, sodium - p-hydroxybenzoate
E	1505		20	MISC	33,36	Ethyl, tri- citrate
E	385		2.5	MISC	28,UPM2(25)	Ethylene diamine tetra-acetate, calcium disodium (CaNa2EDTA)
E	914		A	MISC	33,UPM2(27)	Ethylene, poly- oxidized - wax
E	431		NA	MISC	27	Ethylene, polyoxy- (40) stearate
E	432		10	MISC	29,34,35,UPM2(28)	Ethylene, polyoxy- sorbitan monolaurate (Polysorbate 20)
E	433		10	MISC	29,34,35,UPM2(28)	Ethylene, polyoxy- sorbitan monooleate (Polysorbate 80)
E	434		10	MISC	29,34,35,UPM2(28)	Ethylene, polyoxy- sorbitan monopalmitate (Polysorbate 40)
E	435		10	MISC	29,34,35,UPM2(28)	Ethylene, polyoxy- sorbitan monostearate (Polysorbate 60)
E	436		10	MISC	29,34,35,UPM2(28)	Ethylene, polyoxy- sorbitan tristearate (Polysorbate 65)
E	999		5	MISC	33,UPM2(27)	Extract, quillaia -
E	306		A	MISC	6,13,37,38,39,UPM2(20)	Extract, tocopherol-rich -
E	472	a	NS	MISC	8,13,14,35,40	Fatty, acetic acid esters of mono- and diglycerides of - acids
E	570		NS	MISC	9,UPM2(28)	Fatty, acids
E	470	b	NS	MISC	8,35	Fatty, acids, magnesium salts of -
E	472	c	NS	MISC	8,35,40,UPM2(29)	Fatty, citric acid esters of mono- and diglycerides of - acids
E	472	b	NS	MISC	8,40	Fatty, lactic acid esters of mono- and diglycerides of - acids
E	472	f	NS	MISC	8,14	Fatty, mixed acetic and tartaric acid esters of mono- and diglycerides of - acids
E	472	e	25	MISC	8,14,35	Fatty, mono- and diacetyl tartaric acid esters of mono- and diglycerides of - acids
E	471		NS	MISC	8,11,12,13,14,15,35,37,38,40,UPM2(20,28,29,30)	Fatty, mono- and diglycerides of - acids
E	475		25	MISC	29,35	Fatty, polyglycerol esters of - acids
E	477		25	MISC	30	Fatty, propane 1,2-diol esters of - acids
E	470	a	NS	MISC	8,35,UPM2(28)	Fatty, sodium potassium and calcium salts of - acids
E	473		20	MISC	29,35,UPM2(26,28,29)	Fatty, sucrose esters of - acids
E	472	d	NS	MISC	8,14	Fatty, tartaric acid esters of mono- and diglycerides of - acids
E	479	b	25	MISC	30	Fatty, thermally oxidized soya bean oil interacted with mono- and diglycerides of - acids
E	538		0.025	MISC	31	Ferrocyanide, calcium -
E	536		0.025	MISC	31	Ferrocyanide, potassium -
E	535		MISC	MISC	31	Ferrocyanide, sodium -
E	579		A	MISC	32	Ferrous, gluconate
E	585		A	MISC	32	Ferrous, lactate
E	297		6	MISC	26,UPM2(23)	Fumaric, acid
E	312		0.5	MISC	25,UPM2(23)	Gallate, dodecyl -

Appendix Table 2 (Continued)

E #			ADI SCF mg/kg	DIR	Directives PAGES	NAME / Alphabetically sorted - TABLE 2 -
UPS1(18)	= on page 18 of the 1st. adopted update of the sweeteners Directive					
UPM2(36)	= on page 36 of the 2nd. adopted update of the "*miscellaneous*" Directive					
E	311		0.5	MISC	25,UPM2(23)	Gallate, octyl -
E	310		0.5	MISC	25,UPM2(23)	Gallate, propyl -
E	308		A	MISC	6,13,37,38,39,UPM2(20)	Gamma-tocopherol
E	418		NS	MISC	7,11	Gellan gum
E	425	ii	NS	MISC	UPM2(27,28)	Glucomannane, konjac -
E	578		NS	MISC	9	Gluconate, calcium
E	579		A	MISC	32	Gluconate, ferrous
E	577		NS	MISC	9,36	Gluconate, potassium
E	576		NS	MISC	9	Gluconate, sodium
E	574		NS	MISC	9	Gluconic acid
E	575		NS	MISC	9,13,14,15,40,UPM2(21)	Glucono-delta-lactone
E	623		NS	MISC	32	Glutamate, calcium di-
E	625		NS	MISC	32	Glutamate, magnesium di-
E	624		NS	MISC	32	Glutamate, monoammonium -
E	622		NS	MISC	32	Glutamate, monopotassium -
E	621		NS	MISC	32	Glutamate, monosodium -
E	620		NS	MISC	32	Glutamic, acid
E	472	a	NS	MISC	8,13,14,35,40	Glycerides, acetic acid esters of mono- and di- of fatty acids
E	472	c	NS	MISC	8,35,40,UPM2(29)	Glycerides, citric acid esters of mono- and di- of fatty acids
E	472	b	NS	MISC	8,40	Glycerides, lactic acid esters of mono- and di- of fatty acids
E	472	f	NS	MISC	8,14	Glycerides, mixed acetic and tartaric acid esters of mono- and di- of fatty acids
E	471		NS	MISC	8,11,12,13,14,15,35,37,38,40,UPM2(20,28,29,30)	Glycerides, mono- and di- of fatty acids
E	472	e	25	MISC	8,14,35	Glycerides, mono and diacetyl tartaric acid esters of mono- and - of fatty acids
E	474		20	MISC	29,UPM2(26)	Glycerides, sucro-
E	472	d	NS	MISC	8,14	Glycerides, tartaric acid esters of mono- and di- 1of fatty acids
E	479	b	25	MISC	30	Glycerides, thermally oxidized soya bean oil interacted with mono- and di- of fatty acids
E	422		NS	MISC	7,10,34	Glycerol
E	445		12.5	MISC	29,UPM2(26)	Glycerol, esters of wood rosin
E	475		25	MISC	29,35	Glycerol, poly- esters of fatty acids
E	476		7.5	MISC	30,UPM2(26)	Glycerol, poly- polyricinoleate
E	1518		A	MISC	36,UPM2(27)	Glyceryl triacetate (triacetin)
E	640		A	MISC	9,36	Glycine and its sodium salt
Solvent			25	MISC	34	Glycol, propan-1,2-diol (propylene -)
E	175		A	COLR	25	Gold
E	142		5	COLR	24,28	Green S
E	629		NS	MISC	32	Guanilate, calcium -
E	628		NS	MISC	32	Guanilate, dipotassium -
E	627		NS	MISC	32	Guanilate, disodium -
E	628		NS	MISC	32	Guanilate, potassium, dipotassium -

Appendix Table 2 (Continued)

E #			ADI SCF mg/kg	DIR	Directives PAGES	NAME / Alphabetically sorted - TABLE 2 -
UPS1(18)	= on page 18 of the 1st. adopted update of the sweeteners Directive					
UPM2(36)	= on page 36 of the 2nd. adopted update of the *"miscellaneous"* Directive					
E	626		NS	MISC	32	Guanilic, acid
E	412		NS	MISC	7,11,34,38,40,UPM2(29,30)	Guar gum
E	414		NS	MISC	7,10,15,34,40,UPM2(28)	Gum acacia (gum arabic)
E	418		NS	MISC	7,11	Gum, gellan -
E	412		NS	MISC	7,11,34,38,40,UPM2(29,30)	Gum, guar -
E	416		12.5	MISC	28	Gum, karaya -
E	425	i	NS	MISC	UPM2(27,28)	Gum, konjac -
E	410		NS	MISC	7,11,12,34,38,40,UPM2(30)	Gum, locust bean gum -
E	417		NS	MISC	7	Gum, tara -
E	415		NS	MISC	7,11,12,34,40,UPM2(30)	Gum, xanthan -
E	939		A	MISC	9	Helium
E	959		5	MISC	33,UPM2(27)	Hesperidine, neohesperidine dihydrochalcone (NHDC)
E	959		5	SWET	11,12,UPS1(19)	Hesperidine, neohesperidine dihydrochalcone (NHDC)
E	239		A	MISC	23	Hexamethylene tetramine
E	507		NS	MISC	8,39	Hydrochloric acid
E	503	ii	NS	MISC	8,10,39	Hydrogen, ammonium - carbonate
E	450	vii	70	MISC	27,UPM2(24)	Hydrogen, calcium di- diphosphate
E	170	ii	NS	COLR	6,27	Hydrogen, carbonate, calcium -
E	170	ii	NS	MISC	6,10,13,35,39,UPM2(21)	Hydrogen, carbonate, calcium -
E	352	ii	NS	MISC	7	Hydrogen, malate, calcium -
E	501	ii	NS	MISC	8,10,11,35,39	Hydrogen, potassium - carbonate (potassium bicarbonate)
E	515	ii	NS	MISC	8,35	Hydrogen, potassium - sulphate
E	228		0.7	MISC	19,20,21,22,UPM2(22,23)	Hydrogen, potassium - sulphite
E	262	ii	NS	MISC	6,13,14,15,39	Hydrogen, sodium - acetate (Sodium diacetate)
E	500	ii	NS	MISC	8,10,11,39,UPM2(21)	Hydrogen, sodium - carbonate (Sodium bicarbonate)
E	350	ii	NS	MISC	7,11	Hydrogen, sodium - malate
E	514	ii	NS	MISC	8,35	Hydrogen, sodium - sulfate
E	222		0.7	MISC	19,20,21,22,UPM2(22,23)	Hydrogen, sodium - sulphite
E	227		0.7	MISC	19,20,21,22,UPM2(22,23)	Hydrogen, sulphite, calcium -
E	469		NS	MISC	UPM2(25,28)	Hydrolysed carboxy methyl cellulose, enzymatically -
E	469		NS	MISC	UPM2(25,28)	Hydrolysed carboxy methyl cellulose, enzymatically - - - -
E	527		NS	MISC	9,10	Hydroxide, ammonium -
E	526		NS	MISC	9,10,39	Hydroxide, calcium -
E	528		NS	MISC	9,10	Hydroxide, magnesium -
E	504	ii	NS	MISC	8,13,35,UPM2(21)	Hydroxide, magnesium - carbonate
E	525		NS	MISC	8,10,39	Hydroxide, potassium -
E	524		NS	MISC	8,10,11,39	Hydroxide, sodium -
E	172		NS	COLR	27	Hydroxides, iron oxides and -
E	1442		A	MISC	9,12,35	Hydroxy propyl distarch phosphate

Appendix Table 2 (Continued)

E #			ADI SCF mg/kg	DIR	Directives PAGES	NAME / Alphabetically sorted - TABLE 2 -
UPS1(18)	= on page 18 of the 1st. adopted update of the sweeteners Directive					
UPM2(36)	= on page 36 of the 2nd. adopted update of the "*miscellaneous*" Directive					
E	1440		A	MISC	9,12,35	Hydroxy propyl starch
E	216		10	MISC	16,17,18,19	Hydroxy, propyl p- -benzoate
E	215		10	MISC	16,17,18,19	Hydroxy, sodium ethyl p- -benzoate
E	219		10	MISC	16,17,18,19	Hydroxy, sodium methyl p- -benzoate
E	217		10	MISC	16,17,18,19	Hydroxy, sodium propyl p- -benzoate
E	320		0.5	MISC	25,UPM2(23)	Hydroxyanisole, butylated - (Bha BHA)
E	214		10	MISC	16,17,18,19	Hydroxybenzoate, ethyl p- -
E	218		10	MISC	16,17,18,19	Hydroxybenzoate, methyl p- -
E	463		NS	MISC	8,12,34	Hydroxypropyl cellulose
E	464		NS	MISC	8,12,35	Hydroxypropyl methyl cellulose
E	464		NS	MISC	8,12,35	Hydroxypropyl methyl cellulose
E	321		0.05	MISC	25	Hydroxytoluene, butylated (Bht BHT)
E	132		5	COLR	28	Indigo carmine, indigotine
E	132		5	COLR	28	Indigotine, indigo carmine
E	633		NS	MISC	32	Inosinate, calcium -
E	632		NS	MISC	32	Inosinate, dipotassium -
E	631		NS	MISC	32	Inosinate, disodium -
E	630		NS	MISC	32	Inosinic acid
E	1103		A	MISC	UPM2(20)	Invertase
E	172		NS	COLR	27	Iron oxides and hydroxides
E	444		10	MISC	29	Isobutyrate, sucrose acetate -
E	953		A	MISC	28,34	Isomalt
E	953		A	SWET	6	Isomalt
E	559		1	MISC	31,36,UPM2(26)	Kaolin, aluminium silicate
E	416		12.5	MISC	28	Karaya gum
E	425	ii	NS	MISC	UPM2(27,28)	Konjac glucomannane
E	425	i	NS	MISC	UPM2(27,28)	Konjac gum
E	327		NS	MISC	6,10,11,12,13,14,15,39	Lactate, calcium -
E	585		A	MISC	32	Lactate, ferrous -
E	326		NS	MISC	6,12,13,14,15,39	Lactate, potassium -
E	325		NS	MISC	6,12,13,14,15,39	Lactate, sodium -
E	270		NS	MISC	6,10,11,12,13,14,15,37,38,39,UPM2(20)	Lactic acid
E	472	b	NS	MISC	8,40	Lactic acid esters of mono- and diglycerides of fatty acids
E	966		A	MISC	28,34	Lactitol
E	966		A	SWET	6	Lactitol
E	575		NS	MISC	9,13,14,15,40,UPM2(21)	Lactone, glucono-delta- -
E	482		20	MISC	30	Lactylate, calcium stearoyl-2 -
E	481		20	MISC	30	Lactylate, sodium stearoyl-2- -
E	432		10	MISC	29,34,35,UPM2(28)	Laurate, polyoxyethylene sorbitan mono (Polysorbate 20)

Appendix Table 2 (Continued)

E #			ADI SCF mg/kg	DIR	Directives PAGES	NAME / Alphabetically sorted - TABLE 2 -
UPS1(18)	= on page 18 of the 1st. adopted update of the sweeteners Directive					
UPM2(36)	= on page 36 of the 2nd. adopted update of the *"miscellaneous"* Directive					
E	493		5	MISC	31,35,UPM2(28)	Laurate, sorbitan mono-
E	322		NS	MISC	6,10,11,12,13,14,35,37,38,39,40,UPM2(20,28,29)	Lecithins
E	468		NS	MISC	UPM2(25,28)	Linked, cross - sodium carboxy methyl cellulose
E	468		NS	MISC	UPM2(25,28)	Linked, sodium carboxy methyl cellulose, cross - - - - -
E	180		1.5	COLR	25	Lithorubine BK
E	410		NS	MISC	7,11,12,34,38,40,UPM2(30)	Locust bean gum
E	161	b	NS	COLR	24	Lutein
E	160	d	A	COLR	24,28	Lycopene
E	1105		A	MISC	24	Lysozyme
E	504	i	NS	MISC	8,13,35,UPM2(21)	Magnesium, carbonate
E	511		NS	MISC	8,35	Magnesium, chloride
E	343	ii	70	MISC	UPM2(24)	Magnesium, di- phosphate
E	625		NS	MISC	32	Magnesium, diglutamate
E	625		NS	MISC	32	Magnesium, glutamate, - diglutamate
E	528		NS	MISC	9,10	Magnesium, hydroxide
E	504	ii	NS	MISC	8,13,35,UPM2(21)	Magnesium, hydroxide carbonate
E	530		NS	MISC	9,10	Magnesium, oxide
E	343	i	70	MISC	UPM2(24)	Magnesium, phosphate, mono- -
E	470	b	NS	MISC	8,35	Magnesium, salts of fatty acids
E	553	a i	NS	MISC	31	Magnesium, silicate
E	553	a ii	NS	MISC	31	Magnesium, trisilicate
E	352	i	NS	MISC	7	Malate, calcium -
E	352	ii	NS	MISC	7	Malate, calcium hydrogen -
E	351		NS	MISC	7	Malate, potassium -
E	350	i	NS	MISC	7,11	Malate, sodium -
E	350	ii	NS	MISC	7,11	Malate, sodium hydrogen -
E	296		NS	MISC	6,10,11,39,UPM2(21)	Malic acid
E	953		A	MISC	28,34	Malt, isoma...
E	953		A	SWET	6	Malt, isoma...
E	965		A	MISC	28,34	Maltitol
E	965		A	SWET	6	Maltitol
E	965		A	MISC	28	Maltitol syrup
E	965		A	SWET	6	Maltitol syrup
E	425	ii	NS	MISC	UPM2(27,28)	Mannane, konjac gluco-
E	421		A	MISC	28,34,UPM2(28)	Mannitol
E	223		0.7	MISC	19,20,21,22,UPM2(22,23)	Meta, sodium -bisulphite
E	224		0.7	MISC	19,20,21,22,UPM2(22,23)	Metabisulphite, potassium -
E	353		A	MISC	27	Metatartaric acid
E	466		NS	MISC	8,12,35,UPM2(20,30)	Methyl, carboxy - cellulose + sodium carboxyy methyl cellulose

Appendix Table 2 (Continued)

E #			ADI SCF mg/kg	DIR	Directives PAGES	NAME / Alphabetically sorted - TABLE 2 -
UPS1(18)	= on page 18 of the 1st. adopted update of the sweeteners Directive					
UPM2(36)	= on page 36 of the 2nd. adopted update of the "*miscellaneous*" Directive					
E	461		NS	MISC	8,12,34	Methyl, cellulose
E	468		NS	MISC	UPM2(25,28)	Methyl, cellulose, cross linked sodium carboxy - -
E	468		NS	MISC	UPM2(25,28)	Methyl, cellulose, cross linked sodium carboxy - -
E	469		NS	MISC	UPM2(25,28)	Methyl, cellulose, enzymatically hydrolysed carboxy -
E	469		NS	MISC	UPM2(25,28)	Methyl, cellulose, enzymatically hydrolysed carboxy - -
E	242		A	MISC	23	Methyl, di- dicarbonate
E	242		A	MISC	23	Methyl, di- dicarbonate
E	900		1.5	MISC	32,UPM2(26,28)	Methyl, di- polysiloxane
E	465		NS	MISC	8,12,35	Methyl, ethyl - cellulose
E	464		NS	MISC	8,12,35	Methyl, hydroxypropyl - cellulose
E	218		10	MISC	16,17,18,19	Methyl, p-hydroxybenzoate
E	219		10	MISC	16,17,18,19	Methyl, sodium - p-hydroxybenzoate
E	239		A	MISC	23	Methylene, hexa- tetramine
E	460	i	A	MISC	8,12,34,UPM2(21)	Microcristalline cellulose
E	905		20	MISC	UPM2(27)	Microcristalline wax
E	472	f	NS	MISC	8,14	Mixed acetic and tartaric acid esters of mono- and diglycerides of fatty acids
E	160	a,i	NS	COLR	21,23,24,27	Mixed carotenes
E	466		NS	MISC	8,12,35,UPM2(20,30)	Modified cellulose: carboxy methyl cellulose + sodium carboxyy methyl cellulose
E	468		NS	MISC	UPM2(25,28)	Modified cellulose: cross linked sodium carboxy methyl cellulose
E	469		NS	MISC	UPM2(20,28)	Modified cellulose: enzymatically hydrolyzed carboxy methyl cellulose
E	465		NS	MISC	8,12,35	Modified cellulose: ethyl methyl cellulose
E	463		NS	MISC	8,12,34	Modified cellulose: hydroxypropyl cellulose
E	464		NS	MISC	8,12,35	Modified cellulose: hydroxypropyl methyl cellulose
E	461		NS	MISC	8,12,34	Modified cellulose: methyl cellulose
E	1422		A	MISC	9,12,35,40	Modified starch: acetylated distarch adipate
E	1414		A	MISC	9,12,35,40	Modified starch: acetylated distarch phosphate
E	1451		A	MISC	UPM2(20,28,30)	Modified starch: acetylated oxidized starch
E	1420		A	MISC	9,12,35,40	Modified starch: acetylated starch
E	1412		A	MISC	9,12,35,40	Modified starch: distarch phosphate
E	1442		A	MISC	9,12,35	Modified starch: hydroxy propyl distarch phosphate
E	1440		A	MISC	9,12,35	Modified starch: hydroxy propyl starch
E	1410		A	MISC	9,12,35,40	Modified starch: monostarch phosphate
E	1404		A	MISC	9,12,35,40	Modified starch: oxidized starch
E	1413		A	MISC	9,12,35,40	Modified starch: phosphated distarch phosphate
E	472	e	25	MISC	8,14,35	Mono-, and diacetyl tartaric acid esters of mono- and diglycerides of fatty acids
E	471		NS	MISC	8,11,12,13,14,15,35,37,38,40,UPM2(20,28,29,30)	Mono-, and diglycerides of fatty acids
E	472	a	NS	MISC	8,13,14,35,40	Mono-, and diglycerides, acetic acid esters of - of fatty acids
E	472	c	NS	MISC	8,35,40,UPM2(29)	Mono-, citric acid esters of - and diglycerides of fatty acids
E	472	b	NS	MISC	8,40	Mono-, lactic acid esters of - and diglycerides of fatty acids

Appendix Table 2 (Continued)

E #			ADI SCF mg/kg	DIR	Directives PAGES	NAME / Alphabetically sorted - TABLE 2 -
UPS1(18)	= on page 18 of the 1st. adopted update of the sweeteners Directive					
UPM2(36)	= on page 36 of the 2nd. adopted update of the "*miscellaneous*" Directive					
E	472	f	NS	MISC	8,14	Mono-, mixed acetic and tartaric acid esters of - and diglycerides of fatty acids
E	472	e	25	MISC	8,14,35	Mono-, mono- and diacetyl tartaric acid esters of - and diglycerides of fatty acids
E	472	d	NS	MISC	8,14	Mono-, tartaric acid esters of - and diglycerides of fatty acids
E	479	b	25	MISC	30	Mono-, thermally oxidized soya bean oil interacted with - and diglycerides of fatty acids
E	624		NS	MISC	32	Monoammonium, glutamate
E	333	i	NS	MISC	7,11,12,13,14,35,39,UPM2(20,30)	Monocalcium, citrate
E	341	i	70	MISC	26,35,39,UPM2(24,30)	Monocalcium, phosphate
E	432		10	MISC	29,34,35,UPM2(28)	Monolaurate, polyoxyethylene sorbitan - (Polysorbate 20)
E	493		5	MISC	31,35,UPM2(28)	Monolaurate, sorbitan -
E	343	i	70	MISC	UPM2(24)	Monomagnesium, phosphate
E	433		10	MISC	29,34,35,UPM2(28)	Monooleate, polyoxyethylene sorbitan - (Polysorbate 80)
E	494		5	MISC	31,35,UPM2(28)	Monooleate, sorbitan -
E	434		10	MISC	29,34,35,UPM2(28)	Monopalmitate, polyoxyethylene sorbitan - (Polysorbate 40)
E	495		25	MISC	31,35,UPM2(28)	Monopalmitate, sorbitan -
E	332	i	NS	MISC	7,10,11,12,13,14,15,37,38,39,UPM2(20,29)	Monopotassium, citrate
E	622		NS	MISC	32	Monopotassium, glutamate
E	340	i	70	MISC	26,39,UPM2(24,29)	Monopotassium, phosphate
E	336	i	30	MISC	7,11,13,14,40	Monopotassium, tartrate
E	331	i	NS	MISC	7,10,11,12,13,14,15,37,38,39,UPM2(20,29)	Monosodium, citrate
E	621		NS	MISC	32	Monosodium, glutamate
E	339	i	70	MISC	26,39,UPM2(24,29)	Monosodium, phosphate
E	335	i	30	MISC	7,11,13,14,40	Monosodium, tartrate
E	1410		A	MISC	9,12,35,40	Monostarch, phosphate
E	435		10	MISC	29,34,35,UPM2(28)	Monostearate, polyoxyethylene sorbitan - (Polysorbate 60)
E	491		25	MISC	31,35,UPM2(28)	Monostearate, sorbitan -
E	912		A	MISC	33,UPM2(27)	Montan acid esters
E	235		A	MISC	23	Natamycin
E	959		5	MISC	33,UPM2(27)	Neohesperidine dihydrochalcone (NHDC)
E	959		5	SWET	11,12,UPS1(19)	Neohesperidine dihydrochalcone (NHDC)
E	959		5	MISC	33,UPM2(27)	Nhdc NHDC, neohesperidine dihydrochalcone
E	959		5	SWET	11,12,UPS1(19)	Nhdc NHDC, neohesperidine dihydrochalcone
E	234		0.13	MISC	23,UPM2(23)	Nisin
E	252		5	MISC	24,UPM2(23)	Nitrate, potassium -
E	251		5	MISC	24,UPM2(23)	Nitrate, sodium -
E	249		0.1	MISC	24	Nitrite, potassium -
E	250		0.1	MISC	24	Nitrite, sodium -
E	941		A	MISC	9	Nitrogen
E	942		A	MISC	9	Nitrous oxide
E	160	b	0.065	COLR	21,25,28	Norbixin, annatto, bixin

Appendix Table 2 (Continued)

E #			ADI SCF mg/kg	DIR	Directives PAGES	NAME / Alphabetically sorted - TABLE 2 -
UPS1(18)			= on page 18 of the 1st. adopted update of the sweeteners Directive			
UPM2(36)			= on page 36 of the 2nd. adopted update of the "*miscellaneous*" Directive			
E	634		NS	MISC	32	Nucleotides, calcium 5'-ribo...
E	635		NS	MISC	32	Nucleotides, disodium 5'-ribonucle...
E	635		NS	MISC	32	Nucleotides, sodium, disodium 5'-ribonucle...
E	1450		A	MISC	9,12,35,40,UPM2(30)	Octenyl, starch sodium - succinate
E	311		0.5	MISC	25,UPM2(23)	Octyl gallate
E	479	b	25	MISC	30	Oils, thermally oxidized soya bean - interacted with mono- and diglycerides of fatty acids
E	476		7.5	MISC	30,UPM2(26)	Oleate, polyglycerol polyricin-
E	433		10	MISC	29,34,35,UPM2(28)	Oleate, polyoxyethylene sorbitan mono- (Polysorbate 80)
E	494		5	MISC	31,35,UPM2(28)	Oleate, sorbitan mono-
E	110		2.5	COLR	22,24,28	Orange, sunset yellow FCF, - yellow S
E	232		NE	MISC	23	Ortho, sodium -phenyl phenol
E	231		NE	MISC	23	Orthophenyl phenol
E	529		NS	MISC	9	Oxide, calcium -
E	290		A	MISC	6	Oxide, carbon di-
E	530		NS	MISC	9,10	Oxide, magnesium -
E	942		A	MISC	9	Oxide, nitrous -
E	551		NS	MISC	31,36,UPM2(26,28)	Oxide, silicon di-
E	220		0.7	MISC	19,20,21,22,UPM2(22,23)	Oxide, sulphur di-
E	171		NS	COLR	27	Oxide, titanium di-
E	172		NS	COLR	27	Oxides, iron - and hydroxides
E	914		A	MISC	33,UPM2(27)	Oxidized polyethylene wax
E	1404		A	MISC	9,12,35,40	Oxidized starch
E	1451		A	MISC	UPM2(20,28,30)	Oxidized starch; acetylated - -
E	479	b	25	MISC	30	Oxidized, thermally - soya bean oil interacted with mono- and diglycerides of fatty acids
E	431		NA	MISC	27	Oxyethylene, poly- (40) stearate
E	432		10	MISC	29,34,35,UPM2(28)	Oxyethylene, poly- sorbitan monolaurate (Polysorbate 20)
E	433		10	MISC	29,34,35,UPM2(28)	Oxyethylene, poly- sorbitan monooleate (Polysorbate 80)
E	434		10	MISC	29,34,35,UPM2(28)	Oxyethylene, poly- sorbitan monopalmitate (Polysorbate 40)
E	435		10	MISC	29,34,35,UPM2(28)	Oxyethylene, poly- sorbitan monostearate (Polysorbate 60)
E	436		10	MISC	29,34,35,UPM2(28)	Oxyethylene, poly- sorbitan tristearate (Polysorbate 65)
E	948		A	MISC	9	Oxygen
E	304	i	A	MISC	6,11,13,14,39,UPM2(20,29)	Palmitate, ascorbyl -
E	434		10	MISC	29,34,35,UPM2(28)	Palmitate, polyoxyethylene sorbitan mono- (Polysorbate 40)
E	495		25	MISC	31,35,UPM2(28)	Palmitate, sorbitan mono-
E	160	c	A	COLR	21,23,24,27	Paprika extract, capsanthin, capsorubin
E	131		15	COLR	28	Patent blue V
E	440	i	NS	MISC	7,10,11,12,34,38,40,UPM2(21,30)	Pectins
E	440	ii	NS	MISC	7,10,11,12,34,38,40,UPM2(21,30)	Pectins, amidated -
E	451	ii	70	MISC	27,UPM2(24)	Pentapotassium, triphosphate

Appendix Table 2 (Continued)

E #			ADI SCF mg/kg	DIR	Directives PAGES	NAME / Alphabetically sorted - TABLE 2 -
UPS1(18)	= on page 18 of the 1st. adopted update of the sweeteners Directive					
UPM2(36)	= on page 36 of the 2nd. adopted update of the *"miscellaneous"* Directive					
E	451	i	70	MISC	27,UPM2(24)	Pentasodium, triphosphate
E	231		NE	MISC	23	Phenol, orthophenyl -
E	232		NE	MISC	23	Phenol, sodium orthophenyl -
E	230		NE	MISC	23	Phenyl, bi- , di-
E	231		NE	MISC	23	Phenyl-, ortho- phenol
E	232		NE	MISC	23	Phenyl, sodium ortho- phenol
E	231		NE	MISC	23	Phenyl-, sodium ortho- phenol
E	450	iii	70	MISC	27,UPM2(24)	Phophate, tetrasodium di-
E	1414		A	MISC	9,12,35,40	Phosphate, acetylated distarch -
E	450	vii	70	MISC	27,UPM2(24)	Phosphate, calcium dihydrogen di-
E	452	iv	70	MISC	27,UPM2(24)	Phosphate, calcium poly-
E	341	ii	70	MISC	26,35,39,UPM2(24)	Phosphate, calcium, dicalcium -
E	341	i	70	MISC	26,35,39,UPM2(24,30)	Phosphate, calcium, monocalcium -
E	341	iii	70	MISC	26,35,39,UPM2(24)	Phosphate, calcium, tricalcium -
E	341	ii	70	MISC	26,35,39,UPM2(24)	Phosphate, dicalcium -
E	450	vi	70	MISC	27,UPM2(24)	Phosphate, dicalcium di-
E	450	vii	70	MISC	27,UPM2(24)	Phosphate, dihydrogen, calcium dihydrogen di-
E	343	ii	70	MISC	UPM2(24)	Phosphate, dimagnesium -
E	340	ii	70	MISC	26,39,UPM2(24,29)	Phosphate, dipotassium -
E	450	iv	70	MISC	27,UPM2(24)	Phosphate, dipotassium di-
E	339	ii	70	MISC	26,39,UPM2(24,29)	Phosphate, disodium -
E	450	a	70	MISC	27	Phosphate, disodium di-
E	450	i	70	MISC	27,UPM2(24)	Phosphate, disodium di-
E	1412		A	MISC	9,12,35,40	Phosphate, distarch -
E	450	vii	70	MISC	27,UPM2(24)	Phosphate, hydrogen, calcium dihydrogen di-
E	1442		A	MISC	9,12,35	Phosphate, hydroxy propyl distarch -
E	341	i	70	MISC	26,35,39,UPM2(24,30)	Phosphate, monocalcium -
E	343	i	70	MISC	UPM2(24)	Phosphate, monomagnesium -
E	340	i	70	MISC	26,39,UPM2(24,29)	Phosphate, monopotassium -
E	339	i	70	MISC	26,39,UPM2(24,29)	Phosphate, monosodium -
E	1410		A	MISC	9,12,35,40	Phosphate, monostarch -
E	451	ii	70	MISC	27,UPM2(24)	Phosphate, pentapotassium tri-
E	451	i	70	MISC	27,UPM2(24)	Phosphate, pentasodium tri-
E	1413		A	MISC	9,12,35,40	Phosphate, phosphated distarch -
E	452	ii	70	MISC	27,UPM2(24)	Phosphate, potassium poly-
E	340	ii	70	MISC	26,39,UPM2(24,29)	Phosphate, potassium, dipotassium -
E	340	i	70	MISC	26,39,UPM2(24,29)	Phosphate, potassium, monopotassium -
E	340	iii	70	MISC	26,39,UPM2(24,29)	Phosphate, potassium, tripotassium -
E	101	ii	NS	COLR	22,23,24,27	Phosphate, riboflavine-5'- -

Appendix Table 2 (Continued)

E #			ADI SCF mg/kg	DIR	Directives PAGES	NAME / Alphabetically sorted - TABLE 2 -
UPS1(18) = on page 18 of the 1st. adopted update of the sweeteners Directive						
UPM2(36) = on page 36 of the 2nd. adopted update of the "*miscellaneous*" Directive						
E	541		1	MISC	31,40	Phosphate, sodium aluminium -, acidic
E	452	iii	70	MISC	27,UPM2(24)	Phosphate, sodium calcium poly-
E	452	i	70	MISC	27,UPM2(24)	Phosphate, sodium poly-
E	339	ii	70	MISC	26,39,UPM2(24,29)	Phosphate, sodium, disodium -
E	339	i	70	MISC	26,39,UPM2(24,29)	Phosphate, sodium, monosodium -
E	339	iii	70	MISC	26,39,UPM2(24,29)	Phosphate, sodium, trisodium -
E	450	v	70	MISC	27,UPM2(24)	Phosphate, tetrapotassium di-
E	341	iii	70	MISC	26,35,39,UPM2(24)	Phosphate, tricalcium -
E	340	iii	70	MISC	26,39,UPM2(24,29)	Phosphate, tripotassium -
E	339	iii	70	MISC	26,39,UPM2(24,29)	Phosphate, trisodium -
E	450	ii	70	MISC	27,UPM2(24)	Phosphate, trisodium di-
E	1413		A	MISC	9,12,35,40	Phosphated distarch phosphate
E	442		30	MISC	29,34,UPM2(26)	Phosphatides, ammonium -
E	338		70	MISC	26,37,38,39,UPM2(24)	Phosphoric acid
E	150	a	A	COLR	21,23,24,27	Plain caramel
E	476		7.5	MISC	30,UPM2(26)	Poly, polyglycerol -ricinoleate
E	1200		NS	MISC	9,36	Polydextrose
E	914		A	MISC	33,UPM2(27)	Polyethylene, oxidized - wax
E	475		25	MISC	29,35	Polyglycerol esters of fatty acids
E	476		7.5	MISC	30,UPM2(26)	Polyglycerol polyricinoleate
E	431		NA	MISC	27	Polyoxyethylene (40) stearate
E	432		10	MISC	29,34,35,UPM2(28)	Polyoxyethylene sorbitan monolaurate (Polysorbate 20)
E	433		10	MISC	29,34,35,UPM2(28)	Polyoxyethylene sorbitan monooleate (Polysorbate 80)
E	434		10	MISC	29,34,35,UPM2(28)	Polyoxyethylene sorbitan monopalmitate (Polysorbate 40)
E	435		10	MISC	29,34,35,UPM2(28)	Polyoxyethylene sorbitan monostearate (Polysorbate 60)
E	436		10	MISC	29,34,35,UPM2(28)	Polyoxyethylene sorbitan tristearate (Polysorbate 65)
E	452	iv	70	MISC	27,UPM2(24)	Polyphosphate, calcium -
E	452	iii	70	MISC	27,UPM2(24)	Polyphosphate, calcium sodium -
E	452	ii	70	MISC	27,UPM2(24)	Polyphosphate, potassium -
E	452	i	70	MISC	27,UPM2(24)	Polyphosphate, sodium -
E	452	iii	70	MISC	27,UPM2(24)	Polyphosphate, sodium calcium -
E	900		1.5	MISC	32,UPM2(26,28)	Polysiloxane, dimethyl -
E	432		10	MISC	29,34,35,UPM2(28)	Polysorbate 20, polyoxyethylene sorbitan monolaurate
E	434		10	MISC	29,34,35,UPM2(28)	Polysorbate 40, polyoxyethylene sorbitan monopalmitate (Polysorbate 40)
E	435		10	MISC	29,34,35,UPM2(28)	Polysorbate 60, polyoxyethylene sorbitan monostearate
E	436		10	MISC	29,34,35,UPM2(28)	Polysorbate 65, polyoxyethylene sorbitan tristearate
E	433		10	MISC	29,34,35,UPM2(28)	Polysorbate 80, polyoxyethylene sorbitan monooleate
E	1202		A	MISC	33,36	Polyvinylpolypyrrolidone
E	1201		A	MISC	33,36	Polyvinylpyrrolidone

Appendix Table 2 (Continued)

E #			ADI SCF mg/kg	DIR	Directives PAGES	NAME / Alphabetically sorted - TABLE 2 -
UPS1(18)	= on page 18 of the 1st. adopted update of the sweeteners Directive					
UPM2(36)	= on page 36 of the 2nd. adopted update of the "*miscellaneous*" Directive					
E	124		4	COLR	22,24,28	Ponceau 4R, cochineal red A
E	261		NS	MISC	6,13,14,15,39	Potassium, acetate
E	357		5	MISC	27	Potassium, adipate
E	402		NS	MISC	7,11,12,34,40,UPM2(20)	Potassium, alginate
E	522		1	MISC	31	Potassium, aluminium - sulphate
E	555		1	MISC	31	Potassium, aluminium silicate
E	212		5	MISC	16,17,18,19,UPM2(21,22)	Potassium, benzoate
E	501	i	NS	MISC	8,10,35,39	Potassium, carbonate
E	508		NS	MISC	8,12,35	Potassium, chloride
E	332	i	NS	MISC	7,10,11,12,13,14,15,37,38,39,UPM2(20,29)	Potassium, citrate, mono- citrate
E	332	ii	NS	MISC	7,10,11,12,13,14,15,37,38,39,UPM2(20,29)	Potassium, citrate, tri- citrate
E	450	iv	70	MISC	27,UPM2(24)	Potassium, di- diphosphate
E	628		NS	MISC	32	Potassium, di- guanilate
E	632		NS	MISC	32	Potassium, di- inosinate
E	340	ii	70	MISC	26,39,UPM2(24,29)	Potassium, di- phosphate
E	336	ii	30	MISC	7,11,13,14,40	Potassium, di- tartrate
E	536		0.025	MISC	31	Potassium, ferrocyanide
E	577		NS	MISC	9,36	Potassium, gluconate
E	501	ii	NS	MISC	8,10,11,35,39	Potassium, hydrogen carbonate (Potassium, bicarbonate)
E	515	ii	NS	MISC	8,35	Potassium, hydrogen sulphate
E	228		0.7	MISC	19,20,21,22,UPM2(22,23)	Potassium, hydrogen sulphite
E	525		NS	MISC	8,10,39	Potassium, hydroxide
E	632		NS	MISC	32	Potassium, inosinate, di- inosinate
E	326		NS	MISC	6,12,13,14,15,39	Potassium, lactate
E	351		NS	MISC	7	Potassium, malate
E	224		0.7	MISC	19,20,21,22,UPM2(22,23)	Potassium, metabisulphite
E	332	i	NS	MISC	7,10,11,12,13,14,15,37,38,39,UPM2(20,29)	Potassium, mono- citrate
E	622		NS	MISC	32	Potassium, mono- glutamate
E	340	i	70	MISC	26,39,UPM2(24,29)	Potassium, mono- phosphate
E	336	i	30	MISC	7,11,13,14,40	Potassium, mono- tartrate
E	252		5	MISC	24,UPM2(23)	Potassium, nitrate
E	249		0.1	MISC	24	Potassium, nitrite
E	451	ii	70	MISC	27,UPM2(24)	Potassium, penta- triphosphate
E	340	ii	70	MISC	26,39,UPM2(24,29)	Potassium, phosphate, di- phosphate
E	340	i	70	MISC	26,39,UPM2(24,29)	Potassium, phosphate, mono- phosphate
E	451	ii	70	MISC	27,UPM2(24)	Potassium, phosphate, penta- triphosphate
E	340	iii	70	MISC	26,39,UPM2(24,29)	Potassium, phosphate, tri- phosphate
E	452	ii	70	MISC	27,UPM2(24)	Potassium, polyphosphate
E	283		NS	MISC	24,UPM2(23)	Potassium, propionate

Appendix Table 2 (Continued)

E #			ADI SCF mg/kg	DIR	Directives PAGES	NAME / Alphabetically sorted - TABLE 2 -
UPS1(18)	= on page 18 of the 1st. adopted update of the sweeteners Directive					
UPM2(36)	= on page 36 of the 2nd. adopted update of the "*miscellaneous*" Directive					
E	954		5	SWET	10,11,UPS1(19)	Potassium, saccharin and natrium - and calcium salts
E	470	a	NS	MISC	8,35,UPM2(28)	Potassium, sodium - and calcium salts of fatty acids
E	337		30	MISC	7,11,13,14	Potassium, sodium - tartrate
E	202		25	MISC	16,17,18,19,UPM2(21,22)	Potassium, sorbate
E	515	i	NS	MISC	8,35	Potassium, sulphate
E	450	v	70	MISC	27,UPM2(24)	Potassium, tetra- diphosphate
E	332	ii	NS	MISC	7,10,11,12,13,14,15,37,38,39,UPM2(20,29)	Potassium, tri- citrate
E	340	iii	70	MISC	26,39,UPM2(24,29)	Potassium, tri- phosphate
E	451	ii	70	MISC	27,UPM2(24)	Potassium, triphosphate, penta- triphosphate
E	460	ii	NS	MISC	8,12,34,UPM2(21)	Powdered cellulose
Solvent			25	MISC	34	Propan-1,2-diol (propylene glycol)
E	477		25	MISC	30	Propane 1,2-diol esters of fatty acids
E	405		25	MISC	28,34,UPM2(25,30)	Propane-1,2-diol alginate
E	282		NS	MISC	24,UPM2(23)	Propionate, calcium -
E	283		NS	MISC	24,UPM2(23)	Propionate, potassium -
E	281		NS	MISC	24,UPM2(23)	Propionate, sodium -
E	280		NS	MISC	24,UPM2(23)	Propionic acid
E	310		0.5	MISC	25,UPM2(23)	Propyl gallate
E	216		10	MISC	16,17,18,19	Propyl p-hydroxybenzoate
E	1442		A	MISC	9,12,35	Propyl, hydroxy- distarch phosphate
E	1440		A	MISC	9,12,35	Propyl, hydroxy - starch
E	464		NS	MISC	8,12,35	Propyl, hydroxy- methyl cellulose
E	217		10	MISC	16,17,18,19	Propyl, sodium - p-hydroxybenzoate
Solvent			25	MISC	34	Propylene, propan-1,2-diol (- glycol)
E	1201		A	MISC	33,36	Pyrrolidone, polyvinyl-
E	1202		A	MISC	33,36	Pyrrolidone, polyvinylpoly-
E	999		5	MISC	33,UPM2(27)	Quillaia extract
E	104		10	COLR	22,24,28	Quinoline yellow
E	128		0.1	COLR	25	Red 2G
E	129		7	COLR	22,24,28	Red AC, allura -
E	162		NS	COLR	23,24,27	Red, beetroot - ,betanin
E	124		4	COLR	22,24,28	Red, cochineal - A, ponceau 4R
E	407	a	75	MISC	UPM1(4)	Refined, alternatively - carrageenan
E	101	i	NS	COLR	22,23,24,27	Riboflavine
E	101	ii	NS	COLR	22,23,24,27	Riboflavine-5'-phosphate
E	634		NS	MISC	32	Ribonucleotides, calcium 5'-ribo...
E	635		NS	MISC	32	Ribonucleotides, disodium 5'-ribonucle...
E	635		NS	MISC	32	Ribonucleotides, sodium, disodium 5'-ribonucle...
E	476		7.5	MISC	30,UPM2(26)	Ricin, polyglycerol poly -oleate

Appendix Table 2 (Continued)

E #			ADI SCF mg/kg	DIR	Directives PAGES	NAME / Alphabetically sorted - TABLE 2 -
UPS1(18)	= on page 18 of the 1st. adopted update of the sweeteners Directive					
UPM2(36)	= on page 36 of the 2nd. adopted update of the "*miscellaneous*" Directive					
E	445		12.5	MISC	29,UPM2(26)	Rosin, glycerol esters of wood -
E	954		5	SWET	10,11,UPS1(19)	Saccharin and sodium potassium and calcium salts
E	470	b	NS	MISC	8,36	Salts, magnesium - of fatty acids
E	470	a	NS	MISC	8,35,UPM2(28)	Salts, sodium potassium and calcium - of fatty acids
E	500	iii	NS	MISC	8,39,UPM2(21)	Sesquicarbonate, sodium -
E	904		A	MISC	32,UPM2(26)	Shellac
E	559		1	MISC	31,36,UPM2(26)	Silicate, aluminium - (kaolin)
E	552		NS	MISC	31,36	Silicate, calcium -
E	556		1	MISC	31	Silicate, calcium aluminium -
E	553	a i	NS	MISC	31	Silicate, magnesium -
E	553	a ii	NS	MISC	31	Silicate, magnesium tri-
E	555		1	MISC	31	Silicate, potassium aluminium -
E	554		1	MISC	31	Silicate, sodium aluminium -
E	551		NS	MISC	31,36,UPM2(26,28)	Silicon dioxide
E	174		A	COLR	25	Silver
E	466		NS	MISC	8,12,35,UPM2(20,30)	Sodium, - carboxy methyl cellulose + carboxyy methyl cellulose
E	262	i	NS	MISC	6,13,14,15,39	Sodium, acetate
E	262		NS	MISC	6,13,14,15,39	Sodium, acetates
E	356		5	MISC	27	Sodium, adipate
E	401		NS	MISC	7,11,12,34,40,UPM2(20,30)	Sodium, alginate
E	521		1	MISC	31	Sodium, aluminium - sulphate
E	541		1	MISC	31,40	Sodium, aluminium phosphate, acidic
E	554		1	MISC	31	Sodium, aluminium silicate
E	301		A	MISC	6,11,12,13,14,15,39,UPM2(21,28)	Sodium, ascorbate
E	211		5	MISC	16,17,18,19,UPM2(21,22)	Sodium, benzoate
E	385		2.5	MISC	28,UPM2(25)	Sodium, calcium di- ethylene diamine tetra-acetate (CaNa2EDTA)
E	452	iii	70	MISC	27,UPM2(24)	Sodium, calcium polyphosphate
E	500	i	NS	MISC	8,10,39,UPM2(21)	Sodium, carbonate
E	468		NS	MISC	UPM2(25,28)	Sodium, carboxy methyl cellulose, cross linked -
E	468		NS	MISC	UPM2(25,28)	Sodium, carboxy methyl cellulose, cross linked - - - -
E	331	ii	NS	MISC	7,10,11,12,13,14,15,37,38,39,UPM2(20,29)	Sodium, citrate, di- citrate
E	331	i	NS	MISC	7,10,11,12,13,14,15,37,38,39,UPM2(20,29)	Sodium, citrate, mono- citrate
E	331	iii	NS	MISC	7,10,11,12,13,14,15,37,38,39,UPM2(20,29)	Sodium, citrate, tri- citrate
E	952		7	SWET	9,10	Sodium, cyclamic acid and its - and calcium salts
E	635		NS	MISC	32	Sodium, di- 5'-ribonucleotides
E	450	a	70	MISC	40	Sodium, di- diphosphate
E	450	i	70	MISC	27,UPM2(24)	Sodium, di- diphosphate
E	627		NS	MISC	32	Sodium, di- guanilate
E	631		NS	MISC	32	Sodium, di- inosinate

Appendix Table 2 (Continued)

E #			ADI SCF mg/kg	DIR	Directives PAGES	NAME / Alphabetically sorted - TABLE 2 -
UPS1(18)	= on page 18 of the 1st. adopted update of the sweeteners Directive					
UPM2(36)	= on page 36 of the 2nd. adopted update of the "*miscellaneous*" Directive					
E	339	ii	70	MISC	26,39,UPM2(24,29)	Sodium, di- phosphate
E	335	ii	30	MISC	7,11,13,14,40	Sodium, di- tartrate
E	316		6	MISC	25,UPM2(23)	Sodium, erythorbate
E	215		10	MISC	16,17,18,19	Sodium, ethyl p-hydroxybenzoate
E	535		MISC	MISC	31	Sodium, ferrocyanide -
E	576		NS	MISC	9	Sodium, gluconate
E	640		NS	MISC	9,36	Sodium, glycine and its - salt
E	627		NS	MISC	32	Sodium, guanylate, di- guanilate
E	262	ii	NS	MISC	6,13,14,15,39	Sodium, hydrogen acetate (Sodium, diacetate)
E	500	ii	NS	MISC	8,10,11,39,UPM2(21)	Sodium, hydrogen carbonate (Sodium, bicarbonate)
E	350	ii	NS	MISC	7,11	Sodium, hydrogen malate
E	514	ii	NS	MISC	8,35	Sodium, hydrogen sulfate
E	222		0.7	MISC	19,20,21,22,UPM2(22,23)	Sodium, hydrogen sulphite
E	524		NS	MISC	8,10,11,39	Sodium, hydroxide
E	631		NS	MISC	32	Sodium, inosinate, di- inosinate
E	325		NS	MISC	6,12,13,14,15,39	Sodium, lactate
E	350	i	NS	MISC	7,11	Sodium, malate
E	223		0.7	MISC	19,20,21,22,UPM2(22,23)	Sodium, metabisulphite
E	219		10	MISC	16,17,18,19	Sodium, methyl p-hydroxybenzoate
E	331	i	NS	MISC	7,10,11,12,13,14,15,37,38,39,UPM2(20,29)	Sodium, mono- citrate
E	621		NS	MISC	32	Sodium, mono- glutamate
E	339	i	70	MISC	26,39,UPM2(24,29)	Sodium, mono- phosphate
E	335	i	30	MISC	7,11,13,14,40	Sodium, mono- tartrate
E	251		5	MISC	24,UPM2(23)	Sodium, nitrate
E	250		0.1	MISC	24	Sodium, nitrite
E	232		NE	MISC	23	Sodium, orthophenyl phenol
E	451	i	70	MISC	27,UPM2(24)	Sodium, penta- triphosphate
E	339	i	70	MISC	26,39,UPM2(24,29)	Sodium, phosphate, - mono- phosphate
E	339	ii	70	MISC	26,39,UPM2(24,29)	Sodium, phosphate, di- phosphate
E	451	i	70	MISC	27,UPM2(24)	Sodium, phosphate, penta- triphosphate
E	339	iii	70	MISC	26,39,UPM2(24,29)	Sodium, phosphate, tri- phosphate
E	452	i	70	MISC	27,UPM2(24)	Sodium, polyphosphate
E	470	a	NS	MISC	8,35,UPM2(28)	Sodium, potassium and calcium salts of fatty acids
E	337		30	MISC	7,11,13,14	Sodium, potassium tartrate
E	281		NS	MISC	24,UPM2(23)	Sodium, propionate
E	217		10	MISC	16,17,18,19	Sodium, propyl p-hydroxybenzoate
E	954		5	SWET	10,11,UPS1(19)	Sodium, saccharin and - potassium and calcium salts
E	500	iii	NS	MISC	8,39,UPM2(21)	Sodium, sesquicarbonate
E	1450		A	MISC	9,12,35,40,UPM2(30)	Sodium, starch - octenyl succinate

Appendix Table 2 (Continued)

E #			ADI SCF mg/kg	DIR	Directives PAGES	NAME / Alphabetically sorted - TABLE 2 -
UPS1(18)	= on page 18 of the 1st. adopted update of the sweeteners Directive					
UPM2(36)	= on page 36 of the 2nd. adopted update of the "*miscellaneous*" Directive					
E	481		20	MISC	30	Sodium, stearoyl-2-lactylate
E	514	i	NS	MISC	8,35	Sodium, sulphate
E	221		0.7	MISC	19,20,21,22,UPM2(22,23)	Sodium, sulphite
E	335	ii	30	MISC	7,11,13,14,40	Sodium, tartrate, di- tartrate
E	335	i	30	MISC	7,11,13,14,40	Sodium, tartrate, mono- tartrate
E	450	iii	70	MISC	27,UPM2(24)	Sodium, tetra- diphosphate
E	285		A	MISC	23	Sodium, tetraborate (borax)
E	331	iii	NS	MISC	7,10,11,12,13,14,15,37,38,39,UPM2(20,29)	Sodium, tri- citrate
E	450	ii	70	MISC	27,UPM2(24)	Sodium, tri- diphosphate
E	339	iii	70	MISC	26,39,UPM2(24,29)	Sodium, tri- phosphate
E	451	i	70	MISC	27,UPM2(24)	Sodium, triphosphate, penta- triphosphate
E	203		25	MISC	16,17,18,19,UPM2(21,22)	Sorbate, calcium
E	202		25	MISC	16,17,18,19,UPM2(21,22)	Sorbate, potassium -
E	200		25	MISC	16,17,18,19,UPM2(21,22)	Sorbic acid
E	493		5	MISC	31,35,UPM2(28)	Sorbitan, monolaurate
E	494		5	MISC	31,35,UPM2(28)	Sorbitan, monooleate
E	495		25	MISC	31,35,UPM2(28)	Sorbitan, monopalmitate
E	491		25	MISC	31,35,UPM2(28)	Sorbitan, monostearate
E	432		10	MISC	29,34,35,UPM2(28)	Sorbitan, polyoxyethylene - monolaurate (Polysorbate 20)
E	433		10	MISC	29,34,35,UPM2(28)	Sorbitan, polyoxyethylene - monooleate (Polysorbate 80)
E	434		10	MISC	29,34,35,UPM2(28)	Sorbitan, polyoxyethylene - monopalmitate (Polysorbate 40)
E	435		10	MISC	29,34,35,UPM2(28)	Sorbitan, polyoxyethylene - monostearate (Polysorbate 60)
E	436		10	MISC	29,34,35,UPM2(28)	Sorbitan, polyoxyethylene - tristearate (Polysorbate 65)
E	492		25	MISC	31,35,UPM2(28)	Sorbitan, tristearate
E	420	i	A	MISC	28,34	Sorbitol
E	420	ii	A	MISC	28,34	Sorbitol syrup
E	479	b	25	MISC	30	Soya, thermally oxidized - bean oil interacted with mono- and diglycerides of fatty acids
E	512		A	MISC	31	Stannous chloride
E	1420		A	MISC	9,12,35,40	Starch, acetylated -
E	1451		A	MISC	UPM2(20,28,30)	Starch, acetylated oxidized -
E	1451		A	MISC	UPM2(20,28,30)	Starch, acetylated oxidized -
E	1412		A	MISC	9,12,35,40	Starch, di- phosphate
E	1440		A	MISC	9,12,35	Starch, hydroxy propyl
E	1442		A	MISC	9,12,35	Starch, hydroxypropyl di- phosphate
E	1410		A	MISC	9,12,35,40	Starch, mono- phosphate
E	1404		A	MISC	9,12,35,40	Starch, oxidized -
E	1413		A	MISC	9,12,35,40	Starch, phosphated di- phosphate
E	1450		A	MISC	9,12,35,40,UPM2(30)	Starch, sodium octenyl succinate
E	304	ii	A	MISC	6,11,13,14,39,UPM2(20, ?29?)	Stearate, ascorbyl -

Appendix Table 2 (Continued)

E #			ADI scf mg/kg	DIR	Directives PAGES	NAME / Alphabetically sorted - TABLE 2 -
UPS1(18) = on page 18 of the 1st. adopted update of the sweeteners Directive						
UPM2(36) = on page 36 of the 2nd. adopted update of the *"miscellaneous"* Directive						
E	431		NA	MISC	27	Stearate, polyoxyethylene (40) stearate
E	435		10	MISC	29,34,35,UPM2(28)	Stearate, polyoxyethylene sorbitan mono- (Polysorbate 60)
E	436		10	MISC	29,34,35,UPM2(28)	Stearate, polyoxyethylene sorbitan tri- (Polysorbate 65)
E	491		25	MISC	31,35,UPM2(28)	Stearate, sorbitan mono-
E	492		25	MISC	31,35,UPM2(28)	Stearate, sorbitan tri-
E	482		20	MISC	30	Stearoyl-2-lactylate, calcium -
E	481		20	MISC	30	Stearoyl-2-lactylate, sodium
E	483		20	MISC	30	Stearyl tartrate
E	1450		A	MISC	9,12,35,40,UPM2(30)	Succinate, starch sodium octenyl -
E	363		NS	MISC	28	Succinic acid
E	474		20	MISC	29,UPM2(26)	Sucroglycerides
E	444		10	MISC	29	Sucrose acetate isobutyrate
E	473		20	MISC	29,35,UPM2(26,28,29)	Sucrose esters of fatty acids
E	520		0.3	MISC	31	Sulphate, aluminium -
E	523		1	MISC	31	Sulphate, aluminium ammonium -
E	522		1	MISC	31	Sulphate, aluminium potassium -
E	521		1	MISC	31	Sulphate, aluminium sodium -
E	517		?	MISC	35	Sulphate, ammonium -
E	516		NS	MISC	8,35	Sulphate, calcium
E	515	i	NS	MISC	8,35	Sulphate, potassium -
E	515	ii	NS	MISC	8,35	Sulphate, potassium hydrogen -
E	514	i	NS	MISC	8,35	Sulphate, sodium
E	514	ii	NS	MISC	8,35	Sulphate, sodium hydrogen -
E	150	d	200	COLR	21,23,24,27	Sulphite, ammonia caramel
E	226		0.7	MISC	19,20,21,22,UPM2(22,23)	Sulphite, calcium
E	227		0.7	MISC	19,20,21,22,UPM2(22,23)	Sulphite, calcium hydrogen -
E	150	b	200	COLR	21,23,24,27	Sulphite, caustic - caramel
E	228		0.7	MISC	19,20,21,22,UPM2(22,23)	Sulphite, potassium hydrogen -
E	224		0.7	MISC	19,20,21,22,UPM2(22,23)	Sulphite, potassium metabi-
E	221		0.7	MISC	19,20,21,22,UPM2(22,23)	Sulphite, sodium
E	222		0.7	MISC	19,20,21,22,UPM2(22,23)	Sulphite, sodium hydrogen -
E	223		0.7	MISC	19,20,21,22,UPM2(22,23)	Sulphite, sodium metabi-
E	220		0.7	MISC	19,20,21,22,UPM2(22,23)	Sulphur dioxide
E	513		NS	MISC	8	Sulphuric acid
E	110		2.5	COLR	22,24,28	Sunset yellow FCF, orange yellow S
E	965		A	MISC	28	Syrup, maltitol -
E	965		A	SWET	6	Syrup, maltitol -
E	420	ii	A	MISC	28,34	Syrup, sorbitol
E	553	b	NS	MISC	31,36	Talc

Appendix Table 2 (Continued)

E #			ADI SCF mg/kg	DIR	Directives PAGES	NAME / Alphabetically sorted - TABLE 2 -
UPS1(18)	= on page 18 of the 1st. adopted update of the sweeteners Directive					
UPM2(36)	= on page 36 of the 2nd. adopted update of the "*miscellaneous*" Directive					
E	417		NS	MISC	7	Tara gum
E	334		30	MISC	7,11,13,14,15,40	Tartaric, acid [L(+)-])
E	472	d	NS	MISC	8,14	Tartaric, acid esters of mono- and diglycerides of fatty acids
E	353		A	MISC	27	Tartaric, meta- acid
E	472	f	NS	MISC	8,14	Tartaric, mixed acetic and - acid esters of mono- and diglycerides of fatty acids
E	472	e	25	MISC	8,14,35	Tartaric, mono- and diacetyl - acid esters of mono- and diglycerides of fatty acids
E	354		30	MISC	7,40	Tartrate, calcium -
E	336	ii	30	MISC	7,11,13,14,40	Tartrate, dipotassium -
E	335	ii	30	MISC	7,11,13,14,40	Tartrate, disodium -
E	336	i	30	MISC	7,11,13,14,40	Tartrate, monopotassium -
E	335	i	30	MISC	7,11,13,14,40	Tartrate, monosodium -
E	336	ii	30	MISC	7,11,13,14,40	Tartrate, potassium, dipotassium -
E	336	i	30	MISC	7,11,13,14,40	Tartrate, potassium, monopotassium -
E	337		30	MISC	7,11,13,14	Tartrate, sodium potassium -
E	335	ii	30	MISC	7,11,13,14,40	Tartrate, sodium, disodium -
E	335	i	30	MISC	7,11,13,14,40	Tartrate, sodium, monosodium -
E	483		20	MISC	30	Tartrate, stearyl -
E	102		7.5	COLR	22,24,28	Tartrazine
E	385		2.5	MISC	28,UPM2(25)	Tetra-acetate, calcium disodium ethylene diamine - (CaNa2EDTA)
E	285		A	MISC	23	Tetraborate, sodium - (borax)
E	239		A	MISC	23	Tetramine, hexamethylene -
E	450	v	70	MISC	27,UPM2(24)	Tetrapotassium diphosphate
E	450	iii	70	MISC	27,UPM2(24)	Tetrasodium diphosphate
E	957		A	MISC	33,UPM2(27)	Thaumatin
E	957		A	SWET	11,UPS1(19)	Thaumatin
E	479	b	25	MISC	30	Thermally oxidized soya bean oil interacted with mono- and diglycerides of fatty acids
E	233		0.3	MISC	23,UPM2(23)	Thiabendazole
E	171		NS	COLR	27	Titanium dioxide
E	307		A	MISC	6,13,37,38,39,UPM2(20)	Tocopherol, alpha- -
E	309		A	MISC	6,13,37,38,39,UPM2(20)	Tocopherol, delta- -
E	308		A	MISC	6,13,37,38,39,UPM2(20)	Tocopherol, gamma- -
E	306		A	MISC	6,13,37,38,39,UPM2(20)	Tocopherol-rich extract
E	321		0.05	MISC	25	Toluene, butylated hydroxy- (Bht BHT)
E	413		NS	MISC	7,34	Tragacanth
E	1518		A	MISC	36,UPM2(27)	Triacetate, glyceryl (triacetin)
E	1518		A	MISC	36,UPM2(27)	Triacetin, glyceryl triacetate
E	380		NS	MISC	7	Triammonium, citrate
E	333	iii	NS	MISC	7,11,12,13,14,35,39,UPM2(20)	Tricalcium, citrate
E	341	iii	70	MISC	26,35,39,UPM2(24)	Tricalcium, phosphate

Appendix Table 2 (Continued)

E #			ADI SCF mg/kg	DIR	Directives PAGES	NAME / Alphabetically sorted - TABLE 2 -
UPS1(18)			= on page 18 of the 1st. adopted update of the sweeteners Directive			
UPM2(36)			= on page 36 of the 2nd. adopted update of the *"miscellaneous"* Directive			
E	1505		20	MISC	33,36	Triethyl citrate
E	451	ii	70	MISC	27,UPM2(24)	Triphosphate, pentapotassium -
E	451	i	70	MISC	27,UPM2(24)	Triphosphate, pentasodium
E	340	iii	70	MISC	26,39,UPM2(24,29)	Tripotassium phosphate
E	332	ii	NS	MISC	7,10,11,12,13,14,15,37,38,39,UPM2(20,29)	Tripotassium, citrate
E	553	a ii	NS	MISC	31	Trisilicate, magnesium -
E	331	iii	NS	MISC	7,10,11,12,13,14,15,37,38,39,UPM2(20,29)	Trisodium, citrate
E	450	ii	70	MISC	27,UPM2(24)	Trisodium, diphosphate
E	339	iii	70	MISC	26,39,UPM2(24,29)	Trisodium, phosphate
E	436		10	MISC	29,34,35,UPM2(28)	Tristearate, polyoxyethylene sorbitan - (Polysorbate 65)
E	492		25	MISC	31,35,UPM2(28)	Tristearate, sorbitan
E	153		NS	COLR	21,27	Vegetable carbon
E	1202		A	MISC	33,36	Vinyl, poly - polypyrrolidone
E	1201		A	MISC	33,36	Vinyl, poly - pyrrolidone
E	902		A	MISC	32,UPM2(26)	Wax, candellila
E	903		A	MISC	32,UPM2(26)	Wax, carnauba -
E	905		20	MISC	UPM2(27)	Wax, microcristalline -
E	905		20	MISC	UPM2(27)	Wax, microcristalline -
E	914		A	MISC	33,UPM2(27)	Wax, oxidized polyethylene wax
E	901		A	MISC	32,36,UPM2(26)	White and yellow beeswax
E	445		12.5	MISC	29,UPM2(26)	Wood, glycerol esters of - rosin
E	415		NS	MISC	7,11,12,34,40,UPM2(30)	Xanthan gum
E	967		A	MISC	28,34	Xylitol
E	967		A	SWET	6	Xylitol
E	901		A	MISC	32,36,UPM2(26)	Yellow and white beeswax
E	104		10	COLR	22,24,28	Yellow, quinoline -
E	110		2.5	COLR	22,24,28	Yellow, sunset - FCF, orange - S

B. Additional Sources of Information

1. Full Texts

A CD with the complete texts of the EU additives legislation with the full authorization tables, consolidated texts, and lists of EU food additives is available from the author of this chapter. The layout of the legal texts are represented completely identically as they (bioresco@pandora.be) appear in the *Official Journal*. The format of the texts is pdf (Adobe). Appendix Tables 1 and 2 in EXCEL-97(Microsoft) are also on the same CD.

2. Interesting Internet Addresses

Note that these internet addresses are prone to frequent changes. The URLs indicated below were functional on April 12, 1999.

European institutions with a link to the general search engine:
 http://europa.eu.int/index-en.htm
European Council site:
 http://ue.eu.int/en/summ.htm
European Parliament site:
 http://www.europarl.eu.int/home/default_en.htm
European Commission with a link to the search engine:
 http://www.europa.eu.int/comm/index_en.htm
European Court address:
 http://curia.eu.int/en/index.htm
ECOSOC site:
 http://www.esc.eu.int/
Official Journal of the European Communities:
 http://europa.eu.int/eur-lex/en/oj/index.html
Celex and EU legislative database address:
 http://europa.eu.int/celex/htm/celex_en.htm
 (Paid contract + keyword + PIN required for consultation.)
Scientific Committee for Food of the EU Commission (SCF):
 http://europa.eu.int/comm/food/fs/sc/scf/outcome_en.html

3. Scientific Committee for Food Database

A detailed bibliographic database (in ACCESS-97 and also in EXCEL-97/Microsoft) on the contents of all the reports and of the minutes of the meetings of the SCF can be purchased on a CD from or consulted through the author of this chapter: bioresco@pandora.be

ACKNOWLEDGMENT

Acknowledgment is given to Mrs. S. M. Blackman and Mrs. A. Mannaerts for their assiduity in the preparation of this manuscript.

REFERENCES

1. Council Directive on the approximation of the rules of the Member States concerning the colouring matters authorized in foodstuffs intended for human consumption. (Adopted 23 Oc-

tober 1962.) *Official Journal of the European Communities No. 115* of 11 November 1962: 2645.

2. Council Directive of 18 June 1974 on the approximation of the laws of the Member States relating to emulsifiers, stabilizers, thickeners and gelling agents for use in foodstuffs (74/329/EEC). *Official Journal of the European Communities No. L189* of 12 July 1974:1.
3. Council Directive of 24 July 1973 on the approximation of the laws of the Member States relating to cocoa and chocolate products intended for human consumption (74/241/EEC). *Official Journal of the European Communities No. L228* of 18 August 1973:23.
4. Council Directive of 21 December 1988 on the approximation of the laws of the Member States concerning food additives authorized for use in foodstuffs intended for human consumption (89/107/EEC). *Official Journal of the European Communities No. L40* of 11 February 1989:27.
5. Directive 98/72/EC of the European Parliament and of the Council of 15 October 1998 amending Directive 95/2/EC on food additives other than colours and sweeteners. *Official Journal of the European Communities No. L295* of 04 November 1998:18.
6. Council Directive 88/344/EEC of 13 June 1988 on the approximation of the laws of the Member States on extraction solvents used in the production of foodstuffs and food ingredients. *Official Journal of the European Communities No. L157* of 24 June 1988:28.
7. Council Directive 92/11/EEC of 17 December 1992 amending for the first time Directive 88/344/EEC on the approximation of the laws of the Member States on extraction solvents used in the production of foodstuffs and food ingredients. *Official Journal of the European Communities No. L409* of 31 December 1992:31.
8. Directive 94/52/EC of the European Parliament and of the Council of 7 December 1994 amending for the second time Directive 88/344/EEC on the approximation of the laws of the Member States on extraction solvents used in the production of foodstuffs and food ingredients. *Official Journal of the European Communities No. L331* of 21 December 1994:10.
9. Regulation (EC) No 258/97 of the European Parliament and of the Council of 27 January 1997 concerning novel foods and novel food ingredients. *Official Journal of the European Communities No. L43* of 14 February 1997:1.
10. European Parliament and Council Directive 95/2/EC of 20 February 1995 on food additives other than colours and sweeteners. *Official Journal of the European Communities No. L61* of 18 March 1995:1.
11. Council Directive 88/388/EEC of 22 June 1988 on the approximation of the laws of the Member States relating to flavourings for use in foodstuffs and to source materials for their production. *Official Journal of the European Communities No. L184* of 15 July 1988:61.
12. Council Decision 88/389/EEC of 22 June 1988 on the establishment, by the Commission, of an inventory of the source materials and substances used in the preparation of flavourings. *Official Journal of the European Communities No. L184* of 15 July 1988:67.
13. Commission Directive 91/71/EEC of 16 January 1991 completing Council Directive 88/388/EEC on the approximation of the laws of the Member States relating to flavourings for use in foodstuffs and to source materials for their production. *Official Journal of the European Communities No. L42* of 15 February 1991:25.
14. Commission Directive 91/72/EEC of 16 January 1991 amending Council Directive 79/112/EEC in respect of the designation of flavourings in the list of ingredients on the labels of foodstuffs. *Official Journal of the European Communities No. L42* of 15 February 1991:27.
15. Regulation (EC) No 2232/96 of the European Parliament and of the Council of 28 October 1996 laying down a Community procedure for flavouring substances used or intended for use in or on foodstuffs. *Official Journal of the European Communities No. L299* of 23 November 1996:1.
16. Proposal for a Commission Directive expanding Directive 88/388/EEC on the approximation of the laws of the Member States relating to flavourings for use in foodstuffs and to source

materials for their production. Commission document III/3490/91-EN-Rev3/CG/pb dated 16 March 1992.
17. Draft Commission Directive 96/ . . . /EC laying down the list of food additives for use in flavourings. European Commission document III/5798/96 (1996).
18. Council Directive of 18 December 1978 on the approximation of the laws of the Member States relating to the labelling, presentation and advertising of foodstuffs (79/112/EEC). *Official Journal of the European Communities No. L33* of 8 February 1979:1.
19. Commission Directive 95/31/EC of 5 July 1995 laying down specific criteria of purity concerning sweeteners for use in foodstuffs. *Official Journal of the European Communities No. L178* of 28 July 1995:1.
20. Commission Directive 95/45/EC of 26 July 1995 laying down specific purity criteria concerning colours for use in foodstuffs. *Official Journal of the European Communities No. L226* of 22 September 1995:1.
21. Commission Directive 96/77/EC of 2 December 1996 laying down specific purity criteria on food additives other than colours and sweeteners. *Official Journal of the European Communities No. L339* of 30 December 1996:1.
22. Commission Directive 98/86/EC of 11 November 1998 amending Commission Directive 96/77/EC laying down specific purity criteria on food additives other than colours and sweeteners. *Official Journal of the European Communities No. L334* of 09 December 1998:1.
23. Council Directive of 20 December 1985 concerning the introduction of Community methods of sampling and analysis for the monitoring of foodstuffs intended for human consumption. (85/591/EEC) *Official Journal of the European Communities No. L372* of 31 December 1985: 50.
24. Directive 96/85/EC of the European Parliament and of the Council of 19 December 1996 amending Directive 95/2/EC on food additives other than colours and sweeteners. *Official Journal of the European Communities No. L86* of 28 March 1997:4.
25. Directive 96/83/EC of the European Parliament and of the Council of 19 December 1996 amending Directive 94/35/EC on sweeteners for use in foodstuffs. *Official Journal of the European Communities No. L48* of 19 February 1997:16.
26. European Parliament and Council Directive 94/34/EC of 30 June 1994 amending Directive 89/107/EC on the approximation of the laws of Member States concerning food additives authorized for use in foodstuffs intended for human consumption. *Official Journal of the European Communities No. L237* of 10 September 1994:1.
27. European Parliament and Council Directive 94/35/EC of 30 June 1994 on sweeteners for use in foodstuffs. *Official Journal of the European Communities No. L237* of 10 September 1994:3.
28. European Parliament and Council Directive 94/36/EC of 30 June 1994 on colours for use in foodstuffs. *Official Journal of the European Communities No. L237* of 10 September 1994: 13.
29. Agreement between the European Community and Australia on trade in wine—Protocol—Exchange of letters. 26 January 1994. *Official Journal of the European Communities No. L086* of 31 March 1994:3.
30. Presentation of an application for assessment of a food additive prior to its authorization. Obtainable in English, German and French from the Office for Official Publications of the European Communities, Luxembourg, or from its distribution centers in the Member States. ISBN 92-826-0315-8. (The catalog number of the English edition is CB-57-89-370-EN-C.)
31. Report on adverse reactions to food and food ingredients (expressed on 22 September 1995). Reports of the Scientific Committee for Food 37th Series. Directorate-General for Industry, European Commission, 1997.
32. Draft proposal for a European Parliament and Council Directive amending Directive 79/112/EEC on the approximation of the laws of the Member States relating to the labelling, presentation and advertising of foodstuffs. Directorate-General for Industry, Brussels 16 January 1998, III/5909/97.

33. Draft recommendations for the labelling of foods that can cause hypersensitivity (draft amendment to the general standard for the labelling of prepackaged foods). Codex Alimentarius Commission, 23rd Session, Rome, 28 June–3 July 1999. Report of the 26th session of the Codex Committee on Food Labelling, Ottawa, 26–29 May 1998, ALINORM 99/22. APPENDIX III.
34. Decision No 292/97/EC of the European Parliament and of the Council of 19 December 1996 on the maintenance of national laws prohibiting the use of certain additives in the production of certain specific foodstuffs. *Official Journal of the European Communities No. L48* of 19 February 1997:13.
35. Commission Directive 91/321/EEC of 14 May 1991 on infant formulae and follow-on formulae. *Official Journal of the European Communities No. L175* of 04 July 1991:35.
36. Report on the methodologies for the monitoring of food additive intake across the European Union. Scoop Task Report 4.2. Office for Publications of the European Communities, Luxembourg. 16 January 1998.
37. Commission proposal for a European Parliament and Council Directive amending Directive 79/112/EEC on the approximation of the laws of the Member States relating to the labelling, presentation and advertising of foodstuffs (submitted by the Commission on 10 February 1997). *Official Journal of the European Communities No. C106* of 04 April 1997:5.
38. Commission Directive 93/102/EC of 16 November 1993 amending Directive 79/112/EEC on the approximation of the laws of the Member States relating to the labelling, presentation and advertising of foodstuffs for sale to the ultimate consumer. *Official Journal of the European Communities No. L291* of 25 November 1993:14.
39. Directive 97/4/EC of the European Parliament and of the Council of 27 January 1997 amending Directive 79/112/EC on the approximation of the laws of the Member States relating to the labelling, presentation and advertising of foodstuffs. *Official Journal of the European Communities No. L43* of 14 December 1997:21.
40. Commission Directive 1999/10/EC of 8 March 1999 providing for derogations from the provisions of Article 7 of Council Directive 79/112/EEC as regards the labelling of foodstuffs. *Official Journal of the European Communities No. L69* of 16 March 1999:22.
41. Commission Directive 94/54/EC of 18 November 1994 concerning the compulsory indication on the labelling of certain foodstuffs of particulars other than those provided for in Council Directive 79/112/EEC. *Official Journal of the European Communities No. L300* of 23 November 1994:14.
42. Council Directive 96/21/EC of 29 March 1996 amending Commission Directive 94/54/EC concerning the compulsory indication on the labelling of certain foodstuffs of particulars other than those provided for in Directive 79/112/EEC. *Official Journal of the European Communities No. L88* of 5 April 1996:5.
43. Council Regulation (EC) No. 1139/98 of 26 May 1998 concerning the compulsory indication on the labelling of certain foodstuffs produced from genetically modified organisms of particulars other than those provided for in Directive 79/112/EEC. *Official Journal of the European Communities No. L159* of 3 June 1998:4.
44. Commission Decision of 21 December 1998 on the national provisions notified by the Kingdom of Sweden concerning the use of certain colours and sweeteners in foodstuffs (1999/5/EC). *Official Journal of the European Communities* No. L3 of 07 January 1999:13.
45. Opinion on cyclamic acid and its sodium and calcium salts (expressed on 14 December 1995). Reports of the Scientific Committee for Food 38th series. Directorate-General for Industry, European Commission, 1997.
46. Proposed Draft Standard for the Labelling of and Claims for prepackaged 'Low-Energy' or 'Reduced-Energy' Foods for Special Dietary Uses. Codex Committee on Nutrition and Foods for Special Dietary Uses, 16th Session, Bonn, Bad Godesberg, 29 September–7 October 1989, ALINORM 89/26, Appendix VI, p. 63.
47. Proposed Draft Standard for Labelling and Claims for Prepackaged 'Low Energy' and 'Re-

duced Energy' Foods. Codex Committee on Nutrition and Foods for Special Dietary Uses. CX/FSDU 85/Conference Room Doc. No. 8 Appendix 1.2.

48. Council Directive of 28 July 1981 laying down Community methods of analysis for verifying that certain additives used in foodstuffs satisfy criteria of purity (81/712/EEC). *Official Journal of the European Communities No. L257* of 10 September 1981:1.
49. Corrigendum to European Parliament and Council Directive No. 95/2/EC of 20 February 1995 on food additives other than colours and sweeteners. *Official Journal of the European Communities No. L248* of 14 October 1995:60.
50. Council Directive of 25 July 1978 laying down specific criteria of purity for antioxidants which may be used in foodstuffs intended for human consumption (78/664/EEC). *Official Journal of the European Communities No. L223* of 14 August 1978:30.
51. Council Directive of 25 July 1978 laying down specific criteria of purity for emulsifiers, stabilizers, thickeners and gelling agents for use in foodstuffs (78/663/EEC). *Official Journal of the European Communities No. L223* of 14 August 1978:7.
52. Council Directive of 26 January 1965 laying down specific criteria of purity for preservatives authorized for use in foodstuffs intended for human consumption. (65/66/EEC) *Official Journal of the European Communities No. 22* of 9 February 1965:373/65.
53. Council Directive of 27 June 1967 on the use of certain preservatives for the surface treatment of citrus fruit and on the control measures to be used for the qualitative and the quantitative analysis of preservatives in and on citrus fruit (67/427/EEC). *Official Journal of the European Communities No. 148* of 7 November 1967:1.
54. Commission Regulation (EEC) No. 207/93 of 29 January 1993 defining the content of Annex VI to Regulation (EEC) No. 2092/91 on organic production of agricultural products and indications referring thereto on agricultural products and foodstuffs and laying down detailed rules for implementing the provisions of Article 5(4) thereto. *Official Journal of the European Communities No. L25* of 2 February 1993:5.

8

Regulation of Food Additives in the United States

PETER BARTON HUTT

Covington & Burling, Washington, D.C.

I. INTRODUCTION

For centuries, every recorded civilization has employed some type of regulatory control over the safety of the food supply. All of these previous regulatory systems have involved simple policing of the marketplace. No country attempted to limit new items that could be introduced into the food supply, either from domestic sources or from abroad. New food substances were in fact constantly introduced through new agricultural commodities discovered locally, food introduced from foreign countries, deliberate genetic selection to alter traditional food plants, and functional substances of both natural and synthetic origin added to raw and processed food to permit improvements in the food supply. Enforcement action was available to remove a new food substance from the market if it was found to be harmful, but that very rarely occurred. From earliest recorded history up to the 1950s, this regulatory system worked quite well.

Four decades ago, the United States explored and ultimately adopted a new and untried approach under the Federal Food, Drug, and Cosmetic Act (FD&C Act). Congress divided the food supply into two quite different regulatory categories. The first regulatory category remained subject only to the same policing controls that had been used for centuries. This category included food itself (e.g., raw agricultural commodities) and functional substances added to the food supply that were either approved by the United States Department of Agriculture (USDA) or the Food and Drug Administration (FDA) during 1938–1958 or that were generally recognized as safe (GRAS) for their intended uses. The second regulatory category required the obtaining of premarket approval from FDA prior to marketing for all new food additives that did not fall within the first category and all color additives.

Now we find that half of the system has been a complete success, and the other half is a complete failure. The GRAS process has worked extraordinarily well. The food additive and color additive approval process has broken down completely and requires fundamental reform.

The purpose of this chapter is to explore how these events unfolded, to analyze why half of the new system has been a success and the other half has failed, and to suggest lessons for reforming the food additive approval process to make it work in the future. This chapter does not consider any aspect of the Delaney Clause.

II. HISTORICAL PROSPECTIVE

It is useful to explore the historical antecedents of our current food safety law. They reveal how and why we adopted the new system that was put in place in the 1950s.

A. Ancient History to 1906

Ancient history is filled with examples of concern about, and resulting regulation of, food adulteration. From the dietary laws of Moses to the Roman statutes, through the remarkable laws enacted by Parliament in the Middle Ages, and culminating in the three statutes enacted in England during the third quarter of the 19th century, regulation of food safety has been regarded as an important function of every advanced government (1). In 1266, for example, Parliament prohibited a number of important staple food items if they were so adulterated as to be ''not wholesome for Man's Body'' (2). In the intervening seven centuries, we have been unable to improve upon that remarkably clear, unambiguous, and completely descriptive statutory language. If it remained the law today, it would unquestionably constitute sufficient authority for FDA to take all of the action it in fact does take every day to protect the safety of our food supply.

None of these early regulatory approaches relied upon anything other than government and industry policing the marketplace. There was no required premarket testing, premarket notification, premarket approval, registration of manufacturers, or listing of substances. It was incumbent on the government and the industry guilds to uncover instances of food adulteration and to punish the culprits.

B. 1906 to 1938

Our first national food and drug law, the Federal Food and Drugs Act of 1906 (3), similarly contained no provision for premarket notification, testing, or approval. Section 7 of the 1906 Act declared a food adulterated, and thus illegal, if was shown to ''contain any added poisonous or other added deleterious ingredient which may render such article injurious to health.'' The 1906 Act relied solely upon FDA and its predecessor agencies (4) exercising surveillance over the marketplace and taking adequate enforcement action to assure compliance with this.

But the seeds of premarket approval had already been sewn. As a result of two drug tragedies that occurred in 1901, when contaminated smallpox vaccine caused an outbreak of tetanus in Camden and eleven children were killed in St. Louis because of a contaminated antitoxin, Congress enacted the Biologics Act of 1902 (5). This law required premarket approval of both a new biological product and the establishment in which it was to be manufactured. Although that statute was enacted just four years before the Federal Food and Drugs Act, the 1902 Act was handled by entirely different committees of Congress and thus the concept of premarket approval was not even considered during the enactment of the 1906 Act.

C. 1938 to 1950

No significant change was made in the authority of FDA to regulate food safety when Congress enacted the FD&C Act of 1938 (6). The basic statutory prohibition against added poisonous or deleterious substances remained unchanged from the 1906 Act (7). The 1938 Act, like its predecessor, relied completely upon FDA surveillance and enforcement in the marketplace.

Once again, the potential for premarket approval was lurking in the shadows. As a result of another drug tragedy, which killed more than a hundred people throughout the country in the fall of 1937, Congress included in the 1938 Act a requirement for premarket notification for all new drugs. No one suggested, however, that this approach be expanded to include the ingredients of food or the other products regulated under the new law.

D. The Delaney Committee of 1950 to 1952

During the first half of this century, relatively few functional ingredients were used in food. Following the explosion of food technology during World War II, however, there was a proliferation of functional food ingredients and a revolution in processed food. Representative Frank B. Keefe of Wisconsin was the first to introduce legislation to establish a committee to investigate the use of chemicals in food products. Because Mr. Keefe was in the minority party, and in very poor health, he persuaded a young member of the House of Representatives, James J. Delaney of New York, to sponsor the resolution that was ultimately passed (8) and to serve as chairman of the committee.

The Delaney Committee, as it came to be known, investigated the use of chemicals in food, and later in cosmetics, for two years. In 1952, the Delaney Committee issued its report (9).

The Delaney Committee Report concluded that ''there is a genuine need for the use of many chemicals in connection with our food supply.'' It documented the introduction of new chemicals at an ''ever-increasing rate.'' The Food and Drug Administration testified that approximately 842 chemicals had been used or suggested for use in food, 704 were actually used in food at that time, and the agency could make a definitive determination of safety for only 428 of those substances. The report concluded that chemical additives in food ''raised a serious problem as far as the public health is concerned'' and that existing laws ''are not adequate to protect the public against the addition of unsafe chemicals.'' Finally, ''most witnesses'' were reported to favor a law that would require that ''a chemical or synthetic should not be permitted to be used in the production, processing, preparation, or packaging of food products until its safety for such use has been established'' by the same type of premarket notification then required for new drugs.

Looking back, the language of the Delaney Committee Report is remarkably devoid of the inflammatory rhetoric to which we have become accustomed in the intervening years. No charges were made that the food industry was poisoning the American public. Rather, the committee concluded that the safety of the new ingredients being added to the food supply, and therefore the risk to the consuming public, was unknown and uncertain. This was enough, however, to frighten the public and to bring forth demands for greater consumer protection.

In the mid-1950s, in direct response to the Delaney Committee Report, the National Academy of Sciences (NAS) took a dispassionate look at the use of chemical additives in food processing (10). Noting that the "widely publicized Delaney Committee statement has caused apprehension as to the safety of processed foods," the NAS report pointed out that many substances purposely added to the food supply have been used for centuries, come from natural sources, and serve important functional purposes. The report listed some 600 direct chemical additives for human foods, broken down by functional category.

Not surprisingly, Congress immediately began to consider new legislation. The result was the enactment of three statutes in the next eight years to require premarket evaluation and approval of substances added to the food supply—the Miller Pesticide Amendments of 1954 (11), the Food Additives Amendment of 1958 (12) and the Color Additive Amendments of 1960 (13). These three statutes have dominated food safety issues in the intervening four decades. This chapter focuses on two of those three statutes—the laws regulating food additives and color additives.

III. THE FOOD ADDITIVES AMENDMENT OF 1958

The concept of the 1958 Amendment was quite simple. Congress divided the food supply into four categories. First, it made a distinction between food itself and the substances added to food. Second, it set forth three categories of added substances: (1) added substances approved by FDA or USDA during 1938–1958 (which have come to be called prior-sanctioned substances) (14), (2) added substances that are generally recognized as safe (which have come to be called by the acronym GRAS), and (3) all of the remaining added substances in the food supply, which are neither subject to a prior sanction nor GRAS, which were defined as food additives. For food itself and for prior-sanctioned and GRAS substances, simple policing controls were retained and no premarket approval was required. In contrast, food additives were required to be the subject of premarket testing; a petition had to be submitted to FDA to request the establishment of safe conditions of use; and the promulgation by FDA of a food additive regulation establishing the specific conditions under which the additive can be used in the food supply had to be made (15).

The Food Additives Amendment covered two different kinds of additives—direct additives (incorporated directly into the food supply) and indirect additives (used in various types of food-contact applications, such as packaging and food machinery). The 1958 Amendment also covered additives used in animal feed and pet food. This chapter deals primarily with the subject of food additives and food substances that are directly added to human food.

A. The Transitional Provisions

One of the first policy choices that must be faced in enacting any new regulatory statute is how the statute will be applied to products already on the market. Congress handled that in two different ways under the 1958 Amendment. First, it excluded from the definition of a food additive, and therefore from the requirement of premarket approval, all food and all substances that were subject to a prior sanction or were GRAS. Accordingly, these substances were permanently excluded from premarket approval (except to the extent that a GRAS substance subsequently lost its status as GRAS). Second, for those pre-1958 substances that were not prior-sanctioned or GRAS, and thus properly fell within the definition of a food additive, Congress included a transitional period within which the manufacturers were required to submit and obtain FDA approval of a food additive petition.

The 1958 Amendment became law on September 6, 1958. New Section 409 of the FD&C Act, regulating food additives, became effective six months later. One year after that, all pre-1958 additives were required to have obtained FDA approval and promulgation of a food additive regulation. FDA was authorized to extend that period for an additional year. Thus, all pre-1958 food additives were required to be the subject of food additive regulations promulgated by FDA no later than March 1961.

This schedule proved impossible to meet. In 1961, Congress therefore enacted legislation allowing a further extension for any specific food additive to June 30, 1964 (16). By another law enacted in 1964, this was again extended to December 31, 1965 (17). No further statutory extensions were granted. Accordingly, Congress ultimately provided just slightly over seven years for industry to prepare and to submit to FDA food additive petitions, and for FDA to review these petitions and promulgate food additive regulations, for all of the pre-1958 food additives. During the same time, FDA was required to deal with the prior-sanction and GRAS provisions of the new statute and to handle food additive petitions for post-1958 food additives.

The handling of all of the pre-1958 direct human food additive petitions in seven years reflected a heroic effort by FDA. It required the development and organization of a group of dedicated scientists within FDA, close daily cooperation between FDA and the regulated industry, and a commitment by everyone involved to work through the inevitable problems that arose in a practical and realistic way. No one either in industry or in FDA had the luxury of sitting around for months or years on end, debating theoretical issues. There was a job to be done, and it was done extraordinarily efficiently and with unprecedented success. Not a single additive that was approved during that time has since been removed from the market for lack of safety. The people who participated in that effort, many of whom still are alive today, deserve far greater recognition than they have been given to date.

By the time this transitional process was completed, FDA had promulgated 185 separate direct human food additive regulations, along with an additional 93 indirect food additive regulations, five regulations governing radiation of food, the lists of GRAS substances, and additional lists of prior-sanctioned substances (18). All of these substances were evaluated by FDA, regulations were promulgated, and effective dates were confirmed. It was an extraordinary outpouring of regulatory science. The productivity of FDA at that time in history stands in stark contrast to the inefficiency and paralysis of today.

B. The GRAS List

At the same time that it was laboring to implement the transitional provisions for food additives, FDA also sought to respond to the need to clarify the scope of the GRAS exclusion from the definition of a food additive (19). Based upon the personal knowledge of FDA scientists and a limited survey of academic experts, FDA proposed and promulgated regulations that collectively became known as the GRAS list, consisting of some 450 direct human food GRAS substances (20).

The statue and its legislative history were clear, however, that the FDA GRAS list was not definitive or all-inclusive. Indeed, there was no statutory requirement that FDA even publish a list of GRAS substances (21). The statute excluded GRAS substances from the definition of a food additive, but imposed no requirement that a person interested in marketing a substance that it determined was GRAS either inform FDA of that decision or obtain FDA affirmation that the substance was in fact GRAS. Under the 1958 Amendment, industry has the authority to market a substance that it determines is GRAS without any premarket testing, notification, or approval (22). If FDA determines that a food substance is not GRAS, the agency in turn has the authority to contest the marketing of that substance through all of the informal and formal enforcement provisions under the FD&C Act.

Four decades after its enactment, many important aspects of the scope of the coverage of the 1958 Amendment, and the GRAS exclusion, remain unexplored and uncertain. No court has ever addressed the question whether a food loses its status as a food, and becomes subject to analysis under the 1958 Amendment, when it is combined with other food and food substances in a processed product (23). No court has ever addressed the issue whether deliberate genetic selection, or recombinant DNA technology, to change the characteristics of raw agricultural commodities results in that food losing its status as a food and becoming a food additive. Issues like these, which have major implications for food technology and food policy in our country, are yet to be resolved. As a practical matter, it is the GRAS exclusion that has covered all of these developments, and that has largely made them moot up to this time.

C. The GRAS List Review

In the fall of 1969, Abbott Laboratories announced to FDA that an animal bioassay had shown a combination of saccharin and cyclamate to be carcinogenic. Cyclamate had originally been the subject of a new drug application, but became reclassified as a food substance and was included in the GRAS list promulgated in 1961 (24). FDA promptly removed cyclamate from the GRAS list (25) and, after a futile attempt to restore its status as a drug (26), banned it from the food supply (27). In his consumer message of November 1969, President Nixon ordered FDA to review the GRAS list to assure the American public that the food substances on that list were in fact GRAS and were not subject to safety concerns similar to those for cyclamate (28).

This began another heroic effort, comparable to the seven-year review of pre-1958 transitional food additives. This time, however, FDA turned to an outside organization to do the job (29).

The Food and Drug Administration contracted with the Life Sciences Research Office (LSRO) of the Federation of American Societies for Experimental Biology (FASEB), which established a Select Committee on GRAS Substances (SCOGS) to conduct the GRAS List Review. The Select Committee on GRAS Substances was organized in June 1972. Under strong and able leadership, it held fifty 2-day executive sessions over a 5-year period, and reviewed and prepared substantial GRAS reports on some 400 substances on the FDA GRAS list (30). This was an extremely efficient and effective process. FASEB convened teams of experts who reviewed the published literature, conducted hearings to obtain the views of interested citizens and scientists, considered unpublished data submitted by the regulated industry, and arrived at documented conclusions in well-written reports, all in a very short period of time. By the end of the 1970s, FASEB had completed its job. The remarkable productivity and success of this venture stands alongside the FDA implementation of the transitional provisions of the Food Additives Amendment 15 years earlier.

By this time, however, the ability of FDA to handle its end of the process had begun to disintegrate. As a practical matter, all that FDA had to do was to take each FASEB report, review it to make certain that the agency did not have additional information that would lead to a different result, and publish it in the Federal Register as a proposed GRAS affirmation regulation for public comment. FDA decided, however, that the process should not be that simple and straightforward. The agency concluded that it should also take additional steps to restrict these GRAS affirmation determinations. Initially, the use data on which the GRAS determinations were made were included in the GRAS affirmation regulations themselves, thus severely restricting the scope of the affirmation (31). When even FDA realized that this was unduly restrictive, it adopted a new approach of using those data as guidelines rather than as binding limitations.

Even beyond this technical problem, however, the agency simply lost the capacity to process the FASEB determinations in a timely fashion. In spite of the fact that FASEB completed its work more than 20 years ago, as of today FDA has not only failed to complete its much more limited role in this process, but the end is still not even in sight. It has been a disgrace to the agency. Like the food additive matters that remain unresolved within FDA, this matter received no priority, no management attention, and thus no resolution, until the Institute of Food Technologists and then Congress focused public attention on it in 1995. Recent attempts by FDA to reform the GRAS process are discussed in Section III.F.

D. Direct Human Food Additive Regulations

Once FDA completed its work on pre-1958 transitional food additives, it seems the agency lost heart on food additive matters. Since that time the record of FDA approval of new food additives is appalling. It is useful to review the FDA record on approving new direct human food additives over the past thirty years, from 1970 to the present. Research indicates only eight new direct human food additives during that time:

1. TBHQ, 1972 (32)
2. Aspartame, 1981 (33)
3. Polydextrose, 1981 (34)
4. Acesulfame K, 1988 (35)
5. Gellan gum, 1990 (36)
6. Olestra, 1996 (37)
7. Sucralose, 1998 (38)
8. Sucrose acetate isobutyrate, 1999 (39)

Eight direct human food additives in thirty years (40). That is embarrassing. It represents a serious problem for food technology in the United States. Just at the time when we are learning more about the relationship between diet and health, and new food ingredients could contribute greatly to healthy improvements in our daily diet, we have adopted a regulatory system that has virtually destroyed the incentive to innovate and the ability to get new food additives to the market.

Let us also examine the history of one of the new food additives that languished in the FDA pipeline almost nine years prior to FDA approval: olestra. Information published on olestra shows that it was invented in 1968, the first meetings with FDA were held in the early 1970s, and it was the subject of continuing testing and negotiations with FDA until approved in 1996 (41). The story of the incredible journey of olestra through the food additive quagmire in FDA was so compelling that Congress enacted a statute to permit an extension of the patent term for that compound even after the term of the original patent had expired (42). Nor is that the record. Alitame has been in the FDA food additive pipeline more than thirteen years and still remains unapproved (43).

The lesson taught by the problems encountered by these hapless compounds has been learned by the entire food industry. The food additive approval process in America is dead. It has been killed by FDA. No food company of which I am aware would even consider beginning research today on a new food additive.

The lesson learned is that there is only one type of food substance worth considering: a GRAS substance that can immediately be marketed. If a new food substance cannot be determined to be GRAS, it is simply discarded. Only if it can be regarded as GRAS will it be pursued.

It is easy to be misled about the efficiency of FDA regulation of food additives by counting the *Federal Register* food additive notices during the past 30 years and not looking behind them. The vast majority of these notices relate to indirect food additives that have no possible bearing upon the public health. Indirect food additives are important, serve a highly useful purpose, and should not be given inadequate attention. At the same time, they do not deserve the same degree of scrutiny and FDA priority as new food additives added in significant amounts directly to the food supply. It is apparent that a completely separate and different process, with far less government involvement, is appropriate for the category of indirect food additives. Recent FDA and statutory reforms in the regulation of indirect food additives are discussed in Section III.G.

Once a new direct human food additive is approved, there are always a large number of follow-on amendments that gradually expand its use to additional categories of food and perhaps for additional functional purposes. This is a direct result of the extraordinary conservatism of FDA in reviewing and approving the initial regulation. It is easy to read these and be misled that they represent real work and real progress. In fact, they are a direct reflection of the lack of any realistic evaluation of a new food additive by FDA and an extraordinary amount of sheer busywork as each new food use for an already-approved food additive must be the subject of yet another lengthy petition and review process.

Even as the ability of FDA to handle food additives was disintegrating, the agency announced an initiative—not required by the FD&C Act—to expand its food additive work. In January 1977, FDA presented to Congress an ambitious new program of cyclic reevaluations of previously approved food additives (44). Although FDA continued to pursue plans for the cyclic review for another five years (45), it died a natural death by the mid-1980s. In light of the agency's inability to deal with new food additives, the objective of reviewing prior safety decisions was clearly unachievable. Common sense ultimately prevailed. The agency concluded that prior safety decisions would be reviewed when specific problems arise.

With the development of biotechnology, FDA has been forced to consider how new food substances made through recombinant DNA technology, and traditional plant and animal sources altered through that technology, will be handled under the food additive provisions of the FD&C Act. To its credit, the agency has avoided a rigid response and has instead adopted a very flexible and workable approach. Under the FDA policy announced in 1986 (46) and 1992 (47), FDA clearly stated that the safety of a food or food substance, and not the process by which it is developed, will be the focus of the agency. Food derived from biotechnology will therefore be treated no differently than traditional food, except to the extent that the process of biotechnology adds a substance to the food itself. Those added substances will, in turn, be determined either to be GRAS or to be food additives. FDA has declined to take the rigid view that all such new foods and all such added substances must be reviewed as food additives and has pursued instead a very reasonable process of scientific evaluation based upon each individual food or food substance.

The new food substances created by biotechnology have primarily been handled as GRAS (48), not pursuant to food additive regulations. For the first genetically altered food, a tomato, FDA required a food additive regulation only for the enzyme produced by the marker gene added to the food (i.e., the translation product), but not for the gene or the genetically altered food itself (49).

One of the most serious contributing factors to the decline of the FDA food additive process has been the development of rigid rules for evaluation of food additives. This concept has a long history. Dr. Dale Lindsay, Associate Commissioner for Science in FDA during the late 1960s, was the original proponent of establishing written guidelines for review of food substances, largely to guide the GRAS List Review. Industry strongly opposed this for the reasons that have later turned out to be correct—guidelines may begin as flexible recommendations, but they can all too easily become rigid requirements for which there is no exception. Scientific judgment is too often replaced by rules. Those rules, moreover, are designed to deal with every last possibility. Thus, tests are cascaded upon tests, anticipating any conceivable possibility that might arguably occur in the future. Using cookbook toxicology rather than informed judgment, guidelines can rapidly become a reflection of the lowest common denominator, that is, incorporating every imaginable test, whether or not each test is fully justified from a scientific standpoint for any particular food substance involved (50).

The Lindsay approach was rejected in the late 1960s and early 1970s precisely for these reasons. In June 1980, however, an industry-sponsored Food Safety Council recommended a flexible decision tree approach to determining appropriate testing for food additives (51), and in the early 1980s FDA concluded to prepare its own testing guidelines as an answer to the need for consistent FDA action. The resulting Red Book of food toxicology rules has become exactly what industry and food scientists feared—a rigid compendium of required tests designed more to protect an FDA reviewer from criticism than to protect the consuming public against an unsafe additive. The Red Book was adopted without any opportunity for notice and public comment of the type required for formal regulations (52). A draft of a revised version was made publicly available for comment in 1993 (53), but has never been subject to full public peer review and has not been finalized. Nonetheless, it is as rigidly applied as any provision in the statute or FDA regulations (54).

One of the major failings of FDA, as an institution, is that it cannot admit that it has ever made a wrong decision. There is no better example of this than the ill-fated non-nutritive sweetener cyclamate. Following its ban in 1969, the manufacturer sought repeatedly to obtain reapproval from FDA on the basis of substantial new scientific evidence (55). Cyclamate is now widely regarded as safe, within both the general scientific community and among FDA scientists themselves (56). To admit that the agency had made a mistake in banning it, however, would surely be politically incorrect. Even the recommendations of internal FDA scientists for reapproval have failed to result in returning this substance to the market.

As long as safety evaluation remains solely the domain of FDA, it is extremely difficult, if not impossible, to break down this institutional intransigence. Only if safety evaluation is shared with independent scientists who have no stake in the original decision or in maintaining what FDA perceives as its collective integrity is there any hope that this type of decision will be made solely on scientific grounds.

E. GRAS Affirmation

When FDA established its GRAS affirmation procedures as part of the GRAS List Review in the early 1970s (57), the agency recognized that it did not have the statutory authority either to preclude self-determination of GRAS status or to prevent marketing while the agency reviewed a self-determination of GRAS. In establishing its GRAS affirmation procedure, the agency deliberately and explicitly adopted a policy of recognizing that industry had the legal right to market a new food substance following a self-determination of GRAS, even if the manufacturer then concluded to submit a GRAS affirmation petition to FDA. The purpose of the agency was to encourage submission of GRAS affirmation petitions to FDA in order to assist the agency in its surveillance of the food supply and to settle any safety issues that might arise with the use of new food substances. This was a purposeful tradeoff. FDA gave industry the assurance that it could market products after a self-determination of GRAS, and in return industry began to submit GRAS affirmation petitions in order to demonstrate to potential customers and to FDA that the new substance had been thoroughly evaluated and was in fact GRAS.

While reasonable in concept, FDA implementation of this program broke down almost from the beginning. GRAS affirmation was a process created by FDA, not by statute. There was no statutory deadline for FDA action and thus no priority within FDA. In many instances, FDA seemed to be more intent upon receiving a GRAS affirmation petition than it was upon evaluating and acting upon it. As a result, FDA became simply a dumping ground for GRAS affirmation petitions. The backlog of GRAS affirmation petitions for direct food substances, in various stages of consideration, grew every year. These could undoubtedly have been handled swiftly by FASEB, but the agency was incapable even of delegating this function.

Nonetheless, the program was still a resounding success from the standpoint of public policy, precisely because FDA involvement was not needed to make it work. Self-determination of GRAS permits immediate marketing, whether or not a GRAS affirmation petition is filed and whether or not FDA ever acts on it.

As a general rule, FDA evaluation of GRAS affirmation petitions—although extremely tardy—was reasonable and did not result in taking safe and useful products off the market. In one instance, FDA did deny a GRAS affirmation petition for a non-nutritive sweetener from a natural plant source (58), resulting in the bankruptcy of the small company that sought to market it (59).

Perhaps there is no better example of the success of the GRAS approach than the history of high fructose corn syrup. High fructose corn syrup was the subject of a thorough scientific evaluation and a self-determination of GRAS in the mid-1960s. It was immediately marketed, without asking FDA for an opinion or even informing the agency. Once one manufacturer began to market it, others followed suit. By the early 1970s, it was in widespread use throughout the food industry. It proved to have excellent functional properties and to be an extremely important addition to the food supply.

By the time that FDA realized that high fructose corn syrup had become ubiquitous in the food supply, it was simply too late, as a practical matter, to insist that it be taken off the market and made the subject of a food additive petition. The agency had no basis for questioning safety and would have been in a poor position to contend that it was not

GRAS. The agency therefore requested submission of a GRAS affirmation petition. The industry agreed, a petition was prepared and submitted, and FDA published a notice of filing in 1974 (60). For the next nine years, there was endless discussion about the matter. Finally, in 1983, FDA recognized the obvious and published three separate regulations for this substance (61). Even then the matter was not finished. A further proposal was published in 1988 and was not the subject of a final regulation until eight years later (62).

One should stop to consider what would have happened if high fructose corn syrup had been handled through a food additive petition rather than a self-determination of GRAS. By any standard, this substance is what we now refer to as a macronutrient. It can be consumed at a substantial level in the daily diet, if one selects a diet of processed food sweetened only with this substance (an assumption that FDA routinely makes in evaluating food additives). High fructose corn syrup was a contemporary of olestra. These two macronutrients were subject to initial research and development at the same time, during the mid-1960s. One took the GRAS superhighway to market, and the other followed the meandering food additive path through the FDA woods. One was freely marketed more than 25 years for unlimited food use before the other was allowed to be marketed for very limited food use. In many ways, this simple comparison tells the entire story.

F. Recent Reform for Direct Food Additive and GRAS Substances

By 1995, it was apparent that the FDA GRAS affirmation process was no longer operable. At its June 1995 meeting, the Institute of Food Technologists sponsored a major symposium to analyze food regulatory policy. Three of the papers presented at that symposium focused on the major problems with the current regulation of food additives and GRAS substances in the United States (63). Congress held a hearing (64) to investigate the matter later that month (65) and subsequently issued a report highly critical of the FDA performance (66). A year later the International Society of Regulatory Toxicology and Pharmacology (ISRPT) sponsored a workshop on optimizing the review process of food additive and GRAS petitions (67). The following year, the Institute of Medicine Food Forum held a workshop on the subject, focusing on a major commissioned academic analysis of the entire food additive/GRAS process (68). Although it was difficult to interpret the FDA data presented at the House hearings and the ISRTP workshop, it was clear that the FDA backlog for direct and indirect food additive petitions and GRAS affirmation petitions was very large and increasing each year. As of June 1995, there was a backlog of 295 pending petitions. This backlog was broken down into approximately 70 direct food additive petitions, 150 indirect food additive petitions, and 75 GRAS affirmation petitions. The oldest food additive petition was filed in 1971 and the oldest GRAS affirmation petition was filed in 1972. At the ISRTP workshop a year later, the FDA data presented a slightly improved picture but FDA was quick to point out that the situation was still inadequate.

At the House hearings and the ISRTP workshop, FDA presented proposals for reform in the handling of direct food additives, indirect food additives, and GRAS affirmation petitions. This section of the chapter covers direct food additive and GRAS affirmation petitions. The next section handles indirect food additive and GRAS affirmation petitions.

For direct food additive petitions, FDA presented a plan to Congress that focused on reorganization of the FDA Center for Food Safety and Applied Nutrition (CFSAN) to place the petition review resources under one central manager, set performance goals to review petitions within defined time periods, allocate additional agency resources to reduce the inventory of pending petitions, use external scientific expertise to expedite the review of pending petitions, and eliminate environmental assessment requirements. It is apparent, however, that only one of these reforms—additional agency resources—could make a major impact on the pending backlog. Although FDA has in the past two years received substantial additional funds to implement President Clinton's Food Safety Initiative, the extent to which these resources are being devoted to direct food additive positions is unclear. Thus, although these reforms will produce some changes in the pending backlog, they are, to use FDA's own words, ''differential changes, happening at the margin; they are not sea changes'' (69). None of the reforms proposed by FDA for direct food additives involved any change, of any kind, in the highly conservative scientific requirements imposed by FDA for this class of food substances.

Partly as a response to the findings about the deterioration of the FDA process for review of food petitions, Congress included in the Food and Drug Administration Modernization Act of 1997 (70) a new section (909 of the FD&C Act), explicitly authorizing FDA to enter into contracts with outside experts to review and evaluate any application or submission (including a petition or notification) submitted under the FD&C Act, including those for food additives and GRAS substances. This provided explicit statutory authority for the plan announced by FDA in 1995 to expedite review of food additive and GRAS affirmation petitions. Nonetheless, because the underlying FDA requirements for approval of a direct food additive remain unchanged, and it does not appear that the resources allocated by FDA to this process have substantially increased, this new statutory provision by itself cannot be expected to make a major impact on the pending backlog.

Congressional frustration with the inability of FDA to handle food additive petitions undoubtedly reached its peak when Congress included in the Food and Drug Administration Modernization Act of 1997 an explicit statutory requirement that no later than 60 days following the date of enactment FDA must make a final determination on a pending food additive petition to permit the irradiation of red meat (71). FDA promptly approved the petition (72).

Recognizing that substantial progress on reducing the backlog of direct food additive petitions is unlikely in the near future, FDA sought in 1999 to blunt continuing criticism of its program by announcing that the agency would give ''expedited review'' to any food additive petitions designed to decrease the risk of foodborne illness (73). The agency was careful to point out, however, that this means merely that the expedited petitions would be reviewed first (thus placing other petitions at a disadvantage) and that the standards for review would not change.

For GRAS affirmation petitions for direct food substances, in contrast, FDA chose quite a different approach. In 1997, FDA proposed to replace the entire GRAS affirmation process with a GRAS notification procedure. Under the proposed new procedure, a person would submit a GRAS notification based on the person's determination that a substance is GRAS for its intended use. The GRAS notification would include a detailed summary of the scientific data on which the GRAS determination is based, but not the raw data. Within 90 days of receipt of the notification, FDA would be required to respond in writing either that the agency has no current objection to the notification or that it does have an objection. Although FDA has not yet promulgated a final regulation for this procedure, it has declined to accept new GRAS affirmation petitions, has encouraged those who submitted old GRAS affirmation petitions to convert them to GRAS determination notifications, and has accepted new GRAS determination notifications and taken action on them. In short, FDA is already implementing this proposal.

Without question, the new GRAS determination notification process has the potential for substantially reducing or even eliminating the GRAS affirmation petition backlog very quickly. That potential would be maximized if FDA would agree to delegate primary review of these notifications to FASAB or a comparable organization, and to evaluate only the recommendation made by that outside organization. Even without the benefit of outside expertise, however, the streamlined procedure will result in prompt and efficient FDA decisions if the agency resists the temptation to review these notifications in the same way that it has reviewed GRAS affirmation petitions in the past.

G. Recent Reform for Indirect Food Additive and GRAS Substances

As enacted in 1958, the definition of a food additive broadly includes ''any substance intended for use in producing, manufacturing, packing, processing, preparing, treating, packaging, transporting, or holding food'' (74). Thus, this term has always been understood to include indirect as well as direct food additives. As discussed, the majority of food additive petitions submitted to FDA in the past four decades have been for indirect rather than direct food additives. And if one counts only substances (rather than multiple uses of the same substance), undoubtedly more than 90% of all food additive petitions have been for indirect use. Although these uses are important, they clearly do not need the same level of FDA review as a new direct food additive.

The concept of limiting the number of indirect food additives that would require FDA review was first discussed by industry and agency officials as early as 1968 (75). After 25 years of consideration, a proposal that has come to be called the ''threshold of regulation'' approach was finally published in 1993 (76) and promulgated in 1995 (77). Even then, it contained a classic example of the refusal of FDA to let go of even the most trivial issue. Instead of simply providing that any indirect food additive that met the regulatory criteria for exclusion from regulation could immediately be marketed, FDA has required that this decision, in itself, result in a submission to FDA and an explicit FDA exemption. Indirect food additive petitions have thus been replaced by indirect food additive exemption petitions. No more graphic example of government over-regulation can be found.

By 1995, the FDA backlog of indirect food additive petitions had reached approximately 150. With inadequate resources, a lower priority than direct food additives, and a threshold of regulation policy that suffered from the same deficiencies as indirect food additive petitions themselves, it was apparent that a new approach was needed.

Three approaches have emerged to reduce the resources and time required for FDA handling of petitions for indirect food additives and GRAS substances.

First, industry has simply taken the position that any substance that it determines meets the FDA threshold of regulation policy is also, by that fact alone, GRAS for its intended use. In short, the threshold of regulation policy itself establishes a definition of GRAS status for an indirect food substance. The vast majority of substances that meet the threshold of regulation policy therefore are not submitted to FDA at all. This has been a highly successful result of the threshold of regulation policy, even though FDA unquestionably did not intend this result.

Second, the proposed new GRAS determination notification procedure discussed with respect to direct food substances also applies to indirect food substances. This new procedure therefore has a substantial potential for reducing the FDA backlog for indirect food additive petitions and GRAS affirmation petitions.

Third, and most important, Congress was persuaded to include in the Food and Drug Administration Modernization Act of 1997 an entirely new and different approach for handling indirect food additives (78). Under the 1958 Amendment, no distinction was made between direct and indirect food additives. Under the 1997 Act, however, a new premarket notification procedure is established for indirect food additives to replace the present premarket approval procedure that applies for direct food additives. The new premarket notification procedure represents the result of a successful negotiation between the regulated industry and FDA on this matter.

The new law provides that a manufacturer or a supplier of an indirect food additive may, at least 120 days prior to shipment, notify FDA of the identity and intended use of the substance and of the determination that it is safe for its intended use. All pertinent information must be submitted with this notification. The notification becomes effective 120 days after the date of receipt by FDA, and the substance may be shipped thereafter, unless FDA makes a determination within that period that use of the substance has not been shown to be safe. The Food and Drug Administration may also promulgate regulations prescribing a procedure by which the agency may deem a notification to be no longer effective, based on new information.

Existing indirect food additive regulations remain in effect. The agency may also require that a food additive petition rather than a notification be submitted when the agency determines that this is necessary to provide adequate assurance of safety. It seems likely, however, that a food additive petition will rarely be required.

The legislation initially included user fees to provide adequate resources to FDA for the regulation of indirect food additives. These were deleted by the Conference Committee, which stated that the new procedure is to be implemented by appropriated funds but that ''implementation is to be triggered only when the FDA receives an appropriation sufficient to fund the program'' (79). The agency has issued a report to Congress on the anticipated costs of this program (80) and has held a public meeting to consider implementation of the new procedure (81), but required appropriation has not yet been enacted. If implemented as intended by Congress, this new procedure would substantially reduce FDA resources needed to handle indirect food additive petitions and could eliminate the backlog of these petitions in the near future.

H. Dietary Ingredients in Dietary Supplements

Since the 1920s, when dietary supplements were first marketed, FDA and the dietary supplement industry have been in a continuous regulatory war about both the promotional claims and the safety of these products (82). When FDA sought to impose stricter standards for dietary supplements in the 1960s and early 1970s, Congress enacted the Vitamin–Mineral Amendments of 1976 (83) to overrule the FDA restrictions. The 1976 Amendments, however, made no change in the safety standards applicable to dietary supplements.

Following enactment of the Nutrition Labeling and Education Act of 1990 (84), FDA sought to use this statutory authority to impose new limitations for dietary supplements. Ignoring a congressional invitation to establish separate standards and requirements for dietary supplements (85). FDA instead used the new statutory authority to deny all disease prevention claims for dietary supplements (86), and threatened to impose stringent food additive requirements on important dietary supplement ingredients (87). The dietary supplement industry marshalled its formidable political power and obtained enactment of the Dietary Supplement Health and Education Act of 1994 (88), over the strong objection of FDA. The 1984 Act was the most humiliating defeat for FDA in its entire history. In addition to broadly defining dietary supplements and explicitly authorizing strong new claims for these products, Congress exempted the dietary ingredients in dietary supplement products from the food additive requirements of the FD&C Act and substituted more flexible food safety provisions that place the burden on FDA to demonstrate a lack of safety (89). Old pre-1994 dietary ingredients may be used in dietary supplement products as long as FDA cannot prove that they present a significant or unreasonable risk of illness or injury. New post-1994 dietary ingredients that have not previously been marketed or present in the food supply are required to be subject to a premarket notification submitted to FDA at least 75 days before marketing. FDA may take action against a new dietary ingredient if FDA can prove that the petition shows that there is inadequate information to provide reasonable assurance that the ingredient does not present a significant or unreasonable risk of illness or injury. Thus, for this one category of direct food substances, Congress has repealed the premarket approval requirements of the 1958 Amendment and replaced it with a premarket notification process.

I. The Lack of Market Protection

There are two basic ways for FDA or any other federal agency to regulate new products as they come on the market. The first way is to use a product-specific license, and to require that each new version of the product be the subject of its own petition in order to obtain its own license based upon its own data. This is the way that new drugs have been regulated. When Congress sought to permit increased competition in pharmaceutical products in 1984, it authorized the approval of generic versions of pioneer drugs only after the pioneer drug has exhausted its patent life or a five-year period of market exclusivity (90). For orphan drugs, the period of market exclusivity was established at seven years (91). Thus, a manufacturer could be assured of a period of market protection during which no other company could rely upon its license and data in order to obtain its own marketing approval.

The second means of regulating new products is through a public regulation which, absent a patent, permits competitors to market the product immediately without the need for any type of license or the protection of any period of market exclusivity. This is the approach taken by Congress in 1958 for food additives. The conclusion was that any market protection must come from a patent. Failing that, competitors are free to use the food additive regulation to manufacture and market their own version of the new product.

The economic fallacy in this second approach is obvious. Except in those few instances where strong patent protection is available, there is a substantial disincentive for any person to invest significant resources in research and development on new food substances. Regardless of the potential benefit to public health from new food substances, it is not feasible to recoup any investment made in them because competitors can immediately enter the market at a price that does not need to reflect any investment at all. Not surprisingly, all of the eight new direct food additives listed alone, approved by FDA during the past thirty years, are patented.

If there is a patent for the new food additive, it will, of course, prevent competitors from entering the market until the patent expires. Recognizing that the FDA food additive process erodes the effective patent life for a new patented food additive, Congress provided for patent term extension for food additives as part of the Drug Price Competition and Patent Term Restoration Act of 1984 (92).

For food additives that are unpatented or whose research, development, and FDA review extend up to or beyond the end of patent protection, however, Congress failed to provide any form of nonpatent market exclusivity or other market protection for food additives in the 1984 Act. The food industry is now directly confronted with the reality not only that there is no possible economic justification for investing even one cent in a new food additive for which a patent cannot be obtained, but also that, with horror stories like olestra, even patent protection is illusory. Thus, there is no incentive to innovate in the field of new direct human food additives, however valuable those additives might be to the health of the American public. If one manages to survive the food additive regulatory process, it is likely that the patent will have expired, or be close to expiration, and the enormous investment that has been made in the additive will be lost.

This problem cries out for a legislative solution. Without a statutory period of market exclusivity, American industry cannot be expected to invest in new food additives that cannot be patented, even if the regulatory process itself is reformed. Some form of market protection must be provided in order to assure industry that there will be a reasonable time to recoup the investment before generic competitors are allowed on the market.

We have thus seen a promising new approach to food safety be killed by the dead hand of FDA regulation. The implications are too important for public health to allow this situation to continue. A new system must be found to replace the present one.

IV. THE COLOR ADDITIVE AMENDMENTS OF 1960

Regulation of color additives has a longer history in the United States than regulation of any food additive other than preservatives. These two food substances were the subject of special legislation enacted by Congress in the Color and Preservatives Act of 1900 (93), requiring FDA to investigate their safety even before there was federal legislation authorizing national regulation. Under the 1906 Act, separate provisions were enacted that FDA interpreted to authorize greater control over food colors than any other form of additive (94), even though that authority was on very uncertain legal ground. The 1938 Act explicitly provided for FDA listing of coal tar colors, which are harmless and suitable for use in food and for the batch certification of these colors, the only form of premarket approval for food substances prior to 1958 (95).

Just two years after enactment of the Food Additives Amendment, prompted by a Supreme Court decision interpreting the FD&C Act to require absolute safety for any color additive (96), Congress enacted a new statute to require premarket approval of color additives (97). It established a system similar to that for food additives, but it was different in one critical respect.

The 1960 Amendments, unlike the 1958 Amendment, did not contain a general exclusion for GRAS substances. Instead, it excluded only those GRAS substances that were listed on a formal published FDA GRAS list under the 1958 Amendment (98). As a practical matter, this exclusion has meant nothing. As will be discussed further, this difference accounts for a large part of the failure of the 1960 Amendments. It has meant that every color additive, of any kind, must be the subject of a color additive regulation before it may lawfully be used.

A. The Transitional Provisions

When the investigation of food colors that Congress authorized in the Colors and Preservatives Act of 1900 was published in 1912, there were approximately 80 coal tar color additives used in the general food supply (99). At the time the provisional listing was imposed after the Color Additive Amendments were enacted, this number had been reduced to 12 (100). Today it stands at seven (101).

Like the Food Additives Amendment, the Color Additive Amendments included transitional provisions (102). The 1960 Amendments provided for provisional listing of pre-1960 color additives ''on an interim basis for a reasonable period'' pending the completion of scientific testing that would allow permanent listing under color additive regulations.

The Color Additive Amendments were enacted into law on July 12, 1960. The statute provided a two-and-one-half year transitional period for this testing and promulgation of color additive regulations, but then went on to state that FDA could establish a longer transitional period, without limitation. As the years went by, and FDA testing requirements escalated, the transitional period was repeatedly extended by FDA. In the late 1970s, FDA imposed additional new testing requirements for carcinogenicity, thus forcing a major extension of the time needed to complete testing and to promulgate color additive regulations (103).

The resulting delay provoked litigation that began in the mid-1970s and continued for the next 10 years. In each of these cases, the courts found that the continued provisional listing was lawful under the statute (104).

Today, 40 years after the law was enacted, the transitional provisions are still not yet fully implemented. The FD&C lakes remain provisionally listed (105) and thus have not yet been subject to final FDA action.

There are only two potential explanations for the difference between the remarkably swift and efficient handling of the food additive transitional provisions and the extraordinarily inefficient bungling of the color additive transitional provisions. Both were handled by the same group of people within FDA. First, for food additives, there was a statutory deadline with a ''hammer'' that required definitive FDA action. A food additive that missed the deadline became illegal by operation of law. For color additives, in contrast, there was no statutory deadline and no definitive action required. Second, the food additive decisions were required to be made during an era when FDA was accustomed to making prompt decisions and taking decisive action, before the current paralysis began to set in. By the time that the transitional color additives were ripe for action, the agency's capacity to make decisions had already deteriorated badly.

B. The Lack of a GRAS Exclusion

For a new food substances, the GRAS exclusion has turned out to be the single most important statutory provision. The food additive provisions are basically a dead letter at this point in time. Unless reformed, they will only rarely be used in the future. For color additives, unfortunately, there was only a very limited GRAS exclusion, which has had no impact. The color additive GRAS exclusion explicitly requires that to be exempt from the requirement for a color additive regulation, the coloring substance must be included on a published FDA GRAS list. No such list exists. As a result, all substances intended to color food must be the subject of a color additive regulation.

The lack of a GRAS exclusion under the Color Additive Amendments has put color additives at a distinct disadvantage to food additives. Nor is there any rational scientific or public policy basis for this difference.

Many of the color additives that are not from coal tar sources, and thus are exempt from certification, come from common food products. We simply do not need to waste the time of industry or FDA on color additive regulations for such ordinary substances as beet powder, grape extract, fruit juice, vegetable juice, carrot oil, and paprika (106). The mere existence of the requirement for a color additive petition and regulation for simple compounds like these unnecessarily clogs the regulatory process and stifles product innovation.

C. Implementation of the Color Additive Amendments

The record of color additive technology in the United States in the past 40 years under FDA administration of the Color Additive Amendments is atrocious. Only one new coal tar color additive—FD&C Red No. 40 (107)—has been approved during this time for broad food use. As a practical matter, color additive technology for use in food products in the United States today is not just moribund, it is completely dead.

One new company made a valiant effort to enter this market, during the mid-1970s: the Dynapol Corporation. Recognizing that FDA was beginning to disapprove important food color additives, the company sought a new approach. Focusing on FD&C Yellow Nos. 5 and 6 and on FD&C Red No. 2, the company initially made three polymeric food additives by chemically attaching the colors to nonabsorbable polymer molecules, thus in theory reducing toxicity because they would pass through the gut without being absorbed and metabolized. Five years later this effort came to an abrupt end when, in a rat bioassay, one of these compounds proved to be tumorigenic (108). The company concluded that it would not be feasible to obtain a color additive regulation with these results under the Delaney Clause even if human safety could be shown and thus went out of business.

It is easy to argue that color additives are not a major health priority, and thus that the total destruction of new technology in color additives is not of major concern. The fact remains, however, that color additives are exceedingly important to the entire food and drug supply. Color is essential to the palatability of food products. Drugs are frequently distinguished by their coloring as well as their shape. The entire purpose of many cosmetics is to impart color to the skin. A number of medical devices depend upon coloring as part of their utility and distinctiveness. It would be a drab world, indeed, if color did not exist.

D. The Lack of Market Protection

Unquestionably, the lack of any form of market protection for a new color additive has contributed greatly to the decline of this technology in the United States. The only new color additive approved in the past 40 years, FD&C Red No. 40, was the subject of a patent. It is inconceivable that any company would invest the millions of dollars and years of time necessary to obtain approval of a new color additive under the present law, which permits a competitor to market the identical product immediately, unless there is strong patent protection or the law is changed to provide a substantial period of market exclusivity. Like food additives, color additives were granted patent term restoration in 1984 but not market exclusivity (109).

V. REGULATORY SUCCESS AND FAILURE

Looking back on the past four decades, it is easy to discern areas of success and areas of failure in the regulation of food substances, food additives, and color additives.

A. The GRAS Success

Regulation of GRAS substances under the Food Additives Amendment of 1958 has been an enormous success, largely because it has been implemented outside FDA. It is a classic free market approach. The Food and Drug Administration is not the only access to the market. There are alternative approaches, through equally competent and respected scientific organizations that are much more efficient, less costly, and thus far preferable. Individual companies are free to make their own determinations based upon their own scientific expertise; independent academic experts are available; private companies that specialize in GRAS determinations can be used; and FASEB itself has now, after completing the FDA GRAS List Review, agreed to conduct private GRAS evaluations for the food industry. These compete among each other and with FDA. Under the inexorable rules of competition, in a free enterprise environment, the most effective and efficient organization will ultimately be used.

One must question whether leaving these issues to the free market compromises public health and safety. The evidence over the past four decades provides unequivocal testimony that the public health and safety has been fully protected and not in any way compromised. Not a single food substance that has been added to the food supply under a private GRAS determination based on a thorough and well-documented scientific evaluation since 1958 has been taken off the market by FDA because of a public health or safety problem. The program, in short, has been a complete success.

B. The Food Additive Failure

The failure of the food additive approval process during the past four decades scarcely needs further elaboration. It is a closed process within FDA, not subject to public scrutiny even through an FDA advisory committee, and is solely within the control of the agency. Because there is no other lawful route to the marketplace for a food additive or color additive, manufacturers have no alternative but to do whatever FDA commands. And for the same reason, FDA can demand whatever it wishes, with or without sound scientific justification, without fear of peer review, public scrutiny, or accountability to anyone else. The resulting statistics, documented earlier in this chapter, are the inevitable result.

One might consider retaining such an approach if it produced better results—that is, safer food—than other approaches might achieve. A comparison with the GRAS self-determination approach, however, demonstrates that this is not true. Far more new direct human food substances have been marketed since 1970 after an industry self-determination of GRAS than after FDA approval of a food additive or color additive petition. No product in either category has proved to present a public health hazard. The highly conservative approach of FDA thus contributes nothing more than delay, higher costs to the public for those products that eventually are approved, and a more restricted choice for consumers because the entire process has choked off innovation in this very important field.

VI. LESSONS FOR THE FUTURE

As a nation, we have distrusted one of our most precious heritages—responsible private action within an open and free competitive marketplace—and put our trust instead in a form of control that has repeatedly failed—a government agency with complete and total monopolistic power. We should have known better. No monopoly, no matter how benevolent, no matter how well-intentioned, no matter under what auspices it is established—and regardless of the political party in power—can be efficient, fair, or effective.

Our country has had a longstanding opposition to monopolies of any kind. One does not need to be an economist to understand what happens with a monopoly. Those who enjoy the privilege and protection of a monopoly need not be concerned about efficiency because they lack competition. They need not be concerned about cost because the public has no other choice. Time is of no concern because the public has no other alternative. A monopoly is just a euphemism for a dictatorship. It does what it wants, at its own pace, in its own way, and will brook no interference by others.

That is exactly what we have seen with food additives and color additives. Time is of no consequence; cost is not even considered. Technology languishes, and indeed disappears, because FDA has no incentive to be efficient and fair.

The proof of this analysis lies in the contrast between food additives and color additives on the one hand and GRAS food substances on the other. We have destroyed innovation in the entire color additive industry in this country and have almost achieved that objective for direct human food additives. For GRAS food substances, on the other hand, there is a thriving industry and a highly competitive marketplace. GRAS food substances survive, while food additives and color additives die, precisely because there is a free marketplace for the former and only a government monopoly for the latter.

There are, presumably, three ways that one could change this system. First, FDA could reform itself from within and restore the efficiency and effective regulatory approach taken in the early 1960s. It is highly questionable, however, that this will occur. No monopoly, in all of history, has voluntarily surrendered its power and has established competitive entities. The destruction of monopolies has always been a forced event.

Second, the statute could be amended to require that FDA take action on pending food additive and color additive petitions within the specific statutory time period, or the applications will be deemed to be approved. This is the ''hammer'' approach. It fails, however, to account for the likely FDA response. It is all too easy for FDA simply to issue continual denials of a food additive or color additive petition on the ground that further testing is needed. This is the classic FDA response and accounts for the complete failure of the system as we know it today. It is too much to expect that it would change simply because of the existence of a statutory hammer.

Third, and finally, FDA food additive and color additive regulation could be opened up to free competition, requiring the agency to be judged on a fair comparison with alternative review organizations who have already shown the ability to conduct this type of evaluation in a far more efficient, less time-consuming, and less costly manner. This is the only hope for the future. The market mechanism is the greatest natural form of regulation the world has ever seen. Those who are inefficient go bankrupt. Those who are efficient and effective not only survive, but command the respect and patronage of potential customers.

Let us examine just how this could work. Let us suppose that the statute were amended to permit the manufacturer of a food additive or color additive to seek the review and evaluation of a petition either from FDA or from any other governmental or nongovernmental organization established to conduct these evaluations in the same way that FDA does. Any such outside organization would be required, of course, to meet criteria and standards established by FDA for this type of review. It would be required to apply safety evaluations established by Congress in the statute itself and by FDA in its regulations.

Guidelines, such as the Red Book, would represent acceptable but not mandatory safety evaluation principles, as the FDA regulations provide (110).

Perhaps the best example of an organization that would most assuredly meet those criteria is FASEB. This organization has done this type of work for FDA for thirty years, and FDA has relied upon the FASEB work product repeatedly. It would not take anything more to demonstrate that FASEB is, at this moment, the premier food substance evaluation agency in the United States, respected to a far greater degree than is FDA itself.

The question then becomes whether approval of an organization like FASEB should, in itself, be sufficient to justify immediate marketing. Even though the author remains highly critical of what has happened to the FDA food additive approval process, we must give the agency one more chance to show that it can make a new process like this one work. Following review of a new direct human food additive by an organization like FASEB, the results should be required to be submitted to FDA, to give the agency a limited time within which they can review the decision of FASEB and accept or reject it. Perhaps six months is sufficient for this purpose. If the agency could not take definitive action within six months, the FASEB recommendation would automatically become the decision of the agency.

If this type of approach were to be pursued, it would be essential to incorporate one clear limitation. Once a third party organization like FASEB has made its determination, it would in effect become a rebuttable presumption. FDA would not be permitted, in its review of that recommendation, to veto marketing of the product on the ground that it still needs further testing. The only ground on which marketing could legitimately be stopped by FDA would be if the agency determined that there was a reasonable probability that the additive was, in fact, not safe for its intended use.

This is not a radical approach. It does not dismember FDA or destroy the notion of public protection against unsafe food. It does reject the concept that FDA is the only institution in our country that understands food safety and builds upon the principle that competition, rather than monopoly, is most likely to serve the public interest. It retains in FDA the ultimate authority to protect the public health and the responsibility for enforcement in the marketplace. Public confidence in the food supply will therefore remain undiminished and perhaps even enhanced because two qualified organizations, rather than just one, will have participated in the ultimate decision.

VII. CONCLUSION

It is sad to witness the disintegration of what was once an effective and efficient organization to review food additives and color additives within FDA. It is foolish, however, to believe that the process will be reformed and the organization rebuilt without dramatic change in the statutory provisions. The root causes of the problem are in the statute itself—the unfettered monopoly power of FDA to approve or disapprove a food additive or a color additive without effective time constraints, policy limits, or competition. Unless we want to continue to rely on GRAS food substances and to give up completely on new food additives and color additives, a change in the statute is essential.

Such a change has been proposed in this chapter, and the author believes it would work. It has already been shown to work for GRAS food substances, and there is no meaningful distinction between GRAS substances and those that are regulated as additives. It is basically the only hope for saving what is an otherwise moribund program.

NOTES

1. Peter Barton Hutt and Peter Barton Hutt II, *A History of Government Regulation of Adulteration and Misbranding of Food*, 39 Food Drug Cosmetic Law Journal 2 (January 1984).
2. Ibid. at 14.
3. 34 Stat. 768 (1906).
4. For a history of FDA, see Peter Barton Hutt, *A Historical Introduction*, 45 Food Drug Cosmetic Law Journal 17 (January 1990); Peter Barton Hutt, *The Transformation of United States Food and Drug Law*, 60 Journal of the Association of Food and Drug Officials, No. 3, at 1 (September 1996).
5. 32 Stat. 728 (1902), reenacted in 57 Stat. 682, 702 (1944), now codified in 42 U.S.C. 351.
6. 52 Stat. 1040 (1938), 21 U.S.C. 301 et. seq.
7. The 1938 Act contained new provisions governing the safety of nonadded substances and environmental contaminants, but these did not represent fundamental changes in food safety policy. The statutory prohibition against added poisonous or deleterious substances, which originated in the English Sale of Food and Drugs Act of 1875, 38 & 39 Vict., c.63 (1875), was retained in the 1938 Act unchanged from the 1906 Act. The interpretation of this provision in *United States* v. *Lexington Mill & Elevator Co.*, 232 U.S. 399 (1914) remains the controlling precedent today.
8. 96 Cong. Rec. 8933 (June 20, 1950).
9. H.R. Rep. No. 2356, 82d Cong., 2d Sess. (1952).
10. National Academy of Sciences, *The Use of Chemical Additives in Food Processing*, NAS Pub. No. 398 (1956).
11. 68 Stat. 511 (1954).
12. 72 Stat. 1784 (1958).
13. 74 Stat. 397 (1960).
14. A prior sanction constitutes a permanent exclusion for the sanctioned substance from the definition of a food additive. Although there are important prior sanctions, they obviously play no role in any decision to market a new food substance and therefore are not discussed further in this chapter. See generally Peter Barton Hutt and Richard A. Merrill, *Food and Drug Law: Cases and Materials*, 342–346 (2d ed. 1991).
15. FDA has developed a food additive informational database that is, unfortunately, falsely referred to as ''everything added to food in the United States'' (EAFUS). Not only does this database fail to cover everything that is in fact added to food in the United States, but it does not even cover a significant number of those substances.
16. 75 Stat. 42 (1961).
17. 78 Stat. 1002 (1964).
18. 21 C.F.R. Part 120 (1966).
19. Prior to enactment of the 1958 Amendment, FDA had submitted to Congress a list of representative food substances that FDA regarded as GRAS. ''Food Additives,'' Hearings before a Subcommittee of the Committee on Interstate and Foreign Commerce, House of Representatives, 85th Cong., 461–462 (1958).
20. 23 Fed. Reg. 9511 (December 9, 1958); 24 Fed. Reg. 9368 (November 20, 1959); 25 Fed. Reg. 880 (February 2, 1960); 25 Fed. Reg. 7332 (August 4, 1960); 26 Fed. Reg. 938 January 31, 1961).
21. This statutory policy is reflected in 21 C.F.R. 170.30(d).
22. In one instance, an industry trade association has undertaken a continuing GRAS review of an entire functional field of food substances rather than leaving this matter to FDA. The Flavor and Extract Manufacturers Association has conducted a GRAS review of flavor substances for forty years. John B. Hallagan and Richard L. Hall, *FEMA GRAS—A GRAS Assessment Program for Flavor Ingredients*, 21 Regulatory Toxicology and Pharmacology 422 (June 1995).

23. An attempt by FDA to extend food additive requirements to food substances marketed in a capsule was overturned by the courts. *United States* v. *Two Plastic Drums . . . Black Current Oil*, 984 F.2d 814 (7th Cir. 1992); *United States* v. *29 Cartons of . . . An Article of Food*, 987 F.2d 33 (1st Cir. 1993).
24. 26 Fed. Reg. 938 (January 31, 1961).
25. 34 Fed. Reg. 17053 (October 21, 1969).
26. 34 Fed. Reg. 19547 (December 11, 1969); 34 Fed. Reg. 20426 (December 31, 1969); 35 Fed. Reg. 2774 (February 10, 1970); 35 Fed. Reg. 5008 (March 24, 1970); 35 Fed. Reg. 11177 (July 11, 1970). See ''Cyclamate Sweeteners,'' Hearing Before a Subcommittee of the Committee on Government Operations, House of Representatives, 91st Cong., 2d Sess. (1970); ''Regulation of Cyclamate Sweeteners,'' H.R. Rep. No. 91–1585, 91st Cong., 2d Sess. (1970).
27. HEW News No. 70-42 (August 14, 1970); 35 Fed. Reg. 13644 (August 27, 1970).
28. ''Consumer Protection,'' 5 Weekly Compilation of Presidential Documents 1516 (November 3, 1969).
29. To implement the GRAS List Review, FDA published notices and promulgated procedural regulations establishing criteria for GRAS status and a systematic mechanism for agency handling of FASEB reports. 35 Fed. Reg. 18623 (December 8, 1970); 36 Fed. Reg. 12093 (June 25, 1971); 37 Fed. Reg. 6207 (March 25, 1972); 37 Fed. Reg. 25705 (December 2, 1972); 38 Fed. Reg. 20051, 20053, 20054 (July 26, 1973); 39 Fed. Reg. 34194, 34218 (September 23, 1974); 41 Fed. Reg. 53600 (December 7, 1976).
30. Select Committee on GRAS Substances, *Evaluation of Health Aspects of GRAS Food Ingredients: Lessons Learned and Questions Unanswered*, 36 Federation Proceedings 2519 (October 1977).
31. 38 Fed. Reg. 20044 (July 26, 1973); 39 Fed. Reg. 34173 (September 23, 1974).
32. 36 Fed. Reg. 22617 (November 25, 1971); 37 Fed. Reg. 25356 (November 30, 1972).
33. 42 Fed. Reg. 5921 (March 5, 1973); 39 Fed. Reg. 27317 (July 26, 1974); 40 Fed. Reg. 56907 (December 5, 1975); 44 Fed. Reg. 31716 (June 1, 1979); 46 Fed. Reg. 38285 (July 24, 1981); 46 Fed. Reg. 50947 (October 16, 1981).
34. 44 Fed. Reg. 22816 (April 17, 1979); 46 Fed. Reg. 30080 (June 5, 1981).
35. 47 Fed. Reg. 46139 (October 14, 1982); 53 Fed. Reg. 228379 (July 28, 1988).
36. 51 Fed. Reg. 687 (January 7, 1986); 52 Fed. Reg. 45867 (December 2, 1987); 55 Fed. Reg. 39613 (September 28, 1990). Gellan gum could undoubtedly have been determined to be GRAS, but is included here for purposes of completeness.
37. 52 Fed. Reg. 23606 (June 23, 1987); 61 Fed. Reg. 3118 (January 20, 1996).
38. 52 Fed. Reg. 17475 (May 8, 1987); 63 Fed. Reg. 16417 (April 3, 1998).
39. 52 Fed. Reg. 43927 (September 5, 1991); 64 Fed. Reg. 29949 (June 4, 1999). This ingredient could undoubtedly have been determined to be GRAS, but is included here for purposes of completeness.
40. The 1990 approval of the translation product of the marker gene for the genetically altered tomato is excluded because this could easily have been handled as a GRAS regulation and in any event cannot be regarded as a direct human food additive. See note 49 *infra*.
41. Jennifer Lawrence, *How P&G's Hopes for Food Division's Future Got Mired in FDA Quicksand*, Advertising Age, May 2, 1994, at 16.
42. 107 Stat. 2040 (1993), 35 U.S.C. 156(d)(5); ''Patent Extension Hearing,'' Hearing before the Subcommittee on Patents, Copyrights and Trademarks of the Committee on the Judiciary, United States Senate, 102d Cong., 1st Sess. (1991).
43. 51 Fed. Reg. 34503 (September 29, 1986).
44. ''Food Additives: Competitive, Regulatory, and Safety Problems,'' Hearing Before the Select Committee on Small Business, United States Senate, 95th Cong., 1st Ses. 36 (1977).
45. Alan M. Rulis and Richard J. Ronk, *Cyclic Review—Looking Backward or Looking Forward?*, 36 Food Drug Cosmetic Law Journal 156 (April 1981), Merton V. Smith and Alan

M. Rulis, *FDA's GRAS Review and Priority-Based Assessment of Food Additives*, 35 Food Technology, No. 12, at 71 (December 1981).
46. 51 Fed. Reg. 23309 (June 26, 1986).
47. 57 Fed. Reg. 22984 (May 29, 1992).
48. E.g., 53 Fed. Reg. 3792 (February 9, 1988); 54 Fed. Reg. 40910 (October 4, 1989); 55 Fed. Reg. 10932 (March 23, 1990); 55 Fed. Reg. 10932 (March 23, 1990); 58 Fed. Reg. 27197 (May 7, 1993); 21 C.F.R. 184.1685 (chymosin). See also 53 Fed. Reg. 5319 (February 23, 1988); 53 Fed. Reg. 16191 (May 5, 1988); 54 Fed. Reg. 20203 (May 10, 1989); 55 Fed. Reg. 10113 (March 19, 1990).
49. 56 Fed. Reg. 20004 (May 1, 1991); 57 Fed. Reg. 22772 (May 29, 1992); 58 Fed. Reg. 38429 (July 16, 1993); 59 Fed. Reg. 26647 (May 23, 1994); 59 Fed. Reg. 26700 (May 23, 1994); 21 C.F.R. 173.170.
50. On many occasions, the author has strongly supported realistic and flexible toxicology regulations and guidelines that establish general principles without eliminating individual scientific judgment based upon the facts presented by the data on particular substances. Peter Barton Hutt, *Public Participation in Toxicology Decisions*, 32 Food Drug Cosmetic Law Journal 275 (June 1977). As these regulations and guidelines have developed, however, the elements of flexibility and judgment have been submerged. Rigid rules have unfortunately emerged to take their place.
51. Food Safety Council, *Proposed System for Food Safety Assessment: Final Report of the Scientific Committee of the Food Safety Council* (June 1980).
52. 47 Fed. Reg. 46141 (October 15, 1982).
53. 58 Fed. Reg. 16536 (March 29, 1993); 58 Fed. Reg. 40151 (July 27, 1993).
54. Under 21 C.F.R. 10.90(b)(1) and section 701(h) of the FD&C Act, 21 U.S.C. 371(h), a guideline is defined as an approach that is acceptable to FDA but that is not a legal requirement or that must or even necessarily should be used by industry. The FDA's rigid insistence on use of the Red Book in determining required testing for a food additive directly violates these provisions.
55. See, for example, 45 Fed. Reg. 61474 (September 16, 1980).
56. "Cyclamate Update," FDA Talk Paper No. T89-35 (May 16, 1989); Malcolm Gladwell, *FDA Confirms Cyclamate Ban May End*, Washington Post, May 17, 1989, p. A5.
57. 37 Fed. Reg. 6207 (March 25, 1972); 37 Fed. Reg. 25705 (December 2, 1972).
58. 39 Fed. Reg. 34468 (September 25, 1974); 42 Fed. Reg. 26467 (May 24, 1977).
59. Nathaniel Tripp, *The Miracle Berry: The Fantastic Story of How a Business Venture in Artificial Sweeteners Went Sour*, Horticulture, January 1985, at 58.
60. Six GRAS affirmation petitions were eventually submitted. 39 Fed. Reg. 28310 (August 6, 1974); 41 Fed. Reg. 17953 (April 29, 1976); 41 Fed. Reg. 53545 (December 7, 1976); 42 Fed. Reg. 27298 (May 27, 1977); 42 Fed. Reg. 28601 (June 3, 1977); 46 Fed. Reg. 15953 (March 10, 1981).
61. In 48 Fed. Reg. 5715, 5716 (February 8, 1983), FDA promulgated a regular GRAS regulation for high fructose corn syrup, 21 C.F.R. 182.1866; a GRAS affirmation regulation for the insoluble glucose isomerase enzyme preparations used to make high fructose corn syrup, 21 C.F.R. 184.1372; and a food additive regulation for diethylaminoethylcellulose (DEAE-cellulose) and glutaraldehyde as fixing agents in the immobilization of glucose isomerase enzyme preparations, 21 C.F.R. 173.357.
62. FDA upgraded high fructose corn syrup to the status of affirmed GRAS: 53 Fed. Reg. 44904 (November 7, 1988); 61 Fed. Reg. 43447 (August 23, 1996); 21 C.F.R. 184.1866. An equally egregious example of delay is the food substance Japan wax. This substance was proposed for GRAS affirmation in 47 Fed. Reg. 29965 (July 9, 1982); thirteen years later was reproposed for broader GRAS use in 60 Fed. Reg. 28555 (June 1, 1995); and the matter is not yet concluded.

63. Richard A. Merrill, *Food Additive Approval Procedures Discourage Innovation*, 50 Food Technology, No. 3, at 110 (March 1996); Joseph V. Rodericks, *Safety Assessment of New Food Ingredients*, 50 Food Technology, No. 3, at 114 (March 1996); Peter Barton Hutt, *Approval of Food Additives in the United States: A Bankrupt System*, 50 Food Technology, No. 3 at 118 (March 1996).
64. This was not the first time that Congress reviewed food additive issues under the 1958 Amendment. See, for example, ''Food Additives: Competitive, Regulatory, and Safety Problems,'' Hearings Before the Select Committee on Small Business, United States Senate, 95th Cong., 1st Sess., Parts 1 and 2 (1997); ''Oversight of Food Safety, 1993,'' Hearings Before the Committee on Labor and Human Resources, United States Senate, 98th Cong., 1st Sess. (1983). It was the first time, however, that Congress focused on the substantial deterioration of the FDA petition review process.
65. ''Delays in the FDA's Food Additive Petition Process and GRAS Affirmation Process,'' Hearings Before the Subcommittee on Human Resources and Intergovernmental Relations of the Committee on Government Reform and Oversight, United States House of Representatives, 104th Cong., 1st Sess. (1995).
66. ''The FDA Food Additive Review Process: Backlog and Failure to Observe Statutory Deadline,'' H.R. Rep. No. 104–436, 104th Cong., 1st Sess. (1995).
67. All of the papers presented at the workshop have been published in 24 Regulatory Toxicology and Pharmacology, No. 3, at 213 et. seq. (December 1996).
68. Lars Noah and Richard A. Merrill, *Starting From Scratch?: Reinventing the Food Additive Approval Process*, 78 Boston University Law Review 329 (April 1998).
69. Alan M. Rulis and Laura M. Tarantino, *Food Ingredient Review at FDA: Recent Data and Initiatives to Improve the Process*, 24 Regulatory Toxicology and Pharmacology 224, 226 (December 1996).
70. 111 Stat. 2296 (1997).
71. 111 Stat. 2296, 2353 (1997).
72. 62 Fed. Reg. 64107 (December 3, 1997).
73. 64 Fed. Reg. 157 (January 5, 1999).
74. Section 201(s) of the FD&C Act, 21 U.S.C. 321(s).
75. Food Chemical News, February 19, 1968, p. 3, August 19, 1968, p. 3, and November 18, 1968, p. 3. See L.L. Ramsey, *The Food Additive Problem of Plastics Used in Food Packaging* (November 1969).
76. 58 Fed. Reg. 52719 (October 12, 1993).
77. 60 Fed. Reg. 36582 (July 17, 1995), 21 C.F.R. 170.39.
78. Section 309 of the FDA Modernization Act, 111 Stat. 2296, 2354 (1997), amending Section 409 of the FD&C Act, 21 U.S.C. 348.
79. H.R. Rep. No. 105–399, 105th Cong., 1st Sess. 99 (1997).
80. Letter from HHS Donna E. Shalala to Vice President Al Gore (May 28, 1998).
81. 64 Fed. Reg. 8577 (February 22, 1999).
82. Peter Barton Hutt, *Government Regulation of Health Claims in Food Labeling and Advertising*, 41 Food Drug Cosmetic Law Journal 3 (January 1986).
83. 90 Stat. 401 (1976).
84. 104 Stat. 2353 (1990).
85. Section 403(r)(5)(D) of the FD&C Act, 21 U.S.C. 343(r)(5)(D).
86. 58 Fed. Reg. 33700 (June 18, 1993); 58 Fed. Reg. 53296 (October 14, 1993); 59 Fed. Reg. 395 (January 4, 1994).
87. 58 Fed. Reg. 33690 (June 18, 1993).
88. 108 Stat. 4325 (1994).
89. Sections 402(f) and 413 of the FD&C Act, 21 U.S.C. 342(f) and 350(b).
90. 98 Stat. 1585 (1984); 35 U.S.C. 156; Section 505(j)(4)(D)(ii) of the FD&C Act, 21 U.S.C.

355(j)(4)(D)(ii). See Ellen J. Flannery and Peter Barton Hutt *Balancing Competition and Patent Protection in the Drug Industry: The Drug Price Competition and Patent Term Restoration Act of 1984*, 40 Food Drug Cosmetic Law Journal 269 (July 1985).

91. 96 Stat. 2049 (1983); Section 527(a) of the FD&C Act, 21 U.S.C. 360cc.
92. 35 U.S.C. 156.
93. 31 Stat. 191, 196 (1900).
94. Section 7 of the Federal Food and Drug Act, 34 Stat. 768, 769–770 (1906); ''Dyes, Chemicals, and Preservatives in Foods,'' USDA Food Inspection Decision No. 76 (June 18, 1907); ''Certificate and Control of Dyes Permissible for Use in Coloring Food and Foodstuffs,'' USDA Food Inspection Decision No. 77 (September 16, 1907); ''Certification of Coal-Tar Food Colors,'' USDA Misc. Circ. No. 52 (1925).
95. Section 406(b) of the FD&C Act of 1938, 52 Stat. 1040, 1049 (1938), repealed and replaced by Section 706 as added by the Color Additive Amendments of 1960, now renumbered as Section 721 of the FD&C Act, 21 U.S.C. 379e.
96. *Fleming* v. *Florida Citrus Exchange*, 358 U.S. 153 (1958).
97. 74 Stat. 397 (1960).
98. Section 721(b)(4) of the FD&C Act, 21 U.S.C. 379e(b)(4).
99. B.C. Hesse, *Coal-Tar Colors Used in Food Products*, USDA Bureau of Chemistry Bull. No. 147 (1912).
100. 25. Fed. Reg. 9759 (October 12, 1960).
101. 21 C.F.R. Part 74, Subpart A. In addition, 26 non–coal-tar color additives are approved for direct human food use. 21 C.F.R. Part 73, Subpart A. Two other coal-tar color additives, Orange B and Citrus Red No. 2, are limited to coloring casings on surfaces of frankfurters and sausages and the skins of oranges, respectively. 21 C.F.R. 74.250, 74.302.
102. 74 Stat. 397, 404 (1960). These transitional provisions were not codified in substantive law and thus are not in the FD&C Act or the United States Code.
103. It was the decision to require new chronic toxicity testing for all color additives, based on a new and more demanding protocol than had been required when the studies were originally done years ago, that prompted the short-lived concept of a cyclic review for all food additives. See note 45 *supra*.
104. *Certified Color Manufacturers Association* v. *Mathews*, 543 F.2d 284 (D.C. Cir. 1976), *McIlwain* v. *Hayes*, 690 F.2d 1041 (D.C. Cir. 1982), *Public Citizen* v. *Young*, 831 F.2d 1108 (D.C. Cir. 1987).
105. 44 Fed. Reg. 36411 (June 22, 1979); 61 Fed. Reg. 8372 (March 4, 1996); 21 C.F.R. 81.1.
106. 21 C.F.R. Part 73, Subpart A.
107. 35 Fed. Reg. 10529 (June 27, 1970); 36 Fed. Reg. 6892 (April 10, 1971); 21 C.F.R. 74.340. FD&C Red No. 40 is the only color additive for which FDA has approved lakes as of this time.
108. ''Polymer-Bonded Color Additive Results in Tumors in Rats,'' Food Chemical News, October 13, 1980, at 22.
109. 35 U.S.C. 156.
110. See note 54 *supra*.

9

Nutritional Additives

MARILYN A. SWANSON

National Food Service Management Institute, University, Mississippi

PAM EVENSON

South Dakota State University, Brookings, South Dakota

I. INTRODUCTION

The term nutritional additives can be used to mean the addition of vitamins, minerals, amino acids, fatty acids, as well as other pure chemical compounds to food in order to improve or maintain the nutritional quality of foods. However, manufacturers soon discovered that along with an improvement in nutritional qualities, nutritional additives often provide functional qualities.

The earliest use of nutritional additives was to correct dietary deficiencies. In 1833, the French chemist Boussingault recommended the addition of iodine to table salt to prevent goiter. Salt was first iodized in the United States in 1924 when it was shown that the addition of sodium iodine was effective in preventing goiter. Other examples include vitamin D added to milk and vitamin A added to margarine.

Nutritional additives can be used to restore nutrients to levels found in the food before storage, packaging, handling, and processing. An early example of this is the enrichment of grain products, corn meal, and rice. Another use of nutritional additives is to improve the nutritional status or correct nutritional inferiority in a food that replaces a more traditional nutritional food, an example would be the fortification of breakfast drink substitutes with folacin and vitamin C. With the advent of nutritional labeling and increased public interest in nutritional properties of food, the food industry rapidly recognized that the addition of nutritional additives can be a selling point.

Although it is often thought that the major reason to add nutritional additives to a food supply is to provide nutrients and improve dietary status, nutrients are also added for a variety of other purposes. For example, vitamins C and E may be used for antioxidant properties; beta carotene may be used to provide color. In these cases consumers obtain both a functional and nutrient advantage.

Like all additives, nutritional additives are commercially available in an array of forms such as powders, encapsulated in gelatin, emulsified in oil. The form used depends upon the type of application. Nutrition additives are often protected by protective additives such as antioxidants. Two critical factors in the selection of the form of the additive are stability of the vitamin preparation and its miscibility with the intended food matrix. In order to ensure the second criterion, the point of incorporation of the additive during the food manufacturing process can also be critical.

This chapter is limited to a discussion of nutritional additives as nutrients only, rather than nutrients as pharmaceutical preparations. Included are such nutrients as vitamins, amino acids, fatty acids, minerals, and dietary supplements.

II. VITAMINS

Vitamins are organic compounds that facilitate a variety of biological processes. Since the body cannot manufacture vitamins they must be obtained naturally from foods or added to foods. Vitamins are classified as water soluble or fat soluble depending on their characteristics. Water soluble vitamins show poor solubility in water; they are not stored in body tissues in high amounts and excesses are generally excreted in urine. Fat soluble vitamins are linked with lipid metabolism and are generally insoluble or poorly soluble in water. Since fat soluble vitamins can be stored in the human body it is less important to have a daily supply of these vitamins. Excess amounts of fat soluble vitamins can in some cases lead to toxic effects. The fat soluble vitamins are vitamins A, D, E, and K. Vitamins A, C, and E are also classed as antioxidants because they have the capacity to protect the body from free radicals.

A. Vitamin A

The beginning of the modern nutritional history of vitamin A started in 1913 with the recognition of a growth-stimulating factor in cod liver oil and butter. The active principal component of liver oil is vitamin A. Adequate intake of vitamin A is essential for normal vision, growth, cellular differentiation, reproduction, and integrity of the immune system (1). Olson (2) has extensively reviewed the subject of vitamin A.

Deficiency of vitamin A is commonly seen as night blindness and/or dry and lusterless corneas. Vitamin A is one of the major public health problems in less industrialized countries. In fact, it is the major cause of blindness in the developing world. Inadequate amounts are also associated with protein-calorie malnutrition, low intake of fat, lipid malabsorption syndromes, and febrile diseases (3). Stephensen et al. (4) found that vitamin A, although not normally excreted from the body, was excreted in cases of acute infections with increased risk of developing deficiency. Acute infections of childhood such as measles, respiratory infections, and diarrhea have also been associated with increased risk of developing vitamin A deficiency. Also of concern is the susceptibility of pregnant women to vitamin A depletion because of increased demands during pregnancy and altered dietary habits. Inadequate maternal vitamin A stores may not provide sufficient vitamin A needed for the rapidly growing fetus or satisfactory breast milk concentrations after birth (5).

1. Chemistry

In natural forms, preformed vitamin A, retinol, retinyl esters, and retinaldehydes represent vitamin A. The predominant isomer, all-trans-retinol, possesses maximal vitamin A activity. In most foods, this retinol forms esters with long chain fatty acids, particularly palmitic acid (1).

2. Units and Requirements

Vitamin A has been traditionally expressed in international units (IU). One IU is defined as the amount of vitamin A activity contained in 0.344 μg of all-trans-retinyl acetate, which is equivalent to 0.300 μg of all-trans-retinol (1). However, in 1965 IUs were replaced with the preferred measurement of retinol equivalents (RE), expressed in micrograms of retinol. This unit is defined as the amount of retinol plus the equivalent amount of retinol that can be obtained from the provitamin A carotenoids. Recommended dietary allowances (RDAs) have been redefined based on age, body mass, metabolic activity, and special conditions such as pregnancy and lactation. The RDA for males 11 to 51+ currently is 1000 μg RE (6).

3. Occurrence

Preformed vitamin A is found in foods of animal origin, either in storage areas such as the liver or associated with the fat of milk and eggs (7).

4. Properties

In liquid form, vitamin A is a light yellow to red oil that may solidify on refrigeration. In this form it is very soluble in chloroform and in ether. It is soluble in absolute alcohol and in vegetable oils, but is insoluble in glycerin and in water. In solid form, it may have the appearance of the diluent that has been added to it. It may be nearly odorless or have a mild fishy odor, but it has no rancid odor or taste. In solid form it may be dispersible in water. It is unstable to air and light. Vitamin A should be stored in a cool place in tight containers, preferably under an atmosphere of an inert gas, protected from light (8).

5. Analysis

Several different quantification procedures for vitamin A have been described in literature. The preferred approach is using retinyl acetate, which is converted to retinol by saponification. This approach is preferred because crystalline all-trans-retinyl acetate is commercially available in high purity and is free from cis isomers. High-performance liquid chromatography (HPLC) using UV detection is the most effective for determining vitamin A. Ball (1) explains various HPLC methods for determination of vitamin A compounds in food.

6. Commercial Forms

Vitamin A is available in pure form by chemical synthesis as vitamin A palmitate or the acetate, or recovered from molecularly recovered fish oil (9). Several different commercial preparations are available for fortification: 250 CWS, 250 S, 250 SD, 500, Emulsified RP and Oil (9).

7. Toxicity

In amounts several times higher than the recommended dietary allowance, vitamin A will cause toxicity in humans. With children, signs and symptoms of acute vitamin A toxicity include anorexia, bulging fontanelles, drowsiness, increased intracranial pressure, irritability, and vomiting. Signs and symptoms in adults include abdominal pain, anorexia, blurred vision, drowsiness, headache, hypercalcemia, irritability, muscle weakness, nausea, vomiting, peripheral neuritis, and skin desquamation (10). Toxicity is of greatest concern in persons with compromised liver function, children, and pregnant women. Symptoms disappear when excess intakes are discontinued (7).

B. Carotenoids

Although not vitamins per se, carotenoids serve as a precursor to provide vitamin A (11). This provitamin is converted by the body to vitamin A with various degrees of efficiency. The yellow, red, orange, and violet pigments of carotenoids are largely responsible for the color of many vegetables and fruits. Carotenoids function not only as color and nutrient compounds but also as antioxidants (12). Although carotenoids may have favorable effects and may reduce the risk of some diseases, since carotenoids are not essential nutrients there is no need to use the term ''carotenoid deficiency'' (13).

1. Chemistry

Carotenoids can be divided chemically into two groups: the hydrocarbon carotenoids known as carotenes and the oxygenated derivatives known as xanthophylls. Beta carotene, most likely the most important carotenoid, is made of two molecules of retinol and possesses maximal provitamin A activity (14). There are an estimated 500+ carotenoids found in nature. Approximately 60 carotenoids possess varying levels of provitamin A activity.

2. Units and Requirements

In 1965, an expert committee decided to abandon the international unit measurement for vitamin A in favor of retinol equivalents (14), a purely dietary concept defined as the amount of retinol plus the equivalent amount of retinol that can be obtained from the provitamin A carotenoids. To convert IUs to REs the following equations can be used: 1 RE = 1 μg retinol, and 1 IU = 0.3 μg retinol. However, IUs were based on studies that did not take into account the poor absorption and biological availability of carotenoids in foods. When figuring carotenoids the following equations should be used. In the RE system, 1 μg retinol = 6 μg beta carotene. In the IU system, 1 μg retinol = 2 μg beta carotene (14). There is no separate RDA listing for carotenoids.

3. Occurrence

Provitamin A carotenoids are synthesized exclusively by higher plants and photosynthetic microorganisms. Carotenoids are mainly obtained from plant sources such as carrots, green leafy vegetables, spinach, oranges, and tomatoes. Animal sources include calf liver, whole milk, butter, cheddar cheese, and eggs.

4. Properties

Carotenoids are soluble in most organic solvents but not in water, acids, or alkalies. They are sensitive to oxidation, isomerization, and polymerization when dissolved in dilute solution under light and in the presence of oxygen (13). Beta carotene melts between 176 and 182°C with some decomposition. Optimal storage conditions include a cool place in a tight, light-resistant container under inert gas (15).

5. Analysis

Individual carotenoids have been well characterized using UV and visible adsorption spectroscopy (VIS), infrared spectroscopy, nuclear magnetic resonance spectroscopy, mass spectrometry, and other physical methods (13). Beta carotene, for example, has a wavelength maximum of 450 nm in hexane. Most carotenoids will not fluoresce, but they will form colored complexes with Lewis acids (13).

6. Commercial Forms

Beta carotene is available as red crystals or crystalline powder (CAS: 7236-40-7) (15). Carotenoids are highly sensitive to loss by oxidation and should be protected with appropriate packing materials and storage conditions (12). Most of the beta carotene available commercially is of synthetic origin. These synthetic compounds are identical in every way, both chemically and biologically, to substances isolated from natural sources (13).

7. Toxicity

A yellow, jaundicelike coloration of the skin is evident in individuals who routinely ingest excessively large amounts of carotenoids. This condition may be especially noticeable on the nasolabial folds, the fat pads of the palms of the hands, and the fatty areas of the soles of the feet. However, this condition is completely benign and will slowly disappear once an excess of carotenoid rich foods are removed from the diet (13).

C. Vitamin D

Vitamin D has existed on earth for an estimated 500 million years. The primary function of vitamin D is to maintain serum calcium and phosphorus concentrations in a range that supports cellular processes, neuromuscular function, and bone ossification (16). Vitamin D was found to prevent and cure rickets, a disease associated with malformation of bones. Victims of rickets have traditionally been poor children in industrialized cities where exposure to sunlight has been limited. Individuals in societies where custom requires complete clothing coverage, as well as individuals in areas where sunlight days are limited, may be at greater risk for vitamin D deficiency.

1. Chemistry

There are two common forms of vitamin D: ergocalciferol, known as Vitamin D2, and cholecalciferol, known as vitamin D3. Precursors of vitamin D2 are found in sterol fractions in both animal and plant tissues. Vitamin D3 is produced by the action of sunlight on skin.

2. Units and Requirements

The 1997–1998 *Dietary Reference Intakes* recommend adequate intake (AI) for vitamin D at 5 μg from infancy to age 50, 10 μg for ages 51–70, and 15 μg for over the age of 70 (17).

3. Occurrence

Food sources are mainly animal products in unfortified foods. Saltwater fish such as herring, salmon, and sardines and fish liver oils are good sources. Beginning in the 1930s, milk was fortified with 400 IU (10 μg) of vitamin D2 per quart. Other major sources of fortified foods include butter, margarine, cereals, and chocolate mixes. Vitamin D can be produced in the skin after exposure to sunlight. However, in areas with limited sunlight, air pollution, and large cities where buildings block adequate sunlight, the body's ability to synthesize sufficient amounts of vitamin D may be compromised. Exposure to sunlight must be direct, as windowpane glass absorbs ultraviolet B radiation severely limiting production of cholecalciferol. The elderly, who may have an intolerance for milk or a dislike for milk and other dairy products, may not be obtaining adequate intake levels of vitamin D. Of concern with these individuals is also their age-related decrease in bone mass and their growing susceptibility to falls and fractures (18). Future recommendations for nursing care facilities may include indoor areas with ultraviolet B radiation in order for residents to be able to meet their requirements (19).

4. Properties

Vitamin D is a stable vitamin but is affected by air and by light. Both vitamin D2 and vitamin D3 should be stored in hermetically sealed containers under nitrogen in a cool place and protected from light. Trace metals such as copper and iron will act as prooxidants.

5. Analysis

Assay can be completed using a suitable high pressure liquid chromatograph and comparison to USP Ergocalciferol Reference Standard and USP Cholecalciferol Reference Standard (20).

6. Commercial Forms

Both vitamin D2 (CAS: 50-14-6) and vitamin D3 (CAS: 67-97-0) are white, odorless crystals. Commercially available forms include fat soluble crystals for use in high fat content foods and encapsulated, stabilized versions of the fortificant suitable for use in dry products to be reconstituted with water (21).

7. Toxicity

Excessive amounts of vitamin D are not normal unless there is excessive use of supplemental vitamins. Toxicity includes hypercalcemia, hypercalciuria, anorexia, nausea, vomiting, thirst, polyuria, muscular weakness, joint pains, diffuse demineralization of bones, and general disorientation. Excesses of supplemental vitamin D can lead to massive stimulation of intestinal calcium absorption and bone calcium resorption and ultimately soft tissue calcification and kidney stones. Left unchecked, death will eventually occur (22).

D. Vitamin E

Vitamin E is the major lipid-soluble, membrane-localized antioxidant in humans (23). First discovered in 1922, vitamin E acts in foods to prevent the peroxidation of polyunsaturated fatty acids. In the gut, it enhances the activity of vitamin A by preventing oxidation in the intestinal tract. Vitamin E, at the cellular level, also appears to protect cellular and subcellular membrances from deterioration by scavenging free radicals that contain oxygen. Cell-mediated immunity in healthy elderly subjects treated with 800 mg of vitamin E in the form of DL-α-tocopheryl acetate was found to be significantly improved (24).

Sources of dietary vitamin E are widely available and as a result deficiencies are relatively uncommon. Deficiencies that do occur are usually associated with malabsorption or lipid transport abnormalities. Little vitamin E is transferred across the placenta and newborn infants may have low tissue concentrations. Very low birth weight infants (less than 1.5 kg) should be monitored for their serum vitamin E levels (25). Research continues in regard to the relationship of vitamin E consumption and the risk of coronary disease in women and men. Data from the Nurses' Health Study that began in 1976 suggest that among middle aged women the use of vitamin E supplements are associated with a reduced risk of coronary heart disease (26). The Health Professionals Follow-Up Study in 1986 provided evidence of an association between a high intake of vitamin E and a lower risk of coronary heart disease in men (27). However, further studies are necessary before major public policy recommendations should be made.

1. Chemistry

Eight vitamers of vitamin E occur in nature, four tocopherols and four tocotrienols (28). These compounds consist of substituted hydroxylated ring systems linked to a phytyl side chain. The four major forms are designated alpha, beta, delta, and gamma based on the number and position of methyl groups on the chromanol ring.

2. Units and Requirements

Vitamin E has traditionally been expressed as international units. By international agreement, the 1989 RDAs are listed by milligrams of α-tocopherol equivalents. One milligram of D-α-tocopherol is equivalent to 1.49 IU. The recommended intake of the different forms of vitamin E depends in part on the bioactivity of each form as well as the amount of polyunsaturated fatty acids (PUFAs) consumed. Since intake of PUFAs varies with individuals, the amount of vitamin E needed to balance the minimum requirement for PUFAs is not known, but is estimated to be 3 to 4 mg α-tocopherol (4.5 to 6.0 IU) per day (25). The 1989 recommended dietary allowances for males 11 and older is 10 mg α-tocopherol, and for women 11 and older is 8 mg α-tocopherol, unless pregnant or lactating (29).

3. Occurrence

The richest sources of vitamin E are vegetable oils—including soy bean, corn, cottonseed, and safflower—and products made from these oils such as margarine, shortening, and mayonnaise, as well as wheat germ, nuts, and other grains. Meats, fish, animal fat, and most fruits and vegetables contain little vitamin E.

4. Properties

The acetate ester of α-tocopherol, rather than the free alcohol, is used as a food supplement because of its greater stability (28). Vitamin E is a slightly viscous, pale yellow, oily liquid obtained from molecular distillation of byproducts from vegetable oil refining or by chemical synthesis. The free alcohol form is highly unstable to oxidation and is used in foods as an antioxidant to stabilize the lipid component of foods. Cold water soluble forms have been produced by encapsulation with a suitable matrix (30). Vitamin E should be stored in tight containers blanketed by inert gas and protected from heat and light.

5. Analysis

Current methods of analysis are based on high-performance liquid chromatography. Gas chromatography (GC) methods have also been described for determining vitamin E content of foods, oils, and other substances, but are not routinely used (23).

6. Commercial Forms

The following forms are available as nutrients and/or dietary supplements (31):

DL-α-Tocopherol. CAS: 2074-53-5. Occurs as a yellow to amber, nearly odorless, clear, viscous oil that oxidizes and darkens in air and on exposure to light. Insoluble in water; freely soluble in alcohol; and miscible with acetone, chloroform, ether, fats, and vegetable oils.

D-α-Tocopherol concentrate. Obtained by the vacuum steam distillation of edible vegetable oil products, comprising a concentrated form of d-α-tocopherol. Occurs as brownish red to light yellow, nearly odorless, clear, viscous oil.

Tocopherols concentrate, mixed. Specifications for two types of mixed tocopherols concentration: high-alpha and low-alpha. Obtained by vacuum steam distillation of edible vegetable oil products and differing only in levels of d-tocopherol forms.

D-α-Tocopheryl acetate. CAS: 58-95-7. Obtained by vacuum steam distillation and acetylation of edible vegetable oil products. Occurs colorless to yellow, nearly odorless, clear, viscous oil. May solidify on standing, and melts at about 25°C. Unstable in presence of alkalies.

DL-α-Tocopheryl acetate. CAS:7695-91-2. Colorless to yellow or greenish yellow, nearly odorless, clear, viscous oil. Unstable in presence of alkalies.

D-α-Tocopheryl acetate concentrate. Obtained by vacuum steam distillation and acetylation of edible vegetable oil products. May be adjusted by suitable physical or chemical means. Occurs as light brown to light yellow, nearly odorless, clear, viscous oil.

D-α-Tocopheryl acid succinate. CAS: 4345-03-3. Obtained by vacuum steam distillation and succinylation of edible vegetable oil products. Occurs as a white to off-white crystalline powder having little or no taste or odor. Stable in air but unstable to alkali and to heat. Melts at about 75°C.

7. Toxicity

Vitamin E is relatively nontoxic. However, large intakes of vitamin E might interfere with absorption of Vitamin A and vitamin K. Doses necessary to elicit toxicity far exceed those necessary for nutritional sufficiency. Reported toxic effects in adults have included increased bleeding tendency, impaired immune function, decreased levels of vitamin K–dependent clotting factors, and impairment of leukocyte function.

E. Vitamin K

Vitamin K functions in the liver as an essential cofactor for carboxylase. This enzyme converts specific glutamic acid residues of precursor proteins to a new amino acid, gamma-carboxyglutamic acid (Gla). Vitamin K–dependent blood clotting factor prothrombin (factor II) and factor VII, IX, and X are included in the proteins (32). Vitamin K plays a crucial role in blood clotting activities. Vitamin K is not only associated with coagulation, but also with additional functions in bone, kidney and possibly other tissues (33).

Deficiencies in vitamin K status are relatively uncommon. However, since newborns do not have the necessary menaquinones, due to poor vitamin K placenta transfer, administration of vitamin K to newborns has become common practice. Ferland et al. have demonstrated that subclinical deficiencies in normal human volunteers can occur with dietary restrictions of vitamin K (34). In this study younger subjects were more susceptible to acute deficiency than older subjects. Growing evidence suggests that vitamin K is required for proteins with functions outside of hemostasis. Vitamin K status may influence skeletal fragility and osteoporosis status (35). Further studies need to be conducted in order to further understand the consequences associated with this deficiency.

1. Chemistry

Vitamin K exists in at least three forms, all belonging to a chemical compound group known as quinones. K1 (phylloquinone) occurs in green plants, and K2 (menaquinone) forms as a result of bacterial action in the intestinal tract. K3 (menadione) is the fat soluble synthetic compound (32). Vitamin K is extracted from plant and animal tissues with nonpolar solvents (36).

2. Units and Requirements

The 1989 RDA for vitamin K for men 25 and older is 80 μg, and for women 25 and older is 65 μg. Recommended intake for infants during the first 6 months is 5 μg phylloquinone/day, and for 6 to 12 months is 10 μg phylloquinone/day (37).

3. Occurrence

Green leafy vegetables, especially broccoli, cabbage, turnip greens, and lettuce, have large amounts of vitamin K. Other vegetables, fruits, cereals, dairy products, eggs, and meat contain smaller amounts. A significant amount of vitamin K is formed by the bacterial flora of the human lower intestinal tract.

4. Properties

Vitamin K is fairly resistant to heat. Compounds tend to be unstable in the presence of alkali and light.

5. Analysis

Open column chromatography or thin layer chromatography is used to isolate the vitamin from interfering substances. Since the mid-1970s HPLC has been the method of choice because of its ability to chromatograph the vitamin without the need for derivatization, and because of its nondestructive operation and greater separation and detection selectivity (38). Booth et al. analyzed five different foods using a method that applied a highly sensitive HPLC method and incorporated postcolumn chemical reduction of the quinone followed by fluorescence detection of the hydroquinone form of the vitamin. Their method is currently being used to provide a more accurate and reliable estimation of the phylloquinone content of foods (39).

6. Commercial Forms

Vitamin K (CAS: 84-80-0) is a clear, yellow to amber, very viscous liquid that is stable in air but decomposes when exposed to sunlight. It is slightly soluble in alcohol, and soluble in dehydrated alcohol, in chloroform, in either, and in vegetable oils. It is unsoluble in water (40). Because of its sensitivity to light, vitamin K should be stored in tight, light-resistant containers.

Menadione (K3) is not a natural form. Menadione sodium bisulfite and other water soluble derivatives are synthesized commercially. These are used for animal feed supplements. In excessive amounts, menadione is toxic to infants at excessive dose levels and is not used in human medicine or as a food supplement (38).

7. Toxicity

There is no known toxicity associated with the administration of high doses of vitamin K1. However, administration of vitamin K3 to infants is associated with hemolytic anemia and liver toxicity. Vitamin K1 is now prescribed to prevent hemorrhagic disease in newborns (36).

F. Vitamin C

Vitamin C is the antiscorbutic vitamin. First described during the Crusades and a common plague of early explorers and voyagers, scurvy is now virtually eliminated with the presence of vitamin C in adequate diets. Vitamin C, or ascorbic acid, has many functions as either a coenzyme or cofactor. Because of its ease of losing or taking on hydrogen, it is essential in metabolism. It is well recognized for its role in enhancing the absorption of iron. Vitamin C maintains the integrity of blood vessels. Vitamin C also promotes resistance to infection and, although controversal, has been touted by noted nobel prize recipient Linus Pauling as a cure for the common cold (61).

1. Chemistry

Vitamin C is known as ascorbic acid, ascorbate, or ascorbate monoanion. Currently vitamin C is used as the generic descriptor for all compounds exhibiting qualitatively the biologic activity of ascorbic acid. The biologically active forms are ascorbic acid (AA) and dehydroascorbic acid (DHAA). Since ascorbic acid loses electrons easily it serves as a good reducing agent and is frequently added to commercial food products. Ascorbic acid also serves as an outstanding antioxidant by providing electrons for enzymes or chemical compounds that are oxidants.

2. Units and Requirements

The current RDA is 60 mg/day for adults, based on preventing signs and symptoms of scurvy and to provide adequate reserves. The 1989 RDA is based in part on the concept that urinary excretion of vitamin C indicates that body stores are near saturation (42). During lactation, the requirement for vitamin C is increased 50% or above (43). Other authors suggest an increased requirement for smokers.

Optimal vitamin C ingestion has been found to range from 6 to 750 mg. Levine et al. recommended that daily vitamin C consumption should come from five servings of fruit and vegetables, that best sources are foods, and that prudent vitamin C ingestion should remain at less than 500 mg per day (44).

3. Occurrence

The best sources of vitamin C are fruits and vegetables, preferably acidic and fresh. These sources include citrus fruits, raw leafy vegetables, tomatoes, broccoli, strawberries, cantaloupe, cabbage, and green peppers. Content of ascorbic acid in fruits and vegetables varies with the conditions under which they are grown and degree of ripeness when harvested (45). Even foods collected from the same regions of the country may vary from 10 to 20% in the total vitamin C content (46). Heat and exposure to oxygen can reduce the amount of vitamin C available. Although potatoes are not a good source of vitamin C by themselves, the high level of consumption positions potatoes as providing 20% of the U.S. supply of vitamin C. Ascorbate is also available in vitamin tablets and many multivitamin supplements contain ascorbate.

4. Properties

Ascorbic acid is a stable, odorless, white solid which is soluble in water, slightly soluble in alcohol, and insoluble in organic solvents (47).

5. Analysis

There are approximately a dozen assays for AA and DHAA. Traditionally ascorbic acid has been determined by colorimetric techniques. Care must be taken since these substances can easily degrade during sample procurement, handling, and preparation for assay. For ascorbate, high pressure liquid chromatography with electrochemical detection is the assay of choice. Most common assays use HPLC coupled with an ultraviolet detector (43). Dehydroascorbic acid detection with direct electrochemical is not yet possible.

6. Commercial Form

L-Ascorbic acid (CAS: 50-81-7) is white or slightly yellow crystals or powder. It will gradually darken on exposure to light, and is reasonably stable in air when dry but will rapidly deteriorate in solution in the presence of air. Store in tight light-resistant containers.

7. Toxicity

Since vitamin C is one of the most commonly used supplements in the United States, excessive intake may occur. Symptoms may include diarrhea, abdominal bloating, iron overabsorption, nausea, and kidney stones.

G. Thiamin

Historically thiamin has been known as the vitamin which prevents beriberi. Thiamin is essential in energy transformation and membrane and nerve conduction as well as in the synthesis of pentoses and the reduced coenzyme form of niacin. First identified in 1897 it was not until 1936 that the vitamin was synthesized. There are two major forms, thiamin pyrophosphate (TPP) and thiamin triphosphate (TTP). In the TPP or diphosphate form it is required for the oxidative decarboxylation of pyruvate to acetyl CoA, providing entry of oxidizable substrates into the Krebs cycle for energy (48). Thiamin is most strongly linked with carbohydrate metabolism even though it is also needed to metabolize fats, proteins, and nucleic acids.

Severe thiamin deficiency in humans leads to impairment of the auditory and visual pathways in the brain stem as well as impairment of the endocrine pancreas and the heart (49). In addition, clinical features of thiamin deficiency experienced by the chronic alcoholic includes ataxia and altered memory and alcoholic peripheral neuropathy (50).

1. Chemistry

Also known as vitamin B1, this molecule consists of one pyrimidine and one thiazole ring, linked by a methylene bridge. It is particularly sensitive to sulfites which split the molecule into the pyrimidine and thazole moieties and destroy its biological activity.

2. Units and Requirements

Body stores of thiamin are relatively small and regular intake is necessary, especially since large single doses are poorly absorbed. The 1997–1998 *Dietary Reference Intakes* recommends 1.2 mg and 1.1 mg for adult males and adult females, respectively. Pregnancy recommendations increase to 1.4 mg and lactation to 1.5 mg (51).

3. Occurrence

Food sources include meat, fish, whole cereal grains, fortified cereal and bakery products, nuts, legumes, eggs, yeast, fruits, and vegetables.

4. Properties

Thiamin esters are relatively stable in the dried state if stored at low temperatures in the dark. In solution thiamin is stable at pH 2–4, and TPP is stable at pH 2–6 at low temperature. All thiamin vitamers are unstable at elevated temperatures and under alkaline conditions (52).

5. Analysis

The thiamin vitamers are generally determined as their thiochrome esters and must be released from food matrix first, usually by acid hydrolysis. The fluorimetric thiochrome assay is the most frequently used method. Solid phase chromatography, electrophoresis, and high-performance liquid chromatography have been used recently to separate and quantitate very low concentrations (52).

6. Commercial Forms

Thiamin mononitrate (CAS: 532-43-4) is available as white to yellowish white crystals, or crystalline powder, usually having a slight characteristic odor. It is slightly soluble in alcohol and in chloroform. Thiamin mononitrate should be stored in tight, light-resistant containers. Thiamin hydrochloride (CAS: 67-03-8) is also available as white to yellowish white crystals or crystalline powder (53).

7. Toxicity

Even in high oral doses, thiamin has been found to have no toxic effect other than possibly gastric upset.

H. Riboflavin

Deficiencies of riboflavin are usually in combination with deficiencies of other water soluble vitamins. Intake of riboflavin must be low for several months in order for signs of deficiency to develop. Pellagralike symptoms including skin lesions around the mouth, nose, and ears occur when there is a riboflavin deficiency. There is growing evidence that riboflavin deficiency, along with deprivation of vitamin E, iron, and folate, may be beneficial in the treatment of malaria (54).

1. Chemistry

Riboflavin is a flavin with the flavin ring attached to an alcohol related to ribose. In its pure state it appears as yellow crystals. The free vitamin is a weak base normally isolated or synthesized as a yellowish orange amorphous solid (55). The molecular weight of riboflavin is 376.4, thus 1 mg of riboflavin equals 2.66 μmol.

2. Units and Requirements

The RDA values for riboflavin are given in milligrams. The 1997–1998 *Dietary Reference Intakes* ranges from 0.3 mg for infants to 1.1 mg for adult females and 1.3 mg for adult males (56). The difference between men and women and among various age groups is due to variations in energy intake between sexes. Boisvert et al. (57) found that requirements of riboflavin for the elderly are influenced largely by the macronutrient composition of their diet: the greater the amount of carbohydrates, the lower the dietary requirement. Pregnancy levels are recommended at 1.4 mg and lactation at 1.6 mg.

3. Occurrence

Riboflavin is widely distributed in small amounts. The best sources are milk, cheddar cheese, and cottage cheese. Other good sources include eggs, lean meats, broccoli, and enriched breads and cereals. Cereals and vegetables are the largest sources of riboflavin in developing countries (54).

4. Properties

Vitamin B2 is heat- , oxidation- , and acid-stable but is sparingly soluble in water and disintegrates in the presence of alkali or light, especially ultraviolet. Very little riboflavin is lost in cooking and processing of foods. All vitamin B2 forms fluoresce naturally.

5. Analysis

Because riboflavin is photosensitive, it is imperative that analysis be conducted under subdued light using low actinic glassware (58). Two microbiological and two fluorometric methods are currently recommended for assaying total riboflavin in foods. Fluorometric techniques include a manual and a semiautomated procedure. Overestimates might occur due to presence of interfering artifacts, and the standard method is not suitable for samples of high fat content or active degradative enzyme systems (54).

6. Commercial Forms

The commercial form (CAS: 83-88-5) is a yellow to orange yellow crystalline powder having a slight odor. It melts at about 280°C with decomposition, and its saturated solution is neutral to litmus. When dry it is not affected by diffused light, but in solution light induces deterioration. It is less soluble in alcohol than in water. In ether and in chloroform it is insoluble but very soluble in dilute solutions of alkalies. Store in tight, light-resistant containers. Riboflavin 5′-phosphate sodium (CAS: 130-40-5) is a fine orange yellow crystalline power having a slight odor. It is hygroscopic (59).

7. Toxicity

There are no known effects from toxicity. The capacity of the human gastrointestinal tract to absorb orally administered riboflavin may be less than 20 mg in a single dose. Excess is readily excreted, which is typical of other water soluble vitamins.

I. Niacin

Niacin is a generic term used for nicotinic acid and its derivatives that exhibit the biological activity of niacinamide. These function as a component of the coenzymes nicotinamide adenine dinucleotide (NAD) and nicotinamide adenine dinucleotide phosphate (NADP), which are present in all cells. These coenzymes are essential in oxidation–reduction reactions involved in the release of energy from carbohydrates, fats, and proteins. The coenzyme NAD is also used in glycogen synthesis. Niacin was identified as a result of the search for the cause and cure of pellagra. Clinical signs of pellagra are commonly referred to as the 3 Ds; diarrhea, dementia, and dermatitis. The most characteristic sign is a pigmented rash that develops symmetrically in areas of the skin exposed to sunlight.

1. Chemistry

Niacin is a stable whitish crystalline material. Nicotinic acid and nicotinamide are rapidly absorbed from the stomach and intestine. In humans, the biosynthesis of niacin from tryptophan is an important route for meeting the body's niacin requirement. The efficient conversion of dietary tryptophan to niacin is affected by a variety of nutritional and hormonal factors (60).

2. Units and Requirements

The 1997–1998 *Dietary Reference Intakes* recommendations are given as niacin equivalents (NEs). One niacin equivalent is equal to 1 mg of niacin. In addition the Food and Nutrition Board suggests that 60 mg tryptophan is equivalent to 1 NE. Ranges are from 14 to 16 for adult men and women, respectively, or 6.6 NE per 1000 kcal. Pregnancy dietary reference intake is at 18 NE and lactation at 17 NE (61).

3. Occurrence

Niacin is widely distributed in plant and animal foods. Good sources include meats (including liver), dairy products, cereals, legumes, and seeds. Green leafy vegetables and fish, especially shellfish, as well as enriched breads and cereals, also contain appreciable amounts. The precursor amino acid tryptophan can contribute substantially to niacin intake. Proteins contain approximately 1% of tryptophan, and niacin status could feasibly be adequately sustained with a diet containing greater than 100 g protein (60).

4. Properties

Niacin in general is one of the more stable vitamins. Free forms of the vitamin are white stable crystalline solids. Nicotinamide is more soluble in water, alcohol, and ether than is nicotinic acid. Nicotinic acid's stability is pH independent.

5. Analysis

Niacin vitamers are all UV absorbers. Acid hydrolysis is usually used to estimate the biologically active niacin, while alkaline hydrolysis releases nonbioavailable vitamers and permits determination of the total niacin content. Typical extraction procedures for nicotinamide involve aqueous extractants and the use of dilute acids. High-performance liquid chromatography is the current method of choice for niacin in foods, concentrating on total niacin rather than individual vitamers (62).

6. Commercial Forms

Nicotinic acid, 3-pyridinecarboxylic acid (CAS: 59-67-6) is a white or light yellow crystal or crystalline powder. It is odorless or has a slight odor. It is freely soluble in boiling water and in boiling alcohol and also in solutions of alkali hydroxides and carbonates. It is almost insoluble in ether. Niacinamide (CAS: 98-92-0) is also a white, crystalline powder. It is odorless or nearly so but does have a bitter taste. Its solutions are neutral to litmus. Storage should be in well-closed containers (63).

Niacinamide ascorbate is also available as a nutrient in foods. The combination of ascorbic acid and niacinamide is a lemon yellow colored powder that is odorless or has a very slight odor. It is very soluble in chloroform and in ether and sparingly soluble in glycerin.

7. Toxicity

High doses of niacin may be toxic to the liver. Large doses of nicotinic acid have been shown to reduce serum cholesterol concentrations. However, there are side effects that include flushing of the skin, hyperuricemia, hepatic and ocular abnormalities, and occasional hyperglycemia (XX).

J. Vitamin B6

There are six nutritionally active B6 vitamers occurring in foods. Its original form is pyridoxine (PN), pyridoxal (PL), pyridoxamine (PM), and the corresponding 5′-phosphate esters pyridoxine phosphate (PNP), pyridoxal phosphate (PLP), and pyridoxamine phosphate (PMP) (64). The esters PLP and PMP function primarily in transamination and other reactions related to protein metabolism. The former is also necessary for the formation of the precursor of heme in hemoglobin. Vitamin B6 is essential in the metabolism of tryptophan and its conversion to niacin (65). The immune responses of young and elderly humans are also affected by the nutritional status of vitamin B6 (66). Studies are showing that PLP has a role in hormone modulation; it brings to a second site on the steroid receptor and alters the binding of the receptor to DNA thus decreasing the action of the steroid (67).

Deficiencies of vitamin B6 are relatively rare. However, some medications may interfere with metabolism of vitamin B6. Signs of deficiency in adults include stomatitis, cheilosis, glossitis, irritability, depression, and confusion (68).

1. Chemistry

The three nonphosphate B6 vitamers are 2-methyl-3-hydroxy-5-hydroxy methyl pyridines. Substitutions occur on the 4 position creating the differences between PN, PL, and PM.

2. Units and Requirements

Changes in recommended intakes are now in place with the 1997–1998 *Dietary Reference Intakes* (69): males 14 to 50 at 1.3 mg, males 51 and older at 1.7 mg, females 14 to 18 at 1.2 mg, females 19 to 50 at 1.3 mg, and females older than 51 at 1.5 mg. Pregnancy levels are at 1.9 mg and lactation at 2.0 mg. Recommendations are based in large part on metabolic studies, intake of vitamin B6, and protein intakes of selected population groups (70). In measuring vitamin B6 status, a minimum of three indices should be measured. One should be a direct measure, like PLP in plasma; one should be a short term measure, such as 4-PA; and one should be an indirect measure, such as a tryptophan load test. One should also determine vitamin B6 and protein intake from three-day diet histories (71).

3. Occurrence

Best sources of vitamin B6 are yeast, wheat germ, pork, liver, whole grain cereals, legumes, potatoes, bananas, and oatmeal (65). Plant foods contain primarily the pyridoxine form, and animal products contain primarily the pyridoxal and pyridoxamine forms (68).

4. Properties

Vitamin B6 is considered relatively labile with its degree of lability influenced by pH. All forms are relatively heat stable in acid medium and are heat labile under alkaline conditions. In aqueous solution, most forms are light sensitive.

5. Analysis

Analysis is complicated by several factors, including its six differing structures. Because they are highly light sensitive, care must be given to use subdued lighting and low actinic glassware during extraction and analysis. The sensitivity and specificity of the detection method is critical due to low concentrations at which B6 occurs in foods. All six forms are UV absorbers. High-performance liquid chromatography is most recently being directed at simultaneous determination of the six forms of vitamin B6 (64).

6. Commercial Forms

Vitamin B6 is available commercially in the form of pyridoxine hydrochloride (CAS: 58-56-0) as a colorless or white crystal or a white crystalline powder. It is stable in air and slowly affected by sunlight. It should be stored in light-resistant containers to avoid exposure to sunlight (72).

7. Toxicity

Toxicity, although relatively rare, needs to be understood in relation to the effect of clinical drugs on vitamin B6 metabolism. The possible effects of excess vitamin B6 on drug efficacy should be considered in those cases in which the drug reacts directly with PLP (68). High doses of vitamin B6 have been used in an attempt to treat conditions like premenstrual syndrome and has resulted in a small number of cases of neurotoxicity and photosensitivity.

K. Pantothenic Acid

Pantothenic acid was first synthesized in 1940. As a primary constituent of coenzyme A, it is involved in many areas of cellular metabolism including the synthesis of cholesterol, steroid hormones, vitamin A, vitamin D, and heme A (73). It is involved in the release of energy from carbohydrate and in the degradation and metabolism of fatty acids. Most recently pantothenic acid has been described as a donor of acetate and fatty acid groups to proteins. These modifications of proteins will affect protein localization and activity. Besides its role in molecular synthesis and oxidative degradation, pantothenia (a form of pantothenate) modifies preexisting molecules, in particular proteins (74).

Pantothenic acid is widely available so deficiencies are rare. When there are conditions of severe malnutrition, deficiencies may be detected that mainly affect the adrenal cortex, nervous system, skin, and hair. Insufficiency is characterized by a burning sensation in the soles of feet.

1. Chemistry

A pantoic acid in amide linkage to β-alanine makes pantothenic acid. Pantetheine is formed by addition of β-mercaptoethylamine, which provides the reactive sulfhydryl group, to pantothenic acid (75).

2. Units and Requirements

The 1997–1998 *Dietary Reference Intakes* now provides an adequate intake guideline. For males and females 14 and older adequate intakes are set at 5.0 mg. Pregnancy level is 6.0 mg and lactation 7.0 mg (76). The actual pantothenate content of the average American diet was estimated in one analysis at 5.8 mg/day (77).

3. Occurrence

Pantothenic acid is widely distributed in nature. Rich dietary sources include meat, fish, poultry, eggs, whole grain products, and legumes. Especially high levels of pantothenate are found in royal jelly of bees and in the ovaries of tuna fish and cod.

4. Properties

Pantothenic acid is relatively stable at neutral pH. Strong acids, alkalis, or thermal processing will hydrolyze pantothenic acid. Calcium pantothenate occurs as a slightly hygroscopic, white powder. It is odorless, has a bitter taste, and is stable in air. It is soluble in water and glycerin and practically insoluble in alcohol, in chloroform, and in ether (78).

5. Analysis

Numerous methods are available for analysis of pantothenic acid in foods, including HPLC, nonchromatographic methods, open column chromatography, paper chromatography, thin layer chromatography (TLC), and GC. Analysts are cautioned that food spoilage prior to analysis may lead to inflated pantothenic acid levels (79).

6. Commercial Form

Calcium pantothenate (CAS: 137-08-6) is the calcium salt of the dextrorotatory isomer of panthothenic acid and should be stored in tight containers. Racemic calcium pantothenate (CAS: 6381-63-1) is a mixture of the calcium salts of dextrorotatory and levorotatory isomers of panthothenic acid and also should be stored in tight containers. Calcium chloride double salt of DL- or D-calcium panthothenate (CAS: 6363-38-8) is composed of approximately equal quantities of dextrorotatory or racemic calcium pantothenate and calcium chloride. Dexpanthenol (CAS: 81-13-0) is the dextrorotatory isomer of the alcohol analog of pantothenic acid and occurs as a clear, viscous, somewhat hygroscopic liquid having a slight odor. DL-Panthenol is also available commercially (78).

7. Toxicity

No serious toxic effects are known. Ingestion of excessive amounts may cause diarrhea.

L. Folate

Folate is a generic term for a group of compounds chemically and nutritionally similar to folic acid. Essential to cell division, folate plays an important role in the synthesis of the purines guanine and adenine and the pyrimidine thymine, which are compounds utilized in the formation of nucleoproteins deoxyribonucleic acid (DNA) and ribonucleic acid (RNA) (80). Folate is also essential for the formation of both red and white blood cells and serves as a single carbon carrier in the formation of heme. Low levels of plasma folate and vitamin B12 lead to elevated plasma homocysteine, which has been associated with increased risks of coronary heart disease. Recent compelling studies have shown that a nutritional supplement containing folic acid successfully and markedly reduced the occurrence of neural tube defects (spina bifida or anencephaly) in babies when intervention was administered before conception (81). Another study has shown that large doses of folic acid administered before pregnancy successfully prevented recurrence of neural tube defect in children born to mothers who previously had a child with such a defect (82). Deficiency of folate results in poor growth, megaloblastic anemia, and other blood disorders, elevated blood levels of homocysteine, glossitis, and gastrointestinal tract disturbances.

1. Chemistry

Folate (pteroylglutamate) and folic acid (pteroylglutamic acid) designate any member of the family of pteroylglutamates, or mixtures of them, having various levels of reduction of the pteridine ring, one-carbon substitutions, and numbers of glutamate residues (83). The term *folacin* is no longer the term of choice. Folate occurs in approximately 150 different forms.

2. Units and Requirements

The 1997–1998 *Recommended Dietary Intakes* allowances for folate for men and women 14 and older has been set at 400 μg dietary folate equivalents (DFE). Pregnancy levels are set at 600 μg DFE and lactation levels set at 500 μg DFE. One DFE equals 1 μg food folate, which equals 0.6 μg of folic acid (from fortified food or supplement) consumed with food which equals 0.5 μg of synthetic (supplemental) folic acid taken on an empty stomach. In 1998 the Institute of Medicine recommended that all women of childbearing potential consume 400 mg of synthetic folic acid per day from fortified foods and/or a supplement in addition to food folate (84).

3. Occurrence

Folates are ubiquitous in nature and are present in nearly all natural foods. However, they are highly susceptible to oxidation and folate content in foods and may be destroyed by protracted cooking or processing, such as canning. Best sources are legumes such as kidney beans, lima beans, lentils, fresh dark green leafy vegetables, especially spinach, asparagus, turnip greens, and broccoli.

4. Properties

Crystalline folic acid is yellow. The free acid is almost insoluble in cold water, and the disodium salt is more soluble. Folic acid is destroyed at a pH below 4, but is relatively stable above pH 5 (83).

5. Analysis

Generally analysis of folate is time and labor intensive. Current methods are divided into three broad categories: microbiological (radio-), protein binding, and chromatographic (85). A new method, combined-affinity chromatography and reverse-phase liquid chromatography (affinity/HPLC) has recently been found to be quite rapid and technically simple with a total of 24–30 analyses accomplished within a week by one person (86).

6. Commercial Forms

Folic acid (CAS: 59-30-3) is a yellow or yellowish orange, odorless crystal or crystalline powder. It is insoluble in acetone, in alcohol, in chloroform, and in ether, but dissolves in solutions of alkali hydroxides and carbonates. The pH of a suspension of 1 g in 10 mL of water is between about 4.0 and 4.8 (87). Folic acid should be stored in well-closed light-resistant containers.

7. Toxicity

When ingested in their active forms, folic acid is nontoxic in humans. Being water soluble, excesses tend to be excreted in urine rather than stored in tissue.

M. Vitamin B12

Initially vitamin B12, also known as cobalamin, was identified as the extrinsic factor of food that is effective in the treatment of pernicious anemia. Vitamin B12 is synthesized by bacteria. It is essential for normal function in the metabolism of all cells, especially cells of the gastrointestinal tract, bone marrow, and nervous tissue. It participates, along with folic acid, choline, and methionine, in the transfer of methyl groups in the synthesis of nucleic acids, purines, and pyrimidine intermediates. Vitamin B12 and folate are closely related with each depending on the other for activation. Vitamin B12 removes a methyl group to activate the folate coenzyme. Thus vitamin B12 is involved in the synthesis of DNA and RNA. Vitamin B12 affects myelin formation. With vitamin B12 deficiencies, impaired DNA synthesis results leading to megaloblastic anemia, glossitis, hypospermia, and gastrointestional disorders. Lack of vitamin B12 also results in subacute degeneration of cerebral white matter, optic nerves, spinal cord, and peripheral nerves with resulting symptoms of numbness, tingling, burning of the feet, stiffness, and generalized weakness of the legs (88). Vitamin B12 is found almost exclusively in foods derived from animals. Thus, strict vegetarians (vegans) develop deficiencies because they are not consuming the usual dietary source of meat and meat products (89,90). Deficiencies may take many years to occur since the body reabsorbs much vitamin B12 (90). Deficiencies also produce elevated serum homocysteine levels, which promote heart attacks, thrombotic strokes, and peripheral vascular occlusions (91).

1. Chemistry

The systematic name for vitamin B12 is α-(5,6-dimethyl-benzimidazolyl)-cobamide cyanide. Two cobalamins currently known to be coenzymatically active in humans are methylcobalamin and 5′-deoxyadenoxyl cobalamin (92).

2. Units and Requirements

Recommended dietary allowances as of 1997–1998 for males and females 14 and over is 2.4 μg. Pregnancy level is 2.6 μg and lactation level is 2.8 μg (93). Studies suggest that higher recommendations should be made for elderly due to malabsorption concerns due to atrophic gastritis (94).

3. Occurrence

Usual dietary sources are meat and meat products and to a lesser extent milk and milk products. Richest sources are liver and kidney. Plant products have very little vitamin B12. Fermented soy products and some sea algae do not provide an active form of vitamin B12. Megadoses of vitamin C may adversely affect the availability of vitamin B12 from food (92).

4. Properties

Cyanocobalamin crystals are dark red, and the substance absorbs water. Cobalamins are destroyed by heavy metals and strong oxidizing or reducing agents. Aqueous solutions are neutral with maximum stability at pH 4.5 to 5.0 (95).

5. Analysis

B12 vitamers are all UV absorbers, however their absorption spectra vary from vitamer to vitamer. Detection is complicated by the low concentrations at which they normally occur in foods. Recent developments have concentrated on ligand binding assays for vitamin B12 in foods (96).

6. Commercial Forms

Vitamin B12/cyanocobalamin (CAS: 68-19-9) are dark red crystals or amorphous or crystalline powder. In anhydrous form it is very hygroscopic and when exposed to air it may absorb about 12% of water. It is sparingly soluble in water; soluble in alcohol; and insoluble in acetone, in chloroform, and in ether. Packaging and storage should be in well-closed containers (97).

7. Toxicity

No toxic effects are known.

N. Biotin

Biotin was first isolated in 1936 and synthesized in 1943. First named vitamin H, it was later proved to be the same as a potent growth factor in yeast known as coenzyme R and was renamed biotin. It functions as the coenzyme for reactions involving the addition or removal of carbon dioxide to or from active compounds. These coenzymes are involved in gluconeogenesis, synthesis, and oxidation of fatty acids, degradation of some amino acids, and purine synthesis. The absorption of both dietary biotin and any biotin synthesized by intestinal bacteria is thought to be prevented by dietary avidin, a glycoprotein found in raw egg white. Thus the name ''egg-white injury'' given to the syndrome of severe dermatitis, hair loss, and neuromuscular dysfunction (98).

Prolonged consumption of large quantities of raw egg white and parenteral nutrition without biotin supplementation will lead to biotin deficiency. Biotin deficiency has also been detected in children with severe protein-energy malnutrition (99,100). Deficiencies are relatively rare and can be characterized by muscle aches, skin rashes, mild depression, slight anemia, and increased serum cholesterol (101).

1. Chemistry

Biotin is a bicyclic compound. One ring contains sulfur, termed a tetrahydrothiophene ring, and the other ring contains a ureido group.

2. Units and Requirements

For males and females, ages 19 and above, 1997–1998 adequate intake guidelines state 30 μg biotin. This level is increased to 35 μg for lactation (102).

3. Occurrence

Biotin is widely distributed with major food sources including liver, rice, egg yolks, and vegetables. It may also be found in seeds of many cereals and oilseeds, however it is largely unavailable in this form. Avidin reduces the bioavailability of biotin. Biotin will occur in food as free biotin but is usually bound to protein.

4. Properties

Biotin is soluble in water. It is more soluble in hot water and in dilute alkali and four times more soluble in 95% ethanol than in cold water. It is not soluble in other organic solvents (103). Biotin does not fluoresce nor is it a UV absorber.

5. Analysis

Because biotin is not a UV absorber, nor fluorescent, nor electrochemically active, direct detection is difficult. Refractometry is the only means for direct detection of biotin. Because biotin is predominately protein-bound and is relatively stable, it has to be extracted from food under fairly harsh conditions, autoclaving with sulfuric acid or enzymatic hydrolysis (101). Limited work is being done using HPLC methods.

6. Commercial Forms

cis-Hexahydro-2-oxo-1H-thieno[3,4]imidazole-4-valeric acid and d-biotin (CAS: 58-85-5) is a practically white crystalline powder. It is stable to air and heat. Packaging and storage should be done in tight containers (104).

7. Toxicity

There are no known toxic effects.

O. Choline

Choline was first discovered in 1862 and synthesized in 1866. Choline is found in most animal tissues and is a primary component of the neurotransmitter acetylcholine. It functions with inositol as a basic constituent of lecithin and is a major donor of methyl groups (105). The deposition of fats in the liver is prevented by choline. Choline facilitates the movement of fats into cells (106). Choline can be synthesized from ethanolamine and methyl groups derived from methionine, but most of the time choline comes mainly from dietary phosphatides (107). Diets low in choline can have major consequences that include hepatic, renal, pancreatic, memory, and growth disorders. Zeisel et al. (108) conducted a study with healthy humans and showed that those who consumed a choline-deficient diet for 3 weeks had depleted stores of choline in tissues and developed signs of incipient liver dysfunction. Their study supports the theory that choline may be an essential nutrient for humans when excess methionine and folate are not available. Choline may also be an essential nutrient during long-term TPN (109). However, a second study conducted by Savendahl et al. (105) concluded that prolonged fasting in humans modestly diminished plasma choline, but was not associated with signs of choline deficiency. Further studies are necessary.

1. Chemistry

Choline is a quaternary amine. Two forms are available as nutrient and/or dietary supplements: choline bitartrate, or (2-hydroxyethyl)trimethylammonium bitartrate, and choline chloride, or (2-hydroxyethyl)trimethylammonium chloride.

2. Units and Requirements

The 1997–1998 *Dietary Reference Intakes* adequate intake guidelines for males age 14 and above is 550 mg of choline. Guidelines for females 14 to 18 is set at 400 mg of choline and for females 19 and above is set at 425 mg of choline. Pregnancy guidelines are 450 mg, and lactation guidelines are 550 mg (119).

3. Occurrence

Most common foods contain choline. The rate of an individual's growth determines the relationship between choline and three other nutrients, methionine, folic acid, and vitamin

B12 (108). Richest sources of choline are liver, kidneys, brains, wheat germ, brewers' yeast, and egg yolk. One form of choline, phosphatidylcholine, is often added to processed foods and acts as an emulsifying agent or as an antioxidant.

4. Properties

Choline Chloride is hygroscopic and very soluble in water and in alcohol. Choline bitartrate is freely soluble in water, slightly soluble in alcohol, and insoluble in ether and in chloroform (111).

5. Analysis

Choline and two of its forms found in food, phosphatidylcholine and sphingomyeline, can be measured using a gas chromatography/mass spectrometic assay (109).

6. Commercial Forms

Choline bitartrate (CAS: 87-67-2) is a white, hygroscopic, crystalline powder having an acidic taste. It is odorless or may have a faint trimethylaminelike odor. Packaging and storage should be in tight containers. Choline chloride (CAS: 67-48-1) is colorless or white crystals or crystalline powder, usually having a slight odor of trimethylamine. It should also be packaged and stored in tight containers (112).

7. Toxicity

Excess choline and its relative lecithin can cause short-term discomforts such as gastrointestinal distress, sweating, salivation, and anorexia. Long-term health hazards include injury to the nervous and cardiovascular systems (113).

P. Carnitine

Carnitine is a substance found in skeletal and cardiac muscle and certain other tissues. Carnitine functions as a carrier of fatty acids across the membranes of the mitochondria (114). Other areas of carnitine influence include its role in the metabolism of fuel substrates, its role in modulating the availability of free coenzyme A concentrations, and its role in removing acyl compounds that accumulate to toxic concentrations in cells. Carnitine also stores activated acyl compounds for later use in a variety of metabolic pathways (115). Its actions closely resemble those of amino acids and B vitamins. Because humans can synthesize carnitine it is assumed that it is a nonessential nutrient. Carnitine is synthesized from lysine and methionine through a five-step process in which vitamin B6 and iron play a role. A vitamin B6 deficiency impairs carnitine synthesis (116).

High levels of carnitine concentrations in the liver have been found in morbidly obese patients (117). Low levels of carnitine have been seen in connection with severe fatty liver associated with cirrhosis. Low levels have also been seen in premature infants and neonates (118). Loss of carnitine in body fluids and in tissues are seen in patients with chronic renal failure following hemodialysis, in patients on enteral feedings with protein hydrolysate formulas, and in patients on total parenteral nutrition (118). Carnitine acyltransferase deficiencies have been associated with various metabolic disorders.

1. Chemistry

The structure of carnitine has been shown to be L-b-hydroxy-y-N-trimethyl-aminobutyric acid.

2. Units and Requirements

No U.S. RDA exists for this vitamin.

3. Occurrence

Carnitine levels are high in meat and dairy products. Vegetables, cereals, and fruits contain negligible sources of carnitine (119).

4. Properties

Carnitine can be synthesized with two amino acids, lysine and methionine. It facilitates the transfer of fatty acyl-coenzyme A (CoA) from the cytoplasm across the mitochondria membrane. Once the acyl-CoA is regenerated, carnitine is released to continue the transport process.

5. Analysis

Spectrophotometric methods and radioisotope assays are used to measure carnitine.

6. Commercial Form

There are currently no federal codex monograph specifications for carnitine.

7. Toxicity

Carnitine is a popular supplement among endurance athletes and has been touted as a ''fat burner.'' There is a belief that extra carnitine will help burn more fat and spare glycogen during endurance events. Studies do not support these claims. Excess carnitine may produce diarrhea (120).

Q. Inositol

Inositol is widely distributed in nature and occurs in animals, higher plants, fungi, and bacteria. Inositol, as part of inositol-containing phospholipids, functions as cellular mediators of signal transduction, metabolic regulation, and growth (121). Inositol is important as an intracellular second-messenger precursor. As a natural component of human diets, inositol may be important in the treatment of various psychiatric disorders that are responsive to selective serotonin reuptake inhibitors. Studies have found that inositol is effective in the treatment of depression, panic, and obsessive-compulsive disorders (122–124). Inositol metabolism is altered in several clinical conditions including diabetes, renal disease, and respiratory distress syndrome.

1. Chemistry

The predominant forms are myo-inositol and phosphatidylinositol (PI). The molecular weight of inositol is 180.16 g/mol. Chemical names are 1,2,3,5/4,6-cyclohexanehexol, i-inositol and meso-inositol.

2. Units and Requirements

No U.S. RDA levels have been established.

3. Occurrence

Inositol and its derivatives are widely distributed in nature and occur in animals, higher plants, fungi, and bacteria. Animal products include fish, poultry, meats, and dairy products. Inositol is also present in high concentrations in breast milk.

4. Properties

Inositol occurs as fine, white crystals or as a white crystalline powder. It has a sweet taste, is stable in air, and is odorless. In solutions it is neutral to litmus. It is soluble in water, slightly soluble in alcohol, and insoluble in ether and in chloroform (125).

5. Analysis

Under normal conditions, inositol can be measured accurately using most common optical techniques or by gas chromatography (121).

6. Commercial Forms

Inositol (CAS: 87-89-8) occurs as fine, white crystals or as white crystalline powder. It should be packaged and stored in well closed containers (125).

7. Toxicity

Excessive dietary inositol appears not to be toxic, except in certain clinical situations where inositol metabolism is impaired (121).

III. AMINO ACIDS

Amino acids are the structural parts of proteins. Each amino acid unit contains a base NH_2 group and an acid COOH or carboxyl group. Because they contain both a base and an acid, they are capable of both acid and base reactions in the body. There are 22 amino acids divided into two categories, essential and nonessential. Essential amino acids are those that cannot be adequately synthesized in the body from other amino acids or protein sources and must be provided by diet. The body can synthesize nonessential amino acids. Under certain disease states or special needs states, there are other amino acids that become essential, known as conditionally essential. In addition, these are frequently needed only during early development (126).

Essential	Nonessential	Conditionally essential
Histidine	Alanine	Arginine
Isoleucine	Arginine	Cysteine
Leucine	Asparagine	Glutamine
Lysine	Aspartate	Isoleucine
Methionine	Cysteine	Leucine
Phenylalanine	Glutamate	Taurine
Threonine	Glutamine	Tyrosine
Tryptophan	Glycine	Valine
Valine	Proline	
	Serine	
	Tyrosine	

Deficiencies of both essential and nonessential amino acids can be seen in elderly people with protein-energy malnutrition. Dietary protein may be decreased and there may be an increase in catabolic reactions. These low levels of amino acids may reflect a severe metabolic disturbance (127).

Several observational and intervention studies on humans and experimental studies on animals have been conducted to determine if there is a relationship between dietary protein and blood pressure. However, data results have been inconclusive and contradictory leaving researchers with the unanswered question as to the benefit or harm of dietary protein on blood pressure (128).

According to the American Dietetic Association, high protein diets do not build muscle or burn fat. In fact, diets high in protein may result in missing nutrients from the other food groups: fruits, vegetables, and grains. Very high protein diets can put a strain on one's liver and kidneys and are not a healthy eating plan for life-long health (129). Various types of amino acid solutions are available depending on individual needs. There are conventional solutions available for individuals with normal organ function and special purpose formulas are available, for example, situations such as trauma and liver failure (130).

Amino acids are nonvolatile, crystalline white solids in their pure form. They can decompose at temperatures ranging from 185°C to 342°C. They are all soluble in water to various extents. Glycine is the only amino acid that is not optically active. All are capable of forming salts and these are more cheaply obtained and more stable. The following are FDA approved amino acids for functional use in foods as nutrients (131,132):

DL-Alanine	CAS: 302-72-7
L-Alanine	CAS: 56-41-7
L-Arginine	CAS:74-79-3
L-Arginine monohydrochloride	CAS: 1119-34-2
L-Asparagine anhydrous	CAS: 70-47-3
L-Asparagine monohydrate	CAS: 5794-13-8
DL-Aspartic acid	CAS: 617-45-8
L-Aspartic acid	CAS: 56-84-8
L-Cysteine monohydrochloride	
Monohydrate	CAS:7048-04-6
Anhydrous	CAS: 52-89-1
L-Cystine	CAS: 56-89-3
L-Glutamic acid	CAS: 56-86-0
L-Glutamic acid hydrochloride	CAS: 138-15-8
L-Glutamine	CAS: 56-85-9
Glycine	CAS: 56-40-6
L-Histidine	CAS: 71-00-1
L-Histidine monohydrochloride	
Monohydrate	CAS:5934-29-2
DL-Isoleucine	CAS: 443-79-8
L-Isoleucine	CAS: 73-32-5
DL-Leucine	CAS: 328-39-2
L-Leucine	CAS: 61-90-5
L-Lysine monohydrochloride	CAS: 657-27-2
DL-Methionine	CAS: 59-51-8
L-Methionine	CAS: 63-68-3
DL-Phenylalanine	CAS: 150-30-1
L-Phenylalanine	CAS: 63-91-2

L-Proline	CAS: 147-85-3
DL-Serine	CAS: 302-84-1
L-Serine	CAS: 56-45-1
L-Threonine	CAS: 72-19-5
DL-Tryptophan	CAS: 54-12-6
L-Tryptophan	CAS: 73-22-3
L-Tyrosine	CAS: 60-18-4
L-Valine	CAS: 72-18-4

Food processing offers several dangers to amino acids. Lysine, for example, when in the presence of reducing sugars, may be lost in a treatment of mild heat. In the presence of severe heating conditions, food proteins become resistant to digestion. Lysine and cysteine when exposed to alkali will react together and form lysinoalanine, which is toxic. Methionine is lost when sulfur dioxide is used for oxidation (133).

Analysis of amino acid profiles in foods is a challenge as most amino acids are present as components of proteins. To analyze the individual amino acid, it is necessary to hydrolyze the amide bonds linking the amino acids without destroying the amino acids themselves. Two major categories of analysis exist, free amino acid analysis and amino acid analysis of peptides and proteins. With the creation of HPLC analysis of amino acids in food matrixes, routine procedures can be conducted within a 24-h turnaround period. Recent advances have included the use of microwave heating. According to Baxter (126), with microwave heating, hydrolysis that once took 22 to 24 h can now be completed in minutes, saving considerable analysis time.

IV. FATTY ACIDS

Fatty acids are any of several organic acids produced by the hydrolysis of neutral fat (134). Essential fatty acids, like essential amino acids, cannot be produced by the body and must be provided by diet. Essential fatty acids include linoleic acid (omega-6 family) and alpha-linolenic (omega-3 family). These two fatty acids are parent compounds for other biologically active long-chain polyunsaturated fatty acids (LCPUFAs). Linoleic acid can be converted to gamma-linolenic acid (GLA) and arachidonic acid (AA). Alpha-linolenic acid can be converted to docosahexaenoic acid (DHA) and eicosapentaenoic acid (EPA) (135). These fatty acids are necessary for proper growth, maintenance, and functioning of the body. The LCPUFAs are major essential components of membrane phospholipids.

Of all nutrients, fat is most often linked with chronic diseases. Too much fat or the wrong kind of fat may raise the risks of heart disease, some types of cancer, and obesity. Research is ongoing as to the pros and cons of fatty acids and various diseases. Arachidonic acid and DHA are present in human milk and support exists for adding these acids to formula milk for premature infants. Unexplained failure-to-thrive in infants has been found to be caused by a defect in the mitochondrial metabolism of fat. When low fat diets were given to these infants, improvements were shown and the infants caught up on growth factors. Researchers suspect that either because of genetics or diet, the fatty acid content in regular diets may be causing infants to lose their ability to metabolize fatty acids. Docosahexaenoic acid is depleted in pregnant and lactating women and supplementation may be desired to regain adequate DHA status (136). Women with high-risk pregnancies were given daily dietary supplementation of linoleic acid and calcium during their third trimester and found to have reduced incidence of preeclampsia and improved pregnancy outcome (137). A link between higher consumption of alpha-linolenic acid, such as oil-based salad dressings, has been shown to reduce the risk of fatal ischemic heart disease. Using an original food frequency questionnaire completed by 76,283 women in 1984, a ten-year follow up was conducted. A higher intake of alpha-linolenic acid was associated with a lower relative risk of fatal ischemic heart disease. Two hundred thirty-two cases of fatal ischemic heart disease were documented compared to 597 cases of nonfatal myocardial infarction (138).

Debates continue among professionals as to the role fat plays in breast and prostate cancer. Positive associations have been suggested but have not been supported by large prospective studies where confounding factors are minimized. There is strong evidence that some aspects of foods high in fat increase risk of prostate cancer. Further studies are required of specific dietary fatty acids and their relationship to various malignancies (139).

Over the last five years, there has been growing concern regarding trans fatty acids. Trans fatty acids are formed during the process of hydrogenation. This process of converting liquid oils into solid or semisolid fats dramatically changed the composition of the food supply. Growing numbers of metabolic studies and epidemiological studies are finding that trans fatty acids increase plasma concentrations of low-density-lipoprotein cholesterol and reduce concentrations of high-density-lipoprotein cholesterol relative to the parent natural fat (140). Debate continues regarding the extent of a public health concern related to trans fatty acids consumption. The American Society for Clinical Nutrition and the American Institute of Nutrition in their joint position paper on trans fatty acid were unable to conclude that intake of trans fatty acids were a risk factor for coronary heart disease. However, they did recommend that in order to achieve greatest cholesterol-lowering effects, fats that are sources of unsaturated fatty acids are preferred to fats rich in trans fatty acids. One's primary objective for a healthier diet should be to reduce total fat to less than 30% and to reduce saturated fat to less than 10% of energy (141).

According to the Federal Commercial Codex, linoleic acid is the only fatty acid with monograph specifications. Linoleic acid (CAS: 60-33-3) is the major component in various vegetable oils including cottonseed, soybean, peanut, corn, sunflower seed, safflower, poppy seed, and linseed. This colorless to pale yellow oily liquid is easily oxidized by air. Although insoluble in water, it is freely soluble in ether, soluble in absolute alcohol and in chloroform. Linoleic acid can be mixed with dimethylformamide, fat solvents, and oils. Packaging and storage should be in tight containers (142).

Gas–liquid chromatography (GLC) is used for the analysis of fatty acids. This preferred method of analysis provides very high resolution, high sensitivity, and very good reproducibility in quantitative analyses. The acid hydrolysis–packed column gas–chromatography method satisfies national labeling requirements for the determination of total fat and saturated fat for a large number of foods (143). High-performance liquid chromatography is currently being used for the isolation of specific fatty acid fractions and on a small scale. This type of analysis will pass GLC as technology advances and more highly resolutive capillary columns are developed. Sempore and Bezard provide a detailed history and discussion on the analysis of fatty acids (144).

V. MINERALS AND TRACE MINERALS

Minerals are inorganic elements that retain their chemical identity when in a food product. Minerals can be divided into two groups, the major minerals and the trace minerals. Major minerals are those present in amounts larger than 5 g in the human body. The seven major minerals include calcium, phosphorus, potassium, sulfur, sodium, chloride, and magnesium. Sulfur, however, is not traditionally used as a nutritional additive. There are more than a dozen trace minerals. Some of the most important trace mineral nutritional additives include iron, zinc, copper, iodine, manganese.

Mineral additives are commercially available in one or several salt forms. Some minerals, including iron, are available in the elemental form. Aside from price considerations, the choice of the source depends on three factors:

1. Bioavailability of the mineral in a particular salt form
2. Solubility and/or mixability
3. Potential effects on final product properties

Minerals cannot be destroyed by heat, air, acid, or mixing, and only little care is needed to preserve minerals during food preparation. The ash that remains when a food is burned contains the minerals that were in the food originally.

A. Calcium

Calcium is needed to form bones and to keep bones strong. Bones and teeth are the major storage units of calcium of one's body. Bones are continually torn down and built back as the body works to meet its calcium needs. If calcium needs are not met through dietary intake, the body will pull greater amounts of calcium from bones. Without adequate intake of calcium, use of stored calcium will lead to porous bones and eventually to the crippling bone disease osteoporosis. Dairy products are an excellent source of calcium. Other sources include dark green leafy vegetables, broccoli, spinach, sardines, canned salmon, and almonds. Recommended Dietary Allowances for calcium have been increased to 1300 mg for ages 9–18, 1000 mg for ages 19–50, and 1200 mg for age 50 and older (145).

Researchers have found that it may be beneficial to start efforts to increase calcium intake early in development, prior to the physical changes associated with puberty (146). As children move to adolescence, their awareness of the importance of calcium increases. Educational factors must focus on increasing specific information about daily requirements and calcium content of various dietary sources (147). As the aging process continues, individuals' fears about lactose maldigestion or perceptions of dairy products being fattening should be addressed (148). Education must continue about the risk of osteoporosis and the ability to meet one's needs either through dietary intakes or use of supplemental calcium.

Calcium is commercially available in several forms as a nutrient in foods.

Calcium phosphate, monobasic: anhydrous (CAS: 7758-23-8) and monohydrate (CAS: 10031-30-8). Occurs as white crystals, granules, or granular powder and is sparingly soluble in water and is insoluble in alcohol.

Calcium phosphate, tribasic (CAS: 7758-87-4). This white, odorless, tasteless powder is stable in air. It is insoluble in alcohol and almost insoluble in water. Dissolves readily in dilute hydrochloric and nitric acids.

Calcium acid pyrophosphate (no CAS). This is a fine, white, colorless and acidic powder. It is insoluble in water, but does dissolve in dilute hydrochloric and nitric acids.

Calcium carbonate (CAS: 471-34-1). This fine, white microcrystalline powder is colorless, tasteless and stable in air. The presence of any ammonium salt or carbon dioxide increases solubility in water. The presence of any alkali hydroxide decreases its solubility.

Calcium glycerophosphate (CAS: 27214-00-2). This fine, white, odorless, almost tasteless powder is somewhat hygroscopic. Greater solubility in water at lower temperatures, and citric acid increases solubility in water. Insoluble in alcohol.

Calcium lactobionate (CAS: 5001-51-4). This is a white to cream colored, odorless, free-flowing powder that has a bland taste and readily forms double salts.

Calcium oxide (CAS: 1305-78-8). This comes as hard, white or grayish white masses or granules, or as white to grayish white powder. It is odorless and soluble in glycerin and insoluble in alcohol.

Calcium phosphate, dibasic: anhydrous (CAS: 7757-93-9) and dihydrate (CAS: 7789-77-7). These are white, odorless, tasteless powders that are stable in air.

Calcium pyrophosphate (CAS: 7790-76-3). This is a white, odorless and tasteless powder. It is soluble in dilute hydrochloric and nitric acid and insoluble in water.

Calcium sulfate: anhydrous (CAS: 7778-18-9). This fine, white to slightly yellow white, odorless powder is anhydrous or contains two molecules of water of hydration.

Other commercial forms of calcium exist: calcium silicate, calcium acetate, calcium bromate, calcium chloride, calcium gluconate, calcium hydroxide, and calcium peroxide. However, in the Federal Commercial Codex, their functional use in foods does not include the term nutrient (149,150).

Calcium is determined using either atomic absorption spectrometry or inductively coupled plasma spectroscopy. The calcium content in foods may also be determined by permanganate titration. After initial preparation of calcium to an ash solution, it is then precipitated at a pH of about 4 as the oxalate. The oxalate is then dissolved in sulfuric acid to liberate the oxalic acid. The oxalic acid is then titrated with standard potassium permanganate solution (151).

B. Phosphorus

Following calcium, phosphorus is the second major component of bone and teeth. Besides working with calcium, phosphorus is important as a major regulator of energy metabolism in one's body organs and generates energy in every cell of one's body. Phosphorus also plays an important role in DNA and RNA (152). Most foods contain phosphorus. However, good sources are protein-rich foods including milk, meat, poultry, fish, and eggs. Legumes and nuts are good sources as well. Carbonated beverages also contain phosphorus. With such an abundance of availability to phosphorus, deficiencies are rare. Should deficiency occur, symptoms would include loss of appetite, bone loss, weakness, and pain. Clinical deficiencies may result from those patients on long-term administration of glucose or total parenteral nutrition without sufficient phosphate. Other concerns are those that have excessive use of phosphate-binding antacids, hyperparathyroidism, diabetic acidosis, or alcoholism. Premature infants fed unfortified human milk may also develop hypophosphatemia (153). Calcium and phosphorus can have an inhibitory effect on the absorption of iron from food. However, healthy full-term infants fed calcium-, phosphorus-, and iron-fortified formula were found to have no problems with iron deficiency (154).

With excessive intake of phosphorus, the level of calcium in blood may be reduced. This may be a potentially serious problem in those that already have low calcium intake. Patients with advanced renal failure often become hyperphosphatemic as renal failure progresses. Modifying the protein and phosphorus diets of these patients can be a means to slow the progression of renal insufficiency (155).

The recommended dietary allowances for phosphorus were decreased in 1997. The current recommendations for males and females 9 to 18 years of age is 1250 mg and for males and females 19 years of age and beyond is 700 mg (156).

Commercially phosphorus is available in combination with other minerals: calcium phosphate, calcium pyrophosphate, calcium glycerophosphate, ferric phosphate, ferric pyrophosphate, magnesium phosphate, manganese glycerophosphate, potassium glycerophosphate, sodium phosphate, sodium ferric pyrophosphate, and sodium pyrophosphate (157).

The Food and Drug Administration has a yearly program, the Total Diet Study, that measures mineral content in representative diets of specific age–sex categories. Analysis of phosphorus content was conducted in 1974–1976 and 1978–1979 using colormetric methods, and then, beginning in 1982, inductively coupled plasma spectroscopy was introduced (158). For the purpose of determination of total phosphorus in commercial food oils, graphite furnace atomic absorption provides a rapid and relatively sensitive measure (159).

C. Magnesium

Magnesium is an intracellular cation largely found in bone, followed by muscle, soft tissues, and body fluids. It is important for the part that it plays in more than 300 body enzymes. Enzymes regulate body functions, including producing energy, making body protein, and enabling muscle contractions. Magnesium also plays a part in neuromuscular transmission and activity and works with or against the effects of calcium. Excess magnesium will inhibit bone calcification (160).

Dietary sources high in magnesium include seeds, nuts, legumes, and unmilled cereal grains. Green vegetables are also good sources. Diets high in refined foods, meats, and dairy products are low in magnesium. In the processing of foods like flour, rice, and sugar, magnesium is lost and not returned in the enrichment process.

Because magnesium is relatively common, deficiencies are rare. In cases where the body does not absorb magnesium appropriately, deficiency symptoms of irregular heartbeat, nausea, weakness, and mental derangement may result. Low magnesium intakes or magnesium deficits in the older population have been linked to various disease states, including ischemic heart disease, hypertension, osteoporosis, glucose intolerance, diabetes, and stroke (161). Because there are no clinical manifestations of magnesium deficiency, it is the most overlooked electrolyte alteration among hospital in-patients (162). Magnesium deficiencies are also common in short bowel syndrome where there is binding of calcium and magnesium, as well as with diseases that cause prolonged vomiting or diarrhea (163). Excessive intakes of magnesium do not appear to be of concern unless kidney disease exists (164).

Recommended dietary allowances for males 19 to 30 years old is 400 mg, and from 31 years on is 420 mg. Recommendations for females 19 to 30 years old is 310 mg, and from 31 years on is 320 mg (165).

Commercial forms of magnesium include (166):

Magnesium gluconate: anhydrous (CAS: 3632-91-5) and dihydrate (CAS: 59625-89-7). A white to off-white powder or granulate that is anhydrous, dihydrate, or a mixture of both. Insoluble in ether, sparingly soluble in alcohol, and soluble in water.

Magnesium phosphate, dibasic (CAS: 7782-75-4). A white, odorless, crystalline powder that is soluble in dilute acids. It is insoluble in alcohol and slightly soluble in water.

Magnesium phosphate, tribasic (CAS: 7757-87-1). This white, odorless, tasteless, crystalline powder may contain four, five, or eight molecules of water of hydration. It is soluble in dilute mineral acids and almost insoluble in water.

Magnesium sulfate (CAS: 7487-88-9). This colorless crystal or granular crystalline powder is odorless. May be produced with one or seven molecules of water of hydration or in dried form containing equivalent of about 2.3 waters of hydration. It is readily soluble in water, slowly soluble in glycerine, and sparingly soluble in alcohol.

An integrated analytical scheme involving flame atomic absorption and flame emission spectrometry can be used to determine magnesium elements in foods (167). The Food and Drug Administration's Total Diet Study used atomic absorption spectrometry in their analysis of magnesium in foods for the years 1976–1977 and 1980–1995. Inductively coupled plasma spectroscopy was also used beginning in 1982 (168).

D. Potassium, Sodium, and Chloride

These three minerals are known as electrolytes because of their ability to dissociate into positively and negatively charged ions when dissolved in water. These ions, in delicate balance, help to regulate fluids in and out of body cells.

Potassium, the major cation of intracellular fluid, is present in small amounts in extracellular fluid. With sodium, potassium maintains normal water balance, osmotic equilibrium and acid–base balance. Besides helping to regulate fluids in and out of body cells, potassium is also important, along with calcium, in the regulation of neuromuscular activity. Potassium promotes cellular growth and helps maintain normal blood pressure. Muscle mass and glycogen storage are related to muscle mass. During times of muscle formation an adequate supply of potassium is essential. Muscle contractions require potassium.

A wide range of food sources provide potassium including bananas, whole milk, turkey, haddock, okra, oranges, and tomatoes. Concerns involving too much potassium are rare except in cases where the kidneys are unable to excrete excess, which may lead to heart problems. In cases of excessive vomiting and diarrhea, potassium deficiency may result. Deficiency symptoms may include weakness, appetite loss, nausea, and fatigue. Low potassium (hypokalemia) is one of the most common electrolyte abnormalities encountered in clinical practice (169). Diuretic therapy may lead to hypokalemia. Supplemental potassium administration must be watched carefully to avoid severe hyperkalemia. One of the safest ways to minimize hypokalemia is to insure adequate dietary intake of potassium. There are no established recommended dietary allowance levels for potassium.

Commercial forms of potassium include (170):

Potassium chloride (CAS: 7447-40-7). A colorless, elongated, prismatic or cubical crystal or a white granular powder. It is odorless, has a saline taste, and is stable in air. Potassium chloride containing anticaking free-flowing, or conditioning agents may produce cloudy solutions or dissolve incompletely.

Potassium gluconate: anhydrous (CAS: 299-27-4) and monohydrate (CAS: 35398-15-3). This is an odorless and slightly bitter tasting white or yellowish white, crystalline powder or granules. It is freely soluble in water and in glycerin, slightly soluble in alcohol and insoluble in ether.

Potassium glycerophosphate (CAS: 1319-70-6). This is a pale yellow, syrupy liquid containing three molecules of water of hydration. It may be prepared as a colorless to pale yellow, syrupy solution having a concentration of 50 to 75%.

Potassium iodide (CAS: 7681-11-0). These hexahedral crystals are either transparent and colorless or somewhat opaque and white or a white, granular powder. It is stable in dry air and slightly hygroscopic in moist air.

Both atomic absorption spectrometry and inductively coupled plasma spectroscopy are used to determine potassium in foods (171). Activation analysis without chemical separation was found to be unreliable for determining potassium (172).

Sodium is the major cation of extracellular fluid. As an electrolyte, sodium helps regulate the movement of body fluids in and out of body cells. Sodium also helps muscles relax and helps transmit nerve impulses. Substantial amounts of sodium can be found in bile and pancreatic juices. The skeleton holds 35 to 40% of total body sodium, however this sodium is unexchangeable with sodium in body fluids.

The most common form of sodium is sodium chloride or table salt. Processed foods are high in sodium and only a small amount of sodium occurs naturally in foods. There is no Recommended Dietary Allowance for sodium. For healthy adults, a minimum of 500 mg daily is considered safe and adequate. The Daily Value used in food labeling was set at <2400 mg sodium by the Nutrition Labeling and Education Act of 1994 (174).

In the 1960s sodium restriction began to gain acceptance as a dietary practice to reduce hypertension (175). Studies have indicated that it is not just the sodium intake but may also be confounding factors ingested with sodium that impact blood pressure (176). Diets low in potassium or calcium were found to amplify the effect of high sodium chloride intake on blood pressure. Weinberger (177) found that there is increasing evidence to support the contention that blood pressure is responsive to sodium intake in susceptible healthy individuals and those individuals with hypertension. Advocating modest sodium restriction to those individuals who are likely to be salt sensitive would be advantageous.

Commercial forms of sodium as a nutrient include (170):

Sodium ascorbate (CAS: 134-03-2). A white to yellowish crystalline powder.

Sodium chloride (CAS: 7647-14-5). Generically known as salt. Available in various forms: evaporated salt, rock salt, solar salt, or simply salt. It is transparent to opaque, white crystalline solid of variable particle size. Under humidity of less than 75%, it remains dry but will become deliquescent at higher humidities.

Sodium citrate (CAS: 68-04-2). These colorless crystals or white, crystalline powder are used as the nutrient for cultured buttermilk.

Sodium ferric pyrophosphate (CAS: 1332-96-3). This is a white to tan, odorless powder that is insoluble in water but is soluble in hydrochloric acid.

Sodium gluconate (CAS: 527-07-1). This granular to fine, crystalline powder is white to tan in color. It is soluble in water, sparingly soluble in alcohol, and insoluble in ether.

Sodium phosphate, dibasic (CAS: 7558-79-4). This is anhydrous or contains two molecules of water of hydration. The white, crystalline powder or granules are freely soluble in water and insoluble in alcohol.

Sodium phosphate, monobasic (CAS: 7558-80-7). This occurs in two forms. The anhydrous form is a white, crystalline powder or granules. The hydrated form of one or two molecules of water is white or transparent crystals or granules. Both are odorless and slightly hygroscopic.

Sodium phosphate, tribasic (CAS: 7601-54-9). This occurs as a white odorless crystal or granule or crystalline powder.

Sodium pyrophosphate (CAS: 7722-88-5). This white, crystalline, or granular powder is anhydrous or contains 10 molecules of water. It is soluble in water and insoluble in alcohol. The form containing ten molecules of water will effloresce slightly in dry air.

Atomic absorption spectrometry and inductively coupled plasma spectroscopy are used to determine sodium in foods. Flame photometry will also give high results for sodium.

Chloride is the principal anion of extracellular fluids and is widely distributed in the body. Both chloride and sodium amounts and concentrations are responsible for regulation of extracellular fluids (178). Chloride functions as a regulator of fluids in and out of body cells and as a helper in transmission of nerve impulses. Stomach acid contains chloride which aids in the digestion of foods and absorption of nutrients. Chloride is found mostly in salt made of sodium and chloride. Small amounts of chloride are found in water supplies.

Excessive chloride, like excessive sodium, may play a role in high blood pressure. For those individuals who are sensitive, dietary intake should be monitored. Deficiencies in chloride are rare in healthy adults. Body cells are made up of various ion channels that allow the transmitting of electrolytes back and forth across cell membranes. In cystic fibrosis, the cystic fibrosis transmembrane conductance regulator is mutated and causes a disorder in chloride ion transport (179). Children with cystic fibrosis were found to have chloride deficiencies and metabolic alkalosis. Infants that display unexplained hypochloraemic metabolic alkalosis should have serum electrolyte balances regularly checked (180).

There are no Recommended Dietary Allowances for chloride.

Chloride is commercially available in combination with other minerals as the following forms: calcium pantothenate, calcium chloride double salt (CAS: 6363-38-8), choline chloride (CAS: 67-48-1), manganese chloride (CAS: 7773-01-5), potassium chloride (CAS: 7447-40-7), and sodium chloride (CAS: 7647-14-5). Chloride in foods can be analyzed by either a gravimetric method or a volumetric method (181).

E. Iron

Iron is important as a carrier of oxygen in the hemoglobin of red blood cells. Hemoglobin takes oxygen to body cells where it is used for energy production. The resulting byproduct of energy production, carbon dioxide, is removed by hemoglobin. Iron can exist in different ionic states and therefore can serve as a cofactor to enzymes involved in oxidation–reduction reactions. As energy production proceeds, iron gets recycled protecting against iron deficiency. Iron is also important for its roles in protecting from infections, in converting beta carotene to vitamin A, in helping produce collagen, and in helping make body proteins.

Iron is widely available in foods from a variety of sources, both animal and plant. Iron found in plants is only nonheme iron. Iron found in meat, poultry, and fish contains both heme and nonheme iron in a ratio of about 40 to 60. Heme iron is more rapidly absorbed in the body then nonheme iron. Absorption of nonheme iron can be increased with consumption of vitamin C sources.

Hinderances to nonheme iron absorption can be caused by oxalic acid in spinach and chocolate, phytic acid in wheat bran and legumes, tannins in tea, and polyphenols in coffee (182). These food substances should be limited when trying to absorb iron from plant sources. Recommended Dietary Allowances for males ages 11 to 18 is 12 mg, and for males 19 and older is 10 mg. For females 11 to 50 years old the recommended amount is 15 mg. Over 50 the amount drops to 10 mg largely due to the postmenopausal status of women at this age. Pregnancy raises the recommended amount to 30 mg, and with lactation it is raised to 15 mg (183). Iron is needed most during periods of rapid growth and for women during childbearing years and pregnancy. Prior to menopause, iron is needed to replace that lost from menstrual flow.

Despite its wide availability, iron deficiency is one of the most common nutritional deficiencies, especially among children and women during childbearing years. Deficiencies can be caused by injury, hemorrhage, or illness and can be aggravated by poorly balanced diet containing insufficient iron, protein, folate, vitamin B12, vitamin B6, and vitamin C (184). Anemia occurs as a result of iron deficiency and symptoms include fatigue, weakness, and general poor health.

Excessive iron overload may be caused by hereditary hemochromatosis or transfusion overload. Iron toxicity or poisoning is a short-term disorder that occurs following ingestion of large doses of therapeutic iron. Iron toxicity can lead to severe organ damage and death within hours or days. This is a concern particularly where women in a household are taking iron supplements and children accidently consume large doses of these supplements (185).

Commercial forms of iron include (186):

Ferric ammonium citrate, brown (no CAS). This is a complex salt of undetermined structure, composed of iron, ammonia, and citric acid. It occurs as thin, transparent brown, reddish brown, or garnet red scales or granules or as a brownish yellow powder.

Ferric ammonium citrate, green (no CAS). As with the preceding, this complex salt contains the same undetermined structure but occurs as thin transparent green scales, as granules, as a powder, or as transparent green crystals.

Ferric phosphate (CAS: 10045-86-0). This occurs as an odorless, yellowish white to buff colored powder. It contains from one to four molecules of water of hydration.

Ferric pyrophosphate (CAS: 10058-44-3). This tan to yellowish white, odorless powder is insoluble in water but soluble in mineral acids.

Ferrous fumarate (CAS: 141-01-5). This is an odorless, reddish orange to red-brown powder that may contain small lumps. When crushed these lumps may produce a yellow streak.

Ferrous gluconate (CAS: 299-29-6). This powder or granule may have a slight odor, like that of burned sugar. The powder or granules are fine, yellowish gray or pale greenish yellow.

Ferrous lactate (CAS: 5905-52-2). This greenish white powder or crystal has a distinctive odor. It is sparingly soluble in water and practically insoluble in ethanol.

Ferrous sulfate (CAS: 7720-78-7). This is odorless but has a saline styptic taste and is efflorescent in dry air. It occurs as pale, bluish green crystals or granules.

Ferrous sulfate, dried (no CAS). This is a grayish white to buff colored powder that dissolves slowly in water and is insoluble in alcohol.

Iron, carbonyl (CAS: 37220-42-1). This is an elemental iron produced by the decomposition of iron pentacarbonyl as a dark gray powder.

Iron, electrolytic (no CAS). This is an elemental iron produced by electrodeposition in the form of an amorphous, lusterless, grayish black powder.

Iron, reduced (no CAS). This is produced by a chemical process and is in the form of a grayish black powder. It is lusterless or has not more than a slight luster.

Near-infrared spectroscopic analysis has been used to determine heme and nonheme iron in raw muscle meats. This diffuse reflectance method is a rapid and simple method. Other methods for analysis of iron in foods include inductively coupled argon plasma atomic emission spectrophotometry, Schricker procedure, and Suzuki procedures (187).

F. Zinc

Zinc is second only to iron in its abundance. Zinc assists in the promotion of cell reproduction, tissue growth, and tissue repair. Over 70 enzymes have zinc as a part of them. Zinc is an essential nutrient for normal wound healing (188). Zinc supplementation was found to improve cell-mediated immune response in older populations (189). Zinc is also involved in reactions to either synthesize or degrade major metabolities such as carbohydrates, lipids, proteins, and nucleic acid. Debate continues regarding the role zinc plays with the common cold. Studies have gone both ways, some showing benefits in using zinc gluconate lozenges and other studies showing no benefits. Mossad et al. (190) found that in the doses and form they used, zinc gluconate significantly reduced the duration of the symptoms of the common cold. On the other hand, Macknin et al. (191) found that zinc gluconate lozenges were not effective in treating cold symptoms of children and adolescents. Zinc gluconate does have an unpleasant taste and nausea is a possible adverse effect (192).

Zinc is primarily found in meat, fish, poultry, milk, and milk products. Whole grain products, wheat germ, black-eyed peas, and fermented soybean paste (miso) are also good sources of zinc.

Recommended Dietary Allowances for males 11 years and older is 15 mg, and for females 11 years and older is 12 mg. Zinc requirements are higher for pregnant and lactating women (193).

Deficiencies in zinc can cause retarded growth, loss of appetite, skin changes, and reduced resistance to infections. During pregnancy, zinc deficiencies can cause birth defects (194). High calcium diets have been found to significantly reduce zinc absorption and zinc balance in postmenopausal women (195). Although the interaction between calcium and zinc is not completely understood, those individuals with low calcium levels may also have low zinc levels. Zinc toxicity, although rare, can cause deficiency in copper. Toxic levels can also be harmful to the immune system. Because of conflicting studies and the careful balance of zinc levels, a well-balanced diet including foods rich in zinc and other nutrients should be promoted.

Commercial forms include zinc gluconate, zinc oxide, and zinc sulfate:

Zinc gluconate (CAS: 4468-02-4). A white or nearly white, granular or crystalline powder. Depending on the method of isolation, zinc gluconate can occur as a mixture of various states of hydration.

Zinc oxide (CAS: 1314-13-2). A fine, white, odorless, amorphous powder. Zinc oxide will gradually absorb carbon dioxide from the air.

Zinc sulfate (CAS: 7733-02-2). A colorless, transparent prisms or small needles or granular crystalline powder. It is insoluble in alcohol and its solutions are acid to litmus.

All three commercial forms of zinc should be stored in well-closed containers (196).

Using atomic absorption with a flame mode is found to be efficient and accurate in determination of zinc in foods. This method yielded low interlaboratory coefficients of variation and good recoveries (197).

G. Copper

Copper is involved as a part of many enzymes and helps the body to produce energy in cells. Copper also helps make hemoglobin and is needed to carry oxygen in red blood cells. Studies have found that copper is required for infant growth, host defense mechanisms, bone strength, red and white cell maturation, iron transport, cholesterol and glucose metabolism, myocardial contractility, and brain development (198).

Sources of copper in food are highly variable. Rich sources include organ meats, oysters, seafood, nuts, chocolate, and seeds. Milk is a poor source of copper although human breast milk has a higher content than cow milk. Concentration of copper in breast milk does decrease with the time of lactation. Most infant formulas are fortified with copper (199). Drinking water will also have a variable amount of copper depending on the natural mineral content and pH of the water and the plumbing system.

Deficiencies of copper can result from decreased copper stores at birth, inadequate dietary copper intake, poor absorption, elevated requirements induced by rapid growth, or increased copper losses. Clinical manifestations of copper deficiency are anemia, neutropenia, and bone abnormalities. Copper competes with zinc and iron, therefore high intakes of zinc and/or iron may predispose one to copper deficiency (200). Menkes' kinky hair syndrome is an inherited disorder affecting normal absorption of copper from the intestine and is characterized by growth of sparse, kinky hair (201).

Overt toxicity from dietary copper sources is rare. Wilson's disease is a rare inherited disease which causes copper to be accumulated slowly in the liver and then released and taken in by other parts of the body.

Although there are no established Recommended Dietary Allowance levels for copper, an estimated safe and adequate daily dietary intake (ESADDI) has been established. The ESADDI level for adults is 1.5 to 3 mg/day (202).

Commercial forms of copper include (203):

Copper gluconate (CAS: 527-09-3). This is a fine, light blue powder that is soluble in water and is slightly soluble in alcohol.

Copper sulfate (CAS: 7758-98-7). The blue crystals, crystalline granules or powder function as a nutrient supplement. This chemical will effloresce slowly in dry air.

Atomic absorption in the flame mode can be used efficiently, accurately, with low interlaboratory cooefficients of variation and will provide good results regarding copper content in foods (204).

H. Iodine

Iodine functions as a part of the thyroid hormone, thyroxin. The thyroid regulates the rate that one's body uses energy. It is involved in the regulation of metabolic activities of cells, especially of the brain during fetal and early postnatal life. When requirements are not met, functional and developmental abnormalities can occur. During 1994, it was estimated that some 1.5 billion people in 118 countries were at risk for iodine deficiency, making iodine deficiency one of the world's single most important causes of preventable brain damage and mental retardation (205). In a study of severely iodine deficient (SID) and mildly iodine deficient (MID) male children, the SID children were found to be slower learners than MID children. Also, the rate of improvement in performances was significantly different between the two groups. The SID children were poorly motivated to achieve. Iodine deficiency was found to lead to a range of deficits and reflected developmental disadvantages of the entire community. These iodine deficient areas do not provide children with the necessary supportive sociopsychological environments for learning new skills and various cognitive abilities (206).

Iodine found in foods is rapidly absorbed as iodide. Iodized salt is the main source of this element. Iodine in milk is influenced by the source of animal feed and the sanitizing solutions used in the dairy industry (207). Drinking water from water purification systems used by Peace Corps volunteers in Niger was found to be a possible source of excessive iodine intake (208). Other sources of iodine include saltwater fish, potatoes, spinach, and almonds (209).

Deficiencies of iodine can occur at all stages of development. During pregnancy, infancy, or early childhood, deficiency may lead to endemic cretinism in an infant or child. Cretinism is not reversible. Goiters, the more commonly known iodine deficiency symptom, can be reversed by providing adequate iodine intake.

Iodine can also be toxic. Goiters, thyroiditis, hypothyroidism, and hyperthyroidism may result from excessive iodine in individuals who are salt sensitive, have other thyroid disorders, or have normally low intacts of iodine. Graves' disease is the most common form of hyperthyroidism. Calcium and vitamin D losses may occur in cases of hyperthyroidism, and supplementation with a multivitamin is recommended (210).

The iodine recommended allowance for adults is 150 μg/day. During pregnancy, an additional 25 μg/day is recommended due to the demands of the fetus. An additional 50 μg/day is recommended for lactating women (211). Iodine is available commercially in combination with potassium as potassium iodide. Kelp, a dehydrated seaweed, may be chopped as coarse particles and/or ground for fine powder and provides a salty characteristic taste. Kelp is used as a source of iodine (212).

Reliable determination of iodide in foods is difficult. This difficulty is due to the low levels of iodide in foods and the losses of iodide that occur in sample digestion. In capillary electrophoresis analysis, one can simultaneously determine fluoride, chloride, bromide, iodide, and some oxy-halogenated species with indirect UV detection (213).

I. Manganese

Manganese functions as a part of several enzymes. Besides magnesium, manganese can also activate numerous enzymes. Manganese is associated with the formation of connective and bony tissues, growth and reproduction, and carbohydrate and lipid metabolism (214).

Sources of manganese include nuts, seeds, tea, and whole grains. Small amounts are found in meats, dairy products, and sugary and refined foods. Various dietary components influence the bioavailability of manganese and its absorption, retention, and excretion. These components include iron, phosphorus, phytates, and fiber and also may include calcium, copper, and polyphenolic compounds (215).

Manganese deficiency symptoms include poor reproductive performance, growth retardation, congenital malformations in offspring, abnormal formation of bone and cartilage, and impaired glucose tolerance (216). However, due to homeostatic mechanisms limiting the absorption of manganese from the GI tract, there have been no reported cases of deficiency in humans (217).

The lungs and brain are the primary targets for overexposure to manganese. Fell et al. (218) found cholestatic disease and nervous system disorders associated with high blood concentrations of manganese in patients receiving long-term parenteral nutrition. Manganese madness was first used to describe an initial psychiatric syndrome which included compulsive behavior, emotional lability, and hallucinations that occurred due to the lungs and gastrointestinal tract absorbing too much manganese oxide (217).

No RDA exists for manganese. However, the estimated safe and adequate daily dietary intake for adults is 2.0 to 5.0 mg/day (216).

Commercial forms of manganese include (219):

Manganese chloride (CAS: 7773-01-5). This is available as large, irregular, pink, translucent crystals. Very soluble in hot water and freely soluble in room temperature water.

Manganese gluconate (CAS: 6485-39-8). This is available in either dihydrate or anhydrous form. It is a slightly pink colored powder.

Manganese glycerophosphate (CAS: 1320-46-3). Odorless and nearly tasteless white or pinkish white powder.

Manganese hypophosphite (CAS: 10043-84-2). Odorless and nearly tasteless, pink, granular or crystalline powder. It is stable in air and soluble in alcohol.

Manganese sulfate (CAS: 7785-87-7). This is a pale pink, granular, odorless powder.

Manganese citrate (CAS: 10024-66-5). This is a light pink or pink-white fine, granular solid.

Atomic spectroscopy and neutron activation analysis are two methods believed to have the most sensitivity and greatest potential for accurately measuring manganese (220).

VI. DIETARY SUPPLEMENTS

In 1994, the United States Congress passed the Dietary Supplement Health and Education Act (DSHEA). Through this legislation, dietary supplements have been defined as "a product, other than tobacco, intended to supplement the diet that contains at least one or more of the following ingredients: a vitamin, a mineral, an herb or other botanical, an amino acid, or a dietary substance for use to supplement the diet by increasing the total dietary intake; or a concentrate, metabolite, constituent, or extract or combination of any of the previously mentioned ingredients." Dietary supplements are regulated as a special category of foods rather than as a category of drugs. Because these products are meant to supplement the diet by increasing the total dietary intake of a substance, manufacturers do not have to provide information to the FDA to get their product on the market. Once a dietary supplement is marketed, the responsibility for showing that the product is unsafe falls to the FDA. The FDA cannot restrict the product's use unless the product's dangers can be determined. Individual states, however, can take steps within their jurisdictions to restrict or stop the sale of potentially harmful dietary supplements (221). Food additives, on the other hand, must go through safety studies and premarket approval processes by the FDA prior to marketing.

Three types of claims can be made regarding dietary supplements: nutrient-content claims, disease claims, and nutrition support claims. A nutrient-content claim is based on the level of the nutrient in a food or dietary supplement. For example, a supplement that contains 12 mg of vitamin C could state that it is a ''good source of vitamin C.'' If there is a link between the food and a disease or health-related condition, a disease claim could be made. For example, a product that contains folic acid can make a disease claim that the product decreases the risk of neural tube defects during pregnancy. A nutrient support claim can show a link between the nutrient and a deficiency disease as a result of a lack of the nutrient. For example, vitamin C prevents scurvy.

As of March 1999, dietary supplements must contain a ''Supplement Facts'' panel. Dietary supplements must be labeled as such and be sold in the form of pills, capsules, tablets, gelcaps, liquids, powders, or other forms, and not be represented for use as conventional foods. These products cannot be marketed as the only item in a meal or diet (222).

The dietary supplement industry continues to grow. In 1996, consumers spent an estimated $6.5 billion on supplements, double the expenditures in 1990–1991 (223). The Office of Dietary Supplements is a congressionally mandated office in the Office of the Director, National Institutes of Health (NIH). Through their efforts and those of the USDA Food and Nutrition Information Center, the International Bibliography Information on Dietary Supplements (IBIDS) database are available: http://odp.od.nih.gov/ods/databases/ibids.html.

REFERENCES

Vitamin A

1. Ball, G. 1996. Determination of the fat-soluble vitamins in foods by high-performance liquid chromatography. In: *Handbook of Food Analysis*, Vol. 1, *Physical Characterization and Nutrient Analysis*, Nollet, L. M. L. (Ed.). Marcel Dekker, New York, pp. 602–604.
2. Olson, J. A. 1994. Vitamin A, retinoids, and carotenoids. In: M. E. Shils, J. A. Olson, M. Shike (Eds.), *Modern Nutrition in Health and Disease*, 8th ed., Shils, M. E., Olson, J. A., Shike, M. (Eds.). Williams & Wilkins, Baltimore, pp. 287–307.
3. Olson, J. A. 1996. Vitamin A. In: *Present Knowledge in Nutrition*, 7th ed., Ziegler, E. E., Filer, L. J., Jr. (Eds.). LSI Press, Washington, D.C., pp. 109–119.
4. Stephensen, C. B., Alvarez, J. O., Kohatsu, J., Hardmeier, R., Kennedy, J. I. Jr., Gammon, R. B. Jr. 1994. Vitamin A is excreted in the urine during acute infection. *Am. J. Clin. Nutr.* 60:388–392.
5. Duitsman, P. K., Cook, L. R., Tanumihardjo, S. A., Olson, J. A. 1995. Vitamin A inadequacy in socioeconomically disadvantaged pregnant Iowan women as assessed by the modified relative dose response (MRDR) test. *Nutr. Res.* 15:1263–1276.
6. Food and Nutrition Board. 1989. *Recommended Dietary Allowances*, 10th ed., National Academy Press, Washington, D.C.
7. Mahan, K. L., Escott-Stump, S. 1996. *Krause's Food Nutrition and Diet Therapy*, 9th ed. W. B. Saunders, Philadelphia.
8. Committee on Food Chemicals Codex Food & Nutrition Board. 1996. *Food Chemicals Co-*

dex, 4th ed. Institute of Medicine, National Academy of Sciences. National Academy Press, Washington, D.C.
9. FAO. 1996. Food fortification technology and quality control. FAO technical meeting, Rome. FAO Food and Nutrition Paper 60.
10. Hathcock, J. N., Hattan, D. G., Jenkins, M. Y., McDonald, J. T., Sundaresan, P. R., Wilkening, V. L. 1990. Evaluation of vitamin A toxicity. *Am. J. Clin. Nutr.* 52:183–202.

Carotenoids

11. Mahan, L. K., Escott-Stump, S. 1996. *Krause's Food, Nutrition & Diet Therapy*, 9th ed. W. B. Saunders, Philadelphia.
12. FAO. 1996. Food fortification technology and quality control. FAO technical meeting, Rome. FAO Food and Nutrition Paper 60.
13. Olson, J. A. 1994. Vitamin A, retinoids, and carotenoids. In: *Modern Nutrition in Health and Disease*, 8th ed., Shils, M. E., Olson, J. A., Shike, M. Williams & Wilkins, Baltimore, pp. 287–307.
14. Ball, G. 1996. Determination of the fat-soluble vitamins in foods by high-performance liquid chromatography. In: *Handbook of Food Analysis, Vol. 1, Physical Characterization and Nutrient Analysis*. Marcel Dekker, New York, pp. 602–604.
15. Committee on Food Chemicals Codex Food & Nutrition Board. 1996. *Food Chemical Codex*, 4th ed. Institute of Medicine, National Academy of Sciences. National Academy Press, Washington, D.C.

Vitamin D

16. Holick, M. F. 1994. Vitamin D. In: *Modern Nutrition in Health and Disease*, 8th ed., Shils, M. E., Olson, J. A., Shike, M. (Eds.). Williams & Wilkins, Baltimore, pp. 308–325.
17. Food & Nutrition Board. 1998. *Dietary Reference Intakes*. National Academy Press, Washington, D.C.
18. Webb, A. R., Pilbeam, C., Hanafin, N., Holick, M. F. 1990. An evaluation of the relative contributions of exposure to sunlight and of diet to the circulating concentrations of 25-hydroxyvitamin D in an elderly nursing home population in Boston. *Am. J. Clin. Nutr.* 51: 1075–1081.
19. Holick, M. F. 1995. Environmental factors that influence the cutaneous production of Vitamin D. *Am. J. Clin. Nutr.* 61(Suppl.):638S–645S.
20. Committee on Food Chemicals Codex Food & Nutrition Board. 1996. *Food Chemicals Codex*, 4th ed. Institute of Medicine, National Academy of Sciences. National Academy Press, Washington, D.C.
21. FAO. 1996. Food fortification technology and quality control. FAO technical meeting, Rome. FAO Food and Nutrition Paper 60.
22. Norman, A. W. 1996. Vitamin D. In: *Present Knowledge in Nutrition*, 7th ed., Ziegler, E. E., Filer, L. J., Jr. (Eds.). ILSI Press, Washington, D.C., pp. 120–129.

Vitamin E

23. Sokol, R. J. 1996. Vitamin E. In: *Present Knowledge in Nutrition*, 7th ed., Ziegler, E. E., Filer, L. J., Jr. (Eds). ILSI Press, Washington, D.C., pp. 130–136.
24. Meydani, S. N., Barklund, M. P., Liu, S., Meydani, M., Miller, R. A., Cannon, J. G., Morrow, F. D., Rocklin, R., Blumberg, J. B. 1990. Vitamin E supplementation enhances cell-mediated immunity in healthy elderly subjects. *Am. J. Clin. Nutr.* 52:557–563.

25. Mahan, L. K., Escott-Stump, S. 1996. *Krause's Food, Nutrition and Diet Therapy*, 9th ed., W. B. Saunders, Philadelphia.
26. Stampfer, M. J., Hennekens, C. H., Manson, J. E., Colditz, G. A., Rosner, B., Willett, W. C. 1993. Vitamin E consumption and the risk of coronary disease in women. *New Engl. J. Med.* 328:1444–1449.
27. Rimm, E. B., Stampfer, M. J., Ascherio, A., Giovannucci, E., Colditz, G. A. Willett, W. C. 1993. Vitamin E consumption and the risk of coronary heart disease in men. *New Engl. J. Med.* 328:1450–1456.
28. Ball, G. 1996. Determination of the fat-soluble vitamins in foods by high-performance liquid chromatography. In: *Handbook of Food Analysis, Vol. 1, Physical Characterization and Nutrient Analysis*, Nollet, L. M. L. (Ed.). Marcel Dekker, New York, pp. 602–604.
29. Food and Nutrition Board. 1989. *Recommended Dietary Allowances*, 10th ed. National Academy Press, Washington, D.C.
30. FAO. 1996. Food fortification technology and quality control. FAO technical meeting, Rome. FAO Food and Nutrition Paper 60.
31. Committee on Food Chemical Codex Food & Nutrition Board. 1996. *Food Chemical Codex*, 4th ed. Institute of Medicine, National Academy of Sciences. National Academy Press, Washington, D.C.

Vitamin K

32. Mahan, L. K., Escott-Stump, S. 1996. *Krause's Food, Nutrition, and Diet Therapy*, 9th ed. W. B. Saunders, Philadelphia.
33. Olson, R. E., 1994. Vitamin K. In: *Modern Nutrition in Health and Disease Eighth Edition*, Shils, M. E., Olson J. A., Shike, M. (Eds.). Williams & Wilkins, Baltimore, pp. 342–358.
34. Ferland, G., Sadowski, J. A., O'Brien, M. E. 1993. Dietary induced subclinical vitamin K deficiency in normal human subjects. *J. Clin. Invest.* 91:1761–1768.
35. Binkley, N. C., Suttie, J. W. 1995. Vitamin K nutrition and osteoporosis. *J. Nutr.* 125:1812–1821.
36. Suttie, J. W. 1996. Vitamin K. In: *Present Knowledge in Nutrition*, 7th ed., Ziegler, E. E., Filer, L. J. (Eds.). ILSI Press, Washington, D.C., pp. 137–145.
37. Food and Nutrition Board. 1989. *Recommended Dietary Allowances*, 10th ed. National Academy Press, Washington, D.C.
38. Ball, G. F. M. 1996. In: *Handbook of Food Analysis, Vol. 1 Physical Characterization and Nutrient Analysis*, Nollet, L. M. L. (Ed.). Marcel Dekker, New York, pp. 601–647.
39. Booth, S. L., Davidson, K. W., Sadowski, J. A. 1994. Evaluation of an HPLC method for the determination of phylloquinone (vitamin K1) in various food matrices. *J. Agricult. Food Chem.* 42:295–300.
40. Committee on Food Chemicals Codex, Food and Nutrition Board. 1997. *Food Chemicals Codex: First Supplement to the Fourth Edition*. Institute of Medicine, National Academy of Science. National Academy Press, Washington, D.C.

Vitamin C

41. Pauling, L. 1970. *Vitamin C and the Common Cold*. W. H. Freeman, San Francisco.
42. Food and Nutrition Board. 1989. *Recommended Dietary Allowances*, 10th ed., National Academy Press, Washington, D.C.
43. Levine, M., Rumsey, S., Wang, Y., Park, J., Kwon, O., Xu, W., Amano, N. 1996. Vitamin C. In: *Present Knowledge in Nutrition*, 7th ed., Ziegler, E. E., Filer, L. J. Jr. (Eds.). ILSI Press, Washington, D.C., pp. 146–159.

44. Levine, M., Dhariwal, K. R., Welch, R. W., Wang, Y., Park, J. B. 1995. Determination of optimal vitamin C requirements in humans. *Am. J. Clin. Nutr.* 62(Suppl):1347S–1356S.
45. Mahan, L. K., Escott-Stump, S. 1996. *Krause's Food, Nutrition and Diet Therapy*, 9th ed. W. B. Saunders, Philadelphia.
46. Vanderslice, J. T., Higgs, D. J. 1991. Vitamin C content of foods: sample variability. *Am. J. Clin. Nutr.* 54:1323S–1327S.
47. Jacob, R. A. 1994. Vitamin C. In: *Modern Nutrition in Health and Disease*, 8th ed., Shils, M. E., Olson, J. A., Shike, M. (Eds.). Williams & Wilkins, Baltimore, pp. 432–448.

Thiamin

48. Mahan, L. K. & Escott-Stump, S. 1996. *Krause's Food, Nutrition and Diet Therapy*, 9th ed. W. B. Saunders, Philadelphia.
49. Rindi, G., Patrini, C., Laforenza, U., Mandel, H., Berant, M., Viana, M. B., Poggi, V., Zarra, A. N. F. 1994. Further studies on erythrocyte thiamin transport and phosphorylation in seven patients with thiamin-responsive megaloblastic anaemia. *J. Inherited Metabolic Disease*, 17: 667–677.
50. Rindi, G. 1996. Thiamin. In: *Present Knowledge in Nutrition*, 7th ed. Ziegler, E. E. Filer, L. J., Jr. (Eds.) ILSI Press, Washington, D.C., pp. 160–166.
51. Food and Nutrition Board. 1998. *Dietary Reference Intakes for Thiamin Riboflavin, Niacin, Vitamin B6, Folate, Vitamin B12, Pantothenic Acid, Biotin and Choline*. Committee on the Scientific Evaluation of Dietary Reference Intakes. National Academy Press, Washington, D.C.
52. Russell, L. F. 1996. Water-soluble vitamins. In: *Handbook of Food Analysis*, Vol. 1, Nollet, L. M. L. (Ed.). Marcel Dekker, New York, pp. 649–713.
53. Committee on Food Chemicals Codex Food and Nutrition Board. 1996. *Food Chemicals Codex*, 4th ed. Institute of Medicine, National Academy of Sciences. National Academy Press, Washington, D.C.

Riboflavin

54. Rivlin, R. S., Dutta, P. 1995. Vitamin B2 (riboflavin) relevance to malaria and antioxidant activity. *Nutrition Today*. 30:62–67.
55. McCormick, D. B. 1994. Riboflavin. In: *Modern Nutrition in Health and Disease*, 8th ed. Shils, M. E., Olson, J. A., Shike M. (Eds.). Williams & Wilkins, Baltimore, pp. 366–375.
56. Food & Nutrition Board. 1998. *Dietary Reference Intakes for Thiamin, Riboflavin, Niacin, Vitamin B6, Folate, Vitamin B12, Pantothenic Acid, Biotin and Choline*. Committee on the Scientific Evaluation of Dietary Reference Intakes. National Academy Press, Washington, D.C.
57. Boisvert, W. A., Mendoza, I., Castaneda, C., De Portocarrero, L., Solomons, N. W., Gershoff, S. N., Russell, R. M. 1993. Riboflavin requirement of healthy elderly humans and its relationship to macronutrient composition of the diet. *J. Nutr.* 123:915–925.
58. Russell, L. F. 1996. In: *Handbook of Food Analysis, Vol. 1, Physical Characterization and Nutrient Analysis*, Nollet, L. M. L. (Ed.). Marcel Dekker, New York, pp. 649–713.
59. Committee on Food Chemicals Codex Food & Nutrition Board. 1996. *Food Chemicals Codex*, 4th ed. Institute of Medicine, National Academy of Sciences. National Academy Press, Washington, D.C.

Niacin

60. Jacob, R. A., Swendseid, M. E. 1996. In: *Present Knowledge in Nutrition*, 7th ed. Ziegler, E. E., Filer, L. J., Jr. (Eds.). ILSI Press, Washington, D.C., pp. 184–190.

61. Food & Nutrition Board. 1998. *Dietary Reference Intakes for Thiamin, Riboflavin, Niacin, Vitamin B6, Folate, Vitamin B12, Pantothenic Acid, Biotin and Choline*. Committee on the Scientific Evaluation of Dietary Reference Intakes. National Academy Press, Washington, D.C.
62. Krishnan, P. G., Mahmud, I., Mathees, D. P. 1996. Postcolumn fluorimetric HPLC procedure for determination of niacin content of cereals. *Cereal Chem.* 76(4):512–518.
63. Committee on Food Chemicals Codex Food & Nutrition Board. 1996. *Food Chemicals Codex*. 4th ed. Institute of Medicine, National Academy of Sciences. National Academy Press, Washington, D.C.

Vitamin B_6

64. Russell, L. F. 1996. Water-soluble vitamins. In: *Handbook of Food Analysis, Volume 1 Physical Characterization and Nutrient Analysis*, Nollet, L. M. L. (Ed.). Marcel Dekker, New York, pp. 649–713.
65. Mahan, L. K. & Escott-Stump, S. 1996. *Krause's Food, Nutrition and Diet Therapy*, 9th ed. W. B. Saunders, Philadelphia.
66. Rall, L. C., Meydani, S. N. 1993. Vitamin B6 and immune competence. *Nutr. Rev.* 51:217–225.
67. Compton, M. M., Cidlowski, J. A. 1986. Vitamin B6 and glucocorticoid action. *Endocrine Rev.* 7:140–148.
68. Leklem, J. E. 1994. Vitamin B6. In: *Modern Nutrition in Health and Disease*, 8th ed., Shils M. E., Olson, J. A., Shike M. (Eds.). Williams & Wilkins, Baltimore, pp. 383–394.
69. Food & Nutrition Board. 1998. *Dietary Reference Intakes for Thiamin, Riboflavin, Niacin, Vitamin B6, Folate, Vitamin B12, Pantothenic Acid, Biotin and Choline*. Committee on the Scientific Evaluation of Dietary Reference Intakes. National Academy Press, Washington, D.C.
70. Driskell, J. A. 1994. Vitamin B6 requirements of humans. *Nutr. Res.* 14:293–324.
71. Leklem, J. E. 1990. Vitamin B6: A status report. *J. Nutr.* 120:1503–1507.
72. Committee on Food Chemicals Codex Food & Nutrition Board. 1996. *Food Chemicals Codex*, 4th ed. Institute of Medicine, National Academy of Sciences. National Academy Press, Washington, D.C.

Pantothenic Acid

73. Plesofsky-Vig, N. 1996. Pantothenic Acid. In: *Present Knowledge in Nutrition*, 7th ed., Ziegler E. E., Filer, L. J., Jr. (Eds.) ILSI Press, Washington, D.C., pp. 237–244.
74. Plesofsky-Vig, N., Brambl, R. 1988. Pantothenic acid and coenzyme A in cellular modification of proteins. *Ann. Rev. Nutr.* 8:461–482.
75. Plesofsky-Vig, N. 1994. Pantothenic acid and coenzyme A. In: *Modern Nutrition in Health and Disease*, 8th ed., Shils, M. E., Olson, J. A., Shike, M. (Eds.). Williams & Wilkins, Baltimore, pp. 395–401.
76. Food and Nutrition Board. 1998. *Dietary Reference Intakes for Thiamin, Riboflavin, Niacin, Vitamin B6, Folate, Vitamin B12, Pantothenic Acid, Biotin and Choline*. Committee on the Scientific Evaluation of Dietary Reference Intakes. National Academy Press, Washington, D.C.
77. Tahiliani, A. G., Beinlich, C. J. 1991. Pantothenic acid in health and disease. *Vitamin Hormone*. 46:165–228.
78. Committee on Food Chemicals Codex Food & Nutrition Board. 1996. *Food Chemicals Codex* 4th ed. Institute of Medicine, National Academy of Sciences. National Academy Press, Washington, D.C.
79. Russell, L. F. 1996. Water-soluble vitamins. In: *Handbook of Food Analysis, Vol. 1, Physical*

Characterization and Nutrient Analysis, Nollet, L. M. L. (Ed.). Marcel Dekker, New York, pp. 649–713.

Folate

80. Mahan, L. K., Escott-Stump, S. 1996. *Krause's Food, Nutrition and Diet Therapy*, 9th ed. W. B. Saunders, Philadelphia.
81. Czeizel, A. E. 1995. Folic acid in the prevention of neural tube defects. *J. Pediatr. Gastroenterol. Nutr.* 2:4–16.
82. MRC Vitamin Study Research Group. 1991. Prevention of neural tube defects: results of the medical research council vitamin study. *Lancet* 338:131–137.
83. Herbert, V., Das, K. C. 1994. Folic acid and vitamin B12. In: *Modern Nutrition in Health and Disease*, 8th ed., Shils, M. E., Olson, J. A., Shike, M. (Eds.). Williams & Wilkins, Baltimore, pp. 402–425.
84. Food and Nutrition Board. 1998. *Dietary Reference Intakes for Thiamin, Riboflavin, Niacin, Vitamin B6, Folate, Vitamin B12, Pantothenic Acid, Biotin, and Choline*. Committee on the Scientific Evaluation of Dietary Reference Intakes. National Academy Press, Washington, D.C.
85. Russell, L. F. 1996. Water-soluble vitamins. In: *Handbook of Food Analysis*, Vol. 1, *Physical Characterization and Nutrient Analysis*, Nollet L. M. L. (Ed.). Marcel Dekker, New York, pp. 649–713.
86. Seyoum, E., Selhub, J. 1993. Combined affinity and ion pair column chromatographies for the analysis of food folate. *J. Nutr. Biochem.* 4:488–494.
87. Committee on Food Chemicals Codex Food & Nutrition Board. 1996. *Food Chemicals Codex*, 4th ed. Institute of Medicine, National Academy of Sciences. National Academy Press, Washington, D.C.

Vitamin B_{12}

88. Mahan, L. K., Escott-Stump, S. 1996. *Krause's Food, Nutrition and Diet Therapy*, 9th ed. W.B. Saunders, Philadelphia.
89. Rauma, A.-L., Torronen, R., Hanninen, O., Mykkanen, H. 1995. Vitamin B12 status of long-term adherents of a strict uncooked vegan diet (''living food diet'') is compromised. *J. Nutr.* 125:2511–2515.
90. Herbert, V. 1994. Staging vitamin B12 (cobalamin) status in vegetarians. *Am. J. Clin. Nutr.* 59(Suppl.):1213S–1222S.
91. Nygard, O., Vollset, S. E., Refsum, H., Stensvold, I., Tverdal, A., Nordrehuag, J. E., Ueland, P. M., Kvoale, G. 1995. Total plasma homocysteine and cardiovascular risk profile. *JAMA* 274:1518–1525.
92. Herbert, V. 1996. Vitamin B-12. In: *Present Knowledge in Nutrition*, 7th ed. Ziegler, E. E., Filer, L. J., Jr. (Eds.). ILSI Press, Washington, D.C., pp. 191–205.
93. Food and Nutrition Board. 1998. *Dietary Reference Intakes for Thiamin, Riboflavin, Niacin, Vitamin B6, Folate, Vitamin B12, Pantothenic Acid, Biotin, and Choline*. Committee on the Scientific Evaluation of Dietary Reference Intakes. National Academy Press, Washington, D.C.
94. Herbert, V. 1994. Vitamin B12 and elderly people. *Am. J. Clin. Nutr.* 59:1093–1094 (Letter).
95. Herbert, V., Das, K. C. 1994. Folic acid and vitamin B12. In: *Modern Nutrition in Health and Disease*, 8th ed., Shils, M. E., Olson, J. A., Shike, M. (Eds.). Williams & Wilkins, Baltimore, pp. 402–425.
96. Russell, L. F. 1996. Water-soluble vitamins. In: *Handbook of Food Analysis*, Vol. 1, *Physical Characterization and Nutrient Analysis*, Nollet, L. M. L. (Ed.). Marcel Dekker, New York, pp. 649–713.

97. Committee on Food Chemicals Codex Food & Nutrition Board. 1996. *Food Chemicals Codex*, 4th ed. Institute of Medicine, National Academy of Sciences. National Academy Press, Washington, D.C.

Biotin

98. Mock, D. M. 1996. Biotin. In: *Present Knowledge in Nutrition*, 7th ed., Ziegler, E. E., Filer, L. J., Jr. (Eds.). ILSI Press, Washington, D.C., pp. 220–235.
99. Velazquez, A., Martin-del-Campo, C., Baez, A., Zamudio, S., Quiterio, M., Aguilar, J. L., Perez-Ortiz, B, Sanchez-Ardines, Guzman-Hernandez, J., Casanueva, E. 1989. Biotin deficiency in protein-energy malnutrition. *Eur. J. Clin. Nutr.* 43:169–173.
100. Velazquez, A., Teran, M., Baez, A., Gutierrez, J., Rodriguez, R. 1995. Biotin supplementation affects lymphocyte carboxylases and plasma biotin in severe protein-energy malnutrition. *Am. J. Clin. Nutr.* 61:385–391.
101. Russell, L. F. 1996. Water-soluble vitamins. In: *Handbook of Food Analysis*, Vol. 1, *Physical Characterization and Nutrient Analysis*, Nollet, L. M. L. (Ed.). Marcel Dekker, New York, pp. 649–713.
102. Food and Nutrition Board. 1998. *Dietary Reference Intakes for Thiamin, Riboflavin, Niacin, Vitamin B6, Folate, Vitamin B12, Pantothenic Acid, Biotin, and Choline*. Committee on the Scientific Evaluation of Dietary Reference Intakes. National Academy Press, Washington, D.C.
103. Dakshinamurti, K. 1994. Biotin. In: *Modern Nutrition in Health and Disease*, 8th ed., Shils, M. E., Olson, J. A., Shike, M. (Eds.). Williams & Wilkins, Baltimore, pp. 426–431.
104. Committee on Food Chemicals Codex Food & Nutrition Board. 1996. *Food Chemicals Codex*, 4th ed., Institute of Medicine, National Academy of Sciences. National Academy Press, Washington, D.C.

Choline

105. Savendahl, L., Mar, M.-H. Underwood, L. E., Zeisel, S. H. 1997. Prolonged fasting in humans results in diminished plasma choline concentrations but does not cause liver dysfunction. *Am. J. Clin. Nutr.* 66:622–625.
106. Anderson, K. N., Anderson, L. E., Glanze, W. D. 1994. *Mosby's Medical, Nursing, and Allied Health Dictionary*, 4th ed., Mosby-Year Book, St. Louis.
107. Mahan, L. K., Escott-Stump, S. 1996. *Krause's Food, Nutrition and Diet Therapy*, 9th ed., W. B. Saunders, Philadelphia.
108. Zeisel, S. H., DaCosta, K.-A. Franklin, P. D., Alexander, E. A., Lamont, J. T., Sheard, N. F., Beiser, A. 1991. Choline, an essential nutrient for humans. *FASAB.* 5:2093–2098.
109. Zeisel, S. H. 1994. Choline. In: *Modern Nutrition in Health and Disease*, 8th ed., Shils, M. E., Olson, J. A., Shike, M. (Eds.). Williams & Wilkins, Baltimore, pp. 449–458.
110. Food and Nutrition Board. 1998. *Dietary Reference Intakes for Thiamin, Riboflavin, Niacin, Vitamin B6, Folate, Vitamin B12, Pantothenic Acid, Biotin and Choline*. Committee on the Scientific Evaluation of Dietary Reference Intakes. National Academy Press, Washington, D.C.
111. Committee on Food Chemicals Codex, Food and Nutrition Board. 1996. *Food Chemicals Codex*. 4th ed., Institute of Medicine, National Academy of Science. National Academy Press, Washington, D.C.
112. Committee on Food Chemicals Codex, Food and Nutrition Board. 1996. *Food Chemicals Codex*. 4th ed., Institute of Medicine, National Academy of Science. National Academy Press, Washington, D.C.
113. Whitney, E. N., Rolfes, S. R. 1996. *Understanding Nutrition*, 7th ed., West Publishing, St. Paul, MN.

Carnitine

114. Anderson, K. N., Anderson, L. E., Glanze, W. D. 1994. *Mosby's Medical, Nursing and Allied Health Dictionary*, 4th ed. Mosby-Year Book, St. Louis.
115. Borum, P. R. 1996. Changing perspective of carnitine function and the need for exogenous carnitine of patients treated with hemodialysis. *Am. J. Clin. Nutr.* 64:976–977.
116. Chen, W., Huang, Y.-C., Shultz, T. D., Mitchell, M. E. 1998. Urinary, plasma, and erythrocyte carnitine concentration during transition to a lactoovovegetarian diet with vitamin B6 depletion and repletion in young adult women. *Am. J. Clin. Nutr.* 67:221–230.
117. Harper, P., Wadstrom, C., Backman, L., Cederblad, G. 1995. Increased liver carnitine content in obese women. *Am. J. Clin. Nutr.* 61:18–25.
118. Broquist, H. P. 1994. Carnitine. In: *Modern Nutrition in Health and Disease*, 8th ed., Shils, M. E., Olson, J. A., Shike, M. (Eds.). Williams & Wilkins, Baltimore, pp. 459–465.
119. Mahan, L. K., Escott-Stump, S. 1996. *Krause's Food, Nutrition and Diet Therapy*, 9th ed. W. B. Saunders, Philadelphia.
120. Whitney, E. N., Rolfes, S. R. 1996. *Understanding Nutrition*, 7th ed., West Publishing, St. Paul, MN.

Inositol

121. Aukeman, H. M., Holub, B. J. 1994. Inositol and pyrroloquinoline quinone. In: *Modern Nutrition in Health and Disease* 8th ed., Shils, M. E., Olson, J. A., Shike, M. (Eds.). Williams & Wilkins, Baltimore, pp. 466–472.
122. Fux, M., Levine, J., Aviv, A., Belmaker, R. H. 1996. Inositol treatment of obsessive-compulsive disorder. *Am. J. Psychiatry* 153:1219–1221.
123. Benjamin, J., Levine, J., Fux, M., Aviv, A., Levy, D., Belmaker, R. H. 1995. Double-blind, placebo-controlled, crossover trial of inositol treatment for panic disorder. *Am. J. Psychiatry* 152:1084–1086.
124. Levine, J., Barak, Y., Gonzalves, M., Szor, H., Elizur, A., Kofman, O., Belmaker, R. H. 1995. Double-blind, controlled trial of inositol treatment of depression. *Am. J. Psychiatry* 152:792–794.
125. Committee on Food Chemicals Codex, Food and Nutrition Board. 1996. *Food Chemicals Codex*. 4th ed., Institute of Medicine, National Academy of Science. National Academy Press, Washington, D.C.

Amino Acids

126. Baxter, J. H. 1996. Amino Acids. In: *Handbook of Food Analysis* Vol. 1, *Physical Characterization and Nutrient Analysis*, Nollet, L. M. L. (Ed.). Marcel Dekker, New York, pp. 197–228.
127. Polge, A., Bancel, E., Bellet, H., Strubel, D., Poirey, S., Peray, P., Carlet, C., Magnan de Bornier, B. 1997. Plasma amino acid concentrations in elderly patients with protein energy malnutrition. *Age and Aging* 26:457–463.
128. Obarzanek, E., Velletri, P. A., Cutler, J. A. 1996. Dietary protein and blood pressure. *JAMA* 275:1598–1603.
129. American Dietetic Association. 1999. In the news high-protein/low-carbohydrate diets. http://www.eatright.org/news. Accessed July 5, 1999.
130. Zeman, F. J., Ney, D. M. 1996. *Applications in Medical Nutrition Therapy*, 2nd ed. Prentice Hall, Upper Saddle River, NJ.
131. Committee on Food Chemicals Codex, Food & Nutrition Board. 1996. *Food Chemicals Codex*, 4th ed., Institute of Medicine, National Academy of Sciences. National Academy Press, Washington, D.C.

132. Committee on Food Chemicals Codex, Food & Nutrition Board. 1997. *Food Chemicals Codex: First Supplement to the Fourth Edition*. Institute of Medicine, National Academy of Sciences. National Academy Press, Washington, D.C.
133. Crim, M. C., Munro, H. N. 1994. Proteins and Amino Acids. In: *Modern Nutrition in Health and Disease*, 8th ed., Shils, M. E., Olson, J. A., Shike, M. (Eds.). Williams & Wilkins, Baltimore.

Fatty Acids

134. Andersen, K. N., Anderson, L. E., Glanze, W. D. 1994. *Mosby's Medical, Nursing, and Allied Health Dictionary*, 4th ed. Mosby-Year Book, St. Louis.
135. Mahan, L. K., Escott-Stemp, S. 1996. *Krause's Food, Nutrition & Diet Therapy*, 9th ed., W. B. Saunders, Philadelphia.
136. Rowe, P. M. 1995. Fat of the land and of the lab. *Lancet* 346:46.
137. Walling, A. D. 1998. Linolenic acid and calcium for prevention of preeclampsia. *Am Family Physician*, 58:252.
138. Hu, F. B., Stampfer, M. J., Manson, J. E., Rimm, E. B., Wolk, A., Colditz, G. A., Hennekens, C. H., Wilett, W. C. 1999. Dietary intake of (alpha)-linolenic acid and risk of fatal ischemic heart disease among women. *Am. J. Clin. Nutr.* 69:890.
139. Willett, W. C. 1997. Specific fatty acids and risks of breast and prostate cancer: dietary intake. *Am. J. Clin. Nutr.* 66:1557S–1563S.
140. Ascherio, A., Willett, W. C. 1997. Health effects of trans fatty acids. *Am. J. Clin. Nutr.* 66: 1006S–1010S.
141. ASCN/AIN Task Force on Trans Fatty Acids. 1996. Position paper on trans fatty acids. *Am. J. Clin. Nutr.* 63:663–670.
142. Committee on Food Chemicals Codex, Food and Nutrition Board. 1997. *Food Chemicals Codex*, 4th ed., Institute of Medicine, National Academy of Sciences. National Academy Press, Washington, D. C.
143. Mossoba, M. M., Firestone, D. 1996. New methods for fat analysis in foods. *Food Testing Anal.* 24–32.
144. Sempore, B. G., Bezard, J. A. 1996. Analysis of Neutral Lipids: Fatty Acids. In: *Handbook of Food Analysis*, Vol. 1 *Physical Characterization and Nutrient Analysis*, Nollet, L. M. L. (Ed.). Marcel Dekker, New York, pp. 331–394.

Vitamins and Trace Minerals

145. Food and Nutrition Board. 1998. *Recommended Dietary Allowances*, 11th ed. National Academy Press, Washington, D.C.
146. Abrams, S. A., Grusak, M. A. Stuff, J., O'Brien, K. O. 1997. Calcium and magnesium balance in 9–14-year-old children. *Am J. Clin. Nutr.* 66:1172–1177.
147. Harel, Z. 1998. Adolescents and calcium: what they do and do not know and how much they consume. *JAMA* 279:1678F.
148. Perman, J. A., Dudley, B. S. 1998. Dairy products: try them—you'll like them? *Am. J. Clin. Nutr.* 68:995–996.
149. Committee on Food Chemicals Codex, Food and Nutrition Board. 1996. *Food Chemicals Codex*, 4th ed. Institute of Medicine, National Academy of Sciences. National Academy Press, Washington, D.C.
150. Committee on Food Chemicals Codex, Food and Nutrition Board. 1997. *Food Chemicals Codex First Supplement to the Fourth Edition*. Institute of Medicine, National Academy of Sciences. National Academy Press, Washington, D.C.
151. James, C. S. 1995. *Analytical Chemistry of Foods*. Chapman & Hall, New York.

Phosphorus

152. Larson Duyff, R. 1998. *The American Dietetic Association's Complete Food & Nutrition Guide*. Chronimed Publishing, Minneapolis, MN.
153. Mahan, L. K., Escott-Stump, S. 1996. *Krause's Food, Nutrition, & Diet Therapy*. 9th ed. W. B. Saunders, Philadelphia.
154. Dalton, M. A., Sargent, J. D., O'Connor, GT, Olmstead, E. M., Klein, R. Z. 1997. Calcium and phosphorus supplementation of iron-fortified infant formula: no effect on iron status of healthy full-term infants. *Am. J. Clin. Nutr.* 65:921–926.
155. Boaz, M., Smetana, S. 1996. Regression equation predicts dietary phosphorus intake from estimate of dietary protein intake. *J Am Dietetic Assoc* 96:1268–1270.
156. Food and Nutrition Board. 1998. *Recommended Dietary Allowances*, 11th ed. National Academy Press, Washington, D.C.
157. Committee on Food Chemicals Codex, Food and Nutrition Board. 1996. *Food Chemicals Codex*, 4th ed. Institute of Medicine, National Academy of Sciences. National Academy Press, Washington, D.C.
158. Pennington, J. A. T. 1996. Intakes of minerals from diets and food: is there a need for concern? *J Nutr.* 126:2304S–2308S.
159. Bercowy, GM, Peile, R. 1996. GFAA determination of phosphorus in food oils. *Food Testing Anal.* Aug.–Sept.:10,15,41.

Magnesium

160. Mahan, L. K., Escott-Stump, S. 1996. *Krause's Food, Nutrition, and Diet Therapy*, 9th ed. W. B. Saunders, Philadelphia.
161. Wood, R. J. Suter, P. M., Russell, R. M. 1995. Mineral requirements of elderly people. *Am. J. Clin. Nutr.* 62:493–505.
162. Schuck, P., Gammelin, G., Resch, K. L. 1998. Magnesium and phosphorus. *Lancet* 1474.
163. Nelson, J. K., Moxness, K. E., Jensen, M. D., Gastineau, C. F. 1994. *Mayo Clinic Diet Manual* 7th ed. Mosby-Year Book, St. Louis.
164. Larson Duyff, R. *The American Dietetic Association's Complete Food and Nutrition Guide*. Chronimed Publishing, Minneapolis, MN.
165. Food and Nutrition Board. 1998. *Recommended Dietary Allowances*, 11th ed. National Academy Press, Washington, D.C.
166. Committee on Food Chemicals Codex, Food and Nutrition Board. 1997. *Food Chemicals Codex*. Institute of Medicine, National Academy of Sciences. National Academy Press, Washington D.C.
167. Pomeranz, Y., Meloan, C. E. 1994. *Food Analysis, Theory and Practice*, 3rd ed. Chapman & Hall, New York.
168. Pennington, J. A. T. Intakes of minerals from diets and foods: is there a need for concern? *J Nutr* 126:2304S–2308S.

Potassium, Sodium, and Chloride

169. Gennari, F. J. 1998. Hypokalemia. *New Eng J Med* 339:451–458.
170. Committee on Food Chemicals Codex, Food and Nutrition Board. 1996. *Food Chemicals Codex*. 4th ed. Institute of Medicine, National Academy of Sciences. National Academy Press, Washington D.C.
171. Pennington, JAT. 1996. Intakes of minerals from diets and foods: is there a need for concern? *J Nutr* 126:2304S–2308S.
172. Pomeranz, Y., Meloan C. E. 1994. *Food Analysis, Theory and Practice*, 3rd ed. Chapman and Hall, New York.

173. Mahan, L. K., Escott-Stump, S. 1996. *Krause's Food, Nutrition, and Diet Therapy*. 9th ed. W. B. Saunders, Philadelphia.
174. Duyff, R. L. 1998. *The American Dietetic Association's Complete Food & Nutrition Guide*. Chronimed Publishing, Minneapolis, MN.
175. Engstrom, A., Tobelmann, R. C., Albertson, A. M. 1997. Sodium intake trends and food choices. *Am. J. Clin. Nutr.* 65:704S–707S.
176. Kotchen, T. A., Kotchen, J. M. 1997. Dietary sodium and blood pressure: interactions with other nutrients. *Am. J. Clin. Nutr.* 65:708S–711S.
177. Weinberger, M. H. 1997. More spice on the salt debate. *Arc Inter Med* 157:2407–2409.
178. Luft, F. C. 1996. Salt, water and extracellular volume regulation. In: *Present Knowledge in Nutrition*, 7th ed. Ziegler, E. E., Filer, L. J., Jr. (Eds.). ILSI Press, Washington, D.C. pp. 265–271.
179. Welsh, M. J., Smith, A. E. 1995. Cystic fibrosis. *Scientific American* 273:52–60.
180. Sojo, A. 1995. Chloride deficiency as a presentation or complication of cystic fibrosis. *JAMA* 273:441.
181. Helrich, K. 1990. *Official Methods of Analysis, Vol. 1*, 15th ed. Association of Official Analytical Chemists, Arlington, VA.

Iron

182. Duyff, R. L. 1998. *American Dietetic Association's Complete Food and Nutrition Guide*. Chronimed Publishing, Minneapolis, MN.
183. Food and Nutrition Board. 1989. *Recommended Dietary Allowances*, 10th ed. National Academy Press, Washington, D.C.
184. Mahan, L. K., Escott-Stump, S. 1996. *Krause's Food, Nutrition, and Diet Therapy*, 9th ed. W. B. Saunders, Philadelphia.
185. Yip, R., Dallman, P. R. Iron. 1996. In: *Present Knowledge in Nutrition*, 7th ed., Ziegler, E. E., Filer, L. J., Jr. (Eds). ILSI Press, Washington, D.C., pp. 277–292.
186. Committee on Food Chemicals Codex, Food and Nutrition Board. 1996. *Food Chemicals Codex*, 4th ed. Institute of Medicine, National Academy of Sciences. National Academy Press, Washington, D.C.
187. Hong, J. H., Yasumoto, K. 1996. Near-infrared spectroscopic analysis of heme and nonheme iron in raw meats. *J. Food Comp. Analy.* 9:127–134.

Zinc

188. Andrews, M., Gallagher-Allred, C. 1999. The role of zinc in wound healing. *Advances in Wound Care* 12:137–138.
189. Fortes, C. et al. The effect of zinc and vitamin A supplementation on immune response in an older population. *JAMA* 279:726h (Abstract).
190. Mossad, S. B., Macknin, M. L., Medendorp, S. V., Mason, P. 1996. Zinc gluconate lozenges for treating the common cold. *Ann. Intern. Med.* 125:81–88.
191. Macknin, M. L., Piedmonte, M., Calendine, C., Janosky, J., Wald E. 1998. Zinc gluconate lozenges for treating the common cold in children. *JAMA* 279:1962–1967.
192. Nicoara-Kasti, G. L., Lockey, R. F. 1998. Disease prevention:which nutritional supplements, vitamins, and medications may help? *Consultant* Feb.:397–400.
193. Food and Nutrition Board. 1989. *Recommended Dietary Allowances*, 10th ed. National Academy Press, Washington, D.C.
194. Duyff, R. L. 1998. *American Dietetic Association's Complete Food and Nutrition Guide*. Chronimed Publishing, Minneapolis, MN.
195. Wood, R. J., Zheng, J. J. 1997. High dietary calcium intakes reduce zinc absorption and balance in humans. *Am. J. Clin. Nutr.* 65:1803–1809.

196. Committee on Food Chemicals Codex, Food and Nutrition Board. 1996. *Food Chemicals Codex.* 4th ed. Institute of Medicine, National Academy of Sciences. National Academy Press, Washington DC.
197. Pomeranz, Y., Meloan, C. E. 1994. *Food Analysis, Theory and Practice.* Chapman & Hall, New York.

Copper

198. Olivares, M., Uauy, R. 1996. Copper as an essential nutrient. *Am. J. Clin. Nutr.* 63:791S–796S.
199. Lonnerdal, B. 1996. Bioavailability of copper. *Am. J. Clin. Nutr.* 63:821S–829S.
200. Uauy, R., Olivares, M., Gonzalez, M. 1998. Essentiality of copper in humans. *Am. J. Clin. Nutr.* 67(Suppl):952S–959S.
201. Anderson, K. N., Anderson, L. E. 1998. *Mosby's Medical, Nursing and Allied Health Dictionary*, 5th ed. Mosby-Year Book, St. Louis.
202. Food and Nutrition Board. 1989. *Recommended Dietary Allowances*, 10th ed. National Academy Press, Washington, D.C.
203. Committee on Food Chemicals Codex, Food and Nutrition Board. 1996. *Food Chemicals Codex*, 4th ed. Institute of Medicine, National Academy of Sciences. National Academy Press, Washington, D.C.
204. Pomeranz, Y., Meloan, C. E. 1994. *Food Analysis, Theory and Practice*, 3rd ed. Chapman & Hall, New York.

Iodine

205. Delange, F. 1998. Risks and benefits of iodine supplementation. *Lancet* 351:923–924.
206. Tiwari, B. D., Godbole, M. M., Chattopadhyay, N., Mandal, A., Mithal, A. 1996. Learning disabilities and poor motivation to achieve due to prolonged iodine deficiency. *Am. J. Clin. Nutr.* 63:782–786.
207. Levander, O. A., Whanger, P. D. 1996. Deliberations and evaluations of the approaches, endpoints and paradigms for selenium and iodine dietary recommendations. *J. Nutr.* 126: 2427S–2434S.
208. Khan, L. K., Li, R., Gootnick, D., Peace Corps Thyroid Investigation Group. 1998. Thyroid abnormalities related to iodine excess from water purification units. *Lancet* 352:1519.
209. Duyff, R. L. 1998. *The American Dietetic Association's Complete Food & Nutrition Guide.* Minneapolis, MN: Chronimed Publishing.
210. Schilling, J. S. 1997. Hyperthyroidism: diagnosis and management of graves' disease. *The Nurse Practitioner.* 22:72–95.
211. Food and Nutrition Board. 1989. *Recommended Dietary Allowances.* 10th Ed. Washington, D.C.: National Academy Press.
212. Committee on Food Chemicals Codes, Food and Nutrition Board, Institute of Medicine, National Academy of Sciences. 1996. *Food Chemicals Codex.* 4th Ed. Washington, D.C.: National Academy Press.
213. Sadecka, J., Polonsky, J. 1999. Determination of inorganic ions in food and beverages by capillary electrophoresis. *J. Chromatogr. A.* 834:401–417.

Manganese

214. Mahan, L. K., Escott-Stump, S. 1996. *Krause's Food, Nutrition and Diet Therapy*, 9th ed. W. B. Saunders, Philadelphia.
215. Freeland-Graves, J. H., Turnlund, J. R. 1996. Deliberations and evaluations of the ap-

proaches, endpoints and paradigms for manganese and molybdenum dietary recommendations. *J. Nutr.* 126:2435S–2440S.
216. Food and Nutrition Board. *Recommended Dietary Allowances*, 10th ed. National Academy Press, Washington, D.C.
217. Barceloux, D. G. 1999. Manganese. *J. Toxicol. Clin. Toxicol.* 37:293.
218. Fell, J. M. E., Reynolds, A. P., Meadows, N., Khan, K., Long, S. G., Quaghebeur, G., Taylor, W. J., Milla, P. J. 1996. Manganese toxicity in children receiving long-term parenteral nutrition. *Lancet.* 347:1218–1221.
219. Committee on Food Chemicals Codex, Food and Nutrition Board. 1996. *Food Chemicals Codex*, 4th ed. Institute of Medicine, National Academy of Sciences. National Academy, Press, Washington, D.C.
220. Pomeranz, Y., Meloan, C. E. 1994. *Food Analysis, Theory and Practice*, 3rd ed. Chapman & Hall, New York.

Dietary Supplements

221. Kurtzweil, P. 1999. An FDA guide to dietary supplements. In: *Nutrition*, 11th ed., Cook-Fuller, C. C. (Ed.). Dushkin/McGraw-Hill, Guilford, CT, pp. 27–34.
222. Camire, M. E., Kantor, M. A. 1999. Dietary supplements: nutritional and legal considerations. *Food Tech.* 53:87–96.
223. Kurtzweil, P. 1998. An FDA guide to dietary supplements. *FDA Consumer* 32:28–35.

10

Essential Fatty Acids as Food Additives

DAVID J. KYLE

Martek Biosciences Corporation, Columbia, Maryland

I. INTRODUCTION TO ESSENTIAL FATTY ACIDS

Lipids (oils and fats) are water insoluble macronutrients that provide the most concentrated form of food energy to the body. The major portion of the fat in our diet is in the form of triglyceride, which provides 9.0 kCal/g compared to the 4.0 kCal/g provided by protein or carbohydrate. These lipids are absorbed by the gut and transported to various parts of the body to be oxidized to provide energy in times of need, or to be stored in adipose tissue in times of plenty. Certain lipids, the phospholipids, make up the bilayer membranes surrounding all cells of the body. Lipids are not only structural elements, but they can also be primary and secondary messengers within and between cells and tissues. As a consequence, there are complex biochemical mechanisms controlling synthesis, oxidation, and interconversion of the multitude of different lipids that are obtained from our diet.

It was first suggested that certain fats are essential for life as early as 1929 (Mead, 1982). That is, although the body could synthesise most fats *de novo*, certain fatty acids were required as dietary precursors to the more important functional fatty acids. It was later recognized that two specific fatty acids, linoleic acid (LA) and alpha-linolenic acid (ALA), could not be synthesised *de novo* because of the position of certain double bonds close to the methyl end of the molecule (described in detail in the next section). Thus, these two fatty acids were considered essential for growth. Without them in the diet, animals developed a series of symptoms such as dry, scaly skin, excessive water consumption, reduced growth, infertility, etc., which are now known as classic symptoms of essential fatty acid (EFA) deficiency. We now understand that the symptoms associated with the deficiency are not necessarily due to the absence of LA or ALA, but rather to the resulting absence of the products of LA and ALA metabolism—the long-chain polyunsaturated fatty acids (LC-PUFAs).

Similar to classic vitamin deficiencies, essential fatty acid deficiency can be overcome by only a small amount of dietary LA and ALA, and there are not many cases of essential fatty acid deficiency reported in humans. However, once we go beyond the pathologies of EFA deficiency, we recognize that optimum function of our bodies (not absolute function) is very dependent on the amount, and ratio, of these EFAs and their metabolites. Such a ratio can be, and has been, drastically affected by changes in our diet over the last 100 years (Simopoulos, 1998). It is, therefore, quite appropriate to consider the fortification of foodstuffs with certain dietary fats, as if they were food additives, in order to provide an optimal ratio of these EFAs to maximize health and longevity. Such a rationale has been used for the development of new types of food products referred to as ''functional foods.'' Since this is of growing interest in the food industry, this chapter will discuss the biochemical rationale and functional consequences of the inclusion of EFAs into our diet.

II. CHEMISTRY OF ESSENTIAL FATTY ACIDS

A. Fatty Acid Biochemistry

The most biologically relevant fatty acids are straight chain hydrocarbons (12–22 carbons in length) with a terminal carboxyl group. They are synthesized by a series of enzymatic steps that result in the successive elongation of precursor molecules by two-carbon increments, and can either be fully saturated or dehydrated by the insertion of one to six double bonds at specific locations in the hydrocarbon chain. All fatty acids with multiple double bonds have the double bonds interrupted by a methylene group, and all double bonds are in the *cis* configuration. The position of the double bond is indicated by the number of carbon atoms from the functional (acid) group (e.g., oleic acid has a single cis double bond at the Δ9 position). The standard biochemical nomenclature used in this chapter describes the fatty acid in terms of its carbon chain length, followed by the number of double bonds, and then the position of those double bonds. Oleic acid is, therefore, referred to as C18:1(Δ9), or a fatty acid with 18 carbons and one double bond at the 9 position (Fig. 1).

Nutritionists also use a form of nomenclature which classifies families of fatty acids in terms of the position of the double bond closest to the methyl end of the molecule. Linoleic acid, for example would be chemically described as C18:2(Δ9,12), but also described as an omega-6 (or n-6) fatty acid since the double bond closest to the methyl end of the molecule is six carbons away from that terminal methyl group. This nomenclature is functionally useful because different fatty acid families have significantly different physiological and biochemical effects in the body. The other essential fatty acid, linolenic acid, is chemically described as C18:3(Δ9,12,15), and is nutritionally a part of the omega-3 (n-3) family of fatty acids. The scientific and common names for the principle fatty acids in biology are provided in Table 1.

B. Omega-6 and Omega-3 Fatty Acids

The essential fatty acids, LA and ALA, are the parent molecules of the omega-6 and omega-3 pathways, respectively (Fig. 2). Through the use of stable isotope tracer experiments, it is clear that humans have the ability to synthesize all the fatty acids of the omega-3 and omega-6 pathway from the EFA precursors (Greiner et al., 1997; Salem et al., 1996). The precursors are considered essential because humans do not have the enzymatic

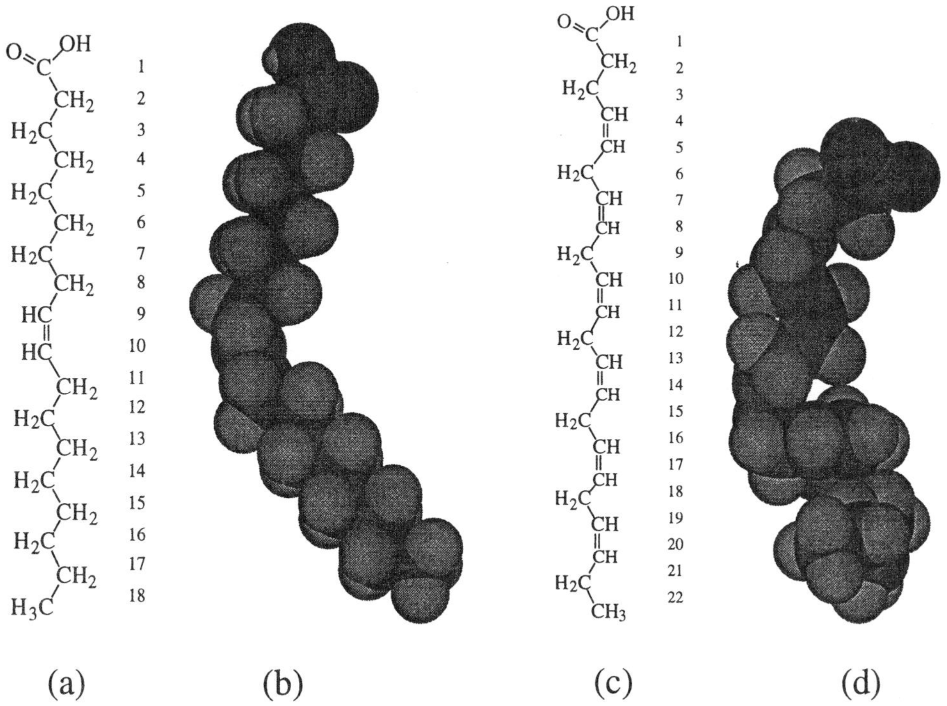

Figure 1 Chemical structure of (a) oleic acid in a stick model; (b) in a space filling model; (c) docosahexaenoic acid in a stick model; and (d) in a space filling model.

capability of desaturating fatty acids toward the methyl terminal of the fatty acyl chain. As shown in Fig. 2 the formation of all members of the omega-3 and omega-6 families of fatty acids require only a Δ6 or a Δ5 desaturation in conjunction with successive elongation steps. The final steps in the conversion of eicosapentaenoic acid (EPA) to docosahexaenoic acid (DHA) was thought to involve the additional elongation of EPA to C22:

Table 1 Scientific and Common Names of the Essential Fatty Acids and Their Common Derivatives

Common name	Scientific name	Chemical notation
Omega-6 family		
linoleic acid (LA)	octadecadienoic acid	C18:2(Δ9,12)
gammalinolenic acid (GLA)	octadecatrienoic acid	C18:3(Δ6,9,12)
dihomogammalinolenic acid	eicosatetraenoic acid	C20:3(Δ8,11,14)
arachidonic acid	eicosatetraenoic acid	C20:4(Δ5,8,11,14)
osbond acid	docosapentaenoic acid	C22:5(Δ4,7,10,13,16)
Omega-3 family		
linolenic acid	octadecatrienoic acid	C18:3(Δ9,12,15)
steriodonic acid	octadecatetraenoic acid	C18:4(Δ6,9,12,15)
timnodonic acid	eicosapentaenoic acid	C20:5(Δ5,8,11,14,17)
cervonic acid	docosahexaenoic acid	C22:6(Δ4,7,10,13,16,19)

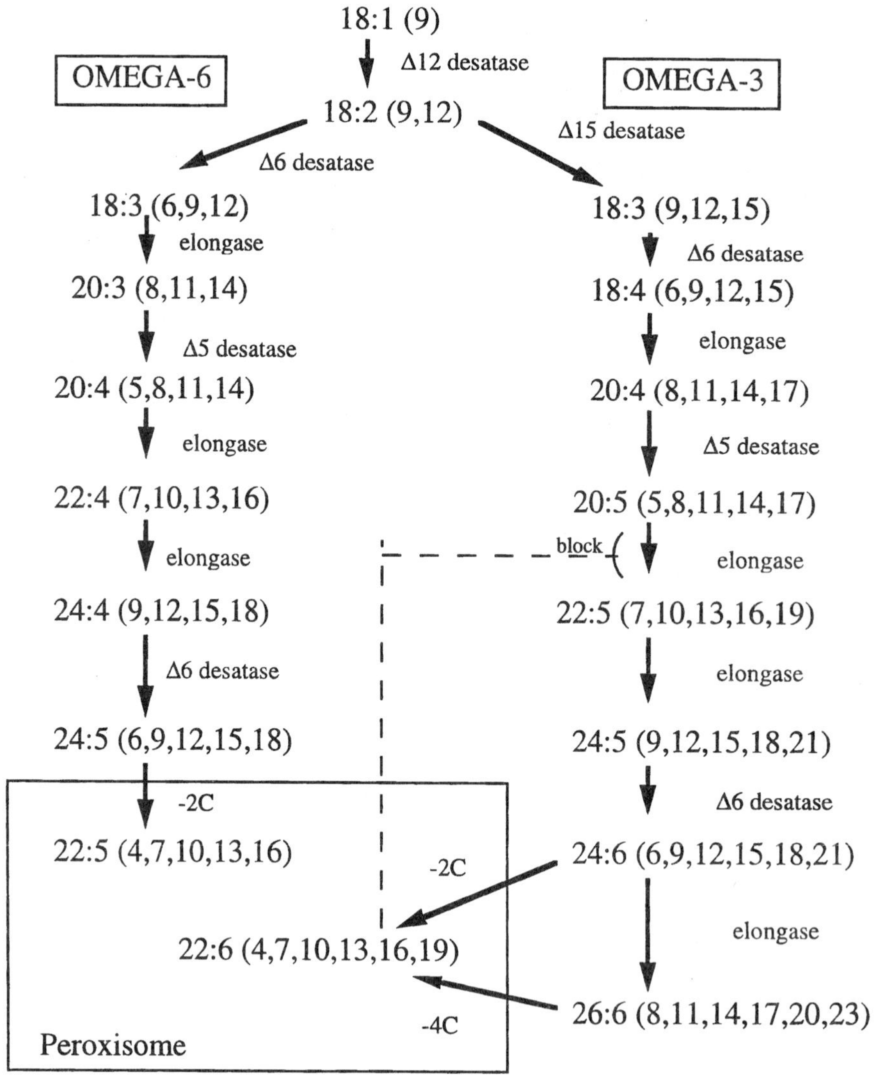

Figure 2 Omega-3 and omega-6 fatty acid biosynthetic pathway.

5(Δ7,10,13,16,19), followed by a Δ4-desaturation to produce DHA [C22:6(Δ4, 7,10,13,16,19)]. Recent studies by Sprecher and colleagues, however, have established that this is not the case (Voss et al., 1991). Rather, a more elaborate pathway involving a progressive elongation of EPA to 22:5(Δ7,10,13,16,19) and then to 24:5(Δ9,12,15,18,21) is followed by a Δ6-desaturation to form 24:6(Δ6,9,12,15,18,21). This fatty acid is then transferred to the peroxisome, where it undergoes one cycle of β-oxidation to form 22: 6(Δ4,7,10,13,16,19), or DHA (Fig. 2). A similar process can occur with the omega-6 pathway to form 22:5(Δ4,7,10,13,16), docosapentaenoic acid (n-6), when an organism is deficient in omega-3 fatty acids.

The twenty carbon fatty acids of the omega-6 and omega-3 families [arachidonic acid (ARA) and EPA, respectively] are the precursors for a family of circulating bioactive molecules called eicosanoids. Both precursors are acted upon by lipoxygenases to form a series of leukotrienes, and by cyclooxygenases to form prostaglandins, prostacyclins, and thromboxanes (Fig. 3). These circulating eicosanoids can affect immune responses, vascular tone, platelet aggregation, and many other cellular functions. In many cases the eicosanoids derived from the omega-6 fatty acids have an opposite effect from those de-

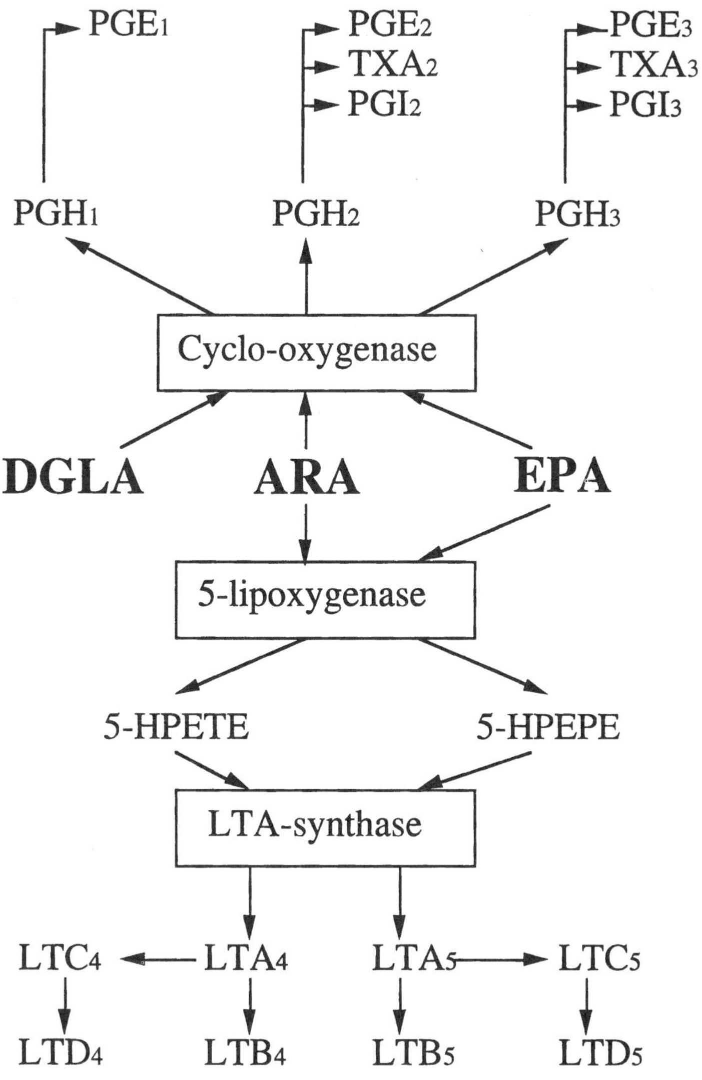

Figure 3 Formation of various eicosanoids from the LC-PUFA precursors: dihomogamma-linolenic acid (DGLA); arachidonic acid (ARA); eicosapentaenoic acid (EPA), prostaglandins (PGEx); thromboxanes (TXAx); prostacyclins (PGIx); leukotrienes (LTAx, LTBx. LTCx, LTDx); 5-hydroxyeicosatetraenoate. (5-HPETE); and 5-hydroxyeicosapentaenoate (5-HPEPE).

rived from the omega-3 fatty acids, and it is, therefore, important to keep the body in a healthy equilibrium between these two fatty acid families. It is because our modern diet has upset this balance that we now consider adding certain components back to the diet to restore this balance.

Docosahexaenoic acid plays a unique role in the body. It is not an eicosanoid precursor and therefore does not feed into the production of prostaglandins, thromboxanes, leukotrienes, or prostacyclins. It is, however, found in massive abundance in the membranes of certain tissues of the central nervous system (Crawford, 1990). Docosahexaenoic acid is the most abundant omega-3 fatty acid of the membranes that make up the grey matter of the brain, and it is found in exceptionally high levels in the synaptic vesicles (Arbuckle and Innis, 1993; Bazan and Scott, 1990), in the retina of the eye (Bazan and Scott, 1990), in cardiac muscle (Gudbjarnason et al., 1978), and certain reproductive tissues (Connor et al., 1995; Zalata et al., 1998). Docosahexaenoic acid, therefore, likely plays a unique role in the integrity or functionality of these tissues.

C. Complex Lipid Forms

Fatty acids are not generally found in the body in the free fatty acid form. Rather, they are found in more complex lipid forms such as triglycerides, phospholipids, sphingomyelin, sterol esters, etc. (Fig. 4). Triglycerides are the primary lipid storage form, and are found in greatest abundance in adipose tissues. Other than triglycerides, the vast majority of fatty acid in the body is found as membrane phospholipid. There are four main phospholipid forms characterized by their polar head groups. They are phosphatidyl choline (PC), phosphatidyl ethanolamine (PE), phosphatidyl inositol (PI), and phosphatidyl serine (PS). All these phospholipid forms are found to different extents in different tissues, and the fatty acyl moieties on the phospholipids help to define the physical and chemical characteristics of the membranes.

III. EFA FUNCTIONS

A. Membrane Protein Boundary Lipids

As noted in the previous section, most fatty acids in the body are present as phospholipids in biological membranes, and the fatty acyl moieties of these phospholipids determine many of the functional and biochemical characteristics of those membranes. Embedded in the lipid matrix of a biological membrane are the proteins that confer specificity to that particular membrane. In the outer segments of the rod cells of the retina, for example, there are a series of pancakelike stacks of membranes in which the retinal-binding protein, rhodopsin, is found (Bazan and Rodriguez de Turco, 1994; Gordon and Bazan, 1990) (Fig. 5). The concentration of rhodopsin in these membranes was recently shown to be dependent on the DHA content of the phospholipids comprising those membranes (Suh et al., 1997). That is, there appears to be a positive correlation between rhodopsin density (a characteristic that should define the light sensitivity of the eye) and DHA content of the retina. Furthermore, if an animal is made omega-3 deficient and the DHA levels are drastically reduced, the DHA is replaced by omega-6 DPA. Under such conditions, the visual acuity of the animal is compromised (Neuringer et al., 1986). Thus, the existence of an additional double bond at the $\Delta 19$ position of an otherwise identical molecule has a dramatic effect on the performance of a specific organ (Salem and Niebylski, 1995; Bloom et al., 1998). In such cases, it is believed that the fatty acid in question must play

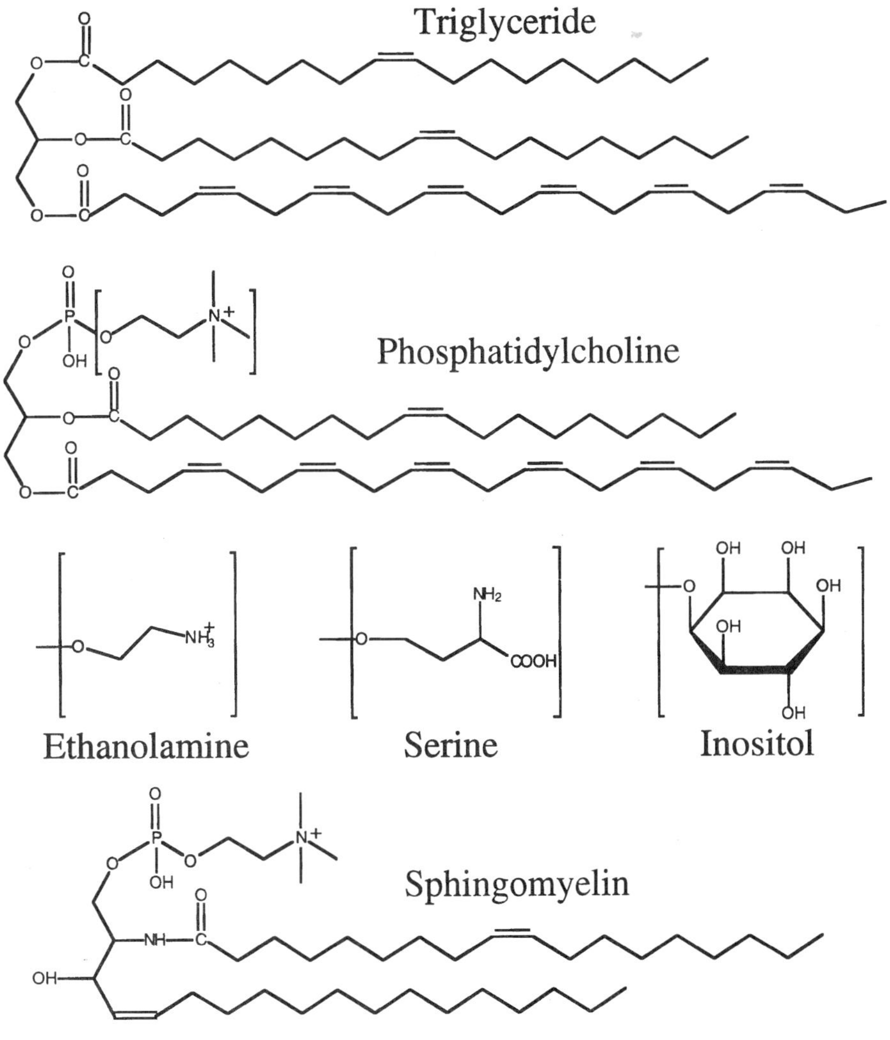

Figure 4 Complex lipid forms.

a pivotal role in the "boundary lipid" of a particular membrane protein such that any alteration in the boundary lipid results in a significant impact of the functioning of that protein. Across 500 million years of evolution DHA—not DPA—has been selected as providing the optimal lipid environment for this photoreceptor pigment protein (Broadhurst et al., 1998).

There are many other examples of how specific fatty acids are required for the optimal function of membrane proteins. This is not unexpected because the orientation of a phospholipid in the membrane is significantly impacted by the degree of unsaturation of its fatty acid moieties. A single *cis*-double bond confers a 37° kink in the orientation of a fatty acid. A second double bond imparts a second bend in the structure and so on. A molecule with six double bonds, such as DHA, can have a helical configuration as a result of successive kinks in the molecule along an axis perpendicular to the plane of the

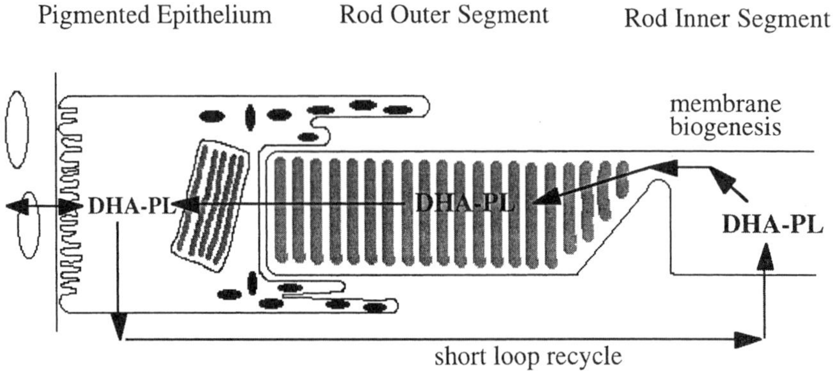

Figure 5 Retinal rod outer segments.

membrane (Fig. 1). Such a molecule has a much larger cross sectional volume than, for example, a simple saturated fatty acid like palmitic or stearic acid and has many very specific packing arrangements that can optimize the function of certain membrane proteins.

B. Eicosanoid Functions

In addition to their role as structural entities affecting the performance of membrane-bound proteins, the twenty carbon metabolites of the essential fatty acids have important roles in the body as circulating eicosanoids (Holman, 1986; Lagarde, 1995; Lands, 1989). The complexity of the biochemical conversions are shown in Fig. 6 for arachidonic acid only. Similar metabolic fates occur in the other twenty-carbon PUFAs such as EPA and dihomo-gamma linolenic acid [DGLA; C20:3(Δ9.12.15)]. These eicosanoids have a relatively high specificity in the body to elicit a biological response, and are therefore under remarkably tight regulation. Prostaglandin E2 (PGE2), for example, is a potent elicitor of platelet aggregation and vasoconstriction (Holman, 1986). The body's response to wounding involves the release of free ARA, the elevation of PGE2, and the subsequent constriction of blood flow around the wound. This is an important, acute survival mechanism. In the longer term, however, it could lead to elevated blood pressure and an increased risk of coronary vascular disease. However, another ARA metabolite, prostacyclin I2, has a strong effect on inhibiting platelet aggregation, so that there may be a considerable latitude of PGE2 levels which provides the organisms with a rapid response mechanism with little, or no, long-term effects.

Many of the eicosanoids have a direct influence on biological responses associated with immune function. These include the inflammatory response as well as induction of macrophages and production of antibodies in response to some challenge to the organism. In general, the omega-6 eicosanoids have been considered as proinflammatory and up-regulators of typical immunological responses. In a recent well-controlled human feeding trial using an ARA-rich triglyceride providing about 1.5 g of ARA per day (over 50 days), both thromboxane B2, and a metabolite of PGI2, were shown to increase, by 41% and 27%, respectively (Ferretti et al., 1997). Although this elicited no measurable change in platelet aggregation or bleeding time (Nelson et al., 1997a), the researchers reported an

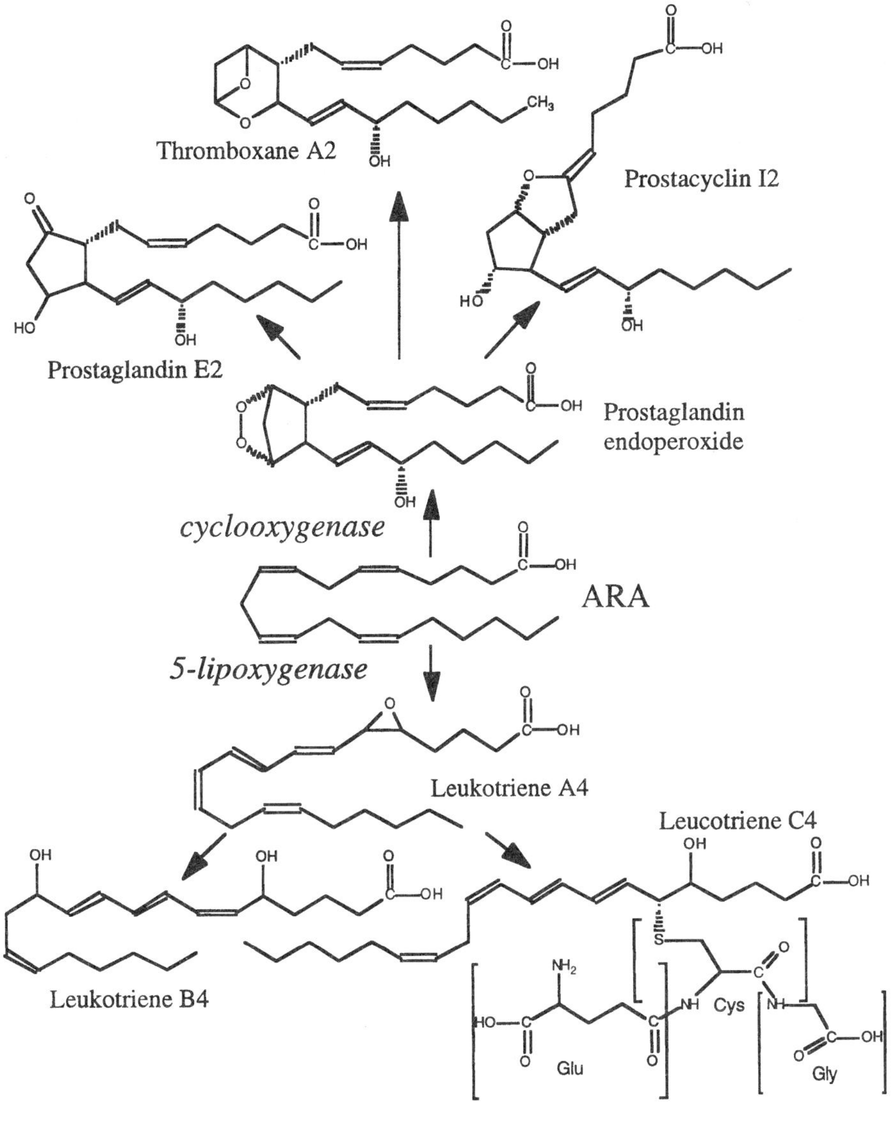

Figure 6 Eicosanoids produced from arachidonic acid and their function.

elevation in antibody titre in response to a heat-killed influenza virus challenge, suggesting an up-regulation of the immune response (Kelley et al., 1997). On the other hand, fish oil feeding trials which provide high levels of the omega-3 eicosanoid precursor EPA, have reported a significant decrease in platelet aggregation and increase in bleeding time (Cobiac et al., 1991; Eritsland et al., 1989; Mark and Sanders, 1994; Ward and Clarkson, 1985), a down-regulation of the immune response (Meydani et al., 1993) and a general anti-inflammatory response (Lee et al., 1991). Such a response is particularly useful in patients with proinflammatory disorders such as rheumatoid arthritis or asthma (Kremer et al., 1990; Lee et al., 1991; Raederstorff et al., 1996).

C. Signalling Pathways

Although EFAs have generally been considered important because of their nutritional value, in recent years it has become clear that these fatty acids may also play an important role in gene regulation (Clarke and Jump, 1994; Sesler and Ntambi 1998). Dietary fats, particularly PUFAs, exhibit a general effect on expression of genes in lipogenic tissues. High levels of PUFAs result in a decrease of activity of liver enzymes involved in lipogenesis, and PUFA restriction appears to induce the expression of lipogenic enzymes. This response does not seem to be specific to either omega-6 or omega-3 LC-PUFAs (Sesler and Ntambi, 1998). Recent studies have identified similar responses in adipose tissue although these may be more class specific. Fatty acid synthetase and lipoprotein lipase expression in adipose tissue of the rat has been shown to be specifically activated by omega-3 LC-PUFAs (Cousin et al., 1993).

Polyunsaturated fatty acid regulation of gene expression in nonlipogenic tissues has also been reported. These genes include Thy-1 antigen on T-lymphocytes, L-fatty acid binding protein, apolipoproteins A-IV and C-III, sodium channel gene in cardiac myocytes, acetyl Co-A carboxylase in pancreatic cells, and steaoryl-CoA desaturase in brain tissue (Gill and Valivety, 1997; Sesler and Ntambi, 1998).

The reported stimulation of superoxide dismutase by LC-PUFAs (Phylactos et al., 1994) represents a mechanism whereby intracellular antioxidants may be stimulated. This may be particularly relevant in neonatology as infants which normally receive a good endogenous supply of DHA from their mother (across the placenta prenatally or via breast milk postnatally) may have a better antioxidant status than those infants fed a formula without supplemental DHA and ARA. Breast-fed infants certainly appear to be protected from certain oxidant-precipitated pathologies such as narcotizing enterocolitis (NEC) (Crawford et al., 1998) and this has been correlated to the higher levels of cupric/zinc superoxide dismutase in the erythrocytes of breast-fed versus formula-fed infants (Phylactos et al., 1995). Such an improvement of intracellular antioxidants may also explain a recent observation that feeding preterm infants with a DHA/ARA-supplemented formula results in an 18% decrease in the incidence of NEC (Carlson et al., 1998).

The molecular mechanisms whereby, essential fatty acids can affect gene expression are still poorly understood. Some studies have suggested that the action of the PUFAs may be both at the level of transcription and others at the stabilization of the mRNA. A cis-acting PUFA responsive element (PUFA-RE) has been proposed to be in the promoter region of all PUFA-regulated genes, and a nuclear factor binding to this element has been demonstrated (Waters et al., 1997). Through an understanding of how dietary EFAs may affect gene regulation, especially with respect to lipogenesis and the production of intracellular antioxidants, we can make better use of supplemental EFAs in our diet to improve long-term health and well being.

D. DHA Function

DHA is somewhat of an enigma in the body. It is an energetically expensive molecule to synthesize, it requires a complicated biochemical route which includes many different gene products, it is prone to oxidative attack because of its abundance of double bonds, and yet is still a major component of many important organ systems in the body. Docosahexaenoic acid is found in close association with membrane proteins of the 7-transmembrane structure (7-Tm), G-protein coupled receptors (i.e., serotonin receptors, acetylcholine receptors, and rhodopsin), and certain ion channels. It is, therefore, believed that DHA may play a greater role than simply as a structural element in many biological membranes.

Leaf and colleagues have recently proposed that LC-PUFAs may play a role in calcium channel regulation (Billman et al., 1997; Xiao et al., 1997). In isolated cardiac myocytes, EPA and DHA are effective in blocking calcium channels, resulting in a restoration of normal rhythmicity to the isolated cells treated with ouabain (Leaf, 1995). In a dog model of sudden cardiac death (induced ventricular arrhythmia), Billman and colleagues (Billman et al., 1994, 1997) have shown that predosing animals with omega-3 LC-PUFAs significantly reduced the number of fatal arrhythmias or ventricular fibrillations. Similar results have been demonstrated with rat (McLennan et al., 1996) and primate models (Charnock et al., 1992). It is now believed that DHA may act as the endogenous calcium channel controlling factor in cardiac cells. The proposal drawn in Fig. 7 suggests such a mechanism in neural tissues, where an elevated intracellular calcium level stimulates a calcium-dependent phospholipase, which in turn cleaves off free DHA from the DHA-rich membranes. The local high concentration of free DHA then closes off the calcium channel, reducing the calcium inflow, and the internal calcium levels drop. Since it is DHA, not EPA that is enriched in neurological and cardiac cells, it is likely that DHA is

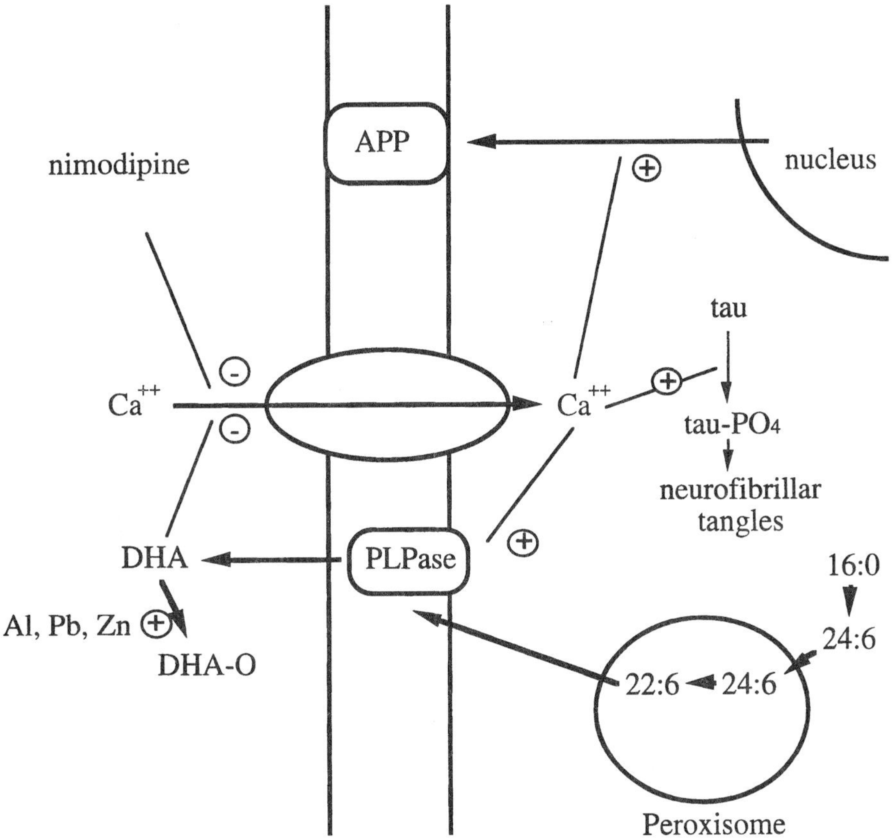

Figure 7 Role of DHA as a modulator of calcium channels.

the active fatty acid in this control mechanism. Indeed, McLennan and colleagues (1996) demonstrated that at low dietary intakes, it is DHA, not EPA, that inhibits ischaemia-induced cardiac arrhythmias in the rat. DHA also represents a safer control mechanism compared to twenty-carbon LC-PUFAs since the release of free eicosanoid precursors may result in a series of unwanted eicosanoid-related responses.

IV. PHYSIOLOGICAL CONSEQUENCES OF LCPUFA DEFICIENCIES

Previous sections have described the biochemistry and functional roles of the essential fatty acids in the body, and it is clear from the definition that they are required for optimal health. We will now consider the physiological consequences of suboptimal levels of EFAs in the body as an introduction to why we should ensure that we receive adequate dietary supplies of these important nutrients.

A. Essential Fatty Acid Requirements of Infants

The tissues with the highest concentration of EFAs in the body are the brain and retina. Moreover, the EFAs of the brain and neurological tissues are almost exclusively DHA and ARA (Crawford et al., 1993). As a result, during the last trimester *in utero* and for the first 2 years of postnatal life, there is a great demand for dietary DHA and ARA to ensure the nominal development of the central nervous system of the infant. Mothers provide this DHA and ARA from their own internal stores, and it is passed to the fetus across the placenta before birth. After birth, it is provided to the baby through its mother's breast milk. The mother's DHA status progressively declines throughout her pregnancy (Al et al., 1995), and the DHA status of an infant from a multiple birth is lower than that of an infant from a singleton birth (Zeijdner et al., 1997).

Infants who are fed infant formulas with no supplemental DHA or ARA have altered blood and brain chemistries compared to infants who are fed breast milk, which provides a natural supplement of both DHA and ARA. The blood of a formula-fed infant will contain less than half of the DHA of the blood of a breast-fed baby (Carlson et al., 1992; Uauy et al., 1992). Brain DHA levels of formula-fed babies can be as much as 30% lower that those of breast-fed babies (Farquharson et al., 1995; Makrides et al., 1994). Thus, feeding an infant a formula without supplemental DHA and ARA as the sole source of nutrition puts that infant into a deficiency state relative to a breast-fed baby. This is especially problematic for the preterm infant whose brain is still developing and who is no longer receiving DHA from the mother across the placenta (Crawford et al., 1998, 1997).

Many studies have attempted to assess the consequences of such a DHA deficiency in the blood and brain of formula-fed infants by comparing the long-term mental outcomes of breast-fed versus formula-fed infants (Florey et al., 1995; Golding et al., 1997; Horwood and Fergusson, 1998; Lanting et al., 1994; Lucas et al., 1992). The vast majority of these studies have clearly established that even after all the confounding data, including socioeconomic status, sex, birth order, etc., have been taken into account, there is still a small but significant advantage (3–4 IQ points) for the breast-fed babies when measured by standard IQ assessment, general performance tests (Anderson et al., 1996), or long-term neurological complications (Lanting et al., 1994).

Recent double-blind and randomized studies have now also compared neurological and visual outcomes of babies fed either formulas containing supplemental DHA or standard infant formulas. In most cases, developmental delays in visual acuity (Birch et al.,

1992b, 1998; Carlson et al., 1996a) or mental acuity (Agostoni et al., 1995; Carlson et al., 1994; Willatts et al., 1998) were detected with the standard formula-fed babies. These differences were overcome in babies who were fed DHA/ARA-supplemented formulas (Table 2). However, the DHA-supplemented formula-fed babies were not better off than the breast-fed babies. Thus, the unsupplemented formula–fed babies may be at a significant disadvantage because of the DHA deficiency in the brain and eyes during this early period of life. The single exception to the above observation (Auestad et al., 1997) may have been due to the very low levels of DHA used in the fortified formulas in that particular study (Table 2).

As the result of the need for supplemental DHA and ARA by infants who are not receiving their mother's milk, several professional organizations have made recommendations that all infant formulas contain supplemental DHA and ARA at the levels normally found in mother's milk (Table 3). Of particular importance was the recommendation by a joint select committee of the Food and Agriculture Organisation and the World Health

Table 2 Clinical Studies Comparing Neurological or Visual Outcomes of Infants Fed Formulas with or Without Supplemental DHA and ARA

Supplementation		Sample size	Effect of DHA/ARA supplementation
Preterm infant studies			
DHA	ARA		
0.33	0.10[1,2]	83	Improved rod ERG function (37 wk PCA[a]); improved visual acuity (57 wk PCA)
0.20	0.00[3,4]	67	Improved visual acuity (4 mo); improved visual information processing (12 mo)
0.20	0.00[5–7]	59	Improved visual acuity (2 mo); improved mental acuity (12 mo); improved visual recognition memory and visual attention (12 mo)
Term infant studies			
DHA	ARA		
0.30	0.44[8]	86	Improved psychomotor development (4 mo)
0.36	0.01[9]	79	Improved visual acuity (16 and 30 wk)
0.12	0.43[10]	58	Improved visual acuity (2 mo)
0.12	0.43[11]	197	No difference
0.25	0.40[12]	44	Improved problem solving ability (10 mo)
0.36	0.72[13]	108	Improved visual acuity (1.5 and 12 mo)
0.32	0.06[14]	54	No significant difference (positive trend) in visual acuity (4 mo)[b]

[a] PCA, postconceptual age.
[b] Formula also contained high levels of EPA (0.4%) and GLA (0.54%).
Sources: [1]Birch et al., 1992a, [2]Hoffman et al., 1993, [3]Carlson et al., 1993, [4]Werkman and Carlson, 1996, [5]Carlson et al., 1994, [6]Carlson and Werkman, 1996, [7]Carlson et al., 1996b, [8]Agostoni et al., 1995, [9]Makrides et al., 1995, [10]Carlson et al., 1996a, [11]Auestad et al., 1997, [12]Willatts et al., 1998, [13]Birch et al., 1998, [14]Horby Jorgensen et al., 1998.

Table 3 Recommendations by Expert Panels for the Supplementation of Infant Formula with DHA and ARA

	BNF[a]	ISSFAL[b]	FAO/WHO[c]
Year	1993	1994	1995
Preterm			
ARA (% of formula fat)	0.30%	1.0–1.5%	1.0%
DHA (% of formula fat)	0.30%	0.5–1.1%	0.8%
Full-term			
ARA (% of formula fat)			0.6%
DHA (% of formula fat)			0.4%
EPA/DHA ratio		> 5:1	10:1

[a] British Nutrition Foundation (Garton, 1992).
[b] International Society for the Study of Fatty Acids and Lipids (ISSFAL, 1994).
[c] Food and Agriculture Organization/World Health Organization (FAO/WHO, 1993).

Organisation which drew on the expertise of nearly 50 researchers in this field, and concluded that adequate dietary DHA was also important for the mother postnatally, prenatally, and even preconceptually (Crawford, 1995).

B. Essential Fatty Acids and Visual Function

Visual tissues are an extension of the tissues of the central nervous system, and in the multilamellar membranes of the retinal rods [rod outer segments (ROS)] we find the highest local concentration of DHA in the body. The phospholipids of the ROS are enriched in DHA to levels 50–60% of the total lipid (Salem et al., 1986). Once again, this tissue is unique in that the visual systems of all animals have selected only one of the EFAs—DHA—as the preferred structural unit for this membrane. The ROS membranes are also rapidly cycled with about 10–20% of the distil portion of the stack being resorbed each day, and new membrane formed at the opposite end. Within these cells there is a complex recycling mechanism in place to reuse the DHA each day for the formation of the new membrane (Anderson et al., 1992; Bazan et al., 1993).

Several pathologies which have a visual dysfunction component also have been shown to be associated with abnormally low levels of circulating DHA. These include retinitis pigmentosa (Hoffman and Birch, 1995; Schaefer et al., 1995), certain peroxisomal disorders (e.g., Zellweger's, Refsum's, and Batten's diseases) (Gillis et al., 1986; Infante and Huszagh, 1997; Martinez, 1990, 1996), long-chain hydroxyacyl-CoA dehydrogenase deficiency (LCHADD) (Gillingham et al., 1997), and even dyslexia (Stordy, 1995). In some of these cases, clinical studies have shown that dietary DHA supplementation has not only improved the EFA status of the patient, but also significantly improved their visual function (Table 4). Individuals with dyslexia generally have poor night vision, and a supplementation study with DHA (and EPA provided as fish oil) has recently shown significant improvements in dark adaptation or night vision (Stordy, 1995). Such a result would be expected if the elevation of DHA in the ROS of such affected patients would also elevate the concentration of rhodopsin in those membranes as has been previously been reported in animal studies (Suh et al., 2000). Supplementation with DHA has resulted in significant improvements in visual function in infants with peroxisomal disorders (Martinez, 1996) and even in children with a degenerative visual function such as LCHADD (Gillingham et al., 1997).

Table 4 Examples of Clinical Conditions Improved with Essential Fatty Acid Supplementation

Condition	Supplementation results
Neurological	
Zelweger's Syndrome	Improved visual and physical outcome[1]; remyelination in brain[2]
Batten's disease	Arrested natural course of disease[3]
Schizophrenia	Significant improvement in schizophrenic symptoms[4]
Alzheimer's disease	Improvement of mental function[5]
Bipolar depression	Increased time periods between manic phases[6]
LCHADD	Significant improvement of vision (VEP)[7]
Dyslexia	Improvement of night vision[8]
ADD/ADHD	Improvement of attention[9]
Cardiovascular	
Elevated triglycerides	Reduction of triglycerides/elevation of HDL[10]
Low HDL	Elevation of HDL[11]
Hypertension	Reduction of blood pressure[12]
Other	
Asthma	Improved forced expiratory volume[13]
Rheumatoid arthritis	Reduced morning stiffness[14]
Diabetic neuropathy	Improvement of nerve conduction velocity[15]; reduction of global
Premenstrual syndrome	symptoms of PMS[16]

[1](Martinez et al., 1993); [2](Martinez and Vazquez, 1998); [3](Bennett et al., 1994); [4](Laugharne et al., 1996); [5](Yazawa, 1996); [6](Stoll, 1998); [7](Gillingham et al., 1997); [8](Stordy, 1995); [9](Stordy, 1998); [10](Davidson et al., 1997); [11](Mori et al., 1994); [12](Bonaa et al., 1990); [13](Dry and Vincent, 1991); [14](Kjeldsen-Kragh et al., 1992); [15](Horrobin, 1991); [16](Oeckerman et al., 1986).

C. Essential Fatty Acids and Neurological Function

The importance of certain essential fatty acids in optimal neurological function is suggested by the high concentration of both DHA and ARA in the tissues of the central nervous system. Within the neuronal cells, DHA and ARA are found in highest concentration in the synaptosomal membranes (Suzuki et al., 1997; Wei et al., 1987; Yeh et al., 1993). When rats are made omega-3 deficient by using a feeding regimen completely devoid of omega-3 EFAs, the brain DHA levels are dramatically reduced, the DHA is replaced with the omega-6 counterpart, n-6 docosapentaenoic acid, and the rats have more difficulty with learning tasks (Fujimoto et al., 1989). It is remarkable how this minor change in the molecule (a double bond at the Δ19 position) can contribute so substantially to the performance of the organism as a whole.

Recent studies using nonhuman primates have also demonstrated that early nutrition and/or social interaction has dramatic long-term behavioral consequences that are manifested in adolescence and adulthood. Higley and coworkers (1996a,b) have shown that infant Rhesus monkeys who are fed their own mother's milk (a dietary supply of DHA and ARA) and who are nurtured by their mothers for the first three months of life have remarkably different developmental outcomes than infants raised with a peer group and fed a formula deficient in DHA and ARA. Formula-fed, peer-reared infants develop more aggressive tendencies in adolescence; they are more depressed and never achieve a very high social rank as adults in a free-living monkey colony. These researchers also demonstrated that the formula-fed monkeys exhibited a lower level of brain serotonin from as early as 14 days after birth to adulthood compared to the breast-fed, DHA/ARA replete infants. How DHA can affect levels of brain serotonin is not well understood, but it has recently been shown that serotonin levels in adult humans are directly related to blood DHA levels (Hibbeln, 1998a).

Attention deficit hyperactivity disorder (ADHD) is also correlated with subnormal serum levels of DHA and ARA. Burgess and coworkers (Stevens et al., 1995) demonstrated that the lower the blood levels of DHA in the ADHD children, the more prevalent were the hyperactivity symptoms. The increase in the incidence of ADHD coincides with the loss of DHA from our diet, and the increasing usage of infant formulas (particularly in the United States) that do not provide supplemental DHA and ARA to an infant early in life. Although it is tempting to propose a causal relationship, the existing data only allow us to conclude that unsupplemented formula feeding is a risk factor for ADHD. McCreadie (1997) also recently showed that formula feeding (lack of early DHA and ARA supplementation) is a significant risk factor for schizophrenia. As in the case of the Rhesus monkeys, it is possible that many long-term behavioral problems may stem from a suboptimal essential fatty acid status (particularly a deficiency of DHA and ARA) at this crucial period for brain development.

Several other neurological pathologies are also related to subnormal levels of DHA in the plasma (Table 4). These include depression (Hibbeln, 1998b; Hibbeln and Salem, 1995; Peet et al., 1998), schizophrenia (Glen et al., 1994; Laugharne et al., 1996) and tardive dyskinesia (Vaddadi et al., 1989). In the latter case, the researchers demonstrated that the symptomology of tardive dyskinesia was most severe in individuals with the lowest DHA levels and that supplementation with omega-3 long chain polyunsaturated fatty acids significantly improved the condition. Preliminary supplementation studies in patients with bipolar disorder (manic depression) using oils rich in DHA and EPA also resulted in a remarkable improvement and significant delay in the onset of symptoms (Stoll, 1998). This result is consistent with the observation by Hibbeln (1998b) that the incidence of major depression seems to be negatively correlated with consumption of fish, the major source of DHA, in a worldwide cross-cultural comparison. A similar correlation can also be drawn between fish (DHA) in the diet and postpartum depression in women.

The brain tissue of patients who were diagnosed with Alzheimer's dementia (AD) contains about 30% less DHA (especially the hippocampus and frontal lobes) compared to similar tissue isolated from pair-matched geriatric controls (Prasad et al., 1998; Soderberg et al., 1991). In a 10-year prospective study with about 1200 elderly patients monitored regularly for signs of the onset of dementia, it was determined that a low serum phosphatidyl choline DHA (PC-DHA) level was a significant risk factor for the onset of senile dementia (Kyle et al., 1998). Individuals with plasma PC-DHA levels less than 3.5% had a 67% greater risk of being diagnosed with senile dementia in the subsequent 10 years than those with DHA levels higher than 3.5%. Furthermore, for women who had a least one copy of the Apolipoprotein E4 allele, the risk of attaining a low minimental state exam (MMSE) score went up 400% if their plasma PC-DHA levels were less than 3.5%.

Clinical studies have supported a role of gamma linolenic acid (GLA) in reducing the symptoms of premenstrual syndrome (PMS) (Oeckerman et al., 1986), and GLA-containing products are marketed in Europe with this indication. Perhaps of equal importance have been the reports that supplemental GLA in the diet of diabetics may significantly improve nerve conduction velocity in individuals with diabetic neuropathy (Horrobin, 1991). Whether GLA is playing a role directly in the neuronal function, or if it is simply acting as a precursor to ARA is not well understood at this time.

Finally, perhaps one of the most profound consequences of EFA deficiency is found in patients with a certain inborn error of metabolism related to peroxisomal dysfunction (Martinez, 1991, 1992). As discussed previously, the last steps of the biosynthesis of DHA require one step of β-oxidation of C24:6(6,9,12,15,18,21), which takes place exclusively in the peroxisome (Fig. 2). Certain peroxisomal diseases including Zellweger's Disease, Refsums's Disease, and neonatal adrenoleukodystrophy, are accompanied by DHA deficiency. The development of these progressive neurodegenerative diseases have been recently shown to be arrested if DHA can be reintroduced into the diet at an early stage (Martinez et al., 1993). Much of the neurodegeneration is manifest in a progressive demyelination, which until now has been thought to be irreversible. Martinez and Vasquez (1998) have recently shown that not only do symptoms improve upon treatment for certain of these patients with DHA, but magnetic resonance imaging data indicate that the brain begins to remyelinate once DHA therapy is initiated.

D. EFAs and Cardiovascular Function

The potential role of essential fatty acids (especially omega-3 fatty acids) in cardiovascular function has been well studied for the last 40 years. Since the first observation by Dyerberg and Bang (1979) that indigenous populations in the Arctic who were consuming large amounts of omega-3 fatty acids (from fish and marine mammals) had a very low incidence of cardiovascular disease, there have been hundreds of clinical studies assessing the effects of fish diets or fish oil pills on cardiovascular outcomes. Several major studies have been completed and many excellent books and reviews on this matter have been written (Chandra, 1989; Harris, 1989, 1997; Simopoulos et al., 1986). Most of these studies have concluded that the main effects of fish oil supplementation include the reduction of triglycerides and an improvement of the HDL/LDL ratio. Furthermore, the effect of EPA on the reduction of platelet aggregation and increase in bleeding time has been viewed as advantageous in some respects, but potentially problematic in others. Fish oil supplementation studies have been difficult to interpret because different researchers have used different types of fish oils with different ratios of DHA to EPA and different levels of endogenous cholesterol. Some studies have even shown negative effects (elevation of total cholesterol and LDL cholesterol) with fish oil consumption (FDA, 1993). Furthermore, it may not be valid to equate whole fish consumption with fish oil consumption since fish generally are a much better source of DHA (DHA/EPA ratio 3–4:1) than are most fish oils (DHA/EPA 0.5:1).

More recently, the antiarrhythmic effect of DHA and other long chain omega-3 fatty acids has been demonstrated. Using a rat model of sudden cardiac death (SCD), McLennan and coworkers (1996) first showed that the frequency of ventricular fibrillation could be significantly reduced by simply predosing the animals with omega-3 LCPUFAs, particularly DHA. Billman and colleagues (1997) demonstrated a similar significant reduction in SCD in a dog model by predosing the animals with DHA/EPA. More directly relating the human condition, Siscovick et al. (1995) followed over 250 paramedic responses to cardiac emergencies in the Pacific Northwest and showed that consumption of at least three fatty fish meals per week (about 200 mg DHA/day) reduced the likelihood of dying from SCD by 50%, although it did not decrease the risk that one would have a cardiac emergency.

Although a general triglyceride-lowering effect has been measured with fish oil (a combination of both EPA and DHA), it was not clear until recently which omega-3 fatty acid was responsible for this effect. Several clinical studies have recently clarified this

Table 5 Effects of DHA Supplementation in the Absence of Any Other Supplemental Omega-3 Fatty Acids in Double-Blind, Placebo-Controlled, Clinical Trials

Measurement	Study 1	Study 2	Study 3	Study 4[a]	Study 5
Sample size	55	24	27	32	6
Dose (g DHA/day)	1.6	1.8	1.2	3.0	6.0
Duration (days)	98	42	42	14	90
Blood DHA	+90%	+250%	+170%	+180%	+344%
Triglyceride	−18%	−17%	−21%	−44%	−17%
tChol	ns	ns	ns	+26%	ns
HDL	+10%	+17%	+6%	+17%	+9%
LDL	ns	ns	ns	ns	ns

[a] This supplementation included 3.4g ARA/day in addition to the DHA.
Sources: Study 1, Agren et al., 1996; Study 2, Conquer and Holub, 1996; Study 3, Davidson et al., 1997; Study 4, Innis and Hansen, 1996; Study 5, Nelson et al., 1997b.

issue by using a single cell algal oil which contains only DHA and no EPA or any other LC-PUFA. Five such studies have now been published (Agren et al., 1996; Conquer and Holub, 1996; Davidson et al., 1997; Innis and Hansen, 1996; Nelson et al., 1997b) and it is clear that DHA alone can reduce serum triglycerides by about 20%, and specifically elevate HDL by about 10% (Table 5). Interestingly, in those studies with the DHA oil, none of the researchers reported any change in platelet aggregation or bleeding time in spite of a relatively high rate of intake (up to 6.0 g DHA/day). Kelley et al. (1998) further showed that the DHA oil supplementation also had little or no effect on various immunological parameters. This is quite unlike fish oils (EPA/DHA combinations), which are classically known for their effect on the down-regulation of the immune response and inflammation. As is the case with the decreased platelet aggregation by EPA, the down-regulation of the immune function may be advantageous for some conditions, but problematic for others.

V. APPLICATIONS OF EFA ADDITION TO FOODS

A. Food Additives, Foods, or Dietary Supplements

Given that the essential fatty acids or, more precisely, their metabolic products, are requirements for the body's optimal function (i.e., they are essential), they don't fit the general description of a food additive. They are not added to food to convey some new or better functionality to the food product; they are foods themselves. However, certain LC-PUFAs are unusual in that they are only needed in a small quantity relative to all dietary calories in order to generate the required function. Furthermore, the two classes of EFAs—the omega-3 family and the omega-6 family—may be presented in the diet in different proportions to one another. The consequences of this unbalancing of the diets of many Western cultures is thought to be the cause of the increase in incidence of many chronic diseases (Simopoulos, 1991).

Our diet has changed drastically in the last 100 years. With the advent of modern agricultural processes, the production of seed oils (corn, soybean, palm, etc.) has become so inexpensive that they now make up a major part of our diet. In the United States, the average intake of fat represents about 35% of dietary calories today, whereas this number was only about 25% of dietary calories at the turn of the century. Since the seed oils are also omega-6 dominant, not only has the fat content increased significantly, but the proportion of omega-6 to omega-3 families of EFAs has drastically increased to about 15:1. This is a long way from the ratio of EFAs thought to be part of the diet of mankind throughout 5 million years of evolution (Broadhurst et al., 1998). In other words, a dramatic change in the dietary ratio of omega-6 to omega-3 EFAs away from the ratio to which our species evolved has occurred in the last 100 years (Simopoulos, 1998). It is for this reason that it is important to return to a much lower ratio of omega-6 to omega-3 EFAs in our diet. This can be done by increasing the levels of omega-3 EFAs in our diet, or decreasing the level of omega-6 EFAs in our diet. Since the latter requires a radical change in our food consumption habits, we should consider improving the omega-3 content of our diet by supplementing our dietary calories with small amounts of omega-3 LC-PUFAs.

B. Not All Omega-3 EFAs Are Created Equal

When we consider supplementing foods with omega-3 fatty acids, the first question that arises is what omega-3 fatty acids do we use? Many of the chronic diseases thought to be associated with the increase in the omega-6 to omega-3 ratio are believed to have their effect by an alteration of the ratios of the omega-6 versus omega-3 eicosanoids. However, they may also result from a deficiency of DHA—the important structural component of neural membranes. Although supplementing with the parent omega-3 EFA—ALA—should result in the elevation of EPA and DHA (the key fatty acids in poising the n-6/n-3 ratio), supplementation with the preformed EPA and DHA themselves is much more effective. Much of the dietary ALA is oxidized for energy, and several groups have shown that it takes about 20 ALA molecules to make (or be equivalent to) one EPA molecule. Likewise, it takes about 10 EPA molecules to make a single DHA. This phenomenon has been referred to as biomagnification.

A biomagnification index is shown in Table 6, which can be used to approximate how much of a particular supplement would be required to get a certain effect. From this data we can see that if one wants to improve the dietary DHA levels by about 100 mg DHA per day, this would require either 100 mg DHA directly, 1000 mg of EPA, or about 20,000 mg of ALA per day. Since the richest, commercially available source of ALA is flaxseed (linseed) oil (ca. 67% ALA), it would require about 30 g of flaxseed oil per day

Table 6 Biomagnification of Parent Essential Fatty Acids into Their Metabolites

Fatty acid	Biomagnification factor
Omega-6	
LA	100
GLA	20
ARA	1
Omega-3	
ALA	200
EPA	10
DHA	1

(about 2 tablespoons) to provide the equivalent of about 100 mg DHA/day. This amount of oil would represent nearly 300 kCal per day, or about 15% of our daily caloric intake as this one fat. Clearly, if we want to amend the functional n-6/n-3 ratio and reduce our dietary fat intake at the same time, it would be more effective to use preformed EPA and DHA than the parent ALA.

C. Technological Hurdles

The principal technological hurdle in amending the diet with EFAs is a consequence of their high degree of unsaturation. Unsaturated fatty acids pose problems in food preparation and processing as they are prone to oxidation, and the oxidation products are organoleptically unacceptable. This becomes more of a problem with the higher degree of unsaturation in the very long chain PUFAs like EPA and DHA. Fish oil, for example, is very rich in highly unsaturated fatty acids and oxidises very quickly to give an unacceptable fishy odor and taste to a product. Even flaxseed (linseed) oil was used as an industrial oil historically because of the sensitivity of ALA to oxidation and the formation of polymers (varnish). As a result, if we want to supplement food with highly unsaturated EFAs, then food process technology must be modified to account for the oxidative sensitivity. Furthermore, such products fortified with these EFAs may not have the same shelf-life as products prepared with saturated fatty acids.

One option recently employed by several manufacturers is to microencapsulate the highly unsaturated oil so that there is a greater oxygen barrier around the oil to prevent oxidation. Typical microencapsulated oils have an oil content of 20–30% by weight, and the microencapsulation matrix can have casein or gelatin as protein components and polydextrose as a carbohydrate component. Microencapsulated oils have a much greater stability than the neat oils, and this should translate into a greater shelf-life for the food product itself.

Although microencapsulation may lend itself to certain food products, it is not acceptable in others. In salad oils, spoonable dressings, and margarines, for example, a neat oil is necessary. In such cases we must look to ways of stabilizing the oil itself. Some oils appear to have an inherently higher degree of stability, which may have to do with the presence of pro-oxidants in the oil itself, the history of the oil processing (many oils carry a ''memory'' of oxidative insults imposed during processing), the total unsaturation index, or even the position of the unsaturated fatty acid on the triglyceride molecule itself. When fish oils are randomized (i.e., the fatty acid moieties are scrambled with respect to their position on the triglyceride), they become more oxidatively stable, whereas the opposite is true for the oils of marine mammals. The principal difference between the two oils is that the DHA molecule is primarily on the sn2 position of the triglyceride (the middle position) in the nonrandomized fish oil and on the sn1 and sn3 positions of the nonrandomized whale oil (Ackman, 1989). It may well be that when the DHA resides on the external positions of the triglyceride, there is a self-stabilization that takes place as a consequence of the overlap of adjacent π bonds of the DHA. This is supported by the fact that a DHA-rich oil produced by a microalgae which has DHA preferentially on positions sn1 and sn3 appears to have about a tenfold greater stability than a typical fish oil on a DHA for DHA basis (Kyle, 1996).

D. Sources of EFAs

If we choose to fortify food with EFAs, prepare functional foods, or provide dietary supplements to improve the EFA status of the consumer, there are several choices of omega-

Table 7 Commercialised Sources of Essential Fatty Acids

Fatty acid	Commercial sources		Levels
Omega-6			
LA	Vegetable oils:	corn[1]	59
		soy[1]	50
		canola[1]	30
GLA	Specialty plant oils:	primrose[1]	9
		borage[1]	22
		black current seed[1]	17
	Single cell oils:	*Mucor*[2]	18
		Mortierella[2]	8
ARA	Single cell oils:	*Mortierella*[3]	48
	Animal:	egg yolk	4[a]
Omega-3			
LNA	Vegetable oils:	soy[1]	9
		canola[1]	7
		flax (linseed)[1]	58
EPA	Animal (fish):	menhaden[4]	10
		salmon[4]	12
		tuna[4]	6
DHA	Animal (fish):	menhaden[4]	13
		salmon[4]	4
		tuna[4]	17
	Animal:	egg yolk	2[a]
	Single cell oils:	*Crypthecodinium*[5]	47
		Schizochytrium[6]	25[b]

[a] Primarily in the form of phosphatidylcholine.
[b] Also contains 13–15% omega-6 DPA.
Sources: [1]Murray, 1996; [2]Ratledge, 1992; [3]Koskelo et al., 1997a; [4]Perkins, 1993; [5]Behrens and Kyle, 1996; [6]Kendrick and Ratledge, 1992.

6 or omega-3 sources that can be used for enrichment. Some of the popular sources of omega-3 and omega-6 fatty acids are shown in Table 7. The choice of the source will be dictated by the sort of benefit the manufacturer wants to convey with that food. For example, if one wants to elevate an individual's DHA status by about 50%, it would take about 200 mg of DHA per day. This could be provided in about 0.5 g DHA-rich algal oil per day, or 60 g flaxseed oil per day (Table 6). Clearly the former provides a much lower caloric load than the latter (5 versus 600 kCal per day). Likewise, the elevation of GLA levels can be easily attained with the consumption of 1000 mg of evening primrose oil, but would require over 20 g of corn oil per day. It is, therefore, important for the manufacturer to first determine what effect is desired before choosing the source of oil that will give that effect.

E. Foods with Supplemental EFAs

Various foods have been supplemented with omega-3 EFAs, particularly in Japan and Southeast Asia. At first, they were indicated as a specialty food, but are now in the mainstream market, including fast and convenience foods. Several companies are marketing specialty drink products in Japan that are enriched in DHA, but at a very low level. Cookies and biscuits have also been produced, but it has proven difficult to disguise the taste of

Table 8 Examples of Food Products Supplemented with Essential Fatty Acids

Category	Product name	Manufacturer	EFA supplementation
Beverage	Recovery	Great Circles (USA)	Fish oil/DHASCO (EPA/DHA)
	UltraCare	UltraBalance (USA)	DHASCO (DHA)
Bread	Live	Irish Pride (Ireland)	Fish oil (EPA/DHA)
	Heartbeat	British bakeries (UK)	Fish oil (EPA/DHA)
	North's Extra	Allied Foods (NZ)	Fish oil (EPA/DHA)
	Flax bread	Natural Ovens (USA)	Flax seed (ALA)
Cereal	Golden Flax	Health Valley (USA)	Flax seed (ALA)
	Flax Plus	Lifestream (USA)	Flax seed (ALA)
Bar	Recovery	Great Circles (USA)	Fish oil/DHASCO (EPA/DHA)
	Prozone	Nutribiotic (USA)	Borage oil (GLA)
Spread	Pact	MD Foods (DK)	Fish oil (EPA/DHA)
	Essential Omega	Spectrum Naturals (USA)	Flax oil (ALA)
	Live	Golden Vale (Ireland)	Fish oil (EPA/DHA)

the fish oil which is the primary source of the EPA and DHA added to many of these foods. In all cases, there appears to be a strong advertising message in the Japanese DHA-containing products to claims of improving mental or visual function. Some products are even advertised as study aids.

In Europe and the United States, the EFA-enriched foods have been mainly restricted to dairy products (butters and margarines) or other spreads (Table 8). In most cases, such products are presented as having important cardiovascular benefits and may be labeled as fish oil products. One morning spread enriched with fish oil is even sold in a container shaped like a heart. Such claims were generally supported by a long list of clinical studies demonstrating the advantages of fish oil consumption on lowering triglycerides.

A common problem to all PUFA-enriched products, however, is shelf-life. Since PUFAs are highly susceptible to oxidation, care must be taken in processing, packaging, and storage of the foods. Since foods in the dairy case are generally refrigerated, these make the best candidates for supplementation. Packaging with low oxygen permeability materials with ''flavor seals'' that are removed by the customer at first use are common to avoid the oxidation and consequential off flavors and odors. An alternative to adding the active PUFAs themselves is to add the precursors—GLA and/or ALA. Although generally less susceptible to oxidation than ARA, EPA, or DHA, these components must be added in a much higher level to provide the same level of effectiveness as the final products, and the overall effect on the shelf-life of the finished product may be no better. This approach has been used in the United States, however, to market bread products enriched with whole flax seeds as a source of ALA. The advantage of this approach is that the flax seeds themselves are not only protected from oxidation by the barrier of the seed coat, but the seeds also contain a large amount of endogenous antioxidants.

VI. REGULATORY STATUS

A. EFAs as Food Additives

In the United States, most EFAs are generally recognized as safe (GRAS) because of their historical presence in the diet, and are therefore exempt from the premarket approval requirements for food additives. An exception to this generality involves the use of EFAs in infant formulas. The use of PUFA supplementation in infant formulas, or any new ingredient to infant formulas, requires a premarket assessment by the Office of Special Nutritionals of the FDA. A recent change in the infant formula regulation in the United States also requires that any new additive to infant formulas to be either a GRAS substance or an FDA-approved food additive. Outside the specific example of infant formula, however, the enrichment of a food product with supplemental PUFAs (either omega-6 or omega-3) generally would not require prior approval of the EFA as a food additive.

B. Novel Foods Regulations

Throughout Europe, sources of PUFAs for use in foods are controlled by the Novel Foods Regulations. Foods, or food products which were not in common use prior to 1997 are considered novel foods by the European Union (EU) and, therefore, require specific toxicological evaluations before they are presented to the marketplace. Infant formula rules in the EU also prohibit the use of any novel foods in infant formulas without extensive premarket evaluations. Certain fish oils, the algal and fungal oils used in formulas throughout Europe (*e.g., Crypthecodinium cohnii* and *Mortierella alpina* oils), borage oil, and primrose oil, were all in use prior to the EU Novel Foods Regulation of 1997 and, therefore, are not considered novel foods. New sources of EFAs introduced to the market after the 1997 agreement, however, would need to be considered as novel foods and will be subject to the Novel Foods Regulation. No such novel foods concept exist in the United States.

C. Health Claim Labeling Regulations

Other than being added directly to foods to provide a ''functional food,'' many PUFAs are provided as dietary supplements, similar to vitamins and minerals. As such, in the United States, these dietary supplements will be subject to the Dietary Supplements Health and Education Act (DSHEA) of 1994. This act provides guidelines for the industry as to how to label such products. Any claims that the product treats, prevents, cures, or mitigates a disease condition will be considered as drug claims and are not allowed. However, language relating the structure and/or function of the supplement in the body may be allowed. Since there is a large base of knowledge on the location in the body and the function of these PUFAs, much can be said describing them in terms appropriate to the DSHEA. These regulations would apply only to the PUFAs provided as dietary supplements, not as whole foods.

Health claims associated with nutrients in the United States are the perview of the Nutritional Labelling and Education Act (NLEA) of 1990. This act provides guidance to the industry indicating that a health claim may be made for a food providing that there is ''significant scientific agreement among experts'' that such health promoting effects are real and FDA has issued a regulation authorizing the claim. In the case of a health claim that omega-3 fatty acids (fish oil) promotes cardiovascular health, the FDA reviewed the literature and ruled in 1993 that such a claim could not be made because there was not substantial agreement in the literature about the specific benefits of fish oil. This was due to some clinical studies concluding that fish oil supplementation reduced cholesterol, some concluded that there was no effect, and some even concluded that fish oil supplementation elevated cholesterol levels (FDA, 1993). The inconsistency in these studies was likely due to a lack of consistency of the source of fish oil used in the various studies.

Not all fish oils are the same. The ratio of EPA to DHA can vary over an order of magnitude among fish oils, and the cholesterol content of fish liver oils is much higher than from fish body oils. It may, therefore, be difficult to ever get a health claim for fish oil since ''fish oil'' is poorly defined. Rather, it may be more appropriate to get a health claim on a specific oil that demonstrates consistent results in various applications and clinical studies. Such was the case in 1998 when Menhaden oil was affirmed as GRAS by the FDA. This affirmation, however, was for Menhaden oil only, and not for any other fish oil since the safety profile of one fish oil may be quite different from another fish oil.

VII. TOXICOLOGY

Essential fatty acids are nutrients that have been in the diet of humans on an evolutionary timescale. They are substituents of the body and are found in fairly high levels in certain tissues. Clearly, the PUFAs themselves are nontoxic under normal dietary circumstances. However, these components are obtained in the diet from a multitude of sources, any of which may carry problematic ''co-travellers,'' so the requirement of toxicological testing is still critical. This poses a difficult problem, however, because very large doses of the macronutrient in question must be used in such safety studies and the researcher must be able to deconvolute the effects of any unknown toxin from the physiological and biochemical consequences of very large doses of a single nutrient. A common problem associated with macronutrient studies, for example, is the development of vitamin and mineral deficiencies as the macronutrient delivered may artificially deplete the diet of these components (Borzelleca, 1992). In order to differentiate true toxicological findings from normal macronutrient findings, one must run a positive as well as a negative control. When testing oils from new sources, therefore, one should include a common food oil whose safety is well established (e.g., corn oil or soy oil) as a positive control for comparative purposes.

A. Safety Data

Many of the sources of ALA, GLA, EPA, DHA, and ARA have been used in food systems or dietary supplements for several years although very few sources used today would be considered traditional foods. As a consequence, many of the PUFA sources are considered safe by implication rather than by thorough and critical safety testing. Flax oil and fish oils for example, have been used commercially for hundreds of years as varnishes or lacquers because of their sensitivity to oxidation and formation of polymers. Primrose and certain algae have not necessarily been in the direct food system until recently. Menhaden oil was recently affirmed by the FDA as GRAS after careful scrutiny of many toxicological studies using this specific fish oil. It is the only fish oil which has been tested with such a high degree of scrutiny and is the only fish oil that is affirmed as GRAS. One must not make the assumption, however, that the oil obtained from Menhaden would be expected to have a similar safety profile to that from, for example, the highly toxic puffer fish (fugu) of the South Pacific. Oils isolated from different species of fish should also be carefully tested and affirmed to be safe before their extensive use as a food additive because in many cases they may represent novel introductions directly into the food chain of man (i.e., never before consumed by man). Such testing would be required by the FDA in order to grant food additive approval for any component.

The same argument is true for any new plant or microbial oil introduced to the food system. Classical toxicological studies have been completed for two microbial oils (the

Table 9 Safety Profile for DHASCO, a New Essential Fatty Acid Supplement

Study	Dose	Results
In vitro toxicology		
AMES mutagenicity		
DHASCO	Up to 5 mg/plate	Nonmutagenic[1]
Whole cells	Up to 5 mg/plate	Nonmutagenic[1]
Forward mutation		
DHASCO	Up to 5 mg/plate	Nonmutagenic[1]
Whole cells	Up to 1 mg/plate	Nonmutagenic[1]
Chromosomal aberration		
DHASCO	Up to 5 mg/plate	Nonclastogenic[1]
Whole cells	Up to 1 mg/plate	Nonclastogenic[1]
In vivo toxicology		
Toxin Bioassay		
DHASCO	10% body wt I.P.	Nontoxic[2]
Whole cells	1.25 g/kg I.P.	Nontoxic[2]
Acute oral dosing		
DHASCO	20 g/kg	Nontoxic[1] $LD_{50} > 20$g/kg
Whole cells	7 g/kg	Nontoxic[2] $LD_{50} > 7$g/kg
28-day subchronic		
DHASCO	Up to 1.25 g/kg/d	Nontoxic[3]
DHASCO*	Up to 13 g/kg/d	Nontoxic[4]
54-day subchronic		
DHASCO	Up to 2 g/kg/d	Nontoxic[2]
90-day subchronic		
DHASCO	Up to 1.25 g/kg/d	Nontoxic[3]
DHASCO*	Up to 13 g/kg/d	Nontoxic[2]
Neurotoxicity		
DHASCO	Up to 1.25 g/kg/d	Not neurotoxic[2]
Developmental toxicology		
DHASCO	Up to 1.25 g/kg/d	Nontoxic[2]

Sources: [1]Boswell et al., 1995; [2]Kyle and Arterburn, 1998; [3]Boswell et al., 1996; [4]Wibert et al., 1997.
* Provided in combination with an arachidonic acid single-cell oil (ARASCO).

DHA-rich oil from the microalgae *Crypthecodinium cohnii* and the ARA-rich oil from the fungi *Mortierella alpina*) (Boswell et al., 1996; Koskelo et al., 1997b). A typical safety profile of these tests is presented in Table 9 for one of these oils. Similar data have not been published for GLA sources including primrose oil, borage oil, black current seed oil, or the fungal producer *Mucor issabelina*. This does not necessarily mean that the products are not safe, since safety may be based on historical use or published data on clinical studies with the products. The objective of toxicity studies is to determine the lowest no adverse effect level. Clinical studies or historical use data using these sources of PUFAs are not generally designed to establish toxicity or safety margins, and they may not be the best measures of safety of a product.

B. Historical Uses

In lieu of a well-established toxicological profile of a particular source of PUFA, one must rely on historical use information. Has this product been consumed by humans or other mammals in the past, or has this product been used in clinical studies where toxicological end points would be recognized? Positive answers to both of these questions gives a better assurance for safety than otherwise, but it must be recognized that these data sources do not establish safety limits. Since these sources do not necessarily require food additive approval from the FDA, there is little control on the safety of such sources. In Europe, the situation is somewhat different as any novel food requires a serious toxicological examination before it enters the food system.

VIII. SUMMARY AND CONCLUSIONS

It is quite clear that the products of essential fatty acid metabolism (i.e., eicosanoids) as well as some longer chain products that have a significant role in the structure of neurological tissues (i.e., ARA and DHA) are important components in the diet. It is also quite clear that the human dietary fats have changed radically in the last 100 years with the advent of high intensity agriculture and the low cost production of omega-6-rich vegetable oils. We are now beginning to realize the long-term effects of such an altered fat composition in the diet and are beginning to recognize the importance of returning the omega-3/omega-6 ratio to that to which our species evolved. One of the best examples of the consequences of our dietary fat manipulation is that of providing babies a fat mixture (in infant formulas) that is completely devoid of the building blocks of the neurological tissues (ARA and DHA). The mid- to long-term consequences of the resulting suboptimal EFA status during these early, developmentally critical first months of life are now being realized. Why should we be surprised when the results of such a dramatic dietary alteration is a visual and neurological developmental delay in infants fed formulas devoid of these important precursors? As the importance of the balance of these fatty acids becomes better understood, there may be a need to fortify our present adult diet with these components as well.

With a healthy, diverse diet there is likely no need to include additives that improve the nutritional quality of the food. However, the absence of such an ideal diet reflects the reality of the situation in the United States. As a consequence, the food industry has already started developing ''functional foods'' as a way of enriching conventional foods with these components. As this market expands in the future, there will be a greater need to consider such fortification, and to ensure the safety of the public consuming these materials, these fortifiers should be well characterized and well tested. Such new food products will be designed to prevent disease and should improve the health and quality of life for all.

REFERENCES

Ackman, R. G. 1989. Problems in fish oils and concentrates. In: *Fats for the Future*, Cambie, R. C. (Ed.). Ellis Horwood, Chichester, pp. 189–204.

Agostoni, C., Trojan, S., Bellu, R., Riva, E., Giovannini, M. 1995. Neurodevelopmental quotient of healthy term infants at 4 months and feeding practice: the role of long-chain polyunsaturated fatty acids. *Pediatr. Res.* 38:262–266.

Agren, J. J., Hanninen, O., Julkunen, A., Fogelholm, L., Vidgren, H., et al. 1996. Fish diet, fish oil and docosahexaenoic acid rich oil lower fasting and postprandial plasma lipid levels. *Eur. J. Clin. Nutr.* 50:765–771.

Al, M. D., van Houwelingen, A. C., Kester, A. D., Hasaart, T. H., de Jong, A. E., Hornstra, G.

1995. Maternal essential fatty acid patterns during normal pregnancy and their relationship to the neonatal essential fatty acid status. *Br. J. Nutr.* 74:55–68.

Anderson, J., Johnstone, B., Kyle, D. J. 1996. Meta-analysis of the effect of breast-feeding on neurodevelopmental scores. In: *AOCS PUFA in Infant Nutrition Meeting*, Barcelona, Spain.

Anderson, R. E., PJ, O. B., Wiegand, R. D., Koutz, C. A., Stinson, A. M. 1992. Conservation of docosahexaenoic acid in the retina. *Adv. Exp. Med. Biol.* 318:285–294.

Arbuckle, L. D., Innis, S. M. 1993. Docosahexaenoic acid is transferred through maternal diet to milk and to tissues of natural milk-fed piglets. *J. Nutr.* 123:1668–1675.

Auestad, N., Montalto, M. B., Hall, R. T., Fitzgerald, K. M., Wheeler, R. E., et al. 1997. Visual acuity, erythrocyte fatty acid composition, and growth in term infants fed formulas with long chain polyunsaturated fatty acids for one year. Ross Pediatric Lipid Study. *Pediatr. Res.* 41: 1–10.

Bazan, N. G., Rodriguez de Turco, E. B. 1994. Review: pharmacological manipulation of docosahexaenoic-phospholipid biosynthesis in photoreceptor cells: implications in retinal degeneration. *J. Ocul. Pharmacol.* 10:591–604.

Bazan, N. G., Scott, B. L. 1990. Dietary omega-3 fatty acids and accumulation of docosahexaenoic acid in rod photoreceptor cells of the retina and at synapses. *Ups. J. Med. Sci. Suppl.* 48:97–107.

Bazan, N. G., Rodriguez de Turco, E. B., Gordon, W. C. 1993. Pathways for the uptake and conservation of docosahexaenoic acid in photoreceptors and synapses: biochemical and autoradiographic studies. *Can. J. Physiol. Pharmacol.* 71:690–698.

Behrens, P. W., Kyle, D. J. 1996. Microalgae as a source of fatty acids. *J. Food Sci.* 3:259–272.

Bennett, M. J., Gayton, A. R., Rittey, C. D. C., Hosking, G. P. 1994. Juvenile neuronal ceroid-lipofuscinosis: developmental progress after supplementation with polyunsaturated fatty acids. *Developmental Medicine and Child Neurology* 36:630–638.

Billman, G. E., Hallaq, H., Leaf, A. 1994. Prevention of ischemia-induced ventricular fibrillation by omega-3 fatty acids. *Proc. Natl. Acad. Sci. U. S. A.* 91:4427–4430.

Billman, G. E., Kang, J. X., Leaf, A. 1997. Prevention of ischemia-induced cardiac sudden death by n-3 polyunsaturated fatty acids in dogs. *Lipids* 32:1161–1168.

Birch, D. G., Birch, E. E., Hoffman, D. R., Uauy, R. D. 1992a. Retinal development in very-low-birth-weight infants fed diets differing in omega-3 fatty acids. *Invest. Ophthalmol. Vis. Sci.* 33:2365–2376.

Birch, E. E., Birch, D. G., Hoffman, D. R., Uauy, R. 1992b. Dietary essential fatty acid supply and visual acuity development. *Invest. Ophthalmol. Vis. Sci.* 1992 Oct. 33:3242–3253.

Birch, E. E., Hoffman, D. R., Uauy, R., Birch, D. G., Prestidge, C. 1998. Visual acuity and the essentiality of docosahexaenoic acid and arachidonic acid in the diet of term infants. *Pediatr. Res.* 44:201–209.

Bloom, M., Linseissen, F., Lloyd-Smith, J., Crawford, M. 1998. Insights from NMR on the functional role of polyunsaturated lipids in the brain. *Proc. of 1998 Enrico Fermi Int. School of Physics*, course #139, ed. B. Maraviglia, p. 527–553, IOS Press 1999.

Bonaa, K. H., Bjerve, K. S., Straume, B., Gram, I. T., Thelle, D. 1990. Effect of eicosapentaenoic and docosahexaenoic acids on blood pressure in hypertension. A population-based intervention trial from the Tromso study. *N. Engl. J. Med.* 322:795–801.

Borzelleca, J. F. 1992. Macronutrient substitutes: safety evaluation. *Regul. Toxicol. Pharmacol.* 16: 253–264.

Boswell, K., Koskelo, E.-K., Carl, L., Kyle, D. J. 1995. Preclinical evaluation of designer oils which are highly enriched in docosahexaenoic and arachidonic acids. *FASEB J.* 9:A475.

Boswell, K., Koskelo, E.-K., Carl, L., Glaza, S., Hensen, D. J., et al. 1996. Preclinical evaluation of single-cell oils that are highly enriched with arachidonic acid and docosahexaenoic acid. *Food Chem. Toxicol.* 34:585–593.

Broadhurst, C. L., Cunnane, S. C., Crawford, M. A. 1998. Rift Valley lake fish and shellfish provided brain-specific nutrition for early *Homo. Br. J. Nutr.* 79:3–21.

Carlson, S. E., Werkman, S. H. 1996. A randomized trial of visual attention of preterm infants fed docosahexaenoic acid until two months. *Lipids* 31:85–90.

Carlson, S. E., Cooke, R. J., Rhodes, P. G., Peeples, J. M., Werkman, S. H. 1992. Effect of vegetable and marine oils in preterm infant formulas on blood arachidonic and docosahexaenoic acids. *J. Pediatr.* 120:S159–S167.

Carlson, S. E., Werkman, S. H., Rhodes, P. G., Tolley, E. A. 1993. Visual-acuity development in healthy preterm infants: effect of marine-oil supplementation. *Am. J. Clin. Nutr.* 58:35–42.

Carlson, S. E., Werkman, S. H., Peeples, J. M., Wilson, W. M. 1994. Long-chain fatty acids and early visual and cognitive development of preterm infants. *Eur. J. Clin. Nutr.* 48:S27–S30.

Carlson, S. E., Ford, A. J., Werkman, S. H., Peeples, J. M., Koo, W. W. 1996a. Visual acuity and fatty acid status of term infants fed human milk and formulas with and without docosahexaenoate and arachidonate from egg yolk lecithin. *Pediatr. Res.* 39:882–888.

Carlson, S. E., Werkman, S. H., Tolley, E. A. 1996b. Effect of long-chain n-3 fatty acid supplementation on visual acuity and growth of preterm infants with and without bronchopulmonary dysplasia. *Am. J. Clin. Nutr.* 63:687–697.

Carlson, S. E., Montalto, M. B., Ponder, D. L., Werkman, S. H., Korones, S. B. 1998. Lower incidence of necrotizing enterocolitis in infants fed a preterm formula with egg phospholipids. *Pediatr. Res.* 44:491–498.

Chandra, R. K. 1989. *Health Effects of Fish and Fish Oils.* ARTS Biomedical, St. John's, Newfoundland, Canada.

Charnock, J. S., McLennan, P. L., Abeywardena, M. Y. 1992. Dietary modulation of lipid metabolism and mechanical performance of the heart. *Mol. Cell. Biochem.* 116:19–25.

Clarke, S. D., Jump, D. B. 1994. Dietary polyunsaturated fatty acid regulation of gene transcription. *Annu. Rev. Nutr.* 14:83–89.

Cobiac, L., Clifton, P. M., Abbey, M., Belling, G. B., Nestel, P. J. 1991. Lipid, lipoprotein, and hemostatic effects of fish vs. fish-oil n-3 fatty acids in mildly hyperlipidemic males. *Am. J. Clin. Nutr.* 53:1210–1216.

Connor, W. E., Weleber, R. G., Lin, D. S., Defrancesco, C., Wolf, D. P. 1995. Sperm abnormalities in retinitis pigmentosa. *J. Invest. Med.* 43:302A.

Conquer, J. A., Holub, B. J. 1996. Supplementation with an algae source of docosahexaenoic acid increases (n-3) fatty acid status and alters selected risk factors for heart disease in vegetarian subjects. *J. Nutr.* 126:3032–3039.

Cousin, B., Casteilla, I., Dani, C., Muzzin, P., Revelli, J. P., Penicaud, L. 1993. Adipose tissues from various anatomical sites are characterized by different patterns of gene expression and regulation. *Biochem. J.* 292:873–876.

Crawford, M. A. 1990. The early development and evolution of the human brain. *Ups. J. Med. Sci.* 48(Suppl):43–78.

Crawford, M. A., Doyle, W., Leaf, A., Leighfield, M., Ghebremeskel, K., Phylactos, A. 1993. Nutrition and neurodevelopmental disorders. *Nutr. Health* 9:81–97.

Crawford, M. A., Costeloe, K., Ghebremeskel, K., Phylactos, A., Skirvin, L., Stacey, F. 1997. Are deficits of arachidonic and docosahexaenoic acids responsible for the neural and vascular complications of preterm babies? *Am. J. Clin. Nutr.* 66:1032S–1041S.

Crawford, M. A., Costeloe, K., Ghebremeskel, K., Phylactos, A. 1998. The inadequacy of the essential fatty acid content of present preterm feeds. *Eur. J. Pediatr.* 157:S23–S27.

Davidson, M. H., Maki, K. C., Kalkowski, J., Schaefer, E. J., Torri, S. A., Drennan, K. B. 1997. Effects of docosahexaenoic acid on serum lipoproteins in patients with combined hyperlipidemia: a randomized, double-blind, placebo-controlled trial. *J. Am. Coll. Nutr.* 16:236–243.

Dry, J., Vincent, D. 1991. Effect of a fish oil diet on asthma: results of a 1-year double-blind study. *Int. Arch. Allergy Appl. Immunol.* 95:156–157.

Dyerberg, J., Bang, H. O. 1979. Haemostatic function and platelet polyunsaturated fatty acids in Eskimos. *Lancet* 2:433–446.

Eritsland, J., Arnesen, H., Smith, P., Seljeflot, I., Dahl, K. 1989. Effects of highly concentrated omega-3 polyunsaturated fatty acids and acetylsalicylic acid, alone and combined, on bleeding time and serum lipid profile. *J. Oslo City Hosp.* 39:97–101.

FAO/WHO 1993. Fats and Oils in Human Nutrition. FAO Food and Nutrition Paper 57. M. Hegsted, Chairman, Publications Division of FAO. Rome, Italy.

Farquharson, J., Jamieson, E. C., Abbasi, K. A., Patrick, W. J., Logan, R. W., Cockburn, F. 1995. Effect of diet on the fatty acid composition of the major phospholipids of infant cerebral cortex. *Arch. Dis. Child* 72:198–203.

FDA. 1993. Food labelling: health claims and label statements: omega-3 fatty acids and coronary heart disease. *Federal Register* 58:2682–2738.

Ferretti, A., Nelson, G. J., Schmidt, P. C., Kelley, D. S., Bartolini, G., Flanagan, V. P. 1997. Increased dietary arachidonic acid enhances the synthesis of vasoactive eicosanoids in humans. *Lipids* 32:435–439.

Florey, C. D., Leech, A. M., Blackhall, A. 1995. Infant feeding and mental and motor development at 18 months of age in first born singletons. *Int. J. Epidemiol.* 24:S21–S26.

Fujimoto, K., Yao, K., Miyazawa, T., Hirono, H., Nishikawa, M., et al. 1989. *The Effect of Dietary Docosahexaenoate on the Learning Ability of Rats,* Vol. I. ARTS Biomedical Publishers and Distributors Limited, St. John's, Newfoundland, Canada, pp. 275–284.

Garton, A. 1992. *Unsaturated Fatty Acids: Nutritional and Physiological Significance; The Report of the British Nutrition Foundation's Task Force.* Chapman & Hall, London.

Gill, I., Valivety, R. 1997. Polyunsaturated fatty acids, Part 1: occurrence, biological activities and applications. *TIBTECH* 15:401–409.

Gillingham, M. B., Mills, M. D., Vancalcar, S. C., Verhoeve, J. N., Wolff, J. A., Harding, C. O. 1997. DHA supplementation in children with long chain 3-hydroxyacyl-CoA dehydrogenase deficiency. *7th International Congress of Inborn Errors of Metabolism:* 108.

Gillis, W. S., Bennett, M. J., Galloway, J. H., Cartwright, I. J., Hosking, G., Smith, C. M. L. 1986. Lipid abnormalities in Batten's Disease. *J. Inherit. Metabol. Dis.* 10:329–332.

Glen, A. I. M., Glen, E. M. T., Horrobin, D. F., Vaddadi, K. S., Spellman, M., et al., 1994. A red cell membrane abnormality in a subgroup of schizophrenic patients: evidence for two diseases. *Schizophrenia Res.* 12:53–61.

Golding, J., Rogers, I. S., Emmett, P. M. 1997. Association between breast feeding, child development and behaviour. *Early Hum. Dev.* 29:S175–S184.

Gordon, W. C., Bazan, N. G. 1990. Docosahexaenoic acid utilization during rod photoreceptor cell renewal. *J. Neurosci.* 10:2190–2202.

Greiner, R. C., Winter, J., Nathanielsz, P. W., Brenna, J. T. 1997. Brain docosahexaenoate accretion in fetal baboons: bioequivalence of dietary alpha-linolenic and docosahexaenoic acids. *Pediatr. Res.* 42:826–834.

Gudbjarnason, S., Doell, B., Oskarsdottir, G. 1978. Docosahexaenoic acid in cardiac metabolism and function. *Acta. Bio. Med. Germ.* 37:777–787.

Harris, W. S. 1989. Fish oils and plasma lipid and lipoprotein metabolism in humans: a critical review. *J. Lipid Res.* 30:785–807.

Harris, W. S. 1997. n-3 fatty acids and serum lipoproteins: human studies. *Am. J. Clin. Nutr.* 65: 1645S–1654S.

Hibbeln, J. R., Linnoila, M., Umhua, J. C., Rawlings, R., George, D. T., Salem, N., Jr. 1998a. Essential fatty acids predict metabolites of serotonin and dopamine in cerebrospinal fluid among healthy control subjects, and early and late-onset alcoholics. *Biol. Psychiatry* 44:235–242.

Hibbeln, J. R. 1998b. Fish consumption and major depression. *Lancet* 351:1213 (letter).

Hibbeln, J. R., Salem, N., Jr. 1995. Dietary polyunsaturated fatty acids and depression: when cholesterol does not satisfy. *Am. J. Clin. Nutr.* 62:1–9.

Higley, J. D., Suomi, S. J., Linnoila, M. 1996a. A nonhuman primate model of type II alcoholism? Part 2. Diminished social competence and excessive aggression correlates with low cerebrospinal fluid 5-hydroxyindoleacetic acid concentrations. *Alcohol Clin. Exp. Res.* 20:643–650.

Higley, J. D., Suomi, S. J., Linnoila, M. 1996b. A nonhuman primate model of type II excessive alcohol consumption? Part 1. Low cerebrospinal fluid 5-hydroxyindoleacetic acid concentrations and diminished social competence correlate with excessive alcohol consumption. *Alcohol Clin. Exp. Res.* 20:629–642.

Hoffman, D. R. Birch, D. G. 1995. Docosahexaenoic acid in red blood cells of patients with X-linked retinitis pigmentosa. *Invest. Ophthalmol. Vis. Sci.* 36:1009–1018.

Hoffman, D. R., Birch, E. E., Birch, D. G., Uauy, R. D. 1993. Effects of supplementation with omega-3 long-chain polyunsaturated fatty acids on retinal and cortical development in premature infants. *Am. J. Clin. Nutr.* 57:807S–812S.

Holman, R. T. 1986. *Essential Fatty Acids, Prostaglandins and Leukotrienes.* Pergamon Press, London.

Horby Jorgensen, M., Holmer, G., Lund, P., Hernell, O., Michaelsen, K. F. 1998. Effect of formula supplemented with docosahexaenoic acid and gamma-linolenic acid on fatty acid status and visual acuity in term infants. *J. Pediatr. Gastroenterol. Nutr.* 26:412–421.

Horrobin, D. F. 1991. Therapeutic effects of gamma-linolenic acid (GLA) as evening primrose oil in atopic eczema and diabetic neuropathy. In: *Health Effects of Dietary Fatty Acids*, Nelson, G. J. (Ed.). American Oil Chemists' Society, Champaign, IL, pp. 234–244.

Horwood, L. J., Fergusson, D. M. 1998. Breastfeeding and later cognitive and academic outcomes. *Pediatr.* 101:e9.

Infante, J. P., Huszagh, V. A. 1997. On the molecular etiology of decreased arachidonic (20:4n-6), docosapentaenoic (22:5n-6) and docosahexaenoic (22:6n-3) acids in Zellweger syndrome and other peroxisomal disorders. *Mol. Cell. Biochem.* 168:101–115.

Innis, S. M., Hansen, J. W. 1996. Plasma fatty acid responses, metabolic effects, and safety of microalgal and fungal oils rich in arachidonic and docosahexaenoic acids in healthy adults. *Am. J. Clin. Nutr.* 64:159–167.

ISSFAL. 1994. Recommendations for the essential fatty acid requirement for infant formula. *ISSFAL Newsletter* 1:4–5.

Kelley, D., Taylor, P., Nelson, G., Mackey, B. 1998. Dietary docosahexaenoic acid and immunocompetence in young healthy men. *Lipids* 33:559–566.

Kelley, D. S. Taylor, P. C., Nelson, G. J., Schmidt, P. C., Mackay, B. E., Kyle, D. J. 1997. Effects of dietary arachidonic acid on human immune response. *Lipids* 32:449–456.

Kendrick, A., Ratledge, C. 1992. Lipids of selected molds grown for production of n-3 and n-6 polyunsaturated fatty acids. *Lipids* 27:15–20.

Kjeldsen-Kragh, J., Lund, J. A., Riise, T., Finnanger, B., Haaland, K., et al. 1992. Dietary omega-3 fatty acid supplementation and naproxen treatment in patients with rheumatoid arthritis. *J. Rheumatol.* 19:1531–1536.

Koskelo, E.-K., Boswell, K., Carl, L., Lanoue, S., C., K., Kyle, D. J. 1997a. High levels of dietary arachidonic acid triglyceride exhibit no subchronic toxicity in rats. *Lipids* 32:397–405.

Koskelo, E.-K., Boswell, K., Carl, L., Lanoue, S., Kelly, C., Kyle, D. 1997b. High levels of dietary arachidonic acid triglyceride exhibit no subchronic toxicity in rats. *Lipids* 32:397–405.

Kremer, J. M., Lawrence, D. A., Jubiz, W., DiGiacomo, R., Rynes, R., et al. 1990. Dietary fish oil and olive oil supplementation in patients with rheumatoid arthritis. Clinical and immunologic effects. *Arthritis Rheum.* 33:810–820.

Kyle, D. J. 1996. Production and use of a single cell oil which is highly enriched in docosahexaenoic acid. *Lipid Technol.* 2:109–112.

Kyle, D. J. Arterburn, L. M. 1998. Single cell oil sources of docosahexaenoic acid: clinical studies. In: *The Return of Omega-3 Fatty Acids into the Food Supply: I. Land Based Animal Food Products and Their Health Effects,* Simopoulis, A. P. (Ed.), *World Review of Nutrition and Diet*, Vol. 83. Karger, Basel; pp. 116–131.

Kyle, D. J., Schaefer, E., Baiser. 1998. Low serum DHA is a risk factor for senile dementia. ISSFAL Conference, Lyon, France.

Lagarde, M. 1995. Biosynthesis and functions of eicosanoids. Recent data. *C. R. Seances Soc. Biol. Fil.* 189:839–851.

Lands, W. E. 1989. n-3 fatty acids as precursors for active metabolic substances: dissonance between expected and observed events. *J. Intern. Med.* 225(Suppl):11–20.
Lanting, C. I., Fidler, V., Huisman, M., Touwen, B. C., Boersma, E. R. 1994. Neurological differences between 9-year-old children fed breast-milk or formula-milk as babies. *Lancet* 344:1319–1322.
Laugharne, J. D., Mellor, J. E., Peet, M. 1996. Fatty acids and schizophrenia. *Lipids* 31:S163–S165.
Leaf, A. 1995. Omega-3 fatty acids and prevention of ventricular fibrillation. *Prostaglandins Leukot. Essent. Fatty Acids* 52:197–198.
Lee, T. H., Arm, J. P., Horton, C. E., Crea, A. E., Mencia-Huerta, J. M., Spur, B. W. 1991. Effects of dietary fish oil lipids on allergic and inflammatory diseases. *Allergy Proc.* 12:299–303.
Lucas, A., Morley, R., Cole, T. J., Lister, G., Leeson-Payne, C. 1992. Breast milk and subsequent intelligence quotient in children born preterm. *Lancet* 339:261–264.
Makrides, M., Neumann, M. A., Byard, R. W., Simmer, K., Gibson, R. A. 1994. Fatty acid composition of brain, retina, and erythrocytes in breast- and formula-fed infants. *Am. J. Clin. Nutr.* 60: 189–194.
Makrides, M., Neumann, M, Simmer, K., Pater, J., Gibson, R. 1995. Are long-chain polyunsaturated fatty acids essential nutrients in infancy? *Lancet* 345:1463–1468.
Mark, G., Sanders, T. A. 1994. The influence of different amounts of n-3 polyunsaturated fatty acids on bleeding time and in vivo vascular reactivity. *Br. J. Nutr.* 71:43–52.
Martinez, M. 1990. Severe deficiency of docosahexaenoic acid in peroxisomal disorders: a defect of delta 4 desaturation? *Neurology* 40:1292–1298.
Martinez, M. 1991. Developmental profiles of polyunsaturated fatty acids in the brain of normal infants and patients with peroxisomal diseases: severe deficiency of docosahexaenoic acid in Zellweger's and pseudo-Zellweger's syndromes. *World Rev. Nutr. Diet* 66:87–102.
Martinez, M. 1992. Abnormal profiles of polyunsaturated fatty acids in the brain, liver, kidney and retina of patients with peroxisomal disorders. *Brain Res.* 583:171–182.
Martinez, M. 1996. Docosahexaenoic acid therapy in docosahexaenoic acid-deficient patients with disorders of peroxisomal biogenesis. *Lipids* 31:S145–S152.
Martinez, M., Vazquez, E. 1998. MRI evidence that docosahexaenoic acid ethyl ester improves myelination in generalized peroxisomal disorders. *Neurology* 51:26–32.
Martinez, M., Pineda, M., Vidal, R., Conill, J., Martin, B. 1993. Docosahexaenoic acid—a new therapeutic approach to peroxisomal-disorder patients: experience with two cases. *Neurology* 43:1389–1397.
McCreadie. 1997. The Nithsdale Schizophrenia Survey 16. Breast-feeding and schizophrenia: preliminary results and hypotheses. *Br. J. Psychiatr.* 178:334–337.
McLennan, P., Howe, P., Abeywardena, M., Muggli, R., Raederstorff, D., et al. 1996. The cardiovascular protective role of docosahexaenoic acid. *Eur. J. Pharmacol* 300:83–89.
Mead, J. F. 1982. The essential fatty acids: past, present and future. *Prog. Lipid Res.* 20:1–6.
Meydani, S. N., Lichtenstein, A. H., Cornwall, S., Meydani, M., Goldin, B. R., et al. 1993. Immunologic effects of national cholesterol education panel step-2 diets with and without fish-derived N-3 fatty acid enrichment. *J. Clin. Invest* 92:105–113.
Mori, T. A., Vandongen, R., Beilin, L. J., Burke, V., Morris, J., Ritchie, J. 1994. Effects of varying dietary fat, fish, and fish oils on blood lipids in a randomized controlled trial in men at risk of heart disease. *Am. J. Clin. Nutr.* 59:1060–1068.
Murray, M. T. 1996. *Encyclopedia of Nutritional Supplements.* Prima Publishing, Rockin, CA.
Nelson, G. J., Schmidt, P. C., Bartolini, G., Kelley, D. S., Kyle, D. J. 1997a. The effect of dietary arachidonic acid on platelet function, platelet fatty acid composition, and blood coagulation in humans. *Lipids* 32:421–425.
Nelson, G. J., Schmidt, P. C., Bartolini, G. L., Kelley, D. S., Kyle, D. 1997b. The effect of dietary docosahexaenoic acid on plasma lipoproteins and tissue fatty acid composition in humans. *Lipids* 32:1137–1146.

Neuringer, M., Connor, W. E., Lin, D. S., Barstad, L., Luck, S. 1986. Biochemical and functional effects of prenatal and postnatal omega 3 fatty acid deficiency on retina and brain in rhesus monkeys. *Proc. Natl. Acad. Sci. U.S.A.* 83:4021–4025.

Oeckerman, P. A., Bachrach, I., Glans, S., Rassner, S. 1986. Evening primrose oil as a treatment of the premenstral syndrome. *Rec. Adv. Clin. Nutr.* 2:404–405.

Peet, M., Murphy, B., Shay, J., Horrobin, D. 1998. Depletion of omega-3 fatty acid levels in red blood cell membranes of depressive patients. *Biol Psychiatry* 43:315–319.

Perkins, E. G. 1993. Nomenclature and classification of lipids. In: *Analysis of Fats, Oils and Derivatives*, Perkins, E. G. (Ed.). American Oils Chemists Society, Champaigne, IL, pp. 1–9.

Phylactos, A. C., Harbige, L. S., Crawford, M. A. 1994. Essential fatty acids alter the activity of manganese-superoxide dismutase in rat heart. *Lipids* 29:111–115.

Phylactos, A. C., Leaf, A. A., Costeloe, K., Crawford, M. A. 1995. Erythrocyte cupric/zinc superoxide dismutase exhibits reduced activity in preterm and low-birthweight infants at birth. *Acta. Paediatr.* 84:1421–1425.

Prasad, M. R., Lovell, M. A., Yatin, M., Dhillon, H., Markesbery, W. R. 1998. Regional membrane phospholipid alterations in Alzheimer's disease. *Neurochem. Res.* 23:81–88.

Raederstorff, D., Pantze, M., Bachmann, H., Moser, U. 1996. Anti-inflammatory properties of docosahexaenoic and eicosapentaenoic acids in phorbol-ester-induced mouse ear inflammation. *Int. Arch. Allergy Immunol.* 111:284–290.

Ratledge, C. 1992. Microbial lipids: commercial realities or acedemic curiosities. In: *Industrial Applications of Single Cell Oils*, Kyle, D. J., Ratledge, C. (Eds.). American Oil Chemists' Society, Champaigne, IL, pp. 1–15.

Salem, N., Jr., Niebylski, C. D. 1995. The nervous system has an absolute molecular species requirement for proper function. *Mol. Membr. Biol.* 12:131–134.

Salem, N., Jr., Wegher, B., Mena, P., Uauy, R. 1996. Arachidonic and docosahexaenoic acids are biosynthesized from their 18-carbon precursors in human infants. *Proc. Natl. Acad. Sci. U.S.A.* 93:49–54.

Salem, N., Jr., Kim, H.-Y., Yergey, J. A. 1986. Docosahexaenoic acid: membrane function and metabolism. In: *Health Effects of Polyunsaturated Fatty Acids in Seafoods* (ed., A. P. Simopoulis, R. R. Kifer and R. E. Martin). Academic Press, Orlando, pp. 263–317.

Schaefer, E. J., Robins, S. J., Patton, G. M., Sandberg, M. A., Weigel-DiFranco, C. A., et al. 1995. Red blood cell membrane phosphatidylethanolamine fatty acid content in various forms of retinitis pigmentosa. *J. Lipid. Res.* 36:1427–1433.

Sesler, A. M., Ntambi, J. M. 1998. Polyunsaturated fatty aid regulation of gene expression. *J. Nutr.* 128:923–926.

Simopoulos, A. P. 1991. Omega-3 fatty acids in health and disease and in growth and development. *Am. J. Clin. Nutr.* 54:438–463.

Simopoulos, A. P. 1998. Overview of evolutionary aspects of omega-3 fatty acids in the diet. In: *The Return of Omega-3 Fatty Acids into the Food Supply: I. Land Based Animal Food Products and Their Health Effects*, Simopoulos, A. P. (Ed.). *World Review of Nutrition and Diet*, Vol. 83. Karger, Basel, pp. 1–11.

Simopoulos, A. P., Kifer, R. R., Martin, R. E. 1986. *Health Effects of Polyunsaturated Fatty Acids in Seafoods.* Academic Press, Orlando.

Siscovick, D. S., Raghunathan, T. E., King, I., Weinmann, S., Wicklund, K. G., et al. 1995. Dietary intake and cell membrane levels of long-chain n-3 polyunsaturated fatty acids and the risk of primary cardiac arrest. *JAMA* 274:1363–1367.

Soderberg, M., Edlund, C., Kristensson, K., Dallner, G. 1991. Fatty acid composition of brain phospholipids in aging and in Alzheimer's disease. *Lipids* 26:421–425.

Stevens, L. J., Zentall, S. S., Deck, J. L., Abate, M. L., Watkins, B. A., et al. 1995. Essential fatty acid metabolism in boys with attention-deficit hyperactivity disorder. *Am. J. Clin. Nutr.* 62: 761–768.

Stoll, A. 1998. Essential fatty acids and bipolar disorder. In: *NIH Workshop on Essential Fatty Acids and Psychiatric Disorders.* Washington, D.C.

Stordy, B. J. 1995. Benefit of docosahexaenoic acid supplements to dark adaptation in dyslexics. Lancet 346:385.

Stordy, J. 1998. Essential Fatty Acids and Attention Deficit Disorder. In: *NIH Workshop on Essential Fatty Acids and Psychiatric Disorders.* Washington, D.C.

Suh, M., Wierzbicki, A. A., Lien, E. L., Clandinin, M. T. 2000. Dietary 20:4 n-6 and 22:6 n-3 modulates the profile of long- and very-long-chain fatty acids, rhodopsin content, and kinetics in developing photoreceptor cells. *Pediatr. Res.* 48:524–530.

Suzuki, H., Manabe, S., Wada, O., Crawford, M. A. 1997. Rapid incorporation of docosahexaenoic acid from dietary sources into brain microsomal, synaptosomal and mitochondrial membranes in adult mice. *Int. J. Vitam. Nutr. Res.* 67:272–278.

Uauy, R., Birch, E., Birch, D., Peirano, P. 1992. Visual and brain function measurements in studies of n-3 fatty acid requirements of infants. *J. Pediatr.* 120:S168–S180.

Vaddadi, K. S., Courtney, P., Gilleard, C. J., Manku, M. S., Horrobin, F. 1989. A double-blind trial of essential fatty acid supplementation in patients with tardive dyskinesia. *Psychiatry Res.* 27: 313–324.

Voss, A., Reinhart, M., Sankarappa, S., Sprecher, H. 1991. The metabolism of 7,10,13,16,19-docosapentaenoic acid to 4,7,10,13,16,19-docosahexaenoic acid in rat liver is independent of a 4-desaturase. *J. Biol. Chem.* 266:19995–20000.

Ward, M. V., Clarkson, T. B. 1985. The effect of a menhaden oil-containing diet on hemostatic and lipid parameters of nonhuman primates with atherosclerosis. *Atherosclerosis* 57:325–335.

Waters, K. M., Miller, C. W., Ntambi, J. M. 1997. Localization of polyunsaturated fatty acid response region in steaoyl-CoA desaturase gene 1. *Biochim. Biophys. Acta.* 1349:33–40.

Wei, J. W., Yang, L. M., Sun, S. H., Chiang, C. L. 1987. Phospholipids and fatty acid profile of brain synaptosomal membrane from normotensive and hypertensive rats. *Int. J. Biochem.* 19: 1225–1228.

Werkman, S. H., Carlson, S. E. 1996. A randomized trial of visual attention of preterm infants fed docosahexaenoic acid until nine months. *Lipids* 31:91–97.

Wibert, G. J., Burns, R. A., Diersen-Schade, D. A., Kelly, C. M. 1997. Evaluation of single cell sources of docosahexaenoic acid and arachidonic acid: a 4-week oral safety study in rats. *Food Chem. Toxicol.* 35:967–974.

Willatts, P., Forsyth, J. S., DiModugno, M. K., Varma, S., Colvin, M. 1998. Effect of long chain polyunsaturated fatty acids in infant formul on problem solving at 10 months of age. *Lancet* 352:688–691.

Xiao, Y. F., Gomez, A. M., Morgan, J. P., Lederer, W. J., Leaf, A. 1997. Suppression of voltage-gated L-type Ca^{2+} currents by polyunsaturated fatty acids in adult and neonatal rat ventricular myocytes. *Proc. Natl. Acad. Sci. U.S.A.* 94:4182–4187.

Yazawa, K. 1996. Clinical experience with docosahexaenoic acid in demented patients. International Conference on Highly Unsaturated Fatty Acids in Nutr and Disease. Barcelona, Spain.

Yeh, Y. Y., Gehman, M. F., Yeh, S. M. 1993. Maternal dietary fish oil enriches docosahexaenoate levels in brain subcellular fractions of offspring. *J. Neurosci. Res.* 35:218–226.

Zalata, A. A., Christophe, A. B., Depuydt, C. E., Schoonjans, F., Comhaire, F. H. 1998. The fatty acid composition of phospholipids of spermatozoa from infertile patients. *Mol. Hum. Reprod.* 4:111–118.

Zeijdner, E. E., van Houwelingen, A. C., Kester, A. D., Hornstra, G. 1997. Essential fatty acid status in plasma phospholipids of mother and neonate after multiple pregnancy. *Prostaglandins Leukot. Essent. Fatty Acids* 56:395–401.

11

Fat Substitutes and Replacers

SYMON M. MAHUNGU

Egerton University, Njoro, Kenya

STEVEN L. HANSEN

Cargill Analytical Services, Minnetonka, Minnesota

WILLIAM E. ARTZ

University of Illinois, Urbana, Illinois

I. INTRODUCTION

The macronutrient composition of the diet can influence hunger, satiety, food intake, body weight, and body composition (Rolls, 1995). Fat, rather than carbohydrates, has been the macronutrient most associated with overeating and obesity. Fat is often consumed in excess because it is highly palatable and provides a high level of energy per given volume of food. Thus, low fat foods and fat substitutes can help reduce fat intake.

This chapter focuses on fat substitutes and fat replacers as additives and/or potential additives in foods. The terms *fat replacer* and *fat substitute* have been differentiated in a paper by Miraglio (1995). In this publication, fat replacer is used to denote an ingredient that replaces some or all the functions of fat and may or may not provide nutritional value; whereas a fat substitute replaces all the functions of a fat with essentially no energy contribution. However, these terms will be used interchangeably in this chapter since they both serve the same primary purpose—a reduction in fat calories in the diet.

Fats contribute to the appearance, taste, mouth-feel, lubricity, texture, and flavor of most food products; they provide essential fatty acids and are carriers of fat soluble vitamins (Akoh, 1995; Artz and Hansen, 1996a,b). The amount and type of fats present in foods determine the characteristics of that food and consumer acceptance. An ideal fat replacer should be completely safe and physiologically inert to achieve a substantial fat and caloric reduction while maintaining the desired functional and sensory properties of a conventional high-fat product (Grossklaus, 1996). Historically, dietary fats and oils have been considered a primary source of energy without regard to the health effects of their specific complement of fatty acids and sterols (Glueck et al., 1994) and currently account for about 38% of the total calories in the diet of Western populations, especially the United States (Akoh, 1995). However, dietary fat intakes greater than 11% of the total caloric intake developed after the domestication of mammals and the subsequent selective breeding of genetically fatter animals (Garn, 1997).

Although there are many nutrition recommendations that remain controversial, there is a consensus among health and nutrition professionals that most Americans should lower their intake of dietary fat and alter its composition (Gershoff, 1995). The relationship of dietary fat and cholesterol to coronary heart disease is supported by extensive and consistent clinical, epidemiologic, metabolic, and animal evidence. Studies strongly indicate that the formation of atherosclerotic lesions in coronary arteries is increased in proportion to the levels of total and low density lipoprotein (LDL) cholesterol in blood, which in turn are increased by diets high in total fats (Glueck et al., 1994). Restriction of fatty foods is an effective means of reducing caloric intake and is consistent with public health goals to reduce the risk of chronic diseases (Borzelleca, 1996; Degraaf et al., 1996). Thus, dietary fat is currently the number one nutrition concern of Americans. In response to the rising consumer demand for reduced-fat foods, the food industry has developed a multitude of nonfat, low fat, and reduced fat versions of regular food products. To generate reduced fat or fat free products that have the same organoleptic characteristics of the regular fat version, food manufacturers frequently employ fat substitutes in the formulation of these foods (Miller and Groziak, 1996). These fat substitutes are made from either carbohydrates, protein, or fat, or a combination of these components. Many of the carbohydrate- and protein-based fat substitutes have either received or would probably receive GRAS (generally recognized as safe) status from the Food and Drug Administration (FDA) (Artz and Hansen, 1996b). In January 1996, the U.S. Food and Drug Administration approved olestra (now Olean) for use in savory snacks (Akoh, 1996, Freston et al., 1997; Zorich et al., 1997). This fat substitute is chemically referred to as sucrose polyester or sucrose fatty acid polyester, whose Code of Federal Regulation (CFR) reference number is CFR 172.867 (CFR, 1997). Other fat substitutes with the potential to partially replace some, but not all, calories from fat are either under development or are already in the market.

Fat substitutes could replace a significant proportion of dietary fat and as such become macronutrient substitutes (Borzelleca, 1996). Hence, the safety of these materials must be established via extensive safety testing prior to FDA approval and introduction into the food supply (Artz and Hansen, 1996b). Appropriate methods of safety evaluation must be used. Traditional methods for the safety evaluation of the food additives are inappropriate, since concentrations of the test materials high enough to provide a 100-fold safety factor cannot be used (Borzelleca, 1996).

II. CHEMISTRY

This section addresses the synthesis and/or preparation and analysis of some of the major fat substitutes in use or those under development that have potential as fat substitutes or fat replacers. While it was intended that most of the fat substitutes would be included in this chapter, the absence of a fat substitute from this chapter implies nothing about the product's utility, safety, or potential. Some of the fat substitutes that will be discussed are those that contain fatty acids attached to a molecule other than glycerol, such as olestra, or where attachment has been modified to reduce susceptibility of the compound to lipase of which the esterified propoxylated glycerols are a good example. The other categories of fat substitutes include the modified starches such as tapioca, the modified proteins such as simplesse, and the modified triglycerides, like salatrim.

A. Synthesis

1. Esterified Propoxylated Glycerols

Fatty acid esterified propoxylated glycerols (EPGs) (ARCO Chemical Company, Newtown Square, PA) were developed for use as fat substitutes. Glycerol is propoxylated with propylene oxide to form a polyether polyol, which is then esterified with fatty acids (Anonymous, 1990a; Arciszewski, 1991; Cooper, 1990; Duxbury and Meinhold, 1991; Dziezak, 1989; Gills, 1988; Hassel, 1993; White and Pollard, 1988, 1989a,b,c). The preferred fatty acids are in the C_{14}–C_{18} range. The resulting triacylglycerol is similar to natural fats in structure and functionality. Fatty acid EPG is a low to noncaloric oil, heat-stable, and only very slightly digestible. The in vivo threshold for nondigestibility occurs when the number of added propylene oxide groups equals 4. Feeding experiments with rats (White and Pollard, 1989b) and mice (White and Pollard, 1989c) indicated no toxicity. Preparation of propoxylated glycerides for use as fat substitutes involved transesterifying propoxylated glycerol with esters of C_{10}–C_{24} fatty acids in a solvent-free, nonsaponifying system (ARCO Chemical Technology Limited Partnership, Wilmington, DE) (Cooper, 1990) to avoid reagents unacceptable in food systems.

2. Fatty Acid Partially Esterified Polysaccharide

The ARCO Chemical Technology Limited Partnership (Wilmington, DE) has patented a fatty acid partially esterified polysaccharide (PEP) that is partially esterified with fatty acids (White, 1990). It is nonabsorbable, indigestible, and nontoxic. Suitable oligo/polysaccharide materials include xanthan gum, guar gum, gum arabic, alginates, cellulose hydrolysis products, hydroxypropyl cellulose, starch hydrolysis products ($n < 50$), karaya gum, and pectin. The degree of esterification is controlled by the length of the acyl ester chain and the total number of hydroxyl groups available for esterification. The preferred level of esterification involves one or more hydroxyl groups per saccharide unit with one or more C_8–C_{24} fatty acids. The preferred fatty acid sources are soybean, olive, cottonseed, and corn oils and tallow and lard. Preparation may involve direct esterification or transesterification, with metal (sodium methoxide, potassium hydroxide, titanium isopropoxide, or tetraalkoxide) catalyzed transesterification preferred due to the charring of saccharides that can occur during direct esterification.

3. Carbohydrate Fatty Acid Esters

The carbohydrate-based fat substitutes include polydextrose, altered sugars, starch derivatives, cellulose, and gums. They can also be made from rice, wheat, corn, oats, tapioca, or potato, and can replace from 50 to 100% of the fat in foods (Glueck et al., 1994).

Reports on the synthesis and analysis of carbohydrate fatty acid esters have been published by several groups (Akoh and Swanson, 1989a,b, 1994; Drake et al., 1994d; Rios et al., 1994). Carbodrate fatty acid polyesters with a degree of substitution (DS: number of hydroxyl groups esterified with long chain fatty-acids) of 4 to 14 are lipophilic, nondigestible, nonabsorbable, fatlike molecules with physical and chemical properties of conventional fats and oils and are referred to as low-calorie fat substitutes (Akoh, 1994a,

1995). Swanson's research group at Washington State University has published much on carbohydrate fatty acid esters synthesized from a variety of carbohydrate sources under a variety of catalytic conditions (Akoh, 1994, 1995; Akoh and Swanson, 1990). The synthesis of several carbohydrate–fatty acid derived fat substitutes, including glucose fatty acid esters, sucrose fatty acid esters, raffinose fatty acid esters, and even larger polysaccharide oligomers, such as stachyose and verbascose fatty acid esters, were reported. These researchers reported the synthesis of novel trehalose octaoleate and sorbitol hexaoleate by a one-stage, solvent-free interesterification of methyl oleate with trehalose octaacetate and sorbitol hexaacetate, respectively, in the presence of 1 to 2.5% sodium metal as catalyst (Akoh and Swanson, 1989b). About 97.5% of trehalose octaoleate and 96.0% of sorbitol hexaoleate were obtained at molar ratios of methyl oleate to trehalose octaacetate of 8:1 and methyl oleate to sorbitol hexaacetate of 6:1 at a synthesis time of 2.5 h and a temperature of 105 to 115°C. Chung et al. (1996) prepared sorbitol fatty acid polyoleate using sodium oleate, methyl oleate and sorbitol in the presence of potassium carbonate as a catalyst. The yields ranged from 89–94%.

Esterified alcohols with more than three ester groups are not hydrolyzed by pancreatic lipases. Oleic acid esters of erythritol, pentaerythritol, adonitol, and sorbitol were prepared by transesterification with an excess of methyl oleate to form complete esters (Mattson and Volpenhein, 1972). The esters formed were erythritol tetraoleate, pentaerythritol tetraoleate, adonitol pentaoleate, and sorbitol hexaoleate. These esters were not susceptible to in vivo lipolysis by lipolytic enzymes of rat pancreatic juice, suggesting potential application as low calorie oils (Akoh and Swanson, 1989b; Mattson and Volpenhein, 1972).

The synthesis of stachyose polyoleate has been reported (Akoh and Swanson, 1989a). Stachyose is a nonreducing heterogeneous tetrasaccharide similar to raffinose with 14 replaceable hydroxyl groups. The stachyose tetradecaacetate (mp 102.2°C) was prepared by acylation with acetic anhydride. In the synthesis of stachyose polyoleate, 1.5 to 2.0% sodium metal was used to promote acyl migrations in the ester–ester interchange reaction. A molar ratio of stachyose tetradecaacetate to methyl oleate of 1:14 gave optimum yields (99.2%) of stachyose polyoleate (DS 12.4) at temperatures of 98 to 110°C for 2.5 h and pressure of 0 to 5 mmHg (Akoh and Swanson, 1989a).

Enzymatic methods for the synthesis of carbohydrate fatty acid esters are discussed in detail by Riva (1994). One of the most promising enzymes tested, particularly for fatty acid esterification of the alkylated glycosides, was a lipase from the yeast *Candida antartica*, which had been immobilized on macroporous resin beads. Mutua and Akoh (1993) have reported the synthesis of glucose and alkyl glycoside fatty acid esters in organic solvents using *Candida antartica* as a catalyst. Since only a limited number of successful enzymatic syntheses have been reported, more research is needed.

4. Sucrose Polyester/Olestra

Probably the most extensively studied and publicized of the low to noncaloric fat substitutes are the sucrose fatty acid polyesters (SPEs). Typically, sucrose fatty acid polyesters are prepared from the reaction of sucrose with long chain fatty acid methyl esters (mostly C16, C18, C18:1, C18:2, and C18:3 fatty acids) (Gardner and Sanders, 1990). Depending upon the reaction conditions, anywhere from one (monoester) to eight (octaester) fatty acids can be attached to the sucrose molecule. Olestra is the generic name for the mixture of hexa-, hepta-, and octaesters of sucrose formed with long chain fatty acids. Digestive enzymes do not release the fatty acids, so olestra is noncaloric and not sweet (Gershoff, 1995). It tastes and behaves like fat and can be used in high-heat applications such as baking and frying. Olestra was approved by the U.S. FDA for use in savory snacks in 1996 (Akoh, 1996).

The review by Akoh (1995) has described various methods that are used to prepare sucrose polyesters. Sucrose polyesters have been prepared (80 to 90% yields) by reacting sucrose octaacetate (SOAC) and methyl palmitate in the presence of 2% sodium or potassium at reaction temperatures of 110 to 130°C for a reaction time of 3 to 6 h under reduced pressure of 15 to 25 mmHg. In this process, a clear melt of SOAC and palmitic methyl ester was formed prior to catalyst addition. This solvent-free method was improved by Volpenhein (1985) by heating a mixture of carbonate; fatty acid methyl ester; 2-methoxy ethyl or benzyl ester; an alkali metal fatty acid soap; and potassium carbonate, sodium carbonate, or barium carbonate as a catalyst to form a homogeneous melt. An excess acyl donor, such as fatty acid methyl ester, was then added to the melt to yield carbohydrate fatty acid polyesters at temperatures from 110 to 180°C and pressures of 0.1 to 760 mmHg. The preferred catalyst was potassium carbonate, and the reaction time for step two was between 2 to 8 h. The molar ratio of fatty acid methyl ester to free sucrose was 12:1. The product formed was a mixture of the higher polyesters, with 36 to 85% octaester content based on sucrose.

Yamamoto and Kinami (1986) used a mixture of sucrose oleate (DS 1.5), molten sucrose, and methyl oleate, a basic catalyst (1 to 10%) such as sodium and potassium carbonate or hydroxide, to form a homogeneous melt at 120 to 180°C under less than 10 mmHg pressure. The length of the reaction varied with the reaction conditions but was generally between 1 to 3 h. The methods of Volpehein (1985) and Yamamoto and Kinami (1986) required molecular distillation at 60 to 150°C (a high-energy process) to remove unreacted fatty acid methyl ester.

Akoh and Swanson (1990) reported an optimized synthesis of SPE that produced yields between 96.6 to 99.8% of the purified SPE based on the initial weight of SOAC. This was a one-stage, solvent-free process involving admixing of SOAC, 1 to 2% sodium metal as catalyst, and fatty acid methyl esters of vegetable oils in a three-neck reaction flask prior to application of heat. Formation of a one-phase melt was achieved 20 to 30 min after heat was applied. High yields of SPE were obtained at temperatures as low as 105°C and synthesis times as short as 2 h by reacting the mixture in a vacuum of 0 to 5 mmHg. Volatile methyl acetate formed as a result of interesterification (simple ester–ester interchange reaction) and was trapped in a liquid nitrogen (−196°C) cold trap, thereby driving the synthesis toward product formation. The advantage of this process is that the acetate groups in SOAC are good leaving and protecting groups against sucrose degradation and caramelization during the SPE synthesis. The preferred molar ratio of fatty acid methyl ester to sucrose octaacetate was at least 8:1 to maximize the yield of SPE. Washing with methanol and hexane removed unreacted fatty acid methyl esters and sucrose fatty acid esters of long chain fatty acid with DS of about 1 to 3.

5. Alkyl Glycoside Fatty Acid Esters

Alkyl glycoside fatty acid esters are nonionic, nontoxic, odorless, and biodegradable compounds with emulsification properties. Direct esterification of reducing sugars such as glucose and galactose often results in excessive sugar degradation and charring. Therefore, alkylation is necessary to convert reducing sugars with reactive C-1 anomeric centers to nonreducing, less reactive, anomeric C-1 centers (Akoh and Swanson, 1989a).

Alkyl glycoside fatty acid esters could be used to replace fat (from 5–95%) in such items as frying oils and Italian salad dressings (Curtice-Burns, Inc., Rochester, NY) (Meyer et al., 1989). Alkyl glycosides can be formed by reacting a reducing saccharide with a monohydric alcohol. Soybean, safflower, corn, peanut, and cotton seed oils are preferred since they contain C_{16}–C_{18} fatty acids that do not volatilize at the temperatures used for interesterification. The preferred alkyl glucosides are the reaction products between glucose, galactose, lactose, maltose, and ethanol and propanol. Blending of unsaturates and saturates (>25% C_{12} or larger) produces a heterogeneous alkyl glycoside fatty acid polyester, which does not induce undesirable anal leakage. Anal leakage may be prevented if the alkyl glycoside fatty acid polyesters have a melting point greater than or equal to 37°C.

Albano-Garcia et al. (1980) reported a solventless synthesis of methyl glucoside esters of coconut fatty acids by reacting a 1:1 to 3:1 molar ratio of methyl glucoside to methyl ester in the presence of 5% anhydrous potassium soap and 0.5% sodium metal as catalysts at temperatures of 140 to 148°C and a pressure of 20 mmHg for 3 to 5 h. Other catalysts utilized with success include 0.5% sodium and 8% potassium soap; 0.5% sodium and 5% potassium soap; 0.5% sodium carbonate and 5% potassium soap; and 0.5% sodium methoxide and 5% potassium soap.

Akoh and Swanson (1989a) have synthesized novel alkyl glycoside polyesters, such as methyl glucoside polyesters, methyl galactoside polyesters, and octyl glucoside polyesters by a solvent free interesterification of the alkyl glycoside tetraacetates with fatty acid methyl esters of long chain fatty acids and vegetable oils. In order to achieve high yields, the alkyl glycosides free hydroxyl groups were first protected by acetylation prior to interesterification. At a molar ratio of alkyl glycoside acetate to fatty acid methyl ester of 1:5, yields of 99.5% were achieved with 2% sodium as the catalyst and heating at 98 to 105°C.

A patent was awarded for the incorporation of the alkyl glycoside fatty acid polyester into food products (Curtice-Burns, Inc., Rochester, NY) (Winter et al., 1990). The fat substitute could replace ''visible fats'' in such products as shortening, margarine, butter, salad and cooking oils, mayonnaise, salad dressing, and confectionery coatings, or ''invisible fats'' in such foods as oilseeds, nuts, dairy, and animal products. Visible fats are defined as the fats and oils that have been isolated from animal tissues, oilseeds, or vegetable sources. Invisible fats are fats and oils that are not isolated from animal or vegetable sources and are consumed along with the protein and carbohydrate components of the sources as they naturally occur. Substitution at 10 to 100% is possible, however less than 100% is preferred, with the ideal in the range of 33–75%. The addition of an anti–anal leakage (AAL) agent is not necessary, especially if the diet has invisible fats containing fatty acids with melting points above 37°C.

Glucosides containing from 1–50 alkoxy groups can be used as fat substitutes at substitution ranges of 10–100% in low calorie salad oils, plastic shortenings, cake mixes, icing mixes, mayonnaise, salad dressings, and margarines (Procter & Gamble Co., Cincinnati, OH) (Ennis et al., 1991). The alkoxylated alkyl glucoside (i.e., propoxylated or ethoxylated methyl glucoside) is esterified with 4 to 7 C_2–C_{24} fatty acids. For example, ethoxylated glucoside tetraoleate was prepared by reacting oleyl chloride with Glucam E-20 (ethoxylated methyl glucoside obtained from reacting methyl glucoside with ethylene oxide).

6. Starch-Based Fat Replacers

There are several starch-derived fat replacers available which are essentially maltodextrins. They are produced upon partial enzymatic or acid catalyzed hydrolysis of starch and are fully digestible. The low dextrose equivalent (DE) maltodextrins have fat binding functional properties, unlike the high DE products. Starches and sugars contain 4 kcal/g, but since they are generally used at concentrations of much less than 100%, the actual caloric content is typically 0.5 to 2 kcal/g. They include Lycadex-100 and Lycadex-200 from Roquette Freres; Crestar SF from Euro Centre Food; N-Lite, Instant N-Oil, and N-Oil II from National Starch Co.; Maltrin M040 from Grain Processing Corp.; Paselli SA-2 from Avebe America, Inc.; Tapiocaline from Tripak; Star Dri, Stellar, and Sta-Slim from A. E. Staley; Amalean I from American Maize Products, and Rice-trin from Zumbro. Typically, a smooth viscous solution or a soft gel is formed when hydrated.

Lycadex is an enzymatically hydrolyzed corn starch–based maltodextrin. The typical concentration used is 20% and it is nongelling. Typical applications include sauces and salad dressings.

Maltrin M040 is a maltodextrin (DE = 5) made from corn starch (Frye and Setser, 1993; Tamime et al., 1994), which is hot water soluble. A solution containing 30–50% solids produces a thermoreversible gel with a bland flavor, smooth mouth feel, and texture similar to hydrogenated oils for butterfat. According to the manufacturer, it can be used for dairy products, frozen foods, sauces, salad dressings, confectionery products, and dry mixes.

Paselli SA2 is a maltodextrin derived from potato starch (Frye and Sester, 1993; Anonymous, 1989). If forms a shiny, white, thermoreversible gel with a smooth, fatlike texture and neutral taste that is also suited for acid formulations. The gel is stable with fat- and oil-containing products. A gel is formed at concentrations above 20% and the gel strength increases with an increase in concentration.

N-Oil is a tapioca-based maltodextrin (Frye and Setser, 1993; Anonymous, 1989). An instantized version, N-Oil II, can be used with foods that either require no heat processing or very modest thermal processing, such as high-temperature, short time thermal processing. N-Oil is heat stable, and it can withstand high temperatures and high shear and acidic conditions. Upon cooling, a solution of hot water soluble N-Oil will develop a texture similar to hydrogenated shortening.

Amalean I is a modified high-amylose corn starch used at a relatively low concentration (8%) compared to the maltodextrins. A gel for fat replacement can be prepared upon heating a paste or slurry to 88–90°C for 3 min. An Amalean II product is available for enhancing batter aeration.

Sta-Slim is a modified starch designed to provide a creamy texture for a wide range of food products, but it is particularly well suited for salad dressings. In contrast to other products, it is usually processed as a warm liquid in the formulation. It is prepared as a slurry at a concentration of 3–20% and heated with agitation to 65–71°C to completely solubolize the starch. It is recommended for use in salad dressings, cheese products, and soups.

Stellar is a microparticulated corn starch gel used in pastries, snack foods, frostings and fillings, cheese products, margarines, meat products, salad dressings, and other selected products (Anonymous, 1991; Frye and Setser, 1993). The water is bound sufficiently such that water migration from the cake to the filling is slowed and staling is retarded. The small, intact starch granules duplicate the mouth feel sensation of fat. The particle aggregates are 3–5 μm in diameter, slightly larger than the protein-based fat replacer products, but approximately the same size as the fat crystals they are designed to replace. The microparticulate character is required to maintain the smooth creamy texture. If it is heated sufficiently to completely gelatinize and disperse the starch (105°C), much of the fat substitute or mimetic functionality is lost permanently. Applications include low fat margarines, salad dressings, soups, confectionery products, baked goods, frostings, fillings, and selected dairy and meat products.

Scientists at American Maize were able to use an unmodified, pregelatinized, waxy starch (waxy starches contain nearly 100% amylopectin) at a concentration of 2% to produce a blueberry muffin with only 3% fat, yet with the textural characteristics comparable to a muffin containing 15% fat (Hippleheuser et al., 1995). The waxy starch assists in providing a uniform cell size, the appropriate crumb structure, and a strong tendency to retain moisture, which extends the shelf life and produces a desirable moist mouth-feel. Since the addition of starch allows an increased moisture content, the potential problem of rapid mold storage should be addressed (0.1% potassium sorbate).

Polydextrose is a low-calorie polymer that can be used to replace some of the fat in a food product. It is formed by the random polymerization of glucose, sorbitol, and citric acid. Sorbitol provides an upper molecular weight limit and reduces the formation of water insoluble material. Most of the linkages are glycosidic and the 1–6 bond in particular, predominates. Litesse, which is a polydextrose-type product, and Veri-Lo, which consists of fat-coated carbohydrate based gel particles, are both from Pfizer.

The fiber-based products include gums, celluloses, hemicelluloses, pectins, β-glucans, and lignins. These are isolated from a wide variety of sources, including cereals, fruits, legumes, nuts, and vegetables. Gums are typically not used as fat substitutes directly, rather they are used at low concentrations (0.1–0.5%) to form gels that increase product viscosity. Agar, alginate, gum arabic, carrageenan, konjac, guar gum, high and low methoxy pectin, xanthan gum, and cellulose derivatives can all potentially be used.

Fiber-based products available include Nutrio-P-fiber from the Danish Sugar Factory; Nutricol derived from konjac flour; P-150 C and P-285 F derived from pea fiber from Grindsted Products; Avicel, a microcrystalline cellulose, and carrageenan are both from FMC. Slendid is a pectin that form gels with fatlike meltability from Hercules. Quaker oatrim is a product of Rhone-Poulenc Food Ingredients derived from oats.

Cellulose derivatives used include α-cellulose, carboxymethyl cellulose, hydroxypropyl cellulose, microcrystalline cellulose, and methyl cellulose (Frye and Setser, 1993). The gels produced from the cellulose derivatives have several desirable fatlike functional properties, which include creaminess, fatlike mouth-feel, stability, texture modification, increased viscosity, and the glossy appearance of high-fat emulsions. FMC, for example, has cellulose-based gels (Penichter and McGinley, 1991) as well as gums which increase emulsion viscosity and, hence, help product stability. This allows one to reduce the oil content because the gel mimics some of the rheological properties associated with high-fat emulsions.

Oatrim is a cold water dispersible amylodextrin rich in β-glucans (Anonymous, 1990b) produced upon partial enzymatic hydrolysis of the starch. It is derived from whole oat or debranned whole oat flour. An oatrim-based gel (25% solids) can function as a fat replacer. In addition, the β-glucan components in the oatrim have a demonstrated hypocholesterolemic effect (Inglett and Grisamore, 1991).

7. Protein-Based Fat Substitutes

The protein-based fat substitute that has received the most attention has been Simplesse, which consists of microparticulated milk and/or egg white proteins, sugar, pectin, and citric acid (Gershoff, 1995). The protein is particulated during a combined pasteurization and homogenization process that produces microparticles of uniform size and spherical shape approximately 1 μm in diameter (Singer and Moser, 1993). The nutritive quality of the protein is essentially unaltered during the preparative processing. Most of the protein-based fat substitutes have a similar basis, that is, the protein is particulated into spherical particles with some combination of heat and shear, with or without added components such as gums, stabilizers, and emulsifiers. The simulated fat produced from an aqueous suspension of microparticulated protein particles of the appropriate size and characteristics have shear thinning characteristics and a smoothness and creaminess similar to fat. Particles with significantly smaller or larger dimensions than 0.5–2 μm do not seem to have the desired sensory characteristics, unless the particles are very soft, hydrated, and compressible, which can extend the size range considerably. While the protein in the fat substitutes contain 4 kcal/g, an aqueous microparticulate protein suspension contains fewer than 4 kcal/g, generally 1–2 kcal. The protein-based substitutes can be used to formulate a variety of products, including cheesecake, puddings, sauces, pie fillings, sour cream, ice cream, cream cheese, mayonnaise, dips, and spreads. Simplesse has been affirmed as GRAS by the FDA for use in several products. The main limitation, which is the limitation for most of the currently approved fat substitutes, is that it cannot be used for heated product applications, particularly frying.

There is more than one form of Simplesse available (Frye and Setser, 1993). One of the newest versions is Simplesse 100, which is a thixotrophic fluid derived from whey protein concentrate. It can be used in a wide variety of dairy and bakery products. It is also available as a readily hydratable powder. Simplesse 300 is a mixture of egg white and milk proteins that can be used in food products that require heating. Of course, it cannot be used as a frying medium.

Once investigators realized that microparticulation could be used to produce a homogeneous suspension of small protein particles that had fatlike properties, other protein-based fat replacer microparticulation processes and systems were examined. Research efforts in this area have been reviewed by Cheftel and Dumay (1993). Membrane processing (ultrafiltration, followed by diafiltration) has been used to concentrate casein micelles 4- to 9-fold to produce a product than can be used to partially or completely replace fat in ice cream, chocolate mousse, dairy product–based spreads, and sauces (Habib and Podolski, 1989). The micelles are reported to function as microparticles with an approximate diameter of 0.1 to 0.4 μm.

Protein precipitation under carefully controlled conditions can produce a microparticulated product that can also be used as a fat substitute (Mai et al., 1990). A dilute solution of water soluble protein (1–5%) is precipitated with heat and/or a change in pH to the isoelectric point of the protein. Starches, gums, emulsifying agents, etc., can be used to enhance product characteristics and prevent extensive aggregation.

A solution of alcohol soluble (70–80% aqueous ethanol) proteins (prolamines from corn, wheat, rice, etc.) can be precipitated by dilution with water to produce a microparticulated spherical protein precipitate (Stark and Gross, 1990). Gums should be used to prevent extensive aggregation. To produce a concentrated protein suspension, ultrafiltration followed by diafiltration can be used. Freeze-drying can be used to produce a powdered precipitate.

One of the first thermomechanical coagulation processes for the microparticulation of protein was described by Singer et al. (1988). A whey protein concentrate is first dispersed in water, then acidified to approximately pH 4, and finally heated (90–120°C) under high shear conditions (500,000 s^{-1}). The resultant dispersion contains small spherical particles (0.1–2.5 μm) and has a creamy and smooth texture. Protein sources other than whey protein can also be used (Cheftel and Dumay, 1993).

In addition to the dual heating/high shear particulation processes used to make Simplesse-type products, extrusion can be used to prepare a microparticulated protein-based fat substitute (Cheftel and Dumay, 1993). The extrusion conditions are very mild, compared to typical extrusion conditions. For example, a whey protein concentrate (20%) within a pH range of 3.5–3.9 (extruder conditions' barrel temperature of 85–100°C and a screw speed of 100–200 rpm) can be extruded to produce a semisolid spread with a smooth and creamy texture. No die is used, so there is no pressure buildup and subsequent expansion. The particles are rather large (12 μm) on the average. If the extrusion is done at a higher pH (4.5–6.8) the protein solubility is reduced, and the percentage of large particles (>20 μm) is much greater and the resultant texture is coarse and grainy. The nitrogen solubility index of the protein remains relatively high: 43–47% for the acid product, and 69–70% for the neutral product. The addition of 0.5% xanthan gum improved the creamy texture and reduced the particle size.

8. Reduced Calorie Fat–Based Fat Replacers

The objective for these products is similar to that for the protein- and carbohydrate-based fat substitutes, a substantial reduction in calories rather than a complete elimination of fat in the product. Examples for fat-based replacers include caprenin, captrin, and salatrim.

Caprenin is a reduced calorie triglyceride (Procter & Gamble) formed by the esterification of three naturally occurring fatty acids: caprylic, capric, and behenic. Since the behenic acid is only partially absorbed, the caprenin contains 5, rather than the normal 9 kcal/g. Caprenin has functional properties similar to cocoa butter and is intended to replace some of the cocoa butter in selected confectionery products. It is digested, absorbed, and metabolized by the same pathways as other triacylglycerols. Captrin, from Stepan Food Ingredients (Anonymous, 1994), is a randomized triglyceride made from linear saturated fatty acids primarily C8 to C10 in length. Some of the proposed uses include baked goods, confections, dairy product analogs, snack foods, and soft candy.

Another fabricated triacylglycerol, similar to caprenin, is Salatrim, which is a triacylglycerol comprised of a mixture of long chain (primarily stearic acid) and short chain (acetic, propionic, and butyric) fatty acids randomly esterified to glycerol (Smith et al., 1994). It contains approximately 5 kcal/g, rather than the 9 kcal/g contained in regular fats and oils. It is a product of the Nabisco Foods Group. A symposium on Salatrim was published in the February 1994 issue of the *Journal of Agriculture and Food Chemistry*. The research reported included structural characterization(s) of the oil, an analysis of the oil in food products, and an extensive series of papers on the metabolism and toxicology of the oil in various animal and human model systems. Salatrim has the same utility that caprenin does as a fat replacer in reduced fat systems and could be used as a cocoa butter substitute in confectionery products and in baked products and filled dairy products. Caprenin and Salatrim are of little use for deep fat frying applications. For example, the smaller molecular weight fatty acids may cause undesirable flavor effects upon hydrolysis.

Salatrim has been prepared by sodium methoxide catalyzed interesterification of saturated long chain fatty acid (LCFA) triglycerides and short chain fatty acid (SCFA) triglycerides (Klemann et al., 1994). The SCFA (triacetin, tripropionic, tributyrin) were mixed with LCFA (hydrogenated canola oil, cotton seed oil, and soybean oil) and the mixture combined with a catalytic amount of sodium methoxide and maintained at a temperature of 100–150°C for 5–60 min. The reaction mixture was cooled, 5% (w/w) water added, the aqueous phase removed, and the organic phase filtered through Tonsil Optimum

105 bleaching clay. The filtrate was subjected to vacuum steam deodorization for sufficient time to remove all of the volatile SCFA triglycerides.

B. Analysis

This section will describe the various chromatographic methods that have been used to purify, characterize and quantitate the fat substitutes and their decomposition products. However, most of the discussion will be on carbohydrate fatty acid esters, especially olestra, esterified propoxylated glycerols, and Salatrim.

1. Sucrose Polyester/Olestra

Various chromatographic techniques for the isolation of highly esterified sucrose polyester oils have been reported by Rios et al. (1994). High-performance size-exclusion chromatography was used to check the purity of the samples, particularly to show that SPEs were free from unreacted fatty acid methyl esters. Thin layer chromatography (TLC), Iatroscan TLC flame ionization detection (FID), and high-performance liquid chromatography (HPLC) on reversed phase were applied for the separation of octa-, hepta-, and hexaesters of sucrose. Pure fractions from the total mixture, obtained by column chromatography, were identified by infrared and nuclear magnetic resonance spectroscopy. When necessary, specific reactions were applied; particularly silylation and lead acetate dichlorofluorescein (in toluene) spray were used to ascertain the degree of esterification of sucrose. The octa-, hepta-, and hexaesters were quantitated by silica column chromatography, TLC/FID, and HPLC on reversed phase. Tallmadge and Lin (1993) developed a liquid chromatographic (LC) method for determining the amount of olestra in lipid samples. Samples were analyzed by reversed phase LC using an evaporative light scattering detector. An octadecylsilane Zorbax column that separates olestra from other lipophilic components was used. Three types of olestra standards (soybean oil–olestra, unheated cottonseed oil–olestra, and heated cottonseed oil–olestra) each were analyzed in soybean oil.

During heating and during deep fat frying, numerous oxidative and thermally induced changes can occur to triacylglycerols (TAGs). The TAG decomposition rate, as well as nonvolatile and volatile decomposition products that occur, are dependent upon the composition of the fatty acids that are present. If the fat-based fat substitutes contain the same fatty acids as found in native fats and oils, they are likely to have similar stability and produce most, if not all, of the same compounds that occur in heated, native TAGs. Some new compounds may be produced as well. These new compounds may or may not affect product flavor, oil stability, and product safety, depending upon the compounds produced and the amount formed.

Reports of heating studies have been published on the sucrose fatty acid polyester or olestra. The first report on olestra, referred to prior to that time as sucrose polyesters, was in 1990 (Gardner and Sanders, 1990). Two samples were heated and compared, the first was a sample of heated olestra, while the second was a heated mixture of olestra and partially hydrogenated soybean oil. The 100% olestra sample was heated at 190°C for six 12-h days in a 15-lb capacity fryer, while the mixture was heated at 185°C for seven 12-h days in a 15-lb fryer as well. Raw, french-style cut potatoes were heated in the mixture through the seven days of heating (49 batches/day at 380 g/batch). As expected, the relative percentage of polyunsaturated fatty acids decreased after heating, while the relative percentage of saturated fatty acids increased after heating.

The dimers in the heated olestra were fractionated with preparative thin layer chromatography, while the dimers contained in the heated mixture of olestra and soybean oil were fractionated with preparative size exclusion chromatography (SEC). Fractions containing suspected dimers were transesterified. The methyl ester fatty acid dimers were separated with capillary gas chromatography and tentatively identified based on their retention times relative to methylated dimer fatty acid standards (Emery Chemical Co., Henkel-Emery Group, Cincinnati, OH) (Gardner and Sanders, 1990). In addition, mass spectral analysis, after gas chromatographic separation, was completed on the methyl ester fatty acid dimers. Polymer linkages identified were very similar to those occurring in thermal oxidized vegetable oils containing the same fatty acids.

Nuclear magnetic resonance (NMR) and infrared (IR) spectra were obtained on the intact olestra monomer and dimer isolates. Plasma desorption mass spectrometry (PDMS) of intact olestra dimers (Gardner and Sanders, 1990) as well as the olestra monomer (Sanders et al., 1992) were completed. During heating there was an increase in the high molecular weight components, as indicated by HPSEC analysis. The PDMS characterization of the high molecular weight olestra fraction isolated with HPSEC indicated that the fraction was a dimer of olestra. In addition, PDMS analysis indicated the presence of an olestra–triglyceride component in the heated olestra–triglyceride mixture. The IR, carbon-13 NMR, and proton NMR spectra of the olestra monomer were virtually identical to those of the olestra dimer, indicating that changes in the sucrose backbone structure did not occur and that the high molecular weight components are olestra dimers. Analysis by GC/MS indicated the presence of dehydro dimers of methyl linoleate in the methylated dimer fatty acid fraction isolated from dimerized olestra and the HPSEC isolated olestra–triglyceride fraction, which indicates the linkages are the same between the olestra dimers and the olestra–triglyceride component. In total, the results strongly suggest that the changes were occurring to the polyunsaturated fatty acids, rather than the sucrose backbone.

In 1992 two articles (Gardner et al., 1992; Henry et al., 1992) were published on the analysis of heated olestra. The first report (Gardner et al., 1992) described in detail an analytical scheme developed for detailed qualitative comparisons of heated fats and oils. The oils (soybean oil, and olestra) were transesterified and the fatty acid methyl esters (FAMEs) were isolated. The FAMEs were separated on the basis of their polarity by adsorption chromatography and solid-phase extraction. Mass spectrometry and infrared spectroscopy were used to identify specific structural components in the compounds separated with capillary GC. The FAMEs (and other components soluble in the hexane phase) were first fractionated with silica gel column chromatography into four fractions: I and II, unaltered FAMEs; III, altered FAMEs; and IV, polar materials. Fraction III, the altered FAMEs, was further fractionated with solid-phase extraction (C18) into III-A, oxidized FAMEs, and III-B, FAME dimers. Gas chromatographic separation of fraction III-A resulted in the detection of FAMEs of various chain lengths with aldehydic, hydroxy, epoxy, and keto functional groups. Gas chromatographic separation of fraction III-B resulted in the detection of three plant sterols and some dimer FAMEs. High-performance liquid chromatographic separation of fraction IV produced components consisting of oxidized di- and triglycerides, monoglycerides, and two very polar FAMEs, one of which was tentatively identified as 9-hydroxyperoxy-10,12-octadecadienoate.

The second report (Henry et al., 1992), referred to as Part 2 by the authors, used the fractionation methods presented in the first paper (Gardner et al., 1992) to provide a detailed qualitative comparison of the transesterified fatty acids from the heated soy oil–based olestra and heated soybean oil. Olestra and soybean oil were used separately to fry potatoes. After frying, the oils were transesterified and fractionated into four fractions based on polarity with the fractionation scheme presented in the paper by Gardner et al. (1992). Analysis indicated that the fatty acid components of both oils undergo similar chemical reactions and changes upon frying, and that the altered FAMEs found in the oils were similar to those found in heated fats and oils by other investigators.

The oxidative stability index (OSI) is commonly used in the food industry to measure oil quality, particularly concerning the expected life of an oil used for deep fat frying. Akoh (1994b) compared eight different oils, including some fat substitutes, with an Omnion OSI instrument. The four vegetable oils examined were crude soybean oil; crude peanut oil; refined, bleached, and deodorized (RBD) soybean oil; and RBD peanut oil. The four oil substitutes analyzed included a soybean oil–derived sucrose polyester, a high oleate/stearate sucrose polyester, a butter fat–derived sucrose polyester, and a methyl glucoside polyester of soybean oil. As determined by OSI, the crude oils were the most stable, second were the RBD oils, and the least stable were the oil substitutes.

Akoh (1994a,1995) has given a detailed account of various methods used for purification and analysis of various carbohydrate-based fat substitutes. These include isolation of methyl glucoside fatty acid esters, purification of raffinose polyester (RPE), analysis of carbohydrates fatty acid polyesters using ^{13}C-NMR and use of infrared spectrometry for structure elucidation of carbohydrate polyesters.

The isolation of methyl glucoside proceeds by dissolving the crude reaction product in a mixture of ethyl acetate and water containing 0.67 mol phosphoric acid per mole of potassium soap as a catalyst. The organic phase is washed with 5% sodium chloride in water, dried over anhydrous sodium sulfate, and filtered.

Raffinose polyester can be synthesized by a solvent-free one-stage process with raffinose undecaacetate and fatty acid methyl esters as substrates. Sodium was added as a catalyst, followed by purification (Akoh and Swanson, 1987). The RPE reaction mixture is neutralized with 1 to 3 mL acetic acid, dissolved in hexane, stirred, and bleached with activated charcoal. The RPE is filtered, washed with methanol, and redissolved in hexane. The methanol removes unreacted FAME. The methanol and hexane are removed by rotary evaporator.

Analysis of carbohydrate fatty acid polyesters by ^{13}C-NMR has been used to determine degree of substitution (DS). Akoh and Swanson (1989) reported quantitative ^{13}C-NMR spectroscopic data calculations on the DS of the saccharide acetate starting material and the saccharide fatty acid polyester products by using chromium acetyl acetonate (NMR relaxation reagent for quantitative NMR analysis). The DS was determined by integrating the anomeric carbon peaks, acetate peaks, bound long chain fatty acid terminal methyl group peaks, and the average intensities of acetate and terminal methyl group peaks compared to the average peak area of the anomeric carbon used as a base.

Infrared spectroscopy has been used for structure elucidation. Depending on the starting material or product, the characteristic functional group absorption bands in the IR spectrum include 3420 to 3500 cm^{-1} (OH), 1740 to 1750 cm^{-1} (ester carbonyl), 1460 to 1470 cm^{-1} (C–H stretch in CH_3 and/or CH_2), and 900 to 920 cm^{-1} (ring vibration). Some spectra may contain low intensity broadbands at 3840 cm^{-1} (weak overtone bands of carbonyl stretch).

2. Esterified Propoxylated Glycerols

Capillary supercritical fluid chromatography has been used to separate and quantitate oligomers of propoxylated glycerols and fatty acid–esterified propoxylated glycerols (fatty acid EPGs, which are heat stable fat-based fat substitutes) (Lu et al., 1993). The propoxylated glycerols analyzed included polyol 382, polyol 550, and polyol 1000; they contained average mole ratios (propylene oxide to glycerol) of 5, 8, and 14, respectively. The oleic acid EPGs included EPG-05, EPG-08, and EPG-14 trioleate; they were prepared from polyol 382, 550, and 1000, respectively. Tridecanoin was used as the internal standard and methylene chloride as the sample solvent. Quantitative analysis of the oligomer distribution for the EPG samples was based on the peak area percent in conjunction with the internal standard. Separations were found to depend primarily upon the molecular weight or the number of the propylene oxide units in each oligomer.

Nuclear magnetic resonance and capillary gas chromatography interfaced with a mass selective detector have been used for the analysis of the nonvolatile components produced during the heating of oleic acid esterified propoxylated glycerol (EPG-08), a fat substitute model compound (Hansen et al., 1997; Hansen, 1995). The heated fat substitute oil sample was fractionated by supercritical fluid extraction and the purity and molecular weights of the fractions estimated by high-performance size exclusion chromatography in combination with light scattering detection. A formula for estimating the molecular weight (MW) based on the retention volume (V_r) was determined using polypropylene glycol triol (PPGT) with MWs of 6000, 4100, and 3000, in addition to triolein, diolein, and monoolein:

$$\log \text{MW} = 6.99 - 0.0991 V_r$$

The response (R_f) of the evaporative light scattering detector to MW was nonlinear, and hence was determined by plotting the area/concentration ratio versus the MW of the standards:

$$R_f = (1.23 \times 10^6) + 30.2\ (\text{MW}) - 0.0220\ (\text{MW})^2$$

The nuclear magnetic resonance spectra recorded for the fractionated fat substitute samples were ^{13}C-NMR, ^{1}H-NMR, distortionless enhancement via polarization transfer (DEPT), and ^{1}H–^{13}C heteronuclear correlation spectroscopy (HETCOR). The samples from NMR analysis were transesterified using a boron trifluoride–methanol mixture for capillary gas chromatography–mass spectrometry analysis (GC-MS). The GC-MS analysis indicated that the fatty acid portion of oleic acid esterified propoxylated glycerol is involved in dimer formation. The ^{13}C and DEPT NMR analysis indicated that the oxypropylene backbone and glycerol backbone of the EPG-08 oleate was not altered significantly, as was the fatty acid component of the molecule. Three groups of peaks specific to the fractionated dimer samples were seen at approximately 107, 67–68, and 23 ppm. The peak at 107 ppm indicates a double bond that may have migrated toward the carbonyl portion of the fatty acid chain, as indicated by the large chemical shift between the olefinics at 129 ppm and the peak at 107 ppm. The peaks at 67–68 ppm were methylene carbons and may indicate that the chemical environment for the α glycerol carbon changed during dimerization. The peak at 23 ppm suggests that a quaternary carbon between C_2 and C_{17} in the fatty acids is involved in the dimerization process. These results suggest that the dimerization occurs in the fatty acid portion of the backbone rather than the oxypropylene portion of the backbone.

The volatile decomposition products formed on heating oleic fatty acid esterified propoxylated glycerol (EPG-08) have been reported (Mahungu, 1994). The volatiles were collected, separated, identified, and quantitated using static headspace–capillary gas chromatography/infrared spectroscopy–mass spectrometry. The major volatiles identified were those associated with oxidation of oleic acid.

3. Salatrim

The analysis of salatrim, which is a ''family'' of triacylglycerols produced by interesterification of highly hydrogenated vegetable oils with triacylglycerols of acetic and/or propionic and/or butyric acids, has been reported (Smith et al., 1994). Some of the techniques reported include proton and ^{13}C-NMR, supercritical fluid chromatography (SFC), high-performance liquid chromatography (HPLC), high-temperature capillary gas chromatography (HTCGC), supercritical fluid extraction (SFE) combined with liquid chromatography–mass spectrometry (PBLC-MS), and molecular modeling.

Henderson et al. (1994) have reported the quantitation and structure elucidation of positional isomers in salatrim using proton and carbon one- and two-dimensional NMR. An interesterified triacylglycerol mixture prepared from an 11:1:1 mole ratio of triacylglycerol/tripropionylglycerol/hydrogenated canola oil was quantitatively characterized using ^{1}H and ^{13}C-NMR experiments. Resonance assignments were facilitated by the combination of ^{1}H–^{1}H and ^{1}H–^{13}C two-dimensional NMR experiments, and species present at 1 mol% or greater were identified. Because of the multiplicity of triacylglycerol species present in the mixture, the experiments were performed in mixtures of perdeuteriobenzene and deuteriochloroform to resolve overlapping resonances. Through the use of one- and two-dimensional ^{1}H and ^{13}C-NMR, the fatty acid resonances for salatrim were identified and assigned to either 2- or 1,3-glycerol positions. The relative quantities of the various triacylglycerol isomers were estimated from the carbonyl and glycerol regions of the ^{13}C spectrum in deuteriochloroform.

Huang et al. (1994a) characterized salatrim by a combination of reversed phase high-performance liquid chromatography and high-temperature capillary gas chromatography in sequence. The salatrim was separated into individual components. An on-line coupling of HTCGC with positive chemical ionization mass spectrometry revealed the structure of the individual components. To obtain detailed quantitative information, a combination of a 5-m OV-1 capillary column and a 25-m methyl 65% phenyl silicone aluminum-clad column was used to determine the quantities of acylglycerols. A total of 55 components were characterized (52 triacylglycerols and 3 diacylglycerols). Huang et al. (1994b) were able to quantitate salatrim in foods by use of supercritical fluid extraction combined with liquid chromatography–mass spectrometry.

The methods of molecular mechanics and semiempirical quantum mechanics have been used to investigate the steric and electrostatic features of salatrim (Yan et al., 1994). Computational studies of the molecules in the gas phase indicated that the hydrophobic interactions between the fatty acid chains were dominant in determining the molecular conformation. Analysis showed that trilaurin and salatrim have similar molecular electrostatic potentials.

III. APPLICATIONS IN FOODS

There are numerous factors that must be considered when selecting a fat substitute, in addition to the obvious and critical sensory quality questions. Is any thermal processing applied to the product? How severe is the thermal processing (pasteurization versus sterilization)? How pH sensitive is the fat substitute? How long will the product be stored, i.e., are there undesirable textural or flavor changes that occur during long-term storage or during some excessively turbulent shipping? Will it be refrigerated? Must it be refrigerated? What are the home preparation steps involved? Is the product microbiologically stable? Are there ''opportunities'' for abuse in the home, i.e., if opened and left on the counter overnight, is food poisoning a possibility?

A comparison of yogurts produced with one of seven selected fat substitutes was made with a control yogurt sample containing milk fat. The overall acceptability and organoleptic quality was comparable to the control for five of the seven fat substitutes; Litesse, N-Oil II, Lycadex-100, Lycadex-200, and Paselli SA2. Interestingly, the overall acceptability, including the flavor and aroma, generally improved after 20 days of storage (Barrantes et al., 1994a,b; Tamime et al., 1996), in contrast to the control. Yogurt made with Simplesse was very similar to the control yogurt made with anhydrous milk fat (Barrantes et al., 1994c), except that the Simplesse-containing product had less sour flavor intensity and more serum separation than the control.

Ice milk samples made with fat mimetics (Simplesse or N-Lite D) and 2.1% milk fat were compared to control ice milk samples containing 4.8% milk fat (Schmidt et al., 1993). Overall, the protein- or Simplesse-containing product was more similar to the control sample than the sample containing the carbohydrate-based fat mimetic. The Simplesse-based ice milk samples had rheological properties that were generally the same as the control. However, the maximum overrun was over 40% greater for the Simplesse containing sample than the control. The maximum overrun for the carbohydrate-based fat mimetic was approximately 60% that of the control.

Milk fat content can be substantially reduced in frozen dairy dessert products with the addition of gums or maltodextrin gels (Frye and Setser, 1993; White, 1993). Low DE maltodextrins, such as Paselli SA2, can be used with frozen dairy desserts, if the maltodextrins are thermostable, cold water soluble, and able to form stable mixtures with other components. Other thermostable, cold water soluble, carbohydrate-based fat substitutes include selected gums, for example, carrageenan, guar gum, and carboxymethylcellulose.

Mono- and diglyceride emulsifiers can be used to replace as much as 50% of the fat in baked goods since they can be used as a 50:50 emulsifier/water mixture (Frye and Setser, 1993; Vetter, 1993). Additional emulsifiers used similarly include sodium stearoyl lactylate, sorbitan monostearate, and polysorbate 60 at high hydration levels. A hydrated blend of emulsifiers developed specifically for fat replacement include stearyl monoglyceridyl citrate, glycol monostearate, and lactylated monoglycerides. Some of the products that can be used to make acceptable frozen dessert products include oatrim, Maltrin 40, and N-Oil. Simplesse, protein hydrolysates, Sta-Slim 143, and sucrose fatty acid esters have been used as fat substitutes in yogurt, sour cream, cream cheese, cheese spread, and frozen dairy desserts. Additional applications are discussed in the chapter by Frye and Setser (1993).

Olestra can be used as a partial fat substitute in shortening, margarines, and frying oils (Hollenback and Howard, 1983; Robbins and Rodriguez, 1984; Roberts, 1984). Cheddar cheese has been prepared with milk fat sucrose polyesters (Crites et al., 1997). No significant differences in moisture, pH, or whey-tritratable acidity were observed between the control cheese and cheeses containing milk fat sucrose polyester. Cheese containing milk fat sucrose polyester contained fat globules that were smaller and more uniform in size compared to a control cheese with no added sucrose polyester.

Simplesse and selected maltodextrins can be used in soft margarine applications for reduced calorie spreads. The product has the mouth-feel, flavor, and spreading properties of a soft margarine (Frye and Setser, 1993). Simplesse has also been used in low fat milks

(Phillips and Barbano, 1997). The addition of 3% Simplesse to skim milk gave milk an appearance that was similar to that of 0.5% fat milk, and the mouth-feel was improved so that it was equivalent to that of milk containing 1% fat. Suspension of titanium dioxide in the skim milk made the milk whiter, which resulted in improved sensory scores for appearance, creamy aroma, and texture.

Gums are used extensively in salad dressings to provide the viscosity associated with high fat products (Quesada and Clark, 1993). Low pH stability by each gum is also required. Maltodextrins can be used, providing better product characteristics than gums (Frye and Setser, 1993). An aqueous 25% maltodextrin solution can replace 30–50% of the fat in salad dressings. Simplesse, as well as the fat-based substitutes, can be used in salad dressings and mayonnaise.

Meat applications for fat substitutes include breakfast sausages, hamburger, hot dogs, gravies, and soups. Some of the products used include Paselli SA2, N-Oil, carrageenan, hydrolyzed vegetable protein, cellulose gum, alginate, and maltodextrins (Frye and Setser, 1993; Keeton, 1994). Carrageenan was used to produce a ground beef patty similar in product characteristics to the high fat product (Egbert et al., 1991). Spices and a mixture of hydrolyzed vegetable protein and salt (1:2) were added (0.375%) to enhance beef flavor intensity and 0.5% carrageenan and 10% water were added to enhance other product organoleptic attributes. That formulation was adapted by McDonalds as their "McLean Deluxe." Others have reported that hydrolyzed soy protein (McMindes, 1991) and oat bran (Pszczola, 1991) can both be effective as the basis for a fat reduction system for ground meat.

Low fat shortbread cookies have been prepared using carbohydrate-based fat substitutes (Sanchez et al., 1995). A combination of carbohydrate-based fat substitutes (Litesse, N-Flate, Rice*Trin, Stellar, or Trim-choice) and emulsifiers (diacetyl-tartaric esters of momnoglycerides, glycerol monostearate, sodium stearoyl-2-lactylate) were used. The principal effects of fat substitutes on shortbread cookie attributes were higher moisture content, greater toughness, and lower specific volume.

One rather creative use of fat substitutes was the suggestion that sucrose fatty acid esters could be used to inhibit lipoxygenase and thereby improve food quality (Nishiyama et al., 1993). There was an increase in the binding strength of the sucrose fatty acid monoester and soybean lipoxygenase-1 (L-1) as the fatty acid carbon chain length was increased from 8 to 12. Thermodynamic analysis of the binding constants indicated that the binding was hydrophobic in character. Sucrose fatty acid esters can also suppress lipase activity and have an antibacterial effect in some cases.

IV. REGULATORY STATUS

The potential appearance of a large number of fat substitutes in the American food supply, varying in composition from those manufactured from food ingredients generally regarded as safe by the Food and Drug Administration to wholly new low- to noncaloric substances, has generally resulted in a requirement that they be evaluated for safety and regulatory acceptability (Gershoff, 1995). The 1958 Food Additives Amendment to the Federal Food, Drug and Cosmetic Act established a premarket approval process for food additives. Guidelines for the safety testing of new food ingredients, referred to as the Redbook, have been published by the FDA (Food and Drug Administration, 1982). Fat substitutes, unlike most food additives that are consumed in very small quantities each day, could replace a substantial proportion of the fat in the diet and may be consumed by some people in several gram quantities each day. Because of this, special considerations must be made in the toxicological and nutritional testing of these substitutes. However, there is resistance to change within the regulatory and governing agencies in spite of the prevailing attitude of the general public toward a healthier lifestyle and diet. Thus, the move toward lower fat foods and the development of fat replacements or substitutes has been a slow process, mostly due to the regulation of fat substitutes.

There are three ways for a manufacturer to obtain regulatory approval for the use of fat substitutes (Glueck et al., 1994). These include: (1) creating a new molecule, (2) applying a new process to make old molecules serve a new purpose, or (3) expanding upon the use of old processes applied to new molecules. If a new molecule is developed, it must be cleared as a food additive (olestra, for example). Millions of dollars and years of delay are involved. If a new process is applied to an old molecule, a generally regarded as safe affirmation petition would probably be required by the FDA. This takes almost as much time and money as the clearance of a food additive. The affirmation process also publicizes the technology and marketing strategies of a company. This makes the protection of confidential information very difficult. Both the food additive clearance and affirmation petition processes are very long. It is possible that by the time the submitted technology is approved it may be obsolete. Each new, improved technology would have to go through the same procedure. The third option, expanding on the use of old processes applied to new molecules, is the most used. The current growth of reduced-fat foods already in the market is proof of this. Existing technologies and uses of familiar molecules are refined and expanded upon. By going this route, companies can avoid the lengthy preclearance requirement of the federal government (Thompson, 1992).

Many of the carbohydrate- and protein-based fat substitutes have either received or would probably receive GRAS status from the Food and Drug Administration. Olestra (Olean), which is a sucrose fatty acid ester developed by Procter & Gamble, was approved by the U.S. Food and Drug Administration in 1996 and given the reference number CFR 172.867 (CFR, 1997). One should also be aware that GRAS status for a compound approved for use as a food additive at low concentrations would not necessarily mean approval for use as a macronutrient substitute.

All of the fat-based macronutrient substitutes are likely to require extensive safety testing prior to Food and Drug Administration approval. Since the carbohydrate-(except olestra) and protein-based fat substitutes approved to date cannot be used for frying, the discussion will include the heat-stable, low- to noncaloric, fat-based fat substitutes in anticipation that one or more will eventually receive FDA approval. Fat-based substitutes are unique in that they contribute little to the caloric content of the food, yet they retain the important functional attributes associated with ''regular'' fat. Early developments in this area have been discussed elsewhere (Harrigan and Breene, 1989; Haumann, 1986; La Barge, 1988).

The widespread use of fat-based fat substitutes could result in a substantial reduction in the amount of fat calories in the U.S. diet, which would have important positive implications in terms of heart disease and other cardiovascular problems. According to the U.S. Surgeon General (1988), a substantial reduction in fat consumption would be the single most important positive change in the American diet. A significant reduction in the percentage of fat in the diet should have a positive effect on cardiovascular disease, arteriosclerosis, obesity, and numerous related health problems. Another potential positive effect, due to the presence of a nondigestive oil substitute in the digestive tract, might be a decrease in absorption of dietary cholesterol, which would reduce the serum cholesterol concentration (Mattson et al., 1979). In contrast, fat-soluble vitamin (Hassel, 1993) and drug absorption can be depressed, and laxative effects could be induced. Some studies have indicated that the caloric intake is increased to compensate for a reduction in fat intake (Beaton et al., 1992; Birch et al., 1993; Caputo and Mattes, 1992; Rolls et al., 1992). However, even if the caloric content of the American diet is not reduced, the use of fat-based fat substitutes should reduce the percentage of fat in the diet, and it should improve food product taste.

There are important questions about the fat-based fat and oil substitutes that must be answered (Borzelleca, 1992; Munro, 1990; Vanderveen, 1994) prior to approval. Until the last few years, regulatory agencies had little experience in the evaluation of macronutrient substitutes, which has compelled them to take a very cautious approach with respect to approval. There are many more limitations to the safety testing of macronutrient substitutes than with food additives used in small concentrations, because the normal level of safety guarantees can not be assured. Luckily, many of the potential fat substitutes are closely related structurally to emulsifiers that have been used for decades, so it may be possible to make inferences about the effect of long-term exposure. A comprehensive and informative text titled *Low-Calorie Foods Handbook*, by Aaron M. Altschul, was published in 1993 by Marcel Dekker. The chapter on regulatory aspects of low-calorie foods was written by John E. Vanderveen of the Food and Drug Administration. He also presented guidelines at the 1994 IFT meeting that would be of particular interest for companies who intend to request regulatory approval for a macronutrient substitute (Vanderveen, 1994). These have been summarized as follows:

1. The exact chemical structure of the substitute must be known and documented. If it is composed of more than one compound, the structure of each of these components must be completely elucidated.
2. The stability should be well documented, especially during food production, storage, and final preparation. If partial degradation occurs, the byproducts should be identified. If any harmful impurities are present, there should be limits established as to the concentration of these components.
3. Detailed exposure assessment for each population type should be included, which would include toxicological data demonstrating that the use of the macronutrient substitute, as well as any of the degradative byproducts, and would show them to not cause harm at least the levels of exposure for the intended use.
4. Additional tests may be required to provide pharmakinetic, metabolic, and/or nutritional data not normally required for other additives.
5. It will be necessary to determine whether the macronutrient substitute is absorbed or not. If it is poorly absorbed, one needs to determine whether large concentrations of the material in the gut affect the morphology, physiology, biochemistry, or normal flora of the gastrointestinal tract.
6. Since some of the normal testing procedures used for the evaluation of macronutrients are either inappropriate or provide insufficient data, an innovative or novel approach may be required. The selection of animal models, experimental protocols, and measurement techniques must be carefully planned to provide data which can be used to accurately assess the safety of the substitute. In addition, consideration of the appropriate controls and control experiments is needed.

7. Sufficient data must be provided to demonstrate that the substitute or its metabolites do not interfere with absorption or metabolism of essential nutrients. If interference does occur, one must determine whether it can be safely offset with nutrient fortification. If the compound is absorbed at a measurable rate, there must be data indicating the impact, if any, on the nutrient status of the animal. In general, traditional assessments of nutritional status should be sufficient for nutrient status assessment.
8. Another area of special concern is the effect of macronutrient substitutes on selected, particularly sensitive, population segments, such as the very young, senior citizens, or those afflicted with certain health problems. One example is aspartame, due to its effect on individuals susceptible to phenylketonuria.
9. The effect of the macronutrient on drug absorption and/or activity should be assessed. Some of the oil substitutes can have a slight laxative or anal leakage effect, at least during the initial stages of consumption until the digestive system has adapted to the product. Since this effect is often greater among children than adults, it may require additional evaluation. In addition, some consumers may not be aware of the relationship between substitute consumption and the laxative effects, as they are with prunes, for example. Special labeling may be needed to address the problem.
10. Postmarket surveillance will be needed in most cases. One long-term potential problem of particular concern could occur due to the consumption of several macronutrient substitutes concurrently, since it is likely that more than one of those materials may be approved in the future. There may be a synergistic or interactive effect on some of factors just discussed. This concern is also likely to limit the number of fat-based fat substitutes that are eventually approved.

The FDA gave safety conditions for use of olestra (CFR, 1997), which are summarized as follows:

1. Olestra is a mixture of octa-, hepta-, and hexaesters of sucrose with fatty acids derived from edible fats and oils or fatty acid sources that are generally recognized as safe or approved for use as food ingredients. The chain lengths of the fatty acids are no less than 12 carbon atoms.
2. Olestra meets the following specifications: the total content of octa-, hepta-, and hexaesters is not less than 97% as measured by size exclusion chromatography; the content of octaester is not less than 70%; the content of hexaester is not more than 1%; the content of pentaester is not more than 0.5%; the unsaturated fatty acid content is not less than 25% and not more than 83%; the content of C_{12} and C_{14} fatty acids is each not more than 1%, and total C_{20} and longer fatty acids are not more 20%; C_{16} and C_{18} make up the remainder with total content not less than 78%; the free fatty acid is not more than 0.5%; the residue ignition (sulfated ash) is not more than 0.5%; total methanol content is not more than 300 ppm; the total heavy metal content (as Pb) is not more than 10 ppm; lead is not more than 0.1 ppm as determined by the atomic absorption spectrophotometric graphite furnace method; water is not more than 0.1%; the peroxide value is not more than 10 meq/kg; the stiffness is not more than 50 kiloPascals/s.
3. Olestra may be used in place of fats and oils in prepackaged ready-to-eat savory (i.e., salty or piquant but not sweet) snacks. In such foods, the additive may be

used in place of fats and oils for frying or baking, in dough conditioners, in sprays, in filling ingredients, or in flavors.

4. To compensate for any interference with absorption of fat-soluble vitamins, the following vitamins shall be added to foods containing olestra: 1.9 milligrams α-tocopherol equivalent per gram olestra; 51 retinol equivalents per gram olestra (as retinyl acetate or retinyl palmitate); 12 IU vitamin D per gram of olestra; and 8 μg vitamin K_1 per gram olestra.
5. The label of a food containing olestra shall bear the following statement: "THIS PRODUCT CONTAINS OLESTRA. Olestra may cause abdominal cramping and loose stools. Olestra inhibits absorption of some vitamins and other nutrients. Vitamins A, D, E, and K have been added."
6. The FDA will review and evaluate all data and information bearing on the safety of olestra received by the agency after the approval of olestra.

These conditions will be very useful to other parties that have applied and/or intend to apply for approval of new fat substitutes.

V. TOXICOLOGY

Fat replacements are so new that independent research results on usage, efficacy, and safety on a broad scale, including public diet modification, are not yet available (Foreyt, 1992). For each new fat substitute, a unique safety evaluation program must be developed, taking into consideration the chemistry and properties of the substance in question. Like other macronutrients, noncaloric fat substitutes need to possess a well-defined product specification that provides assurance of both its chemical stability and its presumed metabolic outcome, and ensures that it is free of potentially toxic impurities.

As with any new food ingredient, fat substitutes must be tested on animals before they are tested on humans. Ordinarily, in animal toxicological tests food additives are fed at dietary levels many times in excess of the levels that will appear in foods for human consumption (Gershoff, 1995). This is done to provoke potential toxic responses and to establish safety factors. Because the amounts of fat substitutes that could appear in the human diet are large relative to other food additives such as colors or flavors, feeding them at such high levels could result in spurious results, because this would require replacing a large part of the nutrients in the diet. Munro (1990) has pointed out that responses that "at first glance may be considered to be of toxicological significance may on further investigation be the result of dietary nutrient imbalance or physiological perturbation induced by the test material when fed at excessive exposure levels." These very-high-level diets could become unpalatable with a resultant poor growth that might be interpreted as a sign of toxicity. Besides growth, measurements of blood and urine chemistry and gross and histologic examination of tissues are often made, and, when appropriate, carcinogenicity, genotoxicity, and reproductive and developmental toxicity testing may also be performed. Even if animal testing proves negative, FDA recognizes that confirmatory studies in humans are an important part of confirming the safety of macronutrient substitutes (Gershoff, 1995).

In toxicological studies, potential effects of fat substitutes that may not be evident in standard toxicological tests also need to be considered based on physiologic effects that may be specific to the chemical or physical properties or the mechanism or site of action of the substitute. There is also a need for confirmatory human studies in normal as well as at-risk populations, such as people with diabetes or compromised gastrointestinal (GI) tracts, or abnormalities that could possibly be caused by the fat substitute under consideration. For nonabsorbable fat substitutes, effects on GI epithelium, colonic microflora ecology, bile acid physiology, pancreatic function, and laxation effects should be considered (Gershoff, 1995; Glueck et al., 1994; Munro, 1990; Vanderveen and Glinsmann, 1992). For fat substitutes that are absorbed, absorption, distribution, metabolism, and elimination of the substitute should be considered. Specifically designed studies for a substitute's potential to alter the absorption or utilization of other dietary nutrients need to be undertaken (Gershoff, 1995; Vanderveen and Glinsmann, 1992). Nutrient–substitute interactions should be identified. Their nature and magnitude should be determined in an appropriate animal model, using defined and extreme testing conditions such as well-characterized and purified diets, exaggerated dietary levels, controlled feeding patterns, and an extensive battery of nutrient status measures. Results from the animal studies can be confirmed and extended in human studies. Dietary levels and consumption frequencies used in these studies should include and exceed as much as is practical the expected real-life usage by the general population as well as subgroups of the population that may have special nutrient demands or unique dietary patterns. During testing, the substitute should be provided in the food forms in which it will be marketed, as part of a controlled diet that provides the recommended dietary allowances, but not excessive levels of nutrients so that potential effects of the substitute on nutrient status are not masked. In addition to the effect on the digestion and absorption of other dietary components, the effect of the fat substitute on the absorption of orally dosed drugs should be considered (Vanderveen and Glinsmann, 1992).

The most exhaustively studied fat substitute has been the heat-stable, non-caloric fat substitute olestra. Olestra is neither hydrolyzed nor absorbed (Mattson and Volpenhein, 1972; Miller et al., 1995). Olestra is not toxic, carcinogenic, mutagenic, or teratogenic and when fed to animals at doses up to 10% of the total diet, there were no noted toxic effects on weight gain, hematology, urinalysis, or tissue pathology (Bergholtz, 1992). Since it is not absorbed, the only organ that olestra contacts is the GI tract. It has no significant effect on gastric emptying, total transit time, fecal water, or pH of pancreas, fecal bile acids, or interohepatic circulation of bile acids. Recently, it was reported that for specific GI symptoms (gas, diarrhea, abdominal cramping), there were no significant differences between humans who consumed olestra and triglyceride chips (Cheskin et al., 1998). Gut microflora do not metabolize olestra under anaerobic conditions, but during waste treatment it is degraded aerobically in sludge-amended soils (Haighbaird et al., 1997).

REFERENCES

Akoh, C. C. 1994a. Synthesis of carbohydrate fatty acid polyesters. In: *Carbohydrate Polyesters as Fat Substitutes*. Akoh, C. C. Swanson, B. G. (Eds.). Marcel Dekker, Inc., New York.

Akoh, C. C. 1994b. Oxidative stability of fat substitutes and vegetable oils by the oxidative stability index method. *J. Am. Oil Chem. Soc.* 71:211.

Akoh, C. C. 1995. Lipid based fat substitutes. *Crit. Rev. Food Sci. Nutr.* 35:405.

Akoh, C. C. 1996. New developments in low calorie fats and oils substitutes. *J. Food Lipids* 3:223.

Akoh, C. C., Swanson, B. G. 1987. One-stage synthesis of raffinose fatty acid polyesters. *J. Food Sci.* 52:1570.

Akoh, C. C., Swanson, B. G. 1989a. Synthesis and properties of alkyl glycoside and stacyose fatty acid polyesters. *J. Am. Oil Chem. Soc.* 66:1295.

Akoh, C. C., Swanson, B. G. 1989b. Preparation of trehalose and sorbitol fatty acid polyesters by interesterification. *J. Am. Oil Chem. Soc.* 66:1581.
Akoh, C. C., Swanson, B. G. 1990. Optimized synthesis of sucrose polyesters: comparison of physical properties of sucrose polyesters, raffinose polyesters and salad oils. *J. Food Sci.* 55:236.
Albano-Garcia, E., Lorica, R. G., Pama, M., de Leon, L. 1980. Solventless synthetic methods for methyl glucoside and sorbitol esters of coconut fatty acids. *Phillip J. Coconut Stud.* 5:51.
Anonymous. 1989. Fats, oils and fat substitutes. *Food Technol.* 43:72.
Anonymous. 1990a. Fat substitute update. *Food Technol.* 44:92.
Anonymous. 1990b. USDA's oatrim replaces fat in many food products. *Food Technol.* 44:100.
Anonymous. 1991. Carbohydrate-based ingredient performs like fat for use in a variety of food applications. *Food Technol.* 45:262.
Anonymous. 1994. Stepan seeks GRAS status for Captrin. *INFORM* 5:1167.
Arciszewski, H. 1991. Fat functionality, reduction in baked foods. *INFORM* 2:392.
Artz, W. E., Hansen, S. L. 1996a. Current development in fat replacers. In: *Food Lipids and Health*, McDonald, R. E., Min, D. B. (Eds.). Marcel Dekker, Inc., New York.
Artz, W. E., Hansen, S. L. 1996b. The chemistry and nutrition of nonnutritive fats. In: *Deep Frying: Chemistry, Nutrition and Practical Applications*, Perkins, E. G., Erickson, M. D. (Eds.). American Oil Chemists' Society, Champaign, IL.
Barrantes, E., Tamime, A. Y., Davies, G., Barclay, M. N. I. 1994a. Production of low-calorie yogurt using skim milk powder and fat-substitute. 2. Compositional quality. *Milchwissenschaft* 49:135.
Barrantes, E., Tamime, A. Y., Sword, A. M. 1994b. Production of low-calorie yogurt using skim milk powder and fat-substitute. 3. Microbiological and organoleptic qualities. *Milchwissenschaft* 49:85.
Barrantes, E., Tamime, A. Y., Muir, D. D., Sword, A. M. 1994c. The effect of substitution of fat by microparticulate whey protein on the quality of set-type, natural yogurt. *J. Soc. Dairy Technol.* 47:61.
Beaton, G. H., Tarasuk, V., Anderson, G. H. 1992. Estimation of possible impact of non-caloric fat and carbohydrate substitutes on macronutrient intake in the human. *Appetite* 19:87.
Bergholtz, C. M. 1992. Safety evaluation of olestra, a non-absorbed fatlike fat replacement. *Crit. Rev. Food Sci. Nutr.* 32:141.
Birch, L. L., Johnson, S. L., Jones, M. B., Peters, J. C. 1993. Effects of a nonenergy fat substitute on children's energy and macronutrient intake. *Am. J. Clin. Nutr.* 58:326.
Borzelleca, J. F. 1992. Macronutrient substitutes: safety evaluation. *Regulatory Toxicol. Pharmacol.* 16:253.
Borzelleca, J. F. 1996. A proposed model for safety assessment of macronutrient substitutes. *Regulatory Toxicol. Pharmacol.* 23:S15.
Caputo, F. A., Mattes, R. D. 1992. Human dietary responses to covert manipulations of energy, fat and carbohydrate in a mid-day meal. *Am. J. Clin. Nutr.* 56:36.
Cheftel, J. C., Dumay, E. 1993. Microcoagulation of proteins for development of creaminess. *Food Rev. Inter.* 9:473.
Cheskin, L. J., Miday, R., Zorich, N., Filloon, T. 1998. Gastrointestinal symptoms following consumption of olestra or regular triglyceride potato chips—a controlled comparison. *JAMA* 279:150.
Chung, H.-Y., Park, J., Kim, J.-H., Kong, U.-Y. 1996. Preparation of sorbitol fatty acid polyesters, potential fat substitutes: optimization of reaction conditions by response surface methodology. *J. Am. Oil Chem. Soc.* 73:637.
Code of Federal Regulations. 1997. Foods and Drugs. Title 21. Parts 170–199. U.S. Government Printing Office, Washington, DC.
Cooper, C. F. 1990. Preparation of propoxylated glycerides as dietary fat substitutes. European Patent 353,928.
Crites, S. G., Drake, M. A., Swanson, B. G. 1997. Microstructure of low-fat cheddar cheese con-

taining varying concentrations of sucrose polyesters. *Lebensmittel-Wissenschaft and Technologie* 30:762.

Degraaf, C., Hulshof, T., Weststrate, J. A., Hautvast, J. G. A. J. 1996. Nonabsorbable fat (sucrose polyester) and the regulation of energy intake and body weight. *Am. J. Physiol.—Regulatory, Integrative, and Comparative Physiology* 39:R1386.

Drake, M. A., Nagel, C. W., Swanson, B. G. 1994. Sucrose polyester content in foods by a colorimetric method. *J. Food Sci.* 59:655.

Duxbury, D. D., Meinhold, N. M. 1991. Dietary fats and oils. *Food Processing* 52:58.

Dziezak, J. D. 1989. Fats, oils, and fat substitutes. *Food Technol.* 43:66.

Ennis, J. L., Kopf, P. W., Rudolf, S. E., van Buren, M. F. 1991. Esterified alkoxylated alkyl glycosides useful in low calorie fat-containing food compositions. Eur. Patent 415,636.

Food and Drug Administration. 1982. Toxicological principles for the safety assessment of direct food additives and color additives used in food. Bureau of Foods. U.S. Government Printing Office, Washington, D.C.

Foreyt, J. P. 1992. Potential impact of sugar and fat substitutes in American diet. *J. Natl. Cancer Inst. Monogr.* 12:99.

Freston, J. W., Ahnen, D. J., Czinn, S. J., Earnest, D. L., Farthing, M. J., Gorbach, S. L., Hunt, R. H., Sandler, R. S., Schuster, M. M. 1997. Review and analysis of the effects of olestra, a dietary fat substitute, on gastrointestinal function and symptoms. *Regulatory Toxicol. Pharmacol.* 26:210.

Frye, A. M., Setser, C. S. 1993. Bulking agents and fat substitutes. In: *Low-Calorie Foods Handbook*. Altschul, A. M. (Ed.), Marcel Dekker, Inc., New York.

Gardner, D. R., Sanders, R. A. 1990. Isolation and characterization of polymers in heated olestra and an olestra/triglyceride blend. *J. Am. Oil Chem. Soc.* 67:788.

Gardner, D. R., Sanders, R. A., Henry, D. E., Tallmadge, D. H., Wharton, H. W. 1992. Characterization of used frying oils. Part 1: isolation and identification of compound classes. *J. Am. Oil Chem. Soc.* 69:499.

Garn, S. M. 1997. From the Miocene to olestra—a historical perspective on fat consumption. *J. Am. Dietetic Assoc.* 97:S54.

Gershoff, S. N. 1995. Nutrition evaluation of dietary fat substitutes. *Nutr. Rev.* 53:305.

Gillis, A. 1988. Fat substitutes create new issues. *J. Am. Oil Chem. Soc.* 65:1708.

Grossklaus, R. 1996. Fat replacers—requirements from a nutritional physiological point of view. *Fett Wissenschaft Technologie* 98:136.

Habib, M., Podolski, J. S. 1989. Concentrated, substantially non-aggregated casein micelles as a fat/cream substitute. NutraSweet Co. European Patent Application 0,334,226.

Haighbaird, S. D., Bus, J., Engelen, C., Hill, R. N. 1997. Biodegradation of noncaloric fat substitutes sucrose polyesters in sewage sludge amended soil. *Chemosphere* 35:413.

Hansen, S. L. 1995. The analysis of non-volatile components of heated fatty acid esterified propoxylated glycerols. Ph.D. Thesis, University of Illinois, Urbana, IL.

Hansen, S. L., Krueger, W. J., Dunn, Jr., L. B., Artz, W. E. 1997. Nuclear magnetic resonance and gas chromatography/mass spectroscopy analysis of the nonvolatile components produced during heating of oleic acid esterified propoxylated glycerol, a fat substitute model compound, and trioleylglycerol. *J. Agric. Food Chem.* 45:4730.

Harrigan, K. A., Breene, W. M. 1989. Fat substitutes: sucrose esters and simplesse. *Cereal Foods World* 34:261.

Hassel, C. A. 1993. Nutritional implications of fat substitutes. *Cereal Foods World* 38:142.

Haumann, B. F. 1986. Getting the fat out. *J. Am. Oil Chem. Soc.* 63:278.

Henderson, J. M., Petersheim, M., Templeman, G. J., Softly, B. J. 1994. Quantitation and structure elucidation of the positional isomers in a triacylglycerol mixture using proton and carbon one- and two-dimensional NMR. *J. Agric. Food Chem.* 42:435.

Henry, D. E, Tallmadge, D. H., Sanders, R. A., Gardner, D. R. 1992. Characterization of used frying oils. Part 2: comparison of Olestra and triglyceride. *J. Am. Oil Chem. Soc.* 69:508.

Hippleheuser, A. L., Landberg, L. A., Turnak, F. L. 1995. A system approach to formulating a low-fat muffin. *Food Technol.* 49:92.

Hollenbach, E. J., N. B. Howard. 1983. Emulsion concentrate for palatable polyester beverage. U.S. Patent 4,368,213.

Huang, A. S., Delano, G. M., Pidel, A., Janes, L. E., Softly, B. J., Templeman, G. J. 1994a. Characterization of triacylglycerols in saturated lipid mixtures with application to salatrim. *J. Agric. Food Chem.* 42:453.

Huang, A. S., Robinson, L. R., Gursky, L. G., Profita, R., Sabidong, C. G. 1994b. Identification and quantitation of salatrim 23CA in foods by the combination of supercritical fluid extraction, particle beam LC-mass spectrometry, and HPLC with light scattering detector. *J. Agric. Food Chem.* 42:468.

Inglett, G. E., Grisamore, S. B. 1991. Maltodextrin fat substitute lowers cholesterol. *Food Technol.* 45:104.

Keeton, J. T. 1994. Low-fat meat products—technological problems with processing. *Meat Science* 36:261.

Klemann, L. P., Aji, K., Chrysam, M. M., D'Amelia, R. P., Henderson, J. M., Huang, A. S., Otterburn, M. S., Yarger, R. G. 1994. Random nature of triacylglycerols produced by the catalyzed interesterification of short- and long-chain fatty acid triglycerides. *J. Agric. Food Chem.* 42: 442.

LaBarge, R. G. 1988. The search for a low-caloric oil. *Food Technol.* 42:84.

Lu, X. J., Myers, M. R., Artz, W. E. 1993. Supercritical fluid chromatographic analysis of the propoxylated glycerol esters of oleic acid. *J. Am. Oil Chem. Soc.* 70:355.

Mahungu, S. M. 1994. Analysis of the volatile components of heated esterified propoxylated glycerols. Ph.D. Thesis, University of Illinois, Urbana, IL.

Mai, J., Breitbar, D., Fischer, C. D. 1990. Proteinaceous material. Unilever, NV. European Patent Application 0,400,714.

Mattson, F. H., Volpenhein, R. A. 1972. Hydrolysis of fully esterified alcohols containing from one to eight hydroxyl groups by the lipolytic enzymes of rat pancreatic juice. *J. Lipid Res.* 13:325.

Mattson, F. H., Glueck, C. J., Jandacek, R. J. 1979. The lowering of plasma cholesterol by sucrose polyester in subjects consuming diets with 800, 300 or less than 50 mg cholesterol per day. *Am. J. Clin. Nutr.* 32:1636.

McMindes, M. K. 1991. Applications of isolated soy protein in low-fat meat products. *Food Technol.* 45:61.

Meyer, R. S., Root, J. M., Campbell, M. L., Winter, D. B. 1989. Low caloric alkyl glycoside fatty acid polyester fat substitutes. U.S. Patent 4,840,815.

Miller, G. D., Groziak, S. M. 1996. Impact of fat substitutes on fat intake. *Lipids* 31:S293.

Miller, K. W., Lawson, K. D., Tallmage, D. H. 1995. Disposition of ingested olestra in the Fischer 344 rat. *Fund. Appl. Toxicol.* 24:229.

Miraglio, A. M. 1995. Nutrient substitutes and their energy values in fat substitutes and replacers. *Am. J. Clin. Nutr.* 62:S1175.

Munro, I. C. 1990. Issues to be considered in the safety evaluation of fat substitutes. *Food Chem. Toxic.* 28:751.

Mutua, L. N., Akoh, C. C. 1993. Synthesis of alkyl glycoside fatty acid esters in nonaqueous media by *Candida sp.* lipase. *J. Am. Oil Chem. Soc.* 70:43.

Nishiyama, J., Shizu, Y., Kuninori, T. 1993. Inhibition of soybean lipoxygenase-1 by sucrose esters of fatty acids. *BioSci. Biotech. Biochem.* 57:557.

Penichter, K. A., McGinley, E. J. 1991. Cellulose gel for fat-free food applications. *Food Technol.* 45:105.

Phillips, L. G., Barbano, D. M. 1997. The influence of fat substitutes based on protein and titanium dioxide on the sensory properties of lowfat milks. *J. Dairy Sci.* 80:2726.

Pszczola, D. E. 1991. Oat-bran-based ingredient blend replaces fat in ground beef and pork sausage. *Food Technol.* 45:60.

Quesada, L. A., Clark, W. L. 1993. Low-calorie foods: General Category. In: *Low-Calorie Foods Handbook*. Altschul, A.M. (Ed.). Marcel Dekker, Inc., New York.

Rios, J. J., Perezcamino, M. C., Marquezrui, G., Dobarganes, M. C. 1994. Isolation and characterization of sucrose polyesters. *J. Am. Oil Chem. Soc.* 71:385.

Riva, S. 1994. Enzymatic synthesis of carbohydrate esters, In: *Carbohydrate Polyesters as Fat Substitutes*. Akoh, C. C., Swanson, B. G. (Eds.). Marcel Dekker, Inc., New York.

Robbins, M. B., Rodriguez, S. S. 1984. Low calorie baked products. U.S. Patent 4,461,782.

Roberts, B. A. 1984. Oleaginous compositions. U.S. Patent 4,446,165.

Rolls, B. J. 1995. Carbohydrates, fat and saiety. *Am. J. Clin. Nutr.* 61:S960.

Rolls, B. J., Pirraglia, P. A., Jones, M. B., Peters, J. C., 1992. Effects of Olestra, a noncaloric fat substitute, on daily energy fan fat intakes in lean men. *Am. J. Clin. Nutr.* 56:84.

Sanchez, C., Klopfenstein, C. W., Walker, C. E. 1995. Use of carbohydrate-based fat substitutes and emulsifying agents in reduced-fat shortbread cookies. *Cereal Chem.* 72:25.

Sanders, R. A, Gardner, D. R., Lacey, M. P., Keough, T. 1992. Desorption mass spectrometry of Olestra. *J. Am. Oil Chem. Soc.* 69:760.

Schmidt, K., Lundy, A., Reynolds, J., Lee, L. N. 1993. Carbohydrate or protein based fat mimicker effects on ice milk properties. *J. Food Sci.* 58:761.

Singer, N. S., Moser, R. H. 1993. Microparticulated proteins as fat substitutes. In: *Low-Calorie Foods Handbook*. Altschul, A. M. (Ed.), Marcel Dekker, Inc., New York, NY.

Singer, N. S., Yamamoto, S., Latella, J. 1988. Protein product base. John Labatt, Ltd. U. S. Patent 4,734,287 and European Patent Application 0,250,623.

Smith, R. E. 1993. Food demands of the emerging consumer: the role of modern food technology in meeting that challenge. *Am. J. Clin. Nutr.* 58:S307.

Smith, R. E., Finely, J. W., Leveille, G. A. 1994. Overview of salatrim, a family of low-calorie fats. *J. Agric. Food Chem.* 42:432.

Stark, L. E., Gross, A. T. 1990. Hydrophobic protein microparticles and preparation thereof. Enzytech, Inc. World Patent Application. WO 90/03123.

Tallmadge, D. H., Lin, P. Y. T. 1993. Liquid chromatographic method for determining the percent of olestra in lipid samples. *J. AOAC Int.* 76:1396.

Tamime, A. Y., Barclay, M. N. I., Davies, G., Barrantes, E. 1994. Production of low-calorie yogurt using skim milk powder and fat-substitute. 1.A review. *Milchwissenschaft* 49:85.

Tamime, A. Y., Barrantes, E., Sword, A. M. 1996. The effect of starch based fat substitutes on the microstructure of set-style yogurt made from reconstituted skimmed milk powder. *J. Soc. Dairy Technol.* 49:1.

Thompson, M. S. 1992. Issues associated with the use and regulation of fat substitutes. *Crit. Rev. Food Sci. Nutr.* 32:123.

U. S. Surgeon General. 1988. The Surgeon General's Report on Nutrition and Health. U. S. Department of Health and Human Services. Publication No. 88-50210. U. S. Government Printing Office, Washington, D.C.

Vanderveen, J. E. 1994. Regulatory status of macronutrient substitutes: what FDA needs to assure safety. Paper No. 15-1. Presented at the Institute of Food Technologists Meeting, Atlanta, GA, July 25–29.

Vanderveen, J. E., Glinsmann, W. H. 1992. Fat substitute: A regulatory perspective. *Ann. Rev. Nutr.* 12:473.

Vetter, J. L. 1993. Low-calorie bakery foods. In: *Low-Calorie Foods Handbook*. Altschul, A. M. (Ed.). Marcel Dekker, Inc., New York.

Volpenhein, R. A. 1985. Synthesis of higher polyol fatty acid polyesters using carbonate catalysts. U. S. Patent, 4,517,360.

White, C. H. 1993. Low-fat dairy products. In: *Low-Calorie Foods Handbook*. Altschul, A. M. (Ed.). Marcel Dekker, Inc., New York.

White, J. F. 1990. Partially esterified polysaccharides (PEP) fat substitutes. U. S. Patent 4,959,466.

White, J. F., Pollard, M. R. 1988. Esterified epoxide-extended polyols as nondigestible fat substitutes of low-caloric value. Eur. Patent 254,547.

White, J. F., Pollard, M. R. 1989a. Non-digestible fat substitutes of low-calorie value. U. S. Patent 4,861,613.

White, J. F., Pollard, M. R. 1989b. Non-digestible fat substitutes of low-calorie value. Eur. Patent 325,010.

White, J. F., Pollard, M. R. 1989c. Low-calorific and non-digestive substitute of fat/oil. China Patent 1,034,572.

Winter, D. B., Meyer, R. S., Root, J. M., Campbell, M. L. 1990. Process for producing low calorie foods from alkyl glycoside fatty acid polyesters. U. S. Patent 4,942,054.

Yamamoto, T., Kinami, K. 1986. Production of sucrose fatty acid polyester, U. S. Patent, 4,611,055.

Yan, Z. Y., Huhn, S. D., Klemann, L. P., Otterburn, M. S. 1994. Molecular modeling studies of triacylglycerols. *J. Agric. Food Chem.* 42:447.

Zorich, N. L., Biedermann, D., Riccardi, K. A., Bishop, L. J., Filloon, T. G. 1997. Randomized, double-blind, placebo-controlled, consumer rechallenge test of olean salted snacks. *Regulatory Toxicol. Pharmacol.* 26:209.

12

Food Additives for Special Dietary Purposes

JOHN M. V. GRIGOR

University of Lincolnshire and Humberside, Grimsby, England

WENDY S. JOHNSON

SHS International Ltd., Liverpool, England

SEPPO SALMINEN

University of Turku, Turku, Finland, and Royal Melbourne Institute of Technology, Melbourne, Australia

I. INTRODUCTION

In recent years, the general public in the ''developed'' world have become increasingly aware of the health issues concerning food. This has been partly due to increased knowledge generated by clinical research and partly to the dissemination of this information and orchestrated lobbying by special interest and media groups. Thus, for example, it is now widely perceived that dietary practices can play a significant role in the prevention of diseases such as hypertension, hyperlipidaemia, and bowel cancer. In response, the food industry has introduced a class of foods referred to as functional foods, or nutraceuticals. Within this classification, some foods are concerned only with the manipulation of nutritional components, such as the reduction in fat in low fat products or the increase in polyunsaturates, whilst others may be considered to play a more positive role. Food additives used in this group are defined as substances with no nutritional value, but which function either to maintain health (e.g., vitamins or minerals) or act as a therapeutic agent (e.g., soluble fibre in noninsulin-dependent diabetes).

Functional foods as defined above are concerned with the dietary needs of the general public as a whole. In contrast, minorities can be identified who require foods specially engineered for their particular medical nutritional condition. Such foods, termed enteral clinical nutrition products, although normally covered by food regulations, will also be supported by a complete portfolio of clinical data, ensuring efficacy in a clinical environment. For example, the medical condition phenylketonuria (PKU), which is a hereditary disease, renders its sufferers unable to effectively metabolize excesses of the amino acid phenylalanine from birth. A normal diet for these sufferers would eventually result in mental retardation. Products have therefore been designed in which all phenylalanine is absent. These are mixtures of protein hydrolysates or synthetic amino acids with vitamins, minerals, carbohydrates, and fats, and are heavily dependent on food additives for their nutritional (U.S. regulations) and technical requirements.

This chapter is not intended to provide a complete review of all the developments and issues within this rapidly changing area, but rather to feature some specific examples. It will highlight the unique role which food additives play in the engineering of clinical nutrition products as well as discuss possible future developments in this area. It will then proceed to discuss the new role which food additives are playing in the functional food area, focusing on therapeutic function and regulatory concerns.

II. FOOD ADDITIVES AND CLINICAL NUTRITION

A. Nutrition

The regulatory environment for food additives is subject to continuous change and evolution; there are also significant differences in approach and detail between U.S. regulations and those developed, for examples, within the European Union. These are the subject of extensive discussion elsewhere in this book. In general, additives as defined by EU law provide a technological rather than nutritional function, whereas under U.S. law components described as additives may have a nutritional element. The long-running debate regarding the ''additive'' status of complex carbohydrates (such as inulin and oligosaccharides) highlights some of the difficulties which have been faced in this context.

B. Palatability

The primary function of a clinical nutrition product is its therapeutic or health maintaining performance. It is also crucially important that the product is designed to optimize compliance by individuals who may have a reduced appetite or interest in food. The main design criterion is to ensure that all nutrients and dietary supplements required for each particular medical condition are added to their functional level as indicated by clinical or physiological data. It is only once these constraints are met that the technologist can intervene to improve palatability and general acceptability. Nutritionally complete products for inborn errors of metabolism and a range of gastrointestinal derangements may have their protein requirements partially or wholly provided by amino acids or by protein hydrolysates. Amino acids have their own individual sensoric characteristics (1,2), while protein hydrolysates often have characteristic bitter tastes.

In an amino acid–based nutritional product, these amino acids in the final, diluted, unflavored product are added well above their threshold level; the taste is therefore predominantly a mixture of their sensoric characteristics. The unique flavor properties of L-glutamic acid (2) and the sulphurous persistent off taste of L-methionine (3) and L-cysteine (2) contribute toward an obnoxious, meaty lingering flavor profile and a predominantly bitter taste. It has also been suggested that a further reason for unpalatability of amino acid aqueous solutions is due to lower alkyl mercapton, such as methyl mercaptons associated with or degraded from L-methionine (4). The inclusion of vitamins and minerals also contributes to the problems by adding a persistent metallic note; such flavor interactions may be additive or even ''enhance'' each other.

There are several ways in which a product of this type can be rendered palatable. (From a clinical point of view this is very important, since dietary compliance is more likely to be achieved if the product is pleasant tasting). These include

1. Addition of flavors and food acids
2. Chemical derivatization, substitution, and purification of key unpleasant tasting amino acids
3. Microencapsulation of key unpleasant tasting components
4. Addition of bitterness inhibitors

1. Addition of Flavors and Food Acids

Most amino acid–containing products are flavored using sweet fruity notes. One reason for this is that the addition of a sweetener helps to mask and reduce the bitter note which predominates in these products. One of the best tasting sweeteners on the market at present is aspartame (L-aspartyl-L-phenylalanine); however, its addition into products for phenylketonurics is forbidden due to the phenylalanine element of the sweetener. In this case saccharin is added, which, with its bitter secondary note,is a far from perfect alternative. There has also been concern, especially in the United States, regarding possible carcinogenic effects of saccharin, based on early animal toxicity data. Intensity of flavor is the main criterion for choice of flavor.

The addition of food acids may also play an important role. Apart from the reported function of enhanced fruity taste (5), food acids also seem to play a pivotal role in improving palatability. One such reason may be their reported ability to chelate metal ions (6). In this respect the acid could act as a shield in preventing the metal ion from reaching flavor receptors. DL-malic acid has been reported to reduce the intensity of off flavors and bitterness in soybean protein hydrolysates (7). It has also been reported that at pH 5.7, 7.5 mg of mercaptons renders formulations containing amino acids unpalatable, whereas at a pH of 3.7, twice the level of mercaptons is required before the formulation becomes unpalatable (4).

The specific taste characteristics of food acids differ considerably: whereas citric acid has an initial burst of acid taste, with intensity descending rapidly with time, DL-malic acid has been reported to have a taste profile which lingers far longer (5). This can be used to the flavorist's advantage: if both acids are added to the product, citric acid will help mask the initial burst of off flavor, while the malic acid will help mask the lingering aftertaste. As reported, the addition of sweetness and acidity will help mask bitterness (4).

2. Chemical Derivatization Substitution, and Purification of Unpleasant Amino Acids

Cysteine and methionine have sulfur-containing side chains resulting in taste profiles which are very odorous and lingering. Protection of these side chains by derivatization can lead to vast taste improvement. One example is N-acetyl-L-methionine, where the acetyl group blocks the sulfur groups passage to flavor receptors. This material has been approved for food use in the United States, where it is classed as an additive. However, it has been reported that other N-acetylated amino acids are poorly utilized in humans owing to restricted acylase capacities (8). Cysteine can be replaced with the dimeric form, cystine, which has a comparatively neutral taste; this may be due to the chemical positioning of the sulphide groups into a disulphide bond.

Glutamic acid has a meaty/acidic flavor, which is considered to have flavor-enhancing properties in some situations. In cases where this is not so, its offensive taste can be removed by substitution with an alternative amino acid which can functionally replace glutamic acid on a nutritional basis. L-glutamine (4) has been offered as an alternative, although glutamine is susceptible to hydrolysis in aqueous solution and is therefore not suitable for liquid formulations (9).

3. Microencapsulation of Key Unpleasant Tasting Components

The coating of functional food additives to change their diffusional properties, thus controlling their release into the food matrix, is not a new idea (10). In clinical nutrition, one approach to improve palatability using encapsulation techniques is to coat offending molecules, such as amino acids, with a hardened fat or wax. Such a coating has to function as a physical barrier preventing the amino acids from contacting taste and flavor receptors. The fat used must therefore be insoluble in an aqueous solution, solid at oral cavity temperatures, and degradable by stomach/intestinal enzymes. The fat matrix must also be impermeable to the guest molecule. The use in clinical nutrition of fat coated amino acids has been limited, partly due to the expense of the raw material and its limited applicability (for example, the difficulty of stabilization in liquid products).

4. Addition of Bitterness Inhibitors

There are many compounds which have been listed as potential bitterness inhibitors. However, as yet no food additive solely functioning as a bitterness inhibitor has been used in clinical nutrition products. This may be due to bitterness inhibitors being specific to one particular sapophore or the strictness in food regulations. Clinical nutrition products could be classified into one category where a strong need for a bitterness inhibitor could be shown.

One particular type of product which has earned a great amount of attention is the chemical group of compounds called cyclodextrins. These synthetic structures are able to form inclusion complexes with ''guest'' molecules. The most common cyclodextrins are α, β, and γ cyclodextrins, which consist of 6, 7, and 8 (1–4) linked α-D-glucosyl residues, respectively. The central cavity of the cyclodextrin is hydrophobic and it is this characteristic which enables hydrophobic guest molecules, such as amino acids (11) to become entrapped within its ring structure.

One compound being used at present as a fruity flavor enhancer has also been reported as having bitterness inhibitor activity. This compound is ethyl maltol/maltol (E636, E635) and is at present being used in some flavoring preparations for synthetic diets in the management of inflammatory bowel disorders. It has been claimed to mask the bitterness associated with B complex vitamins and high intensity sweeteners (12).

a. Protein Hydrolysates. Protein hydrolysates are being used increasingly in clinical nutrition because of their superior absorption by deranged or dissected gut (13), although they are often exceedingly bitter and unpalatable. These are manufactured by the addition of proteolytic enzymes to a protein such as whey or casein, resulting in a mixture of peptides and free amino acids. Peptide size or range can be defined by the extent of hydrolysis and choice of molecular separation techniques. Protein hydrolysates currently available vary considerably in their taste and peptide profile. This highlights the fact that choice of enzymes, reaction times, choice of raw material, column conditions, and packing all affect the final product.

Proteolytic enzymes used for hydrolysate production are either endo- or exopeptidases. Endopeptidases internally break down the protein molecule at specific amino acid sequence points, whereas exopeptidases split individual amino acids from the C- and N-terminal ends of the peptides, and show affinity for hydrophobic amino acids. The bitterness of hydrolysates is partly due to the formation of oligopeptides (1000–5000 daltons), but the hydrophobicity of the oligopeptide also plays a part: increased hydrophobicity leads to increased bitterness. Enzyme combinations have been devised which specifically contain large amounts of exopeptidases to produce less bitter hydrolysates (14). In principle, palatability may be further improved by the application of the plastein reaction, in which amino acids are linked onto peptides, but the cost of this process at present renders it uneconomic.

Currently, there is no community-wide EEC legislation on the use of enzymes for food processing and the regulation of these products as food additives or processing aids varies on a country-by-country basis.

C. Manufacture

The majority of clinical nutrition products are considered most presentable for dietitians/patients in the form of either a ready-made drink or as a powder to be dissolved in water. Due to the inherent insolubility of amino acids such as tyrosine, cystine, histidine, and glutamine (15) there is a need to study ways in which elemental diets and PKU diets can be made more appealing by improving the solubility of individual components without compromising the nutritional efficacy of the product. For example, tyrosine's inherent insolubility (0.045g/100 mL of water at 25°C) causes problems when designing products for PKU patients. Tyrosine is an essential component of the diet as it replaces nutritionally the phenylalanine, which is toxic to PKU sufferers.

More soluble alternatives for tyrosine, glutamine, and histidine can be added. In the case of glutamine, it has been known for some time that this amino acid represents an important fuel for the cells lining the gastrointestinal tract (16), and there is a perceived requirement to supplement glutamine in many enteral feeds. However, as discussed earlier, the possibility of glutamine incorporation into a ready-made drink is limited due to its tendency to undergo quantitative aqueous hydrolysis with formation of cyclic products and ammonia (9), and reported limited solubility. Thus it has been recommended that to avoid the risk of precipitation, glutamine concentrations in feeding solutions should not exceed 1–1.5% (8). Options for amino acid derivatives include use of the hydrochloride form or conversion to the ester. They are claimed by the author to be biologically safe and available for absorption. However, currently this approach is little used. Alternatively soluble peptide could be used. This route is normally very expensive as well as increases the bitterness of the final product. As yet, no satisfactory solution has been offered.

Emulsifiers are also used to stabilize the emulsion of any essential fats added into the formula. The absence of protein, which contributes to emulsion formation in other circumstances, presents particular challenges in amino acid–based systems.

1. Osmolarity

The osmolarity of products designed for clinical applications is a matter of some importance. The use of hydrolyzed protein and small molecular weight components such as simple sugars and amino acids will increase the osmolarity of a product, although it is difficult to measure these accurately using existing equipment, particularly in products containing fat and partially insoluble material.

Although administration of hyperosmolar solutions has been associated with adverse symptoms such as "osmotic" diarrhea, this can often be overcome by appropriate dilution or slowing down of the rate of administration of the feed. It should be borne in mind that the normal digestive process will naturally break down protein to smaller components in the gut, and it may not be simply product osmolarity which is the significant factor in this.

III. FOOD ADDITIVES AS THERAPEUTIC AGENTS—AN ALTERNATIVE FUNCTION

A. Stabilisers and Thickeners

A number of materials used in food manufacture may fulfil more than one function. Typical of these are some of the complex carbohydrates otherwise known both as soluble fibres and as effective stabilizing and thickening agents.

In recent years health aficionados have been promoting the beneficial effects of soluble fibre (18). In normal subjects, soluble dietary fiber can have a beneficial cholesterol-lowering effect, a fact highlighted by manufacturers of foods which are naturally high in soluble fiber. In non–insulin dependent diabetics, an additional postprandial reduction in blood glucose is observed. However, these beneficial effects are dose dependent, and it can be argued that large quantities of food rich in natural fiber would have to be consumed before any real benefits would be shown (The British Diabetic Association have recommended a minimum daily intake of 30 g of dietary fiber (19)).

An alternative approach relies on the use of polysaccharide gums such as guar, pectin, sodium alginate, and konjac glucomannan incorporated at appropriated doses in everyday consumer products such as bread or cereal. Some such "designer high-fiber foods" have been clinically tested for their therapeutic effect and results have been very positive.

1. Guar Gum

This additive is a purified extract from the leguminous plant *Cyamopsis tetragonoloba* (L.) *Taub*. It is composed mainly of the galactomannan polymer, which increases viscosity in the gut, causing a slowing in the absorption of carbohydrates and increased bile production (initiating cholesterol reduction). Work carried out using guar gum incorporated into bread points to an effective dosage of between 6 and 15 g/day (20), which is significantly below the British Diabetic Association's minimum daily intake for dietary fiber, showing that incorporation into food increases the additive's effectiveness.

Guar gum has also been used as a slimming aid; theoretically it swells in the stomach, giving a feeling of satiety and thus reducing the rate of food ingestion (21). However, due to the possibility of esophageal obstruction using guar gum, the Ministry of Agriculture, Fisheries and Foods (UK) in 1988 imposed limits on the use of guar gum tablets and capsules for slimming (22).

2. Konjac Glucomannan

This polysaccharide gum is extracted from the plant called *Amorphophallus konjac* (K) *Koch*, more commonly known as the elephant yam. Manufactured and marketed by Shimizu Chemical Corporation under the trade name Propol, a product containing approximately 90% glucomannan has been added to powdered soup mixes and other food products at a recommended dosage of 1.3 g per meal time. It has been claimed to function as a low-calorie high-fiber ingredient, hypocholesterolemic agent, bulk laxative agent, stabilizer, and fat replacer (23).

B. Lecithins

Lecithin in general is an approved food additive (E322 and FDA approved) for use as an emulsifier in margarine, chocolate, and bakery products. Commercially, it is predominantly extracted from soya beans. Recently, research and marketing have been directed toward the development of lecithin fractions, with their own characteristic phospholipid composition (24). Similarly, in the health food sector lecithin is marketed as a natural emulsifier of fats, rich in choline and inositol. Claims are made which emphasize its role as a facilitator in the metabolism of fat. However, as yet there is no direct evidence which can link lecithin consumption with a reduction in heart disease (25).

IV. FUTURE DEVELOPMENTS

A. Colostrum as an Active Additive or Ingredient for Functional and Clinical Foods

Bovine colostrum is a rich source of biologically active material including growth promoting factors, immunoglobulins and essential nutrients. Colostrum fractionation methods have been developed and several fractions have been clinically tested as new functional ingredients for clinical and special dietary food. Thus, colostrum may provide new possibilities for disease-specific functional foods. Other possibilities include colostrum from hyperimmunized cows.

1. General Properties of Colostrum

Bovine colostrum, milk secreted during the first days after calving, has special nutritional and immunological properties essential for the newborn calf. In addition to the milk proteins and other essential nutrients found in mature milk, colostrum contains growth promoting factors including insulin and insulinlike growth factors, as well as relatively high concentrations of immunoglobulins (especially lgG). Because of its growth promoting activity, colostrum can be used as a serum substitute in mammalian cell cultures (26) to support cell growth. However, colostrum has to be fractionated before its use in cell cultures to remove protein and lipid fractions. Thus, a range of fractionated products is available.

2. Research on Colostrum Derivatives

Possible application for fractions derived from colostrum have been proposed. These include treatment of gastritis (27) and uses in sports nutrition. An in vitro model system to study the potential effects of colostrum derivatives has been developed. Some ongoing projects are listed in Table 1 and current products in Table 2.

The production of colostrum from hyperimmunized cows may enable development of functional foods with disease-specific properties. Examples include the possible treatment of hypercholesterolemia, intestinal infections (such as *Helicobacter pylori*) and prevention of oral *Streptococcus mutans* infection. Also, an immune milk against rheumatoid

Table 1 Current Studies on Bovine Colostrum

Type of study	Subjects	Result
Athletic performance	Finnish Olympic ski team	Enhanced performance, lower creatine kinase activities
Athletic performance	Volunteer athletes	Not completed
Gastritis/immune colostrum	Volunteers	Improved status
Helicobacter pylori gastritis (children) immune colostrum	Children with abdominal complaints and verified *Helicobacter*	Improved status

arthritis has been successfully tested in young arthritis patients in Finland. Colostrum products so far developed include immune colostrum for the treatment of rotavirus diarrhea (Gastrogard™, Australia) and for lowering serum cholesterol levels (Taiwan, New Zealand).

Colostrum from both normal and hyperimmunized cows appears to have properties that are beneficial to health. The desirable properties of colostrum derivatives can be concentrated to enhance the positive effects. It is also possible to combine the effective fractions of colostrum with probiotic lactic acid bacteria. These may facilitate combinations that can be used as functional additives or ingredients for clinical foods and special dietary foods.

B. The Potenttial Use of Probiotics

Almost a hundred years have passed since the introduction of theories on the prolongation of life by modulation of the intestinal ecosystem. The scientific basis for use of probiotic organisms has only recently been firmly established and some sound clinical studies have been published. The physiological and nutritional properties of selected bacterial strains are well characterized, and it is possible to verify that some strains are "probiotic" with documented effects of maintaining and promoting the health of the host when used as part of the daily diet.

Few well-documented probiotic dairy strains are available at present. The most important are listed in Table 3. Specific effects of probiotics include, for example, modulation of diarrhea from various causes, relief of lactose intolerance, and constipation. More recent claims with some strains include enhancement of immune function, vaccine adjuvant effects, reduction of serum cholesterol levels, and changes in colon cancer–related parameters. The immune enhancing effects have been reported in several studies for two strains, *Lactobacillus acidophilus* LCI and *Lactobacillus* GG, which appear to act as immunoadjuvants (28–38). Effects on cholesterol levels remain to be verified since so many confound-

Table 2 Examples of Current Colostrum Products

Name	Type of product	Where marketed
Bioenervi™	Sports nutrition supplement	Finland, Germany, and other countries
Immune colostrum	Cholesterol lowering	Taiwan, New Zealand, United States
Immune colostrum	Rotavirus diarrhea prevention	Australia
HC1, HC3, and AC2	Cell culture medium	Several countries

Table 3 Current Probiotic Bacteria and Their Reported Effects

Strain	Reported effects in clinical studies	Selected references
Lactobacillus acidophilus La-5	Immune enhancer, adjuvant, protection against traveller's diarrhea; balances intestinal microflora	Link-Amstr et al. 1994; Bernet et al. 1994; Bernet et al. 199.
Lactobacillus acidophilus NCFB 1748	Lowering of fecal enzymes, decreasing fecal mutagenicity, prevention of radiotherapy-related diarrhea	Salminen et al. 1987; Salminen et al. 1993
Streptococcus boulardii	Prevention of antibiotic-associated diarrhea; treatment of *Clostridium difficile* diarrhea	Kaila et al., 1993 Siitonen et al., 1991 Isolauri et al., 1992 Salminen et al., 1993 Majamaa et al., 1995 Raza et al., 1995 Kaila et al., 1995 Antila et al., 1994
Lactobacillus johnsonii LA1	Adherence to intestinal cells; balances intestinal microflora; immune enhancement; adjuvant in *H. pylori* treatment	Salminen et al., 1998
Lactobacillus casei Shirota	Prevention of intestinal disturbances, balancing intestinal bacteria, lowering fecal enzyme activities, positive effects on superficial bladder cancer	Asa et al., 1995 Salminen et al., 1993
Streptococcus thermophilus; Lactobacillus bulgaricus	No effect on rotavirus diarrhea; no immune enhancing effect during rotavirus diarrhea; no effect on fecal enzymes; strain-dependent improvement of lactose intolerance symptoms	Majamaa et al., 1995 Goldin et al., 1992
Bifidobacterium lactis Bb-12	Treatment of viral including rotavirus diarrhea; balancing intestinal microflora; alleviation of food allergy symptoms	Saavedra et al., 1994
Lactobacillus reuteri	Shortening of rotavirus diarrhea; colonizing the intestinal tract	Pedrosa et al., 1995

ing factors have contributed to varying extents in different population groups. At the moment, no firm proof on any of the probiotic strains of lactic acid bacteria is available, and more well-defined clinical work is needed.

The future of research on probiotic bacteria will focus on selecting new, more specific strains for the well-being of the host. It may well be that different regions of the gastrointestinal tract benefit from different probiotic bacteria, and maybe the time has come for disease-specific strains. This is particularly true with conditions such as rotavirus diarrhea and gastritis caused by *Helicobacter pylori*. However, the requirements for good clinical studies will become ever more important. Carefully controlled studies on selected strains could result in the development of probiotic bacteria targeted for specific diseases and their prevention. Mixtures of such organisms may be of general therapeutic value as additives in clinical foods.

REFERENCES

1. Birch, G. G., Kemp, S. E. 1989. Apparent specific volumes and tastes of amino acids. *Chem. Senses* 14(2):249.
2. Schiffman, S. S., Clark, T. B., Gagnon, J. 1981. Influence of chirality of amino acids on the growth of perceived taste intensity with concentration. *Physiology and Behaviour* 28:457.
3. Kemp, S. E., Birch, G. G. 1992. An intensity time study of the taste of amino acids. *Chem. Senses* 17(2):151.
4. Winitz, M. 1972. Process for making nutrient composition, U.S. Patent 3,698,912.
5. Kuntz, L. A. 1993. Acid basics: the use and function of food aciducants. *Food Product Design* May:58.
6. Motekaitis, R. J., Martell, A. E. 1984. Complexes of aluminum (III) with hydroxy carboxylic-acids. *Inorg. Chem.* 23:18.
7. Belitz, H., Wieser, H. 1985. Bitter compounds: occurence and structure–activity relationships *Food Rev. Int.* 1(2):271.
8. Furst, P. 1994. New parenteral substrates in clinical nutrition. Part I. Introduction: new substrates in protein nutrition. *Eur. J. Clin. Nutr.* 48:607.
9. Furst, P. Albers, S., Stehle, P. 1990. Glutamine-containing dipeptides in parenteral nutrition *J. Parent. Ent. Nutr.* 14:118.
10. Karel, M., Langer, R. 1988. Controlled release of food additives. ACS Symposium Series 370. *Am. Chem. Soc.*: 177.
11. Pazur, J. 1991. Enzymatic synthesis and use of cyclic dextrins and linear oligosaccharides of the amylodextrin type. Biotechnology of amylodextrin oligosaccharides, Friedman, R. (Ed.). *ACS Symp. Series* 458:51.
12. Murray, P. R., Webb, M. G., Stagnitti, G. 1995. Advances in maltol and ethyl maltol applications. *Food Technol. Int. Eur.*: 53.
13. Clemente, A. 2000. Enzymatic protein hydrolysates in human nutrition. *Trends in Food Science and Technology.* 11:254.
14. Roy, G. M. 1990. The applications and future implications of bitterness reduction and inhibition in food products. *Crit. Rev. Food Sci. Nutr.* 29(2):59.
15. Jakubke, H. D., Jeschkert, H. 1977. *Amino Acids, Peptides and Proteins*. The Macmillan Press Ltd., London.
16. Windmueller, H. G. 1982. Glutamine utilization by the small intestine. *Adv. Enzymol.* 53:201.
18. The British Nutrition Foundation. 1990. *Report of the British Nutrition Foundation's Task Force on Complex Carbohydrates in Foods*. Chapman and Hall, London.
19. Nutrition Subcommittee of the British Diabetic Association's Professional Advisory Committee. 1992. *Diabetic Med.* 9:187.
20. Ellis, P. R. 1994. Polysaccharide gums: their modulation of carbohydrate and lipid metabolism and role in the treatment of diabetes melititus. In: *Gums and Stabilisers for the Food Industry 7*. (Phillips, G. O., Williams, P. A., Wedlock, D. J. (Eds.). Oxford University Press, Cambridge, p. 205.
21. Vinik, A. I., Jenkins, D. J. A. 1988. Dietary fibre in management of diabetes. *Diabetes Care* 11(2):160.
22. Ministry of Agriculture, Fisheries and Food. 1988. Additives in Food: The Emulsifiers and Stabilisers Regulation. Amendment to Regulations S1 1980 No. 1833, HMSO, London.
23. Singhavanish, C. 1992. Glucomannan production and application in health foods. *International Food Ingredients* 6:25.
24. Nieuwenhuyzen, W. 1995. Lecithin fractions for novel foods. *Food Technol. Int. Eur.*: 47.
25. Macrue, R., Robinson, R. K., Sadler, M. J. (Eds.) *Encyclopaedia of Food Science, Food Technology and Nutrition*, Vol. 4. Academic Press, London.
26. Pakkanen, R., Kanttinen, A., Satama, L., Aalto, J. 1992. Bovine colostrum fraction as a serum substitute for the cultivation of mouse hybridomas. *Appl. Microbiol. Biotechnol.* 37:451–456.
27. Korhonen, H., Tarpila, S. Salminen, S. 1994. Immune colostrum in the treatment of gastritis in human volunteers. IDF Annual Meeting, Adelaide, Australia.

13

Flavoring Agents

GABRIEL S. SINKI

Sinki Flavor Technology, Little Falls, New Jersey

ROBERT J. GORDON

Givaudan Roure Inc., Dramptom, Ontario, Canada

I. HISTORY

The history of an industry is important for understanding that particular industry, its roots, its heritage, and its philosophy. A review of history can provide insight into the true nature of the business. The nature of today's flavor industry is in marked contrast to that of the spice and essential oil business of the Middle Ages. This contrast is so great that some strongly claim that there was no flavor industry before the 17th century. Nevertheless, spices and essential oils remain essential ingredients for flavor creation.

In the following discussion, we make a distinction between the two eras, pre-1800 and post-1800, because prior to 1800 there were no simulated flavors. These became available in the 19th century.

A. Before 1800

1. Spices

During the early days of history, people used spices mainly to enhance or modify the flavor of their food. Although some of the herbs and spices were grown in many parts of the world, the most important spices came from the east, especially from India, Ceylon, and the spice islands (Sumatra, Java, Bali, etc.). The use of spices and recognition of their value can be traced as far back as biblical times. The Queen of Sheba offered spices to King Solomon, and Joseph was sold to spice traders.

The technology of utilizing spices progressed slowly. They were originally used mainly whole or ground. It was only in the 18th century that some extraction started, followed by distillation of essential oils.

Thus was born the flavor industry of the 19th century. It borrowed from two technologies: the extraction technique of natural matters by Scheele and synthetic organic chemistry. It appears that in the early stages (1800), the business was handled mainly by druggists. Around the middle of the 19th century, it evolved into the flavor industry. An excellent review entitled "The History of the U.S. Industry" appears in the book *The Fragrance and Flavor Industry*, by Wayne E. Dorland and James A. Rogers, Jr. (3). In this review, one can see that most of today's firms started about 140 years ago. From about 1850 to early 1900, most flavors consisted of a single chemical. Sophisticated flavors at that time constituted a mixture of three or four ingredients, selected from the 50 available chemicals. Over 90% of the raw materials used for flavors were of natural origin. During the 1930s and 1940s, almost any known chemical could be made synthetically at affordable prices, and the flavor industry flourished with artificial flavors. Some manufacturers thought they had reached the pinnacle, while others, persisting in their search for excellence, borrowed from available technologies to reach new horizons. This led to the next era of progress for the flavor industry, the analytical era.

C. The Gas Chromatography Hype

Around the 1950s, the food industry, the largest market for flavors, was searching for better-tasting, more sophisticated flavors. There was a general fatigue with the synthetic nature of the flavors that faintly suggested the copied fruit but were too perfumy and harsh. Some flavor firms had the vision to apply the new technique of chromatography. In the early hype stages of this application, optimists thought that nature's secrets would finally be unlocked. They hoped that food could be analyzed by gas chromatography and that Mother Nature's flavor secrets would be unveiled.

It did not take long for reality to prevail and it was realized that the new analytical tools—the gas chromatograph, the mass spectrometer, nuclear magnetic resonance, etc.—were only excellent tools and would not replace the artist.

In the 1960s and 1970s, following the advent of these new analytical tools, a number of flavor breakthroughs appeared in the market. New aroma chemicals were identified and became available, increasing the number of components in the chemical "library" from the few dozens of the early 1900s to slightly less than 500 by 1963. The 1960s were the golden era of the flavor industry. More than 75% of the business was represented by artificial (synthetic) flavors. Profit margins were good, and the flavor chemists were in the driver's seat. The industry produced a good return for investors, and interest in acquisition started to peak.

In the following decade a new demand for natural flavors forced changes in the flavor industry. This demand provided those blessed with vision and imagination the opportunity to apply the process of biotechnology, with fermentation and enzymology being the main tools.

2. Essential Oils

The term "essential oils" probably originated with Paraceisus Von Hohenheim (1493–1541), who claimed that they were the most sublime extractive, the *quinta essentia* (quintessence). The odoriferous oils and ointments traded in the ancient countries of the Orient, Greece, and Rome were not truly essential oils. It is claimed that these odoriferous oils were produced by simply macerating flowers, roots, etc., in oil. It was when the process of distillation became known in the Middle Ages that the large-scale production of essential oils commenced.

In the second half of the 16th century essential oils began to be widely produced and used, due largely to the publishing of the book *Liger de Arte Distillandi*, by Brunschwig (1500 and 1507).

Essential oils were used predominantly by pharmacists. It was not until the 19th century that they became a factor in the flavor industry. Xavier Givaudan, on the occasion of his company's fiftieth anniversary in 1940, wrote the following:

> My brother Leon, having completed his studies at the Ecole de Chimie Industrielle de Lyon, had the opportunity of meeting Professors Berber and Bouveualt, who were among the pioneers in the field of essential oils. Hearing them speak on the subject (1890), he realized the importance of their work and the vastness of the new field that was opening up.

One can trace the progress of the flavor industry by following the steps of new technology. (See Ref. 1 for more information on essential oils.)

B. After 1800

1. The Chemistry Renaissance

Chemistry progressed from "alchemist methodology" to science slowly but steadily (2):

1608	Succinic acid and benzoic acid were isolated.
1661	Pyroligneous acid was isolated by dry distillation of wood.
1769–1785	Swedish chemist Carl Wilhelm Scheele was the first to apply solvent extraction of plant and animal products to investigate chemical components. Tartaric, citric, malic, gallic, lactic, and uric acids were isolated from natural matters.
1772–1777	Lavoisier (France) was the first to shed some light on the chemical nature of these substance (identification of oxygen and "azote").
1807	Berzelius (Stockholm) was the first to refer to the earlier studies as "organic chemistry" and was the father of the "vital force." Wohler (Germany) was the father of synthetic organic chemistry and the first scientist to refute the hypethesis of the "vital force." He succeeded in preparing synthetic urea.

D. The Natural Trend

It is interesting to note that flavors produced by the industry in the late 19th century were about 90% natural, derived mainly from spices and essential oils. In the 1950s, flavors became about 90% artificial, due to the availability of synthetic chemicals. In the 1980s and 1990s natural flavors comprise roughly 70% of the mix. It is almost as if the industry has gone full circle.

E. Conclusion

As the foregoing review indicates, the flavor industry was developed primarily by entrepreneurs blessed with vision and creativity. It is a service industry that functions well with the entrepreneurial spirit of small to medium-sized organizations. The industry was founded on creativity, and creativity still represents its most crucial element for growth.

As we have seen through its short history, the application of new technologies was the main vehicle in reaching new plateaus.

II. FLAVORS, THEIR NATURE, CREATION, AND PRODUCTION

A. Definition of a Flavor

What is a flavor? A consumer, when describing a flavor, has most of his/her senses at work. His/her description of a flavor might be influenced by a psychological response to the sight (color, shape, appearance, etc.) of the item. Touch and hearing also affect one's judgement of taste and odor. A trained judge or an expert, on the other hand, tries not to be influenced by stimulation of senses other than taste and odor. Hall (4) directed his definition to cover the flavor perception: ''Flavor is the sum of those characteristics of any material taken in the mouth, perceived principally by the senses of taste and smell, and also the general pain and tactile receptors in the mouth, as received and interpreted by the brain.''

The Society of Flavor Chemists (5) formulated in 1969 the following definition of the product itself: ''A flavor is a substance which may be a single chemical entity, or a blend of chemicals of natural or synthetic origin, whose primary purpose is to provide all or part of the particular effect to any food or other product taken in the mouth.''

The International Organization of the Flavor Industry (IOFI) defined flavors from the industry's point of view (6): ''Concentrated preparation, with or without solvents or carriers, used to impart flavor, with the exception of only salty, sweet, or acid tastes. It is not intended to be consumed as such.''

According to the Council of Europe: ''Flavoring is a substance which has predominantly odor-producing properties and which possibly affects the taste.''

B. The Process of Flavor Creation

Flavor chemists' approach to creation of a flavor varies, depending on the nature of the project and their training. The oldest and simplest method is the artistic approach. The second approach is to combine art with scientific know-how. The third approach is to follow nature's footsteps and develop flavors biosynthetically.

1. The Artistic Approach

During their training period, flavor chemists learn the organoleptic qualities of the multitude of raw materials approved for use in flavor compounding. These several components comprise synthetic organic chemicals, natural essential oils, botanical extracts, juices, spices, nuts, herbs, and concentrates of animal origin. The sensory characteristics of these ingredients leave various impressions. Each flavor chemist will have the characteristics of these components and their corresponding food-nuance associations stored in memory, dependent on his or her personal ability and training.

In this approach, the artistic powers of the flavor chemist are at work. The food flavor to be developed—for example, butter—may have its flavor profile described as typical buttery, cooked, cheesy, waxy, creamy, nutty.

At this stage, the flavor chemist tries to associate the flavor profile of the model food with the raw materials that would serve as the building block in the process of flavor reconstitution. Again, each flavor chemist has a unique interpretation and association. A theoretical simplification of four flavor profiles is shown in Tables 1–4. This is the stage of dreaming and planning that precedes preliminary formulations and bench trials.

Table 1 Flavor Profile of Butter and Its Associated Building Blocks as Perceived by Flavor Chemist

Flavor chemist's perception of butter	Associated components as building blocks
Typical buttery	Diacetyl, starter distillate, acetoin, acetyl propionyl
Lactone, cooked	Δ-Decalactone, Δ-dodecalactone, γ-decalactone
Cheesy	Butyric acid, caprioc acid, caprylic acid, capric acid
Waxy	Myristic acid, palmitic acid, dodecanal
Creamy	*cis*-4-Heptenal, methyl amyl ketone
Nutty	2-Hexenal, pyrazines

Table 2 Flavor Profile of Mango and Its Associated Building Blocks as Perceived by Flavor Chemist

Flavor chemist's perception	Associated components as building blocks
Fresh	Acetaldehyde, hexyl butanoate *cis*-3 hexenol
Sweet	Nerol, γ-octalactone, γ-decalactone, γ-ionone
Cooked/juicy	4-Hydroxy-2,5-dimethyl-3(2H)-furanone
Tropical/sulfury	Dimethyl sulfide
Citrus	Linalool, nerol, citronellol, geraniol
Floral	Linalool, nerol, linalyl acetate

Table 3 Flavor Profile of Chicken (Boiled Type) and its Associated Building Blocks as Perceived by Flavor Chemist

Flavor chemist's perception	Associated components as building blocks
Meaty	4-Methyl-5-thiazole ethanol acetate
Cooked	2,3-Butane dithiol, dimethyl disulfide
Sulfury	Hydrogen sulfide
Fatty/oily	2,4-Decadienal, linolenic acid, oleic acid
Skin	2,4-Heptadienal

Table 4 Flavor Profile of Roast Beef and Its Associated Building Blocks as Perceived by Flavor Chemist

Flavor chemist's perception	Associated components of building blocks
Roasted	Trimethyl pyrazine, 2-ethyl-5-methyl-pyrazine, dimethyl sulfide
Meaty	Dimethyl sulfide, 3,5-dimethyl-1,3,4-trithiolane
Fatty	Oleic acid, hexanoic acid
Cooked	Methyl mercaptan, hydrogen sulfide, dimethyl sulfide

We might divide flavors, related to the nature of their formulation, into two categories:

Flavors containing one or two major "characterizing keys" representing the core of the flavor profile
Flavors without a major "characterizing key"

In the process of creating a flavor of the first type, the characterizing key component becomes the cornerstone upon which the artist builds the flavor creation. Examples of this type of flavor and characterizing key chemical are listed in Table 5.

Dealing with the second type, where no major characterizing components are present, flavor creation needs more talent and ingenuity. Examples of this category would be strawberry, blueberry, and chicken. In this case the flavor chemist might use a formulated key of several components as the core of the creation, or just use a total formula that will be adjusted during the process of fine tuning.

It is important to emphasize that this description of the artistic approach is provided as a simple illustration. Each flavor chemist will handle the task in a manner that suits his/her own talent and creativity. Some of the various techniques will be described later in this chapter.

This approach can be called an iterative method, or a trial by error approach. The flavorist produces a trial formulation, evaluates the trial by taste, and then makes revisions based upon their evaluation. This is the traditional type of methodology and has been used by flavorists for many years.

2. The Scientific Approach

The artistic approach is used along with the scientific knowledge. With the introduction of chromatographic analysis in the early 1960s, new horizons were opened that allowed flavor creations that rivaled nature in their quality. The utilization of chromatography, along with various other analytical techniques, has helped the flavor chemist gain a better understanding of nature's process in producing flavors. This knowledge is used in combination with the artistic skills of the flavor chemist to create unique flavors.

Table 5 Example of Flavors of the Type Containing Characterizing Key Chemicals

Flavor	Characterizing key chemical
Anise, fennel	Anethole
Bitter almond	Benzaldehyde
Dill	D-Carvone
Spearmint	L-Carvone
Cassia, Cinnamon	Cinnamaldehyde
Lemon peel, lemongrass	Citral
Cumin	Cuminaldehyde
Tarragon (estragon)	Estragole
Clove, allspice, bay leaf	Eugenol
Eucalyptus	Eucalyptol
Peppermint	L-Menthol
Oil wintergreen, sweet birch, teaberry	Methyl salicylate

It is interesting to note that before 1900 fewer than 100 flavor chemicals were known, and by 1963 there were still fewer than 500 chemicals available for use by the flavor chemist (7). By 1983, after only 20 years of applying modern scientific techniques to flavor creation, the list had expanded to more than 4000 compounds and continues to grow.

Whether these analytical techniques are used to determine an entire flavor profile or simply to identify certain notes or subtleties within a flavor, various analytical practices and precautions must be observed if meaningful results are to be obtained.

a. Analytical Techniques. The analytical techniques (8) have been the subject of numerous articles (some of which will be referenced in this section), which present a broad overview of the subject.

The process of determining the flavor components in a food can be divided into three general phases. These are

1. Isolation of the volatile flavor components from the bulk of the nonvolatile matrix
2. Separation of these components into individual or small groups of components
3. Identification of the flavor components

Prior to analysis, however, most foods must be prepared to a physical form compatible with the extraction techniques to be used. As an example, samples consisting of large particles must have their particle size reduced. Heat produced in operations such as milling and grinding should be kept to a minimum to avoid the formation of artifacts resulting from chemical reactions at elevated temperatures.

ISOLATION OF VOLATILES (9). Once the sample has been reduced to the proper physical form, the volatile flavor components are usually isolated by either distillation or extraction, or a combination of the two.

Since most liquid samples are in an aqueous medium, a simple extraction with an organic solvent in a separatory funnel will often be sufficient, using either polar solvents such as methylene chloride or nonpolar solvents such as ether or pentane. There are also solvents that can be used in special cases, such as isopentane, which can extract the volatile components from a matrix high in ethanol or propylene glycol.

Equipment is also available for the extraction of solid samples, such as ground nuts or seeds. A soxhlet extraction apparatus is commonly used. However, since this method collects the extracted volatiles in boiling solvent, heat-sensitive compounds may be lost or produce artifiacts.

Where significant heat-sensitive materials are present, an extraction in which solvent is continuously recirculated through the sample by means of a pump at ambient temperature is more suitable. Since there is no boiling of the extractant, mixed solvents such as alcohol/water may be used with this method.

Considerable interest has also been reported recently in the extraction process utilizing supercritical CO_2, whereby the sample is placed in a modified Soxhlet extractor with suitable temperature and pressure controls to allow extractions with liquid CO_2.

Distillation is another common method for separating volatile components from the matrix. Since all distillations use heat to some extent, care should be taken if significant amounts of heat-sensitive components are present. A steam distillation is usually employed, which will allow the temperature to rise no higher that 100°C. Vacuum distillations are also commonly used and can be run at lower temperatures.

Once the volatile components have been isolated, it is usually necessary to concentrate them, since they will almost always be contained in a large volume of solvent. If the solvent is water, as is the case with most distillations, then volatiles must be extracted into a lower-boiling organic solvent.

An interesting combination of the distillation and extraction process is found in the Likens-Nickerson apparatus. As with other solvent extracts, the organic phase from the Likens-Nickerson distillation is very dilute in the volatile components of interest. Removal of the majority of the solvent is necessary before they can be examined further. This is usually done under mild heat and vacuum in an apparatus such as a rotary evaporator.

SEPARATION OF THE ISOLATED VOLATILE COMPONENTS. Separation into individual or groups of components is performed. Chromatographic techniques of one form or another are usually used by modern chemists to separate the isolated volatiles into individual components. Column and thin-layer chromatography are generally used for rough separations into groups of components, while gas chromatography (GC) and high-performance liquid chromatography (HPLC) are of greater value in determining individual components. The latter, using a liquid medium, can detect nonvolatile components in an extract as well as the volatile components. For more detailed information on HPLC, the reader is referred to the books by Snyder (10) and Kirkland (11). Gas chromatography is an ideal technique for the study of volatile components (12). In the 20 years since GC became commercially available, it has been widely accepted, and instrument configurations and accessories are available to fill almost any need.

Advances in column technology have produced a wide range of both packed and open tubular columns. The recently introduced bonded phase fused silica columns are the latest to reach the market. In today's instruments, they give excellent separations and quantitative data with a high degree of reproducibility.

A large number of liquid phases, having a wide range of polarities, are available and can accomplish almost any separation.

Highly sensitive detectors have also been developed. The flame ionization and thermal conductivity detectors are the most commonly used by the flavor chemist. Applications are also found for others, such as photoionization and electron capture detectors.

For more detailed information on gas chromatography and its relation to flavor analysis, the reader is referred to the publication by Dickes and Nicholas (13).

HEAD SPACE ANALYSIS (14). The perceived aroma from a food or food product is of strong interest. A technique uniquely suited for use in conjunction with GC has been developed. Using this technique, the volatiles in the head space, the atmosphere surrounding the sample, are collected and injected directly into the gas chromatograph. This can be done either dynamically or statically.

In a dynamic head space analysis, the atmosphere above the sample is continually swept over a vapor trap of activated charcoal or Tenax, for example, where the volatiles are absorbed and concentrated. Then either the trap can be extracted with a solvent such as CS_2 and the extract injected into the chromatograph, or it can be placed in the carrier gas line and heated to introduce the volatiles into the instrument.

A static head space analysis consists of placing the sample in a sealed container, either with or without the application of heat, until an equilibrium between the sample and the surrounding atmosphere is attained. A sample of the head space volatiles is then withdrawn with a syringe through a septum in the container and injected directly into the gas chromatography.

Although both of these techniques can be performed manually, head space analyzers are manufactured as accessories to the chromatograph. These automatically carry out the sampling of the head space volatiles and their injection into the equipment, with a resulting increase in both sensitivity and reproducibility.

IDENTIFICATION OF THE FLAVOR COMPONENTS (15). There are a wide variety of tools for the analyst to identify unknown components. If the general class of the flavor components is known and compared to an accumulated library of standards, retention time data can be used for identification.

Tables of retention indices such as the Kovats index have been published. Kovats indices provide the retention times of individual components compared with those of a homologous series of alkanes or esters. Flavor houses have also built up their own private tables. Comparing the retention times on two dissimilar columns (i.e., polar and nonpolar) with widely different retention properties increases the accuracy of identification.

SNIFF analysis may also be used. In this method, an unknown sample is injected into the gas chromatograph and is separated into the various components. As the components are eluted, a portion of each is sent to the detector as usual. A separate portion is also sent to a sniffing port where the flavorist evaluates the odor of the component as it is released from the gas chromatograph and identifies the component.

Mass spectrometry (MS) is also a widely used tool to identify unknown components. The mass spectrometer is attached to a gas chromatograph which separates the unknown sample into its components. Because compounds fragment in a set pattern according to their chemical structure, the fragmentation pattern or mass spectrum of the unknown component can be identified by comparing it to a library of known spectra. Such comparisons are normally done by computer. With the progress in computers over the past fifteen years, GC-MS systems are available at reasonable costs.

There are commercially available libraries [such as the Wiley and National Institute of Standards and Technology (NIST) libraries] but many of the components contained in such libraries are irrelevant to flavor analysis. Some published compilations such as the TNO/Zeist Library of Volatile Compounds in Food are more useful for flavor analysis, but most flavor companies develop their own libraries, some with over 150,000 components in their database. Other analytical tools such as Raman, nuclear magnetic resonance, and x-ray spectroscopy can also be used; however, the individual components must be isolated and purified before these techniques can be applied.

The reproducibility and accuracy of modern GC instruments coupled with advances in computer technology make GC a key analytical method for the flavor industry.

UTILIZATION OF ANALYTICAL DATA. Regardless of the care taken in the extraction and analysis, the analytical data obtained can rarely be used to formulate a flavor that exactly matches the original target. The reasons for this deviation from a complete analysis include the following:

- The low concentration of total flavor volatiles in food products (For example, 3–5 ppm for tomato, with individual components in the parts per trillion range). This is below the detection limits for routine analysis.)
- The large number of volatile substances in foods, which range from 100 to 700 components.
- The loss of the more volatile components during distillation and concentration.
- Chemical reactions among the various volatiles that produce artifacts.

For these reasons, the analytical data are usually used to form a ''skeleton'' flavor that approximates the original. From this, the flavor chemist's expertise will be called on to modify and round out the formula to produce the desired flavor.

3. The Biosynthetic Approach

In this approach, the biochemist attempts to duplicate nature's biogenetic pathways. Although science has not yet unlocked many of nature's secrets in developing its flavors, some of the known enzymatic and fermentation reactions are crudely exploited to produce building blocks. Table 6 differentiates between the two broad classifications of natural flavors: primary origin and secondary origin. Under primary origin, flavors of purely biological origin, with little human interference, are listed. Secondary origin flavors, mainly produced through technology, are clearly differentiated. Enzymatically modified cheeses, fermented fruits and wine, and cooked and roasted foods are just a few examples.

This area is usually handled by research chemists rather that flavor chemists. In this endeavor, the researcher exploits natural biological processes (16). The primary areas of biotechnology that are being utilized and explored for use in the flavor business are (15,17)

- The product of natural flavor ingredients
- The production of part or complete flavors
- Increased yields of essential oils, oleoresin, and flavor components by the use of enzymes in the processing of natural materials
- Plant cloning for maximum yields of secondary metabolites
- Specific chemical steps to produce economies in the production of expensive flavor chemicals or to support natural claims (e.g., production of natural benzaldehyde from other natural chemicals)
- Modified traditional fermentation systems to produce flavor enhanced bases for direct use or further concentration
- Secondary metabolites from in vitro tissue culture of vegetable cells

Due to the increased demand for natural flavors, research in this area is dramatically increasing. It ranges from preparing single chemicals to producing complex mixtures of aroma products. Table 7 provides some examples. Another route to the production of natural flavors and aroma chemicals is the technique of plant tissue culture. In this process, a part of the plant responsible for the bioproduction of aroma is allowed to reproduce in a manner similar to that of microbial fermentations. Proper selection of plant cells, cultivation medium, nutrients, and hormones will influence yields. Although numerous reports

Table 6 Classification of Nature's Flavors

		Secondary origin	
Source	Primary origin	Biological	Thermal
Botanical	Fruits, vegetables, spices, flowers, nuts	Wine, beer, bread	Coffee, cocoa, caramel
Animal	Fish, beef, chicken, milk	Fermented sausage, cheese	Roast beef, boiled chicken, grilled cheese

Table 7 Examples of Biosynthetic Production of Chemicals

Major product(s) produced in nature	Examples of occurrence	Microorganism applied[a]	Major portion of substrate
Methyl ketones	Cheese	*Penicillium Roqueforti*	Fatty acids
Lactones	Peaches, coconut	*Pityrosporum* species	Lipids
Butyric acid	Butter	*Clostridium Butyricum*	Dextrose
Carveol, carvone, dihydrocarvone, perillyl alcohol	Spearmint and other essential oils	*Pseudomonas* species	Limonene
d-Verbenone d-*cis*-Verbenol		*Aspergillus Niger*	α-Pinene
Cheeselike flavor	Cheese	*Streptococcus* species *Lactobacillus* species	Reconstituted milk
Breadlike flavor	Bread	*Saccharomyces Cerevisiae*	Sugar and milk

[a] Exclusively food microorganisms are applied.

claim major breakthroughs in this area, the few products produced cannot yet compete economically with fermentation and enzymatic processes.

For further details on this subject, readers are advised to review Chapter 19 of the book *Comprehensive Biotechnology*, M. Moo-Young, Editor (Pergamon Press).

4. Thermally Produced Flavors

In this process, the chemist follows the natural route of forming the ''secondary origin'' flavors mentioned in Table 6. The nomenclature for flavors and process flavors produced thermally is now standardized by IOFI and FEMA. The standards established by the industry specify both the ingredients used and the process applied to produce these types of flavors. Ingredients used in their preparation consist of one or more of the following:

- A protein nitrogen source, for example, meat, poultry, dairy products, seafood, and their hydrolysis products
- A carbohydrate source, for example, vegetable, fruit, sugars, and their hydrolysis products
- A fat or fatty source
- Herbs, spices, water, selected vitamins, acid, emulsifiers, and nucleotides
- Flavoring preparations and flavor adjuncts to be added after processing

The prepared mixture is then processed under controlled conditions. The temperature should not exceed 180°C for the specified time.

The general concept of producing flavors thermally is to utilize foodstuffs or constituents of foodstuffs and apply processes comparable to traditional kitchen treatments.

5. General Rules for Flavor Creation

In spite of the increasing number of technological processes and tools, artistry is still the most important factor in the process of flavor creation. Modern flavor chemists utilize all the scientific resources available to them to create the proper balance between art and science to achieve excellence. Attempts to use science only, such as using analytical tools to recreate flavors from data produced, have proven to be fruitless. Reasons for failure of such an approach can be summarized as follows:

Only a few of the hundreds of chemicals identified in foods have sensory significance.
Availability and cost of the total chemicals identified will prove prohibitively costly.
The methodology to obtain qualitative and quantitative identification of trace chemicals is not very accurate.
Synthetic chemicals, utilized to duplicate nature, contain trace impurities that affect the final taste of the compound.

Therefore, technical resources are tools that should be employed to service the flavor chemist. The following will highlight the important factors influencing flavor creation.

a. Flavor Profile of the Target. The flavor target could be any flavor, from a fruit such as a mango to roast beef. Drawing the preliminary flavor profile of this target is an important step in the process of the flavor development and constitutes the foundation upon which the flavor will be created. However, the economic importance of the flavor to be developed will dictate the investment in time allocated for this area. Therefore, depending on the size of the project, the flavor profiling panel can range from one individual (the flavor chemist) up to a team of experts. Moreover, techniques applied in the area can range from simple descriptives up to the sophisticated techniques of ''flavor mapping.''

Currently, flavor evolution has been leaning toward more of the ''fantasy'' type flavors. In this case, there is no target provided. The flavor created will not be the exact or modified image of nature; rather, it will be a combination of pleasant sensations. Tropical punches and fruit mixtures are some examples. In these cases, a model target might have been concocted in the kitchen or left totally to the flavor chemist's imagination.

b. Flavor Descriptive Language. The descriptive language used in the flavor industry, as well as in the food industry, is subjective and nonstandardized. A communication gap exists not only between the industry and its customers (the food scientists), but also among the flavor chemists themselves.

It is meaningless to utilize the multidimensional scaling technique (18) to quantify, for example, ''green,'' while every individual panelist's perception of green is different. Therefore, it is imperative to use a standard descriptive language among flavor chemists and food scientists. Table 8 is a guide chart developed by Givaudan for expert panel communications. In this method, defined chemicals are associated with the descriptive terms. For example, ''green'' is differentiated into several associations of chemicals such as the green of 2-hexenoic acid or 2-ethyl-1-hexanol or *cis*-3-hexenol.

Descriptive language for consumer panels represents another challenge. For details on this subject, readers are advised to refer to the comprehensive work of Moskowitz (19).

c. Sensory Evaluation (20). What will differentiate between widely acceptable flavors and a ''flavorist's flavor'' is the method of sensory evaluation applied during fine tuning prior to the release of the flavor. Minimal use of personal judgment (on the part of the flavorist), the application of proper sensory techniques, and large panels will make the difference in the fine tuning process. Reviewing the flavor profile by utilizing focus groups will bring the right balance of art and science. Consumers should be the ultimate judges of the quality of a flavor.

d. Reproducibility. Since flavors are complex mixtures of various components, the flavor chemists' task is to ensure the compatibility of such mixtures. Some reactions are expected and considered normal aging processes. However, equilibrium should be attained

Table 8 Givaudan's Flavor Descriptive Language

Flavor term	Acids	Alcohols	Aldehydes	Esters	Ethers	Indoles	Ketones	Lactones	Phenols	Pyridines	Pyrazines	Sulfur chemical
Balsamic	Cinnamic acid	Cinnamyl alcohol		Benzyl-cinnamate			Zingerone	a		b	c	d
Burnt									Carvacrol			
Camphoraceous		Borneol		Iso-bornyl acetate	Eucalyptol		Fenchone					
Citrus		Decanol	Decanal	Linalyl acetate								
Earthy		Terpinen-4-ol		Benzyl formate	Methyl eugenol						2-Ethyl-3-methoxy pyrazine	
Fatty	Decanoic acid, dodecanoic acid	Dodecanol	2,4-Decadienal	Butyl butyryl lactate			Acetyl methyl carbinol					
Fecal						Skatole						
Floral		Geraniol	Phenyl acetaldehyde	Benzyl acetate	Methyl isoeugenol		Most ionones, methyl-ionones, methyl naphthyl ketone	Omega-6-hexadecenlactone				
Fruity		Iso-butyl alcohol	Acetaldehyde	Most esters, isoamyl acetate		Skatole	2-Octanone					

Table 8 (Continued)

Flavor term	Acids	Alcohols	Aldehydes	Esters	Ethers	Indoles	Ketones	Lactones	Phenols	Pyridines	Pyrazines	Sulfur chemical
Green (stemmy, leafy)	*trans*-2-Hexenoic acid	2-Ethyl-1-hexanol, *cis*-3-hexenol		*trans*-2-Hexenyl acetate			Geranyl acetone	Gamma-n-Butyro-lactone				
Herbal			Cuminalde-hyde	Isopule-gyl acetate	Estragole		Pipertone					
Minty		L-Menthol		Carvyl acetate			L-carvone					
Musty		Isomenthol		Phenethyl isobuty-rate	Methyl eugenol							
Nutty	Pyruvic acid		2-Hexyl cinnamal-dehyde	Butyl formate			Methyl cyclo-pente-nolone	Gamma-hepta-lactone	2-Acetyl pyridine	2-Acetyl pyrazine		
Oily		Furfuryl alcohol	2,4-Deca-dienal	Ethyl oleate								
Phenolic (medicinal)									Thymol			
Pungent	Acetic acid		2-Hexenal	Ethyl acrylate			Benzo-phenone		*p*-Cresol			

Smoky	Pyroligneous acid								Guaiacol			
Soapy	10-Undecenoic acid											
Spicy			Cinnamaldehyde						Engenol			
Sweet			*trans*-Cinnamaldehyde	Iso-Amyl acetate	Anethole		Methyl acetophenone		Methyl guaiacol (creosol)			
Waxy	Myristic acid	Tetradecanol	2,4-Decadienal	Ethyl myristate								
Woody				Bornyl valerate	Caryophyllene oxide		Methyl ionones					
Garlic												Propyl disulfide,
Onion												Furfuryl methyl sulfide

[a]Most lactones are peach- and coconut-like. Some are more fatty and green than others.
[b]These chemicals are usually dry, harsh, penetrating, and quite diffusive. Mostly used in chocolate, coffee, and tobacco flavors.
[c]These are similar to pyridines, very diffusive and powerful. Used in nut, coffee, and some vegetable flavors.
[d]Powerful and very diffusive. Usually offensive and pungent. Used in onion, garlic, cooked vegetable, chocolate, coffee, and meat flavors.
Source: *Givaudan Corporation*.

in a few days. A slow reaction between components that produce artifacts for extended periods of time should be avoided. As an example, the reaction of benzaldehyde with propylene glycol and ethyl alcohol should be avoided. Any laboratory process should be reproducible in production. Quality assurance starts with a stable, well-formulated compound; no amount of quality control can substitute for good formulation practices.

6. Recent Flavor Creation Technologies

Flavor companies are using computer technology to assist the flavorist in their creative task. Many companies have introduced systems where a flavorist has access to a database of thousands of different flavor preparations. By doing selective searches through this database, the flavorist can retrieve a starting formulation which meets specific parameters (e.g., liquid, natural, kosher, and heat stable). From there, the flavorist can tailor a new formulation that meets a specific customer's needs. By the use of this type of technology, the time to develop a new flavor has been reduced significantly.

There are other tools which have been developed to assist in flavor creation, some commercially available but most developed by flavor companies themselves (21). Some of this equipment allows the flavorist to evaluate almost instantaneously the aroma of blends they make using computer technology even to the extent of including any aroma of a food product—for example, yogurt.

Other equipment is used to accurately measure threshold levels either of individual components or of flavors in finished food products. By using this type of equipment, the flavorist, supported by sensory specialists, can evaluate the different release rates of a flavor in different bases or products and then can adjust formulations so that the same flavor perception can be achieved in different applications (e.g., a beverage and a yogurt).

C. Raw Materials

Components used to compound flavors are either natural or synthetic. Table 9 illustrates the sources of raw materials available to flavorist chemists. The number of synthetic chemicals permitted for use in food flavors changes on a constant basis due to reviews by authorities. At the time of this writing, it can be estimated to be about 1600.

A simple count of natural raw materials could amount to about 600. However, the true count might far exceed this number when all variables are included. Just consider a spice that can be used as is but is also offered in the following forms: roasted oleoresin, extract, tincture, isolate (from the spice), and essential oil. Each of these produces a different taste profile. A tincture, for example, is quite different from a fluid extract of the same

Table 9 Sources of Raw Materials Used in Flavor Compounding

	Natural	
Synthetic[a]	Botanical	Animal
Benzaldehyde Cinnamic alcohol	Fruit and vegetable juice, extract, and distillate Herbs Spices Nuts	Plasma, drippings Seafood byproducts, enzyme-modified cheese, meat extract

[a] See classifications in Table 10 and examples of the groups in Table 11.

spice. Therefore, we can only roughly estimate that the total gamut of raw materials, synthetic and natural, well exceeds 4000 in the United States.

1. Synthetic Chemicals

Only selected chemicals are permitted for use in flavor compounding. Chemicals are evaluated for safety, as will be described in Section V, and approved for use when shown to be innocuous. Chemicals identified in natural materials are not allowed simply on the basis of their occurrence in nature. It is important to note that if such a chemical is proven to be harmful, it will not be permitted (examples: coumarin, safrole, thujone). On the other hand, chemicals not yet found in nature might be permitted when they are proved to be safe (examples: ethyl vanillin, dibenzyl ether, glycol acetate). Table 10 and 11 provide examples of some organic components used in flavors.

2. Natural Raw Materials

Since the flavor industry is primarily concerned with the sensory quality of materials. It is very rare that a spice or any other natural food product is used in its native form. The spice or food is processed to separate the chemical compounds from the neutral matrix (cellulose, fiber, pectin, etc.) The goal is to produce the utmost concentration of aromatic chemicals within the minimum amount of neutral components at the most desirable combination of cost, flavor profile, and stability. In practice, one criterion will usually have to be sacrificed to gain the other advantages required.

a. Methods of Extraction. Botanical or animal raw materials are usually pretreated before being subjected to extraction solvents. Simple grinding is performed with conventional mills or shredders. The goal is to expose as much surface area as possible for efficient extraction. The product at this stage might also be treated to facilitate release of chemical compounds from the cell matrix by maceration or enzyme treatment or other techniques permitted by law. Selection of methods and solvents is usually dependent on the nature of the product and the type of extracts to be produced. The same precautions mentioned in Section II.B.2 on sample preparation for analysis also apply in this area. Raw materials must be carefully handled, avoiding harsh processing, loss of aromatics, and production of artifacts.

(text continues on p. 378)

Table 10 Organic Synthetic Chemicals Used in Flavors

Aromatic		Aliphatic	
Benzenoid	Heterocyclic ring	Cyclic	Acyclic
Phenois	Thiazoles	Lactones	Hydrocarbons
Ethers	Furans		Alcohols
Acetals	Pyrans		Carbonyls
Carbonyls	Thiophenes		Carboxylic acids
Carboxylic acids	Pyrazines		Esters
Esters	Imidazoles		Isoprenoids
Lactones	Pyridines		Sulfur compounds
Sulfur compounds	Pyrroles		Nitrogen compounds
	Oxazoles		
	Thiazoles		

Table 11a Examples of Groups of Synthetic Chemicals and Their Organoleptic Characteristics

	Example				
Subgroup	Structure	Name	FEMA #	Natural occurrence	Organoleptic characteristics
		Aromatic			
Phenols	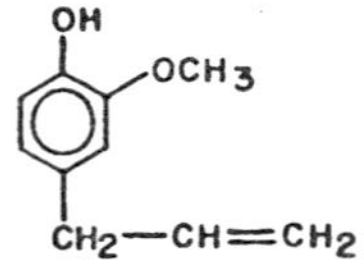	Eugenol	2467	Clove oil, banana, cinnamon leaf oil, cocoa, coffee	Clove-like, spicy
		*p*-Cresol	2337	Ylang-ylang, jasmine, raspberry, cheese, coffee, cocoa	Smoky, medicinal
Ethers	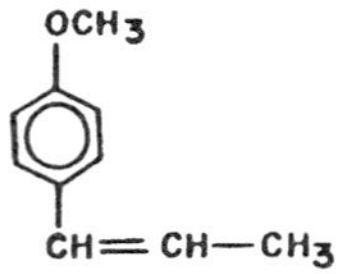	Anethole	2086	Anise, fennel, basil, mint, cheese, tea	Anise odor, sweet, herbaceous
	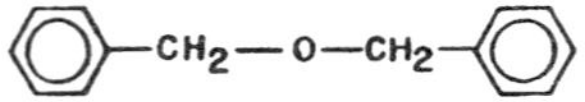	Dibenzyl ether	2371	None reported	Earthy, slightly rosy

Table 11b Examples of Groups of Synthetic Chemicals and Their Organoleptic Characteristics

		Aromatic			
Acetals	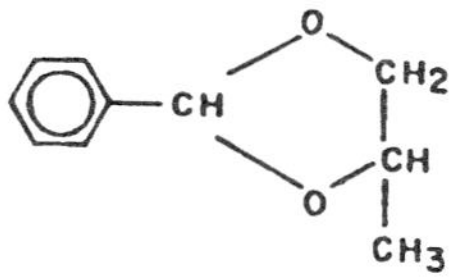	Benzaldehyde propylene glycol acetal	2130	None reported	Weak, almond-like, dirty
	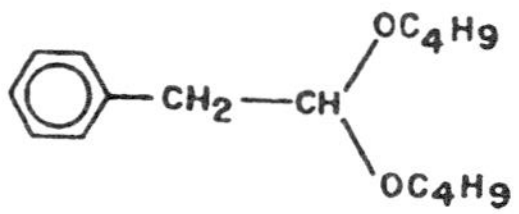	Phenylacetaldehyde diisobutyl acetal	3384	None reported	Sweet, floral, green
Carbonyls	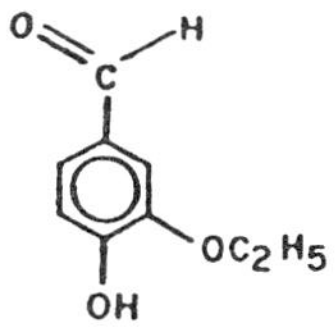	Ethyl vanillin	2464	None reported	Intense vanilla, sweet, creamy

Table 11c Examples of Groups of Synthetic Chemicals and Their Organoleptic Characteristics

Subgroup	Example: Structure	Example: Name	FEMA #	Natural occurrence	Organoleptic characteristics
		Aromatic			
Carbonyls	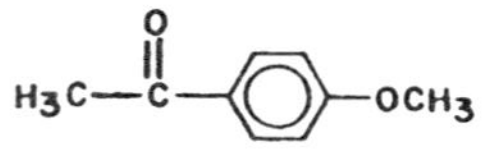	Acetanisole	2005	Anise seed, tomato, tea	Floral, bitter
Esters	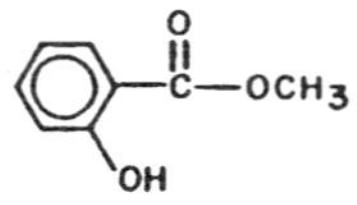	Methyl salicylate	2745	Wintergreen oil, cherry, apple, tomato, wine	Characteristic, wintergreen
	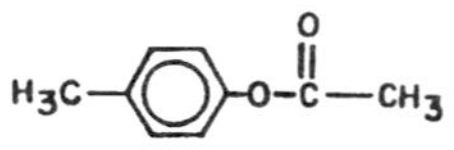	*p*-Tolyl acetate	3073	Cananga oil, ylang-ylang oil	Floral, honey-like
Lactones	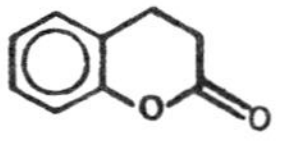	Dihydrocoumarin	2381	Sweet clover	Spicy, vanilla

Table 11d Examples of Groups of Synthetic Chemicals and Their Organoleptic Characteristics

		Aromatic			
Furans and Pyrans	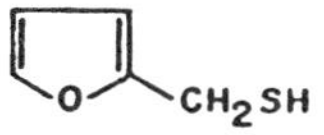	Furfuryl mercaptan	2493	Coffee, beef	Strong, unpleasant, coffee-like
	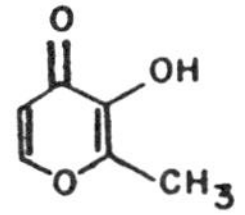	Maltol	2656	Larch trees, pine needles, chicory, roasted malt, strawberry, bread	Sweet, fruity, jam-like
Pyrroles and pyridines	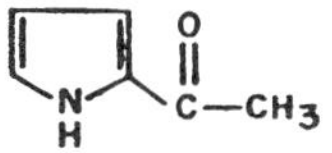	2-Acetyl pyrrole	3202	Bread, cheese, roasted filberts, tobacco, tea	Strong, roasted
		Pyridine	2966	Wood oil, coffee, tobacco	Penetrating, fishy odor, burnt

Table 11e Examples of Groups of Synthetic Chemicals and Their Organoleptic Characteristics

Subgroup	Example: Structure	Example: Name	FEMA #	Natural occurrence	Organoleptic characteristics
		Aromatic			
Sulfur compounds (thiophenes)	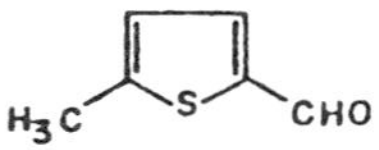	5-Methyl-2-thiophene-carboxaldehyde	3209	Roasted peanuts	Strong, nutty, meaty
Pyrazines	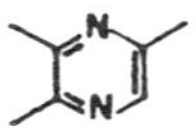	2,3,5-Trimethylpyrazine	3244	Baked goods, coffee, cocoa, peanuts, potatoes	Sweet, roasted peanut
	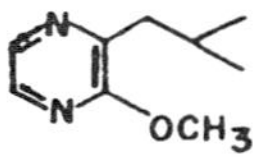	2-Isobutyl-3-methoxy pyrazine	3132	Bell pepper, peas, coffee, potatoes, bread	Powerful, earthy, bell pepper
Thiazoles	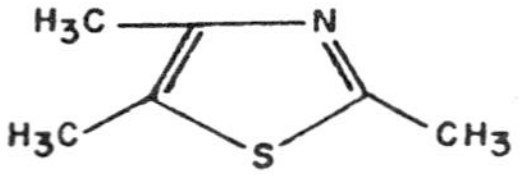	2,4,5-Trimethylthiazole	3325	Potatoes, beef, coffee	Chocolate, nutty, coffee

Table 11f Examples of Groups of Synthetic Chemicals and Their Organoleptic Characteristics

		Aliphatic (cyclic and acyclic)			
Hydrocarbons	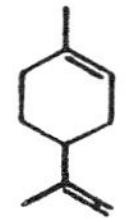	*d*-Limonene	2633	Lemon, orange, mandarin, peppermint	Weak orange or lemon
		α-Pinene	2902	Turpentine, rosemary, lemon, thyme, cheese, nuts	Piney, balsamic
Alcohols		*cis*-3-Hexenol	2563	Apple, orange, raspberry, grapefruit, tea, strawberry	Intense, green odor, leafy
		Decanol	2365	Citrus, ambrette mushroom, wine, apple	Fatty, slightly floral odor

Table 11g Examples of Groups of Synthetic Chemicals and Their Organoleptic Characteristics

	Example				
Subgroup	Structure	Name	FEMA #	Natural occurrence	Organoleptic characteristics
		Aliphatic (cyclic and acyclic)			
Acetals	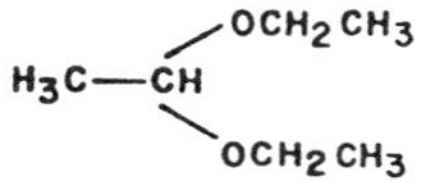	Acetal	2002	Apple, grape, bread, whiskey, rum	Fruity, green
	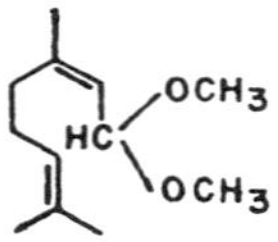	Citral, dimethyl acetal	2305	None reported	Mild, lemon-like odor, oily, green
Carbonyls (ketones and aldehydes)	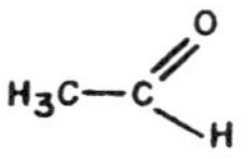	Acetaldehyde	2003	Fruits, tobacco, orange, nuts	Pungent and penetrating
	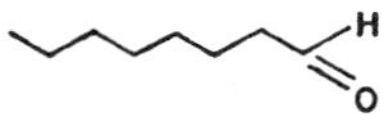	Octanal	2797	Orange, mandarin, grapefruit, rose, beef	Fatty, orange

Table 11h Examples of Groups of Synthetic Chemicals and Their Organoleptic Characteristics

		Aliphatic (cyclic and acyclic)			
Carbonyls (ketones and aldehydes)	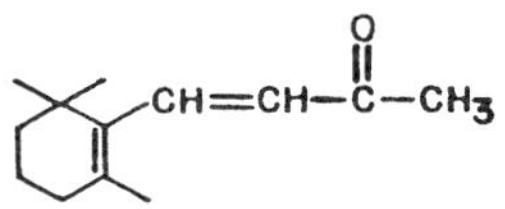	β-Ionone	2595	Raspberry, citrus, tomato, wine	Woody, violet
	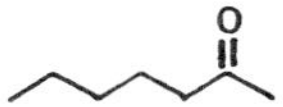	2-Heptanone	2544	Cheese, banana, clove, apple, bread, meat	Fresh, creamy, spicy
Carboxylic acids	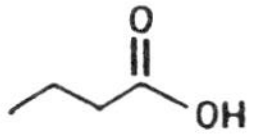	Butyric acid	2221	Dairy products, citronella, bread, strawberry, beef	Rancid, sour milk
	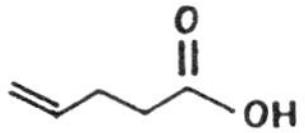	4-Pentenoic acid	2843	None reported	Acrid, caramelic

Table 11i Examples of Groups of Synthetic Chemicals and Their Organoleptic Characteristics

Subgroup	Example: Structure	Example: Name	FEMA #	Natural occurrence	Organoleptic characteristics
	Aliphatic (cyclic and acyclic)				
Carboxylic acids	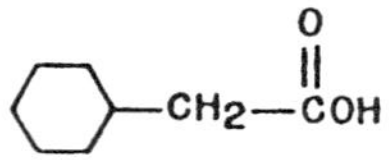	Cyclohexanacetic acid	2347	None reported	Waxy, fatty
Esters	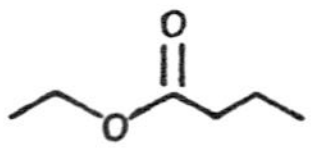	Ethyl butyrate	2427	Strawberry, olive oil, apple, wine, cheese	Fruity, powerful
	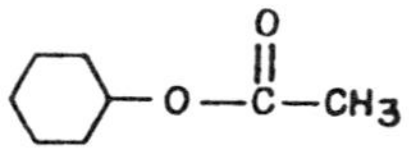	Cyclohexyl acetate	2349	None reported	Fruity, overripe banana, sweet
Lactones	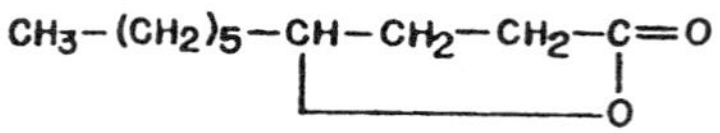	γ-Decalactone	2360	Peach, apricot, strawberry, butter, cheese, meat	Pleasant, fruity, peach-like, creamy

Table 11j Examples of Groups of Synthetic Chemicals and Their Organoleptic Characteristics

		Aliphatic (cyclic and acyclic)			
Lactones	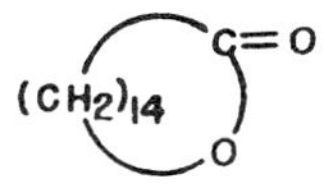	ω-Pentadecalactone	2840	Angelica root	Musk-like
Functionalized isoprenoids	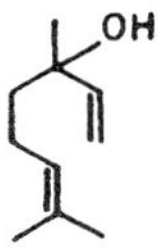	Linalool	2635	Orange, coriander, nutmeg, peach, tomato, beer	Floral, woody, citrusy
	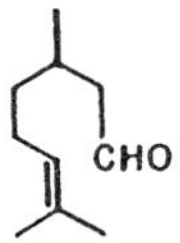	Citronellal	2307	Citronella, lemon, mandarin, grape, cocoa	Floral, citronella, rose-like

Table 11k Examples of Groups of Synthetic Chemicals and Their Organoleptic Characteristics

Subgroup	Example: Structure	Example: Name	FEMA #	Natural occurrence	Organoleptic characteristics
		Aliphatic (cyclic and acyclic)			
Sulfur compounds	$H_3C{-}S{-}CH_3$	Methyl sulfide	2746	Dairy products, meat, peppermint	Unpleasant, cabbage-like
	$CH_2{=}CH{-}CH_2{-}S{-}S{-}CH_2{-}CH{=}CH_2$	Allyl disulfide	2028	Garlic, meat, onion	Characteristic garlic, pungent
	$CH_3{-}SH$	Methyl mercaptan	2716	Meat, cheese, bread	Objectionable, rotting cabbage
	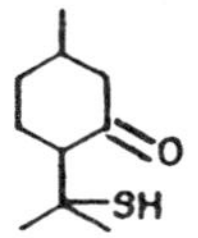	*p*-Mentha-8-thiol-3-one	3177	Buchu leaf	Black currant-like

Table 11I Examples of Groups of Synthetic Chemicals and Their Organoleptic Characteristics

		Aliphatic (cyclic and acyclic)			
Nitrogen compounds	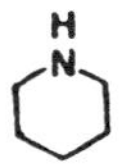	Piperidine	2908	Black pepper, tobacco, bread, meat, fish	Heavy, sweet, animal-like
		Butylamine	3130	Mulberry, cabbage, bread, meat, fish	Ammoniacal

There is a wide array of extraction equipment from which a processor can choose, from simple percolators to sophisticated modified soxhlet extractors and pressurized reactors. Oils and liquid materials are concentrated either by distillation or by liquid–liquid extractors, such as the countercurrent Kerr column and centrifuge extractors.

Although each country has a list of permissible solvents, some extracts are permitted only in conjunction with selected solvents. For example, to produce a household vanilla extract, only a defined percentage of alcohol/water solvent is allowed in the United States. Moreover, other solvents permitted for production of concentrate and oleoresins are regulated, and their trace residue must be controlled. For further information on this subject, readers must refer to local regulations for each country as well as the IOFI publication for the Code of Practice for the Flavor Industry (6).

b. Bases Produced. Except for a few, the hundreds of products thus produced are rarely sold as such to food producers. However, they do constitute the basic building blocks for the creation and production of natural flavors. These products are referred to in the flavor industry by various general terms such as "keys," "bases," or "intermediates."

There are no standard definitions for the following terms pertaining to intermediates used in the flavor industry. The authors have used as a reference Steffen Arctander's book *Perfume and Flavor Materials of Natural Origin* (22). The following definitions refer only to flavoring materials and their terminology.

Absolute. Obtained by alcohol extraction of concrete (see below) of plant materials. Entirely alcohol soluble and usually in liquid form, it is not used extensively in flavors.

Balsam. Natural tree or plant exudate. Semisolid or viscous liquid, insoluble in water and soluble in alcohol.

Concrete. Extract from flowers or plants, obtained by using various types of solvents. Solid or waxy in form and soluble in alcohol.

Distillate. Also known as aroma distillate, referring to the aromatic produced by distillation of extracts or press-cake of fruits. Colorless to pale colored liquid, mostly water soluble.

Essential oil. An oil derived from odorous plants by distillation or expression. The essential oil of a few botanicals such as bitter almond or wintergreen is developed by fermentation or enzymatic reaction. It generally constitutes the odoriferous principles of the plant from which it is produced. Mostly soluble in alcohol and readily soluble in vegetable oil and other essential oils. Essential oils that are produced only by physical expression are called cold-pressed oils.

Extract. Generally an aromatic compound produced by treating a natural raw material with a solvent. This is concentrated either partially or totally by removing all the solvents. Flavor extracts are mainly fluid or semisolid products that are generally soluble in alcohol/water solutions. Some extracts have standards of identity where the aromatic principles as well as solvents are defined (e.g., vanilla). In the flavor industry, the term "fluid extract" refers to a water/alcohol extract, and "solid extract" refers to concentrate made from a fluid extract.

Fold. Indicates the concentration of oils or extracts and refers to the ratio of the quantity of starting material to the quantity of concentrate: for example, a fivefold oil is one in which 10 kg of natural oil has been concentrated to 2 kg. The potency of vanilla extract, expressed in folds, is regulated in the United States and Canada.

Infusion. Rarely used in flavor formulations, an infusion is prepared by refluxing a solvent over raw materials, usually by heat over an extended period of time.

Isolate. A concentrated aromatic compound composed of one isolated ingredient, such as menthol or anethole, derived from natural raw materials.

Oleoresin. Either a natural plant exudate or a concentrated botanical extract, usually prepared by solvent extraction and subsequent total evaporation of the solvent to the pure resinous residue and the volatile and nonvolatile oils derived from the plant. A clear viscous liquid partially soluble in alcohol.

Terpeneless oils. Since most so-called terpeneless oils contain varying amounts of terpenes, this nomenclature should be changed or redefined. It refers to essential oils from which some of the monoterpenes have been removed by solvent extraction and/or distillation processes. Recent studies on the stability of essential oils have demonstrated that the stability is not necessarily achieved by removing all terpenes. There is some evidence indicating that better stability is achieved by reaching a proper balance between oxygenated and nonoxygenated compounds. Sesquiterpeneless oils are essential oils whose sesquiterprenes have also been removed.

Tincture. Alcoholic extract produced directly from botanical raw materials by maceration or percolation without further processing. Heat is not applied. This process stems from original pharmaceutical applications. Tinctures of flavors have no standard for their strength. It is important that the term ''tincture'' be reserved for processes in which the botanical's aroma has been preserved and not destroyed by harsh processing.

D. Production of Flavors

In flavor production, there is no substitute for good manufacturing practices and experienced, conscientious employees for producing finished flavor compounds. The production know-how in this industry is not in the area of sophisticated production equipment but rests, rather, in well-trained staff. The main objective of the production is to ensure the following:

Good maintenance and handling of the multitude of raw materials used in production
Strict adherence to formula and process instructions
Compliance to sanitation and safety rules

For further information, readers are referred to the *Code of Practice for the Flavor Industry* published by IOFI (6).

Due to the diversity of raw materials and formulations, the flavor industry follows the same strict rules that the pharmaceutical industry adheres to. Some of these rules will be discussed in the following pages.

1. General Rules of Good Manufacturing Practices

Incoming raw materials should be kept separate in quarantine from those already approved and passed by quality control.

Only approved QC raw materials and raw material codes should be used. This approval can be by a sticker system or a more elaborate computer system. Complete records should be kept on each lot of raw materials received either on paper or computer file.

All formulas should be compounded by weight, preferably using the metric system. Batch records, including lot number of all ingredients as well as unique batch number for the finished flavor, should be kept for a minimum of two years for ingredient traceablitiy unless local regulations stipulate longer data retention times.

Components of formulas should preferably be single ingredients and not composed of other compounded flavors. However, for the sake of accuracy, components that represent a small percentage should be put together in a separate compound as a key or base.

If aging of raw material or an intermediate is necessary, dates of manufacture and appropriate aging periods should be clearly marked on the containers. An example is the aging of some extracts before using in the flavor compounding.

Raw materials that have separation of oil and water phases or sedimentations should be totally avoided or further treated to ensure homogeneous, uniform components. Quality assurance should never allow a heterogeneous substance to be used in flavor production.

Intermediates and solutions of chemicals (cuts) should be routinely checked by quality assurance personnel. Change of concentrations due to alcohol evaporation or chemical degradation should not be ignored.

Specific raw material naming and QC codes should clearly differentiate similar ingredients. Some conscientious manufacturers use color codes for closely related names—for example, a yellow sticker for ginger oleoresin and a blue one for ginger extract—or use more sophisticated bar code labels and bar code readers.

The flavor formulation must be given to the compounder either in written form or through an advanced electronic system. Complete working instructions must be included with the formulation.

Strict rules of shelf-life have to be adhered to for both raw materials and stocks of finished flavors. Fruit juices that are more than one year old should be discarded even though they might still appear to be normal. Citrus oils have a limited life of less than a year. It is very difficult to reassess the quality of old materials by subjective organoleptic tests.

Formula components and processes must be strictly adhered to by manufacturing personnel; there should be no shortcuts. Accuracy in handling minor components should be well emphasized due to their high impact on the organoleptic quality of the flavor.

In the proper storage of raw materials, their sensitivity to light, temperature, and exposure to oxygen should be taken into consideration. These raw materials should be easily accessible, requiring minimum expenditure of energy and time in handling. The rule of ‘‘first in, first out’’ should be followed for stock rotation.

2. Compounding of Flavors

Generally, flavor manufacturers physically segregate flavor production into two separate areas: one for liquid and semiliquid compounds and another for powdered flavors.

a. Liquid or Semiliquid Flavors. No matter how different the flavor production design is among flavor manufacturers, a good design takes into account the important rules just listed for good manufacturing practices, especially those concerning storage, shelf-life, and accessibility to raw materials. One scheme in which these specifics are considered is the following:

Raw materials are stored under inert gas in stainless steel containers, drums, and tanks, as shown in Fig. 1. Check valves are mounted for each supply line to avoid cross

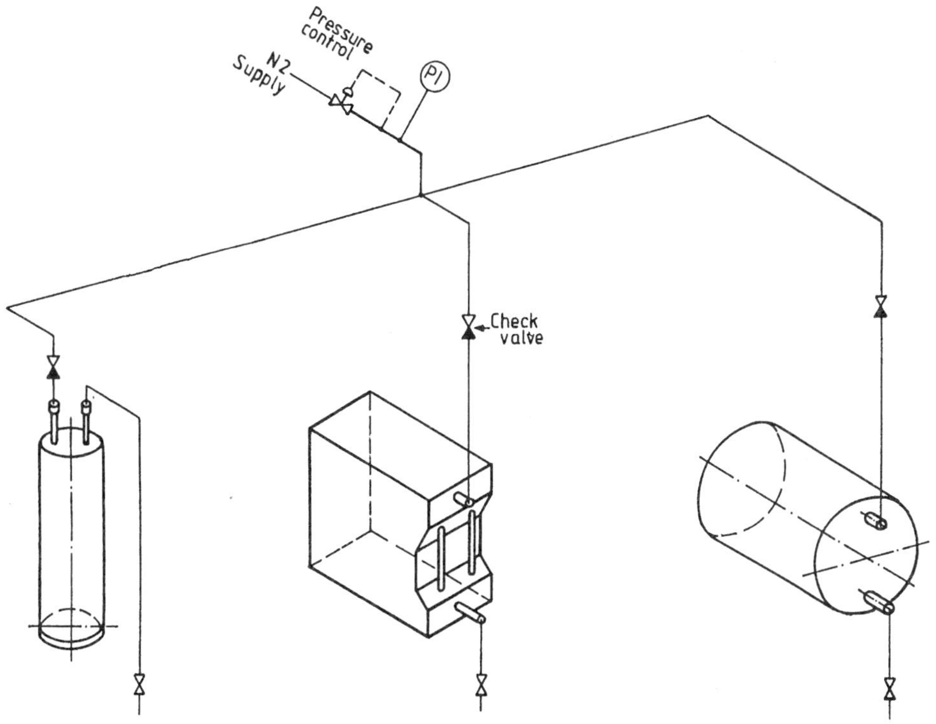

Figure 1 Schematic drawings of flavor raw materials stored under nitrogen.

contamination. The whole system is kept under pressure between 0.5 and 1.5 bar depending on the type of container. Such storage assures a uniform and extended shelf-life.

Ease of access is made possible by the use of portable compounding containers mounted on scales. In this procedure, the compounding vessel is moved around in a one-way route, rather than displacing hundreds of raw materials back and forth from storage areas to production; this can be seen in Figs. 2 and 3, where medium-size and large compounds vessels are used. Moreover, such a system offers added benefits, such as better productivity and control of raw material supplies.

Some of the large producers of spice mixes utilize computer-controlled manufacturing systems. Such computerized compounding is also applicable to the flavor industry. In this case, a flavor formula is fed into the computer, and the components are qualitatively and quantitatively dispensed through the automatically controlled valves, pumps, and scales within the operation.

b. Powder Flavors. Most powdered flavors are produced from liquid or semiliquid concentrates, fixed on an inert carrier at a level of 10–20%. Higher percentages of fix can be made up to 50%; however, for better shelf stability, lower fixation rates should be applied, especially for citrus flavors.

Quite often, incorrect nomenclature is used to designate powdered flavors. As an example, the term ''microencapsulation'' is sometimes used erroneously to refer to spray-

Figure 2 Compounding of medium-size batch flavor. (Courtesy of Givaudan Corporation.)

Figure 3 Compounding of large-size batch flavor. (Courtesy of Givaudan Corporation.)

dried products. The distinction between the various forms of powdered flavors and the process utilized for their production can be illustrated as follows:

Nomenclature	Distribution of flavor in the matrix
Spray-dried	Fine dispersion
Adsorbate	Surface film
Microencapsulation (coacervation)	Coarse dispersion
Spray-chilled distribution	Molecular dispersion

The following paragraphs briefly describe each of these forms of powder.

SPRAY-DRIED (23). Spray-dried flavors that are widely used by the food industry represent the major part of all the dry forms of flavors. In their manufacture, an oil-soluble liquid flavoring preparation is made into an oil-in-water emulsion, the flavoring compound being the disperse phase, water being the continuous phase, and the carrier acting as the emulsifier. Water-soluble flavor compounds are made into a solution or slurry with the carrier. Oil-soluble flavor compounds are made into an oil-in-water emulsion. These preparations are forced through a nozzle or centrifugal device as a mist into a stream of hot gas moving through a chamber. The hot gas, usually air, is heated up to 250°C (450°F) and causes the mistlike feedstock to evaporate and instantly produce dried particles between 10 and 250 micron in size. Elapsed time of operation ranges between 5 and 30 s. Outlet temperature ranges between 71 and 99°C (160–210°F). The spray-drying process is continuous, economical, and reasonably suitable for fixing most volatile aromatics. Today's technology allows a chemical with a boiling point below room temperature, such as acetaldehyde, to be spray-dried. This same technology has resulted in spray-dried citrus oil flavors that are quite stable. A general rule to improve stability is to keep surface oils of the spray-dried flavor at extremely low levels and the rate of fixed oils below 25%.

During the last few years, the matrix of spray-dried flavors, changed from being a single ingredient mostly composed of gum acacia to a complex system of a mixture of two to more carriers. Some of the carriers are acacia, starch hydrolysates, selective hydrocolloids, and simple carbohydrates.

ADSORBATE (24). Adsorbates are produced by plating liquid flavor compounds on solid carriers. Only stable aromatic preparations not prone to oxidation are prepared as adsorbates, due to their large surface exposure to air. Common examples of this type of flavoring preparation are the vanilla-sugar flavors used in the baking industry. These liquid–solid blends exhibit a strong smell of the flavor, have limited shelf-life, are dilute, and are inexpensive. Adsorbates have not found wide applications in food.

MICROENCAPSULATION (25–27). Microencapsulation refers to flavors produced by coacervation of aqueous phase separation. This was largely pioneered by National Cash Register Company, which owns most of the patents. Microencapsulation has found more application, in fields such as paper and drugs. Among the various processes, complex coacervation is the most important. In this process, a dilute aqueous solution of two colloids with opposite electric charge are mixed with a flavor compound to form a homogeneous dispersion of emulsion. The pH is adjusted to an isoelectric point where coacervation is induced and causes the coating material to precipitate around the flavor droplets, thus encapsulating the flavor compound. This produces liquid microcapsules that are solidified by chilling or chemical methods. Mostly gum acacia and gelatin are used in microencapsulation.

One of the drawbacks to this system used to be low flavor loading, but now it is possible to achieve 20 to 30% flavor load, the same range as for spray-drying. The loss of volatiles is less than spray-drying because heat is not involved in the process, but the encapsulate must be kept refrigerated due to the high water content. Also, this technology used to require that the flavor be oil soluble, but improvements have been made.

This technology has been further developed so that it is possible to produce encapsulated flavors by using a blank beadlet made by coacervation, applying a liquid (nonoil) flavor concentrate which is then partitioned into the core of the beadlet. The end result is a dry, free-flowing powdered encapsulated flavor which has up to a 50% flavor load.

Coacervated flavors offer good thermal stability as flavor release is as a result of fracture, that is, when the flavor capsule is broken upon chewing the finished food.

SPRAY-CHILLED. Spray-chilled flavors are very rarely produced by the industry. They are manufactured by using a carrier such as fat, in which a flavor compound has been mixed above the melting point of the fat. The mix is then sprayed into a chamber at room temperature, or in chilled air, thus producing spheroids of spray-chilled flavors.

A quick comparison of the four procedures just described will show that the most economical and practical approach to drying flavors is the spray-drying technique.

E. Quality Assurance and Quality Control

When the quality assurance concept was first introduced, many industries erroneously thought that it was only a sophisticated synonym for quality control (25). It is imperative for the flavor industry to make the distinction and differentiate between the two functions. This task of reproducing the sensory quality of a flavor will be achieved through quality assurance backed up by quality control. Quality assurance uses preventive measures, applied directly to employees and systems, to eliminate problems at their roots. Quality assurance begins with flavor chemists, who must realize that quality starts with their formulations and that whatever process or ingredients are utilized must be exactly reproducible in commercial production. Table 12 describes and differentiates various functions of quality control and quality assurance.

1. Quality Assurance

Quality control is mainly the verification of the quality or raw materials and finished goods against a set of specifications. Quality assurance, on the other hand, is a preventive system where quality is surveyed and monitored right through from product development to ultimate consumption (29). In the flavor industry, quality assurance starts with the formulation of the flavor and deals with all the facets of the production of the flavor to assure excellence in quality. Starting from flavor creation and development, the flavor chemist should consider the following quality assurance points:

Stability of the formulation and compatibility of the ingredients—for example, the reactivity of benzaldehyde with propylene glycol.

Attainment of equilibrium between flavor components within an acceptable period. If aging is required, the necessary holding period should be specified in the process.

Table 12 Differences Between Quality Control and Quality Assurance

	Quality control	Quality assurance
Goals	Screening system—inspection devised to prevent defective products from reaching the marketplace.	A system devised to complement and support quality control, aimed at preventing problems at their roots, to ideally achieve a "zero defects" record.
Fields	Ingredients and products—the inspection, screening and analysis of raw materials and finished goods.	Employees and systems.
End results	Some prevention through the control of ingredients.	Prevention throughout all steps of manufacturing.
	Defects are caught after the fact.	When coupled with an effective quality control program, it is possible to achieve zero defects.
	Defects still exist, and some will pass quality control on to the market.	

Source: Ref. 25.

Specifying and distinctly coding components of the flavor formula—for example, fluid extracts and solid extracts of roasted or unroasted seeds.

Proper accounting of theoretical and actual yields.

Consideration of the practicality of unit weights. Laboratory accuracy is quite different from production accuracy.

Insurance of general compliance with regulations and the specification of only approved ingredients in the formulation.

The inclusion of processing instructions as well as the necessary sequence of addition of the components of the formula.

Verification of reproducibility in scaling up, where necessary.

Proper maintenance of laboratory notebooks or records in which each and every trial is recorded.

Retention of representative samples of flavors compounded by the flavor creation laboratory to compare with the first production runs.

Application of quality assurance to production was discussed in Section II.D.1. Details of other fields of application in the flavor industry are given by Sinki (28).

2. Quality Control

All of the food industry's quality control principles are applicable to the flavor industry. Moreover, the implementation of some of the stringent rules of the pharmaceutical industry have proved to be profitable. Thanks to the application of computers in this area, more can be achieved to ensure quality.

Since the role of flavor is mainly to add organoleptic quality to foods, sensory evaluation should be the most important test for flavor control. The following few lines are devoted to a discussion of quality control of flavors from the point of view of the users rather than the producers.

a. Quality Control of Purchased Flavors. The questions usually raised by users concerning flavor specifications are

1. What will constitute a complete specification?
2. What tests should we apply?

To answer the first question, a specification issued by suppliers should respond to the three basic quality tests: purity, quality, and identity.

Purity. This addresses microbiological and incidental contaminants such as heavy metals.
Quality. A brief description of the organoleptic character of the flavor should be provided. Dosage and taste media should be specified for organoleptic evaluation
Identity. Usually some tests for total aldehydes, volatile oils, etc., are provided. An organoleptic test might also serve for identification. For example, the identity test for a good ingredient such as quinine could be analytically verified. On the other hand, the verification of whether a flavor is strawberry or raspberry is best accomplished organoleptically.

From these notes, it becomes apparent that there is no substitute for sensory evaluation of flavors. Specifications can mention various analytical methods but should always include organoleptic evaluations.

In the 1960s, many flavor users sought the application of gas chromatography to the determination of flavor quality and identity. Such graphs should be correlated with other tests, as they can sometimes be misleading. This can be best exemplified by the fact that poor quality essential oil can sometimes produce excellent GC graph. Moreover, off flavors caused by contamination from minute quantities of potent chemicals are not easily identified by routine QC/GC runs.

The answer to the second question—What tests should be applied?—becomes more evident. The number of physical and chemical tests provided should not superficially judge a specification for a flavor. Emphasis should be on how meaningful or relevant these tests are. For example, a flavor that is mainly an essential oil does not need to have a salmonella test in its microbiological specifications, such as would be required for a dairy flavor.

Tests provided in specifications should address both the nature of the flavor and the end product where the flavor will be used (e.g., completely sugar-free food). The ultimate is the specification worked out in cooperation between suppliers and their customers, taking into consideration the above points.

III. FUNCTION OF FLAVORS AND THEIR UTILIZATION

We shall make an important distinction between the role of flavor and its function. The role of a flavor is simply to impart sensory pleasure to a good, beverage, tobacco, or pharmaceutical, and in doing so it serves a diverse function. It is important to understand the function expected of the flavor so that the type best suited for this function can be delivered.

A. Flavor Functions

Since flavor functions are diverse, we must differentiate these functions, although it must be emphasized that most of the them are intertwined and related. In most applications, however, one function is more important than the others. To better understand how flavors

Table 13 Functional Use of Flavors

Economic	Physiological	Psychological
Simulate	Metabolic response	Nostalgia
Extend	Intestinal absorption	Association
Flavor the unflavored	Appetite and consumption	Intellect/belief, cognitive factors
Modify (cover taste)		Trend
Compensate for flavor losses		Flavor the flavored
Improve shelf-life		

are used by processors Table 13 differentiates among three major categories of function: economic, physiological, and psychological. Many flavor applications perform functions in all three categories, although each category will have a different emphasis. For example, some applications are mainly for pleasure (psychological functions) as exemplified by a hard-boiled candy as opposed to a flavored yogurt positioned for breakfast where the three functions—economic, physiological, and psychological—are applied in differing degrees.

We strongly believe that new foods to be launched in the marketplace must have a raison d'être. The very reason for their innovation should be clearly communicated to consumers by proper positioning and relevant flavor selection. The flavor selected for pleasure in a food positioned as a ''fun snack'' will be more successful when the broad term of pleasure is further reduced to its specifics as shown in Table 13. A flavor profile and character for nostalgic pleasure is quite different from that directed toward pleasure by association.

1. Economic Functions

Obviously, a simulated flavor or an extender must have economic benefits; otherwise it defeats the purpose of its application. Some healthful, nutritional food preparations might have undesirable tastes (e.g., soya and vitamins). Flavors that can modify their taste and make these nutritional, economical food preparations more palatable would be desirable. Another example of this application is the use of flavor to compensate for flavor losses in food processing or to extend freshness during storage.

One interesting concept is known as ''flavor the unflavored'' (copyright, Givaudan Roure). In this instance, the concept is applied to bring variety and palatability to a bland food such as rice or cereal.

2. Physiological Function

Research in this area is still in its infancy. The Monell Chemical Senses Center has a program to study the contribution of taste and smell to nutrition. Various studies on taste and fat digestion provide some indication that taste can alter the metabolic response to a fatty meal (30,31). Research on taste and intestinal absorption of glucose proposes that oral stimulation affects intestinal absorption (32). Monell has recently expanded into the following areas:

1. Chemosensory function and dietary preferences in disease (33)
2. Intravenous feeding and appetite (34)
3. Sodium intake and preference for salty foods (35,36)
4. Saliva composition and taste perception (37,38)

3. Psychological Functions

Although the main role of a flavor is to provide sensory pleasure, psychological analysis of this pleasure can greatly assist in proper flavor selection.

We do not claim that there are rigid boundaries, even within the vertical classification (see Table 13). A nostalgic pleasure can be trendy or both trendy and pleasure-associated. This differentiation is rather conceptual and represents a language developed by Givaudan Roure for use in brainstorming sessions for product innovation.

When attributes such as ''*natural*'' are emphasized in flavors, pleasure derived solely from these attributes is directly dependent on a consumer's belief, intellect, and cognitive factors of how much natural is healthier. Religious belief can also have an important influence on pleasure perception.

Examples of today's trendy flavors are the fantasy types that do not necessarily represent a commonly known food, but rather a novel combination, thus creating a new identity. These are exemplified by the various fruit punches and good hybrids.

The concept of ''flavor the flavored'' is also useful to add variety and increase sensory pleasure to foods that normally need no further flavoring. The addition of flavors to tea and coffee is an example of this concept.

B. Applications of Flavors

Flavors are applied to various products in the food, beverage, tobacco, pharmaceutical, and oral hygiene areas. Lately, some applications have extended to other segments such as the toy industry. Flavored products fall into two categories:

1. *Flavor-dependent.* These are foods and beverages that cannot exist without the application of flavors. Examples are hard-boiled candy, chewing gum, carbonated and nonjuice drinks, gelatin desserts, and powdered artificial beverages.
2. *Flavor-independent.* These are products that can be marketed without flavors or for which flavors are legally prohibited. Examples of the first type of product are crackers, cereals, and nuts. The second type include milk, orange juice, and butter, in which flavor reinforcement is not permitted, unless a new identity is given to the food.

1. Direct and Indirect Flavoring

It is interesting to note that almost every food can be flavored provided that consumers are not misled or deceived. The food industry achieves this, as in the following examples:

1. *New identity.* A food such as milk, butter, or juice can be flavored provided that it is name to convey to consumers its new identity and avoid consumer deception. Examples are blends of butter and margarine.
2. *Indirect flavoring.* Indirect flavorings are preparations that might be added by consumers to their home-cooked meals. Examples include preparations that can be added to meats and poultry in the form of extenders. Salad dressing might also be considered a form of indirect flavoring of home-served salads.

2. Flavor Selection

It is important for flavor users to understand the complexity of flavor selection. It will be helpful to begin by separating flavors from all other additives and placing them in a class by themselves. The various reasons for this differentiation are

1. *Not a commodity.* Some food additives such as sweeteners, spices, colors, and emulsifiers, unless they are in a proprietary mixture, are commodities that can be purchased from alternative suppliers. On the other hand, a flavor such as strawberry is not only unique among individual suppliers, but also must have an identification number to differentiate it from other strawberry flavors produced by the same supplier.
2. *Specificity.* Flavors are very specific to certain applications. For example, the same orange flavor (same identification number and supplier) might taste excellent in some applications but be much less acceptable in others. Flavors in many instances are tailored to suit a complex food system as well as the processing conditions applied.

Therefore, the success of flavor applications is heavily dependent upon and directly proportional to the quality of communication between suppliers and users. In the following section, we briefly highlight the important factors that will assist users in flavor selection and screening.

C. What Does the Flavor Chemist Need to Know from Flavor Users?

In this section, we are not concerned with the extent of communication but rather with its quality. As mentioned earlier, the flavor industry is a service-oriented one that strives for personal contact and mutual trust with its customers. Within the ethics of confidentiality between trusting parties, the following notes are intended as a checklist for both.

1. Marketing Information

In the conceptual stages of development, it is important to share knowledge about the type of consumer product being formulated, its positioning, and its special message to the consumer. This was explained Section III.A.

2. General Information

The following notes briefly list the general information that is usually required for flavor development.

a. Consumer's Product

Type. Details are required on the category and type of food or beverage to be flavored. As an example, if the product is a frozen dessert, it should be specified whether it is sherbert, pudding, or ice cream.

Shelf-life. Expected shelf life within the storage and packaging conditions should be stated.

Type of packaging. The type of container should be identified as being carton, plastic, metal, or glass.

Claims. Will the product have specific health claims? Examples: sodium-free, low cholesterol, sugar-free, fat-free.

Directions. What type of preparation, if any, do consumers require?

Target. Could a model product be developed in a laboratory or kitchen? Is there a similar commercial product available?

Positioning. Information on targeted consumers' age group as well as product market positioning would be very helpful in the process of selecting the proper flavor profile.

Table 14 Flavor Dosage and Most Widely Used Forms Corresponding to Application

Application	Normal dosage range (% in product as consumed		Flavor form				
			Liquid[a]				
	Low	High	W/S	Alc/S	O/S	Powder	Emulsion
Baked goods	0.5	1.5	→		→	→	→
Beverages							
carbonated and still	0.05	0.75	→	→			→
alcoholic	1.0	3.0	→	→			
powdered	0.025	0.40				→	
Cereal	0.25	0.50			→	→	→
Chewing gum	1.0	2.0			→	→	
Condiments	0.1	0.25	→		→	→	→
Confections	0.5	2.00	→	→	→	→	→
Dairy analogues							
cheese and yogurt	0.25	2.00	→		→	→	→
margarine	0.005	0.025	→		→	→	→

Frosting	0.10	0.50	→			→	→
Frozen desserts	0.25	0.75	→				→
Gelatin and puddings	0.05	0.15				→	
Gravies and sauces							
for main meal	0.25	1.00	→		→	→	
Meat, processed	0.10	0.50	→		→		
Snacks	0.50	0.75			→	→	
Toppings and syrups	0.10	0.25	→				→
Dental preparations							
(oral hygiene)	1.0	3.0	→	→			
Pharmaceutical							
chewable tablets	1.0	3.0				→	
elixirs	0.10	0.25	→	→			→
liquid preparations	0.25	0.75	→	→			
Animal foods							
feed	0.10	0.40	→			→	
semimoist pet food	0.25	0.75	→			→	→
regular pet food	0.10	0.25	→		→	→	

[a]W/S = water-soluble; Alc/S = alcohol-soluble; O/S = oil soluble.

b. Product Matrix (Base Recipe)

Type of sugar (sucrose, fructose, glucose, etc.), starches, gums, colors, preservatives, oils, etc. Special ingredients that might be reactive with flavors (proteins, vitamins, acidulants, minerals, stabilizers, oxidizing/reducing agents, etc.)

Active components, if any, as well as pH, viscosity, percent solids, and any other useful information on the base.

c. Processing

Provide information on the equipment to be used such as homogenizer, extruder, etc.

State conditions of pasteurization or sterilization in terms of temperature and holding time.

Indicate whether filling will be at hot or at ambient temperature.

Type of pumps utilized which might affect emulsions or cause foaming.

d. Type of Flavor

Description: fruity, savory, fantasy, etc., fresh, processed, canned, cooked, etc.? Can a model be developed in a kitchen or a laboratory? Can a target profile be well specified? Is there a cost limitation? What type, size, and methodology of sensory panels will be used to judge the flavor?

Claims: natural and/or artificial.

Forms: liquid, powder, emulsion, suspension. (This is usually recommended by the flavor supplier.)

D. Flavor Forms and Dosage

Flavors are supplied in the following forms:

	Forms	Solvents and carriers
Liquid	Water-, alcohol-, or oil-soluble	Alcohol, propylene glycol, triacetin, benzyl alcohol, glycerin, syrup, water, vegetable oil
Powder	Spray-dried, absorbates, or powder mixes	Gum acacia, starch hydrolysates, selective hydrocolloids, simple carbohydrates
Pastes and emulsions	Emulsion of the oil-in-water type	Same ingredients as for powder and liquid

Flavor strength and potency vary considerably. Some flavors are diluted with solvents and carriers, whereas others are compounds of aromatic chemicals without a solvent. Although the flavor dosage is quite varied, there are some industry acceptable ranges, as shown in Table 14.

IV. FLAVOR REGULATIONS

A. The Objectives of Flavor Regulations

The purpose of food laws is to assure safety for the consumer and prevent deception. Labeling regulations require that information be provided on the kind of food, its processing, and the additives contained in it. These principles also apply to flavors. The vast number of flavoring raw materials makes this a difficult task for legislators, resulting in various countries taking different approaches.

1. Flavor Safety

Flavorings are a part of food, and they must be safe for human consumption. Only regulated ingredients that are recognized to be safe are permitted. The following is a review of the types of systems that regulate flavor components.

a. Positive List System. A positive list is composed of flavoring raw materials that are believed to be innocuous and safe for use in food. Positive lists are used mostly to regulate the use of food additives such as preservatives, colorants, and antioxidants. For such additives, the number of products to be controlled is comparatively small. In the case of flavors, however, a positive list is rather voluminous.

The advantage of a positive list system is its convenience in that it indicates exactly what is permitted. Forbidden substances are not listed. The disadvantage of the positive list is that it is voluminous and legally difficult to monitor.

b. Negative List System. A negative list records materials that shall not be used in flavorings because they are known to be harmful and also lists substances for which an upper consumption limit has been fixed.

The advantage of a negative list system is that is usually short, providing a simple means for control of forbidden substances. The main disadvantage is the lack of information on the permitted materials, which are not included in the list. Thus, information on new innocuous substances does not get published.

c. Mixed System. The mixed system was introduced in some European countries to combine the advantages of the positive and negative list systems. The positive list includes approved artificial components (not reported in foods) based on their toxicological appraisal and their past safe use in foods, taking into account their recognizable risks. The negative list indicates forbidden natural flavorings and nature-identical flavoring substances based on proof of adverse effects associated with their intake.

The advantage of the mixed system is the clear listing of forbidden substances, providing adequate consumer protection; it can be enforced in practice without undue cost. New artificial flavoring substances proven to be innocuous can be added to the positive list.

The legal aspects of regulating food additives through a listing system have been extensively discussed by Bigwood and Gerard (39).

The Latin and South American countries mostly follow the mixed system with certain associations to U.S. food laws. In some of these countries, only nature-identical substances that are listed as GRAS by the United States are permitted.

2. Consumer Deception

The purpose of flavoring a food is to add sensory value (pleasure) to that food without altering its identity. Most food regulations are formulated to put a stop to practices where flavor might be used to give false values. The addition of a characterizing flavor is usually not permitted to compensate for the lack of natural flavor due to poor or improper processing. For example, butter or cheese flavors are not permitted to flavor natural butter or cheese, respectively.

3. Consumer Information

The labeling of foods in relation to added flavors is regulated in many countries. The purpose is to label the flavors in conformity with their true nature so that consumers will be properly advised as to the type of flavors used. Although regulations differ greatly from one country to another, the chief objective is to differentiate among the main types of flavors:

Natural flavors
Nature-identical flavors (where applicable)
Artificial flavors
Mixtures of natural flavors and nature-identical or artificial flavors

This is a simplification of terminology. What is considered natural or artificial in one country is not necessarily perceived as such in another.

B. Nomenclature of Flavors

Since all flavors are compounded mixtures, it is imperative to classify the raw materials that are used in the industry. The IOFI defined classes of raw materials according to their origin and nature as follows (40):

1. *Natural aromatic raw materials.* A natural aromatic raw material is a vegetable or animal product used for its flavoring properties, either as such or processed, and is acceptable for human consumption in the form in which it is used; for example, fruit, fruit juice, spices, herbs, balsams, roasted coffee, meat cheese, wine vinegar.
2. *Natural flavoring substances.* A natural flavoring substance is isolated from a natural aromatic raw material by physical methods; for example, citral by fractionation from oil of lemongrass.
3. *Nature-identical flavor substances.* Nature-identical flavoring substances are obtained by synthesis or are isolated through chemical processes from a natural aromatic raw material and are chemically identical to a substance present in natural products intended for human consumption, either processed or not. Examples include vanillin from lignin and citral obtained by chemical synthesis or from oil of lemongrass through its bisulfate derivative.
4. *Artificial flavor substances.* Artificial flavoring substances are substances that have not yet been identified in a natural product intended for human consumption, either processed or not; for example, ethyl vanillin.

Although these definitions for raw materials are acceptable to most countries, the naming of flavors made from such raw materials differs within some countries in their label declaration.

C. Label Declaration of Flavors

Flavors are declared in finished food products according to national food labeling regulations. A brief summary of the flavor labeling regulations for some major countries is given below:

1. Australia

Flavors are considered as a food rather than as food additives. There are references to FEMA, FDA lists, and the Council of Europe book for approved artificial flavoring substances. There is also a negative list of prohibited substances which may not be used in flavors. Flavors may contain designated solvents, carriers, emulsifiers, stabilizers, thickeners, acids, colors, and preservatives, all within prescribed maximum use levels.

When used in standardized food products flavors must, with some exceptions (e.g., cocoa, chocolate, cheese, confectionery, cakes, and other products using flour except breads), be labeled as ''artificially flavored.''

2. Brazil

Flavors are declared as natural, nature-identical, or artificial. There is a short list of approved artificial flavoring ingredients. There is also a negative list of substances considered harmful and not permitted in food. Additionally there is a list of approved food additives with maximum use levels given for each category of food product in which they are permitted.

3. Canada

Canadian Food and Drug Regulations define flavors or essences as natural, naturally fortified (containing at least 51% of the name flavor), or artificial. The term ''imitation'' may be used in place of ''artificial.'' The absence of the words ''artificial'' or ''imitation'' in the description of a flavor indicates that the flavor or essence is of natural origin. There is a list of nineteen standardized extracts, essences, and flavors. The term ''nature-identical'' has no official status in Canada.

There are fifteen tables of various food additives along with the foods in which they are used, the purpose of use and maximum use levels as part of the Canadian Food Regulations, and this includes a table of approved solvents for use in flavors and extracts.

4. Japan

Natural flavorings are regarded as foods and are not directly regulated. All permitted food additives, including nature-identical and artificial flavoring substances are listed as specific chemical compounds and by chemical group. Maximum use levels are stipulated although flavoring substances are listed ''for the purpose of flavor.''

5. Switzerland

All natural and nature-identical flavoring substances are permitted; artificial flavoring substances are permitted if they are on a recognized official list (e.g., FEMA GRAS).

6. United States

a. Natural Flavors. In the United States natural flavors are defined in the Code of Federal Regulations 21CFR101.22(a)(3) as follows:

> The term ''natural flavor'' or ''natural flavoring'' means the essential oil, oleoresin, essence or extractive, protein hydrolysate, distillate, or any product of roasting, heating or enzymolysis, which contains the flavoring constituents derived from a spice, fruit or fruit juice, vegetable or vegetable juice, edible yeast, herb, bark, bud, root, leaf or similar plant material, meat, seafood, poultry, eggs, dairy products, or fermentation products thereof, whose significant

function in food is flavoring rather than nutritional. Natural flavors include the natural essence or extractives obtained from plants listed in other sections of the regulation.

b. Natural With Other Natural Flavor (WONF). This is a class of natural flavorings, containing characterizing and other natural flavoring substances that simulate, resemble, or reinforce the characterizing flavor substance. The criteria here is that the flavor must contain some of the name flavor.

c. Flavor. When the term ''flavor'' is used in the United States, it strictly means that the flavor is natural, since an artificial flavor must always have the adjective ''artificial'' used in its name.

d. Artificial Flavors. Artificial flavorings are made from synthetic components. In the United States, as in Canada, the regulations do not make a distinction for the synthetic components known to occur in nature recognized by the term ''nature-identical'' in Europe.

7. European Community

Flavor Directive 88/388 has been adopted by all member countries for the labeling of flavors. There are six categories of flavorings:

1. Natural flavoring substances
2. Nature-identical substances
3. Artificial (non–nature-identical)
4. Processed flavorings
5. Smoke flavorings
6. Flavoring preparations

The rules for what may be considered as a natural flavor are in Flavor Directive 88/388 and the declaration of a flavor in a food is mandatory as ''flavor.'' A more specific description such as ''nature-identical'' or ''artificial'' is permitted but not mandatory. There are complicated rules concerning the declaration of ''natural flavors.''

At present, the individual member countries are working together to establish a list of approved flavoring substances and their respective natures. This list is expected to be completed by 2004. Until the EU list of approved flavoring substances is complete and in effect in the EU countries, each individual member state is following their respective list of flavoring substances.

D. Flavors and National Food Laws

1. The United States

The use of flavors in food is regulated in the United States by the Food and Drug Administration. Relevant chapters in the Code of Federal Regulations refer to the definition of flavors, labeling provisions, and lists of permitted flavoring materials and substances. In addition to the FDA regulations, the Flavor and Extracts Manufacturers' Association (FEMA) issues regularly updated lists of flavoring substances that have been generally recognized as safe by an independent expert panel.

Founded in 1909, FEMA is well known internationally for its contribution to the industry. It is appropriate to devote a few lines to this association to highlight its various scientific activities:

Safety Evaluation Coordination Committee. This committee directs an extensive safety testing program through its Expert Panel. This is a panel of independent expert pharmacologists and toxicologists who pass on the safety of current or new ingredients and whose unanimous conclusion that a substance is generally recognized as safe is published shortly thereafter. Such data and publications benefit several national and international organizations.

Technical Committee. Through its members, this committee has been instrumental in proofing analytical techniques and adapting methods of establishing standards and detection of adulteration.

2. National Flavor Associations Abroad

Not only do flavor regulations vary from one country to another, but the application for flavors to foods is also regulated in various ways. For precise information on flavor regulations and food laws in a particular country, the reader is advised to contact the respective authorities of that country. The local flavor associations are well prepared to supply information on local food laws and on import/export regulations. The national associations in 21 countries are listed in the appendix at the end of this chapter.

E. International Regulatory Organizations

1. Codex Alimentarius

Shortly after World War II, it became evident that rapidly growing international trade would largely be facilitated by a harmonization of the individual national food laws. One of the first organizations to study this problem was Codex Alimentarius. This organization was founded in 1962 jointly by the United Nations Food and Agriculture Organization (FAO) and the World Health Organization (WHO). In order to establish generally acceptable food standards for a number of internationally traded food items (e.g., meat, frozen food, fats and oils, soups and sauces, cocoa, and chocolate), it maintains international committees on these various food categories. Each of the food committees possesses state-of-the-art expertise in its respective field. In stepwise procedures, specifications for different food items are elaborated and finally adopted. All member countries of the Codex Alimentarius are obligated to accept these international standards, at least for imported food.

As to the needs and technological aspects of food additives, the Codex Alimentarius Committee on Food Additives is consulted. Their objective is to elaborate and to propose standards for food additives in order to harmonize their regulations. With respect to the food additives, the Codex Alimentarius Food Additive Committee has only a consulting function. The final decision as to whether flavors should be permitted in chocolate products rests entirely with the Committee on Cocoa and Chocolate. Concerning flavors, the Food Additive Committee realizes that positive lists for the large number of available flavorings are not practical, and for this reason, and in view of the self-limiting nature of flavors, they have not reached a final decision on how flavorings are to be listed. The delegates of this committee agree, however, that there should be a distinction between natural, nature-identical, and artificial flavoring raw materials.

2. Joint Expert Committee on Food Additives

Safety aspects of individual additives are the exclusive responsibility of the Joint Export Committee on Food Additives (JECFA), which is a joint FAO/WHO program. On request of the Codex Committee on Food Additives, JECFA reviews at regular intervals the safety of additives and establishes values for acceptable daily intakes (ADIs).

3. The Common Market Commission

Within the Common Market countries, the situation is similar. The Commission of the European Community in Brussels accepts the principle of dividing flavoring substances into three groups, natural, nature-identical, and artificial. A final decision has not been reached; still open is the question of what form a future harmonization in the Common Market countries should take. Whether a positive list will be established or the mixed system, as presently used in continental Europe, will be adopted is uncertain. Both options are still under decision. Nightingale (41) pointed out that the introduction of positive list systems in Europe would not be legally possible because of the constitutional provisions of the Treaty of Rome.

4. The Council of Europe

The Council of Europe, with headquarters in Strasbourg, France, announced 20 years ago that a special ad hoc working group would be formed to study the flavor situation. In 1974, this body released a publication entitled *National Flavoring Substances, Their Sources and Added Artificial Flavoring Substances*, referred to as the Blue Book (42), which lists admitted and temporarily admitted flavoring raw materials. For each listed product, some identifying specifications are given. Only natural and artificial flavoring substances are listed. No distinction, however, is made between nature-identical and artificial flavoring substances. Since the Council of Europe has no constitutional power in Europe, this book must be regarded as a state-of-the-art compilation. Its main drawback is the fact that it is not regularly updated, which is a basic requirement for the flavor industry. On the basis of this publication, the Council of Europe has nevertheless proposed a project for a flavor regulation that would distinguish between natural, nature-identical, and artificial flavoring raw materials. For the natural materials, a negative list is envisaged, whereas nature-identical and artificial raw materials would have to be included in a positive list. Due to its liaison with the Common Market Commission in Brussels, the proposal of the Council of Europe has a good chance of becoming official.

5. International Organization of the Flavor Industry

Founded in 1989, IOFI represents the national flavor manufacturer associations of 21 countries. The members of the national associations are either national flavor houses or internationally operating flavor companies with local representatives or manufacturing facilities. Only national flavor associations can become members of IOFI; individual companies are not accepted. The headquarters of IOFI are located in Geneva, Switerland.*

The objectives of IOFI are to encourage a harmonization of the different flavor regulations to assist in formulating flavor regulations in countries that so far have no or insufficient food laws, and to supply information on raw materials, products, and procedures to authorities on a confidential basis. In 1972, a symposium on the safety evaluation of flavoring substances was organized in Geneva. The proceedings of this symposium, in which more than 20 toxicologists of international reputation participated, were compiled in a small brochure, "Basic Features of Modern Flavor Regulation," which IOFI pub-

---

* International Organization of the Flavors Industry, 8, rue Charles Humbert, CH1205 Geneva, Switzerland.

lished in 1976. Within IOFI, several committees work on such matters as analytical procedures, the status of flavoring raw materials, and legal questions.

The technical committee, at regular intervals, edits the IOFI Code of Practice for the flavor industry, which contains, among other features, a regularly updated list of all artificial flavoring substances not yet found in nature permitted for use in the United States.

Today, IOFI enjoys a worldwide reputation as an independent association. It attends the Codex Alimentarius meetings as well as the Council of Europe sessions as an observer. In some Latin American countries and in Europe, many of the existing food laws were introduced with the help of IOFI or are based on its recommendations.

F. Adulteration of Flavors

Food adulteration is as old as humankind. During the days of the Roman Empire, the trade with water and food was strictly controlled, but it was nevertheless common practice to improve the acceptability of food, for example, by the addition of spices or by aeration. In 1524, Martin Luther called all merchants robbers and thieves (43), complaining mainly about measuring practices and the adulteration of food.

Recent scientific advances have assisted in achieving better safety and quality of flavors. The few cases of adulteration are encountered mainly in the area of extending expensive natural flavors with artificial nature-identical or other inexpensive natural ingredients. Such practice has diminished in the last two decades due to advances in analytical techniques that make it possible to detect them. Vanilla extract, as an example, used to be extended with the less expensive extract of St. John's Bread or carob. This type of adulteration stopped long ago due to easy detection of such adulteration by chromatographic analysis. Synthetic vanillin from lignin was also added to boost natural vanillin in vanilla extracts. Thanks to Bricout et al. (44), who applied stable isotopic rations [^{13}C ^{12}C] to differentiate between the two types of vanillin and helped in preventing extensive adulteration with added synthetic vanillin.

Carbon dating analysis, known as ^{14}C analysis (43) has also helped in the determination of the authenticity of some natural materials. Due to this new technique, synthetic benzaldehyde, as an example, is easily differentiated from natural sources (15).

Flavor houses play an important role in ensuring the quality and authenticity of flavors offered to their customers. They work very closely with scientists from various disciplines to help them establish new and reliable analytical methods. Through its associations (IOFI and FEMA), the industry works to establish self-policing procedures and to promote good manufacturing practices and goodwill among its members.

V. FLAVOR SAFETY

A. General Considerations

Flavors are part of food, so they must be wholesome and safe. The concern of consumers and the responsibility of flavor manufacturers make safety aspects topics of utmost interest and importance. Various publications by R. L. Hall reflect this concern (47–51).

Since the authorities simply demand that flavors be innocuous, various approaches have been chosen to verify the safety of flavoring substances. In 1972, a number of toxicologists and pharmacologists from Belgium, Denmark, France, Germany, Italy, Japan, the Netherlands, Switzerland, the United Kingdom, and the United States met in Geneva and reached the following conclusions (52):

Any flavoring substance may be placed, by expert judgement, on a spectrum of confidence regarding safety-in-use, ranging from one extreme to the other. At the extreme of greatest confidence may be placed those substances which fulfill the following description:

- They belong to a group of substances simple in chemical structure and closely related, several of which have been studied toxicologically, and have been found not to possess significant toxicity at levels higher, by a suitable safety factor, than those which could reasonably be encountered in the diet of man.
- They are known, or can, with confidence, be assumed, to be metabolized to safe products or excreted by known mechanisms. By "metabolized to safe products" is implied absence of appreciable tissue accumulation and biotransformation to products that are not considered to constitute any contradiction to the use of the parent compound as a flavoring.
- They are used, or are proposed for use, at defined and very low levels in the diet.
- They occur widely and naturally, in foods, at amounts generally closely related to the present or proposed use.

At the other extreme, where confidence is lacking, are these compounds, which are

- Complex, unusual, or unknown in chemical structure
- Not closely related to substances whose safety has been demonstrated or are related to substances of toxicological concern
- Used at high dietary levels, especially when this use is in special risk groups, such as children
- Not known to occur, and probably do not occur, naturally in food

At the extreme of high confidence, a low priority for testing is, at present, required, although additional requirements arising from knowledge may be imposed at any time. There is also less need for analytical methodology today, except in those instances in which production figures lead to the suspicion that the levels of use, and the varieties of uses, create a total dietary burden that substantially exceeds that attainable from natural sources.

At the other extreme of low confidence, the full range of toxicological and metabolic studies will be required as well as suitable methods for analysis of the substance in food.

An intermediate position of confidence should carry appropriate graduated intermediate evaluation and testing requirements.

The hundreds of different flavoring materials make it difficult to establish levels of confidence for each one of them. To arrive at reasonable and acceptable priority settings, certain assumptions had to be made to make the problem less complex:

Complex mixtures of natural aromatic raw materials that have always been used in foods are regarded as safe for the present time as long as analytical results indicate no presence of foreign materials such as pesticides, heavy metals, or other known or unknown contaminants.

Under these conditions, it can be assumed with good probability that such aromatic raw materials are safe.

Consequently, nature-identical flavoring substances are considered, by some authorities, as being safe, as long as they are not used in food in concentrations significantly higher than in the natural complex mixture.

Artificial flavoring substances are considered by some authorities as potentially hazardous as long as their innocuousness has not been fully established in extensive and suitable tests.

From these considerations, which used to be common thinking 20 years ago, two conclusions became evident: artificial flavoring substances represent the class of highest risk, and some means had to be found to assess the risk involved with natural or nature-identical flavoring substances in order to set priorities for further testing. The latter point, a setting of priorities, was still not possible or attainable on the basis of these recommendations. This situation is well phrased in Chapter 8 of the Council of Europe's book (42): "Indeed, the prolonged use of a substance, with visible harmful effects, does not prevent that substance from contributing to a marked degree towards the morbidity of the population exposed to it."

The safety evaluation of flavorings is as well a topic for JECFA. In its 17th report (53), it stated that the safety of flavorings cannot be evaluated by simply applying the procedures traditionally used for other food additives. In its 20th report (54), JECFA considered the large number of flavoring raw materials and concluded that a priority list for testing should be established. In a later report, JECFA (55) acknowledged that it would not be possible to carry out full toxicological testing with all substances used in the flavor trade.

B. Correlation of Flavor Consumption with Total Food Consumption

Whereas JECFA provided no priority settings, the FAO/WHO Codex Alimentarius Commission (56) concluded that the consumption ratio proposed by J. Stofberg (57) would be a useful tool for determining priorities as to which substances should be investigated first. The consumption ratio is defined as the ratio between the amount of a particular flavoring substance consumed as a natural ingredient of basic and traditional food and the quantity of the same substance consumed as a component of added flavors by the same population and over the same period of time.

If the consumption ratio for a given material is larger than 1, it can be concluded that the intake of this material is predominantly derived from natural sources. If it is zero, we are dealing with an artificial flavoring material that does not occur naturally in food. The proposal to use the consumption ratio as a means to determine priorities for safety testing is still the only logical and generally accepted procedure. Stofberg and Stoffelsma (58,59) have published quantitative data on naturally occurring flavoring substances. The IOFI, its member associations, and the flavor industry support this approach by sponsoring the Dutch CIVO/TNO publication of annual literature reviews on quantitative data of flavoring substances in food, by encouraging independent researchers to collect such data, and by encouraging efforts of the individual flavor companies to compile such information from their existing files or through additional analytical investigation (60).

The next step, as foreseen in the Stofberg proposal, will be to assign priorities to the numerous flavoring substances according to their consumption ratio. Substances with low consumption ratios would then be selected by the application of a protocol similar to the decision tree approach, as proposed by Cramer et al. (61) for appropriate safety testing.

C. Safety Assessment of Flavors

1. Laboratory Studies

There is no scientific foundation for the common assumption that natural foods and flavors are safe. Moreover, there are no available test methods where ''absolute safety'' can be determined by what is being consumed. It is also difficult to foresee the availability of such test methods in the future. The major drawback on toxicological tests is the inaccuracy of extrapolating animal laboratory test results to humans. Not only are humans vastly different in their metabolism compared to test animals, but the differences among humans in their response to various foods and nutrients is another major variable.

Hall (4) defined safety in food ingredients as ''the practical certainty that injury will not result from the substance when it is used in the manner and quantity proposed for its use within the lifetime of the individual.'' Another definition for safety can be indirectly extracted from the U.S. Food Additive Amendment, where an acceptable food additive is defined as follows:

> A substance which is generally recognized among experts qualified by scientific training and expertise to evaluate safety, as having been adequately shown through scientific procedures or expertise based on minimum use of food to be safe under the conditions of its intended use.

Present methodology of high-level toxicological animal testing does not provide conclusions with a reasonable degree of confidence for judging safety for humans. Therefore, the safety of the FEMA GRAS materials was determined by its Expert Panel using judgment based on technical information provided on these materials. On the other hand, some authorities have adopted the position that all flavoring materials are subject to critical reevaluation before their final legal status is confirmed.

Some of the toxicological test methods used to establish the safety of an additive or flavoring material are discussed next.

a. Acute Toxicity. A group of test animals, usually consisting of ten mice, are given a dose of the substance to be tested. The dose levels, consumed by test animals for 2 weeks, are set to establish the median lethal dose, the dose from which half of the animals in the group die. The results are then expressed as LD_{50} in milligrams of the test substance per kilogram of body weight of the test animal. Acute toxicity is usually used for preliminary investigations, and the results provide some information on the metabolic degradation and actions of metabolites of the substance. The need for further testing will be determined by the results of the acute toxicity study.

b. Mutagenicity Tests (62,63). These are usually the second phase of testing, where gene mutations of bacteria or animal cells are observed in vitro (tissue cell cultures). Some investigators believe that if a substance causes mutation, it should be considered unsafe. However, the disagreement in this area among scientists is based on the fact that while it appears that all carcinogens are mutagenic, there is no reason to believe that all mutagens are in fact carcinogens. In some cases, a mutagenicity test is run in vivo.

c. Metabolic and Pharmacokinetics Studies. In this type of experiment, radioisotopes may be used to determine the distribution kinetics and biochemical fate of the test substance. As an example, if an ester is investigated, it should hydrolyze in vivo tests to the corresponding acid and alcohol. Both of which are known to be normal metabolic ingredients. Observation of the test material for its physicochemical or biochemical degradation in the intestinal tract will help to identify metabolites that might provide information as to its potential toxicity.

e. Subchronic Studies. Subchronic studies are used to provide information on dose-response, characteristics, and to determine dosage levels for chronic feeding studies. Normally, subchronic studies are carried out with two animal species over a span of 90 days. Preferably one of the test animals should be a nonrodent.

f. Reproduction and Teratogenicity Studies. Although only rarely used by the flavor industry, these studies include the exposure of test animals for their life span and for two or three generations. The objectives are to determine whether the test substance affects the reproduction of the animal. The test animals are usually rodents.

g. Chronic Toxicity and Carcinogenicity Studies. The objective of these tests is to demonstrate the absence of carcinogenesis induced by the test material in low doses over most of the animal's life span. The test animals are usually rodents. In order to exclude nonspecific chronic toxic effects, it was suggested that a level of 5% of the test material not be exceeded in the feed.

Flavors in dosage levels of up to 5% pose a number of problems. Possibly dosed to their high sensory capabilities, test animals normally reject such food. If the products were encapsulated, the ingestion of at least five times are much carrier material would be required, resulting in a increase to 20% of the total feed.

2. Interpretation and Evaluation of Results

Flavoring materials, before undergoing toxicological tests, are assessed to determine the extent to which toxicity tests are needed. Flavor manufacturers, in preliminary evaluation, examine the following information:

The chemical and physicochemical properties to assess purity
The chemical structure in order to assess its simplicity or complexity and the occurrence of "suspicious" components, such as epoxides
Similarity to other related substances, the toxicity or metabolism of which is known
Maximum use level, consumption data, and the type of food in which the substance will be used
Occurrence in nature
Past experience of food consumption in which this material was applied

Another proposal in risk assessment is to consider the assumption that nature-identical substance might be relatively safer than artificial chemicals. If this assumption is used along with knowledge of the nature of the chemical structure, substances can be classified into four general categories providing a maximum safe dosage in foods as follows:

Safe Maximum Dosage in Food (ppm)

Nature-identical	Simple structure	5
	Complex structure	1
Artificial	Simple structure	1
	Complex structure	0.1

It was also suggested by Givaudan Roure that using a safety factor* of 5000 instead of the traditional factor of 100 in the preliminary toxicological testing of the four categories might provide some insurance of safety or indicate the necessity of further studies using more stringent toxicological testing.

Besides assessing risk and establishing safety for flavoring materials, toxicological data are also used to establish acceptable daily intake data. The ADI is expressed for humans in milligrams of a substance per kilograms of body weight by the Joint Expert Committee of FAO/WHO. An ADI of "no upper limit" is assigned to materials when no risk could be assessed after extensive testing. When sufficient proof is available to exclude possible risk to consumers with incomplete data, a temporary ADI is assigned until more information is available.

The "no effect level" (NEL) is the dosage for which test results have demonstrated that ingested material exhibits no visible harm. However, it should be noted that science might never be capable of assuring absolute safety Human beings are not created with equal metabolisms, and even when a certain type of food is scientifically considered safe, some of the population may have adverse reactions to it. Individuals with no allergy are the exceptions rather than the rule, and even these individuals might have very slow, nonapparent, adverse reactions.

ACKNOWLEDGMENT

The authors wish to thank the following people for their contributions to this chapter: Dr. Wolfgang Schlegel, Messrs. Jerry DiGenova, Guiseppe D'Urso, and Jerome Lombardo, Ms. Anne Prins, and Ms. Judy Booth.

APPENDIX: NATIONAL FLAVOR ASSOCIATIONS

Australia	Flavor and Perfume Manufacturers' Association of Australia Box 3968 G.P.O. Sydney, 2001 Australia Phone 290-0700 Telex 24191
Austria	Fachverband der Nahrungs- und Genussmittel-industrie Oesterreichs Zaunergasse 1-3 A-1037 Vienna, Austria Phone 72 21 21 Telex 011247
Belgium	Groupement des Fabricants d'Essences, Huiles essentielles, Extraits, Produits chimiques aromatiques et Colorants—AROMA 49 Square Marie-Louise B-1040 Brussels, Belgium Phone 230-4090 Telex 23267
Brazil	Associacao Brasileira das Industrias da Alimentacao—ABIFRA AV Brigadeiro Faria de Lima, 1570 Cj. 72 Pinheiros

* Safety factor is a number by which the daily intake of a substance is multiplied to arrive at a dose level for feeding test animals.

	Caixa postal 202529 10452 S. Paula S.P., Brazil Phone 813-5431
Canada	Flavor Manufacturers Association of Canada 885 Don Mills Road, Suite 301 Don Mills, Ontario. M3C 1V9 Phone 510-8036 Fax 510-8043
Columbia	Associacion Nacional de Industriales—ANDI PO Box 4430 Bogota, Columbia Phone 2349 620
Denmark	Essens Fabrikant Fogeningen Gabrdretorv 16 D-1154 Copenhagen K, Denmark Phone 123880
France	Syndicat National des Industries Aromatiques Alimentaires 89 rue du Fanbourg St. Honore F-75008 Paris, France Phone 265-09 65 Telex 64 231 8
Germany	Verband der Deutschen Essenzenindustrie e.V. Meckenheimer Allee 87 D 5300 Bonn 1, Federal Republic of Germany Phone 65 37 11/65 37 29
India	Perfumes & Flavors Association of India—PAFAI 2-B, Court Chambers 35, Sir Vithaldas Thakersey Marg. Bombay 400 020, India Phone 29 58 75
Italy	Fererazione Nazionale dell' Industria Chemica—FEDERCHIMICA (Gruppo Essence Naturali e Sintetiche) Via Accademia 33 1-20131 Milano Phone 263621 Telex 332 488
Japan	Japan Flavor and Fragrance Manufacturer's Association—JFFMA 3F. Nomura Bld. 14-14 Kodenma-Cho. Nihombashi, Chuo-Ku Tokyo 103, Japan Phone 663 2471 Telex 222 26 45
Mexico	Asociacion Nacional de Fabricantes de Productos Aromaticos A.C. Jose M. Rico No. 55 Mexico 12, D.F.
Netherlands	Vereniging van Reuk-en Smaakstoffenfabrikanten-NEA Vlietweg 14 Postbus 411 NL-1400 AK Bussum The Netherlands Phone 31 2159/448911 Telex 43050

Norway	Norske Aromaprodusenters Forening Postboks 6656, Rodelkka Osio, Norway Phone 2 37 87 40
South Africa	The South African Association of Industrial Flavor and Fragrance Manufacturers PO Box 4581 Johannesburg, Union of South Africa Phone 23 57 91
Spain	Asociasion Espanola de Fabricantes de Aromas para Alimentacion—AEFAA San Bernardo, 23, 2° Madrid 28015, Spain Phone 242 16 16 or 242 2019
Sweden	Foreningen Svenska Aromtillverkare Box 5501 S-11485 Stockholm, Sweden Phone 63 50 20
Switzerland	Schweizerische Gesellschaft für Chemische Industrie—SGCI Nordstrasse 15 CH-8035 Zurich, Switzerland Phone 363 10 30 Telex 528 72
United Kingdom	The British Essence Manufacturers' Association—BEMA 6 Catherine Street London, WC2B5JJ, UK Phone 836 24 60 Telex 299 388
United States	Flavor and Extract Manufacturers' Association of the United States—FEMA 900 17th Street N.W. Washington, D.C. 20006 Phone 202 293 5800

REFERENCES

1. Guenther, E. 1948. *The Essential Oils*. Van Nostrad, New York.
2. Fieser, F. L., Fieser, M. 1956. *Organic Chemistry*. Reinhold, New York.
3. Dorland, E. W., Rogers, J. A., Jr., 1977. *The Fragrance and Flavor Industry*. Wayne E. Dorland Co., Medham, NJ.
4. Hall, R.L. 1968. Flavor and flavoring, seeking a consensus of definition. *Food Technol*. 22: 1496.
5. Society of Flavor Chemists. Flavor chemists define flavor two ways. *Food Technol*. 23:1360.
6. IOFI. *The Code of Practice for the Flavor Industry*, 2nd ed. IOFI, Geneva, Switzerland.
7. Weurman, C. 1963. *Lists of Volatile Compounds in Foods*. Division of Nutrition and Food Research, TNO, Zeist, The Netherlands.
8. Moarse, H., Belz, R. 1981. *Isolation, Separation, and Identification of Volatile Compounds in Aroma Research*. Akademie-Verlag, Berlin.
9. Weurman, C. 1969. Isolation and concentration of volatiles in food research. *J. Agr. Food Chem*. 17:370–384.

10. Snyder, L. R. 1968. *Principles of Absorption Chromatography*. Marcel Dekker, New York.
11. Kirkland, J. J. 1971. *Modern Practice of Liquid Chromatography*. Wiley, New York.
12. Keulemans, A.I.M. 1959. *Gas Chromatography*, 2nd ed. Reinhold, New York.
13. Dickes, G. J., Nicholas, P. W. 1966. *Gas Chromatography in Food Analysis*. Butterworth, London.
14. Kepner, R. E., Maarse, H., Strating, J. 1964. Gas chronographic head space techniques for the quantitative determination of volatile components in multicomponent aqueous solitions. *Anal. Chem.* 36:77–82.
15. Acree, T. E., Teranishi, R., 1993. *Flavor Science Sensible Principles and Technologies*, American Chemical Society. Washington, D.C.
16. Sharpell, F. H. 1986. Microbial flavors and fragrances. In: *Comprehensive Biotechnology*, Cooney, C. L., Humphrey, A. E. [Eds.]. Pergamon Press, Oxford, England, pp. 968–981.
17. Ashurst, P. R., *Food Flavorings*, AVI, New York.
18. Drewnowski, A. 1984. New techniques: multidimensional analysis of taste responsibeness. *Int. J. Obesity* 8:599–607.
19. Moskowitz, R. H. 1983. *Product Testing and Sensory Analysis of Foods*. Food and Nutrition Press, Wesport, CT.
20. Amerine, M. A., Pagborn, R. M., Roessler, E. B. 1965. *Principles of Sensory Evaluation of Foods*. Academic, New York.
21. Charalambous, G. 1963. *Food Flavors: Ingredients and Composition*, Elsevier, New York.
22. Arctander, S. 1960. *Perfume and Flavor Materials of Natural Origin*. Stefen Arctander, Elizabeth, NJ.
23. Masters, K. 1972. *Spray Drying*. Leonard Hill Books, London.
24. Fischer, J. J. 1962. Liquid solid blending. *Chem. Eng.* 69(3):83–98.
25. Bokan, J. A. 1973. Microencapsulation of foods and related products. *Food Technol.* 27(11): 34–44.
26. Gatcho, M. H. 1976. *Microcapsules and Microencapsulation Techniques*. Noyes Data Corp., Park Ridge, NJ.
27. Todd, R. D. 1970. Microencapsulations and the flavour industry. *Flavor Ind.* 1:768–771.
28. Sinki, G. S. 1980. The role of quality assurance: how it differs from quality control. *Chem. Times Trends* III:38–42.
29. Wodicka, V. O. 1971. Quality assurance—the state of the art. *Food Technol.* 25(10):29–30.
30. Michael, N., Brand, J. G., Kare, M. R., Carpenter, R. G. 1985. Energy intake, weight gain, and fat deposition in rats fed flavoured, nutritionally controlled diets in a multichoice (''cafeteria'') design. *J. Nutrition* 1447–1458.
31. Ramirez, I. 1985. Oral Stimulation alters digestion of intragastric oil meals in rats. *Am. J. Physiol.* R459–R463.
32. Threatte, R. M., Giduck, S. A., Kling, M. 1986. Oropharyngeal stimulation of glucose absorption from the small intestine in conscious, unrestrained rats. *Fed. Proc.* 45(3):537.
33. Mattes, R. G., Kare, M. R. 1986. Gustatory sequelae of elementary disorders. *J. Digestive Diseases* 4:129–138.
34. Friedman, M. I., Gill, K. M., Rothkopf, M. M., Askanazi, J. 1986. Post-absorptive control of food intake in humans. Presented at IXth International Conference on the Physiology of Food and Fluid Intake. Seattle, WA. *Appetite* 7.
35. Beauchamp, G. K., Bertino, M. 1985. Rats do not prefer salted solid food. *J. Camp Psychol. I* 99:240–247.
36. Bertino, M., Beauchamp, G. K., Eugelman, K. 1982. Long-term reduction in dietary sodium alters the taste of salt. *Am. J. Clin. Nutr.* 36:1134–1144.
37. Christensen, C. M., Navazesh, M., Briehtman, V. J. 1984. Effects of pharmacologic reductions in salivary flow on taste thresholds in man. *Arch. Oral Biol.* 29:17–23.
38. Christensen, C. M. 1986. Role of saliva in human taste perception. In: *Clinical Measurement*

of Taste and Smell, Meiselman, H. L., Rivlin, R. S. [Eds.]. Macmillan, New York, pp. 414–428.

39. Bigwood, E. J., Gerard, A. 1967. *Fundamental Principles and Objectives of a Comparative Food Law*, Vols. I–IV. Karger, Basel, Switzerland.
40. IOFI. 1984. *Code of Practice for the Flavour Industry*, 2nd ed. IOFI, Geneva, Switzerland.
41. Nightingale, W. H. 1978. The influence of legislation on research in flavour chemistry. *Chem. Soc. Rev.* 7(2):195–200.
42. Council of Europe. 1981. Partial agreement in the social and public health field. In: *Natural Flavouring Substances, Their Sources and Added Artificial Flavouring Substances*, 3rd ed. Maisonneuve, Strasbourg, France.
43. Schmauderer, E. 1975. *Studien zur Geschichte der Levensmittewissenschaft*. Steiner Verlag, Wiesbaden, Germany.
44. Bricout, J., Fontes, J.C., Mervilat, L. 1974. Detection of synthetic vanillin in vanilla extracts by isotopic analysis. *J. Assoc. Anal. Chem.* 57:714.
45. *The United States Pharmacopoeia XXI*. 1985. Mack Publishing Co., Easton, PA, p. 1572.
46. Noakes, J. E., Hoffmann, P. G. 1979. Liquid scintillation counting: recent applications and developments. *Proc. Int. Conf.* 2:457–468.
47. Hall, R. L., 1959. Flavoring agents as food additives. *Food Technol.* 13:14.
48. Hall, R. L. 1960. Recent progress in the consideration of flavoring ingredients undere the food additive amendment. *Food Technol.* 14:488.
49. Hall, R. L. Oser, B. L. 1961. Recent progress in the consideration of flavoring ingredients under the food additive amendment. *Food Technol.* 15:20.
50. Hall, R. L., Oser, B. L. 1965. Recent progress in the consideration of flavoring ingredients under the food additive amendment. III. GRAS substances. *Food Technol.* 19:253.
51. Hall. R. L., Oser, B. L. 1970. The safety of flavoring substances. *Flavor Ind.* 45:47–53.
52. IOFI. 1972. *The Safety Evaluation of Flavouring Substances*. Symposium, June 27–28, Geneva, Switzerland,
53. WHO. 1974. 17th Report of the Joint FAO/WHO Expert Committee on Food Additives. *Tech. Rep. Ser.* 539:13.
54. WHO. 1976. 20th Report of the Joint FAO/WHO Expert Committee on Food Additives. *Tech. Rep. Ser.* 599:20.
55. WHO. 1978. 22nd Report of the Joint FAO/WHO Expert Committee on Food Additives. *Tech. Rep. Ser.* 631:11.
56. FAO/WHO. *Food Standards Program*. Alinorm. 83/12A.
57. Stofberg, J. 1981. Setting priorities for the safety evaluation of flavoring materials *Perf. Flav.* 6(4):69.
58. Stofberg, J. Stoffelsma, J. 1980. Consumption of flavoring materials as food ingredients and good additives *Perf. Flav.* 5(7):19.
59. Stofberg, J. 1983. Consumption ratio and food predominance of flavoring materials. *Perf. Flav.* 8(3):61.
60. van Straten, S., Maarse, H., Vischer, C. A. *Volatile Compounds in Food-Quantitative Data*, Vols. 1 and 2. Central Institute for Nutrition and Food Research, TNO, Institute CIVO—Analysis TNO, Zeist, The Netherlands.
61. Cramer, G. M., Ford, R. A., Hall, R. L. 1978. Estimation of toxic hazard, a decision tree approach. *Food Cosmet. Toxicol.* 16:255.
62. Ames, B. N. Choi, E., Yamasaki, E. 1975. Methods for detecting carcinogens and mutagens with the salmonella/mammalian microsome mutagenecity test. *Mut. Res.* 31:347.
63. McCann, J., Choi, E., Yamasaki, E., Ames, B. N. 1975. Detection of carcinogens as mutagens in the salmonella/microsome test: assay of 300 chemicals. *Proc. Natl. Acad. Sci.-U.S.A.* 72: 5135.

14

Flavor Enhancers

YOSHI-HISA SUGITA

Ajinomoto Co., Inc., Tokyo, Japan

I. GENERAL INTRODUCTION

A. Definitions

A *flavor enhancer* is a substance that is added to a food to supplement or enhance its original taste or flavor. The term *flavor potentiator* has also been used with the same meaning.

The most commonly used substances in this category are monosodium L-glutamate (MSG), disodium 5′-inosinate (IMP), and disodium 5′-guanylate (GMP).

B. Historical Background

Dried seaweed for the preparation of soup stock was registered even in the oldest record on foods written on narrow strips of wood in the 8th century in Japan. The seaweed *kombu (Laminaria japonica)* has been an important item of trade from the northern islands to central Japan ever since. It became an offering for the divine service and a sacred gift at the formal engagement ceremony. Dried fermented bonito, *katsuobushi*, has also been used and was thought to be in the same category. (*Kombu* connotes delight, and *katsuobushi* connotes victory.)

Cooks around the world have known how to prepare good soup using vegetables and meat or bones from time immemorial. Ancient Romans loved *Garme*, fermented fish sauce. In Japan soups were prepared with the unique raw materials mentioned. This may be the reason why Professor Kikunae Ikeda tried and succeeded in isolating the essence of ''tastiness'' of soup. In 1908, he isolated glutamic acid from kombu bouillon and named the unique glutamate taste ''umami.'' He suggested that this should be a basic taste independent of the four traditional basic tastes: sweet, sour, bitter, and salty. In China, the word *xianwei*, which represents the taste common in fish and meat, corresponds to umami. The same is true for *savory* in English, *osmazome* in old French, and *gulih* in Indonesian. Glutamic acid was first isolated from gluten (wheat protein) and named after it by Ritthausen in 1866.

Commercialization of glutamate began in 1909 with its isolation from wheat gluten. Today about 640,000 tons (Anonymous, 1993) of MSG are manufactured annually in some 14 countries throughout the world.

In 1913 the investigations of Ikeda's protege Shintaro Kodama into dried bonito led to a second important realization: the discovery that inosinic acid, known since German scientist Justus Freiherr's mid–19th century experiments on beef broth, was another typical umami substance (Kodama, 1913).

Clarification of the full configuration of umami was not achieved until 1960, when Akira Kuninaka recognized the role of 5′-guanylate as another key component. The shiitake mushroom (*Lentinus edodus*) has been used as an invigorant. In 1960, guanylic acid—which in 1898 had been first introduced to the scientific community by British researcher Ivar Bang in his work on pancreatic nucleic acid—was extracted from the broth of the common shiitake mushroom (Kuninaka, 1960).

Many other flavor enhancers are reported and listed in Tables 6 and 7. Ibotenic and tricholomic acids were isolated from the mushrooms *Amanita strobiliforms* and *Tricholoma muscarium*, respectively. However, these two compounds are not commercially produced.

C. Food Occurrence

It is important to note that both compounds comprising umami, that is, glutamate (salts of glutamic acid) and nucleotides, are key components of living organisms.

Glutamate is naturally present in virtually all foods, including meat, fish, poultry, milk (human milk), and many vegetables. It occurs in bound form when linked with other amino acids to form protein, and also in free form when it is not protein bound or in peptides.

Therefore, protein-rich foods such as human milk, cow's milk, cheese, and meat contain large amounts of bound glutamate, while most vegetables contain little. Despite their low protein content, many vegetables, including mushrooms, tomatoes, and peas, have high levels of free glutamate. The glutamate content in foods is shown in Table 1.

It has been noted that glutamate is an important element in the natural and traditional ripening processes that allow the fullness of taste in food to be achieved. Perhaps this is why foods naturally high in glutamate, such as tomatoes, cheese, and mushrooms, have become important to the popular cuisines of the world.

Nucleotides are specifically distributed as shown in Tables 2 and 3. Disodium 5′-inosinate (IMP) is dominant in meat, poultry, and fish, whereas adenosine monophosphate (AMP) is dominant in crustaceans and mollusks; furthermore, almost all vegetables contain AMP.

The GMP content of mushrooms is particularly high, especially in the shiitake species, which is a traditional cooking ingredient in Japan and China.

Table 1 Glutamate Content in Foods

Food	Protein in food[a] (%)	Glutamate in protein (%)	Protein-bound glutamate (g/100 g)[a]	Free glutamate (mg/100 g)
Cow's milk	2.9	19.3	0.560	1.9[b]
Human milk	1.1	15.5	0.170	22[b]
Camembert cheese	17.5[c]	27.4	4.787[c]	390[d]
Parmesan cheese	36.0[c]	27.4	9.847[c]	1400[e]
Cheddar cheese			5.092[f]	
1 month				21.8[g]
8 months				182[g]
Gruyere			5.981[f]	
Appenzel				460[e]
Beaufort				910[e]
Comte				630[e]
Blue cheese			5.189[f]	
Roquefort				1230[e]
Eggs	12.8	12.5	1.600	23[h]
Chicken	22.9	16.1	3.700	44[i]
Beef	18.4	13.5	2.500	33[i]
Pork	20.3	15.7	3.200	23[i]
Green peas	7.4	14.8	1.100	75[j]
Sweet corn	3.3	15.1	0.500	100[k]
Tomato	0.7	37.1	0.260	246[k]
Tomato				
green				20.0[l]
red				143.3[l]
Canned tomato			0.343[m]	202[n]
Tomato juce			0.303[m]	109[n]
Tomato paste			1.510[m]	556[n]
Onion			0.190[m]	102[n]
Potato			0.347[m]	180[n]
Broccoli			0.375[m]	115[n]
Spinach	3.3	9.1	0.300	47[k]

[a] Resources Council, Science and Technology Agency, Japan (1986).
[b] Rassin et al. (1978).
[c] The Glutamate Association.
[d] Giacometti (1979).
[e] Orsan (1993).
[f] USDA (1976).
[g] Weaver (1978).
[h] Maeda et al. (1961).
[i] Maeda et al. (1958).
[j] Suzuki et al. (1976).
[k] Kiuchi and Kondo (1984).
[l] Inaba (1980).
[m] USDA, (1984).
[n] Skurray (1988).

Table 2 Distribution of Nucleotides in Animal Foods

Food	Nucleotides content (mg/100 g)		
	IMP	GMP	AMP
Beef	163		7.5
Pork	186	3.7	8.6
Chicken	115	2.2	13.1
Whale	326	5.3	2.4
Horse mackerel	323	0	7.2
Sweet fish	287	0	8.1
Common sea bass	188	0	9.5
Pilchard	287	0	0.8
Black sea bream	421	0	12.4
Pike mackerel	227	0	7.6
Mackerel	286	0	6.4
Keta salmon	235	0	7.8
Tuna	286	0	5.9
Globefish	287	0	6.3
Eel	165	0	20.1
Dried benito	630–1310	0	trace
Squid	0	0	184
Common octopus	0	0	26
Spiny lobster	0	0	82
Hairy crab	0	0	11
Squilla	26	0	37
Common abalone	0	0	81
Round clam	0	0	98
Common scallop	0	0	116
Short-neck clam	0	0	12

Source: Nakajima et al. (1961); Ohara et al. (1964); Saito (1961); Hashida (1963); Fujita and Hashimoto (1959).

II. CHEMICAL PROPERTIES

A. Structural Relationships

1. Glutamate-Related Substances

The L forms of α-amino dicarboxylates that have four to seven carbons possess taste properties similar to that of L-glutamate (I) (Fig. 1). Likewise, compounds that have the *threo* form and the hydroxy group in the β position, such as DL-*threo*-β-hydroxyglutamate (II), impart a more intense taste than those compounds that have the *erythro* form and the hydroxy group in the γ position. L-Homocystate (III), which has an SO_3H group in the γ position of the L-glutamate molecule, also has the umami taste.

Other amino acid salts that have similar sensory properties are ibotenate (IV), tricholomate (V), and L-theanine (VI); in contrast, α-methyl-L-glutamate (VII), in which the α-hydrogen atom is substituted for a methyl group, is tasteless, and pyrrolidone carboxylic acid (VIII), which is formed by the loss of water from the NH_2 and γ-COOH groups of L-glutamate, has a sour taste (Kaneko, 1938).

Table 3 Distribution of Nucleotides in Vegetable Foods

	Nucleotides content (mg/100 g)		
Food	IMP	GMP	AMP
Asparagus	0	trace	4
Welsh onion	0	0	1
Head lettuce	trace	trace	1
Tomato	0	0	12
Green peas	0	0	2
Cucumber	0	0	2
Japanese radish	trace	0	2
Onion	trace	0	1
Bamboo shoot	0	0	1
Mushroom, shiitake	0	103	175
Dried mushroom, shiitake	0	216	321
French mushroom	0	trace	13
Dried french mushroom	0	trace	190
Mushroom, enokidake	0	32	45
Mushroom, matsutake	0	95	112
Mushroom, syoro	0	9	16
Mushroom, hatsutake	0	85	58
Mushroom, benitengu dake	0	0	trace
Mushroom, naratake	0	0	trace

Source: Hashida (1963,1964).

2. 5′-Nucleotides

The relationship between umami taste and the chemical structure of nucleotides (Fig. 2) has been systematically studied. As shown in Fig. 2, purine ribonucleotides having a hydroxy group on the 6-carbon of the purine ring and a phosphate ester on the 5′-carbon of the ribose moiety impart umami tastes, whereas purine ribonucleotides phosphorylated at C-2′ or C-3′ of the ribose moiety are tasteless. Thus, IMP, GMP, and XMP (disodium xanthylate) have umami tastes. Purine deoxyribonucleotides having a hydroxy group on the C-6 of the purine ring and a phosphate ester on C-5′ of the deoxyribose moiety also have umami tastes, but their taste intensities are weaker than those of ribonucleotides. The phosphate ester linkage on C-5′ of the ribose moiety is necessary for imparting umami taste. The C-5′ phosphate must have both primary and secondary dissociation of the hydroxy group to exhibit umami taste; if the hydroxyl is esterified or amidified, the umami taste is lost. Synthetic derivatives of nucleotides, such as 2-methyl-IMP, 2-ethyl-IMP, 2-*N*-methyl-GMP, 2-methylthio-IMP, and 2- ethylthio-IMP, are known to impart a stronger umami taste in the presence of glutamate than do normal nucleotides (Kuninaka, 1960).

B. Stability

1. Glutamate

Glutamate is not hygroscopic and does not change in appearance or quality during storage. The characteristic taste of glutamate, umami, is a function of its stereochemical molecular

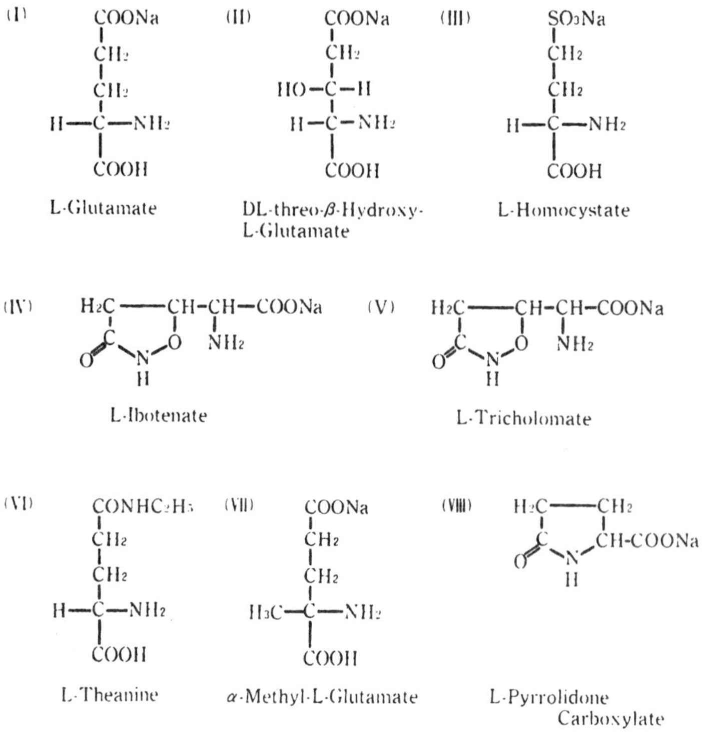

Figure 1 Chemical structure of glutamate-related substances.

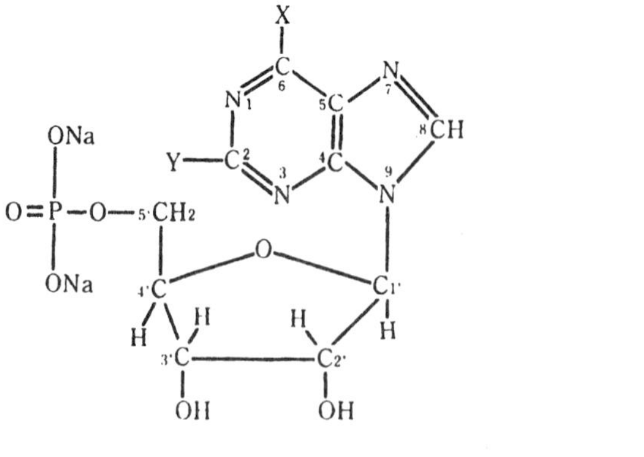

X : OH and Y : H············IMP
X : OH and Y : NH_2·········GMP
X : OH and Y : OH·········XMP

Figure 2 Chemical structures of nucleotides.

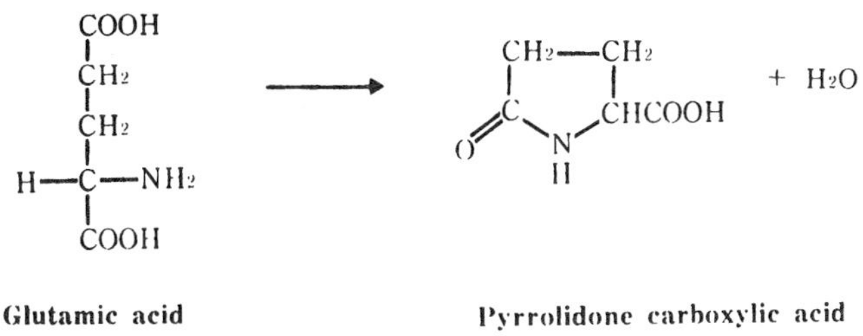

Figure 3 Dehydration of glutamic acid to pyroglutamate.

structure. The D-isomer of glutamate does not possess a characteristic taste or enhance flavors (Yoshida, 1978).

Glutamate is not decomposed during normal food processing or in cooking. In acidic (pH 2.2–4.4) conditions with high temperatures, a portion of glutamate is dehydrated and converted into 5-pyrrolidone-2-carboxylate (Yoshida, 1978).

At very high temperature, glutamate racemizes to DL-glutamate in strong acid or alkaline conditions, but especially in the latter. Maillard (or browning) reactions occur when glutamate is treated at high temperatures with reducing sugars, as is the case with other amino acids (Fig. 3) (Yoshida, 1978).

2. 5′-Nucleotides

IMP and GMP are not hygroscopic. IMP and GMP are stable in aqueous solution, but in acidic solution at high temperature, decomposition of the nucleotides occurs. The ribose linkage of 5′-nucleotides is more labile than the phosphomonoester linkage, and the purine base is completely liberated by heating at 100°C in 1 N HCl.

Enzymatic activity can also have a significant influence on flavor enhancer breakdown and buildup. The phosphomonoester linkage of 5′-nucleotides is easily split by phosphomonoesterases, which are readily found in plant and animal products. From a practical standpoint, these enzymes should be inactivated before the addition of 5′-nucleotide flavor enhancers to foods. Heating or storage below 0°C is usually sufficient to cause inactivation.

C. Manufacturing Process

Glutamate and 5′-nucleotides were originally isolated from natural sources. Even today certain flavor enhancers can be economically isolated from various natural products, but certainly not in the quantities required by the food industry.

Presently the vast majority of commercial MSG is produced through a fermentation process: Most L-glutamic acid producing bacteria are gram-positive, non–spore forming, and nonmotile and require biotin for growth. Among these strains, bacteria belonging to the genera *Corynebacterium* and *Brevibacterium* are in widespread use along with an oleic acid requiring auxotrophic mutant, which was derived from biotin-requiring *Brevibacterium thiogenitalis*.

These bacteria can utilize various carbon sources, such as glucose, fructose, sucrose, maltose, ribose, or xylose, as the substrate for cell growth and L-glutamic acid biosynthesis. For industrial production, starch (tapioca, sago, etc.), cane molasses, beet molasses, or sugar is generally employed as the carbon source.

An ample supply of a suitable nitrogen source is essential for L-glutamic acid fermentation, since the molecule contains 9.5% nitrogen. Ammonium salts such as ammonium chloride or ammonium sulfate and urea are assimilable. The ammonium ion is detrimental to both cell growth and product formation, and its concentration in the medium must be maintained at a low level. The pH of the culture medium is very apt to become acidic as ammonium ions are assimilated and L-glutamic acid is excreted. Gaseous ammonia has a great advantage over aqueous bases in maintaining the pH at 7.0–8.0, the optimum for L-glutamic acid accumulation. It serves as a pH-controlling agent and as a nitrogen source, and solves various technological problems.

Moreover, recent technological innovations, such as genetic recombination, cell fusion, and bioreactor development, are now being applied for further improvement of L-glutamic acid fermentation. Genetic recombination and cell fusion techniques might be useful for the genetic construction of microorganisms with higher production yields or with the capability to assimilate less expensive raw materials such as C_1 compounds and cellulosic materials. Bioreactors packed with L-glutamic acid producing microorganisms are being investigated in an attempt to improve productivity (Hirose et al., 1985).

IMP and GMP are commercially produced by two procedures: (1) degradation of RNA with 5′-phosphodiesterase to form 5′-nucleotides, and (2) fermentation, resulting in the production of nucleosides, which in turn can be phosphorylated into 5′-nucleotides.

MSG, IMP, and GMP occur as colorless or white crystals or as white crystalline powders. They are odorless and dissolve in water readily.

D. Assay Techniques/Analysis for Flavor Enhancers in Food

1. Glutamate

As glutamic acid is the predominant amino acid in most proteins, excessive protein denaturation can result in significantly higher measured glutamic acid levels than can be attributed to actual glutamate addition.

Paper and thin-layer chromatography, amino acid analyzer procedures, gas chromatographic measurement of the trimethylsilyl ether derivative of glutamic acid, and potentiometric titration methods are available. Enzymatic analysis has also been conducted utilizing L-glutamate decarboxylase from pumpkin rind or *Escherichia coli* and L-glutamate dehydrogenase, which catalyzes the conversion of L-glutamate to α-ketoglutarate (Schanes and Schanes, 1946).

In general, glutamate is extracted from the food, preferably under acidic conditions, as free glutamic acid and is subjected to quantitative analysis using liquid column chromatography. A typical analytical method using amino acid analyzer is detailed below (Fujii et al., 1982).

a. Preparation of Sample Solution

WATER-BASED OR SOLUBLE FOOD PRODUCTS. Weigh accurately an amount of sample corresponding to 50–100 mg of glutamic acid and dissolve or disperse it in water. The total volume of the sample solution should be adjusted to 200 mL accurately.

Take about 2 mL of this solution and add to 10 mL of 1 g/dL picric acid solution. For a suspension, perform the following deproteinization operation. For a nonsuspension, use this solution as is.

Deproteinization operation. Take accurately 50 mL of the sample solution, add 150 mL of ethanol, shake well, and filter the solution. Wash the residue three times with 20 mL of 40% ethanol solution each time. Combine both filtrates and washing ethanol solution, concentrate under reduced pressure at 50–60°C to a volume of about 40–45 mL, and then add water to make accurately a volume of 50 mL for sample solution.

PROTEIN FOOD PRODUCTS. Weigh accurately an amount of sample corresponding to 50–100 mg of glutamic acid, place into centrifuge tubes, and add about 5 mL of water at about 100°C for each 1 g of sample. Heat in a water bath for 15 min, cool, centrifuge under refrigeration for about 10 min at about 5000 rpm, and decant the supernatant solution. Repeat the same procedure with 3.3 mL of hot water (100°C).

Repeat this operation three times. Combine all of these separate supernatants quantitatively, add water to a final accurate volume of 200 mL, and filter as needed for sample solution. If the total of the combined supernatants exceeds 200 mL, concentrate under reduced pressure until a volume of approximately 180–190 mL is reached; then add water to an accurate volume of 200 mL, and filter the sample solution if necessary.

Take about 2 mL of this sample solution and add to 10 mL of 1 g/dL picric acid solution. In the case of a suspension, perform the deproteinization operation described under Water-Soluble Foods on the sample solution. If the sample solution is clear, this operation is not needed.

FATTY FOODS. In general, perform the same operation as described under protein food products for the sample solution. However, if there is separation into oil and water layers or if there is marked suspension, perform the following defatting operation for sample solution.

Defatting operation. Place quantitatively all the separate solutions into a separating funnel and add 50 mL ethyl ether or ethyl ether–*n*-hexane (2:1) mixture for each solution. Shake well. Repeat the defatting operation two times and separate out the water layer. Heat the water layer in a water bath until the ethyl ether or *n*-hexane odor disappears, and add water to make an accurate total volume of 200 mL; filter, if necessary.

b. Preparation of Standard Solution. Weigh accurately 127.2 mg of monosodium L-glutamate monohydrate, and dissolve in water to an accurate volume of 100 mL. Take accurately 2 mL of this solution and add citric acid buffer solution (pH 2.2)* to an accurate total volume of 100 mL for the standard solution.

c. Method of Assay

MEASUREMENT CONDITIONS. Using the amino acid automatic analyzer for liquid chromatography, measure under the following conditions:

Packings: gel-form strong cation-exchange resin; average grain size 17 μm; cross-linkage rate 8%
Column dimensions and column temperature: 500 mm × 9 mm, 55°C
Eluent: citric acid buffer (pH 3.25),† 0.6 mL/min

---

* Preparation of citric acid buffer (pH 2.2): Dissolve 19.6 g of sodium citrate dihydrate, 16.6 mL of hydrochloric acid, 0.1 mL of *n*-caprylic acid, 20 mL of thiodiglycol and BRIJ-35 solution in water and adjust to 1000 mL.
† Preparation of citric acid buffer (pH 3.25): Dissolve 7.7 g of sodium citrate dihydrate, 17.9 g of citric acid monohydrate, 7.1 g of sodium chloride, 0.1 mL of *n*-caprylic acid, 5 mL of thiodiglycol, 1 mL of BRIJ-35 solution, and 80 mL of ethyl alcohol in water, and adjust to 1000 mL.

Reaction coil: 20 m × 0.5 mm
Reaction bath temperature: 98°C
Ninhydrin solution‡ flow rate: 0.3 mL/min
Injection volume: 500 μL
Wavelength: 570 nm

PREPARATION OF TEST SOLUTION. Take accurately 10 mL of the sample solution, add hydrochloric acid solution (1 → 6), adjust the pH to 2.2, and make up accurately 100 mL with citric acid buffer (pH 2.2) for the test solution.

DETERMINATION. Calculate the amount of L-glutamic acid by the following formula:

$$\text{Amount of L-glutamic acid (\%)} = \frac{W_S A}{W A_S} \times 0.03145$$

where W_S = milligrams of standard monosodium L-glutamate monohydrate, W = grams of sample, A_S = the L-glutamic acid peak area of the chromatogram obtained using the standard solution, and A = the L-glutamic acid peak area of the chromatogram obtained using the test solution.

$$\text{Amount of monosodium L-glutamate monohydrate (\%)} = \text{amount of L-glutamic acid (\%)} \times 1.272$$

2. 5′-Nucleotides

In the case of 5′-nucleotides, there are numerous procedures that utilize paper and thin layer chromatography and high-performance liquid chromatography. Enzymatic assays specific for 5′-nucleotides are also available.

Thus it can be appreciated that a large number of techniques are available to measure flavor enhancer levels. Perhaps the most important limiting factor associated with all, or at least most, of these techniques is that they are not effective for a wide variety of foods, mainly because of incomplete extractions associated with certain foods or interference due to extraneous compounds. Thus, before a specific technique is chosen for a food, its potential limitations must be evaluated (Maga, 1983).

III. FUNCTION IN FOODS

A. Basic Qualities

1. Umami Taste

The theory of four basic tastes (sweet, sour, salty, and bitter) was proposed by a German psychologist (Henning, 1916), and was accepted for a long time without sufficient scientific data to support it. He explained that all tastes experienced could be made up from the mixture of the four basic tastes, located at the corners of a tetrahedron and located

‡ Preparation of ninhydrin solution: Dissolve 136 g of sodium acetate trihydrate and 25 mL of acetic acid in water and adjust to 250 mL. Add to this solution 750 mL of ethyleneglycol monomethyl ether, pass nitrogen gas for 20 min with stirring. Add 20 g of ninhydrin, pass nitrogen gas for 15 min, and stir well to clear solution. Add 1.7 mL of titanium trichloride, pass nitrogen gas for 10 min, stir well to completely clear solution, press at 0.28 kg/cm^2 in pressure regulator, and preserve.

somewhere on the surface of the tetrahedron. Contrary to this, recent psychometric, biochemical, and neuroelectropysiological studies indicate the following:

Multidimensional scaling analyses of human sensory tests demonstrated that the umami taste is located outside the tetrahedron of the traditional four basic tastes, and the taste quality is distinctly different from those of the other basic tastes (Yamaguchi, 1987). The taste physiological studies in mice also supported that umami taste quality is different from those of the other basic tastes (Fig. 4) (Ninomiya and Funakoshi, 1987).

The taste quality of umami is not produced by mixing any of the other four basic tastes (Yamaguchi, 1987).

Recent electrophysiological studies in primates suggest that a single taste nerve in the chorda tympani is responsible for the taste stimuli of umami substances such as MSG and GMP (Hellekant and Ninomiya, 1991). A study of the monkey cortex demonstrates that umami achieves independence as a taste quality at higher levels of neural processing (Rolls et al., 1994).

Electrophysiological studies suggest that the taste bud receptor site for glutamate is different from those for the traditional four basic tastes (Ohno et al., 1984; Kumazawa and Kurihara, 1990). These electrophysiological data are supported by the

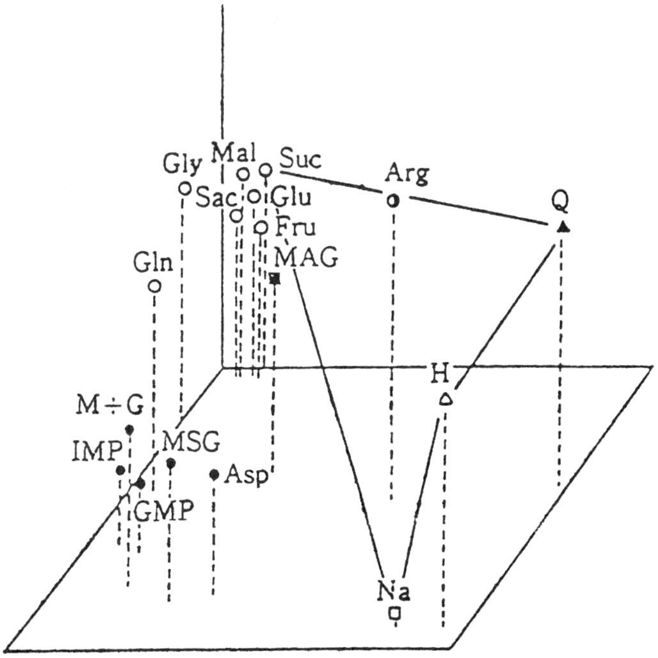

Figure 4 Locations of each of the 17 test stimuli in three-dimensional space obtained through multidimensional scaling based on the data obtained from glossopharyngeal fibers in mice. Suc, sucrose; Sac, Na saccharin; Fru, fructose; Glu, glucose; Mal, maltose; Gly, glycine; Gln, L-glutamine; Arg, L-arginine HCl; MAG monoammonium L-glutamate; Asp, monosodium L-aspartate; MSG, monosodium L-glutamate; IMP, disodium 5′-inosinate; GMP, disodium 5′- guanylate; M+G, a mixture of 0.1M MSG with 0.5mM GMP; Na, NaCl; H, HCl, Q, quinine-HCl. (From Ninomiya and Funakoshi, 1987).

Table 4 Detection Thresholds for Five Taste Substances (in g/dL)

MSG	Sucrose	Sodium chloride	Tartaric acid	Quinine sulfate
0.012	0.086	0.0037	0.00094	0.000049

Source: Yamaguchi and Kimizuka (1979).

most recent molecular biological study on glutamate receptor sites in taste bud cells (Chaudhari et al., 1994).

Umami is one of the primary tastes, independent of the four traditional basic tastes.

2. Taste Thresholds

a. Glutamate. The detection threshold for MSG was as low as 0.012 g/100 mL, or 6.25×10^{-4} M. It was higher than that of quinine sulfate or tartaric acid, lower than that of sucrose, and almost the same as that of sodium chloride at isomolar concentrations (Table 4). Several umami substances have lower thresholds than MSG (Yamaguchi et al., 1979).

b. Nucleotides. The threshold values of IMP and GMP are 0.025 and 0.0125 g/100 mL, respectively (Table 5). The taste threshold for 50:50 blends of GMP and IMP has been reported to be 0.0063%. When they were used in combination with 0.8% MSG, however, the resulting threshold was lowered to 0.000031%, which represents a dramatic synergistic effect.

3. Taste Intensity

The relationship between MSG concentration and the taste intensity of MSG is found to follow a straight line. The slope for MSG is not as steep as for the four basic tastes (Fig. 5) (Yamaguchi, 1979). Moreover, the taste intensity of IMP increases hardly at all, even when its concentration is increased considerably.

The relative taste intensities of amino acid–based derivatives and nucleotides are given in Tables 6 and 7, respectively.

Table 5 Taste Thresholds of Basic Taste Substances

	Absolute threshold[a]				
Solvent	Sucrose (sweet)	Sodium chloride (salty)	Tartaric acid (sour)	Quinine sulfate (bitter)	Glutamate (umami)
Pure water	2.50×10^{-3} M (8.6×10^{-2} %)	6.25×10^{-4} M (3.7×10^{-3} %)	6.25×10^{-5} M (9.4×10^{-4} %)	6.25×10^{-7} M (4.9×10^{-5} %)	6.25×10^{-4} M (1.2×10^{-2} %)
Glutamate, 5×10^{-3} M	1.25×10^{-3} M	6.25×10^{-4} M	1.25×10^{-4} M (1.9×10^{-3} %)	6.25×10^{-7} M	—
IMP, 5×10^{-3} M	1.25×10^{-3} M	6.25×10^{-4} M	2.00×10^{-3} M (3.0×10^{-2} %)	2.50×10^{-6} M (2.0×10^{-4} %)	—

[a] Significant at 5% level.

Source: Yamaguchi and Kimizulea (1979).

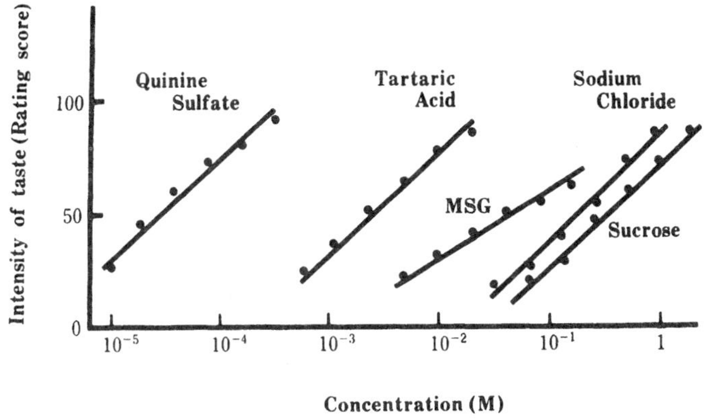

Figure 5 Relationship between concentration and taste intensity.

B. Taste Synergism

1. The Synergistic Effects of Umami Substances

Figure 6 shows that taste intensity of mixtures (IMP and MSG) increases exponentially with their concentration and that the degree of synergism depends upon the ratio of IMP to MSG.

Figure 7 shows the relationship between the intensity of umami and the proportion of IMP in the mixture of MSG and IMP. Because the umami intensities of the samples on both extremes are very weak and almost the same, the curve would have been horizontally linear if the synergistic effect had been absent. The symmetric curve illustrates this remarkable synergistic effect. In this curve, the intensity of umami at its maximum is equivalent to that of 0.78 g/100 mL of MSG alone. The mixture is thus 16 times as strong as that of MSG. This amplification factor is concentration dependent and becomes higher with increasing concentration.

The synergistic effect between MSG and IMP can be expressed by means of the following simple equation:

$$y = u + 1200uv \qquad (1)$$

Table 6 Taste Intensities of Amino Acid Based Derivatives

Substance	Relative umami intensity (α)
Momosodium L-glutamate · H_2O	1
Monosodium DL-*threo*-β-hydroxy glutamate · H_2O	0.86
Monosodium DL-homocystate · H_2O	0.77
Monosodium L-aspartate · H_2O	0.077
Monosodium L-α-amino adipate · H_2O	0.098
L-Tricholomic acid (*erythro* form)[a]	5–30
L-Ibotenic acid[a]	5–30

[a] From Terasaki et al. (1965a,b).
Source: Yamaguchi et al. (1971).

Table 7 Taste Intensities of Nucleotides Having Synergistic Effect on Glutamate Taste

Substance (disodium salt)	Relative potency of umami (β)
5′-Inosinate · 7.5 H_2O	1
5′-Guanylate · 7 H_2O	2.3
5′-Xanthylate · 3 H_2O	0.61
5′-Adenylate	0.18
Deoxy-5′-guanylate · 3 H_2O	0.62
2-Methyl-5′-inosinate · 6 H_2O	2.3
2-Ethyl-5′-inosinate · 1.5 H_2O	2.3
2-Phenyl-5′-inosinate · 3 H_2O	3.6
2-Methylthio-5′-inosinate · 6 H_2O	8.0
2-Ethylthio-5′-inosinate · 2 H_2O	7.5
2-Ethoxyethylthio-5′-inosinate[a]	13
2-Ethoxycarbonylethylthio-5′-inosinate[a]	12
2-Furfurylthio-5′-inosinate · H_2O[a]	17
2-Tetrahydrofurfurylthio-5′-inosinate · H_2O[a]	8
2-Isopentenylthio-5′-inosinate (Ca)[a]	11
2-(β-Methallyl)thio-5′-inosinate[a]	10
2-(γ-Methallyl)thio-5′-inosinate[a]	11
2-Methoxy-5′-inosinate · H_2O	4.2
2-Ethoxy-5′-inosinate[a]	4.9
2-*i*-Propoxy-5′-inosinate[a]	4.5
2-*n*-Propoxy-5′-inosinate[a]	2
2-Allyloxy-5′-inosinate (Ca) · 0.5 H_2O	6.5
2-Chloro-5′-inosinate · 1.5 H_2O	3.1
N^2-Methyl-5′-guanylate · 5.5 H_2O	2.3
N^2, N^2-Dimethyl-5′-guanylate · 2.5 H_2O	2.4
N^1-Methyl-5′-inosinate · H_2O	0.74
N^1-Methyl-5′-guanylate · H_2O	1.3
N^1-Methyl-2-methylthio-5′-inosinate	8.4
6-Chloropurine riboside 5′-phosphate · H_2O	2
6-Mercaptopurine riboside 5′-phosphate · 6 H_2O	3.4
2-Methyl-6-mercaptopurine riboside 5′-phosphate · H_2O	8
2-Methylthio-6-mercaptopurine riboside 5′-phosphate · 2.5 H_2O	7.9
2′,3′-*o*-Isopropylidene 5′-inosinate	0.21
2′,3′-*o*-Isopropylidene 5′-guanylate	0.35

[a] From Imai et al. (1971).
Source: Yamaguchi et al. (1971).

where u and v are the respective concentrations of MSG and IMP in the mixture, and y is the equi-umami concentration of MSG alone (Yamaguchi, 1967).

The synergistic effect can be demonstrated between any combination of substances in Tables 6 and 7, and the intensity of umami can also be expressed by an equation essentially equivalent to Eq. (1). The taste intensities of all substances in Table 6 are always proportional to that of MSG. Therefore, u' g/100 mL of any substance cited in Table 6 can be replaced with $\alpha u'$ g/100 mL of MSG, where α is a constant of proportional-

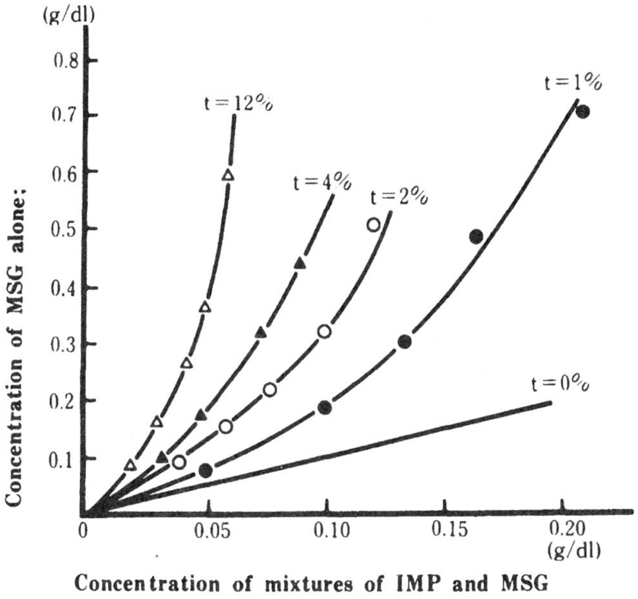

Figure 6 Taste equivalency of mixture of IMP and MSG to MSG alone. *t* represents IMP content (percent) in mixtures.

ity; values for various substances are listed in Table 6. On the other hand, the strength of umami taste of nucleotides in Table 7 is consistently proportional to that of IMP. Hence, v' g/100 mL of any nucleotide is replaceable with $\beta v'$ g/100 mL of IMP. The constants β for all nucleotides are given in Table 7. Therefore, the umami intensity of the mixture of any combination of substances in Tables 6 and 7 can be calculated by substituting $\alpha u'$ for u and $\beta v'$ for v. Because the interrelationships within each series of substances are additive, the intensity of umami of the mixture of two or more different L-α-amino acids

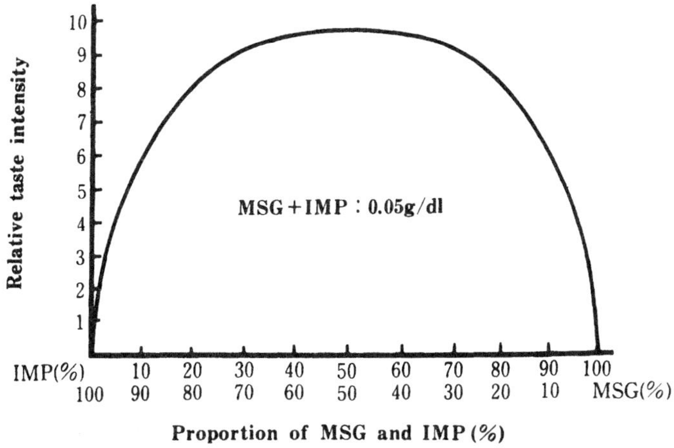

Figure 7 Intensity of umami taste in MSG–IMP mixture.

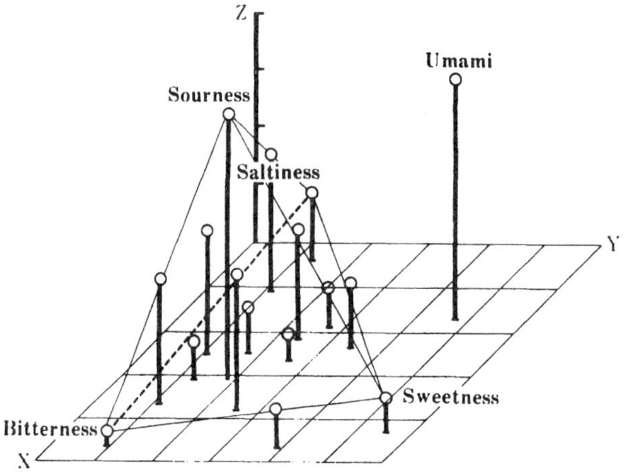

Figure 8 Three-dimensional configuration for five taste stimuli.

and two or more nucleotides can be calculated by substituting the product sums $\sum \alpha_i u_i$ and $\sum \beta_j v_j$ for u and v, respectively, in Eq. (1).

2. The Synergistic Action of Umami in Foods

Multidimensional analysis has shown umami to be present in the taste of natural foods (Yamaguchi, 1987). *Umami* is definitely located outside the tetrahedron of the four basic tastes and is an independent basic taste (Fig. 8). The broths made from animal and fish stocks fall outside the area of the four basic tastes and lie nearer to umami (Fig. 9). This demonstrates that umami is a vitally important element in broth taste composition.

In contrast, broth made from vegetables also contains umami, but some of the taste factors are sweetness or sourness. Thus these broths are distributed widely over the five

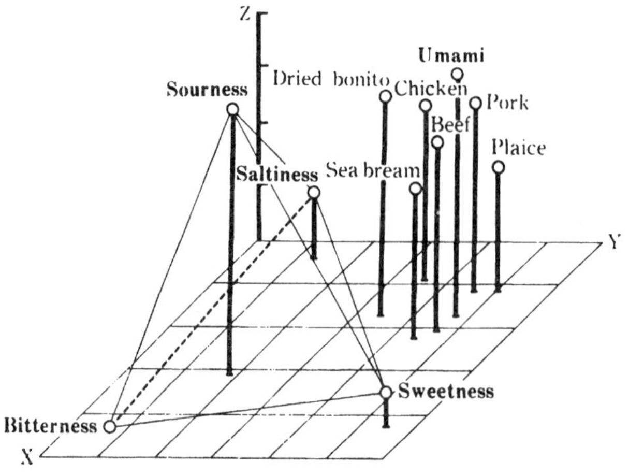

Figure 9 Three-dimensional configuration for meat and fish and the five taste stimuli.

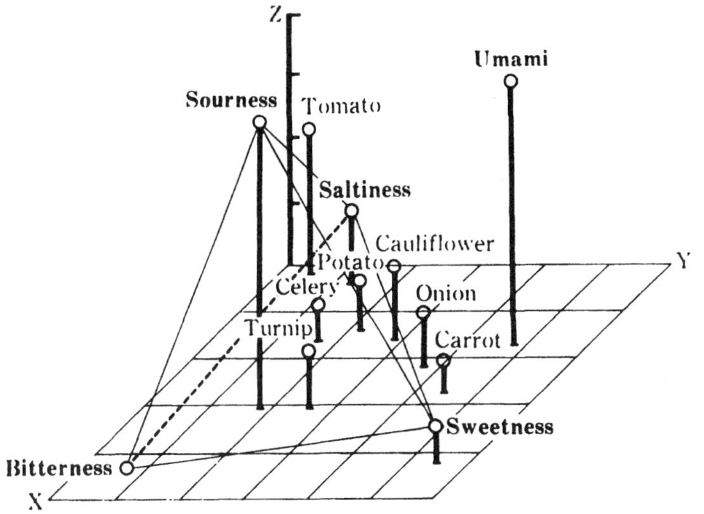

Figure 10 Three-dimensional configuration for vegetable stocks and the five taste stimuli.

taste areas (Fig. 10). However, if a small amount of IMP is added, the tastes of all the broths move in the direction of umami (Fig. 11). This shows the synergistic effect of umami that is brought into existence between the glutamate contained in the vegetables and the added inosinate (Yamaguchi et al., 1967).

C. Taste Physiological Data

The biological significance of the chemoreception of amino acids was reviewed by Kawamura (1987).

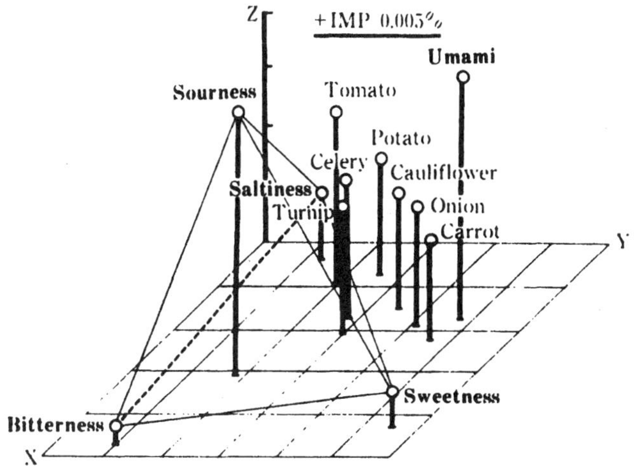

Figure 11 Three-dimensional configuration for vegetable stocks with 0.005% disodium inosinate (IMP) added and the five taste stimuli.

Amino acids are one of the most important classes of nutrients and are potent attractants for many living organisms; organisms as diverse as bacteria, protozoa, annelids, gastropods, crustaceans, insects, fish, amphibians, and mammals all have chemoreceptors sensitive to amino acids. The treatment of carp olfactory epithelium, bullfrog tongue, and rat tongue with pronase E obliterated responses to amino acids in these animals; this showed that receptor molecules for amino acids, including glutamate, are protein. Cross-adaptation experiments utilizing various combinations of amino acids in olfactory and taste systems of the carp explored the receptor sites for various amino acids. Results showed that there are multiple types of receptors corresponding to different amino acids and that the receptor for glutamate is different from those for other amino acids. Evidence suggesting that there is a specific receptor protein for glutamate in chemoreceptor membranes is consistent with reports on receptor potentials in bacteria.

Taste buds on the front of the tongue receive inputs from the chorda tympani (CT) branch of the facial nerve, whereas those on the posterior part of the tongue receive inputs from the glossopharyngeal (GP) nerve. Single-unit nerve responses to glutamate and 5′-nucleotides were recorded in the taste pore, chorda tympani nerve, glossopharyngeal nerve, geniculate ganglion, petrosal ganglion, and cortical neurons in the rat, hamster, mouse, cat, goat, and dog. Results varied somewhat depending on the nerve fiber recorded and the species of experimental animal studied. For example, although some units of the cat geniculate ganglion responded strongly to nucleotides, these same units failed to respond to MSG. In the hamster, MSG-sensitive units of cortical neurons also responded to NaCl; however, the responses of MSG-sensitive units recorded within the hamster taste pore were independent of the responses to any of the four basic taste stimuli.

Differences in responsiveness of CT and GP nerves to various taste stimuli have been observed in mice. Some of the glossopharyngeal nerve fibers of the mouse are especially sensitive to MSG, and these fibers are only slightly sensitive to NaCl or to sucrose. Patterns of suppression of licking in denervated mice suggested that the CT nerve is important for the sodium component of MSG and for discrimination among sugars, whereas inputs from the GP nerve play an important role in discrimination between salty and umami taste components of MSG. Patterns of suppression of licking across various taste stimuli also showed that MSG and GMP elicit similar patterns but that MSG shows the highest similarity to NaCl, especially at high concentrations.

The mechanism of the taste synergism between glutamate and nucleotides has been examined both electrophysiologically and biochemically. Recordings from the rat CT nerve demonstrate the remarkable synergism between MSG and purine-based 5′-nucleotides, which is consistent with and corroborates the psychophysical data. Analysis of data from the rat has suggested that the presence of 5′-nucleotides induces an increase in the affinity (i.e., strength of binding) of glutamate for the receptor sites while not increasing the amount of glutamate bound. However, measurements of binding of L-[^{3}H]glutamate to bovine circumvallate papillae showed that in this system the presence of nucleotides causes a remarkable increase in the maximal capacity of glutamate binding to the papillae without affecting binding affinity. The results were explained in terms of allosteric regulation.

In the dog, large synergism between MSG and GMP or IMP, similar to that in humans, was observed in the chorda tympani nerves. Amiloride, an inhibitor of Na response, suppressed the response to MSG similarly to that observed with NaCl, suggesting that the response to MSG is primarily a salt response. Amiloride did not, however, suppress the response to GMP alone or that induced by the synergism between low concentrations of MSG and GMP, which suggests that these responses are independent of salt (Kumazawa et al., 1991).

Brain neuronal responses to umami taste stimulation were recorded in the rat and monkey. In the nucleus of the solitary tract of the rat, there were no specific responses to umami taste substances. These results were similar to those observed with the CT nerves. Using macaques, recordings were made from 190 taste responsive neurons in the taste cortex and adjoining or orbitofrontal cortex taste area, single neurons were found that were tuned to respond best to MSG. Across the population of neurons, the responsiveness to MSG was poorly correlated with the responsiveness to NaCl, suggesting that the representation to MSG was clearly different from that of NaCl. In addition, cluster analysis and multidimensional scaling analyses confirmed that the neuronal representation of MSG fell outside the space defined by sweet, salt, bitter, and sour. Therefore, in primate taste cortical areas, MSG is as well represented as are the tastes produced by glucose, NaCl, HCl, and quinine (Baylis and Rolls, 1991).

D. Influence on Food Consumption

1. Taste Preferences and Nutritional Status

The perinatal human infant can communicate feelings and emotions by a repertoire of nonverbal communications of facial expressions and actions. When a sour or bitter taste or an unflavored vegetable soup is given, infants respond with expressions and actions indicating displeasure and dissatisfaction. If glutamate is added to the unflavored vegetable soup, however, infants not only detect its presence but also indicate, with their facial expression, enjoyment of the taste in a manner similar to their responses to sweet solutions (Steiner, 1987).

Similar behavioral manifestations have been observed in rats given dietary glutamate (Torii et al., 1983). There is a reproducible correlation between protein level in the diet and selection of taste. If the protein content of the taste material is insufficient for growth, rats do not choose glutamate even when it is available. The preference for glutamate returns when the protein level of the diet is sufficient for rat growth. Even if the quantity of protein is sufficient, the rats do not grow, and they choose sodium chloride rather than glutamate (as in the case of the low protein diet) when lysine, one of the essential amino acids, is deficient. When lysine is supplemented and the diet becomes well balanced both in quality and in quantity, the rats begin to grow and their preference for glutamate appears. It has therefore been suggested that the change in taste preference could be a reaction to an altered physiological state depending upon the nutritional status of the animal (Torii et al., 1985).

To relate neural activity to preference for amino acids, the electrical responses of lateral hypothalamus (LHA) neurons were observed during the application of amino acids to the tongue. The data demonstrated that during an essential amino acid deficiency, more LHA neurons responded to the taste of the deficient amino acid (Tabuchi et al., 1991).

Spontaneously hypertensive rats (SHR) have a tendency to ingest a large amount of sodium chloride and become hypertensive. However, sodium intake declined as the protein content in the diet was increased. When an umami substance (MSG and MSG plus GMP) was available as an aqueous solution, rats selected it, and a drastic decrease in total sodium intake—even up to 70%—was observed when the solutions were offered in conjunction with the diet of higher protein content. Although the predisposition of SHR to hypertension was essentially unchanged, the pathogenesis of atheroma was actually depressed when the diet was sufficient in protein and umami substances were available (Kimura et al., 1984).

2. Physiological Responses to Umami Taste

Sham-feeding experiments with dogs have shown that oral stimulation by MSG produces significant and dose-dependent stimulation of both pancreatic flow and protein output in conscious dogs (Naim et al., 1991).

In the rat, the effects on metabolic parameters such as total and background metabolism and respiratory quotient (RQ) of umami-flavored meal was compared with an unflavored one. MSG added to the meal modified the metabolic parameters closer to those for protein metabolism (Viarouge et al., 1992).

These studies along with a nutritional study by Torii et al support that umami taste stimuli could be a marker of protein intake.

3. MSG and Sodium in the Diet

MSG contains 12.3% sodium, or one-third that of common table salt; as the usual MSG use level is around one-tenth that of salt, the contribution of MSG to dietary sodium is usually one-twentieth to one-thirtieth of that of NaCl. In general, as the salt level in food is reduced, there is a corresponding reduction in food acceptability. By using a small amount of MSG, more than 30% of sodium content may be reduced while maintaining a very palatable and acceptable level of taste (Yamaguchi and Takahashi, 1984b; Chi and Chen, 1992; Altug and Demirag, 1993).

A recent report states that calcium glutamate may be used to improve the acceptability of sodium free diets of children suffering kidney disease (Bellisle et al., 1992).

4. Role of Umami in Food

The role of umami substance in food tastes has been examined with snow crab extract, which was analyzed for amino acids, nucleotides and related compounds, organic bases, sugars, organic acids, and minerals. The extract from the meat of snow crab was reconstructed with about 40 pure chemicals in accordance with the analytical data. A series of organoleptic tests of the reconstructed extract, using omission test methods, revealed that the unique flavor of boiled crab is derived from a rather limited number of components (Gly, Ala, Arg, Glu, IMP, NaCl, KH_2PO_4). The relationship between the taste of a composite extract and characteristics in snow crab meat is shown in Table 8. Fairly high correla-

Table 8 The Relationship Between the Taste of a Composite Extract and Characteristics in Snow Crab Meat

	Sweet (Gly + Ala)	Salty (NaCl + K_2HPO_4)	Umami (Glu + IMP)
Continuity			**
Complexity		*	**
Fullness		*	**
Mildness	*		**
Seafoodlike flavor		*	**
Overall preference		*	**

* $P < 0.05$, ** $P < 0.01$.
Source: Konosu and Yamaguchi (1987).

tions were observed between umami and five descriptors of flavor characteristics and overall preference. Thus when umami was omitted, not only did the continuity, complexity fullness, and mildness diminish, but the characteristic taste of crab itself disappeared, and so did the overall preference. This fact suggests the function of umami substances in flavor-enhancing properties (Konosu and Yamaguchi, 1987).

The role of umami taste in meat flavor was also studied (Kato and Nishimura, 1987). The increase in strength of umami and brothy tastes obtained through conditioning is more evident in pork and in chicken than in beef, which contains less glutamate. The addition of glutamate or glutamate plus IMP to broths enhanced the umami and brothy tastes and showed that umami plays an important role in meat taste.

There is large natural variability in the free glutamate levels in cheese of different varieties. It has been well established that free amino acids content of cheese, including glutamic acid, increases during maturation. The free glutamic acid content of cheese is used as an indicator of maturation (Weaver, 1978) Similarly, the free glutamic acid content of tomato increases during ripening. Glutamate value in green tomato increases during the ripening period (Inaba, 1980; Skurray, 1988). The free glutamate content of ripened tomato is about seven times as high as unripend green tomato.

5. Hedonic Functions and Self-Limiting Properties

Psychometric studies on aqueous solutions of the four basic tastes revealed that three (salty, bitter, and sour) are rated as unpalatable over a wide concentration range in that they received unpleasantness ratings. Only sweetness was given a pleasantness rating. In a similar fashion to salty, sour, or bitter stimuli, umami (MSG in aqueous solution) also had an unpleasantness rating or was rated neutral in acceptability at all concentrations studied (Yamaguchi and Takahashi, 1984a).

Of additional importance is the fact that there is an optimal concentration for MSG added to food. Beyond this most palatable concentration, the palatibility of food decreases. Thus the use of MSG is self-limiting in that overuse decreases palatability (Fig. 12) (Yamaguchi and Takahashi, 1984a).

IV. USE OF GLUTAMATE IN FOODS AND REGULATIONS

A. Food Applications

1. Glutamate

The major use of MSG in cooking around the world is as a flavor enhancer in soups and broths, sauces and gravies, and flavorings and spice blends. MSG is also included in a wide variety of canned and frozen meats, poultry, vegetables, and combination dishes. Results of taste panel studies indicate that a level of 0.1–0.8% by weight in food gives the best enhancement of the food's natural flavor. In home or restaurant cooking, this amounts to about 1–2 teaspoonfuls per kilogram of meat or per 8–12 servings of vegebles, casseroles, soups, etc.

There appears to be some variability from one person to another as to the preferred optimum level of use. Some recipes call for adding MSG during food preparation and then again at the time of serving to "season to taste." Because MSG is readily soluble in water, recipes often call for dissolving it in the aqueous ingredients of products such as salad dressings before they are added to food.

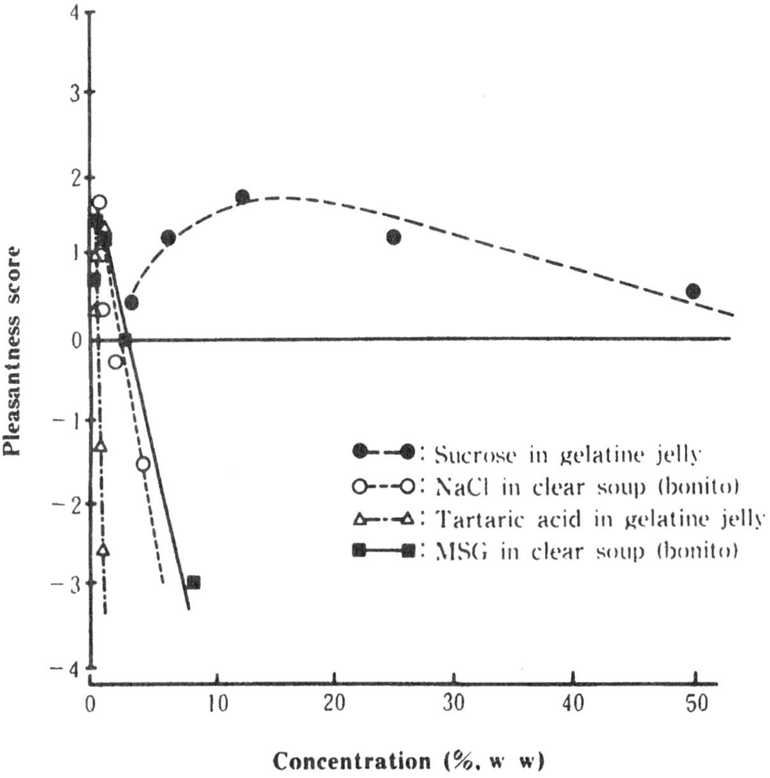

Figure 12 Typical hedonic patterns for four taste substances, showing the self-limiting effect of the amount of MSG (and also NaCl and tartaric acid) to be added to foods. No self-limiting effect is seen with sucrose.

2. 5′-Nucleotides

Generally, the nucleotides are useful in enhancing the flavor of many products containing soups, certain canned meats, fish, vegetables, and vegetable juice (Table 9).

In animal tissues, IMP is produced from ATP, and thus the IMP content of fresh raw meats is usually rather high. However, the IMP content of processed meat products is usually rather low, because raw meats contain phosphomonoesterase, and the IMP that is naturally present in meat is easily lost in processing steps such as thawing, washing, and salting. The 5′-nucleotides added together with glutamate after these steps can be preserved without loss and improve the flavor of the meat products effectively if the phosphomonoesterase is inactivated by heating the products before or immediately after addition of the 5′-nucleotides.

As mentioned previously, the presence of nucleotides enhances the activity of glutamate. Both IMP and GMP can greatly reduce the glutamate requirements of many foods. Currently, MSG, IMP, mixtures of IMP and GMP, and mixtures of MSG, IMP, and GMP are commercially available. In most soups, 4.5–7 kg of a 95% MSG/2.5% IMP/2.5% GMP mixture can replace 45 kg of MSG without appreciably changing the flavor of the products (Titus, 1964). In this application, the nucleotide concentration is reduced to the point where its own taste is not detectable, and yet the synergistic relationship between

Table 9 Processed Foods to Which Flavor Enhancers Are Added and Their Usage Levels

Food	Usage levels MSG (%)	5′-Nucleotides (50:50 IMP and GMP) (%)
Canned soups	0.12–0.18	0.002–0.003
Canned asparagus	0.08–0.16	0.003–0.004
Canned crab	0.07–0.10	0.001–0.002
Canned fish	0.10–0.30	0.003–0.006
Canned poultry, sausage, ham	0.10–0.20	0.006–0.010
Dressings	0.30–0.40	0.010–0.150
Ketchup	0.15–0.30	0.010–0.020
Mayonnaise	0.40–0.60	0.012–0.018
Sausage	0.30–0.50	0.002–0.014
Snacks	0.10–0.50	0.003–0.007
Soy sauce	0.30–0.60	0.030–0.050
Vegetable juice	0.10–0.15	0.005–0.010
Processed cheese	0.40–0.50	0.005–0.010
Dehydrated soups	5–8	0.10–0.20
Soup powder for instant noodles	10–17	0.30–0.60
Sauces	1.0–1.2	0.010–0.030

Source: Maga (1983).

glutamate and the nucleotide is strong enough to give the impression of a much higher concentration of glutamate than is actually present. The amount of this mixture required to give the same flavor effect as 45 kg of MSG varies from product to product but is usually about 7 kg. In one bouillon, only 2.5 kg of the mixture was reported to replace effectively 45 kg of MSG.

Recent developments in processed foods have given rise to many flavoring materials such as hydrolyzed vegetable proteins, hydrolyzed animal proteins, yeast extracts, meat extracts, and vegetable extracts. Hydrolyzed vegetable or animal proteins contain generous amounts of glutamate (approximately 10 g MSG/100 g); meat extracts contain some amount of IMP (approximately 1 g/100 g). GMP-rich yeast extract itself is also available in the market.

Although the nucleotides and glutamate naturally contained in these materials accomplish the synergism, the additional glutamate and nucleotides enhance the quality of processed foods and result in considerable cost savings.

B. Regulations

The use of monosodium glutamate in foods, like that of hundreds of other flavors, spices, and food additives, is subject to a variety of standards and regulations on a worldwide basis. In 1987, the Joint Expert Committee on Food Additives (JECFA) of the Food and Agricultural Organization of the United Nations and the World Health Organization

(FAO/WHO) reviewed and endorsed the safety of glutamate, allocating an acceptable daily intake (ADI) for MSG as ''not specified.''

The previous numerical ADI has been removed; the implied exclusion of use by humans under the age of 12 weeks has also been deleted. This is JECFA's most favorable classification for food additives.

JECFA considered the issue of MSG hypersensitivity and concluded that ''studies have failed to demonstrate that MSG is the causal agent in provoking the full range of symptoms of Chinese Restaurant Syndrome. Properly conducted double-blind studies among individuals who claimed to suffer from the syndrome did not confirm MSG as the causal agent'' (JECFA 1987,1988).

Along with JECFA specifications, various national bodies have also established standards of purity for glutamates. For example, monographs of identity for purity are listed in the U.S. Food Chemicals Codex and the Japanese Standard of Food Additives. In the United States, MSG is included in the GRAS (generally recognized as safe) list of food ingredients by the U.S. Food and Drug Administration, along with salt, pepper, sugar, and vinegar. The Scientific Committee for Foods of the European Community evaluated MSG and gave number E621 as a safe food additive (EL/SCF, 1991). The regulation was published as a Council Directive in 1995 (European Parliament, 1995). In Japan, MSG is a permitted food additive with no limitation.

For calcium and disodium salts of guanylic and inosinic acid, the committee stated ''ADI not specified.'' This means that, on the basis of the available data—toxicological, biochemical, and so forth—an ADI with a numerical limitation is not deemed necessary (JECFA 1974,1975,1993). These salts are listed as food additives permitted for direct addition food for human consumption.

V. BIOCHEMICAL ASPECTS

A. Glutamate

1. Absorption and Metabolism

As glutamate plays a vital role in the metabolism in the body, it is synthesized in the body to meet its demand (endogenicity) and is among those amino acids that are classified as nonessential. It is not necessary to supply these amino acids in the diet. However, the nonessential amino acids do provide twofold benefits to our well-being: they provide important sources of nitrogen and act to supplement or to conserve the essential amino acids, whose supply could otherwise be depleted.

During absorption of both free and protein-bound glutamate, a large quantity of its α-amino nitrogen appears in the portal blood as alanine. This alanine results from the transamination of glutamate to pyruvate, α-ketoglutarate being the other end product. When a large quantity of glutamate is ingested, portal plasma glutamate increases. This elevation results in an increase in the liver's metabolism of glutamate, with the release of glucose, lactate, and glutamate into the systemic circulating plasma. Glutamate originating from protein or in free form is metabolized similarly after absorption (Stegink, 1976; Stegink et al., 1973b,1975).

Glutamate is used by the organism for a wide variety of metabolic processes and plays an important role in nitrogen and energy metabolism (Fig. 13).

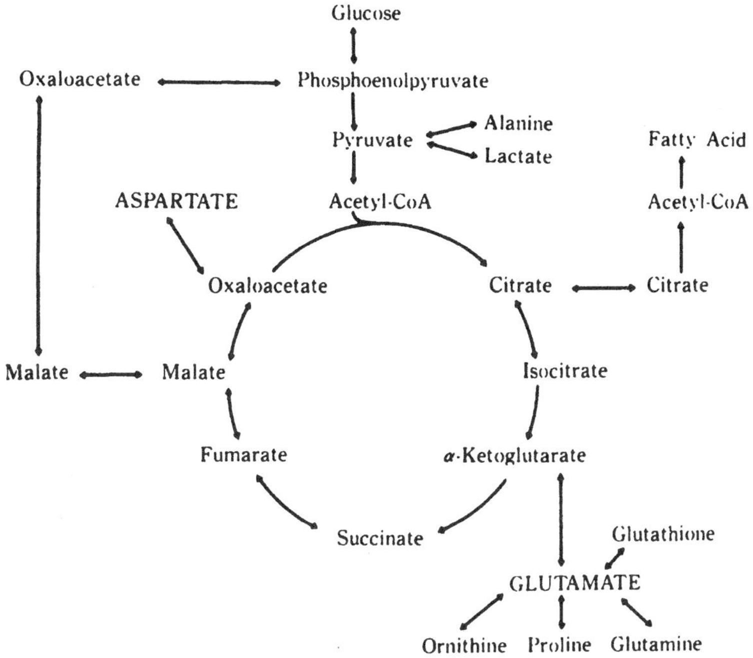

Figure 13 Available pathways of glutamate metabolism. (From Stegink, 1984.)

The normal steps in the metabolism of glutamate include oxidative deamination, transamination, decarboxylation, and amidation. All are well established in mammals (Meister, 1974).

An extensive study on the metabolism of exogenous monododium L-glutamate was carried out in neonatal pigs using ^{14}C-labeled glutamate (Stegink et al., 1973a,b). Despite the enormous quantity of glutamate administered in these studies, a considerable portion did not reach the peripheral circulation but was converted to glucose and lactate by the liver before release. Similar data have been obtained in newborn primates (Boaz et al., 1974; Stegink et al., 1975).

Other studies using different routes of administration and other animal species or organ systems have given results consistent with those of the above studies. The rodent appears to metabolize glutamate somewhat differently from the pig and the primate in that metabolites such as α-keto-glutarate and acetoacetate are found in mouse plasma following administration of glutamate but not in primates (Stegink, 1976).

2. Blood–Brain Barrier

Glutamate is normally present in high concentrations in the central nervous system where it functions as an excitatory neurotransmitter. It is produced from glucose in the brain. It does not have ready access to the brain from circulation or diet. The blood–brain barrier and the very powerful glial and neuronal uptake systems for glutamate help to keep the extracellular concentration of glutamate low in the brain. The dietary consumption of glutamate has not been shown to cause neuropathology in man. (Meldrum, 1993; Fernstrom, 1994).

Pardridge also showed that the normal plasma glutamate level is nearly four times the K_m (0.04 mM) of the transport rate of glutamate to the brain; thus glutamate carriers to the brain are virtually saturated at physiological plasma levels of this amino acid. Furthermore, when rates of glutamate influx are compared to net glutamate release from brain, an active efflux system for glutamate appears to exist (net release is sevenfold greater than the rate of influx) (Pardridge, 1979).

The effects of glutamate administration on the glutamate concentration of regional brain areas have been studied.

Blood–brain and blood–retinal barriers are impermeable to glutamate except when massive doses (greater than 2 g/kg body weight of MSG dissolved in water to yield greater than 20-fold increases over normal plasma glutamate levels) are given to infant mice or rats (Liebschutz et al., 1977).

3. Placental Barrier

Stegink has examined glutamate metabolites following infusion of radioactive MSG into pregnant monkeys. When MSG was infused at a constant rate of 1 g/h (equivalent to 0.16–0.22 g/kg body weight), maternal plasma glutamate levels increased 10- to 20-fold over baseline; fetal plasma glutamate levels, however, remained unchanged. During infusion, 60% of the radioactivity in the maternal circulation was in the form of glutamate. The major radioactive compounds detected in the fetal circulation were glucose and lactate, indicative of rapid metabolism of MSG by the mother. There was essentially no radioactivity present as glutamate in fetal plasma. At higher rates of infusion, maternal plasma levels of glutamate rose to approximately 70 times normal; however, peak fetal plasma glutamate levels increased to less than 10 times normal (Stegink et al., 1975).

Schneider has studied the transfer of glutamate following an in vitro perfusion technique of the human placenta. When the fetal perfusate was recirculated, glutamate was progressively removed from the fetal circulation so that fetal concentrations fell below maternal concentrations (Schneider et al., 1979).

Thus one can conclude that the placenta serves as an effective barrier to the transfer of glutamate by rapidly metabolizing it.

4. Glutamate Levels of Human Breast Milk

Free glutamate concentrations in human milk collected from lactating women ranged from 47 to 128 μmol/dL. This is around 10 times as much free glutamate as is found in cow's milk. To determine if glutamate concentrations could be altered by oral ingestion of MSG, six normal lactating women received 6 g of MSG in capsules with water or with Slender, a liquid, ready-to-eat meal product. Although there was a significant increase in plasma glutamate levels, glutamate levels in milk collected over 12 h following ingestion were not affected. Glutamate levels in milk of these subjects were similar to those noted in the fasting state or in subjects receiving a lactose placebo (Stegink et al., 1972; Baker et al., 1979).

B. 5′-Nucleotides

1. Absorption and Metabolism

Nucleotides are widely distributed throughout the body and have various functions in metabolism. They can be synthesized from nonpurine precursor or obtained from dietary sources (Lagerkvist, 1958).

GMP is apparently degraded enzymatically to uric acid in the rat intestine (Wilson and Wilson, 1962) or converted to allantoin (Kojima, 1974).

When 25 mg/kg body weight of 5′-nucleotides was given orally to male rats, 70–80% was excreted in the urine after 24 h, with most of the remainder being found in the organ-free body. When pregnant rats were given the same dosage, 0.77% 5′-IMP and 0.01% 5′-GMP were found in the fetuses after 24 h.

In humans, administration of 5′-nucleotides induced elevation of uric acid levels in both serum and urine, thus indicating that at least partial degradation had occurred. For example, daily administration of 2.5 g of disodium 5′-inosinate caused a rise in serum uric acid levels from 3.6 to 6.9 mg%, whereas urinary levels increased from 506 to 1100 mg/day. As the intake of nucleotide is around 15 mg/day per person, there is little possibility that this 5′-nucleotide becomes the cause of gout in humans (Kojima, 1974).

2. Pharmacological Effects

Few studies have investigated the pharmacological effects of 5′-nucleotides, except for Kojima's report (Kojima, 1974). Administration of 500 mg/kg of 5′-guanylate to mice induced abnormal floor positions and also slight respiratory depression as well as depression of the avoidance response. These effects were observed about 15 min after administration. With disodium 5′-inosinate, behavioral abnormalities were apparent, but the mice became calm after 60 min. Intravenous injection of 5′-nucleotides (500 mg/kg) did not induce muscle relaxation or alter electroshock-induced convulsion rates. In addition, the analgesic responses of mice to a thermal stimulus were not affected by oral administration of nucleotodes.

5′-Nucleotides administered to rats had no diuretic effect and caused no change in gastric juice secretion. Also 10 mg/kg of disodium 5′-inosinate administered intravenously induced no significant changes in calcium, potassium, or chloride and no apparent change in sodium.

Administration of disodium 5′-inosinate did not affect blood pressure, heart rate, ECG, or blood flow in the hind limbs of anesthetized cats. Disodium 5′-guanylate did not increase blood flow.

VI. TOXICOLOGY

A. General Toxicity

1. Acute and Chronic Toxicity

The acute toxicities of glutamate and 5′-nucleotides are very low (Tables 10 and 11). Apparent LD_{50} values for the oral route are higher than those for the parenteral routes.

MSG (monosodium-L-glutamate), monosodium-DL-glutamate, and L-glutamic acid were added to diets at various concentration levels and fed to mice, rats, or dogs for approximately 2 years. No evidence of carcinogenicity or specific adverse effects of these substances was demonstrated by any of these studies (Ebert, 1979; Owen et al., 1978).

When rats or dogs were fed diets with 5′-nucleotides added (2%) for 2 years, there was no evidence of chronic toxicity or carcinogenicity (Kojima, 1974; Worden et al., 1975).

2. Reproduction and Multigenerational Studies

In a large three-generation study carried out in mice fed diets containing 1 or 4% MSG, average intakes at 1% MSG concentration were between 1500 and 1800 mg/kg body

Table 10 Acute Toxicity Studies of MSG in Mice and Rats

		LD_{50} (g/kg body weight)	
Species	Route	Male	Female
Mice	*po*	17.7	15.4
	sc	8.20	8.40
	ip	6.57	5.70
	iv	3.70	3.30
Rats	*po*	17.3	15.8
	sc	5.58	6.40
	ip	5.70	4.80
	iv	3.30	3.30

Source: Moriyuki and Ichimura (1978).

weight per day (500–600 times the average daily human consumption of MSG). The intake of MSG in dams increased considerably in all groups during lactation. Nevertheless, the pre- and postweanling performance of the young was unaffected (Anantharaman, 1979).

In multigenerational studies including studies on reproductive performance, mice were fed diets containing various levels of MSG. These studies reported no adverse effects of MSG (Semprini et al., 1974; Yonetani and Matsuzawa, 1978; Yonetani et al., 1979).

Table 11 Acute Toxicity Studies of 5′-Nucleotides in Mice and Rats

			LD_{50} (g/kg body weight)	
Species	Route	Compound	Male	Female
Mice	*po*	IMP	>14.4	>14.4
		GMP	>14.4	>14.4
	sc	IMP	5.48	5.63
		GMP	5.05	5.05
	ip	IMP	6.30	6.02
		GMP	6.80	5.01
	iv	IMP	3.95	4.60
		GMP	3.58	3.95
Rats	*po*	IMP	>10.0	>10.0
		GMP	>10.0	>10.0
	sc	IMP	3.90	4.34
		GMP	3.55	3.40
	ip	IMP	5.40	4.85
		GMP	4.75	3.88
	iv	IMP	2.73	2.87
		GMP	2.72	2.85

Source: Kojima (1974).

A three-generation study in rats that were fed up to 2% 5′-nucleotides did not show significant differences in growth and reproduction performance over controls (Kojima, 1974).

3. Teratology

In studies investigating the teratogenicity of MSG, MPG, and monoammonium glutamate in mice, rats, rabbits, and chicken embryos, these three substances were demonstrated to have no teratogenic effects under any of the experimental conditions used (Food and Drug Research Laboratory, 1974a,b; U.S. FDA,1975,1976a,b).

Likewise, studies on the potential for teratogenicity of 5′-nucleotides in pregnant mice, rats, rabbits, and cynomolgus monkeys demonstrated no teratologic effects in fetuses (Kojima, 1974; Kaziwara et al., 1971).

4. Mutagenicity

Glutamates were tested for genotoxicity in a series of Ames tests with and without metabolic activation. No mutagenic activities were shown (Litton Bionetics, 1975a,b, 1977a,b; Ishidate et al., 1984). Neither chromosome aberration tests in a Chinese hamster fibroblast cell line (Ishidate et al., 1984) nor host-mediated assay in rats (Industrial Bio-Test Laboratories, 1973) showed any mutagenic effects.

B. Special Toxicity

Studies on brain lesions in sensitive animal species, utilizing the injection or forced feeding of a huge dose of MSG in high concentrations, indicate that glutamate, like other food components, can induce toxic effects in test animals. However, the levels required to induce toxicity represent amounts and nondietary methods of administration not experienced by humans. When a huge amount of glutamate is given to a test animal species in the same rates, routes, and concentrations as used by humans, no toxic effects are seen even in the most sensitive animal species (Takasaki, 1978a,b).

In studies done on mice using huge dose levels (500 mg/kg body weight) (Olney, 1969; Olney and Ho, 1970; Olney et al., 1972; Lemkey-Johnston and Reynolds, 1972), MSG was shown to produce brain lesion when the compound was injected or administered by forced intubation. Treatment of mice with MSG by nondietary routes of administration, that is, injection or forced intubation, resulted in sufficiently high blood glutamate levels to apparently force enough MSG into the brain to induce toxicity (O'hara and Takasaki, 1979).

In mice, dietary feeding of MSG in amounts and concentrations in great excess of its use by humans (over 12,000 times its maximum anticipated level of use in humans) was without toxic effects to the brain. Nondietary routes of administration—subcutaneous injection or oral forced feeding—were required to induce toxic effects even in highly susceptible fetal, new born, and young rodents (O'hara et al., 1977; O'hara and Takasaki, 1979; Takasaki, 1979; Takasaki and Yugari, 1980; Heywood and Worden, 1979; Daabees et al., 1984; Anantharaman, 1979).

Five independent laboratories found that the monkey showed no brain damage following the injection of huge amounts of MSG in high concentrations. Even the direct injection to the fetus did not cause lesions in the hypothalamus (Abraham et al., 1975; Stegink et al., 1975; Heywood and Worden, 1979; Reynolds et al., 1976, 1979).

Data using microdialysis in freely moving rats showed that oral MSG does not affect striatal extracellular glutamate levels in vivo, even when consumed at very high doses (Bogdanov, 1994).

C. Observations in Humans

An anecdotal report in 1968 triggered curiosity about the so-called Chinese Restaurant Syndrome (CRS), which was associated with transient subjective symptoms of burning, numbness, and a tight sensation in the upper part of the body, reportedly after a meal in a Chinese restaurant (Kwok, 1968). Possible association with food ingredients such as MSG was suggested (Schaumburg et al., 1969). No objective changes in skin temperature, heart rate, ECG, or muscle tone were observed, and no correlation between plasma glutamate levels and symptoms. The symptoms are mild, self-limited and transient, occur 15–20 min after ingestion of a meal, and disappear within 1–2 h with no sequelae (Kenney and Tidball, 1972; Stegink et al., 1979). For assessment of the relevance of this kind of subjective reaction, experimental conditions without double-blind techniques are not suitable (Zanda et al., 1973). Quite a few double-blind placebo-controlled clinical investigations were done and concluded that there was no correlation between MSG added and postprandial claims (Marrs et al., 1978; Tung and Tung, 1980; Byung, 1980; Gore and Salmon, 1980; Kenney, 1986). Kenney suggested that the symptoms arise from irritation of the upper esophagus, and as the concentration is more important than dose, it was unlikely to have these symptoms with the usual level of concentration of glutamate added to food. He further demonstrated that orange juice, coffee, and spiced tomato juice triggered the same kinds of sensations and suggested that ''perhaps the CRS is no more than a manifestation of a common low-grade oesophagitis brought on by American dietary habits'' (Kenney, 1980).

He further challenged self-claimed CRS suffers with MSG in a double-blind clinical test. Six subjects, convinced that they had reacted to MSG, were given on four separate days either a soft drink solution that contained 6 g of MSG or a placebo. Four showed no symptoms with either MSG or the placebo. The other two experienced mild CRS to MSG but reacted also to placebo (Kenney, 1986).

Tarasoff and Kelly completed a randomized double-blind crossover study on 71 healthy subjects. They were given placebos and MSG at doses of 1.5, 3.0, and 3.15 g per person before a standardized breakfast over 5 days. Capsules and specially formulated drinks that can disguise the intrinsic taste of MSG were used as vehicles for placebo and MSG treatments. Subjects mostly had no responses to placebo (86%) and MSG (85%) treatments. Rigorous statistical analysis was performed. Sensations, previously attributed to MSG, did not occur at a significantly higher rate than did those solicited by placebo treatment. This study concluded that CRS is an anecdote applied to a variety of postprandial illness; rigorous and realistic scientific evidence linking the syndrome to MSG could not be found (Tarasoff and Kelly, 1993).

Epidemiological or survey reports using a questionnaire approach were done. Trying to avoid demand-bias results, and selecting an unbiased large-scale population, Kerr made a survey and analyzed 3222 answers. There were no ''definite CRS'' or ''probable CRS,'' and only 1.8% of people had ever experienced ''possible CRS.'' It was found that only 0.19% of the people reported CRS-like symptoms after a meal in a Chinese restaurant (Kerr et al., 1979).

Data from the U.S. Centers for Disease Control (CDC) surveillance of foodborne disease outbreaks show that during the period from 1975 to 1987, there were only 2 or

3 so-called CRS incidences per year. With no outbreak in seven of the thirteen years in the United States, CDC deleted the entry of CRS from the surveillance list (CDC, 1990).

The allegation of asthmatic reaction associated with MSG was denied by Harvard, the NIH group, and Scripps Clinic (Drazen 1990, Manning, 1991, 1992; Germano, 1991, 1993).

REFERENCES

Abraham, R., Swart, J., Golberg, L., Coulston, F. 1975. Electron microscopic observations of hypothalami in neonatal rhesus monkeys (*Macaca mulatta*) after administration of monosodium L-glutamate. *Exp. Mol. Pathol.* 23(2):203.

Altug, T., Demirag, K. 1993. Influence of monosodium glutamate on flavour acceptablity and on the reduction of sodium chloride in some ready made soups. *Chem. Mikorobiol. Technol. Lebensmittel*, 15(5/6):161–164.

Anantharaman, K. 1979. In utero and dietary administration of monosodium L-glutamate to mice: reproductive performance and development in a multigeneration study. In: *Glutamic Acid: Advances in Biochemistry and Physiology*, Filer, L. J., Jr., et al. (Eds.). Raven Press, New York, p. 231.

Anonymous, *Shurui Shokuhinn Tohkeigeppo* June, 1994:38.

Baker, G. L., Filer, L. J., Jr., Stegink, L. D. 1979. Factors influencing dicarboxylic amino acid content of human milk. In: *Glutamic Acid: Advances in Biochemistry and Physiology*, Filer, L. J., Jr., et al. (Eds.). Raven Press, New York, p. 111.

Baylis, L. L., Rolls, E. T. 1991. Responses of neurons in the primate taste cortex to glutamate. *Physiol. Behavior*, 49(5):973–979.

Bellisle, F., Dartois, A. M., Broyer, M. 1992. Two studies on the acceptablility of calcium glutamate as a potassium-free sodium substitute in children. *J. Renal Nutr.* 2(3) (Suppl. 1, July):42–46.

Boaz, D. P., Stegink, L. D., Reynolds, W. A., Filer, L. J., Pitkin, R. M., Brummel, M. C. 1974. Monosodium glutamate metabolism in the neonatal primate. *Fed. Proc.* 33:651.

Bogdanov, M. B., Wurtman, R. J. 1994. Effects of systemic or oral *ad libitum* monosodium glutamate administration on striatal glutamate release, as measured using microdialysis in freely moving rats. *Brain Res.* (in press).

Byung, S. M. 1980. Human development: human serum glutamate levels of Koreans. *Nutr. Rep. Int.* 22(5):697.

CDC 1990. Foodborn disease outbreaks, 5 year summary, 1983–1987 CDC Surveillance Summaries, MMWR 39/No. SS-1:15–57. (Definitions from 1979 summary, data from 1983 summary attached.).

Chaudhari, N., Yang, H., Lamp, C. and Roper, S. 1994. A metabotoropic glutamate receptor is expressed in rat taste bud cells. *Abstracts of AChemS 1994.*

Chi, S. P., Chen, T. C. 1992. Predicting optimum monosodium glutamate and sodium chloride concentrations in chicken broth as affected by spice addition. *J. Food Processing Preservation*, 16:313–326.

Daabees, T. T., Andersen, D. W., Zike, W. L., Filer, L. J., Stegink, L. D. 1984. Effect of meal components on peripheral and portal plasma glutamate levels in young pigs administered large doses of monosodium L-glutamate. *Metabolism* 33:58.

Ebert, A. G. 1979. The dietary administration of L-monosodium glutamate, DL-monosodium glutamate and L-glutamic acid to rats. *Toxicol. Lett.* 3:71.

EC/SCF. 1991. First series of food additives of various technological functions. *Report of the Scientific Committee for Food* (25th Series), Commission of the European Communities, pp. 16–17.

European Parliament. 1956. Council Directive No. 95/2/EC of 20 February 1995 on food additives other than colours and sweeteners.

Fernstrom, J. D. 1994. Dietary amino acids and brain function. *J. Am. Diet. Assoc.* 94(1):72.

Food and Drug Research Laboratories. 1974a. Teratologic evaluation of FDA 71–69 (monosodium glutamate) in mice, rats and rabbits. Prepared for Food and Drug Administration, PB-234 865. Waverly, New York.

Food and Drug Research Laboratories. 1974b. L-(+)-Glutamic acid: teratologic evaluation of FDA 73–58 (monopotassium glutamate) in mice and rats. Prepared under FDA contract No. 223-74-2176. Waverly, New York.

Fujii, M. et al. 1982. L-Glutamic acid and L-sodium glutamate. In: *Method of Analysis for Food Additives in Foods*, Ministry of Health and Welfare of Japan (Ed.). Koudansha, Tokyo, p. 401.

Fujita, T., Hashimoto, Y. 1959. Inosinic acid content of foodstuffs II. Katsuobushi (dried bonito). *Bull. Jap. Soc. Sci. Fish* 25:312.

Garattini, S. 1979. Evaluation of the neurotoxic effects of glutamic acid. In: *Nutrition and the Brain*, Vol. 4, Wurtman, R. J., Wurtman, J. J. et al. (Eds.). Raven Press, New York, p. 79.

Germano, P., Cohen, S. G., Hahn, B., Metcalfe, D. D. 1991. An evaluation of clinical reactions to monosodium glutamate (MSG) in asthmatics using a blinded, placebo-controlled challenge. *J. Allerg. Clin. Immunol.* 87(1):Abstracts 177.

Germano, P., Cohen, S. G., Hibbard, V., Metcalfe, D. D. 1993. Assessment of bronchial hyperactivity by methacholine challenge (MTC) in asthmatics before and after monosodium glutamate (MSG) administration. *J. Allergy Clin. Immunol.* 91:340.

Giacometti, T. 1979. Free and bound glutamate in natural products. In: *Glutamic Acid: Advances in Biochemistry and Physiology*, Filer, L. J., Jr., et al. (Eds.). Raven Press, New York, p. 25.

Gore, M. E., Salmon, P. R. 1980. Chinese restaurant syndrome: fact or fiction? [letter.] *Lancet* 1(8162):251.

Hashida, D. 1963. Studies on nucleic acid related substances in food-stuffs. I. Distribution of 5′-nucleotides in vegetables. *Hakko Kogyo Zasshi* 41:420. (In Japanese.)

Hashida, D. 1964. Annual Meeting of Society of Fermentation Technology Japan.

Hellekant, G., Ninomiya Y. 1991. On the taste of umami in chimpanzee. *Physiol. Behavior*, 49(5): 927–934.

Henning, H. 1916. Die Qualitatenreige des geschmacks. *Z. Psychol.* 74:203–219.

Heywood, R., Worden, A. N. 1979. Glutamate toxicity in laboratory animals. In: *Glutamic Acid: Advances in Biochemistry and Physiology*, Filer, L. J., Jr., et al. (Eds.). Raven Press, New York, p. 203.

Hirose, Y., Enei, H., Shibai, H. 1985. L-Glutamic acid fermentation. In: *Comprehensive Biotechnology: The Principles, Applications and Regulations of Biotechnology in Industry, Agriculture and Medicine*, Moo-Young, M., et al. (Eds.). Pergamon, Oxford, p. 593.

Imai, K., Marumoto, R., Kobayashi, K., Yoshida, Y., Toda, J., Honjo, M. 1971. *Chem. Pharm. Bull.* 19:576.

Inaba, A., Yamamoto, T., Ito, T., Nakamura, R. 1980. Changes in the concentrations of free amino acids and soluble nucleotides in attached and detached tomato fruits during ripening. *J. Japan Soc. Hort. Sci.* 49(3):435–441.

Industrial Bio-Test Laboratories. 1973. Host mediated assay for detection of mutations induced by Ac'cent brand monosodium-L-glutamate. Northbrook, ILI:1.

Ishidate, M., Sofuni, T., Yoshikawa, K., Hayashi, M., Nohmi, T. 1984. Primary mutagenicity screening of food additives currently used in Japan. *Food Chem. Toxicol.* 22(8):623.

JECFA. 1974. Evaluation of certain food additives. *Eighteenth Report of the Joint FAO/WHO Expert Committee of Food Additives*, WHO Technical Report Series 557, FAO Nutrition Meeting Report Series 54, p. 14.

JECFA. 1975. Toxicological evaluation of some food colours, enzymes, flavour enhancers, thickening agents, and certain food additives. *Eighteenth Report of the Joint FAO/WHO Expert Committee on Food Additives*, WHO Technical Report Series 557, FAO Nutrition Meeting Report Series 54, pp. 14–37.

JECFA. 1987. Evaluation of certain food additives and contaminants. *Thirty-First Report of the Joint FAO/WHO Expert Committee on Food Additives*, Technical Report Series 759, pp. 29–31.

JECFA. 1988. Toxicological evaluation of certain food additives. The 31st Meeting of the Joint FAO/WHO Expert Committee on Food Additives. *WHO Food Additives Series* 22:97–161.

JECFA. 1993. Evaluation of certain food additives and contaminants. *Forty-First Report of the Joint FAO/WHO Expert Committee on Food Additives*, WHO Technical Report Series 837, pp. 13–14.

Kaneko, T. 1938. Relationship between taste and structure of α-amino acids. *J. Chem. Soc. Japan* 59:433.

Kato, H., Nishimura, T. 1987. Taste components and conditioning of beef, pork and chicken. In: *Umami: A Basic Taste*, Kawamura, Y., Kare, M. R. (Eds.). Marcel Dekker, New York, p. 289.

Kawamura, Y. 1987. Recent developments in umami research. In: *Umami: A Basic Taste*, Kawamura, Y., Kare, M. R. (Eds.). Marcel Dekker, New York, p. 637.

Kawamura, Y., Kare, M. R. (Eds.). 1987. *Umami: A Basic Taste*. Marcel Dekker, New York.

Kaziwara, K., Mizutani, M., Ihara, T. 1971. *J. Takeda Res. Lab.* 30:314.

Kenney, R. A. 1980. Chinese restaurant syndrome. *Lancet* 1(Feb. 9):311.

Kenney, R. A. 1986. The Chinese restaurant syndrome: an anecdote revisited. *Food Chem. Toxicol.* 24(4):351.

Kenney, R. A., Tidball, C. S. 1972. Human susceptibility to oral monosodium L-glutamate. *Am. J. Clin. Nutr.* 25(2):140.

Kerr, G. R., Wu-Lee, M., El-Lozy, M., McGandy, R., Stare, F. J. 1979. Food-symptomatology questionnaires: risks of demand-bias questions and population-biased surveys. In: *Glutamic Acid: Advances in Biochemistry and Physiology*. Filer. L. J., et al. (Eds.). Raven Press, New York, p. 375.

Kimura, S., Komai, M., Yokomukai, Y., Torii, T. 1984. Effects of dietary protein levels and umami on the palatability to saltiness in rats. *Chem. Senses* 9:79.

Kiuchi, T., Kondo, Y. 1984. The study of free amino acids and related compounds in vegetable foods. *Rep. Hiroshima Womens Univ.* 20:65.

Kodama, S. 1913. Separation of inosinic acid (Japanese). *J. Tokyo Chem. Soc.* 34:751.

Kojima, K. 1974. Safety evaluation of disodium 5′-inosinate, disodium 5′-guanylate and disodium 5′-ribonucleotide. *Toxicology* 2:185.

Konosu, S., Yamaguchi, Y. 1987. Role of extractive components of boiled crab in producing the characteristic flavor. In: *Umami: A Basic Taste*. Kawamura, Y., et al. (Eds.). Marcel Dekker, New York, p. 235.

Kumazawa, T., Kuirhara, K. 1990. Large synergism between monosodium glutamate and 5′-nucleotides in canine taste nerve responses. *Am. J. Physiol.* 259:R420–R426.

Kumazawa, T., Nakamura, M., Kurihara, K. 1991. Canine taste nerve responses to umami substances. *Physiol. Behavor*, 40(5):875–881.

Kuninaka, A. 1960. Studies on taste of ribonucleic acid derivatives. *Nippon Nogeikagaku Kaishi* 34:489.

Kurihara, K. 1987. Recent progress in the taste receptor mechanism. In: *Umami: A Basic Taste*, Kawamura, Y., Kare, M. R. (Eds.). Marcel Dekker, New York, p. 3.

Kwok, H. M. 1968. Chinese-restaurant syndrome. *N. Engl. J. Med.* 4:796.

Lagerkvist, Y. 1958. Biosynthesis of guanosine 5′-phosphate. II. Amination of xanthosin 5′-phosphate by purified enzyme from pigeon liver. *J. Biol. Chem.* 233:143.

Lemkey-Johnston, N., Reynolds, W. A. 1972. Incidence and extent of brain lesions in mice following ingestion of monosodium glutamate (MSG). *Anat. Rec.* 172:354.

Liebschutz, J., Airoldi, L., Brownstein, M. J., Chinn, N. G., Wurtman, R. J. 1977. Regional distribution of endogenous and parenteral glutamate, aspartate and glutamine in rat brain. *Biochem. Pharmacol.* 26:443.

Litton Bionetics. 1975a. Mutagenic evaluation of compound, FDA 73–58. 000997-42-2. Monopotassium glutamate. U.S. Department of Commerce, National Technical Information Service. PB-254 511.

Litton Bionetics. 1975b. Mutagenic evaluation of compound, FDA 75–11. 007558-63-6. Monoammonium glutamate. FCC. U.S. Department of Commerce, National Technical Information Service, PB-254 512.

Litton Bionetics. 1977a. Mutagenic evaluation of compound. FDA 75–59, L-glutamic acid, HCl. U.S. Department of Commerce, National Technical Information Service, PB-266 892.

Litton Bionetics. 1977b. Mutagenic evaluation of compound. FDA 75-65, L-glutamic acid. FCC: U.S. Department of Commerce, National Technical Information Service, PB-266-889.

Maeda, S., Eguchi, S., Sasaki, H. 1958. The content of free L-glutamic acid in various foods. *J. Home Econ. (Japan)* 9:163.

Maeda, S., Eguchi, S., Sasaki, H. 1961. The content of free L-glutamic acid in various foods (Part 2). *J. Home Econ.* 12:105.

Maga, J. A. 1983. Flavor potentiators. *CRC Crit. Rev. Food Sci. Nutr.* 18:231.

Manning, M. E., Stevenson, D. D. 1991. Pseudoallergic drug reactions. *Immunol. Allerg. Clin. N.A.*, 11(3):659.

Manning, M. E., Stevenson, D. D., Mathison, D. A. 1992. Reactions to aspirin and other nonsteroidal anti-inflammatory drugs. *Anaphylaxis and Anaphylactoid Reactions* 12(3):611.

Marrs, T. S., Salmona, M., Garattini, S., Murston, D., Matthews, D. M. 1978. The absorption by human volunteers of glutamic acid from monosodium glutamate and from a partial enzymic hydrolysate of casein. *Toxicology* 11:101.

Meister, A. 1974. Glutamine synthetase of mammals. In: *The Enzymes*, Vol. 10, Boyer, P. D. (Ed.). Academic, New York, p. 699.

Meldrum, B. 1993. Amino acids as dietary excitotoxins: a contribution to understanding neurodegenerative disorders. *Brain Res.* 18:293.

Moriyuki, R., Ichimura, M. 1978. Acute toxicity of monosodium L-glutamate in mice and rats. *Oyo Yakuri (Pharmacometrics)* 15:433.

Naim, M., Ohara, I., Kare, M. R., Levinson, M. 1991. Interaction of MSG taste with nutrition: perspectives in consummatory behavior and digestion. *Physiol. Behavior*, 49(5):1019–1024.

Nakajima, N., Ichikawa, K., Kamata, M., Fujita, E. 1961. Food chemical studies on 5′-ribonucleotides. Part I. On the 5′-ribonucleotides in foods. (1) Determination of the 5′-ribonucleotides in various stocks by ion exchange chromatography. *Nihon Nogeikagaku Kaishi* 35:797.

Ninomiya, Y., Funakoshi, M. 1987. Qualitative discrimination among ''umami'' and the four basic taste substances in mice. In: *Umami: A Basic Taste*, Kawamura, Y., Kare, M. R. (Eds.). Marcel Dekker, New York, p. 365.

Ohara, M., Maeda, S., Komata, Y., Matsuno, T. 1964. Fifth Kanto Branch Congress, Agr. Chem. Soc. Japan.

O'hara, Y., Takasaki, Y. 1979. Relationship between plasma glutamate levels and hypothalamic lesions in rodents. *Toxicol. Lett.* 4:499.

O'hara, Y., Iwata, S., Ichimura, M., Sasaoka, M. 1977. Effect of administration routes of monosodium glutamate on plasma glutamate levels in infant, weanling and adult mice. *J. Toxicol. Sci.* 2:281.

Ohno, T., Yoshii, K., Kurihara, K. 1984. Multiple receptor types for amino acids in the carp olfactory cells revealed by quantitative cross-adaptation method. *Brain Res.* 310:13–21.

Oldendorf, W. H. 1971. Brain uptake of radiolabeled amino acids, amines and hexoses after arterial injection. *Am. J. Physiol.* 221:1629.

Olney, J. W. 1969. Brain lesions, obesity and other disturbances in mice treated with monosodium glutamate. *Science* 164:719.

Olney, J. W., Ho, O. L. 1970. Brain damage in infant mice following oral intake of glutamate, aspartate or cysteine. *Nature* 227:609.

Olney, J. W., Sharpe, L. G., Feigin, R. D. 1972. Glutamate-induced brain damage in infant primates. *J. Neuropathnol. Exp. Neurol.* 31:464.

Orsan. 1993. IGTC Letter to FASEB on September 3, 1993, p. 2. (FDA Docket No. 92N-0391.)

Owen, G., Cherry, C. P., Prentice, D. E., Worden, A. N. 1978. The feeding of diets containing up to 10% monosodium glutamate to beagle dogs for 2 years. *Toxicol. Lett.* 1:217.

Pardridge, W. M. 1979. Regulation of amino acid availability to brain: selective control mechanisms for glutamate. In: *Glutamic Acid: Advances in Biochemistry and Physiology*. Filer, L. J., Jr., et al. (Eds.). Raven Press, New York, p. 125.

Rassin, D. R., Sturman, J. A., Gaull, G. E. 1978. Taurine and other free amino acids in milk of man and other mammals. *Early Human Dev.* 2:1.

Resources Council, Science and Technology Agency, Japan. 1986. Standard tables of food composition in Japan. Amino acid composition of foods.

Reynolds, W. A., Butler, V., Lemkey-Johnston, N. 1976. Hypothalamic morphology following ingestion of aspartame or MSG in the neonatal rodent and primate: a preliminary report. *J. Toxicol. Environ. Health* 2(2):471.

Reynolds, W. A., Lemkey-Hohnston, N., Stegink, L. D. 1979. Morphology of the fetal monkey hypothalamus after in utero exposure to monosodium L-glutamate. In: *Glutamic Acid: Advances in Biochemistry and Physiology*, Filer, L. J., Jr., et al. (Eds.). Raven Press, New York, p. 217.

Rolls, E. T., Critchley, H. D., Wakeman, E. A., Mason, R. 1994. Responses of neurons in the primate's taste cortex to the glutamate ion and to inosine 5′-monophosphate. *Chem. Senses* (in press).

Saito, T. 1961. Adenosine triphosphate and the related compounds in the muscles of aquatic animals. *Bull. Jap. Soc. Sci. Fish* 27:461.

Schanes, O., Schanes, S. 1946. Glutamic acid decarboxylase of high plants. III. Enzymatic determination of L-(+)-glutamic acid. *Arch. Biochem.* 11:445.

Schaumburg, H. H., Byck, R., Gerstl, R., Mashman, J. H. 1969. Monosodium glutamate: its pharmacology and role in the Chinese restaurant syndrome. *Science* 163:826.

Schneider, H., Moehlen, K. H., Challier, J. C., Dancis, J. 1979. Transfer of glutamic acid across the human placenta perfused in vitro. *Br. J. Obstet. Gynecol.* 86:299.

Semprini, M. E., D'Amicis, A., Mariani, A. 1974. Effect of monosodium glutamate on fetus and newborn mouse. *Nutr. Metab.* 16:276.

Skurray, G. R., Pucar, N. 1988. L-glutamic acid content of fresh and processed foods. *Food Chem.* 27:177–180.

Stegink, L. D. 1976. Absorption, utilization, and safety of aspartic acid. *J. Toxicol. Environ. Health* 2:215.

Stegink, L. D. 1984. Aspartate and glutamate metabolism. In: *Aspartame: Physiology and Biochemistry*, Stegink, L. D., et al. (Eds.). Marcel Dekker, New York, p. 47.

Stegink, L. D., Filer, L. J., Baker, G. L. 1972. Monosodium glutamate: effect on plasma and breast milk amino acid levels in lactating women. *Proc. Soc. Exp. Biol. Med.* 140:836.

Stegink, L. D., Brummel, M. C., Boaz, D. P., Filer, L. J. 1973a. Monosodium glutamate metabolism in the neonatal pig: conversion of administered glutamate into other metabolites in vivo. *J. Nutr.* 103:1146.

Stegink, L. D., Filer, L. J., Baker, G. L. 1973b. Monosodium glutamate metabolism in the neonatal pig: effect of load on plasma, brain, muscle and spinal fluid free amino acid levels. *J. Nutr.* 103:1138.

Stegink, L. D., Reynolds, W. A., Filer, L. J., Pitkin, R. M., Boaz, D. P., Brummel, M. C. 1975. Monosodium glutamate metabolism in the neonatal monkey. *Am. J. Physiol.* 299:246.

Stegink, L. D., Reynolds, W. A., Filer, L. J., Baker, G. L., Daabees, T. T., Pitkin, R. M. 1979.

Comparative metabolism of glutamate in the mouse, monkey, and man. In: *Glutamic Acid: Advances in Biochemistry and Physiology*, Filer, L. J., Jr., et al. (Eds.). Raven Press, New York, p. 85.

Steiner, J. E. 1987. What the neonate can tell us about umami. In: *Umami: A Basic Taste*, Kawamura, Y., Kare, M. R. (Eds.). Marcel Dekker, New York, p. 97.

Suzuki, T., Kurihara, Y., Tamura, S. 1976. Amino acid content of fruits, vegetables and processed goods and similarily between their amino acid patterns. *Rep. Nat. Food Res. Inst.* 31:42.

Tabuchi, E., Ono, T., Hishijo, H., Torii, K. 1991. Amino acid and NaCl appetite and LHA neuron responses of lysine-deficient rat. *Physiol. Behavior*, 49(5):951–964.

Takasaki, Y. 1978a. Studies on brain lesions after administration of monosodium L-glutamate in the diet. *Toxicol. Lett.* 4:205.

Takasaki, Y. 1987b. Studies on brain lesion by administration of monosodium L-glutamate to mice. I. Brain lesions in infant mice caused by administration of monosodium L-glutamate. *Toxicology* 9(4):293.

Takasaki, Y. 1979. Protective effect of mono and disaccharides on glutamate induced brain damage in mice. *Toxicol. Lett.* 4:205.

Takasaki, Y., Yugari, Y. 1980. Protective effect of arginine, leucine and pre injection of insulin on glutamate neurotoxicity in mice. *Toxicol. Lett.* 5(1):39.

Tarasoff, L., Kelly, M. F. 1993. Monosodium L-glutamate; a double-blind study and review. *Food Chem. Toxicol.* 31(12):1019.

Terasaki, M., Fujita, I., Wada, S., Nakajima, T., Yokobe, T. 1965a. Studies on the taste of tricholomic acid and ibotenic acid. Part I. *J. Jap. Soc. Food Nutr.* 18:172.

Terasaki, M., Wada, S., Takemoto, T., Nakajima, T., Fujita, E., Yokobe, T. 1965b. Studies on the taste of tricholomic acid and ibotenic acid. Part II. *J. Jap. Soc. Food Nutr.* 18:222.

Titus, D. S. 1964. The nucleotide story: applications presented at *Symposium of Flavor Potentiation*, Boston, March 18.

Torii, K., Mimura, T, Yugari, Y. 1986. Preference for umami, sweet and salty taste in rats fed diets containing various amounts and quality of protein. *Chem. Senses* 9:80.

Torii, K., Mimura, T., Yugari, Y. 1985. Effects of dietary protein on the taste preference for amino acids in rats. In: *Interaction of the Chemical Senses with Nutrition*, Academic, New York 47.

Tung, T. C., Tung, K. S. 1980. Serum free amino acid levels after oral glutamate intake in infant and adult humans. *Nutr. Rep. Int.* 22:431.

Umami Information Center. 1986. The World of Umami (video). Umami Information Center, Tokyo.

U.S. Department of Agriculture. 1974. Composition of Foods: Dairy and Egg Products. Agriculture Handbook No. 8-1.

U.S. Department of Agriculture. 1984. Composition of Foods: Vegetables and Vegetable Products. Agriculture Handbook No. 8-11.

U.S. FDA. 1975. Investigation of the toxic and teratogenic effects of GRAS substances on the developing chicken embryo. Unpublished report. Sept. 9.

U.S. FDA. 1976a. Investigation of the toxic and teratogenic effects of GRAS substances on the developing chicken embryo. Unpublished report. Apr. 20.

U.S. FDA. 1976b. Investigation of the toxic and teratogenic effects of GRAS substances on the developing chicken embryo. Unpublished report. Sept. 9.

Virarouge, C., Caulliez, R., Nicolaidis, S. 1992. Umami taste of monosodium glutamate enhances the thermic effect of food and affects the respiratory quotient in the rat. *Physiol. Behavior* 52: 879–884.

Weaver, J. C., Kroger, M. 1978. Free amino acid rheological measurements on hydrolyzed lactose cheddar cheese during ripening. *J. Food Sci.* 43:579–583.

Wilson, D. W., Wilson, H. C. 1962. Studies in vitro of the digestion and absorption of purine ribonucleotides by the intestine. *J. Biol. Chem.* 237:1643.

Worden, A. N., Rivett, K. J., Edwards, D. B., Street, A. E., Newman, A. J. 1975. Long-term feeding study on disodium 5′-ribonucleotide on reproductive function over three generations in the rat. *Toxicology* 3:349.

Yamaguchi, S. 1967. The synergistic taste effect of monosodium glutamate and disodium 5′-inosinate. *J. Food. Sci.* 32:473.

Yamaguchi, S. 1979. The umami taste. In: *Food Taste Chemistry*, Boudreau, J. C. (Ed.). American Chemical Society, Washington, D. C.

Yamaguchi, S. 1987. Fundamental properties of umami in human taste sensation. In: *Umami: A Basic Taste*, Kawamura, Y., Kare, M. R. (Eds.). Marcel Dekker, New York, p. 41.

Yamaguchi, S., Kimizuka, A. 1979. Psychometric studies on the taste of monosodium glutamate. In: *Glutamic Acid: Advances in Biochemistry and Physiology*, Filer, L. J., Jr., et al. (Eds.). Raven Press, New York, p. 35.

Yamaguchi, S., Takahashi, C. 1984a. Hedonic functions of monosodium glutamate and four basic taste substances used at various concentration levels in single and complex systems. *Agr. Biol. Chem.* 48:1077.

Yamaguchi, S., Takashashi, C. 1984b. Interactions of monosodium glutamate and sodium chloride on saltiness and palatability of a clear soup. *J. Food Sci.* 49:82.

Yamaguchi, S., Yoshikawa, T., Ikeda, S., Ninomiya, T. 1971. Measurement of the relative taste intensity of some L-α-amino acids and 5′-nucleotides. *J. Food Sci.* 36(6):846.

Yonetani, S., Matsuzawa, Y. 1978. Effect of monosodium glutamate on serum luteinizing hormone and testosterone in adult male rats. *Toxicol. Lett.* 1:207.

Yonetani, S., Ishii, H., Kirimura, J. 1979. Effect of dietary administration of monosodium L-glutamate on growth and reproductive functions in mice. *Oyo Yakuri (Pharmacometrics)* 17:143.

Yoshida, T. 1978. L-Monosodium glutamate (MSG). In: *Encyclopedia of Chemical Technology*, Vol. 2. Wiley, New York, p. 410.

Yoshii, K., Yokouchi, C., Kurihara, K. 1986. Synergistic effects of 5′-nucleotides on rat taste response to various amino acids. *Brain Res.* 367:45.

Zanda, G., Franciosi, P., Tongnoni, G., Rizzo, M., Standen, S. M., Morselli, P. L., Garattini, S. 1973. A double blind study on the effects of monosodium glutamate in man. *Biomedicine* 19: 202.

15

Sweeteners

SEPPO SALMINEN

University of Turku, Turku, Finland, and Royal Melbourne Institute of Technology, Melbourne, Australia

ANJA HALLIKAINEN

National Food Administration, Helsinki, Finland

I. INTRODUCTION

A. Theory of Sweetness

Sweetness is one of the most important taste sensations for humans and for many animal species as well. Sweet compounds almost universally induce a positive hedonic response in humans, and this response, which is found in the neonate, is often thought to be inborn. There is scarcely any area of food habits today that does not in some way involve the sweet taste. The importance of sweetness is reflected in the world production of sugar, which rose from 8 million tons in 1900 to 70 million tons in 1970. No other agricultural product has shown a similar increase in production during the same period.

Sucrose is not consumed only for its sweetness. It also has many functional properties in foods that make it useful as a bulking agent, texture modifier, mouth-feel modifier, and preservative. Sucrose additionally offers an important energy source for many food fermentations.

At present little is known about the basic mechanisms of the sweet taste. Numerous attempts have been made to develop a theory of sweetness. It has been suggested by many investigators that the stereochemical basis of sweetness can be verified with the aid of simple model sugars and their deoxy derivatives (Birch, 1980). The original hypothesis by Schallenberger (1963) and Shallenberger and Acree (1967) suggesting that hydrogen bonding might explain the taste properties of sugars has withstood many tests over the years. This theory explains the stereochemical fit that is thought to be responsible for the sweetness quality (Schallenberger, 1973). However, even today the complex research on various aspects of sweetness has not been able to answer the question of sweetness perception completely, but there may be a genetic background (Davenport 2001).

The sweetness of individual sweeteners is usually measured in model systems and compared to that of sucrose. The sweetness of individual sweeteners and mixtures can therefore also be estimated and expressed as the concentration of the equisweet reference sugar (usually glucose or sucrose).

The increased sweetness obtained by mixing natural and synthetic sweeteners has been of economic and nutritional interest. The unpleasant aftertastes of many artificial or nonnutritive sweeteners have also stimulated the development of sweetener mixtures to reduce these aftertastes. A review of the synergism between sweeteners and the factors affecting it has been written by Hyvönen (1980).

B. Sugar Substitutes in Foods

To nutritionists and many consumers sugar is not a satisfactory food. It compares unfavorably with many other foods in nutritional value. It is generally considered bad for our teeth, and it has been related to many diseases. Most of all, nutrition specialists have been worried about the decreasing nutrient density of diets high in refined sugar.

For nutritional and health reasons there has been a growing desire in most Western countries to utilize sweeteners other than sucrose. Consumers are urged to control their energy intake to avoid obesity, and, in addition to reducing fat consumption it is usually recommended that sugar intake be reduced. According to Hyvönen (1980) the main reason for alternative sweetening of foods today appears to be people's desire to reduce their energy intake. However, disorders in carbohydrate metabolism and dental caries are also problems that lead to the use of sugar substitutes. Because caries is still one of the most common diseases in the world, the dental profession has been forced to recommend decreases in sugar consumption. This growing pressure to use noncaloric or artificial sweeteners rather than carbohydrate-based sweeteners extends to the search for noncariogenic or even anticariogenic sugar substitutes.

The use of artificial sweeteners gives rise to a variety of problems in food technology due to some basic differences between them and the carbohydrate sweeteners. Nonnutritive sweeteners are usually not carbohydrate based and therefore have different chemical and physical properties. Often nonnutritive sweeteners also have flavor characteristics that differ from those of carbohydrate sweeteners and are intensely sweet compared to carbohydrate sweeteners. These properties often influence the cost of food manufacturing because the resulting dietetic or special dietary foods are expected to be as acceptable as those with carbohydrate sweeteners.

Recent doubts about the safety of nonnutritive sweeteners have resulted in the development of caloric sucrose substitutes for use as sweet food ingredients. While nonnutritive sweeteners include a range of natural products and some synthetic chemicals, caloric sweeteners or sucrose substitutes are usually carbohydrates or carbohydrate derivatives.

The major difference between the nonnutritive and nutritive sweeteners, in addition to the energy content, is the amount of sweetener needed. Therefore, for the purpose of this chapter all sweeteners are divided into two groups: the nonnutritive sweeteners and the nutritive or bulk sweeteners (Table 1). Nonnutritive sweeteners are here defined in the same manner as in the U.S. Code of Federal Regulations (Title 21, Section 170.6). Thus nonnutritive sweeteners are substances that have less than 2% of the caloric value in an equivalent unit of sweetening capacity.

Table 1 Some Common Nutritive and Nonnutritive Sweeteners with their European E-Codes

Nutritive sweeteners		Nonnutritive sweeteners	E-Code
Glucose		*Commonly used*	
Fructose		Saccharin	E 954
Invert sugar		Cyclamates	E 952
Saccharose		Aspartame	E 951
Polyols		Acesulfame K	E 950
Hydrogenated glucose syrups		*Others*	
Lactitol	(E 966)	Thaumatin	E 957
Maltitol	(E 965)	Stevioside	
Mannitol	(E 421)	Neuhesperidine	E 959
Sorbitol	(E 420)	Monellin	
Xylitol	(E 967)	Miraculin	
		Dulcin	
		Sucralose	

The aim of this chapter is to provide guidelines for the use and safety of nonnutritive and bulk sweeteners. Information will be given on the source and availability, toxicology, and safety aspects of the sweetener, as well as its regulatory position and its application to foods. It is assumed that new sweeteners will continue to be developed in the future. However, sucrose and the main nonnutritive sweeteners saccharin, cyclamates, aspartame, and acesulfame K will probably hold a strong role as sweeteners in the near future.

II. NONNUTRITIVE SWEETENERS

A. Saccharin

1. History

Saccharin was first synthesized in the United States in 1879 by two chemists (Remsen and Fahlberg, 1879). Essentially the same production method is still used by many manufacturers. Initially saccharin was used as an antiseptic and a preservative, but its potential as a food sweetening agent was soon established. Saccharin was first introduced as a food additive in the United States in 1900, and immediately concern over its safety arose. In Europe the use of saccharin increased during the two world wars due to the lack of sugar. Since World War II the consumption of saccharin has steadily increased due to the widespread acceptance of special dietary and dietetic foods even though its safety has repeatedly been questioned (National Research Council/National Academy of Sciences, 1978). Only recently, since the introduction of aspartame, has a small decline in the increase in saccharin consumption taken place.

2. Chemistry

Saccharin is a general name used for saccharin, sodium saccharin, and calcium saccharin. The molecular formula of saccharin is $C_7H_5NO_3S$, and the structural formula is presented in Fig. 1. Chemically saccharin is 1,2-benzisothiazol-3(2H)-one-1,1-dioxide and its sodium or calcium salt.

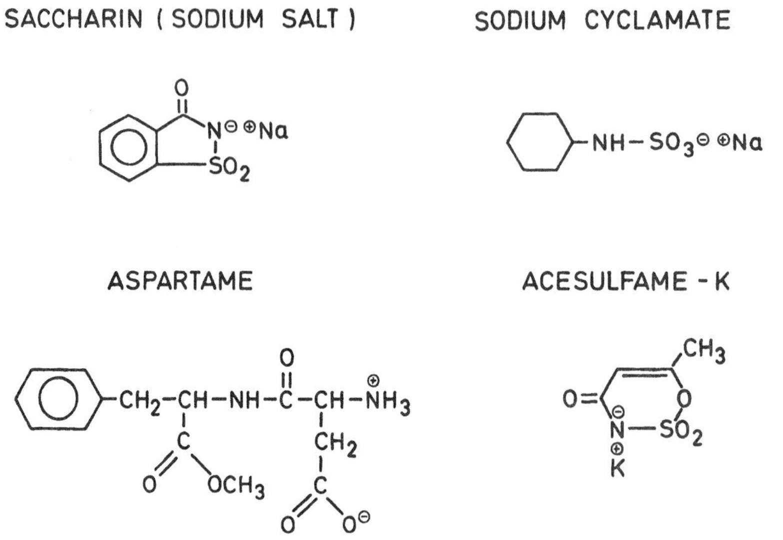

Figure 1 The chemical structures of the most important nonnutritive sweeteners.

Saccharin can be produced by two methods, either from toluene and chlorosulfonic acid (Remsen and Fahlberg, 1879) or from methyl anthranilate (National Research Council/National Academy of Sciences, 1978). Saccharin and sodium saccharin do not occur in nature. Manufactured saccharin and sodium saccharin are white crystalline powders that are stable at both high temperatures (up to 300°C) and low temperatures. [This stability has been questioned by Kroyer and Washüttl (1982).] Both compounds are soluble in water and ethanol. In dilute aqueous solutions they are about 300 times sweeter than a solution containing an equal concentration by weight of sucrose.

3. Technological Properties

Saccharin and its calcium and sodium salts are very stable under almost all food processing conditions, and they also have a long shelf life (Table 2). It remains to be further studied whether the reactions of saccharin and many food components are as significant as indicated by Kroyer and Washüttl (1982). Saccharin is at present the only noncaloric sweetener that appears to be stable during cooking and baking of food products and can be utilized in most drugs, special dietary products, and cosmetics. However, saccharin tends to have a slight to moderate metallic or bitter aftertaste. This aftertaste can be masked by the use of lactose or by combining saccharin with aspartame or other sweeteners. When combined with other sweeteners, saccharin usually has a synergistic effect on the sweetness and thereby the total amount of noncaloric sweeteners can be reduced (Hyvönen, 1980).

4. Intake

In the United States the major uses of saccharin include soft drinks, tabletop sweeteners, and dietetic foods. Similar usage patterns are also found in most European countries. Saccharin and sodium saccharin have other uses in cosmetics and pharmaceuticals. It is estimated that, despite all the adverse publicity about its potential hazards to health, some 70 million Americans are fairly regular users of saccharin. About 60% is consumed in

Table 2 Some Properties of Nonnutritive Sweeteners

Sweetener	Sweetness in relation to sucrose	Aftertaste	Stability		ADI[a] (mg/kg body weight)
			In solution	During heating	
Acesulfame K	150×	Very slight, bitter	Stable	Stable	0–9
Aspartame	180×	Prolonged sweetness	Not stable in acid conditions	Unstable, sweetness may disappear	40
Cyclamate	30–60×	Chemical flavor	Relatively stable	Relatively stable	0–7
Saccharin	300×	Bitter metallic	Stable in pH < 2.0	Relatively stable	2.5
Stevioside	100–300×	Bitter	Relatively stable	Relatively stable	Not acceptable
Talin	200–2500×	Licorice-like	Relatively stable	Stable at neutral to low pH	Not specified
Sucralose	600×	—	Stable	Stable	0–15

[a] WHO and European Union Scientific Committee on Food.

soft drinks, 20% in other drinks and foods, and 20% as tabletop sweeteners (Higginbotham, 1983). The average consumption of sodium saccharin in the United States has been reported to be 7.1 mg/day for the whole population. For saccharin-consuming population groups the intake was 25 mg/day (National Academy of Sciences, 1979). In Finland the average per capita saccharin consumption has been reported at a level of 15 mg/day and in diabetics 14 mg/kg body weight per day (Virtanen et al., 1988). In other Scandinavian countries the intakes appear to be similar (Penttilä et al., 1988). Two reviews on the history and usage of saccharin are available (Cranmer, 1980; Oser, 1985).

5. Toxicology of Saccharin

a. Metabolism and Short-Term Toxicity. Saccharin itself is absorbed slowly but almost completely from the gut after oral administration, and it is rapidly excreted in the urine as unchanged saccharin. The remaining saccharin is excreted unchanged in feces (Renwick, 1983,1985). Therefore, metabolites are not likely to cause toxic effects. Saccharin does not increase the urinary excretion of dietary oxalate in mice, and therefore oxalate crystals are not likely to be involved with saccharin toxicity or bladder effects (Salminen and Salminen, 1986a). Although most mutagenicity studies are negative, saccharin has been indicated to pose mutagenic properties at least in the host-mediated assay. This test proved that highly purified saccharin itself was not mutagenic, but the urine of mice treated with saccharin exhibited mutagenic activity in *Salmonella typhimurium* TA 100 strain (Batzinger et al., 1977). However, most studies on mutagenicity have indicated that sodium saccharin is nonreactive to DNA and inactive as a gene mutagen in vitro (Ashby, 1985).

b. Long-Term Toxicity. Saccharin has been the subject of a very large number of long-term toxicity tests. In some long-term rat studies an induction of a higher incidence of bladder cancer was found among rats consuming saccharin. This incidence has been observed in several long-term rat studies conducted by the U.S. Food and Drug Administration, the Wisconsin Alumni Research Foundation (Tisdel et al., 1974), and the International Research and Development Corporation (1983). All these studies indicated that high dietary levels (5–7.5%) increase the incidence of urinary bladder tumors in rats. In the latest studies the effect was clearest, and even a dose-response relation was observed (Carlborg, 1985). However, a possibility of a threshold dose level for the bladder effects has recently been suggested (Carlborg, 1985). Numerous additional studies have been conducted on the bladder carcinogenicity and cocarcinogenicity of saccharin. The results appear to support weak carcinogenicity and possible cocarcinogenicity or promotion effect of saccharin. A review of some of these studies has been completed by Cranmer (1980) and more recently by Cohen (1985).

c. Epidemiological Studies on Saccharin. The safety of saccharin has been assessed in many epidemiological studies involving both normal subjects and diabetics (Hoover and Strasser, 1980; Morrison and Buring, 1980; Wynder and Stellman, 1980). Most studies in many population groups including diabetics have failed to demonstrate any statistical evidence of an association between human bladder cancer and saccharin consumption (Morgan and Wong, 1985). However, most studies are inadequate in many ways and cannot determine exact saccharin consumption figures. On the other hand, diabetics, who are significant users of saccharin and other nonnutritive sweeteners, form a special group with different dietary habits than the general population. Therefore, results from epidemiological saccharin studies in diabetic populations may not be relevant for the general population. A review of the studies summarizing relative risks has been recently published (Morgan and Wong, 1985).

The idea that saccharin may be a weak carcinogen in the rat appears to be evident on the basis of numerous animal studies. At present, the linkage between animal studies and human studies is difficult for risk evaluation purposes, but it appears that saccharin is unlikely to present a cancer risk at present average exposure rates. However, since alternative nonnutritive sweeteners are currently available it is easy to utilize these whenever technological and consumer preference aspects require saccharin substitution with other sweeteners.

6. Regulatory Status

The Joint FAO/WHO Expert Committee on Food Additives (JECFA) has established an acceptable daily intake (ADI) of 2.5 mg/kg body weight. Saccharin is currently being used almost worldwide. However, restrictions of use to only a limited number of products apply in most countries. In 1977, the U.S. Food and Drug Administration proposed a ban on saccharin in the United States. The proposal resulted in a congressional moratorium on the saccharin ban pending further toxicity studies. This moratorium has been extended several times to allow more studies to be completed. Studies have thus accumulated and as new sweeteners have become available the use of saccharin has decreased.

B. Cyclamates

1. History

Sodium cyclamate was synthesized in 1937, but it was first produced commercially in the United States in 1950. Now cyclamates are produced in many countries including Japan, Germany, Spain, Taiwan, and Brazil (IARC, 1980). Due to regulatory decisions the use of cyclamates decreased in the 1960s. Recently, after the allocation of a new acceptable daily intake value by JECFA, the use has started to increase again.

2. Chemistry

Cyclamates is a group name used for the following compounds: cyclamic acid, sodium cyclamate, and calcium cyclamate. The molecular formula for calcium cyclamate is $C_{12}H_{24}CaN_2O_6S_2 \cdot 2H_2O$; and the structural formula is presented in Fig. 1.

Cyclamates are chemically synthesized products that are not found in nature. They are synthesized from cyclohexylamine by sulfonation of various chemicals (chlorosulfonic acid, sulfamic acid) followed by neutralization with hydroxides (IARC, 1980).

Cyclamates are stable at both high and low temperatures. They provide a sweet taste that is 30 times sweeter than sugar (Table 2). Cyclamates are easily soluble in water and can be used as noncaloric sweeteners in most foods, including soft drinks, confections, desserts, and processed fruits and vegetables. Cyclamates have a synergistic sweetening effect when combined with saccharin. Sodium and calcium cyclamates have been used mainly in the form of the 10:1 cyclamate/saccharin mixture.

3. Toxicology

a. Metabolism of Cyclamate. Cyclamate is absorbed partially from the intestine, and a variable amount is converted to cyclohexylamine by microorganisms in the large bowel.

In humans the conversion of cyclamate is thought to be limited but may vary markedly from person to person. The maximum amount of cyclamate that is theoretically available for metabolism is 60%, which is also the maximum conversion of cyclamate to cyclohexylamine. Ninety percent of the population will convert less than 1% of ingested cyclamate to cyclohexylamine (Renwick and Williams, 1972; WHO, 1982). Both cyclamate and cyclohexylamine can cross the placental barrier, thereby affecting the fetus.

After cyclamate administration in some nonhuman mammals, *N*-hydroxycyclohexylamine has been found in urine. Cyclohexanol and cyclohexane have also been reported as trace metabolites of cyclamate in rats, rabbits, guinea pigs, monkeys, and humans, probably arising from metabolism of cyclohexylamine (Renwick and Williams, 1972).

b. Short-Term Toxicity. Although many short-term tests have been conducted, there have been no assays for mammalian cell DNA damage and gene mutation for cyclamate and no DNA damage test for cyclohexylamine.

Tests for gene mutations in bacteria have been negative for cyclamate and cyclohexylamine. Positive tests include some short-term tests like mammalian cytogenetic tests. In vitro studies have shown tumor-promoting properties for cyclamate (IARC, 1980; FDA, 1985; National Research Council, 1985).

c. Carcinogenicity and Epidemiological Studies. In two carcinogenicity studies, cocarcinogenic and tumor-promoting activities for cyclamate were observed. In one study cyclamate was incorporated into the bladders of mice, and in the other cyclamate was fed to rats after *N*-methyl-*N*-nitrosourea had been instilled into the bladder. However, these studies were of problematic design, and the results have been questioned (DeSesso, 1987).

In one bioassay, rats dosed with a 10:1 cyclamate/saccharin mixture were reported to develop urinary bladder cancer. Two later attempts to duplicate the results of the initial study failed to show carcinogenicity. Numerous other studies demonstrate that even after ingestion of high doses of cyclamate throughout the lifetime of laboratory animals, cyclamate does not cause cancer.

Epidemiological studies in humans indicate that there is suggestive evidence that the use of cyclamate/saccharin mixtures may be associated with a small increase in the risk of bladder cancer. In addition to carcinogenicity data there are some adverse effects observed in laboratory animals such as testicular atrophy in animals exposed to cyclohexylamine. The clarification of these adverse effects as well as sone other questions give rise to the need for further studies. Epidemiological monitoring should be continued, comparing heavy and long-term users. The promoter or cocarcinogenic role of cyclamate should also be elucidated (National Research Council, 1985).

Cyclamates themselves appear not to be carcinogenic, and epidemiological studies support this conclusion (IARC, 1980; FDA, 1985; National Research Council, 1985).

4. Use and Intake

There is little information about the use of cyclamates in different countries. Therefore it is not known whether the proposed acceptable daily intake is exceeded by some consumer groups. However, data from the United States before the ban of cyclamate in 1977 are available and indicate that the greatest intake is from beverages and sugar substitutes. According to this information it has been calculated that a person who is a 90th percentile consumer of cyclamate in all products would have a daily consumption of cyclamate of less than 1500 mg (MRCA, 1985).

5. Regulatory Status of Cyclamates

The Joint FAO/WHO Expert Committee on Food Additives established in 1967 a temporary acceptable daily intake of 50 mg/kg body weight for total cyclamates. This was withdrawn in 1970, and a temporary ADI of 4 mg/kg body weight expressed as cyclamic acid was recommended in 1977 (WHO, 1978). The most recent European assessment raises the ADI to 9 mg/kg body weight.

In the United States cyclamate was approved for use as a nonnutritive sweetening agent in 1950. In 1969, the FDA banned the use of cyclamate in food because its safety was questioned. The data on carcinogenicity were reexamined in 1976 by a National Cancer Institute Committee, in 1984 by the FDA Cancer Assessment Committee, and in 1985 by the National Research Council Committee. All the committees have concluded that experimental and epidemiological evidence does not indicate that cyclamate is carcinogenic. However, cyclamates were not approved for food use in the United States.

In 1984 the Commission of the European Communities issued a report on sweeteners and established an ADI value for cyclamate (Table 2).

More than 40 countries in Europe, Asia, North and South America, and Africa have approved cyclamate as nonnutritive sweetening agent.

C. Aspartame

1. History

Aspartame was discovered accidentially in the G. D. Searle laboratories by J. M. Schlatter in the early 1960s. Since the discovery safety studies on aspartame have been carefully conducted by Searle laboratories and by many other independent research laboratories. In the early 1980s aspartame was approved in many countries as an alternative sweetener to saccharin and cyclamate.

2. Chemistry

Chemically, aspartame is the methyl ester of *L*-aspartyl-*L*-phenylalanine. Aspartame is produced from the amino acids phenylalanine and aspartic acid. Preliminary amino acids can be produced by fermentation.

Aspartame is an odorless white crystalline powder and has a clean sweet taste. It is slightly soluble in water and sparingly soluble in alcohol. The sweetening potency of aspartame is 150–200 times that of sucrose (Table 2). Aspartame provides 4 kcal/g. Under certain moisture, temperature, and pH conditions, the *O*-methyl ester bond is hydrolyzed, forming the dipeptide aspartylphenylalanine and methanol. Alternatively, methanol may be eliminated by the cyclization of aspartame to form its diketopiperazine (DKP), which in turn can be hydrolyzed to aspartylphenylalanine and then ultimately to aspartate and phenylalanine (Fig. 2). When these compounds are formed in food products, a loss of sweetness is perceived (Stegink and Filer, 1984; Homler, 1984). However, the stability of aspartame in dry products is good. In addition to sweetening, aspartame also enhances some food flavors (Homler, 1984).

3. Technological Properties

Aspartame provides sugarlike sweetness both in foods and as a tabletop sweetener, but it is not suitable for all foods or food processes. A good review has been compiled by

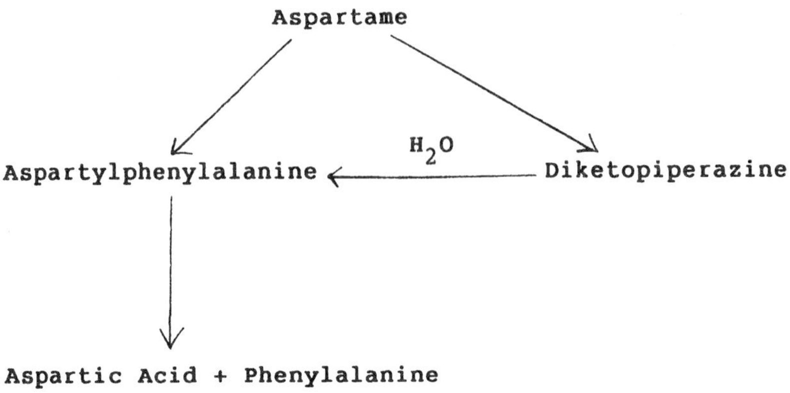

Figure 2 Typical chemical reactions by which aspartame is converted to nonsweet compounds. [According to Homler (1984).]

Stegink and Filer (1984). Aspartame is an excellent sweetener for dry products like powdered drinks or tabletop sweeteners. At high temperature or low pH, aspartame is gradually hydrolyzed and aspartylphenylalanine and methanol are produced. Since these compounds are not sweet, a loss of sweetness is observed in foods that have an extremely low pH or in foods heated for long periods. Aspartame can be easily used in chewing gum, instant coffee, and tea. Aspartame is also suitable for the sweetening of most soft drinks, dairy products such as yogurts and ice cream, and dessert mixes (Homler, 1984). Soft drink manufacturers often increase the stability of aspartame by raising the product pH slightly and controlling the inventory. Soft drinks are usually sweetened with a saccharin/aspartame mixture that further increases the stability. Notable sweetness differences are observed only after an excessive storage time and a 40% loss in aspartame concentration (Anonymous, 1986).

In most cases sugar cannot simply be replaced by aspartame due to the functional differences between the two. Therefore, a complete reformulation of the product is often needed. Aspartame provides a clear sugarlike sweetness, but the physical and functional properties have to be supplied by suitable bulking agents and carbohydrates. For instance, polydextrose can be utilized as a bulking agent in many aspartame-sweetened foods. In some applications such as powdered beverage mixes, bulk reduction can be beneficial, reducing the packaging and shipping costs (Homler, 1984).

4. Intake

The metabolism of aspartame yields, on a weight per weight basis, is approximately 50% phenylalanine, 40% aspartic acid, and 10% methanol. It has been calculated that daily replacement of all sweeteners with aspartame (using a sweetener ratio of 180:1) would increase the phenylalanine intake by 12% and aspartic acid intake by 5%. It is clear that aspartame is not likely to alter daily phenylalanine and aspartic acid intake (3.6 and 6.8 g, respectively) appreciably. The potential extreme intakes of aspartame have been calculated and estimated using various methods. The highest daily intakes have been estimated for young children (10 mg/kg body weight) and diabetics (8 mg/kg body weight), while the ADI value for aspartame is 40 mg/kg body weight (ILSI-NF, 1986). In Finland the actual aspartame intake of diabetics was found to be 1.2 mg/kg body weight (Virtanen et al., 1988). It appears that even at high aspartame product consumption the intakes remain well below the ADI value.

5. Toxicology

a. Metabolism. Aspartame can be absorbed and metabolized in two major ways. In both cases large doses of aspartame release aspartate, phenylalanine, and methanol to the portal blood, and these compounds are metabolized and/or excreted.

Phenylalanine makes up over half of the aspartame molecule and is an essential component of body proteins that cannot be synthesized by mammals. The metabolism and nutritional roles of phenylalanine and tyrosine are linked (Stegink and Filer, 1984). Some ingested aspartame is transaminated to glutamate. There are different interactions between glutamate and aspartate metabolism.

b. Long-Term and Carcinogenicity Studies. Some food sweetened with aspartame might contain the diketopiperazine derivative (DKP) at levels up to 5% of the amount of aspartame added. Therefore DKP has also been subjected to intensive toxicological testing (WHO, 1980).

Extensive data have been reported to support the safety of aspartame. Mutagenicity and reproduction studies do not indicate any adverse effects. Mouse data obtained from chronic feeding studies were negative. The results of one long-term rat study with aspartame were consistent with an increased incidence of intracranial neoplasms in the treated animals. However, the increase was not dose or sex related, and the overall incidence was within that previously observed in untreated animals of the same strain. Furthermore, no such increase was seen in two subsequent long-term studies with aspartame. The DKP study results were negative (WHO, 1980).

c. Other Safety Issues Related to Aspartame. Regarding the aspartic acid moiety of aspartame, there was concern that aspartame, either alone or in combination with glutamate, may cause focal brain lesions and endocrine disorders. Both glutamate, which is widespread in the food supply, and aspartame in high concentrations have shown neurotoxic effects on rodents. Although the metabolism of aspartame in humans is similar to that of phenylalanine and aspartic acid, clinical studies involving both adults and children have indicated no untoward effects at levels higher than can be expected for the sweetener in a normal diet.

Phenylketonuria (PKU) is a human genetic disorder of phenylalanine metabolism. In PKU phenylalanine is metabolized poorly and accumulates in blood and tissues. Grossly elevated plasma phenylalanine levels are associated with harmful health effects.

Aspartame in combination with dietary carbohydrates might alter brain neurotransmitter activity. In rats the administration of glucose and aspartame (200 mg/kg) by gavage increased brain levels of tyrosine and phenylalanine and decreased brain serotonin concentration. Plasma phenylalanine concentrations were also increased in fasted humans, who received large doses of aspartame and confections containing sucrose. However, a nine-month study on infant monkeys showed neither abnormalities nor physical or behavioral impairment when they were removed from high aspartame/phenylalanine diets (Council of Scientific Affairs, 1985).

Plasma phenylalanine concentration normally ranges from 6 to 12 μ mol/dL. The toxic threshold for plasma phenylalanine is 100 μ mol/dL for normal persons including infants, and 50 μmol/dL for pregnant women. Clinical studies have shown that only a loading dose of 200 mg/kg aspartame increased plasma phenylalanine levels to 50 μ mol/dL. This is equivalent to 24 L of aspartame-sweetened beverage. However, because individuals with phenylketonuria need to control their phenylalanine intake, products containing aspartame should usually be labeled with the statement: ''Phenylketonurics: contains phenylalanine'' (WHO, 1980).

Approximately 10% by weight of aspartame is converted to methanol. Therefore the potential toxicity of markedly elevated blood methanol concentration must also be reconsidered. Ingestion of large doses of the substance by special population groups that may have increased sensitivity (due to infancy, pregnancy, lactation, heterozygosity for PKU) has been specially observed in toxicology studies.

The FDA has reported that in clinical studies, measurable blood levels of methanol have not been detected until loading doses of aspartame exceeded the projected 99th percentile exposure level of 34 mg/kg (Council on Scientific Affairs, 1985).

Available evidence suggests that normal consumption of aspartame is safe because consumption of aspartame from foods is far below any suspected toxic levels. These data do not provide evidence for serious adverse health effects, although certain individuals might have an unusual sensitivity to the product.

6. Regulatory Status

The FAO/WHO Joint Expert Committee on Food Additives has recommended an acceptable daily intake of 40 mg/kg body weight for humans for aspartame. Simultaneously an ADI value of 7.5 mg/kg body weight was given for diketopiperazine.

In 1981 aspartame was approved in the United States for use in tabletop sweeteners and dry beverage mixes. Further testing of the safety of aspartame was recommended. Later in 1983 aspartame was approved for use in beverages and many other foods (*Federal Register*, 1981; Stegink and Filer, 1984).

Aspartame has been approved for use in most European countries, in the United States, and in Canada.

D. Acesulfame K

1. History

Acesulfame K is one of the most recently introduced nonnutritive sweeteners. It was developed by Hoechst Company in West Germany in 1967 and has only recently been recommended for use in foods in several countries.

2. Chemistry

Acesulfame K is a name utilized for the potassium salt of 6-methyl-1,2,3-oxathiazine-4(3)-one-2,2-dioxide. The composition of acesulfame K is $C_4H_4NO_4KS$. The compound is freely soluble in water and forms a neutral solution. Acesulfame K is not hygroscopic, and it decomposes during heating at temperatures over 235°C (Table 2). The molecular weight of the compound is 201.2. At room temperature acesulfame K is an intensely sweet (150–200 times sweeter than sucrose), white, odorless, crystalline powder.

3. Food Technological Uses

Major uses will probably include soft drinks, tabletop sweeteners, and chewing gum. In pharmaceutical areas, mouthwashes and toothpastes may be included (Higginbotham, 1983). At present, more food applications are being developed.

4. Toxicology

Acesulfame K was first evaluated by JECFA in 1981, but some shortcomings were found in long-term carcinogenicity studies and therefore no acceptable daily intake value was allocated (WHO, 1981). Also, mouse carcinogenicity studies were found not to meet the requirements of the expert group, although the study was carried out according to good laboratory practice at the time of completion. Additional studies showed that acesulfame K is not mutagenic and not carcinogenic in the rat. The present toxicological information appears to be complete and the safety of acesulfame K established.

5. Regulatory Status

An acceptable daily intake of 0–9 mg/kg body weight has been allocated for acesulfame K (WHO, 1983). In Britain the Food Additives and Contaminants Committee (1982) recommended the use of acesulfame K in various foods; and Switzerland, Germany, Ireland, Denmark, and Sweden have approved some food uses. In some countries it has been accepted for toothpastes and mouthwashes.

E. Thaumatin

1. Chemistry

Thaumatin (Thalin) is a macromolecular protein sweetener with a molecular weight of around 22,000. The major protein constituents of the sweetener consist of the normal amino acids except for histidine, which is absent. The extensive disulfide cross-linking confers thermal stability and resistance to denaturation. The tertiary structure of the polypeptide chain gives thaumatin its sweet character. Cleavage of just one disulfide bridge results in a loss of sweet taste (Iyengar, 1979).

Thaumatin is purified from the fruit of the West African perennial plant *Thaumatococcus danielli*. A small amount of organic nonprotein impurity remains in the commercial products. This consists principally of the arabinogalactan and arabinoglucuronoxylan polysaccharides, both of which are normal constituents of plant gums.

Thaumatin is 2000–3000 times sweeter than sucrose. It is stable in freeze-dried form. Degradation is unlikely to occur in acidic materials. The protein structure is unstable when baked or broiled (Table 2).

2. Use and Intake

Although the primary organoleptic property of thaumatin is sweetness, its main use is as a flavor enhancer (Higginbotham, 1983). It acts synergistically with saccharin, acesulfame K, and stevioside. The major applications include chewing gum, savory flavor, dairy products, animal feeds, and pet foods.

Potential daily intake estimates based on American food consumption figures indicate that the maximum daily intake would be less than 2 mg per capita (about 0.03 mg/kg/day).

3. Toxicology

a. Metabolism. Thaumatin is completely metabolized to its constituent amino acids before absorption. The measured digestibility of thaumatin has been at least as great as that of egg albumin.

b. Other Toxicological Studies. In short-term studies rats and dogs were unaffected by thaumatin treatments. Teratology studies did not show any abnormalities in treated groups compared with controls. Mutagenicity studies in vitro and in vivo, Ames salmonella tests, and dominant lethality were negative. No anaphylactic antibodies or adverse effects in blood were detected in human volunteers consuming thaumatin (>100 mg/day) for several weeks (WHO, 1983).

It has been demonstrated that thaumatin is not allergenic, mutagenic, or teratogenic. Both short-term tests and clinical human exposure studies, some at exaggerated levels, showed no adverse effects. However, long-term studies have not been conducted. It has been questioned whether sufficient data exist for the safety assessment of thaumatin. In any case thaumatin has a long history as a sweetening agent in West Africa and has been used for many years in Japan without any reported reactions.

4. Regulatory Status

In 1983 thaumatin was evaluated by JECFA (WHO, 1983). Specifications were prepared, but an ADI could not be established without data from long-term studies and adequate studies in humans. In 1985 an ADI of ''not specified'' was allocated for thaumatin, indicating that it does not represent any hazard to health.

The Commission of the European Communities has accepted thaumatin for use as a sweetener in foods. Thaumatin was first permitted in Japan in 1979 as a natural food. In the United States and Switzerland it is permitted in chewing gum as a flavor enhancer. In the United Kingdom the use of thaumatin in food is not restricted, and in Australia it is permitted as a flavor enhancer.

F. Sucralose

Sucralose is the generic name of a relatively new intense sweetener made from ordinary sugar. Sucralose was first discovered in 1976, and it is a unique sweetener as it is made from ordinary sugar. It is a thrichloro derivative of the C-4 epimer galactosucrose which is not broken down during its passage through the gastrointestinal tract and thus does not provide calories. Sucralose tastes like sugar, but it is about 600 times sweeter. However, the taste profile is similar to sucrose and it can be used for almost all applications where sucrose is used. The sweetness does not react with food components or other ingredients and sucralose has good water solubility. Sucralose has excellent product stability even under high temperatures and it can be used in a broad range of food products.

1. Safety

A large number of studies have proven that sucralose is safe for human consumption. Sucralose does not break down in the gastrointestinal tract or accumulate in fatty tissues (McLean et al., 2000). Sucralose is also noncariogenic. It is currently evaluated by several regulatory bodies and it has received approval in Canada, Australia, and Russia. The ADI for sucralose has been set at a level of 15 mg/kg/day.

G. Other Nonnutritive Sweeteners

In addition to the traditional and extensively studied nonnutritive sweeteners, a growing number of new compounds have been suggested as sugar substitutes. Some of these are listed in Table 1. Many new developments are plant extracts. Monellin is derived from a noncultivated African plant, *Dioscoreophyllum cummiusii*, by enzymatic treatment. This sweetener is a protein composed of two polypeptide chains. Relative to sucrose it is 150–300 times sweeter, but it is reported to have a licorice flavor. Monellin is not very stable in food processing (Anonymous, 1986).

Glycyrrhizin is a terpene glycoside extracted from licorice root. It consists of the salts of glycyrrhizic acid and is reported to be 50–180 times sweeter than sucrose. However, the sweetness is perceived slowly, and it it followed by licorice aftertaste. Glycyrrhizin is mainly used in Japan for soy products to control saltiness (Crossby and Wingard, 1979; Anonymous, 1986). In the United States glycyrrhizin is used as a flavoring agent.

Neosugar is a fructo-oligosaccharide produced enzymatically from sucrose. It has 40–60% of the sweetness of sucrose and is claimed not to be extensively metabolized in the gastrointestinal tract (Anonymous, 1986). Neosugar represents a new model of developing oligosaccharides that are not metabolized in the body. Some oligosaccharides may be metabolized by specific types of intestinal bacteria to produce desirable metabolites and to promote desired types of intestinal bacteria. However, more research is needed to prove the safety and usefulness of these compounds (Anonymous, 1985).

Phyllodulcin is 200–300 times sweeter than sucrose. It is a 3,4-dihydroxy-isocoumarin compound with a licoricelike flavor (Crossby, 1976).

Miraculin is a glycoprotein derived from the African plant *Richardella dulcifica*. Miraculin is actually not a sweetener but a taste-modifying glycoprotein. In sour or tart foods miraculin exhibits a sweet taste that may last for several hours (Janelm et al., 1985).

A recent development in the plant extract field is hernandulcin, which is a sweet plant extract from the Mexican herb *Lippia dulcis*. Hernandulcin has a sweetness of 1000 times that of sucrose (Compadre et al., 1985).

Dulcin is a synthetic sweetener that is chemically *p*-phenetolcarbamide. It has been reported to cause cancer in laboratory animals and is therefore not accepted for food use in most countries (WHO, 1968).

Stevioside is a widely used high-intensity sweetener in Japan, where the fact that it is a plant extract allows it to be included in the class of natural food additives. Stevioside can be extracted from the leaves of *Stevia rebaudiana*, which is cultivated in Japan, Korea, and some South American countries. The sweetness of stevioside varies according to the plant source and the food in which it is used. Relative sweetness values varying between 100 and 300 have been quoted in the literature (Table 2).

Stevioside is used in Japan in soft drinks, candy, and chewing gum. It has also been used in many sugar-free or diebetic foods either alone or in combination with other nonnutritive sweeteners. However, little information exists about the toxicological properties of stevioside, and therefore it has not been approved for food use in most European or North American countries. The Joint FAO/WHO Expert Committee of Food Additives has not evaluated stevioside, but the European Commission considered stevioside unacceptable for food additive use (Commission of the European Communities, 1984).

Most nonnutritive sweeteners described here are either not generally accepted or are approved for food use in only a few countries. Japan is often an exception and allows the use of natural compounds in foods. In Europe and the United States these compounds have only a very limited use at present.

III. NUTRITIVE SWEETENERS

A. Fructose

1. History

Fructose is a hexose monosaccharide that is one of the most commonly occurring natural sugars. It is often called fruit sugar or levulose. Free fructose is found in almost all fruits and berries and in most vegetables. Fructose was earlier very difficult to produce and therefore remained unavailable for food use until the late 1960s.

Crystalline fructose has been commercially available for only a decade, and therefore it is not commonly known in all parts of the food industry. However, its use has increased significantly due to its properties in diabetic and other special dietary foods. Also the interest in natural sweeteners has centered on fructose. Therefore, fructose may have significant applications in dietetic and health foods in the future.

2. Properties and Uses

Commercially crystalline fructose is produced from sucrose by inversion or from glucose by enzymatic isomerization. One of the major advantages of the use of crystalline fructose is its relative sweetness, which is about 1.5 times that of sucrose. Fructose is the only carbohydrate with a sweetness higher than that of sucrose on a weight basis. Usually smaller amounts of fructose are needed for the same sweetness in food products, and this often results in reduction of energy content. Fructose is also more slowly absorbed than glucose or sucrose, and it does not stimulate insulin per se. Fructose does not increase serum triglycerides in normal subjects. Therefore, fructose has been recommended as a sweetener for diabetics and is presently included in the diabetic diet recommendations in many countries. However, one must take into account the calories in fructose.

Results of clinical dental trials on fructose have indicated that fructose is significantly less carcinogenic than sucrose (Scheinin and Mäkinen, 1975). Later studies appear to have verified the findings as reviewed by Grenby (1983).

Fructose is readily soluble in water, and it is sweetest in cold solution. Fructose also tends to enhance the flavor of fruits and berries, which makes it suitable as a sweetener for fresh fruits, iced tea, and lemonade. As a ketose-type sugar, fructose undergoes Maillard reactions with amino groups at even lower temperatures than sucrose; therefore, browning reactions in baked or cooked foods can be obtained at lower temperatures. The rationale for fructose use has been extensively reviewed by Doty (1980) and Würsch and Daget (1987).

In a food technological sense, fructose has its major applications in dietetic and diabetic foods, where its slow absorption and greater sweetness are of value. Good applications for fructose can be found in presweetened cerals, canned fruits, cakes, biscuits, and ice cream. Fructose is also ideal for powdered juices and frozen products.

3. Regulatory Status

The regulatory status of fructose in the United States has been reviewed by Frattali (1982), particularly with respect to its use in special dietary foods. Fructose is a natural sugar that has been tested in many clinical situations as indicated by Scheinin and Mäkinen (1975). No side effects have been observed, and therefore most countries consider fructose an ingredient rather than an additive. There are usually no restrictions on fructose use.

B. Xylitol

1. Chemistry

Xylitol is a pentitol that can be found in most fruits and berries as well as vegetables (Washüttl et al., 1973). Commercially xylitol is produced from xylan-containing plant material by acid hydrolysis, hydrogenation, and purification. Xylitol can also be produced by microbiological methods (Aminoff et al., 1978). At room temperature xylitol is equisweet with sucrose and therefore twice as sweet as sorbitol and three times as sweet as mannitol (Aminoff et al., 1978; Mäkinen, 1978). Xylitol is a sweetener with the same caloric content as sucrose. It is supplied as a colorless, nonhygroscopic crystal and has a caloric content equal to sucrose (Table 3). Xylitol is shown to be a noncariogenic sweetener, and it is also suitable for diabetic and dietetic foods (Mäkinen, 1978; Alanen et al., 2000).

2. Technology

Interest in xylitol is primarily due to its properties and potential uses as a noncariogenic sugar substitute (Mäkinen, 1978). Xylitol can be incorporated successfully into a wide range of foods. Usually it is possible to produce confections, sweet snacks, chocolate, and chewing gum using the common recipes with xylitol replacing sucrose (Aminoff et al., 1978; Voirol, 1978). The main application of xylitol appears to be in confectionery, especially in sugar-free products and noncariogenic chewing gum. Xylitol can also be utilized in diabetic and dietetic foods provided its energy value is taken into account (Mäkinen, 1978). Xylitol has also been suggested for functional food use for the prevention of acute otitis media in children (Uhari et al., 1996). A review on the safety and efficacy of xylitol as a functional food ingredient is available (Salminen and Ahukas, 2000).

3. Toxicology

The toxicological properties of xylitol have been extensively studied, and all tests have proved xylitol to be safe at dietary levels likely to be consumed. Xylitol is absorbed slowly, and therefore a major part of the substance may be metabolized by intestinal microorganisms (Salminen et al., 1986). Otherwise xylitol is metabolized by the liver. Salminen et al. (1982, 1986) have shown that xylitol can be consumed without ill effects when the single dose does not exceed 30 g. A 30-g dose does not cause significant increases in blood glucose or insulin secretion (Salminen et al., 1982). Larger doses are often likely to cause diarrhea; this may be due to the intestinal bacteria, which are not able to metabolize large xylitol doses (Salminen et al., 1986). The toxicological studies include studies on carcinogenicity, mutagenicity, and teratogenicity. All these studies have indicated that xylitol is safe for food use. A comprehensive review is available from the World Health Organization (WHO, 1983).

4. Regulatory Status

The Joint FAO/WHO Expert Committee on Food Additives has established an acceptable daily intake of ''not specified'' for xylitol (WHO, 1983). This means that on the basis of the available data, the potential daily intakes do not represent a hazard to health.

The regulatory status of xylitol varies from country to country. In Europe, xylitol is used as a sweetener and food additive according to the directive on sweetening agents.

Table 3 Some Properties of Polyol Sweeteners

Polyol	E code	Synonyms	Sweetness (sucrose = 100)	Melting point (°C)	Solubility at 25°C (g/100 g H_2O)	Impact on blood sugar	Laxative effect[a]	ADI[b] (year of evaluation)
Xylitol	E 967	Xylit	90–100	93–94.5	64	Very low	++	Not specified (1985)
Sorbitol	E 420	Glucitol	50–60	93–112	72	Low	++	Not specified (1982)
Mannitol	E 421	Mannit, mannose sugar	50–60	165–168	18	Low	+++	Not specified (1987)
Lactitol	E 966	Lactit	30–40	94–97	149[c]	None	+	Not specified (1983)
Maltitol	E 965	Maltit	80–90	—	Easily soluble	Low	++	Not specified (1985)
Isomaltitol	E 953	Isomalti	50	—		None	+++	Not specified (1985)

[a] Gradual adaptation to large xylitol, sorbitol, mannitol, and lactitol doses occur in most humans during prolonged consumption; initial laxative threshold varies between 20 and 40 g/day.
[b] WHO.
[c] As a monohydrate.

C. Sorbitol

1. Chemistry

Sorbitol is a six-carbon sugar alcohol that was originally found in the berries of mountain ash. It occurs in many fruits and vegetables (Washüttl et al., 1973). Sorbitol has the same steric configuration as glucose, and it is chemically synthesized from glucose or dextrose for commercial use. Sorbitol has about half the sweetness of sucrose. Some properties of sorbitol are listed in Table 3.

2. Technological Properties

Sorbitol is used mainly as a sweetening agent for dietetic foods, where it combines moderate sweetening power, specific flavor characteristics, and pleasant viscosity in liquids. It is used in sugar-free candies and chewing gum and in diabetic foods. In mixtures with other sugars, sorbitol modifies the crystallization properties of foods. When added to syrups containing sucrose it reduces crystal deposition during storage. Sorbitol also has some uses as a humectant and stabilizer, and it can be used as a substitute for glycerol. Small amounts of sorbitol have been added to low-calorie drinks to mask the aftertaste of saccharin and to provide the normal mouth-feel (Grenby, 1983). Typical products in Europe include confectionery, pastilles, diabetic jam and cookies, ice cream, chocolates, and pastries.

3. Toxicology and Safety

The toxicology of sorbitol has been reviewed recently by WHO. The Joint FAO/WHO Expert Group on Food Additives has given sorbitol an ADI of "not specified," which means that no health hazards are foreseen (WHO, 1982). Since sorbitol is absorbed slowly, foods sweetened with sorbitol are thought to be suitable for diabetics provided that the calories are taken into account. However, large amounts of sorbitol can cause flatulence, diarrhea, and abdominal distension. Gradual addition of sorbitol into the diet may increase the tolerance of the individual (Salminen et al., 1985a,b).

4. Regulatory Status

In most countries sorbitol is approved as an ingredient when used for sweetening purposes. In the United States sorbitol is permitted for use as a food additive by the U.S. Food and Drug Administration. Sorbitol is also considered GRAS for use as a nutrient and dietary supplement (FASEB/SCGOS, 1973a).

D. Mannitol

1. Chemistry

Mannitol is a hexitol that is stereoisomeric to sorbitol. It is commonly found naturally in some plant foods, including beets, celery, olives, and seaweed. Mannitol has about 0.4–0.5 the sweetness of sucrose, and its properties are fairly similar to those of sorbitol (Table 3). Only the solubility of mannitol is poor compared to sorbitol. Mannitol is produced from sucrose or dextrose and can also be obtained as a byproduct of some fermentations.

2. Technological Properties

Mannitol is used in sugar-free dietary foods, sugar-free chewing gum, sweets, and ice cream. In addition to its use as a sweetener, mannitol can be used as a texturizing agent, anticaking agent, or humectant. Sometimes it is used in breakfast cereals and frostings. At present mannitol is used mainly in sugar-free chewing gums as a sweetening agent and for dusting the chewing gum sticks.

3. Toxicological Studies and Safety

Mannitol is slowly absorbed from the intestinal tract and may cause diarrhea and flatulence. In experimental animals an adaptation to mannitol can be seen (Salminen et al., 1985b). In humans a laxative effect is observed after intakes of 20–30 g of mannitol. Toxicity studies have not indicated any adverse effects other than diarrhea. Therefore mannitol is considered safe for use in foods. Mannitol is also on the U.S. FDA GRAS list. An evaluation of its health effects has been conducted (FASEB/SCOGS, 1973b). An acceptable daily intake of ''not specified'' has been allocated for mannitol (WHO, 1986).

E. Lactitol

1. Chemistry

Lactitol is a disaccharide alcohol [(4-*O*-β-D-galactopyranosyl)-D-glucitol] produced by the hydrogenation of lactose or lactulose. Lactitol has been known since 1912, and its manufacture was mentioned by Aminoff in 1974, but attention has been directed to its properties only lately (Linko et al., 1980). Lactitol has a low sweetness (Table 3), and therefore its major use is not as a sweetener. More recent biological research cited by WHO (1983) indicates that lactitol may have a lower energy value than other carbohydrate sweeteners.

Commercially available lactitol has a molecular weight of 344 and crystallizes as a colorless and odorless monohydrate with a pleasant mild sweetness (Saijonmaa et al., 1978; Linko et al., 1980). Since lactitol does not have a carbonyl group it cannot undergo Maillard reactions. In general use, lactitol is more stable than lactose. Its relative sweetness is about 50% that of glucose (Linko et al., 1980). Like lactulose lactitol also promotes the amount of bifidobacteria and lactic acid bacteria in the human colon. Thus, lactitol can be considered a bifidogenic factor or a prebiotic substance (Ballongue et al., 1997; Salminen and Salminen, 1997).

2. Technological Uses

Lactitol as well as most polyols can be applied to special dietary foods that can be consumed by diabetics provided that the calories are taken into account. Lactitol can be used as a sweetener in most foods, but due to its low sweetness it is not very attractive. However, it may have other uses as a sweet bulking or texturizing agent in the future because of its low energy value (WHO, 1983). The products that are manufactured with lactitol have excellent palatability, and no unpleasant aftertaste is associated with lactitol. Lactitol also has uses as an ingredient for the pharmaceutical industry. Lactitol can also be utilized instead of lactulose for some therapeutic applications and special dietary foods (Patil et al., 1987) and as a prebiotic in functional foods (Salminen and Salminen, 1997).

3. Toxicology and Safety

Lactitol has gone through extensive toxicological studies. All required toxicology studies have been completed, including long-term and carcinogenicity studies. Also, short studies on the gut microflora changes have been completed (Salminen and Salminen, 1986b).

These studies indicate that apart from diarrhea after consumption of large lactitol doses no toxicologically significant adverse effects have been noted (WHO, 1983). The EEC Scientific Committee on Food has accepted lactitol and most other polyols for use in food. However, it was pointed out that laxation may occur at high intakes.

4. Regulatory Status

Since lactitol has been made commercially available only recently, most countries have not classified it as an ingredient or a food additive. JECFA has given lactitol an ADI of "not specified," indicating that it is considered safe for food use (WHO, 1983). Food additive and ingredient petitions have been filed in most countries; lactitol has been approved for special dietary foods in the Netherlands, and a lactitol preparation is used as a pharmaceutical in Switzerland.

F. Lactulose

1. Chemistry

Lactulose (4-*O*-β-D-galactopyranosyl)-D-fructose is a keto analog of lactose. Unlike lactose, it resists the hydrolytic action of intestinal β-galactosidases and therefore is not absorbed from the small intestine. Lactulose is mostly available in syrup containing 67% lactulose, but recently a crystalline form has also been produced. Lactulose has a relative low sweetness, but it is stable in most foods.

2. Technological Properties

Lactulose can be utilized in most liquid foods, but due to its low sweetness it does not have many food applications. In addition it has laxative properties that prevent its use in most common foods. Only special dietary foods that are intended for people suffering from constipation have utilized lactulose as a sweetener. It has been claimed that lactulose in infant foods and formulas enhances the development of intestinal flora containing lactic acid bacteria and bifidobacteria (*Bifidobacterium bifidum*) that mimic the flora of breast-fed infants (Mendez and Olano, 1979). Lactulose may also act as a nonabsorbable substrate for colonic bacteria, thereby causing other favorable changes in the microflora and making lactulose suitable for special dietary foods in this area. Lactulose also has a number of applications in pharmaceutical preparations. During the fermentation short chain fatty acids (e.g., acetic, lactic, and butyric acids) are formed with consequent lowering of the colon pH and modification of the microflora of the intestinal contents. It has been reported that lactulose promotes the growth of lactobacilli and especially *Lactobacillus acidophilus* in the colon. In addition to lactulose also the ingestion of dairy products fermented with lactic acid bacteria and lactitol have been shown to lower colonic pH and to relieve abnormalities in intestinal transit (Ballongue, 1997; Salminen and Salminen, 1997).

3. Toxicology

The toxicity of lactulose has not been studied according to present guidelines. However, lactulose has been utilized for over 30 years in the treatment of chronic constipation and portal-systemic encephalopathy, and the patients have received large lactulose doses for long treatment periods. It is apparent that microflora changes in the intestinal tract occur during lactulose ingestion (Salminen and Salminen, 1997; Salminen et al., 1988). However, apart from meteorism and occasional diarrhea no other harmful effects have been observed.

4. Regulatory Status

In most countries lactulose is utilized as a pharmaceutical preparation, and occasionally it is used in special dietary foods.

G. Hydrogenated Glucose Syrups

1. Chemistry

There are a number of hydrogenated glucose syrups that consist of 2–8% sorbitol, 50–55% maltitol, 15–20% maltotriol, and 20–30% hydrogenated tri- or heptasaccharides.

Hydrogenated glucose syrups are usually marketed as syrups containing about 75% solids that do not crystallize at high concentrations. No brown pigments are formed in the Maillard reaction because there are no aldehyde groups. Hydrogenated glucose syrup is used as a substitute for glucose, sucrose, or sorbitol. It has reduced cariogenicity compared to glucose, sucrose, or sorbitol and reduced laxative effect (Food Additives and Contaminants Committee, 1982).

2. Toxicology

Short-term studies in the rat and dog together with metabolic data, mutagenicity data, and human tolerance studies do not give any reason to doubt the safety of hydrogenated glucose syrups as sweetening agents. However, no long-term studies are available. In the light of evidence that hydrogenated glucose syrup breaks down to glucose and sorbitol, it is possible to consider their safety according to the results on sorbitol (WHO, 1980a,b).

3. Regulatory Status

An acceptable daily intake (ADI) of ''not specified'' has been allocated to hydrogenated glucose syrups (WHO, 1985).

H. Maltitol

1. Chemistry

The maltitol molecule, 4-*O*-β-D-glucopyranosyl-D-glucitol, consists of a glucose and a sorbitol unit linked 1,4 (1,4-glucosyl-glucitol).

Maltitol is produced by enzymatic hydrolysis of starch (potato or corn) to obtain a high maltose syrup, which is hydrogenated to the corresponding high maltose syrup, from which crystalline maltitol is obtained. Both liquid and crystalline maltitol are used. They are soluble in water and are very stable both at different pH conditions and thermally. The sweetening power for crystalline maltitol is 0.9 and for liquid 0.6 (sucrose = 1).

2. Metabolism

Maltitol is more slowly hydrolyzed to glucose and sorbitol by maltase than maltose, which is a natural substrate. More than 50% of the ingested dose remains unsplit and enters the large intestine, where it is fermented by microflora.

3. Toxicity

Maltitol has a low acute toxicity by oral administration ($LD_{50} > 24$ g/kg body weight). Maltitol is not mutagenic. Teratogenic studies have also been negative. Subscute and long-term toxicological studies with maltitol in various animal species failed to reveal any differences in behavior and biochemical patterns between control and maltitol-fed animals. Only a moderately decreased growth rate was observed in fed animals due to the reduced caloric intake (Yamasaki et al., 1973a,b; Shimpo et al., 1977).

I. Isomalt

1. Chemistry

Isomalt is also called hydrogenated isomaltulose or hydrogenated palatinose. It is an equimolar mixture of 6-*O*-β-D-glucopyranosyl-D-glucitol and 1-*O*-β-D-glucopyranosyl-D-mannitol. Isomalt is produced by the enzymatic trans-glucosidation of sucrose to isomaltulose followed by dehydrogenation. It is about 0.5 times as sweet as sucrose. It is stable in acid and alkaline media under conditions normally occurring in food production (Siebert, 1975).

2. Use

Isomalt can be used as a sugar substitute in confectioneries, chewing gum, soft drinks, and desserts. It is claimed to be less cariogenic than sucrose and less laxative than sorbitol or xylitol.

3. Toxicology

Only 50% of isomalt is metabolized in humans. It is broken down initially in the gastrointestinal tract to form sorbitol, mannitol, and glucose. Some metabolic and tolerance studies in animals are available. The only effect seen in short-term studies in the rat and dog is a dose-related increase in bilirubin level in the study of the rat. There are no long-term studies available on isomalt. Long-term studies have not been considered necessary because the breakdown products are glucose, sorbitol, and mannitol (WHO, 1981).

4. Regulatory Status

Isomalt was evaluated by the WHO expert group in 1981. Safety documentation was inadequate with respect to long-term feeding studies. A temporary ADI of 0–25 mg/kg body weight was allocated. In 1985 an ADI of ''not specified'' was allocated (WHO, 1985).

J. Fructose Syrups

Fructose syrups should not be misunderstood to be similar to crystalline fructose (Doty, 1980). The amount of fructose in fructose syrup products varies from 40 to 90%. Relative sweetness compared to 15% saccharose solution is related to fructose content. In 40–90% fructose-containing products, the relative sweetness is 100–160. Fructose syrups are very hygroscopic, and the tendency to absorb moisture increases with increasing amounts of fructose. Fructose syrups are used to reduce saccharose intake. Their main use is in products with very high water content, such as drinks and juices (Young and Long, 1982).

IV. CHOICE OF SWEETENER

Sweetness is important for food acceptability, but in a food sweetness is seldom tasted alone. Sweet tastes are almost always mixed with a variety of complex flavors, and sweetness is also affected by the texture and aroma of the food. Therefore, the overall properties of a food determine the amount and choice of sweeteners. When considering the choice of sweeteners, one of the most important aspects is naturally the degree or intensity of sweetness. However, nutritive sweeteners especially have many other functions in food systems. Sucrose or sugars have a great influence on the texture and bulking of many foods. These include baked foods, sugar confectionery, soft drinks, and dairy foods. In soft drinks, sugars produce viscosity and a pleasant mouth-feel, which are difficult to mimic in products sweetened with nonnutritive sweeteners. In preserves and jams as well as marmalades, sugars affect texture, sweetness, and bulk properties. Additionally they have osmotic effects and thereby help preserve the product and extend its shelf-life. An additional property is the ability to absorb water in baked products to help retain the moist texture. In many applications nutritive sweeteners also offer an energy source for food fermentations. The most important properties of sweeteners in foods are summarized in Table 4.

The general properties of nutritive polyol sweeteners are summarized in Table 3. The substitution of sugars with polyols is often easily accomplished. In many cases sorbitol and xylitol can be used in place of sucrose in foods. Polyols in general (except for lactitol) also have humectant properties. All polyols act as bulking agents, and their use for bulking and texturizing may reduce the energy content of foods. This is especially true for lactitol and mannitol, which have relatively low sweetness. In terms of reduced cariogenicity, xylitol offers an excellent substitute for most common sugars. Fructose, on the other hand, also appears to be a less cariogenic alternative to sucrose and sorbitol. An added advantage of polyols is their suitability for special dietary foods and diabetic foods. Fructose, xylitol, and lactitol appear to be especially suitable sugar substitutes for diabetics provided their calories are taken into account.

The properties of nonnutritive sweeteners are summarized in Table 2. Most nonnutritive sweeteners are suitable sugar substitutes for special dietary foods and foods manufactured for diabetics. The sweeteners in this group have a relatively intense sweetness, and therefore only small amounts of these compounds are needed in food products. This results in the need of bulking agents in many food products sweetened with the compounds in this group. Until recently the choice of bulking agents has been limited. However, now polydextrose is a good low-calorie bulking agent suitable to many baked foods and dairy

Table 4 Properties of Sweeteners in Foods

Sweetener, functional ingredient, prebíotic
Bulking agent, texturing agent, viscosity modifier
Preservative, antimicrobial
Substrate for fermentations and fermentation processes
Humectant (water adsorption modifier)
Freezing point modifier
Crystallization modifier
Noncariogenic or anticariogenic agent
Mouth-feel modifier

foods such as ice cream. Additionally, lactitol can be utilized as a low-calorie bulking agent with a relatively low sweetness.

The manufacture of jams and marmalades sweetened with nonnutritive sweeteners requires the use of pectins or other thickening agents to obtain normal viscosity. Often antimicrobials are also needed to obtain a shelf-life similar to that of sucrose jams.

Other uses for nonnutritive sweeteners include pharmaceutical preparations (sweetness to cover the flavor of the product), tabletop sweeteners, and nonnutritive sweetener preparations for home use. For these purposes the stability and taste properties are most important, as indicated in Tables 2 and 3. In some cases, noncariogenic properties are important, as in pharmaceutical preparations for children.

Another important factor is the price of the sweetener. Nonnutritive sweeteners usually appear to be relatively cheap compared to the intensity of sweetness they produce. However, if bulking agents or other special measures are required, the cost may increase significantly. Table 5 summarizes the relative costs of some sweeteners as reported in Scandinavia by the Swedish National Food Administration (Hallström and Janelm, 1985). Also included in Table 5 are some suggestions of products for which particular sweeteners are suitable.

In conclusion, for the practical choice of sweetener, at least the following aspects should be considered: regulatory acceptability, availability, price for sweetening equivalent (including other ingredient and additive costs), nutritional value, sensory characteristics, and functional properties in the food system. After deciding the general direction for the development of a product and choice of sweetener, product optimization may be of further help. Würsch and Daget (1987) have reviewed both the overall problems of sweetness in product development and product optimization, which was introduced by Sidel and Stone (1983).

The ideal sweetener has been described by the Calorie Control Council (1985). According to their definition it should have the same sweetness as sucrose or greater sweetness than sucrose. In addition it should be odorless, colorless, stable,and readily soluble in food systems. It should be functional and economically feasible. It should also be noncaloric or its use should produce reduced caloric content in foods. The ideal sweetener should additionally be noncariogenic and nontoxic. At present we do not have an ideal sweetener,

Table 5 Price of Various Sweeteners on a Sweetness Equivalent Basis in Scandinavia in 1985 and Suitable Food Uses

Sweetener	Price/sweetness equivalent (USD)	Suitable food uses
Saccharin	0.01–0.02	Soft drinks, tabletop sweeteners, dessert mixes, yogurt
Cyclamates	0.15	Soft drinks, tabletop sweeteners
Aspartame	0.25–0.85	Soft drinks, dry foods, ice cream, yogurt, fruit juices, tabletop sweeteners
Acesulfame K	0.12	Soft drinks, tabletop sweeteners
Sorbitol	1.0–2.0	Special dietary foods, diabetic foods
Xylitol	4.2–6.1	Noncariogenic candy, chewing gum, dietetic foods, pharmaceuticals
Sucrose	0.17	All foods
Fructose	0.28	Almost all foods

Source: Halström and Janelm, 1985.

but many sweeteners are approaching these goals. We can also achieve almost ideal combinations by mixing two or more sweeteners.

REFERENCES

Alanen, P., Isokangas, P., Gutmann, K. 2000. Xylitol candies in caries prevention: results of a field study in Estonian children. *Community Dent. Oral Epidemiol.* 3:218–224.

Aminoff, C., Vanninen, E., Doty, T. E. 1978. The occurrence, manufacture and properties of xylitol. In: *Xylitol*, Counsell, J. N. (Ed.). Applied Science Publishers, London.

Anonymous. 1985. Meosugar—a fructo-oligosaccharide nonnutritive sweetener. *Nutr. Rev.* 43:155–157.

Anonymous. 1986. Alternatives to cane and beet sugar. *Food Technol.* 40:116–128.

Ashby, J. 1985. The genotoxicity of sodium saccharin and sodium chloride in relation to their cancer-promoting properties. *Food Chem. Toxic.* 23:507–519.

Ballongue, J., Schumann, C., Quignon, P. 1997. Effects of lactulose and lactitol on colonic microflora and enzymatic activity. *Scand. J. Gastroenterol.*, 32, Suppl. 222:41–44.

Batzinger, R. P., Ou, S. L., Bueding, E. 1977. Saccharin and other sweeteners: mutagenic properties. *Science* 198:944–946.

Birch, G. G. 1980. Theory of sweetness. In: *Carbohydrate Sweeteners in Food and Nutrition*, Koivistoinen, P., Hyvönen, L. (Eds.). Academic Press, London.

Butchko, H. H., Kotsonis, F. N. 1989. Aspartame: review of recent research. *Comments Toxicol.* 3(4):253–278.

Calorie Control Council. 1985. *Sweet Choices.* Atlanta.

Carlborg, F. W. 1985. A cancer assessment for saccharin. *Food Chem. Toxicol.* 23:499–506.

Cohen, S. M. 1985. Multi-stage carcinogenesis in the urinary bladder. *Food Chem. Toxicol.* 23:521–528.

Comprande, C., Pezzutto, J., Kinghorn, A. 1985. Hernandulcin: an intensely sweet compound discovered by review of ancient literature. *Science* 227:417–418.

Council on Scientific Affairs. 1985. *Aspartame, review of safety issues. JAMA* 254:400–402.

Crammer, M. 1980. *Saccharin: A Report.* American Drug Research Institute. Scherr, G. H. (Ed.). Pathotox Publishers.

Crossby, G. A. 1976. New sweetnesses. *CRC Crit. Rev. Food Sci. Nutr.* 7:300–349.

Crossby, G. A., Wingard, R. E. 1979. A survey of less common sweetness. In: *Development of Sweeteners*, Vol. 1, Hough, M., Parker, K., Vlitos, A. (Eds.). Applied Science Publishers, London.

Davenport, R. J. 2001. New gene may be key to sweet tooth. *Science*, 292:5517.

DeSesso, J. 1987. The Winter Toxicology Forum, pp. 141–154.

Doty, T. 1980. Fructose: rationale for traditional and modern use. In: *Carbohydrate Sweeteners in Foods and Nutrition*, Hyvönen, L., Koivistoinen, P. (Eds.). Academic Press, London.

FASEB/SCOGS. 1973a. *Evaluation of Health Aspects of Sorbitol as a Food Ingredient.* Report No. 9. NTIS PB 221–951.

FASEB/SCOGS. 1973b. *Evaluation of the Health Aspects of Mannitol as a Food Ingredient.* Report No. 10. MTIS PB 221–953.

FDA. 1985. FDA Talk Paper. NAS Report on Cyclamate. U.S. Food and Drug Administration, Bethesda.

Federal Register. Part IV. Department of Health and Human Services, July 24, 1981. Docket No. 75F-0555. *Aspartame: Commission's Final Decision.*

Frattali, V. P. 1982. Fructose—a regulatory perspective. In: *Food Carbo hydrates*, Lineback, D., Inglett, G. (Eds.). Avi, Westport, CT.

Grenby, T. H. 1983. Nutritive sucrose substitutes and dental health. In: *Developments in Sweeteners*,

Vol. 2, Grenby, T., Parker, K., Lindley, M (Eds.). Applied Science Publishers, London, pp. 51–88.

Hallström, H., Janelm, A. 1985. Sötningmedel förr och nu: från honung till aspartam. *Vår Föda* 37:434–441.

Higginbotham, J. D. 1983. Recent Developments in nonnutritive sweeteners. In: *Developments in Sweeteners*, Vol. 2, Grenby, T., Parker, K., Lindley M. (Eds.). Applied Science Publishers, London, pp. 119–158.

Homler, B. E. 1984. Aspartame; implications for the food scientist. In: *Aspartame, Physiology and Biochemistry*, Stegink, L. D., Filer, L. J. (Eds.). Marcel Dekker, New York, pp. 247–262.

Hoover, R. N., Strasser, P. H. ±980. Artificial sweeteners and human bladder cancer. Preliminary results. *Lancet* 1:837–839.

Hyvönen, L. 1980. Synergism between sweeteners. In: *Carbohydrate Sweeteners in Foods and Nutrition*, Koivistoinen, P., Hyvönen, L. (Eds.). Academic, London.

IARC. 1980. *IARC Monographs* 22.

ILSI-NF. 1986. *International Aspartame Workshop Proceedings*. November 1986, Marbella, Spain.

International Research and Development Corporation. 1983. *Evaluation of the Dose-Response and In Utero Exposure of Saccharin in the Rat*, Mattawan, Michigan.

Iyengar, R. B. 1979. The complete amino acid sequence of the sweet protein thaumatin I. *Eur. J. Biochem.* 96:193.

Janelm, A., Larsson, A., Nilsson, I., Ericsson, B. 1985. Faktablad om sötningsmedel. *Vår Föda* 37: 527–555.

Jobe, P. C., Dailey, J. W. 1993. Aspartame and seizures, review article. *Amino Acids* 4:197–235.

Kroyer, G., Washuttl, J. 1982. Interactions of artificial sweeteners with food additives. In: *Recent Developments in Food Analysis*. Baltes, W., Czedik-Eysenberg, P., Pfannhouser, W. (Eds.). Verlag Chemie, Weinheim.

Lavin, P. T., Sanders, P. G., Mackey, M. A., Kotsonis, F. N. 1994. Intense sweeteners use and weight change among women: a critique of the Stellman and Garfinkel study. *J. Am. College Nutr.* 13(1):102–105.

Linko, P., Saijonmaa, T., Heikonen, M., Kreula, M. 1980. Lactitol. In: *Carbohydrate Sweeteners in Foods and Nutrition*, Koivistoinen, P., Hyvönen, L. (Eds.). Academic, London.

Mäkinen, K. K. 1978. Biochemical principles of the use of xylitol on medicine and nutrition with special consideration of dental aspects. *Experientia*, Suppl. 30, pp. 1–160.

McLean, B., Shephard, N., Merritt, R., Hildick-Smith, G. 2000. Repeated dose study of sucralose in human subjects. *Food Chem. Toxicol.*, 38:Suppl 2:S123–S129.

Mendez, A., Olano, A. 1979. Lactulose. A review of some chemical properties and application in infant nutrition and medicine. *Dairy Sci. Abstr.* 41:531–535.

Metcalfe, D. D. 1987. Proceedings of the Sixth International Food Allergy Symposium, Boston 75.

Morgan, R. W., Wong, O. 1985. A review of epidemiological studies on artificial sweeteners and bladder cancer. *Food Chem. Toxicol.* 23:529–533.

Morrison, A., Buring, J. E. 1980. Artificial sweeteners and cancer of the lower urinary tract. *New Engl. J. Med.* 302:537–539.

MRCA. 1985. Market Research Corporation of America. *MRCA IV*. National Academy of Sciences. 1979. *The 1977 Survey of Industry on the Use of Food Additives*. National Academy of Sciences, Washington, D.C.

National Research Council/National Academy of Sciences. 1978. *Saccharin: Technical Assessment of Risks and Benefits*, Report No. 1, Committee for a Study on Saccharin and Food Safety Policy, Washington, D.C.

National Research Council. 1985. *Evaluation of Cyclamate for Carcinogenicity*. National Academy Press, Washington, D.C.

Oser, B. L. 1985. Highlights in the history of saccharin toxicology. *Food Chem. Toxicol.* 23:535–542.

Patil, D., Grimble, G., Silk, D. 1987. Lactitol—a new hydrogenated lactose derivative: intestinal absorption and laxative threshold in human subjects. *Br. J. Nutr.* 57:195–199.
Penttilä, P. L., Salminen, S., Niemi, E. 1988. Estimates on the intake of food additives in Finland. *Lebensm. Unters. Forsch.* 186:11–15.
Remsen, I., Fahlberg, C. 1879. On the oxidation of substitution products of aromatic hydrocarbons. IV. On the oxidation of orthotoluenesulphamide. *J. Am. Chem. Soc.* 1:426–438.
Renwick, A. G. 1983. The fate of nonnutritive sweeteners in the body. In: *Developments in Sweeteners*, Vol. 2, Grenby, T., Parker, K., Lindley, M. (Eds.). Applied Science Publishers, London, pp. 179–224.
Renwick, A. G. 1985. The disposition of saccharin in animals and man—a review. *Food Chem. Toxicol.* 23:429–435.
Renwick, A. G., Williams, R. T. 1972. The fate of cyclamate in man and other species. *Biochem. J.* 129:869.
Saijonmaa, T., Heikonen, M., Kreula, M., Linko, P. 1978. Preparation and characterization of milk sugar alcohol, lactitol. *Milchwissenschaft* 33:733–736.
Salminen, S., Ahokas, J. 2000. Assessment of safety and efficacy of functional foods. In: *Essentials of Functional Foods*, Schmidt, M., Lalouza, T. (Eds.). Aspen Publishers, Maryland.
Salminen, E., Salminen, S. 1986a. Urinary excretion of orally administered oxalic acid in saccharin and o-phenylphenol fed NMRI mice. *Urol. Int.* 41:88–90.
Salminen, E., Salminen, S. 1986b. Lactulose and lactitol induced caecal enlargement and microflora changes in mice. *Proc. EuroFood Toxicol.* II:313–317.
Salminen, S., Salminen, E. 1997. Lactulose, lactic acid bacteria, intestinal microecology and mucosal protection. *Scand. J. of Gastro.*, 32, Suppl. 222:41–44.
Salminen, S., Salminen, E., Marks, V. 1982. The effects of xylitol on the secretion of insulin and gastric inhibitory polypeptide in man and rats. *Diabetologia* 22:480–482.
Salminen, S., Salminen, E., Bridges, J. W., Marks, V. 1985a. The effects of sorbitol on the gastrointestinal flora in rats. *Z. Ernährungswiss.* 25:91–95.
Salminen, S., Salminen, E., Bridges, J. W., Marks, V., Koivistoinen, P. 1985b. Urinary excretion of orally administered oxalic acid in sorbitol and mannitol fed CD-1 mice. In: *Renel Heterogeneity and Target Cell Toxicity*, Bach, P., Lock, E. (Eds.). Wiley, Chichester, England.
Salminen, S., Salminen, E., Koivistoinen, P., Bridges, J., Marks, V. 1986. Gut microflora interactions with xylitol in the mouse, rat, and man. *Food Chem. Toxicol.* 23:985–990.
Salminen, E., Elomaa, I., Minkkinen, J., Vapaatalo, H., Salminen, S. 1988. Preservation of intestinal integrity during radiotherapy using live *Lactobacillus acidophilus* cultures. *Clin. Radiol.* 39, 435–437.
Scheinin, A., Mäkinen, K. K. 1985. Turku sugar studies. *Acta Odont. Scand.* 33(Suppl. 70).
Schallenberger, R. S. 1963. Hydrogen bonding and the varying sweetness of sugars. *J. Food Sci.* 28:584–589.
Schallenberger, R. S. 1973. Sugar Structure and taste. *Advan. Chem. Ser.* 117:256–263.
Schallenberger, R. S., Acree, T. E. 1967. Molecular theory of sweet taste. *Nature* 216:480–482.
Shimpo, K., Togashi, H., Yokoi, Y., Futsiwara, S., Katada, H., Hiraga, K., Tanabe, T. 1977. Long-term toxicity of Malti, a sweet material consisting in maltitol, with special reference to tumorgenic activity in rats. *J. Toxicol. Sci.* 2:417–432.
Sidel, J. L., Stone, H. 1983. An Introduction to optimization research. *Food. Technol.* 37:36–38.
Siebert, G. 1975. Studies on isomaltitol. *Nutr. Metab.* 18 (Suppl. 1):191–196.
Stegink, L. D., Filer, L. J. (Eds.). 1984. *Aspartame: Physiology and Biochemistry*. Marcel Dekker, New York.
Stellman, S. D., Grafinkel, L. 1986. Artificial sweetener use and one-year weight change among women. *Prev. Med.* 15:195–202.
Tisdel, M. O., Nees, P. O., Harris, D. L., Derse, P. H. 1974. Long-term feeding of saccharin in rats. In: *Sweeteners*, Inglett, C. E. (Ed.). Avi, Westport, CT.

Uhari, M., Kontiokari, T., Koskela, M., Niemelä, M. 1966. Xylitol chewing gum in prevention of acute otitis media: double-blind, randomized trial. *Brit. Med. J.*, 313:1180–1184.
Voirol, F. 1978. The value of xylitol as an ingredient in confectionery. In: *Xylitol*, Counsell, J. N. (Ed.). Applied Science Publishers, London, pp. 11–20.
Washüttl, S., Riederer, P., Bancher, E. 1973. Qualitative and quantitative study of sugar alcohols in several foods. *J. Food Sci.* 38:1262–1267.
WHO. 1968. *WHO/Food Add*/68.33.
WHO. 1978. *Food Additive Series*, Vol. 13.
WHO. 1980a. *Food Additive Series*, Vol. 14.
WHO. 1980b. *Food Additive Series*, Vol. 15.
WHO. 1981. *Food Additive Series*, Vol. 16.
WHO. 1982. *Food Additive Series*, Vol. 17.
WHO. 1983. *Food Additive Series*, Vol. 18.
WHO. 1985. *Food Additive Series*, Vol. 19.
WHO. 1987. *Food Additive Series*, Vol. 20.
WHO. 1987. *Food Additive Series*, Vol. 21.
WHO. 1988. *Food Additive Series*, Vol. 22.
Wilson, J. 2000. Lunch eating behaviour of preschool children: effect of age, gender, and type of beverage served. *Phys. and Behavior*, 70:27–33.
Wynder, E. L., Stellman, S. D. 1980. Artificial sweetener use and bladder cancer: a case control study. *Science (N.Y.)* 207:1214–1216.
Würsch, P., Daget, N. 1987. Sweetness in product development. In: *Sweetness*, Dobbing, J. (Ed.). Springer-Verlag, New York, pp. 247–260.
Yamasaki, M., Tanabe, K., Kimishima, K. 1973a. Acute toxicity of Malbit in mice. *J. Yonago Med. Assoc.* 24:1.
Yamasaki, M., Tanabe, K., Matsumoto, Y. 1973b. Toxicological studies of Malbit in rats. *J. Yonago Med. Assoc.* 24:38.
Young, L. S., Long, J. E. 1982. Manufacture, use and nutritional aspects of 90% high fructose corn sweeteners. In: *Chemistry of Food and Beverages: Recent Developments*. Academic, London, pp. 195–210.

16

Synthetic Food Colorants

JOHN H. THORNGATE III

University of California, Davis, California

I. INTRODUCTION

The impact of the physical world on our lives is, quite literally, colored by our perceptions. Through the mechanisms of our various sensory receptors we convert such environmental stimuli as pressure waves into sound, chemical structure into taste and smell, and electromagnetic radiation into sight. Indeed, the last sense is of immense importance to us; humans are decidedly visual creatures. Vision not only represents the most complex sensory system in the human body (Mason and Kandel, 1991), but also dominates the central processing unit itself, the human brain (Sekular and Blake, 1985).

Of those qualities associated with vision, color has perhaps the most immediate impact. Lindberg (1938) outlined a number of early psychological experiments which examined the relative impact of color, and devised tests to measure ''color attitude,'' that is, the prevalence of response to color over form. With regard to food, color has often been stated, usually apocryphally, to be of preeminent sensory importance (Noonan, 1972; IFT, 1986; Newsome, 1990; von Elbe and Schwartz, 1996), although Clydesdale (1993) notes that color's true role remains ''elusive and difficult to quantify.''

Whether or not color is indeed the most important sensory characteristic discerned is academic; color's practical importance to food acceptability and flavor has long been recognized. Foods and beverages have been ''artificially'' colored since ancient times. Pliny, in Book XIV, Section viii of his *Naturalis Historia*, mentions that unscrupulous dealers in Narbonne artificially colored their wines through the use of smoke or aloe (Pliny, 1945). Pliny also notes the practice of adding gypsum (calcium sulfate dihydrate) to wine ostensibly to ''soften any roughness,'' although besides modulating the acidity gypsum would also reduce the red color (Book XIV, Section xxiv). In the same section Pliny mentions the Greek habit of adding seawater to wine, which, as Amerine and Singleton (1977) note, would increase the color's brightness.

Scully (1995) discusses the importance of appearance to the medieval cook, for whom color, especially yellow, was prized. Saffron and egg yolk were the colorants of preference, although gold leaf was also (infrequently) used. It is interesting to note that the color yellow was also central to an early edict against color adulteration; in 1396 the Parisian government prohibited the coloring of butter (NAS/NRC, 1971). Unfortunately not all coloring agents were harmless; medieval cooks had also learned that verdigris (copper oxides) imparted an excellent green color (Scully, 1995).

The increasing use of toxic colorants and other adulterants led Fredrick Accum to publish *A Treatise on Adulterations of Food and Culinary Poisons* in 1820, in which he decried, among other practices, the coloring of hedge leaves with verdigris to resemble green tea and the artificial greening of pickles and candies with copper salts (Accum, 1820). Accum's criticisms, while internationally sensationalistic, did little to spur immediate reform (Hutt and Hutt, 1984); ironically, the use of toxic colorants accelerated following Sir William Henry Perkin's synthesis of mauve from the coal-tar derivative methylanaline in 1856. This synthetic purple dye sparked not only the birth of the modern organic chemical industry, but also the quest for and use of synthetic colorants by a food industry which was turning increasingly to processed foods in order to feed a burgeoning industrial middle class (Tannahill, 1988).

The food industry favored the synthetic dyes over animal-, plant-, or mineral-based colorants due to their more consistent hues, dye strength, and stability (NAS/NRC, 1971). Only four years following Perkin's initial discovery the French were coloring wine with the coal tar derivative fuchsine (now known as Magenta I) (Marmion, 1979); by 1886 the United States Congress had approved the use of synthetic coloring for butter and by 1896 for cheese (Marmion, 1979). By 1900 approximately 80 dyes were being used as food colorants in the United States (Noonan, 1972).

II. U.S. GOVERNMENT REGULATION

Accum's book initiated a flurry of investigations and reports on food adulteration in the United States; Hutt and Hutt (1984) list a number of treatises published in the mid- to late-1800s which served to prompt state governments increasingly into legislating the safety of the food supply. Consideration of this issue reached a national climax with the National Board of Trade's 1879 competition to award $1000 to the person submitting the best draft of a food adulteration act; the competition winner, announced in 1880, was one G. W. Wigner of England (Hutt and Hutt, 1984). Wigner [cited in Hutt and Hutt (1984)] outlined seven specific elements which defined food adulteration, the first and sixth being

> If any substance, or any substances, has, or have been mixed with it, so as to reduce, or lower, or injuriously affect its quality, strength, purity, or true value.
>
> If it be colored, or coated, or polished, or powdered, whereby damage is concealed, or it is made to appear better than it really is, or of greater value.

This increased awareness was however insufficient to compel Congress into legislating adulteration at the national level; although bills were introduced as early as 1879 Congress chose instead (not atypically) to further investigate the issue (Hutt and Hutt, 1984). The positive result of this inaction was that Congress began appropriating funds for the Department of Agriculture's Division of Chemistry (later Bureau of Chemistry) to investigate the adulteration of foods (in 1889) and the specific use of colorants (in 1900). The latter action was, to quote Marmion (1979), "The first effective step taken by the government to check such practices." Based upon this research, the Secretary of Agriculture released a Food Inspection Decision in 1906, the first such pertaining to the use of colorants: the coal tar derivative Martius Yellow was declared unsafe for foods (Marmion, 1979).

Concurrent with the food preservatives research conducted by Dr. Harvey W. Wiley's Poison Squad in the Bureau of Chemistry, Dr. Bernhard C. Hesse was investigating the suitability of the available coal tar dyes (some 700) as food colorants (NAS/NRC, 1971). Hesse's literature search and subsequent physiological studies, along with Wiley's work, led to the passage of the Pure Food and Drug Act of 1906, the first comprehensive federal legislation regulating additives (IFT, 1986). Hesse's findings were specifically incorporated into the Act through the Food Inspection Decision of 1907, which recognized seven dyes as meeting the requisite suitability standards. These are listed in Table 1.

Hesse's criteria for assessing the suitability of dyes for food use were incorporated into the 1906 legislation; these included the requirements that all dyes be manufactured under strict control, and tested via human and animal physiological studies before being certified as suitable for food use (NAS/NRC, 1971). The certification process was, however, a voluntary process overseen by the Secretary of Agriculture (NAS/NRC, 1971); nevertheless, the dye industry quickly saw the advantages to marketing certified colorants, and by 1908 the first voluntary certified dye was available (Marmion, 1979). Under this process ten additional dyes were certified for food use in the years prior to 1938 (see Table 2), although two of these (Sudan I and Butter Yellow) were almost immediately delisted for causing contact dermatitis in dye workers (NAS/NRC, 1971).

While the groundwork had been laid with the Act of 1906, the question of enforcement remained problematic. In 1907 the Secretary of Agriculture created the Board of Food and Drug Inspection, charged with conducting hearings on alleged violations, and in 1908 President Roosevelt commissioned the Referee Board of Consulting Scientific Experts to rule on specific issues pertaining to food adulteration (Hutt and Hutt, 1984). The USDA Bureau of Chemistry, however, remained the enforcing agency of record. (Wiley had chaired the Board of Food and Drug Inspection.) In 1927 the Bureau evolved into the Food, Drug, and Insecticide Administration, which within three years became the Food and Drug Administration. Hutt (1996) notes that the 1906 Act required the Bureau to

Table 1 Colorants Permitted for Use in Foods Following FID No. 76, July 1907

Common name	1938 FDA nomenclature	Color index number	Year delisted
Ponceau 3R	FD&C Red No. 1	16155	1961
Amaranth	FD&C Red No. 2	16185	1976
Erythrosine	FD&C Red No. 3	45430	—
Indigotine	FD&C Blue No. 2	73015	—
Light Green SF	FD&C Green No. 2	42095	1966
Napthol Yellow S	FD&C Yellow No. 1	10316	1959
Orange 1	FD&C Orange No. 1	14600	1956

Source: Marmion (1979).

Table 2 Additional Colorants Permitted for Use in Foods, 1907–1938

Common name	1938 FDA nomenclature	CI number	Year listed	Year delisted
Tartrazine	FD&C Yellow No. 5	19140	1916	—
Sudan I	—	12055	1918	1918
Butter Yellow	—	—	1918	1918
Yellow AB	FD&C Yellow No. 3	11380	1918	1959
Yellow OB	FD&C Yellow No. 4	11390	1918	1959
Guinea Green B	FD&C Green No. 1	42085	1922	1966
Fast Green FCF	FD&C Green No. 3	42053	1927	—
Brilliant Blue FCF	FD&C Blue No. 1	42090	1929	—
Ponceau SX	FD&C Red No. 4	14700	1929	1976
Sunset Yellow FCF	FD&C Yellow No. 6	15985	1929	—

Source: NAS/NRC (1971); Marmion (1979).

enforce the legislation solely in response to malfeasance; the Act mandated no premarket approval process other than the voluntary certification schema.

As early as 1917 the Bureau of Chemistry had noted the shortcomings of the Act of 1906, specifically with regards to various economic repercussions of food adulteration (Hutt and Hutt, 1984). A bill was eventually put before the Senate in 1933 which became upon its passage the Federal Food, Drug and Cosmetic Act of 1938. From the standpoint of food colorants the 1938 Act was notable for making certification mandatory and for creating three specific certified categories: colors suitable for foods, drugs, and cosmetics (FD&C); colors suitable for drugs and cosmetics (D&C); and colors suitable for externally applied drugs and cosmetics (Ext. D&C). The fifteen food colorants then listed (the original seven of Hesse in addition to the eight which had been successfully listed in the intervening years) were again subjected to toxicological testing; in 1940 those colorants were again deemed suitable for use, although subject to the 1938 Act's provisions regarding specifications, uses and restrictions, labeling, and certification (refer to 21 C.F.R. §74, 1996).

Following passage of the Food, Drug and Cosmetic Act, and prior to the amending legislation of 1958, four additional dyes obtained certification as food colorants (see Table 3). However, in the early 1950s three incidences involving excessively applied colorants in popcorn and candies prompted the FDA to begin a more rigorous toxicological testing program (NAS/NRC, 1971; Noonan, 1972). As a consequence FDA delisted FD&C Red

Table 3 Additional Colorants Permitted for Use in Foods, 1939–1956

Common name	FDA nomenclature	CI number	Year listed	Year delisted
Napthol Yellow S[a]	FD&C Yellow No. 2	10316	1939	1959
Oil Red XO	FD&C Red No. 32	12140	1939	1956
Orange SS	FD&C Orange No. 2	12100	1939	1956
Benzyl Violet 4B	FD&C Violet No. 1	42640	1950	1973

[a] Potassium salt.
Source: NAS/NRC (1971); Marmion (1979).

No. 32, and FD&C Orange Nos. 1 and 2 in 1956. Although manufacturers petitioned the FDA to consider the safety of the colorants in the context of proper usage, FDA instead chose to interpret the "harmless and suitable" standard of the 1938 Act as "harmless *per se*," meaning absolutely harmless at all levels. Lower court challenges to this interpretation culminated in the 1958 Supreme Court case *Flemming v. Florida Citrus Exchange*, in which the Supreme Court ruled that the FDA did not have the authority to set quantity limitations; that is, a zero toxicity criterion was established (Noonan, 1972; Taylor, 1984). This led to the subsequent delisting of the four FD&C Yellow dyes, Nos. 1, 2, 3, and 4 in 1959.

It was clear that the "zero toxicity" criterion of the Flemming case would eventually disallow the use of any synthetic colorant; the FDA was being forced to delist colorants regardless of their true capacity to endanger public health (Noonan, 1972). In response the color manufacturing industry, through the Certified Color Industry Committee, and the FDA helped to initiate legislation which became the Color Additives Amendments of 1960. The Amendments established an alternative safety standard to the 1938 Act; however, in order to ensure public health the Amendments called for premarket approval of all colorants, now referred to as "color additives." Color additives encompassed a wider set of colorants than the synthetic coal tar derivatives regulated by the 1938 Act; a color additive was defined as "a dye, pigment or other substance made . . . or . . . derived from a vegetable, animal, mineral or other source and that, when added or applied to a food, drug, or cosmetic or to the human body or any part thereof, is capable . . . of imparting a color thereto" [21 C.F.R. §70.3(f), 1996]. Both synthetic and natural colorants were subject to premarket safety evaluation, but certification (as a guarantee of purity) was only required for synthetic colorants (Newsome, 1990).

Additionally, the Color Additives Amendments prescribe how the FDA determines the safety of color additives. With regards to the safety standard, the Color Additives Amendments replaced the "harmless *per se*" standard with a general safety clause based upon a premarket evaluation of the color additive's safety; "safety" was not explicitly defined in the statute, but was interpreted as "convincing evidence that establishes with reasonable certainty that no harm will result from the intended use of the color additive" [21 C.F.R. §70.3(i), 1996]. As Taylor (1984) notes, the burden of proof of an additive's safety was now incumbent upon the manufacturer, and that proof had to be evidentiary in nature. The factors which have to be considered by FDA in the safety determination include exposure, cumulative effects, pertinent safety factors, and availability of practicable analytical detection methodology (Bachrach, 1984); the approved process for determining safety is detailed in the FDA publication *Toxicological Principles for the Safety Assessment of Direct Food Additives and Color Additives Used in Food* (the so-called Redbook).

If the color additive is known or suspected of causing cancer, however, the Color Additives Amendments invoke a Delaney clause similar to that of the Food Additives Amendment of 1958. The Delaney clause explicitly states that a color additive "shall be deemed unsafe, and shall not be listed for any use . . . if the additive is found . . . to induce cancer" [21 U.S.C. §376(b)(5)(B), 1976]. Though this is frequently misinterpreted as yet again meaning "harmless *per se*," the Delaney clause was not intended represent an absolute ban; in 1984 the Supreme Court upheld this reasoning in the case of *Scott v. Food and Drug Administration*, regarding the presence of a carcinogenic impurity in D&C Green No. 5 (Bachrach, 1984). It was argued that the presence of *para*-toluidine, a known carcinogen in D&C Green No. 5 should invoke the Delaney clause and cause the dye to be delisted. The FDA, however, had previously interpreted their regulatory responsibility thusly:

> [If data suggest that] the color additive, including its components or impurities, induces cancer in man or animal, the Commissioner shall determine whether . . . cancer has been induced and whether the color additive, including its components or impurities, was the causative substance. If it is his judgment that the data do not establish these facts, the cancer clause is not applicable; and *if the data considered as a whole establish that the color additive will be safe under the conditions that can be specified in the applicable regulation, it may be listed for such use.* (21 C.F.R. §70.50, 1996; emphasis added)

This interpretation, the basis for the so-called constituents policy, was upheld by the court.

The primary difficulty associated with the Color Additive Amendments was not the inclusion of the Delaney clause, but rather the failure of the Amendments to allow previously listed colorants as having generally recognized as safe (GRAS) status. As Hutt (1996) notes, this meant that all color additives, regardless of previous status or type, were subjected to the regulatory requirements of the Amendments; this is the basis of Title II in the Color Additive Amendments, which allowed provisional listing status for then-current colorants until their safety and suitability could be ascertained. Unfortunately, as Hutt (1996) also notes, Congress did not establish a statutory deadline for this provisional status; the courts have instead extended the transitional period from the original $2^1/_2$ years until the present day. Much of the delay was due to increased testing requirements imposed by the FDA, especially with respect to carcinogenicity. Only two new certified colors have been permanently listed in the years following the passage of the Color Additive Amendments: FD&C Red No. 40 and its alumina lake. The ten certified color additives currently permanently listed for use in foods in the United States are presented in Table 4; the lakes of FD&C Blue No. 1, Blue No. 2, Green No. 3, Yellow No. 5, and Yellow No. 6 continue to have provisional status.

The Color Additive Amendments also saw the creation of the class of uncertified colors (colors exempt from certification): the natural-source and nature-identical (synthetic

Table 4 Certified Color Additives Permanently Listed for Use in Foods

Common name	FDA nomenclature	CI number	EEC number	Year (re)listed
Brilliant Blue FCF	FD&C Blue No. 1	42090	E133	1969
Indigotine	FD&C Blue No. 2	73015	E132	1983
Fast Green FCF	FD&C Green No. 3	42053	—	1982
Erythrosine	FD&C Red No. 3	45430	E123	1969
Allura Red AC	FD&C Red No. 40	16035	E129	1971
Allura Red AC Lake	FD&C Red No. 40 Lake	16035	—	1994
Tartrazine	FD&C Yellow No. 5	19140	E102	1969
Sunset Yellow FCF	FD&C Yellow No. 6	15985	E110	1986
Citrus Red No. 2[a]	—	12156	—	1959
Orange B[b]	—	19235	—	1966

[a] Restricted to the coloring of orange skins, not to exceed 2 ppm by weight.
[b] Restricted to the coloring of casings or surfaces of frankfurters and sausages, not to exceed 150 ppm by weight.
Source: 21 C.F.R. 74 (1996); Marmion (1979); Jukes (1996); von Elbe and Schwartz (1996).

Table 5 Color Additives Exempt from Certification Permanently Listed for Use in Foods

Color additive	CI number	EEC number	Restrictions[a]
Annatto extract	75120	E160b	
β-Apo-8′-carotenal	40820	E160e	nte 15 mg[b]
Beets, dehydrated (beet powder)		E162	
Canthaxanthin	40850	E161g	nte 30 mg[b]
Caramel		E150	
β-Carotene	75130	E160a	
Carrot oil			
Cochineal extract; carmine	75470	E120	
Cottonseed flour, toasted, partially defatted			
Ferrous gluconate		E579	Ripe olives
Fruit juice			
Grape color extract		E163	Nonbeverages
Grape skin extract (enocianina)		E163	Beverages
Iron oxide, synthetic	77499	E172	Sausage casings
Paprika		E160c	
Paprika oleoresin			
Riboflavin		E101	
Saffron	75100		
Titanium dioxide	77891	E171	nte 1%[c]
Turmeric	75300		
Turmeric oleoresin	75300	E100	
Vegetable juice			

[a] nte—not to exceed.
[b] Per pound or pint.
[c] By weight.
Source: 21 C.F.R. §73 (1996); Marmion (1979); Rayner (1991); Jukes (1996), von Elbe and Schwartz (1996).

dyes which are identical to natural pigments) colors. The FDA does not recognize any category of colorant as being "natural," as the addition of any colorant to food results in an artificially colored product (Newsome, 1990). As noted previously, these colorants are subject to the same safety standard as certified colors, but do not require chemical purity certification. The color additives exempt from certification are listed in Tables 5 (for direct consumables) and 6 (for animal feeds).

III. INTERNATIONAL GOVERNMENT REGULATION

Only five of the certified food color additives allowed for use in the United States are permitted for use by the European Economic Community. As von Elbe and Schwartz (1996) note, the disparate range of colorants allowed and disallowed by various countries has created a formidable trade barrier to the food industry. Norway, for instance, does not allow any synthetic food colorants in food manufacture. Table 7 [after von Elbe and Schwartz (1996)] reviews current regulations in the EEC, Canada, Japan, and the United States.

Table 6 Color Additives Exempt from Certification Permanently Listed for Use in Foods

Color additive	CI number	EEC number	Animal
Algae meal, dried			Chickens
Astaxanthin			Salmonid fish
Canthaxanthin	40850	E161g	Chickens
Corn endosperm oil			Chickens
Iron oxide, synthetic	77499	E172	Cats, dogs
Tagetes (Aztec marigold) meal and extract	75125	E161b	Chickens
Ultramarine blue	77007		General

Source: 21 C.F.R §73 (1996); Marmion (1979); Rayner (1991); von Elbe and Schwartz (1996).

Table 7 Synthetic Color Additives Allowed for Use in Food Worldwide

Common name	FDA nomenclature	CI number	EEC number	Countries permitting[a]
Allura Red AC	FD&C Red No. 40	16035	E129	C, US
Brilliant Blue FCF	FD&C Blue No. 1	42090	E133	C, EEC,[b] J, US
Erythrosine	FD&C Red No. 3	45430	E127	C, EEC, J, US
Fast Green FCF	FD&C Green No. 3	42053	—	US
Indigotine	FD&C Blue No. 2	73015	E132	C, EEC, J, US
Sunset Yellow FCF	FD&C Yellow No. 6	15985	E110	EEC, US
Tartrazine	FD&C Yellow No. 5	19140	E102	C, EEC,[c] J, US
Amaranth	(FD&C Red No. 2)[th]	16185	E123	C, EEC[e]
Brilliant Black BN		28440	E151	EEC[ef]
Brown FK		—	E154	EEC[d]
Carmoisine		14710	E122	EEC[efg]
Chocolate Brown HT		20285	E155	EEC[f]
Green S		44090	E142	EEC[efg]
Patent Blue V		42051	E131	EEC[f]
Ponceau 4R		16255	E124	EEC,[e] J
Quinoline Yellow	D&C Yellow No. 10	47005	E104	EEC[f]
Red 2G		18050	E128	EEC[d]
Yellow 2G		18965	E107	EEC[d]

[a] C—Canada, EEC—European Economic Community, J—Japan, US—United States.
[b] Permitted only in Denmark, Ireland, and the Netherlands.
[c] Permitted only in Ireland, and the Netherlands.
[d] Permitted only in Ireland.
[e] Not permitted in Finland.
[f] Not permitted in Portugal.
[g] Not permitted in Sweden.
[h] Delisted.
Source: Rayner (1991); Jukes (1996); von Elbe and Schwartz (1996).

IV. COLOR CHEMISTRY

The perceptual property of color is induced by two broad mechanisms: either white light is selectively interacted with by matter and thus decomposed into its constituent wavelengths, or else nonwhite light is directly emitted by some source (Nassau, 1987). Nassau (1987) proceeds to list fifteen specific physicochemical mechanisms whereby color is produced; from the standpoint of the food scientist, however, those mechanisms involving electron transitions between molecular orbitals are the most important. These molecular orbital transitions are largely responsible for the color associated with organic compounds, whether synthetic or natural in origin.

It is the bond conjugation within the organic molecule which is responsible for color; delocalization of the π-bonding electrons lowers their excitation energies, allowing them to absorb light (Nassau, 1987). Extensive conjugation, or the presence of electron donor and acceptor groups within the molecule, serves to shift the absorption of light to the lower energies (that is, longer wavelengths) comprising the visible spectrum (wavelengths of 400 to 750 nm). Those wavelengths absorbed will be dependent on the existence of molecular orbital levels separated by energy, hc/λ, where h is Planck's constant, c is the velocity of light, and λ the wavelength of the absorbed radiation; the incident radiation will cause an electron to shift to the higher energy ($+ hc/\lambda$) orbital. This harmless interaction of photon with electron defines what wavelengths are visible to us; lower energy light ($\lambda >$ 750 nm) induces only small vibrational changes (at most perceived as heat) whereas higher energy light ($\lambda <$ 400 nm) will ionize matter (that is, knock out electrons completely) (Nassau, 1987).

Note, however, what we properly see as color are not the wavelengths of light absorbed, but rather the remainder of the incident light reflected (or in the case of transparent objects or solutions, the remainder of light transmitted) following absorption (Pavia et al., 1979). These perceptual colors are termed "complementary" to those wavelengths absorbed (see Table 8).

A. Colorants Subject to Certification

The certified colors discussed previously are all organic dyes, and may be grouped into the following five classes based upon general chemical structure: monoazo (FD&C Yellow

Table 8 Relationship Between Absorbed Color and Observed (Complementary) Color

Wavelength absorbed (nm)	Color of absorbed light	Color of observed light
400	Violet	Yellow
450	Blue	Orange
500	Blue-green	Red
530	Yellow-green	Red-violet
550	Yellow	Violet
600	Orange-red	Blue-green
700	Red	Green

Source: Pavia et al. (1979).

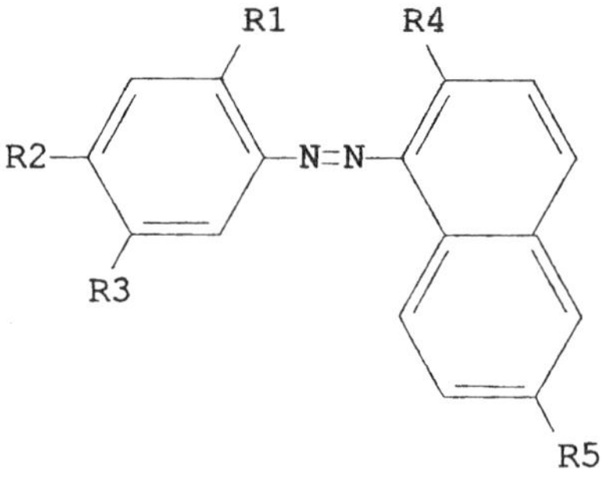

FD&C Yellow No. 6	R1=H; R2=SO_3Na; R3=H; R4=OH; R5=SO_3Na
FD&C Red No. 40	R1=OCH_3; R2=SO_3Na; R3=CH_3; R4=OH; R5=SO_3Na
Citrus Red No. 2	R1=OCH_3; R2=H; R3=OCH_3; R4=OH; R5=H

Figure 1 Monoazo colorants.

No. 6, FD&C Red No. 40, Citrus Red No. 2; see Fig. 1), pyrazolone (FD&C Yellow No. 5, Orange B; see Fig. 2), triphenylmethane (FD&C Blue No. 1, FD&C Green No. 3; see Fig. 3), indigoid (FD&C Blue No. 2; see Fig. 4), and xanthene (FD&C Red No. 3; see Fig. 5) (Marmion, 1979). Tables 9 and 10 summarize their respective chemical properties and stabilities. As is evident from the tables, the dyes have varying degrees of stability dependent upon their chemical structure. The monoazo and pyrazolone structures are subject to SO_2 decolorization through HSO_3^- addition to the nitrogens, resulting in the colorless hydroazo sulfonic acids (von Elbe and Schwartz, 1996); although data were not available for Citrus Red No. 2 and Orange B, their structures indicate that they too would be

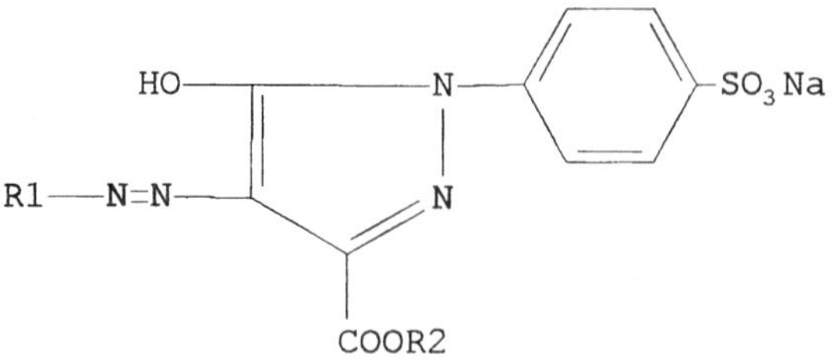

FD&C Yellow No. 5	R1= 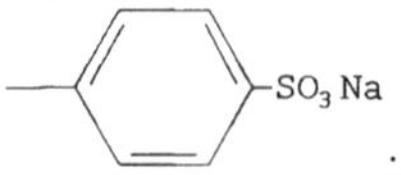 ; R2=Na
Orange B	R1= 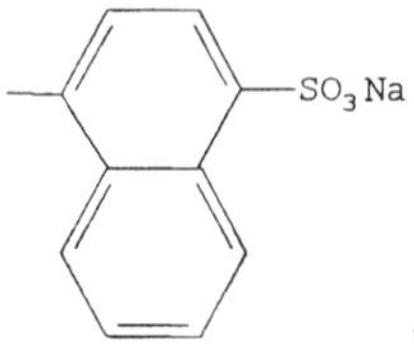 ; R2=C_2H_5

Figure 2 Pyrazolone colorants.

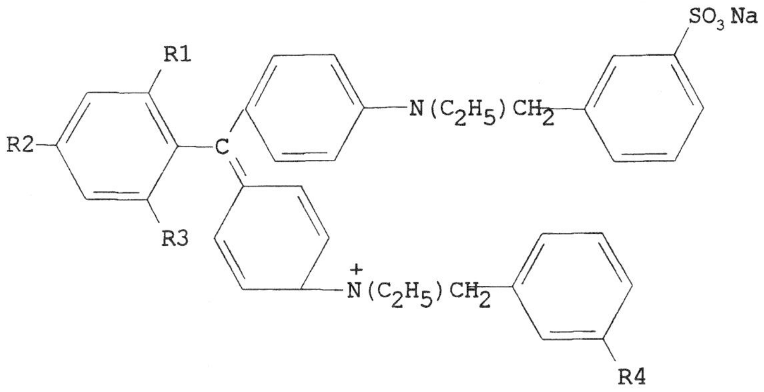

FD&C Blue No. 1	R1=H; R2=H; R3=SO_3^-; R4=SO_3Na
FD&C Green No. 3	R1=SO_3Na; R2=OH; R3=H; R4=SO_3^-

Figure 3 Triphenylmethane colorants.

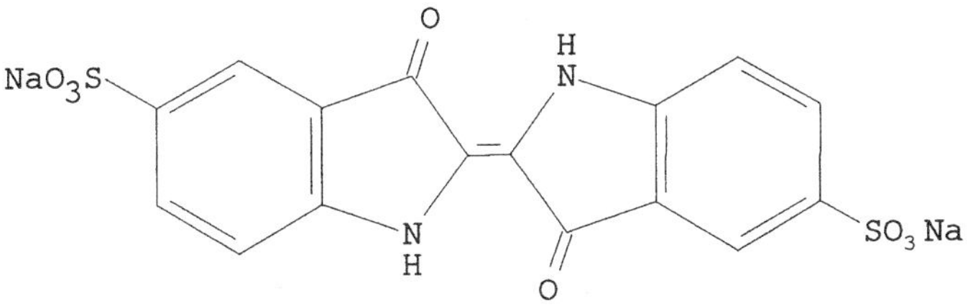

Figure 4 Indigoid colorant FD&C Blue No. 2.

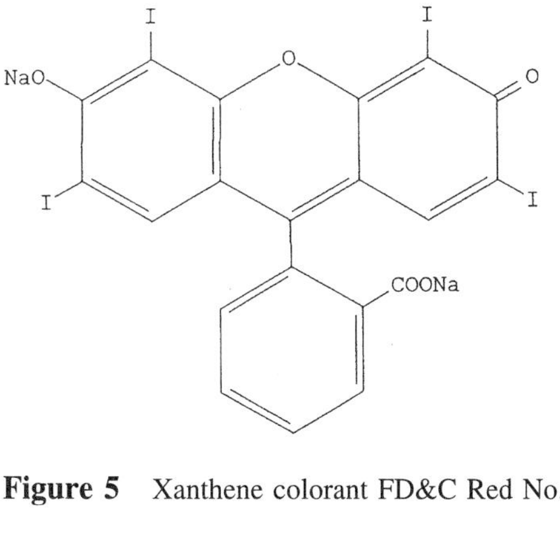

Figure 5 Xanthene colorant FD&C Red No. 3.

Table 9 Chemical Data and Properties of the U.S. Certified Colorants

FDA nomenclature	Chemical classification	Empirical formula	Molecular weight	Solubility (g/100 mL, 25°C)[a]			
				H_2O	EtOH	Gly	Glycol
FD&C Yellow No. 6	Monoazo	$C_{16}H_9N_4O_9S_2Na_3$	452.36	19.0	IN	20.0	2.2
FD&C Red No. 40	Monoazo	$C_{18}H_{14}N_2O_8S_2Na_2$	496.42	22.0	0.01	3.0	1.5
Citrus Red No. 2	Monoazo	$C_{18}H_{16}N_2O_3$	308.34	IN	VSS	VSS	VSS
FD&C Yellow No. 5	Pyrazolone	$C_{16}H_9N_4O_9S_2Na_3$	534.36	20.0	IN	18.0	7.0
Orange B	Pyrazolone	$C_{22}H_{16}N_4O_9S_2Na_2$	590.49	22.0[b]	na	na	na
FD&C Blue No. 1	Triphenylmethane	$C_{37}H_{34}N_2O_9S_3Na_2$	792.84	20.0	0.15	20.0	20.0
FD&C Green No. 3	Triphenylmethane	$C_{37}H_{34}O_{10}N_2S_3Na_2$	808.84	20.0	0.01	20.0	20.0
FD&C Blue No. 2	Indigoid	$C_{16}H_8N_2O_8S_2Na_2$	466.35	1.6	IN	1.0	0.1
FD&C Red No. 3	Xanthene	$C_{20}H_6O_5I_4Na_2$	879.86	9.0	IN	20.0	20.0

[a] H_2O—water EtOH—ethanol, Gly—glycerine, Glycol—propylene glycol, IN—insoluble, VSS—very slightly soluble, na—data not available.
[b] At 77°C.
Source: NAS/NRC (1971); Marmion (1979).

Table 10 Stability Data for the U.S. Certified Colorants

FDA nomenclature	Original hue	Stability to[a]								
		pH 3	pH 5	pH 7	pH 8	Light	Heat	Acid	Base	SO_2
FD&C Yellow No. 6	Reddish	VG	VG	VG	VG	M	VG[b]	VG	M	F
FD&C Red No. 40	Yellowish red	VG	VG	VG	VG	VG	VG[c]	VG	F	VG
Citrus Red No. 2	Orangish red	na	na	na	na	na	na	na	na	na
FD&C Yellow No. 5	Lemon yellow	VG	VG	VG	VG	G	VG[c]	VG	P	F
Orange B	Orangish red	na	na	na	na	M	na	na	na	na
FD&C Blue No. 1	Greenish blue	M	G	G	G	F	VG[c]	VG	VP	G
FD&C Green No. 3	Bluish green	M	G	G	M	F	F[b]	VG	VP	G
FD&C Blue No. 2	Deep blue	F	F	P	VP	VP	F[c]	VP	P	VP
FD&C Red No. 3	Bluish pink	IN	IN	VG	VG	P	G[c]	IN	VP	IN

[a] Acid—10% acetic acid, Base—10% sodium hydroxide, SO_2—250 mg/L, VG—very good, G—good, M—moderate, F—fair, P—poor, VP—very poor, IN—insoluble, na—data not available.
[b] To 205°C.
[c] To 105°C.
Source: Marmion (1979); Newsome (1990); Rayner (1991).

bleached by SO_2). The notable exception is the monoazo FD&C Red No. 40; the electron donor groups on the aromatic rings apparently inhibit the nitrogens from serving as nucleophilic sites, thus forestalling HSO_3^- addition. Note that the lack of sulfonic acid groups on Citrus Red No. 2 prevent it from being water soluble.

The triphenylmethane colorants FD&C Blue No. 1 and FD&C Green No. 3 are very similar in structure, differing only in the presence of a hydroxyl group in FD&C Green No. 3. These dyes are highly subject to alkali decolorization due to the formation of colorless carbinol bases (von Elbe and Schwartz, 1996).

FD&C Blue No. 2 (indigoid) is highly susceptible to oxidation by ultraviolet light and fades rapidly (Newsome, 1990). FD&C Red No. 3, the sole xanthene, is also very unstable to light. As Noonan (1972) notes, however, this instability is product dependent, with FD&C Blue No. 2 performing well in candies and baked goods, and FD&C Red No. 3 performing well in retorted products.

Pure dye concentration in these colorants is strictly controlled by the FDA. FD&C Blue Nos. 1 and 2, FD&C Green No. 3, and FD&C Red 40 must contain no less than 85% pure dye; Orange B, FD&C Red No. 3, and FD&C Yellow Nos. 5 and 6 must contain no less than 87% pure dye; and Citrus Red No. 2 must contain no less than 98% pure dye (21 C.F.R. §74, 1996). While the coloring power of a colorant, the tinctorial strength, is directly proportional to the pure dye concentration, variations of a few percent have little practical significance (von Elbe and Schwartz, 1996). Furthermore, the tinctorial strength is ultimately an intrinsic property of the dye's chemical structure (i.e., the extinction coefficient); the coloring power is thus best manipulated through optimization of such parameters as the physical form of the dye used and the carrying vehicle (Marmion, 1979).

The colorants listed in Table 9 all exhibit (with the exception of Citrus Red No. 2) some degree of water solubility. However, not all foodstuffs have an adequate moisture content to insure complete dissolution. Previously oil-soluble dyes were used, but these were found to present health hazards, and the last four oil-soluble dyes (FD&C Yellow Nos. 1, 2, 3, and 4) were delisted in 1959 (Noonan, 1972). Although nonaqueous solvents may be used to disperse the water-soluble dyes (e.g., glycerine and propylene glycol), an alternative method is to precipitate the aluminum salt of a dye onto an insoluble substrate; these colorants are known as *lakes* and are physically dispersed in a food as a pigment rather than being adsorbed or dissolved in the matrix. The dye content of the lake may range from 1 to 45%, although they usually contain 10 to 40% (Newsome, 1990). The only dye substratum currently approved for food use is alumina hydrate.

The tinctorial strength of lakes is not proportional to their dye content; particle size, however, greatly affects the color intensity, as the smaller the particle size, the more complete the dispersion and the greater the reflective surface area (Dziezak, 1987; von Elbe and Schwartz, 1996). Lakes have better light, chemical, and thermal stabilities than their associated dyes (due in part to their lower dye content) (Dziezak, 1987); however, the complexity of preparing the substratum and extending the dyes increases the cost of lakes (Meggos, 1984). Furthermore, the energy required to properly disperse the lakes is high; improper dispersion, however, will result in particle-clumping, which is perceptually evidenced as speckling (Dziezak, 1987).

B. Colorants Exempt from Certification

There has been an increasing demand on the part of consumers for natural-source colorants, due in large part to the perception that foods thus colored are more wholesome and of better quality (Wissgott and Bortlik, 1996). However, natural-source colorants are far less stable to heat, light, or pH, and the colorants themselves may impart extraneous flavors (Moore, 1991). Furthermore, natural-source colorant production is not easily scaled up to meet industrial demand, even with in vitro techniques replacing whole-plant cultivation (Wissgott and Bortlik, 1996). This makes natural-source colors more expensive; red and yellow colorants may cost 100 times more than their synthetic counterparts to deliver the same tinctorial strength (Riboh, 1977).

While natural-source could refer to any colorant derived from animal, vegetable, or mineral sources, it is most often taken to mean ''derived from plant sources.'' Of the 356 colorant patents on natural sources filed in the years 1969 through 1984, 63% were of plant origin (Francis, 1987); of the twenty-two colorants exempt from certification for use in foods, thirteen are of plant origin and three are nature-identical (see Table 5). The non–plant-derived colorants, however, are not to be discounted; this group includes such important colorants as caramel, cochineal, *Monascus* derivates, and titanium dioxide. See Lee and Khng (this volume) for information regarding the chemistry and properties of the colorants exempt from certification.

V. COLOR ANALYSIS

Color analysis may be performed through either one of two approaches: the chemical quantification of the colorant compounds or the assessment of the resulting color, the latter being conducted either instrumentally or with human observers. The impetus underlying these two approaches differs greatly, however. In the former case the desire is to closely monitor the concentration of the color additive used (especially in the case of the certified colorants) in order to insure the health and safety of the consumer populace. Though all of the synthetic colorants have been subject to thorough toxicological assessment (and many show minimal toxicity), high concentrations of certain colorants (e.g., FD&C Red No. 40 and FD&C Yellow No. 5) may cause adverse reactions in some persons (Davidson, 1997). Such risks necessitate that the concentration of these additives, and their impurities, be closely monitored (see also Section VIII). Color assessment, however, is concerned with the perceptual impact of the colorants used, and is thus complicated by the implicit requirement of a human judgment.

A. Chemical Analysis

The methods for analyzing the individual colorants follow standard organic analysis techniques, the development of more sophisticated techniques having paralleled the development of new analytical instrumentation (Yeransian et al., 1985). While titrametric and gravimetric methods are allowed for determining pure dye content of color additives (Bell, 1990), spectrophotometric methods have been listed in the *AOAC Official Methods of Analysis* since 1960. The speed, ease, and efficacy of the spectrophotometric methods make them of particular value, as does their minimal (on the order of ng) sample requirement (Marmion, 1979).

Spectrophotometric methods often prove inadequate, however, in the analysis of real samples due to the overlapping of spectral absorption maxima (Capitán-Vallvey et al., 1997). These separation difficulties have been surmounted through the use of specific chromatographic (Puttemans et al., 1981; Maslowska, 1985; Patel et al., 1986; Karovicová et al., 1991; Oka et al., 1994), electroanalytical (Fogg et al., 1986; Ni et al., 1996), or absorption (Capitán et al., 1996) procedures and through the use of multivariate-calibration techniques (e.g., partial least squares regression analysis) (Ni et al., 1996; Capitán-Vallvey et al., 1997). High performance liquid chromatography (HPLC) in particular has received much attention for the separation of colorants (Puttemans et al., 1981); Gennaro, Abrigo, and Cipolla have reviewed the use of HPLC in the identification and determination of dyes and their impurities (Gennaro et al., 1994).

It is often the case, however, that isolating colorants from the sample matrix is more problematic to the chemist than is the problem of resolving individual colorant components (Greenway et al., 1992). Matrix isolation techniques have typically depended upon one of three general methodologies: leaching, solvent–solvent extraction, or active substrate absorption (Marmion, 1979); the presence of high-affinity binding agents such as proteins, however, necessitates removal of the interfering matrix (e.g., through precipitation) (Puttemans et al., 1984). As Marmion (1979) notes, no one method is applicable to all sample matrices, thus the chemist requires comprehensive knowledge to optimize conditions on a sample-specific basis.

B. Visual Colorimetry

Quantification of colorants does not, it should be emphasized, serve as a specification of the color. As Little and Mackinney (1969) noted, "the measurement of the light-modifying properties of an object does not qualify as a measurement of color." Color is an exclusively human perceptual phenomenon; as such any method for assessing color depends at some point upon human response.

Visual assessments of color have typically depended upon the comparison of the sample color to that of reference standards or to a color atlas (Billmeyer, 1988). Fortunately for that segment of the population devoid of color vision deficits, color is perceived in a far more uniform manner than are other sensations such as taste or aroma (Clydesdale, 1977). Biases can still affect the assessment, however, either through psychological preferences or through lack of control of the viewing conditions (Mabon, 1993). The latter is especially problematic when comparing samples to a color order system such as the

Munsell Book of Color [see Billmeyer (1987) for a survey of the common color order systems], which presupposes that a standard illuminant will be used for viewing (Billmeyer, 1988). Billmeyer (1988) notes that while color comparison represents a straightforward enough task, the difficulty of control may preclude accurate color assessment.

C. Instrumental Colorimetry

The subjectivity of human response, coupled to the relative constancy of human color perception, make instrumental measures of color both desirable and feasible. These assessments depend upon rigorously defined color spaces, notably CIE-LAB and CIE-LUV. However, as Billmeyer (1988) cautions, these color spaces were derived for perceptual matching purposes and not for absolute color specification. A discussion of instrumental colorimetry being beyond the scope of this article, the reader is referred to Mackinney and Little (1962), Francis and Clydesdale (1975), or Hutchings (1994).

VI. COLOR EFFECTS IN FOODS

As was noted prior, the impact of colorant usage in foods has been known since ancient times. Controlled studies of the effects of color on flavor and food acceptance, however, have only been conducted during the past 70 years. One of the earliest studies was that of Moir (1936), who noted that ''Many people find that their sense of taste is affected by the colour of the foodstuff to be tasted. They may give a totally wrong answer where delicate flavours are concerned as a result of being misled by their eyes.''

Moir used both inappropriately colored jellies (yellow vanilla, green orange, amber lime, and red lemon) and brown-colored sponge cookies (one flavored with cocoa, the other with vanilla) to determine people's abilities to correctly ascertain flavors. With regard to the jellies, only 1 person out of 60 correctly identified all four flavors; the majority of persons were only able to identify two or fewer flavors correctly.

Hall (1958), in a widely cited study, reported a similar experiment in which subjects were presented with appropriately, inappropriately, and white-colored sherbets. Similar to Moir's findings, inappropriately or white-colored sherbets were poorly identified as to their true flavors. Hall concluded that flavor, unless ''outstandingly good or bad'' was ''outweighed in importance by other factors, particularly by visual factors.'' Hall's conclusions are seemingly the source of the notion that color is of preeminent sensory importance, a notion reflected in the apocryphal report of Wheatley (cited in Kostyla and Clydesdale, 1978) that diners were sickened by inappropriately colored food [a report relegated to the status of ''food folklore'' by Hutchings (1994)].

It is important, therefore, to note that Schutz (1954) reached a quite different conclusion in his study of color on orange juice preference. When subjects visually assessed preference, an orange-colored juice was preferred over a yellow-colored juice. However, when subjects assessed preference based upon tasting the products, there was no statistically significant difference in preference. Schutz concluded that food preference based upon color alone was not indicative of true food preference: ''We can conclude that although people may prefer one color juice to another on the basis of appearance alone, when they actually taste the juices, the color variable becomes insignificant in determining their preference ratings.''

Hall (1958) had also obtained hedonic data, and found that within a flavor appropriately colored sherbets were preferred over white-colored sherbets, which were in turn preferred over the inappropriately colored sherbets. The difference between the two studies is that Hall used colors outside of the normal range of expectation, whereas Schutz (1954) did not. As Schutz noted, "We may conjecture that if we had colored the frozen juice blue we would have found a lower preference rating." Therefore an alternative hypothesis regarding color and flavor would be that, within the range of normal color expectations, preference is dependent in a complex fashion upon both color and flavor, and only in the artificial situation of inappropriate color does color take precedence over flavor, relegating flavor to a minor role unless flavor also varies inappropriately (i.e., extremely).

This alternative hypothesis is supported by the research of Tuorila-Ollikainen (Tuorila-Ollikainen, 1982; Tuorila-Ollikainen et al., 1984). Tuorila-Ollikainen and coworkers found that the pleasantness of carbonated soft drinks depended primarily upon the flavor and sweetness; color had minimal influence. Clydesdale (1993) raised the objection that this lack of a color effect was due to the soft drinks not being identified by flavor, pointing out that in an earlier study (Tuorila-Ollikainen, 1982) color had been found to have a positive effect on soft drink pleasantness. However, this was only true when the product flavor was identified; when the flavor was unspecified color had no effect on pleasantness. Indeed, the colorless soft drink was rated as more pleasant than the appropriately colored sample.

Tuorila-Ollikainen's studies further demonstrate the importance of context upon the magnitude of color's effect upon acceptability judgments. Martens and coworkers (1983) also found that color played a significant role in preference of black currant juice when the color was modified from the expected color to a darker, more brown and less green juice. Similar results were obtained by Du Bose et al. (1980), who found that color affected acceptability when judgments were made within the context of the appropriateness of the color to a specified flavor; however, overall acceptability was more closely correlated to a product's flavor than a product's color.

Finally, in addition to influencing flavor identification and product acceptability, color may also affect both real and perceived nutritional value (Bender, 1981). Yellow-orangish fruits and vegetables are accurately identified as being high in carotenes; however, with the exception of true whole-wheat bread, brown color is typically a poor indicator of nutritional value (e.g., brown eggs are no more nutritious than white eggs; brown sugar is no more nutritious than white sugar).

VII. COLORANT APPLICATIONS

As Dziezak (1987) notes, colorants are added to consumable products for the sole purpose of enhancing the visual appeal. Possible reasons underlying the need for enhancement include (NAS/NRC, 1971; IFT, 1986)

1. Correcting for natural variations in food or ingredient colors
2. Correcting for color changes during storage, processing, packaging,or distribution
3. Emphasizing associated flavors or preserving unique identifying characteristics
4. Protecting flavor and vitamins from photodegradation

Colorants represented a $245 million business in 1991, with certified colorants representing some 37% of the market total (TPC Business Research Group Report, 1992). At that time the growth projection for certified colorants was half that for natural colorants; the reason for this conservative projection was anticipated negative consumer perception regarding synthetic colorants. As it turned out, in 1994 certified colorants did indeed account for an even smaller percentage of the market (~33%), with the total dollar value remaining flat at $90 million (Fitzgerald, 1996).

Of the certified colorants, FD&C Red No. 40 remains the predominant dye used in the United States; Henkel (1993) estimated 1992 production at ~1,361,000 kg, or 454,000 kg more than the next leading colorant, FD&C Yellow No. 5. In 1982 Red No. 40, Yellow No. 5, and Yellow No. 6 comprised ~90% of the total certified color production (Marmion, 1984), a figure similar to that reported by the CCIC (1968) for Red No. 2, Yellow No. 5, and Yellow No. 6 production. Typical applications for the certified colorants are presented in Table 11.

The beverage industry is the largest user of certified colorants (NAS/NRC, 1971), with most fruit-flavored beverages utilizing certified colorants (Noonan, 1972) in the range of 5–150 mg/kg (Dziezak, 1987). Candies, confections, and dessert powders (e.g., gelatins) are the next largest users of certified colorants (CCIC, 1968); whereas lakes are not suitable for beverages, they are often used to produce the striping effect in hard candies, as the lakes do not bleed or migrate (Dziezak, 1987). Lakes are also the colorants of choice for compound coated candies to prevent color specking (Noonan, 1972). The major categories of processed foods utilizing certified colorants are presented in Table 12.

Meggos (1994) outlined the key decisions to be made in determining the optimal colorant for a specific application:

Table 11 Common Applications of the FD&C Certified Colorants

Colorant	Common Name	Hue	Applications
FD&C Blue No. 1	Brilliant Blue FCF	Bright greenish-blue	Bakery, beverages, condiments, confections, dairy products, extracts, icings, jellies, powders, syrups
FD&C Blue No. 2	Indigotine	Deep royal blue	Baked goods, cereal, cherries, confections, ice cream, snack foods
FD&C Green No. 3	Fast Green FCF	Sea green	Baked goods, beverages, cherries, confections, dairy products, ice cream, puddings, sherbet
FD&C Red No. 3	Erythrosine	Bluish pink	Baked goods, confections, dairy products, fruit cocktail cherries, snack foods
FD&C Red No. 40	Allura Red AC	Yellowish red	Beverages, cereals, condiments, confections, dairy products, gelatins, puddings
FD&C Yellow No. 5	Tartrazine	Lemon yellow	Bakery, beverages, cereals, confections, custards, ice cream, preserves
FD&C Yellow No. 6	Sunset Yellow FCF	Orange	Bakery, beverages, cereals, confections, dessert powders, ice cream, snack foods

Source: Moore (1991); FDA/IFICF (1993).

Table 12 Certified Colorant Use and Concentration by Processed Food Category

Category	Total colorant purchased (kg)[a]	Colorant concentration (mg/kg)	
		Average	Range
Beverages (liquid and powdered)	255,569	75	5–200
Candies and confections	91,938	100	10–400
Dessert powders	85,050	140	5–600
Bakery goods	80,693	50	10–500
Sausage (surface)	74,705	125	40–250
Cereals	48,027	350	200–500
Dairy products	41,759	30	10–200
Snack foods	15,677	200	25–500

[a] These figures represent sales data for the first nine months of 1967.
Source: CCIC (1968).

1. What is the target shade?
2. What are the physical/chemical attributes of the food?
3. Are there any nontechnical marketing requirements that must be met?
4. In what countries will the finished product be marketed?
5. What processing will the food undergo, both by the processor and the consumer?
6. What type of packaging will be used for the finished food?
7. What will the storage conditions be for the finished food?

As Meggos (1994) notes, answering these questions will usually constrain the choices to one or two colorants, the final selection then being determined in pilot plant studies. Additional product specific information may be found in Noonan (1972) and Dziezak (1987).

VIII. SAFETY

As discussed previously (Section II) the passage of the Pure Food and Drug Act of 1906 formalized the requirements that all synthetic dyes be manufactured under strict control and undergo voluntary certification; certification of synthetic colorants became mandatory with the passage of the Food, Drug and Cosmetic Act of 1938. Finally, the Color Additive Amendments of 1960 required additional testing using modern techniques, and also invoked the Delaney Clause regarding carcinogenicity. It was clearly recognized from Hesse's time that, even though the primary role of food colorants is cosmetic (their only nutritional contribution being indirect, in the preservation of light-degradable vitamins), colorants nonetheless can have a significant impact on human health.

A. Food Intake Estimates

The Certified Color Industry Committee (now the International Association of Color Manufacturers) estimated that the maximum daily per capita consumption of food colorants was on the order of 50 mg, with an average daily intake of some 15 mg (CCIC, 1968); Marmion (1984) estimated an approximate daily per capita consumption value of 30 mg. Both of these values, it should be noted, were based upon sales or manufacturing data, as opposed to actual studies of dietary intake. Given that the average concentration of certified colorant in carbonated beverages is ~75 mg/L (see Table 12), one can (355 mL) of soda alone would deliver 26 mg of dye, making these values appear low.

The National Academy of Sciences published a report in 1979 in which the 1977 intake patterns of some 12,000 individuals were analyzed for food additives consumption [''The 1977 Survey of Industry on the Use of Food Additives: Estimates of Daily Intake,'' cited in Newsome (1990)]. The NAS/NRC participants reported a total daily intake of ~325 mg colorant/kg body weight; however, this reflects data for the 99th percentile of the population (meaning that 99% of the population sampled was estimated to have intakes equal to or below this value; these values are thus leveraged by extreme consumers). NAS estimates that actual intakes are approximately 20% of the self-reported amounts (Newsome, 1990), thus a corrected average daily consumption of food colorants could be on the order of 65 mg/kg body weight for the 99th percentile. Even this value seems excessive, however (for example, a 70 kg man would be consuming 4.5 g of food colorant daily).

Louekari et al. (1990) point out that the mean value of consumption may be 5- to 40-fold less than that of the 99th percentile; assuming a 40-fold difference would suggest an overall adjusted, corrected consumption value of 1.6 mg/kg, or ~115 mg per day for a 70-kg man. This value is on the same order as that suggested by Louekari et al.'s data (indicating a total mean intake of ~110 mg/person). This estimate, higher than CCIC's or Marmion's, most likely reflects the increasing percentage of processed food in the modern diet.

B. Toxicological Considerations

Given the ubiquity of colorants in the diet, their inherent safety is, of course, of preeminent concern. In order to safeguard the populace, the concept of an acceptable daily intake (ADI) level was introduced by FDA scientists in the 1950s (Rodricks, 1996). ADIs are most commonly established through extensive animal studies; the current Redbook requirements for toxicological testing include [FDA Redbook, cited in IFT (1986)]:

1. One subchronic feeding study, of 90 days duration, in a nonrodent species, usually the dog
2. Acute toxicity studies in rats
3. Chronic feeding studies in at least two animal species (one with in utero exposure), lasting at least 24–30 months
4. One teratology study
5. One multigeneration reproduction study using mice
6. One mutagenicity test

Such testing establishes no adverse effect dietary levels for each specific colorant; these levels are then adjusted by a safety factor (typically 100) to arrive at ADIs (Kokoski et al., 1990). Daily intake levels for the certified colorants and their respective ADIs are presented in Table 13. Once ADIs are obtained they must perforce be compared to the estimated daily intake (EDI); if the EDI is sufficiently below the ADI the additive is deemed safe for consumption. Specific toxicological summaries are presented in the IFT Expert Panel on Food Safety and Nutrition's Scientific Status Summary (IFT, 1986).

Table 13 Comparative Daily Intake Data for the Certified Colorant Dyes

Colorant	Average daily intake (mg/kg)[a]	Adjusted, corrected daily intake (mg/kg)[b]	Acceptable daily intake (mg/kg)[c]
FD&C Blue No. 1	16	0.08	12.5
FD&C Blue No. 2	7.8	0.04	5.0
FD&C Green No. 3	4.3	0.02	12.5[d]
FD&C Red No. 3	24	0.12	0.05[e]
FD&C Red No. 40	100	0.50	Not available
FD&C Yellow No. 5	43	0.22	7.5
FD&C Yellow No. 6	37	0.19	5.0

[a] NAS/NRC data (Newsome, 1990); values for the 99th percentile of the population.
[b] Corrected for estimated over-reporting and adjusted to represent the mean (see text).
[c] FAO/WHO data (Vettorazzi, 1981).
[d] FAO/WHO (1987).
[e] FAO/WHO (1989).

C. Hypersensitivity and Hyperactivity

Tartrazine (FD&C Yellow No. 5) has long been linked with hypersensitivity in certain individuals. In addition, tartrazine was thought to provoke bronchospasms in asthmatics and aspirin-intolerant persons (IFT, 1986). Newer studies indicated, however, that the incidence of tartrazine hypersensitivity was rare (fewer than 1 out of 10,000 people being susceptible) (FDA/IFICF, 1993) and that tartrazine responses in asthmatic or aspirin-sensitive individuals were exceptional and idiosyncratic (IFT, 1986). More recently, Sunset Yellow FCF (FD&C Yellow No. 6) was proposed for label declaration by the FDA because of putative hypersensitivity, although the International Association of Color Manufacturers charged that the supporting studies were seriously flawed (Food Chemical News, 1995). The passage of the Nutrition Labeling and Education Act of 1990 made this issue moot, as all certified color additives are legally required to be listed on the ingredient label (Segal, 1993).

With regard to hyperactivity, B. F. Feingold first hypothesized in the early 1970s that ingestion of certified colorants promoted hyperactivity in children. While Feingold claimed a 50% improvement in patients on total elimination diets, subsequent studies either found no differences or obtained inconclusive results (IFT, 1986). The FDA has categorically stated that there is "no evidence that food color additives cause hyperactivity or learning disabilities in children" (FDA/IFICF, 1993). Interestingly, a recent double-blind, placebo-controlled study once again linked tartrazine ingestion with irritability, restlessness, and sleep disturbances in some children (Rowe and Rowe, 1994).

REFERENCES

Accum, F. 1820. *A Treatise on Adulterations of Food and Culinary Poisons*. Abraham Small, Philadelphia.

Amerine, M. A., Singleton, V. L. 1977. *Wine: An Introduction*, 2nd ed. University of California Press, Berkeley, CA.

Bachrach, E. E. 1984. D & C Green No. 5: judicial review of the constituents policy. *Food Drug Cosmetic Law Journal* 39(3):299–305.

Bell, S. 1990. Color additives. In: *Official Methods of Analysis: Food Composition, Additives, Natural Contaminants* Vol. 2, 15th ed., Helrich, K., (Ed.). Association of Official Analytical Chemists, Arlington, VA, pp. 1115–1136.

Bender, A. E. 1981. The appearance and the nutritional value of food products. *J. Human Nutr.* 35(3):215–217.

Billmeyer, F. W., Jr. 1987. Survey of color order systems. *Color Research and Application* 12(4): 173–186.

Billmeyer, F. W., Jr. 1988. Quantifying color appearance visually and instrumentally. *Color Research and Application* 13(3):140–145.

Capitán, F., Capitán-Vallvey, L. F., Fernández, M. D., de Orbe, I., Avidad, R. 1996. Determination of colorant matters mixtures in foods by solid-phase spectrophotometry. *Anal. Chim. Acta* 331(1/2):141–148.

Capitán-Vallvey, L. F., Fernández, M. D., de Orbe, I., Vilchez, J. L., Avidad, R. 1997. Simultaneous determination of the colorants Sunset Yellow FCF and Quinoline Yellow by solid-phase spectrophotometry using partial least squares multivariate calibration. *Analyst* 122(4):351–354.

Certified Color Industry Committee. 1968. Guidelines for good manufacturing practice: use of certified FD&C colors in food. *Food Technol.* 22(8):14–17.

Clydesdale, F. M. 1977. Color measurement. In: *Current Aspects of Food Colorants*, Furia, T. E. (Ed.). CRC Press, Cleveland, OH, pp. 1–17.

Clydesdale, F. M. 1991. Color perception and food quality. *J. Food Quality* 14(1):61–74.

Clydesdale, F. M. 1993. Color as a factor in food choice. *Crit. Rev. Food Sci. Nutr.* 33(1):83–101.

Davidson, P. M. 1997. Food additives. In: *Encyclopedia Britannica*, Vol. 19, ''Food Processing,'' Chicago, pp. 402–405.

DuBose, C. N., Cardello, A. V., Maller, O. 1980. Effects of colorants and flavorants on identification, perceived flavor intensity, and hedonic quality of fruit-flavored beverages and cake. *J. Food Sci.* 45(5):1393–1399,1415.

Dziezak, J. D. 1987. Applications of food colorants. *Food Technol.* 41(4):78–88.

FAO/WHO Expert Committee on Food Additives. 1987. Toxicological evaluation of certain food additives and contaminants. Joint FAO/WHO Expert Committee on Food Additives. 29th Meeting, Geneva, Switzerland, 1985. Cambridge University Press, New York.

FAO/WHO Expert Committee on Food Additives. 1989. Toxicological evaluation of certain food additives and contaminants. Joint FAO/WHO Expert Committee on Food Additives. 33rd Meeting, Geneva, Switzerland, 1989. Cambridge University Press, New York.

Fitzgerald, P. 1996. Natural food colorants. *Chemical Marketing Reporter* 249(26):SR24.

Fogg, A. G., Barros, A. A., Cabral, J. O. 1986. Differential-pulse adsorptive stripping voltammetry of food and cosmetic synthetic colouring matters and their determination and partial identification in tablet coatings and cosmetics. *Analyst* 111(7):831–835.

Food and Drug Administration/International Food Information Council Foundation. 1993. Food Color Facts. http://vm.cfsan.fda.gov/~lrd/coloradd.txt.

Food Chemical New. 1995. Allergic reactions to Yellow 6 not supported by evidence: IACM. *Food Chemical News* 37(35):18.

Francis, F. J. 1987. Lesser-known food colorants. *Food Technol.* 41(4):62–68.

Francis, F. J., Clydesdale, F. M. 1975. *Food Colorimetry: Theory and Applications.* AVI Publishing Co., Westport, CT.

Gennaro, M. C., Abrigo, C., Cipolla, G. 1994. High-performance liquid chromatography of food colours and its relevance in forensic chemistry. *J. Chromatogr.* 674(1/2):281–299.

Greenway, G. M., Kometa, N., Macrae, R. 1992. The determination of food colours by HPLC with on-line dialysis for sample preparation. *Food Chem.* 43(2):137–140.

Hall, R. L. 1958. Flavor study approaches at McCormick & Company, Inc. In: *Flavor Research and Acceptance.* Reinhold Publishing Corp., New York, pp. 224–240.

Henkel, J. 1993. From shampoos to cereal: seeing to the safety of color additives. *FDA Consumer* 27(10):14–21.

Hutchings, J. B. 1994. *Food Colour and Appearance*. Blackie Academic & Professional, London.

Hutt, P. B. 1996. Approval of food additives in the United States: a bankrupt system. *Food Technol.* 50(3):118–128.

Hutt, P. B., Hutt, P. B., II. 1984. A history of government regulation of adulteration and misbranding of food. *Food Drug Cosmetic Law Journal* 39(1):2–73.

IFT Expert Panel on Food Safety and Nutrition. 1986. Food colors. *Food Technol.* 40(7):49–56.

Jukes, D. 1996. Food Additives in the European Union. The Department of Food Science and Technology, The University of Reading, UK. Food Law page. http://www.fst.rdg.ac.uk/people/ajukesdj/l-addit.htm.

Karovicová, J., Polonský, J., Príbela, A., Simko, P. 1991. Isotachophoresis of some synthetic colorants in foods. *J. Chromatogr.* 545(2):413–419.

Kokoski, C. J., Henry, S. H., Lin, C. S., Ekelman, K. B. 1990. Methods used in safety evaluation. In: *Food Additives*, Branen, A. L., Davidson, P. M., Salminen, S. (Eds.). Marcel Dekker, New York, pp. 579–616.

Kostyla, A. S., Clydesdale, F. M. 1978. The psychophysical relationship between color and flavor. *CRC Crit. Rev. Food Sci. Nutr.* 10(3):303–321.

Lee, Y.-K., Khng, W.-P. 1997. Natural color additives. In: *Food Additives*, 2nd ed., Branen, A. L., Davidson, P. M., Salminen, S. (Eds.). Marcel Dekker, New York.

Lindberg, B. J. 1938. *Experimental Studies of Colour and Non-Colour Attitude in School Children and Adults*. Levin & Munksgaard, Copenhagen.

Little, A., Mackinney, G. 1969. The sample as a problem. *Food Technol.* 23(1):25–28.

Louekari, K., Scott, A. O., Salminen, S. 1990. Estimation of food additive intakes. In: *Food Additives*, Branen, A. L., Davidson, P. M., Salminen, S. (Eds.). Marcel Dekker, New York, pp. 9–32.

Mabon, T. J. 1993. Color measurement of food. *Cereal Foods World* 38(1):21–25.

Mackinney, G., Little, A. C. 1962. *Color of Foods*. AVI Publishing Co., Westport, CT.

Marmion, D. M. 1979. *Handbook of U.S. Colorants for Foods, Drugs, and Cosmetics*. John Wiley & Sons, New York.

Marmion, D. M. 1984. *Handbook of U.S. Colorants for Foods, Drugs, and Cosmetics*, 2nd ed. John Wiley & Sons, New York.

Martens, M., Risvik, E., Schutz, H. G. 1983. Factors influencing preference—a study on black currant juice. In: *Research in Food Science and Nutrition, Vol. 2: Basic Studies in Food Science*, McLoughlin, J. V., McKenna, B. M. (Eds.). Boole Press Ltd., Dublin, pp. 193–194.

Maslowska, J. 1985. A new chromatographic method for the separation of food dye mixtures on thin MgO layers. *Chromatographia* 20(2):99–101.

Mason, C., Kandel, E. R. 1991. Central visual pathways. In: *Principles of Neural Science*, 3rd ed., Kandel, E. R., Schwartz, J. H., Jessell, T. M. (Eds.). Elsevier, New York, pp. 420–439.

Meggos, H. N. 1984. Colors—key food ingredients. *Food Technol.* 38(1):70–74.

Meggos, H. N. 1994. Effective utilization of food colors. *Food Technol.* 48(1):112.

Moir, H. C. 1936. Some observations on the appreciation of flavour in foodstuffs. *Chem. Ind.* 55: 145–148.

Moore, L. 1991. The natural vs. certified debate rages on. *Food Eng.* 63(8):69–72.

Nassau, K. 1987. The fifteen causes of color: the physics and chemistry of color. *Color Research and Application* 12(1):4–26.

National Academy of Sciences/National Research Council. 1971. *Food Colors*. National Academy Press, Washington D.C.

Newsome, R. L. 1990. Natural and synthetic coloring agents. In: *Food Additives*, Branen, A. L., Davidson, P. M., Salminen, S. (Eds.). Marcel Dekker, New York, pp. 327–345.

Ni, Y., Bai, J., Ling, J. 1996. Simultaneous adsorptive voltammetric analysis of mixed colorants by multivariate calibration approach. *Anal. Chim. Acta* 329(1/2):65–72.

Noonan, J. 1972. Color additives in food. In: *Handbook of Food Additives*, 2nd ed., Furia, T. E. (Ed.). CRC Press, Cleveland, OH, pp. 587–615.

Oka, H., Ikai, Y., Ohno, T., Kawamura, N., Hayakawa, J., Harada, K., Suzuki, M. 1994. Identification of unlawful food dyes by thin-layer chromatography–fast atom bombardment mass spectrometry. *J. Chromatogr.* 674(1/2):301–307.

Patel, R. B., Patel, M. R., Patel, A. A., Shah, A. K., Patel, A. G. 1986. Separation and determination of food colours in pharmaceutical preparations by column chromatography. *Analyst* 111(5): 577–578.

Pavia, D. L., Lampman, G. M., Kriz, G. S., Jr. 1979. *Introduction to Spectroscopy*. Saunders College Publishing, Philadelphia.

Pliny. 1945. Book XIV. In: *Natural History, Vol. IV, Books XII–XVI*, Rackham, H., translator. Harvard University Press, Cambridge, MA, pp. 185–285.

Puttemans, M. L., Dryon, L., Massart, D. L. 1981. Ion-pair high performance liquid chromatography of synthetic water-soluble acid dyes. *J. Assoc. Off. Anal. Chemists* 64(1):1–8.

Puttemans, M. L., de Voogt, M., Dryon, L., Massart, D. 1984. Extraction of organic acids by ion-pair formation with tri-*n*-octylamine. Part 7. Comparison of methods for extraction of synthetic dyes from yogurt. *J. Assoc. Off. Analytical Chemists* 68(1):143–145.

Rayner, P. 1991. Colours. In: *Food Additive User's Handbook*, Smith, J. (Ed.). Blackie (AVI), New York, pp. 89–113.

Riboh, M. 1977. Natural colors: what works . . . what doesn't. *Food Eng.* 49(5):66–72.

Rodricks, J. V. 1996. Safety assessment of new food ingredients. *Food Technol.* 50(3):114,116–117.

Rowe, K. S., Rowe, K. J. 1994. Synthetic food coloring and behavior: a dose response effect in a double-blind, placebo-controlled, repeated-measures study. *J. Pediatrics* 125(5, pt. 1):691–698.

Schutz, H. G. 1954. Color in relation to food preference. In: *Color in Foods*, Farrell, K. T., Wagner, J. R., Peterson, M. S., Mackinney, G. (Eds.) National Academy of Sciences–National Research Council, Washington, D.C., pp. 16–21.

Scully, T. 1995. *The Art of Cookery in the Middle Ages*. The Boydell Press, Woodbridge, England.

Segal, M. 1993. Ingredient labeling: what's in a food? *FDA Consumer* 27(3):14–18.

Sekular, R., Blake, R. 1985. *Perception*. Knopf, New York.

Tannahill, R. 1988. *Food in History*, 2nd ed. Crown Publishers, New York.

Taylor, M. R. 1984. Food and Drug Administration regulation of color additives—overview of the statutory framework. *Food Drug Cosmetic Law Journal* 39(3):273–280.

TPC Business Group Research Report. 1992. *Food Additives: U.S. Products, Applications, Markets*. Technomic Publishing Co., Lancaster, PA.

Tuorila-Ollikainen, H. 1982. Pleasantness of colourless and coloured soft drinks and consumer attitudes to artificial food colours. *Appetite* 3(4):369–376.

Tuorila-Ollikainen, H., Mahlamäki-Kultanen, S., Kurkela, R. 1984. Relative importance of color, fruity flavor and sweetness in the overall liking of soft drinks. *J. Food Sci.* 49(6):1598–1600,1603.

Vettorazzi, G., Ed. 1981. *Handbook of International Food Regulatory Toxicology, Vol. 2: Profiles*. SP Medical & Scientific Books, New York.

von Elbe, J. H., Schwartz, S. J. 1996. Colorants. In: *Food Chemistry*, 3rd ed., Fennema, O.R. (Ed.). Marcel Dekker, New York, pp. 651–722.

Wissgott, U., Bortlik, K. 1996. Prospects for new natural food colorants. *Trends Food Sci. Technol.* 7(9):298–302.

Yeransian, J. A., Sloman, K. G., Foltz, A. K. 1985. Food. *Anal. Chem.* 57(5):278R–315R.

17

Natural Color Additives

YUAN-KUN LEE

National University of Singapore, Singapore

HWEE-PENG KHNG

PSB Corporation Pte. Ltd., Singapore

I. INTRODUCTION

Food colorants play a very important role in enhancing the aesthetic appeal of food. The practice of adding colors to food dates back to ancient times. One of the earliest records of the use of food colorants is that of the coloring of wine as early as 400 BC. Until the discovery of the first synthetic dye in 1856 by Sir William Henry Perkins, mankind had been relying on pigments extracted from plants, animals, and minerals as food colorants. Since then, due to the superior properties of synthetic dyes with respect to tinctorial strength, hue, and stability as well as their easy availability and low cost, synthetic dyes are used more extensively than the natural extracts in the coloring of food.

Unfortunately, stringent safety tests required for the synthetic food colorants have led to the prohibition of the use of some of the synthetic colors in food due to the discovery of possible toxic substances in them. On top of that, the less stringent tests necessary for the use of natural colorants and the increase in the demand for natural ingredients by increasingly health conscious consumers have led food manufacturers to take another look into the use of natural food colorants. This has also resulted in a proliferation of interest in the development of natural food colorants, as can be seen from the vast number of patents filed in recent years (1).

Due to the vast diversity of naturally occurring pigments that are suitable for use as food colorants, this chapter cannot fully review all the natural food colorants. Instead, it will focus on some of the common natural food colorants that are of biological origin and which are being used at the moment. Some novel but naturally occurring coloring materials, such as those of microbial origin and plant tissue culture, will also be examined. There will also be a short discussion on the legislation, toxicology, and future aspects of the natural food colorants.

II. NATURAL FOOD COLOURS OF BIOLOGICAL SOURCES

A. Pigments from Plant Sources

The plant kingdom, with its multitude of colors, generates vast interest among many researchers and is most widely studied as a major source of food colorants. Flavonoids, carotenoids, and chlorophyll are the major contributors to the natural colors of most plants, with betalines and curcumin playing a minor yet significant role.

1. Flavonoids

Anthocyanin, chalcone, and flavones belong to a group of compounds collectively known as flavonoids.

a. Anthocyanin. Anthocyanins are basically glycosides of anthocyanidins (aglycones). The six major anthocyanidins are illustrated in Fig. 1. The sugar moieties are usually attached to the anthocyanidins via the 3-hydroxyl or 5-hydroxyl positions and to a lesser extent the 7-hydroxyl position. The anthocyanin sugars may be simple sugars—the most common being glucose, galactose, rhamnose, and arabinose—or complex sugars such as rutinose and sambubiose (2). These sugar moieties may be acylated, the most common of which being phenolic acids such as coumaric acid and caffeic acids, and to a lesser extent p-hydroxybenzoic, malonic, and acetic acids (2,3).

Anthocyanins are the most established food colorants and may be found in a wide variety of edible plant materials, such as the skin of red apples, plums, and grapes, in

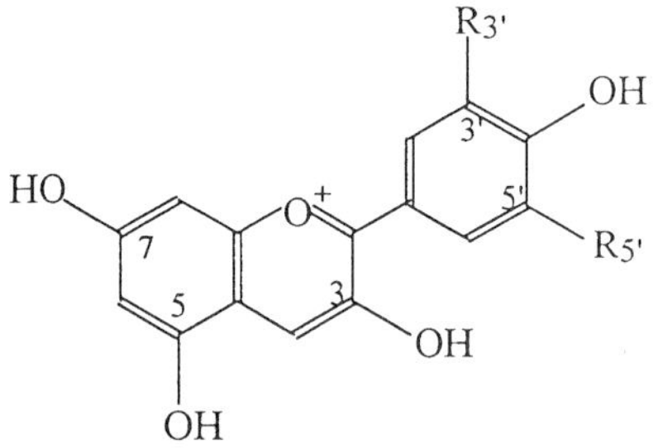

Pelargonidin : $R_{3'}=R_{5'}=H$
Cyanidin : $R_{3'}=OH$, $R_{5'}=H$
Peonidin : $R_{3'}=OMe$, $R_{5'}=H$
Delphinidin : $R_{3'}=R_{5'}=OH$
Petunidin : $R_{3'}=OMe$, $R_{5'}=OH$
Malvidin : $R_{3'}=R_{5'}=OMe$

Figure 1 Structure of some major anthocyanins.

addition to strawberries, red cabbage, and shiso (*Perilla ocimoidis* Varcripsa) leaves and blueberries. Concentrated or spray-dried juices of cranberries, raspberries, and elderberries have also been reported to be used as food colorants in certain food products (4). Meanwhile, extraction of anthocyanins are being carried out on the flowers of *Tibouchina grandiflora* (5) and *Clitoria ternata* (6), while the skin of black currant, elderberry, choke berry, and bilberry (7) are also being looked into as possible sources for the commercial extraction of anthocyanins. Extraction of anthocyanin generally involves the use of an alcoholic solvent (8). One of the conventional methods of extracting anthocyanin from macerated plant materials involves the use of low boiling point alcohols (such as methanol, ethanol, and *n*-butanol) which have been acidified with mineral acids like HCl.

The major source of anthocyanins is still the grape skin. It is, therefore, not surprising that nearly all the commercially available anthocyanins, known under the generic name of enocyanina, are obtained from the grape skin and other byproducts of the vine industry. Currently, Europe (in particular, Italy, France, and Germany) is the main producer of commercial anthocyanin using grape skins, producing about 50 tons annually. This is followed closely by the United States.

Application of anthocyanins in food is restricted due to their ability to participate in a number of reactions, resulting in its decolorization. These include reactions with ascorbic acids, oxygen, hydrogen peroxide, and sulfur dioxide to form colorless compounds; formation of complexes with metal ions and proteins; and hydrolysis of the sugar moieties to form unstable anthocyanidins. Anthocyanins are also sensitive to pH, being more stable at low pH (4). In addition, the anthocyanin colors vary with changes in the pH: at pH 1 and below, the anthocyanin pigment gives an intense red but becomes colorless or purple when the pH is increased to between 4 and 6. Meanwhile, the pigment turns a deep blue when the pH is between 7 and 8. Further increase in pH sees the anthocyanin pigment turning from blue to green and then to yellow. Such variation in color has been attributed to structural transformation in response to changes in pH, as illustrated in Fig. 2 (8). On the other hand, anthocyanins, which are only soluble in water or polar organic solvent, are fairly stable to heat. Studies on the effect of temperature on anthocyanin have indicated that the stability is dependent on the structure of anthocyanin, with the sugar moiety playing a significant role (9,10).

The stability of anthocyanin in the lower pH range means that anthocyanins are best suited for use in food of low pH. Anthocyanins are currently being used to provide a natural red or blue coloring for foodstuffs. Successful application of the anthocyanins includes the coloring of canned fruit, fruit syrups, yogurt, and soft drinks. Commercial anthocyanins have also been used to intensify the color of wine.

In view of the considerable consumption of anthocyanin, toxicological as well as mutagenic studies of the pigment have been carried out. Timberlake (11) reviewed the studies that have been done and concluded that anthocyanin is neither toxic nor mutagenic. On the other hand, anthocyanins were found to have beneficial therapeutic properties and would, therefore, find increasing application in not just the food area but in the medical field as well.

b. Chalcone. The chalcones are water-soluble pigments (Fig. 3) extracted from petals of safflower (*Carthamas tinctorius*). Red carthamin, safflor yellow A, and safflor yellow B are the three chalcones that have been isolated and identified in the safflower extract. At the moment, chalcone does not have many applications in the food industry. This is attributed to the very little information available on the stability of the color under various

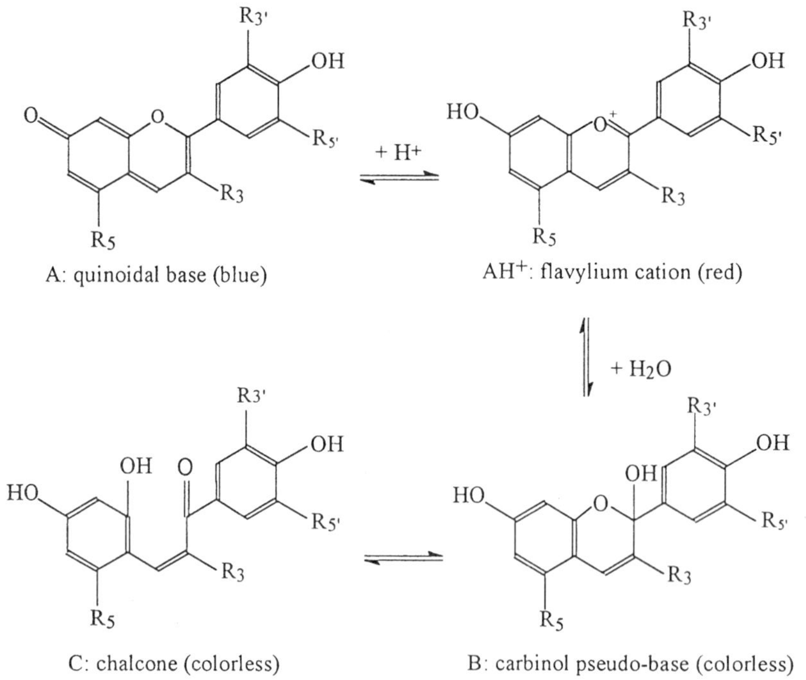

Figure 2 Structural transformation of anthocyanin.

conditions. However, it has been noted that chalcone color is relatively insensitive to pH changes, light, and microbial degradation. Other observations regarding changes in the color of chalcone indicate that it is affected by heating and contact with metal. Due to such limitations, applications of chalcone so far are restricted to foodstuffs such as noodles, yogurts, and fruit juices (e.g., pineapple juice).

2. Carotenoids

Carotenoids are noted for their great diversity and distribution. They can be found not only in plants (e.g., carrots, tomatoes, and capsicum) but also in bacteria, fungi, algae, and animals. To date over 500 carotenoids have been isolated and identified (12). The general structure of the carotenoids comprises a C_{40} hydrocarbon chain made up of eight isoprenoid units. Naturally occurring carotenoids exist mainly as the more stable *trans*

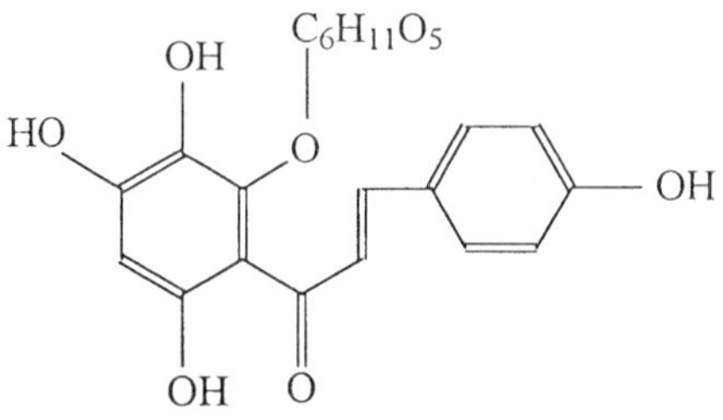

Figure 3 Structure of Safflower Yellow.

isomers. Diversity in the carotenoids arises as a result of additional substituents at both terminals of the hydrocarbon chain (Fig. 4). The extent of the double bonds in the hydrocarbon chain as well as the stereochemistry of the molecule affects the molecule's absorption spectrum and determines the color displayed by the different carotenoid compounds. A minimum of seven conjugated double bonds in the tetraterpenoid molecule is required before the carotenoid compound may have perceivable color (13). The presence of numerous double bonds in the carotenoid molecule also causes the compound to be very prone to oxidation, especially in the presence of light, enzymes, metals, and lipid hydroperoxides (14). Such reactions are believed to promote *trans–cis* isomerisation. Generally, however, carotenoids are relatively stable over a wide pH range and are fat soluble.

Beta carotene is the most abundant carotenoid in nature, particularly in plant materials. It is the major coloring principle in carrot and as well as palm oil seed extracts. The extracts are oil soluble and impart a yellow color to foods; they find applications in dairy products, cakes, soup, and confectionery. It is also a known fact that beta carotene is a precursor of vitamin A while possessing antioxidation properties which may help in the prevention of cancer and other diseases. This has resulted in the incorporation of beta carotene in health products, such as functional or nutraceutical beverages, with increasing usage being predicted in the future.

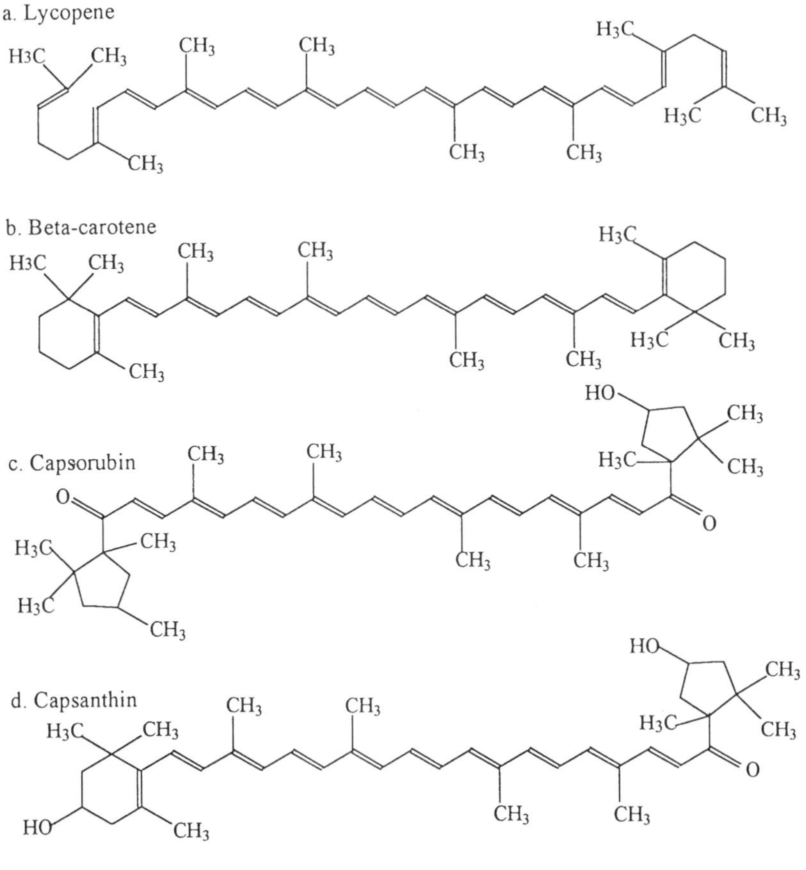

Figure 4 Structures of some common carotenoids.

Although present in a lesser amount than beta carotene, annatto, saffron, and gardenia extracts are the more commonly used carotenoids for coloring foodstuffs. Paprika, tomato, carrot, and palm oil seed have also been utilized for the extraction of carotenoids. The carotenoids are used to provide orange and yellow colors in food, particularly in fat-based food products.

a. Annatto. Annatto is an orange-yellow colored carotenoid derived from the pericarp of the seeds belonging to the shrub *Bixa orellana*, which can be found in tropical countries such as Brazil, Mexico, Peru, Jamaica, and India. It is estimated that every year about 7000 tons of the annatto seed are used for the production of the food colorant worldwide, with the main market being the United States and Western Europe.

Unlike other carotenoids, annatto is quite stable to pH changes and exposure to air, but moderately stable to heat. It is, however, unstable when exposed to strong light. The annatto color precipitates in acidic conditions and is also decolorized by sodium dioxide, as in the case of anthocyanin. Annatto is basically a mixture of two compounds, bixin and norbixin. Bixin, which is the mono-methyl ester of a dicarboxylic carotenoid, is the major component in the mixture (Fig. 5). It is also a fat-soluble compound which is extracted from annatto seeds with edible oil (14). Bixin, whose tinctorial strength is comparable to that of beta carotene, gives rise to an orange color which finds uses in dairy and fat-based food products, such as margarine, cheese, creams, and baked goods. The fat-soluble bixin is also used in conjunction with other food colorants to produce various shades of color. For example, it may be used with paprika oleoresin to give a redder shade in processed cheese. Alternatively, bixin may be combined with tumeric oleoresin to provide a much more yellow shade when required.

Alkaline hydrolysis of bixin yields the free acid norbixin (Fig. 5), which is also the minor component found in annatto. The water-soluble norbixin may also be extracted from the annatto seeds using aqueous alkali (14). Applications of norbixin include smoked fish, cheese, baked goods, meat products (e.g., frankfurter sausages), snack foods, and sugar confectionery.

b. Saffron. Saffron is one of the earliest food additives used by man. It is a water-soluble extract obtained from the stigma of the flowers of *Crocus sativus*. This plant is the major source for the commercial production of the pigment. The same pigment may also be obtained from the flowers of *C. albifloris*, *C. lutens*, *Cedrela toona*, *Nyctasthes arbortristes*, *Verbascum phlomoides*, and *Gardenia jasminoides*.

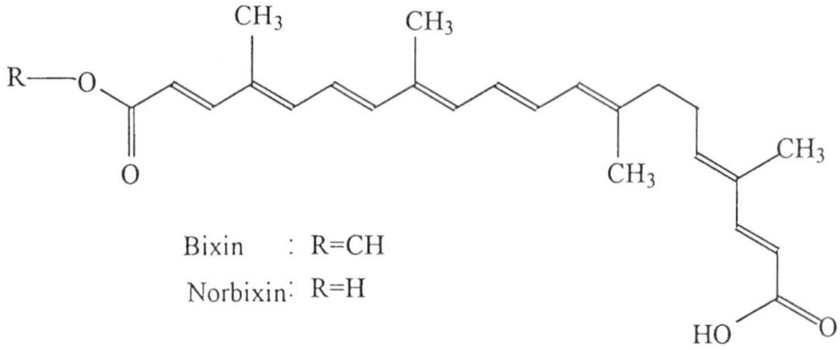

Figure 5 Structure of annatto pigments.

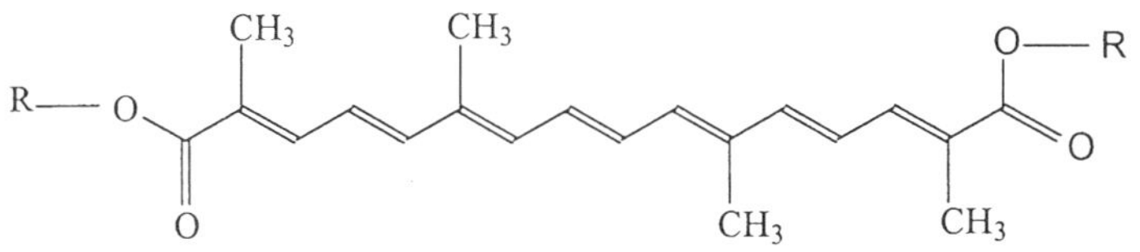

Crocin : R=gentiocide
Crocetin : R=H

Figure 6 Structure of saffron pigments.

Many studies have been carried out on the stability of saffron (15). Unlike the annatto extract, saffron extract is rather sensitive to pH changes and is prone to oxidation. It is, however, moderately resistant to heat. The saffron extract is made up of water-soluble crocin and fat-soluble crocetin. The major component of the saffron extract is crocin, which is the digentiobioside ester of crocetin (Fig. 6). Crocetin, like bixin, is a dicarboxylic carotenoid. In addition to crocin and crocetin, zeaxanthin, beta carotene and certain flavoring compounds (mainly picrocrocin and safranal) are also found to be present in saffron extract. The flavoring compounds impart a distinct spicy flavor, thus restricting the usage of saffron extract as a food colorant.

Generally, it takes about 140,000 stigma from the *Crocus* flowers to produce about 1 kg of saffron powder. Coupled with the high cost of production, it makes saffron one of the most expensive food colorants (at about US$1,000 per kilogram). In view of that, it is used sparingly not only as a colorant but as a spice as well. That is why it is usually added to foodstuffs, such as curry products, soups, meat, and certain confectionery goods, where a spicy flavor is desirable while at the same time to enhance the yellow color of these products.

c. Other Carotenoid Pigments. Lycopene is one of the numerous carotenoids that are being assessed for use as new food colorants. This carotenoid can be found in watermelon and red grapefruit and has an intense red color. However, tomato is the major source of this pigment. The apparent stability of lycopene in tomatoes during standard industrial tomato processing has prompted investigation into its potential as a new commercial natural colorant (16). Studies have shown that lycopene is soluble in aqueous solution and to a certain extent in nonpolar solution. Potential application of lycopene includes beverages, confectionery, boiled sweets, bread, and cakes.

Paprika extract is an orange-red oil-soluble extract obtained from the red pepper *Capsicum annum*. The color is insensitive to light but stable at high temperature. The main carotenoids found in this extract are capsanthin and capsorubin as well as beta carotene (Fig. 4). Like the saffron extract, it also contains some flavoring compounds which impart a characteristic pungency to the food. Paprika's characteristic color has seen use in coloring sauces, confectionery, salad dressing, meat products, sausages, and baked goods.

3. Betalains

Centrospermae, the plant order to which beet belongs, is the only group of plants known to produce betalains. Betalains can be divided into two classes of pigments, namely, betacyanins and betaxanthin. Betacyanin refers to the red pigment that may be extracted from the red beetroot *Beta vulgaris*. The major component in this class of pigment is betanin

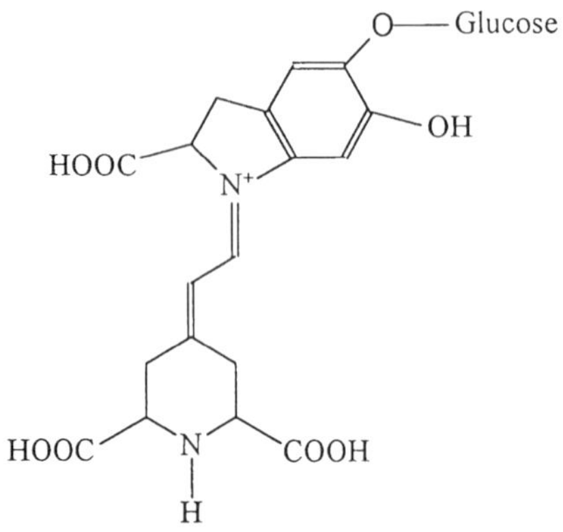

Figure 7 Betanin, a major component of betacyanin, which is extracted from the red beet root.

(Fig. 7). Betaxanthin, on the other hand, refers to the yellow pigment obtained from the yellow beet root *Beta vulgaris* var. *lutea*. The major components found in this class of pigment are vulgaxanthine I and II (Fig. 8).

Due to the high level of betalains found in beetroot, the plant is deemed a valuable source of food colorant. Commercial production of betalains involves a countercurrent liquid/solid extraction process. This is followed by an aerobic fermentation, generally with *Candida utilis*, to remove the large amount of sugar present (17,18).

Much work has been done on both betanin and vulgaxanthine in order to determine their suitability as food colorants. The stability of betanin with respect to pH, temperature, light, and air was studied by von Elbe et al. (19). Studies on the degradation rates of vulgaxanthine I with regard to pH, temperature, and oxygen were also carried out by Singer et al. (16). Both studies established that betanin and vulgaxanthine are most stable between pH 4.0 and 6.0. Both pigments are quite sensitive to air and relatively heat labile. Betanin was also found to be sensitive to light, since in the presence of light the degradation rate of the pigment was found to be increased by 15.6 $\pm$ 0.5% (20). As a result, both betanin and vulgaxanthine can only be used in foodstuffs with a short shelf-life as well as in those food products that do not undergo prolonged heat treatment.

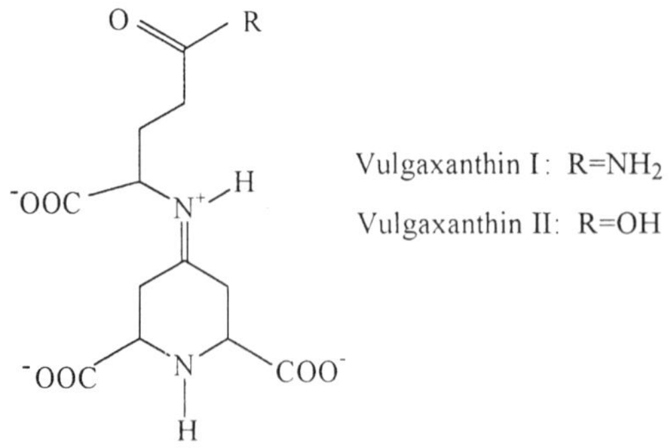

Figure 8 Vulgaxanthin I and II, major components of betaxanthin, which is extracted from the yellow beet root.

Nevertheless, application studies of betanin in selected food have suggested the presence of a protective effect on betanin from light and oxygen. This was evident based on data collected regarding color change in sausages, protein-gel, and soy protein containing added betanin (21). The protective effect has been attributed to the protein system present in the food. As a result, betanin is used mainly in food products with a high protein content, such as poultry meat sausages, soya protein products, gelatin dessert, and dairy products like yogurt and ice cream.

4. Chlorophyll

Chlorophyll is the green pigment found in all green plants as well as green alga. The pigment is responsible for the photosynthetic process in plants. Chlorophylls a and b are the two main types of chlorophyll pigment found in nature. The former is a bluish green pigment, while the latter is yellowish green in color. In addition, related pigments known as bacteriochlorophylls are found in photosynthetic bacteria.

The chlorophyll is a porphyrin pigment, made up of four pyrrole rings joined together via methine linkages (Fig. 9). There is also a magnesium atom within the center of the porphyrin structure, held in position by two covalent and two coordinate bonds. The magnesium can be easily released from the molecule through acid-catalyzed hydrolysis to give phaeophytin. However, its stability is increased when chlorophyll is subjected to hydrolysis in an alkaline condition. In addition to the magnesium atom, a 20-carbon monounsaturated alcohol, phytol, is also associated with the porphyrin molecule. This phytol side chain is responsible for the hydrophobic nature of the chlorophyll. Removal of phytol through hydrolysis yields chlorophylide, which has increased solubility in polar solvents.

The self-renewing nature of the sources of chlorophyll has generated much commercial interest due to its economic value. Commercial production of chlorophyll for use as a food colorant dates back to the 1920s. The current commercial output of chlorophyll stands at the estimated figure of 11×10^8 tons per year. The United Kingdom is thought to be the largest producer, accounting for about one-third of the world's output. Three-quarters of the plant materials that are used for chlorophyll extraction are of aquatic origin with the rest being terrestrial plants. However, most of the chlorophyll that is being used as a food colorant is obtained from land plants. Lucerne, alfalfa, and nettles are some of the popular plant materials being used. The chlorophyll pigments are usually extracted from dried plant materials using aqueous solvents, such as chlorinated hydrocarbons and

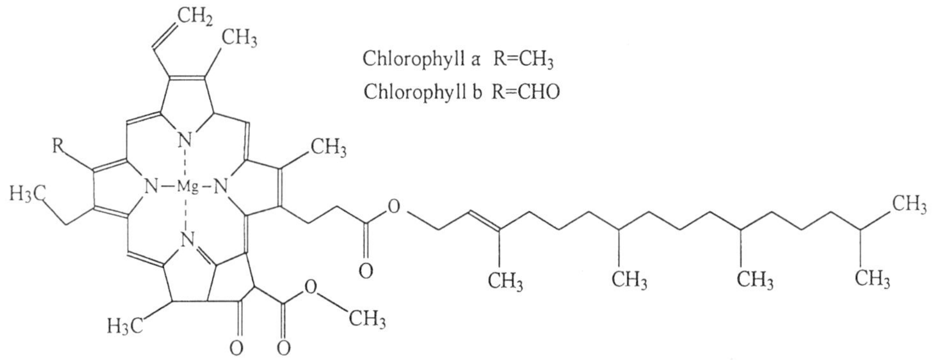

Figure 9 The pyrrole rings of chlorophyll.

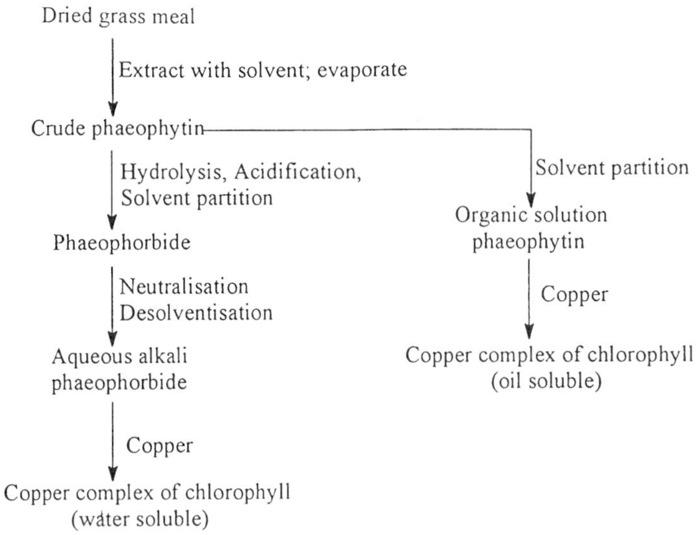

Figure 10 Commercial production of chlorophyll.

acetone (22). The resulting phaeophytin extract is then further processed to give a more stable copper complex. The general scheme for the commercial production of chlorophyll is as shown in Fig. 10.

Both the oil-soluble and water-soluble forms of chlorophyll are commercially available in the form of the stable copper complex. Both forms of commercial chlorophyll are relatively stable toward light and heat. However, unlike the water-soluble chlorophyll, the oil-soluble form is not very stable in acids and alkalis. A major portion of the commercial chlorophyll is used in the food industry for coloring dairy products, edible oil, soups, chewing gum, and sugar confectionery. It is mainly added to fat-based food, particularly canned products, confectionery, and pet foods. The pharmaceutical and cosmetic industries also apply chlorophyll to some of their products.

5. Miscellaneous Plant Pigments

Turmeric is a fluorescent yellow colored extract obtained from the root of the curcuma plant, with *Curcuma longa* being the important commercial source. Traditional use of tumeric involves grinding the tuber into powder and adding it to the food as a spice rather than as a coloring agent. The tumeric extract actually comprises three pigments: curcumin, demethoxycurcumin, and bisdemethoxycurcumin. The major pigment is curcumin, which is insoluble in water. However, it has been reported that a water-soluble complex may be obtained by reaction the pigment with metals such as zinc chloride (22). The major disadvantage of using tumeric or curcumin is that they impart a characteristic odor and sharp taste to the foodstuff to which they are applied. On top of that, curcumin is not stable in the presence of light and alkaline conditions. Consequently, uses of tumeric or curcumin extracts are rather restricted. However after a deodorization process, the odorless commercial extract generally finds application in food products such as soups, mustard, pickles, confectionery, and canned products. Its acid stability has also found application in salad dressing and some other food products with low pH.

B. Pigments from Microbial Sources

Microorganisms are known to produce a variety of pigments, namely, chlorophyll, carotenoids, and some unique pigments, and are, therefore a promising source of food colorants. Use of microorganisms as a source of food colorants also has the added advantages of rapid growth and ease of control.

Of all the microorganisms, the Monascus pigments and algae are perhaps the most widely studied for their potential as a source of food colorants.

1. Monascus Pigments

Monascus pigments have been used in the Orient, particularly China, Japan, Indonesia, and the Philippines, to color food for centuries. Chinese red rice wines, red soybean cheese, pickled vegetables, fish, and salted meats are some examples of Oriental food colored red by the Monascus pigment. The main source of the Monascus pigments is the fungus *Monascus purpureus*, and the traditional method of pigment production involves the growth of the fungus on solid medium such as steamed rice (23). The resulting mass is then dried and ground to powder to be used as a colorant.

However, Lin (24) observed that *Monascus* sp. was able to produce the pigment in a submerged culture. Rice powder (24) and tapioca starch (25) were found to be the suitable carbon source for maximum pigment production with the optimum ratio of carbon source to nitrogen source being 5.33:7.11 (24). Since then, many studies have been devoted to the optimization of cultural conditions for maximum pigment yield, particularly the method of submerged fermentation of the fungus (26–29). Novel methods of pigment production using roller bottles and solid–liquid state fermentation were also suggested.

The Monascus pigments are essentially a mixture of pigments whose major components are the orange monascorubrin and rubropunctatin, the red monascorubramine and rubropunctamine, as well as the yellow monascin and ankaflavin. Their structures are shown in Fig. 11.

The native Monascus pigments generally exist in the orange-yellow insoluble form, which is bound to the membrane of the fungal mycelial. Extraction of the pigments would involve the use of organic solvents, such as methylene chloride and methanol, thereby making the pigments undesirable for use in food. Attempts in producing water-soluble pigments involve the reaction of the extracted pigments with water-soluble proteins to form the water-soluble red pigments. Solubilization of the Monascus pigments has been attributed to the formation of a complex that involves the substitution of an oxygen atom by the nitrogen of the amino group belonging to the water-soluble protein through a ring opening and a Schiffs rearrangement reaction, as shown in Fig. 11. Moll et al. (31) recommended the use of chitosan to produce a water-soluble red colorant, as it is not metabolized by man and is abundant in nature (31). A method for the direct aqueous extraction of water-soluble red Monascus pigments has also been developed by Chen (32). The Monascus pigments are relatively stable to heat treatment and are able to withstand the autoclave process. The pigments are also stable in pH ranging from 3 to 10. The pigments may be freeze-dried or spray-dried and may be considered for use in protein-rich foods such as meat (33), sausages (34), processed seafoods, milk, and baked goods.

2. Algal Pigments

Beside chlorophyll, of which Chlorella is one of the major producers, algae are also known to produce a group of pigments known as biliprotein or phycobiliprotein. This group of

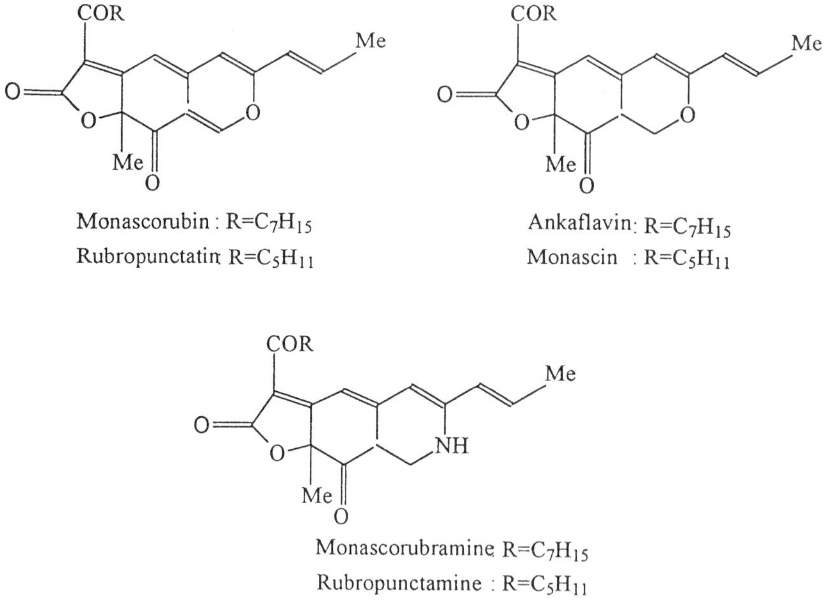

Figure 11 Structural variation of the different Monascus pigments.

pigments is produced mainly by the red algae (Rhodophyta), blue-green algae (Cyanophyta), and the cryptomonad algae (Cryptophyta). The biliproteins may further be divided into two groups: the red phycoerythrins and the blue phycocyanins. The bilin portion of the biliprotein is made up of four pyrrole rings linked together forming an open chain. The tetrapyrrole is, in turn, bound to an apoprotein by means of one or two thioether linkages (Fig. 12).

Both phycoerythrins and phycocyanins are soluble in water. Studies carried out by Arad et al. (35) showed that the biliproteins are more stable in pH ranging from 5 to 9 and tend to precipitate at lower pH. However, stability of the pigments at low pH could be achieved by subjecting the pigments to hydrolytic reaction using proteolytic enzymes such as pronase (35). While pigments extracted from the normal algae are relatively sensitive to heat, pigments obtained from thermophilic algae are quite stable to heat.

At the moment, a blue colorant extracted from the blue-green algae *Spirulina platensis*, known as ''Lina blue,'' is being produced commercially by the Japanese company Dainippon Ink and Chemicals Inc. The Spirulina, with a protein content of 60 to 70%, has been a part of the African and Mexican diets for centuries. With its increasing popularity, clinical studies have been carried out by Dainippon Ink and Chemicals Inc. The studies include acute toxicity and pharmacological studies, as well as chronic toxicity of Spirulina and effects of Spirulina on photoallergic sensitivity. So far, no adverse effects on Spirulina have been reported.

Uses of the colorant in chewing gum, soft drinks, alcoholic drinks, and fermented milk products such as yogurt, have been patented in Japan. Other applications of the biliprotein includes confectioneries, candied ices, and sherbets.

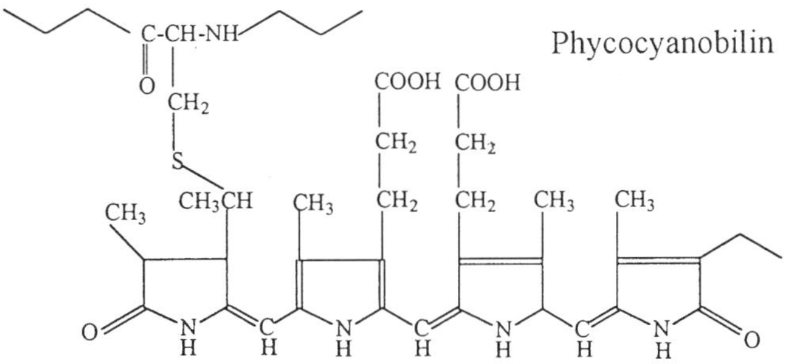

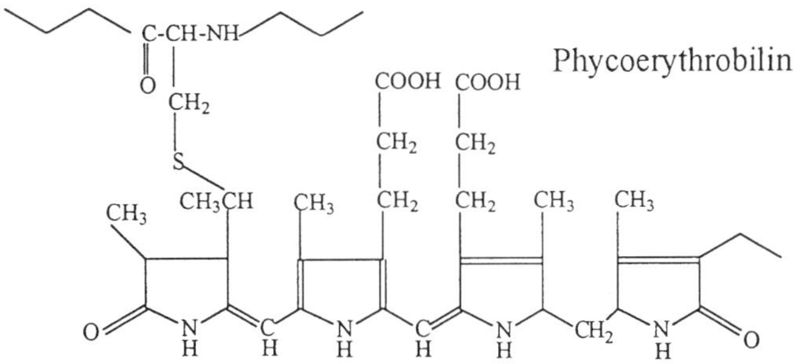

Figure 12 The biliproteins found in alga.

C. Pigments from Animal and Insect Sources

1. Cochineal

Cochineal is an anthraquinone which forms part of a group of pigments known as the quininoid pigments. It has been used for centuries as a red colorant, mainly for dyeing textiles. The cochineal pigment actually refers to various red pigments obtained from female coccid insect. Different species of coccid insect are associated with different cochineal colors. The most well known cochineal pigment is carminic acid which comes from the female *Dactylopius coccus Costa*, a parasite of the cactus plants belonging to the Opuntia and Nopalea genera, particularly *N. cochenillifera*. The latter is a native of Central and South America. Other cochineal-related pigments that are also extracted from insects are

1. Armenian red—obtained from the coccid insect *Porphyrophyra hameli*, which is found growing on the lower stems and roots of grasses found in wet and alkaline regions of Azarbaijan and Armenia.
2. Kermes—extracts of *Kermes lilcis*, an insect which grows on species of Quercus.
3. Polish cochineal—extracted from the insect *Margarodes polonicus*, which inhabits the roots of *Scleranthus perennis*, a plant native to Central and Eastern Europe.
4. Lac dyes (laccaic acid)—produced by the lac insect, otherwise known as *Laccifera lacca*, which can be found in the trees *Schleichera oleosa, Zizyphus maure-*

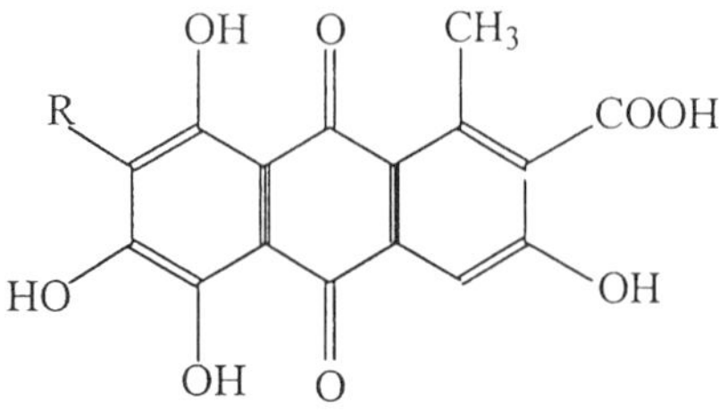

Figure 13 Carminic acid.

tania, and *Butea monosperma* in the Indian subcontinent, China, and Malaysia. It has been used traditionally by the Chinese to color agar.

In most cases, the female insects are harvested by hand when they reach sexual maturity. The insects collected are then dried in the open air. Conventional extraction of the cochineal pigment involves the treatment of the dried bodies with hot water. However, treatment of the ground insect bodies with proteolytic enzymes in a suitable surfactant has been proposed. This is then followed by a purification step using ion-exchange chromatography. Better extraction yield of the pigment has been reported with this method (36).

The principal cochineal pigment that is being used industrially is carminic acid (Fig. 13). On its own, carminic acid has little intrinsic color at pH 7. However, it is able to complex with various metals to produce a bright red color. Aluminium is the metal usually used in the commercial preparation of the carminic acid complex. The resulting product is known as carmine. The intense red color of carmine makes it a popular coloring agent for jams, syrups, preserves, confectionery, and baked goods. By varying the ratio of carminic acid to aluminium, one can also obtain a range of colors, ranging from pale ''strawberry'' to near ''black currant'' (36). The soluble form of carmine is obtained through the treatment of the complex with ethanol, while the nonsoluble form is obtained by the addition of calcium salt to the final solution.

2. Heme Pigments

Like chlorophyll, heme also possesses four pyrrole rings. However, it differs from chlorophyll in that the central magnesium atom has been replaced by and iron atom. Heme is most abundant throughout the animal kingdom where it tends to associate with proteins forming complexes, such as the myglobin in the muscle and the hemoglobin in the blood (Fig. 14). It functions as an oxygen carrier within the animal bodies. When the central iron atom is oxidized, as in oxygenated blood, the complex is bright red in color. Upon heating, however, the oxygen atom which is loosely bound will be lost, giving rise to a brownish color, characteristic of cooked meat.

Various extraction methods of heme from the protein complexes have been reported. Generally, it involves the use of a mixture of organic solvent and acid, such as ether and acetic acid or ethyl acetate and acetic acid. A yield of up to 80% for bulk preparations of heme using a mixture of acetone and acetic acid has been reported (37). In order to preserve the red color in heme pigment, other ligands are used to substitute the less stable oxygen atom. Ligands suggested include imidazole, S-nitrosocysteine, carbon monoxide, various amino acids, and nitrite (38). Alternatively, the central iron atom in the heme structure may be replaced with a more stable metal.

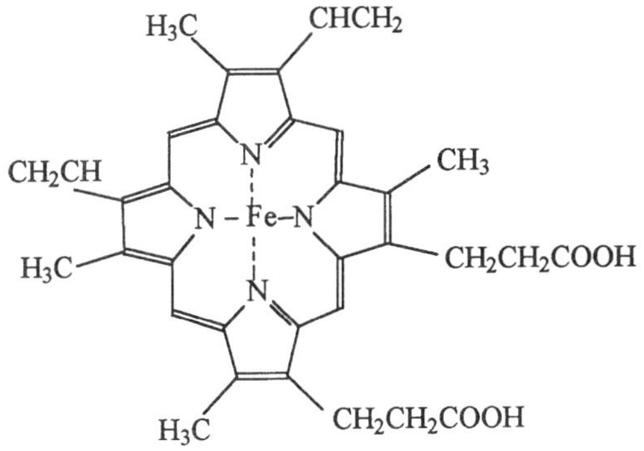

Figure 14 Structure of heme.

A preliminary toxicity study carried out on animals had indicated that the heme pigments are harmless. Nevertheless, the characteristic color of the heme pigments had restricted their use to food products in which the color of cooked meat is desired, for example, in sausages and meat analogs.

II. ANALYTICAL METHODS

Assaying of the natural colorants is made more difficult due to the presence of a mixture of the pigments and their derivatives in the extract. In addition, suitable standards are usually not available for quantifying purposes. Nevertheless, the same analytical methods, namely, spectrophometric and chromatographic assays, that are used for the analysis of the synthetic pigments may also be applied to the analysis of the natural colorants.

A. Spectrophotometric Analysis

This is the most fundamental means of analysis, involving the measurement of light absorption by the sample at its maximum wavelength, usually in the ultraviolet and visible light region. Scanning and automatic baseline adjustment are some of the recent developments in spectrophotometers that facilitate the spectrophotometric assay of the natural colorants. Anthocyanins, annatto, chlorophyll, and Monascus pigment are some of the natural pigments that have been assayed spectrophotometrically. The use of mathematically derived equations in some cases, such as for chlorophyll, has resulted in the accurate quantification of the total and individual chlorophylls (39).

B. Chromatographic Analysis

Unlike spectrophotometric analysis, chromatographic analysis enables one to separate and identify the individual components in a mixture, which is common in natural pigments, with their mixture of derivatives. Paper chromatography and thin-layer chromatography are widely used because of their cost-effectiveness. Thin-layer chromatography, in particular, has been used for the analysis of annatto and chlorophyll (13,39). Unfortunately, both paper chromatography and thin-layer chromatography are rather time consuming. In addition, thin-layer chromatography involves the use of inorganic absorbents, such as silica gel, which might cause chemical alternation in some of the organic compounds.

A more rapid and sensitive method would be high-performance liquid chromatography (HPLC). This quantitative method has a high resolution with a short analysis time. Coupled to a photodiode array detector, one can easily characterize the spectral properties of the sample. Hong et al. (40) described a systematic approach in the use of HPLC and an on-line photodiode array detector for the separation and detection of anthocyanins in cranberry, roselle, and strawberry.

III. LEGISLATION

In the United States, legislation concerning the use of colors in foods is covered by the Federal Food Drug & Cosmetic Act of 1938 and the Color Additives Amendments Act of 1960. Under the American legislation, food colorants are divided in to two main groups: certified and uncertified colors. Natural food colorants, which have a history of safe use in food, are classified as uncertified colors and are thus exempted from certification (Table 1). However, clearance from the FDA would still be required before any of the natural pigments could be used in food products. Generally, the colorants must be accorded the GRAS (generally recognized as safe) status by the commissioner before they are approved for use. A complete listing of natural food colorants with GRAS status may be found in the CFC handbook (41). On top of that, one must still follow a set of guidelines regarding the applications of these colorants (42).

The development of food colorant legislation in the rest of the world followed the same pattern as that in the United States. Nevertheless, the list of permitted natural food

Table 1 Uncertified Natural Color Additives Permanently Listed for Use in Food and Feed in the United States (1993)

Color additive	Restriction
Algal meal	In chicken feed only
Annatto extract	General colorant (ξ73.30, ξ73.1030, ξ73.2030)
β-Apo-8′-carotenal	15 mg/lb
β-Carotene	—
Beet powder	General food colorant (ξ73.40, ξ73.260)
Canthaxanthin	30 mg/lb 4.4 mg/lb (chicken feed)
Caramel	GRAS (ξ182.1235, ξ73.85, ξ73.1085)
Carrot oil	GRAS (ξ182.20, ξ73.300)
Cochineal extract (carmine)	In chicken feed only
Corn endosperm oil	—
Partially defatted, cooked, or toasted cottonseed flour	—
Fruit juice	—
Grape color extract	For nonbeverage foods only
Grape skin extract (enocianina)	For still and carbonated drinks and ales, beverage bases, and alcoholic beverages with specific restriction (ξ73.170)
Paprika and paprika oleoresin	GRAS (ξ182.20)
Saffron	GRAS (ξ182.10, ξ182.20)
Tumeric and tumeric oleoresin	GRAS (ξ184.1351)
Vegetable juice	—

Table 2 EEC Scientific Committee for Food: Classification of Food Colors

Group 1 Colors with established ADI[a]	Group 2 Colors with temporary ADI[a]	Group 3 Colors with no assigned ADI[a] (natural source)	Group 4 Colors prohibited in food
β-Apo-88′ carotenal	Amaranth	Anthocyanins,	Alkanet
β-Apo-88′ carotenic acid ethyl ester	Annatto extracts	Beet red	Allura Red AC
β-Carotene	Azorubine (Carmoisine)	Chlorophyll	Black 7984
Chlorophyllin copper complexes	Brilliant Black BN	Curcumin	Burnt Umber
Erythrosine	Brilliant Blue FCF	Lycopene	Chocolate Brown FB
Indigotine	Brown FK	Mixture of α-, β-, τ-tocopherols	Chrysoine S
Iron oxides and hydrated iron oxides	Caramel	Carotenes	Fast Red E
Red 2G	Chocolate Brown HT	Xanthophylls	Fast Yellow AB
Sunset Yellow FCF	Food Green S		Indanthrene Blue RS
Tartrazine	Patent Blue V	*For some alcoholic beverages*: cochineal and carminic acid	Orange G
	Ponceau 4R		Orange GGN
	Quinoline Yellow		Orange RN
	Yellow 2G		Orchil and Orcein
			Ponceau 6R
			Scarlet GN
			Violet 6B

[a] ADI = Acceptable Daily Intake (mg/kg).

colorants differ from country to country. For example, besides those natural colorants listed in the CFC handbook, the EEC also approves the use of other natural pigments in food (43). These include curcumin, erythrosine, chlorophyll, and lycopene. This discrepancy is due to the absence of definite specifications of many natural pigments as a result of large variation in natural sources.

An attempt is therefore being made to create a worldwide permitted list of food colorants. This involves rationalization of food colors legislation by the joint FAO/WHO Codex Alimentarius Commission. Since 1975, the EEC Scientific Committee for Food (SCF) has also undertaken a toxicological review of all permitted colors of member states. Based on the data collected, natural as well as synthetic colors are divided into four groups based on the concept of acceptable daily intake (ADI) (Table 2).

IV. FUTURE PROSPECTS

In a 1987 survey carried out by the Commission of the European Communities, 63 experts from 9 countries were interviewed. It was predicted that by the year 2000 about 50% of the synthetic additives would be replaced by natural additives, with the bulk being synthetic colors and flavors (44).

At the moment, the use of synthetic food colorants is still very much favored by most food manufacturers. One of the reasons being the high cost of natural colorants due to the low yield as compared to the synthetic food colorants. The superior properties of the latter, such as the greater tinctorial strength and stability, also contribute to the greater demand for synthetic colorants at the moment. Other major obstacles of the natural food colorants are their variability due mainly to plant variety as well as changes in weather and soil conditions and the uncertainty in their supply. Seasonal variation of the pigment sources also means that a constant supply would be difficult.

Those pigments obtained from plant materials are usually secondary metabolites produced by the plant cells. With the emergence of biotechnology, it is believed that plant tissue culture and fermentation could offer possible solutions (45,46) to the problems encountered in the production of natural food colorants. Both methods offer several advantages over the extraction of pigment from whole plants: They are more reliable and predictable as the processes are not prone to seasonal variation. The quality of the pigments are improved as they are no longer affected by strain variation and changes in weather conditions.

Natural pigments obtained by tissue culture include anthocyanin, anthraquinone, and shikonin. However, until recently, commercial production via tissue culture has been plagued with several setbacks. The main obstacle being the relatively low yield by such methods, thus rendering it economically nonviable. The low yield is due mainly to the slow growth rate of the plant tissue culture and the low production potential of the parent plant (47,48). Nowadays, with better understanding of the biosynthetic pathway of these natural pigments within the plant, improved reactor design, and advances in genetic engineering, one can actually improve the yield of these pigments.

Currently shikonin, a red dye, is produced commercially for cosmetics use by the Japanese company Mitsui Petrochemical Ltd. by plant tissue culture. At present one can optimize plant tissue culture cells to produce 845 times more shikonin than that from the plant roots (46,49). Meanwhile, Pirt et al. developed a bioreactor that is capable of cultivating large amount of microalgae for the production of beta carotene for use as a food colorant (50). In comparison to the traditional open ponds, the use of the bioreactors was

Table 3 Properties of Major Natural Food Colors

Color (pH of solution)	Name of pigment	Solubility			Stability					Tinctorial strength
		Water	Alcohol	Fat	Light	Heat	Salt	Microbe	Metal	
Yellow	Gardenia yellow	VG	G	VP	P	P	G	G	G	VG
Yellow	Safflower yellow	VG	G	VP	G	P	G	G	P	P
Yellow-orange	Annatto	P	G	VP	G	P	G	G	P	P
Yellow-orange	Carrot (β-carotene)	VP	P	VG	P	G	G	G	G	VP
Yellow-orange	Paprika	VP	P	VG	P	G	G	G	G	VP
Orange (acid)-red (neutral)-red purple (alkaline)	Cochineal	VG	G	VP	VG	VG	VG	GG	VP	P
Orange (<3)-red	Beet root	VG	G	VP	P	VG	G	G	G	P
Orange red-red (<3 not soluble)	Monascus red	G	VG	VP	P	G	G	G	G	VG
Orange red (acid)-red (neutral)-red purple (alkaline)	Lac	P	VG	VP	VG	VG	VG	G	VP	P
Red (3)-red purple (4–6)-blue (8)	Cora	VG	G	VP	G	G	G	G	VP	
Red (3)-red purple (4–6)-blue (8)	Berry	VG	G	VP	G	G	G	G	VP	P
Red (3)-red purple (4–6)-blue (8)	Shiso	VG	G	VP	G	G	G	G	VP	P
Purple red (acid)-red purple (alkaline)	Shikon	P	VG	—	VG	VG	VG	G	VP	P
Purple red (acid)-red purple (alkaline)	Red cabbage	VG	G	VP	G	G	G	G	VP	P
Purple red (3)-red purple (4–6)-blue (8)	Grape skin	VG	G	VP	G	G	G	G	VP	P
Purple red (3)-red purple (4–6)-blue (8)	Grape juice	VG	G	VP	G	G	G	G	VP	P
Brown	Cacao	VG	G	VP	VG	VG	VG	G	G	VG
Brown (.3 not soluble)	Kaoliang	VG	G	VP	VG	VG	VG	G	G	VG
Green (.5 brown)	Chlorophyll	VG	G	VG	VP	G	G	G	G	P
Green (alkaline)-brown	Tumeric	P	VG	G	P	P	P	G	P	VG
Blue	Gardenia blue	VG	P	VG	P	G	G	G	G	P
Blue	Spirulina	VG	G	VP	P	G	G	G	G	P

Notes: VG = very good; G = good; P = poor; VP = very poor.

able to reduce the cost of biomass recovery since a smaller volume of liquid was being processed.

Table 3 summarizes the properties of major natural food colorants. It is obvious that no single colorant is stable at the whole range of food preparation and processing conditions. Industrial users would have to choose a suitable colorant based on the physical and chemical properties of the food product and the methods for processing and preserving the product.

To summarize, there is a growing trend toward the use of natural colorants for food. The shift will not occur overnight but will rather be a slow process. The major obstacle is the economics of these natural colorants. Much work is still needed especially in increasing the product yield, before the natural food colorants can be economically viable.

REFERENCES

1. Francis, F. J. 1987. Lesser known food colours. *Food Technol.* 41(4):62.
2. Timberlake, C. F., Bridle, P. 1980. Anthocyanins. In: *Developments in Food Colors*, Walford, J. (Ed.). Applied Science Publishers, London, p. 115.
3. Francis, F. J. 1992. A new group of food colorants. *Trends Food Sci. Technol.* 3(2):27.
4. The Herald Organization. 1993. *Science, Applications & Marketing of Natural Colors for Foods, Confectionery & Beverages*, Trimester International S.I.C. Publishing, Hamden, CT.
5. Bobbio, F. O., Bobbio, P. A., Degaspari, C. H. 1985. Anthocyanins from Tibouchina grandiflora. *Food Chem.* 18:153.
6. Srivastava, B. K., Pande, C. S. 1977. Anthocyanins from the flowers of Clitoria ternata. *Plant Med.* 32:138.
7. Wilskajeszka, J. 1991. Anthocyanins as natural food colorants. *Intl. Food Ingredients.* 3:10.
8. Timberlake, C. F. 1980. Anthocyanins—occurrence, extraction and chemistry. *Food Chem.* 5:69.
9. El-Fadeel, M. G. A. 1993. Anthocyanins and Betacyanins as natural red colors. Proceedings from the Natural Colors for the Food Industry Conference and Exhibition, Ismailia, Egypt, pp. 40–48.
10. Dougall, D. K., Baker, D. C., Gakh, E., Redus, M. 1997. Biosynthesis and stability of Monoacylated Anthocyanin. *Food Technol.* 51(11):69.
11. Timberlake, C. F. 1988. The biological properties of Anthocyanins. *Natcol. Quaterly Information Bulletin.* 1:4.
12. Weedon, B. C. L., Moss, G. P. 1995. *Carotenoids*, Vol. 1A: *Isolation and Analysis.* Briton, G., Liauen-Jensen, S., Pfander, H. (Eds.). Birkhäuser, Basel, p. 27.
13. Rodriguez-Amaya, D. B. 1993. Carotenoids—properties and determination. In: *Encyclopaedia of Food Science, Food Technology and Nutrition.* Macrae, R., Robinson, R. K., Sadler, S. J. (Eds.). Academic Press, London, p. 707.
14. Preston, H. D., Rickard, M. D. 1980. Extraction and chemistry of Annatto. *Food Chem.* 5: 47.
15. Francis, F. J. 1996. Less common natural colorants. In: *Natural Food Colorants.* Hendry, G. A. F., Houghton, J. D. (Eds.). Blackie Academic and Professional, Glasgow, p. 325.
16. Nir, Z., Hartal, D., Raveh, Y. 1993. Lycopenes from tomatoes—a new commercial natural carotenoid. *Intl. Food Ingredients* 6:45.
17. Harmer, R. A. 1980. Occurrence: chemistry and application of betanin. *Food Chem.* 5:81.
18. Adams, J. P., von Elbe, J. H., Amundson, C. H. 1976. Production of a betacyanine concentrate by fermentation of red beet juice with *Candida utilis. J. Food Sci.* 41:78.
19. von Elbe, J. H., Maing, I. Y., Amundson, C. H. 1974. Color stability of betanin. *J. Food Sci.* 39:334.

20. Singer, J. W., von Elbe, J. H. 1980. Degradation rates of vulgaxanthine I. *J. Food Sci.* 45: 489.
21. von Elbe, J. H. Klement, J., Amundson, C. H., Cassens, R. G., Lindsay, R. C. 1974. Evaluation of betalain pigments as sausage colorants. *J. Food Sci.* 39:128.
22. Humphrey, A. M. 1980. Chlorophyll. *Food Chem.* 5:57.
23. Chinese soy sauce, pastes and related fermented foods. In: *Handbook of Indigenous Fermented Foods*, 1978. Steinkraus, K. H. (Ed.). Marcel Dekker, Inc., New York, p. 433.
24. Lin, C. F. 1973. Isolation and cultural conditions of *Monascus* sp. for the production of pigment in a submerged culture. *J. Ferment. Technol.* 51(6):407.
25. Lee, Y. K., Chen, D. C., Chauvatcharin, S., Seki, T., Yoshida, T. 1995. Production of *Monascus* pigment by a solid-liquid state culture method. *J. Ferment. Bioeng.* 79(5):516.
26. Yoshimura, M., Yamanaka, S., Mitsugi, K., Hirose, Y. 1975. Production of Monascus-pigment in a submerged culture. *Agr. Biol. Chem.* 39(9):1789.
27. Shepherd, D., Carels, M. S. C. 1979. Red pigment production. U.S. Patent 4,145,254.
28. Su, Y. C., Huang, J. H. 1980. Fermentative production of Anka-pigments (Monascus-pigments). *Proc. Natl. Sci. Counc. ROC* 4(2):201.
29. Lee, Y. K., Lim, B. L., Ng, A. L., Chen, D. C. 1994. Production of polyketide pigments by submerged culture of *Monascus*: effects of substrates limitation. *A.-P. J. Mol. Biol. Biotech.* 2(1):21.
30. Mak, N. K., Fong, W. F., Wong-Leung, Y. L. 1990. Improved fermentative production of Monascus pigments in roller bottle culture. *Enzyme Microb. Technol.* 12:965.
31. Moll, H. R., Farr, D. R. 1976. Red pigment and process. U.S. Patent 3,993,789.
32. Cheng, D. C. 1994. Fermentation production of red pigment by *Monascus*. M.S. Thesis, Department of Microbiology, National University of Singapore.
33. Nagawa, N., Watanabe, S. 1979. Meat-coloring agent. *Chem. Abstr.* 79:11374.
34. Wasilewski, S. 1979. An attempt to use monascorubrin as sausage colorant. *Proceedings Eur. Meet. Meat Res. Workers* 25(12.2):1.
35. Arad, S. M., Yaron, A. 1992. Natural pigments from red microalgae for use in foods and cosmetics. *Trends Food Sci. Technol.* 3(4):92.
36. Lloyd, A. G. 1980. Extraction and chemistry of cochineal. *Food Chem.* 5:91.
37. Houghton, J. D. 1996. Haems and bilins. In: *Natural Food Colorants*, Hendry, G. A. F., Houghton, J. D. (Eds.). Blackie Academic and Professional, Glasgow, p. 157.
38. Taylor, A. J. 1984. Natural colors in food. In: *Developments in Food Colors*, Walford, J. (Ed.). Elsevier Applied Science Publishers, London.
39. Daood, H. G. 1993. Chlorophyll. In: *Encyclopaedia of Food Science, Food Technology and Nutrition*, Macrae, R., Robinson, R. K., Sadler, S. J. (Eds.), Academic Press, London, p. 904.
40. Hong, V., Wrolstad, R. E. 1990. Use of HPLC separation/photodiode array detection for characterization of Anthocyanins. *J. Agric. Food Chem.* 38(3):709.
41. CFR. 1993. Listing of color additives exempt from certification. *Code of Federal Regulations—Food and Drugs*, Vol. 21, Part 73.
42. CFR. 1993. Substances generally recognised as safe. *Code of Federal Regulations—Food and Drugs*, Vol. 21, Part 182.
43. Agra Europe. 1997. Additives. In: *EuroMonitor: European Union Legislation on Foodstuffs.* Agra Europe, Kent, UK.
44. Paulus, K. 1989. Necessary actions of the industry for successful food technology in the year 2000. In: *Proceedings of the International Minisymposium on Food Technology in the Year 2000*, Lindroth, S., and Ryynänen, S. S. I., (Eds.), Helsinki, Finland, p. 67.
45. Ilker, R. 1986. In-vitro pigment production: an alternative to color synthesis. In: *Modern Technologies in Food Colorants.* Proceedings of the 46th Annual Meeting of the Institute of Food Technologists Symposium, Texas.
46. Stafford, A. 1991. The manufacture of food ingredients using plant cell and tissue cultures. *Trends Food Sci. Technol.* 2(5):116.

47. Dörnenburg, H., Knorr, D. 1997. Challenges and opportunities for metabolite production from plant cell and tissue cultures. *Food Technol.* 51(11):47.
48. Frenkel, D. H., Dorn, R., Leustel, T. 1997. Plant tissue culture for production of secondary metabolites. *Food Technol.* 51(11):56.
49. Singh, G. 1997. Reactor design for plant cell culture of food ingredients and additives. *Food Technol.* 51(11):62.
50. Pirt, J. 1989. Mass transformation. *Food Flavorings Ingredients Processing & Packaging* 11(5):34.

18

Antioxidants

J. BRUCE GERMAN

University of California, Davis, California

I. INTRODUCTION

A. Lipids in Foods

The majority of fat in foods is present as saturated, monounsaturated, or polyunsaturated species of fatty acids esterified into triacylglycerides, also known as triglycerides. As such, fat provides a significant caloric density to the human diet. In addition, dietary fat in the form of triacylglycerols has been shown to exert effects on a variety of metabolic systems. These effects depend on the relative quantity of total fat in the diet, the presence of other dietary constituents, the fatty acid composition, and the arrangement of fatty acids on the glycerol. These arrangements are named according to the IUPAC stereochemical numbering system as the sn-1, sn-2, or sn-3 position of the triacylglycerol.

Lipids are the food molecules most susceptible to oxidative free radical reactions. This instability is due to their content of polyunsaturated fatty acids (PUFA) and includes the esters of glycerol with fatty acids, triacylglycerols, and phospholipids. Triacylglycerols play a major storage role in both plants and animals. Adipose tissue is the main storage form of fat in animals, in which specific cells, adipocytes, are the primary depository of triacylgycerides stored in phospholipid-delimited droplets within the cytoplasm. In those plants in which triacylglycerides are a storage form in seeds, the seed oils are stored as droplets in the cytoplasm of seed endosperm cells.

Phospholipids make up a smaller proportion of lipids than triacylglycerols in most food commodities, but they are compounds of structural importance to final foods. The unique amphiphilic properties of phospholipids, the major lipids in biological membranes, cause them to spontaneously form lamellae or align at the interface between lipid and aqueous environments. Phospholipids, as the most abundant surfactants of biological tissues in both plants and animals, also facilitate the formation of dispersed triacylglycerides as emulsions in foods derived from fat and oil-containing commodities. Phospholipids are used as functional ingredients for their emulsifying properties and are valuable to the preparation of food emulsions such as mayonnaise.

Both phospholipids and triacylglycerides are assembled from glycerol and activated fatty acids within the endoplasmic reticulum of cells. There is considerable variation in the fatty acid composition of both phospholipids and triacylglycerides depending on genetics, cell type, and, in animals, the diet. Certain structural principles appear to be retained within both plant and animal phospholipids. The three positions on the glycerol moiety are not esterified randomly, and if the sn-3 position is assigned to the phosphoryl moiety, the sn-1 and sn-2 positions tend to be acylated by saturated and unsaturated fatty acids, respectively. Thus, on average, biological membranes tend to be made up of approximately one-half saturated and one-half unsaturated fatty acids. It is likely a requirement that PUFA occupy the sn-2 position to maintain the simple structural properties of the liquid crystal state of the phospholipids in a viable cellular membrane. The presence of PUFA as an integral component of the phospholipids surrounding every cell and cellular compartment renders the membranes and foods containing them unstable to oxidation. Living organisms possess a variety of preventive measures to maintain the oxidative stability of the membrane, however many of these measures are lost on processing.

II. OXIDATION CHEMISTRY

The mechanism(s) of oxidative reactions leading to decreased quality of processed foods are the same mechanisms described for lipid oxidation in general chemistry. Lipid oxidation is a multistep, multifactorial process, and in foods the variables encompassed include individual fatty acid susceptibility, molecular structure of lipids, physical state of lipids, initiation reactions, hydroperoxide (ROOH) decomposition catalysts (e.g., metals), presence of oxidized lipids, and the amounts and selectivities of antioxidants present. The reader interested in a detailed discussion of oxidation chemistry is referred to a recent book by Frankel (1998) that provides an excellent presentation of the many aspects of the field of lipid oxidation.

As a simple consequence of their chemistry, thermodynamic equilibrium strongly favors the net oxidation of reduced, carbon-based biomolecules. The kinetic stability of all biological molecules in an oxygen-rich atmosphere results from the unique spin state of the unpaired electrons in ground state molecular (triplet) oxygen in the atmosphere. This property renders atmospheric oxygen kinetically inert to reduced, carbon-based biomolecules. Hence, reactions between oxygen and proteins, lipids, polynucleotides, carbohydrates, etc., proceed at slow rates unless they are catalyzed. Nevertheless, unsaturated fatty acids will oxidize autocatalytically in the presence of oxygen once initiated into a free radical chain reaction (Kanner et al., 1987). The basic reactions of autocatalytic lipid oxidation are summarized in Fig. 1. The ability of the peroxy radical, $ROO^{\bullet}$, to participate as an initiating, single-electron oxidant drives the very destructive self-perpetuating oxidation of unsaturated lipids. The free radicals generated rapidly propagate and interact directly with various targets and also yield ROOH. These ROOH are readily attacked by reduced metals, leading to a host of decomposition products. Some of these products cause further damage; some, formed through self-propagating reactions, are themselves free radicals; thus oxidation is reinitiated. A large volume of literature points to several key participants in the reaction course (Frankel, 1991; Porter et al., 1995).

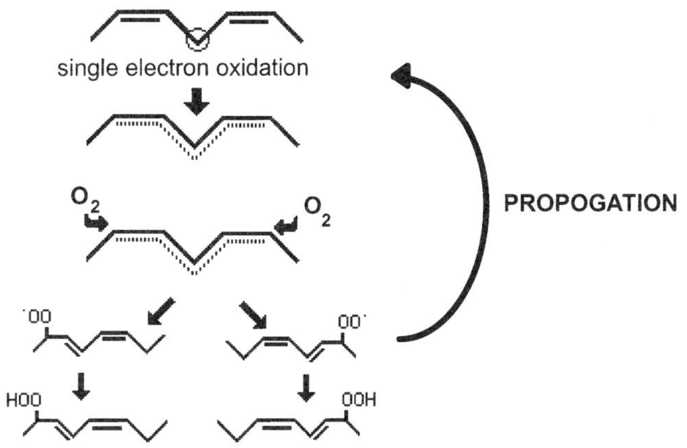

Figure 1 Initiation and propagation of single-electron oxidation of a polyunsaturated unsaturated lipid.

III. OXIDATION OF POLYUNSATURATED OILS

Initiators of oxidation eliminate the reactive impediments imposed by the spin restrictions of ground state oxygen by converting stable organic molecules, RH, to free radical-containing molecules, R$^\bullet$. Oxygen reacts readily with such species to form the peroxy radical, ROO$^\bullet$. Initiators of lipid oxidation are relatively ubiquitous, are primarily single-electron oxidants, and include trace metals, ROOH cleavage products, and light. A risk of food systems that contain PUFA is that these molecules are oxidized by the ROO$^\bullet$ species to yield another free radical, R$^\bullet$, and a lipid ROOH. This effectively sets up a self-propagating free radical chain reaction, $R^\bullet + O_2 \rightarrow ROO^\bullet \rightarrow ROOH + R^\bullet$, that can lead to the complete consumption of PUFA in a free radical chain reaction (Kanner et al., 1987). The ability of the peroxy radical to act as an initiating, single-electron oxidant drives the destructive and self-perpetuating reaction of PUFA oxidation. The products of the initial chain reaction stage of lipid oxidation are thus lipid ROOH. These ROOH are readily detected by a variety of colorimetric assays, and all such assays can provide a relatively accurate estimate of the concentration of ROOH in a food or oil sample, typically referred to as the peroxide value. ROOH, though formed stoichiometrically by the free radical chain reaction, are not the most important products of oxidation reactions in terms of food quality spoilage. Most lipid ROOH are organoleptically undetectable, and even when present as a detectable fraction of the lipid material, have little if any ostensible effect on the structure or functions of the molecules on which they are present. However, the ROOH are themselves highly susceptible to a homolytic cleavage reaction catalyzed by reduced metals, including ferrous (Fe^{++}) and cuprous (Cu^{+}) ions. The cleavage reaction products are highly unstable radicals that result in a large number of breakdown products yielded from the original lipid molecule. Some of these breakdown products are volatile molecules with very low flavor thresholds and are ostensibly responsible for the distinctive off-flavors of rancidity. Their potency as flavorants is responsible for the first and most important loss of quality of lipid-containing foods.

As a result of the largely free radical driven formation and breakdown of food lipid ROOH, short chain aldehydes, ketones, and alcohols are released (Frankel, 1991). These volatile compounds are sensed by receptors in the olfactory epithelia and normally perceived as off-flavors. The presence and accumulation of volatile off-flavors are the most important results of oxidation of foods that lead to quality losses. These off-flavors are so distinctive and familiar as rancidity that their presence alone ultimately leads to rejection as a result of the perceived rancidity of the fat or food. The voluminous literature that has gradually unraveled this chemistry in food lipids, and more recently in biological lipids, points to several key participants in the reactions. Initiators of lipid oxidation are relatively ubiquitous, primarily single-electron oxidants including trace metals, peroxides, and light (Frankel, 1998). These initiators tend to actively promote initiation of the radical chain reactions of lipid oxidation.

The overall rates of oxidation are affected by the ease of hydrogen abstraction afforded by the abundance of double bonds on fatty acids. Thus, oxidation of oils tends to increase dramatically with the content of PUFA. As the number of double bonds in a molecule increases, the number of methylenic hydrogens increases, which increases the rate of oxidation. The relative rate of oxidation as a function of number of double bonds has important practical consequences on the stability of edible oils. The relative oxidation rates for fatty acids with one, two, three, and four double bonds are 1, 50, 100, and 200, respectively. Additionally, the number of double bonds dramatically increases the number of possible breakdown pathways and the overall accumulation of products.

IV. SAFETY AND HEALTH IMPLICATIONS OF OXIDIZED LIPIDS

ROOH produced in foods are themselves odorless. However, carbonyl compounds formed upon decomposition can impart to the food undesirable off-flavors. These products, including short chain aldehydes, ketones, and alcohols (Frankel, 1991), as well as radicals, can compromise health in a number of ways. (1) The direct oxidation of susceptible molecules can result in loss of normal biological function. For example, oxidation of membrane lipids alters membrane integrity, promotes red blood cell fragility and membrane leakage. The oxidation of proteins results in loss of enzyme catalytic activity and/or regulation. (2) Reaction of some of these products leads to adduct formation with loss of native functions of specific molecules. The oxidative modification of the apoB molecule on low-density lipoprotein (LDL) prevents uptake by the LDL receptor and stimulates uptake by the scavenger receptor. (3) Oxidation can cleave DNA, and cause point, frame shift, deletion, and base damage. This oxidative cleavage impairs or destroys normal functionality. (4) Oxidative reactions can liberate signal molecules or analogs that elicit inappropriate responses such as the activation of platelet aggregation and down-regulation of vascular relaxation by leukotoxin and eicosanoid analogs.

The susceptibility and overall rate of oxidation of a lipid molecule, whether in food or within the body, is related to the number of double bonds on the fatty acids. The rate of oxidation is determined by the ease of hydrogen abstraction. An increase in the number of double bonds increases the oxidation rate. These relative rates of oxidation as a function of number of double bonds may be important to rates of deterioration of various components in foods as well as biological molecules in vivo. Foods high in PUFA require more antioxidants to prevent oxidation and rancidity (Fritsche and Johnston, 1988). Consumption by animals, including humans, of foods with high amounts of PUFA appears to increase the antioxidant requirement to prevent tissue damage (Muggli, 1989). The molecular basis for this increased requirement is not known. Frequently, reports of increased in vivo oxidative damage are based on crude measures of lipid oxidation such as thiobarbituric acid–reacting substances. The thiobarbituric acid assay does not distinguish oxidation among different dietary fats as it responds differently to the same amount of oxidation in PUFA with different numbers of double bonds. A diet enriched in highly unsaturated fatty acids would appear to increase the tendency to oxidation and increase the incidence of oxidation-associated chronic degenerative diseases; however, this has not been observed. In fact, studies have shown that replacement of diets high in saturated fat with highly unsaturated fat diets frequently reduces atherosclerosis, thrombosis, and other chronic diseases (Keen et al., 1991). Thus, when considering the myriad effects of food fats on subsequent oxidation within tissues, it is also critical to understand all of the various biochemical and metabolic consequences. Because oxidative processes are initiated through a variety of chemical and enzymatic reactions (Kanner et al., 1987), the inhibition of oxidative biochemical and metabolic pathways by a variety of antioxidants may significantly alter their net effect on oxidative damage to tissues.

With the increasing interest in the public health implications of edible oils and the possibility that oxidation products of particular lipids may have uniquely detrimental properties, some emphasis has been placed on analyzing fats for the presence of oxidation products. Foremost among these are the oxides of cholesterol (Esterbauer, 1993; Morin and Peng, 1989; Sevanian et al., 1991). How much of the toxicity of cholesterol oxides in vivo is related to cholesterol oxides from ingested food is not known. Overall, the overt toxicity of all oxidized lipids ingested in foods appears to be remarkably low (Esterbauer, 1993). Nevertheless, the long-term influence of lipid peroxides and oxides of cholesterol, etc., may be deleterious, especially toward the development of chronic disease (Halliwell, 1993). In ongoing research attempting to elucidate the basis of LDL modification and the accumulation of atherosclerotic plaque, many studies have measured cholesterol oxides as the index of LDL oxidation. This is a useful but not particularly sensitive barometer of the deterioration of LDL particles; the cholesterol molecule is not readily oxidized, hence the accumulation of the oxidation products of cholesterol readily confirms extensive decomposition of the particle. At the point at which LDL contain oxides of cholesterol, they can be readily shown to be taken up by the so-called scavenger receptor on macrophage. This then is consistent with the conversion of native LDL to modified and atherogenic LDL (Steinberg et al., 1989). This cholesterol oxide index of oxidation may have been overly interpreted as reflecting the fact that oxides of cholesterol are the most toxic and important components leading to the atherogenicity of LDL, and this has been further translated to foods that contain cholesterol in general. Although cholesterol oxides are toxic to a variety of cells and biochemical processes, it is not yet clear that particular oxides of cholesterol found in foods are uniquely toxic or that their presence in foods contributes inordinately to chronic disease. Nevertheless, research that is beginning to catalog the abundance of oxidized lipids in foods and edible oils has emphasized the need to stabilize oils against oxidation and has increased the rate of disappearance of animal-derived, cholesterol-containing cooking fats (Fontana et al., 1992; Morin and Peng, 1989; Zhang and Addis, 1990), all with arguably highly beneficial results. The potential adverse effect of this research focus on cholesterol oxides as uniquely deleterious agents is that it may shift emphasis away from dietary antioxidants and the broader protection of susceptible in vivo targets such as lipoproteins. Ongoing research suggests that such protection is involved in attenuating the risk of atherosclerosis and other degenerative diseases (Ames, 1989; Ames et al., 1993; Esterbauer, 1993; Halliwell, 1993).

V. MECHANISMS OF OXIDATION INHIBITION

The overall response of living organisms to oxidation is highly sophisticated, in keeping with the obvious potential catastrophic consequences to cell membranes. There are many complex biochemical pathways that have evolved to prevent, delimit, or respond to oxidation: (1) Within tissues, oxidation is isolated and compartmentallized within subcellular organelles such as mitochondria and peroxisomes. (2) Activated oxidants are scavenged by redox-active phenolics, including the lipid-soluble species α-tocopherol and ubiquinone, and water-soluble hydrogen donors such as ascorbate, uric acid, polyphenolics, various flavonoids and their polymers, amino acids, and protein thiols. (3) Various enzymes reduce reactive oxygen and reactive oxygen–containing molecules such as peroxides. Catalase, superoxide dismutases, and peroxidases all detoxify active-oxygen species in vivo and in vitro after harvest. (4) Enzymatic repair systems are ubiquitous in cells. For example, proteases, lipases, ribonucleases, etc., constantly turn over cellular constituents, and degradative enzymes often have higher affinities for modified molecules. Substrate affinities of synthetic enzymes discriminate against oxidized forms of lipids, proteins, and nucleotides. This discriminatory process removes damaged molecules from the living cell; however, these enzymes typically remain active to various extents and their hydrolytic activities can pose significant problems to the integrity of tissues post-harvest. The complexity and interdependence of these antioxidative protection systems is highly dependent on the structural integrity of tissues, and the disruption associated with food processing abolishes many of the oxidant defenses that are present in natural tissues. As a result, the major antioxidant protection remaining in biomaterials destined to be foods are the redox-active free radical–scavenging phenolics. These phenolic molecules are also most effective when added back to food materials as either natural or synthetic antioxidants. The mechanisms by which this class of antioxidants acts to slow oxidation will therefore be discussed in greater detail. Importantly, antioxidant is a broad classification for molecules that may act prior to or during a free radical chain reaction, at initiation, propagation, termination, decomposition, or the subsequent reaction of oxidation products with sensitive targets. Antioxygenic compounds are broadly defined as those molecules that can participate in any of the protective strategies. However, differences between antioxidant molecules, especially in the stage or site of activity at which they act, are not trivial and influence various aspects of the efficacy of a given compound to act as a net antioxidant or protectant. How and where different molecules act can affect the rate of oxidation, its reaction course, the products formed, and the final targets damaged.

A. Metal Inactivation

The oxidation reactions of polyunsaturated lipids are most successfully prevented by avoiding initiation. Because metals are the single most important contributors to the initiation reactions in most foods, the most effective inhibitors of oxidation initiation are the metal chelators. Metal chelators are structurally diverse, from simple polyacids such as ethylenediaminetetraacetic acid (EDTA) to slightly more elaborate peptides such as carnosine and even a variety of proteins such as lactoferrin and ceruloplasmin. The fundamental property of these molecules that renders them effective inhibitors of the initiating activities of metals is their ability to bind metals and either completely occupy their liganding orbitals or alter their redox potential so as to prevent them from participating in the redox chemistry necessary to form initiating radicals. These structural considerations are not all met by just any compound that binds a metal. Ironically, some proteins actually activate metals to promote their oxidative properties. The most stunning example is lactoferrin, which binds and inactivates two moles of iron per mole of protein, but subsequent moles of iron bind to activating sites on the protein, and at high iron to lactoferrin ratios oxidation is increased. There are normally sufficient chelating species in living tissues (plant or animal) to inactivate all of the metals present. However, processing frequently isolates the various tissue components, permitting metals to accumulate in foods in excess of the capabilities to inactivate them. Those chelators occurring naturally within the commodity that are lost can be either added explicitly during processing or their net benefits replaced by alternative chelating choices (Frankel, 1989).

B. Inhibition of Hydroperoxide Formation

The alkyl radical $R^{\bullet}$ is too reactive in an oxygen-rich environment for any competing species to successfully re-reduce $R^{\bullet}$ to RH before oxygen adds to form the peroxy radical $ROO^{\bullet}$. At this point, however, $ROO^{\bullet}$ is a relatively stable free radical that reacts relatively slowly with targets such as PUFA. This is the most widely accepted point of action for free radical-scavenging antioxidants such as the phenolic tocopherol. Tocopherol can reduce $ROO^{\bullet}$ to ROOH with such ease that tocopherol is competitive with biologically sensitive targets such as unsaturated lipids, RH, even at 10,000-fold lower concentration. The tocopheroxyl radical $A^{\bullet}$ is in general a poor oxidant and reacts significantly more slowly than $ROO^{\bullet}$. Conversion of $ROO^{\bullet}$ to ROOH and formation of $A^{\bullet}$ effectively impart a kinetic hindrance on the propagating chain reaction. The tocopheroxyl radical can be either re-reduced by reductants such as ascorbate, dimerized with another radical, or be oxidized further to a quinone. These free radical–scavenging functions of tocopherol are well documented (Buettner, 1993).

VI. NATURAL ANTIOXIDANTS

A. Tocopherol and Nonessential Polyphenols as Antioxidants

Tocopherols are the most active chain-breaking antioxidants, and there is an explicit dietary requirement for tocopherols as vitamin E. Tocopherols occur to varying extents in most foods unless they are removed by specific processes during manufacture. Loss of the free radical–scavenging properties of tocopherol is believed to be the basis for its essentiality and the pathologies associated with its deficiency. That the basis for the essentiality of tocopherol lies in its ability to prevent oxidative damage raises an important nutritional question, "Are these actions also provided by nonessential polyphenolics present in plants and foods derived from them?" Many phytochemicals that are present in the food supply have been implicated as being capable of interfering with and inhibiting free radical chain reactions of lipids. Plant phenolics inhibit lipid ROOH formation catalyzed by metals, radiation, and heme compounds (Buettner, 1993; Hanasaki et al., 1994), and also scavenge peroxy, alkoxy, and hydroxy radicals and singlet oxygen (Hanasaki et al., 1994; Laughton et al., 1991; Tournaire et al., 1993). Tocopherol in oxidizing lipid systems are spared by flavonoids (Terao et al., 1994; Jessup et al., 1990). If α-tocopherol (Fig. 2) is an essential antioxidant that acts where no other compound can, the sparing effect of nonessential antioxidants may be one of their most important actions.

B. Inhibition of Hydroperoxide Decomposition, Alkoxyl Radical Reduction, and Aldehyde Scavenging

Antioxidant activity is not limited to prevention of ROOH formation. ROOH are not damaging to foods or biological molecules, but their presence is an indication that oxidation

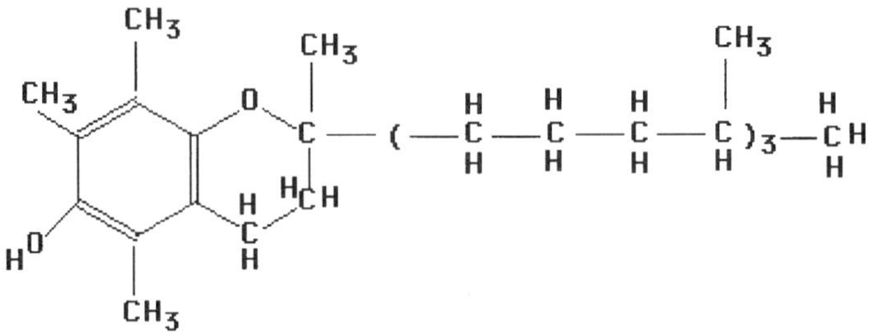

Figure 2 α-Tocopherol.

has occurred. Although ROOH are not directly damaging, their decomposition by reduced metals generates the reactive hydroxyl radical $HO^{\bullet}$ or the alkoxyl radical $RO^{\bullet}$. These strongly electrophilic oxidants react with and oxidize virtually all biological macromolecules. The alkoxy radical typically fragments the parent lipid molecule and liberates electrophilic aldehydes, hydrocarbons, ketones, and alcohols. Both the highly reactive hydroxy and alkoxy radicals and the electrophilic aldehydes liberated with their reduction react readily with polypeptides (proteins) and polynucleotides (DNA). Thus, additional antioxidant actions include preventing ROOH decomposition, reducing alkoxyl radicals, or scavenging the electrophilic aldehydes. The efficacy of different antioxidants varies during this phase of the oxidation process. Even tocopherol isomers differ with respect to their ability to prevent decomposition of ROOH (Huang et al., 1994, 1995). Plant phenolics vary in their ability to interrupt a free radical chain reaction, with differences detectable among different lipid systems, oxidation initiators, and other antioxygenic components.

As discussed, the chemistry of free radical oxidations is multistage and complex. Oxidation is not a single catastrophic event. There is no single initiating oxidant that generates all free radicals; there are a great many sources of single electrons. Similarly, there is no single reactive product of oxidation; there are classes of products, many of which are both selectively and broadly damaging. Free radicals and their products react with virtually all biological molecules, and there is no single defense against all targets of oxidative damage. Thus, a variety of mechanisms have evolved to prevent or respond to oxidative stresses and free radicals and their products at one or more of the many steps of oxidation.

C. Natural Antioxidants Added to Foods

Soluble chain-breaking antioxidants used in foods include ascorbate, as either the naturally occurring free acid or as synthetic ascorbate, and various soluble and insoluble ester forms. The acid form is an excellent electron donor in foods. This is the principle property that makes it an excellent antioxidant at low concentrations, but at high concentrations its ability to reduce metal initiators can actually lead to a prooxidant effect (Frankel, 1989).

In response to a perceived desire by consumers for less chemically processed food ingredients, several naturally occurring, chain-breaking antioxidants are being introduced to accomplish essentially the same effects as those of substituted phenols such as butylated hydroxy hydroxyanisole (BHA) and butylated hydroxytoluene (BHT) (Aruoma et al., 1992). These natural antioxidants are primarily extracts of herbs or plant materials with inordinately high concentrations of particular polyphenolics with good electron-donating and chain-breaking properties. Rosemary extract is perhaps the most widely known of these natural antioxidants, although a variety of novel materials are being pursued.

Because once initiated, lipid oxidation of PUFA tends to proceed autocatalytically, breaking the chain reaction is a critical step in affording stability. This stability is afforded both in vivo and in food lipids by the presence of scavengers of the peroxy radical oxidant. The abundance of scavengers of oxidants and peroxyl radicals is therefore a critical variable limiting the progress of lipid oxidation.

VII. SYNTHETIC ANTIOXIDANTS IN FOODS

Many antioxidants have been evaluated for use in foods as preservatives. Whereas natural antioxidants, (e.g., vitamin E) do not withstand processes such as frying and baking, synthetic antioxidants can survive these processes. Four synthetic antioxidants are particularly widespread in their use in foods: BHA, BHT, propyl gallate, and 2-(1,1-dimethylethyl)-1,4-benzenediol, also known as tertiarybutyl hydroquinone (TBHQ). Although a number of synthetic substances have been studied as food antioxidants, relatively few are approved and listed in the Code of Federal Regulations (Code of Federal Regulations, 1998a) by the Food and Drug Administration for use in the United States. The limits on types of antioxidants and allowable concentrations vary from country to country, which can be a problem in international marketing. Among the food preservatives that are allowed to be directly added to foods are the following antioxidants: Anoxomer, BHA, BHT, ethoxyquin, 4-hydroxymethyl-2,6-di-*tert*-butylphenol, TBHQ, and 2,4,5-trihydroxybutyrophenone (THBP) (Code of Federal Regulations, 1998b). The phenolic antioxidants are quite effective and are often used at concentrations of less than 0.01% (Rajalakshmi and Narasimhan, 1996). Other compounds that function as antioxidants are approved for direct addition to food and are on the list of food additives affirmed as generally recognized as safe (GRAS). Among the most prevalent of these substances are propyl gallate, lecithin, gum or resin guaiac, and glycine. Substances classified as antioxidants, when migrating from food-packaging material (limit of addition to food, 0.005%), include BHA, BHT, dilauryl thiodipropionate, distearyl thiodipropionate, gum guaiac, nordihydroguairetic acid, propyl gallate, thiodipropionic acid, and 2,4,5-trihydroxy butyrophenone (Code of Federal Regulations, 1998b).

A. Synthetic Antioxidants Added Directly to Food

Commercial antioxidants are prepared as solids or blends of liquid. The blends of antioxidants are solubilized and thus are more readily added to foods during processing. Antioxidant mixtures are prepared in solvents such as propylene glycol, which is odorless, tasteless, and inert. BHT, however, is insoluble in propylene glycol, and carriers such as fat or the fat derivatives mono- and diglycerides can be used. Solvents such as alcohol or acetic acid are often used as antioxidant carriers they are usually removed later by a combination of heat and vacuum (Chipault, 1962). The commercial antioxidants usually are mixtures of phenolic antioxidants and synergists.

1. Anoxomer

Anoxomer is a polymeric antioxidant that is prepared by condensation polymerization of divinylbenzene (*m*- and *p*-) with *tert*-butylhydroquinone, *tert*-butyl-phenol, hydroxyanisole, *p*-cresol, and 4,4′-isopropylidenediphenol. This polymeric antioxidant must meet the following specifications (Code of Federal Regulations, 1998c): not less than 98.0% purity; total monomers, dimers, and trimers below MW 500 not to exceed 1%; phenol content, not less than 3.2 meq/g and not more than 3.8 meq/g; and heavy metals, not more than 10 ppm lead, 3 ppm arsenic or 1 ppm mercury. Anoxomer may be safely used as an antioxidant in food at a level of not more than 5000 ppm based on fat and oil content of the food.

2. Butylated Hydroxyanisole

Butylated hydroxy hydroxyanisole (MW, 180.24 melting point, 48–55°C; boiling point 264–270°C) is a mixture of 2-*tert*-butyl-4-methoxyphenol and 3-*tert*-butyl-4-methoxyophenol, with the 3-isomer being 90% or more of the mixture (Fig. 3). The molecule is a ''hindered'' phenol, and the *tert*-butyl group *ortho* or *meta* to the hydroxyl group serves to suppress antioxidant activity. The steric hindrance is probably responsible for the relative ineffectiveness of BHA in vegetable oils because the tertiary butyl group interferes with the antioxidant activity of the phenolic structure. The steric hindrance may account for the carry-through effect of BHA in fats used in baked foods. BHA is commonly used in combination with other primary antioxidants, such as gallates, to take advantage of synergistic effects. BHA has a strong phenolic odor that becomes particularly noticeable when an oil treated with this antioxidant is subjected to high temperatures such as in baking or frying operations. This water-insoluble, white, waxy solid is soluble in fats and oils. BHA is often produced in tablet form to prevent caking. BHA effectively controls the oxidation of animal fats, but is a relatively ineffective antioxidant in most vegetable oils. It is a particularly effective antioxidant for use in palm kernel and coconut oils, which are typically used in cereal and confectionery products. BHA provides good carry-through, which is the ability to be added to food, survive processing, and remain stable in food, especially in baked products. It is the most effective of all food-approved antioxidants for protecting the flavor and color of essential oils. BHA is also added to packaging materials, either being added directly to the wax used in making waxed inner liners or applied to the packaging as an emulsion.

BHA exhibits antioxidant properties and synergism with acids, BHT, propyl gallate, hydroquinone, methionine, lecithin, thiodipropionic acid, etc. This antioxidant is used especially in foods, and the American Meat Institute Foundation proposed an antioxidant mixture known as AMIF-72 that contains 20% BHA, 6% propyl gallate, and 4% citric acid in propylene glycol (Budavari, 1989).

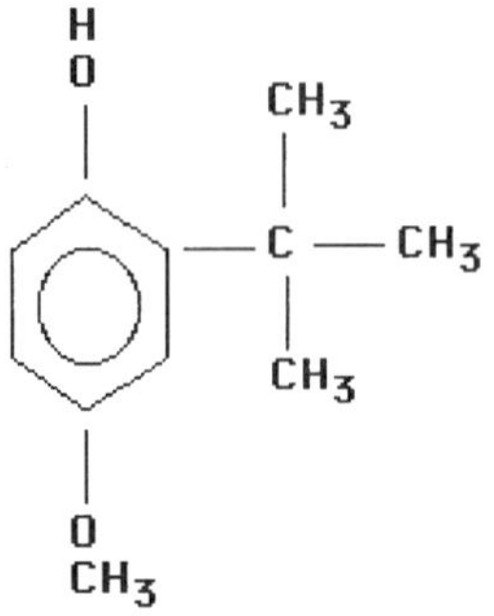

Figure 3 Butylated hydroxyanisole.

The food additive BHA, alone or in combination with other antioxidants permitted in food for human consumption, may be safely used in or on specified foods as follows (Code of Food Regulations, 1998d):

1. Total BHA must assay at 98.5% minimum, with a minimum melting point of 48°C.
2. BHA may be used alone or in combination with BHT, as an antioxidant in foods, as follows: food (total BHT and BHA, ppm): dehydrated potato shreds (50); active dry yeast (1000 BHA only); beverages and desserts prepared from dry mixes (2 BHA only); dry breakfast cereals (50); dry diced glazed fruit (32 BHA only); dry mixes for beverages and desserts (90 BHA only); emulsion stabilizers for shortenings (200); potato flakes (50); potato granules (10); and sweet potato flakes (50).
3. To assure safe use of the additive BHA: (1) The label of any market package of the additive shall bear, in addition to the other information required, the name of the additive. (2) When the additive is marketed in a suitable carrier, the label shall declare the percentage of the additive in the mixture. (3) The label or labeling of dry mixes for beverages and desserts shall bear adequate directions for use to provide that beverages and desserts prepared from the dry mixes contain no more than 2 ppm BHA.

3. Butylated Hydroxytoluene

Butylated hydroxytoluene (2,6-di-*tert*-butyl-*p*-cresol; 2,6-bis(1,1-dimethylethyl)-4-methylphenol; MW, 220.34; melting point, 70°C; boiling point, 265°C) is a water-insoluble, white, crystalline solid antioxidant that is more soluble in food oils and fats than is BHA (Fig. 4). BHT is soluble in toluene, methanol, ethanol, isopropanol, methyl ethyl ketone, acetone, Cellosolve, petroleum ether, benzene, and most other hydrocarbon solvents. BHT is an antioxidant for foods, animal feed, petrol products, synthetic rubbers, plastics, animal and vegetable oils, and soaps. While it is effective in animal fats, BHT is not as effective in vegetable oils. BHT is frequently used in combination with BHA in foods because the two antioxidants are synergistic in their actions. The oxidative reactions of nuts and nut products are very responsive to this combination of antioxidants.

BHT is noted for its high-temperature stability and its carry-through effect in fats or shortenings in baked food. However, BHT is less effective than BHA because of the greater steric hindrance presented by two tert-butyl groups surrounding the hydroxyl group

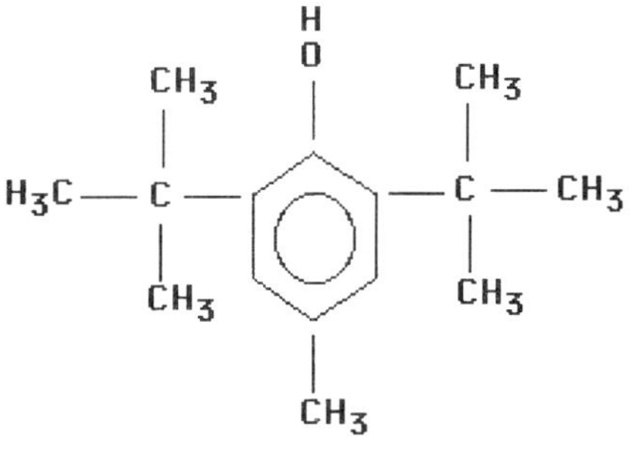

Figure 4 Butylated hydroxytoluene.

in BHT. BHT is important as a food antioxidant because it is readily soluble in glycerides, is insoluble in water and is susceptible to loss by volatilization and distillation under, certain food processing conditions, such as frying. One negative aspect of BHT, is that it may give a yellow coloration due to the formation of stilbenequinone in the presence of iron. Because of its relatively low antioxidant effectiveness in vegetable oils, BHT is often used in combination with other primary antioxidants.

The food additive BHT, alone or in combination with other permitted antioxidants, may be safely used in or on specified foods as follows (Code of Federal Regulations, 1998e):

1. Total BHT must assay at 99% minimum.
2. BHT may be used alone or in combination with BHA, as an antioxidant in foods, as follows: food (total BHT and BHA, ppm)—dehydrated potato shreds (50); dry breakfast cereals (50); emulsion stabilizers for shortenings (200); potato flakes (50); potato granules (10); and sweet potato flakes (50).
3. To assure safe use of the additive BHT: (1) The label of any market package of the additive shall bear, in addition to the other information required, the name of the additive. (2) When the additive is marketed in a suitable carrier, the label shall declare the percentage of the additive in the mixture.

Reische et al. (1998) postulated that the synergistic mechanism of BHA and BHT involves interactions of BHA with peroxy radicals to produce a BHA phenoxy radical. The BHA phenoxy radical can then abstract a hydrogen from the hydroxyl group of BHT; BHT thus replenishes hydrogen to BHA, which regenerates its effectiveness. According to Belitz and Grosch (1987), the BHT radical then can react with a peroxy radical, thus acting as a chain terminator.

4. Ethoxyquin

Ethoxyquin (1,2-dihydro-6-ethoxy-2,2,4-trimethylquinoline; MW, 217.30; boiling point, 123–125°C) is a yellow liquid antioxidant used principally in animal feed and as an antidegradation agent for rubber. When used in oil, ethoxyquin is primarily in the form of a radical. Dimerization of the radical will inactivate the antioxidant (Reische et al., 1998). The Code of Federal Regulations states that ethoxyquin may be safely used as an antioxidant for preservation of color in the production of chili powder, paprika, and ground chili at levels not to exceed 100 ppm (Code of Federal Regulations, 1998f). Although approved for limited use in human foods, ethoxyquin has long been used effectively as an antioxidant in feeds used for poultry and swine production, mainly for the protection of carotenoids. Commercial antioxidant mixtures for feeds contain both chelators and at least two effective antioxidants to prevent oxidative rancidity. Protection of feeds allows producers to achieve optimal growth rate, reproductive efficiency, and liveability in intensively produced poultry. The recommendation of the American Soybean Association (URL http://www.pacweb.net.sg/asa) is that antioxidants be incorporated into a program comprising good manufacturing practices with appropriate quality control over ingredients, adequate rotation of inventory, and consumption of rations within one week of manufacture. The FDA has established tolerances for residues of ethoxyquin in or on edible products of animals as follows: 5 ppm in or on the uncooked fat of meat from animals except poultry, 3 ppm in or on the uncooked liver and fat of poultry, 0.5 ppm in or on the uncooked muscle meat of animals, 0.5 ppm in poultry eggs, and zero in milk.

Extensive research has been done on the effects of feeding exotic birds ethoxyquin at 20% greater concentration than is used in normal feeds over a 5-year period and on feeding dogs over an 11-year period; dogs are the species most sensitive to potential effects of ethoxyquin. A Kaytee Avian Research Center (2000) report states that ''To date, no ethoxyquin-related tissue changes have occurred, even in the 11 year feeding group. This is the only test ever conducted on ethoxyquin use in psittacine species and is now one of the longest and largest tests conducted in any specie. There is no legitimate reason to believe that any of the commercial antioxidants are a significant risk compared to the risk of unprotected food products.''

5. 4-Hydroxymethyl-2,6-di-tert-butylphenol

The food additive 4-hydroxymethyl-2,6-di-*tert*-butylphenol may be safely used in food in accordance with the following proscribed conditions (Code of Federal Regulations, 1998g): (1) This food additive must have a solidification point of 140–141°C. (2) As an antioxidant, it may be used alone or in combination with other permitted antioxidants. (3) The total amount of all antioxidants added to such food shall not exceed 0.02% of the oil or fat content of the food, including the essential oil content of the food.

6. 2-(1,1-Dimethylethyl)-1,4-Benzenediol

2-(1,1-Dimethylethyl)-1,4-benzenediol (TBHQ; MW, 166.22; boiling point, 300°C; melting point, 126.5–128.5°C), also known as tertiary butylhydroquinone (Fig. 5), is the most recently developed major phenolic antioxidant for food use. TBHQ is a white to light tan crystalline solid that effectively increases oxidative stability (shelf-life) of polyunsaturated food fats and oils. Features that make this antioxidant favorable are its moderate solubility (5–10%) in fats and oils, its slight (1%) water solubility and its lack of discoloration with metals, such as iron. TBHQ is the best antioxidant for protecting frying oils against oxidation, and it provides good carry-through to the finished product. However, TBHQ is relatively ineffective in baking applications. The response of vegetable oils to treatment with TBHQ is generally greater than with other approved primary antioxidants. This compound filled a need for an antioxidant for polyunsaturated oils, such as safflower seed oil. TBHQ also improves the color and stability of hydrogenated fats. TBHQ used in combination with citric acid further enhances its stabilizing properties, primarily in vegetable oils, shortenings, and animal fats. However, combination of TBHQ with propyl gallate is not permitted.

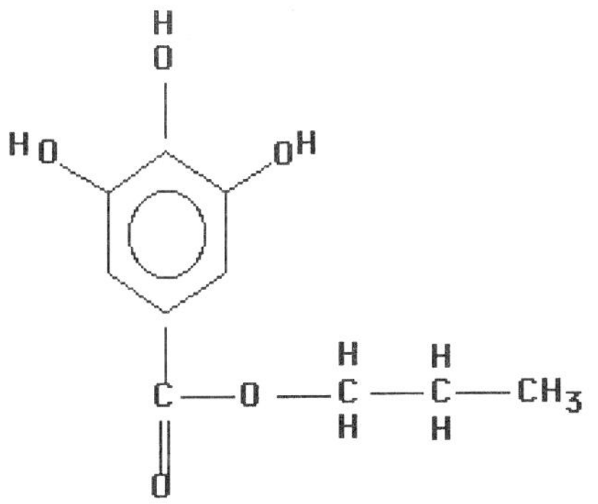

Figure 5 Tertiary butylhydroquinone (TBHQ).

TBHQ can be safely used as an antioxidant in food in accordance with the following prescribed conditions (Code of Federal Regulations, 1998h): (1) The additive must have a melting point of 126.5–128.5°C. (2) TBHQ may be used as an antioxidant alone or in combination with BHA and/or BHT. (3) The total antioxidant content of a food containing the additive must not exceed 0.02% of the oil or fat content of the food, including the essential oil content of the food.

7. 2,4,5-Trihydroxybutyrophenone

2,4,5-Trihydroxybutyrophenone (THBP; MW, 196; melting point, 149–153°C) is a tan powder that is slightly soluble in water, moderately soluble in fats, soluble in alcohol, propylene glycol, and paraffin. This antioxidant can brown in the presence of metals. The food additive THBP may be safely used in food in accordance with the following conditions (Code of Federal Regulations, 1998i): (1) The additive must have a melting point of 149–153°C. (2) THBP can be used as an antioxidant alone or in combination with other permitted antioxidants. (3) The total antioxidant content of a food containing TBHQ must not exceed 0.02% of the oil or fat content of the food, including the essential oil content of the food.

B. Antioxidants Generally Recognized as Safe

1. Propyl Gallate

Propyl gallate (3,4,5-trihydroxybenzoic acid propyl ester; MW, 212.20; melting point, 150°C; decomposes above 148°C) is a white to light gray, crystalline antioxidant that is partially soluble in water, alcohol, ether, vegetable oils, and lard (Fig. 6). Propyl gallate is used as an antioxidant in foods, fats, oils, ethers, emulsions, waxes, and transformer oils. This antioxidant is used to prevent rancidity in meat products such as rendered fats or pork sausage. The low oil solubility of propyl gallate makes this antioxidant difficult to incorporate into fats and oils, and its solubility in water makes it more likely to complex with iron and iron salts, which causes dark discoloration in some applications. Propyl gallate is usually used with citric acid to eliminate this unappealing discoloration. The low melting point of propyl gallate renders it ineffective at temperatures greater than 190°C for frying. As a result of heat lability, propyl gallate provides little or no carry-through protection in many heat-processed foods. This is especially true when the foods are alkaline, such as in some baked goods. Propyl gallate is synergistic with BHA and

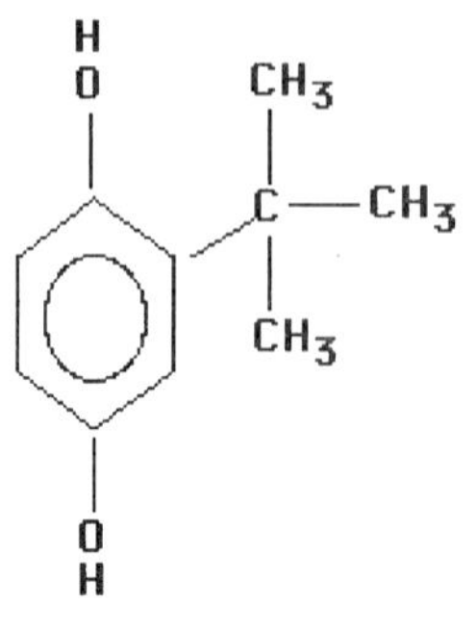

Figure 6 Propyl gallate.

BHT, and the combined effects provide improved storage stability and carry-through protection. Some countries allow the use of longer-chain esters, octyl and dodecyl gallates.

The Code of Federal Regulations (1998j) specifies propyl gallate use as follows: (1) Propyl gallate is the n-propylester of 3,4,5-trihydroxybenzoic acid. The natural occurrence of propyl gallate has not been reported. Propyl gallate is commercially prepared by esterification of gallic acid with propyl alcohol followed by distillation to remove excess alcohol. (2) The ingredient meets the specifications of the *Food Chemicals Codex*, 3d ed., 1981. (3) The ingredient is used as an antioxidant. (4) The ingredient is used in food at levels not to exceed good manufacturing practice, which results in a maximum total content of antioxidants of 0.02% of the fat or oil content, including the essential (volatile) oil content, of the food.

2. Gum Guaiac

Gum guaiac, also known as resin guaiac (melting point, 85–90°C) is insoluble in water but freely soluble in alcohol, chloroform, ether, alkalis. The antioxidant properties of gum guaiac were first described by Grettie (1933) in the early 1930s and was the first antioxidant to be approved for use in lard (Chipault, 1962). Gum guaiac is more effective as a stabilizer of animal fats than of vegetable oils. Gum guaiac is a relatively weak antioxidant, and it has not been used to any large extent in recent years. This substance is generally recognized as safe when used in edible fats or oils in accordance with good manufacturing or feeding practice and can be used at 0.1% (equivalent antioxidant activity, 0.01%).

3. Nordihydroguaiaretic Acid

Nordihydroguaiaretic acid (NDGA) is the chemical 4,4′-(2,3-dimethyl-1,4-butanediyl) bis[1,2-benzenediol]. NDGA was first isolated from the creosote bush, and it occurs with gums, resins, and waxes on the surface of leaves. This white, crystalline solid has a melting point of 184–185°C and is slightly soluble in water and dilute acid. Lundberg et al. (1944) first described its antioxidant properties, and it has been used as an antioxidant in foods, especially animal fats (Chipault, 1962). Now, however, food containing NDGA is deemed to be adulterated and in violation of the act based upon an order published in the *Federal Register* of April 11, 1968 (33 FR 5619).

C. Synergistic Antioxidants

Several considerations are basic to the use of phenolic antioxidant formulations in food fats and oil (Sherwin, 1989). Antioxidants are combined to take advantage of their different types of effectiveness. Specific combinations avoid or minimize solubility or color problems presented by the individual antioxidants; combinations permit better control and accuracy of application; combinations enable more complete distribution or solution of antioxidants and chelating agents in fats and oils; some combinations of antioxidants are more convenient to handle than individual antioxidant compounds; and some provide synergistic effects offered by some antioxidant combinations.

Over three decades ago, Chipault (1962) reviewed the then current information about antioxidant synergists. It was clear that at the time some compounds, which by themselves have very little effect on the oxidation of fats, may enhance or greatly prolong the antioxygenic action of primary antioxidants. The mechanisms of synergism vary, and while part of the activity of synergists is due to their inactivation of prooxidant metals, they may also function by inhibiting the decomposition of peroxides.

Many low molecular weight hydroxy acids or amino acids exhibit synergistic activity. Among synergistic antioxidants are substituted mercaptopropionic acids, such as 3,3-thiodipropionic acid, phospholipids, citric acid, ascorbic acid, and phosphoric acids. As stated, mixtures of primary antioxidants, such as propyl gallate and BHA and mixtures of BHA and BHT, are used synergistically in some food systems.

1. 3,3-Thiodipropionic Acid

3,3-Thiodipropionic acid (3,3′-thiobis[propanonic acid]; MW, 178.20; melting point, 134°C) is an antioxidant that is freely soluble in hot water, alcohol, and acetone. In addition to being used as a food antioxidant, it is added to plasticizers, lubricants, soap products, and polymers of ether. Thiodipropionic acid is a slightly effective antioxidant when used alone in lard, but it is a powerful synergist when used in combination with BHA (Kraybill et al., 1949).

2. Lecithin

Phospholipids can function as either antioxidants or prooxidants. Most of the antioxidant effect noted appears to result from the chelation of metal ions. Commercial lecithin is a naturally occurring mixture of the phosphatides of choline, ethanolamine, and inositol, with smaller amounts of other lipids. It is isolated as a gum following hydration of solvent-extracted soy, safflower, or corn oils. Lecithin can be bleached by hydrogen peroxide and benzoyl peroxide and dried by heating. Lecithin can be used in food with no limitation other than current good manufacturing practice.

3. Citric Acid

Citric acid, 2-hydroxy-1,2,3-propane-tricarboxylic acid (MW, 192.12), is highly soluble in water and primarily insoluble in fats. Citric acid is a natural antioxidant that is active in vegetable oils but without effect in lard, lard esters, or the purified, distilled esters of cottonseed oil. Although citric acid can counteract the prooxidant effect of iron, it can act as a synergist in the presence of a phenolic synthetic antioxidant when no metallic accelerators are present. At least two free carboxylic groups are necessary for antioxidative potency (Chipault, 1962). Although decomposed by heat, the thermal decomposition products are also good synergists. Citric acid is listed among the specific substances affirmed as GRAS.

4. Ascorbic Acid

Ascorbic acid (3-oxo-L-gulofuranolactone; MW, 176.12) is a crystalline substance that decomposes near 160°C. This natural antioxidant, which is extremely insoluble in fats, was first used as an antioxidant to improve the stability of mayonnaise. The synergistic antioxidant effect of ascorbic acid can be ascribed partly to the binding of metal ions. The free acid acts as a synergist with most phenolic antioxidants, but not with gallic acid (Filer et al., 1944). Ascorbic acid is among the substances affirmed as GRAS.

VIII. TOXICOLOGY OF ANTIOXIDANTS

The toxicological aspects of antioxidants used in the food industry have been reviewed in great detail by Madhavi and Salunkhe (1995).

REFERENCES

Ames, B. N. 1989. Endogenous oxidative DNA damage, aging, and cancer. *Free Radic. Res. Commun.* 7:121–128.

Ames, B. N., Shigenaga, M. K., Hagen, T. M. 1993. Oxidants, antioxidants, and the degenerative diseases of aging. *Proc. Nat. Acad. Sci. U.S.A.* 90:7915–7922.

Aruoma, O. I., Halliwell, B., Aeschbach, R., Loligers, J. 1992. Antioxidant and pro-oxidant properties of active rosemary constituents: carnosol and carnosic acid. *Xenobiotica* 22:257–268.

Belitz, H. D., Grosch, W. 1987. *Food Chemistry*. Springer-Verlag, New York, pp. 128–200.

Budavari, S. (Ed.). 1989. *The Merck Index*, 11th ed., Merck & Co., Rahway, NJ, p. 238.

Buettner, G. R. 1993. The pecking order of free radicals and antioxidants: lipid peroxidation, alpha-tocopherol, and ascorbate. *Arch. Biochem. Biophys.* 300:535–543.

Chipault, J. R. 1962. Antioxidants for use in foods. In: *Antioxidation and Antioxidants*, Vol. II, Lundberg, W. O. (Ed.) Interscience Publishers, New York, pp. 477–542.

Code of Federal Regulations. 1998a. Title 21—Food and Drugs, Ch. 1 (4-1-98 edition), Part 172—Food Additives Permitted for Direct Addition to Food for Human Consumption, Subpart B—Food Preservatives.

Code of Federal Regulations. 1998b. Title 21, Vol. 3, Parts 170–199, Ch. 1 (4-1-98 edition), Part 181—Prior-Sanctioned Food Ingredients, Subpart B—Specific Prior-Sanctioned Food Ingredients, Sec. 181.24—Antioxidants.

Code of Federal Regulations. 1998c. Title 21, Ch. 1 (4-1-98 Edition), Part 172—Food Additives Permitted for Direct Addition to Food for Human Consumption; Subpart B—Food Preservatives, Sec. 172.105—Anoxomer.

Code of Food Regulations. 1998d. Title 21, Ch. 1 (4-1-98 Edition), Part 172—Food Additives Permitted for Direct Addition to Food for Human Consumption; Subpart B—Food Preservatives, Sec. 172.110—BHA.

Code of Federal Regulations. 1998e. Title 21, Ch. 1 (4-1-98 Edition), Part 172—Food Additives Permitted for Direct Addition to Food for Human Consumption, Subpart B—Food Preservatives, Sec. 172.120—BHT.

Code of Federal Regulations. 1998f. Title 21, Ch. 1 (4-1-98 Edition), Part 172—Food Additives Permitted for Direct Addition to Food for Human Consumption, Subpart B—Food Preservatives, Sec. 172.140—Ethoxyquin.

Code of Federal Regulations. 1998g. Title 21, Ch. 1 (4-1-98 Edition), Part 172—Food Additives Permitted for Direct Addition to Food for Human Consumption, Subpart B—Food Preservatives, Sec. 172.150—4-Hydroxymethyl-2,6-di-*tert*-butylphenol.

Code of Federal Regulations. 1998h. Title 21, Ch. 1 (4-1-98 Edition), Part 172—Food additives permitted for direct addition to food for human consumption; Subpart B—Food Preservatives, Sec. 172.185—TBHQ.

Code of Federal Regulations. 1998i. Title 21, Ch. 1 (4-1-98 Edition), Part 172—Food Additives Permitted for Direct Addition to Food for Human Consumption, Subpart B—Food Preservatives, Sec. 172.190—THBP.

Code of Federal Regulations. 1998j. Title 21, Ch. 1 (4-1-98 Edition); Part 184—Direct Food Substances Affirmed as Generally Recognized as Safe, Subpart B—Listing of Specific Substances Affirmed as GRAS, Sec. 184.1660—Propyl gallate.

Esterbauer, H. 1993. Cytotoxicity and genotoxicity of lipid-oxidation products. *Am. J. Clin. Nutr.* 57(5, Suppl.):779S–786S.

Filer, L. J., Mattil, K. F., Longenecker, H. E. 1944. Antioxidant losses during the induction period of fat oxidation. *Oil Soap* 21:289–292.

Fontana, A., Antoniazzi, F., Cimino, G., Mazza, G., Trivellone, E., Zanone, B. 1992. High-resolution NMR detection of cholesterol oxides in spray dried egg yolk. *J. Food Sci.* 57:869–872.

Frankel, E. N. 1989. The antioxidant and nutritional effects of tocopherols, ascorbic acid and beta-carotene in relation to processing of edible oils. *Bibl. Nutr. Dieta* 43:297–312.

Frankel, E. N. 1991. Recent advances in lipid oxidation. *J. Sci. Food Agric.* 54:495–511.

Frankel, E. N. 1998. *Lipid Oxidation*, The Oily Press, Dundee, Scotland.

Fritsche, K. L., Johnston, P. V. 1988. Rapid autoxidation of fish oil in diets without added antioxidants. *J. Nutr.* 118:425–426.

Grettie, D. P. 1933. Gum guaiac A new anti-oxidant for oils and fats. *Oil Soap* 10:126–127.

Halliwell, B. 1993. The role of oxygen radicals in human disease, with particular reference to the vascular system. *Haemostasis* 23(Suppl. 1):118–126.

Hanasaki, Y., Ogawa, S., Fukui, S. 1994. The correlation between active oxygen scavenging and antioxidative effects of flavonoids. *Free Radic. Biol. Med.* 16:845–850.

Huang, S.-W., Frankel, E. N., German, J. B. 1994. Antioxidant activity of alpha- and gamma-tocopherols in bulk oils and in oil-in-water emulsions. *J. Agric. Food Chem.* 42:2108–2114.

Huang, S.-W., Frankel, E. N., German, J. B. 1995. Effects of individual tocopherols and tocopherol mixtures on the oxidative stability of corn oil triglycerides. *J. Agric. Food Chem.* 43:2345–2350.

Jessup, W., Rankin, S. M., De Whalley, C. V., Hoult, J. R., Scott, J., Leake, D. S. 1990. Alpha-tocopherol consumption during low-density lipoprotein oxidation. *Biochem. J.* 265:399–405.

Kanner, J., German, J. B., Kinsella, J. E. 1987. Initiation of lipid oxidation in biological systems. *Crit. Rev. Food Sci. Nutr.* 25:317–364.

Kaytee Avian Research Center. 2000. Antioxidants and preservatives. Ask the Experts FAQ, September 19, 2000, www.kaytee.com/experts/faq/925841935.html.

Keen, C. L., German, J. B., Mareschi, J. P., Gershwin, M. E. 1991. Nutritional modulation of murine models of autoimmunity. *Rheum. Dis. Clin. North Am.* 17:223–234.

Kraybill, H. R., Dugan, L. R., Jr., Beadle, B. W., Vibrans, F. C., Swartz, V., Rezabek, H. 1949. Butylated hydroxyanisole as an antioxidant for animal fats. *Am. Oil Chemists' Soc.* 26:449–453.

Laughton, M. J., Evans, P. J., Moroney, M. A., Hoult, J. R., Halliwell, B. 1991. Inhibition of mammalian 5-lipoxygenase and cyclo-oxygenase by flavonoids and phenolic dietary additives. Relationship to antioxidant activity and to iron ion-reducing ability. *Biochem. Pharmacol.* 42:1673–1681.

Lundberg, W. O., Halvorson, H. O., Burr, G. O. 1944. The antioxidant properties of nordihydroguaiaretic acid. *Oil Soap* 21:33–35.

Madhavi, D. L., Salunkhe, D. K. Antioxidants. 1995. Antioxidants. In: *Food Additive Toxicology*, Magu, J. A., Tu, A. T. (Eds.). Marcel Dekker, New York, pp. 89–177.

Morin, R. J., Peng, S. K. 1989. The role of cholesterol oxidation products in the pathogenesis of atherosclerosis. *Ann. Clin. Lab. Sci.* 19:225–237.

Muggli, R. 1989. Dietary fish oils increase the requirement for vitamin E in humans. In: *Health Effects of Fish and Fish Oils*, Chandra, R. K. (Ed.). ARTS Biomedical Publishers and Distributors, St. John's, Newfoundland, Canada, 201–210.

Porter, N. A., Caldwell, S. E., Mills, S. A. 1995. Mechanisms of free radical oxidation of unsaturated lipids. *Lipids* 30:277–290.

Rajalakshmi D., Narasimhan, S. 1996. Food antioxidants: sources and methods of evaluation. In: *Food Antioxidants: Technological, Toxicological, and Health Perspectives*, Madhavi, D. L., Deshpande, S. S., Sallunkhe, D. K. (Eds.). Marcel Dekker, New York, pp. 65–157.

Reische, D. W., Lillard, D. A., Eitenmiller, R. R. 1998. Antioxidants. In: *Food Lipids: Chemistry, Nutrition, and Biotechnology*, Akoh, C. C., Min, D. B. (Eds.). Marcel Dekker, New York, pp. 423–448.

Sevanian, A., Berliner, J., Peterson, H. 1991. Uptake, metabolism, and cytotoxicity of isomeric cholesterol-5,6-epoxides in rabbit aortic endothelial cells. *J. Lipid Res.* 32:147–155.

Sherwin, E. R. 1989. Antioxidants. In: *Food Additives*, Branen, A. L., Davidson, P. M., Salminen, S. (Eds.). Marcel Dekker, Inc., New York, pp. 139–193.

Steinberg, D., Parthasarathy, S., Carew, T. E., Khoo, J. C., Witztum, J. L. 1989. Beyond cholesterol. Modifications of low-density lipoprotein that increase its atherogenicity. *New Eng. J. Med.* 320:915–924.

Terao, J., Piskula, M., Yao, Q. 1994. Protective effect of epicatechin, epicatechin gallate, and quercetin on lipid peroxidation in phospholipid bilayers. *Arch. Biochem. Biophys.* 308:278–284.

Tournaire, C., Croux, S., Maurette, M. T., Beck, I., Hocquaux, M., Braun, A. M., Oliveros, E. J. 1993. Antioxidant activity of flavonoids: efficiency of singlet oxygen (1 delta g) quenching. *J. Photochem. Photobiol. B, Biol.* 19:205–215.

Zhang, W. B., Addis, P. B. 1990. Prediction of levels of cholesterol oxides in heated tallow by dielectric measurement. *J. Food Sci.* 55:1673–1675.

19

Antibrowning Agents

GERALD M. SAPERS, KEVIN B. HICKS, and ROBERT L. MILLER

Agricultural Research Service, U.S. Department of Agriculture, Wyndmoor, Pennsylvania

I. INTRODUCTION

A. Browning as a Problem

Many plant foods are subject to degradative reactions during handling, processing, or storage, collectively described as browning reactions, that result in the formation of brown, black, gray, or red colored pigments (Nichols, 1985; Feinberg et al., 1987). Such reactions are generally grouped into two categories: enzymatic browning and nonenzymatic browning. Examples of the former include browning of cut apples or potatoes, while examples of the latter include browning of shelf-stable, pasteurized juices and dehydrated vegetables.

Enzymatic browning results from the oxidation of polyphenols to quinones, catalyzed by the enzyme polyphenol oxidase (E.C. 1.14.18.1 and E.C. 1.10.3.1; also known as PPO, tyrosinase, *o*-diphenol oxidase, and catechol oxidase), and subsequent further reaction and polymerization of the quinones. This discoloration is generally a problem with raw fruit and vegetable products rather than blanched or thermally processed products since enzymes would be inactivated in the latter. Enzymatic browning of raw commodities may result from physiological injury; senescence; pre- or postharvest bruising; disruption of the fruit or vegetable flesh by peeling, coring, slicing, or juicing; tissue disruption from freeze–thaw cycling; and tissue disruption by bacterial growth. The occurrence of enzymatic browning can limit the shelf-life of fresh-cut fruits and salad vegetables, fresh mushrooms, prepeeled potatoes, and other fresh products of commercial importance (Huxsoll et al., 1989). This problem has held back the development and commercialization of fresh-cut fruits such as sliced apples. Enzymatic browning also may be a problem with some dehydrated and frozen fruits and vegetables (Shewfelt, 1986; Hall, 1989). In addition to causing discoloration, enzymatic browning reactions in fruit and vegetable products also can result in loss of ascorbic acid (vitamin C) through reaction with quinones. Enzymatic browning is usually controlled by blanching, where applicable (McCord and Kilara, 1983; Hall, 1989; Ma et al., 1992); acidification; and application of sulfites (which are now subject to regulatory constraints with a number of commodities) or sulfite substitutes such as ascorbic acid or cysteine. These substitutes are generally less effective than sulfites.

Nonenzymatic browning reactions may result from the classic Maillard reaction between carbonyl and free amino groups, i.e., reducing sugars and amino acids (Hodge, 1953), which produces melanoidin pigments in a wide variety of foods including dairy, cereal, fruit, and vegetable products (Labuza and Schmidl, 1986; Handwerk and Coleman, 1988). Such discolorations generally occur in products that are subjected to heat and/or prolonged storage. Nonenzymatic browning can be minimized by avoidance of excessive exposure to heat, control of moisture content in dehydrated products, and application of sulfites. Other nonenzymatic browning reactions will be discussed in Section II.

B. Regulatory Issues and Other Constraints

Sulfites have been associated with occurrence of allergic reactions, in some cases severe, with some individuals who are asthmatic. Consequently, the U.S. Food and Drug Administration has banned the use of sulfites in certain raw fruit and vegetable products. Other products such as wine and packaged dehydrated fruits and vegetables must be labeled to indicate the presence of sulfites.

While a number of sulfite substitutes exist, there is a strong reluctance on the part of some food processors to use them since a label declaration would normally be required. Many consumers seek "natural" ingredients and might be expected to reject products in which "chemicals" are used as additives. Various plant extracts and other natural products have been found to possess antibrowning activity, but there is no assurance that such products are free of toxicants, and some form of regulatory approval would probably be required before they could be used.

C. Scope of Review

Because of the economic importance of browning reactions in foods, restrictions in the use of sulfites, and limitations in the efficacy of conventional sulfite substitutes, the search for more effective alternatives has been a very active area of research. In this chapter, we will briefly review the chemistry of browning reactions in foods and the use of sulfites as browning inhibitors. However, the main focus will be on alternatives to sulfites, especially developments in the area of new antibrowning agents and methods of application. We also will address specific browning control issues with certain commodities such as fresh-cut fruits, prepeeled potatoes, mushrooms, and fresh juices.

II. CHEMISTRY OF BROWNING REACTIONS IN FOODS

A. Enzymatic Browning

In this section, we will provide an overview of the reactions in foods that result in browning or other related dark discolorations. The sequence of reactions classified as enzymatic browning is usually initiated by the hydroxylation of monophenolic compounds to *o*-diphenols in the presence of atmospheric oxygen and PPO. The *o*-diphenols then undergo further oxidation, also catalyzed by PPO, to *o*-quinones. The highly reactive quinones

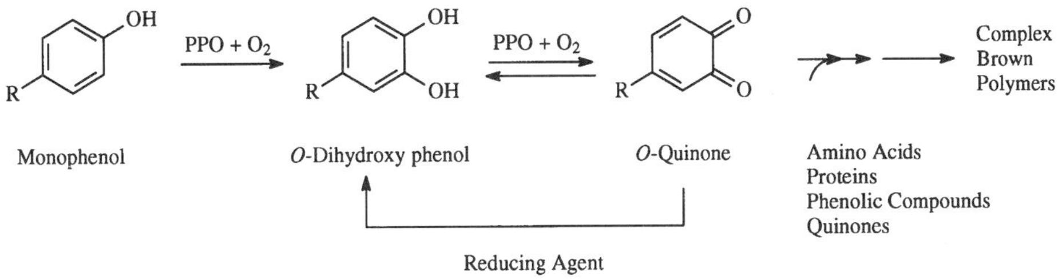

Figure 1 Enzymatic browning reaction, showing action of reducing agents as browning inhibitors. (From Sapers, 1993.)

condense and react nonenzymatically with various other compounds, including phenolic compounds and amino acids, to produce pigments of indeterminate structure (Fig. 1). The subject of enzymatic browning has been reviewed by many authors, including Mayer and Harel (1979), Vamos-Vigyazo (1981, 1995), McEvily et al. (1992), Walker (1995), and Martinez and Whitaker (1995).

Enzymatic browning reactions occur in many plants, usually when plant cells are disrupted so that endogenous substrates and PPO, which are normally compartmentalized, become mixed in the presence of oxygen. Each plant species contains a characteristic pattern of phenolic compounds that are substrates of PPO. Numerous papers and reviews describing these compounds have been published (Maga, 1978; Mayer and Harel, 1979; Vamos-Vigyazo, 1981; Gross, 1981; Herrmann, 1989; Friedman, 1997). Some of the more important substrates in fruits and vegetables are catechins, chlorogenic acid, 3,4-dihydroxyphenylalanine (DOPA), and tyrosine (Fig. 2).

The optimum pH for PPO activity is between 5 and 7. PPO is relatively heat labile, and PPO-catalyzed reactions can be inhibited by acids, halides, phenolic acids, sulfites, chelating agents, reducing agents such as ascorbic acid and dithiothreitol, quinone couplers such as cysteine, and substrate binding compounds such as polyvinylpolypyrollidone (PVPP) and β-cyclodextrin. With some of these compounds, the inhibitory effect is directly on PPO, but in most cases the effect is to inhibit the browning reaction by removing substrates or blocking further reaction of intermediates. A detailed examination of PPO inhibitors and mechanisms of inhibition is beyond the scope of this review since the subject has been addressed elsewhere (Pifferi et al., 1974; Mayer and Harel, 1979; Walker and McCallion, 1980; Vamos-Vigyazo, 1981; Dudley and Hotchkiss, 1989; Ferrar and Walker, 1996). However, we will examine a number of applications of PPO inhibitors as browning inhibitors for fruit and vegetable products.

B. Nonenzymatic Browning

Nonenzymatic browning via the Maillard reaction between carbonyl and amino groups (Fig. 3) represents a complex sequence of nonenzymatic reactions that ultimately result in the formation of nitrogenous polymers and copolymers of variable composition (Hodge, 1953). In addition to discolorations associated with pigment formation, flavor changes may result from the formation of Strecker degradation aldehydes and other volatile compounds (Whitfield, 1992). Participation of free amino groups in browning reactions may result in losses of essential amino acids and reduced protein digestibility (O'Brien and Morrissey, 1989). Nonenzymatic browning reactions also can result in a loss of nutrients such as

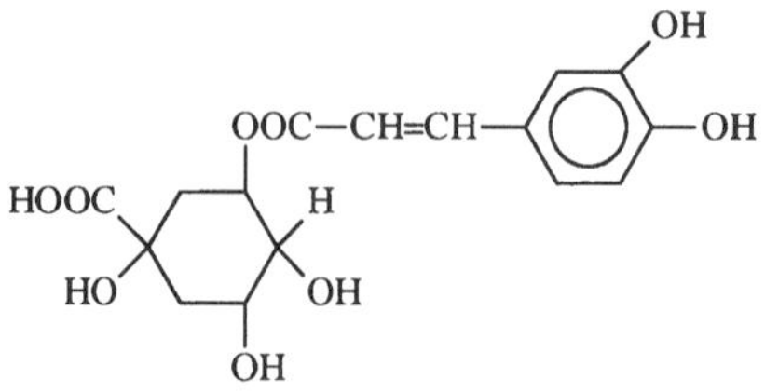

Chlorogenic Acid

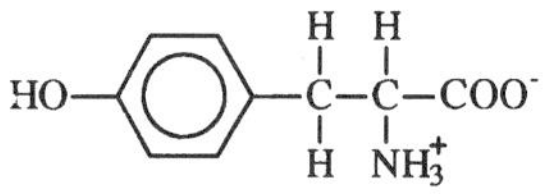

Tyrosine

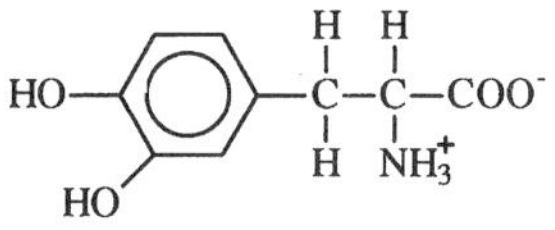

DOPA

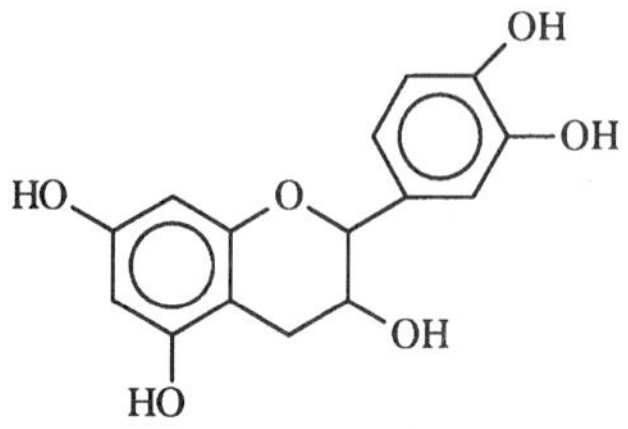

Catechin

Figure 2 Common substrates of polyphenol oxidase in fruits and vegetables.

ascorbic acid, which may become oxidized to dehydroascorbic acid, a highly unstable intermediate that reacts further via aldol condensation or reaction with amino groups to form brown pigments (Kacem et al., 1987; Wong and Stanton, 1989; Löscher et al., 1991). Browning due to anaerobic degradation of ascorbic acid is very important in processed fruit juices enriched with vitamin C. Browning also may result from sugar degradation (Lee and Nagy, 1988). These reactions have been the subject of numerous reviews (Waller and Feather, 1983; Handwerk and Coleman, 1988; Namiki, 1988) and will not be discussed further.

Other nonenzymatic reactions may result in brown or other dark discolorations. The ''after-cooking darkening'' reaction induced by heat during cooking or steam-peeling of potatoes is attributed to formation of iron complexes of chlorogenic acid and is controlled by addition of chelating agents such as sodium acid pyrophosphate (Smith, 1987a). Phenolic compounds can undergo nonenzymatic oxidation that results in browning (Cilliers and Singleton, 1989). We have observed browning in mushrooms treated with 0.01% sodium hypochlorite which oxidizes polyphenols to yield pigments similar to those produced by the enzymatic polyphenol oxidation (Choi and Sapers, 1994a).

Nonenzymatic browning reactions in fruit and vegetable products depend on product composition (Wong and Stanton, 1989; Kennedy et al., 1990) and pH (Wedzicha and Goddard, 1988; O'Brien and Morrissey, 1989) and is usually associated with exposure to heat during processing and storage (Nagy et al., 1990), exposure to oxygen (Kacem et

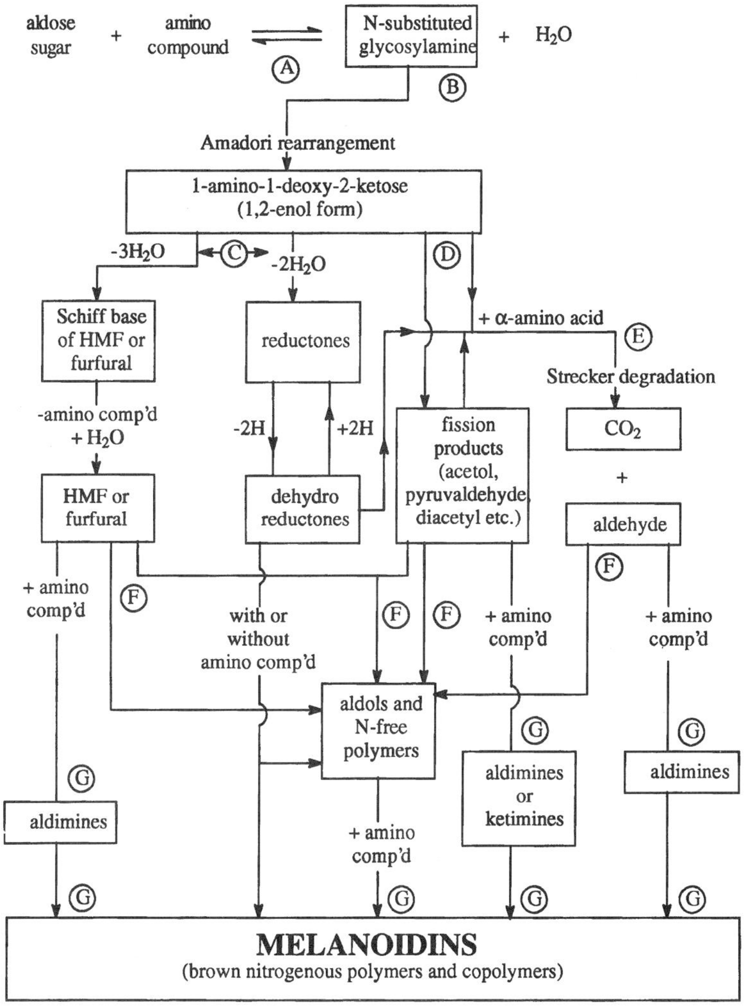

Figure 3 Nonenzymatic browning reaction (Maillard reaction). A, sugar-amine condensation; B, Amadori rearrangement; C, sugar dehydration; D, sugar fragmentation; E, Strecker degradation of amino acid moiety; F, aldol condensation; G, aldehyde–amine polymerization and formation of heterocyclic nitrogen compounds. (From Hodge, 1953.)

al., 1987), and insufficient moisture reduction in dehydrated products (Monsalve et al., 1990). As indicated previously, discoloration and other quality loss due to nonenzymatic browning can be prevented by treatment of fruits and vegetables with sulfites (Bolin and Steele, 1987). If sulfiting is not permitted, there are few options for controlling browning other than reduced exposure to high temperatures, better control of water activity (Labuza and Saltmarch, 1981), and in some cases removal of reducing sugars by treatment with glucose oxidase (Low et al., 1989) or manipulation of raw material storage conditions, e.g., potatoes for chipping (Smith, 1987b). Sulfhydryl-containing amino acids show some capacity to inhibit nonenzymatic browning (Friedman and Molnar-Perl, 1990), but such treatments must have regulatory approval and cannot introduce atypical flavors to treated products. Bolin and Steele (1987) reported that cysteine was ineffective in controlling browning of dried apple.

III. SULFITES AS BROWNING INHIBITORS

A. Treatment of Foods with Sulfites

Sulfites are unique in their ability to perform a number of useful functions as food additives—control of both enzymatic and nonenzymatic browning, suppression of microbial growth, and bleaching (Taylor et al., 1986). They have been used since antiquity for these and other purposes. In the case of enzymatic browning, sulfites act as PPO inhibitors and also react with intermediates to prevent pigment formation (Sayavedra-Soto and Montgomery, 1986). Sulfites inhibit nonenzymatic browning by reacting with carbonyl intermediates, thereby blocking pigment formation (Wedzicha, 1987).

Treatment conditions vary widely. Sulfite may be applied as sulfur dioxide; sulfurous acid; or sodium (or potassium) sulfite, bisulfite, or metabisulfite. Treatment levels vary widely, but treatment residues usually do not exceed several hundred ppm, although some products may contain 1000 ppm (Taylor et al., 1986). Among the products that are treated with sulfites are dehydrated fruits and vegetables, prepeeled potatoes, fresh grapes, and wine. Maximum levels of 300, 500, and 2000 ppm have been proposed for fruit juices, dehydrated potatoes, and dried fruit, respectively (FDA, 1988b).

B. Safety and Regulatory Issues

Sulfite residues in foods have been responsible for some severe allergic reactions in susceptible individuals, usually asthmatics. Fatal anaphylactic reactions have been reported (Taylor et al., 1986). The FDA has restricted use of sulfites in certain categories of foods where there is no means of alerting sensitive consumers to their presence (FDA, 1986). Fruit and vegetable products that are consumed raw and sold unlabeled in salad bars, restaurants, or from bulk containers so that the consumer cannot be alerted to their presence fall in this category (FDA, 1986). However, a ban on sulfiting prepeeled potatoes, a product usually sold to the food service industry (FDA, 1990), was vigorously opposed by the prepeeled potato industry on the ground that there were no effective substitutes, and the ban was eventually overturned in the courts on a technicality (Anonymous, 1990). FDA established labeling requirements for foods containing sulfites and affirmed the GRAS status of sulfiting agents in 1988 (FDA, 1988a,1988b).

Fear that use of sulfites as browning inhibitors for fruit and vegetable products might be restricted prompted the food industry to seek alternatives. During the 1980s, this was an area of great research activity, and a number of new browning inhibitors were developed (Sapers and Hicks, 1989). In recent years, the number of papers in this area has diminished somewhat, but there is still commercial interest in developing improved sulfite substitutes for important commodities such as fresh-cut apples and potatoes. Use of benign ingredients as browning inhibitors and treatments that leave no residues have been especially prominent. In part this is driven by the reluctance of consumers to buy products containing ''chemicals'' of any kind.

In the remainder of this chapter, we will review the range of antibrowning agents that have been considered, including some of the newer developments in this area.

IV. ALTERNATIVES TO SULFITES

The search for alternatives to sulfite has been complicated by the fact that sulfites are extremely potent as browning inhibitors, inexpensive to use, and multifunctional, exhibiting antimicrobial activity as well as anti-browning activity. Ideally, sulfite substitutes should exhibit similar properties. Moreover, sulfite substitutes should be safe and free of regulatory constraints. Unfortunately, most alternatives to sulfites do not meet these criteria, so that the choice of agent must represent a compromise. Furthermore, since individual anti-browning agents may be deficient in some respects, most alternatives to sulfite represent combinations of agents that act synergistically or at the very least have an additive effect. Even so, there is no ''magic bullet'' to control browning in commercial use or under development. Perhaps the eventual development of genetically engineered commodities lacking PPO or with reduced levels of this enzyme will eliminate the need for browning inhibitors. For the foreseeable future, however, there will be a need for safe and cost-effective anti-browning agents.

A. Conventional Alternatives to Sulfite

1. Ascorbic Acid–Based Formulations

Ascorbic acid (vitamin C) has been used as an antibrowning agent for more than five decades and is still the most widely used alternative to sulfiting agents. This may be an outgrowth of the common kitchen practice of using lemon juice to delay browning during food preparation. The earliest scientific studies were reported by Tressler and DuBois (1944) and Esselen et al. (1945). These pioneering investigators added ascorbic acid or its isomer erythorbic (*d*-isoascorbic) acid to syrups or dips to control browning of fresh sliced and frozen apples and peaches. In most respects, ascorbic and erythorbic acids are similar in activity as antioxidants (Borenstein, 1965) and browning inhibitors (Sapers and Ziolkowski, 1987). However, the latter compound does not have vitamin C activity. In some cases, an organic acid such as citric acid and a firming agent such as calcium chloride were added along with the ascorbic acid. To improve the uptake of these agents and at the same time remove oxygen from the product void spaces, vacuum infiltration was sometimes used in conjunction with browning inhibitor treatment (Guadagni, 1949). More recently, Sapers et al. (1990) reported that the ''water-logging'' effects seen in vacuum-infiltrated fresh apples could be avoided by infiltrating ascorbate or erythorbate solutions under pressure.

The chemical basis for the efficacy of ascorbic acid treatments is the ability of ascorbic acid to reduce quinones, produced by PPO-catalyzed oxidation of polyphenols, back to dihydroxy polyphenols. As long as quinones do not accumulate, further reactions leading to pigment formation are avoided. When the added ascorbic acid is depleted quinones will accumulate, and browning will result. Thus, the primary effect of ascorbic acid is as an inhibitor of the enzymatic browning reaction, not as an inhibitor of PPO per se. However, ascorbic acid does have some direct inhibitory effect against PPO (Vamos-Vigyazo, 1981). Dehydroascorbic acid, the oxidation product of ascorbic acid that is formed during quinone reduction, can itself undergo nonenzymatic browning, leading to product discoloration. We have seen such discolorations during refrigerated storage of fresh apple and pear juices that were treated with ascorbic acid during juicing as a temporary means to suppress browning and then rapidly filtered or centrifuged to remove particulate-bound PPO. This treatment permanently eliminates the capacity of juices to undergo enzymatic browning (See Section V) (Sapers, 1991).

During the 1980s, many ascorbic acid–based browning inhibitor formulations were marketed. A survey conducted by the National Restaurant Association in 1986 identified 13 suppliers of such products. These formulations usually contained ascorbic or erythorbic acid (or their sodium salts) in combination with adjuncts such as citric acid or some other acidulant, a calcium salt for firming, a phosphate such as sodium acid pyrophosphate (a chelating agent), sodium chloride, cysteine, and a preservative such as sodium benzoate or potassium sorbate (Sapers, 1993). Different formulations usually were provided for apples, potatoes, and salad vegetables. Use levels for these products varied over a wide range, and there appeared to be no consensus as to optimal ascorbic acid levels. These products were claimed to provide a shelf-life of 4–7 days under refrigeration, far less than could be achieved with sulfites.

A longer shelf-life could be achieved by shipping peeled potatoes in a preservative solution after treatment with browning inhibitors (Santerre et al., 1991). However, the logistics of this approach are highly unfavorable. Another approach to shelf-life extension was use of vacuum packing after browning inhibitor treatment (Langdon, 1987). However, rapid browning would ensue once the vacuum was broken. Vacuum packing of raw potatoes might be hazardous because of the potential ability of *Clostridium botulinum* to grow and produce toxin in peeled, partially cooked potatoes under anaerobic conditions (Tamminga et al., 1978). A recent study by Juneja et al. (1998) demonstrated growth without concurrent spoilage by *Listeria monocytogenes* in inoculated vacuum-packed potatoes under conditions of temperature abuse.

Research on ascorbic acid–2-phosphates and fatty acid esters as stabilized forms of ascorbic acid showed that these derivatives exhibited some advantages over ascorbic acid for control of browning in cut apples (Sapers et al., 1989b), cut potatoes (Sapers and Miller, 1992), and fresh fruit juice (Sapers et al., 1989b). However, no commercial interest in this technology was forthcoming, probably because of the need for regulatory approval.

At the present time, major suppliers of ascorbic acid or erythorbic acid–based browning inhibitor formulations are Monsanto Chemical Company (Snow Fresh™), EPL Technologies, Inc. (Potato Fresh™), and Mantrose-Haeuser Co., Inc. (NatureSeal™). Numerous companies supply the ingredients from which such formulations can be made.

2. Cysteine

The ability of cysteine to inhibit enzymatic browning is well established and has been used commercially for a number of years (Cherry and Singh, 1990). This alternative to sulfites is a key ingredient of browning inhibitor formulations for apples and prepeeled potatoes supplied by EPL Technologies, Inc. (Brereton and Sapers, 1997). Cysteine is also a component of browning inhibitor treatments developed by Senesi and Pastine (1996) and by Gunes and Lee (1997). Cysteine reacts with quinone intermediates, formed by PPO-catalyzed oxidation of polyphenols, to yield stable, colorless compounds, thereby blocking pigment formation (Dudley and Hotchkiss, 1989). Cysteine also directly inhibits the enzyme (Robert et al., 1996). We have observed that cysteine-based inhibitors are particularly effective against browning of apple core tissue. However, under some conditions, cut pears and potatoes treated with cysteine show pink discolorations. At high treatment levels, a noticeable sulfury odor can result. Studies by Molnar-Perl and Friedman (1990 a,b) and Friedman (1994) suggest that reduced glutathione and N-acetylcysteine are nearly as effective as sulfites in controlling browning of apple, potato, and fresh fruit juices. However, only cysteine is approved for food use.

3. 4-Hexylresorcinol

This PPO inhibitor is used commercially to control discoloration of unpeeled shrimp (Everfresh™) and is highly effective as a browning inhibitor for some fruits and vegetables (McEvily et al., 1991; Monsalve-Gonzalez et al., 1993; Luo and Barbosa-Canovas, 1997). While 4-hexylresorcinol has a long history of human consumption (Frankos et al., 1991), it is only approved for use on shrimp (where the application is to the peel, which is not usually eaten). Regulatory approval for other applications where it might be consumed in quantity is uncertain. We have found that it is particularly effective against core browning in fresh-cut pears (Sapers and Miller, 1998). However, 4-hexylresorcinol caused darkening and tissue breakdown in fresh mushrooms at concentrations greater than 50 ppm (Sapers et al., 1994).

B. Other PPO Inhibitors

Various PPO inhibitors have been proposed as alternatives to sulfites, but they lack regulatory approval at this time. Walker and Wilson (1975) investigated inhibition of apple PPO by a number of phenolic acids. Walker (1976) reported that cinnamic acid at concentrations greater than 0.5 mM prevented browning of Granny Smith apple juice for 7 hours. Experiments carried out with cinnamate in our laboratory with cut apples also showed short-term browning inhibitor activity, but the treatment appeared to induce browning after 24 hours at ambient temperature (Sapers et al., 1989b). Benzoate also induced browning in cut apples after prolonged storage.

Kojic acid [5-hydroxy-2-(hydroxymethyl)-γ-pyrone] has been considered for use as an alternative to sulfite. This compound can reduce quinones to polyphenols, thereby preventing browning by the same mechanism as ascorbic acid (Chen et al., 1991). It is not clear whether this compound acts as a true PPO inhibitor. The regulatory status of kojic acid is in doubt because of reported mutagenic activity (Wei et al., 1991).

Tong and Hicks (1991, 1993) reported that carrageenans and other sulfated polysaccharides showed browning inhibitor activity in apple juice and diced apples; citric acid acted synergistically with these compounds in inhibiting browning. The mechanism of browning inhibition by the sulfated polysaccharides is not known. Xu et al. (1993) reported that maltodextrin inhibited browning of ground apple.

Natural PPO inhibitors have been found in honey (Oszmianski and Lee, 1990), pineapple (Lozano de Gonzalez et al., 1993; Wen and Wrolstad, 1998), fig latex (McEvily, 1991), and a large number of botanical products (Choi et al., 1997). In some cases, the natural origin of such products has been equated with safety. This view is probably misguided since their purity and composition may be unknown, and consumption patterns of these products as browning inhibitors may be substantially different than their use as herbs or traditional medicinal agents. In any event, FDA approval would be required if the products were to be isolated, concentrated, and applied to foods as browning inhibitors.

C. Complexing Agents

In studies carried out in our laboratories and in France, β-cyclodextrin was found to be an effective browning inhibitor for fruit and vegetable juices (Sapers et al., 1989b; Hicks et al., 1990; Billaud et al., 1995; Hicks et al., 1996). Treatment with β-cyclodextrin did not control browning of cut apple or potato (Sapers et al., 1989b). β-cyclodextrin has been shown to form inclusion complexes with chlorogenic acid and presumably other PPO substrates, thereby effectively removing them from contact with PPO in fresh juices (Irwin et al., 1994). Although there has been commercial interest in this antibrowning agent, regulatory approval has not been forthcoming. Fruit and vegetable juices treated with insoluble polymerized forms of β-cyclodextrin resist browning indefinitely (Hicks et al., 1996). Use of such forms of this compound should avoid regulatory issues associated with treatment residues.

PVPP, an insoluble product used as a fining agent for juices, can bind the polyphenol substrates of PPO and thus prevent browning (Van Buren, 1989). PVPP can be separated from juices by decanting or centrifugation so that there are no treatment residues. Studies in our laboratory have shown that this agent is highly effective in preventing browning of tropical fruit juice blends (Essa and Sapers, 1998).

Sodium acid pyrophosphate (SAPP) is widely used as a chelating agent in potato products. Its primary purpose is to prevent after-cooking darkening by chelating ferrous ion, thus preventing formation of dark-colored iron-chlorogenic acid complexes (Feinberg et al., 1987). We have used it in browning inhibitor formulations for prepeeled potatoes (Sapers and Miller, 1995).

Another complexing agent, a proprietary polyphosphate marketed as Sporix, was studied in our laboratory as a potential sulfite substitute. This product is highly acidic and is reputed to be a powerful chelating agent (Friedman, 1986; Gardner et al., 1991). We found that this product was highly effective in controlling browning of cut apple and apple juice when used in combination with ascorbic acid (Sapers et al., 1989b). However, further studies with Sporix were curtailed since the supplier was unable to obtain regulatory approval for its use.

While polyphosphates and cyclodextrins each provide browning inhibition in fresh juices, use of these complexing agents in combination leads to even more effective (synergistic) results (Hicks et al., 1996). Combinations of β-cyclodextrin with either SAPP, Sporix, sodium hexametaphosphate, or phytic acid resulted in greater browning inhibition than predicted from the effects of each component tested separately. This synergistic effect allows greater inhibition of browning and use of lower levels of additives to achieve control of browning.

V. SPECIAL PROBLEMS IN CONTROL OF ENZYMATIC BROWNING

The efficacy of browning inhibitor treatments for fruit and vegetable products sometimes depends on factors other than the antibrowning agent and application conditions used. In some cases, commodity condition plays a key role in determining treatment efficacy. In other cases, the treatment may exert an adverse effect on some quality attribute, necessitating treatment modification. In all cases, one should take a holistic approach in developing browning inhibitor treatments so that treatments are effective and avoid unforeseen adverse consequences. The following examples illustrate some of these complicating factors and strategies used to circumvent them.

A. Fresh-Cut Apples

In earlier work, we found that apple plugs and slices responded better to treatment with neutral solutions of sodium ascorbate or sodium erythorbate than to treatment with the respective acids (Sapers, 1988; Sapers et al., 1990). Laboratory studies with fresh-cut apples demonstrated effective control of enzymatic browning by application of dips containing 2–4% sodium erythorbate and 0.1–0.2% calcium chloride. Products were packaged in films that produced a modified atmosphere with reduced oxygen and elevated carbon dioxide during storage. The modified atmosphere suppressed both browning and microbial spoilage. Typically, products had a shelf-life of 2 weeks at 4°C, based on absence of browning and microbial spoilage.

Two factors complicated this treatment. Apple slices prepared by hand in the laboratory were generally free of residual core tissue and showed uniform response to treatment. However, field tests of the treatment with commercially cored and sliced apples containing residual core demonstrated that the core tissue did not respond as well to antibrowning agents as the flesh portion of slices. Thus, slices with residual core tissue showed conspicuous browning when most slice surfaces were still free of discoloration. Research is in progress to address this problem by use of alternative antibrowning agents and modified treatment conditions (Brereton and Sapers, 1997).

The second complicating factor was the development of atypical, pineapplelike odors during modified atmosphere storage of treated products. We attribute this quality defect to the development of near-anaerobic conditions within the packages that caused apple slices to undergo anaerobic respiration and generate ethanol (Zhuang et al., 1998). We speculate that the ethanol reacted with endogenous esters on the cut surfaces of slices to produce some atypical ethyl esters by transesterification. The resulting change in volatile composition was perceived as pineapplelike. This problem could be addressed by using a packaging material with greater oxygen permeability so that an anaerobic condition would not occur during normal product storage (Brereton and Sapers, 1997). However, this represents a tradeoff since the greater oxygen concentration within the package resulted in more browning and a shorter shelf-life.

B. Fresh-Cut Pears

In many respects, fresh-cut pears are similar to apples in their response to browning inhibitor treatments (Sapers and Miller, 1998). Neutral sodium ascorbate or erythorbate treatments with added calcium chloride were more effective than acidic treatments. Residual core tissue on slices did not respond as well to antibrowning treatments as the flesh portion. However, with fresh-cut d'Anjou and Bartlett pear slices, blackening of the cut edge of the skin during storage was even more conspicuous. Residual core browning could be controlled by addition of 50–100 ppm 4-hexylresorcinol to the browning inhibitor dip.

By far the most important factor affecting browning of fresh-cut pears was raw material ripeness. We found that slightly under-ripe pears responded better than fully ripe pears to application of browning inhibitor dips in combination with modified atmosphere packaging. The less ripe fruit showed less browning of cut surfaces, residual core tissue, and the cut edge of the skin than fully ripe pears. The slightly under-ripe fresh-cut pears probably suffered less tissue damage than the riper fruit during slicing. Use of slightly under-ripe fruit resulted in slightly lower flavor levels but better retention of firm texture during storage.

C. Pre-Peeled Potatoes

In studies with prepeeled potatoes, we found that enzymatic browning could be controlled by treatment with ascorbic acid–2-phosphates (Sapers and Miller, 1992) or by surface digestion with sodium hydroxide prior to application of an ascorbic acid–based browning inhibitor (Sapers and Miller, 1993). Good control of browning also could be obtained when acidic browning inhibitor solutions containing ascorbic and citric acids were applied to peeled potatoes at 45–55°C (Sapers and Miller, 1995). However, such treatment resulted in case-hardening of potato surfaces which interfered with mashing and slicing. Similar effects, although not as severe, were observed when peeled potatoes were treated with conventional acidic sulfite substitutes. We believe that this effect was due to activation of pectinmethylesterase during treatment which partially demethylated pectin, creating new binding sites for calcium or magnesium ions. Subsequent cross-linking of pectin by these ions would result in the atypical surface firming that we observed (Sapers et al., 1997).

The textural abnormality could not be addressed by reducing the treatment temperature or by using a neutral browning inhibitor treatment since these steps would have greatly reduced browning inhibitor efficacy. However, by buffering the browning inhibitor solution, a compromise treatment was devised that yielded an extended product shelf-life without the textural defect (Martin et al., 1997).

D. Washed Fresh Mushrooms

Another example of a factor interfering with browning inhibitor application was seen with washed, fresh mushrooms. It is well-known that washing predisposes fresh mushrooms to premature spoilage by *Pseudomonas tolaasii*, which produces a dark, sunken lesion on mushroom surfaces called ''brown'' or ''bacterial blotch.'' Washing with sulfite solution avoided this problem because of sulfite's antimicrobial activity and, at the same time, inhibited browning of the mushroom surface. However, sulfites can no longer be used on fresh mushrooms. Washing with chlorinated water sometimes caused mushroom darkening (Choi and Sapers, 1994a) or induced purple discolorations on mushroom surfaces, attributed to the reaction between L-DOPA oxidation products and sinapic acid (Choi and Sapers, 1994b). We were unable to find an effective alternative to sulfite for controlling browning of fresh mushrooms until we addressed the microbiological spoilage problem first. We found that a wash in 5% hydrogen peroxide to reduce the bacterial load prior to application of a sodium erythorbate spray inhibited browning with no shelf-life penalty due to microbial spoilage (Sapers et al., 1994, 1995).

These examples clearly show the fallacy of looking at browning inhibitor treatments in isolation. One must consider the tendency of the commodity and even specific cultivars to undergo browning; commodity maturity or ripeness; raw material condition; and peripheral effects of treatment on product quality or susceptibility to microbial spoilage. Therefore, antibrowning agents should be tested with a broad range of raw material types, prepared in a form representative of commercial practice, stored under realistic conditions, and evaluated for overall quality as well as efficacy in controlling browning.

E. Fresh Juices

The growing popularity of minimally processed fruit and vegetable products is reflected in the appearance of supermarket juice bars and other retail establishments selling beverages based on fresh fruit homogenates (''smoothies'') and juice blends. Such products may contain active enzymes, including PPO, which could affect product color with some commodities such as pear, apple, and mango. Recent concern over the microbiological safety of unpasteurized cider will undoubtedly restrict production and marketing of some products, but several nonthermal pasteurization treatments under development (i.e., membrane filtration, ultra-high pressure, pulsed electric fields, UV-pasteurization, high intensity pulsed light) may provide a means of safely preserving fresh juice products. In most cases, these treatments do not inactivate enzymes, and products derived from commodities containing sufficient PPO may undergo browning during storage. Browning inhibition treatments entailing use of additives may not be acceptable to the fresh juice market.

We have developed technology for controlling browning in fresh juices without use of additives that may be applicable to this situation. In some unfiltered juices such as apple and pear, PPO is bound to particulate fractions and can be removed by filtration or centrifugation to yield a nonbrowning product (Sapers, 1991). Removal of PPO can be enhanced by addition of insoluble fining agents such as diatomaceous earth, bentonite, or chitosan (Sapers, 1991, 1992). Alternatively, an insoluble adsorbent or complexing agent such as polymerized β-cyclodextrin (Hicks et al., 1990, 1996) or PVPP (Essa and Sapers, 1998) could be added to remove PPO substrates from the juice. These treatments leave no residues in the juice but might adsorb or otherwise remove components important to juice quality such as flavor compounds, pigments, and suspended solids that contribute to desirable cloud. These potential problems as well as treatment efficacy, cost, and regulatory status must be considered in determining the feasibility of technology to control browning in fresh juices without use of additives.

F. Lettuce

Harvested lettuce is subject to enzymatic browning of the cut stem (butt discoloration). Peiser et al. (1998) demonstrated that wounding of lettuce during cutting induced production of phenylalanine ammonia lyase (PAL) and subsequent formation of phenolic substrates of PPO by the phenylpropanoid pathway, which, upon oxidation, caused browning. Application of acetic acid to stem tissue completely inhibited PAL activity and the production of wound-induced phenolics, thereby preventing browning (Tomas-Barberan et al., 1997). Castaner et al. (1997) found that acetic and propionic acid treatments were effective in reducing butt discoloration during storage and commercial handling. Heat shock treatments also suppressed PAL activity; synthesis of chlorogenic acid, dicaffeoyl tartaric acid, and isochlorogenic acid; and subsequent browning (Loaizavelarde et al., 1997). These studies demonstrate the potential value of treatments that interfere with the formation of PPO substrates as an alternative to the direct inhibition of PPO or addition of reducing or complexing agents that block pigment formation.

G. Use of Nonbrowning Cultivars

Control of enzymatic browning without use of antibrowning agents might be achieved in some commodities by use of raw materials that are deficient in PPO. Some cultivars are naturally deficient in PPO or PPO substrates so that they are less subject to browning than more common cultivars (Coseting and Lee, 1987; Sapers and Douglas, 1987; Sapers et al., 1989a). Nonbrowning cultivars might be developed through breeding programs (Woodwards and Jackson, 1985) or genetic engineering (Bachem et al., 1994; Mooibroek et al., 1996). However, such cultivars should be equivalent to conventional cultivars in quality, processability, disease resistance, yield, and other important characteristics. Whether the cost of eliminating the browning trait by genetic engineering can be justified remains to be seen.

VI. CONCLUSIONS

Recent advances in antibrowning agents and improved application methods, used in conjunction with antimicrobial treatments and modified atmosphere packaging, have yielded large improvements in control of browning in fresh-cut and other minimally processed fruit and vegetable products. However, whether such improvements can be translated into shelf-life extension will depend on the retention of other quality attributes such as flavor and texture as well as on the suppression of microbial spoilage. Of course, we must always be alert to the possibility that suppression of spoilage organisms might create a niche for human pathogens.

Technical achievements in development of antibrowning agents are only part of the story. Regulatory hurdles, labeling issues, lack of competitive advantage, and cost may limit the commercialization of promising treatments. Thus, there is a continuing need for research to develop effective antibrowning treatments that meet the requirements of technical and economic feasibility, safety, and consumer acceptance.

REFERENCES

Anonymous. 1990. Court overturns sulfite ban on fresh potatoes. *Food Chemical News*, August 6, p. 60.

Bachem, C. W. B., Speckmann, G.-J. van der Linde, P. C. G., Verheggen, F. T. M., Hunt, M. D., Steffens, J. C., Zabeau, M. 1994. Antisense expression of polyphenol oxidase genes inhibits enzymatic browning in potato tubers. *Bio/Technol.* 12:1101.

Billaud, C., Regaudie, E., Fayad, N., Richard-Forget, F., Nicolas, J. 1995. Effect of cyclodextrins on polyphenol oxidation catalyzed by apple polyphenol oxidase. In: *Enzymatic Browning and Its Prevention*, ACS Symposium Series 0097–6156, 600, American Chemical Society, Washington, D.C. pp. 295–312.

Bolin, H. R., Steele, R. J. 1987. Nonenzymatic browning in dried apples during storage. *J. Food Sci.* 52:1654.

Borenstein, B. 1965. The comparative properties of ascorbic acid and erythorbic acid, *Food Technol.* 19:1719.

Brereton, E., Sapers, G. M. 1997. Unpublished data. Eastern Regional Research Center, Wyndmoor, PA.

Castaner, M., Gil, M. I., Artes, F. 1997. Organic acids as browning inhibitors on harvested baby lettuce and endive. *Z. Lebensm. Unters. Forsch.* 205:375.

Chen, J. S., Wei, C.-I., Marshall, M. R. 1991. Inhibition mechanism of kojic acid on polyphenol oxidase. *J. Agric. Food Chem.* 39:1897.

Cherry, J., Singh, S. S. 1990. Discoloration preventing food preservative and method. U.S. patent 4,937,085.

Choi, S.-W., Sapers, G. M. 1994a. Effects of washing on polyphenols and polyphenol oxidase in commercial mushrooms (*Agaricus bisporus*). *J. Agric. Food Chem.* 42:2286.

Choi, S.-W., Sapers, G. M. 1994b. Purpling reaction of sinapic acid model systems containing L-DOPA and mushroom tyrosinase. *J. Agric. Food Chem.* 42:1183.

Choi, S.-W., Kim, H.-J., Chang, E.-J., Sapers, G. M. 1997. Inhibition of tyrosinase activity by plant extracts. *Foods Biotechnol.* 6:44.

Cilliers, J. J. L., Singleton, V. L. 1989. Nonenzymatic autoxidative phenolic browning reactions in a caffeic acid model system. *J. Agric. Food Chem.* 37:890.

Coseteng, M. Y., Lee, C. Y. 1987. Changes in apple polyphenoloxidase and polyphenol concentrations in relation to degree of browning. *J. Food Sci.* 52:985.

Dudley, E. D., Hotchkiss, J. H. 1989. Cysteine as an inhibitor of polyphenol oxidase. *J. Food Biochem.* 13:65.

Essa, H. A. A., Sapers, G. M. 1998. Unpublished data. Eastern Regional Research Center, Wyndmoor, PA.

Esselen, W. B., Jr., Powers, J. J., Woodward, R. 1945. *d*-Isoascorbic acid as an antioxidant. *Ind. Eng. Chem.* 37:295.

FDA. 1986. Sulfiting agents; revocation of GRAS status for use on fruits and vegetables intended to be served or sold raw to consumers. *Fed. Reg.* 51(131):25021.

FDA. 1988a. Sulfiting agents in standardized foods; labeling requirements. *Fed. Reg.* 53(243): 51062.

FDA. 1988b. Sulfiting agents; affirmation of GRAS status. *Fed. Reg.* 53(243):51065.

FDA. 1990. Sulfiting agents; revocation of GRAS status for use on ''fresh'' potatoes served or sold unpackaged and unlabeled to consumers. *Fed. Reg.* 55(51):9826.

Feinberg, B., Olson, R. L., Mullins, W. R. 1987. Prepeeled potatoes. In: *Potato Processing*, 4th ed., Talburt, W. F., Smith, O. (Eds.). AVI–Van Nostrand Reinhold, New York, p. 697.

Ferrar, P. H., Walker, J. R. L. 1996. Inhibition of diphenol oxidases: a comparative study. *J. Food Biochem.* 20:15.

Frankos, V. H., Schmitt, D. F., Haws, L. C., McEvily, A. J., Iyengar, R., Miller, S. A., Munro, I. C., Clydesdale, F. M., Forbes, A. L., Sauer, R. M. 1991. Generally recognized as safe (GRAS) evaluation of 4-hexylresorcinol for use as a processing aid for prevention of melanosis in shrimp. *Reg. Toxicol. Pharmacol.* 14:202.

Friedman, M. 1994. Improvement in the safety of foods by SH-containing amino acids and peptides. A review. *J. Agric. Food Chem.* 42:3.

Friedman, M. 1997. Chemistry, biochemistry, and dietary role of potato polyphenols—a review. *J. Agric. Food Chem.* 45:1523.

Friedman, M. and Molnar-Perl, I. 1990. Inhibition of browning by sulfur amino acids. 1. Heated amino acid-glucose systems. *J. Agric. Food Chem.* 38:1642.

Friedman, S. 1986. Personal communication. International Sourcing, Inc., South Ridgewood, NJ.

Gardner, J., Monohar, S., Borisenok, W. S. 1991. Method and composition for preserving fresh peeled fruits and vegetables. U.S. patent 4,988,523.

Gross, G. G. 1981. Phenolic acids. In: *The Biochemistry of Plants*, Vol. 7. Academic Press, New York, pp. 301–316.

Guadagni, D. G. 1949. Syrup treatment of apple slices for freezing preservation. *Food Technol.* 3: 404.

Gunes, G., Lee, C. Y. 1997. Color of minimally processed potatoes as affected by modified atmosphere packaging and antibrowning agents. *J. Food Sci.* 62:572.

Hall, G. C. 1989. Refrigerated, frozen and dehydrofrozen apples. In: *Processed Apple Products*, Downing, D. D. (Ed.). AVI–Van Nostrand Reinhold, New York, p. 239.

Handwerk, R. L., Coleman, R. L. 1988. Approaches to the citrus browning problem. A review. *J. Agric. Food Chem.* 36:231.

Herrmann, K. 1989. Occurrence and content of hydroxycinnamic and hydroxybenzoic acid compounds in foods. *Crit. Rev. Food Sci. Nutr.* 28:315.

Hicks, K. B., Sapers, G. M., Seib, P. A. 1990. Process for preserving raw fruit and vegetable juices using cyclodextrins and compositions thereof. U.S. patent 4,975,293.

Hicks, K. B., Haines, R. M., Tong, C. B. S., Sapers, G. M., El-Atawy, Y., Irwin, P. L., Seib, P. A. 1996. Inhibition of enzymatic browning in fresh fruit and vegetable juices by soluble and insoluble forms of beta-cyclodextrin alone or in combination with phosphates. *J. Agric. Food Chem.* 44:2591.

Hodge, J. E. 1953. Dehydrated foods: chemistry of browning reactions in model system. *J. Agric. Food Chem.* 1:928.

Huxsoll, C. C., Bolin, H. R., King, A. D., Jr. 1989. Physicochemical changes and treatments for lightly processed fruits and vegetables. In: *Quality Factors of Fruits and Vegetables: Chemistry and Technology*, Jen, J. J. (Ed.). ACS Symposium Series 405, American Chemical Society, Washington, D.C., p. 203.

Irwin, P. L., Pfeffer, P. E., Doner, L. W., Sapers, G. M., Brewster, J. D., Nagahashi, G., Hicks, K. B. 1994. Binding geometry, stoichiometry, and thermodynamics of cyclomalto-oligosaccharide (cyclodextrin) inclusion complex formation with chlorogenic acid, the major substrate of apple polyphenol oxidase. *Carbohydrate Research* 256:13.

Juneja, V. K., Martin, S. T., Sapers, G. M. 1998. Control of *Listeria monocytogenes* in vacuum-packaged pre-peeled potatoes. *J. Food Sci.* 63:911.

Kacem, B., Cornell, J. A., Marshall, M. R., Shireman, R. B., Matthews, R. F. 1987. Nonenzymatic browning in aseptically packaged orange drinks: effect of ascorbic acid, amino acids and oxygen. *J. Food Sci.* 52:1668.

Kennedy, J. F., Rivera, Z. S., Lloyd, L. L., Warner, F. P., Jumel, K. 1990. Studies on non-enzymatic browning in orange juice using a model system based on freshly squeezed orange juice. *J. Sci. Food Agric.* 52:85.

Labuza, T. P., Saltmarch, M. 1981. The nonenzymatic browning reaction as affected by water in foods. In: *Water Activity: Influence on Food Quality*, Rockland, L. B., Stewart, G. F. (Eds.). Academic Press, New York, p. 605.

Labuza, T. P., Schmidl, M. K. 1986. Advances in the control of browning reactions in foods. In: *Role of Chemistry in the Quality of Processed Food*, Fennema, O. R., Chang, W. H., Lii, C. Y. (Eds.). Food & Nutrition Press, Westport, CT, p. 65.

Langdon, T. T. 1987. Prevention of browning in fresh prepared potatoes without the use of sulfating agents. *Food Technol.* 41(5):64.

Lee, H. S., Nagy, S. 1988. Relationship of sugar degradation to detrimental changes in citrus juice quality. *Food Technol.* 42(11):91.

Loaizavelarde, J. G., Tomasbarbera, F. A., Saltveit, M. E. 1997. Effect of intensity and duration of heat-shock treatments on wound-induced phenolic metabolism in iceberg lettuce. *J. Am. Soc. Hort. Sci.* 122:873.

Löscher, J., Kroh, L., Westphal, G., Vogel, J. 1991. L-Ascorbic acid—a carbonyl component of non-enzymatic browning reactions. 2. Amino-carbonyl reactions of L-ascorbic acid. *Z. Lebensm. Unters. Forsch.* 192:323.

Low, N., Jiang, Z., Ooraikul, B., Dokhani, S., Palcic, M. M. 1989. Reduction of glucose content in potatoes with glucose oxidase. *J. Food Sci.* 54:118.

Lozano de Gonzalez, P. G., Barrett, D. M., Wrolstad, R. E., Durst, R. 1993. Enzymatic browning inhibited in fresh and dried apple rings by pineapple juice. *J. Food Sci.* 58:399.

Luo, Y., Barbosa-Canovas, G. V. 1997. Enzymatic browning and its inhibition in new apple cultivars slices using 4-hexylresorcinol in combination with ascorbic acid. *Food Sci. Technol. Int.* 3: 195.

Ma, S., Silva, J. L., Hearnsberger, J. O., Garner, J. O., Jr. 1992. Prevention of enzymatic darkening in frozen sweet potatoes [*Ipomoea batatas* (L.) Lam.] by water blanching: relationship among darkening, phenols, and polyphenol oxidase activity. *J. Agric. Food Chem.* 40:864.

Maga, J. A. 1978. Simple phenol and phenolic compounds in food flavor. *CRC Crit. Rev. Foods Sci. Nutr.* 10:323.

Martin, S. T., Sapers, G. M., Miller, R. L. 1999. Process for inhibiting enzymatic browning and maintaining textural quality of fresh peeled potatoes. U.S. Patent 5,912,034.

Martinez, M. V., Whitaker, J. R. 1995. The biochemistry and control of enzymatic browning. *Trends Food Sci. Technol.* 6:195.

Mayer, A. M., Harel, E. 1979. Polyphenol oxidases in plants. *Phytochemistry* 18:193.

McCord, J. D., Kilara, A. 1983. Control of enzymatic browning in processed mushrooms (*Agaricus bisporus*). *J. Food Sci.* 48:1479.

McEvily, A. J. 1991. Method of preventing browning of food utilizing protease free latex extracts, particularly from figs. U.S. patent 4,981,708.

McEvily, A. J., Iyengar, R., Otwell, S. 1991. Sulfite alternative prevents shrimp melanosis. *Food Technol.* 45(9):80.

McEvily, A. J., Iyengar, R., Otwell, S. 1992. Inhibition of enzymatic browning in foods and beverages. *Crit. Rev. Food Sci. Nutr.* 32:253.

Molnar-Perl, I., Friedman, M. 1990a. Inhibition of browning by sulfur amino acids. 2. Fruit juices and protein-containing foods. *J. Agric. Food Chem.* 38:1648.

Molnar-Perl, I., Friedman, M. 1990b. Inhibition of browning by sulfur amino acids. 3. Apples and potatoes. *J. Agric. Food Chem.* 38:1652.

Monsalve, A., Powers, J. R., Leung, H. K. 1990. Browning of dehydroascorbic acid and chlorogenic acid as a function of water activity. *J. Food Sci.* 55:1425.

Monsalve-Gonzalez, A., Barbosa-Canovas, G. V., Cavalieri, R. P., McEvily, A. J., Iyengar, R. 1993. Control of browning during storage of apple slices preserved by combined methods. 4-hexylresorcinol as anti-browning agent. *J. Food Sci.* 58:797.

Mooibroek, H., van de Rhee, M., Rivas, C. S., Mendes, O., Werten, M., Huizing, H., Wichers, H. 1996. Progress in transformation of the common mushroom, *Agaricus bisporus*. In: *Mushroom Biology and Mushroom Products*, Royse, D. (Ed.). Pennsylvania State University Press, University Park, PA.

Nagy, S., Lee, H., Rouseff, R. L., Lin, J. C. C. 1990. Nonenzymatic browning of commercially canned and bottled grapefruit juice. *J. Agric. Food Chem.* 38:343.

Namiki, M. 1988. Chemistry of Maillard reaction: recent studies on the browning reaction mechanism and the development of antioxidants and mutagens. *Adv. Food Res.* 32:115.

Nichols, R. 1985. Post-harvest physiology and storage. In: *The Biology and Technology of the Cultivated Mushroom*, Flegg, P. B., Spencer, D. M., Wood, D. A. (Eds.). John Wiley & Sons, New York, p. 195.

O'Brien, J., Morrissey, P. A. 1989. Nutritional and toxicological aspects of the Maillard browning reaction in foods. *Crit. Rev. Food Sci. Nutr.* 28:211.

Oszmianski, J., Lee, C. Y. 1990. Inhibition of polyphenol oxidase activity and browning by honey. *J. Agric. Food Chem.* 38:1892.

Peiser, G., Lopezgalvez, G., Cantwell, M., Saltveit, M. E. 1998. Phenylalanine ammonia lyase inhibitors control browning of cut lettuce. *Postharvest Biology & Technology* 14:171.

Pifferi, P. G., Baldassari, L., Cultrera, R. 1974. Inhibition by carboxylic acids of an *o*-diphenyl oxidase from *Prunus avium* fruits. *J. Sci. Food. Agric.* 25:263.

Ponting, J. D., Jackson, R. 1972. Pre-freezing processing of Golden Delicious apple slices. *J. Food Sci.* 37:812.

Robert, C., Richard-Forget, F., Rouch, C., Pabion, M., Cadet, F. 1996. A kinetic study of the inhibition of palmito polyphenol oxidase by L-cysteine. *Intl. J. Cell Biol.* 28:457.

Santerre, C. R., Leach, T. F., Cash, J. N. 1991. Bisulfite alternatives in processing abrasion-peeled Russet Burbank potatoes. *J. Food Sci.* 56:257.

Sapers, G. M. 1988. Unpublished data. Eastern Regional Research Center, Wyndmoor, PA.

Sapers, G. M. 1991. Control of enzymatic browning in raw fruit juice by filtration and centrifugation. *J. Food Processing Preservation* 15:443.

Sapers, G. M. 1992. Chitosan enhances control of enzymatic browning in apple and pear juice by filtration. *J. Food Sci.* 57:1192.

Sapers, G. M. 1993. Browning of foods: control by sulfites, antioxidants and other means. *Food Technol.* 47(10):75.

Sapers, G. M., Douglas, F. W., Jr. 1987. Measurement of enzymatic browning at cut surfaces and in juice of raw apple and pear fruits. *J. Food Sci.* 52:1258.

Sapers, G. M., Hicks, K. B. 1989. Inhibition of enzymatic browning in fruits and vegetables. In: *Quality Factors of Fruits and Vegetables: Chemistry and Technology*, Jen, J. J. (Ed.). ACS Symp. Series No. 405, American Chemical Society, Washington, D.C., p. 29.

Sapers, G. M., Miller, R. L. 1992. Enzymatic browning control in potato with ascorbic acid–2-phosphates. *J. Food Sci.* 57:1132.

Sapers, G. M., Miller, R. L. 1993. Control of enzymatic browning in pre-peeled potatoes by surface digestion. *J. Food Sci.* 58:1076.

Sapers, G. M., Miller, R. L. 1995. Heated ascorbic/citric acid solution as browning inhibitor for pre-peeled potatoes. *J. Food Sci.* 60:762.

Sapers, G. M., Miller, R. L. 1998. Browning inhibition in fresh-cut pears. *J. Food Sci.* 63:342.

Sapers, G. M., Ziolkowski, M. A. 1987. Comparison of erythorbic and ascorbic acids as inhibitors of enzymatic browning in apple. *J. Food Sci.* 52:1732.

Sapers, G. M., Douglas, F. W., Jr., Bilyk, A., Hsu, A.-F., Dower, H. W., Garzarella, L., Kozempel, M. 1989a. Enzymatic browning in Atlantic potatoes and related cultivars. *J. Food Sci.* 54:362.

Sapers, G. M., Hicks, K. B., Phillips, J. G., Garzarella, L. G., Pondish, D. L., Matulaitis, R. M., McCormack, T. J., Sondey, S. M., Seib, P. A., El-Atawy, Y. S. 1989b. Control of enzymatic browning in apple with ascorbic acid derivatives, polyphenol oxidase inhibitors, and complexing agents. *J. Food Sci.* 54:997.

Sapers, G. M., Garzarella, L., Pilizota, V. 1990. Application of browning inhibitors to cut apple and potato by vacuum and pressure infiltration. *J. Food Sci.* 55:1049.

Sapers, G. M., Miller, R. L., Miller, F. C., Cooke, P. H., Choi, S.-W. 1994. Enzymatic browning control in minimally processed mushrooms. *J. Food Sci.* 59:1042.

Sapers, G. M., Cooke, P. H., Heidel, A. E., Martin, S. T., Miller, R. L. 1997. Structural changes related to texture of pre-peeled potatoes. *J. Food Sci.* 62:797.

Sapers, G. M., Miller, R. L., Choi, S.-W. 1995. Mushroom discoloration: new processes for improving shelf life & appearance. *Mushroom News* 43(3):7.

Sayavedra-Soto, L. A., Montgomery, M. W. 1986. Inhibition of polyphenoloxidase by sulfite. *J. Food Sci.* 51:1531.

Senesi, E., Pastine, R. 1996. Pre-treatments of ready-to-use fresh cut fruits. *Industrie Alimentari* 35:1161.

Shewfelt, R. L. 1986. Flavor and color of fruits as affected by processing. In: *Commercial Fruit Processing*, 2nd ed., Woodroof, J. G., Luh, B. S. (Eds.). AVI Publishing, Westport, CT, p. 481.

Smith, O. 1987a. Effect of cultural and environmental conditions on potatoes for processing. In: *Potato Processing*, 4th ed., Talburt, W. F., Smith, O. (Eds.). AVI–Van Nostrand Reinhold, New York, p. 73.

Smith, O. 1987b. Transport and storage of potatoes. In: *Potato Processing*, 4th ed., Talburt, W. F., Smith, O. (Eds.). AVI–Van Nostrand Reinhold, New York, p. 203.

Tamminga, S. K., Beumer, R. R., Keijbets, M. J. H., Kampelmacher, E. H. 1978. Microbial spoilage and development of food poisoning bacteria in peeled, completely or partly cooked vacuum packed potato. *Archiv Lebensmittelhygiene* 29:215.

Taylor, S. L., Higley, N. A., Bush, R. K. 1986. Sulfites in foods: uses, analytical methods, residues, fate, exposure assessment, metabolism, toxicity, and hypersensitivity. *Adv. Food Res.* 30:1.

Tomas-Barberan, F. A., Gil, M. I., Castaner, M., Artes, F., Saltveit, M. E. 1997. Effect of selected browning inhibitors on phenolic metabolism in stem tissue of harvested lettuce. *J. Agric. Food Chem.* 45:583.

Tong, C. B. S., Hicks, K. B. 1991. Sulfated polysaccharides inhibit browning of apple juice and diced apples. *J. Agric. Food Chem.* 39:1719.

Tong, C. B. S., Hicks, K. B. 1993. Inhibition of enzymatic browning of raw fruit and/or vegetable juice. U. S. patent 5,244,684.

Tressler, D. K., DuBois, C. 1944. No browning of cut fruit when treated by new process. *Food Industries* 16(9):701.

Vamos-Vigyazo, L. 1981. Polyphenol oxidase and peroxidase in fruits and vegetables. *CRC Crit. Rev. Food Sci. Nutr.* 15:49.

Vamos-Vigyazo, L. 1995. Prevention of enzymatic browning in fruits and vegetables: a review of principles and practice. In: *Enzymatic Browning and Its Prevention*, ACS Symposium Series 0097-6156, 600, American Chemical Society, Washington, D.C., pp. 49–62.

Van Buren, J. P. 1989. Causes and prevention of turbidity in apple juice. In: *Processed Apple Products*, Downing, D. D. (Ed.). AVI–Van Nostrand Reinhold, New York, p. 239.

Walker, J. R. L. 1976. The control of enzymatic browning in fruit juices by cinnamic acids. *J. Food Technol.* 11:341.
Walker, J. R. L. 1995. Enzymatic browning in fruits: its biochemistry and control. In: *Enzymatic Browning and Its Prevention*, ACS Symposium Series 0097-6156, 600, American Chemical Society, Washington, D.C., pp. 8–22.
Walker, J. R. L., McCallion, R. F. 1980. The selective inhibition of *ortho*- and *para*-diphenol oxidases. *Phytochemistry* 19:373.
Walker, J. R. L., Wilson, E. L. 1975. Studies on the enzymic browning of apples. Inhibition of apple *o*-diphenol oxidase by phenolic acids. *J. Sci. Food Agric.* 26:1825.
Waller, G. R., Feather, M. S. 1983. *The Maillard Reaction in Foods and Nutrition*. ACS Symposium Series 215, American Chemical Society, Washington, D.C.
Wedzicha, B. L. 1987. Review: chemistry of sulphur dioxide in vegetable dehydration. *Intl. J. Food Sci. Technol.* 22:433.
Wedzicha, B. L., Goddard, S. J. 1988. The dissociation constant of hydrogen sulfite ion at high ionic strength. *Food Chem.* 30:67.
Wei, C. I., Fernando, S. Y., Huang, T. S. 1991. Mutagenicity studies of kojic acid. Proceedings of the 15th Annual Conference, Tropical and Subtropical Fisheries Technology Conference of the Americas, Dec. 2–5. Florida Sea Grant Program SGR-105:464.
Wen, L., Wrolstad, R. E. 1998. Phenolics composition of pineapple juice concentrate. Abstract 34-B-3. Presented at 1998 Annual Meeting of Institute of Food Technologists, June 20–24, Atlanta.
Whitfield, F. B. 1992. Volatiles from interactions of Maillard reactions and lipids. *Crit. Rev. Food Sci.* 31:1.
Wong, M., Stanton, D. W. 1989. Nonenzymatic browning in kiwijuice concentrate systems during storage. *J. Food Sci.* 54:669.
Woodwards, L., Jackson, M. T. 1985. The lack of enzymic browning in wild potato species, series Longipedicellata and their crossability with *Solanum tuberosum*. *Z. Pflanzenzüchtg.* 94:278.
Xu, Q., Chen, Y. J., Nelson, P. E., Chen, L. F. 1993. Inhibition of the browning reaction by maltodextrin in freshly ground apples. *J. Food Process. Preserv.* 16:407.
Zhuang, H., Brereton, E., Sapers, G., Romig, W., Hotchkiss, A., Revear, L. 1998. Developing a quantitative GC-SPME method for study of volatile formation of fresh-cut apples in polymeric bags. Abstract 20A-37. 1998 Annual Meeting of Institute of Food Technologists, June 20–24, Atlanta.

20

Antimicrobial Agents

P. MICHAEL DAVIDSON

University of Tennessee, Knoxville, Tennessee

VIJAY K. JUNEJA

U.S. Department of Agriculture, Wyndmoor, Pennsylvania

JILL K. BRANEN

University of Illinois, Urbana, Illinois

I. INTRODUCTION

Humans have attempted to preserve food products from the detrimental effects of microorganisms since prehistoric times. Processes such as heating, drying, fermentation, and refrigeration have been used to prolong the shelf-life of food products. Some chemical food preservatives, such as salt, nitrites, and sulfites, have been in use for many years, however some have seen extensive use only recently. One of the reasons for the increased use of chemical preservatives has been the changes in the ways foods are produced and marketed. Today, consumers expect foods to be available year-round, to be free of foodborne pathogens, and to have a reasonably long shelf-life. While some improvements have been made using packaging and processing systems to preserve foods without chemicals, today antimicrobial food preservatives still play a significant role in protecting the food supply.

In selecting a food antimicrobial agent, several factors must be taken into consideration (Branen, 1993). First, the antimicrobial spectrum of the compound to be used must be known. This, along with knowledge of the bioburden of the food product, will allow the use of correct antimicrobial agent for the microorganism(s) of concern. Second, the chemical and physical properties of both the antimicrobial and the food product must be known. Such factors as pK_a and solubility of the antimicrobial and the pH of the food will facilitate the most efficient use of an antimicrobial. Third, the conditions of storage of the product and interactions with other processes must be evaluated to ensure that the antimicrobial will remain functional over time. Fourth, a food must be of the highest microbiological quality initially if an antimicrobial is to be expected to contribute to its shelf-life. None of the antimicrobials discussed in this section is able to preserve a product that is grossly contaminated. In most cases, while food antimicrobials will extend the lag phase or inactivate low numbers of microorganisms, their effects can be overcome. With rare exceptions, food antimicrobials are not able to conceal spoilage of a food product, i.e., the food remains wholesome during its extended shelf-life. Because food antimicrobials are generally bacteriostatic or fungistatic, they will not preserve a food indefinitely. Depending upon storage conditions, the food product will eventually spoil or become hazardous. Finally, the toxicological safety and regulatory status of the selected compound must be known.

This chapter will focus on food antimicrobials approved by regulatory agencies for use in foods as direct additives. For each antimicrobial or class of antimicrobials, characteristics of the compound(s), the antimicrobial spectrum, the antimicrobial effectiveness in foods, the mechanism of action, applications, regulations, and toxicology are discussed. The detail of the discussions varies depending upon research available and importance of the compounds.

II. DIMETHYL DICARBONATE

Dimethyl dicarbonate (CH_3–O–O–C–O–C–O–O–CH_3) is a colorless liquid which is slightly soluble in water. The compound is very reactive with substances including water, ethanol, alkyl and aromatic amines, and sulfhydryl groups (Ough, 1993a). The use of dimethyl dicarbonate (DMDC) as an antimicrobial was reviewed by Ough (1993a). The primary target microorganisms for DMDC are yeasts including *Saccharomyces, Zygosaccharomyces, Rhodotorula, Candida, Pichia, Torulopsis, Torula, Endomyces, Kloeckera*, and *Hansenula*. Terrell et al. (1993) evaluated sulfur dioxide, sorbic acid, and DMDC for ability to act as preservatives against yeast spoilage in grape juice. Grape juice was inoculated with yeast at 2200 and 20,000 CFU/mL and fermentation monitored at 21 or 31°C. DMDC at 0.8 mM was most effective in suppressing fermentation for all inoculation levels and temperatures. When added to tomato juice stored at 5 or 20°C, DMDC was highly effective in inactivating molds and yeasts (Bizri and Wahem, 1994). It was more effective than sorbate/benzoate in controlling aerobic plate counts in tomato juice acidified to pH 3.7 and stored at 5 or 20°C. The compound is also bactericidal at 30–400 μg/mL to a number of species including *Acetobacter pasteurianus, E. coli, Pseudomonas aeruginosa, Staphylococcus aureus*, several *Lactobacillus* species, and *Pediococcus cerevisiae* (Ough, 1993a). DMDC has been shown to be bactericidal against *E. coli* O157:H7 in apple cider (Fisher and Golden, 1998). Molds are generally more resistant to DMDC than yeasts or bacteria. van der Riet and Pinches (1991) evaluated DMDC against the heat resistant mold, *Byssochlamys fulva*, in apple and strawberry juices. They found that DMDC decreased viable *Byssochlamys fulva* ascospores when it was added at 24-h intervals during storage of the fruit juices at 25–30°C. They described the treatment as a modified tyndallization process using DMDC rather than heat as the lethal agent.

DMDC may be used in wine, teas, carbonated and noncarbonated nonjuice beverages (e.g., sports drinks), carbonated and noncarbonated fruit-flavored or juice beverages. It is approved by the United States Food and Drug Administration (FDA) (21 CFR 172.133) as an inhibitor of yeast for the following: (1) in wine, dealcoholized wine, and low alcohol wine that has less than 500 yeast CFU/mL at ≤200 ppm (μg/mL); (2) in ready-to-drink teas (<500 yeast CFU/mL) at ≤250 ppm; (3) in carbonated or noncarbonated, nonjuice-containing (≤1%), flavored or unflavored beverages containing added electrolytes at ≤250 ppm; and (4) in carbonated, dilute beverages containing juice, fruit flavor, or both, with juice content ≤50% at ≤250 ppm. The label of the product to which compound is added must indicate that "dimethyl dicarbonate" is added.

III. LYSOZYME

Lysozyme (1,4-β-N-acetylmuramidase; EC 3.2.1.17) is a 14,600-Da enzyme present in avian eggs, mammalian milk, tears and other secretions, insects, and fish. While tears contain the greatest concentration of lysozyme, dried egg white (3.5%) is the commercial source (Tranter, 1994). The enzyme catalyzes hydrolysis of the β-1,4 glycosidic bonds between N-acetylmuramic acid and N-acetylglucosamine of the peptidoglycan of bacterial cell walls. This causes cell wall degradation and lysis in hypotonic solutions. Lysozyme is stable to heat (80°C for 2 min). It is inactivated at lower temperatures when the pH is increased. It has an optimum temperature for activity of 55–60°C but has ca. 50% activity at 10–25°C (Inovatech, 2000). Yang and Cunningham (1993) studied the effect of pH, ionic strength, and other antimicrobial substances on the lytic ability of lysozyme on *Micrococcus lysodeikticus*. The compound remained stable for over 30 days at pH 7.0 or ionic strength <0.1. Activity was reduced at pH 9.0 and ionic strength >0.14. The compound was stable or relatively stable (>70% activity retained) for 30 days to 1.0% NaCl, 100 μg/mL sodium nitrite, 4.0% ethanol, 0.1% sodium benzoate, 0.3% calcium propionate, 0.1% potassium sorbate, or 0.1% propyl paraben. In 0.5% EDTA, 50% of the activity was lost and no activity was detected in the presence of 0.5% lactic acid, 4% acetic acid, or 100 μg/mL chlorine.

Lysozyme is most active against gram positive bacteria most likely because the peptidoglycan of the cell wall is more exposed than in gram negative bacteria. The enzyme has been shown to be inhibitory to *Clostridium botulinum, Clostridium thermosaccharolyticum, Clostridium tyrobutyricum, Bacillus stearothermophilus, Bacillus cereus, Micrococcus lysodeikticus*, and *Listeria monocytogenes* (Vakil et al., 1969; Carminiti et al., 1985; Duhaiman, 1988; Hughey and Johnson, 1987; Hughey et al., 1989). The enzyme has shown potential for use as an antimicrobial with EDTA to control the growth of *Listeria monocytogenes* in vegetables, but was less effective in refrigerated meat and soft cheese products (Hughey et al., 1989). At 1000 μg/mL, lysozyme alone was effective against only two of four strains of *Listeria monocytogenes* in milk and gave less than 1 log reductions for both (Carminati and Carini, 1989). Wang and Shelef (1991) showed that lysozyme was the primary antimicrobial compound in egg albumen, but that activity was enhanced by ovotransferrin, ovomucoid, and alkaline pH. Johansen et al. (1994) suggested that low pH (5.5) caused increased inhibition of *Listeria monocytogenes* by lysozyme because the organism had a slower growth rate allowing enzymatic hydrolysis of the cell wall to exceed the cell proliferation rate. Variation in susceptibility of gram positive bacteria is likely due to the presence of teichoic acids and other materials that bind the enzyme and the fact that certain species have greater proportions of 1,6 or 1,3 glycosidic linkages in the peptidoglycan which are more resistant than the 1,4 linkage (Tranter, 1994). For example, four strains of *Listeria monocytogenes* were not inhibited by lysozyme alone, but when EDTA was added growth inhibition resulted (Hughey and Johnson, 1987; Payne et al., 1994). Hughey and Johnson (1987) hypothesized that the peptidoglycan of the microorganism may be partially masked by other cell wall components and EDTA enhanced penetration of the lysozyme to the peptidoglycan.

Lysozyme is less effective against gram negative bacteria due to their reduced peptidoglycan content (5–10%) and presence of the outer membrane of lipopolysaccharide (LPS) and lipoprotein (Wilkins and Board, 1989). Gram negative cell susceptibility can be increased by pretreatment with chelators (e.g., EDTA). In addition, gram negative cells may be sensitized to lysozyme if the cells are subjected to pH shock, heat shock, osmotic shock, drying, freeze–thaw cycling, and trisodium phosphate (Wilkins and Board, 1989; Ray et al., 1984; Tranter, 1994; Carneiro de Melo et al., 1998). Samuelson et al. (1985) found that EDTA plus lysozyme was inhibitory to *Salmonella* Typhimurium on poultry. In contrast, no inhibition was demonstrated with up to 2.5 mg/mL EDTA and 200 μg/mL lysozyme in milk against either *Salmonella* Typhimurium or *Pseudomonas fluorescens* in a study by Payne et al., 1994. It was theorized that there may be a significant influence of the food product on activity of lysozyme and EDTA. According to Inovatech (2000), lysozyme is effective against spoilage and pathogenic bacteria in beer/wine, bread, canned foods, cheeses, meat, and rice. Ibrahim et al. (1996) found that lysozyme with reduced enzymatic activity through heating at 80°C, pH 6.0 exhibited strong bactericidal activity against gram negative and gram positive bacteria suggesting action independent of catalytic function.

The minimum inhibitory concentration of lysozyme against most fungi tested, including *Candida, Sporothrix, Penicillium, Paecilomyces*, and *Aspergillus*, was >9,530 μg/mL in potato dextrose agar at pH 5.6 (Razavi-Rohani and Griffiths, 1999a). Only *Fusarium graminearum* PM162 (1,600 μg/mL) and *A. ochraceus* MM184 (3260 μg/mL) were less than the maximum concentration evaluated. However, when combined with an equivalent concentration EDTA, lysozyme was inhibitory to all species of fungi except *Candida lipolytica* 1591 (MIC > 9530 μg/mL) and *Candida parapsilosis* NCPF 3207 (MIC = 960 μg/mL) at ≤500 μg/mL.

Lysozyme is one of the few naturally occurring antimicrobials approved by regulatory agencies for use in foods. In Europe, lysozyme is used to prevent gas formation (''blowing'') in cheeses such as edam and gouda by *Clostridium tyrobutyricum* (Wasserfall et al., 1976; Carini and Lodi, 1982). Cheese manufacturers using egg white lysozyme for this purpose add a maximum of 400 mg/L. A tentative final rule (*FR* 1998. 63:12421–12426) listing egg white lysozyme as a ''direct food substance affirmed as generally recognized as safe'' (21 CFR 184.1550) was published by FDA in 1998. The enzyme is allowed to be used in cheeses to prevent gas formation. Lysozyme is used to a great extent in Japan to preserve seafood, vegetables, pasta, and salads. Lysozyme has been evaluated for use as a component of antimicrobial packaging (Padgett et al., 1998).

Since egg whites have been used for food since the beginning of recorded history, there is little concern by regulatory agencies about the toxicity of lysozyme. However, there exists the potential for allergenicity to the protein.

IV. NATAMYCIN

Natamycin was first isolated in 1955 from a culture of *Streptomyces natalensis*, a microorganism found in soil from Natal, South Africa (Anonymous, 1991). The generic name ''natamycin,'' which is approved by the World Health Organization, is synonymous with ''pimaricin,'' a name used in earlier literature.

Natamycin ($C_{33}H_{47}NO_{13}$; MW, 665.7 Da) is a polyene macrolide antibiotic. Like many polyene antibiotics, natamycin is amphoteric, possessing one basic and one acidic group. Natamycin has low solubility in water (30–100 mg/L) and polar organic solvents and is practically insoluble in nonpolar solvents (Anonymous, 1991). Solubility ranges in other solvents include (mg/L): methanol, 2–15; ethanol, 0.04–1.2; n-butanol, 0.05–0.12; chloroform, 0.01–0.013 (Brik, 1981). Raab (1967) reported the isoelectric point of natamycin as 6.5.

Natamycin is active against nearly all molds and yeasts, but has no effect on bacteria or viruses. Most molds are inhibited at concentrations of natamycin from 0.5 to 6 μg/mL while some species require 10–25 μg/mL for inhibition. Most yeasts are inhibited at natamycin concentrations from 1.0 to 5.0 μg/mL. Ray and Bullerman (1982) reported that 10 μg/mL natamycin inhibited aflatoxin B_1 production of *Aspergillus flavus* by 62.0% and eliminated ochratoxin production by *A. ochraceus*. The same level of natamycin inhibited penicillic acid production by *Penicillium cyclopium* by 98.8% and eliminated patulin production of *P. patulum*. The inhibitory effect of natamycin was reported to be greater against mycotoxin production than mycelial growth. Gourama and Bullerman (1988) studied the effect of natamycin on growth and mycotoxin production (penicillic acid) by *Aspergillus ochraceus* OL24 in yeast extract sucrose (YES) medium and olive paste. Natamycin at 20 μg/mL delayed onset of growth, inhibited sporulation and reduced mycelial weight of *A. ochraceus* at 15, 25, and 35°C. Penicillic acid production by *A. ochraceus* at all temperatures was inhibited by 10 μg/mL natamycin. This was in contrast to potassium sorbate which, at sublethal concentrations, caused stimulation of penicillic acid production. Growth initiation was delayed and penicillic acid production inhibited by natamycin in olive paste. The authors concluded that natamycin could provide protection against fungal growth and mycotoxin formation in olives.

Several factors affect the stability and resulting antimycotic activity of natamycin, including pH, temperature, light, oxidants, and heavy metals. While pH has no apparent effect on antifungal activity, it does influence stability of the compound. After 3-week storage at 30°C, 100% of natamycin activity is retained at pH 5–7, while ca. 85% remains at pH 3.6 and only about 75% at pH 9.0 (Anonymous, 1991). In the pH range of most food products, natamycin is very stable. Under normal storage conditions, temperature has little effect on natamycin activity when in neutral aqueous suspension. Little or no decrease in activity occurs after several days at 50°C or a short time at 100°C (Brik, 1981). In contrast, dilute solutions of natamycin are less stable and susceptible hydrolysis (Anonymous, 1991). Irradiation due to sunlight, contact with certain oxidants (e.g., organic peroxides and sulfhydryl groups), and heavy metals all adversely affect stability of natamycin solutions or suspensions (Brik, 1981; Anonymous, 1991). van Rijn et al. (1999) reported that complexing natamycin to one or more proteins, such as whey, or amino acids increased that antifungal activity of the compound. Further, they theorized that this increased activity was due to improved availability due to improved solubility. The complexed natamycin is also less susceptible to hydrolysis.

All microorganisms which are susceptible to polyene macrolide antibiotics contain sterols, while resistant microorganisms do not (Hamilton-Miller, 1973). The mode of action of polyene macrolide antibiotics therefore is binding to ergosterol and other sterol groups of the fungal cell membrane. Ergosterol is a naturally occurring sterol that can be found at concentrations of up to 5% (dry weight) in strains of *Saccharomyces* (Hamilton-Miller, 1974). Other sterols found associated with the fungal cell membrane include 24,28 dehydroergosterol and cholesterol (Hamilton-Miller, 1974). Generally, binding of natamycin by sterols causes inhibition of ergosterol biosynthesis and distortion of the cell membrane with resultant leakage (Hamilton-Miller, 1973). While inhibition of glycolysis and respiration by polyene macrolide antibiotics can be demonstrated, they are considered to be secondary to cell membrane effects (Hamilton-Miller, 1973). Ziogas et al. (1983) studied 17 natamycin-resistant mutants of *Aspergillus nidulans*. Some mutants contained no ergosterol while others had reduced levels of the compound compared to the wild type. Ergosterol-deficient mutants were most resistant to natamycin (ca. 14–16 μg/mL) compared to the wild type (ca. 2 μg/mL), but grew at much slower rate than the wild type.

Nilson et al. (1975) determined the effect of natamycin in comparison with mycostatin on the shelf-life of cottage cheese stored at 4.4, 10.0, or 15.6°C. The compounds were added through curd wash water or in the cheese dressing. The cottage cheese was inoculated with *Aspergillus niger* or *Saccharomyces cerevisiae* or was uninoculated. Natamycin (100 μg/mL) added in the wash water was effective in increasing the days to spoilage of uninoculated cottage cheese by 13.6, 7.7, and 6.3 days over the control when stored at 4.4, 10.0, and 15.6°C, respectively. Cottage cheese inoculated with *A. niger* stored at the same temperatures had increased days to spoilage of 12.7, 6.0, and 4.3 days, and samples inoculated with *S. cerevisiae* had increased shelf-life of 10.3, 6.3, and 3.7 days, respectively. Adding natamycin to the cottage cheese dressing was even more effective in extending shelf-life. At 4.4, 10.0, and 15.6°C, the inoculated and uninoculated cottage cheese had increased days to spoilage ranges of 20.4–26.7, 9.7–12.3, and 2.6–5.0, respectively. Natamycin was found to be slightly more effective that mycostatin. *Aspergillus niger* was found to be the most sensitive of the two microorganisms used in the inoculation studies. Lück and Cheeseman (1978) found that 500 or 1000 μg/mL concentration of natamycin delayed mold growth on cheese for up to 6 months but did not prevent it completely. Verma et al. (1988) evaluated the effectiveness of natamycin against sorbic acid, benzoic acid, and nystatin in inoculated and uninoculated butter and cheese. All four preservatives were effective in reducing fungal growth on uninoculated butter and cheese samples stored 30 days at 7°C compared to the controls. Against *Aspergillus terreus, Trichoderma harzianum, Penicillium janthinellum*, or *Saccharomyces* sp. inoculated on butter or cheese and stored at 7°C for up to 30 days, nystatin was the most effective antifungal agent followed by natamycin. Lodi et al. (1989) found that natamycin was effective in preserving seven types of Italian cheeses with no detrimental effect on ripening. The antimycotic has been used to some extent as a butter preservative by being applied to the wrapper in small concentrations (Anonymous, 1991).

In addition to cheese, early work with natamycin suggested its use to inhibit fungal growth on fruits and meats. Ayres and Denisen (1958) investigated several antifungal agents including nystatin, rimocidin, ascosin, candidan, and natamycin (Myprozine®) for their potential in extending the shelf-life of berries. Strawberries, raspberries, and cranberries were dipped in solutions containing 0, 5, 10, 20, 50, and 100 μg/mL of each of the antibiotics. The berries were then stored at 5 ± 3°C for various periods. Natamycin and rimocidin were the most effective of the antifungal agents tested. Natamycin at 10–50 μg/mL decreased fungi on strawberries after 3–5 days storage and 50 μg/mL maintained fungal counts at equal to or less than initial count for 9 days. Natamycin (100 μg/mL) was also effective in prolonging the shelf-life of raspberries by 4 days. On cranberries, 10 μg/mL decreased the viable yeast count for 14 days of storage. When sprayed on raspberries and strawberries in the field, 50 μg/mL lowered the percentage of deterioration of the fruits during storage compared to controls. Shirk and Clark (1963) investigated the effectiveness of natamycin against yeast spoilage of orange juice. At 20 μg/mL, natamycin immediately reduced viable yeasts in uninoculated and inoculated (*Saccharomyces cerevisiae*) samples and eliminated viable yeast cells within 1 week of storage at 2.5–4°C. No spoilage was detected in inoculated or uninoculated natamycin-treated samples after 8 weeks storage, whereas after 1 week storage inoculated control samples were spoiled. The uninoculated control was judged to be palatable after 7 weeks. In a second study, natamycin at 5 μg/mL and sorbic acid at 1000 μg/mL eliminated viable yeast cells in orange juice inoculated with natural contaminants and stored for up to 12 weeks at 2.5–4°C. The uninoculated control sample spoiled in approximately 4 weeks and the inoculated control in 1 week. The authors concluded that natamycin was of potential use in orange juice because it was an effective antimycotic and did not cause off-flavors, while sorbic acid did.

Ayres et al. (1956) evaluated several antibiotics including chlortetracycline, oxytetracycline, tetracycline, streptomycin, neomycin, mycostatin, aerosporin, ascosin, rimocidin, and natamycin against spoilage microflora of raw cut-up chicken. They added the antibiotics at 10 ppm to chill water (1.7°C) and dipped the chicken for 2 h. Fungi were enumerated on acidified malt agar incubated at 30°C for 4 days. Rimocidin and natamycin, alone and in combination with chlortetracycline (all at 10 ppm), were the most effective antifungal agents and inhibited yeast growth on chicken stored 12–15 days at 4.4°C. There was a 2 log reduction in yeast counts at day 12 compared to the control (untreated). In contrast, there was little effect of the two antifungal compounds on total microflora (nutrient agar) when used in combination with chlortetracycline as a dip. Natamycin is an effective inhibitor of fungi on the surface of sausage (van Rijn et al., 1999).

Ticha (1975) investigated natamycin for use in the baking industry. Natamycin at 100 ppm was found to be effective against 5 molds, including *Aspergillus flavus* isolated from bakery products. The compound was effective in preventing growth of molds and yeasts in quarg fillings and icings at 0.05% and cream fillings at 0.01%. Natamycin at 50 ppm was effective in inhibiting *A. parasiticus* growth and toxin production on raw peanuts by 99% after 11 days (Gelda et al., 1974).

In the United States, natamycin is approved for use in cheese making as a mold spoilage inhibitor (21 CFR 172.155). Natamycin may be applied to the surface of cuts and slices of cheese by dipping or spraying an aqueous solution containing 200–300 ppm. The regulation concerning natamycin specifies that it may be applied to cuts and slices of cheese only if the cheese standard allows for use of ''safe and suitable'' antimycotics.

Regulations in the Netherlands limit natamycin concentrations on cheese surfaces to ≤ 2 mg/dm^2 (surface) and ≤ 1 mm in depth (Daamen and Van den Berg, 1985). The acceptable daily intake allowed for natamycin (FAO/WHO Expert Committee on Food Additives) is 0.3 mg/kg body weight/day (Smith and Moss, 1985).

The intravenous route is the path by which polyene macrolide antibiotics are most toxic and oral administration is least toxic. There is apparently no absorption of up to 500 mg/day natamycin from the human intestinal tract after 7 days administration (Brik, 1981). Levinskas et al. (1966) carried out a study to determine the acute and chronic toxicity of natamycin. The single oral dose LD_{50} for natamycin in the male rat was found to be 2.73 g/kg (1.99–3.73 g/kg) and 4.67 g/kg (3.0–7.23 g/kg) for female rats. The oral LD_{50} for fasted male albino rabbits was 1.42 g/kg (0.46–4.39 g/kg). The single dermal dose LD_{50} was estimated at >1.25 g/kg. For rats, no signs of toxicity occurred after large single doses and no gross lesions could be detected related to natamycin ingestion. Rabbits which died had congested and hemorrhagic gastric mucosa. In a 3 month feeding study with rats, Levinskas et al. (1966) found that animals fed 8000 ppm natamycin had body weights which averaged 54–67% of the control group (no natamycin). At 2000 ppm, animal weight averaged 85% of the control group. Natamycin had no apparent effect on body organs nor did it produce any lesions.

Oral administration of natamycin to rats for 2 years at up to 1000 ppm did not have an effect on survival of the animals (Levinskas et al., 1966). The diet containing natamycin was judged to have had no adverse effects on food utilization, reproductive performance, neoplasms, or other lesions above that encountered in controls. Dogs were fed diets containing 125, 250, or 500 ppm natamycin for 2 years (Levinskas et al., 1966). Oral administration of 500 ppm resulted in a slight decrease in body weight. No significant hematologic abnormalities nor significant lesions could be attributed to the consumption of natamycin (Levinskas et al., 1966). Nausea, vomiting, and diarrhea in humans given natamycin orally in doses exceeding 1000 mg/day (Brik, 1981), and vomiting and diarrhea in dogs fed 5000 ppm natamycin (Levinskas et al., 1966), have been reported.

V. NISIN

Nisin was first recognized in 1928 by Rogers and Whittier and was later isolated, characterized, and named by Mattick and Hirsh (1947). The compound is a peptide produced by a strain of the dairy starter culture *Lactococcus lactis* ssp. *lactis*. The structure and amino acid content of nisin was determined by Gross and Morell (1971). The molecular weight of nisin is 3500 (Gross and Morell, 1967); however, it usually occurs as a dimer with a molecular weight of 7000 (Jarvis et al., 1968). Nisin is a 34 amino acid peptide which contains the unusual amino acids dehydroalanine, dehydrobutyrine, lanthionine, and β-methyl-lanthionine (Delves-Broughton and Gasson, 1994). The latter amino acids are found in other inhibitory peptides produced by gram positive bacteria. The group of antimicrobial peptides containing these amino acids are called "lantibiotics" (Delves-Broughton and Gasson, 1994). A variant of nisin, called nisin Z, is produced by some strains of *L. lactis* ssp. *lactis* and contains an asparagine in place of a histidine at residue 27.

The solubility of the compound is dependent upon the pH of the solution. The solubility at pH 2.2 is 56 mg/mL, at pH 5.0 the solubility is 3 mg/mL, and the solubility at pH 11 is 1 mg/mL (Liu and Hansen, 1990). Nisin solution at pH 3 retains the greatest activity (97.5%) following heating at 115°C for 20 min (Davies et al., 1998). At pHs above and below that level, activity is reduced. At pH 7.0, inactivation occurs at room temperature. Nisin stability during storage is variable. Hirsch et al. (1951) noted rapid degradation of nisin in Swiss cheese after 8 days, and after 20 days very little was detected. Nisin degradation has also been reported in meat slurries and pasteurized process cheese spread at elevated temperature storage (Rayman et al., 1981; Delves-Broughton, 1990). Nisin is readily inactivated by proteolytic enzymes at pH 8. Nisinase from *Streptococcus thermophilus*, *Lactobacillus plantarum*, other lactic acid bacteria, and certain *Bacillus* species inactivate nisin (Koop, 1952; Alifax and Chevalier, 1962; Jarvis, 1967).

Nisin by itself has a narrow spectrum affecting only gram positive bacteria, including *Alicyclobacillus, Bacillus, Clostridium, Desulfotomaculum, Enterococcus, Lactobacillus, Leuconostoc, Listeria, Pediococcus, Staphylococcus*, and *Sporolactobacillus* (Table 1) (Hawley, 1957; Aplin & Barrett Ltd. no date). It does not generally inhibit gram negative bacteria, yeasts, or molds. *Clostridium botulinum* types A, B, and E were evaluated for their nisin susceptibility in brain heart infusion (BHI) broth and cooked meat medium (CMM) (Scott and Taylor, 1981a). Nisin concentrations necessary to inhibit the organism

Table 1 Susceptibility of Various Bacteria to Nisin

Microorganism	Susceptible strains/no. strains tested	MIC (μg/mL)	Reference	Method
Streptococci groups A–M exc. C,D,L	17/17	0.00625–0.1	Mattick and Hirsch, 1947	1
Pneumococcus	4/4	0.00625–2.5	Mattick and Hirsch, 1947	1
Staphylococcus	33/33	ND[a]	Laukova, 1995	2
Staphylococcus pyogenes	6/6	2.5	Mattick and Hirsch, 1947	1
Enterococcus	26/26	ND	Laukova, 1995	2
Neisseria	3/3	0.05–1.25	Mattick and Hirsch, 1947	1
Bacillus	6/6	0.05–0.1	Mattick and Hirsch, 1947	1
Clostridium	8/8	0.00625–2	Mattick and Hirsch, 1947	1
Corynebacterium	12/12	0.1–3	Mattick and Hirsch, 1947	1
Actinomyces	6/6	0.025–.25	Mattick and Hirsch, 1947	1
Mycobacterium tuberculosis	6/6	2.5–12.5	Mattick and Hirsch, 1947	1
Erysipelothrix monocytogenes	3/3	0.05	Mattick and Hirsch, 1947	1
Listeria monocytogenes	9/9 (TSA)	18.5–2950	Benkerroum and Sandine, 1988	3
Pediococcus	30/30 (GMA and TJA)	ND	Radler, 1990a	4
Leuconostoc	18/18 (GMA and TJA)	ND	Radler, 1990a	4
Lactobacillus	31/35 (GMA)	ND	Radler, 1990a	4

[a] ND, not determined. Author established susceptibility to nisin but did not specifically determine the MIC.

Methods:

1. Determined by broth dilution assay in 1% dextrose–0.3% Lemco–0.5% peptone broth. Units are estimated because experiments were done before units for nisin were standardized.
2. Determined by well diffusion assay on nutrient agar with wells containing 250 μg nisin. Susceptible strains showed a zone of inhibition around wells.
3. Determined by agar dilution assay on tryptic soy agar (TSA) or MRS agar.
4. Determined by well diffusion assay on Grape Must Agar (GMA), pH 4.5, and Tomato Juice Agar (TJA), pH 6.0. Wells contained 0.25, 2.5, and 10 μg/mL nisin. Resistant strains had no zone of inhibition around wells with 10 μg/mL nisin.

in BHI were 200, 80, and 20 μg/mL for types A, B, and E, respectively. The concentration required to inhibit *C. botulinum* in CMM was not determined, as it was beyond the highest concentrations tested for types A (>200 μg/mL) and B (>80 μg/mL). It was theorized that the higher levels required in CMM were due to binding of the nisin by meat particles. Rose et al. (1999) using matrix-assisted laser desorption/ionization time-of-flight mass spectrometry demonstrated that nisin is likely inactivated in raw meat by an enzymatic reaction with glutathione. In contrast, *Clostridium sporogenes* was inhibited in a meat system (pork) by 5–75 μg/mL nisin at pH 6.5–6.6 (Rayman et al., 1981). Strains of psychrotrophic *Bacillus cereus* were evaluated for their resistance to nisin (up to 50 μg/mL) in BHI broth or trypticase soy broth at 8 or 15°C (Jaquette and Beuchat, 1998). Tolerance of *Bacillus cereus* vegetative cells and spores was dependent upon temperature and pH. As the temperature was increased from 8 to 15°C and the pH from 5.53 to 6.57, more strains were tolerant to up to 50μg/mL nisin. At pH 5.53 and 15°C, all vegetative cells of all 8 strains tested were inhibited by 50μg/mL and only 1 grew at 10μg/mL. At pH 6.07 and 15°C, spores of all 6 strains tested were inhibited by 10 μg/mL. Laukova (1995) found that several strains of *Staphylococcus* (including *S. aureus*) and *Enterococcus* were sensitive to nisin. In general, *Enterococcus* was more susceptible than *Staphylococcus*. Radler (1990a) reported that the majority of 83 strains of lactic acid bacteria, including *Pediococcus, Leuconostoc*, and *Lactobacillus*, were sensitive to 0.25–10 μg/mL nisin in grape must agar and tomato juice agar. Nisin has been shown by a number of researchers to be inhibitory to *Listeria monocytogenes* (Benkerroum and Sandine, 1988; Harris et al., 1991; Ukuku and Shelef, 1997).

The spectrum of activity of nisin can be expanded to include gram negative bacteria when it is used in combination with chelating agents, such as ethylenediamine tetraacetic acid (EDTA) or trisodium phosphate, heat, and freezing (Delves-Broughton and Gasson, 1994; Carneiro de Melo et al., 1998). Stevens et al. (1992) reported that 20 mM of chelators such as ethylenebis tetraacetic acid (EGTA), citrate, and phosphate enhanced activity of nisin against gram negative bacteria, but none were as effective as EDTA. Schved et al. (1996) found that 20 mM maltol and ethyl maltol, which are flavor enhancers and have chelating properties, enhanced the activity of nisin (0.2–0.8 μg/mL) against two *E. coli* strains in cell buffer. Carneiro de Melo et al. (1998) demonstrated that *Campylobacter jejuni, Escherichia coli*, and *Salmonella* Enteritidis exposed to sublethal concentrations of trisodium phosphate (0.5–5mM) had increased susceptibility to 1 μM nisin in deionized water and dried on the surface of chicken skin.

Nisin activity is affected by a number of environmental factors, including pH, inoculum size, and interaction with food components (Somers and Taylor, 1972; Scott and Taylor, 1981a,b; Benkerroum and Sandine, 1988; Rayman et al., 1984; Harris et al., 1991; Ukuku and Shelef, 1997). Activity generally increases with decreasing pH and decreased initial numbers of microorganisms. The presence of food components such as lipids and protein influence nisin activity (Scott and Taylor, 1981a). Nisin was less active against *L. monocytogenes* in milk (Jung et al., 1992) and ice cream (Dean and Zottola, 1996) with increasing fat concentrations. This was probably due to binding of nisin to fat globules (Jung et al., 1992); this binding was overcome by adding emulsifiers (e.g., Tween 80). Thomas et al. (1998) demonstrated that sucrose fatty acid esters (sucrose palmitate and sucrose stearate) enhanced the activity of nisin against *Bacillus cereus, Listeria monocytogenes, Lactobacillus plantarum*, and *Staphylococcus aureus*, but had no effect on the activity of nisin against any gram negative bacteria.

The cytoplasmic membrane of vegetative cells is the primary site of action of nisin. The primary mechanism of nisin is believed to be the formation of pores in the cytoplasmic membrane which result in depletion of proton motive force and loss of cellular ions, amino acids, and ATP (Crandall and Montville, 1998). Ruhr and Sahl (1985) reported that the addition of nisin to susceptible cells (*Staphylococcus cohnii, Bacillus subtilis, Micrococcus luteus, Streptococcus zymogenes*) resulted in a rapid loss of intracellular amino acids and Rb^+, a K^+ analog. In addition, nisin inhibited proline and glutamine uptake by membrane vesicles of susceptible cells. The loss of ions resulted in a rapid decrease in the membrane potential, $\Delta\Psi$. The effects of nisin in depleting proton motive force components, $\Delta\Psi$ and ΔpH, were confirmed in *L. monocytogenes* (Bruno et al., 1992) and *C. sporogenes* (Okereke and Montville, 1992). Okereke and Montville (1992) reported that nisin treatment also results in a decrease in intracellular reserves of ATP. Winkowski et al. (1994) also reported efflux of ATP from nisin-treated *L. monocytogenes* cells but it only accounted for 20% of the loss of intracellular ATP. Therefore these researchers suggest that ATP hydrolysis also occurs within cells, perhaps as an attempt to maintain the proton motive force.

The effects of nisin on energized cells is consistent with a pore-forming mechanism (Winkowski et al., 1994). Sahl et al. (1987) detected nisin-induced pores of 0.2–1 nm diameter in black lipid membranes (artificial membranes with induced $\Delta\Psi$), which would allow the passage of solutes as big as 500 Da but would exclude larger compounds. Garcera et al. (1993) found that nisin caused leakage of fluorescein from liposomes indicating that pores allowing passage of low molecular weight compounds were formed. Garcera et al. (1993) suggested that nisin may induce pores in membranes by a "barrel-stave" mechanism consisting of three steps: (1) nisin monomers bind to a target membrane, (2) nisin inserts into the membrane, (3) nisin molecules aggregate to form a barrel-like hole around a water-filled pore. The nisin molecules may also aggregate before inserting into the membrane.

Several studies have shown that nisin-induced dissipation of the proton motive force is more effective against energized cells rather than starved or low energy cells (Ruhr and Sahl, 1985; Gao et al., 1991), indicating that nisin activity may be dependent upon a membrane potential. Kordel et al. (1989) also reported the need for a membrane potential in order for nisin to span liposome membranes. In further studies of nisin mechanism, Gao et al. (1991) reported that either a membrane potential and/or a pH gradient (with a more alkaline interior pH) is necessary in order for nisin to increase permeability of model liposomes made from various phospholipids. The $\Delta\Psi$ needed for dissipation of proton motive force was lower at lower external pH. Therefore it is possible that as the pH gradient increases, the dependence of nisin activity on $\Delta\Psi$ may be lower.

In most cases nisin is sporostatic rather than sporocidal (Delves-Broughton et al., 1996). Hitchens et al. (1963) determined that nisin inhibits postgermination swelling to prevent spore outgrowth. Morris et al. (1984) reported the existence of sulfhydryl groups in the membrane of germinated *B. cereus* spores. Sulfhydryl agents such as S-nitrosothiols inhibit outgrowth of germinated spores by interacting with the sulfhydryl groups in the membrane. At low concentrations, nisin inhibited spore outgrowth and competed with sulfhydryl agents for membrane sulfhydryl groups, indicating that the membrane sulfhydryl groups may be the target for nisin. Morris et al. (1984) hypothesized that the dehydroalanine (DHA) groups on nisin may react with the membrane sulfhydryl groups. In light of the results of Hansen (1994), where alterations of the 5-dehydroalanine group of subtilin eliminated activity against spores, it is becoming clearer that the DHA groups may in fact play a very important role in the sporostatic activity of nisin.

The application of nisin as a food preservative has been studied extensively (Marth, 1966; Lipinska, 1977; Hurst, 1981; Hurst and Hoover, 1993). Nisin was first used as a food preservative by Hirsch et al. (1951). Nisin-producing starter cultures were used to prevent ''blowing'' of Swiss-type cheese caused by *Clostridium tyrobutyricum* and *Clostridium butyricum*. This was followed by a similar application for the preservation of processed Swiss-type cheese (McClintock et al., 1952). Later, Hawley (1957) recommended the addition of a nisin-containing skim milk powder. Somers and Taylor (1987) studied the use of nisin to prevent *C. botulinum* outgrowth in process cheese spread formulated to have higher than normal moisture content and/or lower salt content. Nisin was an effective antibotulinal agent at 12.5–250 μg/g. The higher nisin levels allowed for the safe formulation of cheese spreads with higher moisture content and lower salt concentration. Delves-Broughton (1990) reported that nisin levels of 6 to 12.5 μg/g controlled non-*Clostridium botulinum* spoilage in process cheese.

The use of nisin to prevent growth of spoilage and pathogenic bacteria in fluid milk has also been investigated. Heinemann et al. (1965) reported that the time required for heating canned chocolate milk at 250°F could be reduced from 12.0 to 3.3 min and still prevent growth of added *Bacillus stearothermophilus* or *Clostridium sporogenes* PA 3679 spores with 2.5 μg/mL nisin. Jung et al. (1992) studied the efficacy of nisin against *L. monocytogenes* growth and the effect of fat and emulsifiers. They reported that nisin (0.25 and 1.25 μg/mL) reduced *L. monocytogenes* counts by 4.0 log CFU/mL after 24 h in skim milk, but, as fat content was increased, activity of nisin decreased. A similar study by Dean and Zottola (1996) found that nisin decreased *L. monocytogenes* cells to undetectable levels in 3% and 10% fat ice cream stored at −18°C. Nisin did not kill cells as quickly in the 10% fat ice cream, but after 3 months the effect of nisin was the same in both the 3% and 10% fat samples.

Nisin has been recommended for use in canned vegetable products to prevent the outgrowth of *Clostridium botulinum* when less severe sterilization conditions are desired or required (Campbell et al., 1959; Denny et al., 1961). In addition, Delves-Broughton et al. (1992) demonstrated that nisin may be used to increase the shelf-life of pasteurized liquid whole eggs. They found that 5 μg/mL nisin added to eggs prior to pasteurization increased the refrigerated shelf-life by 11–14 days. In addition, nisin prevented growth of the pathogen *B. cereus*, which grew well in pasteurized eggs with no added nisin.

Nisin has been shown to have potential benefit in some meat products. Jarvis and Burke (1976) demonstrated that 400 mg/kg nisin, in conjunction with 0.1% sorbic acid and 2.5% (w/w) polyphosphate retarded spoilage of fresh sausage at 5°C. Chung et al. (1989) dipped meat in a nisin solution (250 μg/mL) for 10 min at room temperature, then inoculated the surface of the meat and allowed 10 min for attachment of bacteria. Nisin significantly ($p < 0.05$) lowered the number of *Lactococcus lactis*, *S. aureus* and *L. monocytogenes* attached to the meat. However nisin had no effect on attachment by gram negative bacteria including *S.* Typhimurium, *Serratia marcescens*, and *Pseudomonas aeruginosa*. Scannel et al. (1997) investigated the use of nisin combined with sodium lactate on *S. aureus* and *Salmonella* Kentucky in fresh pork sausage. They found that 2% lactate combined with 12.5 μg/g nisin was superior to nisin alone at controlling growth of total aerobes, *S. aureus*, and *S.* Kentucky in sausage stored at 4°C for 10 days. Budu-Amoaka et al. (1999) found that heating canned lobster in brine at 60 or 65°C for 5 or 2 min, respectively, in combination with 25 μg/g nisin reduced *Listeria monocytogenes* by 3–5 logs. They proposed that using nisin could reduce the commercial thermal process for this product (13–18 min at 65.5°C) with equivalent lethality and reduced drained weight loss.

Nisin has been suggested as an adjunct to nitrite in cured meats for the purpose of preventing the growth of clostridia (Caserio et al., 1979; Holley, 1981). Calderon et al. (1985) found that 12–24 μg/g nisin added to bacon containing 50 μg/g nitrite increased the shelf-life by one day compared to bacon containing 50 and 150 μg/g nitrite only. They predicted that 18 μg/g nisin would increase the shelf-life of bacon at 5°C by one week. Rayman et al. (1981) showed that 75 μg/g nisin was more effective than 150 μg/g nitrite at inhibiting outgrowth of *C. sporogenes* spores (3.0 log CFU/g) in pork, beef, and turkey meat slurries. Although nisin appears to be more effective than nitrites at preventing the growth of some pathogenic and spoilage microorganisms in cured meats, it is yet to be shown to prevent *C. botulinum* growth in cured meats. Nisin is effective against *C. botulinum* in TPYG broth, but it is ineffective in cooked meat medium (Scott and Taylor, 1981a; Somers and Taylor, 1972). Therefore it would not be wise to replace some or all of the nitrites in cured meat with nisin until studies confirm that nisin will prevent *C. botulinum* growth in meat systems.

Lactic acid bacteria are common spoilage microorganisms in beer and wine. Since nisin is effective against most lactic acid bacteria but is inactive against yeasts, there is potential use for nisin in alcoholic beverages to prevent growth of spoilage lactic acid bacteria. Ogden et al. (1988) recommended nisin for preservation of beer and ale against spoilage by lactic acid bacteria. Radler (1990b) added nisin (2.5 and 25 μg/mL) to grape musts artificially contaminated with lactic acid bacteria. He reported that 25 μg/mL nisin inhibited growth of the lactic acid bacteria and had no effect on the fermentation of the musts, or on the composition and tastes of the wines. Choi and Park (2000) used nisin at 100 IU/mL to inhibit lactobacilli responsible for spoilage of kimchi, traditional Korean fermented vegetables. Nisin has been evaluated for use as a component of antimicrobial packaging (Ming, et al., 1997; Padgett et al., 1998).

Nisin is approved in many countries. In the United States, nisin is approved as a "nisin preparation" (21 CFR 184.1538) with a content of not less than 900 IU/mg. It is approved to inhibit outgrowth of *Clostridium botulinum* spores and toxin formation in pasteurized cheese spreads and pasteurized process cheese spreads (21 CFR 133.175); pasteurized cheese spread with fruits, vegetables, or meats (21 CFR 133.176); pasteurized process cheese spread (21 CFR 133.179); and pasteurized process cheese spread with fruits, vegetables, or meats (21 CFR 133.180). It has a maximum use level of 250 ppm of nisin in the finished product. In addition, nisin is approved for use in liquid eggs, salad dressings, and sauces.

Nisin has generally been considered nontoxic. The oral LD_{50} in mice is 6950 mg/kg body weight, which is similar to that of common salt (Hara et al., 1962). Nisin was found to have no effect on animals in studies of subchronic or chronic toxicity, reproduction, or sensitization (*FR* 1988, 53:11247). It has also been shown that nisin does not produce cross-resistance in microorganisms to therapeutic antibiotics (*FR* 1988, 53: 11247).

Bacteriocins with potential for use in foods have been demonstrated to be produced by strains of *Carnobacterium, Lactobacillus, Lactococcus, Leuconostoc, Pediococcus*, and *Propionibacterium*. Many of these compounds could potentially be used as food antimicrobials, but at the present time none are approved in the United States to be added to foods in their purified form. One approach to using these compounds has been to grow bacteriocin-producing starter cultures in a medium such as whey, nonfat dry milk, or dextrose. The fermentation medium is then pasteurized and spray-dried, which kills the starter culture but retains the active antimicrobial. These products act as antimicrobial additives but are generally considered GRAS and may be listed as ''cultured whey'' or ''cultured nonfat dry milk'' on the food label. Examples of such products are Microgard®, Alta™, and Perlac™. Alta at 0.1–1.0% was shown to decrease the growth rate of *Listeria monocytogenes* on vacuum-packaged smoked salmon stored at 4 or 10°C (Szabo and Cahill, 1999). Degnan et al. (1994) inoculated fresh blue crab (*Callinectes sapidus*) with a three strain mixture of *L. monocytogenes* (ca. 5.5 log CFU/g) and washed with various fermentation products (2000–20,000 arbitrary units [AU]/mL of wash) and stored at 4°C. Counts of *Listeria monocytogenes* decreased 0.5–1.0 log with Perlac or MicroGard and 1.5–2.7 logs with Alta.

VI. NITRITES

Nitrite salts (KNO_2 and $NaNO_2$) have been used in meat curing for many centuries. Meat curing utilizes salt, sugar, spices, and ascorbate or erythorbate in addition to nitrite. The reported contributions of nitrite to meat curing include characteristic color development, flavor production, texture improvement, and antimicrobial effects (IFT, 1987). The specific contribution of nitrite to the antimicrobial effects of curing salt was not recognized until the late 1920s (NAS/NRC, 1981), and evidence that nitrite was an effective antimicrobial agent came even later (Steinke and Foster, 1951). Nitrites are white to pale yellow hygroscopic crystals that are quite soluble in water and liquid ammonia but much less so in alcohol and other solvents.

The primary use for sodium nitrite as an antimicrobial is to inhibit the growth and toxin production of *Clostridium botulinum* in cured meats. In association with other components in the curing mix—such as salt, ascorbate, and erythorbate—and pH, nitrite exerts a concentration-dependent antimicrobial effect on the outgrowth of spores from *C. botulinum* and other clostridia. The use of nitrites in cured meat products to control *C. botulinum* has been studied extensively. An in-depth review of these studies is beyond the scope of this discussion, and the reader is referred to Holley (1981), Roberts and Gibson (1986), and Tompkin (1993).

The effectiveness of nitrite is dependent upon several environmental factors. Nitrite is most inhibitory to bacteria at an acidic pH. A tenfold increase in the inhibitory effect of nitrite against *C. botulinum* was found when the pH was reduced from 7.0 to 6.0 (Roberts and Ingram, 1966). Nitrite is more inhibitory under anaerobic conditions (Castellani and Niven, 1955; Lechowich et al., 1956; Buchanan and Solberg, 1972). Temperature, salt concentration, and initial inoculum size also significantly influence the antimicrobial role of nitrite (Riemann et al., 1972; Roberts, 1975; Roberts et al., 1966, 1976; Genigeorgis and Riemann, 1979). Ascorbate and isoascorbate enhance the antibotulinal action of nitrite most likely by acting as reducing agents (Roberts et al., 1991).

Nitrite has been shown to have variable effects on microorganisms other than *C. botulinum. Clostridium perfringens* growth in laboratory medium at 20°C was inhibited by 200 μg/mL nitrite and 3% salt or 50 μg/mL nitrite and 4% salt at pH 6.2 (Gibson and Roberts, 1986b). *Listeria monocytogenes* growth was inhibited for 40 days at 5°C by 200 ppm sodium nitrite with 5% NaCl in vacuum packaged and film wrapped smoked salmon (Pelroy et al., 1994b). Fecal streptococci showed growth in the same medium at 20°C with 400 μg/mL and 6% salt present (Gibson and Roberts, 1986b). In a similar study,

Gibson and Roberts (1986a) tested nitrite and salt against *Salmonella* sp. and Enteropathogenic *E. coli*. *Salmonella* showed visible growth within a week at 20°C in the presence of 400 μg/mL nitrite and 4% salt. Significant inhibition by salt and nitrite was achieved only at lower temperatures (10 or 15°C) and at pH 5.6 or 6.2. *Escherichia coli* was more resistant than *Salmonella*. Inhibition was demonstrated only at the extremes of pH (5.6), salt (6%), nitrite (400 μg/mL), and temperature (10°C). Other researchers have also demonstrated microbial resistance to nitrite among species of *Salmonella* (Castellani and Niven, 1955; Rice and Pierson, 1982), *Lactobacillus* (Castellini and Niven, 1955; Spencer, 1971), *Clostridium* (*perfringens*) and *Bacillus* (Grever, 1974). In TSB at pH 5.0, 4 strains of *Escherichia coli* O157:H7 were completely inhibited in the presence of 200 μg/mL nitrite at 37°C (Tsai and Chou, 1996). At higher pH values or lower temperatures, inhibition and/or inactivation of the microorganism was diminished. Korenekova et al. (1997) found that 100 mg/kg of nitrite added to milk inhibited the growth of yogurt culture.

The mechanism of nitrite inhibition of microorganisms has been studied for 50 years (Roberts and Gibson, 1986; Woods et al., 1989; Tompkin, 1993). The inhibitory effect of nitrite on bacterial sporeformers is apparently due to inhibition of outgrowth and during cell division (Cook and Pierson, 1983; Tompkin, 1978; Genigeorgis and Riemann, 1979). Nitrite does not inhibit spore germination to a significant extent. Ingram (1939) first postulated that nitrite inactivated enzymes associated with respiration. Since that time, nitrate has been found to affect a variety of enzymes and enzyme systems. Nitrite was shown to inhibit active transport, oxygen uptake, and oxidative phosphorylation of *Pseudomonas aeruginosa* by oxidizing ferrous iron of an electron carrier, such as cytochrome oxidase, to the ferric form (Rowe et al., 1979). Nitrite inhibited the active transport of proline in *Escherichia coli* but not group translocation by the phosphoenol–pyruvate : phosphotransferse system (Yarbrough et al., 1980). The growth of *Clostridium sporogenes* and *C. botulinum* was inhibited by nitrite through interference with the phosphoroclastic system (Woods et al., 1981; Woods and Wood, 1982). Inhibition is due to the reaction of nitric oxide with the nonheme iron of pyruvate–ferredoxin oxidoreductase (Woods et al., 1981). Nitrite also inhibits the iron-sulfur enzyme, ferredoxin, of *C. botulinum* and *C. pasteurianum* (Carpenter et al., 1987). McMindes and Siedler (1988) reported that nitric oxide was the active antimicrobial principle of nitrite and that pyruvate decarboxylase may be an additional target for growth inhibition by nitrite. Roberts et al. (1991) also confirmed inhibition of the phosphoroclastic system and found that ascorbate enhanced inhibition. In addition, they showed that other iron-containing enzymes of *Clostridium botulinum* were inhibited including other oxidoreductases and the iron-sulfur protein, hydrogenase. It has been suggested that inhibition of clostridial ferridoxin and/or pyruvate–ferroxin oxidoreductase is the ultimate mechanism of growth inhibition for clostridia (Carpenter et al., 1987; Tompkin, 1993). These observations are substantiated by the fact that the addition of iron to meats containing nitrite reduces the inhibitory effect of the compound (Tompkin et al., 1978). Chelating agents like sodium ascorbate, EDTA, and polyphosphate enhance the antibotulinal action of nitrite.

The mechanism of inhibition against nonsporeforming microorganisms may be different than for sporeformers. Nitrite was shown to inhibit active transport, oxygen uptake, and oxidative phosphorylation of *Pseudomonas aeruginosa* by oxidizing ferrous iron of an electron carrier, such as cytochrome oxidase, to the ferric form (Rowe et al., 1979). Muhoberac and Wharton (1980) and Yang (1985) also found inhibition of cytochrome oxidase of *Pseudomonas aeruginosa. Streptococcus faecalis* and *S. lactis*, bacteria that do not depend on active transport or cytochromes, were not inhibited by nitrite (Rowe et al., 1979). Woods et al. (1989) theorized that nitrites inhibited aerobic bacteria by binding the heme iron of cytochrome oxidase.

In a bacteriological medium, the inhibitory effect of nitrite is enhanced tenfold after heating due to a substance known as ''Perigo inhibitor'' (Perigo et al., 1967). The Perigo inhibitor is formed at 105°C or higher, which exceeds the temperatures normally used in the processing of cured meats. Its antibacterial activity is also neutralized by meat particles. Perigo inhibitor is formed in culture medium only when sulfhydryl groups and iron are present (Holley, 1981).

As stated, the use of nitrites in meats produces characteristic cured meat color, contributes to flavor development, inhibits *C. botulinum* growth and neurotoxin production, and retards rancidity. Meat products that may contain nitrites include bacon, bologna, corned beef, frankfurters, luncheon meats, ham, fermented sausages, shelf-stable canned cured meats, and perishable canned cured meat (e.g., ham). Nitrite is also used in a variety of fish and poultry products. The concentration of nitrite used in these products is specified by the U.S. FDA and USDA regulations. Sodium nitrate is used in certain European cheeses to prevent spoilage by *Clostridium tyrobutyricum* or *C. butyricum* (Tompkin, 1993).

Regulations for nitrite in (9 CFR 318.7) in bacon allow one of the following: (1) 120 ppm sodium nitrite or 148 ppm potassium nitrite plus 550 ppm sodium erythorbate or isoascorbate, (2) 100 ppm sodium nitrite or 123 ppm potassium nitrite plus 550 ppm sodium erythorbate or isoascorbate if a demonstration of adequate process control is met, or (3) 40–80 ppm sodium nitrite or 49–99 ppm potassium nitrite plus 550 ppm sodium erythorbate or isoascorbate plus 0.7% sucrose and a lactic acid bacterial culture (*Pediococcus*). The use of nitrite in other products is limited to a maximum residual level of 200 ppm. The input levels for nitrite are as follows: 2 lb/100 gal of pickle, 1 oz/100 lb (625 ppm) of meat for dry cured products, and 0.25 oz/100 lb (156 ppm) of chopped meat.

Use of nitrites in certain types of smoked and cured fish is specified in 21 CFR 172.160; 21 CFR 172.1170; 21 CFR 172.175; and 21 CFR 172.177. Presently, the level of nitrites allowed is a maximum of 10 ppm in smoked cured tuna fish and 200 ppm (input not to exceed 500 ppm) in smoked cured sablefish, salmon, shad, cod roe, and in home-curing mixtures. The level in smoked chub has been fixed at 100–200 ppm.

The lethal dose of nitrites in humans is 32 mg/kg body weight or 2 g (Burden, 1961) and 4–6 g (Wagner, 1956). Prolonged ingestion of sodium nitrite or sodium nitrate has been shown to cause methemoglobinemia, that is, excessive production of abnormal hemoglobin, especially in infants (NAS/NRC, 1981). Exposure to nitrites has been implicated as a causative agent of a variety of diseases in human beings and in animals. The major adverse effect is the possible induction of cancer. Studies conducted at the Massachusetts Institute of Technology have indicated that nitrites increase the incidence of lymphoma in rats fed 250–2000 ppm nitrite in their food or water (Newberne, 1979). The induction of cancer was considered to be a direct effect of nitrite on the lymphocytes and was independent of nitrosamine formation. A study of Newberne's data by the FDA (1980), however, disputed his histopathological findings and concluded that no demonstration of nitrite-induced tumors could be found. At one time there was concern over the reaction of nitrites with secondary amines to form nitrosamines or with substituted amides to form nitrosamides, both of which are known to cause cancer in many animal species (NAS/NRC, 1981; IFT, 1987). Epidemiological studies indicated a possible link between exposure to high levels of nitrites and a high incidence of stomach and esophageal cancer. However, in none of these studies was the exposure of nitrites or *N*-nitroso compounds actually measured in the individuals who developed cancer (NAS/NRC, 1981).

VII. ORGANIC ACIDS

Organic acids are commonly used by food manufacturers as antimicrobial preservatives or acidulants in a variety of food products. This is due to their solubility, flavor, and low toxicity.

Many factors influence the effectiveness of organic acids as antimicrobials, including hydrophobicity. However, the most important factor in the use of these compounds is undoubtedly the pH of the food. Early research demonstrated that the activity of organic acids was related to pH and that the undissociated form of the acid was primarily responsible for its antimicrobial activity (Cruess and Irish, 1932; Fabian and Wadsworth, 1939; Levine and Fellers, 1940). Many subsequent studies on organic acids have yielded similar findings concerning the effects of pH on activity (Ingram et al., 1956; Doores, 1993). In selecting an organic acid for use as an antimicrobial food additive, both the use pH and the pK_a of the acid must be taken into account. The use of organic acids is generally limited to foods with pH < 5.5., since most have a pK_a in the range of 3–5 (Doores, 1993).

The mechanism(s) by which organic acids inhibit microorganisms has been studied extensively. There is little evidence that the organic acids influence cell wall synthesis in prokaryotes or that they significantly interfere with protein synthesis or genetic mechanisms. Instead, organic acids more likely act at the cytoplasmic membrane level. The undissociated form of the organic acid penetrates the cell membrane lipid bilayer, and once inside the cell it dissociates because the cell interior has a higher pH than the exterior. Since bacteria must be able to maintain an internal pH near neutrality, protons generated from dissociation of the organic acid must be extruded to the exterior. Therefore, since protons generated by the organic acid inside the cell must be extruded using energy in the form of ATP, a constant influx of these protons will eventually deplete cellular energy. It must be noted that this same phenomenon could take place due to interference with membrane permeability as well. Sheu and Freese (1972) observed that short-chain organic acids altered the structure of the cell membrane. They hypothesized that this interfered with the regeneration of ATP by uncoupling the electron transport system or inhibiting the transport of metabolites into the cell. Sheu et al. (1972) determined that short-chain organic acids acted as uncouplers of amino acid carrier proteins from the electron transport system. As proof, they showed that transport of L-serine and other L-amino acids was inhibited in membrane vesicles of *Bacillus subtilis* when exposed to acetate and other fatty acids. Sheu et al. (1975) determined that organic acids inhibited active transport by interfering with the proton motive force (PMF) of the cell membrane. Eklund (1985) evaluated the effect of sorbic acid and parabens on the ΔpH and $\Delta\Psi$ components of the PMF in *E. coli* membrane vesicles. Both compounds eliminated the ΔpH but did not significantly affect the $\Delta\Psi$ component of the PMF so that active transport of amino acids continued. Eklund (1989) concluded that neutralization of the PMF and subsequent transport inhibition was not the sole mechanism of action of the parabens. To summarize, the organic acids and their esters have a significant effect on bacterial cytoplasmic membrane, interfering with the transport of metabolites and maintenance of membrane potential.

A. Acetic Acid and Acetate Salts

Acetic acid (CH_3COOH; pK_a = 4.75; MW, 60.05 Da), the major component of vinegar, and its salts are widely used in foods as acidulants and antimicrobials. Acetic acid is more effective against yeasts and bacteria than against molds (Ingram et al., 1956). Only acetic, lactic, and butyric acid–producing bacteria are markedly tolerant to the acid (Baird-Parker, 1980; Doores, 1993).

The activity of acetic acid varies with food product, environment, and microorganism. Woolford (1975a) studied the antimicrobial role of acetic acid at pH 4, 5, and 6. At pH 6, it was observed that *Bacillus, Clostridium*, and gram negative bacteria were inhibited more than lactic acid bacteria, yeast, molds, and other gram positive bacteria. When the pH was reduced to 5, gram positive bacteria were more inhibited than lactic acid bacteria, yeast, and molds. At pH 4, the concentration of acetic acid required for inhibition was considerably reduced. The foodborne pathogen, *Staphylococcus aureus* was inactivated 90 and 99% in 12 h at pH 5.2 and 5.0, respectively, by acetic acid (Minor and Marth, 1970). Against *Listeria monocytogenes*, Sorrells et al. (1989) demonstrated that, at an equivalent pH and temperature, acetic acid was a more effective antimicrobial than lactic acid, malic acid, or citric acid. However, on an equimolar basis, malic and citric acids were more effective. One percent acetic acid inactivated *Pseudomonas aeruginosa* by > 99.9% in 1 h (Hedberg and Miller, 1969). Brackett (1987) found that acetic acid was one of the least effective organic acids against the growth of *Yersinia enterocolitica*. In contrast, Karapinar and Gonul (1992a) showed that acetic acid completely inhibited the growth of *Yersinia enterocolitica* at 0.156% (v/v) in microbiological media at 22°C for 48 h. Entani et al. (1998) found that 0.1% (w/v) acetic acid from vinegar was bacteriostatic to multiple strains of *E. coli* O157:H7, *Salmonella* Enteritidis, *Salmonella* Typhimurium, and *Aeromonas hydrophila* after 4 days at 30°C on the surface of nutrient agar. The same concentration of acetic acid did not inhibit growth of *Bacillus cereus* or *S. aureus*. Enterohemorrhagic *E. coli* strains were less susceptible to bactericidal concentrations (2.5%) of acetic acid than an enteropathogenic strain. Finally, Entani et al. (1998) showed that stationary phase cells of *E. coli* O157:H7 were more resistant to acetic acid than log phase cells.

Acetic acid was inhibitory at pH 3.5 to *A. niger* and *Rhizopus nigricans* (Kirby et al., 1937). At 0.8–1% and pH 3.5, acetic acid inhibited growth of *S. cerevisiae* var. *ellipsoideus* and *Penicillium glaucum* (Cruess and Irish, 1932). A concentration greater than 4% was required to inhibit their growth at pH 7.0.

Acetic acid and its salts have shown variable success as antimicrobials in food applications. Acetic acid was used to increase the shelf-life of poultry when added to cut-up chicken parts in cold water at pH 2.5 (Mountney and O'Malley, 1965). Addition of acetic acid at 0.1% to scald tank water used in poultry processing decreased the D_{52} of *Salmonella* Newport, *Salmonella* Typhimurium, and *Campylobacter jejuni* five- to tenfold (Okrend et al., 1986). Increasing the acetic acid to 1.0% caused inactivation of all three microorganisms. In contrast, Lillard et al. (1987) found that 0.5% acetic acid in the scald water had no significant effect on *Salmonella*, total aerobic bacteria, or *Enterobacteriaceae* on unpicked poultry carcasses. At 1.2% as a 10-s dip for beef, acetic acid reduced microflora such as *S. Typhimurium, Shigella sonnei, Y. enterocolitica, E. coli, Pseudomonas aeruginosa*, and *Streptococcus faecalis* by an average of 65% (Bell et al., 1986). The compound has been used as a spray sanitizer at 1.5–2.5% on meat carcasses and as an effective antimicrobial dip for beef, lamb, and catfish fillets (Anderson et al., 1988; Bala and Marshall, 1998). Use of 2% acetic acid resulted in reductions in viable *E. coli* O157:H7 on beef after 7 days at 5°C (Siragusa and Dickson, 1993). Acetic acid was the most effective antimicrobial in ground roasted beef slurries against *E. coli* O157:H7 growth in comparison with citric or lactic acid (Abdul-Raouf et al., 1993). Acetic acid added at 0.1% to bread dough inhibited growth of 6 log CFU/g rope-forming *Bacillus subtilis* in wheat bread (pH 5.14) stored at 30°C for >6 days (Rosenquist and Hansen, 1998). In brain heart infusion broth (BHI), 0.2% acetic acid at pH 5.1 or 0.1% at pH 4.8 inhibited rope-forming strains of *B. subtilis* and *Bacillus licheniformis* for >6 days at 30°C (Rosenquist and Hansen, 1998). In a study by Post et al. (1985), shrimp, shrimp puree, tomato puree, and shrimp and tomato puree were acidified to 4.2 and 4.6 with acetic acid. No significant *Clostridium botulinum* growth or toxin production occurred in the products after 8 weeks at 26°C. In a frankfurter emulsion slurry, acetic acid (0.3–1.16%) caused increased inactivation rates of *Bacillus stearothermophilus* and *B. coagulans* at pH 4.6, 121°C and pH 4.2, 105–110°C, respectively (Lynch and Potter, 1988). Karapinar and Gonul (1992b) demonstrated that dipping parsley contaminated with 7.0 log CFU/g of *Y. enterocolitica* into the 2% (v/v) acetic acid or 40% (v/v) vinegar solution for 15 min was bactericidal against the microorganism. Delaquis et al. (1999) utilized gaseous acetic acid at 242 μL/L of air to inactivate up to 3–5 log CFU/g *Escherichia coli* O157:H7, *Listeria monocytogenes*, and *Salmonella* Typhimurium inoculated on mung bean seeds for sprouting. The procedure did not effect germination of the seeds.

Sodium acetate is an inhibitor of rope-forming *Bacillus* in baked goods and of the molds, *Aspergillus flavus, A. fumigatus, A. niger, A. glaucus, Penicillium expansum*, and *Mucor pusillus* at pH 3.5–4.5 (Glabe and Maryanski, 1981). It is useful in the baking industry because it has little effect on the yeast used in baking. Al-Dagal and Bazaraa (1999) found that whole or peeled shrimp dipped in a 10% sodium acetate (w/w) solution for 2 min had extended microbiological and sensory shelflife compared to controls.

Sodium diacetate (pK_a = 4.75) is effective at 0.1–2.0% in inhibiting mold growth in cheese spread (Doores, 1993). In BHI broth adjusted to pH 5.4, 0.45% (32 mM) sodium diacetate was shown to be inhibitory to *L. monocytogenes, E. coli, Pseudomonas fluorescens, Salmonella* Enteritidis, and *Shewanella putrefaciens*, but not *S. aureus, Yersinia enterocolitica, Pseudomonas fragi, Enterococcus faecalis*, or *Lactobacillus fermentis* after 48 h at 35°C (Shelef and Addala, 1994). In addition, 0.2% or 0.3–0.4% (21–28 mM) sodium diacetate suppressed growth by the natural microflora of ground beef after storage at 2°C for 14 days or 5°C for up to 8 days, respectively (Ajjarapu and Shelef, 1999; Shelef and Addala, 1994). Degnan et al. (1994) showed a 2.6 log decrease in viable *L. monocytogenes* in blue crab meat washed with 2M sodium diacetate after 6 days at 4°C. In contrast, Ajjarapu and Shelef (1999) found no effect of sodium diacetate at 0.2% on survival of *Escherichia coli* O157:H7 in ground beef stored at 2 or 10°C.

Dehydroacetic acid has a high pK_a of 5.27 and is therefore active at higher pH values. It is inhibitory to bacteria at 0.1–0.4% and fungi at 0.005–0.1% (Doores, 1993).

In the United States, acetic acid is GRAS (21 CFR 184.1005) for use as a pickling agent in (maximum amounts shown in parentheses) baked goods (0.25%), cheeses (0.8%), condiments and relishes (9.0%), dairy product analogs (0.8%), fats and oils (0.5%), gravies and sauces (3.0%), and meats (0.6%). Other foods may contain up to 0.15%. Sodium acetate (21 CFR 184.1721) is available for use in breakfast cereals (0.007%), fats and oils (0.1%), hard candy (0.15%), jams and jellies (0.12%), meats (0.12%), soft candy (0.2%), and snack foods, soup mixes, and sweet sauces (all 0.05%). Calcium acetate (21 CFR 184.1185) is approved for cheese (0.02%), gelatin (0.02%), snack foods (0.06%), and sweet sauces (0.15%). Sodium diacetate is also GRAS (21 CFR 184.1754) for use in baked goods (0.4%); cheese spreads, gravies, and sauces (0.25%); meats (0.1%); candy (0.1%); and soup mixes (0.05%). Acetic acid may also be added to meat products (9 CFR 318.7).

B. Benzoic Acid and Benzoates

Benzoic acid is found naturally in apples, cinnamon, cloves, cranberries, plums, prunes, strawberries, and other berries. Sodium benzoate (144.1 Da) is a stable, odorless, white granular or crystalline powder that is soluble in water (66.0 g/100 mL at 20°C) and ethanol (0.81 g/100 mL at 15°C). Benzoic acid (122.1 Da), also called phenylformic acid, occurs as colorless needles or leaflets and is much less soluble in water (0.27% at 18°C) than sodium benzoate. For the latter reason the salt is preferred for use in most foods.

The primary uses of benzoic acid and sodium benzoate are as antimycotic agents. Most yeasts and molds are inhibited by 20–2000 μg benzoic acid per mL at pH 5.0 (Baird-Parker, 1980; Chipley, 1993). One yeast that it is particularly resistant is *Zygosaccharomyces bailii*. Wind and Restaino (1995) found that up to 0.3% sodium benzoate did not prevent spoilage of a salsa mayonnaise by *Z. bailii*. While some bacteria associated with food poisoning are inhibited by 1000–2000 μg/mL undissociated acid, the control of many spoilage bacteria requires much higher concentrations (Chipley, 1993). Benzoates are most effective at pH 2.5–4.0 and significantly lose effectiveness at > pH 4.5 (Lloyd and Drake, 1975; Chichester and Tanner, 1981). The antimicrobial spectrum of benzoic acid against selected microorganisms is shown in Table 2. Rajashekhara et al. (1998) demonstrated that sodium benzoate reduced the $D_{85°C}$ value of the heat resistant mold *Neosartorya fischeri* from 69.3 and 63.5 min to 50.1 and 27.6 min in grape and mango juices, respectively.

The undissociated from of benzoic acid ($pK_a = 4.19$) is the most effective antimicrobial agent. Rahn and Conn (1994) reported that the compound was 100 times as effective in acid solutions as in neutral solutions and that only the undissociated acid had antimicrobial activity. Macris (1975) studied the effect of benzoic acid on *Saccharomyces cerevisiae*. A rapid uptake of benzoic acid by the yeast was observed, reaching saturation in about 2 min. Only the undissociated form was taken up by the cells. When cells were subjected to elevated temperatures, a decrease in the rate of uptake was observed above 60°C. This indicated an irreversible heat inactivation of the uptake process, similar to that observed for enzymatic inactivation. Therefore, protein compounds may be involved in the uptake of this preservative.

The target of benzoic acid in the microbial cell has not been completely elucidated but is likely similar to other organic acids. For example, Freese (1978) suggested that benzoic acid destroyed the proton motive force of the cytoplasmic membrane by continuous transport of protons into the cell causing disruption of the transport system. Benzoates also inhibit enzymes in bacterial cells such as those controlling acetic acid metabolism and oxidative phosphorylation (Bosund, 1962), α-ketoglutarate and succinate dehydrogenases in the citric acid cycle (Bosund, 1962), lipase production by *Pseudomonas fluorescens* (Anderson et al., 1980) and trimethylamine-*N*-oxide reductase activity (57% inhibition at 2.5 mM, pH 6.0) of *Escherichia coli* (Kruk and Lee, 1982). In fungi, aflatoxin production by a toxigenic strain of *Aspergillus flavus* (Uraih and Chipley, 1976; Uraih et

Table 2 Minimum Inhibitory Concentrations of Benzoic Acid Against Growth of Selected Microorganisms

Microorganism	pH	Minimum inhibitory concentration (μg/mL)
Bacteria, Gram Positive		
Bacillus cereus	6.3	500
Lactobacillus	4.3–6.0	300–1800
Listeria monocytogenes	5.6, 4°C; 5.6, 21°C	2000; 3000
Micrococcus	5.5–5.6	50–100
Streptococcus	5.2–5.6	200–400
Bacteria, Gram Negative		
Escherichia coli	5.2–5.6	50–120
Pseudomonas	6.0	200–480
Molds		
Aspergillus	3.0–5.0	20–300
Aspergillus parasiticus	5.5	>4000
Aspergillus niger	5.0	2000
Byssochlamys nivea	3.3	500
Cladosporium herbarum	5.1	100
Mucor racemosus	5.0	30–120
Penicillium	2.6–5.0	30–280
Penicillium citrinum	5.0	2000
Penicillium glaucum	5.0	400–500
Rhizopus nigricans	5.0	30–120
Yeasts		
Candida krusei	—	300–700
Debaryomyces hansenii	4.8	500
Hansenula	4.0	180
Pichia membranefaciens	—	700
Rhodotorula	—	100–200
Saccharomyces bayanus	4.0	330
Zygosaccharomyces bailii	4.8	4500
Zygosaccharomyces rouxii	4.8	1000

Sources: Chipley, 1993; El-Gazzar and Marth, 1987; El-Shenawy and Marth, 1988; Jermini and Schmidt-Lorenz, 1987; Marwan and Nagel, 1986; Roland et al., 1984.

al., 1977; Chipley and Uraih, 1980) and 6-phosphofructokinase activity (Francois et al., 1986) were inhibited.

Sodium benzoate is used as an antimicrobial in carbonated and still beverages (0.03–0.05%), syrups (0.1%), cider (0.05–0.1%), margarine (0.1%), olives (0.1%), pickles (0.1%), relishes (0.1%), soy sauce (0.1%), jams (0.1%), jellies (0.1%), preserves (0.1%), pie and pastry fillings (0.1%), fruit salads (0.1%), and salad dressings (0.1%), and in the storage of vegetables (Chipley, 1993). It is also used in pharmaceutical preparations, toiletries, and cosmetics. Kimble (1977) reviewed the use of these compounds for the preservation of fish by means of dips or germicidal ices. Zhao et al. (1993) showed that 0.1% benzoic acid was effective in reducing viable *E. coli* O157:H7 in apple cider (pH 3.6–4.0) by 3–5 logs in 7 days at 8°C. Kasrazadeh and Genigeorgis (1994,1995) found that 0.3% sodium benzoate added to queso fresco cheese had an inhibitory effect on *Escherichia coli* O157:H7 and *Salmonella*.

Benzoic acid and sodium benzoate were the first antimicrobial compounds permitted in foods by the FDA (Jay, 1992). Benzoates are GRAS preservatives (21 CFR 184.1021; 21 CFR 184.1733; 9 CFR 318.7) up to a maximum of 0.1%. In most countries of the world, the maximum permissible use concentration is 0.15–0.25%.

Evidence has shown that benzoates have a low order of toxicity for animals and humans (FAO/WHO, 1962). In humans, toxic doses of 6 mg/kg were reported upon intradermal administration (Sax, 1979). However, oral administration of sodium benzoate at rates of 5–10 g for several days had no adverse effect on health (Dakin, 1909). The reason for low toxicity is that humans and animals have an efficient detoxification mechanism for benzoate. The compounds are conjugated with glycine in the liver to form hippuric acid, which is then excreted in the urine (Chipley, 1993). Griffith (1929) reported that this mechanism removed 66–95% of the benzoic acid from individuals ingesting large quantities. It was also suggested that the remaining benzoate was detoxified through conjugation with glucuronic acid. Sensitization reactions to benzoate have been reported in some individuals (Michaelsson and Juhlin, 1973).

Sodium benzoate had no teratogenic activity when administered orally (Minor and Becker, 1971). No carcinogenic effect was observed when rats were given 1–2% (Sodemoto and Enomoto, 1980), or 5% (Hartwell, 1951) sodium benzoate, orally. Sodium benzoate has been found to be nonmutagenic (Njagi and Gopalan, 1980).

C. Lactic Acid and Lactates

Lactic acid ($pK_a = 3.79$) is a primary end-product of the lactic acid bacteria and serves to assist in preservation of many fermented dairy, vegetable, and meat products. It is used as a food additive primarily for pH control and flavoring. The antimicrobial activity of the compound is variable. For example, in cold pack cheese formulated with lactic and acetic acid and inoculated with *Salmonella* Typhimurium, no increase in destruction of the organism was found (Park et al., 1970). In contrast, lactic acid was found to be four times as effective as malic, citric, propionic, and acetic acids in inhibiting growth of *Bacillus coagulans* in tomato juice (Rice and Pederson, 1954). Lactic acid inhibited sporeforming bacteria at pH 5.0 but was much less effective against yeasts and molds (Woolford, 1975b). *Staphylococcus aureus* was inactivated by 90 and 99% in 12 h at pH 4.9 and 4.6, respectively, by lactic acid (Minor and Marth, 1970). Based upon molar concentration, pH, and activity of undissociated acid, lactic acid was one of the most effective organic acids against the growth of *Yersinia enterocolitica* (Brackett, 1987). Addition of 5.0% lactic acid in ground beef patties stored at 4°C for 10 days reduced coliforms by 3.9 logs (Podolak et al., 1996b). Oh and Marshall (1993) reported that there was little interaction between lactic acid and ethanol against *L. monocytogenes*. The MIC value of lactic acid alone was 0.5%, but was lower when 1.25% ethanol was combined with 0.25% lactic acid. When 2.5% ethanol was combined with 0.25% lactic acid, the extent of inhibition was not more than that of the most active single compound alone.

Smulders et al. (1986) and Snijders et al. (1985) advocated the use of lactic acid for surface decontamination of fresh meats, slaughter byproducts, and poultry. Lactic acid at 1–2% was reported to reduce *Enterobacteriaceae* and aerobic plate counts by 0.3–2.7 log on beef, veal, pork, and poultry. The compound also delayed growth of spoilage microflora during long-term storage of products. Visser et al. (1988) confirmed the antimicrobial effectiveness of 2.0% (v/v) L-lactic acid on fresh veal tongues and demonstrated that the compound decreased the growth rate of spoilage organisms during vacuum-packaged storage at 3°C.

Brewer et al. (1991) reported that 2 or 3% sodium lactate added to fresh pork sausage delayed microbial deterioration, pH decline, development of sour and off-flavors by 7–10 days at 4°C and appeared to protect red color of products. Lamkey et al. (1991) reported that 3% (w/w) sodium lactate was effective in maintaining low microbial numbers and the shelf-life was extended by more than 2 weeks when fresh pork sausage was stored at 4°C. Sodium lactate at 3 or 4% prolonged shelf-life of refrigerated beef roasts up to 84 days (Papadopoulos et al., 1991). Blom et al. (1997) added 2.5% sodium lactate plus 0.5% sodium acetate to vacuum-packaged sliced ham or sausage inoculated with *Listeria monocytogenes* and stored at 4 or 9°C for 5 weeks. The antimicrobials inhibited the microorganism for the 5 week period at 4°C.

Psychrotrophic and coliform populations in 40% fat fresh pork sausage patties were retarded with 2% potassium lactate (Bradford et al., 1993). Shelef and Yang (1991) showed that 4% (w/w) potassium lactate in beef corresponding to a concentration of 416 mM in the water phase, was more effective in controlling growth of *L. monocytogenes* than was 3% (w/w) NaCl, corresponding to a concentration of 684 mM in the water phase.

Sodium lactate (1.8–5.0%) inhibits *Clostridium botulinum, Clostridium sporogenes,* and *Listeria monocytogenes* in various meat products (Anders et al., 1989; Maas et al., 1989; Shelef and Yang, 1991; Unda et al., 1991; Chen and Shelef, 1992; Stillmunkes et al., 1993; Weaver and Shelef, 1993; Pelroy et al., 1994a; Hu and Shelef, 1996). The antimicrobial effect of sodium lactate against *L. monocytogenes, Salmonella spp., Yersinia enterocolitica* increases with decreasing pH values (5.7–7.0). However, for *S. aureus* the MIC value of sodium lactate did not change with pH (Houtsma et al., 1996). In contrast, 1.8% sodium lactate alone or with 0.2% sodium diacetate had no effect on survival of *Escherichia coli* O157:H7 in ground beef at 2 or 10°C (Ajjarapu and Shelef, 1999).

Presumably lactic acid applied as antimicrobial to foods functions similarly to other organic acids and has a primary mechanism involving disruption of the cytoplasmic membrane proton motive force (Eklund, 1989). In contrast, there has been some controversy concerning the mechanism of lactate salts used at high concentrations. These salts have been shown to have little effect on product pH. Therefore, most of the lactate remains in the less effective anionic form. Initially it was thought that the high concentration of the salts may reduce water activity sufficiently to inhibit microorganisms (Loncin, 1975; Debevere, 1989). However, Chen and Shelef (1992) and Weaver and Shelef (1993) using cooked meat model systems and liver sausage, respectively, containing lactate salts up to 4% concluded that water activity reduction was not sufficient to inhibit *Listeria monocytogenes.* In addition, Papadopoulos et al. (1991) reported that water activity was not lowered in cooked beef top rounds injected with various levels of sodium lactate. Inhibition may be due to the presence of sufficient undissociated lactic acid, possibly in combination with a slightly reduced pH and water activity, to cause inhibition of some microorganisms.

Lactic acid was probably one of the first acids used in food due to its wide distribution in nature as a product of fermentation. The U.S. FDA has approved lactic acid as GRAS for miscellaneous and general purpose usage (21 CFR 184.1061) with no limitation upon the concentration used. It may not be used in infant foods and formulas. The U.S. Department of Agriculture allows lactic acid in meat products at the lowest concentration necessary for the intended purpose (9 CFR 318.7).

D. Propionic Acid and Propionates

Up to 1% propionic acid (pK_a = 4.87) is produced naturally in Swiss cheese by *Propionibacterium freudenreichii* ssp. *shermanii*. The use of propionic acid and propionates has been directed primarily against molds. Some yeasts and bacteria, particularly gram negative strains, may also be inhibited. The activity of propionates depends upon the pH of the substance to be preserved. In an early study, O'Leary and Kralovec (1941) demonstrated that the microorganism in bread dough that causes rope formation, *Bacillus subtilis* (*mesentericus*), was inhibited at 0.19% at pH 5.8 and 0.16% at pH 5.6. In a similar study, another strain of *B. subtilis* was inhibited by propionates at pH 6.0 (Woolford, 1975b). Propionates (0.1–5.0%) have been found to retard the growth of *S. aureus, Sarcina lutea, Proteus vulgaris, Lactobacillus plantarum, Torula (Candida)*, and *Saccharomyces cerevisiae* var. *ellipsoideus* for up to 5 days (Woolford and Anderson, 1945). The minimum inhibitory concentration of undissociated propionic acid against three *Bacillus* species, *E. coli*, and *Staphylococcus aureus* ranged from 0.13% to 0.52% and for the yeast *Candida albicans*, 0.29%. Chung and Goepfert (1970) found that propionic acid was the most effective inhibitor of *Salmonella* serotypes Anatum, Senftenberg, and Tennessee among acetic, adipic, citric, lactic, propionic, and tartaric acids. *Listeria monocytogenes* growth was inhibited at 13, 21, and 35°C by 0.3% propionic acid at pH 5.0 and totally inhibited at 4°C (El-Shenawy and Marth, 1989). Propionic acid and propionates at 8–12% retard mold growth on the surface of cheese and butter (Ingle, 1940; Deane and Downs, 1951). The activity of propionic acid against *Aspergillus* and *Fusarium* was enhanced by the presence of EDTA (Razavi-Rohani and Griffiths, 1999b).

Early research on the mechanism of inhibition of microorganisms by propionic acid showed that sodium propionate inhibition of *E. coli* was overcome by the addition of β-alanine (Wright and Şkeggs, 1946). However, this potential interference with β-alanine synthesis is probably not a universal mechanism as a reduction in inhibitory effect was not observed with *B. subtilis, Pseudomonas*, or *Aspergillus clavatus*. Some strains of *Penicillium*, however, may grow in nutrient media containing over 5% propionic acid (Heseltine, 1952). The primary mode of action of propionic acid is likely similar to that of the other short chain organic acids discussed, i.e., acidification of the cytoplasm and inhibition of an unspecified function by the undissociated acid.

Propionic acid and propionates are used as antimicrobials in baked goods and cheeses. Propionates may be added directly to bread dough because they have no effect on the activity of baker's yeast. There is no limit to the concentration of propionates allowed in foods but amounts used are generally less than 0.4% (Robach, 1980).

In the United States, propionic acid and calcium and sodium propionates are approved as GRAS (21 CFR 184.1081; 21 CFR 184.1221; and 21 CFR 184.1784). No upper limits are imposed except for products which come under standards of identity. A limit of 0.32% is allowed in flour in white bread and rolls, 0.38% in whole wheat products, and 0.3% in cheese products (Robach, 1980).

E. Sorbic Acid and Sorbates

Sorbic acid and its potassium, calcium, or sodium salts are collectively known as sorbates. Sorbic acid was first isolated in 1859 by A. W. Hoffman, a German chemist, from the berries of the mountain ash tree (rowanberry) (Sofos and Busta, 1993). The structure of sorbic acid was determined around 1880, and it was first synthesized by O. Doebner in 1900. It was not until ca. 1940, however, that the antimicrobial preservative power of sorbic acid was discovered. In 1945, a U.S. patent was awarded to C. M. Gooding and Best Foods, Inc. (Gooding, 1945). They recognized sorbic acid as an effective fungistatic agent for food. Since that time, sorbates have been used in foods as effective inhibitors of fungi, including those that produce mycotoxins, and certain bacteria (Robach, 1980; Sofos and Busta, 1981,1993). Sorbic acid (pK_a = 4.75) is a trans–trans, unsaturated monocarboxylic fatty acid (CH_3–CH=CH–CH=CH–COOH), which is slightly soluble in water (0.15 g/100mL at 20°C). The potassium salt of sorbic acid is highly soluble in water (58.2 g/100 mL at 20°C).

As with other organic acids, the antimicrobial activity of sorbic acid is greatest in the undissociated state. The effectiveness of the compound is greatest as the pH decreases below 6.5 (Anonymous, 1999). Eklund (1983) determined the minimum inhibitory concentrations of dissociated and undissociated sorbic acid against several strains of bacteria and a yeast. Both forms were shown to cause inhibition, but the undissociated acid was 10–600 times more effective than the dissociated acid. At pH > 6, however, the dissociated acid was responsible for >50% of the inhibition observed. Sorbates are reported to be more effective against spoilage microorganisms than propionates or benzoate at the same pH (Anonymous, 1999).

Food-related yeasts inhibited by sorbates include species of *Brettanomyces, Byssochlamys, Candida, Cryptococcus, Debaryomyces, Endomycopsis, Hansenula, Oospora, Pichia, Rhodotorula, Saccharomyces, Schizosaccharomyces, Sporobolomyces, Torulaspora, Torulopsis*, and *Zygosaccharomyces* (Anonymous, 1999; Sofos and Busta, 1993). Food-related mold species inhibited by sorbates belong to the genera *Alternaria, Ascochyto, Aspergillus, Botrytis, Cephalosporium, Fusarium, Geotrichum, Gliocladium, Helminthosporium, Humicola, Mucor, Penicillium, Phoma, Pullularia* (*Auerobasidium*), *Rhizopus, Sporotrichum*, and *Trichoderma* (Anonymous, 1999; Sofos and Busta, 1993). Sorbates have been found to inhibit the growth of yeasts and molds in cucumber fermentations, high-moisture dried prunes (Nury et al., 1960), and cheeses (Smith and Rollin, 1954), on the surface of country-style hams (Baldock et al., 1979), in broth culture (Deak et al., 1970), and in Mexican hot sauce (Flores et al., 1988). Rajashekhara et al. (1998) demonstrated that potassium sorbate reduced the $D_{85°C}$ value of the heat-resistant mold *Neosartorya fischeri* from 69.3 and 63.5 min to 25.1 and 29.4 min in grape and mango juices, respectively.

Certain species of yeast are more resistant, and some acquire a resistance to sorbates. Pitt (1974) reported that *Zygosaccharomyces* (*Saccharomyces*) *bailii*, a preservative-tolerant yeast, was not inhibited by sorbic acid at 0.06% in 10% glucose. Deak and Novak (1972) reported that sorbic acid suppressed yeast metabolism and growth at high concentrations but was metabolized by yeast at low intracellular concentrations. Warth (1977) further studied the ability of *Z. baillii* to resist the antimicrobial activity of sorbic acid. Data indicated that, while the yeast accumulated the acid at a rate expected from conditions of concentration and pH, it was capable of transporting the sorbic acid from the cells. This transport system required energy as was demonstrated by the increased efflux in the presence of glucose. Further, Warth (1977) suggested that metabolism of sorbic acid played little, if any, part in resistance. Bills et al. (1982) showed that *Saccharomyces rouxii* that were exposed to 0.1% sorbate became resistant to subsequent exposure to the compound and that this resistance was not influenced by the presence of sucrose. Certain mold species are also resistant to sorbic acid, which results in occasional spoilage of foods. Melnick et al. (1954) reported that high initial mold populations were able to degrade sorbic acid in cheese. Liewen and Marth (1985) found that a sorbic-resistant *Penicillium roqueforti* strain grew in yeast extract sucrose broth and yeast extract/malt extract broth with 6000 and 9000 μg/mL, respectively. Degradation of sorbates is through a decarboxylation reaction carried out in the fungal mycelium and is accompanied by the formation of 1,3-pentadiene, which may produce a kerosenelike or hydrocarbonlike odor (Marth et al., 1966; Liewen and Marth, 1985).

The inhibitory effect of sorbates may be lethal as well as static (Melnick et al., 1954; Costilow et al., 1955). Pederson et al., (1961) showed that yeasts die slowly in fruit juices treated with sorbates, while Harada et al. (1968) found only fungistatic effects. Przybylaski and Bullerman (1980) reported that conidia of *Aspergillus parasiticus* lost viability in the presence of sorbate. Bullerman (1984) studied the effect of potassium sorbate on growth and patulin production by strains of *Penicillium patulum* and *P. roqueforti* isolated from cheese. Potassium sorbate at 0.05, 0.10, and 0.15% delayed initiation of growth, prevented spore germination, and decreased the rate of growth of *P. patulum* in potato dextrose broth at 12°C. *Penicillium roqueforti* was affected less. Similar results were noted with *A. parasiticus* and *Aspergillus flavus* in yeast extract sucrose broth (Bullerman, 1983). Potassium sorbate reduced or prevented the production of the mycotoxins patulin by *P. patulum* and of aflatoxin B_1 by *A. parasiticus* and *A. flavus* for up to 70 days at 12°C (Bullerman, 1983,1984). Potassium sorbate has been found to inhibit markedly the growth and patulin production of *P. expansum* and *Byssochlamys nivea* in grape and apple juice (Lennox and McElroy, 1984; Roland et al., 1984; Roland and Beuchat, 1984). In contrast, Yousef and Marth (1981) found that sorbate delayed mold growth but did not inhibit biosynthesis of aflatoxin by *A. parasiticus*. The ability to synthesize aflatoxin was greater in the early stages of growth and then decreased as mold growth progressed. In a subsequent study, Rusul and Marth (1987) showed that 0.2% (pH 5.5) or 0.05% (pH 4.5) potassium sorbate completely inhibited *A. parasiticus* growth and toxin production for 3 days in a glucose–yeast extract salts medium. However, growth and toxin production were nearly normal by day 7 of incubation. Tsai et al. (1984) reported that potassium sorbate was more effective against injured spores of *A. parasiticus* and that inhibition was dependent on the concentration of sorbate and pH. The activity of potassium sorbate as an antifungal agent has been shown to be enhanced by the presence of EDTA and vanillin (Razavi-Rohani and Griffiths, 1999b; Matamoros-León et al., 1999).

Sorbate has been found to inhibit the growth of *Salmonella, Clostridium botulinum*, and *Staphylococcus aureus* in cooked, uncured sausage (Tompkin et al., 1974); *S. aureus* in bacon (Pierson et al., 1979); *Pseudomonas putrefaciens* and *P. fluorescens* in TSB (Robach, 1978,1979); *Vibrio parahaemolyticus* in crab meat and flounder homogenates (Robach and Hickey, 1978); *S. aureus* and *E. coli* in poultry (Robach, 1980; Robach and Sofos, 1982); *Yersinia enterocolitica* in pork (Myers et al., 1983); and *Escherichia coli* O157:H7 and *Salmonella* in queso fresco cheese (Kasrazadeh and Genigeorgis, 1994,1995). Doell (1962) reported that sorbate at concentrations as low as 0.075% inhibited *Salmonella* Typhimurium and *E. coli*. Sorbates also inhibited *S.* Typhimurium in laboratory media and in milk and cheese (Park et al., 1970; Park and Marth, 1972). Uljas and Ingham (1999) developed processes that were effective in reducing *Escherichia coli* O157:H7 and *Salmonella* Typhimurium DT104 by 5 logs in apple cider at pH 3.3–4.1. Sorbic acid (0.1%) in combination with 12 h at 25°C plus freeze-thawing (FT = 48 h at −20°C, 4 h at 4°C), 6 h at 35°C, or 4 h at 35°C plus FT all resulted in a 5 log reduction of the two pathogens in apple cider at pH 4.1. Potassium sorbate at 0.5% inhibited growth and histamine production by *Proteus morgani* and *K. pneumoniae* strains in a TSB fortified with histidine. The inhibition was effective for 215 h and 120 h at 10°C and 32°C, respectively (Taylor and Speckhard, 1984). Potassium sorbate inhibited anaerobic growth of *S. aureus* more than growth under aerobic conditions in an agar–meat model system and was more inhibitory when lactic acid was added (Smith and Palumbo, 1980). Larocco and Martin (1987) demonstrated no inhibition or injury of *S. aureus* MF-31 by potassium sorbate (0.3%) or sorbate in combination with salt at pH 6.3. Microorganisms isolated from seafood show varying degrees of sensitivity to sorbate (Chung and Lee, 1982). Potassium sorbate (0.1%) added to scallops (pH 6.3–6.5) resulted in more rapid *Clostridium botulinum* toxin development than in controls (Fletcher et al., 1988). It was postulated that this might be due to inhibition of microflora competitive to *C. botulinum.*

Curran and Knaysi (1961) found that 0.01–0.1% sorbic acid did not affect the germination of *Bacillus subtilis* spores. In contrast, Gould (1964) demonstrated that germination and outgrowth of six *Bacillus* species was decreased by sorbate at pH 6.0. Smoot and Pierson (1981) reported that potassium sorbate was a strong inhibitor of bacterial spore germination at pH 5.7 but much less so at pH 6.7.

Reports in the 1950s concluded that sorbates had either stimulatory or no effect on clostridia (York and Vaughn, 1954; Hansen and Appleman, 1955). Later studies concluded that sorbates acted as an anticlostridial agent in cured meat products (Sofos et al., 1979b; Robach and Sofos, 1982; Sofos and Busta, 1983). The contradictory results of the studies are most likely due to the fact that in the 1950s, media with pH values approaching 7.0 were used. This was optimal for growth of clostridia but not optimal for the activity of sorbates, since only about 0.6% of sorbic acid is in the undissociated form at pH 7.0. Potassium sorbate, sufficient to give an undissociated sorbic acid concentration of 250 mg/L in culture medium at pH 5.5–7.0, retarded the growth of proteolytic strains of *Clostridium botulinum* from spores and vegetative cells (Blocher et al., 1982; Blocher and Busta, 1983). Sorbate has been shown to prevent spores of *C. botulinum* from germinating and forming toxin in poultry frankfurters and emulsions (Sofos et al., 1979c,d,1980a; Huhtanen and Fienberg, 1980) as well as beef, pork, and soy protein frankfurter emulsions and bacon (Ivey et al., 1978; Sofos et al., 1979a,1980b). Roberts et al. (1982) found that 0.26% (w/v) potassium sorbate significantly decreased *C. botulinum* toxin production in a model cured meat system. The effect of the sorbate was greatest at 3.5% sodium chloride, a pH less than 6.0, and low storage temperatures. Lund et al. (1987) did a systematic study to determine the concentration of potassium sorbate necessary to inhibit a single *C. botulinum* cell in culture medium. They reported that 1000 mg/L (0.1%) at pH 5.0 would reduce the probability of growth of *C. botulinum* at 30°C in 14 days to 1 in 10^8. They also concluded that 0.2% sorbic acid alone in meat products at pH 6.5 would not significantly inhibit *C. botulinum* vegetative cells.

Research on the effect of sorbic acid on catalase-producing microorganisms is contradictory. Several studies showed that sorbic acid inhibited catalase-positive microorganisms and could be used as a selective agent for lactic acid bacteria and clostridia (Phillips and Mundt, 1950; York and Vaughn, 1954,1955). However, Emard and Vaughn (1952) reported that some catalase-positive strains of *S. aureus* grew as well in the presence of sorbate as catalase-negative *Lactobacillus*. Sorbic acid at 0.05–0.10% inhibited both growth and acid production of catalase-negative organisms such as *Streptococcus thermophilus* and *Lactobacillus delbrueckii* ssp. *bulgaricus* (Hamden et al., 1971). Costilow et al. (1955) reported that two strains of *Pediococcus cerevisiae* (weakly catalase-positive) were found to be as tolerant to sorbic acid as catalase-negative strains.

The mechanism by which sorbic acid inhibits microbial growth may partially be due to its effect on enzymes. Melnick et al. (1954) postulated that sorbic acid inhibited dehydrogenases involved in fatty acid oxidation. Addition of sorbic acid resulted in the accumulation of β-unsaturated fatty acids that are intermediate products in the oxidation of fatty acids by fungi. This prevented the function of dehydrogenases and inhibited metabolism and growth. Sorbic acid has also been shown to be a sulfhydryl enzyme inhibitor. These enzymes are very important in microorganisms and include fumarase, aspartase, succinic dehydrogenase, and yeast alcohol dehydrogenase (Whitaker, 1959; Martoadiprawito and Whitaker, 1963; York and Vaughn, 1964). York and Vaughn (1964) suggested that sorbate reacted with sulfhydryl enzymes through an addition reaction with the thiol groups of cysteine. Rehm (1963) indicated that the activity of sorbic acid against *Aspergillus niger* was increased by cysteine. Whitaker (1959) suggested that sorbate activity is due to the formation of stable complexes with sulfhydryl-containing enzymes through a thiohexenoic acid derivative. Sorbate therefore inhibits enzymes either by formation of a covalent bond between sulfur of the essential sulfhydryl group or the ZnOH of the enzyme and the (α and/or β) carbons of the sorbate ion. Other suggested mechanisms for sorbate have involved interference with enolase, proteinase, and catalase, or through inhibition of respiration by competitive action with acetate in acetyl CoA formation (Azukas et al., 1961; Troller, 1965; Dahl and Nordal, 1972). Kouassi and Shelef (1995) found that sorbate inhibited activation of the hemolytic activity of listeriolysin O of *Listeria monocytogenes* by reacting with cysteine. Freese (1978) suggested that lipophilic acids, such as sorbic acid, interfere with transport across the cytoplasmic membrane. Eklund (1985) demonstrated that sorbic acid eliminated the Δ-pH component of the proton motive force (PMF) in *E. coli* membrane vesicles but did not affect the Δ-ψ component. Ronning and Frank (1987) also showed that sorbic acid reduced the cytoplasmic membrane proton gradient and consequently the PMF. They concluded that the loss of PMF inhibited amino acid transport, which could eventually result in the inhibition of many cellular enzyme systems. The varied proposed mechanisms indicate that researchers disagree regarding the exact mechanism(s) of sorbate action on microorganisms. It might be concluded that no single mechanism holds true under all conditions.

The methods of sorbate addition available include direct addition into the product, dipping, spraying, dusting, or incorporation into packaging. Sorbates are used in many food products including artificially sweetened confections, cakes and cake mixes, cheeses, diet drinks, doughnuts, dried fruits, fruit drinks, fudges, icing, jams, jellies, mayonnaise, olives, orange juice, packaged fresh salads, pet foods, pickles, pies and pie fillings, relishes, salad dressings, sausage casings, sour cream, wine, and yogurt. Use concentrations for sorbates vary considerably (Table 3). In the United States, sorbic acid and potassium sorbate are considered GRAS (21 CFR 182.3089; 21 CFR 182.3225; 21 CFR 182.3640; and 21 CFR 182.3795). They may be used in more than 90 food products having standards of identity, and their use may be requested in any food product that allows preservatives. The maximum concentration of sorbic acid is set at 0.2% in pasteurized blended cheese; pasteurized process cheese, cheese food, and cheese spread; and cold pack cheese, cheese food, and cheese spread (21 CFR 133). A maximum of 0.3% is allowed in other cheeses (21 CFR 133). Potassium sorbate at 0.05–0.08% is used in olives, sauerkraut, pickled cucumbers, and sweet relish. It does not interfere with the lactic acid fermentation necessary for the production of these products. In wines it may be used as a replacement or adjunct to sulfur dioxide as a sterilizing and preservative agent to prevent the occurrence of secondary fermentation. The maximum level of sorbic acid or its salt has been set at 0.1% in wine or in materials for the production of wine, and it is not necessary to declare its presence on the label. United States standards of identity (21 CFR 150.141 and 21

Table 3 Food Products in Which Sorbates Are Used and Typical Use Concentrations

Food product	Use concentration (%)
Beverage syrups	0.1
Cakes and icings	0.05–0.1
Cheese and cheese products	0.2–0.3
Cider	0.05–0.1
Fruits, dried	0.02–0.05
Fruit drinks	0.025–0.075
Margarine	0.1
Pie fillings	0.05–0.1
Pet food, semimoist	0.1–0.3
Salad dressings	0.05–0.1
Salads, prepared vegetable	0.05–0.1
Wine	0.02–0.04

Source: Anonymous, 1999.

CFR 150.161) for artificially sweetened jams, jellies, and preserves permit the use of sorbic acid and its salts to maximum levels of 0.1%. The compound is generally added during the cooling cycle. Potassium sorbate may be used to protect dry sausages from mold spoilage by dipping the casing, either before or after stuffing (9 CFR 318.7). The use of sorbic acid and sorbates are permitted in all countries of the world for the preservation of a variety of foods.

Sorbic acid is considered one of the least harmful antimicrobial preservatives, even at levels exceeding those normally used in foods (Lueck, 1980; Sofos and Busta, 1981). The LD_{50} for sorbates in rats ranges from 7.4 to 10.5 g/kg body weight. Rats fed sorbic acid at 10% in feed for 40 days had no ill effects (Lueck, 1980). When the feeding period was increased to 120 days, the growth rate and liver weight increased (Demaree et al., 1955). Sorbic acid at 5% in the diet of rats did not affect health after 1000 days (except for two tumors in 100 rats) (Lang, 1960b). When rats were given sorbic acid (10 mg/100 mL) or potassium sorbate (0.3%) in drinking water or 0.1% levels in the diet, no tumors were observed after 100 weeks. The growth of tumors was not seen when mice were fed 40 mg sorbic acid per kilogram body weight (Shtenberg and Ignat'ev, 1970). Repeated subcutaneous administration of sorbic acid in peanut oil or water in rats produced sarcomas at the site of injection. Sarcomas produced locally in this manner were not considered a valid index of carcinogenicity (Gangolli et al., 1971). When sorbic acid is used for cosmetic and pharmaceutical products, it may irritate the mucous membranes, and in highly sensitive individuals it may cause skin irritation (Lueck, 1980).

F. Other Organic Acids

While citric acid generally is not used as an antimicrobial, it has been shown to possess activity against some molds and bacteria. Reiss (1976) found that 0.75% citric acid slightly reduced growth and greatly reduced toxin production by *Aspergillus parasiticus*. With *Aspergillus versicolor*, growth was inhibited at the same level, but toxin production was prevented by 0.25% citric acid. In contrast, 0.75% citric acid did not influence the growth or toxin production of *Penicillium expansum* (Reiss, 1976). Citric acid was observed to be more inhibitory to *Salmonella* than lactic or hydrochloric acids (Subramanian and Marth, 1968). In a related study, Thomson et al. (1967) found that as little as 0.3% citric acid could reduce the level of viable *Salmonella* on poultry carcasses. Shrimp, shrimp puree, tomato puree, and shrimp and tomato puree acidified to pH 4.2 and 4.6 with citric acid showed no significant growth or toxin production by *C. botulinum* after 8 weeks at 26°C (Post et al., 1985). Minor and Marth (1970) showed that *Staphylococcus aureus* was inhibited 90% and 99% in 12 h at pH 4.7 and 4.5, respectively, by citric acid. Citric acid was found to be particularly inhibitory to flat-sour organisms isolated from tomato juice, but the inhibition was pH-dependent (Murdock, 1950). Xiong et al. (1999) reported that citric acid in lemon juice could be used to prepare homemade mayonnaise with raw eggs that was free of *Salmonella* enteritidis PT4. Both the concentration of lemon juice and time of storage influenced inactivation of the *Salmonella* serovar.

The mechanism of inhibition by citrate has been theorized to be related to its ability to chelate metal ions. Branen and Keenan (1970) were the first to suggest that inhibition may be due to chelation, in studies with citrate against *Lactobacillus casei*. Chelation was also indicated as the reason for inhibition of *S. aureus* (Rammell, 1962) and *Arthrobacter pascens* (Imai et al, 1970) by citrate. In contrast, Buchanan and Golden (1994) found that while undissociated citric acid was inhibitory against *Listeria monocytogenes*, the dissociated molecule protected the microorganism. They theorized that this protection was due to chelation by the anion. The U.S. FDA has classified citric acid as GRAS for miscellaneous and general purpose use (21 CFR 182.1033).

Fumaric acid has been used to prevent the occurrence of malolactic fermentation in wines (Ough and Kunkee, 1974) and as an antimicrobial agent in wines (Pilone, 1975). Esters of fumaric acid (monomethyl, dimethyl, and ethyl) at 0.15–0.2% have been tested as a substitute or adjunct for nitrate in bacon. Fumaric acid esters (0.125%) retarded swelling and toxin formation in canned bacon inoculated with *Clostridium botulinum* for up to 56 days at 30 °C (Huhtanen, 1983). Dymicky et al. (1987) studied the structure–activity relationships of *n*-mono-alkyl, di-alkyl, and methyl *n*-alkyl fumarates in a microbiological medium against *C. botulinum* 62A. The *n*-mono-alkyl fumarates (C_1–C_{18}) were found to have the greatest activity, with minimum inhibitory concentrations ranging from 6.2 to 400 μg/mL. Fungal growth in tomato juice was inhibited with the use of 0.2% methyl and ethyl fumarates or 0.05% dimethyl and diethyl fumarates. Fumaric acid esters were also found to inhibit mold growth on bread (Huhtanen et al., 1981; Huhtanen and Guy, 1984). At 1% as a sanitizer on lean beef, fumaric acid reduced *L. monocytogenes* and *E. coli* O157:H7 by approximately 1 log after 5 s at 55°C (Podolak et al., 1996a). Addition of 5.0% fumaric acid in ground beef patties stored at 4°C for 10 days resulted in less than a 1.5 log increase in aerobic and psychrotrophic microorganisms and fecal coliforms (Podolak et al., 1996b).

Many other organic acids, including adipic, caprylic, malic, succinic, and tartaric have been evaluated for their antimicrobial properties. The antimicrobial activity of adipic, malic, and tartaric acids may be attributed only to their ability to reduce pH. Malic, citric, and tartaric acids did not have a statistically significant effect on the thermal resistance of spores of *Alicyclobacillus acidoterrestris* spores in a model fruit juice system (Pontius et al., 1998). Caprylic acid has been shown to have variable effects on microorganisms (Kabara et al., 1972). The compound has been found to be inhibitory to *Vibrio parahaemolyticus* (Beuchat, 1980), *Escherichia coli*, and *Shigella* spp. (Nakamura and Zangar, 1968). In contrast to other organic acid antimicrobials, Woolford (1975b) found that as the pH was decreased, the concentration of caprylic acid required for inhibition was not markedly lowered. Succinic acid at 3% or 5% was found to be effective in reducing the microbial level on chicken carcasses; however, the appearance of the product was adversely effected (Cox et al., 1974).

VIII. PARABENS

Alkyl (methyl, ethyl, propyl, butyl, and heptyl) esters of *p*-hydroxybenzoic acid are collectively known as the "parabens." Sabalitschka and coworkers (Prindle, 1983) first described antimicrobial action of parabens in the 1920s. Esterification of the carboxyl group of benzoic acid allows the molecule to remain undissociated up to pH 8.5 versus benzoic acid with a pK_a of 4.2. While the pH optimum for antimicrobial activity of benzoic acid is 2.5–4.0, the parabens are effective at pH 3–8 (Aalto et al., 1953; Chichester and Tanner, 1972). They are effective at both acidic and alkaline pH levels. The molecular weights of various esters are: methyl, 152.14; ethyl, 166.17; propyl, 180.21; butyl, 196.23; and heptyl, 236.21. The solubility of parabens in ethanol increases from methyl to heptyl ester, but the water solubility is inversely related to the alkyl chain length.

The antimicrobial activity of *p*-hydroxybenzoic acid esters is, in general, directly proportional to the chain length of the alkyl component (Table 4). Parabens are generally more active against molds and yeast than against bacteria. Against bacteria, they are more effective against gram positive than gram negative bacteria.

Moir and Eyles (1992) compared the effectiveness of methyl paraben and potassium sorbate on the growth of four psychrotrophic foodborne bacteria: *A. hydrophila, Listeria monocytogenes, Pseudomonas putida*, and *Yersinia enterocolitica*. At pH 5, there was little difference in inhibition between the antimicrobials. At pH 6 however, methyl paraben was effective at a lower concentration than potassium sorbate for all pathogens except *A. hydrophila*, where the two were equal. Robach and Pierson (1978) investigated the effect of methyl and propyl paraben on toxin production of *Clostridium botulinum* NCTC 2021. At 100 μg/mL of methyl and 100 μg/mL of propyl paraben, toxin formation was prevented, while 1200 μg/mL methyl and 200 μg/mL propyl were necessary for growth inhibition. Reddy and Pierson (1982) and Reddy et al. (1982) determined the effect of methyl, ethyl, propyl and butyl parabens on growth and toxin production of ten *Clostridium botulinum* strains (5 Type A, 5 Type B). In TYG medium at pH 7.0 and 37°C, 1000 μg/mL methyl paraben blocked growth and toxin formation for only 1 day. Ethyl and propyl paraben, at the same concentration, prevented growth and toxin production for the maximum incubation time of 7 days. Butyl paraben, as might be expected, was most effective and prevented growth and toxin production for 7 days at 200 μg/ml. Inhibition of *C. botulinum* by the parabens in food systems is less than in laboratory media (Davidson, 1993). Propyl paraben added to vacuum-packaged sliced ham or sausage inoculated with *Listeria monocytogenes* and stored at 4 or 9°C for 5 weeks was ineffective in controlling the microorganism (Blom et al., 1997).

As with bacteria, inhibition of fungi increases as the alkyl chain length of the parabens increases. Jermini and Schmidt-Lorenz (1987) evaluated ethyl paraben against osmotolerant yeasts at various water activities and pH levels. They found that the concentration of ethyl paraben necessary for inhibition was a function of initial number of yeast cells present. At 600 μg/mL ethyl paraben, the time to initiate growth was approximately 15, 12, 5, and 2–3 days for 10^2, 10^3, 10^4, and 10^5 cells at a_w 0.900 and pH 4.8. Inhibition by parabens was also a function of pH, with more acid environments requiring less ethyl

Table 4 Minimum Inhibitory Concentrations of Methyl, Propyl, and Heptyl Esters of *p*-Hydroxybenzoic Acid Against Growth and End-Product Production of Selected Microorganisms

	Minimum inhibitory concentration (μg/mL)		
Microorganism	Methyl	Propyl	Heptyl
Bacteria, gram positive			
Bacillus cereus	2000	125–400	12
Bacillus subtilis	2000	250	—
Clostridium botulinum Type A	1000–1200	200–400	—
Clostridium botulinum toxin production	100	100	—
Clostridium perfringens	500[a]	—	—
Lactococcus lactis	—	400	12
Listeria monocytogenes	>512	512	—
Staphylococcus aureus	4000	350–500	12
Bacteria, gram negative			
Aeromonas hydrophila, protease secretion	—	>200	—
Enterobacter aerogenes	2000	1000	—
Escherichia coli	2000	400–1000	—
Klebsiella pneumoniae	1000	250	—
Pseudomonas aeruginosa	4000	8000	—
Pseudomonas fragi	—	4000	—
Pseudomonas fluorescens	2000	1000	—
Salmonella Typhi	2000	1000	—
Salmonella Typhimurium	—	>300	—
Vibrio parahaemolyticus	—	50–100	—
Fungi			
Aspergillus flavus	—	200	—
Aspergillus niger	1000	200–250	—
Byssochlamys fulva	—	200	—
Candida albicans	1000	125–250	—
Penicillium chrysogenum	500	125–250	—
Rhizopus nigricans	500	125	—
Saccharomyces bayanus	930	220	—
Saccharomyces cerevisiae	1000	125–200	100

[a] 3:1 methyl:propyl.

Sources: Aalto et al., 1953; Bargiota et al., 1987; Davidson, 1993; Dymicky and Huhtanen, 1979; Jurd et al., 1977; Klindworth et al., 1979; Lee, 1973; Marwan and Nagel, 1986; Moustafa and Collins, 1969; Robach and Pierson, 1978; Veugopal, 1984.

paraben for inhibition. Little effect of water activity or type of humectant was observed. Of the genera tested, *Zygosaccharomyces bailii* was most resistant, requiring 900 μg/mL at 25°C, a_w 0.900 and pH 4.8. Other yeasts evaluated included *Torulaspora delbrueckii, Z. rouxii, Z. bisporus*, and *Debaryomyces hansenii* with minimum inhibitory concentrations of 700 μg/mL, 700 μg/mL, 400 μg/mL, and 400 μg/mL, respectively. They concluded that the concentration of ethyl paraben required to preserve a product from the effect of osmotolerant yeast for 30 days at 25°C and a_w 0.795–0.980 was 900 μg/mL or 400 μg/mL at a pH of 4.8 or ≤4.0, respectively. Thompson (1994) evaluated butyl, propyl, ethyl and methyl parabens, alone and in combination, against multiple strains of mycotoxigenic *Aspergillus, Penicillium*, and *Fusarium*. The most effective parabens were the propyl and butyl esters with minimum inhibitory concentrations of 1.0–2.0 mM in potato dextrose agar. Combinations of the various parabens were determined to have synergistic activity against the mold species.

The mechanism by which the parabens inhibit microorganisms is most likely related to their effects on the cytoplasmic membrane. Freese et al. (1973) demonstrated that parabens inhibited serine uptake and ATP production in *Bacillus subtilis*. They concluded that the parabens were capable of inhibiting both membrane transport and the electron transport system. Eklund (1985) found that parabens eliminated the ΔpH of the cytoplasmic membrane of *E. coli*. The compounds did not significantly affect the $\Delta\Psi$ component of the proton motive force. He concluded that neutralization of the proton motive force and subsequent transport inhibition was not the sole mechanism of action of the parabens. Bargiota et al. (1987) postulated that mediation of the effectiveness of parabens as antimicrobials was due to cellular lipid components. Juneja and Davidson (1992) altered the lipid composition of *L. monocytogenes* by growth in the presence of added fatty acids (C14:0, C18:0 or C18:1). Growth of *L. monocytogenes* in the presence of exogenously added C14:0 or C18:0 fatty acids increased the resistance of the cells to TBHQ and parabens. However, growth in the presence of C18:1 led to increased sensitivity to the antimicrobial agents. Results indicated that, for *L. monocytogenes*, a correlation existed between lipid composition of the cell membrane and susceptibility to antimicrobial compounds.

To take advantage of their respective solubility and increased activity, methyl and propyl parabens are normally used in a combination of 2–3:1 (methyl:propyl). The compounds may be incorporated into foods by dissolving in water, ethanol, propylene glycol, or the food product itself. Methyl and propyl paraben (3:1) at the 0.03–0.06% level may be used to increase the shelf-life of fruit cakes, pie crusts, pastries (nonyeast), icing, toppings, and fillings such as fruit, jellies, and creams. Methyl and propyl paraben (2:1) at 0.03–0.05% may be used in soft drinks. A combination of esters at 0.03–0.06% is recommended for marinated, smoked, or jellied fish products. A combination of esters at 0.05–0.1% is used in flavor extracts. Approximately 0.05% methyl and propyl paraben (2:1) are used to preserve fruit salads, juice drinks, sauces, and fillings. Methyl paraben at 0.05–0.1% or a combination of methyl and propyl paraben has been used for gelatinous coatings or jellied foods. Methyl and propyl paraben (2:1) may be used as preservatives in jams and jellies (0.07%), in salad dressings (0.1–0.13%), and in wines (0.1%) (Davidson, 1993).

The FDA considers methyl (21 CFR 184.1490) and propyl (21 CFR 184.1670) parabens as GRAS, with a total addition limit of 0.1%. Both methyl and propyl parabens are permitted as antimycotics in food packaging material (21 CFR 121.2001). Heptyl paraben is permitted in fermented malt beverages (beer), noncarbonated soft drinks, and fruit-based beverages at a maximum of 12 ppm (21 CFR 172.145). In the United Kingdom, methyl, ethyl, and propyl parabens are permitted in food. In many countries, the butyl ester is allowed for use in foods.

Parabens are known to have low toxicity. They are rapidly hydrolyzed, conjugated in the body, and excreted in the urine (Jones et al., 1956; FAO/WHO, 1967). Methyl paraben has been shown to be noncarcinogenic in rats fed 2–8% in the diet (Matthews et al., 1956). Methyl paraben at 1% administered intraperitoneally to mice was shown to be noncarcinogenic. Similar results were observed by subcutaneous administration of 1.0 mL/week through the life span (WHO, 1974). Ethyl paraben at 2% levels in feed also has been shown to be noncarcinogenic in rats (WHO, 1974).

Parabens have been observed to have a local anesthetic effect. The action increases with the number of carbon atoms in the alkyl group (Adler-Hradecky and Kelentey, 1960; WHO, 1974). The effect of a 0.1% solution of methyl paraben is similar to that of a 0.05% procaine solution. Propyl or ethyl paraben at 0.05% concentration also have been reported to have a local anesthetic effect on the buccal mucosa (Bubnoff et al., 1957).

Parabens in foods have been reported to cause dermatitis of unknown etiology (Epstein, 1968). Cross-sensitization phenomenon among the parabens has been observed in individuals sensitive to *p*-hydroxybenzoate. Matthews et al. (1956) reported that 0.1% methyl or propyl paraben solution applied to skin produced no skin reaction.

IX. PHOSPHATES

Phosphates are used extensively in food processing. Some phosphate compounds, including sodium acid pyrophosphate (SAPP), tetrasodium pyrophosphate (TSPP), sodium tripolyphosphate (STPP), sodium tetrapolyphosphate, sodium hexametaphosphate (SHMP), and trisodium phosphate (TSP), have demonstrated variable levels of antimicrobial activity in foods. There are over 30 phosphate salts used in food products and their functions include buffering or pH stabilization, acidification, alkalization, sequestration or precipitation of metals, formation of complexes with organic polyelectrolytes (e.g., proteins, pectin, and starch), deflocculation, dispersion, peptization, emulsification, nutrient supplementation, anticaking, antimicrobial preservation, and leavening (Ellinger, 1981).

Gram positive bacteria appear to be generally more susceptible to phosphates than gram negative bacteria. Post et al. (1968) found that 0.1% sodium hexametaphosphate (SHMP) was effective against many gram positive bacteria, while gram negative bacteria grew at 10% SHMP. Kelch (1958) tested commercial mixtures of tetrasodium pyrophosphate (TSPP), sodium acid pyrophosphate (SAPP), sodium tripolyphosphate (STPP), and SHMP with and without heat against gram positive bacteria. *Staphylococcus aureus* and *Streptococcus faecalis* were completely inhibited in nutrient medium plus heat (50°C). Without heat, susceptibility was variable. The MICs of food grade phosphates added to early exponential phase cells of *Staphylococcus aureus* ISP40 8325 in a synthetic medium were determined to be 0.1% for sodium ultraphosphate and sodium polyphosphate glassy and 0.5% for sodium acid pyrophosphate, sodium tripolyphosphate, and tetrasodium pyrophosphate (Lee et al., 1994a). A mixture of TSPP (15%)–STPP (70%)–SHMP (15%) was an effective inhibitor of *Bacillus subtilis, Clostridium sporogenes*, and *Clostridium bifermentans* at 0.5%. In a similar study, Jen and Shelef (1986) tested seven phosphate derivatives against the growth of *Staphylococcus aureus* 196E. Only 0.3% SHMP (with a phosphate chain length n of 21) and 0.5% STPP or SHMP ($n = 13$; 15) were effective growth inhibitors. Magnesium reversed the growth-inhibiting effect. Wagner and Busta (1985) found that while 0.4% SAPP had no effect on the growth of *Clostridium botulinum* 52A, the compound delayed or prevented toxicity to mice. It was theorized that toxin inhibition was due to binding of the toxin molecule or inactivation of the protease responsible for protoxin activation. Gould (1964) showed that 0.2–1.0% SHMP permitted germination of *Bacillus* spores but prevented outgrowth. Zaika and Kim (1993) found that 1% sodium polyphosphates inhibited lag and generation times of *Listeria monocytogenes* in BHI broth, especially in the presence of NaCl.

Phosphate derivatives also have antimicrobial activity in food products. In several studies, SAPP, SHMP, or polyphosphates have been shown to increase the effectiveness of the curing system (nitrite–pH–salt) against *Clostridium botulinum* (Ivey and Robach,

1978; Nelson et al., 1980; Roberts et al., 1981; Wagner and Busta, 1983). Other studies have reported variable results. Post et al. (1968) found that a 10% sodium tetrapolyphosphate dip was effective in preserving cherries against the growth of several mold species including *Penicillium, Rhizopus*, and *Botrytis*. SHMP, STPP, and tetrasodium pyrophosphate were less effective. In cheese and pasteurized process cheese products, phosphates serve several roles. Tanaka (1982) showed that phosphate along with sodium chloride, water activity, water content, pH, and lactic acid interacted to prevent the outgrowth of *Clostridium botulinum* in pasteurized process cheese. Ebel et al. (1965) reported preservative action of condensed phosphates on fish and inhibitory activity against *Staphylococcus aureus* and *Bacillus subtilis* in a broth medium. Phosphate salts have also been shown to have varying antimicrobial activity against rope-forming *Bacillus* in bread and *Salmonella* in pasteurized egg whites (Tompkin, 1983).

Trisodium phosphate (TSP) at levels of 8–12% has been reported to reduce pathogens, especially *Salmonella*, on poultry (Giese, 1992). Kim et al. (1994) found a 1.6–1.8 log reduction in *S*. Typhimurium on postchill chicken carcasses using a 10% TSP dipping treatment. Wang et al. (1997) reported that 10% TSP reduced *S. typhimurium* attached to 38.5 cm^2 chicken skin by 1.6 to 2.3 log. In addition, Slavik et al. (1994) evaluated a 10% TSP dipping of chicken carcasses inoculated with *Campylobacter jejuni* and reported a 1.5 log reduction. A treatment of beef surfaces with TSP reduced the level of attached *E. coli* O157:H7 and *S*. Typhimurium by 0.51–1.39 logs, respectively (Kim and Slavik, 1994). Lillard (1994) determined that while 10% TSP appeared to reduce viable *Salmonella* by 2 logs, it was a function of the high pH of the (11–12) of the system. Waldroup (1995) and Waldroup et al. (1995) concluded that TSP had no effect on pathogens or indicator microorganisms on poultry and may actually allow greater survival of *Listeria monocytogenes* than non-TSP processes. A trisodium phosphate (10 or 15%) dip (37°C) of 15 s inactivated >5 log CFU/cm^2 *Salmonella* Montevideo inoculated on the surface of tomatoes, but was only capable of reducing the microorganism by 2 logs in the tomato core tissue (Zhuang and Beuchat, 1996). The pH of the dip solutions was 11.8–12.6.

Several mechanisms for bacterial inhibition by polyphosphates (Sofos, 1986). The ability of polyphosphates to chelate metal ions appears to play an important role in the antimicrobial activity of these compounds. The presence of magnesium has been shown to reverse inhibition of gram positive bacteria by antimicrobial phosphates (Post et al., 1968; Jen and Shelef, 1986; Lee et al., 1994b). With some phosphates, calcium and iron are also effective in reversing inhibition (Jen and Shelef, 1986; Lee et al., 1994b). Knabel et al. (1991) stated that the chelating ability of polyphosphates was responsible for growth inhibition of *Bacillus cereus*, *Listeria monocytogenes*, *Staphylococcus aureus*, *Lactobacillus* and *Aspergillus flavus*. The target metal chelated was dependent upon the microorganism. Orthophosphates had no inhibitory activity against any of the microorganisms and have no chelating ability. Further, Knabel et al. (1991) reported that inhibition was reduced at lower pH due to protonation of the chelating sites on the polyphosphates. They concluded that polyphosphates inhibited gram positive bacteria and fungi by removal of essential cations from binding sites on the cell walls of these microorganisms. It has been suggested that polyphosphates may also interfere with RNA function or metabolic activities of cells (Ellinger, 1981; Sofos, 1986).

Excessive intake of phosphates may decrease the availability of calcium, iron, and other minerals; however, no adverse effects have been reported with moderate doses (Lauerson, 1953; Lang, 1959). Nephrocalcinosis was reported when rats were fed a diet containing 1% of a mixture of diphosphates and polyphosphates (Van Esch et al., 1957).

At 2.5% of the diet, the mixture caused anemia characterized by a decrease in erythrocytes, and at 5% retarded growth rate and decreased fertility resulted (Van Esch et al., 1957). Life span was not significantly altered with up to 2.5% phosphates (Van Esch et al., 1957). Problems in humans are likely to occur when high levels of phosphates are consumed, as in large quantities of soft drinks (Jacobs, 1959).

X. SULFITES

Sulfur dioxide and its various salts claim a long history of use dating back to times of the ancient Greeks (Ough, 1993b). They have been used extensively as antimicrobials and to prevent enzymatic and nonenzymatic discoloration in a variety of foods (Wedzicha, 1981). The earliest recorded food use of sulfur dioxide was the treating of wines in ancient Rome (IFT, 1975). The use of sulfur dioxide as a food preservative was reported in the literature in the 17th century by Evelyn (1664), who suggested that carts should be filled with cider that contained sulfur dioxide (produced by burning sulfur).

The salts of sulfur dioxide include (formula; solubility in g/L at temperature specified): potassium sulfite (K_2SO_3; 250, 20°C), sodium sulfite (Na_2SO_3; 280, 40°C), potassium bisulfite ($KHSO_3$; 1000, 20°C), sodium bisulfite ($NaHSO_3$; 3000, 20°C), potassium metabisulfite ($K_2S_2O_5$; 250, 0°C), and sodium metabisulfite (NaS_2O_5; 540, 20°C) (Ough, 1993b).

The most important factor impacting the antimicrobial activity of sulfites is pH. Sulfur dioxide and its salts set up a pH-dependent equilibrium mixture when dissolved in water:

$$SO_2 \cdot H_2O \leftrightarrows HSO_3^- + H^+ \leftrightarrows SO_3^{2-} + H^+$$

Aqueous solutions of sulfur dioxide theoretically yield sulfurous acid (H_2SO_3), however evidence indicates that the actual form is more likely $SO_2 \cdot H_2O$ (Gould and Russell, 1991). As the pH decreases, the proportion of $SO_2 \cdot H_2O$ increases and the bisulfite (HSO_3^-) ion concentration decreases. The pK_a values for sulfur dioxide, depending upon temperature, are 1.76–1.90 and 7.18–7.20 (Rose and Pilkington, 1989; Gould and Russell, 1991; Ough, 1993b). The inhibitory effect of sulfites is most pronounced when the acid or $SO_2 \cdot H_2O$ is in the undissociated form (Hailer, 1911). Therefore the most effective pH range is <4.0. King et al. (1981) proved this when they found that undissociated $H_2SO_3(SO_2 \cdot H_2O)$ was the only form active against yeast and that neither HSO_3^- nor SO_3^{2-} had antimicrobial activity. Similarly, $SO_2 \cdot H_2O$ was shown to be 1000, 500, and 100 times more active than HSO_3^- or SO_3^{2-} against *E. coli*, yeast, and *Aspergillus niger*, respectively (Rehm and Wittman, 1962). Increased effectiveness at low pH likely is due to the fact that unionized sulfur dioxide can pass across the cell membrane in this form (Rahn and Conn, 1944; Ingram et al., 1956; Rose and Pilkington, 1989).

Sulfites, especially as the bisulfite ion, are very reactive. These reactions not only determine the mechanism of action of the compounds, they also influence antimicrobial activity. For example, sulfites form addition compounds (α-hydroxysulfonates) with aldehydes and ketones. These addition compounds are in equilibrium in solution with free sulfite ions, resulting in the formation of a thiol (R–SH) and S-substituted thiosulfates ($R{-}SSO_3$) (Means and Feeney, 1971). It is generally agreed that these bound forms have much less or no antimicrobial activity compared to the free forms. For example, Ough (1993b) reported that addition compounds with sugars completely neutralized the antimicrobial activity of sulfites against yeast. However, Stratford and Rose (1985) did demonstrate that pyruvate–sulfite complexes retained some activity against *Saccharomyces cerevisiae*.

Sulfurous acid inhibits yeast, molds, and bacteria; however, yeasts and molds are generally less sensitive to sulfur dioxide than bacteria (Hailer, 1911). As antimicrobials, sulfites are used primarily in fruit and vegetable products to control three groups of microorganisms: spoilage and fermentative yeasts and molds on fruits and fruit products (e.g., wine), acetic acid bacteria, and malolactic bacteria (Ough, 1993b).

Sulfur dioxide is fungicidal even in low concentrations against yeast and mold. The inhibitory concentration range of sulfur dioxide against yeasts is as follows (μg/mL): *Saccharomyces*, 0.1–20.2; *Zygosaccharomyces*, 7.2–8.7; *Pichia*, 0.2; *Hansenula*, 0.6; *Candida*, 0.4–0.6 (Rehm and Wittmann, 1962). Goto (1980) determined that wild yeasts of the general *Saccharomyces* and *Torulopsis* were the most tolerant to sulfur dioxide in grape juice. Roland et al. (1984) and Roland and Beuchat (1984) compared the effectiveness of sodium benzoate, potassium sorbate, and sulfur dioxide against *Byssochlamys nivea* growth and patulin production in grape and apple juices. Sulfur dioxide at 25–100 μg/mL was the most effective inhibitor (based upon concentration) of both growth and toxin production. Parish and Carroll (1988) found that sulfur dioxide had greater antimicrobial effectiveness alone than in combination with benzoate or sorbate against *Saccharomyces cerevisiae* Montrachet 522. Similarly, Knox et al. (1984) showed no synergistic antimicrobial interactions with combinations of sulfur dioxide, potassium sorbate, and butylated hydroxyanisole against *Saccharomyces cerevisiae* in grape or apple juice.

Against some bacteria, low concentrations of sulfur dioxide (1–2 μg/mL) are bacteriostatic, while only high concentrations are bactericidal. This may be due to differential uptake of sulfur dioxide by fungi and bacteria. Sulfur dioxide at 1–10 μg/mL is capable of inhibiting most lactic acid bacteria in fruit products at pH 3.5 or less (Wibowo et al., 1985). Tompkin et al. (1980) found that addition of sodium metabisulfite as a source of sulfur dioxide delayed *C. botulinum* outgrowth in perishable canned comminuted pork when it was temperature-abused at 27°C. Sulfurous acid and an excess of aldehyde in a medium inhibited growth of *Lactobacillus hilgardii* and *Leuconostoc mesenteroides* isolated from wine (Fornachon, 1963). Sulfur dioxide is reported to be more inhibitory to gram negative bacteria such as *E. coli* and *Pseudomonas* than to gram positive bacteria (Roberts and McWeeny, 1972). Banks and Board (1982) tested several genera of *Enterobacteriaceae* isolated from sausage for their metabisulfite sensitivity. The microorganisms tested and the concentration of free sulfite (μg/mL) necessary to inhibit their growth at pH 7.0 were as follows: *Salmonella* sp., 15–109; *E. coli*, 50–195; *Citrobacter freundii*, 65–136; *Yersinia enterocolitica*, 67–98; *Enterobacter agglomerans*, 83–142; *Serratia marcescens*, 190–241; and *Hafnia alvei*, 200–241.

The most likely targets for inhibition by sulfites include disruption of the cytoplasmic membrane, inactivation of DNA replication, protein synthesis, inactivation of membrane-bound or cytoplasmic enzymes, or reaction with individual components in metabolic pathways. Cell damage may result from interaction with SH groups in structural proteins and interactions with enzymes with SH groups in structural proteins and interactions with enzymes, cofactors, vitamins, nucleic acids, and lipids. The sensitivity of enzymes with SH groups is a primary inhibitory effect against NAD-dependent reactions (Pfleiderer et al., 1956). Sulfur dioxide also reacts with end products or intermediate products and inhibits enzyme chain reactions. Sulfur dioxide cleaves essential disulfide linkage in proteins and induces changes in the molecular confirmation of enzymes. This modifies the enzyme active site or destroys the coenzymes. It destroys the activity of thiamine and thiamine-dependent enzymes by cleavage and produces cytotoxic effects by cross-linking individual nucleic acid residues or nucleic acid residues and proteins. It also damages cell metabolism and membrane function by peroxidizing lipids. One or more of these factors may result in microbial death or inhibition (Hammond and Carr, 1976).

Sulfur dioxide is used to control the growth of undesirable microorganisms in soft fruits, fruit juices, wines, sausages, fresh shrimp, and acid pickles, and during extraction of starches. It is added to expressed grape juices used for making wines to inhibit molds, bacteria, and undesirable yeasts. The concentration of sulfur dioxide used depends on the cleanliness, maturity, and general condition of the grapes, but 50–100 ppm (μg/mL) is generally used (Amerine and Joslyn, 1960). At appropriate concentrations, sulfur dioxide does not interfere with wine yeasts or with the flavor of wine. During fermentation, sulfur dioxide also serves as an antioxidant, clarifier, and dissolving agent. The optimum level of sulfur dioxide (50–75 ppm) is maintained to prevent post fermentation changes by microorganisms. Sulfur dioxide, at 0.01–0.2%, is used as a temporary preservative in fruit products. Residues from the final products are removed by heat or vacuum (Lueck, 1980). Sulfur dioxide is not only used as an antimicrobial, but also has other functions such as protection against oxidative, enzymatic, and nonenzymatic browning reactions and inhibition of chemically induced color losses. Sulfur dioxide used as a solution in water is very effective and controls the growth of *Botrytis*, *Cladosporium*, and other molds on soft fruits (Roberts and McWeeny, 1972). It is used extensively in preserving strawberries, raspberries, and gooseberries after picking for jam production. In this way, jam production may be spread over the year rather than concentrated in the harvesting season. Sulfur dioxide solution in water is used to sanitize equipment.

In some countries, sulfites may be used to inhibit the growth of microorganisms on fresh meat and meat products (Kidney, 1974). Sulfur dioxide restores a bright color, but may give a false impression of freshness. Sulfite or metabisulfite added in sausages is effective in delaying the growth of molds, yeast, and salmonellae during storage at refrigerated or room temperature (Ingram et al., 1956). Banks and Board (1982) showed that 600 μg/g sodium metabisulfite prevented the growth of species of *Enterobacteriaceae* at 4, 10, and 15°C, but not at 22°C in fresh sausage. *Salmonella* did not grow in sulfited sausage except at 25°C.

The U.S. FDA considers sulfur dioxide and several sulfite salts as GRAS (21 CFR 182). However, sulfites cannot be used in meats, food recognized as a source of thiamine, or on fruits or vegetables intended to be served raw to consumers, sold raw to consumers, or presented to consumers as fresh. They are allowed in fruit juices and concentrates, dehydrated fruits and vegetables, and in wine. In other countries, sulfites may be permitted in meats, meat products, poultry, poultry products, and seafood (Chichester and Tanner, 1981). The maximum level of sulfur dioxide allowed in wine was set at 350 mg/L by the regulating body for the U.S. alcoholic beverage industry, the Bureau of Alcohol, Tobacco and Firearms of the Department of the Treasury. The amount of sulfites used in food products is dictated by good manufacturing practice.

The oral LD_{50} for rats is 1000–2000 mg sulfur dioxide per kilogram body weight. The LD_{50} for rabbits and cats was determined as 600–700 and 450 mg sulfur dioxide/kg body weight, respectively. In dogs and human beings, fatal poisoning is not possible because sulfur dioxide induces vomiting (Lang, 1960a). The toxic effect of sulfur dioxide in humans is variable. Some persons may tolerate up to 50 mg/kg body weight, while others have headache, nausea, and diarrhea (Schroeter, 1966). Sodium bisulfite fed at 0.5–2% in feed to rats had an injurious effect on the nervous system, reproductive organs, bone tissue, kidneys, and other visceral organs within 12 months (Fitzhugh et al., 1946). Inhalation of sulfur dioxide at concentrations higher than 33 mg/L in air may cause death due to pulmonary dysfunction (Amadur, 1980). Postmortem lesions include pulmonary edema, lung hemorrhage, and visceral congestion. Symptoms observed before death include coughing, lacrimation, and sneezing. Sulfites binding with certain nucleotides may cause point mutations. Such mutagenicity was observed in a lambda phage of *E. coli* with a dose of 3 M sodium bisulfite at pH 5.6 for 37°C for 1.5 h (Hayatsu and Miura, 1970). Mutation was induced in cytosine–guanine pairs at specific sites (Mukai et al., 1970). That sulfites destroy thiamine has been known since 1935 (Williams et al., 1935). Symptoms of thiamine deficiency accompanied by decreased urinary excretion of thiamine are observed (Til et al., 1972a,b). Growth inhibition in rats fed 0.1% or more of sulfite was due to thiamine destruction (Fitzhugh et al., 1946). Bhagat and Lockett (1964) observed that feeds containing 0.6% sodium metabisulfite produced two types of toxic effects in rats. Feeds stored for 7 weeks resulted in vitamin B_1 (thiamine) deficiency symptoms. However, feeds stored for 3–4 months produced diarrhea and growth retardation that were not reversed by thiamine administration. Sulfites elicit allergenic responses in certain individuals, especially steroid-dependent asthmatics (Stevenson and Simon, 1981). This led to the ban on the use of sulfites on raw fruits and vegetables to be consumed fresh.

REFERENCES

Aalto, T. R., Firman, M. C., Rigier, N. E. 1953. *p*-Hydroxybenzoic acid esters as preservatives. I. Uses, antibacterial and antifungal studies, properties and determination. *J. Am. Pharm. Assoc. (Sci. Ed.)* 42:449.

Abdul-Raouf, U. M., Beuchat, L. R., Ammar, M. S. 1993. Survival and growth of *Escherichia coli* O157:H7 in ground, roasted beef as affected by pH, acidulants, and temperature. *Appl. Environ. Microbiol.* 59:2364–2368.

Adler-Hradecky, C., Kelentey B. 1960. The toxicity and local analgesic effect of *p*-hydroxybenzoic acid ester. *Arch. Int. Pharmacodyn.* 128:135.

Ajjarapu, S., Shelef, L. S. 1999. Fate of pGFP-Bearing *Escherichia coli* O157:H7 in ground beef at 2 and 10°C and effects of lactate, diacetate and citrate. *Appl. Environ. Microbiol.* 65:5394–5397.

Al-Dagal, M. M., Bazaraa, W. A. 1999. Extension of shelf life of whole and peeled shrimp with organic acid salts and bifidobacteria. *J. Food Prot.* 62:51–56.

Alifax, R., Chevalier, R. 1962. Etude de la nisinase produtie par *Streptococcus thermophilus. J. Dairy Res.* 29:233.

Amadur, M. O. 1980. Air pollutants. In: *Toxicology: The Basic Sciences of Poisons*, Doull, J., Klaassen, C. D., Amadur, M. O. (Eds.). Macmillan, New York, p. 608.

Amérine, M. A., Joslyn, M. A. 1960. Commercial production of table wines. *Calif. Agric. Exp. Sta. Bull.* 639:143.

Anders, R. J., Cerveny J. G., Milkowski, A. L. 1989. Method for delaying *Clostridium botulinum* growth in fish and poultry. U.S. Patents 4,798,729, Jan. 17, and 4,888,191, Dec. 19.

Anderson, M. E., Huff, H. E., Naumann, H. D., Marshall, R. T. 1988. Counts of six types of bacteria on lamb carcasses dipped or sprayed with acetic acid at 25°C or 55°C and stored vacuum packaged at 0°C. *J. Food Prot.* 51:874.

Andersson, R. E., Bodin, H. G., Snygg, B. G. 1980. The effect of some food preservatives on growth, lipase production and lipase activity of *Pseudomonas fluorescens. Chem. Mikrobiol. Technol. Lebensm–Food Chem. Microbiol Tech.* (Nurnberg) 6(6); 161–164.

Anonymous. 1991. Delvocid®. Technical Bulletin. Gist-brocades Food Ingredients, Inc. King of Prussia, PA.

Anonymous. 1999. Sorbic acid and potassium sorbate for preserving freshness. Public. ZS-1D. Eastman Chemical Co., Kingsport, TN.

Aplin & Barrett, Ltd. no date Nisaplin® Technical Data. Aplin & Barrett Ltd., Trowbridge, UK.

Ayres, J. C., Denisen, E. L. 1958. Maintaining freshness of berries using selected packaging materials and antifungal agents. *Food Technol.* 12:562.

Ayres, J. C., Walker, H. W., Fanelli, M. J., King, A. W., Thomas, F. 1956. Use of antibiotics in prolonging storage life of dressed chicken. *Food Technol.* 10:563.

Azukas, J. J., Costilow, R. N., Sadoff, H. L. 1961. Inhibition of alcoholic fermentation by sorbic acid. *J. Bacteriol.* 81:189.

Baird-Parker, A. C. 1980. Organic acids. In: *Microbial Ecology of Foods*, Vol. I. *Factors Affecting Life and Death of Microorganisms.* International Commission on Microbiological Specifications for Foods. Academic Press New York, p. 126.

Bala, M. F. A., Marshall, D. L. 1998. Organic acid dipping of catfish fillets: effect on color, microbial load, and *Listeria monocytogenes. J. Food Prot.* 61:1470–1474.

Baldock, J. D., Frank, P. P., Graham, P. P., Ivey, F. J. 1979. Potassium sorbate as a fungistatic agent in country ham processing. *J. Food Prot.* 42:780.

Banks, J. G., Board, R. G. 1982. Sulfite-inhibition of enterobacteriaceae including *Salmonella* in British fresh sausage and in culture systems. *J. Food Prot.* 45:1292.

Bargiota, E., Rico-Munoz, E., Davidson, P. M. 1987. Lethal effect of methyl and propyl parabens as related to *Staphylococcus aureus* lipid composition. *Intl. J. Food Microbiol.* 4:257.

Bell, M. F., Marshall, R. T., Anderson, M. E. 1986. Microbiological and sensory tests of beef treated with acetic and formic acids. *J. Food Prot.* 49:207.

Benkerroum, N., Sandine, W. E. 1988. Inhibitory action of nisin against *Listeria monocytogenes. J. Dairy Sci.* 71:3237–3245.

Beuchat, L. 1980. Comparison of anti-vibrio activities of potassium sorbate, sodium benzoate and glycerol and sucrose esters of fatty acids. *Appl. Environ. Microbiol.* 39:1178.

Bhagat, B., Lockett, M. F. 1964. The effect of sulphite in solid diets on the growth of rats. *Food Cosmet. Toxicol.* 2:1.

Bills, S. Restaino, L., Lenovich, L. M. 1982. Growth response of an osmotolerant sorbate-resistant yeast, *Saccharomyces rouxii* at different sucrose and sorbate levels. *J. Food Prot.* 45:1120–1124.

Bizri, J. N., Wahem, I. A. 1994. Citric acid and antimicrobials affect microbiological stability and quality of tomato juice. *J. Food Sci.* 59:130–134.

Blocher, J. C., Busta, F. F. 1983. Influence of potassium sorbate and reduced pH on the growth of vegetative cells of four strains of type A and B *Clostridium botulinum. J. Food Sci.* 48:574.

Blocher, J. D., Busta, F. F., Sofos, J. N. 1982. Influence of potassium sorbate and pH on ten strains of type A and B *Clostridium botulinum. J. Food Sci.* 47:2028.

Blom, H., Nerbrink, E., Dainty, R., Hagtvedt, T., Borch, E., Nissen, H., Nesbakken, T. 1997. Addition of 2.5% lactate and 0.25% acetate controls growth of *Listeria monocytogenes* in vacuum-packed, sensory acceptable servelat sausage and cooked ham stored at 4°C. *Intl. J. Food Microbiol.* 38:71–76.

Bosund, L. 1962. The action of benzoic and salicyclic acids on the metabolism of microorganisms. *Adv. Food Res.* 11:331.

Brackett, R. E. 1987. Effect of various acids on growth and survival of *Yersinia enterocolitica. J. Food Prot.* 50:598.

Bradford, D. D., Huffman, D. L., Egbert, W. R., Jones, W. R. 1993. Low-fat pork sausage patty stability in refrigerated storage with potassium lactate. *J. Food Sci.* 58:488–491.

Branen, A. L. 1993. Introduction to use of antimicrobials. In: *Antimicrobials in Foods*, 2nd ed. Davidson, P. M., Branen, A. L. (Eds.). Marcel Dekker, New York, p. 1.

Branen, A. L., Keenan, T. W. 1970. Growth stimulation of *Lactobacillus casei* by sodium citrate. *J. Dairy Sci.* 53:593.

Brewer, S. M., McKeith, F., Martin, S. E., Dallmierm, A. W., Meyer, J. 1991. Sodium lactate effects on shelf-life, sensory and physical characteristics of fresh pork sausage. *J. Food Sci.* 56: 1176.

Brik, H. 1981. Natamycin. In: *Analytical Profiles of Drug Substances*, Flory, K. (Ed.). Academic Press, New York, p. 513.

Bruno, M. E. C., Kaiser, A., Montville, T. J. 1992. Depletion of proton motive force by nisin in *Listeria monocytogenes* cells. *Appl. Environ. Microbiol.* 58:2255–2259.

Bubnoff, M., von Schnell, D., Vogt-Moykeff, J. 1957. Pharmacology of benzoic acid, *p*-chlorobenzoic acid, *p*-hydroxybenzoic acid, and their esters. II. *Arzneimittel Forsch.* 7:340.

Buchanan, R. L., Solberg, M. 1972. Interaction of sodium nitrite, oxygen, and pH on growth of *Staphylococcus aureus. J. Food Sci.* 37:81.

Buchanan, R. L., Golden, M. H. 1994. Interaction of citric acid concentration and pH on the kinetics of *Listeria monocytogenes* inactivation. *J. Food Prot.* 57:567–570.

Budde, C. C. L. G. 1904. Ein neues verfahren zur sterilsierung der milch. *Tuberculosis* (Leipzig) 3:94.

Budu-Amoako, E., Ablett, R. F., Harris, J., Delves-Broughton, J. 1999. Combined effect of nisin and moderate heat on destruction of *Listeria monocytogenes* in cold-pack lobster meat. *J. Food Prot.* 62:46–50.

Bullerman, L. B. 1983. Effects of potassium sorbate on growth and aflatoxin production by *Aspergillus parasiticus* and *Aspergillus flavus. J. Food Prot.* 46:940.

Bullerman, L. B. 1984. Effects of potassium sorbate on growth and patulin production by *Penicillium patulum* and *Penicillium roqueforti. J. Food Prot.* 47:312.

Burden, E. H. W. J. 1961. The toxicology of nitrates and nitrites with particular reference to the potability of water supplies. *Analyst* 86:429.

Calderon, C., Collins-Thompson, D. L., Usborne, W. R. 1985. Shelf-life studies of vacuum-packaged bacon treated with nisin. *J. Food Prot.* 48:330–333.

Campbell, L. L., Sniff, E. E., O'Brien, R. T. 1959. Subtilin and nisin as additives that lower the heat-process requirements of canned foods. *Food Technol.* 13:462.

Carini, S., Lodi, R. 1982. Inhibition of germination of clostridial spores by lysozyme. *Industria-del-Latte* 18:35–48.

Carminati, D., Carini, S. 1989. Antimicrobial activity of lysozyme against *Listeria monocytogenes* in milk. *Microbiologie Aliment Nutr.* 7:49–56.

Carminati, D., Neviani, E., Mucchetti, G. 1985. Activity of lysozyme on vegetative cells of *Clostridium tyrobutyricum*. Latte 10:194–198.

Carneiro de Melo, A. M. S., Cassar, C. A., Miles, R. J. 1998. Trisodium phosphate increases sensitivity of gram-negative bacteria to lysozyme and nisin. *J. Food Prot.* 61:839–844.

Carpenter, C. E., Reddy, D. S. A., Cornforth, D. P. 1987. Inactivation of clostridial ferredoxin and pyruvate-ferredoxin oxidoreductase by sodium nitrite. *Appl. Environ. Microbiol.* 53:549.

Caserio, G., Ciampella, A., Gennari, M., Barluzzi, A. M. 1979. Research on the employ of nisin in cooked sausage. *Ind. Aliment.* (Italy) 18:1.

Castellani, A. G., Niven, C. F. 1955. Factors affecting the bacteriostatic action of sodium nitrite. *Appl. Microbiol.* 3:154.

Chen, N., Shelef, L. A. 1992. Relationship between water activity, salts of lactic acid, and growth of *Listeria monocytogenes* in a meat model system. *J. Food Prot.* 55:574–578.

Chichester, D. F., Tanner, F. W. 1981. Antimicrobial food additives, In: *Handbook of Food Additives*, 2nd ed. Furia, T. W. (Ed.). CRC Press, Cleveland, p. 115.

Chipley, J. R. 1993. Sodium benzoate and benzoic acid. In: *Antimicrobials in Foods*, 2nd ed. Davidson, P. M., Branen, A. L. (Eds.). Marcel Dekker, New York, p. 11.

Chipley, J. R., Uraih, N. 1980. Inhibition of *Aspergillus* growth and aflatoxin release by derivatives of benzoic acid. *Appl. Environ Microbiol.* 40:352.

Choi, M. H., Park, Y. H. 2000. Selective control of lactobacilli in kimchi with nisin. *Lett. Appl. Microbiol.* 30:173–177.

Chung, K., Dickson, J. S., Crouse, J. D. 1989. Effects of nisin on growth of bacteria attached to meat. *Appl. Environ. Microbiol.* 55:1329–1333.

Chung, K. C., Goepfert, J. M. 1970. Growth of *Salmonella* at low pH. *J. Food Sci.* 35:326–328.

Chung, Y. M., Lee, J. S. 1982. Potassium sorbate inhibition of microorganisms isolated from seafood. *J. Food Prot.* 45:1310.

Cook, F. K., Pierson, M. D. 1983. Inhibition of bacterial spores by antimicrobials. *Food Technol.* 37:115.

Costilow, R. N., Ferguson, W. E., Ray, S. 1955. Sorbic acid as a selective agent in cucumber fermentations. I. Effect of sorbic acid on microorganisms associated with cucumber fermentations. *Appl. Microbiol*, 3:341.

Cox, N. A., Mercuri, A. J., Juven, B. J., Thomson, J. E., Chew, V. 1974. Evaluation of succinic acid and heat to improve the microbial quality of poultry meat. *J. Food Sci.* 39:985.

Crandall, A. D., Montville, T. J. 1998. Nisin resistance in *Listeria monocytogenes* ATCC 700302 is a complex phenotype. *Appl. Environ. Microbiol.* 64:231–237.

Cruess, W. V., Irish, J. H. 1932. Further observations on the relation of pH value to toxicity of preservatives to microorganisms. *J. Bacteriol.* 23:163.

Curran, H. R., Knaysi, G. 1961. Survey of fourteen metabolic inhibitors for their effect on endospore germination in *Bacillus subtilis. J. Bacteriol.* 82:793.

Daamen, C. B. G., Van den Berg, G. 1985. Prevention of mould growth on cheese by means of natamycin. *Voedingsmiddelentechnologie* 18:26; *Dairy Sci. Abstr.* 1985. 49:3768.

Dahl, H. K., Nordal, J. 1972. Effect of benzoic acid and sorbic acid on the production and activities of some bacterial proteinases. *Acta Agric. Scand.* 22:29.

Dakin, H. D. 1990. The fate of sodium benzoate in the human organism. *J. Biol. Chem.* 7:103.

Davidson, P. M. 1993. Parabens and Phenolic compounds. In: *Antimicrobials in Foods*, 2nd ed. Davidson, P. M., Branen, A. L. (Eds.). Marcel Dekker, New York, p. 263.

Davies, E. A., Bevis, H. E., Potter, R., Harris, J., Williams, G. C., Delves-Brougton, J. 1998. Research note: the effect of pH on the stability of nisin solution during autoclaving. *Lett. Appl. Microbiol.* 27:186–187.

Deak, T., Novak, E. K. 1972. Assimilation of sorbic acid by yeasts. *Acta Aliment.* 1:87.

Deak, T., Tliske, M., Novak, E. K. 1970. Effects of sorbic acid on the growth of some species of yeast. *Acta Microbiol. Acad. Sci. Hung.* 17:237.

Dean, J. P., Zottola, E. A. 1996. Use of nisin in ice cream and effect on the survival of *Listeria monocytogenes. J. Food Prot.* 59:476–480.

Deane, D., Downs, P. A. 1951. Flexible wrappers for cheddar cheese. *J. Dairy Sci.* 23:509.

Debevere, J. M. 1989. The effect of sodium lactate on the shelflife of vacuum-packed coarse liver pate. *Fleischwirtschaft* 69:223–224.

Degnan, A. J., Kaspar, C. W. Otwell, W. S., Tamplin M. L., Luchansky, J. B. 1994. Evaluation of lactic acid bacterium fermentation products and food-grade chemicals to control *Listeria monocytogenes* in blue crab (*Callinectes sapidus*) meat. *Appl. Environ. Microbiol.* 60:3198–3203.

Delaquis, P. J., Sholberg, P. L., Stanich, K. 1999. Disinfection of mung bean seed with gaseous acetic acid. *J. Food Prot.* 953–957.

Delves-Broughton, J. 1990. Nisin and its uses as a food preservative. *Food Tech.* 44:100–112,117.

Delves-Broughton, J., Gasson, M. J. 1994. Nisin. In: *Natural Antimicrobial Systems and Food Preservation*, Dillon, V. M., Board, R. G. (Eds.). CAB International, Wallingford, UK, pp. 99–132.

Delves-Broughton, J., Williams, G. C., Wilkinson, S. 1992. The use of the bacteriocin, nisin, as a preservative in pasteurized liquid whole egg. *Lett. Appl. Microbiol.* 15:133–136.

Delves-Broughton, J., Blackburn, P., Evans, R. J., Hugenholtz, J. 1996. Applications of the bacteriocin, nisin. *Antonie van Leeuwenhoek* 69:193–202.

Demaree, G. E., Sjogren, D. W., McCashland, B. W., Cosgrove, R. P. 1955. Preliminary studies on the effect of feeding sorbic acid upon the growth, reproduction, and cellular metabolism of albino rats. *J. Am. Pharm. Assoc.* (*Sci. Ed.*) 44:619.

Denny, C. B., Sharpe, L. E., Bohrer, C. W. 1961. Effects of tylosin and nisin on canned food spoilage bacteria. *Appl. Microbiol.* 9:108.

Doell, W. 1962. The antimicrobial action of potassium sorbate. *Arch. Lebensmittelhyg.* 13:4.

Doores, S. 1993. Organic acids. In: *Antimicrobials in Foods*, 2nd ed. Davidson, P. M., Branen, A. L. (Eds.). Marcel Dekker, New York, p. 95.

Duhaiman, A. S. 1988. Purification of camel milk lysozyme and its lytic effect on *Escherichia coli* and *Micrococcus lysodeikticus*. *Comp. Biochem. Physiol.* 91B:793.

Dymicky, M., Huhtanen, C. N. 1979. Inhibition of *Clostridium botulinum* by *p*-hydroxybenzoic acid n-alkyl esters. *Antimicrob. Agents Chemother.* 15:798.

Dymicky, M., Bencivengo, M., Buchanan, R. L., Smith, J. L. 1987. Inhibition of *Clostridium botulinum* 62A by fumarates and maleates and relationship of activity to some physiochemical constants. *Appl. Environ. Microbiol.* 53:110.

Ebel, J. P. Dirheimer, G., Stahl, A., Muller-Fetter, S. 1965. Biochemistry of inorganic polyphosphates. IV. Antibacterial effects of condensed phosphates. *Colloq. Intl. Centre Natl. Rech. Sci.* (Paris), pp. 289–299.

Eklund, T. 1980. Inhibition of growth and uptake processes in bacteria by some chemical food preservatives. *J. Appl. Bacteriol.* 48:423.

Eklund, T. 1983. The antimicrobial effect of dissociated and undissociated sorbic acid at different pH levels. *J. Appl. Bacteriol.* 54:383.

Eklund, T. 1985. The effect of sorbic acid and esters of *p*-hydroxybenzoic acid on the protonmotive force in *Escherichia coli* membrane vesicles. *J. Gen. Microbiol.* 131:73.

Eklund, T. 1989. Organic acids and esters. In: *Mechanisms of Action of Food Preservation Procedures*, Gould, G. W. (Ed.). Elsevier, London, pp. 161–200.

El-Gazzar, F. E., Marth, E. H. 1987. Sodium benzoate in the control of growth and aflatoxin production by *Aspergillus parasiticus*. *J. Food Prot.* 50:305.

El-Shenawy, M. A., Marth, E. H. 1989. Behavior of *Listeria monocytogenes* in the presence of sodium propionate. *Intl. J. Food Microbiol.* 8:85.

Ellinger, R. H. 1981. Phosphates in food processing. In: *Handbook of Food Additives*, 2nd ed., Furia, T. E. (Ed.). CRC Press, Cleveland.

Emard, L. O., Vaughn, R. H. 1952. Selectivity of sorbic acid media for the catalase negative lactic acid bacteria and clostridia. *J. Bacteriol.* 63:487.

Entani, E., Asai, M., Tsujihata, S. Tsukamoto, Y., Ohta, M. 1998. Antibacterial action of vinegar against food-borne pathogenic bacteria including *Escherichia coli* O157:H7. *J. Food Prot.* 61: 953–959.

Epstein, S. 1968. Paraben sensitivity: subtle trouble. *Ann. Allergy* 26:185.

Evelyn, J. 1664. Pomona, or an appendix concerning fruit trees. In: *Relation to Cider and Several Ways of Ordering It*, Supplement; *Aphorisms Concerning Cider*. Beale, P. J. M., Allestry, J. A. (Eds.). London.

Fabian, F. W., Wadsworth, C. K. 1939. Experimental work on lactic acid in preserving pickles and pickle products. I. Rate of penetration of acetic and lactic acid in pickles. *Food Res.* 4:499.

FAO/WHO. 1962. Evaluation of the toxicity of a number of antimicrobials and antioxidants. sodium benzoate. *Tech. Rep. Ser.* No. 228. FAO/WHO, Geneva.

FAO/WHO. 1967. Specifications for the identity and purity of food additives and their toxicological evaluation: some emulsifiers and stabilizers and certain other substances. *Tech. Rep. Ser.* No. 373. FAO/WHO, Geneva.

FDA. 1980. Re-evaluation of the pathology findings of studies on nitrite and cancer: histological lesions in Sprague-Dawley rats. Final report submitted by Universities Associated for Research and Education in Pathology, Inc. U.S. Food and Drug Administration, Dept. of Health and Human Services, Washington, D.C.

Fisher, T. L., Golden, D. A. 1998. Survival of *Escherichia coli* O157:H7 in apple cider as affected by dimethyl dicarbonate, sodium bisulfite, and sodium benzoate. *J. Food Sci.* 63:904–906.
Fitzhugh, O. G., Knudson, L. F., Nelson, A. A. 1946. The chronic toxicity of sulfites. *J. Pharmocol. Exp. Ther.* 86:37.
Fletcher, G. C., Murrell, W. G., Statham, J. A., Stewart, B. J., Bremner, H. A. 1988. Packaging of scallops with sorbate: an assessment of the hazards from *Clostridium botulinum. J. Food Sci.* 43:349.
Flores, L. M., Palomar, L. S., Roh, P. A., Bullerman, L. B. 1988. Effect of potassium sorbate and other treatments on the microbial content and keeping quality of a restaurant-type Mexican hot sauce. *J. Food Prot.* 51:4.
Fornachon, J. C. M. 1963. Inhibition of certain lactic acid bacteria by free and bound sulfur dioxide. *J. Sci. Food Agric.* 14:857.
Francois, J., Van Schaftingen, E., Hers, H. G. 1986. Effect of benzoate on the metabolism of fructose 2,6-biphosphate in yeast. *Eur. J. Biochem.* 154:141–145.
Freese, E. 1978. Mechanism of growth inhibition by lipophilic acids. In: *The Pharmacological Effect of Lipids*, Kabara, J. J. (Ed.). American Oil Chemists Society, Champaign, IL, p. 123.
Freese, E., Sheu, C. W., Galliers, E. 1973. Function of lipophilic acids as antimicrobial food additives. *Nature* 241–321.
Gangolli, S. D., Grasso, P., Goldberg, L., Hooson, J. 1972. Protein binding by food colourings in relation to the production of subcutaneous sarcoma. *Food Cosmet. Toxicol.* 10:449.
Gao, F. H., Abee, T. L., Koonings, W. N. 1991. Mechanism of action of the peptide antibiotic nisin in liposome and cytochrome C oxidase containing proteoliposomes. *Appl. Environ. Microbiol.* 57:2164–2170.
Garcera, M. J. G., Elferink, M. G. L., Driessen, A. J. M., Konings, W. N. 1993. In vitro pore-forming activity of the lantibiotic nisin: role of protonmotive force and lipid composition. *Eur. J. Biochem.* 212:417–422.
Gelda, C. S., Mathur, A. D., Stersky, A. K. 1974. The retarding effect of the antifungal agent pimaricin on the growth of *Aspergillus parasiticus. IV, Intl. Congr. Food Sci. Technol.* 9a:261.
Genigeorgis, C., Riemann, H. 1979. Food processing and hygiene. In: *Food-Borne Infections and Intoxications*, 2nd ed., Riemann, H., Bryan, F. L. (Eds.). Academic Press, New York, p. 613.
Gibson, A. M., Roberts, T. A. 1986a. The effect of pH, water activity, sodium nitrite and storage temperature on the growth of enteropathogenic *Escherichia coli* and salmonellae in laboratory medium. *Intl. J. Food Microbiol.* 3:183.
Gibson, A. M., Roberts, T. A. 1986b. The effect of pH, sodium chloride, sodium nitrite and storage temperature on the growth of *Clostridium perfringens* and faecal streptococci in laboratory medium. *Intl. J. Food Microbiol.* 3:195.
Giese, J. 1992. Experimental process reduces *Salmonella* on poultry. *Food Technol.* 46(4):112.
Glabe, E. F., Maryanski, J. K. 1981. Sodium diacetate: an effective mold inhibitor. *Cereal Foods World* 26:285.
Gooding, C. M. 1945. Process of inhibiting growth of molds. U.S. Patent 2,379,294, June 26.
Goto, S. 1980. Changes in the wild yeast flora of sulfited grape musts. *J. Inst. Enol. Vitic.* 15:29.
Gould, G. W. 1964. Effect of food preservatives on the growth of bacteria from spores. In: *Microbial Inhibitors in Food*, Molin, N. (Ed.). Almqvist and Miksell, Stockholm, p. 17.
Gould, G. W., N. J. Russell. 1991. Sulphite, p. 72–88 In: *Food Preservatives*, Russell, N. J., Gould, G. W. (Eds.). Blackie and Son Ltd., Glasgow.
Gourama, H., Bullerman, L. B. 1988. Effects of potassium sorbate and natamycin on growth and penicillic acid production by *Aspergillus ochraceus. J. Food Prot.* 51:139.
Grever, A. B. G. 1974. Minimum nitrite concentrations for inhibition of clostridia in cooked meat products. In: *Proceedings of the International Symposium on Nitrite in Meat Products, 1973*, Krol, B., Tinbergen, B. J. (Eds.). Wageningen Center for Agricultural Publishing and Documentation, Wageningen, The Netherlands, pp 103–109.

Griffith, W. H. 1929. Benzoylated amino acids in the animal organism. IV. A method for the investigation of the origin of glycine. *J. Biol. Chem.* 82:415.

Gross, E., Morell, J. L. 1967. The presence of dehydration in the antibiotic nisin and its relationship to activity. *J. Am. Chem. Soc.* 89:2791.

Gross, E., Morell, J. L. 1971. The structure of nisin. *J. Am. Chem. Soc.* 93:4634–4635.

Hailer, E. 1911. Experiments on the properties of free sulfurous acid of sulfites and a few complex compounds of sulfurous acid in killing germs and retarding theri development. *Arb. Kais. Gesundh.* 36:297; *Chem. Abstr.* 5:1805 (1911).

Hamden, I. Y., Deane, D. D., Kunsman, J. E. 1971. Effect of potassium sorbate on yogurt cultures. *J. Milk Food Technol.* 34:307.

Hamilton-Miller, J. M. T. 1973. Chemistry and biology of the polyene macrolide antibiotics. *Bact. Rev.* 37:166.

Hamilton-Miller, J. M. T. 1974. Fungal sterols and the mode of action of the polyene antibiotics. *Adv. Appl. Microbiol.* 17:109.

Hammond, S. M., Carr, J. C. 1976. The antimicrobial activity of SO_2—with particular reference to fermented and non-fermented fruit juices. *Soc. Appl. Bacteriol. Symp. Ser.* No. 5:89.

Hansen, J. N. 1994. Nisin as a model food preservative. *CRC Crit. Rev. Food Sci. Nutr.* 34:69–93.

Hansen, J. D., Appleman, M. D. 1955. The effect of sorbic, propionic, and caprioic acid on the growth of certain Clostridia. *Food Res.* 20:92.

Hara, S., Yakazu, K., Nakakawaji, K., Takenchi, T., Kobyashi, T., Sata, M., Imai, Z., Shibuya, T. 1962. An investigation of toxicity of nisin. *Tokyo Med. Univ. J.* 20:175.

Harada, K., Hizuchin, R., Utsumi, I. 1968. Studies on sorbic acid. IV. Inhibition of the respiration in yeasts. *Agric. Biol. Chem.* 32:940.

Harris, L. J., Fleming, H. P., Klaenhammer, T. R. 1991. Sensitivity and resistance of *Listeria monocytogenes* ATCC 19115, Scott A, and UAL500 to nisin. *J. Food Prot.* 54:836–840.

Hartwell, J. L., 1951. *Survey of Compounds Which Have Been Tested for Carcinogenic Activity*, 2nd ed., Public Health Service, National Institutes of Health, Bethesda, MD.

Hawley, H. B. 1957. Nisin in food technology. *Food Manuf.* 32:370.

Hayatsu, H., Miura, A. 1970. The mutagenic action of sodium bisulfite. *Biochem. Biophys. Res. Commun.* 39:156.

Hedberg, M., Miller, J. K. 1969. Effectiveness of acetic acid, betadine, amphyll, polymyxin B, colistin and gentamicin against *Pseudomonas aeruginosa. Appl. Microbiol.* 18:854.

Heinemann, B., Voris, L., Stumbo, C. R. 1965. Use of nisin in processing food products. *Food Technol.* 19:160–164.

Heseltine, W. W. 1952. Sodium propionate and its derivatives as bacteriostatics and fungistatics. *J. Pharm. Pharmacol.* 4:577.

Hirsch, A., Grinsted, E., Chapman, H. R., Mattick, A. T. R. 1951. A note on the inhibition of an anaerobic sporeformer in swiss-type cheese by a nisin producing *Streptococcus. J. Dairy Res.* 18:205.

Hitchens, A. D., Gould, G. W., Hurst, A. 1963. The swelling of bacterial spores during germination and outgrowth. *J. Gen. Microbiol.* 30:445–453.

Holley, R. A. 1981. Review of the potential hazard from botulism in cured meats. *Can. Inst. Food Sci. Technol. J.* 14:183.

Houtsma, P. C., Wit, J. C., Rombouts, F. M. 1996. Minimum inhibitory concentration (MIC) of sodium lactate and sodium chloride for spoilage organisms and pathogens at different pH values and temperatures. *J. Food Prot.* 59:1300–1304.

Hu, A. C., Shelef, L. A. 1996. Influence of fat content and preservatives on the behavior of *Listeria monocytogenes* in beaker sausage. *J. Food Safety* 16:175–181.

Hughey, V. L., Johnson, E. A. 1987. Antimicrobial activity of lysozyme against bacteria involved in food spoilage and food-borne disease. *Appl. Environ. Microbiol.* 53:2165–2170.

Hughey, V. L., Wilger, R. A., Johnson, E. A. 1989. Antibacterial activity of hen egg white lysozyme against *Listeria monocytogenes* Scott A in foods. *Appl. Environ. Microbiol.* 55:631–638.

Huhtanen, C. N. 1983. Antibotulinal activity of methyl and ethyl fumarates in comminuted nitrite-free bacon. *J. Food Sci.* 48:1574.

Huhtanen, C. N., Feinberg, J. 1980. Sorbic acid inhibition of *Clostridium botulinum* in nitrite-free poultry frankfurters. *J. Food Sci.* 45:453.

Huhtanen, C. N., Guy, E. J. 1984. Antifungal properties of esters of alkenoic and alkynoic acids. *J. Food Sci.* 49:281.

Huhtanen, C. N., Guy, E. J., Milnes-McCaffrey, L. 1981. Antifungal properties of fumaric acid esters. 41st Annual Meeting, Institute of Food Technologists, Atlanta, GA.

Hunter, D., Segel, I. H. 1973. Effect of weak acids on amino acid transport by *Penicillium chrysogenum*. Evidence for a proton or charge gradient as the driving force. *J. Bacteriol.* 113:1184.

Hurst, A. 1981. Nisin. *Adv. Appl. Microbiol.* 27:85.

Hurst, A., Hoover, D. G. 1993. Nisin. In: *Antimicrobials in Foods*, 2nd ed. Davidson, P. M., Branen, A. L. (Eds.). Marcel Dekker, New York, p. 369.

Ibrahim, H. R. Higashiguchi, S., Koketsu, M., Juneja, L. R., Kim, M., Yamamoto, T., Sugimoto, Y., Aoki, T. 1996. Partially unfolded lysozyme at neutral pH agglutinates and kills gram-negative and gram-positive bacteria through membrane damage mechanism. *J. Agric. Food Chem.* 44:3799–3806.

IFT, Expert Panel on Food Safety and Nutrition and Committee on Public Information. 1975. Sulfites as food additives: a scientific status summary. *Food Technol.* 29:117.

IFT, Expert Panel on Food Safety and Nutrition and Committee on Public Information. 1987. Nitrate, nitrite and nitroso compounds in foods. *Food Technol.* 41:127.

Imai, K., Banno, I., Ujima, T. 1970. Inhibition of bacterial growth by citrate. *J. Gen. Appl. Microbiol.* 16:479.

Ingle, J. D. 1940. Some preliminary observations on the effectiveness of propionates as mold inhibitors on diary products. *J. Dairy Sci.* 23:509.

Ingram, M. 1939. The endogenous respiration of *Bacillus cereus*. II. The effect of salts on the rate of absorption of oxygen. *J. Bacteriol.* 24:489.

Ingram, M., Ottoway, F. J. H., Coppock, J. B. M. 1956. The preservative action of acid substances in food. *Chem. Ind.* (London) 42:1154.

Inovatech. 2000. Inovapure product description. Canadian Inovatech Inc., Abbotsford, BC, Canada.

Ivey, F. J., Robach, M. C. 1978. Effect of sorbic acid and sodium nitrite on *Clostridium botulinum* outgrowth and toxin production in canned comminuted pork. *J. Food Sci.* 43:1782.

Ivey, F. J., Shaver, K. J., Christiansen, L. N., Tompkin, R. B. 1978. Effect of potassium sorbate on toxigenesis by *Clostridium botulinum* in bacon. *J. Food Prot.* 41:621.

Jacobs, M. B. 1959. Acids and acidulation. In: *Manufacture and Analysis of Carbonated Beverages*. Chemical Publishing Co., New York, p. 68.

Jaquette, C. B., Beuchat, L. R. 1998. Combined effects of pH, nisin, and temperature on growth and survival of psychrotrophic *Bacillus cereus*. *J. Food Prot.* 61:563–570.

Jarvis, B. 1967. Resistance to nisin and production of nisin-inactivating enzymes by several *Bacillus* species. *J. Gen. Microbiol.* 47:33.

Jarvis, B., Burke, C. S. 1976. Practical and legislative aspects of the chemical preservation of food. In: *Inhibition and Inactivation of Vegetative Microbes*, Skinner, F. A., Hugo, W. B. (Eds.). Academic, New York, p. 345.

Jarvis, B., Jeffcoat, J., Cheeseman, G. C. 1968. Molecular weight distribution of nisin. *Biochem. Biophys. Acta* 168:153.

Jay, J. M. 1992. *Modern Food Microbiology*, 5th ed., Chapman and Hall, New York.

Jen, C. M. C., Shelef, L. A. 1986. Factors affecting sensitivity of *Staphylococcus aureus* 196E to polyphosphate. *Appl. Environ. Microbiol.* 52:842.

Jermini, M. F. G., Schmidt-Lorenz, W. 1987. Activity of Na-benzoate and ethyl paraben against osmotolerant yeasts at different water activity values. *J. Food Prot.* 50:920.

Johansen, C., Gram, L., Meyer, A. S. 1994. The combined inhibitory effect of lysozyme and low pH on growth of *Listeria monocytogenes*. *J. Food Prot.* 57:561–566.

Jones, P. S., Thigpen, D., Morrison, J. L., Richardson, A. P. 1956. *p*-Hydroxybenzoic acid esters as preservatives. III. The physiological disposition of *p*-hydroxybenzoic acid and its esters. *J. Am. Pharm. Assoc.* (Sci. Ed.) 45:268.

Juneja, V. K., Davidson, P. M. 1992. Influence of altered fatty acid composition on resistance of *Listeria monocytogenes* to antimicrobials. *J. Food Prot.* 56:302–305.

Jung, D., Bodyfelt, F. W., Daeschel, M. A. 1992. Influence of fat and emulsifiers on the efficacy of nisin in inhibiting *Listeria monocytogenes* in fluid milk. *J. Dairy Sci.* 75:387–393.

Jurd, L., King, A. D., Mihara, K., Stanely, W. L. 1971. Antimicrobial properties of natural phenols and related compounds. 2. Obtusastyrene. *Appl. Microbiol.* 21:507.

Kabara, J. J., Swieczkowski, D. M., Conley, A. J., Truant, J. P. 1972. Fatty acids and derivatives as antimicrobial agents. *Antimicrob. Agents Chemother.* 2:23.

Karapinar, M., Gonul, S. A. 1992a. Effects of sodium bicarbonate, vinegar, acetic and citric acids on growth and survival of *Yersinia enterocolitica*. *Intl. J. Food Microbiol.* 16:343–347.

Karapinar, M., Gonul, S. A. 1992b. Removal of *Yersinia enterocolitica* from fresh parsley by washing with acetic acid or vinegar. *Intl. J. Food Microbiol.* 16:261–264.

Kasrazadeh, M., Genigeorgis, C. 1994. Potential growth and control of *Salmonella* in Hispanic type soft cheese. *Intl. J. Food Microbiol.* 22:127–140.

Kasrazadeh, M., Genigeorgis, C. 1995. Potential growth and control of *Escherichia coli* O157:H7 in soft hispanic type cheese. *Intl. J. Food Microbiol.* 25:289–300.

Kelch, F. 1958. Effect of commercial phosphates on the growth of microorganisms. *Fleischwirtschaft* 10:325.

Kidney, A. J. 1974. The use of sulfite in meat processing. *Chem. Ind.* (London) 1974:717.

Kim, J. W., Slavik, M. F. 1994. Trisodium phosphate (TSP) treatment of beef surfaces to reduce *Escherichia coli* O157:H7 and *Salmonella typhimurium*. *J. Food Safety* 59:20–22.

Kim, J. W., Slavik, M. F., Pharr, M. D., Raben, D. P., Lobsinger, C. M., Tsai, S. 1994. Reduction of *Salmonella* on post-chill chicken carcasses by trisodium phosphate (Na_3PO_4) treatment. *J. Food Safety* 14:9–17.

Kimble, C. E. 1977. Chemical food preservatives. In: *Disinfection, Sterilization, and Preservation*, 2nd ed. Block, S. S. (Ed.). Lea and Febiger, Philadelphia, p. 834.

King, A. D., Ponting, J. D., Sanshuck, D. W., Jackson, R., Mihara, K. 1981. Factors affecting death of yeast by sulfur dioxide *J. Food Prot.* 44:92.

Kirby, G. W., Atkin, L., Frey, C. N. 1937. Further studies on the growth of bread molds as influenced by acidity. Cereal Chem. 14:865.

Klindworth, K. J., Davidson, P. M., Brekke, C. J., Branen, A. L. 1979. Inhibition of *Clostridium perfringens* by butylated hydroxyanisole. *J. Food Sci.* 44:564.

Knabel, S. J., H. W. Walker, P. A. Hartman. 1991. Inhibition of *Aspergillus flavus* and selected gram-positive bacteria by chelation of essential metal cations by polyphosphates. *J. Food Prot.* 54:360–365.

Knox, T. L., Davidson, P. M., Mount, J. R. 1984. Evaluation of selected antimicrobials in fruit jusices as sodium metabisulfite replacements or adjuncts. 44th Annual Meeting, Institute of Food Technologist, Anaheim, CA.

Koop, J. S. 1952. Strains of *Lactobacillus plantarum* which destroy the antibiotic made by *Streptococcus lactis*. *J. Neth. Milk Dairy* 6:223.

Kordel, M., Schuller, F., Sahl, H. 1989. Interaction of the pore-forming, peptide antibiotics Pep5, nisin and subtilin with non-energized liposomes. *FEBS Lett.* 244:99–102.

Korenekova, B., Kottferova, J., Korenek, M. 1997. Observation of the effects of nitrites and nitrates on yogurt culture. *Food Res. Intl.* 30:55–58.

Kouassi, Y., Shelef, L. A. 1995. Listeriolysin O secretion by *Listeria monocytogenes* in the presence of cysteine and sorbate. *Lett. Appl. Microbiol.* 20:295–299.
Kruk, M., Lee, J. S. 1982. Inhibition of *Escherichia coli* trimethylamine-*N*-oxide reductase by food preservatives. *J. Food Prot.* 45:241.
Lamkey, J. W., Leak, F. W., Tule, W. B., Johnson, D. D., West, R. L. 1991. Assessment of sodium lactate addition to fresh pork sausage. *J. Food sci.* 56:220.
Lang, K. 1959. Summary report of the survey on the demand for phosphate and the harm caused by a large phosphate supply. *Z. Lebensmitt.-Untersuch.* 100:450.
Lang, K. 1960a. Die physiologischen Wirkungen von schwefliger Saure. Heft 31. Der Schriftenreihe des Bundes fur Lebensmittelrecht und Lebesmittelkunde. B. Behr's Verlag, Hamburg.
Lang, K. 1960b. Tolerance to sorbic acid. *Arzneimittelforsch.* 10:997.
Larocco, K. A. and Martin, S. E. 1987. Effects of potassium sorbate alone and in combination with sodium chloride on growth of *Staphylococcus aureus* MF-31. *J. Food Prot.* 50:750.
Lauersen, F. 1953. Summary of a survey report on the health considerations of the use of phosphoric acid and primary phosphates in soft drinks. *Z. Lebensmitt.-Untersuch.* 96:418.
Laukova, A. 1995. Inhibition of ruminal staphylococci and enterococci by nisin *in vitro. Lett. Appl. Microbiol.* 20:34–36.
Lechowich, R. V., Evans, J. B., Niven, C. F. 1956. Effect of curing ingredients and procedures on the survival and growth of staphylococci in and on cured meats. *Appl. Microbiol.* 4:360.
Lee, J. S. 1973. What seafood processors should know about *Vibrio parahaemolyticus. J. Milk Food Technol.* 36:405.
Lee, R. M., Hartman, P. A., Olson, D. G., Williams, F. D. 1994a. Bacteriocidal and bacteriolytic effects of selected food-grade phosphates, using *Staphylococcus aureus* as a model system. *J. Food Prot.* 57:276–283.
Lee, R. M., Hartman, P. A., Olson, D. G., Williams, F. D. 1994b. Metal ions reverse the inhibitory effects of selected food-grade phosphates in *Staphylococcus aureus. J. Food Prot.* 57:284–288.
Lennox, J. E., McElroy, L. J. 1984. Inhibition of growth and patulin synthesis in *Penicillium expansum* by potassium sorbate and sodium propionate in culture. *Appl. Environ. Microbiol.* 48:1031.
Levine, A. S., Fellers, C. R. 1939. The inhibiting effect of acetic acid with sodium chloride and sucrose on microorganisms. *J. Bacteriol.* 39:17.
Levine, A. S., Fellers, C. R. 1940. Action of acetic acid on food spoilage microorganisms. *J. Bacteriol.* 39:499.
Levinskas, G. J., Ribelin, W. E., Shaffer, C. B. 1966. Acute and chronic toxicity of pimaricin. *Toxicol. Appl. Pharmacol.* 8:97.
Liewen, M. B., Marth, E. H. 1985. Growth of sorbate-resistant and -sensitive strains of *Penicillium roquefortii* in the presence of sorbate. *J. Food Prot.* 48:525.
Lillard, H. S. 1994. Effect of trisodium phosphate on salmonellae attached to chicken skin. *J. Food Prot.* 57:465–469.
Lillard, H. S., Blankenship, L. C., Dickens, J. A., Craven, S. E., Shackelford, A. D. 1987. Effect of acetic acid on the microbiological quality of scalded picked and unpicked broiler carcasses. *J. Food Prot.* 50:112.
Lipinska, E. 1977. Nisin and its applications In: *Antibiotics and Antibiosis in Agriculture*, Woodbine, M. (Ed.). Butterworths, London, p. 103.
Liu, W., Hansen, J. N. 1990. Some chemical and physical properties of nisin, a small-protein antibiotic produced by *Lactococcus lactis. Appl. Environ. Microbiol.* 56:2551–2558.
Lloyd, A. G., Drake, J. J. P. 1975. Problems posed by essential food preservatives. *Br. Med. Bull.* 31:214.
Lodi, R. Todesco, R., Bozzetti, V. 1989. New applications of natamycin with different types of Italian cheese. *Microbiol. Aliment. Nutr.* 7:81.
Loncin, M. 1975. Basic principles of moisture equilibria. In: *Freeze Drying and Advanced Food*

Technology, Goldblith, S. A. Rey, L., Rothmayr, W. W. (Eds.). Academic Press, New York, p. 599.

Lück, H. Cheeseman, C. E. 1978. Mould growth on cheese as influenced by pimaricin or sorbate treatments. *S. Afr. J. Dairy Technol.* 10:143.

Lueck, E. 1980. *Antimicrobial Food Additives.* Springer-Verlag, Berlin.

Lund, B. M., George, S. M., Franklin, J. G. 1987. Inhibition of type A and type B (proteolytic) *Clostridium botulinum* by sorbic acid. *Appl. Environ. Microbiol.* 53:935.

Lynch, D. J., Potter, N. N. 1988. Effects of organic acids on thermal inactivation of *Bacillus stearothermophilus* and *Bacillus coagulans* spores in frankfurter emulsion slurry. *J. Food Prot.* 51: 475.

Mass, M. R., Glass, K. A., Doyle, M. P. 1989. Sodium lactate delays toxin production by *Clostridium botulinum* in cook-in-bag turkey products. *Appl. Environ. Microbiol.* 55:2226–2229.

Macris, B. J. 1975. Mechanism of benzoic acid uptake by *Saccharomyces cerevisiae. Appl. Microbiol.* 30:503.

Marth, E. H. 1966. Antibiotics in foods—naturally occurring, developed and added. *Residue Rev.* 12:65.

Marth, E. H., Capp, C. M., Hasenzah., L., Jackson, H. W., Hussong, R. V. 1966. Degradation of potassium sorbate by pencillium species. *J. Dairy Sci.* 49:1197.

Martoadiprawito, W., Whitaker, J. R. 1963. Potassium sorbate inhibition of yeast alcohol dehydrogenase. *Biochem. Biophys. Acta* 77:536.

Marwan, A. G., Nagel, C. W. 1986. Quantitation determination of infinite inhibition concentrations of antimicrobial agents. *Appl. Environ. Microbiol.* 51:559.

Matamoros-León, B., Argaiz, A., López-Malo, A. 1999. Individual and combined effects of vanillin and potassium sorbate on *Penicillium digitatum, Penicillium glabrum*, and *Penicillium italicum* growth. *J. Food Prot.* 62:540–542.

Matthews, C., Davidson, J., Bauer, E., Morrison, J. G., Richardson, A. P. 1956. *p*-Hydroxybenzoic acid esters as preservatives. II. Acute and chronic toxicity in dogs, rats, and mice. *J. Am. Pharm. Assoc.* (Sci. Ed.) 45:260.

Mattick, A. T. R., Hirsch, A. 1947. Further observation on an inhibitor (nisin) from lactic streptococci. *Lancet* ii:5.

McClintock, M., Serres, L., Marzolf, J. J., Hirsch, A., Mocquot, G. 1952. Action inhibtrice des streptocoques producteurs de nisine sur le developpement des sponiles anaerobics dans le fromage de Gruyere fondu. *J. Dairy Res.* 19:187.

McMindes, M. K., Siedler, A. J. 1988. Nitrite mode of action: inhibition of yeast pyruvate decarboxylase (E.C. 4.1.1.1) and clostridial pyruvate:oxidoreductase (E.C. 1.2.7.1) by nitric oxide. *J. Food Sci.* 53:917.

Means, G. E., Feeney, R. E. 1971. *Chemical Modification of Proteins.* Holden-Day, San Francisco.

Melnick, D., Luckmann, F. H., Gooding, C. M. 1954. Sorbic acid as a fungistatic agent for foods. VI. Metabolic degradation of sorbic acid in cheese by molds and the mechanism of mold inhibition. *Food Res.* 19:44.

Michaelsson, G., Juhlin, L. 1973. Urticaria induced by preservatives and dye additives in food and drugs. *Br. J. Dermatol.* 88:525.

Ming, X., Weber, G. H., Ayres, J. W., Sandine, W. E. 1997. Bacteriocins applied to food packaging materials to inhibit *Listeria monocytogenes* on meats. *J. Food Sci.* 62:413–415.

Minor, J. L., Becker, B. A. 1971. A comparison of the teratogenic properties of sodium salicylate, sodium benzoate, and phenol. *Toxicol. Appl. Pharmacol.* 19:373.

Minor, T. E., Marth, E. H. 1970. Growth of *Staphylococcus aureus* in acidified pasteurized milk. *J. Milk Food Technol.* 33:516.

Moir, C. J., Eyles, M. J. 1992. Inhibition, injury, and inactivation of four psychrotrophic foodborne bacteria by the preservatives methyl *p*-hydroxybenzoate and potassium sorbate. *J. Food Prot.* 55:360–366.

Mountney, G. J., O'Malley, J. 1985. Acids as poultry meat preservatives. *Poultry Sci.* 44:582.

Morris, S. L., Walsh, R. C., Hansen, J. N. 1984. Identification and characterization of some bacterial membrane sulfhydryl groups which are targets of bacteriostatic and antibiotic action. *J. Biolog. Chem.* 259:13590–13594.

Moustafa, H. H., Collins, E. B. 1969. Effects of selected food additives on growth of *Pseudomonas fragi*. *J. Dairy Sci.* 52:335.

Muhoberac, B. B., Wharton, D. C. 1980. EPR study of heme–NO complexes of ascorbic acid-reduced *Pseudomonas* cytochrome oxidase and corresponding model complexes. *J. Biol. Chem.* 255:8437–8442.

Mukai, F., Hawryluk, I., Shapiro, R. 1970. The mutagenic specificity of sodium bisulfite. *Biochem. Biophys. Res. Commun.* 39:983.

Murdock, D. I. 1950. Inhibitory action of citric acid on tomato juice flat sour organisms. *Food Res.* 15:107.

Myers, B. R., Edmondson, J. E., Anderson, M. E., Marshall, R. T. 1983. Potassium sorbate and recovery of pectinolytic psychrotrophs from vacuum-packaged pork. *J. Food Prot.* 46:499.

Nakamura, M., Zangar, M. J. 1968. Effect of fatty acids on *Shigella*. *Proc. Mont. Acad. Sci.* 28: 51.

NAS/NRC. 1981. *The Health Effects of Nitrate, Nitrite and N-Nitroso Compounds*. Committee on Nitrite and Alternative Curing Agents, National Research Council, National Academy Press, Washington, D. C.

Nelson, K. A., Busta, F. F., Sofos, J. N., Allen, C. E. 1980. Effect of product pH and ingredient forms in chicken franfurter emulsions on *Clostridium botulinum* growth and toxin production. 40th Annual Meeting, Institute of Food Technologists, New Orleans, LA.

Newberne, P. M. 1979. Nitrite promotes lymphoma incidence in rats. *Science* 204:1079.

Nilson, K. M., Shahani, K. M., Vakil, J. R., Kilara, A. 1975. Pimaricin and mycostatin for retarding cottage cheese spoilage. *J. Dairy Sci.* 58:668.

Njagi, G. D. E., Gopalan, H. N. B. 1980. Mutagenicity testing of some selected food preservatives, herbicides, and insecticides. 2. Ames test. *Bangladesh J. Bot.* 9:141.

Nury, F. S., Miller, M. W., Brekke, J. E. 1960. Preservative effect of some antimicrobial agents on high moisture dried fruits. *Food Technol.* 14:113.

Ogden, K. M., Weites, J., Hammond, J. R. M. 1988. Nisin and brewing. *J. Inst. Brew.* 94:233.

Oh D.-H., Marshall, D. L. 1993. Antimicrobial activity of ethanol, glycerol monolaurate or lactic acid against *Listeria monocytogenes*. *Intl. J. Food Microbiol.* 20:239–246.

Okereke, A., Montville, T. J. 1992. Nisin dissipates the proton motive force of the obligate anaerobe *Clostridium sporogenes* PA 3679. *Appl. Environ. Microbiol.* 58:2463–2467.

Okrend, A. J., Johnston, R. W., Moran, A. B. 1986. Effect of acetic acid on the death rates at 52°C of *Salmonella newport, Salmonella typhimurium*, and *Campylobacter jejuni* in poultry scald water. *J. Food Prot.* 49:500.

O'Leary, D. K., Kralovec, R. D. 1941. Development of *Bacillus mesentericus* in bread and control with calcium acid phosphate or calcium proprionate. *Cereal Chem.* 18:730.

Ough, C. S. 1993a. Dimethyl dicarbonate and diethyl dicarbonate. In: *Antimicrobials in Foods*, 2nd ed. Davidson, P. M., Branen, A. L. (Eds.), Marcel Dekker, New York, pp. 343–368.

Ough, C. S. 1993b. Sulfur dioxide and sulfites. In:*Antimicrobials in Foods*, 2nd ed. Davidson, P. M., Branen, A. L. (Eds.). Marcel Dekker, New York, p. 137.

Ough, C. S., Kunkee, R. E. 1974. The effect of fumaric acid on malolactic fermentation in wines from warm areas. *Am. J. Enol. Vitic.* 25:188.

Padgett, T., Han, I. Y., Dawson, P. L. 1998. Incorporation of food-grade antimicrobial compounds into biodegradable packaging films. 61:1330–1335.

Papadopoulos, L. S., Miller, R. K., Acuff, G. R., Vanderzant, C., Cross, H. R. 1991. Effect of sodium lactate on microbial and chemical composition of cooked beef during storage. *J. Food Sci.* 56: 341.

Parish, M. E., Carroll, D. E. 1988. Effects of combined antimicrobial agents on fermentation initiation by *Saccharomyces cerevisiae* in a model broth system. *J. Food Sci.* 53:240.

Park, H. S., Marth, E. H. 1972. Inactivation of *Salmonella typhimurium* by sorbic acid. *J. Milk Food Technol.* 35:532.

Park, H. S., Marth, E. H., Olson, N. F. 1970. Survival of *Salmonella typhimurium* in cold-pack cheese food during refrigerated storage. *J. Milk Food Technol.* 33:383.

Payne, K. D., Rico-Munoz, E., Davidson, P. M. 1989. The antimicrobial activity of phenolic compounds against *Listeria monocytogenes* and their effectiveness in a model milk system. *J. Food Prot.* 52:151.

Payne, K. D., Davidson, P. M., Oliver, S. P. 1994. Comparison of EDTA and apo-lactoferrin with lysozyme on the growth of foodborne pathogenic and spoilage bacteria. *J. Food Prot.* 57:62–65.

Pederson, M., Albury, N., Christensen, M. D. 1961. The growth of yeasts in grape juice stored at low temperature. IV. Fungistatic effect of organic acids. *Appl. Microbiol.* 9:162.

Pelroy, G. A., Holland, P. J., Eklund, M. W. 1994a. Inhibition of *Listeria monocytogenes* in cold-process (smoked) salmon by sodium lactate. *J. Food Prot.* 57:108–113.

Pelroy, G. A., Peterson, M. E., Paranjpye, R., Almond, J., Eklund, M. W. 1994b. Inhibition of *Listeria monocytogenes* in cold-process (smoked) salmon by sodium nitrite and packaging method. *J. Food Prot.* 57:114–119.

Perigo, J. A., Whiting, E., Bashford, T. E. 1967. Observations on the inhibition of vegetative cells of *Clostridium sporogenes* by nitrite which has been autoclaved in a laboratory medium discussed in the context of sublethally processed cured meats. *J. Food Technol.* 2:377.

Pfleiderer, G., Jeckel, D., Wieland, T. 1956. Uber die Eingirkung von sulfit auf einige DPN hydrierende enzyme. *Biochem. Z.* 328:187.

Phillips, G. F., Mundt, J. O. 1950. Sorbic acid as an inhibitor of scum yeast in cucumber fermentations. *Food Technol.* 4:291.

Pierson, M. D., Smoot, L. A., Stern, N. J. 1979. Effect of potassium sorbate on growth of *Staphylococcus aureus* in bacon. *J. Food Prot.* 42:302.

Pierson, M. D., Smoot, L. A., Vantassell, K. R. 1980. Inhibition of *Salmonella typhimurium* and *Staphylococcus aureus* by butylated hydroxyanisole and the propyl ester of *p*-hydroxybenzoic acid. *J. Food Prot.* 43:191.

Pilone, G. J. 1975. Control of malo-lactic fermentation in table wines by addition of fumaric acid. In: *Lactic Acid Bacteria in Beverages and Foods*. Carr, J. G., Cutting, C. V., Whiting, G. C. (Eds.). Academic Press, London, p. 121.

Pitt, J. I. 1974. Resistance of some food yeasts to preservatives. *Food Technol. Austral.* 26:238.

Podolak, R. K., Zayas, J. F., Kastner, C. L., Fung, D. Y. C. 1996a. Inhibition of *Listeria monocytogenes* and *Escherichia coli* O157:H7 on beef by application of organic acids. *J. Food Prot.* 59:370–373.

Podolak, R. K., Zayas, J. F., Kastner, C. L., Fung, D. Y. C. 1996b. Reduction of bacterial populations on vacuum-packaged ground beef patties with fumaric and lactic acids. *J. Food Prot.* 59:1037–1040.

Pontius, A. J., Rushing, J. E., Foegeding, P. M. 1998. Heat resistance of *Alicyclobacillus acidoterrestris* spores as affected by various pH values and organic acids. *J. Food Prot.* 61:41–46.

Post, F. J., Coblentz, W. S., Chou, T. W., Salunhke, D. K. 1968. Influence of phosphate compounds on certain fungi and their preservative effect on fresh cherry fruit (*Prunus cerasus* L.). *Appl. Microbiol.* 16:138.

Post, L. S., Amoroso, T. L., Solberg, M. 1985. Inhibition of *Clostridium botulinum* type E in model acidified food systems. *J. Food Sci.* 50:966.

Prindle, R. F. 1983. Phenolic compounds, p. 197. In: *Disinfection, Sterilization, and Preservation*, 3rd ed., Block S. S. (Ed.). Lea and Febiger, Philadelphia, p. 197.

Przybylaski, K. S., Bullerman, L. B. 1980. Influence of sorbic acid on viability and ATP content of conidia of *Aspergillus parasiticus*. *J. Food Sci.* 45:385.

Raab, W. 1967. Pimaricin ein Antibiotikum gegen Pilze und Trichomonaden. *Arzneimittel-Forschung* 17:538.

Radler, F. 1990a. Possible use of nisin in winemaking. I. Action of nisin against lactic acid bacteria and wine yeasts in solid and liquid media. *Am. J. Enol. Vitic.* 41:1–6.

Radler, F. 1990b. Possible use of nisin in winemaking. II. Experiments to control lactic acid bacteria in the production of wine. *Am. J. Enol. Vitic.* 41:7–11.

Rahn, O., Conn, J. E. 1944. Effect of increase in acidity on antiseptic efficiency. *Ind. Eng. Chem.* 36:185.

Rajashekhara, E., Suresh, E. R., Ethiraj, S. 1998. Thermal death rate of ascospores of *Neosartorya fischeri* ATCC 200957 in the presence of organic acids and preservatives in fruit juices. *J. Food Prot.* 61:1358–1362.

Rammell, C. G. 1962. Inhibition by citrate of the growth of coagulase-positive staphylococci. *J. Bacteriol.* 84:1123.

Ray, B., Johnson, C., Wanismail, B. 1984. Factors influencing lysis of frozen *Escherichia coli* cells by lysozyme. *Cryo-Letters* 5:183–190.

Ray, L. L., Bullerman, L. B. 1982. Preventing growth of potentially toxic molds using antifungal agents. *J. Food Prot.* 45:953.

Rayman, M. K., Aris, B., Hurst, A. 1981. Nisin: a possible alternative or adjunct to nitrite in the preservation of meats. *Appl. Environ. Microbiol.* 41:375–380.

Razavi-Rohani, S. M., Griffiths, M. W. 1999a. The antifungal activity of butylated hydroxyanisole and lysozyme. *J. Food Safety* 19:97–108.

Razavi-Rohani, S. M., Griffiths, M. W. 1999b. Antifungal effects of sorbic acid and propionic acid at different pH and NaCl conditions. *J. Food Safety* 19:109–120.

Reddy, N. R., Pierson, M. D. 1982. Influence of pH and phosphate buffer on inhibition of *Clostridium botulinum* by antioxidants and related phenolic compounds. *J. Food Prot.* 45:925.

Reddy, N. R., Pierson, M. D., Lechowich, R. V. 1982. Inhibition of *Clostridium botulinum* by antioxidants, phenols and related compounds. *Appl. Environ. Microbiol.* 43:835.

Rehm, H. J. 1963. Einfluss chemischer verbindungen auf die antimikrobielle konservierungsstoffwirkung. I. Einfluss verschiedener stoffgruppen auf die konservierungsstoffwirkung gegen *Aspergillus niger. Z. Lebensm. Untersuch. Forsch.* 118:508.

Rehm, H. J., Whittman, H. 1962. Beitrag Zur Kenntnis der antimikrobiellen wirkung der Schwefligen Saure I. Ubersicht uber einflussnehmende Factoren auf die antimikrobielle wirking der Schwefligen Saure. *Z. Lebens. Untersuch. Forsch.* 118:413.

Reiss, J. 1976. Prevention of the formation of mycotoxins in whole wheat bread by citric acid and lactic acid. *Experientia* 32:168.

Rice, A. C., Pederson, C. S. 1954. Factors influencing growth of *Bacillus coagulans* in canned tomato juice. II. Acidic constituents of tomato juice and specific organic acids. *Food Res.* 19: 124.

Rice, K. M., Pierson, M. D. 1982. Inhibition of *Salmonella* by sodium nitrite and potassium sorbate in frankfurters. *J. Food Sci.* 1615–1617.

Riemann, H., Lee, W. H., Genigeorgis, C. 1972. Control of *Clostridium botulinum* and *Staphylococcus aureus* in semipreserved meat products. *J. Milk Food Technol.* 35:514.

Robach, M. C. 1978. Effect of potassium sorbate on the growth of *Pseudomonas flueorescens. J. Food Sci.* 43:1886.

Robach, M. C. 1979. Influence of potassium sorbate on growth of *Pseudomonas putrefaciens. J. Food Prot.* 42:312.

Robach, M. C. 1980. Use of preservatives to control microorganisms in food. *Food Technol.* 34: 81.

Robach, M. C., Hickey, C. S. 1978. Inhibition of *Vibrio parahaemolyticus* by sorbic acid in crab meat and flounder homogenates. *J. Food Prot.* 41:699.

Robach, M. C., Pierson, M. D. 1978. Influence of *para*-hydroxybenzoic acid esters on the growth and toxin production of *Clostridium botulinum* 10755A. *J. Food Sci.* 43:787.

Robach, M. C., Sofos, J. N. 1982. Use of sorbate in meat products, fresh poultry and poultry products: a review. *J. Food Prot.* 45:374.

Roberts, A. C., McWeeny, D. J. 1972. The uses of sulfur dioxide in the food industry. A review. *J. Food Technol.* 7:221.

Roberts, T. A. 1975. The microbiological role of nitrite and nitrate. *J. Sci. Food Agric.* 26:1775.

Roberts, T. A., Gibson, A. M. 1986. Chemical methods for controlling *Clostridium botulinum* in processed meats. *Food Technol.* 40:163.

Roberts, T. A., Gracia, C. E. 1973. A note on the resistance of *Bacillus* spp. faecal streptococci and *Salmonella typhimurium* to an inhibitor of *Clostridium* spp. formed by heating sodium nitrite. *J. Food Technol.* 8:463.

Roberts, T. A., Ingram, M. 1966. The effect of sodium chloride, potassium nitrate and sodium nitrite on the recovery of heated bacterial spores. *J. Food Technol.* 1:147.

Roberts, T. A., Gilbert, R. L., Ingram, M. 1966. The effect of sodium chloride on heat resistance and recovery of heated spores of *C. sporogenes* (PA 3679/52). *J. Appl. Bacteriol.* 29:549.

Roberts, T. A., Jarvis, B., Rhodes, A. C. 1976. Inhibition of *Clostridium botulinum* by curing salts in pasteurized pork slurry. *J. Food Technol.* 11:25.

Roberts, T. A., Gibson, A. M., Robinson, A. 1981. Factors controlling the growth of *Clostridium botulinum* types A and B in pasteurized cured meats. II. Growth in pork slurries prepared from high pH meat (pH ranges 6.3–6.8). *J. Food Technol.* 16:267.

Roberts, T. A., Gibson, A. M., Robinson, A. 1982. Factors controlling the growth of *Clostridium botulinum* types A and B in pasteurized cured meats. *J. Food Technol.* 17:267.

Roberts, T. A., Woods, L. F. J., Payne, M. J., Cammack, R. 1991. Nitrite. In: Food Preservatives, Russell, N. J., Gould, G. W., (Eds.). Blackie and Son Ltd., Glasgow, pp. 89–111.

Roland, J. O., Beuchat, L. R. 1984. Biomass and patulin production by *Byssochlamys nivea* in apple juice as affected by sorbate, benzoate, SO_2 and temperature. *J. Food Sci.* 49:402.

Roland, J. O., Beuchat, L. R., Worthington, R. E., Hitchcock, H. L. 1984. Effects of sorbate, benzoate, sulfur dioxide and temperature on growth and patulin production by *Byssochlamys nivea* in grape juice. *J. Food Prot.* 47:237.

Ronning, I. E., Frank, H. A. 1987. Growth inhibition of putrefactive anaerobe 3679 caused by stringent-type response induced by protonophoric activity of sorbic acid. *Appl Environ. Microbiol.* 53:1020.

Rose, A. H., Pilkington, B. J. 1989. Sulphite. In: *Mechanisms of Action of Food Preservation Procedures*, Gould G. W. (Ed.). Elsevier Applied Science London, pp. 201–224.

Rose, N. L., Sporns, P., Stiles, M. E., McMullen, L. M. 1999. Inactivation of nisin by glutathione in fresh meat. *J. Food Sci.* 64:759–762.

Rosenquist, H., Hansen, Å. 1998. The antimicrobial effect of organic acids, sour dough and nisin against *Bacillus subtilis* and *B. licheniformis* isolated from wheat bread. *J. Appl. Microbiol.* 85:621–631.

Rowe, J. J., Yabrough, J. M., Rake, J. B., Eagon, R. G. 1979. Nitrite inhibition of aerobic bacteria. *Curr. Microbiol.* 2:51.

Ruhr, E., Sahl, H. 1985. Mode of action of the peptide antibiotic nisin and influence on the membrane potential of whole cells and on cytoplasmic and artificial membrane vesicles. *Antimicrob. Agents Chemother.* 27:841–845.

Rusul, G., Marth, E. H. 1987. Growth and aflatoxin production by *Aspergillus parasiticus* NRRL 2999 in the presence of potassium benzoate or potassium sorbate at different initial pH values. *J. Food Prot.* 50:820.

Sahl, H. G., Kordel, M., Benz, R. 1987. Voltage-dependent depolarization of bacterial membranes and artificial lipid bilayers by the peptide antibiotic nisin. *Arch. Microbiol.* 149: 120–124.

Samuelson, K. J., Rupnow, J. H., Froning, G. W. 1985. The effect of lysozyme and ethylenediaminetetraacetic acid on *Salmonella* on broiler parts. *Poultry Sci.* 64:1488–1490.

Sax, M. I. 1979. *Dangerous Properties of Industrial Materials*, 5th ed. Van Nostrand–Reinhold, New York.

Scannel, A. G. M., Hill, C., Buckley, D. J., Arendt, E. K. 1997. Determination of the influence of organic acids and nisin on shelf-life and microbiological safety aspects of fresh pork sausage. *J. Appl. Microbiol.* 83:407–412.

Schroeter, L. C. 1966. *Sulfur dioxide: Applications in Foods, Beverages and Pharmaceuticals*. Pergamon, Oxford.

Schved, F., Pierson, M. D., Juven, B. J. 1996. Sensitization of *Escherichia coli* to nisin by maltol and ethyl maltol. *Lett. Appl. Microbiol.* 22:189–191.

Scott, V. N., Taylor, S. L. 1981a. Effect of nisin on the outgrowth of *Clostridium botulinum* spores. *J. Food Sci.* 46:117–120.

Scott, V. N., Taylor, S. L. 1981b. Temperature, pH, and spore load effects on the ability of nisin to prevent the outgrowth of *Clostridium botulinum* spores. *J. Food Sci.* 46:121–126.

Shelef, L. A., Addala, L. 1994. Inhibition of *Listeria monocytogenes* and other bacteria by sodium diacetate. *J. Food Safety* 14:103–115.

Shelef, L. A., Yang, Q. 1991. Growth suppression of *Listeria monocytogenes* by lactates in broth, chicken and beef. *J. Food Prot.* 54:282–287.

Sheu, C. W., Freese, E. 1972. Effects of fatty acids on growth and envelope proteins of *Bacillus subtilis. J. Bacteriol.* 111:516.

Sheu, C. W., Konings, W. N., Freese, E. 1972. Effects of acetate and other short-chain fatty acids on sugars and amino acid uptake of *Bacillus subtilis. J. Bacteriol.* 111:525.

Sheu, C. W., Saloman, D., Simmons, J. L., Sreevalsan, T., Freese, E. 1975. Inhibitory effects of lipophilic acids and related compounds on bacteria and mammalian cells. *Antimicrob. Agents Chemother.* 7:349.

Shibasaki, I. 1969. Antimicrobial activity of alkyl esters of *p*-hydroxybenzoic acid. *J. Ferment Technol.* 47:167.

Shirk, R. J., Clark, W. L. 1963. The effect of pimaricin in retarding the spoilage of fresh orange juice. *Food Technol.* 17:1062.

Shtenberg, A. J., Ignat'ev, A. D. 1970. Toxicological evaluation of some combinations of food preservatives. *Food Cosmet. Toxicol.* 8:369.

Siragusa, G. R., Dickson, J. S. 1993. Inhibition of *Listeria monocytogenes, Salmonella typhimurium* and *Escherichia coli* O157:H7 on beef muscle tissue by lactic or acetic acid contained in calcium alginate gells. *J. Food Safety* 13:147–158.

Slavik, M. F., Kim, J. W., Pharr, M. D., Raben, D. P., Tsai, S., Lobsinger, C. M. 1994. Effect of trisodium phosphate on *Campylobacter* attached to post-chill chicken carcasses. *J. Food Prot.* 57:324–326.

Smith, D. P., Rollin, N. J. 1954. Sorbic acid as a fungistatic agent for foods. VII. Effectiveness of sorbic acid in protecting cheese. *Food Res.* 19:59.

Smith, J. E., Moss, M. O. 1985. *Mycotoxins: Formation, Analysis and Significance*. Wiley, New York.

Smith, J. L., Palumbo, S. A. 1980. Inhibition of aerobic and anaerobic growth of *Staphylococcus aureus* in a model sausage system. *J. Food Safety* 4:221.

Smoot, L. A., Pierson, M. D. 1981. Mechanisms of sorbate inhibition of *Bacillus cereus* T and *Clostridium botulinum* 62A spore germination. *Appl. Environ. Microbiol.* 42:477.

Smulders, F. J. M., Barendsen, P., van Logtestjin, J. G., Mossel, D. A. A., Van der Marel, G. M. 1986. Review: lactic acid: considerations in favour of its acceptance as a meat decontaminant. *J. Food Technol.* 21:419.

Snijders, J. M. A., van Logtestjin, J. G., Mossel, D. A. A., Smulders, F. J. M. 1985. Lactic acid as a decontaminant in slaughter and processing procedures. *Vet. Quart.* 7:277.

Sodemoto, Y., Enomoto, M. 1980. Report of carcinogenesis bioassay of sodium benzoate in rats: absence of carcinogenicity of sodium benzoate in rats. *J. Environ. Pathol. Toxicol.* 4:87.

Sofos, J. N. 1986. Use of phosphates in low-sodium meat products. *Food Technol.* 40(9):52–68.

Sofos, J. N., Busta, F. F. 1981. Antimicrobial activity of sorbates. *J. Food Prot.* 44:614.

Sofos, J. N., Busta, F. F. 1983. Alternatives to the use of nitrite as an antibotulinal agent. *Food Technol.* 34:244.

Sofos, J. N., Busta, F. F. 1993. Sorbates In: *Antimicrobials in Foods*, 2nd ed. Davidson, P. M., Branen, A. L. (Eds.). Marcel Dekker, New York, p. 49.

Sofos, J. N., Busta, F. F., Allen, C. E. 1979a. *Clostridium botulinum* control by sodium nitrite and sorbic acid in various meat and soy protein formulations. *J. Food Sci.* 44:1662.

Sofos, J. N., Busta, F. F., Allen, C. E. 1979b. Botulism control by nitrite and sorbate in cured meats. A review. *J. Food Prot.* 42:379.

Sofos, J. N., Busta, F. F., Allen, C. E. 1979c. Sodium nitrite and sorbic acid effects on *Clostridium botulinum* spore germination and total microbial growth in chicken frankfurter emulsions during temperature abuse. *Appl. Environ. Microbiol.* 34:1103.

Sofos, J. N., Busta, F. F., Bhothipaksa, K., Allen, C. E. 1979d. Sodium nitrite and sorbic acid effects on *Clostridium botulinum* toxin formation in chicken frankfurter-type emulsions. *J. Food Sci.* 44:668.

Sofos, J. N., Busta, F. F., Allen, C. E. 1980a. Influence of pH on *Clostridium botulinum* control by sodium nitrite and sorbic acid in chicken emulsions. *J. Food Sci.* 45:7.

Sofos, J. N., Busta, F. F., Bhothipaksa, K., Allen, C. E., Robach, M. C., Paquette, M. W. 1980b. Effects of various concentrations of sodium nitrite and potassium sorbate on *Clostridium botulinum* toxin production in commercially prepared bacon. *J. Food Sci.* 45:1285.

Somers, E. B., Taylor, S. L. 1972. Further studies on the antibotulinal effectiveness of nisin in acidic media. *J. Food Sci.* 46:1972–1973.

Somers, E. B., Taylor, S. L. 1987. Antibotulinal effectiveness of nisin in pasteurized process cheese spreads. *J. Food Prot.* 50:842–848.

Sorrells, K. M., Enigl, D. C., Hatfield, J. R. 1989. Effect of pH, acidulant, time, and temperature on growth and survival of *Listeria monocytogenes. J. Food Prot.* 52:571–573.

Spencer, R. 1971. Nitrite in curing: microbiological implications. Proceedings of the 17th European Meeting of Meat Research Workers Conference, Bristol, England.

Steinke, P. D. W., Foster, E. M. 1951. Botulism toxin formation in liver sausage. *Food Res.* 16: 477.

Stevens, K. A., Sheldon, B. W., Klapes, N. A., Klaenhammer, T. R. 1992. Effect of treatment conditions on nisin inactivation of gram-negative bacteria. *J. Food Prot.* 55:763–766.

Stevenson, D. D., Simon, R. A. 1981. Sensitivity to ingested metabisulfites in asthmatic individuals. *J. Allergy Clin. Immunol.* 68:26.

Stillmunkes, A. A., Prabhu, G. A., Sebranek, J. G., Molins, R. A. 1993. Microbiological safety of cooked beef roasts treated with lactate, monolaurin or gluconate. *J. Food Sci.* 58:953–958.

Stratford, M., Rose, A. H. 1985. Hydrogen sulphide production from sulphite by *Saccharomyces cerevisiae. J. Gen. Microbiol.* 131:1417–1424.

Subramanian, C. S., Marth, E. H. 1968. Multiplication of *Salmonella typhimurium* in skim milk with and without added hydrochloric, lactic and citric acids. *J. Milk Food Technol.* 31:323.

Szabo, E. A., Cahill, M. E. 1999. Nisin and ALTA™ 2341 inhibit the growth of *Listeria monocytogenes* on smoked salmon packaged under vacuum or 100% CO_2. *Lett. Appl. Microbiol.* 28: 373–377.

Tanaka, N. 1982. Challenge of pasteurized process cheese spreads with *Clostridium botulinum* using in-process and post-process inoculation. *J. Food Prot.* 45:1044.

Taylor, S. L., Speckhard, M. W. 1984. Inhibition of bacterial histamine production by sorbate and other antimicrobial agents. *J. Food Prot.* 47:508.

Terrell, F. R., Morris, J. R., Johnson, M. G., Gbur, E. E., Makus, D. J. 1993. Yeast inhibition in grape juice containing sulfur dioxide, sorbic acid, and dimethyldicarbonate. *J. Food Sci.* 58: 1132–1134.

Thomas, L. V., Davies, E. A., Delves-Brougton, J., Wimpenny, J. W. T. 1998. Synergist effect of sucrose fatty acids on nisin inhibition of gram-positive bacteria. *J. Appl. Microbiol.* 85:1013–1022.

Thompson, D. P. 1994. Minimum inhibitory concentration of esters of *p*-hydroxybenzoic acid (paraben) combinations against toxigenic fungi. *J. Food Prot.* 57:133–135.

Thomson, J. E., Banwart, G. J., Sanders, D. H., Mercuri, A. J. 1967. Effect of chlorine, antibiotics, ß-propiolactone, acids and washing on *Salmonella typhimurium* on eviscerated fryer chickens. *Poult. Sci.* 46:146.

Ticha, J. 1975. A new fungicide, pimaricin, and its application in the baking industry. *Mlynsko-Pekarensky Prumysl* 21:225; *Food Sci. Technol. Abst.* 8:163.

Til, H. P., Feron, V. J., DeGroot, A. P. 1972a. The toxicity of sulphite. I. Long-term feeding and multigeneration studies in rats. *Food Cosmet. Toxicol.* 10:291.

Til, H. P., Feron, V. J., DeGroot, A. P., Van der Wal, P. 1972b. The toxicity of sulphite. I. Short- and long-term feeding studies in pigs. *Food Cosmet. Toxicol.* 10:463.

Tompkin, R. B. 1978. The role and mechanism of the inhibition of *C. botulinum* by nitrite—is a replacement available? Proceedings of the 31st Annual Reciprocal Meats Conference, Storrs, CT, p. 135.

Tompkin, R. B. 1983. Indirect antimicrobial effects in foods: phosphates. *J. Food Safety* 6:13.

Tompkin, R. B. 1993. Nitrite. In: *Antimicrobials in Foods*, 2nd ed. Davidson, P. M., Branen, A. L. (Eds.). Marcel Dekker, New York, p. 191.

Tompkin, R. B., Christiansen, L. N., Shaparis, A. B., Bolin, H. 1974. Effect of potassium sorbate on salmonellae, *Staphylococcus aureus*, *Clostridium perfringens*, and *Clostridium botulinum* in cooked uncured sausage. *Appl. Microbiol.* 28:262–264.

Tompkin, R. B., Christiansen, L. N., Shaparis, A. B. 1978. The effect of iron on botulinal inhibition in perishable canned cured meat. *J. Food Technol.* 13:521–527.

Tompkin, R. B., Christiansen, L. N., Shaparis, A. B. 1980. Antibotulinal efficacy of sulfur dioxide in meat. *Appl. Environ. Microbiol.* 39:1096–1099.

Tranter, H. S. 1994. Lysozyme, ovotransferrin and avidin. In: *Natural Antimicrobial Systems and Food Preservation*, Dillon, V. M., Board, R. G. (Eds.). CAB International. Wallingford, UK, pp. 65–97.

Troller, J. A. 1965. Catalase inhibition as a possible mechanism of the fungistatic action of sorbic acid. *Can. J. Microbiol.* 11:611.

Tsai, W. Y. J., Shao, K. P. P., Bullerman, L. B. 1984. Effects of sorbate and propionate on growth and aflatoxin production of sublethally injured *Aspergillus parasiticus*. *J. Food Sci.* 49:86.

Tsai, S.-H., Chou, C.-C. 1996. Injury, inhibition and inactivation of *Escherichia coli* O157:H7 by potassium sorbate and sodium nitrite as affected by pH and temperature. *J. Sci. Food Agric.* 71:10–12.

Ukuku, D. O., Shelef, L. A. 1997. Sensitivity of six strains of *Listeria monocytogenes* to nisin. *J. Food Prot.* 60:867–869.

Uljas, H. E., Ingham, S. C. 1999. Combinations of intervention treatments resulting in 5-$\log_{10}$-unit reductions in numbers of *Escherichia coli* O157:H7 and *Salmonella typhimurium* DT104 organisms in apple cider. *Appl. Environ. Microbiol.* 65:1924–1929.

Unda, J. R., Mollins, R. A., Walker, H. W. 1991. *Clostridium sporogenes* and *Listeria monocytogenes*: survival and inhibition in microwave-ready beef roasts containing selected antimicrobials. *J. Food Sci.* 56:198–205.

Uriah, N., Chipley, J. R. 1976. Effects of various acids and salts on growth and aflatoxin production by *Aspergillus flavus*. *Microbios* 17:51.

Uriah, N., Cassity, T. R., Chipley, J. R. 1977. Partial characterization of the action of benzoic acid on aflatoxin biosynthesis. *Can J. Microbiol.* 23:1580.

Vakil, J. R., Chandan, R. C., Parry, R. M., Shahani, K. M. 1969. Susceptibility of several microorganisms to milk lysozymes. *J. Dairy Sci.* 52:1192–1197.

van der Riet, W. B., Pinches, S. E. 1991. Control of *Byssochlamys fulva* in fruit juices by means of intermittent treatment with dimethyldicarbonate. *Lebensmittel Wissenschaft Technologie* 24: 501–503.

Van Esch, G. J., Vink, H. H., Wit, S. J., Van Genderen, H. 1957. Physiological effects of polyphosphates. *Arzneimittelforschung* 7:172.

van Rijn, F. T. J., Stark, J., Geijp, E. M. L. 1999. Antifungal complexes. U.S. patent 5,997,926.

Verma, H. S., Yadav, J. S., Neelakantan, S. 1988. Preservative effect of selected antifungal agents on butter and cheese. *Asian J. Dairy Res.* 7:34.

Veugopal, V., Pansare, A. C., Lewis, N. F. 1984. Inhibitory effect of food preservatives on protease secretion by *Aeromonas hydrophila. J. Food Sci.* 49:1078.

Visser, I. J. R., Koolmees, P. A., Bijker, P. G. H. 1988. Microbiological conditions and keeping quality of veal tongues as affected by lactic acid decontamination and vacuum packaging. *J. Food Prot.* 51:208.

Wagner, H. J. 1956. Vergiftung mit pokelsalz. *Arch. Toxikol.* 16:100.

Wagner, M. K. 1986. Phosphates as antibotulinal agents in cured meats. A review. *J. Food Prot.* 49:482.

Wagner, M. K., Busta, F. F. 1983. Effect of sodium acid pyrophosphate in combination with sodium nitrite or sodium nitrite/potassium sorbate on *Clostridium botulinum* growth and toxin production in beef/pork frankfurter emulsions. *J. Food Sci.* 48:990.

Wagner, M. K., Busta, F. F. 1985. Inhibition of *Clostridium botulinum* 52A toxicity and protease activity by sodium acid pyrophosphate in media systems. *Appl. Environ. Microbiol.* 50:16.

Waldroup, A. 1995. Evaluating reduction technologies. *Meat Poult.* 41(8):10.

Waldroup, A., Marcy, J., Doyle, M., Scantling, M. 1995. TSP: A market survey. *Meat Poult.* 41(12): 18–20.

Wang, C., Shelef, L. 1991. Factors contributing to antilisterial effects of raw egg albumen. *J. Food Sci.* 56:1251–1254.

Wang, W. C., Li, Y., Slavik, M. F., Xiong, H. 1997. Trisodium phosphate and cetylpyridinium chloride spraying on chicken skin to reduce attached *Salmonella typhimurium. J. Food Prot.* 60:992–994.

Warth, A. D. 1977. Mechanism of resistance of *Saccharomyces bailii* to benzoic, sorbic, and other weak acids used as food preservatives. *J. Appl. Bacteriol.* 43:215–230.

Wasserfall, F., Voss, E., and Prokopek, D. 1976. Experiments on cheese ripening: the use of lysozyme instead of nitrite to inhibit late blowing of cheese. *Kiel. Milchwirtsch. Forschungsber.* 238:3–16.

Weaver, R. A., Shelef, L. A. 1993. Antilisterial activity of sodium, potassium or calcium lactate in pork liver sausage. *J. Food Safety* 13:133–146.

Wedzicha, B. L. 1981. Sulphur dioxide: the reaction of sulphite species with food components. *Nutr. Food Sci.* 72:12.

Whitaker, J. R. 1959. Inhibition of sulfhydryl enzymes with sorbic acid. *Food Res.* 24:37.

WHO. 1974. *p*-Hydroxybenzoic, ethyl, methyl, propyl esters. In: *Toxicological Evaluation of Some Food Additives Including Anticaking Agents, Antimicrobials, Antioxidants, Emulsifiers and Thickening Agents*, World Health Organization, Geneva, p. 89.

Wibowo, D., Eschenbruch, R., Davis, C. R., Fleet, G. H., Lee, T. H. 1985. Occurrence and growth of lactic acid bacteria in wine. A review. *Am. J. Enol. Vitic.* 36:302.

Wilkins, K. M., Board, R. G. 1989. Natural antimicrobial systems. In: *Mechanisms of Action of Food Preservation Procedures*, Gould, G. W. (Ed.). Elsevier Applied Science, London, pp. 285–362.

Williams, R. R., Waterman, R. E., Keresztesy, J. C., Buchman, E. R. 1935. Studies of crystalline vitamin B_1. III. Cleavage of vitamin with sulfite. *J. Am. Chem. Soc.* 47:536.

Wind, C. E., Restaino, L. 1995. Antimicrobial effectiveness of potassium sorbate and sodium benzoate against *Zygosaccharoymyces bailii* in a salsa mayonnaise. *J. Food Prot.* 58:1257–1259.

Winkowski, K., Bruno, M. E. C., Montville, T. J. 1994. Correlation of bioenergetic parameters with cell death in *Listeria monocytogenes* cells exposed to nisin. *Appl. Environ. Microbiol.* 60:4186–4188.

Woods, L. F. J., Wood, J. M. 1982. The effect of nitrite inhibition on the metabolism of *Clostridium botulinum. J. Appl. Bacteriol.* 52:109.

Woods, L. F. J., Wood, J. M., Gibbs, P. A. 1981. The involvement of nitric oxide in the inhibition of the phosphoroclastic system in *Clostridium sporogenes* by sodium nitrite. *J. Gen. Microbiol.* 125:399.

Woolford, E. R., Anderson, A. A. 1945. Propionates control microbial growth in fruits and vegetables. *Food Ind.* 17:622.
Woolford, M. K. 1975a. Microbiological screening of the straight chain fatty acids (C_1–C_{12}) as potential silage additives. *J. Sci. Food Agric.* 26:219.
Woolford, M. K. 1975b. Microbiological screening of food preservatives, cold sterilants and specific antimicrobial agents as potential silage additives. *J. Sci. Food Agric.* 26:229.
Wright, L. D., Skeggs, H. R. S. 1946. Reversal of sodium propionate inhibition of *Escherichia coli* with β-alanine. *Arch. Biochem.* 10:383.
Xiong, R., Xie, G., Edmondson, A. S. 1999. The fate of *Salmonella enteritidis* PT4 in home-made mayonnaise prepared with citric acid. *Lett. Appl. Microbiol.* 28:36–40.
Yang, T. 1985. Mechanism of nitrite inhibition of cellular respiration in *Pseudomonas aeruginosa. Curr. Microbiol.* 12:35–40.
Yang, T. S., Cunningham, F. E. 1993. Stability of egg white lysozyme in combination with other antimicrobial substances. *J. Food Prot.* 56:153–156.
Yarbrough, J. M., Rake, J. B., Eagon, R. G. 1980. Bacterial inhibitory effects of nitrite: inhibition of active transport, but not of group translocation, and of intracellular enzymes. *Appl. Environ. Microbiol.* 39:831.
York, G. K., Vaughn, R. H. 1954. Use of sorbic acid enrichment media for species of Clostridium. *J. Bacteriol.* 68:739.
York, G. K., Vaughn, R. H. 1955. Resistance of *Clostridium parabotulinum* to sorbic acid. *Food Res.* 20:60.
York, G. K., Vaughn, R. H. 1964. Mechanisms in the inhibition of microorganisms by sorbic acid. *J. Bacteriol.* 88:411.
Yousef, A. E., Marth, E. H. 1981. Growth and synthesis of aflatoxin by *Aspergillus parasiticus* in the presence of sorbic acid. *J. Food Prot.* 44:736.
Zaika, L. L., Kim, A. H. 1993. Effect of sodium polyphosphates on growth of *Listeria monocytogenes. J. Food Prot.* 56:577–580.
Zhao, T., Doyle, M. P., Besser, R. E. 1993. Fate of enterohemorrhagic *Escherichia coli* O157:H7 in apple cider with and without preservatives. *Appl. Environ. Microbiol.* 59:2526–2530.
Zhuang, R.-Y., Beuchat, L. R. 1996. Effectiveness of trisodium phosphate for killing *Salmonella montevideo* on tomatoes. *Lett. Appl. Microbiol.* 22:97–100.
Ziogas, B. N., Sisler, H. D., Lusby, W. R. 1983. Sterol content and other characteristics of pimaricin-resistant mutants of *Aspergillus nidulans. Pesticide Biochem. Physiol.* 20:320.

21

pH Control Agents and Acidulants

STEPHANIE DOORES

The Pennsylvania State University, University Park, Pennsylvania

I. INTRODUCTION

The effective use of organic acids in food products is governed in part by the desired end product. Acidulants contribute a variety of functional properties that enhance the quality of food. The proper selection of an acid depends on the property or combination of properties of the desired acid as well as cost. Acids are most commonly used for flavor and tartness, and the buffering ability of the salts of some acids can modify and smooth out these characteristics. Acidulants all share a characteristic sour flavor, however the degree of flavor profile varies significantly depending on the final pH of the product (Hartwig and McDaniel, 1995). These differences can be used in the formulation of foods with specific sensory qualities that may be unrelated to any antimicrobial effects of the individual acids.

In their synergistic capacity, acids used in conjunction with antioxidants prevent rancidity or complex with heavy metals that might initiate oxidation or browning reactions. Acids stabilize color, reduce turbidity, change melt characteristics, prevent splattering, or enhance gelling. They can also act as leavening or inversion agents, emulsifiers, nutrients, and supplements (Gardner, 1966).

A major use of acidulants is as an antimicrobial agent and this use is receiving much attention today in a desire to achieve safer food products. With many foods, the incorporation of acids into the product at sufficiently high levels ensures a commercially sterile product. The target levels necessary for this purpose, however, can overwhelm the sensory properties of the food and thereby prevent the use of an acid. When used with other preservation processes, such as refrigeration or heating, acids extend shelf-life for almost indefinite periods from an antimicrobial standpoint. The judicious use of an acidulant for its antimicrobial properties is counterbalanced by the desired sensory characteristics that predicate the amount of acid that is utilized. It is in the product development and safety testing phases that the addition of the acid with its inherent properties can be evaluated to produce a product desirable to the consumer.

There are a number of excellent reference texts that contain information concerning acidulants, especially with descriptions of chemical and physical characteristics and toxicology. Among these works are those of Ash and Ash (1995), Burdock (1997), Igoe and Hui (1996), Lewis (1989), and Maga and Tu (1995).

II. INTERNATIONAL REGULATORY USE IN FOODS

Various international agencies regulate the addition of all acids, salts, and derivatives of acids to foods. Limits or conditions for acidulant use vary not only by country, but also for each food product category. *Food Additives Tables* (Bigwood et al., 1975; Fondu et al., 1980) and their supplements list regulations for 19 foreign countries for close to 50 product categories that incorporate acidulants into foodstuffs. The use or prohibition on use of the organic acids discussed in this chapter is constantly changing. For this reason, and because of the shear volume of such information, the reader is referred to the tables referenced in this review. Regulations regarding the use of acidulants in foods in the United States are presented for each acidulant under the heading ''Regulatory Use in Foods.''

III. CHEMICAL ANALYSIS AND ASSAY

A variety of factors dictates the effective use of acidulants not only in the food product, but also by the physical properties of the acidulant. Information on each of the acids discussed in this chapter is given in the *Food Chemicals Codex* (National Research Council, 1996). Supplements to the *Codex* are issued periodically to update contents. For each acid, the *Codex* provides its chemical description, specifications for identification and purity, guidelines for good manufacturing practices, general test methods, and notes on functional use in foods. Physical and chemical properties of the acids are presented in Table 1.

The *Official Methods of Analysis of the Association of Official Analytical Chemists International* are used as the basis for conducting many assays for the various acids in foods (Cunniff, 1995; Helrich, 1990; Williams, 1984). Assays for organic acids in foodstuffs are generally carried out for three reasons: (1) the organic acid is a natural component of the product and therefore the assay is a quantitative or qualitative measure of wholesomeness or lack of adulteration or is a confirmation of the standard of identity for that product; (2) the acid is not normally present in the food product or is present at levels lower than that normally detected in a standard assay and serves as a measure of adulteration either through addition of the acid or fermentation of the substrate to form acidic byproducts; or (3) the organic acid is added to achieve a desired effect as regulated by good manufacturing practice.

In general, methods in common depend on the ease of separation of a specific acid from the food, the food product itself, and the desire for quantitative or qualitative results. For example, it is desirable to know the levels of acetic and propionic acids in breads and cakes, eggs, and seafood; citric, malic, and tartaric acids in wines, nonalcoholic beverages, fruit, and fruit products; lactic acid in wines, eggs, milk and milk products, fruits, and canned vegetables. Other food products assayed for specific organic acid levels include

Table 1 Physical and Chemical Properties of Organic Acids

	Acid (chemical name)	Molecular weight	Formula	Solubility	p*K*
A.	Acetic (monocarboxylic acid) (ethanoic acid)	60.05	$C_2H_4O_2$	Water, alcohol, glycerol	4.75
B.	Acetate Salts Calcium acetate	158.17	$C_4H_6CaO_4$	Water, sl. alcohol	
	Potassium acetate	98.14	$C_2H_3KO_2$	Water, alcohol	
	Sodium acetate	82.03	$C_2H_3NaO_2$	Water, alcohol	
C.	Dehydroacetic acid (methylacetopyranone)	168.15	$C_8H_8O_4$	Sl. water, sl. alcohol	5.27
D.	Sodium diacetate (sodium hydrogen diacetate)	142.09	$C_4H_7NaO_4$	Water	4.75
E.	Adipic (1,4-butane dicarboxylic acid) (hexanedioic)	146.14	$C_6H_{10}O_4$	Sl. water, alcohol	4.43 5.41
F.	Ascorbic acid	176.12	$C_6H_8O_6$	water, sl. alcohol	4.17 11.57
G.	Caprylic (octanoic acid)	144.21	$C_8H_{16}O_2$	Sl. water, alcohol	4.89
H.	Citric (2-hydroxy 1,2,3-propanetricarboxylic acid) (β-hydroxytricarballylic acid)	192.13	$C_6H_8O_7$	Water, alcohol	3.14 4.77 6.39
	Calcium citrate	498.44	$C_{12}H_{10}Ca_3O_{14}$	Sl. water	
	Potassium citrate	324.42	$C_6H_5K_3O_7H_{20}$	Water	
	Sodium citrate	258.07	$C_6H_5Na_3O_7$	Water	
I.	Fumaric (2-butenedioic acid) (*trans*-1,2-ethylenedicarboxylic acid)	116.07	$C_4H_4O_4$	Sl. water, alcohol	3.30 4.44
	Sodium stearyl fumarate	390.54	$C_{22}H_{39}NaO_4$	Almost insoluble, water	
J.	Lactic acid (2-hydroxypropionic acid) (2-hydroxypropanoic acid) (1-hydroxyethane 1-carboxylic acid)	90.08	$C_3H_6O_3$	Water, alcohol	3.08
	Calcium lactate	218.22	$C_6H_{10}CaO_6H_2O$	Water	
K.	Malic (1-hydroxy 1,2-ethanedicarboxylic acid) (hydroxy succinic acid) (hydroxybutanedioic acid)	134.09	$C_4H_6O_5$	Water, alcohol	3.4 5.11
L.	Propionic acid (propanoic acid)	74.08	$C_3H_6O_2$	Water, alcohol	4.87
	Calcium propionate	186.22	$C_6H_{10}CaO_4$	Water	
	Sodium propionate	96.07	$C_3H_5NaO_2$	Water, alcohol	
M.	Succinic (1,4-butanedioic acid)	118.09	$C_4H_6O_4$	Water, alcohol	4.16 5.61
	Sodium succinate	162.05	$C_4H_4Na_2O_4$	Water	
N.	Tartaric (2,3-dihydroxysuccinic acid)	150.09	$C_4H_6O_6$	Water, alcohol	2.98 4.34
	Potassium acid tartrate	188.18	$C_4H_5KO_6$	Water, alcohol	
	Sodium potassium tartrate	282.23	$C_4H_4KNaO_6.4H_2O$	Water	
	Sodium tartrate	194.05	$C_4H_4Na_2O_6$	Water	

Source: Budavari, 1996.

cheese (dehydroacetic, citric, and tartaric), dried milk (citric), maple syrup (malic), baking powders (tartaric), and eggs (succinic).

Enzymatic analysis for acetic, citric, dehydroacetic acid, isocitric, L- and D-lactic, L-malic, and succinic acids can be conducted to identify the acid (Bergmeyer, 1983). Acids can also be quantitated using thin layer chromatography (Tijan and Jansen, 1971), gas chromatography (Martin et al., 1971; Tsuji et al., 1986), high-performance liquid chromatography (HPLC) (Bouzas, 1991; Marsili et al., 1981), and titrimetric or colorimetric methods (Nollett, 1996).

IV. MODE OF ACTION OF ACIDS AS ANTIMICROBIAL AGENTS

In 1932, Cruess and Irish suggested that the preservative effect of acids was greatly dependent on the pH resulting from their addition and that the undissociated acids were the true preserving agents (Cruess and Irish, 1932). Fabian and Wadsworth (1939b) agreed with Cruess and Irish and suggested that pH alone was not a sufficient indicator of the preservative effect of an acid. In their studies, any two acids equilibrated to the identical pH achieved different inhibitory effects. In the same year, Hoffman and coworkers (1939) proposed that acids with fewer than seven carbons were more effective antimicrobially at lower pH levels, whereas acids with 9–12 carbons were more effective at neutrality and above. Work by Levine and Fellers (1940) demonstrated that the antimicrobial action occurring with acetic acid was seen at a higher pH level than that achieved by hydrochloric acid, indicating that some mechanism besides hydrogen ion concentration (pH) alone was responsible for inhibition. This inhibition is apparently also dependent upon the undissociated molecule.

In the 1970s, considerable research on the action of organic acids on the microbial cell focused at the molecular level. It was known that the mode of action of an acid was related to the undissociated portion of the molecule. This action was thought to be more important than any external change in pH brought about by the addition of acids. Fatty acids incorporated into the plasma membrane at a physiological pH of the microorganism would be expected to be almost completely dissociated due to the dissociation constants of the acid (Larsson et al., 1975). Hunter and Segel (1973) hypothesized that membrane transport processes were inhibited and the proton gradient was somehow affected. The undissociated form of a weak acid penetrates rapidly to the interior of the cell because of its lipid solubility and could discharge the gradient by diffusing through the plasma membrane and dissociate internally. Dissociated forms of weak acids, on the other hand, could not be absorbed by microorganisms to any great extent.

Sheu et al. (1972) discovered that acetic acid acts as an uncoupler of amino acid carrier proteins from the electron transport system to which they are coupled by protein interaction or cation gradient. The uncoupling effect inhibits amino acid transport noncompetitively. This may come about by acetic acid reacting with the cellular membrane or its proteins by altering the membrane structure or uncoupling electron transport from the proteins responsible for ATP regeneration or transport (Sheu and Freese, 1972).

Reynolds (1975) proposed that the undissociated portion of the molecule penetrates the cell membrane, dissociates owing to the pH differential, and then causes general protein denaturation. Freese and Levine (1978) found that acids act as proton shuttles that effectively uncouple the transmembrane proton gradients from energy-coupling circuits. This action was reversible. In practical situations, however, prolonged contact of an acid with a microorganism can lead to decreased viability and death. In addition, some microorganisms can utilize acids in anabolic reactions that lower their effectiveness and account, at least partially, for some differences among microbial species in their sensitivity to acids. For further details in this area, the reader is referred to Doores (1983,1993).

V. ORGANIC ACIDS

Numerous studies compare several organic acids for their antimicrobial effect. The results of these studies are usually presented within the microbial function section of the acid providing the most inhibitory effect.

A. Acetic Acid

1. Microbial Function

One of the primary functional uses of acetic acid in food has been that of an acidulant. The microcidal activity has been attributed to a lowering of the pH below that needed for optimal growth. In addition, acetic acid is an excellent antimicrobial additive that has far-ranging bactericidal and fungicidal properties. As with all acids, different organisms can display varying tolerances dependent upon the type, concentration, and pH of the acid when formulated in either a broth or foodstuff system. Lactic, acetic, propionic, and butyric acid–producing bacteria are among the most tolerant of the bacterial groups to the growth-inhibitory aspects of acetic acid or other organic acids. This resistance to the detrimental properties of acids may be due in part to the production by the bacteria of one or several of these acids in their metabolism. In general, bacteria normally found associated with fermented products, which tend to be low-pH foods, will have greater tolerance to acids.

Acetic acid has been used primarily to limit bacterial and yeast growth rather than mold growth, however some molds are sensitive to this acidulant. Acetic acid has been shown to be an effective antifungal agent at pH 3.5 against the black bread molds, particularly *Aspergillus niger* and *Rhizopus nigrificans* (Kirby et al., 1937). Buchanan and Ayres (1976) promoted the surface application of acetic acid or propionic acid by either spraying, dipping, or flotation as an effective means for preventing the growth and aflatoxin production by *Aspergillus flavus* and *A. parasiticus*. The growth of *Saccharomyces ellipsoideus* and *Penicillium glaucum* was inhibited at pH 3.5 in apple juice containing 0.8–1.0% acid, but when the pH was adjusted to 7.0 a concentration of greater than 4.0% acid was required for inhibition (Cruess and Irish, 1932).

In screening various acids for their antimicrobial activity for use as silage additives, Woolford (1975b) found that acetic acid, at a pH level of 6.0, inhibited the growth of *Bacillus*, *Clostridium* spp., and gram negative bacteria more than gram positive and lactic acid bacteria, yeasts, and molds. When the pH level was lowered to 5.0, *Bacillus*, *Clostridium*, spp., and both gram negative and gram positive bacteria were more inhibited than the lactic acid bacteria, yeasts, and molds. However, when the pH level was further reduced to 4.0, the minimum inhibitory concentration for all organisms was reduced significantly.

In comparative studies using *Bacillus cereus*, *Salmonella aertrycke*, *Staphylococcus aureus*, *Saccharomyces cerevisiae*, and *Aspergillus niger* as test organisms, the microstatic or microcidal levels of acetic acid was determined after a 15-min contact period. *Staphylococcus aureus* was inhibited at pH 5.0 (destroyed at pH 4.9), followed by *B. cereus* and *S. aertrycke* at pH 4.9 (pH 4.5), *A. niger* at pH 4.1 (pH 3.9), and *S. cerevisiae* at pH 3.9 (Levine and Fellers, 1939). In addition, it was found that *A. niger* could utilize acetic acid as an energy source, thereby decreasing its effectiveness.

In experiments using *S. aureus* as a test organism, Nunheimer and Fabian (1940) rated the antimicrobial effectiveness of acids on their attendant dissociation constants. Tartaric acid, which has one of the highest dissociation constants, showed a weaker inhibitory action (pH 3.92) than acetic acid, which has the lowest dissociation constant (pH 4.59). This resulted in the conclusion that the effectiveness of organic acids was related to the undissociated portion of the molecule. In their study, acetic, lactic, citric, malic, tartaric, and hydrochloric acids were thus rated in descending order of bacteriostatic effectiveness. In other studies by Minor and Marth (1970), the relationship between pH levels and destruction was explored. As the pH level was adjusted downward with acetic acid to 5.2, 5.0, and 4.9, the destruction of *S. aureus* increased from 90 to 99 and 99.9%, respectively. Lower pH values of 4.6, 4.0, and 3.3, respectively, were required for hydrochloric acid to achieve the same killing effect.

In studies determining the synergistic effect of temperature and pH on the inhibition of *Yersinia enterocolitica*, growth inhibition was dependent upon the acidulant and temperature used. When used in combination with temperature, lower pH levels of the acids were required to limit growth of *Y. enterocolitica* as the temperature increased. Acetic acid was more effective than lactic, which in turn was more effective than citric acid in producing this effect (Adams et al., 1991; Brocklehurst and Lund, 1990; Little et al., 1992). This same effect was also seen with another psychrotrophic pathogen, *Listeria monocytogenes* (Ahmad and Marth, 1989; El-Shenawy and Marth, 1989b; Sorrells et al., 1989).

Adams and Hall (1988) were able to show a weakly synergistic growth inhibitory effect between acetic and lactic acids with *Salmonella enteritidis* and *E. coli*. Although these acid mixtures might be expected to occur naturally during fermentations by heterofermentative bacteria, this apparent synergy has not been demonstrated in other foods. These authors believe that the synergistic action is brought about by lactic acid. This association may help to explain the apparent stability that occurs in natural fermentations, e.g., sauerkraut manufacture. Sauerkraut manufacture is a fermentation process brought about by a natural succession of microbes that inhabit the surface of cabbage leaves. The succession typically begins with growth of *Leuconostoc mesenteroides* followed by *Lactobacillus plantarum*, a more acid-tolerant organism. McDonald et al. (1990) showed that growth of *L. mesenteroides* ceased when the internal pH of the cells reached 5.4–5.7, whereas growth was stopped in *L. plantarum* at pH 4.6–4.8.

Buffered acidulant systems (brain heart infusion broth, pH 4.8, 22°C) containing 1% acetic acid and 1% other organic acid (lactic, malic, tartaric or citric) were far more effective in reducing numbers of *Staphylococcus aureus*, *Salmonella blockley*, and *Escherichia coli* than unbuffered systems. This same acidulant mix was bacteriostatic to *Streptococcus faecalis*. This finding is important in the manufacture of medium acid foods containing more than one organic acid, such as mayonnaise (Debevere, 1988).

Levine and Fellers (1940) suggested that sublethal concentrations of acetic acid would enhance reduction of microbial load during thermal processing. This effect was seen by Ito et al. (1976) in their work on pickle manufacture and processing. *Clostridium botulinum* could not grow or produce toxin when inoculated at levels of 1 million spores per cucumber if preserved in brine supplemented with as little as 0.9% acetic acid and 12°Brix followed by a standard thermal process for pickles. Not only was acetic acid found to penetrate cucumbers used in the production of pickles faster than lactic acid (Fabian and Wadsworth, 1939a), but the rate of acid production was rapid enough to prevent growth of *C. botulimum* even under the acid and Brix conditions stated above (Ito et al., 1976).

The effect of organic acids on reducing microbial loads at various sites during poultry processing has been investigated, particularly in the last decade. A combination of acidulant (0.1% concentration of acetic acid) and temperature (52°C) in the scald tank water reduced decimal reduction times for *Salmonella newport*, *Salmonella typhimurium*, and *Campylobacter jejuni*. When the concentration of acid was raised to 1.0%, however, immediate death resulted (Okrend et al., 1986). In studies by Lillard and coworkers (1987), it was found that the addition of 0.5% acetic acid (pH 3.6) to scald tank water reduced total aerobic plate counts by over 99% and *Enterobacteriaceae* counts to nondetectable levels, confirming the observations of Okrend et al. (1986). Both studies suggested that the addition of acetic acid to scald water was an effective means of eradication of potential foodborne pathogens. Although the addition of acid might limit cross-contamination of carcasses in scald water, it would not significantly reduce microbial loads on the carcasses.

A second potential intervention step in poultry processing is the acidification of the chill water at the end stage of processing. Broilers were immersed for 10 min in 0.6% acetic solution acid (pH 3.0) and immediately sampled. Aerobic plate counts were not significantly reduced because of acid treatment; however, *Enterobacteriaceae* counts were significantly by 0.71 log MPN/mL (Dickens et al., 1994). In a second experiment, Dickens and Whittemore (1994) exposed the broilers to the same 10-min, prechill acid treatment, but at two concentrations of acetic acid (0.3 and 0.6%) with and without the use of air injection to agitate the chill water. Again, aerobic plate counts were unaffected by the treatments, but *Enterobacteriaceae* counts were significantly reduced by 0.86 log MPN/mL for the 0.3% acid and 2.35 log MPN/mL for the 0.6% acid solutions. Air injection did not affect reduction of these counts. There was no significant difference in texture or sensory characteristics between the treatments although the skin of the 0.6% acetic acid–treated carcasses was darkened or yellowed (Dickens et al., 1994). Water pockets occurred under the skin of chicken carcasses with air-agitated samples (Dickens and Whittemore, 1994).

In another study, cut-up chicken parts were added to chill water adjusted with organic acids to pH 2.5 (Mountney and O'Malley, 1965). Acetic acid was more effective antimicrobially than adipic, succinic, citric, fumaric, and lactic acids. However, the addition of the acids to these pH levels not only produced off-odor defects, but also produced a hard and leathery skin on the chicken.

A variety of acid treatments has been used to control of microbial load associated with meat products (Doores, 1993). Numerous organic acid combinations, containing from 0.6 to 3.0% acetic acid have been used to rinse beef, pork and lamb carcasses, or meat tissue. Treatments spanned varying amounts of time with concomitant decrease in microbial loads up to 3 logs. There can be adverse effects resulting in discoloration of meat.

Kotula and Thelappurate (1994) compared acetic acid and lactic acid solutions (0.6 or 1.2%) as a dip for rib-eye steaks. Acids were applied for 20 or 120 s at a temperature of 1–2°C and stored for up to 9 days. Although total plate counts were significantly lower for beef dipped in a 1.2% acetic acid solution for 120 s compared with a water-dipped control, only a 0.8 log reduction was achieved at day 1. Similar results were seen with *E. coli* counts for the same parameters with a 0.7 log reduction. For lactic acid–treated samples, total counts and *E. coli* counts were significantly reduced, but only by 0.4 log for the same parameters. A residual effect was noted for the lactic acid–treated tissue in that microbial counts were still significantly decreased compared to the control tissue after 9 days of storage, but this effect was not seen with acetic acid–treated tissue. Acid-treated samples were lighter in color due to leaching of the pigment during immersion, but the shear values, moisture content, and sensory analysis were not affected by acid treatment.

Dickson (1992) contaminated lean and beef tissue surfaces with *Salmonella typhimurium* followed by treatment with 2% acetic acid. *Salmonella typhimurium* was reduced by 0.5 to 0.8 log cfu/cm^2, however this was not significantly different from the controls. *S. typhimurium* reductions were greater when the bacteria were attached to fatty tissue, possibly because this tissue retained less moisture (20%) than lean tissue (75%), and the reduced water activity may have enhanced the antimicrobial effects. It was noted that the use of acetic acid as a rinse for beef tissue did lead to sublethal injury of bacterial cells. An increase in organic material, such as rumen fluid, dirt, or manure, led to less effective reduction of *Salmonella typhimurium*.

Raw produce is now being subjected to acid rinses in some production processes in the hope of reducing high levels of spoilage organisms and, on occasion, foodborne pathogens. Because produce is typically refrigerated and only minimally processed, these foods can be the carriers of psychrotrophic pathogens, such as *Yersinia enterocolitica*. Parsley, inoculated with ~10^7 cfu/g *Y. enterocolitica* was dipped in acetic acid solutions (1, 2, or 5%) or vinegar (30, 40, or 50%) for 15 or 30 min. *Yersinia enterocolitica* was not detectable after exposure to 2 and 5% acetic acid or 40 and 50% vinegar for 15 and 30 min (Karapinar and Gönül, 1992).

The production of new food products oftentimes calls for a reassessment of the level of acids used for pH adjustment of the traditional food. Reduced-calorie mayonnaise (RCM) manufactured with 0.1, 0.3, 0.5, or 0.7% acetic acid in the aqueous phase and adjusted to pH 4.0 with hydrochloric acid was compared to cholesterol-free (egg yolk–free), reduced-calorie mayonnaise (CFM) manufactured with 0.3 and 0.7% acetic acid. Both formulations were inoculated with ~10^8 cfu/mL of an 8-strain mixture of *Salmonella* spp. or a 6-strain cocktail of *Listeria monocytogenes* and held at 23.9°C. *Salmonella* spp. were inactivated within 48 h in both RCM and CFM formulated with 0.7% acetic acid, the commonly used acid in RCM, with reduction to undetectable levels in 1 and 2 weeks with 0.5 and 0.3% acetic acid formulations, respectively. Destruction of *Salmonella* occurred a little more quickly in CFM than RCM most likely because of the absence of egg yolks in CFM and the presence of egg white, which has recognized antimicrobial properties. *Listeria monocytogenes*, although more resistant than *Salmonella*, was reduced by 4 logs in both mayonnaises within 72 h in formulations containing 0.7% acetic acid and reduced to undetectable levels in RCM within 10 days and in CFM within 14 days compared to 2 days for *Salmonella*. It is unlikely that these products would contain such high initial levels of either *Salmonella* spp. or *L. monocytogenes* especially with the use of pasteurized egg products, therefore the standard use of 0.7% acetic acid, pH 4.0 in the aqueous phase of CFM and RCM would still produce a microbiologically sound product (Glass and Doyle, 1991).

Newer products are now being preserved by a number of antimicrobial compounds used in a multiple barrier approach to preservation. Nisin (an approved bacteriocin) and acetic, lactic, citric acids, or glucono-delta lactone at 0.05 and 0.5% concentrations were studied using *Bacillus* spores heated at pasteurization temperatures of 65 and 95°C and recovered at 12, 20, and 30°C (Oscroft et al., 1990). Although each individual acid influenced the heat resistance of *Bacillus* spores differently, acetic acid was the most destructive acid at all combinations of parameters. As the incubation temperature of recovered spores was lowered, germination and outgrowth of *Bacillus* spores were more restricted. Nisin, in combination with organic acids, displayed synergistic effects in increasing the destruction of the organism. It was found that acetic acid was the most effective acid, followed by glucono-delta lactone, lactic, and citric acid when spores were heated at 95°C for 15 min and acetic, lactic, glucono-delta lactone, and citric acids when temperatures were reduced to 65°C for 60 min. Foods, particularly precooked, chilled, ready-to-eat products are especially susceptible to spoilage because of indigenous microbial populations. A combination of these parameters could be used to extend the shelf-life of these products.

Hedberg and Miller (1969) found that exposure of *Pseudomonas aeruginosa* to a 1.0% concentration of acetic acid for 1 h was effective in reducing the population by 99.999%, and they substantiated its use as a decontaminating agent for hospital equipment.

2. Regulatory Use in Foods

Acetic acid is generally recognized as safe when used in accordance with good manufacturing practice (21 CFR 184.1005). It can be used as a curing and pickling agent, a pH control agent, flavor enhancer, flavoring agent and adjuvant, solvent, and vehicle. Acetic acid can be commonly found in such products as marinades, vinaigrettes, mustard, catsup, salad dressings, sauces, canned fruits, and mayonnaise. The acid is also used in pickled products such as pickled sausages and pigs feet where a combination of acidity, pungent vinegar odor, and sour taste are desirable. Acetic acid can be used for a multitude of purposes in meat and poultry (9 CFR 318.7, 381.147) and to alter the acidity of milk and cheese products, canned fruits and vegetables (21 CFR 131.111, 131.125, 131.136, 131.144, 133.123, 133.124, 133.147, 133.169, 133.170, 133.171, 133.173, 133.174, 133.176, 133.178, 133.179, 133.180; part 145; part 155). It also assists in carmelization (21 CFR 73.85) and in the manufacture of hydroxylated lecithin (21 CFR 172.814).

Commercial vinegar contains not less than 4% acetic acid. When acetic acid is coupled with sodium bicarbonate, carbon dioxide is released as a leavening agent. Acetic acid can also be used as a sanitizing agent on food processing equipment and utensils (21 CFR 178.1010) and as a boiler water additive at levels not to exceed good manufacturing practice (21 CFR 173.310, 184.1005). The maximum levels of acetic acid that are recommended for various food product categories are given in Table 2.

3. Toxicology

Acetic acid causes a variety of adverse reactions in humans ranging from allergic-type symptoms such as mouth sores (Tuft and Ettelson, 1956), cold sensitivities (Wiseman and Adler, 1956),and epidermal reactions (Weil and Rogers, 1951) to death (Palmer, 1932). Persons consuming acetic acid in high concentrations at a low pH have exhibited burned lips, stomach, and intestinal mucosa, corroded lung tissue, and subsequent pneumonia resulting from inhalation of vapors (Gerhartz, 1949). Acidosis and renal failure, reduction of clotting efficiency, and interference in blood coagulation are also reported (Paar et al., 1968; Fin'ko, 1969).

In short-term animal studies, acetic acid added to the drinking water of rats at levels up to 0.2% appeared to have no effect on growth, appetite, or fluid consumption. At the 0.5% feeding level, however, rats lost approximately 2.6% of their body weight, though none of them died (Sollman, 1921). At the 10% level, ulcers appeared in the gastric mucosa, whereas a 20% concentration was required for the same effect in cats (Okabe et al.,

Table 2 Maximum Levels of Organic Acids Recommended for Various Food Product Categories in the United States

	Acid (%)							
Food product category	Acetic[a]	Acetate[b], calcium	Acetate[c], sodium	Diacetate[d], sodium	Adipic[e]	Caprylic[f]	Malic[g]	Succinic[h]
Baked goods, baking mixes	0.25	0.2		0.4	0.05	0.13		
Beverages, nonalcoholic					0.005		3.4	
Breakfast cereal			0.007					
Cheese	0.8	0.02				0.04		
Chewing gum	0.5						3.0	
Condiments, relishes	9.0				5.0			0.084
Dairy product analogs	0.8				0.45			
Fats, oils	0.5		0.5	0.1	0.3	0.005		
Frozen dairy desserts					0.0004	0.005		
Gelatin, pudding, filling		0.2			0.55	0.005	0.8	
Grain products and pastas			0.6					
Gravies and sauces	3.0			0.25	0.1			
Hard candy/cough drops			0.15				6.9	
Jams, jellies (commercial)			0.12				2.6	
Meat products	0.6		0.12	0.1	0.3	0.005		0.0061
Processed fruits, juices							3.5	
Snack food			0.6	0.05	1.3	0.016		
Soft candy			0.2	0.1		0.005	3.0	
Soup mixes (commercial)			0.05	0.05				
Sweet sauces		0.15	0.05					

Source: 21 CFR [a]184.1005 ($\leq$0.15% for all other food categories); [b]184.1185 ($\leq$0.0001%); [c]184.1721; [d]184.1754; [e]184.1009 ($\leq$0.02%); [f]184.1025 ($\leq$0.001%); [g]184.1069 ($\leq$0.7%); [h]184.1091.

1971). Histamine was released in the rat gastric mucosa in response to as little as 0.6% concentration of the acid (Johnson, 1968).

Other short-term studies dealing with exposure to low levels of acetic acid vapors have shown dose-response effects. Rats exposed to levels of acid at 0.01 mg/m^3 showed no ill effects, but at levels of 0.2 or 5 mg/m^3 rats displayed muscle imbalance. For comparative purposes, humans do not detect acetic acid until levels reach 0.6 mg/m^3 (Takhirov, 1969).

Acetic acid occurs naturally in plant and animal tissues. In humans, acetic acid is involved in fatty acid and carbohydrate metabolism and is found as acetyl CoA. The Food and Agriculture Organization (FAO), recognizing that acetic acid is a normal constituent of foods, has set no limit on the acceptable daily intake for humans (FAO/WHO, 1973).

B. Acetate Salts

1. Microbial Function

Derivatives of acetic acid in the form of the calcium, sodium, or potassium salts are substituted for acetic acid in certain formulations. These salts, sometimes used interchangeably with the free acid form, have the same antimicrobial properties at the same pH values as confirmed by Hoffman et al. (1939). Growth and aflatoxin production of *Aspergillus parasiticus* was inhibited by a 1.0% concentration of sodium acetate at pH 4.5. At lower acid concentrations of 0.6 or 0.8%, growth and toxin formation were decreased 70 and 90%, respectively (Buchanan and Ayres, 1976).

Sodium acetate (SA) at concentrations of 0.75 and 1% significantly reduced the initial aerobic plate counts of catfish fillets by 0.6 to 0.7 log when stored at 4°C compared to the nontreated fillets. Counts continued to decrease over the shelf-life of the products and shelf-life was increased by 6 days. The combination of 0.3 to 0.7% sodium acetate with 0.4% monopotassium phosphate (MKP) more effectively prolonged shelf-life to 12 days compared to SA alone, although MKP alone had no effect. Sodium acetate alone or SA-MKP treated fillets were not significantly different in odor from fresh controls up to 9 days, but did appear brownish and watery (Kim et al., 1995).

Sodium acetate combined with 10% potassium sorbate and 10% phosphates extended the shelf-life of pork chops to 10 weeks (Mendonca et al., 1989).

2. Regulatory Use in Foods

Calcium and sodium acetates are generally recognized as safe when used in accordance with good manufacturing practice. Calcium acetate can be used as a firming agent, pH control agent, processing aid, sequestrant, texturizer, stabilizer, and thickener (21 CFR 184.1185). It is also approved as a stabilizer in sausage casings, an antirope agent in bakery products, and when the salts migrate from food packaging material into food (21 CFR 181.29). Sodium acetate can be used as a pH control agent, flavoring agent and adjuvant (21 CFR 184.1721), and as a boiler water additive for food-grade steam (21 CFR 173.310). Sodium acetate is limited at levels not to exceed 2 oz/100 lb of artificially flavored fruit jelly, preserves, and jams (21 CFR 150.141, 150.161). The maximum levels of sodium and calcium acetate that are recommended for various food product categories are given in Table 2.

Other forms of acetate have specific use requirements. Ethyl acetate can be used in food in accordance with good manufacturing practice as a solvent in the decaffeination of coffee and tea (21 CFR 173.228) and in inks for marking fruits and vegetables (21

CFR 73.1). α-Tocopherol acetate is generally recognized as safe when used in accordance with good manufacturing practice as a nutrient (21 CFR 182.8892).

Derivatives of acetates can be used as synthetic flavoring substances and adjuvants but should be used in a minimum quantity required to produce the desired effect (21 CFR 182.60). Polyvinyl acetate can be used in a chewing gum base (21 CFR 172.615) and as a diluent in color additive mixtures for marking foods (21 CFR 73.1). Trisodium nitriloacetate can be added to boiler feed water at levels not to exceed 5 ppm, but it cannot be used where steam will be in contact with milk and milk products (21 CFR 173.310).

Acetates are also approved for many nonfood or indirect uses in situations in which they may come into contact with food. Derivatives of acetates can be used as components of adhesives, resinous and polymeric coatings, (21 CFR part 175), paper and paperboard packaging material (21 CFR 181.30), and cotton and cotton fabrics in dry food packaging (21 CFR 182.70).

There is no set limit in the acceptable daily intake for humans for calcium and sodium acetate (FAO/WHO, 1963,1973).

3. Toxicology

Toxicological data on the salts of acetic acid are not as widely available as data for the acid form. A short-term study involving chickens fed diets supplemented with 5.44% sodium acetate (4% acetic acid) showed reduced growth rate, depressed appetite, and increased mortality rate compared with chickens fed control diets. The effect was attributed to the sodium rather than the acid moiety (Waterhouse and Scott, 1962). No toxicological data were available for calcium acetate nor are reports of acute toxicity to sodium acetate in humans available (FAO/WHO, 1963).

Acetate is a normal intermediate metabolite in humans. Acetate turns over rapidly, which accounts for low endogenous reserves of this component in humans (FAO/WHO, 1975). Therefore, no limit has been set on the acceptable daily intake for humans (FAO/WHO, 1973). Ethyl acetate is limited to 0–25 mg/kg per day.

C. Dehydroacetic Acid

1. Microbial Function

Dehydroacetic acid (DHA) is unique in that, because of its high dissociation constant, it is a more effective inhibitory agent in high pH ranges than other acids. DHA was inhibitory to *Saccharomyces cerevisiae*, *Enterobacter aerogenes*, and *Lactobacillus plantarum* at levels of 0.025, 0.3, and 0.1%, respectively, whereas the sodium salt was inhibitory at twice the levels for the same species (Wolf, 1950). Sodium dehydroacetate was more than twice as effective as sodium benzoate against *S. cerevisiae* at pH 5.0 and 25 times more effective against *Penicillium glaucum* and *A. niger* (Banwart, 1989).

2. Regulatory Use in Foods

Dehydroacetic acid and its sodium salt are generally recognized as safe when used in accordance with good manufacturing practice. They may be used as a preservative for cut or peeled squash only. No more than 65 ppm, expressed as dehydroacetic acid, however can remain as a residual in or on the prepared squash (21 CFR 172.130). Other uses of DHA feature its incorporation as a component of adhesives (21 CFR 175.105).

3. Toxicology

In a short-term study of dehydroacetic acid fed to male rats at levels below 300 mg/kg/day, no differences in weight gain were noted between treated and control groups. Beginning with increased intake levels above 300 mg/kg/day, severe weight loss and internal organ damage were seen. Longer-term studies (2 years) on the chronic effects of DHA at levels up to 0.1% showed no differences in growth, mortality, appearance, hematology, or histopathology compared with controls. No toxic effects were seen in monkeys fed DHA at the rate of five times per week for 1 year at levels up to 100 mg/kg, although growth and organ changes were noted when levels were increased to 200 mg/kg. Prolonged dermal contact with 10% DHA dissolved in butyl carbitol acetate or its sodium salt dissolved in water did not irritate rabbit skin (Spencer et al., 1950).

D. Diacetate Salts

1. Microbial Function

Sodium diacetate has been used primarily as a fungistat to retard mold growth in cheese spreads at levels of 0.1–2.0% and in malt syrups at the 0.5% level. Butter wrappers may also be treated with sodium diacetate at the same concentrations to prevent surface mold growth (Chichester and Tanner, 1972). At levels of 0.05–0.4% at pH 3.5, *Aspergillus flavus*, *A. fumigatus*, *A. niger*, and *Penicillium expansum* were inhibited in feeds and silage. The same strains and *Mucor pusillus* were inhibited at levels of 0.15–0.5% at pH 4.5 (Glabe and Maryanski, 1981).

Sodium diacetate has been used in the baking industry to inhibit bread molds and *Bacillus mesentericus* (*subtilis*), the ''rope-forming'' bacterium. Its usefulness lies in its effectiveness against this bacterium while having little effect on *Saccharomyces* spp., the yeast used in breadmaking (Glabe and Maryanski, 1981).

Shelf-life of chickens was extended about 4 days when held at 2°C by using a surface application of 1–3 mg/cm^2 (2–6 g per chicken) sodium diacetate powder. Sodium diacetate dissolved in the surface moisture of the carcass, dissociated to form acetic acid and sodium acetate, and produced a surface pH of about 4.8. *Enterobacteriaceae* counts dropped 1000-fold over 6 days at 2°C. Sensory characteristics of roasted chickens were not affected even at levels up to 8 g per chicken (Moye and Chambers, 1991).

Schlyter et al. (1993) used the multiple barrier concept to limit growth of *Listeria monocytogenes* in turkey mixtures. Slurries of uncured turkey breast, prepared with sodium diacetate (0.1, 0.3, 0.5%), sodium lactate (2.5%), and pediocin (5000 AU/mL) were pasteurized for 10 min at 68°C. Mixtures were inoculated with *L. monocytogenes* at a level of ~4.5 log cfu/mL slurry and incubated at 25 or 4°C and sampled up to 7 and 42 days, respectively. Sodium diacetate was determined to be listericidal over the 7-day period at 25°C only at the 0.5% concentration (0.54 log reduction) in contrast to 0.3 and 0.5% concentrations when slurries were held at 4°C (0.59 log reduction). The initial pH levels were 6.0, 5.5, and 5.2 at 0.1, 0.3, and 0.5% diacetate. Lactate and pediocin enhanced the listericidal properties of diacetate suggesting some type of synergistic effect rather than a reduction in pH.

In an effort to control the growth of *L. monocytogenes* in crab meat, crab was washed with 2 M sodium diacetate. Levels of *L. monocytogenes* were reduced by 2.6 log during storage at 4°C for 6 days compared with 0.8 log reduction using 4 M sodium acetate. A

1 M sodium lactate solution reduced populations within 2 days, followed by an increase of ~0.5 log within 6 days (Degnan et al., 1994).

Sodium diacetate was also shown to be effective in limiting growth of *Listeria monocytogenes*, *Escherichia coli*, *Pseudomonas fluorescens*, *Salmonella enteritidis*, *Shewanella putrefaciens*, *Pseudomonas fragi*, *Yersinia enterocolitica*, and *Enterococcus fecalis*; *Lactobacillus fermentus* and *Staphylococcus aureus* were not inhibited (Shelef and Addala, 1994). Sodium diacetate was more effective than acetic acid in the pH range of 5.0–6.0 with the minimum inhibitory concentrations decreasing with decreasing temperature.

2. Regulatory Use in Foods

Sodium diacetate is generally recognized as safe when used in accordance with good manufacturing practice. Calcium diacetate can be used as a sequestrant in edible oils (21 CFR 182.6197). Sodium diacetate can be used also as antimicrobial agent, pH control agent, flavoring agent, and adjuvant (21 CFR 184.1754). It is approved for use in cheese spreads, malt syrups, salad dressings, pickled products, butter, and wrapping materials. The maximum levels of sodium diacetate recommended for various food product categories are listed in Table 2.

Sodium acetate and diacetate plus malic acid can be used in the manufacture of synthetic dry vinegars (Chichester and Tanner, 1972).

3. Toxicology

No toxicity data are available for the diacetates, although it is assumed that they would be metabolized in a fashion similar to that of acetic acid (FAO/WHO, 1962). No reports of acute toxicity to sodium diacetate in humans are available (FAO/WHO, 1963). There are no reports of this salt occurring naturally in plants or animals (FAO/WHO, 1975), but it is assumed that it is metabolized in the same way as acetic acid (FAO/WHO, 1962). Sodium diacetate is limited unconditionally for humans to a level of 15 mg/kg daily (FAO/WHO, 1973).

E. Adipic Acid

1. Microbial Function

No data have appeared in the literature to suggest that adipic acid has any unusual antimicrobial properties other than a pH effect. Microorganisms susceptible to low pH levels would most likely be affected by this acid.

2. Regulatory Use in Foods

Adipic acid is generally recognized as safe when used in accordance with good manufacturing practice. It can be used as flavoring, leavening, and pH control agents (9 CFR 318.7; 21 CFR 131.111, 131.136, 131.144, 166.110, 184.1009). The maximum levels of adipic acid recommended for various food product categories are shown in Table 2.

Derivatives of adipic acid have found use as components of adhesives, resinous and polymeric coatings (21 CFR part 175), and plasticizers for food packaging material (21 CFR 178.3740, 181.27). Adipic acid or adipates can be used in closures with sealing gaskets for food containers (21 CFR 177.1210), as a component of articles intended for repeated use (21 CFR 177.2420), and in rubber articles not to exceed 30% by weight of the rubber product (21 CFR 177.2600).

Adipic acid imparts a slowly developing smooth, mild taste essential in supplementing foods with delicate flavors. The solubility of adipic acid is four to five times greater at room temperature than that of fumaric acid, another commonly used wetting agent. It is practically nonhygroscopic, a property that has the advantage of prolonging the shelf-life of powdered products (e.g., drinks, cake mixes, instant pudding) in which it is incorporated. It is used in beverages, lozenges, canned vegetables, and in gelatin and pudding products to improve set and to maintain acidities within the pH 2.5–3.0 range. It is an excellent, slow-acting leavening agent that supports the even release of carbon dioxide in baked goods and can be used as an alternative to tartaric acid in commercial baking powders (Gardner, 1966). It improves the melting characteristics and texture of process cheese and cheese spreads and aids in the gelling process of imitation jams and jellies. Diethyl adipate can enhance the whippability of egg whites (Gardner, 1972). It is also known to prevent rancidity in edible oils, although it is not routinely used. When used at a 1% concentration, adipic acid in combination with tetrasodium pyrophosphate, decreases viscosity in milk solids suspensions in products like evaporated milk (Burdock, 1997).

3. Toxicology

In short-term feeding studies on male rats, no negative effects were seen with adipic acid in comparison with the control groups. A dosage of 800 mg/day for 35 days, decreased growth, produced diarrhea, and altered behavior. No clear effect was seen with offspring from treated female rats fed 40 mg/day for 28 days (Lang and Bartsch, 1953). When rats were fed protein-deficient diets and dosed at 400 mg/day, significant inhibition of growth occurred, but only slight anemia was noted compared with controls.

Long-term feeding studies (90 days) involving male rats fed up to 5.0% adipic acid and female rats fed at a 1.0% level displayed no differences from the controls in pathology, tumor rate, or survival rate. Weight gains for treated rats fed 3.0 or 5.0% adipic acid, however, were significantly lower for the treated rats than for the controls (Horn et al., 1957). Sodium adipate fed at the 5.0% level decreased the rate of weight gain in rats (Anonymous, 1943). Teratological studies involving mice, rats, and hamsters fed doses of adipic acid at 263, 288, and 205 mg/kg body weight, respectively, showed no differences from the controls. No effect was seen on maternal or fetal survival rate compared with control groups (Food and Drug Research Laboratories, 1973b).

Adipic acid is limited conditionally for humans to a level of 5 mg/kg body weight daily on a free acid basis (FAO/WHO, 1966).

F. Ascorbic Acid

1. Microbial Function

Although ascorbic acid is included in this chapter by virtue of its acidic nature, it is not used solely as an acidulant nor does it share the same type of antimicrobial properties with other organic acids.

One of the most important antimicrobial advantages of ascorbic acid or its salts is its use in conjunction with nitrite to inhibit *Clostridium botulinum* growth in cured meats. Ashworth and Spencer (1972) added 0.1% sodium ascorbate as a reducing agent for pork slurry and found that its inclusion increased the inhibitory effect of nitrite. *Bacillus cereus* did not grow when inoculated into uncooked sausage in the presence of 500 mg/kg sodium isoascorbate and 200 mg/kg sodium nitrite and incubated for 48 h at 20°C. Sodium isoascorbate alone had no inhibitory properties (Raevuori, 1975).

Tompkin and coworkers (1978a,b,1979) showed that the enhanced antibotulinal effect of nitrate was not due to the antioxidant properties of sodium ascorbate or sodium isoascorbate but rather to a sequestration of metal ions in the meat. The addition of sodium isoascorbate at a level of 0.02% in conjunction with 50 μg/g of sodium nitrite gave an anitbotulinal effect comparable to that of 156 μg/g of sodium nitrite alone. Furthermore, when the product was abused at the time of manufacturer, isoascorbate was more effective as an antibotulinal agent because of its sequestering properties. Its inhibitory effect decreased when it was used at excessive levels and when the product was refrigerated before abuse. In the latter instance, ascorbate actually hastened the depletion of residual nitrite and, hence, led to increased spoilage.

A secondary, but no less important, effect was demonstrated by Fiddler and coworkers (1973a,b). They found that during the curing process of frankfurters, sodium nitrite reacted with dimethylamine to form the nitrosamine, *n*-nitrosodimethylamine (NDMA), a recognized mutagen. The addition of sodium ascorbate or ascorbic acid markedly inhibited the formation of NDMA. It appeared that these two acids competed with the reductants for the NO_2^-, thereby making it less available for the nitrosation of the secondary amine. Mottram et al. (1975) observed similar results with the addition of 0.1% ascorbic acid in the bacon-curing process. Sen and coworkers (1976) were able to reduce the formation of nitrosopyrrolidine by dipping the bacon in 1000 ppm ascorbate just before frying. The substitution of ascorbyl palmitate for sodium ascorbate resulted in a further reduction from the control. As a result of these studies, they suggested the use of ascorbate or its derivatives during manufacture.

During the controversy surrounding the formation of nitrosamines in cured meats and the proposed reduction of nitrite levels in cured products, sorbic acid was viewed favorably as a substitute for nitrite or at least as an adjunct in the process to lower the amount of nitrite needed for safety purposes. Sorbic acid was able to inhibit in vitro the formation of NDMA from its precursors, dimethylamine and nitrite, about as effectively as ascorbic acid (Tanaka et al., 1978). Sorbic acid, however, was weaker than ascorbic acid in preventing the formation of *n*-nitrosomorpholine. Ascorbic acid or cysteine destroyed chemically a pyrrole mutagen isolated as a reaction product between sorbic acid and sodium nitrite as determined by the Ames assay in conjunction with HPLC (Osawa et al., 1980).

Ascorbic acid, sodium nitrite, and sodium nitrate were studied extensively in heated or unheated pork slurries at low and high pH ranges in various concentrations of the reactants. The addition of isoascorbate significantly reduced toxin production of *Clostridium botulinum* and counteracted any increase in toxin production attributed to polyphosphate (Roberts et al., 1981a,b).

Farkas and Andrássy (1993) used 0.1% ascorbic acid to reduce the pH of a minced meat product from pH 6.0 to 5.2–5.3. The vacuum-packaged product was irradiated with 2 kGy dose from a ^{60}Co source and stored for 4 weeks at 0–2°C. To simulate temperature abuse, some packages were transferred to 10°C after 2 weeks of storage at the lower temperature. The irradiation dose resulted in a 2-log reduction independent of the pH effect on total counts and a 4-log reduction of *Enterobacteriaceae*. Lactic acid bacteria predominated during prolonged storage with a further 1-log reduction when ascorbic acid was added.

2. Regulatory Use in Foods

Ascorbic acid and its calcium and sodium salts are generally recognized as safe when used in accordance with good manufacturing practice (21 CFR 182.3013, 182.3189, 182.3731). Ascorbic acid can be used as a dough conditioner (21 CFR 137.105, 137.200), acidulant, and color preservative to retard dark spots on shrimp (21 CFR 161.175); to preserve color in canned apricots, fruit cocktail, peaches, and asparagus (21 CFR 145.115, 145.116, 145.135, 145.136, 145.170, 145.171, 155.200); and to preserve color and flavor in juice and wine (27 CFR 24.246). Ascorbic or erythorbic acid can be used as an antioxidant preservative not to exceed 150 ppm in canned applesauce (21 CFR 145.110). Ascorbic acid can also be used as a preservative in artificially sweetened fruit jelly, preserves, and jams at levels not to exceed 0.1% by weight of finished food (21 CFR 150.141, 150.161).

As an adjunct to curing systems, ascorbate or isoascorbate reduces nitrite, forming dehydroascorbic acid and nitric oxide, which react with myoglobin under reducing conditions to yield nitrosomyoglobin, the usual red pigment associated with cured meat products. Ascorbic acid accelerates color development and promotes color uniformity and stability in poultry products (9 CFR 381.147).

When ascorbic acid is used as an antioxidant, it is preferentially oxidized in place of other substrates. Ascorbyl palmitate, formed by combining ascorbic acid with palmitic acid, produces a fat-soluble antioxidant that supports other antioxidants in polyphase food systems (21 CFR 166.110, 182.3149). In conjunction with other antioxidants, e.g., BHA, BHT, and propylgallate, ascorbic acid can retard fat rancidity in meat emulsion systems such as bologna. The stability of ascorbic acid depends upon pH, copper and iron content, exposure to oxygen, and temperature.

Ascorbic acid and its sodium, calcium, and ferrous salts are used as a nutrient supplement (21 CFR 184.1307a). Where there is a need to preserve the vitamin content as in fortified foods, the D isomer of isoascorbic acid (erythrobate) is incorporated with ascorbic acid. Because it is more rapidly oxidized than ascorbic acid, the D isomer protects ascorbic acid from oxidation, but it has no vitamin activity itself (Gardner, 1966). The addition of an acidulant such as citric or malic acid with ascorbic acid should not only inhibit oxidation but also reduce vitamin loss. In multivitamin preparations, nicotinamide–ascorbic acid complex serves as a source of two vitamins (21 CFR 172.315).

3. Toxicology

Ascorbic acid exists in nature in its reduced form or as L-dehydroascorbic acid in its readily oxidized form, with biological activity confined to the L isomer (Gardner, 1972).

Plants and all mammals except humans, monkeys, and guinea pigs can synthesize ascorbic acid, and these three animals require external sources such as citrus fruits and vegetables in their daily diet.

G. Caprylic Acid

1. Microbial Function

Caprylic acid has been ineffective as an antimicrobial compound against many gram positive and gram negative bacteria or yeasts in concentrations as high as 7.8 µmol/mL (Kabera et al., 1972). When compared with acetic or propionic acid at pH 6.0, the minimum inhibitory concentration (MIC) for caprylic acid was 4 to 23 times lower. Unlike other acids, e.g., acetic or propionic, which increase antimicrobial effectiveness as the pH level is lowered, the MIC of caprylic acid was not substantially lowered for bacteria, yeasts, or molds as the pH level was decreased (Woolford, 1975b).

The MIC for caprylic acid toward *Vibrio parahaemolyticus* in tryptic soy broth was 100 μg/mL (Beuchat, 1980). A comparable vibriostatic concentration was reduced to 40 g/mL when a caprylic acid derivative, sucrose caprylate, was used. This compound does not act synergistically with potassium sorbate or sodium benzoate to inhibit growth of *V. parahaemolyticus* nor was there an additive effect. *Escherichia coli* can be inhibited with only 0.3% caprylic acid when grown in a chemically defined medium. This concentration was far more effective than acetic acid at 1.5% and propionic acid at 1.8%, which did not significantly reduce the population. Caprylic acid was also more inhibitory to *Shigella* spp. than acetic or propionic acids (Nakamura and Zangar, 1968).

2. Regulatory Use in Foods

Caprylic acid is generally recognized as safe when used in accordance with good manufacturing practice. It can be used as a flavoring agent and adjuvant (21 CFR 184.1025). The maximum levels of caprylic acid recommended for various food product categories are listed in Table 2.

Other uses can be found for derivatives of caprylic acid. Cobalt, iron, and manganese caprylates are classified as drying agents when migrating from food packaging material (21 CFR 181.25). Caprylic acid can also be used in sanitizing solutions on food processing equipment or utensils (21 CFR 178.1010).

Caprylic acid imparts a sweatlike odor or a cheesy and buttery taste in foods. It appears naturally in conjunction with dairy-based flavors.

3. Toxicology

Since caprylic acid occurs normally in various foods and is metabolized as a fatty acid, no limit has been set on the acceptable daily intake for humans.

H. Citric Acid

1. Microbial Function

Citric acid and its salts have been investigated for their effects on inhibition of bacteria, yeasts, and molds. Sorrells et al. (1989) reported that on an equal molar basis, citric acid was more inhibitory than lactic acid, followed by acetic acid. Murdock (1950) found that citric acid was particularly inhibitory to flat-sour organisms isolated from tomato juice, and this inhibition appeared to be related to the inherent pH of the product. Citric acid, rather than acetic or lactic acids, was also shown to have an effect on the inhibition of thermophilic bacteria (Fabian and Graham, 1953), *Salmonella typhimurium* (Subramanian and Marth, 1968), lactic acid bacteria such as *Streptococcus agalactiae* (Sinha et al., 1968), and *S. anatum* and *S. oranienburg* (Davis and Barnes, 1952). As little as 0.3% citric acid has been shown to be particularly effective in decreasing native levels of salmonellae on poultry carcasses (Thomson et al., 1967).

Minor and Marth (1970) reported that *Staphylococcus aureus* was inhibited 90 and 99% within a 12-h period at pH levels of 4.7 and 4.5, respectively. When compared with citric acid at the 90 and 99% inhibition levels, acetic acid was inhibitory at much higher pH values of 5.2 and 5.0, respectively.

In addition to the pH-lowering effects of citric acid, a secondary inhibitory effect is attributed to the chelation of essential minerals. Branen and Keenan (1970), in their work with *Lactobacillus casei*, found that sodium citrate in concentrations of 12–18 μmol/mL of broth, was stimulatory to *L. casei*, whereas concentrations greater than 30 μmol/mL were inhibitory. It was believed that inhibition might have been more attributable to chelation of essential metal ions by citrate rather than inherent acid inhibition per se. Chelation was also believed to be an influencing factor in the inhibition of growth of *S. aureus* (Rammell, 1962) and *Arthrobacter pascens* (Imai et al., 1970).

The effects of citric acid were shown to limit growth and toxin production of molds such as *Aspergillus parasiticus* and *A. versicolor*. At levels of 0.75%, the growth of *A. parasiticus* was only slightly affected, however toxin production was inhibited. With *A. versicolor*, growth was inhibited at that level but toxin production was stopped at the 0.25% level (Reiss, 1976).

Conner et al. (1990) found that *L. monocytogenes* was inhibited at pH 5.0 by propionic acid, 4.5 for acetic and lactic acids, and 4.0 for citric and hydrochloric acids when added to trypticase soy yeast extract broth. The effect was temperature dependent in that survival of *L. monocytogenes* decreased to undetectable levels within 1–3 weeks at 30°C whereas at 10°C *L. monocytogenes* was still surviving after 11–12 weeks in media adjusted with acetic, citric, and propionic acids and for 6 weeks in media containing HCl or lactic acid. Temperature dependency also played a role in inhibition of *L. monocytogenes* by citric acid (Cole et al., 1990). Minimum pH values for growth of *L. monocytogenes* were 4.66 at 30°C, 4.36 at 10°C, and 4.19 at 5°C.

Citric acid was also effective in reducing mesophilic and thermophilic spoilage in canned mushrooms (Beelman et al., 1989; Okereke et al., 1990a,b). Mushroom extract, adjusted with citric acid to pH 6.7 (control), 6.22, 5.34, and 4.65—and asparagus adjusted with citric acid or glucono-delta lactone to pH levels of 4.5, 4.8, 5.1, and 5.4 with pH 5.9—were inoculated with *Clostridium sporogenes* PA 3679 spores and heated at 110, 115, 118, and 121°C. The heat resistance of *C. sporogenes* was decreased with decreasing pH levels, but not affected by higher processing temperatures. It was suggested that the mushroom or asparagus extracts may have some sort of protective effect on *C. sporogenes*, thereby leading to its survival regardless of the change in pH (Ocio et al., 1994; Silla-Santos et al., 1992).

Citric acid dips for produce not only retard some spoilage, but may also act as a chelator of metal ions responsible for enzymatic browning reactions. Cabbage and carrots were pretreated with the addition of a reduced calorie dressing or with a dip of 0.2 or 1.0% citric acid for 5 or 30 min. The products were packed with or without modified atmosphere and stored at 4°C for 10–21 days. Samples pretreated with 1% citric acid displayed significantly lower total numbers, coliforms, and lactic acid bacteria counts than samples dipped with the lower concentration of citric acid. Samples containing the reduced calorie dressing were even lower in microbial numbers than acid-dipped vegetables (Eytan et al., 1992).

Houben and Krol (1991) were able to reduce the thermal process of canned liver paste by the addition of citric acid and citrate. Pastes were inoculated with $\sim 10^6$ to 10^7 *Bacillus* spores/g and $\sim 10^4$ *Clostridium sporogenes* spores/g and heated to a F_0 of 0.05, 0.3, and 0.85. Product samples were formulated using 0.14% citric acid (CA), 0.29% CA, 0.31% CA–2.0% trisodium citrate (TCA), and 0.31% CA–2.0% TCA–0.1% potassium sorbate. Microbially stable products were achieved with a F_0 of 0.3 and addition of 0.14% CA (pH 5.69). The addition of potassium sorbate did not enhance stability.

2. Regulatory Use in Foods

Citric acid is generally recognized as safe with no limitations when used in accordance with good manufacturing practice (21 CFR 184.1033). Citric acid can be used as an acidifying agent in dairy products (21 CFR 131.111, 131.136, 131.144, 133.123, 133.124, 133.125, 133.129, 133.147, 133.169, 133.170, 133.171, 133.173, 133.174, 133.176, 133.178, 133.179, 133.180), as an acidulant in canned fruits (21 CFR Part 145), as a pH control agent in canned prune juice (21 CFR 146.187), artificially sweetened fruit jelly, preserves, and jams (21 CFR 150.141, 150.161), canned corn (21 CFR 155.130, 155.131), and a flavor agent in canned tuna (21 CFR 161.190), and dressings (21 CFR 169.140, 169.150). It can also assist in caramelization (21 CFR 73.85) and adjustment of acidity in and stabilization of grape wines and citrus fruit, juice, and wine (27 CFR 24.182, 24.244, 24.246). Citric acid, sodium citrate, and isopropyl citrate can be used for a multitude of purposes in meat and poultry products (9 CFR 318.7, 381.147).

Citric acid is one of the major acidulants in carbonated beverages, imparting a refreshing, tangy citrus flavor (Gardner, 1966) that enhances fruit flavors, such as lemon, lime, orange, and berry. It is used commercially as a synergist for antioxidants and retardant of browning reactions. Citric acid is more hygroscopic than adipic or fumaric acids, which can create storage problems in powdered products. Among its other uses, citric acid is used as a plasticizer and emulsifier to provide texture to process cheese and to enhance melting, to reduce heat-processing requirements by lowering the pH, and to control acidity in pectin or alginate gel formation.

The calcium (21 CFR 184.1195), sodium (21 CFR 184.1751), and potassium (21 CFR 184.1625) salts of citric acid are generally recognized as safe and can be used in nonalcoholic beverages, dairy product analogs, infant formulas, fish, meat, milk, and poultry products at levels not to exceed good manufacturing practice. Calcium citrate can be added to French dressing in an amount not greater than 25% of the weight of the acids in the vinegar (21 CFR 169.115). Sodium and potassium citrate can be used to acidify artificially sweetened fruit jelly, preserves, and jams at a level not to exceed 2 oz/100 lb finished product (21 FR 150.141, 150.161). Sodium, calcium, and potassium citrate can be used as an emulsifier in pasteurized process cheese (21 CFR 133.169, 133.173). Potassium citrate acts as a pH control agent and sequestrant in the treatment of citrus wines (27 CFR 24.246). Calcium citrate can be used as necessary in gelling of these products (21 CFR 150.141, 150.161). Manganese citrate can also be used as a nutrient supplement (21 CFR 184.1449).

Sodium citrate (trisodium citrate) and tartrate are used as chelating agents with phosphate buffers to prepare noncaking, meat-salt mixtures (Gardner, 1966). Sodium citrate improves the whipping properties in cream and prevents feathering of cream and nondairy whiteners. It emulsifies and solubilizes proteins in processed cheese and prevents precipitation of solids during storage in evaporated milk. In dry soups, it improves rehydration that reduces cooking time. It is used as a sequestrant in puddings and a complexing agent for iron, calcium, magnesium, and aluminum. Usage levels range from 0.1% to 0.25%. Potassium citrate may be substituted for sodium citrate where there is a desire to reduce the sodium content. The salts also modify sharp flavors and prevent products from having too much tartness.

Ammonium citrate can be used as a flavor enhancer, pH control agent in nonalcoholic beverages and cheeses with no limitation other than good manufacturing practice (21 CFR 184.1140). Isopropyl citrate can be used as an antioxidant, sequestrant, solvent, and vehicle in margarine, nonalcoholic beverages, and fats and oils with no limitations (21 CFR 184.1386), however, it is limited to 0.02% in margarine (21 CFR 166.110).

Monoisopropyl citrate; mono-, di-, and tristearyl citrate; and triethyl citrate can be used as plasticizers in food ingredients (21 CFR 181.27). Stearyl monoglyceridyl citrate can be used in or with shortenings containing emulsifiers (9 CFR 318.7). Monoglyceride citrate can be used in antioxidant formulations for addition to oils and fats not to exceed 200 ppm of the combined weight of the additive and fat or oil (21 CFR 172.832). Triethyl citrate is used as a flavoring agent, solvent, and vehicle and surface active agent. Stearyl citrate functions as an emulsifier and antioxidant in nonalcoholic beverages, fats, and oils (21 CFR 184.1851) and as a preservative in margarine at a level of 0.15% (21 CFR 166.110).

Ferric ammonium citrate (21 CFR 184.1296), ferric citrate (21 CFR 184.1298), and ferrous citrate (21 CFR 1307c) can be used as nutrient supplements and in infant formulas. An iron–choline–citrate complex may be used as a source of iron in foods for special dietary use (21 CFR 172.370). Iron ammonium citrate can be used as an anticaking agent in salt for human consumption when the level of the additive does not exceed 25 ppm in the finished product (21 CFR 172.430).

Derivatives of citric acid can be used as indirect food additives as a component of adhesives, as plasticizers in resinous and polymeric coatings (21 CFR Part 175), and as a component of paper and paperboard in contact with dry food (21 CFR 176.180).

Citric acid fermentation in conjunction with lactic acid fermentation produces diacetyl and other flavor and aroma compounds in dairy products. Citric acid can be used at levels up to 0.15% by weight of the cultured milk used and as a flavor precursor; sodium citrate can also be used for the same purpose in the equivalent amount to citric acid (21 CFR 131.112, 131.138, 131.146). Sodium citrate may be added prior to the culturing of sour cream as a flavor precursor (21 CFR 134.160).

3. Toxicology

Short-term feeding studies with male rats showed lowered intake of food and subsequent reduced weight gains and slight blood chemistry abnormalities at citric acid levels up to 5.0% (Yokotani et al., 1971; Horn et al., 1957). Rabbits fed 7.7% sodium citrate (5% citric acid) for 60 days showed no abnormal blood or urine chemistries or histological changes (Packman et al., 1963).

Longer-term feeding studies with rats fed quantities up to 5% citric acid showed no significant changes from control groups (Bonting and Jansen, 1956; Horn et al., 1957). Cramer and coworkers (1956) investigated the relationship of the intake of sodium citrate and citric acid in vitamin D–free diets containing low levels of phosphorus but adequate levels of calcium. When vitamin D was not given, the citrate completely prevented the absorption of calcium. There was no effect on the formation of dental caries when rats were fed 1.5, 4.5, or 12.0 g citric acid per kilogram of food in a noncariogenic diet. In the highest level given, however, the molars showed more wear and attrition than did molars of the control group (Dalderup, 1960).

Gruber and Halbeisen (1948) suggested that several of the symptoms produced with high levels of citric acid resembled signs of calcium deficiency. Gomori and Gulyas (1944) gave 320–1200 mg/kg body weight of sodium citrate to dogs, which left the blood calcium levels unchanged but at the same time increased the urine calcium. Gruber and Halbeisen (1948) concluded that when citrate salts were administered the acid part of the molecule was responsible for toxicity rather than the citrate moiety.

Citric acid is found in all animal tissues as an intermediate in the Krebs cycle, and therefore no limit has been set on the acceptable daily intake for humans for either the acid (FAO/WHO, 1973) or salt (FAO/WHO, 1963). The acceptable daily intake for consumption of calcium or sodium citrate is not specified and is 0–100 ppm for monoglyceride citrate.

I. Fumaric Acid

1. Microbial Function

Fumaric acid has been used as an antimicrobial agent (Pilone, 1975) in the prevention of malolactic fermentation in wines and for adding acidity to wines (Ough and Kunkee, 1974).

Esters of fumaric acid, notably the mono- and di- forms of the methyl and ethyl derivatives, have been evaluated at the 0.15 or 0.2% level as a replacement for nitrite in bacon. Huhtanen et al. (1981a) were able to retard swelling in canned bacon inoculated with *Clostridium botulinum* and increase shelf-life from 4–6 days up to 8 weeks by using these ester compounds. Fumaric acid alone under the same conditions was only slightly inhibitory. When added to tomato juice, fungal growth was also inhibited with these compounds at the 0.2% level for methyl and ethyl fumarates, and at the 0.05% level for dimethyl and diethyl fumarates. Esters of fumaric acid have also been found to be effective in the prevention of mold growth on bread (Huhtanen et al., 1981b). Furthermore, dimethyl and diethyl fumarates were effective at a concentration of 20 mg in a 2.6-L dessicator jar in arresting mold development on the surface of bread (Huhtanen and Guy, 1984). Then *n*-monoalkyl fumarates and malates esterified with C_{15}–C_{18} alcohols showed significant antimicrobial activity. Methyl *n*-alkyl fumarates demonstrated lower activity and di-*n*-alkyl fumarates were almost inactive (Dymicky et al., 1987).

Fumaric acid at a concentration of 0.5% was superior to acetic, citric, malic, or tartaric acids in inactivating *Talaromyces flavus* during heating at 80°C. Lethality also increased as the pH was decreased from 5.0 to 2.5 (Beuchat, 1988). *Neosartorya fischeri* also displayed the same behavior with a 2% concentration and heating at 82°C (Conner and Beuchat, 1987).

2. Regulatory Use in Foods

Fumaric acid and its salts are generally recognized as safe when used in accordance with good manufacturing practice. Fumaric acid can be used to acidify milk (21 CFR 131.111, 131.136, 131.144), to enhance flavor in canned fruits and wine (21 CFR Part 145; 27 CFR 24.182, 24.244, 24.246), and in artificially sweetened fruit jelly, preserves, and jams (21 CFR 150.141, 150.161). Fumaric acid can be used to accelerate color fixing in cured, comminuted meat and poultry products at a level of 0.065% of the weight of the product before processing (9 CFR 318.7, 381.147). Fumaric acid is limited at 25 lb/1000 gal wine such that the fumaric acid content of finished wine does not exceed 3.0 g/L (27 CFR 240.1051).

Sodium stearyl fumarate can be used as a dough conditioner in yeast-leavened bakery products at levels not to exceed 0.5% by weight of the amount of the flour used, as a conditioning agent in dehydrated potatoes and processed cereals for cooking at levels not to exceed 1% by weight, in starch- or flour-thickened foods at levels not to exceed 0.2% by weight of food, and as a stabilizing agent in non–yeast-leavened bakery products at levels not to exceed 1% by weight of the flour used (21 CFR 172.826).

Ferrous fumarate is used as a nutrient supplement with no limitations other than good manufacturing practice. It may also be used in infant formula (21 CFR 172.350, 184.1307d).

Derivatives of fumaric acid can be used as components of adhesives (21 CFR 175.105), resinous and polymeric coatings (21 CFR 175.300), paper and paperboard in contact with dry food (21 CFR 176.180) and in single-use and repeated-use food contact surfaces (21 CFR 177.2420).

Fumaric acid imparts a sour taste to food products and is one of the most acidic of the solid acids. Its low solubility in cold water, however, is a limiting factor in its widespread use. Conversely, its low rate of moisture absorption aids in extending the shelf-life of powdered products (Gardner, 1966). A cold-soluble form of fumaric acid can be attained in admixture with 0.3% dioctyl sodium sulfosuccinate and 0.5% calcium carbonate (Gardner, 1972). Fumaric acid is used in fruit drinks, in gelatin desserts, where it increases the gel strength of gelatin, in pie fillings and biscuit doughs, and in wines. In addition to supplying acidity to the product, fumaric acid displays some antioxidant properties in synergistic combinations with BHA and BHT in fat-containing foods (Gardner, 1972). It can eliminate excessive hardening and rubbery texture in alginate-based desserts.

3. Toxicology

In short-term feedings trails, weanling rats fed fumaric acid at the 1.5% level showed increased mortality (Fitzhugh and Nelson, 1947); however, rabbits administered sodium fumarate at the 6.9% level (5.0% fumaric acid) showed no abnormalities in blood and urine chemistries (Packman et al., 1963).

Long-term studies involving the feeding of 1.0% fumaric acid or 1.38% sodium fumarate to rats and 1.0% to guinea pigs showed no effect on growth rate, hemoglobin formation, or skeletal abnormalities. Humans fed fumaric acid at a level of 500 mg/day for 1 year also showed no negative results (Levey et al., 1946).

Fumaric acid is found as an intermediate in the Krebs cycle. It is limited unconditionally for humans to a level of 6 mg/kg body weight daily and conditionally to 6–10 mg/kg daily (FAO/WHO, 1966,1974).

J. Lactic Acid

1. Microbial Function

Lactic acid has been used more extensively for its sensory qualities than its antimicrobial properties, although lactate compounds are showing promise as effective rinses for muscle foods to remove microbial loads. For extensive reviews of previous uses of lactic acid/lactate sprays, the reader is referred to Doores (1993). For an excellent review of the antimicrobial nature of sodium lactate, see Shelef (1994).

Lactic acid sprays have been effective in the range of 1–1.25% for lowering microbial loads on veal carcasses (Smulders and Woolthus, 1983), but 2% concentrations led to discoloration (Woolthus and Smulders, 1985). Rinse solutions combining acetic acid and lactic acid coupled with higher temperatures and packaging under vacuum also provided increased destruction of microbial loads in beef carcasses and extended shelf-life (Acuff et al., 1987; Anderson and Marshall, 1990; Anderson et al., 1992; Smulders and Woolthus, 1985). Hot-boned, subprimal meat was injected with one of three treatments: 0.3 M calcium chloride solution (CAL; pH 5.1), 0.3 M lactic acid solution (LAC; pH 3.0), or a combination (1:1) CAL/LAC (COM; pH 2.5) at a volume equal to 10% of the initial subprimal weight. Subprimal cuts that were chilled (1°C) or injected with deionized water served as the two controls. Significantly higher aerobic plate counts were seen with hot-boned cuts over 10 and 17 days postmortem compared to cold-boned cuts, and aerobic plate counts of meats injected with LAC or COM were significantly lower than CAL. COM improved tenderness in muscle cuts, but control cuts retained more desirable flavor profile (Eilers et al., 1994).

Siragusa and Dickson (1992) studied the antimicrobial effect of lactic acid contained within a calcium alginate gel applied to the surface of beef tissue. The alginate application was effective in reducing populations of *Listeria monocytogenes* on beef tissue, but non-coated tissue became dehydrated during the 7-day study. Acid treatments alone were only 50% as effective as acid/alginate treatments. In a second study, acetic and lactic acids immobilized in calcium alginate gel were applied to beef tissue and held for 7 days at 5°C. Levels of *Listeria monocytogenes* and *Salmonella typhimurium* were not significantly decreased. Lactic acid significantly reduced levels of *L. monocytogenes* compared to lactic acid/gel combination, but there was no difference with acetic acid alone or in combination with the gel. There was no difference in either lactic or acetic acids alone or in combination with calcium alginate gel in reducing populations of *Salmonella typhimurium* or *Escherichia coli*.

A 1% concentration of lactic acid (pH 2.8) at 55°C had little effect on the aerobic plate counts taken of the surface of pork carcasses. *Salmonella* spp. or *Listeria* spp. were not recovered, nor were sensory characteristics affected (Prasai et al., 1992). By increasing the concentration of lactic acid to 2%, numbers of *Salmonella* spp. and *Campylobacter* spp. were reduced immediately and remained lower 24 h after slaughter (Epling et al., 1993). Van Netten et al. (1994) developed a pork skin model system to determine the effect of lactic acid decontamination on microbial counts and changes in predominance of the microflora. Five treatments were delivered to the pork skins: water-treated (control), or exposure at 21°C for 120, 180, 240, or 360 s to 2% lactic acid. Not only did the microbial counts decrease as the time of decontamination increased, but the population of microorganisms shifted. Lactic acid was effective in decreasing the number of *Enterobacteriaceae* as well as other gram negative mesophilic and psychrotrophic spoilage organisms. Elimination of these groups shifted the population to gram positive bacteria and yeasts. Psychrotrophic, gram negative bacteria were the most sensitive to lactic acid, followed by mesophilic *Enterobacteriaceae*, psychrotrophic gram positive bacteria, lactobacilli and yeasts were the least sensitive.

Salts of lactic acid are equally as effective in muscle food systems. Pork liver sausage was formulated with sodium, potassium, or calcium lactate at concentrations of 2, 3, or 4% and inoculated with $\sim 10^4$–10^5 *Listeria monocytogenes* per gram of sausage. Samples were stored up to 50 days at 5°C and up to 10 days at 20°C. Numbers of *L. monocytogenes* increased to $>10^8$ cfu/g after 50 days at 5°C in sausages not containing lactates, but were reduced with 2 or 3% sodium lactate and 2% potassium lactate. Calcium lactate (2%) reduced populations below the inoculum level. At the abuse temperature of 20°C, 4% sodium or potassium lactate was required to reduce *L. monocytogenes*, while only 3% calcium lactate was needed to reduce populations at this same temperature. Addition of 2% sodium chloride enhanced the listericidal activity of the lactates (Weaver and Shelef, 1993).

Poultry carcasses have also been successfully decontaminated with lactic acid using dips or sprays (van der Marel et al., 1988; Zeitoun and Debevere, 1990). Lactic acid (1%) added to both chill water (0–1.1°C, pH 2.8) and scald water (54°C/2 min) reduced the bacterial level of broilers artificially contaminated with *Salmonella typhimurium* to almost nondetectable numbers. Lactic acid added to scald water alone had minimal effect on reducing the numbers of contaminated birds. The number of *Salmonella*-positive birds was also reduced as a function of time of the dip (Izat et al., 1990a). Lactic acid added to broiler chill water resulted in the development of a brown coloration most likely due to blood coagulation. In an effort to reduce carcass discoloration, lower levels of lactic acid (0.25%, pH 2.88 or 0.5%, pH 2.62) were combined with propylene glycol (20%) in chill water. Salmonellae were eliminated from broiler carcasses after a 1-h exposure, however lactic acid promoted discoloration and propylene glycol contributed an objectionable flavor (Izat et al., 1990b).

Sodium lactate at levels up to 2.5% was effective in delaying growth of *Clostridium botulinum* and producing an organoleptically acceptable product (Mass et al., 1989). Sodium and potassium lactate at levels of 5% were effective in delaying growth of *Listeria monocytogenes* in broth systems and could be used interchangeably (Shelef and Yang, 1991).

Gill and Newton (1982) suggested that the inhibitory effect of lactic acid on gram negative psychrotrophs resulted from a decrease in pH rather than the effect of the undissociated acid. Dubos (1950) concluded that lactic acid had primarily a bacteriostatic effect against *Mycobacterium tuberculosis*, which increased as the pH was decreased. This inhibition was affected by the addition of serum or other components to the medium, supporting the premise that constituents in the medium can affect the antimicrobial activity of the organic acid.

Cold pack cheese, a food product normally manufactured with lactic acid, was compared with the same product manufactured with acetic acid. When the product was inoculated with *Salmonella typhimurium*, it was found that neither acid increased the destruction of the organisms (Park et al., 1970).

Lactic acid was found to be four times more effective than malic, citric, propionic, and acetic acids at inhibiting growth of *Bacillus coagulans* in tomato juice, the causative agent in flat-sour spoilage (Rice and Pederson, 1954). The minimum inhibitory concentration (MIC) for spore-forming bacteria at pH 6.0 ranged from 31 to 63 mM compared with greater than 250 mM for yeasts and molds. When the pH level was lowered to 5.0, the MIC for spore-forming bacteria was reduced to 6–8 mM, however the MIC needed for yeasts and molds did not change (Woolford, 1975a). Minor and Marth (1970) found that *Staphylococcus aureus* was inhibited 90 and 99% within a 12-h period at pH levels of 4.9 and 4.6, respectively.

Sodium lactate is purported to lower water activity, and thereby some of its antimicrobial activity may be attributed to a combined acid/low water activity synergy. The inhibiting effect appears to be temperature dependent with lower temperatures being more antibacterial. Sodium lactate (5% w/w) was effective in retarding growth of *Streptococcus faecalis*, *Staphylococcus aureus*, and *Salmonella typhimurium* in a model medium at pH 7.25 and a water activity of 0.958, compared to media supplemented with sodium chloride to the same water activity level. This effect is also noted in MRS medium at 30°C for the lactic acid bacteria, *Leuconostoc mesenteroides* and *Lactobacillus casei* var. *rhamnosus*. Lactate had little effect on *E. coli* (de Wit and Rombouts, 1994).

2. Regulatory Use in Foods

Lactic acid is generally recognized as safe when used in accordance with good manufacturing practice. It can be used as antimicrobial and pH control agents, curing and pickling agent, flavor enhancer, flavoring agent and adjuvant, solvent, and vehicle (21 CFR 184.1061). Lactic acid can be used as an acidifying agent in dairy products, except in infant foods and formulas (21 CFR 131.111, 131.136, 131.144, 133.123, 133.124, 133.125, 133.129, 133.147, 133.169, 133.170, 133.171, 133.173, 133.174, 133.176, 133.178, 133.179, 133.180) and in grape wines (27 CFR 24.182). It also aids in the emulsification of hydroxylated lecithin (21 CFR 172.814). Lactic acid and its calcium, potassium, and sodium salts can be used for a multitude of purposes in meat and poultry products (21 CFR 318.7, 381.47).

Calcium lactate is generally recognized as safe with no limitation when used in accordance with good manufacturing practice. It can be used as a firming agent, flavor enhancer, flavoring agent or adjuvant, leavening agent, nutrient supplement, and thickener (21 CFR 184.1207). The solubility of calcium lactate can vary with the form of the additive, i.e., monohydrate, trihydrate, or pentahydrate (Igoe and Hui, 1996). As a stabilizing agent, calcium lactate combines with pectin to produce the less soluble form, calcium pectate, thus maintaining the structural integrity of fruits and vegetables processed by canning. Calcium lactate can be used in artificially sweetened fruit jelly, preserves, and jams as a gelling agent (21 CFR 150.141, 150.161). Calcium lactate also serves to maintain foam volume by increasing protein extensibility in meringues, angel food cakes, and whipped toppings.

Potassium lactate can be used as a flavor enhancer, flavoring agent or adjuvant, humectant and pH control agent at levels not to exceed good manufacturing practice (21 CFR 184.1639).

Sodium lactate is used in sponge cake and Swiss roll to produce a tender crumb and to reduce staling. It is a protein plasticizer in biscuits and replaces sodium chloride in frankfurter-type sausages and as a dehydrating salt or humectant in uncured hams.

Ferrous lactate is generally recognized as safe for when used in accordance with good manufacturing practice. It can be used as a nutrient supplement (21 CFR 184.1311). It is also used as a color fixative for ripe olives.

Lactic acid has a mild, creamy odor with a pleasant, sour taste. It is useful in ensuring clarity in the brine of Spanish olives (Gardner, 1972). In salad dressings and marinades, the synergistic combination of lactic and acetic acids is especially inhibitory to heterofermentative lactobacilli, more so than either acid separately.

3. Toxicology

Lactic acid has been found to be lethal in infants who consumed milk that had been acidified with various quantities of lactic acid. Hemorrhaging, gangrenous gastritis, and esophageal burning resulted in these cases (Pitkin, 1935; Trainer et al., 1945; Young and Smith, 1944). With this particular acid, deleterious effects are associated with the use of isomeric forms. This appears to be especially important in the nutrition of premature infants. Infant groups fed milk acidified with the D(−) or DL form of lactic acid developed acidosis, lost weight, became dehydrated, and vomited. Because of this finding, the L(+) form is recommended for use in feeding premature infants (Ballabriga et al., 1970). This recommendation is also supported by the Food and Agriculture Organization of the United Nations (FAO/WHO, 1973,1974). Seymour et al. (1954) found that acetic acid substituted for lactic acid as an acidulant in infant formulas could be an effective preservative with none of the deleterious side effects.

A cariogenic diet supplemented with lactic acid in the drinking water (40 mg/100 mL) or food (45.6 mg/100 g) did not affect the growth rate of hamsters. Although enamel decalcification was apparent, no difference was noted in the extent of carious lesions among the groups (Granados et al., 1949).

Since lactic acid is a normal constituent of food and an intermediate metabolite in humans, there is no set limit on the acceptable daily intake for humans (FAO/WHO, 1973). There is also no ADI for calcium lactate. Neither D (−) nor DL lactic acid should be used in infant food (FAO/WHO, 1973).

K. Malic Acid

1. Microbial Function

No unusual antimicrobial action is attributed to malic acid other than that associated with pH effects (Banwart, 1989). Many of the studies comparing the efficacy of using one organic acid over another have tested malic acid. In studies by Nunheimer and Fabian (1940) on the inhibitory effectiveness of organic acids against *Staphylococcus aureus*, malic acid was found to be inhibitory at pH 3.98, a level between acetic acid at pH 4.59 and tartaric acid at pH 2.94.

2. Regulatory Use in Foods

Malic acid is generally recognized as safe when used in accordance with good manufacturing practice (21 CFR 184.1069). It can be used as a flavor enhancer in canned fruits (21 CFR Part 145); flavoring agent and adjuvant and pH control agent (21 CFR 131.111, 131.136, 131.144, 184.1069), and to increase effectiveness of antioxidants (21 CFR 381.147). Malic acid is limited to 25% of the weight of the acid of the vinegar in French dressings (21 CFR 169.115). It can be used as an acidulant in artificially sweetened fruit jelly, preserves, jams (21 CFR 150.141, 150.161), and grape wines and apples, apple juice, and apple wine (27 CFR 24.182, 24.246). The maximum levels of malic acid recommended for various food product categories are given in Table 2.

Malic acid imparts a smooth, tart taste with no burst of flavor and is commonly used in soft drinks, dry mix beverages, ices, hard candies, baked goods, canned foods, puddings, jams, jellies, and preserves. Less malic acid than citric acid is required to impart the same degree of acidity. It provides excellent antibrowning properties in fruits and acts as a synergist with antioxidants (Gardner, 1972). The Internal Revenue Service has specified the addition of malic acid to volatile fruit flavor concentrates containing 6–15% alcohol to render these products nonpotable during transportation from manufacturer to winery (Gardner, 1966).

3. Toxicology

Long-term studies involving rats fed up to 5000 ppm malic acid in their diet showed no change in growth rate, histology, or blood and urine chemistries. At levels of 50,000 ppm, food consumption and growth were significantly decreased (Hazelton Laboratories, 1971a); however, dogs fed the same levels showed no adverse effects using the same protocol (Hazelton Laboratories, 1971b). At levels of 1000 and 10,000 ppm, rats showed no differences in mating and reproduction as compared with the controls (Hazelton Laboratories, 1970).

The L(+) isomer of malic acid occurs naturally in foods, while the DL form does not and is found as an intermediate in the Krebs cycle. DL-Malic acid is limited to 100 mg/kg body weight as an acceptable daily intake for humans (FAO/WHO, 1967).

L. Propionic Acid

1. Microbial Function

Most research using propionic acid and its salts as microbial inhibitors has been directed against fungi although some bacterial species have also been studied.

The addition of propionic acid at levels of 8–12% retarded mold growth on the surface of cheese and butter (Deane and Downs, 1951; Ingle, 1940). In contrast, a 5% calcium propionate solution adjusted to pH 5.5 with lactic acid was found to be as effective as a 10% unacidified solution in preventing surface mold on butter (Olson and Macy, 1940,1945), in which a decrease of pH from 5.0 to 4.0 increased inhibition (Heseltine, 1952b; Woolford, 1975b). Fungistatic pH effects were also noted with yeasts and molds.

The interaction of concentration and pH levels was also demonstrated using *B. mesentericus (subtilis)*, which causes rope formation in bread. At pH 5.8, 0.188% calcium propionate prevented rope formation, whereas at pH 5.6 only 0.156% was needed (O'Leary and Kralovec, 1941). *Salmonella* species were inhibited from initiating growth at pH 5.5 for propionic, 5.4 for acetic, 5.1 for adipic, 4.6 for succinic, 4.4 for lactic, 4.3 for fumaric and malic, 4.1 for tartaric, and 4.05 for citric acids, indicating a particular sensitivity of this organism to propionic acid (Chung and Goepfert, 1970).

The sodium salts of propionic acid have also shown antimicrobial properties. Sodium propionate retarded growth of *Staphylococcus aureus*, *Sarcina lutea*, *Proteus vulgaris*, *Lactobacillus plantarum*, *Torula* spp., and *Saccharomyces ellipsoideus* for 5 days at levels of 0.1–5.0% when incorporated into food products (Wolford and Anderson, 1945). Sodium propionate inhibition of *E. coli* was reversed, however, by addition of β-alanine to the medium, an antimicrobial phenomenon not found in *B. subtilis, Pseudomonas* spp., *Aspergillus cravats*, or *Trichophyton mentagrophytes* (Wright and Skeggs, 1946).

Tryptose broth containing sodium propionate at concentrations of 0.05, 0.15, and 0.30% was adjusted to pH 5.0 or 5.6 with acetic, tartaric, lactic, or citric acids. Broths were inoculated with *Listeria monocytogenes* and incubated at 13 and 35°C. As would be expected, growth was inhibited to a greater extent with the higher concentration of propionate (El-Shenawy and Marth, 1992; Ryser and Marth, 1988).

2. Regulatory Use in Foods

Propionic acid and its calcium and sodium salts are generally recognized as safe when used in accordance with good manufacturing practice (21 CFR 184.1081). The salts are classified as antimycotics when migrating from food packaging material and as an antimicrobial agent and flavoring agent (21 CFR 181.23). Propionic acid is limited at 1.0%, 0.32% in flour in white bread rolls, 0.38% in whole wheat, and 0.3% in cheese products.

Calcium propionate can be used in baked goods, cheeses, confections and frostings, gelatins, puddings and fillings, jams and jellies with no limitation other than good manufacturing practice (21 CFR 184.1221). It can be used as an antimicrobial agent with no activity against yeasts and limited activity against bacteria, yet is effective against molds. Its activity is optimal at pH 5.0 with decreasing effectiveness above pH 6.0. Calcium and sodium propionates can be used as a preservative at a level not to exceed 0.1% by weight of the finished product in artificially sweetened fruit jelly, preserves, and jams.

Calcium propionate is used extensively as a mold and rope inhibitor in bread products. Mold spores are normally killed during the baking process but can cause problems with postbaking contamination. Mold growth can occur under the wrapper at the proper temperature in a humid environment. In addition to its fungistatic property when used for this purpose, the calcium moiety provides nutritional enrichment to the product. The sodium salt form may be preferred if it is found that the calcium salt interferes with the chemical leavening process in the bread product (Chichester and Tanner, 1972).

Sodium propionate can be used as an antimicrobial and flavoring agent (21 CFR 184.1784) in baked goods, nonalcoholic beverages, cheeses, confections and frostings, gelatins, puddings and fillings, jams and jellies, meat products, and soft candy. It is more soluble than the calcium salt.

Derivatives of propionates can be used as flavoring agents in foods (21 CFR 172.515), as components of resinous and polymeric coatings (21 CFR 175.300), and in closures with sealing gaskets for food containers (21 CFR 177.1210). Cellulose acetate propionate is approved for use as an adhesive (21 CFR 175.105) and in hot melt strippable food coatings (21 CFR 175.230).

Sulfur derivatives of propionate can be used as antioxidants when migrating from food packaging material at levels not to exceed 0.005% (21 CFR 181.24). Thiodipropionic acid and dilauryl thiopropionate are generally recognized as safe for use in food when used in accordance with good manufacturing practice such that the total content of antioxidant does not exceed 0.02% of the fat or oil content of the food (21 CFR 182.3109, 182.3280).

Propionic acid has a pungent, sour odor and taste and is limited in use by its sensory properties (Gardner, 1972). They are naturally developed as a byproduct in the fermentation of milk to produce Swiss cheese in concentrations as high as 1%.

3. Toxicology

In short-term studies using calcium and sodium salts of propionic acid, no differences in growth or mortality, hematology patterns, or histopathology were seen (Graham et al., 1954; Graham and Grice, 1955; Hara, 1965; Harshbarger, 1942). Sodium propionate administered daily to a human male in oral form in concentrations up to 6 g/kg body weight had no toxic effects but was reported to have some local antihistaminic activity (Heseltine, 1952a). In one study, however, albino rats fed propionic acid at concentrations of 50 cm^3/kg rice for 100 days showed umbilicate or warty lesions on the stomach (Mori, 1953).

Propionic acid is a normal constituent of food and an intermediary metabolite in humans and ruminants, and therefore no limit has been set for the acid or salt forms of propionate on the acceptable daily intake for humans (FAO/WHO, 1973).

M. Succinic Acid

1. Microbial Function

Like malic acid, succinic acid has not been used extensively in food products for its antimicrobial activity. Cox et al. (1974) have found that succinic acid at levels of 3 or 5% at 60°C are as effective as acetic acid in decreasing the microbial load on chicken carcasses. Despite the antimicrobial nature of the compound, however, this concentration adversely affected the appearance of the product. Thomson et al. (1967) preinoculated *Salmonella typhimurium* on chicken carcasses and sprayed them with 1% succinic acid to achieve a 60% reduction of numbers.

2. Regulatory Use in Foods

Succinic acid is generally recognized as safe when used in accordance with good manufacturing practice (21 CFR 184.1091). It can be used as both a flavor enhancer and a pH control agent in acidified milk (21 CFR 131.111, 131.136, 131.144). The maximum levels of succinic acid recommended for various food product categories are listed in Table 2.

Derivatives of succinic acid can be used as flavoring agents or in combination with paraffin as a protective coating for selected fruits and vegetables (21 CFR 172.275). Succinylated gelatin can be used in the manufacture of capsules for flavoring substances at levels not to exceed 4.5–5.5% of the content of the gelatin and 15% of the total weight of capsule and flavoring compound (21 CFR 172.230). Succistearin can be used as an emulsifier in or with shortenings and edible oils intended for use in cakes, cake mixes, fillings, icings, pastries, and toppings in accordance with good manufacturing practice (21 CFR 172.765).

Dioctylsodium sulfosuccinate can be used as a wetting agent in fumaric acid–acidulated foods, processing aid in sugar manufacturing, solubilizing agent on gums and hydrophilic colloids, a diluent in color additive mixtures (21 CFR 73.1) or emulsifying agent for cocoa manufacturing (21 CFR 163.117, 172.520, 172.810), with stabilizers in evaporated milk, skim milk, and cream cheese (21 CFR 131.130, 131.132, 133.133, 133.134, 133.162, 133.178) and in sanitizing solutions (21 CFR 178.1010).

Succinylated monoglycerides can be used as emulsifiers in liquid and plastic shortenings and as dough conditioners in bread not to exceed 0.5% by weight of the flour (21 CFR 172.830).

Succinic anhydride is used as a dehydrating agent for the removal of moisture from foods, and it imparts a stability to dry mixes. It can be used as a starch modifier at a level up to 4%. It also aids in the controlled release of carbon dioxide during leaving (Gardner, 1966).

Derivatives of succinic acid can be used as components of paper and paperboard in contact with aqueous and fatty foods (21 CFR 176.170) or dry foods (21 CFR 176.180). They can also be used as components of single and repeated use food contact surfaces (21 CFR 177.1200) and in closures with sealing gaskets for food containers (21 CFR 177.1210).

Succinic acid is odorless and has a sour, acid taste with a slow taste buildup. It is a nonhygroscopic acidulant that can extend shelf-life of dessert powders without damaging flavors. It readily combines with proteins in modifying the plasticity of doughs and aids in the production of edible fats with desired thermal properties (Gardner, 1972).

3. Toxicology

Short-term studies involving rats injected subcutaneously with 0.5 mg succinic acid daily for 60 days with gradually increasing doses up to 2.0 mg/day at 4 weeks showed no abnormal changes when compared with controls. No abnormalities of development were noted in chick embryos when comparable dosages were administered into the air sac (Dye et al., 1944).

Succinic acid occurs naturally in small amounts in fruits and as an intermediate in the Krebs cycle. Therefore, no limit has been set on the acceptable daily intake for humans.

N. Tartaric Acid

1. Microbial Function

The antimicrobial use of tartaric acid is limited.

2. Regulatory Use in Foods

Tartaric acid and its salts are generally recognized as safe when used in accordance with good manufacturing practice (21 CFR 184.1099). Tartaric acid can be used as a firming, flavoring, and pH control agent, flavor enhancer, and humectant. Tartaric acid is an acidifying agent in acidified milks (21 CFR 131.111, 131.136, 131.144), artificially sweetened fruit jelly, preserves, and jams (21 CFR 150.141, 150.161), and grape wines (27 CFR 24.182). Tartaric acid can be used for a multitude of purposes in meat and poultry products (9 CFR 318.7, 381.47).

Potassium acid tartrate (i.e., potassium bitartrate, cream of tartar) can be used as an anticaking, leavening, antimicrobial, and pH control agent, formulation and processing aid, humectant, stabilizer and thickener, and surface-active agent (21 CFR 184.1077). There are no limitations for use other than good manufacturing practice, and it is often used in baked goods, confections, frostings, gelatins, puddings, hard and soft candies, jams, and jellies. It acts as a chemical leavening agent in baking powder that functions to release carbon dioxide more quickly as the temperature rises.

Sodium tartrate and sodium potassium tartrate can be used as an emulsifier and a pH control agent in cheeses, fats, and oils (21 CFR 133.169, 133.173, 184.1801, 184.1804). Sodium tartrate is limited at a level not to exceed 3.0% of cheese products. Sodium tartrate, sodium potassium tartrate and sodium acid tartrate are limited at 2 oz/100 lb of finished product or artificially sweetened fruit jelly, preserves, and jams (21 CFR 150.141, 150.161).

Choline bitartrate can be used as a nutrient (21 CFR 182.8250).

Diacetyl tartaric acid esters of mono- and diglycerides of edible fats or oils or edible fat-forming fatty acids can be used as an emulsifier (9 CFR 318.7; 21 CFR 184.1101) at a level of 0.5% in oleomargarine. It can be used in baked goods, baking mixes, nonalcoholic beverages, confections and frostings, dairy product analogs, fats, and oils.

Tartaric acid has a strong, tart taste and is the most soluble of all acidulants. It is widely used in grape- and lime-flavored beverages as an acidulant and as a flavor enhancer. Tartaric acid salts are used to control the degree of acidity in soft drinks (Gardner, 1972). Tartaric acid acts synergistically with antioxidants to prevent rancidity, and it prevents discoloration in cheese (Gardner, 1972).

3. Toxicology

In short-term studies, rabbits were fed 7.7% sodium tartrate (5.0% acid), and in long-term studies 21-day-old weanling rats were fed tartaric acid at levels up to 1.2%. Results showed no significant changes in any parameters when compared with control rats (Fitzhugh and Nelson, 1947; Horn et al., 1957; Packman et al., 1963). Subcutaneous administration of 0.25–1.0 g of D-tartaric acid as the sodium salt to rabbits produced blood chemistry changes. In teratogenic testing no abnormalities were observed in mice, rats, hamsters, or rabbits at levels of 274, 181, 225, and 215 mg/kg body weight, respectively (Food and Drug Research Laboratories, Inc., 1973a).

Tartaric acid as the L(+) form is limited unconditionally for humans up to a level of 30 mg/kg body weight (FAO/WHO, 1973).

REFERENCES

Acuff, G. R., Vanderzant, C., Savell, J. W., Jones, D. K., Griffin, D. B., Ehlers, J. G. 1987. Effect of acid decontamination of beef subprimal cuts on the microbiological and sensory characteristics of steaks. *Meat Sci.* 19:217–226.

Adams, M. R., Hall, C. J. 1988. Growth inhibition of food-borne pathogens by lactic and acetic acids and their mixtures. *Intl. J. Food Sci. Technol.* 23:287–292.

Adams, M. R., Little, C. L., Easter, M. C. 1991. Modelling the effect of pH, acidulant and temperature on the growth rate of *Yersinia enterocolitica*. *J. Appl. Bacteriol.* 71:65–71.
Ahamad, N., Marth, E. H. 1989. Behavior of *Listeria monocytogenes* at 7, 13, 21, and 35°C in tryptose broth acidified with acetic, citric, or lactic acid. *J. Food Prot.* 52:688–695.
Anderson, M. E., Marshall, R. T. 1990. Reducing microbial populations on beef tissues: concentration and temperature of an acid mixture. *J. Food Sci.* 55:903–905.
Anderson, M. E., Marshall, R. T., Dickson, J. S. 1992. Efficacies of acetic, lactic and two mixed acids in reducing numbers of bacteria on surfaces of lean meat. *J. Food Safety* 12:139–147.
Anonymous. 1943. The toxicity of adipic acid. Monograph on adipic acid, NAS/NRC Questionnaire. Informatics, Inc., Rockville, MD.
Ash, M., Ash, I. 1995. *Handbook of Food Additives*. Gower Publishing, Hampshire, England.
Ashworth, J., Spencer, R. 1972. The perigo effect in pork. *J. Food Technol.* 7:111–124.
Ballabriga, A., Conde, C., Gallart-Catala, A. 1970. Metabolic response of prematures to milk formulas with different lactic acid isomers or citric acid. *Helv. Paediatr. Acta* 25:25–34.
Banwart, G. J. 1989. *Basic Food Microbiology*, 2nd ed. Van Nostrand–Reinhold, New York.
Beelman, R. B., Witowski, M. E., Doores, S., Kilara, A., Kuhn, G. D. 1989. Acidification process technology to control thermophilic spoilage in canned mushrooms. *J. Food Prot.* 52:178–183.
Bell, M. F., Marshall, R. T., Anderson, M. E. 1986. Microbiological and sensory tests of beef treated with acetic and formic acids. *J. Food Prot.* 49:207–210.
Bergmeyer, H. U. 1983. *Methods of Enzymatic Analysis*, 3rd ed. Verlag Chemie, Deerfield Beach, FL.
Beuchat, L. 1980. Comparison of anti-*Vibrio* activities of potassium sorbate, sodium benzoate and glycerol and sucrose esters of fatty acids. *Appl. Environ. Microbiol.* 39:1178–1182.
Beuchat, L. R. 1988. Influence of organic acids on heat resistance characteristics of *Talaromyces flavus* ascospores. *Intl. J. Food Microbiol.* 6:97–105.
Bigwood, E. J., Fondu, M., Art, G., van Gindertael, W. (Eds.). 1975. *Food Additives Tables*. Elsevier Scientific, New York.
Bonting, S. L., Jansen, B. C. P. 1956. The effect of a prolonged intake of phosphoric acid and citric acid in rats. *Voeding* 17:137–148.
Bouzas, J., Kantt, C. A., Bodyfelt, F., Torres, J. A. 1991. Simultaneous determination of sugars and organic acids in cheddar cheese by high-performance liquid chromatography. *J. Food Sci.* 56: 276–278.
Branen, A. L., Keenan, T. W. 1970. Growth stimulation of *Lactobacillus casei* by sodium citrate. *J. Dairy Sci.* 53:593–597.
Brocklehurst, T. F., Lund, B. M. 1990. The influence of pH, temperature and organic acids on the initiation of growth of *Yersinia enterocolitica*. *J. Appl. Bacteriol.* 69:390–397.
Buchanan, R. L., Jr., Ayres, J. C. 1976. Effect of sodium acetate on growth and aflatoxin production by *Aspergillus parasiticus* NRRL 2999. *J. Food Sci.* 41:128–132.
Budavari, S. 1996. *The Merck Index*, 12th ed. Merck & Co., White House Station, NJ.
Burdock, G. A. 1997. *Encyclopedia of Food and Color Additives*. CRC Press, Boca Raton, FL.
CFR. 1997. Code of Federal Regulations Title 9, Title 21, Title 27, Office of the Federal Register, U.S. Government Printing Office, Washington, DC.
Chichester, D. F., Tanner, F. W., Jr. 1972. Antimicrobial food additives. In: *Handbook of Food Additives*, 2nd ed. Vol. 1, Furia, T. E. (Ed.). CRC Press, Cleveland, OH, pp. 115–184.
Chung, K. C., Goepfert, J. M. 1970. Growth of *Salmonella* at low pH. *J. Food. Sci.* 35:326–328.
Cole, M. B., Jones, M. V., Holyoak, C. 1990. The effect of pH, salt concentration and temperature on the survival and growth of *Listeria monocytogenes*. *J. Appl. Bacteriol.* 69:63–72.
Conner, D. E., Beuchat, L. R. 1987. Heat resistance of ascospores of *Neosartorya fischeri* as affected by sporulation and heating medium. *Intl. J. Food Microbiol.* 4:303–312.
Conner, D. E., Scott, V. N., Bernard, D. T. 1990. Growth, inhibition, and survival of *Listeria monocytogenes* as affected by acidic conditions. *J. Food Sci.* 53:652–655.

Cox, N. A., Mercuri, A. J., Juven, B. J., Thomson, J. E., Chew, V. 1974. Evaluation of succinic acid and heat to improve the microbial quality of poultry meat. *J. Food Sci.* 39:985–987.

Cramer, J. W., Porrata-Doria, E. I., Steenbock, H. 1956. A rachitogenic and growth-promoting effect of citrate. *Arch. Biochem. Biophys.* 60:58–63.

Cruess, W. V., Irish, J. H. 1932. Further observations on the relation of pH value to toxicity of preservatives to microorganisms. *J. Bacteriol.* 23:163–166.

Cunniff, P. (Ed.) 1995. *Official Methods of Analysis of the Association of Official Analytical Chemists International*, 16th ed. AOAC International, Arlington, VA.

Dalderup, L. M. 1960. The effects of citric and phosphoric acids on the teeth. *J. Dent. Res.* 39: 420–421.

Davis, F., Barnes, L. A. 1952. Suppression of growth of *Salmonella anatum* and *Salmonella oranienburg* by concentration variation of energy sources in a synthetic basal medium. *J. Bacteriol.* 63:33–38.

Deane, D., Downs, P. A. 1951. Flexible wrappers for cheddar cheese. *J. Dairy Sci.* 34:767–775.

Debevere, J. M. 1988. Effect of buffered acidulant systems on the survival of some food poisoning bacteria in medium acid media. *Food Microbiol.* 5:135–139.

Degnan, A. J., Kaspar, C. W., Otwell, W. S., Tamplin, M. L., Luchansky, J. B. 1994. Evaluation of lactic acid bacterium fermentation products and food-grade chemicals to control *Listeria monocytogenes* in blue crab (*Callinectes sapidus*) meat. *Appl. Environ. Microbiol.* 60:3198–3203.

de Wit, J. C., Rombouts, F. M. 1990. Antimicrobial activity of sodium lactate. *Food Microbiol.* 7: 113–120.

Dickens, J. A., Whittemore, A. D. 1994. The effect of acetic acid and air injection on appearance, moisture pick-up, microbiological quality, and *Salmonella* incidence on processed poultry carcasses. *Poult. Sci.* 73:582–586.

Dickens, J. A., Lyon, B. G., Whittemore, A. D., Lyon, C. E. 1994. The effect of an acetic acid dip on carcass appearance, microbiological quality, and cooked breast meat texture and flavor. *Poult. Sci.* 73:576–581.

Dickson, J. S. 1992. Acetic acid action on beef tissue surfaces contaminated with *Salmonella typhimurium*. *J. Food Sci.* 57:297–301.

Doores, S. 1983. Organic acids. In: *Antimicrobials in Foods*, Branen, A. L., Davidson, P. M. (Eds.). Marcel Dekker, New York, pp. 75–108.

Doores, S. 1993. Organic acids. In: *Antimicrobials in Foods*, 2nd ed., Branen, A. L., Davidson, P. M. (Eds.). Marcel Dekker, New York, pp. 95–136.

Dubos, R. J. 1950. The effect of organic acids on mammalian tubercle bacilli. *J. Exp. Med.* 92: 319–332.

Dye, W. S., Overholser, M. D., Vinson, C. G. 1944. Injections of certain plant growth substances in rat and chick embryos. *Growth* 8:1–11.

Dymicky, M., Bencivengo, M., Buchanan, R. L., Smith, J. L. 1987. Inhibition of *Clostridium botulinum* 62A by fumarates and maleates and relationship of activity to some physicochemical constants. *Appl. Environ. Microbiol.* 53:110–113.

Eilers, J. D., Morgan, J. B., Martin, A. M., Miller, R. K., Hale, D. S., Acuff, G. R., Savell, J. W. 1994. Evaluation of calcium chloride and lactic acid injection on chemical, microbiological and descriptive attributes of mature cow beef. *Meat Sci.* 38:443–451.

El-Shenawy, M. A., Marth, E. H. 1989. Inhibition or inactivation of *Listeria monocytogenes* by sodium benzoate together with some organic acids. *J. Food Prot.* 52:771–776.

El-Shenawy, M. A., Marth, E. H. 1992. Behavior of *Listeria monocytogenes* in the presence of sodium propionate together with food acids. *J. Food Prot.* 55:241–245.

Epling, L. K., Carpenter, J. A., Blankenship, L. C. 1993. Prevalence of *Campylobacter* spp. and *Salmonella* spp. on pork carcasses and the reduction effected by spraying with lactic acid. *J. Food Prot.* 56:536–537,540.

Eytan, O., Weinert, I. A. G., McGill, A. E. J. 1992. Effect of salad dressing and citric acid dip on storage quality of shredded cabbage and carrots packed under modified atmospheres. *Lebensm. Wiss. U. Technol.* 25:445–450.

Fabian, F. W., Graham, H. T. 1953. Viability of thermophilic bacteria in the presence of varying concentrations of acids, sodium chloride and sugars. *Food Technol.* 7:212–217.

Fabian, F. W., Wadsworth, C. K. 1939a. Experimental work on lactic acid in preserving pickles and pickle products. I. Rate of penetration of acetic and lactic acid in pickles. *Food Res.* 4: 499–509.

Fabian, F. W., Wadsworth, C. K. 1939b. Experimental work on lactic acid in preserving pickles and pickle products. II. Preserving value of acetic acid and lactic acid in the presence of sucrose. *Food Res.* 4:511–519.

FAO/WHO. 1962. Evaluation of the Toxicity of a Number of Antimicrobials and Antioxidants. 6th Report of the Joint Food and Agriculture Organization of the United Nations/World Health Organization Expert Committee on Food Additives. WHO Technical Report Series No. 228. FAO Nutrition Meetings Report Series No. 31.

FAO/WHO. 1963. Specifications for the Identity and Purity of Food Additives and Their Toxicological Evaluation: Emulsifiers, Stabilizers, Bleaching and Maturing Agents. 7th Report of the Joint Food and Agriculture Organization of the United Nations/World Health Organization Expert Committee on Food Additives. WHO Technical Report Series No. 281. FAO Nutrition Meetings Report Series No. 35.

FAO/WHO. 1966. Specifications for the Identity and Purity of Food Additives and Their Toxicological Evaluation: Some Antimicrobials, Antioxidants, Emulsifiers, Stabilizers, Flour-Treatment Agents, Acids and Bases. 9th Report of the Joint Food and Agriculture Organization of the United Nations/World Health Organization Expert Committee on Food Additives. WHO Technical Report Series No. 339. FAO Nutrition Meetings Report Series No. 40.

FAO/WHO. 1967. Specifications for the Identity and Purity of Food Additives and Their Toxicological Evaluation: Some Emulsifiers and Stabilizers and Certain Other Substances. 10th Report of the Joint Food and Agriculture Organization of the United Nations/World Health Organization Expert Committee on Food Additives. WHO Technical Report Series No. 313. FAO Nutrition Meetings Report Series No. 43.

FAO/WHO. 1973. Toxicological Evaluation of Certain Food Additives with a Review of General Principles and of Specifications. 17th Report of the Joint Food and Agriculture Organization of the United Nations/World Health Organization Expert Committee on Food Additives. WHO Technical Report Series No. 539. FAO Nutrition Meetings Report Series No. 53.

FAO/WHO. 1974. Evaluation of Certain Food Additives. 18th Report of the Joint Food and Agriculture Organization of the United Nations/World Health Organization Expert Committee on Food Additives. WHO Technical Report Series No. 557. FAO Nutrition Meetings Report Series No. 54.

FAO/WHO. 1975. Specifications for the Identity and Purity of Some Food Colours, Flavour Enhancers, Thickening Agents and Certain Other Food Additives. 18th Report of the Joint Food and Agriculture Organization of the United Nations/World Health Organization Expert Committee on Food Additives. WHO Technical Report Series No. 557. FAO Nutrition Meetings Report Series No. 54B.

Farkas, J., Andrássy, É. 1993. Interaction of ionising radiation and acidulants on the growth of microflora of a vacuum-packaged chilled meat product. *Intl. J. Food Microbiol.* 19:145–152.

Fiddler, W., Pensabene, I., Kushnir, I., Piotrowski, E. G. 1973a. Effect of frankfurter cure ingredients on N-nitrosodimethylamine formation in a model system. *J. Food Sci.* 38:714–715.

Fiddler, W., Pensabene, J. W., Piotrowski, E. G., Doerr, R. C., Wasserman, A. E. 1973b. Use of sodium ascorbate or erythrobate to inhibit formation of N-nitrosodimethylamine in frankfurters. *J. Food Sci.* 38:1084.

Fin'ko, L. N. 1969. Blood coagulation changes in acetic acid poisoning. *Klin. Med.* (Moscow) 47: 84–87.

Fitzhugh, O. G., Nelson, A. A. 1947. The comparative chronic toxicities of fumaric, tartaric, oxalic and maleic acids. *J. Am. Pharm. Assoc.* 36:217–219.

Fondu, M., van Gindertael-Zegers de Beyl, H., Bronkers, G., Carton, P. (Eds.) 1980. *Food Additives Tables*. Elsevier Scientific, New York.

Food and Drug Research Laboratories, Inc. 1973a. NAS/NRC questionnaire on tartaric acid, Contract 71–75. Food and Drug Administration, Rockville, MD.

Food and Drug Research Laboratories, Inc. 1973b. Teratological evaluation of FDA 71-50 (Adipic acid) in mice, rats and hamsters. Food and Drug Administration, Rockville, MD.

Freese, E., Levine, B. C. 1978. Action mechanisms of preservatives and antiseptics. In: *Developments in Industrial Microbiology*, Vol. 19, Underkofler, L. A. (Ed.). American Institute of Biological Sciences, Washington, D.C., p. 207.

Gardner, W. H. 1966. *Food Acidulants*. Allied Chemical Corporation, New York.

Gardner, W. H. 1972. Acidulants in food processing. In: *Handbook of Food Additives*, 2nd ed., Vol. 1, Furia, T. E. (Ed.). CRC Press, Cleveland, OH, pp. 225–270.

Gerhartz, H. 1949. Changes in the liver in a case of acetic acid poisoning and their significance for the timely measurement of liver regeneration and cirrhotic scar formation. *Virchows Arch. Pathol. Anat. Physiol.* 316:456–475.

Gill, C. O., Newton, K. G. 1982. Effect of lactic acid concentration on growth on meat of gram-negative psychrotrophs from a meatworks. *Appl. Environ. Microbiol.* 43:284–288.

Glabe, E. F., Maryanski, J. K. 1981. Sodium diacetate: an effective mold inhibitor. *Cereal Foods World*. 26:285–289.

Glass, K. A., Doyle, M. P. 1991. Fate of *Salmonella* and *Listeria monocytogenes* in commercial, reduced-calorie mayonnaise. *J. Food Prot.* 54:691–695.

Gomori, G., Gulyas, E. 1944. Effect of parenterally administered citrate on the renal excretion of calcium. *Proc. Soc. Exp. Biol. Med.* 56:226–228.

Graham, W. D., Grice, H. C. 1955. Chronic toxicity of bread additives to rats. Part II. *J. Pharm. Pharmacol.* 7:126–134.

Graham, W. D., Teed, H., Grice, H. C. 1954. Chronic toxicity of bread additives to rats. *J. Pharm. Pharmacol.* 6:534–535.

Granados, H., Glavind, J., Dam, H. 1949. Observations on experimental dental caries. III. The effect of dietary lactic acid. *J. Dent. Res.* 28:282–287.

Gruber, C. M., Halbeisen, W. A. 1948. A study on the comparative toxic effects of citric acid and its sodium salts. *J. Pharmacol. Exp. Ther.* 94:65–67.

Hara, S. 1965. Pharmacological and toxic actions of propionates. Examination of general pharmacological actions and toxicity of sodium and calcium propionates. *Chem. Abstr.* 62:977f.

Harshbarger, K. E. 1942. Report of a study on the toxicity of several food preserving agents. *J. Dairy Sci.* 25:169–174.

Hartwig, P., McDaniel, M. R. 1995. Flavor characteristics of lactic, malic, citric, and acetic acids at various pH levels. *J. Food Sci.* 60:384–388.

Hazelton Laboratories 1970. Two generation reproduction study—Rats X-5120 (malic acid). Project 165–128. Final report submitted to Specialty Chemicals Division. Allied Chemical Corporation, Buffalo, NY.

Hazelton Laboratories. 1971a. 24 month dietary administration—Rats. Project No. 165–126 (malic acid). Final report submitted to Specialty Chemicals Division. Allied Chemical Corporation, Buffalo, NY.

Hazelton Laboratories. 1971b. 104 week dietary administration—Dogs. X-5120 (malic acid). Final report submitted to Specialty Chemicals Division. Allied Chemical Corporation, Buffalo, NY.

Hedberg, M., Miller, J. K. 1969. Effectiveness of acetic acid, betadine, amphyll, polymyxin B, colistin and gentamicin against *Pseudomonas aeruginosa*. *Appl. Microbiol.* 18:854–855.

Helrich, K. (Ed.) 1990. *Official Methods of Analysis of the Association of Official Analytical Chemists*, 15th ed. Association of Official Analytical Chemists, Arlington, VA.

Heseltine, W. W. 1952a. A note on sodium propionate. *J. Pharm. Pharmacol.* 4:120–122.

Heseltine, W. W. 1952b. Sodium propionate and its derivatives as bacteriostatic and fungistatics. *J. Pharm. Pharmacol.* 4:577–581.

Hoffman, C., Schweitzer, T. R., Dalby, G. 1939. Fungistatic properties of the fatty acids and possible biochemical significance. *Food Res.* 4:539–545.

Horn, H. J., Holland, E. G., Hazelton, L. W. 1957. Safety of adipic acid as compared with citric and tartaric acid. *J. Agric. Food Chem.* 5:759–761.

Houben, J. H., Krol, B. 1991. Effect of citric acid and citrate on the shelf stability of canned liver paste processed at reduced F_0 values. *Meat Sci.* 30:185–194.

Huhtanen, C. N., Guy, E. J. 1984. Antifungal properties of esters of alkenoic and alkynoic acids. *J. Food Sci.* 49:281–283.

Huhtanen, C. N., Dymicky, M., Trenchard, H. 1981a. Methyl and ethyl esters of fumaric acid as substitutes for nitrite for inhibiting *Clostridium botulinum* spore outgrowth in bacon. Presented at 41st Annual Meeting of the Institute of Food Technologists, Atlanta, GA, June 7–10.

Huhtanen, C. N., Guy, E. J., Milnes-McCaffrey, L. 1981b. Antifungal properties of fumaric acid esters. Presented at the 41st Annual Meeting of the Institute of Food Technologists. Atlanta, GA, June, 7–10.

Hunter, D., Segel, I. H. 1973. Effect of weak acids on amino acid transport by *Penicillium chrysogenum*: evidence for a proton or charge gradient as the driving force. J. Bacteriol. 113: 1184–1192.

Igoe, R. S., Hui, Y. H. 1996. *Dictionary of Food Ingredients*, 3rd ed. Chapman and Hall, New York.

Imai, K., Banno, I., Iijima, T. 1970. Inhibition of bacterial growth by citrate. *J. Gen. Appl. Microbiol.* 16:479–487.

Ingle, J. D. 1940. Some preliminary observations on the effectiveness of propionates as mold inhibitors on dairy products. *J. Dairy Sci.* 23:509.

Ito, K., Chen, J. K., Lerke, P. A., Seeger, M. L., Unverferth, J. A. 1976. Effect of acid and salt concentration in fresh-pack pickles on the growth of *Clostridium botulinum* spores. *Appl. Environ. Microbiol.* 32:121–124.

Izat, A. L., Colberg, M., Thomas, R. A., Adams, M. H., Driggers, C. D. 1990a. Effects of lactic acid in processing waters on the incidence of salmonellae on broilers. *J. Food Qual.* 13:295–306.

Izat, A. L., Hierholzer, R. E., Kopek, J. M., Adams, M. H., McGinnis, J. P., Reiber, M. A. 1990b. The use of propylene glycol and/or lactic acid in chill water for reducing salmonellae on broilers. *J. Food Proc. Pres.* 14:369–374.

Johnson, L. R. 1968. Source of the histamine released during damage to the gastric mucosa by acetic acid. *Gastroenterology*. 54:8–15.

Kabera, J. J., Swieczkowski, D. M., Conley, A. J. Truant, J. P. 1972. Fatty acids and derivatives as antimicrobial agents. *Antimicrob. Agents Chemother.* 2:23–28.

Karapinar, M., Gönül, S. A. 1992. Removal of *Yersinia enterocolitica* from fresh parsley by washing with acetic acid or vinegar. *Intl. J. Food Microbiol.* 16:261–264.

Kim, C. R., Hearnsberger, J. O., Vickery, A. P., White, C. H., Marshall, D. L. 1995. Extending shelf-life of refrigerated catfish fillets using sodium acetate and monopotassium phosphate. *J. Food Prot.* 58:644–647.

Kirby, G. W., Atkin, L., Frey, C. N. 1937. Further studies on the growth of bread molds as influenced by acidity. *Cereal Chem.* 14:865–878.

Kotula, K. L., Thelappurate, R. 1994. Microbiological and sensory attributes of retail cuts of beef treated with acetic and lactic acid solutions. *J. Food Prot.* 57:665–670.

Lang, K., Bartsch, A. R. 1953. Concerning the metabolism and tolerability of adipic acid. *Biochem. Z.* 323:462–468.

Larsson, K., Norén, B., Odham, G. 1975. Antimicrobial effect of simple lipids and the effect of pH and positive ions. *Antimicrob. Agents Chemother.* 8:733–736.

Levey, S., Lasichak, A. G., Brimi, R., Orten, J. M., Smyth, C. J., Smith A. H. 1946. A study to determine the toxicity of fumaric acid. *J. Am. Pharm. Assoc. Sci. Ed.* 35:298–304.

Levine, A. S., Fellers, C. R. 1939. The inhibiting effect of acetic acid with sodium chloride and sucrose on microorganisms. *J. Bacteriol.* 39:17.

Levine, A. S., Fellers, C. R. 1940. Action of acetic acid on food spoilage microorganisms. *J. Bacteriol.* 39:499–514.

Lewis, R. J., Sr. 1989. *Food Additives Handbook.* Van Nostrand–Reinhold, New York.

Lillard, H. S., Blankenship, L. C., Dickens, J. A., Craven, S. E., Shackelford, A. D. 1987. Effect of acetic acid on the microbiological quality of scalded picked and unpicked broiler carcasses. *J. Food Prot.* 50:112–114.

Little, C. L., Adams, M. R., Easter, M. C. 1992. The effect of pH, acidulant and temperature on the survival of *Yersinia enterocolitica. Lett. Appl. Microbiol.* 14:148–152.

Maas, M. R., Glass, K. A., Doyle, M. P. 1989. Sodium lactate delays toxin production by *Clostridium botulinum* in cook-in-bag turkey products. *Appl. Environ. Microbiol.* 55:2226–2229.

Maga, J. A., Tu, A. T. 1994. *Food Additive Toxicology.* Marcel Dekker, New York.

Marsili, R. T., Ostapenko, H., Simmons, R. E., Green, D. E. 1981. High performance liquid chromatography determination of organic acids in dairy products. *J. Food Sci.* 46:52–57.

Martin, G. E., Sullo, J. G., Schoeneman, R. L. 1971. Determination of fixed acids in commercial wines by gas–liquid chromatography. *J. Agric. Food Chem.* 19:995–998.

McDonald, L. C., Fleming, H. P., Hassan, H. M. 1990. Acid tolerance of *Leuconostoc mesenteroides* and *Lactobacillus plantarum. Appl. Environ. Microbiol.* 56:2120–2124.

Mendonca, A. F., Molina, R. A., Kraft, A. A., Walker, H. W. 1989. Microbiological, chemical, and physical changes in fresh, vacuum-packaged pork treated with organic acids and salt. *J. Food Sci.* 54:18–21.

Minor, T. E., Marth, E. H. 1970. Growth of *Staphylococcus aureus* in acidified pasteurized milk. *J. Milk Food Technol.* 33:516–520.

Mori, K. 1953. Production of gastric lesions in the rat by the diet containing fatty acids. *Gann* 44: 421–425.

Mottram, D. S., Patterson, R. L. S., Rhodes, D. N., Gough, T. A. 1975. Influence of ascorbic acid and pH on the formation of N-nitrosodimethylamine in cured pork containing added dimethylamine. *J. Sci. Food Agric.* 26:47–53.

Mountney, G. J., O'Malley, J. 1965. Acids as poultry meat preservatives. *Poult. Sci.* 44:582–586.

Moye, C. J., Chambers, A. 1991. An innovative technology for salmonella control and shelf life extension. *Food Australia.* 43:246–249.

Murdock, D. I. 1950. Inhibitory action of citric acid on tomato juice flat-sour organisms. *Food Res.* 15:107–113.

Nakamura, M., Zangar, M. J. 1968. Effect of fatty acids on *Shigella. Proc. Mont. Acad. Sci.* 28: 51–57.

National Research Council. 1996. *Food Chemicals Codex*, 4th ed. Committee on Codex Specifications, Food and Nutrition Board, National Research Council, National Academy Press, Washington, D.C.

Nollet, L. M. L. (Ed.) 1996. *Handbook of Food Analysis.* Marcel Dekker, New York.

Nunheimer, T. D., Fabian, F. W. 1940. Influence of organic acids, sugars, and sodium chloride upon strains of food poisoning staphylococci. *Am. J. Public Health* 30:1040–1049.

Ocio, M. J., Sánchez, T., Fernandez, P. S., Rodrigo, M., Martinez, A. 1994. Thermal resistance characteristics of PA 3679 in the temperature range of 110–121°C as affected by pH, type of acidulant and substrate. *Intl. J. Food Microbiol.* 22:239–247.

Okabe, S., Roth, J. L. A., Pfeiffer, C. J. 1971. Differential healing periods of the acetic acid ulcer model in rats and cats. *Experientia* 27:146–148.

Okereke, A., Beelman, R. B., Doores, S. 1990a. Control of spoilage of canned mushrooms inoculated with *Clostridium sporogenes* PA 3679 spores by acid blanching and EDTA. *J. Food Sci.* 55: 1331–1333,1337.

Okereke, A., Beelman, R. B., Doores, S., Walsh, R. 1990b. Elucidation of the mechanism of the acid-blanch and EDTA process inhibition of *Clostridium sporogenes* PA 3679 spores. *J. Food Sci.* 55:1137–1142.

Okrend, A. J., Johnston, R. W., Moran, A. B. 1986. Effect of acetic acid on the death rates at 52°C of *Salmonella newport, Salmonella typhimurium* and *Campylobacter jejuni* in poultry scald water. *J. Food Prot.* 49:500–503.

O'Leary, D. K., Kralovec, R. D. 1941. Development of *B. mesentericus* in bread and control with calcium acid phosphate or calcium propionate. *Cereal Chem.* 18:730–741.

Olson, J. C., Macy, H. 1940. Propionic acid and its calcium and sodium salts as inhibitors of mold growth. *J. Dairy Sci.* 23:509–510.

Olson, J. C., Jr., Macy, H. 1945. Propionic acid, sodium propionate and calcium propionate as inhibitors of mold growth. I. Observations on the use of propionate-treated parchment in inhibiting mold growth on the surface of butter. *J. Dairy Sci.* 28:701–710.

Osawa, T., Ishibashi, H., Namiki, M., Kada, T. 1980. Desmutagenic actions of ascorbic acid and cysteine on a new pyrrole mutagen formed by the reaction between food additives; sorbic acid and sodium nitrite. *Biochem. Biophys. Res. Comm.* 95:835–841.

Oscroft, C. A., Banks, J. G., McPhee, S. 1990. Inhibition of thermally-stressed *Bacillus* spores by combinations of nisin, pH and organic acids. *Lebensm. Wiss. U. Technol.* 23:538–544.

Ough, C. S., Kunkee, R. E. 1974. The effect of fumaric acid on malolactic fermentation in wines from warm areas. *Am. J. Enol. Vitic.* 25:188–190.

Paar, D., Helmsoth, V., Werner, M., Bock, K. D. 1968. Haemostatic failure due to consumption of coagulation factors in acute acetic acid poisoning. *Ger. Med. Mon.* 13:421–424.

Packman, E. W., Abbott, D. D., Harrisson, J. W. E. 1963. Comparative subacute toxicity for rabbits of citric, fumaric and tartaric acids. *Toxicol. Appl. Pharmacol.* 5:163–167.

Palmer, A. A. 1932. Two fatal cases of poisoning by acetic acid. *Med. J. Aust.* 1:687.

Park, H. S., Marth, E. H., Olson, N. F. 1970. Survival of *Salmonella typhimurium* in cold-pack cheese food during refrigerated storage. *J. Milk Food Technol.* 33:383–388.

Pilone, G. J. 1975. Control of malo-lactic fermentation in table wines by addition of fumaric acid. In: *Lactic Acid Bacteria in Beverages and Food*, Carr, J. G., Cutting, C. V., Whiting, G. C. (Eds.). Academic Press, London, p. 121.

Pitkin, C. E. 1935. Lactic acid stricture of the esophagus. *Ann. Otol. Rhinol. Laryngol.* 44:842–843.

Prasai, R. K., Acuff, G. R., Lucia, L. M., Morgan, J. B., May, S. G., Savell, J. W. 1992. Microbiological effects of acid decontamination of pork carcasses at various locations in processing. *Meat Sci.* 32:413–423.

Raevuori, M. 1975. Effect of nitrite and erythrobate on growth of *Bacillus cereus* in cooked sausage and in laboratory media. *Zentralbl. Bakteriol. Hyg. I, Abt. Orig. B.* 161:280–287.

Rammell, C. G. 1962. Inhibition by citrate of the growth of coagulase-positive staphylococci. *J. Bacteriol.* 84:1123–1124.

Reiss, J. 1976. Prevention of the formation of mycotoxins in whole wheat bread by citric acid and lactic acid. *Experientia* 32:168–169.

Reynolds, A. E., Jr. 1975. The mode of action of acetic acid on bacteria. *Diss. Abstr.* B35:4935–4936.

Rice, A. C., Pederson, C. S. 1954. Factors influencing growth of *Bacillus coagulans* in canned tomato juice. II. Acidic constituents of tomato juice and specific organic acids. *Food Res.* 19: 124–133.

Roberts, T. A., Gibson, A. M., Robinson, A. 1981a. Factors controlling the growth of *Clostridium botulinum* types A and B in pasteurized cured meats. I. Growth in pork slurries prepared from ''low'' pH meat (pH range 5.5–6.3). *J. Food Technol.* 16:239–266.

Roberts, T. A., Gibson, A., and Robinson, A. 1981b. Factors controlling the growth of *Clostridium botulinum* types A and B in pasteurized cured meats. II. Growth in pork slurries prepared from ''high'' pH meat (pH range 6.3–6.9). *J. Food Technol.* 16:267–281.

Ryser, E. T., Marth, E. H. 1988. Survival of *Listeria monocytogenes* in cold-pack cheese food during refrigerated storage. *J. Food Prot.* 51:615–621.

Schlyter, J. H., Glass, K. A., Loeffelholz, J., Degnan, A. J., Luchansky, J. B. 1993. The effects of diacetate with nitrite, lactate, or pediocin on the viability of *Listeria monocytogenes* in turkey slurries. *Intl. J. Food Microbiol.* 19:271–281.

Sen, N. P., Donaldson, B., Seaman, S., Iyengar, J. R., Miles, W. F. 1976. Inhibition of nitrosamine formation in fried bacon by propyl gallate and L-ascorbyl palmitate. *J. Agric. Food Chem.* 24: 397–401.

Seymour, C. F., Taylor, G., Welsh, R. C. 1954. Substitution of vinegar for lactic acid as bactericidal agent in infant milk mixtures. *Am. J. Dis. Child.* 88:62–66.

Shelef, L. A. 1994. Antimicrobial effects of lactates: a review. *J. Food Prot.* 57:445–450.

Shelef, L. A., Addala, L. 1994. Inhibition of *Listeria monocytogenes* and other bacteria by sodium diacetate. *J. Food Safety* 14:103–115.

Shelef, L. A., Yang, Q. 1991. Growth suppression of *Listeria monocytogenes* by lactates in broth, chicken, and beef. *J. Food Prot.* 54:283–287.

Sheu, C. W., Freese, E. 1972. Effects of fatty acids on growth and envelope proteins of *Bacillus subtilis. J. Bacteriol.* 111:516–524.

Sheu, C. W., Konings, W. N., Freese, E. 1972. Effects of acetate and other short-chain fatty acids on sugar and amino acid uptake of *Bacillus subtilis. J. Bacteriol.* 111:525–530.

Silla-Santos, M. H., Nuñez Kalasic, H., Casado Goti, A., Rodrigo Enguidanos, M. 1992. The effect of pH on the thermal resistance of *Clostridium sporogenes* (PA 3679) in asparagus purée acidified with citric acid and glucono-δ-lactone. *Intl. J. Food Microbiol.* 16:275–281.

Sinha, D. P., Drury, A. R., Conner, G. H. 1968. The *in vitro* effect of citric acid and sodium citrate on *Streptococcus agalactiae* in milk. *Indian Vet. J.* 45:805–810.

Siragusa, G. R., Dickson, J. S. 1992. Inhibition of *Listeria monocytogenes* on beef tissue by application of organic acids immobilized in a calcium alginate gel. *J. Food Sci.* 57:293–296.

Siragusa, G. R., Dickson, J. S. 1993. Inhibition of *Listeria monocytogenes, Salmonella typhimurium* and *Escherichia coli* O157:H7 on beef muscle tissue by lactic or acetic acid contained in calcium alginate gels. *J. Food Safety* 13:147–158.

Smulders, F. J. M., Woolthus, C. H. J. 1983. Influence of two levels of hygiene in the microbiological condition of veal as a product of two slaughtering/processing sequences. *J. Food Prot.* 46: 1032–1035.

Smulders, F. J. M., Woolthus, C. H. J. 1985. Immediate and delayed microbiological effects of lactic acid decontamination of calf carcasses—influence on conventionally boned versus hot-boned and vacuum-packaged cuts. *J. Food Prot.* 48:838–847.

Sollman, T. 1921. Studies of chronic intoxication on albino rats. III. Acetic and formic acids. *J. Pharmacol. Exp. Ther.* 16:463–474.

Sorrells, K. M., Enigl, D. C., Hatfield, J. R. 1989. Effect of pH, acidulant, time, and temperature on the growth of *Listeria monocytogenes. Food Prot.* 52:571–573.

Spencer, H. C., Rowe, V. K., McCollister, D. D. 1950. Dehydroacetic acid. I. Acute and chronic toxicity. *J. Pharmacol. Exp. Ther.* 99:57–68.

Subramanian, C. S., Marth, E. H. 1968. Multiplication of *Salmonella typhimurium* in skim milk with and without added hydrochloric, lactic and citric acids. *J. Milk Food Technol.* 31:323–326.

Takhirov, M. T. 1969. Hygienic standards for acetic acid and acetic anhydride in air. *Hyg. Sanit.* 34:122–125.

Tanaka, K., Chung, K. C., Hayatsu, H., Kada, T. 1978. Inhibition of nitrosamine formation *in vitro* by sorbic acid. *Food Cosmet. Toxicol.* 16:209–215.

Thomson, J. E., Banwart, G. J., Sanders, D. H., Mercuri, A. J. 1967. Effect of chlorine, antibiotics, β-propiolactone, acids and washing on *Salmonella typhimurium* on eviscerated fryer chickens. *Poult. Sci.* 46:146–151.

Tijan, G. H., Jansen, J. T. A. 1971. Identification of acetic, propionic, and sorbic acids in bakery products by thin layer chromatography. *J. Assoc. Off. Anal. Chem.* 54:1150–1151.

Tompkin, R. B., Christiansen, L. N., Shaparis, A. B. 1978a. Antibotulinal role of ascorbate in cured meat. *J. Food Sci.* 43:1368–1370.

Tompkin, R. B., Christiansen, L. N., Shaparis, A. B. 1978b. Enhancing nitrite inhibition of *Clostridium botulinum* with isoascorbate in perishable canned cured meat. *Appl. Environ. Microbiol.* 35:59–61.

Tompkin, R. B., Christiansen, L. N., Shaparis, A. B. 1979. Isoascorbate level and botulinal inhibition in perishable canned cured meat. *J. Food Sci.* 44:1147–1149.

Trainer, J. B., Krippaehne, W. W., Hunter, W. C., Lagozzino, D. A. 1945. Esophageal stenosis due to lactic acid. *Am. J. Dis. Child.* 69:173–175.

Tsuji, S., Tonogai, Y., Ito, Y. 1986. Rapid determination of mono-, di- and tri-isopropyl citrate in foods by gas chromatography. *J. Food Prot.* 49:914–916.

Tuft, L., Ettelson, L. N. 1956. Canker sores from allergy to weak organic acid (citric and acetic). *J. Allergy* 27:536–543.

van der Marel, G. M., van Logtestijn, J. G., Mossel, D. A. A. 1988. Bacteriological quality of broiler carcasses as affected by in-plant lactic acid decontamination. *Intl. J. Food Microbiol.* 6:31–42.

Van Netten, P., Huis In'T Veld, J. H. 1994. The effect of lactic acid decontamination on the microflora on meat. *J. Food Safety* 14:243–257.

Waterhouse, H. N., Scott, H. M. 1962. Effect of sex, feathering, rate of growth and acetates on chick's need for glycine. *Poult. Sci.* 41:1957–1960.

Weaver, R. A., Shelef, L. A. 1993. Antilisterial activity of sodium, potassium or calcium lactate in pork liver sausage. *J. Food Safety* 13:133–146.

Weil, A. J., Rogers, H. E. 1951. Allergic reactivity to simple alipatic acids in man. *J. Invest. Dermatol.* 17:227–231.

Williams, S. 1984. *Official Methods of Analysis of the Association of Official Analytical Chemists*, 14th ed. A.O.A.C., Arlington, VA.

Wiseman, R. D., Adler, D. K. 1956. Acetic acid sensitivity as a cause of cold urticaria. *J. Allergy* 27:50–56.

Wolf, P. A. 1950. Dehydroacetic acid: a new microbiological inhibitor. *Food Technol.* 4:294–297.

Wolford, E. R., Anderson, A. A. 1945. Propionates control microbial growth in fruits and vegetables. *Food Ind.* 17:622–624.

Woolford, M. K. 1975a. Microbiological screening of food preservatives, cold sterilants and specific antimicrobial agents as potential silage additives. *J. Sci. Food Agric.* 26:229–237.

Woolford, M. K. 1975b. Microbiological screening of the straight chain fatty acids. (C_1–C_{12}) as potential silage additives. *J. Sci. Food Agric.* 28:219–228.

Woolthus, C. H. J., Smulders, F. J. M. 1985. Microbial decontamination of calf carcasses by lactic acid sprays. *J. Food Prot.* 48:832–837.

Wright, L. D., Skeggs, H. R. 1946. Reversal of sodium propionate inhibition of *Escherchia coli* with β-alanine. *Arch. Biochem.* 10:383–386.

Yokotani, H., Usui, T., Nakaguchi, T., Kanabayashi, T., Tanda, M., Aramaki, Y. 1971. Acute and subacute toxicological studies of TAKEDA-citric acid in mice and rats. *J. Takeda Res. Lab.* 30:25–31.

Young, E. G., Smith, R. P. 1944. Lactic acid: A corrosive poison. *JAMA* 125:1179–1181.

Zeitoun, A. A. M., Debevere, J. M. 1990. The effect of treatment with buffered lactic acid in microbial decontamination and on shelf life of poultry. *Intl. J. Food Microbiol.* 11:305–312.

22

Enzymes

ARUN KILARA

Arun Kilara Worldwide, Northbrook, Illinois

MANIK DESAI

The Pennsylvania State University, University Park, Pennsylvania

1. INTRODUCTION

Enzymes are stereochemical- and substrate-specific organic biocatalysts that occur in all living beings. They are natural substances without which life itself could not proceed. Enzymes are colloidal substances that are proteinaceous and, being catalysts, they are not consumed in a reaction but merely speed up the rate of a reaction. Although enzymes are proteins containing several hundred amino acid residues, only a few residues are involved directly in substrate binding or reaction catalysis. The nature of the reaction that is catalyzed by a given enzyme is dependent on the identity of the amino acids that constitute the active site. Flexibility and ''plasticity'' of enzyme structure also appears to play a role in substrate specificity (Bone et al., 1989). Substrate specificity implies that they act only on certain types of molecules. Enzymes exhibit various types of specificity like group, bond, stereo, and absolute specificity. For example, protease acts on proteins and not on carbohydrates: furthermore among proteins only certain types of bonds may be affected. Porteases are capable of acting on esters of amino acids. In stereochemical specificity, an enzyme acting on D-glucose will not act on L-glucose because the latter is stereochemically distinct from the former. Another example of this type is trimethylamine-N-oxide demethylase, which forms only dimethyl amine and formaldehyde from trimethylamine-N-oxide, whereas the chemically catalyzed reaction forms a mixture of trimethylamine, diethylamine, and formaldehyde (Parkin and Hultin, 1986).

Glucose kinase exhibits absolute specificity by exchanging phosphate only between ATP and glucose, while alcohol dehydrogenase exhibits group specificity because it oxidizes several primary alcohols.

Among other unique properties of enzymes are that they are active within specific ranges of temperature and pH at a certain specific substrate concentration, are affected by compounds that can speed up their action (such compounds are called activators), and can be hindered by the presence of certain types of molecules called inhibitors. These specificities are conferred through the complex structure of enzyme molecules.

Enzymes, being ubiquitous in living things, can be selectively isolated from plants, animals, and microorganisms. They are of interest here because of the technological advantages conferred to foods by their use as catalysts. Therefore, for ethical and legal reasons a select group of plants, animals, and microorganisms are considered suitable raw materials for isolation of these biocatalysts.

Successful commercialization of enzymatic processes depends on one or more of the following: (1) their use must produce a better quality product than traditional processes in which enzymes are not used, (2) it should be more economical to produce products with enzymes than without, or (3) enzyme-catalyzed reactions must enable the manufacture of products not previously possible because of a lack of adequate raw materials. Therefore, enzymic modifications can produce at least three different categories of products: those that simulate traditional products, those that simulate traditional products but in addition improve cost-effectiveness or product traits, and new products.

A developing area in enzymology is the production of catalytic antibodies (Lerner and Tramatano, 1987). The idea is to design ''smart enzymes'' that can catalyze specific reactions. In the history of enzymology, enzymes were first recognized by Payen and Persoz in 1833. They found a thermolabile substance which converts starch into sugar during alcohol production. The enzyme is known as amylase.

Enzymes are of universal occurrence in all biological materials. They are discovered, obtained, and extracted in a pure form from various sources of plants, animals, and microorganisms and are called the native enzymes. The other group of protein catalysts are xenozymes; these are foreign enzymes which are created by chemical modification techniques, by random mutations, and by genetic and protein engineering. The most common enzymes from plant sources are papain, a protease from the latex of *Carica papaya*, with an estimated tonnage of 100 tons enzyme matter per annum, and malt amylase, which is used extensively in brewing and whose production is estimated to be at 10,000 tons of enzyme matter per annum (Aunstrup, 1977). Other known plant sources of enzymes are pineapple, which yields bromelain, and fig latex, which is the source of ficin. These enzymes, however, are not extensively used in commerce. Enzymes from animal sources are proteases, e.g., bovine or porcine pepsin and porcine trypsin are produced at 15 and 5 tons of enzyme matter per year, respectively. The latter two enzymes are, however, not used to any significant extent in the food industry, other than as rennet extenders. Other enzymes from animal sources are pregastric esterase and lipase preparations obtained from kid and lamb used in the Italian cheese industry (Kilara, 1985a). No accurate production figures are available.

The third living source from which enzymes are isolated and used in the food industry are microorganisms, including bacteria, yeasts, molds, actinomyces, and protozoa. These microorganisms can be divided into three groups: those traditionally used in food fermentation, microorganisms accepted as harmless food contaminants, and those that do not belong in the first two of these classes (see Section IX.A).

Microbial enzymes are more popular than those derived from plant or animal sources because of their ease of production, their variety, their stability, and their safety. Recent techniques in biotechnology also permit improvement in yields of enzymes and facilitate the production of unconventional enzymes from conventional sources. Higher plants accumulate wastes in vacuoles, and these wastes can be toxic; microorganisms do not store their wastes. Wherever possible, nonsporeforming strains of microorganisms are used to minimize the hazards of toxin production. From the standpoint of ease of isolation, extracellular enzymes (enzymes secreted into the growth media) rather than intracellular enzymes (enzymes contained within cells) are preferred. Advances in downstream processing techniques may, however, overcome the disadvantages. Commonly used genera for the production of enzymes are *Rhizopus*, *Aspergillus*, *Endothia*, *Bacillus*, *Mucor*, *Candida*, *Micrococcus*, *Morteirella*, *Streptomyces*, *Actinoplanes*, and *Kluyveromyces*. Considering that in nature there are many thousands of genera, this list is selective and restrictive.

II. ENZYME NOMENCLATURE

In a world in which scientific information is generated and transmitted efficiently, nomenclature or taxonomy assumes great importance, since it helps avoid confusion over what is being studied by scientists. In naming enzymes, it was once customary for the person who isolated and characterized the enzyme to also name it. As the number of enzymes being isolated and studied increased, this system of nomenclature led to confusion because two enzymes were sometimes given the same name or the same enzyme was given two names. Though enzyme names often ended with ''ase,'' there were notable exceptions. For example, trypsin, rennet, and papain are all proteases. Some other enzyme names described the phenomenon rather than a reaction, e.g., reverse transcriptase. Then there were colloquial names, e.g., proteases are enzymes acting on proteins; lipases act on lipids; and carbohydrases act on carbohydrates. Whether such action on a substrate is involved in hydrolysis, synthesis, or group transfer is not clear from such vague descriptions.

The International Union of Biochemistry formed a committee on nomenclature and classification of enzymes (Nomenclature Committee, 1978). This committee came up with a system of nomenclature that should preferably be followed when identifying enzymes. In this system an enzyme is identified by the prefix EC followed by four numerals, separated by periods. The first numeral refers to the class of enzyme. The six main classes of enzymes are as follows: 1, oxidoreductases; 2, transferases; 3, hydrolases; 4, lyses; 5, isomerases; and 6, ligases. For example, an enzyme designated 3.x.x.x is a hydrolase.

The second number in the designation refers to the subclass of the enzyme. For example, in the class of hydrolases, 3.1.x designates a hydrolase (3) acting on ester bonds (1), whereas a 3.2.x is a hydrolase acting on glycosyl compounds, and 3.4.x is a hydrolase acting on peptide bonds (a peptide hydrolase). The third number in the system pertains to the sub-subclass of the enzymes. For example. 3.4.1.x indicates a hydrolase (3) that acts on peptide bonds (4), with a sub-subclass designation 1 indicating that this peptide hydrolase acts on α-amino peptide and is an amino acid hydrolase, previously and variously called an amino peptidase. The fourth number designates the sequence of the enzyme in the sub-subclass, for example, 3.2.1.23 signifies a hydrolase (3) that acts on glycosyl compounds (2) of sub-subclass 1 with a sequence number 23, indicating β-D-galactoside galactohydrolase. This enzyme colloquially is called lactase or β-galactosidase. In spite of the efforts by the International Union of Biochemistry to encourage the use of systematic names, researchers tend to use trivial or officially unrecognized names as cited for lactase or β-galactosidase. Over 2300 enzymes are cataloged by the International Union of Biochemistry Enzyme Commission (IUB) in 1984.

III. ENZYME ASSAYS

To determine whether an enzyme is present and the quantities in which it is present, a test called an assay is run. The activity of an enzyme is the velocity per milliliter or milligram of enzyme, where velocity is defined as the number of molecules of substrate transformed per unit of time. Activity is therefore the number of substrate molecules transformed per unit of time per milliliter or milligram of enzyme. Activity is expressed in units, where a unit is defined as the amount of enzyme required to transform 1 μmol of substrate per minute under the conditions of the assay. The last phrase, ''under the conditions of the assay,'' is a disclaimer that accounts for pH, temperature, ionic strength, presence of cosolvents, etc. The International Union of Biochemistry recommends the use of the Katal as a standard unit. A Katal of enzyme activity represents that amount of enzyme capable of transforming 1 mol of substrate per second and conforms with other SI units of measurement. Its use, however, has not been prevalent because of its tendency to generate smaller numbers than those scientist prefer to use. For example, 1 unit of enzyme, where a unit converts 1 μmol of substrate per minute, would correspond to 6×10^{-7} Katal, or 0.6 μKatals.

There are two main events in an enzyme reaction, and these happen simultaneously. During an enzyme reaction, the substrate concentration diminishes as it is converted to product, and the product concentration increases as the substrate is utilized. Therefore, the velocity of enzyme-catlyzed reaction can be monitored by measuring either the rate of disappearance of substrate or the rate of appearance of products. These differences may be manifested as absorbance, fluorescence, optical rotation, pH, electrical conductivity, viscosity, or volume, any of which can form the basis of assay. Numerous recipes for the assay of enzymes have been published (Bergmeyer, 1974; Barman. 1969; Colwick and Kaplan, 1955). In addition, enzyme manufacturers provide information on assay methods. The U.S. Food and Drug Administration (FDA) generally follows procedures for assay recommended by the Food Chemical Codex and Nutrition Board of the National Research Council of the National Academy of Sciences in Washington, D.C. (NAS/NRC, 1981).

To study the action of amylase on starch, there are four different assay procedures in common use. Due to hydrolysis there is an appearance of reducing groups, decrease in viscosity, and an increase in maltose, glucose, or dextrins. The degree of hydrolysis is also determined by comparing the iodine color of a standard. When amylase hydrolyzes the α-1,4 glucosidic linkages of the polysaccharide, there is a rapid loss in viscosity and iodine color formation decreases. One dextrinizing unit (DU) is the quantity of amylase that will hydrolyze soluble starch in the presence of an excess of β-amylase at the rate of 1 g/h at 30°C. Bacterial amylases are assayed in a similar manner but in the absence of amylase, and 1 bacterial amylase unit (BAU) is defined as the quantity of enzyme that will dextrinize 1 mg of starch per minute under the conditions of the assay.

Color measurements are done by comparing colors with standards. Note that 1 BAU is defined differently from 1 DU and the assay conditions vary with respect to the presence of β-amylase.

Malt is characterized by its diastase activity, or diastatic power (DP), and is different from assays for amylases. Amylase activity in malt is measured by hydrolysis of starch occuring in a 30-min period at pH 4.6 and 20°C. The hydrolysis is quantitated titrimetrically by the reducing groups generated. One unit of diastatic power expressed as DP degrees (°DP) is that amount of enzyme contained in 0.1 mL of a 5% solution of sample enzyme preparation that will produce sufficient reducing sugars to reduce 5mL of Fehling's solution when the sample is incubated with 100 mL of substrate for 1 h at 20°C.

Catalase reacts with chemically reactive oxygen species. It participates in two types of reactions: (1) a catalytic reaction and (2) a peroxide reaction. Catalase activity can be measured by various methods such as spectrophotometric, titrimetric, polarographic, and manometric methods.

Various assays given for commercial catalase preparations convert a rate of absorbance change to a rate of concentration change at a set wavelength. Catalase activity is determined as Baker units (BU). The assay is an exhaustion method that quantitates the breakdown of hydrogen peroxide and the breakdown of catalase by the peroxide under specified conditions. One Baker unit of catalase activity is that amount of enzyme required to decompose 266 mg of hydrogen peroxide under the conditions of the assay. The assay is also performed by gas elecrode method, using phosphate buffer of pH 7.0 saturated with air and addition of hydrogen peroxide and catalase solution in a specific proportion at 25°C. The rate of increase in O_2 saturation in terms of molar concentration is recorded.

$$2\ H_2O_2 \rightarrow 2\ H_2O + O_2$$

Titrimetric activity is also calculated by reaction of hydrogen peroxide with standard potassium permanganate solution in acid.

Cellulase activity is measured in terms of reduction in viscosity of a reaction compound. Cellulose is converted to glucose by the action of a group of enzymes. The viscosity depends upon molecular weight of cellulose which reduces from 400 to 300 centipoise (cP) at pH 5.0 due to action of this enzyme. Manning (1981) has developed the methodology for measuring the viscosity change due to action of endocellulase on carboxymethylcellulose. One unit of cellulase activity (CA) is defined as that quantity of enzyme required to reduce the viscosity of 200 g of 5% solution of cellulose gum (Hercules Type 7-LF) from 400 to 300 cP at 35 ± 0.1°C, pH 5.0, in 1 h.

β-Glucanase activity determination is based on a 15-min hydrolysis lichenin substrate at 40°C and pH 6.5. As reducing groups are generated during hydrolysis, the increase in reducing power is quantitated spectrophotometrically. One β-glucanase unit (BGU) is that quantity of enzyme capable of liberating 1 mol of reducing sugar per minute under conditions of the assay. β-Glucanase assay is also described by Wood and Weisz (1987). This enzyme is mainly found in barley and oats. To 25 mL of MES buffer pH 5.5 was added 12.5 mg β-glucan and 125 mg agarose, and suspension was heated to boiling. The hot suspension was poured into petripalates and wells of 4 mm diameter were cut into gel. To each well 10 μL enzyme solution were added. Congo red was layered on petridishes and incubated for 18 h. The unreacted subtrate gave a purple red color sourrounding a clear zone of enzyme activity.

Glucoamylase or amyloglucosidase hydrolyzes starch, liberating one glucose molecule at a time. The enzyme sample is allowed to hydrolyze a cornstarch hydrolysate under controlled conditions of time, temperature, pH, and substrate concentration. The enzyme is assayed by measuring the reducing sugars resulting from the enzymatic hydrolysis of starch. A1-mL enzyme sample is mixed with an equal volume of soluble starch (30 g/L) in 0.1 M sodium acetate buffer (pH 4.8) and incubated at 30°C. The amount of reducing sugar produced is determined with glucose as the standard. One unit glucoamylase of activity is defined as the amount of enzyme which liberates 1 μmol of reducing sugars per minute.

Glucose isomerase is the major enzyme used in the high-fructose corn syrup industry. Activity of glucose isomerase is generally determined using an immobilized enzyme preparation. The rate of conversion of glucose to fructose is measured using a plug flow or packed bed reactor. Enzyme assays are highly dependent upon the bed configuration, flow rate, and viscosity of the solutions. Glucose concentration is 45% w/w inlet pH 7.0–8.5 at 25°C (varying depending on the source of enzyme), reaction temperature 60°C, magnesium concentration 4×10^{-3} M. One unit of glucose isomerase with the column method (GLcU) is defined as the amount of enzyme that converts glucose to fructose at an initial rate of 1 μmol/min under specified conditions. The conversion of glucose to fructose is monitored polarimetrically by measuring the change in specific rotation. In the automated assay, enzyme reacts with D-glucose and the dialyzed reaction mixture is mixed with acid carbazole. Carbazole gives a colored product only with D-fructose.

Glucose oxidase is an oxidoreductase whose activity is measured by an oxygen electrode based on the oxygen uptake in the presence of excess substrate, excess air, and excess catalase. One glucose oxidase unit (GOU) is defined as the amount of enzyme that will cause an uptake of 10 mm^3 of oxygen per minute in a Warburg manometer in the presence of excess air and excess catalase and with a substrate containing 3.3% glucose monohydrate in 0.1 M phosphate buffer at pH 5.9 with 0.4% sodium dehydroascorbate (Scott, 1953). Glucose oxidase can also use O-benzoquinone as the hydrogen acceptor. 1 mL of benzoquinone (0.1% in water), 0.9 mL citrate buffer (pH 5.0), and 2 mL of glucose are mixed and kept at 25°C; 0.05 mL of enzyme is added and absorbance at 290 nm is recorded at 2-min intervals. The increase in absorbance is due to an increase in concentration of hydroquinone, which has a molar absorptivity of 2.31×10^3 I.M. cm^{-1}.

Hemicellulase activity is measured by the ability of the enzyme to break interior glycosidic bonds of a defined locust bean gum (Type D-200, Meer Corporation) substrate at pH 4.5 and 40°C. The corresponding reduction in viscosity is measured using a viscometer. One hemicellulase unit (HCU) is defined as that amount of enzyme that will produce a relative fluidity change of 1 over a period of 5 min in a locust bean substrate under specified conditions. The viscometer recommended is a Cannon–Fenske type size 100 apparatus.

Invertase is an enzyme used in invert sugar manufacture, and its activity is measured by 30-min hydrolysis of sucrose at 20°C and pH 4.5. The degree of hydrolysis is measured by a change in optical rotation of the solution with a polarimeter. One invertase unit is defined as the quantity of enzyme that will hydrolyze 77% of the sucrose applied under the conditions of the assay.

Lactase (β-galactosidase) activity is based on a 15 min assay involving the hydrolysis of a synthetic substrate O-nitrophenyl–β-D-galactopyranoside (ONPG) at 37°C at a specified pH, dependent on the enzyme source. Substrate hydrolysis is monitored spectrophotometrically by measuring the intensity of yellow color at 420 nm. The yellow color is the O-nitrophenol liberated due to hydrolysis. One lactase unit (LaU) is defined as the amount of enzyme that hydrolyzes 1 μmol of ONPG per minute at 37°C under assay conditions.

Lipase hydrolyzes triacylglycerol liberating free fatty acids, which can be titrated. The assay is based on a 5-min hydrolysis of olive oil emulsion at pH 6.5 and 30°C. The released fatty acids are titrated with standard base. One unit of lipase activity is defined as that amount of enzyme required to liberate 1 μmol of acid per minute from an emulsified substrate under conditions of the assay.

Lipase/esterase (forestomach) activity is assayed using a soluble substrate such as tributyrin, and the liberated acids are continuously titrated in a pH-stat. One lipase unit (LFU) is the activity that releases 1.25 μmol of butyric acid per minute under conditions of the assay.

Milk-clotting activity is often determined for enzyme preparations suitable for cheesemaking. Expression of a milk-clotting unit is complex because it is a ratio. In a general sense it is the number of milliliters of milk that can be clotted in 40 min at 35°C by 1 mL of a coagulant.

Pepsin activity is measured as pepsin units. One pepsin unit is defined as that quantity of enzyme that digests 3000 times its weight of coagulated egg albumin under conditions of the assay.

Plant proteolytic activity depends on a 60-min hydrolysis of casein at pH 6.0 and 40°C. Unhydrolyzed substrate is precipitated with trichloroacetic acid, and the concentration of the solubilized protein is determined spectrophotometrically at 280 nm. One plant proteolytic unit is the quantity of enzyme that liberates 1 mg of tyrosine per hour under conditions of the assay.

The determination of proteolytic activity of bacterial proteases (PC) is similar to the assay for plant proteolytic activity and relies on measuring solubilized casein expressed as tyrosine. One unit of PC is the amount of enzyme that will liberate 1.5 mg of tyrosine equivalent per minute under the conditions of the assay.

Proteolytic activity of fungal enzyme is also similar to these methods, but instead of casein hemoglobin is used as the substrate, and therefore units are expressed as hemoglobin units on tyrosine basis (HUT). A 30-min hydrolysis of hemoglobin at pH 4.7 and 40°C is followed by the precipitation of insolubilized hemoglobin at 280 nm. One HUT of proteolytic activity is that amount of enzyme that produces in 1 min under assay conditions a hydrolysate whose absorbance at 275 nm is the same as that of a solution containing 1.10 mg/mL of tyrosine in 0.006 N hydrochloric acid.

Fungal proteolytic activity is expressed in terms of spectrophotometric acid protease units (SAPU) based on a 30-min assay of proteases with Hammersten casein at pH 3.0 and 37°C. Solubilized casein is determined spectrophotometrically at 280 nm after TCA precipitation of insoluble casein. One SAP unit of enzyme activity is that amount of enzyme that will liberate 1 μmol of tyrosine per minute under conditions of the assay.

Trypsin activity is determined by using the synthetic substrate N-benzoyl-L-arginine ethyl ester hydrochloride. One U.S. Pharmacopoeia (USP) unit of trypsin is that amount of enzyme that will cause a change in absorbance of 0.003/min under conditions of the assay.

These assay scenarios cover all the enzymes deemed important for use in food processing. If assays do not function or do not give expected values, several factors may need to be checked. These include

- Type of unit of activity being measured
- Variation in conditions of the assay, such as pH, temperature, and ionic strength
- Presence of inhibitors and/or activators
- Presence of enzyme "poisons"
- Improper calculations
- Improper enzyme and/or substrate concentrations being used in the assay

For example, proteolytic activity can be measured in any number of ways, as discussed. The number of units generated by these different assay procedures for any given enzyme can vary considerably. Lactase obtained from *Kluyveromyces lactis* assayed using ONPG as the substrate yielded 40,000 units per gram of material, but when assayed against lactose only 15,000 units/g material was reported (Kilara et al., 1977). In each case a unit was the amount of substrate being hydrolyzed per unit time, and in terms of number of O-glycosyl linkages broken, the numbers vary. The other factors affecting assay values are self-explanatory.

IV. FUNCTIONAL ASPECTS OF ENZYMES

Enzymes have been defined as powerful catalysts. The catalytic efficiency of enzymatic reactions approaches 10^8–10^{11}. Catalytic factors are obtained by dividing the rate of a reaction catalyzed by an enzyme by the rate of the same reaction in the absence of the enzyme. In enzyme kinetics, V_{max}/K_m ratios are commonly used to assess the catalytic power of enzymes. Calculations have shown that one enzyme molecule can transform as many as 10^6 substrate molecules per minute. These speeds are achieved at low temperature, under mild pH conditions, and in a solvent that is inexpensive, safe, and abundant, namely, water. Another advantage of enzymes is the wide variety in nature that one can select from. A similar smorgasbord of chemical catalysts would not be easy to compile. As discussed under nomenclature, there is a sub-subclass of enzyme for every type of chemical reaction conceivable. For example, not only can enzymes participate in isomerization reactions, but they can do so in racemic and epimeric modes, facilitate cis-trans isomerization, or cause intramolecular oxidation/reduction or group transfers. Further, among the racemases and epimerases there are specific enzymes that act only on amino acids and their derivatives, hydroxy acids and their derivatives, carbohydrates and their derivatives, or other compounds. One of the exciting prospects of enzyme technology is the creation of ''synzymes''—synthetic enzymes that can catalyze reactions not necessarily observed in nature. These synzymes can be viewed as enzyme analogs. Enzymes also promote ''clean'' reactions, that is, the tendency to form undesirable byproducts is minimized. Chemical catalysts, on the other hand, have a propensity to produce undesirable byproducts. This dedication of enzymes to act on very few substrates to produce specific products is what is termed substrate specificity, but substrate specificity is not absolute and must be addressed in the context of reaction conditions such as solvent, temperature, and pH. Industrially important enzymes are not normally used to convert their physiological substrates. For example, glucose isomerase used in the production of high-fructose corn syrup actually preferentially acts on xylose, its physiological substrate (Bucke, 1977). The specificity of enzymes can also be extended to stereospecificity, or the ability to distinguish between chiral carbon atoms. Enzymes have one other advantage in that they can be synthesized or degraded in rather short order, providing an attractive method for gross control over reactions. Further, binding of small molecules can modify enzyme activities, thereby offering finer control elements in reactions.

Thermodynamic laws, applicable to all reactions, do not make exceptions for enzymes. A catalyst, whether chemical or biochemical, speeds up the rate of reaction but does not alter the equilibrium constant. In order for a bond to be made, broken, or transformed, energy is required. Energy is consumed during the formation of bonds and released during the breaking of the bonds. Energy needed for the bond brakeage has to be obtained from molecular collisions. This energy will be stored until a transition state is formed, and the energy input is called free energy of activation. After this free energy activation is achieved, the reaction proceeds spontaneously, leading to bond breakage. Enzymes lower the free energy of activation, and therefore reactions proceeds faster.

Enzymes, being catalysts, provide new reaction pathways in which (1) and the slowest step (called the rate-limiting or rate-determining step) has a lower free energy of activation than the rate-determining step of the uncatalyzed reaction, and (2) all transition state energies in the catalyzed pathway are lower than the highest one of the uncatalyzed pathway. In an enzyme-catalyzed reaction a third possibility is the creation of the environment in which the free energy of the product is lowered. The formation of transition complexes between the enzyme and the substrate takes 10^{-9}–10^{-6} s, and as long as fresh substrate is available the concentration of free enzyme can be maintained in a dynamic equilibrium.

Rate constants are very sensitive to the free energy of binding of substrate. A change in only 2 kcal/mol between two competing substrates will result in a 97:3 ratio between their reaction rates. Enzymes do not alter the overall free energy of the reaction, but only decrease the free energy of the transition state complex. Jencks (1975) demonstrated that the slowness of non–enzyme-catalyzed and uncatalyzed reactions in free solution is due to a loss of transitional and rotational entropy resulting from the formation of the transition state complex. This explanation, however, does not apply to intramolecular reactions and enzyme-catalyzed reactions because of the formation of enzyme–substrate complexes. Further proof of this explanation was provided by showing that substrate concentration of 10 M would suffice for enzyme-catalyzed reactions, whereas for the same reaction to proceed in free solution, reactant concentration of 10^7–10^9 M would be necessary.

All known enzymes are proteins. Proteins are polymeric entities that exhibit structural hierarchy. The monomeric units are amino acids joined together by amide linkages or peptide bonds. There are 20 amino acids in nature from which enzyme molecule can be built. Sequence of amino acids can be varied to obtained differences in proteins. The long chain of amino acids does not remain extended but rather takes up a definite three-dimensional structure, which is further stabilized by intramolecular crosslinks and interactions. For example, two cysteine residues can combine to form a disulfide bond, and often this facilitates the formation of "loops." Similarly, hydrogen bonds, hydrophobic interactions, and ionic bonds all contribute to holding the coil in place. In some enzymes, polypeptide chains may be linked together in what are called quaternary, or subunit, structure.

Enzymes are specific, and even small changes in the structure of a substrate molecule lead to the inability of the enzyme to convert the compound to products. Therefore, it was suggested that the binding of an enzyme to its substrate is analogous to a lock and key, with the key fitting only one lock. The area where the substrate is bound and transformed is called the active site, and specific residues of amino acids in the active site region are collectively called the active center. The size of the active site is not set and is perhaps dependent on the functional size of the substrate molecule. For example, degradation of a polymer like amylose may require a larger active site than the hydrolysis of maltose to glucose. The groups involved in the active center are generally not sequentially proximal to one another, i.e., they are not near neighbors in terms of the primary structure of the protein. The highest order in structure imposed by secondary, but more often tertiary and quaternary, structure bring the group in the active center into spatial proximity. Therefore, in an enzyme the majority of the amino acid residues are involved in keeping the few necessary residues in close spatial proximity. If this spatial orientation is distorted by such factors as heat, pH, or solvent effects, the catalyst nature of the enzyme will be diminished or even lost. Such distortion can be very minute, say 0.1 Å (100 picometers).

The catalytic efficiency of an enzyme may result from (1) proximity and orientation effects, (2) catalysis by distortion, (3) catalysis by acid–base reactions, and (4) nucleophilic and electrophilic catalysis (charge relay). The contribution of proper positioning of substrate, with respect to the active center, to the overall catalytic efficiency is hard to assess exactly, but approximations are possible. Binding of substrate to enzyme converts an intramolecular reaction of second or higher orders to an intramolecular or first order reaction. On theoretical grounds, orientation and proximity effects can account for 5.5×10^4 enhancement of reaction rates in enzymes-catalyzed biomolecular reactions and 1×10^{22} for enzyme-catalyzed intermolecular reactions based on 10^{-3} M concentration of reactants (Koshland, 1960). Proximity alone is not adequate to explain catalytic efficiencies; anchoring a substrate to the enzyme is another critical process. Such a binding process has to decrease the entropy. The closer the ground state resembles the transition state, the more positive the entropy and the faster will be the reaction rate (Bender et al., 1964).

The second factor, catalysis by distortion, is an elegant theory postulated to play an important role in the action of such enzymes as lysozyme, trypsin, and chymotrypsin. This hypothesis is not verifiable due to experimental limitations. It suggests that the association of enzyme and substrate places a strain on the electronic structure of the substrate.

The third factor is catalysis by acid–base reactions. Acids are defined as substances that donate proteins, while bases are proton acceptors. Many amino acids in protein chains act as acids or bases. Some of these side chains and residues are α-carboxyl groups of the C-terminus peptide, carboxyl groups of the N-terminus and lysyl residues, respectively. The sulfhydryl group of cysteine, the phenolic hydroxyl of tyrosine, and the guanidino group of arginine all act as general bases when they take up protons but also have the ability to give up protons.

A fourth mechanism postulated to operate in enzyme catalysis is dependent on the nucleophilic and electrophilic properties of the amino acids residues in the enzyme molecules. A nucleophile is a group that has a strong tendency to donate a pair of electrons. If the rate of the reaction depends in whole or in part on a step that involves donation of an electron pair from the catalyst to the substrate, it is termed a nucleophilic catalysis. Conversely, if the reaction rate is dependent in whole or in part on a step involving acceptance of an electron pair from the substrate by the end product, it is termed an electrophilic catalysis.

All these briefly discussed mechanisms are only postulations, and often more than one mechanism is operational in catalysis by a single enzyme. Therefore, a thorough understanding of the mechanism of catalysis may not be as important to the successful application of enzyme in food processing as it is to the use of the other food additives.

VI. MANUFACTURE OF COMMERCIAL ENZYMES

For the most part, this discussion focuses on microbial enzymes rather than on enzymes from plant and animal sources. Plant and animal strains, breeds, and cultivars are not selected specifically for enzyme production. Rather, enzymes are byproducts recovered from plants and animals selected, bred, or propagated for other reasons. Additionally, strain and breed selection in plants and animals is a difficult, time-consuming, tedious process. Microorganisms, on the other hand, do not pose these limitations, and strain selection and the propagation of selected strains are rapidly achieved because of their short generation times (Aunstrup, 1980).

Microorganisms can be isolated from such natural sources as soil, compost, and rotting wood or can be obtained from culture collections. Enrichment cultures (e.g., using the substrate of interest as a sole carbon source or nitrogen source) select organisms from a mixed microbial population. These selected organisms can further be stressed by mutation, causing agents (mutagens) to create mutant strains. The frequency of mutation is dependent on the nature of the mutagen, its concentration, and the inherent susceptibility of particular strains to mutae. Strategies for strain improvement are targeted to improve the yield of the desired product. Elimination of undesirable traits may be a secondary benefit obtained during mutation. Increase in enzyme yields by mutation is achieved through a number of mechanisms. Enzyme production may depend on the presence of an inducer molecule in the desired medium. It may also be inhibited by the products of reaction or other breakdown products of the medium constituents—a phenomenon called catabolite repression. When circumvented by mutation, maximal enzyme production can be achieved. Recent techniques of recombinant DNA have provided means of achieving increased enzyme production in some instances.

The ultimate objectives in strain selection are to isolate microbial strains that grow on inexpensive media and give a high, constant yield of enzyme in a short time. The ideal strain should produce a minimal amount of secondary metabolites and secondary enzyme activities in the growth media. Recovery of the produced enzyme should be simple and inexpensive and should lead to a stable product that can be handled safely. The process for producing the enzyme must be safe to the plant operators, and the effluents should not be hazardous.

The selected strain is propagated and maintained in such a manner that the mutated properties are retained. For commercial manufacture of enzymes, the strains are grown either as submerged cultures or as semisolid fermentation products. Numerous factors that affect the performance of cultures during fermentation have been discussed by Frost (1986). Growth of the culture is synonymous with product secretion in fermentations. The product has to be isolated and partially purified. In such processes, the first operation is to separate insolubles from solubles. Insolubles in submerged fermentation may be cell mass, but in semisolid fermentation this can include unutilized growth media constituents in addition to the cell mass. For enzymes secreted into the growth medium the insoluble material is a byproduct, but in instances where the enzyme may be located within the cells, the insoluble material is of interest and the supernatant is considered to be a byproduct (Frost, 1986).

Endocellular enzymes are isolated by rupturing the cells using such methods as ball milling, freeze-thawing, pH, temperature, or osmotic shocking, high-pressure homogenization, or cell lysis with lysozyme. The resulting materials have to be separated from the cell debris, leaving cell-free extracts. Exocellular enzymes have already been made cell-free by separation from the growth medium.

Cell-free extracts can be purified by a series of consecutive fractionation and concentration steps including ion exchange chromatography, reverse osmosis, or ultrafiltration; or the protein can be precipitated by treatment with organic solvents, polyethylene glycol, dextran, polyacrylic acid, or ammonium sulfate. Precipitation of the protein is one method by which enzyme can be separated from carbohydrates in solution.

Enzymes in crude form can then be dried to obtain powders, or they can be manufactured as a concentrated liquid. Enzymes for use in food processing have to meet several other safety guidelines. For example, microbiological specifications such as freedom from *Salmonella* and *Escherichia coli* can be met by filtering enzyme solutions through 0.45 μm or smaller filters and by preventing contamination in the remaining operations as well. For drying enzymes, not only should the product be free from bacteria, but it should also be nonhygroscopic and dust-free. For either type of preparation, the material has to be protected against microbial growth during prolonged storage. Glycerol or propylene glycol, benzoate and its derivatives, and sugar alcohols are used for this purpose. These chemicals reduce water activity, thereby restricting microbial growth, and act as diluents to standardize activities of enzyme preparation from batch to batch. The various categories of additives have been listed by Schwimmer (1983). Most additives serve several functional roles in enzyme preparations, eventually leading to stability of the enzyme.

VII. USE OF ENZYMES IN THE FOOD INDUSTRY

In a historical sense, enzymes were used in food preparation long before their "discovery." Wrapping meat in certain leaves as a tenderizing aid for preparation or cooking serves as one example. Various fermented foods consumed around the world—alcoholic and fermented beverages, cheeses, fermented vegetables, fermented oilseed products, and fermented cereal and grain products—exemplify the use of enzymes in the fermentation or preservation of food. With the advent of modern enzymology, however, isolated and partially purified enzymes have gained increasing use in numerous synthetic, degradative, and analytical reaction.

The successful use of enzymes in any industrial biochemical process is contingent on several factors:

1. The process should be simple and be applicable under conditions that minimize growth of microbial contaminants while reducing viscosity.
2. Enzymes used commercially should be inexpensive, readily available, and approved for use by regulatory agencies.
3. Commercial enzymes must be amenable to use at high substrate concentration and must also be highly active.
4. The interaction of such process parameters as pH, time of residence, temperature, and substrate concentration should be well understood in order to achieve process optimization.

Market statistics on the use of enzymes vary depending on the group that generates the report. In general, 59% of enzymes used are proteases, 28% carbohydrases, and 3% lipases, and all other enzymes combined account for 10% of those used. A majority of the proteases are used in the detergent industry, but rennets account for 10% of all other enzymes used, and all of the rennet usage is attributed to cheese manufacture. Among carbohydrases, amylases, isomerases, and pectinases are important, and these enzymes are used almost exclusively in the food industry. Therefore, one estimate accounts for usage of enzymes as follows: in 1985 the total enzyme market in the United States was \$185 million, and the food sector used \$55 million worth, while the beverage industry used \$53 million worth of enzymes. On the same scale, cleaning products used \$25 million worth of enzymes, and all other categories of enzyme use accounted for \$49 million (Cianci, 1986). The food and beverage sectors used over 58% of the market value of enzymes. Cianci (1986) also reported that the \$185 million enzyme sales could be categorized into \$88 million for the protease market, while rennets accounted for \$35 million and papain for \$3 million. The carbohydrase market similarly could be subdivided into glucose isomerase, \$28 million; glucoamylase, \$21 million; amylase, \$11 million; and pectinases, \$4 million. Whereas, only about 43% of the protease market could be attributed to the food industry, over 90% of the carbohydrases are used by the food and beverage industries. The total U.S. enzyme market was estimated to be \$260 million in 1990, and of this \$68 and \$65 million will be consumed by the beverage and food industries, respectively,

Table 1 Number of Enzymes Identified

Enzyme type	Number of enzymes identified
Oxidoreductase	650
Transferase	720
Hydrolase	636
Lyase	255
Isomerase	120
Ligase	80

representing a total growth of nearly 28% in the beverage sector and 18% in the food sector over a 5-year period. Table 1 lists the number of enzymes identified (Roberts et al., 1995).

A wide variety of uses of enzymes in the food industry can be cataloged. The following subsections provide only some representative example. The reader is referred to a series of references to obtain specific details (Wiseman, 1983,1985; Reed, 1975; Birch et al., 1981; Godfrey and Reichelt, 1983; Fogarty, 1983; Schwimmer, 1983).

A. Hydrolysis of Proteins

It is probable that in the future the production of enzymatic hydrolysates from soybean, fish, meat, and microbial protein will become increasingly important. The factors that affect the degradation of proteins by enzymes include enzyme specificity, protein denaturation, substrate and enzyme concentration, pH, ionic strength, temperature, and the presence of inhibitors. Native proteins are not generally susceptible to degradation by proteolytic enzymes, as their compact conformation makes potentially susceptible bonds inaccessible (Robinson and Jencks, 1965). Denaturation unfolds proteins and makes the peptide bonds accessible to proteolytic cleavage.

Enzymes may be used to hydrolyze proteins in order to improve their functional properties (Whitaker, 1977; Kilara, 1985b) Proteolytic enzymes from *Bacillus licheniformis* and *B. subtilis* have been used to modify soy proteins (Peterson, 1981). Unmodified soy proteins are insoluble at their isoelectric pH of 3–5. Enzymatic hydrolysis improves its usefulness.

Hydrolysis also increases emulsifying and whipping capacities, with the maximum effect being observed at 4 and 5% degree of hydrolysis, respectively. Prior enzyme hydrolysis can give up to a tenfold increase in the volume of whipping capacities. Proteolytic enzymes can be used to reduce the viscosity of protein solutions and improve handling.

Excessive proteolytic action on certain proteins generates bitter peptides. Proteins with a high proportion of amino acids with hydrophobic side chains has a greater tendency to produce bitter hydrolysates than other proteins (Ney, 1971). It has been shown that bitterness in soy protein increases with the degree of hydrolysis. A balance has to be struck so that high yields of soluble protein are obtained without developing excessive bitterness.

The following procedure may be used to produce acceptable soy protein hydrolysates. An 8% solution of soy protein at pH 8 may be hydrolyzed at 50–55°C using alkaline Bacillus proteinase (Petersen, 1981). When the degree of hydrolysis is 9–10%, acid is added to reduce the pH to 4, stop the reaction, and precipitate unhydrolyzed protein. It may also be possible to remove bitter peptides from protein hydrolysates by using enzymes to resynthesize proteinlike polymers called plasteins (Fujimaki et al., 1977).

1. Milk-Clotting Enzymes

Traditionally the enzyme used to clot milk in the formation of cheeses was calf rennet (EC 3.4.4.3) obtained as a saline extract from the abomasum or fourth stomach of unweaned calves. The calf enzyme quickly clots milk at pH 6.7 and has been the most intensively studied of the enzymes that coagulate milk protein (Foltmann, 1971). As the slaughter of young animals has decreased, calf rennet has become increasingly difficult to obtain. This has led to the use of rennet substitutes. Some rennet substitutes consist of mixtures of rennet and pepsin. In the last decade milk-clotting enzymes of microbial origin have become important in the manufacture of cheese. The preparations that have become established in cheese manufacture include enzymes from *Mucor meihei*, *M. pusillus*, and *Endothia parasitica* (Sternberg, 1976). Proteases are also produced by *A. oryzae* and other *Aspergillus* species. These enzymes are known by Koji and it is widely used in brewing of soy sauce and mainly in the baking industry.

In the enzymatic clotting of milk an especially labile bond between Phe 105 and 106 in k-casein micelles is cleaved to yield two peptides (Dalgleish, 1982). One peptide, comprising the amino acid residues 106–169 and referred to as the glycomacropeptide, is soluble and diffuses away from the micelle. The second peptide comprises the amino acid residues on the micelle. Progressive hydrolysis leads to changes in the micelle structure that eventually lead to aggregation. Both the processes, enzymatic hydrolysis of k-casein and the subsequent aggregation, occur simultaneously. Calcium ions are necessary for the precipitation of para-casein to occur (Sardinas, 1976).

In addition, the hydrolysis of k-casein is dependent on pH ionic strength and temperature. Prior heating of milk reduces the susceptibility of k-casein to the action of rennet (Morrissey, 1969; Wilson and Wheelock, 1972). Rennet preparation from Mucor are more affected by variations in calcium ions than calf rennet (Nelson, 1975).

All milk-clotting enzymes are proteolytic enzymes, and their activity is characterized by an increase in soluble nitrogenous compounds in addition to the aggregation of casein micelles to form a gel structure. All are aspartate proteases and depend for their activity on the presence of two reactive aspartate residues at the active sites of the enzymes. Enzymes from different sources appear to share many features of primary structure catalytic activity and three-dimensional structure.

The pH optima for the enzymes are between 1.5 and 5 depending on the enzyme and type of substrate. The acid pH optima arise from the presence of the two aspartate residues that are essential for activity. One of the aspartate residues is protonated, and the second one is ionized. The aspartic acid residues are reactive toward specific reagents. Diazoacetyl norleucine methyl ether and 1,2-epoxy-3-(p-nitro-phenoxy) propane bind to the different aspartate residues and are also inhibited by pepstatin, a peptide from Streptomyces.

In contrast to microbial enzymes, gastric rennets are secreted as inactive proteins that require activation by removal of a peptide of about 40 amino acid residues from the amino end of the enzyme precursor (Asato and Rand, 1977). Microbial enzymes are produced in active form. Rennets of microbial origin have been found to clot milk in a manner that is generally similar to clotting by calf rennet. However, cheese made with calf rennet is the standard for taste, flavor, consistency, and texture. All commercial preparations of microbial rennets are mixtures containing, apart from the milk-clotting activity, enzymes that may be damaging to the cheese-making process. The ratio of milk-clotting activity to proteolytic activity is lower for microbial enzymes than for calf rennin (Aunstrup, 1980).

For milk to be clotted satisfactorily, the clotting enzyme must attack at or close to the Phe 105–Met 106 bond in k-casein. The cleavage destroys the stabilizing properties of k-casein and allows curd to be formed. The capacity for general cleavage must be low in order to avoid nonspecific production of soluble peptides. Excessive proteolytic activity can result in extensive breakdown of milk protein, reduced yield, soft texture, and bitter taste. Compared to calf rennet, microbial enzymes tend to have more unwanted proteolytic activity, which produces more nonprotein nitrogen. It may be necessary to use high-temperature scalding to stop enzyme activity.

In some situations the thermal stability of microbial rennets is a disadvantage. Residual enzymes in whey can continue to break down protein and limit the use of whey in foods (Harper and Lee, 1975). At a particular pH, microbial rennets are more resistant to heat treatment in whey than calf rennet (Hyslop et al., 1975). The different proteolytic specificities of microbial rennets may change the maturation characteristics of cheeses to give unfamiliar off-flavours (Cheeseman, 1981). There is a greater degree of variability with batches of microbial rennets due to fermentation and other conditions than with calf rennet.

2. Proteases Used in Meat Tenderizing

Proteolytic enzymes from pineapple, fig, and *Aspergillus* spp. have been used to tenderize meat (Lawrie, 1974). Papain, isolated from the latex and fruit of Carica papaya, is perhaps the most widely used. Like the other plant proteases, papain is a sulfhydryl protease and will degrade myofibril and connective tissue proteins. Its pH optimum is 5–9. The enzyme is extremely stable even at elevated temperature (Glazer and Smith, 1971).

Papain has little effect at room temperature but will act when the meat is cooked. It shows optimal activity above 50°C. Undenatured collagen is resistant, but over 50°C collagen fibers are loosened, with maximum solubilization occurring at 60–65°C. The enzyme is probably not inactivated completely until about 90°C, but it is usually inactivated by oxidizing agents and by exposure to air.

Papain, in an inactive stabilized form, may be injected into the circulatory systems of animals before slaughter (Dransfield and Etherington, 1981). This ensures an even distribution of enzymes but results in reduced value for the offal, which contains high enzyme levels. Inactivation is effected by oxidizing cysteinyl groups at the active site to form disulfide bridges. The inactivated enzyme remains in this form in the live animal and is excreted if the animal is not slaughtered. In slaughtered animals, continuing glycolysis depletes oxygen levels and produces a relatively reducing environment; under these conditions the papain is reactivated.

Conversion is slow in chilled meat but proceeds rapidly on warming. The action of bromelin and ficin is similar to that of papain. The enzymes are inactivated by several reagents such at Hg^{2+}, parachloromercuribenzoate, iodoacetate, dibromoacetate, and other reagents that react with sulfhydryl groups (Glazer and Smith, 1971).

3. Proteases in Baking and Brewing

The major proteins involved in baking processes are glutenins and gliadins. Both the protein structures are very complex. The major source of enzymes used in the process are fungal proteases. Fungal proteases particularly increase the texture and elasticity of dough. Proteases preparations from *Aspergillus oryzae* have been used to break down flour proteins to increase extensibility and workability, of the dough (Samuel, 1972). Proteinases and peptidases improve flavor by splitting the protein chains and forming amino acids res. There is no adverse effect that occurs by increase in concentration of fungal proteases, while there is a structural decomposition due to excess amounts of addition of bacterial or plant proteases. So the excessive proteolytic degradation produces sticky doughs of poor loaf characteristics. The action of fungal proteases on gluten is inhibited by salt. The proteases show little effect on doughs containing 2% salt (Samuel, 1972).

Fermentation and aging follows the filtration of beer. During storage, cloudiness sometimes develops in beer caused by peptides and polyphenolic procyanidins and some metal ions (Hough et al, 1982). In 1910, a cash award was offered by the United States Brewmasters Association to find a solution for this particular problem. Wallerstein in 1911 came up with a solution by using such proteases as bromelin, papain, and pepsin to stabilize and chillproof beer. Papain is mainly used to degrade the protein to low molecular weight peptides. The affinity of papain for the soluble protein in beer must be very specific, as there is very little proteolysis that occurs. The process must be carefully controlled so that enough polypeptides of sufficient size remain to entrap carbon dioxide in the formation of a head on the beer.

4. Assay for Proteinases

Because their protein substrates are not a homogeneous group and because different peptide bonds are hydrolyzed, the activity of proteolytic enzymes is not easy to measure. Several methods have been used (Ward, 1983). Casein, hemoglobin, or other protein and synthetic peptides or esters can be used as substrates.

In general, the substrate is treated with enzyme for a suitable time. After inactivating the enzyme, the amount of product is measured. This may be done by precipitation of the unhydrolyzed protein with trichloroacetic acid and measuring the absorbance of the supernatant at 280 nm. With milk-clotting enzymes the time required for a given amount of enzyme to start clotting a standard milk sample is a more useful measure of activity than a general measurement of proteolytic activity.

B. Starch-Degrading Enzymes

Starches from various sources contain about 80% amylopectin and 20% amylose. Amylose is a polymer in which glucose units number 300–400 and they are joined through α-1,4 glycosidic bonds. In amylopectin, chains of α-1,4–linked glucose units are joined through β-1,6 glycosidic bonds. Several enzymes can hydrolyze the bonds in starch molecules such as α and β amylases and glucoamylases. Alpha and beta amylases hydrolyze amylose completely to maltose, while glucoamylase hydrolyzes amylose completely to glucose (Figs. 1 and 2). The investigation of amylolytic enzymes is widespread due to applications in industry, such as conversion of starch to different sugars and use as a food sweetners.

1. Alpha Amylase (α-1,4-D-Glucan Glucanohydrolase)

Alpha amylase is obtained from plants, animals, as well as microorganisms. Pancreatic α-amylases hydrolyze starches of indigested food to oligosaccharides. Bacterial as well as fungal amylases have commercial importance (Fogarty, 1983; Crueger and Crueger, 1977). The amylases produced from bacteria such as *Bacillus amyloliquefaciens*, *B. licheniformis*, and *B. stearothermophilus* are endoenzymes that hydrolyze internal α-1,4 glycosidic linkages in amylose, amylopectin, and glycogen. Amylose is hydrolyzed to maltotriose. Maltotriose is hydrolyzed only slowly by amylase to produce glucose and

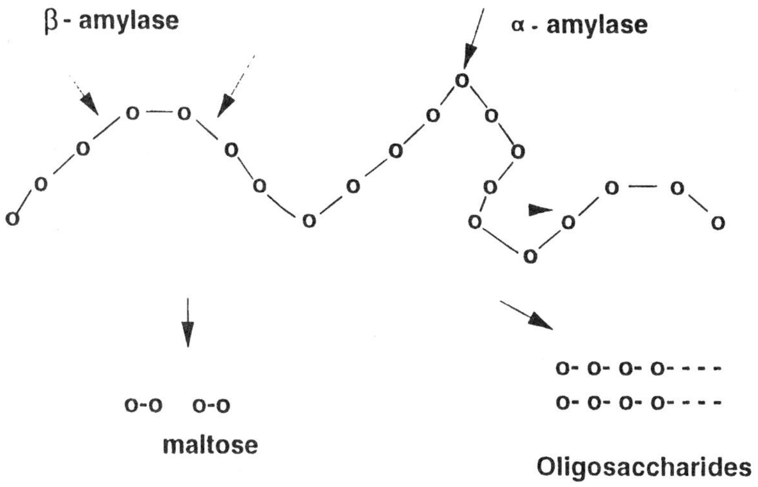

Figure 1 Amylose, a straight chain polymer.

maltose. As amylases cannot attack α-1,6 bonds, amylopectin is hydrolyzed to glucose, maltose, and various limit dextrins that consist of four or more glucose units. The nature of the dextrins formed depends upon the amylase used.

These enzymes are stable in a pH range of 5.5–8.0. If the pH is decreased below the optimum range, these enzymes may be inactivated. Above the pH range formation of byproducts increases, which results in a loss of product yield and an increase in refining costs.

Calcium has a profound effect on enhancing the thermostability of α-amylases. In the presence of calcium ions the enzymes are resistant to inactivation at elevated tempera-

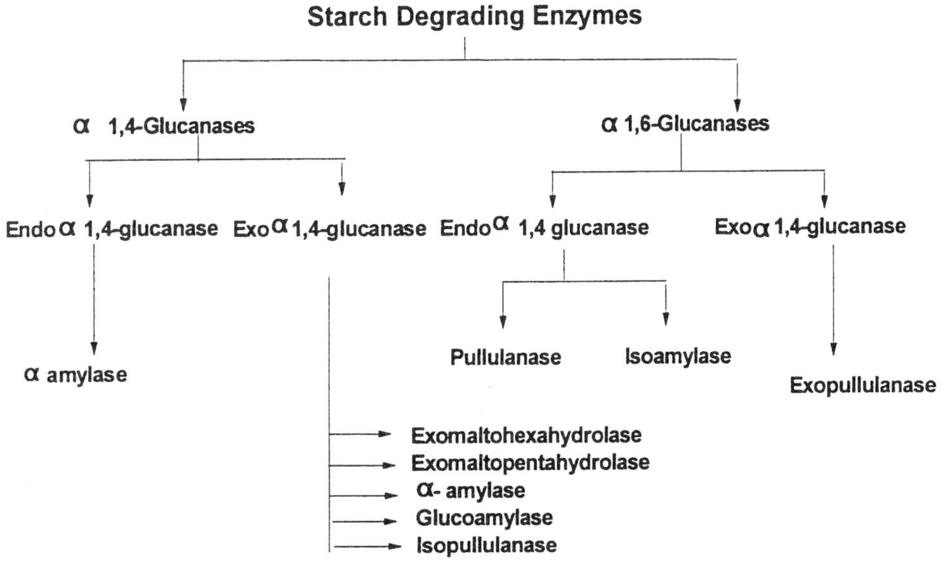

Figure 2 Starch-degrading enzymes.

tures, extremes of pH, treatment with urea, and exposure to proteolytic enzymes. Ca^{2+} binds to the protein very tightly and helps in maintaining the secondary and tertiary structure of the molecule (Stein and Fischer, 1958). Generally 100–200 ppm of calcium is sufficient to enhance the stability of enzymes at high temperatures. The enzymes from *B. licheneiformis* can be used at 110°C for a short duration.

Alpha amylases are used for the production of maltose syrups. The enzymatic process is preferable to acid hydrolysis, because the latter may produce undesirable byproducts such as 5-hydroxymethyl 1-2-furfuraldehyde and anhydroglucose compounds. But some pretreatment methods are required before the enzymatic hydrolysis of starch. The starch is exposed to temperatures above 60°C with subsequent addition of α-amylase from *B. subtilis*, which results in the swelling and disruption of particles due to high temperature and an enzymatic hydrolysis of glycosidic bonds. This process thins the starch slurry. The thinned starch slurry contains 30–40 % solids. The reaction further proceeds in at 103–107°C at pH 6.5. Temperature is lowered to 95°C, and the reaction milleu is held for 1–2 h. This liquefaction process results in a product with a dextrose equivalent of 10–20 and provides a partially hydrolyzed mixture of reduced viscosity (Norman, 1981). These derivatives have commercial value in the food industries.

The pH of the liquefied starch is adjusted to 5, and the temperature is lowered to 50°C. Fungal β-amylase is added to the mixture, and the reaction is allowed to proceed for 24–48 h. When the desired extent of hydrolysis has occurred, the crude syrup is filtered and decolorized with activated carbon. Ion exchange is then used to remove inorganic ash. The overall process is summarized in Fig. 3. The high maltose syrups produced by this process show a reduced tendency to crystallize and are relatively nonhygroscopic. They find use in the manufacture of candy and frozen desserts and in the baking industry.

It has been observed that the most effective amylases are normally thermostable. Bacteria such as Bacillus species, actinomycetes like micromonospora, thermomonospora, and fungi such as Humicola, Mucor, etc., are excellent test organisms for the study of amylase (Table 2). Heat-labile enzymes, such as the amylase from *A. oryzae*, cannot be

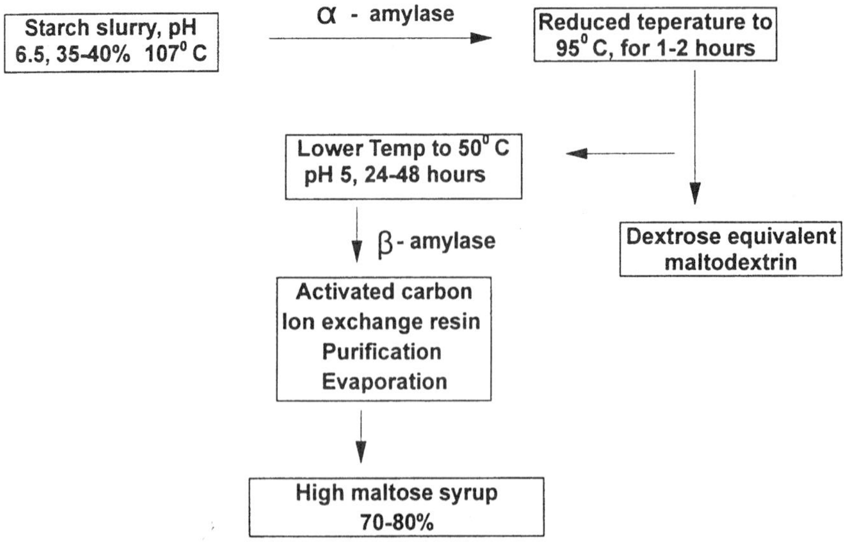

Figure 3 Enzymatic hydrolysis of starch to yield high maltose syrups.

Table 2 Properties of α-Amylase from Three Microbial Sources

	Enzyme source		
Property	*B. amyloliquefaciens*	*B. licheniformis*	*A. oryzae*
Optimum pH	5–9	7–9	4–7
Optimum temperature	70°C	90°C	55°C
End products	Maltotriose	Maltose, maltotriose	Maltose, maltotriose
Isoelectric pH	5.2	5.2	—

used in the liquefaction process because of the elevated temperatures that are required to bring about starch gelatinization. The *A. oryzae* enzyme can be used to hydrolyze liquefied starch to maltose and maltotriose (Barfoed, 1976). Generally the liquefaction of starch is carried out in a batch process using free enzyme, but continuous processes with immobilized α-amylase to reduce the processing costs have also been attempted (Yi-Hsu-J. et al., 1995).

2. Beta-Amylase (EC 3.21.2.2; β-1,4-D-Glucan Maltohydrolase)

Beta amylase is found primarily in microorganisms and such higher plants as barley, wheat, rye, oats, sorghum, soybeans, and sweet potatoes. Microbial β-amylases have been obtained from *B. megaterium* (Higashihara and Okada, 1974) and *B. cereus* (Takasaki, 1976). Plant β-amylases are sulfhydryl enzymes and are sensitive to heavy metal ions and oxidizing agents (Rowe and Weitl, 1962). The bacterial enzymes have higher heat resistance and higher pH optima than their plant counterparts. Beta amylases have higher molecular weight than α-amylases.

In the reaction mechanism of action of β-amylase, maltose splits off from amylose. Enzymes have three specific groups on their active sites (X, A, B) which help in binding and transforming the substrate. The X group recognizes and interacts with the hydroxyl group of the fourth carbon atom of the nonreducing end of the polysaccharide chain. Then the other glucosidic linkages of substrate positioned at A and B groups form an enzyme–substrate complex. In the enzyme–substrate complex, the imidazole group donates a hydrogen to the glycosidic oxygen and forms an intermediate, maltosylenzyme, while the carboxyl group helps in regenerating the enzyme.

3. Glucoamylase (EC 3.2.1.3; α-1,4-D-Glucan Glucohydrolase)

The complete hydrolysis of starch to glucose is generally accomplished by fungal glucoamylase. *A. niger*, *A. awamori*, *A. oryzae*, and *Rhizopus oryzae* are good sources of glucoamylase. The most important commercial application of glucoamylase is the formation of syrups of up to 96% glucose. Such syrups may be used for making crystalline glucose (Kingman, 1969). Glucoamylase hydrolyzes the α-1,4 linkage from the reducing end of amylose, amylopectin, and glycogen to yield glucose residues. The total conversion of starch to glucose can be achieved at 60°C and pH 4.0 and the reaction time is 3–4 days. After an incubation the crude syrup can be vaccum filtered and purified with activated carbon and ion exchange resins (Norman, 1981).

The low pH reduces unwanted isomerization reactions to fructose and other sugars, and also restricts the contamination of microorganisms during the process. The rate of the hydrolysis depends upon the size of the molecule and the order of α-1,4 and α-1,6 linkages. Glucoamylase also acts on the reverse reaction where dextrose molecules are combined to form maltose and isomaltose. The reversion of dextrose specifically involves the condensation of a β anomer of D-glucopyranose with either an α- or β-D-glucose molecule in the presence of glucoamylase.

$$\text{D-Glucose} + \beta\text{-D-Glucose} \rightarrow \text{Disaccharide} + H_2O$$

The high dextrose equivalent syrups produced by glucoamylase are used in the brewing, baking, soft drink, canning, and confectionery industries. Glucoamylase is important in light beer production. Under optimum conditions glucoamylase is capable of transforming at least 95% of dextrins to glucose, which in turn can be easily fermented by yeasts during beer production. To measure glucoamylase activity in commercial preparations, different units are used by different manufacturers, but the most common units are comparable in magnitude. A unit may be grams of glucose produced from soluble starch per hour at pH 4.2 and 60°C.

4. Pullulanases

Prolonged action of α-amylases on starch produces limit dextrins that cannot be hydrolyzed further. Hydrolysis with β-amylases also stops at α-1,6 branch points and produces limit dextrins (Robyt and Whelan, 1976). Pullulanases produced by *Aerobasidium pullulans* specifically hydrolyze the α-1,6 glycosidic bonds found in the fungal carbohydrate pullulan. Pullulanase will also cleave the α-1,6 links found in amylopectin and limit dextrins provided that there are at least two glucose residues on either side of the α-1,6 bond.

Pullulanases are produced by *Bacillus cereus*, *Klebsiella* spp, *Streptococcus mites*, and *Escherichia* intermedia, but the enzyme from *Klebsiella* has been the most investigated (Norman, 1981). The pullulanases of industrial importance are produced by *K. pneumoniae* and *B. cereus*. The enzymes attack α-1,6 linkages randomly to yield maltotriose and branched maltotriose oligosaccharide. The oligosaccharide may eventually be broken down to maltotriose (Fig. 4). It appears that the primary use of pullulanases in the food industry will be the production of maltose and maltose syrups. The use of these enzymes in combinations yields products in excess of 80% maltose.

5. Glucose Isomerase (EC 5.3.1.5)

It is generally accepted that the glucose isomerase used commercially is actually D-xylose isomerase (Antrim et al., 1979; Bucke, 1983, Hemmingsen, 1979). These enzymes normally catalyze the conversion of D-xylose to xylulose. They also act on glucose and other substrates but at much lower rates.

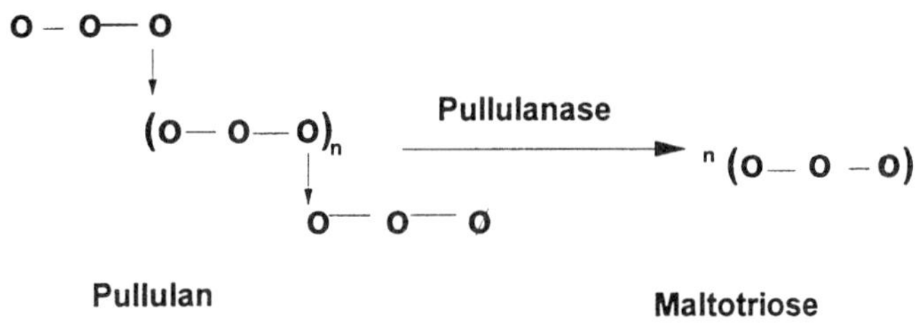

Figure 4 Action of pullulanase on pullulan.

The conversion of glucose to fructose can be effected chemically at high temperatures and in alkaline conditions. In dilute alkaline solutions, reducing sugars are transformed into mixtures of aldoses and ketoses. This is an example of a class of reactions known as the Lobry de Bruyn–Alberda van Eckenstein transformation (Fig. 5). In such procedures, however, it is difficult to obtain 40% fructose without the formation of unmetabolizable products such as psicose and other objectional products. These unwanted materials result in reduced sweetness and may contribute to poor color and off-flavor (Antrim et al., 1979).

Glucose isomerases have been prepared from Lactobacillus (Yamanaka, 1968), Streptomyces, and *Bacillus coagulans* (Yashimura, 1966). The enzymes are heat stable and operate in the range of 45–65°C. The optimal temperatures for activity range from 45°C for the enzymes from *L. brevis* (Yamanaka, 1968) to 90°C for the *Actiniplanes missouriensis* enzyme (Scallet et al., 1974), with most having optima around 65°C. Stability at elevated temperatures is desirable, not only because the reaction rate is increased, but also because microbial contamination is reduced.

Glucose isomerase requires such metal ions such as magnesium, cobalt, manganese, or chromium for activity (Antrim et al., 1979). The enzymes are inhibited by copper, zinc, nickel (Scallet et al., 1974), calcium (Aschengreen, 1975), silver, and mercury ions (Takasoki et al., 1969). Inactivation by calcium ions may be due to competition with the required magnesium ions for the enzyme's active site, and inactivation by mercury suggests the presence of important thiol groups. These enzymes are also inhibited by sugar alcohols, especially xylitol (Scallet et al., 1974; Young et al., 1975; Yamanaka and Takahara, 1977). As trishydroxymethylaminomethane is an inhibitor, it should not be used as a buffer when glucose isomerase are being studied (Danno, 1970).

Glucose isomerases catalyze the production of an equilibrium mixture containing 55–60% fructose; the actual proportions vary slightly with temperature. As it would take

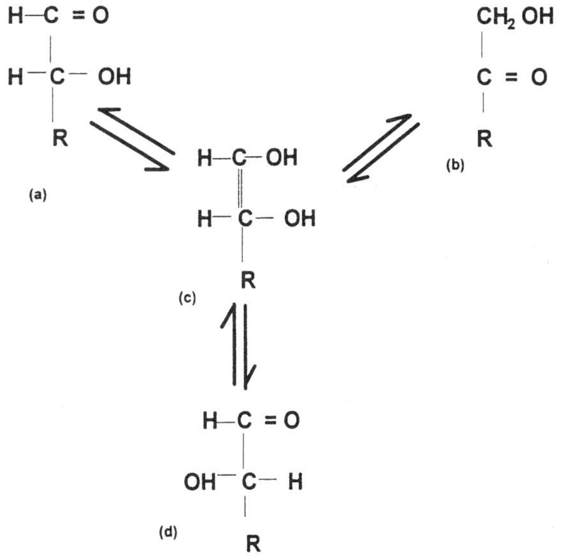

Figure 5 Lobry de Bruyn–Alberda van Eckenstein transformation that converts glucose to fructose.

too long to reach equilibrium, producers usually aim for a mixture containing 42% fructose. Typically the feedstock contains 93% glucose produced by the action of amylase and glucoamylase (Bucke, 1981).

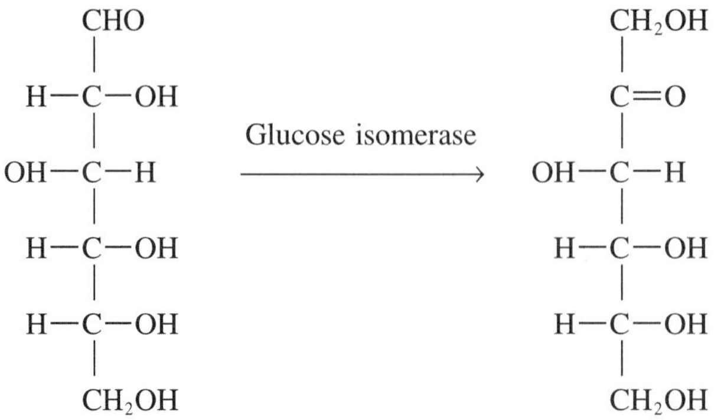

D—Glucose D—Fructose

Because calcium ions are inhibitory to glucose isomerase, it is necessary to deionize syrups after saccharification to remove the calcium ions that are required for α-amylase activity and are always present in starches. Appropriate amounts of magnesium ions are then added for the isomerization step. Alternatively an α-amylase such as that from *Bacillus licheniformis*, which is not dependent on calcium for activity, could be used in the saccharification process (Aschengreen, 1975; Hollo et al., 1975).

Glucose isomerase is an expensive enzyme. It would not have been commercially practicable to use this enzyme were it not for the development of solid supports. The activity of glucose isomerases is measured in terms of the amount of enzyme required to convert 1 μmol or 1 g of glucose to fructose per given time (Antrim et al., 1979). The product fructose is determined spectrophotometrically using cystein carbazole or other ketone-condensing reagents (Fig. 6) (Takasaki, 1966; Lloyd et al., 1972).

C. Pectin-Degrading Enzymes

Along with celluloses and hemicelluloses, pectic substances are the major component of cell wall in higher plants. These plant polysaccharides play an important role in maintaining cell wall structure. The pectic enzymes generally termed as pectinases act on pectic substances. The enzyme sources are commonly derived from Aspergillus, Rhizopus, and some species of Penicillium. The pectin obtained from plant juices, saps, apple pomace, tomatoes, citrus, and beet pulp acts as an inducer during the production of pectic enzymes by submerged fermentation. Their use in the food industry is to improve the filtration rate, increase extraction of fruit pulp, reduce turbidity, and provide clarification. Pectic enzymes reduce haze of grape juice during the wine-making process. Use of pectic enzymes also promotes faster aging of wine. Pectinases also play an important role in coffee and tea fermentation (Jones and Jones, 1984). There are three different types of pectic enzymes; pectinesterase, lyase, and polygalacturonase.

1. Pectinesterase (EC 3.1.1.1.1)

Pectin esterases de-esterify pectin to produce pectic acid and methanol. The enzymes are specific for the methyl ester of pectic acid and will not attack the methyl ester of polymannuronic acid (alginic acid) or gum tragacanth. The ethyl esters of pectic acid are

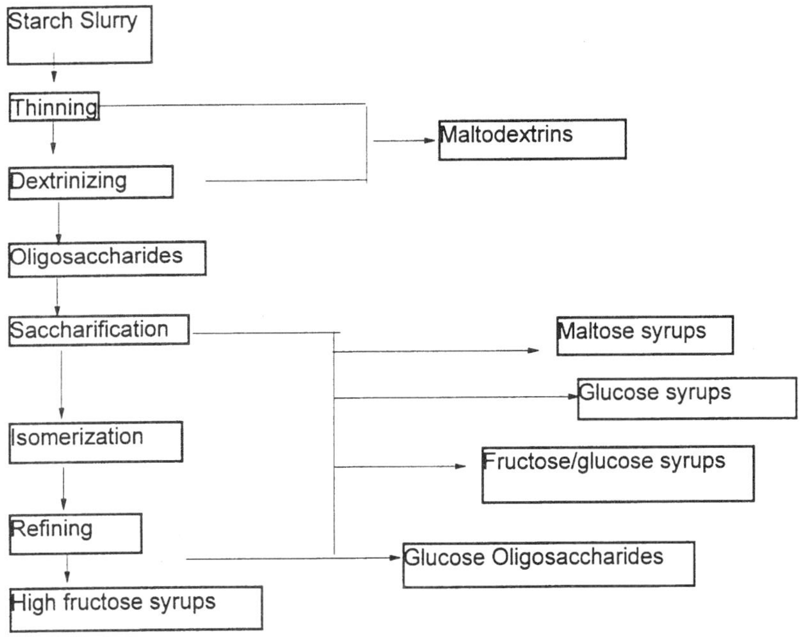

Figure 6 Degradation of starch.

hydrolyzed at reduced rates (Macdonnell et al., 1950). The action of pectinesterase does not proceed to completion but stops at a degree of esterification that has been found to vary between 0.4 and 11%. For substrates containing less than 10 galacturonic acid units, the reaction rate of pectinesterase from orange decreases with substrate chain length until no activity is observed with the triethyl ester of the trimer (McCready and Seegmiller, 1954). There is some support for the hypothesis that pectinesterases act on methyl ester groups that are adjacent to free carboxyl groups (Solms and Denel, 1975). Some microbial enzymes have alkaline pH optima.

The presence of univalent and divalent metal ions enhances the activity of pectinesterases severalfold. The effect is especially pronounced with enzymes of plant origin. Although the enzymes have been reported to have an unusual resistance to chemicals, inhibition by anionic detergents (McColloch and Kertesz, 1947), sucrose, glycerol, and D-glucose (Change et al., 1965) has been reported. They are also inhibited by polygalacturonates with a degree of polymerization greater than 8 (Thermote et al., 1977).

Pectinesterase activity can be measured by determining the carboxyl groups that are formed in pectin by the enzyme (Colle and Wood, 1961). The amount of methanol released can also be measured (Holden, 1945). Methanol may be converted to methyl nitrite, which can be analyzed by gas-liquid chromatography (GLC) (Bartolome and Hoff, 1972). Activity can also be followed by monitoring pH changes in the pH range 7.5–7. [^{14}C]-Methyl pectin has been used as a substrate in pectinesterase assays (Kauss et al., 1969). After enzyme deeesterification, the pectin is precipitated with methanol. The [^{14}C] methol in the supernatant can then be determined in a scintillation counter. One unit of pectinesterase is defined as the amount of enzyme that liberates 1 μmol free carboxyl group (or methanol) per minute under specific assay conditions.

2. Pectin- and Pectate-Depolymerizing Enzymes

Depolymerizing enzymes cleave the α-1,4 glycosidic bonds in pectin and pectin acid. There are two main classes of pectin/pectic acid–depolymerizing enzymes; polygalacturonases and lyases. Polygalacturonases are sometimes referred to as pectin or pectic acid hydrolases.

a. Polygalacturonases (EC 3.2.1.15). Polygalacturonases hydrolyze internal bonds in pectic acid, resulting in rapid reduction in viscosity (Kelly and Fogarty, 1978; Rombouts and Pilnik, 1980). The enzymes are specific for high molecular weight pectic acid and show reduced activity with increased esterification. Activity with oligogalacturonates as substrate decreases with decreasing chain length, and little action is observed with the trimer and dimer (Rexova-Baenkova, 1973). Thus when pectic acid is hydrolyzed by endopolygalacturonases, monomers, dimers, and trimers of galacturonic acid accumulate as end products (Mount et al., 1970). Polygalacturonases are again divided based on the nature of substrate and the activity of enzyme. Enzymes of plant, fungal, and bacterial origin have been described (Rombouts and Pilnik, 1980).

The pH optima of the enzymes are generally in the region 4–5.5 However, the enzyme from *Corticium rolfsii* has an optimum pH of 2.5 (Kaji and Okada, 1969). The pH optima appear to depend on the degree of polymerization of the substrate (Barash and Eyal, 1970).

Exopolygalacturonases (EC 3.2.1.67) hydrolyze the terminal α-1,4 bonds in pectates at the nonreducing end to release galacturonic acid monomers. Several enzymes from plants have been described (Riov, 1975; Pressey and Avants, 1976; Bartley, 1978). Although exopolygalacturonases prefer high molecular weight pectates, they will, unlike the endoenzyme, accept digalacturonic acid as a substrate. The activity of exoenzymes depends on the degree of substrate polymerization (Pressey and Avants, 1973; Mills, 1966). The activity of exopolygalacturonases results in only a gradual reduction in viscosity.

b. Lyases. Highly esterified polymethylgalacturonic acid is directly attacked by pectin lyases (EC 4.2.2.10). Most pectin lyases are of fungal origin. All are endoenzymes and cleave highly esterified pectin at random with a rapid decrease in viscosity. Only glycosidic bonds adjacent to methyl ester groups are split in an eliminative mechanism. This results in the formation of products with a double bond between C—4 and C—5; conjugation of this double bond of the carboxyl group on C—5 leads to light absorption with a maximum at 235 nm. Because of their preference for highly esterified pectin, pectin lyases can be used in the processing of fruit juices that contain highly esterified pectin (Ishii and Yokotsuka, 1975). The pH optima for the enzyme vary from 5.2 to 8.7 (Fogarty and Kelly, 1983). The activity of most pectin lyases is enhanced in the presence of calcium ions. The extent of stimulation depends on the pH and degree of esterification of the substrate (Edstrom and Phaff, 1964).

Endopectate lyases (EC 4.2.2.2) randomly split partially or completely de-esterified pectin chains. They are produced by fungi and bacteria, and their pH optima is in the range 8–10. The pH optimum varies with the chain length of the substrate (Ward and Fogarty, 1972). These enzymes require calcium for activity. Although pectates are generally good substrates, some enzymes prefer substrates with a degree of esterification of 21–44%. The activity of endopectate lyases decreases with decrease in the chain length of the substrate (Atallah and Nagel, 1977).

Exopectate lyases (EC 4.2.2.9) liberate unsaturated dimers from the reducing end of pectic acid. The pH optima for the enzymes are in the range 8–9.5. The enzymes prefer pectate over pectins, and completely esterified pectins are not accepted as substrates. The smallest substrate that can be degraded is the trimer.

Enzymatic treatment of soft fruit pulp facilitates pressing and improves juice and anthocyanin pigment yields (Neubeck, 1975). Pectin-degrading enzymes have been used to degrade highly esterified apple pectin and increase juice yields (DeVos and Pilnik, 1973). The enzyme-treated juices have a slightly higher methanol content than juices obtained by pressing. The health hazard of such low methanol levels has not been assessed but may be of no consequence.

When pectinases are added to viscous and turbid freshly pressed fruit juices, the cloudy material agglomerates and forms flocs. It has been suggested that the enzymes act by dissolving away the negatively charged pectin coating of cloud particles. Electrostatic interaction of the destabilized cloud material would then result in coagulation (Endo, 1965; Yamasaki et al., 1967). Such clarification and removal of pectin is essential for juices that have to be concentrated. The process is often carried out at elevated temperature to increase the rate of reaction and discourage microbial growth and unwanted fermentation (Grampp, 1977).

Pectinases can be used to recover juices from pulp that is sieved out of freshly pressed juice (Braddock and Kesterton, 1976). Pectic enzymes can also be used in conjugation with mechanical processes to produce macerated suspensions of loose cells from fruit and vegetables. Enzyme treatment can also be used to lower the viscosity of such products. When pectinases are used together with cellulases, effective liquefaction of fruit and vegetable products can be achieved.

The activity of pectin-depolymerizing enzymes may be determined by measuring the rate of increase of reducing groups by titration with hypoiodite. After the oxidation of reducing carbonyl groups with iodine, unreacted iodine is measured by titration with sodium thiosulfate (Phaff, 1960; Mills and Tuttobello, 1961). Reducing groups may also be determined colorimetrically with 3,5-dinitrosalicylic acid-phenol reagent (Borel et al., 1952) or cuprous reagent (Somogyi, 1952; Spiro, 1966; Milner and Avigad, 1967). With highly esterified substrates, alkaline conditions should be avoided because under these conditions these substrates may be split by β elimination, leading to high values for reducing group content (Albersheim et al., 1960a,b). For esterified substrate the methods of Launer and Tomimatsu (1959) are suitable.

The activity of exoenzymes can be monitored by measuring the amount of D-galacturonate degraded in the internal positions. The activity of endoenzymes can also be followed by measuring the decrease in substrate viscosity (Pressey and Avants, 1973; Wimborne and Richard, 1978; Mills and Tuttobello, 1961). As viscosities of pectic acid materials are influenced by temperature, pH, and ionic strength, these factors must be carefully controlled. Viscosity changes can be used to distinguish between exoenzymes and endoenzymes. A greater proportion of glycosidic bonds need to be cleaved by exoenzymes to achieve the same reduction in viscosity effected by endoenzymes.

In commercial applications enzymes may be compared by determining the minimal amounts of an enzyme required to perform a given task (Bauman, 1981). The double bond introduced by the action of lyases leads to maximal absorption at 235 nm (Albersheim et al, 1960a,b) or 230 nm (Starr and Moran, 1962). Increase in absorbance at the two wavelengths can therefore be used to follow lyase activity (Saio and Kaji, 1980; Rombouts et al., 1978; Kelly and Fogarty, 1978). This method cannot be used when the unsaturated galacturonic acid monomer is produced. The unsaturated monomer rearranges to 4-deoxy-L-threo-5-hexoseulose uronic acid, which does not absorb at 235 nm. This product, however, may be measured by the triobarbituric acid method.

D. Cellulases

Cellulose is a linear polymer of a simple sugar or saccharide which does not occur in pure form in any natural resource. Most forms of cellulose which are available in nature contain about 10% by weight noncellulosic polysaccahrides, proteins, and mineral elements. Most cellulosic materials are generally found in association with hemicellulose, pectin, and lignin. In cellulose, glucose units are joined by β-1,4 glycosidic bonds. The degree of polymerization varies from 15 to 14,000, but most chains are 3000 glucose units long (Cowling and Brown, 1969). The linear polymers are held together by hydrogen bonding to form crystalline fibrils. Crystalline regions may be interspaced with less-or-dered amorphous regions. Although cellulolytic enzymes from bacteria, fungi, plants, and invertebrates have been described (Whitaker, 1971), only enzymes of microbial origin have industrial potential. Several bacteria—cellulomonas, clostridium, acetobutlylcum cellovibrio, *Clostridium thermomonospora*—and fungi like *Trichoderma reesei*, *Penicillium fusicolsum*, and *A. niger* are good producers of this enzymes.

Three major types of enzyme activity are now generally accepted as acting in concert or in sequence to bring about the degradation of cellulose (Enari, 1983). Endoglucanase (EC 3.2.14; 1,4-β-D-glucan-4-glucanohydrolase) hydrolyzes β-1,4 glycosidic bonds in a random fashion. Cellobiose is not a substrate for this enzyme. Otherwise the enzyme does not appear to be highly specific for any substrate.

Cellobiohydrolase (EC 3.2.14.1; 4-β-glucan-4-glucanohydrolase) cleaves off cellobiose units from the nonreducing ends of cellulose polymers. Substituted celluloses and cellobiose are not hydrolyzed. The third enzyme, β-D-glucoside glucohydrolas (EC 3.2.1.21), hydrolyzes cellobiose and oligosaccharides derived from cellulose to yield glucose.

The association of cellulose with hemicellulose and lignin in most cellulose sources makes the commercial degradation of cellulose difficult as the native cellulose is highly resistant to hydrolysis. The associated materials have to be removed and the crystalline structure destroyed if the enzymatic hydrolysis is to be improved. Treatment with alkali, acid, or steam and milling treatments may be used to prepare cellulosic materials for enzymatic hydrolysis (Enari, 1983).

Several methods are available to determine cellulose-degrading activity (Goksyr and Eriksen, 1980; Enari, 1983). A variety of substrates including cotton fiber, filter paper, and cellulose derivatives such as carboxymethylcellulose, hydroxyethylcellulose, and Avicel have been reported in the literature.

Alpha-glucosidase activity can be measured using cellobiose, saiicin, or p-nitrophenyl-α-D-glucoside as substrate (Selby and Maitland, 1967). The reactions may be followed by determining the reducing sugars produced or by measuring changes in viscosity or absorbance of light.

For commercial purposes it is the total solubilizing effect that is of interest. The amount of reducing sugars present is not a reliable measure of enzyme activity. Total solubilizing effect may be conveniently measured using dyed substrates such as Avicel, Solka floc, and filter paper. The activity of endoglucanases is generally determined by following the reduction in the viscosity of treated carboxymethylcellulose. The activity of cellobiohydrolase may be determined by measuring the reducing sugars formed from a cotton or Avicel substrate.

E. Lipases

Lipases bring about the hydrolysis of insoluble triacylglycerols to produce glycerol and fatty acids. The hydrolysis of triglycerides is reversible and triglycerides may be formed from free fatty acids and glycerol. Their preference for emulsified substrates distinguishes lipases from esterases, which attack substrates in solution. Solid triglycerides are poor substrates for lipases (Suigara and Isobe, 1975). Simple alkyl esters are hydrolyzed by lipases but at reduced rates. The substrate specificity of various lipases has been reviewed (Shahani, 1975; MacRae, 1983).

A group of enzymes will hydrolyze triglycerides at positions 1 and 3 only to produce 1,2-diglycerides and 2-monoglycerides by migration of the fatty acid moiety. Because of this migration, the complete hydrolysis of some triglycerides can be achieved even though no lipases are known to release fatty acids from the 2 position.

Lipases from *Candidum*, *Candida cylindracae (Candida rugośa)*, *Corynebacterium acnes*, *Chromobacterium viscosum*, *Penicillium cyclopium*, and *Humicola lanuginosa* have been reported to be nonspecific, whereas lipases from *Aspergillus niger*, *Mucor javanicus*, *Rhizopus*, and *Pseudomonas fragi*, and pancreatic lipase, show 1,3 specificity (MacRae, 1983). Most microbial enzymes do not appear to show specificity for fatty acids. However, an enzyme from *G. candidum* preferentially releases fatty acids with a cis double bond in position 9 (Jensen, 1974). A lipase from *P. cyclopium* has been shown to hydrolyze the partial glycerides diolein and monolein faster than the triglyceride (Okumura et al., 1976).

Lipases may be used instead of sodium or sodium methoxide to effect interesterification reactions. Because the lipase reaction is reversible, the triglycerides may be hydrolyzed and resynthesized. Resynthesis to yield a product with rearranged fatty acids occurs under conditions of reduced water. With nonspecific enzymes the interesterified products are similar to those obtained by chemical means. With 1,3-specific enzymes, interesterification is limited to positions 1 and 3.

Mixtures of free acids and glycerides may be used for interesterification reactions. There is thus the possibility of producing novel triglycerides, some of which may have desirable properties. Using such procedures it is possible to introduce unsaturated fatty acids such as linoleic acid into saturated fats.

The hydrolysis of milk fat can result in rancid flavors in milk, cream, and other dairy products (Arnold et al., 1975). Controlled lipolysis with enzymes specific for short-chain fatty acids makes it possible to develop desirable flavors (Nelson, 1972). The prepared fat substrate is combined with the enzyme preparation and the mixture is blended into a stable emulsion. After incubation at a suitable, temperature to achieve the desirable extent of lipolysis, the enzyme can be inactivated by pasteurization. Enzymes specific for short-chain fatty acids have also found application in generating flavors in Italian blue and cheddar cheese (Moskovitz et al., 1977)

The activity of lipases may be determined using emulsified triglycerides. Soluble glycerides are not suitable because they are hydrolyzed slowly by lipases and may be hydrolyzed faster by other esterases. Triolein, the best substrate, is too expensive; it may be replaced with less expensive olive oil. Emulsion of the oil stabilized with gum arabic or other emulsion stabilizers are treated with enzyme. The release of fatty acids inhibits glyceride hydrolysis. For this reason calcium is added to the reaction mixture to precipitate fatty acids as calcium salts. After stopping the reaction with organic solvent, the fatty acid product is titrated using thymol blue as an indicator. The release of fatty acids can also be followed with a pH-stat.

VIII. WORLDWIDE REGULATIONS GOVERNING THE USE OF ENZYMES

The authors are more familiar with U.S. laws and regulations regarding enzymes, and views on other countries are not necessarily derived from first-hand knowledge.

A. United States of America

The Food and Drug Administration of the U.S. Department of Health and Human Services has promulgated regulations pertaining to foods, which are published as the Code of Federal Regulations (CFR). This code is divided into 50 titles that represent broad subject areas that come under the scrutiny of federal regulations. Title 21 of CFR pertains to food and drugs. Each title is divided into chapters, which usually bear the name of the issuing agency. Each chapter is subdivided into parts covering specific regulatory areas. Thus most regulations pertaining to foods are 21 CFR 100–199 (Title 21 of the Code of Federal Regulations, Parts 100–199). Part 173 is ''Secondary Direct Food Additives Permitted in Foods for Human Consumption.'' Subpart B of Part 173 specifically deals with enzyme preparations and microorganisms. Subpart B has nine specific items as follows:

173.110 Amyloglucosidase derived from *Rhizopus niveus*
173.120 Carbohydrase and cellulase derived from *Aspergillus niger*
173.130 Carbohydrase derived from *Rhizopus oryzae*
173.135 Catalase derived from *Micrococcus lysodiekticus*
173.140 Esterase–lipase derived from *Mucor miehei*
173.145 α-Galactosidase derived from *Morteirella vinaceae* var. *raffinoseutilizer*
173.150 Milk-clotting enzymes, microbial
173.160 *Candida guilliermondii*
173.165 *Candida lypolytica*

These enzymes are considered direct food additives.

Another group of enzymes are listed in 21 CFR 184, ''Direct Food Substances Affirmed as Generally Recognized as Safe'' (GRAS). 21 CFR 184.1027 is mixed carbohydrase and protease enzyme product; 184.1372 pertains to insoluble glucose isomerase products; 184.1388 addresses lactase enzyme preparation from *Kluyveromyces lactis*; 184.1585 is papain; 184.1685 is rennet (animal-derived). These lists do not account for some other enzymes that are used in food processing, such as pectinases, which are covered under 21 CFR 101, which deals with food labeling. Subpart F covers exemption from food labeling requirements and specifically under 1.2(ii) makes provisions for processing aids. Processing aids further defined as

1. Substances that are added to a food during the processing of such food but are removed in some manner from the food before it is packaged in its finished form

2. Substances that are added to a food during processing that are converted into constituents normally present in the food and do not significantly increases the amount of the constituents naturally found in the food
3. Substances that are added to a food for their technical or functional effect in the finished food at insignificant levels and do not have any technical or functional effect in that food

Item 3 is perhaps most fitting for enzymes such as pectinases in the sense that they are used to reduce viscosity and are then destroyed. Tenderization of meat can also be classified in this category or even perhaps under item 2.

Roland (1981) reported that the GRAS Review Branch of the FDA had received 11 GRAS petitions dealing with enzymes of which only one was processed as a food additive. One petition in particular (3-0016) submitted by the Ad Hoc Enzyme Technical Committee covered a large number of microbial animal and plant enzymes, and even after 7 years no action had been taken. A summary of the current status of enzymes in the United States is presented in Table 3.

The Enzyme Technical Association (ETA) is carrying on a dialogue with the FDA in its quest to seek GRAS status for microbial enzymes. Enzymes listed in 21 CFR often specify not only genus and species of microorganisms but also varieties or variants. This position, if adopted by FDA into new regulations, would narrow the definition of acceptable microbes and exclude certain species from the pre-1958 use category. The FDA's understanding of enzymes would view all microbial enzyme preparations as mixtures of carbohydrases, proteases, and lipases. Further, the status of newer enzymes produced as a result of genetic manipulation of microorganism is an issue that the FDA is studying. One final note on GRAS status and what it means; the GRAS list reflects FDA opinions as to which substances are generally recognized as safe, but the final authority as to GRAS status does not rest with the FDA but with "qualified scientists." The Federal Food, Drug and Cosmetic Act does not define the level of scientific training and experience needed to qualify as an expert. Similarly the nature of scientific information used by experts in making a GRAS designation is not specified. Further, the experts need not inform the FDA of the evidence supporting the designation and the agency need not review and approve the evidence before the substances is used in food. Therefore, if the FDA refuses to affirm an enzyme as GRAS, the food industry can still use the enzyme under its own criteria for GRAS. This in no manner prevents the FDA from initiating legal action or enforcement action to stop the use of a non–FDA-approved GRAS enzyme.

B. Canada

Enzymes are treated as food additives and therefore require approval. Approval is obtained by filing a petition in accordance with procedures in Section B.16.002 of the Food and Drug Regulations. The enzymes currently approved (Denner, 1983) are listed in Table 4. Maximum level of use of enzymes should follow good manufacturing practice. The law also specifies enzyme, source, and permitted uses.

C. European Economic Community

The Association of Microbial Food Enzyme Producers (AMFEP) is the EEC's equivalent of the U.S. Ad Hoc Enzyme Technical Committee. This organization has formulated general standards for enzyme regulation (Anonymous, 1983).

Table 3 Enzymes Permitted for Food Use by the United States Food and Drug Administration

Enzyme	Source	Regulatory status
α-Amylase	*Aspergillus niger*	GRAS
	A. oryzae	GRAS
	Rhizopus oryzae	GRAS
	Bacillus stearothermophilus	Petitioned for GRAS
	B. subtilis	GRAS
	B. licheniformis	184.1027
	Barley malt	GRAS
β-Amylase	*Barley malt*	GRAS
Cellulase	*Aspergillus niger*	GRAS
	Trichoderma reesei	Petitioned for GRAS
α-Galactosidase	*Morteirella vinaceae* var. *raffinoseutilizer*	173.145
Glucoamylase or amyloglucosidase	*Aspergillus niger*	GRAS
	A. oryzae	GRAS
	Rhizopus oryzae	173.145
	R. niveus	173.110
Invertase	*Saccharomyces cerevisiae*	GRAS
Lactase	*A. niger*	GRAS
	A. oryzae	GRAS
	Kluyveromyces fragilis	GRAS
	Candida pseudotropicalis	Petitioned for GRAS
Pectinase	*Aspergillus niger*	GRAS
	Rhizopus oryzae	173.130
Glucose isomerase (immobilized preparation)	*Streptomyces rubiginosus*	184.1372
	Actinoplanes missouriensis	184.1372
	Streptomyces olivaceus	184.1372
	S. olivochromogenes	184.1372
	Bacillus coagulans	184.1372
	Arthrobacter globiformis	Petitioned for GRAS
	Streptomyces murinus	Petitioned for GRAS
Catalase	*Micrococcus lysodeikticus*	173.130
	A. niger	GRAS
	Bovine liver	GRAS
Glucose oxidase	A. niger	GRAS
Bromelain	pineapples	GRAS
Ficin	figs	GRAS
Papain	Papaya	GRAS
Glucose oxidase	*Aspergillus niger*	GRAS
Bromelain	Pineapples	GRAS
Ficin	Figs	GRAS
Papain	Papaya	GRAS
Milk-clotting enzyme	*Endothia parasitica*	173.150(a)(1)
	Bacillus cereus	173.150(a)(2)
	Mucor pusillus Lindt	173.150(a)(3)
	M. miehei Cooney et Emerson	173.150(a)(4)
Rennet	Ruminant fourth stomach (abomasum)	GRAS
Pepsin	Porcine and bovine stomach	GRAS
Trypsin	Porcine, bovine pancrease	GRAS
Esterase/lipase	*Mucor miehei*	173.140
	Bacillus licheniformis	184.1027
	Aspergillus niger	GRAS
	Bacillus subtilis	GRAS
Lipase	Calf, kid, and lamb pancreatic tissue	GRAS
	Aspergillus niger	GRAS
	A. oryzae	GRAS
β-Glucanase	*Aspergillus niger*	GRAS

Source: Denner (1983).

Table 4 Enzymes Permitted for Use in Foods of Canada

Enzymes	Source	Permitted uses
Bovine rennet	Aqueous extract	Cheese, cottage cheese, cream cheese, cream cheese with named additives, cream cheese spread, cream cheese spread with named additives
Bromelain	Pineapples	Ale, beer, light beer, malt liquor, porter, stout; bread, flour, whole wheat flour, edible collagen; sausage casings, hydrolyzed animal, milk, and vegetable proteins, meat cuts; meat tenderizing, preparations; pumping pickle for curing beef cuts
Ficin	Fig tree latex	Same uses as bromelain except not used in bread, flour, whole wheat flour, or pumping pickles
Papain	Fruit of papaya	Same as for bromelain and also for beef before slaughter and precooked (instant) cereals
Milk-coagulating enzyme	*Mucor miehei*, *M. pusillus*, *Endothia parasitica*	Cheese, cottage cheese, sour cream, emmentaler (Swiss) cheese
Pepsin	Glandular layer of porcine stomach	Ale, beer, light beer, malt liquor, porter, stout; cheese, cottage cheese, cream cheese, cream cheese with named ingredients, cream cheese spread, cream cheese spread with named ingredients, defatted soy flour, precooked (instant) cereals
Protease	*Aspergillus oryzae*, *A. niger*, *Bacillus subtilis*	Same as for bromelain and also dairy-based flouring preparations, distillers mash, industrially spray-dried cheese powder, precooked (instant) cereals, unstandardized bakery foods
Rennet	Aqueous extracts from fourth stomach of calves, kids, or lambs	Same as for bovine rennet and in unstandardized milk-based dessert preparations
Lipase	*Aspergillus niger*, *A. oryzae*; edible forestomach tissue of calves, kids, or lamb; animal pancreatic tissue	Dairy-based flavoring preparations, liquid and dried egg white, Romano cheese
Lipoxidase	Soybean whey or meal	Bread, flour
Pancreatin	Pancrease of the hog or ox	Liquid and dried egg white, precooked (instant) cereals, starch used in the production of dextrins, maltose, dextrose, or glucose, or glucose solids
Catalase	*Aspergillus niger*, *Micrococcus lysodeikticus* bovine liver	Soft drinks, egg albumin
Glucose oxidase	*Aspergillus niger*	Soft drinks, liquid whole egg, egg white, and liquid egg yolk destined for drying
Cellulase	*Aspergillus niger*	Distillers mash, liquid coffee concentrate, spice extracts, natural flavor and color extractives

Table 4 Continued

Enzymes	Source	Permitted uses
Hemicellulase	*Bacillus subtilis*	Ale, beer, light beer, malt liquor, porter, corn for degerming distillers mash destined for vinegar manufacture, unstandardized bakery products
Pentosanase	*Aspergillus niger*	Ale, beer, light beer, malt liquor, porter, stout, corn for degerming distillers mash, mash destined for vinegar manufacture, unstandardized bakery products
Pectinase	*Aspergillus niger*, *Rhizopus oryzae*	Cider; wine; distillers mash; juice of named fruits; natural flavor and color extractives; skin of citrus fruits destined for jam, marmalade, and candied fruit production; vegetable stock for use in soup manufacture
Glucanase	*Aspergillus niger*, *Bacillus subtilis*	Same as for pentosanase
Invertase	*Saccharomyces*	Soft-centered and liquid-centered confections, unstandardized baking foods
Lactase	*Aspergillus niger*, *Rhizopus oryzae*, *Saccharomyces*	Lactose-reducing enzyme preparations, milk destined for use in ice cream mix
Glucoamylase (amyloglucosidase maltase)	*Aspergillus niger*, *A. oryzae*,	Ale, beer, light beer, malt liquor, porter, stout, bread, flour, whole wheat flour, chocolate syrups, distillers mash, precooked (instant) cereals, starch used in the production of dextrins, maltose, dextrose, glucose (glucose syrups), glucose solids, unstandardized bakery products
		Distillers mash
		Glucose solids, unstandardized bakery products
	Rhizopus niveus	Distillers mash
	Rhizopus delemar	Mash destined for vinegar making
	Multiplici sporus	Brewers mash, distillers mash, mash for vinegar making, starch used in production of dextrins, maltose, dextrose, glucose (glucose syrups), glucose solids (dried glucose syrups)

Another advisory trade group is the Association of Manufactures of Animal/Plant-Derived Food Enzymes (AMAFE), which unlike AMFEP concerns itself with only enzymes from nonmicrobial sources. Like the Food Chemical Codex in the United States the EEC organization also relies on data submitted and compiled by the FAO/WHO Joint Expert Committee on Food Additives (JECFA, 1981). The JECFA data reflect both FCC and AMAFE/AMFEP advice on enzymes. Individual EEC member states have different regulatory views on enzymes, and these are discussed subsequently.

1. Belgium

This country considers enzymes as processing aids, and authorization is needed for their use. The list of permitted enzymes is narrow; an attempt was made to enlarge and revise this list by royal decree, but this attempt was unsuccessful.

2. Denmark

This largest producer of enzymes in the world requires that 6 months prior to the use of any enzyme a notification to and registration with the Danish National Food Institute be completed. The director of the Institute can approve the use within a 6-month period, but if at the end of this period the director has not responded to the notification, then approval is automatically assumed. The procedures for notification and registration of enzyme products were first published in 1974 and are being revised.

3. Federal Republic of Germany

Enzymes are considered to be food additives. No authorization and sanctions are required, except for cheese, fruit juices, and wine.

4. France

Decrees are necessary for using food additives such as enzymes. In addition memos are issued periodically declaring the ''toleration'' of specific enzymes not sanctioned by prior decree. Applications for the use of enzymes must be filed with Service de la Repression des Frandes.

5. Greece, Ireland, and the Netherlands

These countries have no specific controls on the use of enzymes.

6. Italy

Enzymes are considered to be processing aids, and their use in beer and wine is controlled by decree.

7. United Kingdom

Enzymes are deemed to be processing aids, and as such there are no controls. The Food Additive and Contaminants Committee has recommended that controls be instituted for the use of enzymes. A prior sanction for certain enzymes and sources was also recommended (Table 5).

The list in Table 5 does not consider lipooxygenase, catalase from *Micrococcus lysodeikticus*, hemicellulases, ficin, β-amylase, pentosanase, or esterases as recommended for food use. Additionally, certain microorganisms such as *Penicillium funiculosum*, *P. lilacinum*, *P. emersonii*, *Klebsiella aerogenes*, and *Streptomyces fradiae*, which are not recognized as safe organisms in the United States or Canada, would be permitted for use in the United Kingdom.

8. Japan

Additives can be classified as natural or synthetic. If a naturally occurring compound is modified or synthesized, then it becomes a synthetic additive. For example, citric acid in a lemon would be a natural additive; sodium citrate obtained from fermentation and

Table 5 Food Additives and Contaminants Committee Recommended Prior Sanctioned Enzymes and Sources of Such Enzymes in the United Kingdom

Enzymes	Sources
Acid proteinase (including pepsin and chymosin)	Porcine gastric mucosa; abomasum of calf or lamb; adult bovine abomasum; *Mucor miehei*, *M. pusillus*
α-Amylase	Porcine or bovine pancreatic tissue, *Aspergillus niger*, *A. oryzae*, *Bacillus licheniformis*, *B. subtilis*
Bromelain	*Ananas bracteatus*, *A. comosus*
Catalase	Bovine liver, *Aspergillus niger*
Cellulase	*Aspergillus niger*, *Trichoderma viride*
Dextranase	*Penicillium funiculosum*, *P. lilacinum*
Endothia carboxyl proteinase	*Endothia parasitica*
Fructofuranosidase (invertase)	*Saccharomyces cerevisiae*
Galactofuranosidase (invertase)	*Aspergillus niger*
Endo-1,3(4)-β-D-glucosidase (laminarinase)	*Aspergillus niger*, *Bacillus subtilis*, *Penicillum emersonii*
Glucose isomerase	*Bacillus coagulans*
Glucose isomerase (immobilized)	*Bacillus coagulans*, *Streptomyces olivaceous*
Glucose oxidase	*Aspergillus niger*
Exo-1,4-D-glucosidase (glucoamylase)	*Aspergillus niger*
Exo-1, 4-D-glucosidase (immobilized)	*Aspergillus niger*
Neutral proteinase	*Aspergillus oryzae*, *Bacillus subtilis*
Papain/chymopapain	*Carica papaya*
Pectin esterase	*Aspergillus niger*
Pectin lyase	*Aspergillus niger*
Polygalacturonase	*Aspergillus niger*
Pullulanase	*Klebsiella aerogenes*
Serine proteinase (including ltrypsin)	Porcine or bovine pancratic tissue, *Bacillus licheniformis*, *Streptomyces fradiae*
Triacylglycerol lipase	Edible oral forestomach tissues of the calf, kid, or lamb; porcine or bovine pancreatic tissue

neutralization is the same as chemically synthesized citric acid and therefore is considered synthetic. If an additive falls into the natural category, it needs no approval and listing.

The Food sanitation Investigation Council in accordance with food sanitation laws is involved in the approval process. In order for a compound to be approved the applicant must show that

1. The additive is demonstrably safe.
2. The additive is advantageous to consumer in that it satisfies one or more of the following conditions: (a) it is necessary in the manufacturing process, (b) it helps to maintain the nutritional value of the food, (c) it prevents or minimizes degradation of food, and (e) it reduces the price to the consumer.
3. The additive can be shown to be either superior to or more useful than existing additives.

4. The additive can be chemically identified by analysis in the food as sold to the consumer.

If the additive (1) could disguise poor quality in either the basic food or manufacturing process, (2) impairs nutritional quality, (3) acts as curative or therapeutic, or (4) causes an improvement that could be duplicated by modifying the manufacturing process, it will be rejected.

Nutritional aspects of food additives receive more explicit attention, and there is a list of specific permitted nutritional additives. Once an additive passes preliminary scrutiny, the next step is to determine toxicological safety at proposed levels of use. Such testing must be done at two authentic research institutes in Japan or at such institute if reliable supporting evidence is available from foreign literature sources.

IX. TOXICOLOGY

The purpose of this section is to describe the tests that enzymes are subjected to in order to determine their safety. Many factors are involved in determining safety.

A. Safety of the Source Organism and Its Metabolites

The safety of the source organism should be of prime importance. If the enzyme is derived from nontoxicogenic, non–antibiotic-producing strain of bacterium, yeast or mold normally present in edible food, it would seem reasonable to expect that it would be safe. In the selection of organisms for producing enzymes, parent strains are subjected to genetic and environmental pressures to enhance enzyme synthesis. A technique called mutation is often used. If an organism is safe or nontoxic, it is unlikely that it will be converted to a toxic organism upon mutation.

Second, factors that favor optimal growth may promote toxin production. In other growth environments toxin production may be suppressed to levels below detectable amounts. Therefore absolute toxicity or nontoxicity cannot be resolved. What is commonly done is that the organism is grown under a variety of growth conditions and tested for toxin, and a negative test is taken as reasonable proof of the unlikelihood of the organism being toxic.

Pathogenicity is a clearly demonstrable trait. If an enzyme-producing organism is pathogenic, it will not be used in commercial enzyme production.

Carcinogenicity and mutagenicity are two other traits that are examined. An organism and its enzyme that promotes tumor growth or causes mutation would be deemed unsafe. If current good manufacturing practices are followed, then sources of enzymes approved by the FDA in the United States are neither carcinogenic nor mutagenic.

Teratogenic and reproductive effects of enzymes in food have been shown to be inconsequential. Pariza and Foster (1983) report a study in which four generations of rats were fed rennet from *Mucor pusillus* and no ill effects were observed.

The AMEFP classifies microorganisms used for enzyme production into three groups: (1) microorganisms that have traditionally been use in food or in food processings, (2) microorganisms that are accepted as harmless contaminants present in food, and (3) microorganisms that are not included in groups 1 or 2. Antibacterial activity is to be tested with six organisms and their classifications according to AMEFT are listed in Table 6.

Table 6 Classification of Microorganisms According to Their Use in Food Enzyme Manufacture

Group characteristics	Microorganisms
1. Microorganisms traditionally used in food processing	*Bacillus subtilis* (including strains *mesentericus*, *innatto*, and *amyloliquifaciens*); *Aspergillus niger* (including strains *awamori*, *foetidus*, *saitoe*, and *usami*); *A. oryzae* (including strains *sojae* and *effeses*); *Mucor javanicus*; *Rhizopus arrhizus*; *R. oligosporus*; *R. Oryzae*; *Saccharomyces cerevisiae*; *Kluveromyces fragilis*; *K. lactis*; *Leuconostoc oenus*
2. Microorganisms that are accepted as harmless	*Bacillus stearothermophilus*; *B. licheniformis*; *B. coagulans*; *B. megaterium*; *B. circulans*; *Klebsiella aerogenes*
3. Microorganisms not included in groups A or B	*Mucor miechei*; *M. pusillus*; *Endothia parasitica*; *Actinoplanes missouriensis*; *Streptomyces albus*; *Bacillus cereus*; *Trichoderme reesei (viride)*; *Penicillium lilacium*; *P. emersonii*; *Sporotrichum dimorphosporum*; *Sterptomyces olivaceus*; *P. simplicissium*; *P. funiculosum*

B. Safety During Growth of the Organism

Pure culture techniques are essential in large-scale cultivation of microorganisms in order to ensure that contaminants do not grow and compromise the integrity of the end product. The AMEFP had formulated good manufacturing practices that recommend the checking of ''controlled microbial purity.'' This term recognizes the limitations of commercial sterility and the cost associated with asepsis versus antisepsis. Samples for microbial purity are drawn from stock culture, seed culture before transfer to the fermenter, fermentation liquid at regular intervals during the fermentation, and final sample before transfer to the recovery plant. The uninoculated medium used for seed culture and in the fermenter should be checked for absence of microbial growth. The check for purity can be visual inspection, i.e., macroscopic and microscopic observations. Here such phenotype characteristics as color (especially molds), Gram's stain, spore formation, and cell morphology (clusters, chains, single cells) are observed to ascertain if the traits are what are accepted as ''normal'' fermentation. Streaking of plates also offers one method of quantitating the number of contaminants and enables the isolation of colonies followed by further tests to identify the genus and species of the contaminants.

Submerged broth fermentations may be amenable to all of the controls mentioned, but semisolid fermentations do not offer a convenient method of obtaining representative samples for microbiological analysis. In such cases the absence of undesired growth is assessed solely by visual inspection. Another index of significant contamination can be the fermentation process parameters, such as enzyme activity, pH, absorbance/turbidity, and oxygen consumption. Often these parameters, if understood, monitored, and controlled appropriately, may signal contamination problems in advance of visual observations. If the batch is contaminated, it is discarded.

C. Safety Testing of Enzyme Preparations

Two different viewpoints exist in this area: the AMFEP philosophy and the U.S. FDA approach. The AMFEP recommends that microorganisms classified as group 1 do not require routine testing, whereas microorganisms in groups 2 and 3 should be routinely tested for acute oral toxicity with rats and mice as test animals, for subacute oral toxicity with rats over a 4-week period, and for oral toxicity with rats over a 3-month span, as well as for in vitro mutagenicity using the Ames test. In addition, microorganisms in group 3 must be tested for pathogenicity to rats. In vivo mutagenicity to mice and hampsters, toxicity studies on the final food, carcinogenicity to rats, and effects on fertility and reproduction may be performed under exceptional conditions.

The U.S. FDA places the burden of proof on the manufacturer and/or user of the substance. General guidelines are as follows:

An additive whose molecular structure is not associated with any known toxic potential and whose presence in the daily diet is less than 0.05 ppm would cause a low level of concern. Tests likely to be associated with this concern are a 28-day continuous feeding study in rodents and a genetic toxicity screen.

An additive whose molecular structure is not associated with any known toxic potential but which is present in the daily diet at levels of 0.05–1.0 ppm would cause an intermediate level of concern. Tests such as a 90-day (subchronic) feeding study in a rodent species, a subchronic feeding study in a nonrodent species, a reproductive study with a teratology phase in the rodent, and a genetic toxicity screen may be required.

An additive with a presence in the daily diet of 1 ppm or greater receives the highest level of concern and requires lifetime feeding studies for carcinogenicity in two rodent species and a reproductive study with a teratology phase in the rodent, and a genetic toxicity screen may be required.

The FDA requires submission of data to demonstrate that the microorganisms are nonpathogenic and nontoxic to humans and animals and do not produce antibiotics, with multigeneration feeding studies at three levels (1.25, 2.50, and 5.0% of diet) on a rodent and reproductive study with in utero exposure with teratology. Further, at least 20 litters per generation per level each for reproductive and teratological changes are considered to be minimally acceptable for toxicology studies. For histological assessment, a minimum of 20 surviving animals per dose per sex is considered acceptable. In order to accomplish this, studies begin with 25–30 males and 25–30 females per dose to ensure that 20 litters from F_0 are called F_1, and 50 females per dose are selected. Of these, 25 males and 25 females are used for the growth and development phase of testing and undergo histological examination after the delivery of the F_2a litter. Use of 25 animals ensures that 20 animals per sex per dose will be available for histology. The remaining 25 males and 25 females of the F_1 generation can be used to produce 20 litters for the F_2b generation. The F_2b generation has to be delivered by cesarean section for teratological analysis. Half of these fetuses should undergo viceral and the other half skeletal examination.

These toxicity tests require (1) the use of a hardy, outbred strain with a known and stable background incidence of tumors; (2) the use of two; and preferably three, test dose levels; and (3) termination of trails when total tumor incidence in controls is 20–25%. Since these tests are expensive ($500,000 to $750,000 each) and rodents have a 2- to 3-year lifetime, it is advisable that preliminary in vitro screening be accomplished.

In addition, JECFA, AMFEP, and FDA have general provisions for contaminants and their limits. These are listed in Table 7.

Table 7 Contaminants in Microbial Enzymes and Specification Limits

Contaminant	Specification limit
Arsenic	3 mg/kg
Lead	10 mg/kg
Heavy metals	50 mg/kg
Aflatoxin	5 mg/kg
Other mycotoxins	Tolerances to be established as methods to monitor become available
Pseudomonas	Absent using U.S. Pharmacopeia method (*P. aeruginosa*)
Salmonella	0 per 25 g sample
Coliforms	30/g using a 10-g sample
Antibiotic activity	Absent using JECFA/Food Chemical Codex method

D. Safety of Immobilized Enzyme Preparations

Immobilization of enzymes refers to techniques that permit the containment of enzymes and their separation from the reaction milieu in a manner that renders them reusable. Immobilization enzymes posed several questions that differed from earlier concerns for enzymes. The main concern of regulatory officials was the possibility of the release of enzyme from the immobilized matrix. The possible presence of immobilized enzyme in food streams and the safety of reagents used in achieving immobilization were other areas of concern.

A number of mechanisms may operate in degrading the immobilized enzyme. Some examples are

1. Degradation of enzyme or cross-linker by microorganisms
2. Protein displacement of adsorbed or cross-linked enzyme
3. Ionic displacement of enzyme
4. Adverse pH environment
5. Degradation due to enzymes normally present in foods
6. Adverse cleaning or sanitation producers by plant personnel
7. Particle attrition due to recycle cleaning in plug-floor reactors or in continuously stirred reactors

To minimize these concerns, several practices can be adopted. Sources of the enzymes should be restricted to species suitable for production of soluble enzymes.

The carrier on which the enzyme is immobilized should be inert and generally acceptable for use in food manufacture; it should have mechanical strength, thermal stability, and chemical durability.

The immobilization procedure should be safe and tested as safe in use. Toxicity studies of immobilized enzymes should be carried out.

Leakage tests should be performed to demonstrate that leakages of immobilizing agent, matrix, enzyme, and immobilized enzymes are not problems. Such tests can only be done if suitable assay methods are available.

Product specifications for soluble enzymes should also apply to immobilized enzymes.

These issues of safety and toxicity have been discussed from the point of view of the additive per se. The physical state of the additives and the safety of the workplace in which these enzymes are used are important from an environmental toxicology viewpoint. Enzymes are proteins, and proteins are potent allergens to sensitive individuals. It is generally recommended that enzyme products provide the following information on the label: (1) trade name and generic name; (2) enzyme sources; (3) enzyme activity; (4) contents by volume or weight; (5) major ingredients by percentage to include diluents, stabilizers, preservatives, colors, and salt compositions; (6) manufacturer's name and addresses; (7) country of origin; (8) batch number and ''use-by'' date or manufacturing data ; and (9) warning notices. In many instances this information and more is contained in enzyme data sheets supplied by manufacturers to their clients. The caution notice can be worded as follows:

> Product contains enzymes which may cause allergic reaction in sensitive individuals. Handle the product with care and in accordance with instructions given by the manufacturer's technical literature. Avoid contact with eyes, skin, and mucus membranes. In case of accidental spillage, wash with large volume of water. In case of accidental contact with skin or eyes, wash immediately with tap water.

Typical compositions of enzymes are given in Table 8. One of the enzymes is a liquid, another a powdered preparation. Colors commonly used are annatto or caramel; preservatives commonly used are benzoic acids and its esters; odor masks are cumin, oil of anise, or oil of dill; and standardizing diluents can be lactose, starch, dextrins, and sodium chloride.

Hypersensitivity disease can vary in severity of response and can exhibit either macro or micro effects. Immune responses can be respiratory, ophthalmic or dermatological. These responses have been described in detail by Farrow (1981). Treatment of symptoms may be as simple as removal of the allergen in the environment of the sensitive individuals or the required various medications. For the prevention of such problems the enzyme manufacturer can build in features that allow a margin of safety.

There is also the probability that legislative action in Europe and elsewhere will require the use of liquid enzymes only. Liquid enzymes can be concentrated and measured conveniently in closed pumping systems. Further, because of the increased viscosity of

Table 8 Typical Composition of Liquid and Dry Enzyme Preparations

	Enzyme preparation (% w/w)	
Constituent	Liquid	Powder
Enzyme-active protein	0.5–5.0	1–8
Inactive protein	2–5	1–5
Standardizing diluent	—	90^+
Electrolytes	16–18	2–5
Color	0.05–1.0	Not added
Preservatives	0.2–1.0	Not added
Flavor masks	0.05–0.02	0.1–0.3
Water	90^+	3–5

liquid concentrates, splashing is less of a problem and minimizes the tendency toward aerosolization. Increased concentration increases product stability and reduces storage and freight costs.

Dry enzymes are agglomerated by several techniques that produce encapsulated or aggregated powders. Generally powders in which particle sizes are smaller than 10 μm pose problems of dust and attendant allergies. It is therefore the aim of dry enzyme manufacturers to produce large particles of 500 μm and no particle smaller than 200 μm. Enzymes can be embedded in spheres of waxy materials consisting of a nonionic surfactant by means of spray-cooling or the prilling process. The marumerizer process is another important method by which enzyme is mixed with a filler, binder, and water and then extruded and subsequently formed into spheres in a marumerizer.

General safety precautions should be exercised during the handling of enzymes:

Aviod inhalation of powdered enzymes and contact with eyes, skin, and mucus membranes.
Wear suitable protective clothing, masks, goggles, and gloves.
Work in well-ventilated environments free from drafts.
Avoid contact of liquid enzymes with skin and mucus membranes.
Warn and train all personnel handling enzymes.
Provide safety showers and maintain records of personnel affected by hypersensitivity reactions

Prevent hypersensitized people, asthmatics, and sufferers from dermatitis, hay fever, and eczema from handling enzymes.

Health risks should be kept in perspective in relation to other industrial chemicals. Enzymes are key to the future of industrial food processing, and adoption of precautionary, inexpensive practices can greatly minimize the hazards associated with the use of enzymes.

REFERENCES

Albersheim, P., Neukom, H., Devel, H. 1960a. Splitting of pectin chain molecules in neutral solution. *Arch. Biochem. Biophys.* 90:46–51.

Albersheim, P., Neukom, H., Devel, H. 1960b. Uber de bildung von ungesattigten Abbau producten durch ein perkinabbauendes enzym. *Helv. Chim. Acta* 43:1422–1426.

Anonymous. 1983. *General Standards for Enzyme Regulations.* The Association of Microbial Food Enzyme Procedures, Brussels, pp. 1–14.

Antrim, R. L., Colilla, W., Schnyder, B. J. 1979. Glucose isomerase production of high fructose corn syrup. In: *Applied Biochemistry and Bioengineering*, Vol. 2, *Enzyme Technology*, Wingard, L. B., Jr., Katchalski-Katzir, E., Goldstein, L. (Eds.). Academic, New York, pp. 98–156.

Arnold, R. G., Shahani, K. M., Divivewdi, B. K. 1975. Application of lipolytic enzymes to flavour development in dairy foods. *J. Dairy Sci.* 58:1127–1143.

Asato, N. R., Rand, A. G., Jr. 1977. Activation studies of multiple forms of prochymosin (prorennin). *Biochem. J.* 167:429–434.

Aschengreen, N. H. 1975. Production of glucose/fructose syrup. *Process Biochem.* 10(4):17–19.

Atallah, M. T., Nagel, C. W. 1977. The role of calcium ions in the activity of an endo-pectic acid lyses on oligogalacturonides. *J. Food Biochem.* 1:185–206.

Aunstrup, K. 1977. Industrial approach to enzyme production. In: *Biotechnological Applications of Proteins and Enzymes*, Bohak, Z., Sharon, N. (Eds.). Academic, New York, pp. 39–49.

Aunstrup, K. 1980. Proteinases. In: *Microbial Enzymes and Bioconversions*, Rose, A. H. (Ed.). Academic, New York, pp. 50–112.

Barash, I., Eyal, Z. 1970. Properties of apolygalactuonase produced by *Geotrichum candidum*. *Phytopathology* 60:27–30.

Barfoed, H. C. 1976. Enzymes in starch processing. *Cereal Foods World* 21:588–604.

Barman, T. E. 1969. *Enzyme Handbook*, Vols. 1 and 2. Springer Verlag Chemie, New York.

Bartley, I. M. 1978. Exo-polygalaturonase of apple. *Phytochemistry* 17:213–216.

Bartolome, L. C., Hoff, J. E. 1972. Gas chromatographic methods for the assay of pectin methyl esterases, free methanol and methoxy groups in plant tissues. *J. Agr. Food Chem.* 20:262–266.

Baurnan, J. W. 1981. Application of enzymes in fruit juice technology. In: *Enzymes and Food Processing*, Birch, G. G., Blakebrough, N., Parker. K. J. (Eds.). Applied Science Publishers, New York, pp. 129–148.

Bender, M. L., Clement, G. R., Kedzy, F. J., Heck, H. D. 1964. The correlation of pH (pD) dependence and stepwise mechanism of α-chyumotrypsin-catalysed reaction. *J. Am. Chem. Soc.* 86: 3680–3690.

Bergmeyer, H. U. 1974. *Enzymatic Methods of Analysis*. Springer Verlag, Dusseldorf.

Birch, G. G., Blackbrough, N., Parker, K. J. 1981. *Enzymes and Food Processing*. Applied Science Publishers, London.

Bone, R., Silen, J. L., Agard, D. A. 1989. Structural plasticity brodens the specificity of an engineered protease. *Nature* (London) 339:191–195.

Borel, E., Hostettler, F., Devel, H. 1952. Quantitative zuckerbestimmungmit 3.5-dinitrosalicylsaure und phenol. *Helv. Chim. Acta* 35:115–120.

Braddock, R. J., Kesterton, J. W. 1976. Enzyme to reduce viscosity and increase recovery of soluble solids from citrus pulp washing operations. *J. Food Sci.* 41:82–85.

Bucke, C. 1977. Industrial glucose isomerase. In: *Topics in Enzyme and Fermentation Biotechnology*, Vol. 1, Wiseman, A. (Ed.). Halsted, New York, pp. 147–171.

Bucke, C. 1981. Enzymes in fructose manufacture. In: *Enzymes and Food Processing*, Birch, G. G. Blakebrough, N., Parker, K. J. (Eds.). Applied Science Publishers, New York, pp. 51–72.

Bucke, C. 1983. Glucose transforming enzymes. In: *Microbial Enzymes and Biotechnology*, Fogarty, W. M. (Ed.). Applied Science Publishers, New York, pp. 93–130.

Change, L. W. S., Morita, L. L., Yamamota, H. Y. 1965. Papaya pectinesterase inhibition by sucrose. *J. Food Sci.* 30:218–222.

Cheeseman, G. C. 1981. Rennet and cheesemaking. In: *Enzymes and Food Processing*, Birch, G. G., Blakebrough, N., Parker, K. J. (Eds.). Applied Science Publishers, London, pp. 195–212.

Cianci, J. 1986. The market outlook for enzymes. In: *World Biotechnology Report 1986*, Vol. 2, Part 3. Proceedings of Online Conference, San Fransisco, pp. 55–71.

Colle, M., Wood, R. K. S. 1961. Pectic enzymes and phenolic substances in apples rotted by fungi. *Ann. Bot.* 25:435–452.

Colwick, S. P., Kaplan, N. O. 1955. Methods in Enzymology. Academic, New York.

Cowling, E. B., Brown, W. 1969. Cellulases and their applications. *Adv. Chem. Ser.* 95:152.

Crueger, W., Crueger, A. 1977. *Biotechnology—A Textbook of Industrial Microbiology*. Sinauer Associates, Sunderland, MA.

Dalgliesh, D. G. 1982. The enzymatic coagulation of milk. In: *Development in Dairy Chemistry*, Vol. 1, *Proteins*, Fox, P. F. (Ed.). Applied Science Publishers, London, pp. 157–187.

Danno, G. 1970. Studies on D-glucose isomerization enzyme from *Bacillus coagulans* strain NH-88. Part V. *Agr. Biol. Chem.* 34:1805–1814.

Denner, W. H. B. 1983. The legislative aspects of the use of industrial enzymes. In: *Industrial Enzymology: The Application of Enzymes in Industry*, Godfrey, T., Reichelt, J. (Eds.). Nature Press, New York, pp. 111–137.

DeVos, L., Pilnik, W. 1973. Proteolytic enzymes in apple juice extraction. *Process Biochem.* 8:18–19.

Dransfield, E., Etherington, D. 1981. Enzymes in the tenderization of meat. In: *Enzymes and Food Processing*, Birch, G. G., Blakebrough, N., Parker, K. J. (Eds.). Applied Science Publishers, London, pp. 177–194.

Edstrom, R. D., Phaff, H. J. 1964. Elimination cleavage of pectin and of oligogalacturonide methyl esters by pectin transeliminase. *J. Biol. Chem.* 239:2409–2415.

Emeri, G. H., Gum, E. K., Lang, J. A., Liu, T. H., Brown, R. D. 1974. Cellulases. In: *Food Related Enzymes*, Whitaker, J. R. (Ed.). Adv. Chem. Ser. Vol. 136. American Chemical Society, Washington, D.C., pp. 79–133.

Enari, T. M. 1983. Microbial Cellulases. In: *Microbial Enzymes and Biotechnology*, Fogarty, W. M. (Ed.). Applied Science Publishers, London, pp. 183–224.

Endo, A. 1965. Studies on pectolytic enzymes of molds. 16. Mechanism of enzyme clarification of apple juice. *Agric. Biol. Chem.* 29:229–233.

Farrow, R. I. 1981. Enzymes: health and safety considerations. In: *Enzymes and Food Processing*, Birch, G. G., Blakebrough, N., Parker, K. J. (Eds.). Applied Science Publishers, New York, pp. 261–274.

Fogarty, W. M. (Ed.). 1983. *Microbial Enzymes and Biotechnology*. Applied Science Publishers, London.

Fogarty, W. M., Kelly, C. T. 1983. Pectic Enzymes. In: *Microbial Enzymes and Biotechnology*, Fogarty, W. M. (Ed.). Applied Science Publishers, London, pp. 131–182.

Foltman, B. 1971. The biochemistry of prorennin and rennin. In: *Milk Proteins: Chemistry and Molecular Biology*, Vol. II, McKenzie, A. (Ed.). Academic, New York, pp. 236–265.

Frost, G. M. 1986. Commercial production of enzymes. In: *Developments in Food Proteins*, Vol. 4, Hudson, B. J. F. (Ed.). Applied Science Publishers, London, pp. 57–134.

Fruton, J. S. 1971. Pepsin. In: *The Enzymes*, Vol. 3, Boyer, P. D. (Ed.). Academic, New York, pp. 119–134.

Fujimaki, M., Arai, S., Yamashita, M. 1977. Enzymatic protein degradation and resynthesis for protein improvement. In: *Food Protein: Improvement Through Chemical and Enzymatic Modification*, Feeney, R. E., Whitaker, J. R. (Eds.). Advances in Chemistry Series, Vol. 160. American Chemical Society, Washington, D.C., pp. 156–175.

Glazer, A. N., Smith, E. L. 1971. Papain and other sulfahydrl proteases. In: *The Enzymes*, Vol. 3, Boyer, P. D. (Ed.). Academic, New York, pp. 502–545.

Godfrey, T., Reichelt, J. 1983. *Industrial Enzymology: The Application of Enzymes to Industry*. Nature Press, New York.

Goksyr, J., Eriksen, J. 1980. Cellulases. In: *Microbial Enzymes and Bioconversion*, Vol. 5, Rose, A. H. (Ed.). Academic, New York, pp. 283–326.

Grampp, E. 1977. Hot clarification process improves production of apple juice concentrate. *Food Technol.* 31:38–41.

Hehre, E. J., Okada, G., Genghof, D. S. 1969. Configurational specificity: unappreciated key to understanding enzymic reversion and de novo glycosidic bond syntheses. *Arch. Biochem. Biophys.* 135:75–89.

Harper, W. J., Lee, C. R. 1975. Residual coagulants in whey. *J. Food Sci.*, 40:282–284.

Hemmingsen, S. H. 1979. Developments of immobilized glucose isomerase for industrial application. In: *Applied Biochemistry and Bioengineering*, Vol. 2, *Enzyme Technology*, Wingard, L. B., Jr., Katchalski-Katzi, E., Goldstein, L. (Eds.). Academic, New York, pp. 157–184.

Higashihara, M., Okada, S. 1974. Studies on amylase of *Bacillus megatarium* Strain No. 32. *Agri. Biol. Chem.* 38:1023–1029.

Holden, M 1945. Acid producing mechanism in minced leaves. *Biochem. J.* 39:172–178.

Hollo, J., Laszlo, E., Hosehke, A. 1975. Einige probleme der herstellung fructosehaltiger sirupe ous starke. Starke 27:232–235.

Hough, J. S.; Briggs, D. E.; Stevens, R., Young, T. W. 1982. Malting and Brewing Science, Vol. II, Chapman and Hall, London, pp. 826–828.

Hyslop, D. B., Swanson, A. M., Lund, D. B. 1975. Heat inactivation of milk clotting enzymes. *J. Dairy Sci.* (Abstract) 58:795.

Iizuka, H., Mireki, S. 1978. Studies on the genus Monascus. II. Substrate specificity of two glycoamylases. *J. Gen. Appl. Microbiol.* 24:185–192.

Ishii, S., Yokotsuka, T. 1975. Purification and properties of pectin lyase from *Aspergillus ponicus*. *Agr. Biol. Chem.* 39:313–321.

JECFA. 1981. Specification for identity and purity of carrier solvents, emulsifiers and stabilizers, enzyme preparations, flavouring agents, food colors, sweetening agents and other additives. Food and Nutrition paper No. 19, 25th Session of the Joint Expert Committee on Food Additives, FAO/WHO, Rome, pp. 215–218.

Jencks, W. P. 1975. Binding energy, specificity and enzymic catalyses: the Circe effect. *Adv. Enzymol.* 43:318–340.

Jensen, R. G. 1974. Characteristics of the lipase from the mol *Geotrichum candidum*. A review. *Lipids* 9:149–157.

Jones, K. L., Jones, S. E. 1984. Fermentations involved in the production of cocoa, coffee and tea. *Prog. Ind. Microbiol.* 19:411–456.

Kaji, A., Okada, T. 1969. Purification and properties of an unusually acid-stable endo polygalaturonase produced by *Cortium rolfsii*. *Arch. Biochem. Biophys.* 131:203–209.

Kauss, H, Swanson, A. L., Arnold, R., Odzuk, W. 1969. Biosynthesis of pectic substances: localization of enzymes and products in a lipid membrane complex. *Biochim. Biophys. Acta* 192:55–61.

Kelly, C. T., Fogarty, W. M. 1978. Production and properties of polygalacturonase lyase by an alkalophilic microorganism *Bacillus* sp RK 9. *J. Microbiol.* 24:1164–1172.

Kertesz, Z. I., McColloch, R. J. 1950. Enzymes acting on pectic substances. *Adv. Carbohydrate Chem.* 5:79–101.

Kilara, A. 1985a. Enzymes modified lipid food ingredients. *Process Biochem.* 20(3):35–45.

Kilara, A. 1985b. Enzymes modified lipid food ingredients. *Process Biochem.* 20(5):149–158.

Kilara, A., Shahani, K. M., Wagner, F. W. 1977. Preparation and properties of immobilized lactase. *Lebens. Wiss Technol.* 10:84–88.

Kingman, W. G. 1969. Crystalline dextrose manufacture. *Process Biochem.* 4:19–21.

Koshland, D. E. 1960. The active site and enzyme action. *Adv. Enzymol.* 22:45–98.

Launer, H. F., Tominatsu, Y. 1959. Reaction of sodium chloride with various polysaccharides: rate studies and aldehyde group determinations. *Anal. Chem.* 31:1569–1574.

Lawrie, P. A. 1974. *Meat Science*. Pergamon, Oxford.

Lee, E. Y. C., Whelan, W. J. 1971. Glycogen and starch debranching enzymes. In: *The Enzymes*, Vol. 5. Boyer, P. E. (Ed.). Academic, New York, pp. 192–228.

Lerner, R. A., Tramontano, A. 1987. Antibodies as enzymes. *Trends Biochem. Sci.* 12:427–430.

Lloyd, N. E., Khaleeluddin, K., Lamm, W. R. 1972. Automated method for the determination of D-glucose isomerase activity. *Cereal Chem.* 49:544–553.

McColloch, R. J., Kertesz, Z. I. 1947. Pectic Enzymes. VIII. A comparison of fungal pectin methylesterase with that of higher plants, especially tomatoes. *Arch. Biochem. Biophys.* 13:217–229.

McComb, E. A., McCready, R. M. 1952. Calorimetric determination of pectic substances. *Anal. Chem.* 24:1630–1632.

McCready, R. M., Seegmiller, C. G. 1954. Action of pectic enzymes on oligogalacturonic acids and some of their derivatives. *Arch. Biochem. Biophys.* 50:440–450.

Macdonnel, L. R., Jang, R., Jansen, E. F., Lineweaver, H. 1950. The specificity of pectin esterases from several sources with some notes on purification of orange pectinesterase. *Arch. Biochem. Biophys.* 28:260–273.

MacRae, A. R. 1983. Extracellular microbial lipase. In: *Microbial Enzymes and Biotechnology*, Fogarty, W. M. (Ed.). Applied Science Publishers, London, pp. 225–250.

Manning, K. 1981. Improved viscometric assay for cellulase. *J. Biochem. Biophys. Methods* 5:189–202.

Mills, P. J. 1966. The pectic enzymes of *Aspergillus niger*. *Biochem. J.* 99:557–561.

Mills, P. J., Tuttobello, R. 1961. The pectic enzymes of *Aspergillus niger*. *Biochem. J.* 79:57–64.

Milner, Y., Avigad, G. 1967. A copper reagent for the determination of hexuronic acids and certain ketohexoses. *Carbohydrate Res.* 4:359–360.

Morrissey, P. A. 1969. The rennet hysteresis of heated milk. *J. Dairy Res.* 36:333–341.
Moskovitz, A. J., Shen, T., West, I. R., Cassagene, R., Feldman, L. I. 1977. Properties of esterase produced by *Mucor miehei* to develop flavor in dairy products. *J. Dairy Sci.* 60:1260–1265.
Mount, M. S., Bateman, D. F., Basham, H. G. 1970. Induction of electrolyte loss, tissue maceration and celular death of potato tissue by an endopolygalacturonase trans-eliminase. *Phytopathalogy* 60:924–931.
NAS/NRC. 1981. *Food Chemical Codex*, 3rd ed. National Academy of Sciences, National Research Council, Food and Nutrition Board, Committee on Codex Specifications, National Academy Press, Washington, D.C.
Nelson, J. H. 1972. Enzymatically produced flavors for fatty systems. *J. Am. Oil. Chem. Soc.* 49: 559–562.
Nelson. J. H. 1975. Impact of new milk clotting enzymes on cheese technology. *J. Dairy Sci.* 58: 1739–1750.
Neubeck, C. E. 1975. Fruits, fruit products and wine. In: *Enzymes in Food Processing*, Reed, G. (Ed.). Academic, New York, pp. 397–442.
Ney, K. H. 1971. Predictions of bitterness of peptides from their amino acid composition. *Z. Lebensm. Unters. Forsch.* 147:64–68.
Nomenclature Committee of the International Union of Biochemists on the Nomenclature and Classification of Enzymes. 1978. Enzyme Nomenclature Recommendations. Academic Press, New York.
Norman, B. E. 1981. New developments in starch syrup technology. In: *Enzymes and Food Processing*, Birch, G. G., Blakebrough, N., Parker, K. J. (Eds.). Applied Sciences Publishers, London, pp. 15–50.
Okumura, S., Iwai, M., Tsujikayo, Y. 1976. Positional specificities of four kinds of lipases. *Agr. Biol. Chem.* 40:655–660.
Palmer, T. J. 1975. Glucose syrups in foods and drinks. *Process Biochem.* 10(12):19–20.
Pariza, M. W., Foster, E. M. 1983. Determining the safety of enzymes in food processing. *J. Food Prot.* 46:453–468.
Parkin, K. L., Hultin, H. O. 1986. Characterization of trimethylamine-N-oxide demethylase demethylase activity from fish muscle microsomes. *J. Biochem.* 100:77–86.
Pazur, J. H., Kleppe, K. 1962. The hydrolysis of α-D-glucosidases by amyloglucosidase from *Aspergillus niger. J. Biol. Chem.* 237:1002–1006.
Petersen, B. D. 1981. Impact on the enzymic hydrolysis process on recovery and use of proteins. In: *Enzymes and Food Processing*, Birch, G. G., Blakrbrough, N., Parker, K. J. (Eds). Applied Science publishers, London, pp. 149–176.
Phaff, H. J. 1960. α-1,4-polygalacturomide glycanohydrolase (endo-olygalacturonase) from *Saccharomyces fragilis. Methods Enzymol.* 8:636–641.
Pressey, R., Avants, J. K. 1973. Two forms of polygalactoronase in tomatoes. *Biochem. Biophys. Acta* 309:363–369.
Pressey, R., Avants, J. K. 1976. Pear polygalacturonases. *Phytochemistry* 15:1349–1351.
Reed, G. 1975. Enzymes in Food Processing. Academic, New York.
Reese, E. T., Siu, R. G. H., Levinson, H. S. 1950. The biological degradation of soluble cellulose derivatives and its relationship to the mechanism of cellulose hydrolysis. *J. Bacteriol.* 59:485–497.
Rexova-Baenkova, I. 1973. The size of substrate binding site of an *Aspergillus niger* extracellular endopolygalacturonases. *Eur. J. Biochem.* 39:109–115.
Riov, J. 1975. Polygalacturonase activity in citrus fruit. *J. Food Sci.* 40:201–202.
Roberts, S. M.; Turner, N. J.; Willetts, A. J., Turner, M. K. 1995. The interrelationships between enzymes and cells, with particular reference to whole-cell biotransformation using bacteria and fungi. In: *Introduction to Biocatalysis Using Enzymes and Microorganisms.* Cambridge University Press, Cambridge. pp. 34–78.

Robinson, D. R., Jencks, W. P. 1965. The effect of compounds of the urea-ammonia class on the activity coefficient of acetyltetra—glycine ethyl ester and related compounds. *J. Am. Chem. Soc.* 8:2462–2470.

Robyt, J. F., Whelan, W. J. 1968. The α-amylases. In: *Starch and Its Derivatives*, 4th ed., Radley, J. A. (Ed.). Chapman and Hall, London, pp. 477–497.

Roland, J. F. 1981. Regulation of food enzymes. *Enzyme Microbiol. Technol.* 3:105–110.

Rombouts, F. M., Pilnik, W. 1980. Pectic Enzymes. In: *Microbial Enzymes and Bioconversions*, Rose, A. H. (Ed.). Academic, New York, pp. 228–272.

Rombouts, F. M., Spannsen, C. H., Visser, J., Pilnik, W. 1978. Purification and some characterstics of pectate lyase from *Pseudomonas fluorescence* GK-5. *J. Food Biochem.* 2:1–22.

Rowe, A. W., Weitl, E. D. 1962. The inhibition of β-amylose by ascorbic acid, *Biochem. Biophys. Acta* 65:245–251.

Saio, M., Kaji, A. 1980. Another pectate lyase produced by *Streptomyces nitrosporen. Agr. Biol. Chem.* 44:1345–1349.

Samuel, M. 1972. *Bakery Technology and Engineering.* Avi, Westport, CT.

Sardinas, J. L. 1976. Calf rennet substitutes. *Process Biochem.* 11(4):10–17.

Scallet, B. L., Shieh, K., Ehrenthal, I., Slapshak, L. 1974. Studies in the isomerization of D-glucose. *Staerke* 26:405–408.

Schwimmer, S. 1983. *Source Book of Food Enzymology.* Avi. Westport, CT.

Scott, D. 1953. Glucose conversion in preparation of albumin solids by glucose oxidase–catalase reaction system. *J. Agr. Food Chem.* 1:727–730.

Selby, K., Maitland, C. C. 1967. The cellulase of *Trichoderma viridie. Biochem. J.* 104:716–724.

Shahani, K. M. 1975. Lipases and esterases. In: *Enzymes in Food Processing*, Reed, G. (Ed.). Academic, New York, pp. 182–221.

Solms, J., Denel, H. 1975. Uber den mechanismus dee enzymatischen verseifung von pektinstoffers. *Helv. Chim. Acta* 38:321–329.

Somogyi, M. 1952. Notes on sugar determination. *J. Biol. Chem.* 195:19–23.

Spiro, R. G. 1966. Analysis of sugars found in glycoproteins. *Methods Enzymol.* 8:3–26.

Starr, M. P., Moran, F. 1962. Eliminative split of pectic substances by phytopathogenic soft rot bacteria. *Science* 135:920–921.

Stein, E. R., Fischer, A. H. 1958. The resistance of α-amylases towards proteolytic attack. *J. Biol. Chem.* 232:867–879.

Sternberg, M. 1976. Microbial rennents. *Adv. Appl. Microbiol.* 20:135–153.

Suigara, M., Isobe, M. 1975. Effects of temperature and state of substrate on the rate of hydrolysis of glycerides by lipase. *Chem. Pharm. Bull.* 23:681–683.

Takasaki, Y. 1966. Studies on sugar isomerizing enzyme. Production and utilization of glucose isomerase from *Streptomyces* sp. *Agr. Biol. Chem.* 30:1247–1253.

Takasaki, Y. 1976. Production and utilization of β-amylase and pullulanase from *Bacillus cereus* var. *mycoides. Agr. Biol. Chem.* 40:1515–1522.

Takasaki, Y., Kosugi, Y., Kanbayashi, A. 1969. Streptomyces glucose isomerase. In: *Fermentation Advances*, Perlman, D. (Ed.). Paper presented at 3rd International Fermentation Symposium, 1968. Academic Press, New York.

Thermote, F., Rombouts, F. M., Pilnik, W. 1977. Stabilization of cloud in pectinesterase active orange juice by pectic acid hydrolysates. *J. Food Biochem.* 1:15–34.

Ward, O. P. 1983. Microbial proteinases in Stabilization of cloud in pectinesterase active orange juice by pectic acid hydrolysates. *J. Food Biochem.* 1:15–34.

Ward, O. P., Fogarty, F. M. 1972. Polygalacturonase lyase of a Bacillus species associated with increase in permeability of Sitka spruce (*Picea sitithesis*). *J. Gen. Microbiol.* 73:439–446.

Whitaker, D. R. 1971. Cellulases. In: *The Enzymes*, Vol. 5, Boyer, P. D. (Ed.). Academic, New York, pp. 273–289.

Whitaker, J. R. 1977. *Principles of Enzymology for the Food Sciences.* Marcel Dekker, New York.

Wilson, G. A., Wheelock, J. V. 1972. Factors affecting the actions of rennin in heated milk. *J. Dairy Res.* 39:413–419.

Wimborne, M. P., Richard, P. A. O. 1978. Pectinolytic activity of *Saccharomyces fragilis* cultured in controlled environments of pectin ester content and pectin methylesterase activity. *Anal. Biochem.* 39:418–428.

Wiseman, A. 1983. *Principles of Biotechnology*. Blackie and Son, Glasgow, Scotland.

Wiseman, A. 1985. *Handbook of Enzyme Biotechnology*. Wiley, New York.

Wood, P. J., Siddiqui, I. R. 1971. Determination of methanol and its application to measurements of pectin ester content and pectin methylesterase activity. *Anal. Biochem.* 39:418–428.

Wood, P. J., Weisz, J. 1987. Detection and assay of (1-4)-β-D-glucanase, (1-3)-β-D-glucanase, (1-3) (1-4)-β-D-glucanase, and xylanase based on complex formation of substrate with Congo red. *Cereal Chem.* 64:8–15.

Yamanaka, K., Takahara, N. 1977. Purification and properties of D-xylose isomerase from *Lactobacillus xylosus. Agr. Biol. Chem.* 34:1805–1814.

Yamasaki, M., Kato, A., Chu, S. Y., Arima, K. 1967. Pectic enzymes in the clarification of apple juice. Part II. The mechanism of clarification. *Agr. Biol. Chem.* 31:552–560.

Yashimura, H. 1966. Studies on D-glucose isomerizing activity of D-xylose grown cells from *Bacillus coagulans* strain HN-68. *Agr. Biol. Chem.* 30:1015–1023.

Yi-Hsu-Ju, Chen, W. J., Cheng, K. L. 1995 Starch hydrolysis using α-amylase immobilized on a hollow fiber reactor. *Enz. Microbiol. Technol.* 17:685–688.

Young, J. M., Schray, K. J., Mildvan, A. S. 1975. Proton Magnetic relaxation studies of the interaction of D-xylose and xylitol with D-xylose isomerase. *J. Biol. Chem.* 250:9021–9027.

Zittan, I., Poulsen, P. B., Hemmingsen, H. 1975. Sweetzyme—a new immobilized glucose isomerase. *Staerke* 27:236–241.

23

Emulsifiers

SYMON M. MAHUNGU

Egerton University, Njoro, Kenya

WILLIAM E. ARTZ

University of Illinois, Urbana, Illinois

I. INTRODUCTION

This chapter focuses on emulsifiers, one category in the general class of compounds called surface-active agents, and generally excludes the macromolecular stabilizers and any of the other components involved in emulsion stabilization. In addition, the discussion will be limited to food emulsifiers, rather than the use of emulsifiers for nonfood products such as cosmetics, paints, or drug delivery systems. The macromolecular emulsifiers and stabilizers will be discussed very briefly, even though many biological macromolecules, such as proteins, gums, and starch, have significant emulsifying and/or stabilizing capabilities in food systems.

The definition of an emulsion has continued to evolve since the 1930s. Becher (1957) developed an elaborate definition from several previous authors. ''An emulsion is a heterogeneous system, consisting of at least one immiscible liquid intimately dispersed in another in the form of droplets, whose diameter, in general, exceeds 0.1 μm. Such systems possess a minimal stability, which may be accentuated by such additives as surface-active agents, finely divided solids, etc.'' Freiberg et al. (1969) elucidated the significance of liquid crystalline phases on emulsion stability. This was reflected in the IUPAC-IUB (1972) definition of an emulsion: ''In an emulsion, liquid droplets and/or liquid crystals are dispersed in a liquid.'' Sharma and Shah (1985) defined both micro- and macroemulsions, differentiating them on the basis of size and stability. Macroemulsions were defined as ''mixtures of two immiscible liquids, one of them being dispersed in the form of fine droplets with (a) diameter greater than 0.1 μm in the other liquid. Such systems are turbid, milky in color and thermodynamically unstable.'' Microemulsions were defined as ''clear thermodynamically stable dispersions of two immiscible liquids.'' The dispersed range consists of small droplets in the range of 100–1000 Å.''

Food macroemulsions are unstable systems, even with the addition of emulsifiers. Emulsifiers are added to increase product stability and attain an acceptable shelf-life. The function of an emulsifier is to join together oily and aqueous phases of an emulsion in a homogeneous and stable preparation (Waginaire, 1997). The main characteristic of an emulsifier is that it contains in its molecule two parts. The first part has a hydrophilic affinity, while the second has a lipophilic affinity. Emulsifiers are generally classified as anionic emulsifiers, cationic emulsifiers, amphoteric emulsifiers, and nonionic emulsifiers.

Emulsifier selection is based upon final product characteristics, emulsion preparation methodology, the amount of emulsifier added, the chemical and physical characteristics of each phase, and the presence of other functional components in the emulsion. Food emulsifiers have a wide range of functions. The most obvious is to assist stabilization and formation of emulsions by the reduction of surface tension at the oil–water interface. The most common examples are mayonnaise and margarine. An additional function is that of alteration of the functional properties of other food components. An example of this is the use of emulsifiers in bakery products such as a crumb softener and dough conditioner. A third function is to modify the crystallization of fat, e.g., the reduction of bloom in certain candy products.

II. EMULSIFIER CHEMISTRY

Schuster (1985) has written a comprehensive text on emulsifiers with extensive discussions on the current theories of emulsion formation and emulsifier chemistry, function, and analysis. He included discussions covering all the emulsifiers used in Germany and the United States, plus applications in a wide variety of products.

A. Synthesis and Structure

Food emulsifiers can be categorized (Table 1) on the basis of several characteristics including origin, either synthetic or natural; potential for ionization, nonionic versus ionic; hydrophilic/lipophilic balance (HLB); and the presence of functional groups.

1. Lecithin and Lecithin Derivatives

The primary source of lecithin, the only naturally occurring emulsifier used in any significant quantity in the food industry, is soybeans. Soybean oil contains anywhere from 1–

Table 1 Food Emulsifier Categories

Lecithin and lecithin derivatives
Glycerol fatty acid esters
Hydroxycarboxylic acid and fatty acid esters
Lactylate fatty acid esters
Polyglycerol fatty acid esters
Ethylene or propylene glycol fatty acid esters
Ethoxylated derivatives of monoglycerides
Sorbitan fatty acid esters
Miscellaneous derivatives

3% phospholipids in the crude oil (Haraldsson, 1983). Other, less significant, sources include corn, sunflower, cottonseed, rapeseed, and eggs. Lecithin is obtained by an aqueous extraction of the oil extracted from soybeans. Phase separation occurs upon hydration of the phospholipids and the two phases are separated by centrifugation (Flider, 1985). The crude extract, after water removal, contains about 35% triglycerides and smaller amounts of nonphospholipid materials. Extraction with acetone is used to produce an oilfree lecithin. The term "lecithin" has been used to describe both phosphatidylcholine and mixtures of phospholipids. Current recommendations by IUPAC-IUB (1977) (International Union of Pure and Applied Chemistry–International Union of Biochemistry) suggest the use of 3-sn-phosphatidyl-choline rather than lecithin to describe 1,2-diacyl-sn-glycero-3-phosphatidylcholine. However, a commercial soybean-derived lecithin preparation contains several different phospholipids, primarily phosphatidylcholine, phosphatidylethanolamine, and phosphatidylinositol (Hurst and Martin, 1984). The structures are shown in Fig. 1. Over 90% of the phospholipids in soybean lecithin are these three (Scholfield, 1981). Several papers from a symposium on lecithin presented at the 1980 American Oil Chemists' Society meeting in New York were published in the 1980 October issue of the *Journal of the American Oil Chemists' Society*. The composition of

```
R1-C(O)-O-CH2
          |
R2-C(O)-O-CH
          |
          CH2-O-P(O)-OCH2CH2N+(CH3)3
                   |
                   O-
```

(a) *Phosphatidylcholine*

```
R1-C(O)-O-CH2
          |
R2-C(O)-O-CH
          |
          CH2-O-P(O)-OCH2CH2NH+3
                   |
                   O-
```

(b) *Phosphatidylethanolamine*

```
R1-C(O)-O-CH2
          |
R2-C(O)-O-CH
          |
          CH2-O-P(O)-O-C6H11O5
                   |
                   O-
```

(c) *Phosphatidylinostol*

Figure 1 The primary phospholipids reported in commercial lecithin, where R_1 and R_2 are fatty acids.

lecithin from a variety of plant sources (soybeans, corn, sunflower, and cottonseed) was discussed.

Commercial lecithin preparations can be treated or modified chemically to provide a product with altered functional characteristics. Treatment with either hydrogen peroxide or benzoyl peroxide will produce a lighter colored product. The chemical modification of lecithin by reaction with hydrogen peroxide plus lactic or acetic acid and water will produce a hydroxylated product. Hydroxylation occurs at the double bonds (Schmidt and Orthoefer, 1985), altering lecithin such that its hydrophilic character is increased. The result is a product with improved oil-in-water (O/W) emulsifying properties relative to unmodified lecithin (Prosise, 1985).

Triglycerides are soluble in acetone, whereas phospholipids are not. Therefore, the greater the percentage of acetone-insoluble material, the greater the phospholipid content in crude lecithin (Prosise, 1985). Because of this, one of the primary criteria for the evaluation of lecithin is the percentage of acetone-insoluble material. Lecithin is also evaluated on the basis of several other parameters (*Food Chemicals Codex*, 1981) including acid value (an indication of free fatty acids), hexane insoluble matter (an indication of fibrous material), water, peroxide value, and metallic impurities. Individual phospholipids in soy lecithin can be quantitated using HPLC (Hurst and Martin, 1984). The emulsifying properties of native, purified, and modified lecithin can be determined based on a method developed by Von Pardun (1982). First, an emulsion is formed utilizing lecithin. Next, after placing the emulsion under moderate stress, the half-life of the system is measured. Alternatively, the rate of phase separation, creaming, flocculation or coalescence can be determined based on a visual observation combined with droplet size determination (Rydhag and Wilton, 1981).

The Food Chemicals Codex (1981) contains information on nearly all the food emulsifiers discussed in this chapter. Included in this information is a description of the emulsifier, the requirements in terms of specifications for each emulsifier and a detailed description of each assay used to evaluate either the emulsifier specifications or quality parameters. The information contained in the Food Chemicals Codex is a result of the work by the Committee of Codex Specifications of the Food and Nutrition Board of the National Research Council. The Food and Nutrition Board is an advisory group in the area of food and nutrition. It is not a regulatory agency of any government even though some governments may adopt the Food Chemicals Codex for their use.

2. Mono- and Diglycerides

Mono- and diglycerides are the most commonly used food emulsifiers (Fig. 2). They consist of esters synthesized via catalytic transesterification of glycerol with triglycerides, with the usual triglyceride source of hydrogenated soybean oil. Mono- and diglycerides are also synthesized directly from glycerol and fatty acids under alkaline conditions. Molecular distillation is used to prepare a purified product containing up to approximately 90% monoglyceride. Zlatanos et al. (1985) prepared monoglycerides from the reaction of glycidol (2,3-epoxy-1-propanol) and carboxylic acids with a yield in excess of 90%. Advantages of the process included the synthesis of difficult-to-produce monoglycerides and a good potential for continuous processing. Campbelltimperman et al. (1996) have prepared mono- and diglycerides from a butterfat fraction by chemical glycerolysis, while Pastor et al. (1995) reported the enzymatic preparation of mono- and distearin by glycerolysis of ethylstearate and direct esterification of glycerol in the presence of a lipase from *Candida antarctica*.

$HO\text{-}CH_2$
|
$HO\text{-}CH$
|
$CH_2\text{-}OC(O)\text{-}(CH_2)_{16}\text{-}CH_3$

Monoglyceride

$HO\text{-}CH_2$
|
$HO\text{-}C\text{-}OC(O)\text{-}(CH_2)_{16}\text{-}CH_3$
|
$CH_2\text{-}OC(O)\text{-}(CH_2)_{16}\text{-}CH_3$

Diglyceride

Figure 2 Mono- and diglycerides.

Several tests are used for characterizing commercial sources of mono- and diglycerides, including total monoglycerides, hydroxyl value, iodine value, and the saponification value (Food Chemicals Codex, 1981). With the monoesters, the fatty acid can be attached at either the alpha or beta positions, likewise with the diglycerides.

3. Hydroxycarboxylic and Fatty Acid Esters

To produce an emulsifier with increased hydrophilic character relative to monoglycerides, small organic acids are esterified to monoglycerides (Fig. 3). Some of the acids used are acetic, citric, fumaric, lactic, succinic, and tartaric. Succinylated monoglycerides are synthesized from succinic anhydride and distilled monoglycerides (Larsson, 1976). They are used by the baking industry as dough conditioners and crumb softeners. Acetic acid esters of mono- and diglycerides are synthesized from fatty acids plus acetic anhydride or by transesterification. The product is lipid soluble and water insoluble. Functions in food include control of fat crystallization and improvement of aeration properties of high-fat foods. They are often added to shortenings or cake mixes.

H
|
$H\text{-}C\text{-}OR_1$
|
$H\text{-}C\text{-}OR_2$
|
$H\text{-}C\text{-}OR_3$
|
H

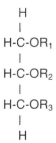

Figure 3 Organic acid ester of monoglyceride, where at least one R is a short chain organic acid, for example, acetic acid.

```
O    H  O
||   |  ||
RC-O-[C - C]n-O-Na
     |
     CH3
```

Figure 4 Sodium stearoyl-2-lactylate, where n normally averages 2 and R is a fatty acid moiety.

To synthesize other acid esters, citric acid esters of mono- and diglycerides, glycerol is esterified with a mixture of citric acid and fatty acids. It can also be prepared by the direct esterification of citric acid with glyceryl monooleate (Food Chemicals Codex, 1981). The product is hot water and lipid soluble. Functions in food include emulsification, anti-spattering agent in margarine, improvement of bakery product characteristics, a fat replacement in high fat foods, and a synergist and solubilizer for antioxidants.

Diacetyl tartaric acid esters of monoglycerides (DATEM) are synthesized from diacetyl tartaric acid anhydride and monoglycerides (Krog and Lauridsen, 1976). The emulsification properties of DATEM depend primarily upon the type of fatty acid and the percentage of esterified tartaric acid. Emulsifier quality is based upon results from the analyses for tartaric and acetic acid, acid value, total fatty acids, saponification value, and metallic residue (Food Chemicals Codex, 1981).

Lactic acid esters of mono and diglycerides consist of a mixture of lactic and fatty acid esters of glycerin. The emulsifier is dispersible in hot water. Important qualitative parameters include the percentage of monoglycerides, total lactic acid, acid value, free glycerin, and the amount of water (Food Chemicals Codex, 1981).

4. Lactylate Fatty Acid Esters

Polymeric lactic acid esters of monoglycerides (Fig. 4) are also available, commonly known as sodium or calcium stearoyl-2-lactylates. Typically, there are two lactic acid groups per emulsifier molecule. To produce the emulsifier a mixture of the fatty acid, polylactic acid, and calcium or sodium carbonate is heated at about 200°C for about 1 h with agitation in an inert atmosphere (Krog and Lauridsen, 1976). The calcium salt is less dispersible in water than sodium stearoyl-2-lactylate.

5. Polyglycerol Fatty Acid Esters

Polyglycerol esters of fatty acid (Fig. 5) are also used in food products, primarily in baked goods. They consist of mixed partial esters synthesized from the reaction of polymerized glycerol with edible fats. Polyglycerols will vary in degree of glycerol polymerization with an average specified. The source of fatty acids as well as the degree of polymerization

```
                OR2
                |
R1-[O-CH2-C-CH2]n-OR3
                |
                H
```

Figure 5 Polyglycerol esters of fatty acids, where R_1, R_2, and R_3 are each either a fatty acid and/or a hydrogen and where the average value of n is greater than 1.

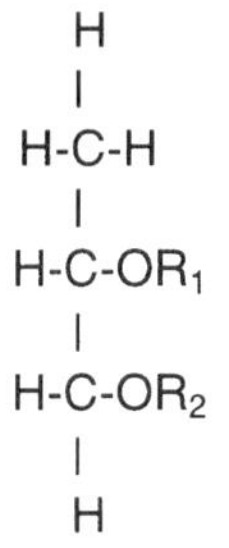

Figure 6 Propylene glycol esters of fatty acids, where R_1 and R_2 represent either a fatty acid and/or a hydrogen and where at least one R represents a fatty acid.

can vary, providing a wide range of emulsifiers, from hydrophilic to very lipophilic (Food Chemicals Codex, 1981).

6. Polyethylene or Propylene Glycol Fatty Acid Esters

Fatty acids can be esterified directly to polyethylene glycol ethers (Fig. 6) (Meffert, 1984) or by enzymatic preparation, which allows better control of the reaction (Bhattacharyya et al., 1984). Shaw and Lo (1994) reported the production of propylene glycol fatty acid (C_{12}, C_{14}, C_{16}, C_{18}, and $C_{18:1}$) monoesters by lipase catalyzed reactions. Propylene glycol monoesters of docosahexaenoic acid and eicosapentaenoic acid, which are water-in-oil (W/O) emulsifiers useful in the food industry, have been synthesized by lipase-catalyzed esterification (Liu and Shaw, 1995). The HLB of the emulsifier is altered by adjusting the degree of ethoxylation. Fatty acid polyglycol esters are good O/W emulsifiers (Maag, 1984).

7. Ethoxylated Derivatives of Monoglycerides

Ethoxylated mono- and diglycerides are produced from the reaction of several moles of ethylene oxide and mono- or diglycerides under pressure (Meffert, 1984). Ethoxylation of monoglycerides results in a product that is much more hydrophilic relative to monoglycerides (Rusch, 1981).

Polyoxyethylene monoglycerides may contain as many as 40 moles of ethylene oxide per mole of monoglyceride (Schuster, 1985). The end product of the synthesis is actually a mixture with a distribution range and peak (Becher, 1967); therefore, lots often vary among manufacturers.

8. Sorbitan Fatty Acid Esters

Polyoxyethylene sorbitan esters are synthesized by the addition, via polymerization, of ethylene oxide to sorbitan fatty acid esters. These nonionic hydrophilic emulsifiers (Fig. 7) are very effective antistaling agents and, thus, are used in a wide variety of bakery products (Schuster and Adams, 1984). These emulsifiers are much more widely known as the polysorbates, e.g., polysorbate 20, 60, and 80. Polysorbate 20, 60, and 80 utilize lauric, stearate, and oleate, respectively, for the fatty acid portion of the molecule (Food Chemicals Codex, 1981). Polysorbate 60 is a monostearate, while polysorbate 65 is a tristearate.

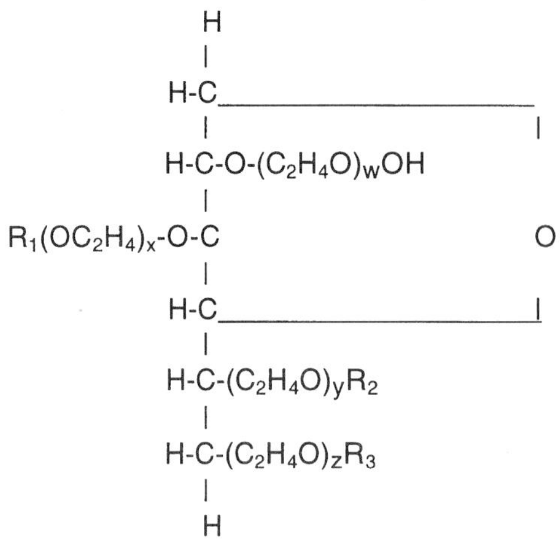

Figure 7 Polysorbates, where w + x + y + z = 20 (approximately) and Rs represent a single fatty acid and hydrogens for polysorbate 20, 40, 60, and 80. For polysorbate 65, each R represents a stearic acid moiety. The fatty acids are lauric, palmitic, stearic, and oleic acid for polysorbate 20, 40, 60, and 80, respectively.

9. Miscellaneous Derivatives

Fatty acids can be esterified directly to compounds other than glycerol, for example, sugar alcohols, like sorbitol, mannitol, and maltitol, and sugars, like sucrose, glucose, fructose, lactose, and maltose (Torrey, 1983).

Sorbitol or sorbitan esters are formed from 1,4-anhydro-sorbitol and fatty acids (Meffert, 1984). Typically, the emulsifier consists of a mixture of stearic and palmitic acid esters of sorbitol and its mono- and dianhydrides (Fig. 8). Ethoxylated derivatives can also be prepared by the addition of several moles of ethylene oxide to the sorbitan monoglyceride ester, and, depending on the number of moles of ethylene oxide added, have a wide range in HLB.

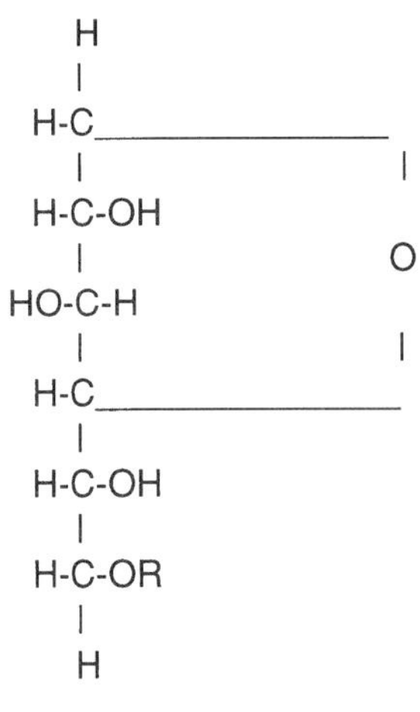

Figure 8 Sorbitan stearate, where R represents a fatty acid moiety, for example, stearic acid, oleic acid, lauric acid, or palmitic acid.

$$\begin{array}{r} CH_2OR_2\qquad\qquad\quad \\ |\qquad\qquad\qquad\quad \\ R_1OCH_2\text{-}C_5H_7O_4\text{-}O\text{-}C_4H_4O_3\text{-}CH_2OR_3 \end{array}$$

Figure 9 Sucrose fatty acid esters, where at least one of either R_1, R_2, or R_3 represents a fatty acid and the reminder may represent a fatty acid or a hydrogen; the degree of substitution is 1–3.

Lactitol (the hydrogenation product of lactose) palmitate is synthesized by direct esterification at a temperature of approximately 160°C (van Velthuijsen, 1979). The product mixture can be separated with silica gel thin layer chromatography (TLC) with chloroform–acetic acid–methanol–water (80:10:8:2, v/v) as the eluent. The esters containing one or two fatty acid groups are effective emulsifiers. Lactitol esters containing at least 4 moles of fatty acid per mole of lactitol could potentially be used as a low-calorie fat substitute, since hydrolysis occurs to only a minor degree in humans.

Akoh (1994) has reported the enzymatic synthesis of acetylated glucose fatty acid esters. Two immobilized lipases from *Candida anarctica* (SP 382) and *Candida cylindraceae* catalyzed the synthesis of novel acetylated glucose fatty acid esters with glucose pentaacetate and Trisun 80 (80% oleic) vegetable oil or methyl oleate as substrates in organic solvents. The incorporation of oleic acid onto the glucose ranged from 30–100%. It was possible to catalyze the synthesis of glucose fatty acid esters with free glucose as the sugar substrate. Other researchers have reported the synthesis of a novel nonionic surfactant, dialkyl glucosylglutamate from delta-gluconolactone, glutamic acid, and alkyl alcohols (Tsuzuki et al., 1993).

Sucrose fatty acid esters (Fig. 9) can be synthesized using a variety of solvents or by direct esterification (Wei, 1984). The first description of a practical commercial process for the preparation of sucrose esters of fatty acids was reported by Osipow et al. in 1956. Enzymatic synthesis of carbohydrate esters of fatty acids has also been reported for the esters of sucrose, glucose, fructose, and sorbitol with oleic and stearic acid (Seino et al., 1984) and fatty acid esters of fructose (Arcos et al., 1998). The report by Arcos et al. indicates that these workers were able to enzymatically prepare three different 1,6-diacyl fructofuranoses. At low temperatures (5°C) the synthesis produces quantitative yields of the diesters by simple addition of the original sugar to a solution of the fatty acid in a solvent (acetone) which is accepted by the European Economic Commission (EEC) for use in the manufacture of additives. By varying the degree of esterification, the HLB, and hence the functionality, can be controlled. Sucrose monoesters have an HLB value greater than 16, while the triesters have an HLB value less than 1. Monoesters are particularly useful for the stabilization of O/W emulsions (Maag, 1984), whereas diesters are best for W/O emulsions. With esterification equal to or greater than 5 moles of fatty acid per mole of sucrose, the emulsification properties of sucrose fatty acid esters are lost (Wei, 1984). But, at that degree of esterification, the sucrose fatty acid polyester can be used as a low-calorie fat replacement since it is neither digestible nor absorbable.

The consistency of both O/W and W/O emulsions can be affected with the addition of ethylene or propylene glycol monostearate. The most common ethylene and propylene glycol esters used as emulsifiers are the monostearate and monopalmitate.

B. Analysis

Quantitation of emulsifiers is difficult, largely for three reasons: (1) the similarities in structure among many of the emulsifiers, (2) most commercial sources of emulsifiers are quite heterogeneous, and (3) due to their nature, emulsifiers can be difficult to extract from foods that contain either protein or starch (Baur, 1973). Baur (1973) listed analyses for 17 different groups of emulsifiers, which included recommended solvents for extraction, chromatographic techniques for separation, and quantitation procedures.

Dieffenbacher and Bracco (1978) developed a thin layer chromatographic method for the detection of a mixture of several emulsifiers. They utilized a chloroform–methanol extraction in combination with a column chromatography cleanup. Thin layer chromatography was used for separation and quantitation. Three different solvent systems were used along with a variety of spray reagents for detection. With a combination of R_f values plus specific and nonspecific detection sprays, 14 different emulsifies were detected.

Gas–liquid chromatography (GLC) analysis for the silylated derivatives of both the polyglycerols and their fatty acid esters (Sahasrabudhe, 1967) and the lactylated monoglycerides (Neckerman and Noznick, 1968) are available. Soe (1983) separated several different emulsifier groups using GLC. Emulsifiers separated included monoglycerides, lactic acid esters of monoglycerides, acetic acid esters of monoglycerides, and propylene glycol esters of fatty acids. Each of the emulsifier groups analyzed contained esters of several different fatty acids. The AOAC INTERNATIONAL has developed a gas chromatographic technique for the analysis of monoglycerides and diglycerides, which has been adopted as an IUPAC/AOCS/AOAC method (Firestone, 1994).

Sudraud et al. (1981) separated mixtures of emulsifiers with high-performance liquid chromatography (HPLC). Analysis of acetylated tartaric acid esters of mono- and diglycerides indicated a complex mixture containing at least 12 different components. Garti and Aserin (1981) utilized partition chromatography HPLC to separate and quantitate polyglycerol esters of fatty acids. Schlegelmilch et al. (1995) have developed an HPLC method for the quantitation of free polyglycerol in polyglycerol caprate emulsifiers, while Martin (1995) has reported the use of an HPLC having an evaporative light scattering detector for the analysis of nonionic emulsifiers.

Sucrose fatty acid esters can by determined with HPLC using a reverse phase column and either a methanol–water solvent or a methanol–isopropanol solvent (Kaufman and Garti, 1981). Tsuda and Nakanishi (1983) and Gupta et al. (1983) have developed methods for the analysis of sucrose mono- and diesters using GLC. The AOAC recommends an HPLC method for the quantitation of sucrose mono- and diesters, but it may be difficult to obtain good peak separation when fatty acid esters of glycerol are present (Gupta et al., 1983).

Daniels et al. (1982) developed an analysis for a mixture of polyoxyethylated stearic acid esters of sorbitol and its anhydrides (polysorbate 60) in salad dressings. The procedure utilized a combination of colorimetric analysis and silica gel TLC plus column chromatography.

Buschmann and Hulskotter (1997) have developed a titration procedure for low-ethoxylated nonionic emulsifiers. An ionic group is introduced into the molecule by a derivatization reaction. The reaction product is then determined by conventional titration methods for anionic emulsifiers without any modification. This method has been used for the analysis of other nonionic emulsifiers such as sorbitan esters and ethoxylated fatty acid amides.

Wheeler (1979) developed an analysis for the determination sodium stearoyl-2-lactylate in flour and flour blends utilizing a chloroform extraction, thin layer chromatography for purification, and a colorimetric procedure for quantitation.

Artz and Myers (1994) developed a supercritical fluid method for extraction, separation, and quantitation of emulsifiers. Selected emulsifiers which included acetylated monoglycerides, lactylated monoglycerides, hexaglycerol distearate, triglycerol mono- and dioleate, and decaglycerol decaoleate were separated with capillary supercritical fluid chromatography on a 25% cyanopropyl stationary phase with a mobile phase of CO_2 at 100 to 150°C. Samples of acetylated monoglycerides were placed in a supercritical fluid extraction cell on a glass bead bed and extracted for 15 min at 50°C at 340, 408, and 680 atm with CO_2. Acetylated monoglycerides added during twin-screw extrusion of corn starch were extracted from the extrudate for 15 and 45 min at pressures of 544 and 646 atm with 0–5% methanol in supercritical carbon dioxide. The percent acetylated monoglyceride extracted after 45 min at 120°C and 646 atm was 60%. Recently, fatty alcohol ethers, nonionic surfactants which are used as general purpose emulsifiers, were separated by use of water-modified carbon dioxide mobile phase in capillary supercritical fluid chromatography (Pyo et al., 1996). Much greater peak intensities were observed in the chromatogram with on-line modified mobile phase than with pure CO_2.

Berchter et al. (1997) reported the use of the matrix assisted laser desorption and ionization time of flight mass spectrometry (MALDI-TOF-MS) for the analysis of nonvolatile highly polar or high-mass substances, especially emulsifiers.

C. Kinetics of Emulsification

This section briefly discusses the kinetics of emulsifiers with respect to the adsorption and desorption kinetics of emulsifiers at hydrophobic and hydrophilic surfaces, as well as micelle lifetimes. Whereas nonionic emulsifiers are adsorbed as submonolayers or monolayers at hydrophobic surfaces, surface micelles or bilayers are formed at hydrophilic surfaces (Tiberg, 1996). Thus, the need for a better understanding of the adsorption/desorption phenomena of emulsifiers and macromolecules at surfaces becomes evident in the desire to control and affect surface properties in a range of technical and biological processes. Depending upon whether the hydrophilic or the hydrophobic segment of the emulsifier has stronger tendency to be adsorbed at the surface, different molecular arrangements are expected at the surface. If the water-soluble component interacts more favorably with the surface and the hydrophobic segments are sufficiently large to trigger self-association, surface micelles or various bilayer type aggregates can be expected at the surface. If, on the other hand, the hydrophobic part is adsorbed more strongly at the surface, it should result in the formation of monolayer or submonolayer structures at the surface.

The behavior of nonionic emulsifiers has been studied using ellipsometry. This technique is used to measure polarization changes that occur using oblique reflection of a polarized light beam from a surface. The polarization changes are very sensitive to the presence of a thin film or a layer of adsorbed molecules at the surface. The difference in the polarization state between the incident and the reflected light is described by the measured ellipsometric angles. From these measurements, adsorption isotherms of emulsifiers can be plotted and determined. Dynamic interfacial tension is another widely used technique for probing emulsifier adsorption/desorption kinetics interfaces. After the interface is disturbed, the interfacial tension is measured as a function of time to monitor emulsifier adsorption at the interface.

1. Nonionic Emulsifiers at Hydrophilic Surfaces

When the emulsifier concentration is substantially less than the critical micelle concentration (CMC), the emulsifiers are adsorbed as unimers to the surface. However, as the emulsifier interfacial concentration due to this small adsorption exceeds a certain threshold concentration, surface micellation begins, resulting in a large increase in concentration in a very narrow region. The bulk concentration at which the aggregation process starts is called the critical surface aggregation concentration (CSAC) (Levitz et al., 1984; Lindheimer et al., 1990; Tiberg et al., 1994a). The cooperative isotherm obtained is characteristic of emulsifiers with weak interaction energies between the head groups and the surface. Below the CSAC, the adsorbed-layer thickness is always very small. Above CSAC, the adsorbed-layer thickness jumps to stable plateau values roughly corresponding to twice the emulsifier length, while the adsorbed amount generally increases more slowly. This indicates that the emulsifiers form discrete surface aggregates, with a well-defined thickness. The whole adsorption (or surface aggregation) process occurs below the CMC, since the surface aggregates have a lower free energy than the corresponding bulk structures. The surface excess increases as the ratio of lipophilic to hydrophilic groups of the emulsifiers increases. This effect is mainly related to the fact that emulsifiers assemble into surface aggregates of different in-plane dimensions. Thus, surface aggregates adsorbed on hydrophilic surfaces exhibit properties that are closely related to the corresponding bulk structures. The adsorbed layer thickness reaches steady state values at low surface coverages. The thickness remains unaffected during desorption indicating that surface micelles are present at hydrophilic surfaces from relatively low to high coverages. The adsorption process is controlled by diffusive transport of monomers and micelles to the interface through a stagnant layer. The absorption process can be described quantitatively using the following equation, which describes the rates of adsorption and desorption under steady state conditions (Tiberg et al., 1994b):

$$d\Gamma/dt = \frac{D_{\mathrm{mon}}}{\delta}\left[\left(c_{\mathrm{mon,s}} - c_{\mathrm{mon,s}}\right) + \frac{ND_{\mathrm{mic}}}{\delta}\left(c_{\mathrm{mic,b}} - c_{\mathrm{mic,s}}\right)\right]$$

where N is the mean aggregation number of the micelles, D_{mon} and D_{mic} are the diffusion coefficients (in $\mathrm{m^2/s}$) of emulsifier monomers and micelles, respectively, δ is the stagnant layer thickness, and Γ is the emulsifier adsorption (in $\mathrm{kmol/m^2}$); $c_{\mathrm{mon,b}}$, $c_{\mathrm{mon,s}}$, $c_{\mathrm{mic,b}}$, and $c_{\mathrm{mic,s}}$, are the concentrations (in $\mathrm{kmol/m^3}$) of monomers and micelles in the bulk and just outside the surface, respectively. The thickness of the stagnant layer δ can be calculated from the equation

$$\frac{d\Gamma}{dt} = -\frac{D_{\mathrm{mon}}}{\delta}(\mathrm{CSAC})$$

The emulsifier activities, both in bulk solution and just outside the adsorbed layer, are approximately constant, the latter through a local equilibrium existing between the adsorbed aggregates and the solution in the immediate vicinity of the adsorbed aggregates.

2. Nonionic Emulsifiers on Hydrophobic Surfaces

The adsorption on hydrophobic surfaces gives a simple Langmuirian type isotherm, where the saturation adsorption is reached around the CMC. The maximum plateau adsorption increases with the length of the hydrocarbon chain and/or a reduction in the size of the hydrophilic group. The isotherm is fairly well represented by a Langmuir isotherm with the general form

$$\Gamma = \frac{\Gamma_p C_b}{C_b + k}$$

where k is the chemical potential difference between adsorbed and bulk surfactants, C_b is the bulk concentration of the emulsifier, and Γ is the emulsifier adsorption (in kmol/ m^2).

The adsorption on a hydrophobic surface is a noncooperative process involving the formation of surface aggregates. As with the case on hydrophilic surfaces, the adsorbed layer thickness on hydrophobic surfaces increases in a steplike fashion with concentration as the CSAC is exceeded. The hydrocarbon region of the adsorbed layer is free of water. Thus, the emulsifier head has roughly the same volume independent of the surface coverage, indicating an increase of the emulsifier concentration, and hence surface coverage, resulting in an average tilt of the hydrophilic group toward the surface.

The kinetics at a hydrophobic surface differs from that at a hydrophilic surface. At a hydrophobic surface, the surface excess and the adsorbed layer thickness evolves in similar manner with time, whereas at a hydrophilic surface they exhibit different time dependencies. The adsorption at a particularly bulk concentration is much faster to the hydrophobic than the hydrophilic surface. However, desorption goes much faster from a hydrophilic surface than a hydrophobic surface. At the hydrophobic surface, the emulsifiers become increasingly tilted toward the surface with adsorption time and surface coverage. The concentration of the monomer emulsifiers in the immediate vicinity of the adsorbed layer, $C_{\mathrm{mon,s}}(\Gamma)$ at a particular Γ value is equal to the inverse isotherm $C_b(\Gamma)$, and $C_{\mathrm{mic,s}}(\Gamma) = 0$, except when $\Gamma \approx \Gamma_p$. The adsorption rate when C_b is much less than the CMC is given by the following equation:

$$\frac{d\Gamma}{dt} = \frac{D_{\mathrm{mon}}}{\delta}\left(C_b - \frac{k\Gamma}{\Gamma_p - \Gamma}\right)$$

When the bulk surfactant concentration quickly adjusts to $C_b \approx 0$, the desorption rate is given by the following equation:

$$\frac{d\Gamma}{dt} = \frac{D_{\mathrm{mon}}}{\delta}\frac{k\Gamma}{\Gamma_p - \Gamma}$$

In general, nonionic emulsifiers are adsorbed as submonolayers or monolayers at hydrophobic surfaces, whereas surface micelles or bilayer type aggregates are formed above a critical concentration on hydrophilic surfaces. At hydrophobic surfaces, adsorbed emulsifiers are tilted toward the surface at low coverages. The degree of hydration of the headgroups is relatively independent of the surface coverage. An increasing surface excess at a hydrophilic surface is accompanied by an increase in the number of surface micelles and in some cases a substantial in-plane growth of the surface micelles, resulting in the formation of bilayer type structures. The kinetics of adsorption and desorption are strongly dependent on the surface properties. These processes are diffusion limited at both hydrophobic and hydrophilic surfaces. Thus, the rates are determined by the stagnant layer thickness, the diffusion coefficients of monomers and micelles, and the concentration gradients over the stagnant layer.

3. Ionic Emulsifiers

The similarity of the physical structure (i.e., short hydrocarbon tail and small headgroup) of a typical ionic emulsifier to that of a nonionic emulsifier that exhibits fast local exchange between solution and interface suggests that ionic emulsifiers should exchange at similar rates. However, small nonionic emulsifiers have sorption kinetics in which diffusion to the interface is the limiting transport resistance, while ionic emulsifiers undergo transport that is diffusion limited, kinetic limited, and mixed diffusion–kinetic (Bonfillon and Langevin, 1993; Joos et al., 1992; MacLeod and Radke, 1994a). In the presence of other ions (such as sodium and chloride ions), the ionic emulsifier must overcome the electrostatic repulsion from the interface. Hence, the rate of transport of an ionic emulsifier to the interface is slower than that of an equivalent nonionic emulsifier (MacLeod and Radke, 1994b). However, the kinetics of emulsifier reorientation and adsorption onto the interface are not substantially different between ionic and nonionic emulsifiers of similar sizes and shapes. For nonionic emulsifiers, kinetic limitations are manifested early, while diffusion-limited transport occurs later when the diffusion boundary layer thickness has grown sufficiently large (MacLeod and Radke, 1994a). However, for ionic emulsifiers, charge accumulation on the interface slows the rate of diffusion to the interface. Thus, the time scale for kinetic-limited transport is less for an ionic emulsifier than for an equivalent nonionic emulsifier (MacLeod and Radke, 1994b). At low electrolyte concentrations, differences between counterion and emulsifier ion diffusion coefficient influence the rate of emulsifier transport to the interface, but at high electrolyte concentrations, the counterion diffusion coefficient has no effect on the rate of emulsifier transport.

4. Micellar Lifetime

Micelles are in a dynamic equilibrium state with monomers in solution. The micelles are dissociated into monomers and monomers are associated into micelles continuously. The multiequilibrium, stepwise association process of micelle formation can be described as:

$$A_1 + A_1 = A_2$$
$$A_1 + A_2 = A_3$$
$$A_1 + A_3 = A_4$$
$$\vdots$$
$$A_1 + A_{n-1} = A_n$$

where A_1 denotes the emulsifier monomer and A_n represents the micellar aggregate with an aggregation number of n. These micellar kinetics have been studied by stopped-flow, temperature-jump, pressure-jump, and ultrasonic absorption methods since Aniansson developed a theoretical model to describe kinetic micellization (Aniansson et al., 1976). There are two relaxation processes: The first one is the fast relaxation process with relaxation time (τ_1, in the microsecond range), which is associated with the fast exchange of monomers between micelles and bulk aqueous phase. This process is considered as the collision between emulsifier monomers and micelles. The second relaxation time (τ_2, in the millisecond range) is related to the micelle dissociation kinetics (Oh and Shah, 1994). The average micellar lifetime is given by the following expression:

$$\frac{1}{\tau_2} = \frac{n^2}{A_1 R}\left[1 + \left(\frac{\sigma^2}{n}\right)a\right]^{-1}$$

$$T(\mathrm{m}) = \frac{\tau_2 n a}{1 + \left(\frac{\sigma^2}{n}\right)a} \tag{1}$$

where $T(\mathrm{m})$ is the average micellar lifetime, τ_2 is the second relaxation time, n is the aggregate number, σ is the distribution width of the distribution curve of micellar sizes, and $a = (A_{tot} - A_1)/A_1$. A_{tot} and A_1 are the total emulsifier concentration and mean monomer concentration, respectively. When the concentration of an emulsifier is much greater than the critical micelle concentration, the micellar lifetime is approximately equal to $n\tau_2$. During dynamic processes such as emulsification, the interface between two phases expands rapidly. The dynamic surface tension of the micellar solutions determines the characteristics of these dynamic processes. However, the dynamic surface tension is influenced by the rate of diffusion of emulsifier monomers from the bulk liquid to the interface (Tiberg, 1996). Thus, relatively unstable micelles can increase significantly the diffusion rate of monomers upon a rapid breakdown of micelles into emulsifier monomers.

Generally, emulsion systems have minimal thermodynamic stability and tend to phase separate. Conventional emulsions or macroemulsions are thermodynamically unstable and turbid systems. This phenomena is even more pronounced in double emulsions, which are termed "emulsions of emulsions." Double emulsions are thermodynamically unstable systems with a strong tendency for coalescence, flocculation, and creaming. Most double emulsions consist of relatively large droplets, cannot withstand storage regimes, and have a strong tendency to release entrapped matter in an uncontrolled manner (Garti, 1997). However, it is possible to have emulsion systems with ultralow interfacial tensions which are thermodynamically stable. These emulsions are called microemulsions, which are thermodynamically stable dispersions produced by using emulsifiers able to reduce the interfacial energy to values close to zero (De Buruaga et al., 1998; Perrin and Lafuma, 1998). Both oil-in-water (direct) and water-in-oil (inverse) microemulsions can be produced and are transparent. Their structural entities are much smaller than the wavelength of light, which is the reason for their transparency (Perrin and Lafuma, 1998).

The primary driving force for phase separation is droplet interfacial free energy (Opawale and Burgess, 1998). During emulsification, a lot of energy is required to disperse one liquid into another as small droplets. The interfacial area is greatly increased during this dispersion process. The work done to expand the interfacial area is given by the following expression:

$$W = \gamma(\Delta A)$$

where γ is the dynamic interfacial tension and ΔA is the increase in the interfacial area. For a constant W, a greater value of γ yields a smaller ΔA. Thus, the emulsion droplet size increases as the interfacial tension increases. The dynamic interfacial tension between oil and water during the emulsification process is determined by how effectively the emulsifier monomers adsorb into the interface from the bulk solution (from the water phase for O/W emulsions, from the oil phase for W/O emulsions). Most of the emulsifier molecules exist as the micellar form in solution. Therefore, the effectiveness of emulsifier monomer adsorption depends on the rate of micellar breakdown because micelles themselves cannot adsorb at the interface.

III. EMULSIFIER FUNCTION AND MECHANISM

A. Emulsion Formation

The first step in the formation of a stable emulsion is dispersion of one liquid phase in another liquid phase. A critical factor in that emulsification process is the formation of a monomolecular layer at the lipid/water interphase by the emulsifier. During emulsion formation there is a large increase in surface area (up to several thousandfold), which is dependent upon the number and size of the droplets. To form and disperse these droplets, a substantial amount of energy or work must be supplied. Since emulsifiers reduce the surface tension, the addition of emulsifiers reduces the amount of work that must be done to form the emulsion. The most common method of emulsion formation is the application of mechanical energy via vigorous agitation.

The emulsifier is first dissolved in the aqueous or organic phase depending on the solubility of the emulsifier and on the type of emulsion desired. Next, sufficient agitation to cause surface deformation and large droplet formation is applied during the addition of one phase to the other. The next step is disruption of the droplets. To form a stable emulsion and prevent coalescence, sufficient emulsifier must be available to adsorb at the aqueous/organic interphase. The emulsifier lowers the Laplace pressure, which facilitates droplet deformation and disruption (Walstra, 1983). After droplet formation, the emulsifier partitions into the interphase of the aqueous/organic system stabilizing the emulsion. Droplet size, which is directly related to the emulsification procedure, is also dependent on the amount of emulsifier added, the type of emulsifier, and the emulsification temperature.

There are several possible methods for emulsion formation and a wide range of equipment is available for emulsion formation. These methods include shaking, stirring, and injection, and the use of colloid mills, homogenizers, and ultrasonics. An excellent summation on the mathematical evaluation of the process of emulsification and the relationship of the factors involved is given by Walstra (1983).

One problem of key importance is the scale-up from laboratory to pilot plant to manufacturing (Lynch and Griffin, 1974). It is important to closely simulate manufacturing conditions during pilot plant preparation of the emulsion, particularly if this is not possible on a laboratory scale. On a laboratory scale, the use of a blender to prepare an emulsion may result in the application of more energy and a much faster rate than is possible in a manufacturing situation. If at all possible, laboratory equipment as well as pilot plant equipment should be of the same design as the production equipment. A motor-driven propeller or a hand-driven homogenizer would provide a much more realistic simulation of actual manufacturing conditions than a blender.

Equipment manufacturers can provide a wide range of equipment capable of emulsification. The main types of equipment are stirrers (propeller and turbine types), colloid mills, homogenizers, and ultrasonic mixers. Stirrers are typically used to produce either coarse emulsions or as a premixer for some other type of emulsifying equipment. However, high-speed mixers are available that, according to one manufacturer (Arde Barinco, Mahwah, NJ), work quite well for the production of stable food emulsions without the need for additional emulsification. Product examples included French dressing, margarine, butter sauce, and a flavor emulsion. Similar equipment, including laboratory scale equipment, is available from Greerco (Hudson, NH), among others.

Colloid mills emulsify based on the shearing action imparted to the liquid by a high-speed rotor moving within a fixed stator. The stator and rotor are separated by a very small gap (Becher, 1957). Clearance between the rotor and stator may be as small as 0.001 in. Many manufacturers, like Chemicolloid Laboratories (Garden City Park, NY), utilize grooved rotors and stators in their design, along with an adjustable clearance between the rotor and stator. Greeco offers a colloid mill with grooving and an adjustable gap that they recommend over their high speed mixer for emulsions of smaller droplet size, like mayonnaise. The gap is adjusted depending upon the type of emulsion. One disadvantage of a very small gap size is reduced throughput. APV Gaulin (Everett, MA), a colloid mill manufacturer, suggests that the colloid mill is best suited to high-viscosity products with high ratios of oil to water, like mayonnaise. With mayonnaise, moderately small droplet sizes are best. With too small a droplet size and the concomitant increased surface area, the total surface area of the emulsion will likely exceed the capacity of the added emulsifier, significantly reducing product stability.

Another type of colloid mill with a rapidly spinning rotor, but minus the stator, is manufactured by Cornell (Springfield, NJ). Material enters a reduced pressure chamber and a film of product is impounded at the center of a rapidly spinning disc. The thin film formed on the surface of the spinning disc is forced outward by centrifugal force, which reduces the film thickness even further. Due to the characteristics of the product and spinning disc, substantial shear forces are developed that greatly reduce droplet size. An example of a food product that would best utilize this equipment design is salad dressings.

A familiar piece of emulsifying equipment is the homogenizer due to its widespread use for the homogenization of milk. Homogenizers emulsify by forcing the product through a small orifice under substantial pressure. The most likely mechanism for droplet size reduction is the dual effect of cavitation and turbulence. As the orifice size is decreased and the pressure is increased, the size of the dispersed phase droplets is decreased. In order to utilize the homogenizer most effectively, one manufacturer suggests that the product entering the homogenizer should be premixed so the particle size is less than 20 μm in diameter. Another possibility is the use of either multistage homogenizers or multiple passes through a single homogenizer to insure small, uniform droplets. The homogenizer is, of course, recommended for milk homogenization. It is also recommended for processing dispersions such as catsup and tomato sauce, for producing flavor and beverage emulsions, and for the production of frozen whipped toppings.

One piece of equipment of somewhat unusual design is the Hydroshear manufactured by APV Gaulin. The equipment utilizes a unique chamber design to subject the product to high shear. The chamber has a double conical shape. The fluid is forced tangentially into the middle of the chamber. The fluid moves in a spiral motion from the outside to the inside of the chamber. Due to the conical shape, the velocity of the product layers increases as the radius decreases and the chamber height increases. Differences in velocities between concentric layers generates regions of very high shear. The emulsified product is then discharged through small openings in each cone apex. Droplet size, although smaller on the average than that obtained with colloid mills, is greater than that possible with homogenizers. It is recommended for reduction of droplet size in preparation for a single-pass homogenizer. Energy input for this equipment is substantially less than for a homogenizer.

Ultrasonic emulsification is the treatment of liquids with high-frequency vibrations to produce high-intensity cavitation. Sound waves move through the liquid, compressing and stretching it, which results in cavity or bubble formation within the liquid. Upon collapse of the bubbles tremendous shear forces are generated. The ultrasonic process can be influenced by controlling static pressure, temperature, amplitude of vibration and flow rates.

According to Branson (Danbury, CT) and Sonic (Stratford, CT), two of several ultrasonic equipment manufacturers, there are both limitations and advantages to ultrasonic emulsifying equipment. The equipment is not particularly effective for large volumes or for highly viscous products. Advantages include lower capital and operating costs, no premixing requirement, lower maintenance, and easier cleaning relative to homogenizers.

B. Emulsion Stability

The following discussion on emulsion stability is rather abbreviated and the reader interested in a comprehensive treatment is referred to the review by Tadros and Vincent (1983) and Friberg et al. (1990). They provide an excellent theoretical treatment of both the destabilization and stabilization mechanisms involved with emulsions.

There are several phenomena that can cause emulsion destabilization. Each is affected by the presence of emulsifiers. To best understand the mechanism of emulsion formation and ultimately the forces stabilizing emulsions, the mechanisms of destabilization should be understood.

1. Mechanisms of Destabilization

Emulsion destabilization can be due to one or all of five possible mechanisms; flocculation, coalescence, sedimentation or creaming, Ostwald ripening, and phase inversion (Tadros and Vincent, 1983).

a. Flocculation. The adherence of droplets to form aggregates or clusters and the buildup of these aggregates is referred to as flocculation. It occurs when the attractive forces between the droplets exceeds that of the repulsive forces, without a breakdown in the structural integrity of the interfacial film surrounding the droplets. These attractive forces are primarily long-range London–van der Waals forces and electrostatic forces. Once flocculated, the droplets sediment faster to the bottom (or rise faster to the top, cream) than drops of the original size (Friberg et al., 1990).

b. Coalescence. When aggregates or flocculates of the dispersed phase combine to form a single, larger drop the phenomena is referred to as coalescence. Coalescence is really a reflection of the nature of the interfacial film on the surface of the droplet. A strong, stable film on the surface of the droplet, due to addition of the correct concentration of the appropriate emulsifier, will minimize this type of destabilization. When coalescence occurs, the integrity of the interfacial film is lost and droplets in close contact combine, with the result of a reduction in the number of droplets. The ultimate effect is the formation of a single ''drop,'' the sedimentation becomes faster, and the emulsion separates into two layers.

c. Changes in Droplet Concentration. Droplet concentration can increase preferentially in either the top or bottom portion of the emulsion depending upon the relative density of the two phases. Sedimentation and creaming are two examples of this phenomena. Reducing the average droplet size and adding an emulsifier will substantially reduce the rate at which this occurs.

d. Ostwald Ripening. If the two phases forming the emulsion are not totally immiscible and there are differences in droplet size within the emulsion, larger droplets will form at the expense of smaller droplets due to a process known as Ostwald ripening. Ostwald ripening is always a factor since variation in initial droplet size always occurs in macroemulsions and both phases are never completely immiscible. The driving force for Ostwald ripening is the difference in chemical potential between droplets of different sizes. Equilibrium will only exist when all droplets are the same size, which really means a single ''drop'' or the presence of two continuous and separate phases.

e. Phase Inversion. The viscosity of an emulsion will gradually increase as more and more of a given phase is added until a critical volume is reached. If more of that same phase is added, exceeding the critical volume, the emulsion will invert, i.e., the discontinuous phase will become the continuous phase.

2. Mechanisms of Stabilization by Emulsifiers

There are several factors, some of which are dependent on the emulsifiers and stabilizers added, involved in emulsion stabilization (Nawar, 1986). The first is the reduction of interfacial tension by the emulsifiers. Next is repulsion between droplets due to similar electrical charges on the surface of the droplets. A third is the formation of mesophases or liquid-crystalline phases which will provide the most stable configuration for a specified set of conditions. A fourth is the addition of macromolecules or particulate material which can substantially increase emulsion viscosity and stability. An increase in viscosity of the continuous phase adds to the kinetic stability. However, without a concurrent energy barrier, viscosity will have a small effect on stabilization. Viscosity enhancers increase the stability of the energy barrier (Friberg et al., 1990).

Emulsion stability is also dependent upon the conditions under which the emulsion is formed. This includes not only the constituents of the emulsion, but the emulsifier concentration, the emulsion temperature, and the physical state (crystalline versus fluid) of the emulsifier (Friberg and Mandell, 1969). Even the order of addition of the constituents is an important factor. Addition of lecithin to the lipid phase prior to the addition of the aqueous phase can substantially alter droplet size, liquid crystal formation, and emulsion stability (Friberg et al., 1976). Another contributing factor is the nature of the internal and continuous phases. Both affect emulsion stability. Two types of emulsions, those prepared with unsaturated emulsifiers and unsaturated oil and those prepared with saturated emulsifiers and saturated oil, were more stable that those prepared with emulsifiers and oil of intermediate or mixed saturation (Garti and Remon, 1984).

a. Interfacial Tension. The reduction of interfacial tension through the addition of emulsifiers is a key factor in emulsion formation. It allows emulsion formation with considerably less energy input than would be required without the presence of an emulsifier (Becher, 1983). Once the interfacial film consisting of emulsifier is formed, it acts as an effective barrier to droplet coalescence (Schuster and Adams, 1984). It has been found that, in the presence of emulsifiers, the droplet interface may acquire viscoelastic properties which are important in the prevention of coalescence (Joanne et al., 1994, Williams et al., 1997). A strong interaction between the hydrophilic portion of the emulsifier and the aqueous phase leads to a large reduction in the surface tension of the water (Boyle, 1997). According to Schuster and Adams (1984), this also effects the type of emulsion formed. A weak interaction between water and the hydrophilic portion of the emulsifier molecule will favor a W/O emulsion, while a strong interaction will favor an O/W emulsion.

b. Electrical Charge. Ionic emulsifiers provide an additional mechanism for emulsion stabilization relative to nonionic emulsifiers, through ion–ion and ion–solvent interactions (Gunnarsson et al., 1980). In addition, the introduction of charged groups on the surface of the emulsion droplets increases the repulsive forces between droplets (Larsson and Krog, 1973). Ionic emulsifiers will form an electrically charged double layer in the aqueous solution surrounding each oil droplet (Nawar, 1986). The explanation for the stability of emulsions due to charge separation relies heavily on the DLVO theory (Friberg et al., 1990; Krog et al., 1985; Petrowski, 1976). According to this theory, if the net repulsive interaction between droplets, due to a combination of electrostatic repulsion and the attractive forces due to van der Waals interactions is greater than the kinetic energy of the droplets, the emulsion will be stable (Nawar, 1986).

c. *Liquid Crystal Stabilization.* Macroemulsions, although thermodynamically unstable, can attain rather long-term stability, strongly suggesting an intermediate stability level. This was attributed to the formation of a liquid-crystalline state by the emulsifier (Friberg and Mandell, 1969; Friberg et al., 1969). Jansson and Friberg (1976) indicated the presence of a liquid-crystalline state reduces the rate of coalescence even if droplet flocculation occurs. Nakama et al. (1997) obtained a stable W/O emulsion without coalescence of the water droplets that contained a substantial amount of water (90%) using a lauroamphoglycinate (LG) and oleic acid (OA) mixture. The X-ray diffraction patterns and the strong hydrophobicity showed that the equimolar complex composed of LG, OA, and water was a liquid crystal with a reversed phase hexagonal structure. The reversed hexagonal liquid crystal was capable of solubilizing a certain amount of liquid paraffin in its alkyl group parts while maintaining its hexagonal structure.

An emulsifier film can actually exist in several different mesophases or liquid-crystalline states (Schuster and Adams, 1984). Conversion between physical states by the emulsifier at the oil/water interface is referred to as lyotropic mesomorphism. This phenomenon occurs when an emulsifier/water mixture is heated to a sufficient temperature so that the hydrocarbon chains liquify and, simultaneously, water penetrates between layers of the polar groups (Flack, 1983). The result is the formation of liquid-crystalline structures referred to as either lamellar, cubic, or hexagonal. Various structural arrays can exist, including alternating films, spheres, and cylinders. The resultant structure is dependent upon several factors including emulsifier molecular structure, the concentration ratio of the emulsifier to water, temperature, ionic strength, and pH (Schuster and Adams, 1984). Each parameter affecting stability is closely related to the others. Altering one, even slightly, may alter the type of mesophase formed and the emulsion stability. Kwon and Rhee (1996) investigated the emulsifying capacity of coconut protein as a function of salt, phosphate, and temperature. They reported that between pH 4.0 and 5.0, protein nitrogen solubilities of coconut flour (CF) and coconut protein concentrate (CPC) in water were lower than those in salt solutions. In salt solutions, the nitrogen solubility was lowest at pH 1, and increased steadily as the pH was increased from 3 to 6. Increased phosphate addition increased emulsifying capacity, while an increase in temperature decreased emulsifying capacity. Other workers (Taiwo et al., 1997) studied the influence of temperature and additives on the adsorption kinetics of food emulsifiers. They found that the presence of salt and greater temperatures reduced the interfacial tension of egg york solutions, while the interfacial tension of whey protein concentrate solutions increased with salt addition. Increased temperature caused the equilibrium interfacial tension of both emulsifiers to decrease and attain equilibrium more quickly.

Another concentration-dependent phenomenon that can occur is critical micelle formation. At low concentrations, strongly hydrophilic emulsifiers will disperse or dissolve completely in water. If sufficient emulsifier is added to an aqueous solution, strongly hydrophilic emulsifiers will form micelles, a type of liquid-crystalline microstructure (Schuster and Adams, 1984). This entropy-driven association occurs to minimize water interaction with the hydrophobic portion of the emulsifier. The emulsifier concentration at which the micelle formation occurs is the critical micelle forming concentration (CMC). It is dependent upon the structure of the hydrophobic portion of the emulsifier molecule and, to a lesser degree, the temperature, ionic strength, and pH. Micellar structures are very small, 0.005 to 0.01 μm in diameter.

d. Stabilization by Macromolecules and Finely Divided Solids. Emulsion stability can be increased by the addition of macromolecules like gums and protein. Tharp (1982) found that colloids, such as xanthan gum, carboxy methyl cellulose, and guar gum, significantly increased emulsion stability. At a constant emulsifier and colloid concentration, emulsion stability was enhanced by increased emulsification temperature, increased degree of shear, and increased pH, in the pH range of 3–6. Colloids act by either increasing the viscosity or by partitioning into the oil/water interface and providing a physical barrier to coalescence (Nawar, 1986).

Tadros and Vincent (1983) have provided a detailed theoretical discussion on emulsion stability due to added macromolecules and finely divided solids. Small, finely divided solids can adhere to the surface of a lipid droplet, stabilizing the emulsion by forming a physical barrier to coalescence, through steric hindrance. The type of emulsion formed, as well as the emulsion stability, is dependent upon the relative ability of the two phases to wet the particles (Nawar, 1986). This can be greatly influenced by the addition of emulsifiers. An example of stabilization due to finely divided solids is the stabilization effects of amorphous silica. Addition of amorphous silica to an O/W emulsion containing an emulsifier substantially increased both emulsion stability and viscosity (Villota, 1985).

To evaluate emulsion stability and thereby characterize the potential of an emulsifier, the rate at which the combined destabilization phenomena occur must be determined. These rates can be determined from the changes in the size and distribution of the oil droplets with time. There are several methods available for this determination. According to Trumbetas et al. (1978), nuclear magnetic resonance (NMR) analysis can provide a better indication of stability than HLB values. Samples were evaluated that contained a mixture of polyoxyethylene 20 sorbitan monostearate, sorbitan monostearate, lipid, and water. Another method, used to evaluate the effects of processing on emulsion stability, was centrifugation (Tornberg and Hermansson, 1977). Frenkel et al. (1982) evaluated the stability of W/O emulsion by monitoring turbidity at 400 and 800 nm. They stated that the method was suitable for determination of the required HLB, the amount and type of emulsifier, and the fraction of water. Tung and Jones (1981) examined the microstructures of mayonnaise and salad dressing with light microscopy, transmission electron microscopy, and scanning electron microscopy (SEM), while SEM was used to determine lipid droplet sizes. Chemical fixation and critical point dehydration were effective in providing a method for observation of undiluted samples.

Sherman (1971) recommended that before an accelerated method for the testing of emulsion stability was used, a rigorous evaluation to determine the correlation between the accelerated method and actual storage conditions must be done. Those procedures, based on assumptions about the effects of either alterations in temperature or centrifugation rather than empirical analysis, may not be valid. He recommends, for simple emulsions, either a determination of the rate of change in particle size after a limited time under actual storage conditions or the phase inversion temperature analysis developed by Shinoda's group (Shinoda, 1969; Shinoda and Sagitani, 1978; and Shinoda et al., 1980).

C. Emulsifier Selection

Perhaps the most important factor in preparing an emulsion is the selection of the appropriate emulsifier. Several methodologies have been developed to assist in such an endeavor

(Shinoda and Kunieda, 1983). These include the HLB system of Griffin (1949, 1954), the H/L numbers (Moore and Bell, 1956), the water number of Greenwald et al. (1956), the phase inversion temperature (PIT) of Shinoda (1967), and the emulsion inversion point (EIP) (Marszall, 1975). Even with the best of methods selection can be very difficult, except perhaps for the few foods that are relatively straightforward emulsions, such as mayonnaise and margarine. Often one of the best sources of information for the food technologist will be the emulsifier manufacturer. There are several companies that have considerable expertise and can provide excellent advice in this area.

Several parameters should be considered during emulsifier selection (Nash and Brickman, 1972). These parameters include (1) approval of the emulsifier by the appropriate government agency, (2) desired functional properties, (3) end product application, (4) processing parameters, (5) synergistic effect of other ingredients, (6) home preparation, and finally (7) cost.

Obviously, before an emulsifier can be used in a food product it must be approved by the appropriate regulatory agency. Assuming this criterion is met, the most important considerations would be both the required functional properties of the selected emulsifier and the application. Delineating the required functional properties such as emulsification, starch complexation, and crystallization control, and the specific end product application are the two major factors in emulsifier selection. An exact determination of these two parameters should focus attention on a limited number of emulsifiers. The processing methodology and equipment available in the processing facility could further limit the range of emulsifiers that are of potential use. It is at this stage that an ingredient supplier(s) could begin to provide helpful assistance.

By far, the most widely used rule for the selection of food emulsifiers is the HLB number, published by Griffin (1949,1954). The HLB index, called the hydrophile–lipophile balance, is based upon the relative percentage of hydrophilic to lipophilic groups within the emulsifier molecule. Griffin assigned values ranging from 1 to 20 (Waginaire, 1997). Lower HLB values indicate a more lipophilic emulsifier, while higher values indicate a more hydrophilic emulsifier. Emulsifiers with HLB numbers in the 3–6 range are best for W/O emulsions, whereas emulsifiers with HLB numbers in the range of 8–18 are best for O/W emulsions. Depending upon the application and the types of oils to be emulsified there is an optimal HLB.

An equation developed by Griffin (1954) can be used to determine the HLB number for several types of nonionic emulsifiers, particularly the ethoxylated alcohols and the polyhydric fatty acid esters (Tadros and Vincent, 1983). To determine the HLB for the fatty acid ester type emulsifiers, Griffin (1949,1954) used the equation

$$\text{HLB} = 20\left(1 - \frac{S}{A}\right)$$

where A is the acid number and S is the saponification number of the ester.

For the polysorbate type of emulsifier the HLB value can be determined from the equation

$$\text{HLB} = \frac{E + P}{5}$$

where E is the weight percentage of oxyethylene and P is the weight percentage of polyhydric alcohol.

Davies (1957) was able to determine the HLB values for a large number of emulsifiers based on the difference in the number of hydrophilic and lipophilic groups.

However, there are several factors that reduce the utility of the HLB selection system. One factor is that the HLB system does not work well for ionic emulsifiers, a problem further complicated by the fact that the charge varies with pH. Another factor is that commercial preparations usually contain two or more emulsifiers. These emulsifiers can have a significant synergistic effect (Rosen and Hua, 1982) that make it very difficult to apply the HLB system (Nash and Brickman, 1972). Another limitation is due to the fact that the HLB system is based upon the molecular structure of the emulsifier and does not take into account the combined oil/aqueous phase/emulsifier system (Marszall, 1981). In addition, once the appropriate emulsifier has been selected, the correct concentration cannot be determined from the HLB value. The concentration required is really a function of the droplet size. The smaller the droplet, the greater the surface area and, therefore, the greater the amount of emulsifier that is required for monolayer coverage of each droplet. In spite of these limitations the HLB is still the most widely used index of emulsifier functionality. An extensive and updated listing of the HLB literature was compiled by Becher (1985). The large number of references in the listing (several hundred) attest to the continued use and wide acceptance of the HLB concept.

Shinoda (1968) developed a method for the evaluation of emulsifiers based on the phase inversion temperature, which is also referred to as the HLB temperature. The temperature at which an emulsion inverts is dependent on the specific emulsifier employed (Sagitani and Shinoda, 1973,1975) and, to a lesser degree, on the lipid phase. The stability of an emulsifier system is highly dependent on the temperature (Wilton and Friberg, 1971). Emulsion stability is also related to the temperature at which phase inversion occurs (Parkinson and Sherman, 1972; Shinoda and Sagitani, 1978). Emulsions are relatively stable when stored at temperatures 25–60°C below the PIT (Shinoda, 1969). So the higher the PIT, the less the rate of droplet coalescence and the more stable the emulsion at a given storage temperature. Phase inversion temperature is really a characteristic of the complete emulsion system rather than just the emulsifier. The PIT can be determined by monitoring the change in conductivity of an emulsion with a change in temperature (Tsutsumi et al., 1978). Phase inversion will cause a large change in conductivity. For example, with a W/O emulsion a large increase in conductivity indicates the continuous phase has converted from oil to aqueous.

Often, a mixture of two or more emulsifiers are used, since blending allows the formation of an emulsifier mixture with any HLB bounded by the HLB values of the individual emulsifiers. Although blends of hydrophilic and lipophilic emulsifiers with largely different HLB values have been suggested, stable emulsions are usually only obtained if the difference in HLB values between the two emulsifiers is not too great (Shinoda et al., 1980). The assumption that HLB values for different emulsifiers are additive is really not valid for emulsifiers that differ substantially in HLB value (Marszall, 1981).

Lynch and Griffin (1974) outlined a product development methodology for the use of emulsifiers in foods. Once the emulsifier has been selected, based on many of the same factors discussed by Nash and Brickman (1972), nine steps for product formulation are listed. The first is to group the components together based on solubility. Next, calculate an approximate HLB for the combined oil phase based on the emulsion type. Third, prepare emulsifier mixtures with high and low HLB values that closely bracket the HLB value as determined from the second step. Fourth, dissolve all the oil-soluble components at a temperature about 5–10°C above the melting point of the highest melting ingredient. Fifth, dissolve all the water-soluble components, exclusive of the salts and acids and a small portion of the water. Sixth, heat the aqueous phase to a temperature 3–5°C greater than the oil phase temperature. Seventh, with agitation, add the oil phase to the aqueous phase. Next, dissolve in the remaining components (salts, acids, water, etc.). Finally, critique the stability of the emulsion to determine (1) a minimum room temperature storage time, (2) number of times the emulsion can survive a freeze–thaw cycle, (3) time of storage at 40 and 50°C, and finally (4) the degree of separation or creaming.

IV. APPLICATIONS IN FOODS

Emulsifiers have a wide range of functional properties in addition to the obvious one: stabilization of food emulsions. Table 2 is a list of functional properties of food emulsifiers compiled from a variety of sources including product brochures from several emulsifier manufacturers.

Table 2 Functional Properties of Food Emulsifiers

Functions	Product examples
Emulsification, water-in-oil emulsions	Margarine
Emulsification, oil-in-water emulsions	Mayonaise
Aeration	Whipped toppings
Whippability	Whipped toppings
Inhibition of fat crystallization	Candy
Softening	Candy
Antistaling	Bread
Dough conditioner	Bread dough
Improve loaf volume	Bread
Reduce shortening requirements	Bread
Pan-release agent	Yeast-leavened and other dough and batter products
Fat stabilizer	Food oils
Antispattering agent	Margarine and frying oils
Antisticking agent	Caramel candy
Protective coating	Fresh fruits and vegetables
Surfactant	Molasses
Control viscosity	Molten chocolate
Improved solubility	Instant drinks
Starch complexation	Instant potatoes
Humectant	Cake icings
Plasticizer	Cake icings
Defoaming agent	Sugar production
Stabilization of flavor oils	Flavor emulsions
Promotion of "dryness"	Ice cream
Freeze–thaw stability	Whipped toppings
Improve wetting ability	Instant soups
Inhibition of sugar crystallization	Panned coatings

A. Cereal-Based Products

With bakery products, emulsifiers not only improve final product characteristics, but also facilitate processing. A review by Schuster and Adams (1984) and another by Stampfli and Nersten (1995) covered research on emulsifiers with respect to baked goods and included an excellent discussion on the mechanism of emulsification, the interaction of emulsifiers with starch, and a wide range of applications.

First of all, emulsifiers can function as dough conditioners. Advantages attributed to dough conditioners include (1) improved tolerance to variations in flour and other ingredient quality, (2) doughs with greater resistance to mixing and mechanical abuse, (3) better gas retention resulting in lower yeast requirements, shorter proof times, and increased baked product volume, (4) increased uniformity in cell size, a finer grain, and a more resilient texture, (5) stronger sidewalls, (6) reduced shortening requirements, and (7) improved slicing (Dubois, 1979; Rusch, 1981; Brummer et al., 1996). Gawrilow (1979) developed a mathmetical model to predict the effect of added emulsifiers on bread quality. Several different types of bakery products, in addition to bread produced via conventional methods, benefit from added emulsifiers, including white bread produced by a highly mechanized process, buns and rolls, yeast-raised sweet rolls and doughnuts, plus variety breads that contain a significant portion of nonwheat flour components (Schuster and Adams, 1984; Stampfli and Nersten, 1995).

Certain emulsifiers also function as crumb softeners, another type of dough conditioner. The soft crumb, a characteristic of freshly baked bread, can be retained longer if the appropriate emulsifiers are added. Crumb firming, associated with staling and starch retrogradation, can be delayed for 2–4 days with the addition of emulsifiers (Knightly, 1968). Emulsifiers are thought to reduce the rate of water migration and hence, staling, by complexing with the starch and adsorbing onto the starch surface (Pisesookbunterng and D'Appolonia, 1983). In fact, the amount of lipid complexation that occurs with amylose correlates well with the crumb softening effect in bread (Krog and Jensen, 1970; Lagendijk and Pennings, 1970).

Emulsifiers can form complexes with amylose that are difficult to dissociate. Investigation of the complex revealed the nonpolar portion of the emulsifier was inserted into helical portions of amylose. Raman spectroscopic analysis of a monostearin–amylose complex indicated there are approximately three turns of the amylose helix per stearic acid chain (Carlson et al., 1979). The ability of amylose to form these complexes is dependent on the formation of a helical structure. Kim and Robinson (1979) suggested that the helical conformation of amylose is the primary factor in the binding of emulsifiers to amylose. Amylopectin, which does not readily form a helical complex, does not bind lipid. Osman et al. (1961) found each of the several emulsifiers investigated, except diglycerides, formed amylose complexes. In addition, formation of the lipid complex reduced the amount of iodine that could bind to the starch. The inclusion of emulsifiers into helical sections of amylose is one of the key reasons starch gelatinization is affected by the addition of emulsifiers (Schuster and Adams, 1984), although there is some effect due to starch granule coating by the emulsifier and emulsifier interaction with the wheat proteins.

There are other explanations as to why crumb softening occurs. The best emulsifiers, in terms of amylose complexation, were the distilled hydrogenated monoglycerides, the stearoyl-2-lactylates, and the succinylated monoglycerides (Krog, 1971), while the acetylated monoglycerides and propylene glycol esters of palmitic and stearic acid did not form complexes with amylose. Interestingly, both of these emulsifiers, acetylated monoglycerides and propylene glycol esters, did not form any kind of aqueous dispersions or micellar solutions. Since the most effective starch complexing emulsifiers exhibited lyotrophic mesomorphism, it suggested that a high degree of molecular freedom in the aqueous phase is essential for the initial association with amylose (Krog, 1971). Also, the greater complexation associated with the saturated monoglycerides relative to the unsaturated, plus the limited internal dimensions of the amylose helix, indicated that there is a significant steric factor involved. Other investigators (MacRitchie, 1977; Larsson, 1980) have attributed the effect of emulsifiers on the delay of crumb firming to the ability of emulsifiers to function as foam stabilizers. According to this hypothesis, polar lipids concentrate in the aqueous phase of the dough surrounding the gas bubbles in ordered, mesophase type structures. Emulsifiers function similarly, supporting the wheat lipids in their foam stabilizing function.

Lorenz et al. (1982) evaluated several emulsifiers for their effect on crumb hardness. Of the eleven commercial sources of emulsifiers tested, four were most effective at 10 and 25°C: propylene glycol mono- and diesters, ethoxylated mono- and diglycerides, polyoxyethylene sorbitan monostearate, and succinylated monoglycerides. All of the emulsifiers tested, except hydroxylated lecithin, were equally effective at 40 and 50°C.

Most investigators have found that lecithin has the same characteristics as other dough conditioners on bakery products; increased water absorption, reduced mixing time, improved machinability, and longer freshness (Schuster and Adams, 1983). Hydroxylation improved lecithin functionality (Pomeranz et al., 1968), whereas addition of hydrogenated lecithin reduced loaf volume severely. Lecithin has been used to improve moisture tolerance, to insure uniform suspension of ingredients, as a wetting agent to provide rapid wetting of powdered formulations, and to facilitate product release from pans or molds. Lecithin can also increase dough and bread yields, improve loaf volume, improve crumb grain, and retard staling. Although lecithin does reduce the rate of staling in bread, it was less effective than several other emulsifiers based on loaf volume. Loaf volume, as determined by the optimized baking test, can be used to determine emulsifier effectiveness in bread products (Schuster and Adams, 1984). It is a good indirect test for staling since it correlates well with the delayed onset of crumb firming. The addition of lecithin to baked goods will also improve fat dispersion, thereby reducing air cell size (Van Nieuwenhuyzen, 1981). Only limited lecithin is used in the United States in bread products, although it is widely used in Europe.

Lecithin can improve product characteristics of modified bread products and nonbread bakery products. Addition of lecithin improved loaf volume, crumb grain, crumb color, and freshness retention in bread supplemented with soy protein (Adler and Pomeranz, 1959). Matz (1968) indicated that the functional characteristics of cracker and cookie dough is favorably altered by the addition of lecithin. Lecithin can alter shortening requirements in cookies (Pomeranz, 1985) and substantially improve the handling properties of the dough during cookie production. Waffles with added lecithin were stronger, crisper, less soggy, and retained freshness longer than waffles without added lecithin. Pan release after baking was improved if lecithin was added to cake mixes. Lecithin can also reduce the amount of eggs and/or shortening required. Lecithin addition improved color, texture, grain, and keeping quality in cakes and other sweet dough products. Lecithin can also improve the wettability of cake mixes. Phospholipids, treated with phospholipase to alter functionality, have been utilized in wheat dough to improve product characteristics (Ohta et al., 1984).

The deleterious effect of wheat bran on bread quality can be offset by the addition of emulsifiers and vital wheat gluten. Lecithin (0.5%) was as effective as shortening (3.0%) and slightly less effective than 0.5% each of ethoxylated monoglycerides, sucrose monopalmitate, and diacetyl tartaric esters for improving whole wheat bread quality (Pomeranz, 1985). The quality of Zwieback and chapatties is also improved with the addition of lecithin (Ebeler and Walker, 1983).

Any emulsifier, including lecithin, is reduced in efficiency if it is not dispersed in the product properly. A mixture of hydrophilic and lipophilic emulsifiers provides the best dispersibility for bakery products (Pomeranz, 1985).

Emulsifiers can affect both starch viscosity and gelatinization (Morad and D'Appolonia, 1980; Krog and Lauridsen, 1976). Krog (1973) found that monoglycerides caused the largest increase in gelatinization temperature, followed by sodium stearoyl-2-lactylate, calcium stearoyl-2-lactylate, and then diacetyl tartaric acid esters of monoglycerides. The binding of monostearin to starch was temperature dependent. The temperature dependence varied depending upon the starch source (von Lonkhuysen and Blankestijn, 1974). At 30°C, wheat starch binds a greater percentage of the added monoglycerides than does cassava starch; while at 90°C the reverse is true.

Both sodium stearoyl-2-lactylated and a mixture of mono- and diglycerides were effective in restoring freshness (in terms of crumb softness) upon reheating (Pisesookbunterng et al., 1983). Saturated monoglycerides relative to unsaturated monoglycerides had a greater effect on reducing moisture loss rates in cake during reheating (Cloke et al., 1984).

The alpha-crystalline form of monoglycerides is the most active in terms of functionality. Cooley (1965) indicated that it improved mixing tolerance and crumb grain and resulted in a softer bread. The monoglyceride hydrate is more active than the nonhydrate. Unlike lecithin, the saturated monoglyceride hydrate is more active than the unsaturated monoglyceride hydrate (Schuster and Adams, 1984). Hydrated emulsifiers can be utilized in a variety of bakery products. They impart both improved dough conditioning and antistaling properties (Gawrilow, 1980; Wittman et al., 1984).

In many cases emulsifier activity can be synergistic, rather than additive in nature. For example, a mixture of monoglyceride and ethoxylated mono- and diglycerides has superior properties over any of the individual components.

Diacetyl tartaric acid esters of monoglycerides (DATEM), a widely used emulsifier in Europe for bread products, appears to act as a dough strengthener and a crumb softener. It also improves loaf volume and delays staling (Hoseney et al., 1976; Rogers and Hoseney, 1983; Lorenz, 1983) and shows synergistic activity with monoglycerides (MGL) on gassing power (Armero and Collar, 1996). Lorenz (1983) suggested that the use of DATEM may increase as the sale of variety breads increases in the United States, since DATEM is particularly useful in European type hearth breads, which are not cooked in pans. Addition of diacetyl tartaric acid esters of mono- and diglcyerides improves the distribution of shortening within the dough. Another advantage is that the proof volume of weak doughs is affected to a greater degree than that of strong doughs. While the cost of DATEM relative to monoglycerides has delayed extensive use, the price difference has decreased (Walker, 1983). DATEM has a more pronounced effect than monoglycerides on crumb softness, loaf volume, and texture, which offsets some of the price difference.

Pomeranz et al. (1959a,b) and Hoseney et al. (1972) found that sucrose fatty acid esters improved loaf volume, crumb grain, and crumb softness in bread such that the deleterious properties of added soy isolate were counteracted. The effectiveness of the sucrose esters was dependent upon the degree of substitution and the structure of the fatty acids (Chung et al., 1981a,b). The palmitate and stearate esters were the most effective. Sucrose esters added to wheat flour strengthened the dough and increased the mixing tolerance (Watson and Walker, 1986). Sucrose esters with an HLB of 8 were a much better replacement for flour lipids than the sucrose esters with an HLB of 1, while sucrose esters with an HLB of 14 replaced both the flour lipids and added shortening (Chung et al., 1976).

Schuster and Adams (1984) summarized the theories of Aidoo (1972) and Aidoo and Tsen (1972) on the interaction of nonwheat protein, emulsifiers, gliadin, and glutenin. Soy protein interaction with gliadin is hydrophobic in nature, while the glutenin–soy protein interaction is hydrophilic in nature. The consequence of soy protein addition is the formation of a gliadin–soy protein–glutenin complex that impairs the functional properties of the wheat proteins. With the addition of emulsifier in the presence of soy protein, a gliadin–emulsifier–glutenin complex is formed, which competes with the soy–wheat protein complex, reducing the deleterious effects of the soy protein. Several different protein sources including milk solids, oilseed meal, peanut protein, legume meal, sunflower, fava bean, and field pea have been utilized to supplement wheat protein. In each case emulsifiers were used to offset at least part of the detrimental effects due to nonwheat proteins (Schuster and Adams, 1984). Mesallam et al. (1979) evaluated the effect of emulsifiers on the substitution of 10% of the wheat flour with corn flour in bread. Loaf volume, specific volume, and bread quality were improved with the addition of each emulsifier tested. Tsen (1974) evaluated several emulsifiers including sucrose fatty acid esters, fatty esters of polyalkoxylated glycosides, the stearoyl-2-lactylates, ethoxylated monoglycerides and glycolipids. All were effective in improving the baking performance of wheat flour fortified with defatted soy flour. Tsen et al. (1975b) used a low-cost extruder to prepare full-fat soy flour for supplementation in wheat flour. With the addition of sodium stearoyl-2-lactylate, they were able to make a bread with an acceptable loaf volume. Fiber reduced the gas retention of bread by altering the crumb structure (Pomeranz et al., 1977). Emulsifiers were used to counteract at least part of the effect of added fiber on bread quality. Schuster and Adams (1983) found that 12 of 14 emulsifiers evaluated were able to compensate for the reduction in loaf volume caused by the addition of cellulose.

Due to ease of handling, many bakers prefer powdered emulsifiers. Many of the dough conditioners are available in powdered form (Suggs, 1983), usually as a mixture (Koizumi et al., 1984; Suggs and Buck, 1980; Jackson, 1979). Campagne et al. (1983) developed a powdered sucroglyceride on a casein and maltodextrin support. An emulsifier mixture in the form of beads is also available, specifically a blend of propylene glycol monoesters and acetylated monoesters that is premixed in the oil by the baker as needed. There is a wide range of baked goods, other than bread, with improved characteristics due to added emulsifiers. Emulsifiers can be used to prevent, inhibit or control starch gelation in macaroni, spaghetti, and snack foods and maintain cooked rice in a dry and nonsticky condition (Anonymous, 1983).

Cake volume can be increased upon addition of several different emulsifiers, including diacetyl tartaric acid esters of monoglycerides, glycerol fatty acid esters, and calcium stearoyl-2-lactylate. Emulsifier addition to cake mixes facilitates increased substitution of sucrose with high-fructose corn syrup (Hartnett, 1979). Fructose causes starch gelatinization earlier during baking, relative to sucrose, reducing cake volume, grain size, textural quality, and shelf-life. Emulsifiers increase the gelatinization temperature offsetting the effect of the high-fructose corn syrup. Emulsifiers have three primary functions in cake: (1) to facilitate air incorporation, (2) to disperse shortening in smaller particles to allow the maximum number of air cells, and (3) to improve moisture retention (Painter, 1981; Flack, 1983). Flack (1983) indicated the most commonly used emulsifiers in cake include monoglycerides, lactylated monoglycerides, and polysorbates. Seward and Warman (1984) developed a lipophilic emulsifying system for cake mixes that contained monoglycerides, propylene glycol monoesters, polyglycerol esters, and lactylated monoglycerides. Morgan et al. (1980) developed a similar mixture, minus the polyglycerol esters.

Aerated batters are used to produce products like sponge and angel food cake. Emulsifiers provide faster whipping rates, improved structure, and greater volume, with the added advantage of supplementing part of the egg that is needed (Flack, 1983). With an increase in HLB from 1 to 15, sucrose fatty acid esters improved high-ratio white layer cake volume and softness (Ebeler and Walker, 1984). According to Ebeler and Walker (1984), there were several indications that the sucrose esters delayed both pasting and gelatinization. Hsu et al. (1980) found that emulsifier addition to cake mix reduced both the rate of moisture loss and the rate of temperature increase in the temperature range associated with starch gelatinization during baking. Several different mixtures of surfactants were effective aerating agents, including monoglycerides, polyglycerol esters, lactylated monoglycerides, and sorbitan monostearate (Johnston, 1973; Seibel et al., 1980). The amount of emulsifier added (Schuster and Adams, 1984) can affect emulsifier effectiveness; for example, the addition of an inadequate amount can even have a deleterious effect.

Cookie characteristics (volume, top grain, and, particularly, spread ratio) can be improved with the addition of the appropriate emulsifier. Tsen et al. (1973) found that several emulsifiers affected cookie characteristics favorably, including sodium stearoyl fumarate, sodium stearoyl lactylate, sucrose fatty acid esters, polysorbates, sorbitan fatty acid esters, and ethoxylated monoglycerides. Tsen et al. (1975a) also found that emulsifiers could alter the effect of hard wheat flours such that hard wheat flour could be substituted for soft wheat flour. Like bread, the effect of nonwheat protein addition to cookies can be minimized by the addition of emulsifiers. Cookie shelf-life can also be increased with addition of emulsifiers (Hutchison et al., 1977). Emulsifier addition to cookie mix can improve the spread ratio, cooking surface release, and texture of cookies (Rusch, 1981). However, Breyer and Walker (1983) found that the addition of sucrose esters had a negligible effect on cookie quality.

The interaction of emulsifiers with starch alters the effect of heat on starch. Puddings, for example, were said to have increased freeze–thaw stability and increased resistance to retort processing, while starch-based sauces have improved freeze–thaw stability and increased stability during holding at high temperatures (Rusch, 1981).

Flat bread products are consumed extensively in the Middle East, North Africa, and the Indian subcontinent. Like other bread products, the flat breads will stale upon storage. Addition of sodium stearoyl-2-lactylate and monoglycerides reduced the degree of firmness upon storage of flat breads (Maleki et al., 1981). The addition of sucrose esters to chapaties, a flat bread product, allowed the substitution of 40% of the whole wheat flour with nonglutenous flour (e.g., corn, sorghum, millet, and soy) with only a slight loss in quality (Ebeler and Walker, 1983). This would facilitate soy flour fortification of chapaties to improve the nutritional quality of the product.

B. Dairy Products

1. Ice Cream

Ice cream is both a frozen foam and an emulsion. Protein and polar lipids (lecithin) found in milk function as surfactants in ice cream. However, these naturally occurring components are usually supplemented with additional emulsifiers to make ice cream. Emulsifiers in ice cream improve fat dispersion, facilitate fat–protein interactions, control fat agglomeration, facilitate air incorporation, impart dryness to formed products, confer smoother texture due to smaller ice crystals and air cells, increase resistance to shrinkage, reduce whipping time, and improve melt-down (Flack, 1983; Walker, 1983). Bayer et al. (1997) recently reported that a polysorbate 80 blend with monoglycerides and diglycerides, 40% alpha-monoglyceride, 70% alpha-monoglyceride, and lecithin increased the consistency of viscosity of low fat ice cream mix and reduced whipping times and ice crystal sizes. These emulsifiers were found to increase stability to heat shock and improved the body and texture of low fat ice cream.

Emulsifiers are particularly useful with formed specialty products like ice cream sandwiches, slices, and factory-filled cones. Emulsifiers coat milk fat, improving the ability of the protein film to surround the air cells in the ice cream. The results are improved whippability, smoother texture and body, drier appearance, and slower melt-down. The smoother texture is favored since both ice crystal and air cell sizes are reduced. Two types of emulsifiers are used: (1) mono- and diglycerides and (2) polyoxethylene derivatives of sugar alcohol fatty acid esters, typically polyoxethylene sorbitan monostearate and monooleate.

Ice cream is an O/W emulsion, but the monoglycerides used in ice cream would be classified as W/O emulsifiers due to their low HLB values, suggesting their function in ice cream is not simply emulsification (Berger, 1976). The monoglycerides improve foamability, impart solidity, and improve the shape-retaining characteristics of the frozen product. The improved foamability is due to interaction of the emulsifier with milk proteins. Berger (1976) found that the interfacial tension between water and oil in the presence of 0.1% casein was substantially reduced when monoglycerides were added. This same effect did not occur upon addition of Tween 60. Berger (1976) suggested that a complex between casein and the monoglycerides is formed altering the tertiary structure of casein such that partial denaturation occurred.

The improved shape-retaining characteristics and solidity attributed to emulsifiers is likely due to the partial destabilization of the fat emulsion (Berger, 1976) and agglomeration of the fat globules (Keeney and Kroger, 1974). Both the monoglycerides and the polyoxethylene sorbitan fatty acid esters are effective. Berger (1976) suggested a series of mesophase transitions by the emulsifier at the O/W interphase that would account for the fat destabilization effect. The polyoxethylene derivatives of the sorbitan fatty acid esters are particularly effective in imparting stiffness to ice cream immediately upon leaving the freezer. The added emulsifiers are added not to stabilize the mix, but to impart desirable end product qualities. Selection of a suitable emulsifier should be based on HLB since optimum product characteristics were dependent on emulsifier HLB. The optimum emulsifier HLB was 16 (Govin and Leeder, 1971). Lin and Leeder (1974) characterized ice cream production as a de-emulsification process due to a combination of agitation, freezing, and the emulsifier.

Monoglycerides differ in functionality depending upon the fatty acid moiety (Keeney and Kroger, 1974). The shorter chain fatty acids are more effective than the longer chain fatty acids for imparting stiffness. Unsaturated fatty acids are also more effective, than saturated fatty acids for imparting stiffness. Oleic acid monoglycerides have good shape-retaining characteristics, while stearic acid monoglycerides have improved stability to oxidation relative to oleic acid esters. Because of this, monoglycerides available commercially are often a mixture of both oleic and stearic acid esters (Torrey, 1983).

Flack (1985) found that mono- and diglycerides, citric acid esters of monoglycerides, lactic acid esters of monoglycerides, and distilled monoglycerides affected the whippability and foam stability of dairy whipping cream that had been pasteurized at 80°C and homogenized at 1000 psi. Lactic acid esters produced a stable foam with high overrun that was substantially different from a fresh whipped cream, while a mixture of citric acid esters and distilled monoglycerides produced a foam similar to that of fresh whipped cream.

Pearce and Harper (1982) developed a procedure for the evaluation of emulsion stability for liquid coffee whiteners upon addition to hot coffee. They found that if the emulsion was particularly unstable, coagulation occurred. However, with the addition of the correct mixture of emulsifiers, feathering could be completely eliminated.

Cheese yield can be increased with the addition of lecithin. Lecithin was added in quantities of 0.001–0.066% by weight of the milk (Bily, 1981).

C. Candy Products

The elimination of ''bloom,'' i.e., the transition of fat crystals from the alpha and beta′ configuration to the less desirable beta configuration, is a key reason for the addition of emulsifiers to candy products. Emulsifiers can be used as crystal structure modifiers in mixtures of triglycerides (Anonymous, 1983). Certain emulsifiers can also be used to control product viscosity in cream fillings and in chocolates. The conversion of the alpha form of stearic acid (Garti et al., 1982a) and tristearin (Garti et al., 1982b) to the beta form can be partially inhibited by the addition of sorbitan esters (Spans) and ethoxylated sorbitan esters (Tweens). Garti et al. (1986) also studied the effect of Spans and Tweens on polymorphic transitions in cocoa butter, suggesting that these emulsifiers might be effective in preventing chocolate bloom. Emulsifiers are often used with both semisweet and milk chocolate.

Emulsifiers aid in processing by reducing the weep or exudate that occurs with heavy sugar pastes during processing (Riedel, 1985). Rubenstein et al. (1980) prepared a marshmallow-based frozen confectionary product with 0.2 to 0.8% emulsifier with an HLB between 3 and 9.

Ogawa et al. (1979) used emulsifiers (sucrose fatty acid esters, sorbitan fatty acid esters, glycerol fatty acid esters, or propylene glycol fatty acid esters) to improve shelf stability of center-filled chewing gum. The emulsifiers are added to the flavored liquid center at 0.01 to 0.5% by weight. Polyglycerol esters are used in coatings for confectionary products (Herzing and Palamidis, 1984).

D. Miscellaneous

Emulsifiers have been utilized in the production of meat analog products. Howard (1980) used emulsifier mixtures selected from the group consisting of polyglycerol fatty acid monoesters, monoacylglycerol esters of dicarboxylic acids, sucrose esters of fatty acids, polyol monoesters of fatty acids, and phospholipids to prepare meat analogs. Haggerty and Corbin (1984) used phospholipids to prepare filled meat products containing substantial amounts of both meat and texturized soybean protein.

Emulsifiers are used in the formulation of flavor emulsions, although the available literature is rather limited. Chilton and Peppard (1978) evaluated several mixtures of Spans and Tweens to determine the optimum for the stabilization of hop oil emulsions. For those interested in flavor applications, recommendations should be solicited from companies that sell emulsifiers.

There is a wide range of dairy substitute and bakery products, where emulsifiers are used to facilitate formulation. These products include low calorie margarines, filled cream products, whipped margarines, synthetic milks and creams, coffee whiteners, whipped cream products, and stable foamed frozen emulsions (Torrey, 1984; Eydt, 1994; La Guardia, 1994). Rule (1983) developed an alcohol and dairy based product that contained 0.14–0.7% emulsifiers selected from a group consisting of polyglycerol esters of fatty acids, ethoxylated fatty acid esters, and sugar esters such that the HLB was greater than 10.

Margarine exists as a W/O emulsion for only a short period of time prior to chilling, at which time the emulsion is converted to a dispersion of water in a semisolid fat phase (Krog et al., 1985). Upon solidification, product stability is greatly enhanced, since coalescence is essentially eliminated. Emulsifiers fulfil three functions in margarine: (1) assistance in emulsion formation, (2) modification of crystal structure in the vegetable fat, and (3) antispattering. Typically, a mixture of lecithin or citric acid monoglycerides and monoglycerides are used. A stable emulsion during frying reduces the coalescence of water droplets into large drops. This facilitates gradual water evaporation, rather than explosive evolution, during frying. An additional explanation may be that the lecithin sludge formed during frying serves as nuclei for formation of small vapor bubbles (Van Nieuwenhuyzen, 1981). To reduce sandiness and ''oiling out'' in margarine due to recrystallization, emulsifiers can be added (Krog et al., 1985). The emulsifiers prevent recrystallization by slowing down the polymorphic transition rates (Dorota, 1996; Paola et al., 1996). The emulsifiers 1-monostearin, sorbitan tristearate, and sugar monostearate hinder the β' to β transformation of tristearin (Paola et al., 1996). The most effective in preventing recrystallization in tristearin were sorbitan monostearate and citric acid esters of monoglycerides. However, to be effective in margarine, the emulsifier must also be highly soluble in the oil phase. Other emulsifiers used include polysorbates and polyglyerol esters of fatty acids (Walker, 1983).

To produce a low calorie imitation dairy product, emulsifiers are usually added to the product (Gilmore et al., 1980). The caloric content can be significantly reduced by replacement of the fat with diglycerides. For low fat margarine containing approximately 40% fat, a mixture of saturated and unsaturated monoglycerides and lecithin has been used successfully (Lefebvre, 1983). Madsen (1976) recommended a mixture of 25% saturated and 75% unsaturated monoglycerides for the production of low calorie margarines. This provided the best combination of emulsion stability and melt-down in the mouth. Similarly, emulsifiers were used for fat replacement in a low fat, butter-flavored liquid spread (Bosco and Sledzieski, 1980). The stability of milk fat and water emulsions are dependent upon the amount of emulsifier added (Titus and Mickle, 1971).

Mayonnaise is an O/W emulsion containing a high percentage of oil (>70%). Due to the large percentage of fat, coalescence, rather than creaming, is the primary problem (Jaynes, 1985). Different protein sources are utilized because of their effectiveness in reducing coalescence. The manufacturing procedure for mayonnaise is very critical since the high percentage of oil favors a W/O emulsion rather than an O/W emulsion (Krog et al., 1985). Lecithin, contained in the added egg yolk, is usually the only emulsifier added. However, the vegetable oil may contain emulsifiers added to inhibit crystallization, e.g., ethoxylated sorbitan monooleate and monostearate (Anonymous, 1983). To inhibit cloud formation in salad dressings and salad oils, emulsifiers can also be added.

To improve wetting characteristics in instant cocoa drinks, the powder is agglomerated. Lecithin will facilitate agglomeration during spray-drying. During the spray-drying process the hydrophobic portion of the emulsifier will dissolve in the cocoa butter, orienting the hydrophilic portion of the phospholipid toward the surface of the particle (Van Nieuwenhuyzen, 1981). The increased affinity of the cocoa powder for water aids dispersion and wetting.

Emulsifiers are often used in peanut butter. The addition of an emulsifier to peanut butter can inhibit oil phase separation.

V. TOXICOLOGY AND WORLDWIDE REGULATIONS CONCERNING USE

There are two sources of international food standards, the Codex Alimentarius standards under the auspices of the Food and Agricultural Organization/World Health Organization (FAO/WHO) on a worldwide basis and the European Economic Community (EEC) directives applicable to the member states within the European Economic Community (Gnauck, 1978). In the United States the Food and Drug Administration (FDA) is the primary source of regulation on food, particularly for food additives.

Recommendations for the Codex Alimentarius Committee come from the Joint Expert FAO/WHO Committee on Food Additives (JEFCA), for the EEC from the Scientific Committee on Food (SCF), and in the United States from within the FDA. Opinions from the SCF and the JEFCA are provided to legislative bodies in interested countries either in Europe or elsewhere in the world. The Scientific Committee on Food is a committee of scientists nominated by the Commission of the European Communities to advise the Commission:

> on any problem relating to the protection of the health and safety of persons arising from the consumption of food, and in particular on the composition of food, processes which are liable to modify food, the use of food additives and other processing aids, as well as the presence of contaminants. (Haigh, 1978)

Evaluations of emulsifiers and recommendations for their limiting concentrations in food and their acceptable daily intake (ADI) are then provided to the Commission by the SCF.

The Joint FAO/WHO Codex Alimentarius Commission, which consists of member nations and associate members of FAO and/or WHO, was established to implement the Joint FAO/WHO Food Standards Program (FAO/WHO, 1979). The Joint FAO/WHO Food Standards Program was established to promulgate international standards for foods in order, to protect the health of consumers, to ensure fair practices in the food trade, to promote coordination of all food standards work undertaken by international governmental and nongovernmental organizations, to determine priorities and initiate and guide the preparation of draft standards through and with the aid of appropriate organizations, to finalize standards, and, after acceptance by governments, to publish them in a Codex Alimentarius either as regional or worldwide standards. The Joint FAO/WHO Codex Committee on Food Additives, an intergovernmental subsidiary body, advises the Codex Alimentarius Commission on all matters relating to food additives based on the views and recommendations from the Joint Expert FAO/WHO Committee on Food Additives. Evaluation of the available scientific data and the resultant recommendations are provided by the Joint FAO/WHO Expert, Committee on Food Additives. The Committee is composed of experts who serve in a personal capacity. The Committee establishes ADIs and specifications of identity or purity for food additives, including, of course, food emulsifiers.

The JEFCA recommended that food additives be used only after authorization by the appropriate authorities and that legal control be based on a system of permitted or positive lists. Two key criteria, technological efficacy and safety, must be satisfied before a substance can be formally accepted as a food additive.

Principles of evaluation and procedures for the toxicological analysis of food additives were dealt with in four separate Committee reports (JEFCA, 1973,1994). The first stage of a toxicological evaluation is the collection of relevant data, while the second is the interpretation and assessment of the data in order to arrive at a decision about the acceptability of the substance as a food additive.

If sufficient data exist, an acceptable daily intake is recommended. The JEFCA may make one of several recommendations: a specific amount in milligrams per kilogram of body weight; an upper limit of ''not specified;'' ''decision postponed'' (pending clarification of matters related to technological use); ''no ADI allocated'' (in the absence of sufficient data for established safety or of adequate specifications); or ''not to be used'' (where there is sufficient information on which to base such a decision). The JEFCA definition of the acceptable daily intake for man, expressed on a body weight basis, ''is the amount of a food additive that can be taken daily in the diet, even over a lifetime, without risk'' (JEFCA, 1973,1994). In the Seventeenth Report, Annex 3 (JEFCA, 1973), the general procedures for the testing of food additives are provided. For example, recommendations are given for acute toxicity studies, biochemical studies, short-term studies, long-term toxicity, and several additional special studies including embryotoxicity, teratogenicity, and carocinogenicity.

No acute or chronic toxic effects have been observed with either lecithin or mono- and diglycerides at normal dosage levels (JECFA, 1974). There was also no evidence of acute or chronic toxic effects with fatty acid salts (JEFCA, 1970). Similarly, there were no observable acute and chronic toxicological effects from acetic, citric, lactic, tartaric, mixed tartaric, or acetic and diacetyltartaric acid esters of mono- and diglycerides (JEFCA, 1974). Sucrose fatty acid esters have shown no toxic effects at a practical dose level. In addition, both long-term and short-term studies indicate no significant organ changes and no adverse effects due to sucrose fatty acid esters. No toxic effects were seen in rats due to high doses of calcium stearoyl-2-lactylate over a short term, while the short-term studies on sodium stearoyl-2-lactylate were conflicting in terms of growth rates and relative organ weights with rats.

Both polyoxethylene (polysorbate 20) sorbitan monolaurate and polyoxethylene (polysorbate 20) sorbitan monooleate have an oral LD_{50} level in excess of 30 g/kg for rats. Monkeys fed polyoxethylene (polysorbate 20) sorbitan monolaurate for 17 months at 1 g/day showed no significant histological abnormalities. Two percent of polysorbate 80 incorporated in rat diets showed no abnormalities attributable to the diet. The same study over a 3-year period showed no evidence of abnormalities in growth rates and gross or histological analysis. Polyoxethylene (polysorbate 20) sorbitan monopalmitate (polysorbate 40) did not show any acute or chronic toxicological effects in rats. Neither polyoxethylene (polysorbate 20) sorbitan monostearate (polysorbate 60) nor polyoxethylene (polysorbate 20) sorbitan tristearate (polysorbate 65) showed any acute or chronic toxicological effects in test animals (JEFCA, 1974). Five percent levels in the diet of rats showed no significant toxicological effects for both polyoxethylene (polysorbate 8) stearate and polyoxethylene (40) stearate. Neither sorbitan monolaurate, or monooleate, or monopalmitate showed any toxicological effects in rats over a 2-year period at the 5% level. The absence of toxicity for sorbitan monostearate and tristerate has also been established. Stearyl tartrate has shown no long- or short-term toxicity based on work done with animal studies. Most of this toxicological information has also been summarized in the EFEMA (1976). Annex 4 lists the ADI for several food additives including emulsifiers.

The FDA issues "Proposed Rules" in the *Federal Register* and solicits input from the public regarding the proposed regulation changes. After due consideration, the proposed changes are issued as Final Rules, initially in the Federal Register and finally in the Code of Federal Regulations (CFR).

The CFR (1997) list concentrations allowed in food, while the Codex Alimentarius Commission (1995) lists both allowable concentrations in foods and the acceptable daily intakes. Table 3 was compiled from both these sources. The limitations imposed by the CAC are based on toxicological evaluations provided by the Joint FAO/WHO Expert Committee on Food Additives published in FAO Nutrition Meetings Reports. The limitations set for concentrations in food in the United States are published yearly in the CFR, while the limitations promulgated by the CAC (1979) are published in the *Guide to the Safe Use of Food Additives*. For specific applications the appropriate reference should be consulted.

In 1974 the Council of the European Communities issued Council Directive No. 74/329 to facilitate trade among countries within the European community. Food emulsifiers, as defined in the directive, are "those substances which, when added to a foodstuff, make it possible to form or maintain a uniform dispersion of two or more immiscible substances." The emulsifiers, as authorized in 1974 by the European Economic Council (EEC), and as amended in 1978, 1980, and 1985, are those emulsifiers listed in Table 4 with an EEC number. A second category, Annex II, published in 1980 by the EEC, is that of emulsifiers that may be authorized for use by member states for a limited time.

In the United States a list of the emulsifiers approved for use has been published in the Code of Federal Regulations, Part 172 of Title 21 (1997). The CFR covers food emulsifiers under two categories: those permitted only under specific conditions and those generally recognized as safe. The emulsifiers covered are in the Code of Federal Regulations, Title 21, Part 182, subpart E. Included in this table is lecithin, covered in Part 184 but also listed as GRAS. Both of these groups are listed in Table 4. As one can see from Table 4, there are several emulsifiers that are allowed in the United States but not in Europe and vice versa.

Most of the approved uses for food additives as food emulsifiers as listed in the CFR (1997) are in Table 5. The CFR also discusses the limiting concentrations for the emulsifiers and the methods approved for their production. A list of the approved uses for food emulsifiers as determined by the Codex Alimentarius Committee can be found

Table 3 Limitations for Emulsifiers List from the U.S. Code of Federal Regulations and the Codex Alimentarius Commission

		CAC[b]	
	CFR[a]	ML[c]	ADI[d]
Lecithin	GRAS[e]	5–20 g/kg[f]	Not specified[g]
Na phosphate derivatives of mono- and diglycerides of fatty acids	GRAS	—	Not specified
Hydroxylated lecithin[h]		—	No ADI[i]
Na, K, and Ca salts of fatty acids[j]		—	Not specified
Mono- and diglycerides of fatty acids	GRAS	2.5–15 g/kg	Not specified[k]
Acetic acid esters of mono- and diglycerides of fatty acids[l,m]		10–20 g/kg (oil)	Not specified[k]
Lactic acid esters of mono- and diglycerides of fatty acids[l,m]		10–20 g/kg (oil)	Not specified[k]
Citric acid esters of mono- and diglycerides of fatty acids[l]	200 ppm	10–20 g/kg (oil)	Not specified[k]
Mixed acetic and tartaric acid esters of mono- and diglycerides of fatty acids[l]		10–20 g/kg (oil)	Not specified[k]
Mono- and diacetyltartaric acid esters of mono- and diglycerides of fatty acids	GRAS	10–20 g/kg (oil)	0–50 mg/kg[n]
Tartaric acid esters of mono- and diglycerides of fatty acids[l]		—	Not specified[k]
Sucrose esters of fatty acids[h,m]		10–20 g/kg	0–2.5 mg/kg
Sucroglycerides	—	10–20 g/kg	0–2.5 mg/kg
Polyglycerol esters of fatty acids[m]		5–20 g/kg	0–25 g/kg
Na stearoyl-2-lactylate	0.2–0.5%[p]	20 g/kg	0–20 mg/kg
Ca stearoyl-2-lactylate	0.05–0.5%[p]	20 g/kg	0–20 mg/kg
Stearyl citrate	—	—	0–50 mg/kg
Polyoxyethylene (20) sorbitan monolaurate (polysorbate 20)[e]	—	—	0–25 mg/kg
Polyoxyethylene (20) sorbitan monopalmitate (polysorbate 40)[q]	—	—	0–25 mg/kg
Polyoxyethylene (20) sorbitan monostearate (polysorbate 60)[q]	0.32 ≥ 1.0%	—	0–25 mg/kg
Polyoxyethylene (20) sorbitan tristearate (polysorbate 65)[q]	0.1–0.32%	15–20 g/kg	0–25 mg/kg
Polyoxyethylene (20) sorbitan monooleate (polysorbate 80)[q]	0.1–1.0%	20 g/kg (oil)	0–25 mg/kg
Polyoxyethylene (8) stearate	—	—	0–25 mg/kg[r]
Polyoxyethylene (40) sorbitan	—	—	0–25 mg/kg[r]
Ammonium phosphatides	—	—	0–30 mg/kg
Thermally oxidized soybean oil interacted with mono- and diglycerides of fatty acids	—	—	0–0.3 mg/kg
Lactylated fatty acid esters of glycerol and propylene glycol[l]	—	—	Not specified[k]
Dioctyl sodium succinate	25 ppm (cocoa fat)	5 g/kg	0–0.1 mg/kg
Stearyl monoglyceridyl citrate[j]	—	—	Not specified[k]
Stearyl propylene glycol hydrogen succinate (succistearin)[h]	—	—	0–25 mg/kg
Copolymer condensates of ethylene oxide and propylene oxide	0.5% (flour)	—	0–25 mg/kg
Methyl glucoside–coconut oil ester	320 ppm (molasses)	—	—
Oxystearin	0.125% (oil)	—	—
Sodium lauryl sulfate	125–1000 ppm (egg whites) 25 ppm (fruit juice drink)	—	—

Sodium stearoyl fumarate	0.5–1.0% (bakery products) 0.2% (starch thickened foods)	—	—
Succinylated monoglycerides	3.0% (shortening) 0.5% (flour)	—	—
Ethoxylated mono- and diglycerides	0.2–0.5% (flour, dairy substitutes, etc.)	—	—
Polyglycerol polyricinoleate	—	5–20 g/kg	0–7.5 mg/kg
Propylene glycol esters of fatty acids[l]	—	20 g/kg	0–25 mg/kg
Propylene glycol alginate[p] (limited by GMP[t])	0.3–1.1%	5–10 g/kg	0–25 mg/kg
Sorbitan esters of monostearate	0.4–1.0%	10–20 g/kg	0–25 mg/kg[u]
Sorbitan esters of tristearate	—	10–20 g/kg	0–25 mg/kg[u]
Sorbitan esters of monopalmitate	—	10–20 g/kg	0–25 mg/kg[u]
Sorbitan esters of monolaurate	—	—	0–25 mg/kg[u]
Sorbitan esters of monooleate	—	—	0–25 mg/kg[u]

[a] Code of Federal Regulation gives the concentrations allowed in food.
[b] Codex Alimentarius Commission.
[c] ML is the designation for the maximum limit, which is product specific.
[d] Acceptable daily intake (ADI) is in milligrams of emulsifier per kilogram of body weight.
[e] Generally recognized as safe.
[f] Grams of emulsifier per kilogram of food product.
[g] An ADI of ''not specified'' is allocated to substances which do not constitute a toxicological hazard to man except under the most gross conditions of overexposure.
[h] Must be used in accordance with good manufacturing practices (GMP).
[i] No ADI indicates no ADI allocated.
[j] Must be used as outlined in the Code of Federal Regulations.
[k] As sum of the total mixed glycerol fatty acids and acetic, citric, lactic, and tartaric acids, provided that the total amount of emulsifier does not exceed 100 mg/kg.
[l] As sum of the total glycerol fatty acids and acetic, citric, lactic, and tartaric acids, provided that the total additive intake of tartaric acid does not exceed 30 mg/kg.
[m] Must not be used in excess of the amount reasonably required to produce its effect in food.
[n] Milligrams of emulsifier per kilogram of body weight.
[o] Calculated as the palmitate ester.
[p] Concentration is dependent on both the food products the emulsifier is used in and the other emulsifiers used (refer to CFR, Title 21).
[q] As the sum of the polyoxyethylene (20) sorbitan esters.
[r] As the sum of polyoxyethylene (20) sorbitan and polyoxyethylene (8) stearate.
[s] As propylene glycol.
[t] The additive in question is self-limiting in food for technological, organoleptic, or other reasons and that, therefore, the additive need not be subject to legal maximum limits. It also means that the food additives must be used according to good manufacturing practice and in accordance with the *General Principles for the Use of Food Additives*.
[u] As the sum of all the sorbitan esters.

Table 4 Emulsifiers Authorized for Use by the EEC

Emulsifier designation	EEC no.	CFR no.
Lecithins	E322	184.1400
Sodium phosphate derivatives of mono- and diglycerides of edible oils	—	182.4521
Stearyl monoglyceride citrate	—	172.755
Stearoyl propylene glycol hydrogen succinate (succistearin)	—	172.765
Copolymer condensates of ethylene oxide and propylene oxide	—	172.808
Hydroxylated lecithins	—	172.814
Methyl glucoside–coconut oil ester	—	172.816
Oxystearin	—	172.818
Sodium lauryl sulfate	—	172.822
Sodium stearoyl fumarate	—	172.826
Na, K, and Ca salts of fatty acids	E470	172.863
Succinylated monoglycerides	—	172.830
Ethoxylated mono- and diglycerides	—	172.834
Mono- and diglycerides of fatty acids	E471	182.4505
Acetic acid esters of mono- and diglycerides of fatty acids	E472(a)	172.828
Lactic acid esters of mono- and diglycerides of fatty acids	E472(b)	172.852
Citric acid esters of mono- and diglycerides of fatty acids	E472(c)	172.832
Tartaric acid esters of mono- and diglycerides of fatty acids	E472(d)	—
Mono- and diacetyltartaric acid esters of mono- and diglycerides of fatty acids	E472(e)	182.4101
Mixed acetic and tartaric acid esters of mono- and diglycerides of fatty acids	E472(f)	—
Sucrose esters of fatty acids	E473	172.859
Sucroglycerides	E474	—
Polyglycerol esters of fatty acids	E475	172.854
Propane-1,2-diol esters of fatty acids (or propyleneglycol esters of fatty acids)	E476	172.856
Propylene glycol alginate	—	172.858
Na stearoyl-2-lactylate	E481	172.846
Ca stearoyl-2-lactylate	E482	172.844
Stearyl tartrate	E483	—
Polyoxyethylene (20) sorbitan monolaurate (polysorbate 20)	Annex II	172.515
Polyoxyethylene (20) sorbitan monopalmitate (polysorbate 40)	Annex II	172.515
Polyoxyethylene (20) sorbitan monostearate (polysorbate 60)	Annex II	172.836
Polyoxyethylene (20) sorbitan tristearate (polysorbate 65)	Annex II	172.838
Polyoxyethylene (20) sorbitan monooleate (polysorbate 80)	Annex II	172.840
Polyoxyethylene (8) stearate	Annex II	—
Polyoxyethylene (40) sorbitan	Annex II	—
Ammonium phosphatides	Annex II	—
Thermally oxidized soybean oil interacted with mono- and diglycerides of fatty aids	Annex II	—
Lactylated fatty acid esters of glycerol and propylene glycol	—	172.850
Lactylic esters of fatty acids	—	172.848
Dioctyl sodium sulfosuccinate	Annex II	172.810
Polyglycerol polyricinoleate	Annex II	—
Sorbitan esters of monostearate	Annex II	172.842
Sorbitan esters of tristearate	Annex II	—
Sorbitan esters of monolaurate	Annex II	—
Sorbitan esters of monooleate	Annex II	—
Sorbitan esters of monopalmitate	Annex II	—

Table 5 Approved Uses for Food Additives as Emulsifiers

Additive	Approved uses or functions
Stearyl monoglyceridyl citrate	Emulsion stabilizer in shortenings
Succistearin oil	Shortenings and edibles used in baked products, e.g., pastries
Copolymer condensates of ethylene oxide and propylene oxide	Yeast-leavened products and flavor concentrates
Dioctyl sodium sulfosuccinate	Emulsifier in cocoa fat for noncarbonated beverages
Hydroxylated lecithin[a]	
Methyl glucoside–coconut oil ester	Surfactant in molasses
Oxystearin	Inhibition of crystallization of vegetable oils
Sodium lauryl sulfate	To facilitate product formulation with egg whites
Sodium stearoyl fumarate	Dough conditioner
Acetylated monoglycerides[a]	
Succinylated monoglycerides	Emulsifier in shortenings and as a dough conditioner
Monoglyceride citrate	Used in antioxidant formulations added to fats and oils
Ethoxylated mono- and diglycerides	Emulsifier in pan-release agents, in yeast-leavened products, cakes, whipped vegetable toppings, icings, frozen desserts, nondairy creamers
Polysorbate 60	Whipped toppings, cake mixes, icings, confectionery products, shortenings, nondairy creamers, nonstandard dressings, protective vegetable coatings, chocolate flavored coatings, coloring formulations, and as a dough conditioner
Polysorbate 65	Ice cream, cakes, whipped vegetable toppings, and nondairy creamers
Polysorbate 80	Ice cream, cakes, whipped vegetable toppings, and nondairy creamers
Sorbitan monostearate	Confectionary coatings, cakes, nondairy creamers, whipped toppings, protective coatings for fruits and vegetables
Calcium stearoyl-2-lactylate	Yeast-leavened bakery products, whipped toppings, dehydrated potatoes
Sodium stearoyl-2-lactylate	Waffles, pancakes, cake icings and fillings, toppings, nondairy creamers, dehydrated potatoes, snack dips, imitation cheeses, sauces, and gravies
Lactylic esters of fatty acids	Bakery mixes, bakery products, cake icings, fillings, toppings, dehydrated fruit juices, vegetable oil emulsions, frozen desserts, shortening, pancake mixes, precooked rice, and puddings
Lactylated fatty acid esters of glycerol and propylene glycol	Emulsifier, plasticizer, or surfactant
Glyceryl-lacto fatty acid esters	Emulsifier or plasticizer
Polyglycerol fatty acid esters	Whipped toppings
Propylene glycol mono and diesters of fats and fatty acids[a]	
Propylene glycol alginate	Confections and frostings, baked goods, cheeses, fats and oils, gelatins, puddings, gravies, sweet sauces, condiments, relishes[a,b]
Sucrose fatty acid esters	Baked goods, baking mixes, protective coatings in fresh fruits and vegetables[a]
Salts of fatty acids[a]	

[a] The additive may be used in amounts not in excess of that reasonably required to produce the desired physical or technical effect.

[b] Although a polymer, this additive was included because of its many approved uses as an emulsifier.

in the *General Principles for the Use of Food Additives* (Codex Alimentarius Commission, 1995).

REFERENCES

Adler, L., Pomeranz, Y. 1959. Use of lecithin in production of bread containing defatted soya-flour as a protein supplement. *J. Sci. Food Agric.* 10:449.

Aidoo, E. S. 1972. High-protein bread: interactions of wheat proteins and soy proteins with surfactants in doughs and in model systems. Ph.D. Dissertation. Kansas State University, Manhattan, KS.

Aidoo, E. S., Tsen, O. K. 1973. Influence of surfactants or soy proteins on the extractability, gel filtration and disk electrophoretic patterns. (Abstract) *Cereal Sci. Today* 18:302.

Akoh, C. C. 1994. Enzymatic synthesis of acetylated glucose fatty acid esters in organic solvent. *J. Am. Oil Chem. Soc.* 71:319.

Aniansson, E. A. G., Wall, S. N. Almgren, M., Hoffmann, H., Kielmann, I., Ulbrich, W., Zana, R., Lang, J., Tondre, C. 1976. Theory of the kinetics of micellar equilibria and quantitative interpretation of chemical relaxation studies of micellar solutions of ionic surfactants. *J. Phys. Chem.* 80:905.

Anonymous. 1983. Emulsifiers for the food industry. *Food, Flavour, Ingred., Process. Packag.* 6:27.

Arcos, J. A., Bernabe, M., Otero, C. 1998. Quantitative enzymatic production of 1,6-diacylfructofuranoses. *Enzyme Microbial Technol.* 22:27.

Armero, E., Collar, C. 1996. Antistaling additives, flour type and sourdough process effects on functionality of wheat doughs. *J. Food Sci.* 61:299.

Artz, W. E., Myers, M. R. 1994. Supercritical fluid extraction and chromatography of emulsifiers. *J. Am. Oil Chem. Soc.* 72:219.

Baer, R. J., Wolkow, M. D., Kasperson, K. M. 1997. Effect of emulsifiers on the body and texture of low fat ice cream. *J. Dairy Sci.* 80:3123.

Baur, F. J. 1973. Analytical methodology for emulsifiers used in fatty foods: a review. *J. Am. Oil Chem. Soc.* 50:85.

Becher, P. 1957. *Emulsions: Theory and Practice*. ACS Monograph No. 135. Rheinhold, New York.

Becher, P. (Ed.) 1983. *Encyclopedia of Emulsion Technology*, Vol. I: *Basic Theory*. Marcel Dekker, New York.

Becher, P. 1984. Hydrophile–lipophile balance: history and recent developments. Langmuir lecture, 1983. *J. Dispers. Sci. Tech.* 5:81.

Becher, P. 1985. *Encyclopedia of Emulsion Technology*, Vol. 2: *Applications*. Marcel Dekker, New York.

Berchter, M., Meister, J., Hammes, C. 1997. MALDI-TOF-MS—a new analytical technique for the characterizing products on the base of renewable resources. *Fett.-Lipid* 99:384.

Berger, K. G. 1976. Ice cream. In: *Food Emulsions*, Friberg, S. (Ed.). Marcel Dekker, New York.

Bhattacharyya, D. N., Krishnan, S., Kelkar, R. Y., Chikale, S. V. 1984. Preparation and emulsifying properties of polyethyleneglycol (1500) diesters of fatty acids. *J. Am. Oil Chem. Soc.* 61:1925.

Bily, R. R. 1981. Addition of lecithin to increase yield of cheese. U.S. patent 4,277,503. In: *Food Additives: Recent Developments*, Johnson, J. C. (Ed.). Food Technology Review No. 58. Noyes Data Corporation, Park Ridge, NJ.

Bonfillon, A., Langevin, D. 1993. Viscoelasticity of monolayers at oil–water interfaces. *Langmuir* 9:2172.

Bosco, P. M., Sledzieski, W. L. 1981. Low fat liquid spreads. U.S. patent 4,292,333. In: *Food Additives: Recent Developments*, Johnson, J. C. (Ed.). Food Technology Review No. 58. Noyes Data Corporation, Park Ridge, NJ.

Boyle, E. 1997. Monoglycerides in food systems—current and future uses. *J. Food Technol.* 51:52.

Breyer, L. M., Walker, C. E. 1983. Comparative effect of various sucrose–fatty acid esters on bread and cookies. *J. Food Sci.* 48:955.

Brummer, J. M., Mettler, E., Seibel, W., Pfeilsticker, K. 1996. Studies on the optimization of emulsifier and hydrocolloid effects in wheat doughs and wheat breads. *Lebensmittel-Wissenschaft and Technologie* 29:106.

Buschmann, N., Hulskotter, F. 1997. Titration procedure for the low ethoxylated nonionics surfactants. *Tenside, Surfactants, Detergents* 34:8.

Campagne, J.-C., Chollet, J., Redien, P. Phone Poulenc Industries. 1983. Sucroglyceride on a support. U.S. patent 4,380,555.

Campelltimperman, K., Choi, J. H., Jimenez-Flores, R. 1996. Mono- vand diglycerides prepared by chemical glycerolysis from a butterfat fraction. *J. Food Sci.* 61:44.

Carlson, T. L.-G., Larsson, K., Dinhnguyen, N., Krog, N. 1979. A study of the amylose–monoglyceride complex by Raman spectroscopy. *Staerke* 31:222.

Chilton, H. M., Peppard, T. L. 1978. Suitable emulsifiers for hop oil emulsions. *J. Inst. Brew.* 84: 177.

Chung, O. K., Pomeranz, Y., Goforth, D. R, Shogren, M. D., Finney, K. F. 1976. Improved sucrose esters in breadmaking. *Cereal Chem.* 53:615.

Chung, O. K., Shogren, M. D., Pomeranz, Y., Finney, K. F. 1981a. Defatted and reconstituted wheat flours. VII. The effect of 0–12% shortening (flour basis) in breadmaking. *Cereal Chem.* 58:69.

Chung, H. Seib, P. A., Finney, K. F., Magoffin, C. D. 1981b. Sucrose monoesters and diesters in breadmaking. *Cereal Chem.* 58:164.

Cloke, J. D., Davis, E. A., Gordon, J. 1984. Water loss during reheating of fresh and stored cakes made with saturated and unsaturated monoglycerides. *Cereal Chem.* 61:371.

Code of Federal Regulations. 1997. Title 21. Food and Drugs. Parts 170–199. Office of the Federal Register, National Archives and Records Administration. U.S. Government Printing Office, Washington, D.C.

Codex Alimentarius Commission. 1995. *General Principles for the Use of Food Additives*, Vol. 1A. Food and Agriculture Organization of the United Nations. World Health Organization, Rome.

Cooley, J. A. 1965. Role of shortening in continuous dough processing. *Bakers Dig.* 39:37.

Council of European Communities. 1974. Council Directive No. 74/329 on the approximation of the laws of the member states relating to emulsifiers, stabilizers, thickeners and gelling agents for use in foodstuffs. *Official J. European Communities* 17:No. L189/1.

Council of European Communities. 1978. Council Directive No. 78/612 amending for the first time Directive 74/329/EEC on the approximation of the laws of the Member States relating to emulsifiers, stabilizers, thickeners and gelling agents for use in foodstuffs. *Official J. European Communities* 21:No. L197/22.

Council of European Communities. 1980. Council Directive No. 80/597 amending for the second time Directive 74/329/EEC on the approximation of the laws of the Member States relating to emulsifiers, stabilizers, thickeners and gelling agents for use in foodstuffs. *Official J. European Communities* 23:No. L155/23.

Council of European Communities. 1985. Council Directive No. 85/6 amending for the third time Directive 74/329/EEC on the approximation of the laws of the Member States relating to emulsifiers, stabilizers, thickeners and gelling agents for use in foodstuffs. *Official J. European Communities* 28:No. L2/21.

Daniels, D. H., Warner, C. R., Selim, S. 1982. Determination of polysorbate 60 in salad dressings by colorimetric and thin layer chromatographic techniques. *J. Assoc. Off. Anal. Chem.* 65:162.

Davies, J. T. 1957. A quantitative kinetic theory of emulsion type. I. Physical chemistry of the emulsifying agent. *Proc. Int. Cong. Surface Activ. 2nd, London* 1:426.

De Buruaga, A. S., Capek, I., De La Cal, J. C., Asua, J. M. 1998. Kinetics of the photoinitiated inverse microemulsion polymerization of 2-methacryloyl oxyethyl trimethyl ammonium chloride. *J. Polymer Sci.* 36:737.

Dieffenbacher, A., Bracco, U. 1978. Analytical techniques in food emulsifiers. *J. Am. Oil Chem. Soc.* 55:642.

Dorota, J. 1995. The influence of temperature on interactions and structures in semisolid fats. *J. Am. Oil Chem. Soc.* 72:1091.

Dubois, D. K. 1979. Dough strengtheners and crumb softeners. I. Definition and classification. *Res. Dep. Tech. Bull.*, Vol. 1, Issue 4. American Institute of Baking, Manhattan, KS.

EFEMA (European Food Emulsifier Manufacturers' Association). 1976. *Monographs for Emulsifiers for Foods.* EFEMA, Brussels, Belgium.

Ebeler, S. E., Walker, C. E. 1983. Wheat and composite flour chapaties: effects of soy flour and sucrose-ester emulsifiers. *Cereal Chem.* 60:270.

Ebeler, S. E., Walker, C. E. 1984. Effects of various sucrose fatty acid ester emulsifiers on high-ratio white layer cakes. *J. Food Sci.* 49:380.

Eydt, A. J. 1994. Formulating reduced-fat foods with polyglycerol ester emulsifiers. *Food Technol.* 48:82.

FAO/WHO. 1979. *Guide to the Safe Use of Food Additives.* Issued by the Secretariat of the Joint FAO/WHO Standards Programme, FAO, Rome.

Firestone, D. 1994. Gas chromatographic determination of monoglycerides and diglycerides in fats and oils—summary of collaborative study. *J. AOAC Intl.* 77:677.

Flack, E. A. 1983. The use of emulsifiers to modify the texture of processed foods. *Food, Flavour, Ingred., Process. Packag.* 5:32.

Flack, E. A. 1985. Foam stabilization of dairy whipping cream. *Dairy Industries Intl.* 50:35.

Flider, F. J. 1985. The manufacture of soybean lecithins. In: *Lecithins*, Szuhaj, B. S., List, G. R. (Ed.). American Oil Chemistry Society, Champaign, IL.

Food Chemicals Codex. 1981. Committee of Codex Specifications of the Food and Nutrition Board, National Research Council. National Academy Press, Washington, D.C.

Frenkel, M., Shwartz, R., Garti, N. 1982. Turbidity measurements as a technique for evaluation of water-in-oil emulsion stability. *J. Dispersion Sci. Technol.* 3:195.

Friberg, S. 1976. Emulsion stability. In: *Food Emulsions*, Friberg, S. (Ed.). Marcel Dekker, New York.

Friberg, S., Mandell, L. 1969. Phase equilibria and their influence on the properties of emulsions. *J. Am. Oil Chem. Soc.* 47:149.

Friberg, S., Mandell, L., Larsson, M. 1969. Mesomorphous phases, a factor of importance for the properties of emulsions. *J. Colloid Interface Sci.* 29:155.

Friberg, S., Jansson, P. O., Cederberg, E. 1976. Surfactant association structure and emulsion stability. *J. Colloid Interface Sci.* 55:614.

Friberg, S. E., Goubran, R. F., Kayali, I. H. 1990. Emulsion stability. In: *Food Emulsions*, Friberg, S. E. (Ed.). Marcel Dekker, New York.

Garti, N. 1997. Double emulsions—scope, limitations and new achievements. *Colloids and Surfaces A: Physicochemical and Engineering Aspects* 123–124:233.

Garti, N., Aserin, A. 1981. Analyses of polyglycerol esters of fatty acids using high performance liquid chromatography. *J. Liquid Chrom.* 4:1173.

Garti, N., Remon, G. F. 1984. Relationship between nature of vegetable oil, emulsifier and the stability of W/O emulsion. *J. Food Technol.* 19:711.

Garti, N., Wellner, E., Sarig, S. 1982a. Crystal structure modifications of tristearin by food emulsifiers. *J. Am. Oil Chem. Soc.* 59:181.

Garti, N., Wellner, E., Sarig, S. 1982b. Effect of food emulsifiers on crystal structure and habit of stearic acid. *J. Am. Oil Chem. Soc.* 58:1058.

Garti, N., Schlichter, J., Sarig, S. 1986. Effect of food emulsifiers on polymorphic transitions of cocoa butter. *J. Am. Oil Chem. Soc.* 63:230.

Gawrilow, I. 1977. Utilization of mathematical models to characterize functional properties of selected emulsifiers in continuous mix bread. *J. Am. Oil Chem. Soc.* 54:397.

Gawrilow, I. 1980. Hydrated stable fluent shortening containing solid phase emulsifier components. SCM Corporation. U.S. patent 4,226,894.

Gilmore, C., Miller, D. E., Zielinski, R. J. 1980. Low calorie containing imitation products. U.S.

patent 4,199,608. In: *Food Additives: Recent Developments*, Johnson, J. C. (Ed.). Food Technology Review No. 58. Noyes Data Corporation, Park Ridge, NJ.

Gnauck, D. 1978. International food legislation and its acceptance in the Federal Republic of Germany. In: *Chemical Toxicology of Food*, Galli, C. L., Paoletti, R., Vettorazzi, G. (Eds.). Elsevier, New York.

Govin, R., Leeder, J. G. 1971. Action of emulsifiers in ice cream utilizing the HLB concept. *J. Food Sci.* 36:718.

Greenwald, H. L., Brown, G. L., Fineman, M. N. 1956. Determination of the hydrophilic–lipophilic character of surface-active agents and oils by a water titration. *Anal. Chem.* 28:1693.

Griffin, W. C. 1949. Classification of surface-active agents by ''HLB.'' *J. Soc. Cosmetic Chemists* 1:31.

Griffin, W. C. 1954. Calculation of HLB values of non-ionic surfactants. *J. Soc. Cosmetic Chemists* 5:249.

Gunnarsson, G., Jonsson, B., Wennerstrom, H. 1980. Surfactant association into micelles. An electrostatic approach. *J. Phys. Chem.* 84:3114.

Gupta, R. K., James, K., Smith, F. J. 1983. Sucrose esters and sucrose ester/glyceride blends as emulsifiers. *J. Am. Oil Chem. Soc.* 60:862.

Haggerty, J. A., Corbin, D. D. 1984. Method for preparing meat-in-sauce, meat-in-gravy and meat filling. Central Soya Co. U.S. patent 4,472,448.

Haigh, R. 1978. The activities of the Scientific Committee for Food of the Commission of the European Communities. In: *Chemical Toxicology of Food*, Galli, C. L., Paoletti, R., Vettorazzi, G. (Eds.). Elvsevier, New York.

Haraldsson, G. 1983. Degumming, dewaxing and refining. *J. Am. Oil Chem. Soc.* 60:203A.

Hartnett, D. J. 1979. Emulsifier systems facilitate ingredient substitutes in cakes. *Food Prod. Develop.* 13:60.

Herzing, A. G., Palamidis, N. 1984. Confectionary coating compositions containing polyglycerol ester emulsifiers. SCM Corp. U.S. patents 4,464,411.

Hoseney, R. C., Kinney, K. F., Shogren, M. O. 1972. Functional breadmaking and biochemical properties of wheat flour components. IX. Replacing total free lipid with synthetic lipids. *Cereal Chem.* 49:366.

Hoseney, R. C., Hsu, K. H., Ling, R. S. 1976. Use of diacetyl-tartaric acid esters of monoglycerides. *Bakers Dig.* 59:28.

Howard, N. B. 1980. Meat analog compositions. The Proctor and Gamble Co. U.S. patent 4,226,890.

Hsu, E. E., Gordon, J., Davis, E. A. 1980. Water loss rates and scanning electron microscopy of model cake systems made with different emulsification systems. *J. Food Sci.* 45:1280.

Hurst, J. W., Martin, R. A. 1984. The analysis of phospholipids in soy lecithin by HPLC. *J. Am. Oil Chem. Soc.* 61:1462.

Hutchison, P. E., Baiocchi, F., Del Vecchio, A. J. 1977. Effect of emulsifiers on the texture of cookies. *J. Food Sci.* 42:399.

IUPAC. 1972. *Manual on Colloid and Surface Science*. International Union of Pure and Applied Chemistry. Butterworths, London.

IUPAC-IUB Commission on Biochemical Nomenclature. 1977. *Lipids* 12:455.

Jackson, C. A. 1979. Powdered hydrated emulsifiers and their method of preparation. Southland Corp. U.S. patent 4,159,952.

Jansson, P.-O., Friberg, S. 1976. Van der Waals potential in coalescing emulsion drops with liquid crystals. *Mol. Cryst. Liq. Cryst.* 34:75.

Janssen, J. J. M., Boon, A., Agterof, W. G. M. 1994. Influence of dynamic interfacial properties on droplet breakup in simple shear flow. *AICHE J.* 40:1929.

Jaynes, E. N. 1985. Applications in the food industry: II. In: *Encyclopedia of Emulsion Technology*, Vol. 2: *Applications*, Becher, P. (Ed.). Marcel Dekker, New York.

JEFCA. 1970. Toxicological evaluation of some food colours, emulsifiers, stabilizers, anti-caking agents and certain other substances. Joint FAO/WHO Expert Committee on Food Additives. FAO Nutrition Meetings Report Series 46A, WHO/Food Additives/70.36.

JEFCA. 1973. Seventeenth Report of the Joint FAO/WHO Expert Committee on Food Additives: Toxicological evaluation of certain food additives with a review of general principles and of specifications. FAO Nutrition Meeting Report Series No. 53, WHO Technical Report Series No. 539, World Health Organization, Geneva.

JEFCA. 1974. Toxicological evaluation of some food additives including anticaking agents, antimicrobials, antioxidants, emulsifiers and thickening agents. Joint FAO/WHO Expert Committee on Food Additives. FAO Nutrition Meeting Report Series No. 53A, WHO Food Additives Series No. 5, World Health Organization, Geneva.

JEFCA. 1994. Summary of Evaluations. Joint FAO/WHO Expert Committee on Food Additives, Rome.

Johnston, N. F. 1973. Continuous cake production using hydrated surfactants. *Bakers Dig.* 47:30.

Joos, P., Fang, J. P., Serrien, G. 1992. Comments on some dynamic surface tension measurements by dynamic bubble pressure method. *J. Colloid Interface Sci.* 151:144.

Kaufman, V. R., Garti, N. 1981. Analysis of sucrose fatty acid esters composition by HPLC. *J. Liquid Chrom.* 4:1195.

Keeney, P. G., Kroger, M. 1974. Frozen dairy products. In: *Fundamentals of Dairy Chemistry*, 2nd ed., Webb, B. H., Johnson, A. H., Alford, J. A. (Eds.). AVI Publishing, Westport, CT.

Kim, Y. J., Robinson, R. J. 1979. Effect of surfactants on starch in a model system. *Staerke* 31: 293.

Knightly, W. H. 1968. The role of surfactants in baked goods. In: *Surface-active lipids in foods.* Monograph No. 32, Soc. Chem. Industries, London.

Koizumi, Y., Yamada, K., Sakka, H., Yuuda, M., Yamaguchi, T. 1984. Emulsifier composition and quality improvement method of starch containing foods. Riken Vitamin Co. U.S. patent 4,483,880.

Krog, N. 1971. Amylose complexing effect on food grade emulsifiers. *Staerke* 23:206.

Krog, N. 1973. Influence of food emulsifiers on pasting temperatures and viscosity of various starches. *Staerke* 25:22.

Krog, N. 1981. Theoretical aspects of surfactants in relation to their use in breadmaking. *Cereal Chem.* 58:158.

Krog, N., Lauridsen, J. B. 1976. Food emulsifiers and their associations with water. In: *Food Emulsions*, Friberg, S. (Ed.)., Marcel Dekker, New York.

Krog, N., NyboJensen, B. 1970. Interaction of monoglycerides in different physical states with amylose and their anti-firming effects in bread. *J. Food Technol.* 5:77.

Krog, N. J., Riisom, T. H., Larsson, K. 1985. Applications in the Food Industry: I. In: *Encyclopedia of Emulsion Technology*. Vol. 2: *Applications*, Becher, P. (Ed.). Marcel Dekker, New York.

Kwon, K. S., Rhee, K. C. 1996. Emulsifying capacity of coconut protein as a function of salt, phosphate, and temperature. *J. Am. Oil Chem. Soc.* 73:1669.

Lagendijk, J., Pennings, H. J. 1970. Relation between complex formation of starch with monoglycerides and the firmness of bread. *Cereal Sci. Today* 15:354.

La Guardia, M. K. 1994. High-performance fat systems in baked products. *Cereal Foods World* 39:147.

Larsson, K. 1976. Food emulsifiers and their association with water. In: *Food Emulsion*, Stiberg, F. (Ed.). Marcel Dekker, New York.

Larsson, K. 1980. Technical effects in cereal products of lipids, naturally present and additives. In: *Cereals for Food and Beverages: Progress in Cereal Chemistry and Technology*, Inglett, G. E., Munck, L. (Eds.). Academic Press, London.

Larsson, K., Krog, N. 1973. Structural properties of the lipid–water gel phase. *Chem. Phys. Lipids* 10:177.

Lefebvre, J. 1983. Finished product formulation. *J. Am. Oil Chem. Soc.* 60:247A.

Levitz, H., Van Damme, H., Keravis, D. 1984. Fluorescence decay study of the adsorption of nonionic surfactants at the solid–liquid interface. 1. structure of the adsorption layer on a hydrophilic solid. *J. Phys. Chem.* 88:2228.

Lin, P.-M., Leeder, J. G. 1974. Mechanism of emulsifier action in an ice cream system. *J. Food Sci.* 39:108.

Lindheimer, M., Keh, E., Zaini, S., Partyka, S. 1990. Interfacial aggregation of nonionic surfactants onto silica gel: calorimetric evidence. *J. Colloid Interface Sci.* 138:83.

Liu, K. J., Shaw, J. F. 1995. Synthesis of propylene glycol monoesters of docosahexaenoic acid and eicosapentaenoic acid by lipase-catalysed esterification in organic solvents. *J. Am. Oil Chem. Soc.* 72:1271.

Lorenz, K. 1983. Diactyl tartaric acid esters of monoglycerides (DATEM) as emulsifiers in breads and buns. *Bakers Digest* 57:6.

Lorenz, K, Dilsaver, W., Kulp, K. 1982. Comparative efficiencies of bread crumb softeners at varied bread storage temperatures. *J. Am. Oil Chem. Soc.* 59:484.

Lynch, M. J., Griffin, W. C. 1974. Food emulsions. In: *Emulsions and Emulsion Technology*, Lissant, K. J. (Ed.). Marcel Dekker, New York.

Maag, H. 1984. Fatty acid derivatives: important surfactants for household, cosmetic and industrial purposes. *J. Am. Oil Chem. Soc.* 61:259.

MacLeod, C. A., Radke, C. J. 1994a. Surface exchange kinetics at air/water interface from the dynamic tension of growing liquid drops. *J. Colloid Interface Sci.* 166:73–78.

MacLeod, C. A., Radke, C. J. 1994b. Charge effects in the transient adsorption of ionic surfactants at fluid interfaces. *Langmuir* 10:3555.

MacRitchie, F. 1977. Flour lipids and their effects in baking. *J. Sci. Food Agric.* 28:53.

Madsen, J. 1976. Ingredient requirements, development techniques for low calorie table spreads. *Food Prod. Develop.* 10(4):72.

Maleki, M., Vetter, J. L., Hoover, W. J. 1981. The effect of emulsifiers, sugar, shortening and soya flour on the staling of barbari flat bread. *J. Sci. Food Agric.* 32:1209.

Marszall, L. 1981. The effective hydrophile–lipophile balance of nonionic surfactant mixtures. *J. Dispersion Sci. Technol.* 2:443.

Martin, N. 1995. Analysis of nonionic surfactants by HPLC using evaporative light scattering detector. *J. Liquid Chromatogr.* 18:1173.

Matz, S. A. 1968. *Cookie and Cracker Technology*. AVI Publishing, Westport, CT.

Meffert, A. 1984. Technical uses of fatty acid esters. *J. Am. Oil Chem. Soc.* 61:255.

Mesallam, A. S. F., Salem, A. E, Mohasseb, Z., Zoueid, M. E. 1979. Effect of certain kinds of emulsifiers on the mixing and baking of wheat and corn flour admixture. Alexandria. *J. Agric. Res.* 27:381.

Moore, C. D., Bell, M. 1956. Nonionic surface-active agents. *Soap, Perfumery Cosmetics* 29:893.

Morad, M. M., D'Appolonia, B. L. 1980. Effect of baking procedure and surfactants on pasting properties of bread crumb. *Cereal Chem.* 57:141.

Morgan, J. E., DelVecchio, A. J., Brooking, B. L., Laverty, D. M. 1980. Emulsifier system and cake mix containing same. The Pillsbury Co. U.S. patent 4,242,366.

Nakama, Y., Shiojima, Y., Takeshita, Y. 1997. Complex formation in a ternary system composed of lauroamhoglycinate, oleic acid and water. *J. Am. Oil Chem. Soc.* 74:803.

Nash, N. T., Brickman, L. M. 1972. Food emulsifiers-science and art. *J. Am. Oil Chem. Soc.* 49: 457.

Nawar, W. W. 1985. Lipids. In: *Food Chemistry*, Fennema, O. (Ed.). Marcel Dekker, New York.

Neckerman, E. F., Noznick, P. P. 1968. Gas–liquid chromatographic analysis of lactoylated monoglycerides. *J. Am. Oil Chem. Soc.* 45:845.

Ogawa, K., Tezuka, S., Terasawa, M., Iwata, S. 1979. Center-filled chewing gum. Lotte Co., Ltd. U.S. patent 4,157,402.

Oh, S. G., Shah, D. O. 1994. Micellar lifetime: its relevance to various technological processes. *J. Dispersion Sci. Technol.* 15:297.

Ohta, S., Inoue, S., Torigoe, T., Kobayashi, M. 1984. Emulsifiers comprising lysophosphatidic acid or a salt thereof and processes for making a dough containing same. Kyowa Hakko Kogyo Kabushiki Kaisha. U.S. Patent 4,478,866.

Opawale, F. O., Burgess, D. J. 1998. Influence of interfacial properties of lipophilic surfactants on water-in-oil emulsion stability. *J. Colloid Interface Sci.* 197:142.

Osipow, L. I., Snell, F. D., Marra, D. York, W. C. 1956. Methods of preparation. Fatty acid esters of sucrose. *Indust. Eng. Chem.* 48:1459.

Osman, E. M., Leith, S. J., Fles, M. 1961. Complexes of amylose with surfactants. *Cereal Chem.* 38:449.

Painter, K. A. 1981. Functions and requirements of fats and emulsifiers in prepared cake mixes. *J. Am. Oil Chem. Soc.* 58:92.

Paola, E., Amelia, D., Francois, D. 1996. Polymorphism of stabilized and nonstabilized tristearin, pure and in the presence of food emulsifiers. *J. Am. Oil Chem. Soc.* 73:187.

Parkinson, C., Sherman, P. 1972. Phase inversion temperature as an accelerated method for evaluating emulsion stability. *J. Colloid Interface Sci.* 41:328.

Pastor, E., Otero, C., Ballesteros, A. 1995. Enzymatic preparation of mono- and distearin by glycerolysis of ethylstearate and direct esterification of glycerol in the presence of a lipase from candida antarctica (NOVOZYM 435). *Biocatalysis and Biotransformation* 12:147.

Pearce, R. J., Harper, W. J. 1982. A method for the quantitative evaluation of emulsion stability in coffee whiteners. *J. Food Sci.* 47:680.

Perrin, P., Lafuma, F. 1998. Low hydrophobically modified poly(acrylic acid) stabilizing macroemulsions: relationship between copolymer structure and emulsion properties. *J. Colloid Interface Sci.* 197:317.

Petrowski, G. E. 1976. Emulsion stability and its relation to foods. *Adv. Cereal Chem.* 22:309.

Pisesookbunterng, W., D'Appolonia, B. L. 1983. Bread staling studies. I. Effect of surfactants on moisture migration from crumb to crust and firmness values of bread crumb. *Cereal Chem.* 60:298.

Pisesookbunterng, W., D'Appolonia, B. L., Kulp, K. 1983. Bread staling studies. II. The role of refreshening. *Cereal Chem.* 60:301.

Pomeranz, Y. 1985. Lecithin in baking. In: *Lecithins*, Szuhaj, B. S., List, G. R. (Eds.). American Oil Chemists' Society, Champaign, IL.

Pomeranz, Y., Shogren, M. D., Finney, K. F. 1959a. Improving breadmaking properties with glycolipids. I. Improving soy products with sucroesters. *Cereal Chem.* 46:503.

Pomeranz, Y., Shogren, M. D., Finney, K. F. 1959b. Improving breadmaking properties with glycolipids. II. Improving various protein-enriched products. *Cereal Chem.* 46:512.

Pomeranz, Y., Shogren, M. D., Finney, K. F. 1968. Natural and modified phospholipids: effects on bread quality. *Food Technol.* 22:897.

Pomeranz, Y., Shogren, M. D., Finney, K. F., Bechtel, D. B. 1977. Fiber in breadmaking: effects on functional properties. *Cereal Chem.* 54:25.

Prosise, W. E. 1985. Commercial lecithin products: food use of soybean lecithin. In: *Lecithins*, Szuhaj, B. F., List, G. R. (Eds.). American Oil Chemists' Society, Champaign, IL.

Pyo, D. J., Lee, K., Lee, H., Kim, H., Lee, M. L. 1996. New analytical method using polar modifiers in supercritical fluid chromatography and its application to the separation of fatty acid alcohol ethers. *Bulletin of the Korean Chemical Society* 17:496.

Riedel, H. R. 1985. Treatment of toffees and various fillings with a natural anti-oxidant emulsifier. *Confectionery Production* 65:678.

Rogers, D. E., Hoseney, R. C. 1983. Breadmaking properties of DATEM. *Bakers Digest* 57:12.

Rosen, M. J., Hua, X. Y. 1982. Synergism in binary mixtures of surfactants. II. Some experimental data. *J. Am. Oil Chem. Soc.* 59:582.

Rubenstein, I. H. Maryland Cup Corporation. 1980. Marshmallow variegate for frozen confections and frozen confections containing same. U.S. patent 4,189,502.

Rule, C. E. 1983. Cream/alcohol-containing beverages. SCM Corp. U.S. patent 4,419,378.

Rusch, D. T. 1981. Emulsifiers: uses in cereal and bakery foods. *Cereal Foods World* 26:110.

Rydhag, L., Wilton, I. 1981. The function of phospholipids of soybean lecithin in emulsions. *J. Am. Oil Chem. Soc.* 58:831.
Sagitani, H., Shinoda, K. 1973. Example of emulsifier selection for a W/O type emulsion by the HLB-temperature method. *Yukagaku* 22:438.
Sagitani, H., Shinoda, K. 1975. The selection of emulsifier in W/O type emulsions by HLB-temperature (PIT) system. *Yukagaku* 24:171.
Sahasrabudhe, M. R. 1967. Chromatographic analysis polyglycerols and their fatty acids. *J. Am. Oil. Chem. Soc.* 44:376.
Schlegelmilch, F., Siemanowski, W., Wiebrock, L. 1995. Development of an HPLC method for determining the free polyglycerol content of polyglycerol caprates. *Fett Wissenschaft Technologie* 97:408.
Schmidt, J. C., Orthoefer, F. T. 1985. In: *Lecithins*, Szuhaj, B. S., List, G. R. (Eds.). AOCS Monograph 12. American Oil Chemists' Society, Champaign, IL.
Scholfield, C. R. 1981. Composition of soybean lecithin. *J. Am. Oil Chem. Soc.* 58:889.
Schuster, G. 1985. *Emulgatoren fur Lebensmittel.* Springer-Verlag, New York.
Schuster, G., Adams, W. 1983. Emulgatoren als Zusatzstoffe fur Lebensmittel. III. Ensatz der im Anhang I der EG-Richtlinie genannten Emulgatoren in Lebensmittel. 2. Emulgatoren in Brot und Kleingeback. *Intl. J. Food Technol. Food Eng.* (ZFL) 33:362.
Schuster, G., Adams, W. F. 1984. Emulsifiers as additives in bread and fine baked products. *Adv. Cereal Sci.* 6:139.
Seibel, W., Ludewig, H. G., Bretschneider, F. 1980. Untersuchungen mit Aufschlagmitten bei Biskuit und Sandmassen. *Getreide Mehl Brot* 34:298.
Seino, H., Uchibori, T., Nishitani, T., Inamasu, S. 1984. Enzymatic synthesis of carbohydrate esters of fatty acid (I) esterification of sucrose, glucose, fructose and sorbitol. *J. Am. Oil. Chem. Soc.* 61:1761.
Seward, L. O., Warman, B. 1983. Cake mix containing a lipophilic emulsifier system. The Proctor and Gamble Co. U.S. patent 4,419,377.
Sharma, M. K., Shah, D. O. 1985. Introduction to macro- and microemulsions. In: *Macro- and Microemulsions: Theory and Practice*, Shah, D. O. (Ed.). ACS Symposium Series. American Chemical Society, Washington, D.C.
Shaw, J. F., Lo, S. 1994. Production of propylene glycol fatty acid monoesters by lipase catalysed reactions in organic solvents. *J. Am. Oil Chem. Soc.* 71:715.
Sherman, P. 1971. Accelerated testing of emulsion stability. *Soap, Perfumery and Cosmetics.* 44: 693.
Shinoda, K. 1968. Comparison between the PIT system and the HLB-value system for emulsifier selection. *Proc. Intl. Congr. Surface Activity* (5th, Barcelona) 3:275.
Shinoda, K. 1969. The stability of O/W emulsions as functions of temperature and the HLB of emulsifiers: the emulsification by PIT method. *J. Colloid Interface Sci.* 30:258.
Shinoda, K., Kunieda, H. 1983. Phase properties of emulsions: PIT and HLB. In: *Encyclopedia of Emulsion Technology*, Vol. 1: *Basic Theory*, Becher, P. (Ed.). Marcel Dekker, New York.
Shinoda, K., Sagitani, H. 1978. Emulsifier selection in water/oil type emulsions by the hydrophile–lipophile balance—temperature system. *J. Colloid Interface Sci.* 64:68.
Shinoda, K., Yoneyama, T, Tsutsumi, H. 1980. Evaluation of emulsifier blending. *J. Dispersion Sci. Technol.* 1:1.
Soe, J. B. 1983. Analysis of monoglycerides and other emulsifiers by gas chromatography. *Fette Seifen Anstichmittel* 85:72.
Stampfli, L., Nersten, B. 1995. Emulsifiers in bread making. *Food Chem.* 52:353.
Sudraud, G., Coustard, J. M., Retho, C., Caude, M., Rosset, R., Hagemann, R., Gaudin, D., Vierlizier, H. 1981. Analytical and structural study of some food emulsifiers by high-performance liquid chromatography and off-line mass spectrometry. *J. Chromatogr.* 204:397.
Suggs, J. L. 1983. The benefit of powdered emulsifiers. *Cereal Foods World* 28:715.

Suggs, J. L., Buck, D. F. 1980. Emulsifiers for baked goods. Eastman Kodak Co. U.S. patent 4,229,480.

Szuhaj, B. F., List, G. R. (Eds.). 1985. *Lecithins*. AOCS Monograph 12. American Oil Chemists' Society, Champaign, IL.

Tadros, T. F., Vincent, B. 1983. Liquid/liquid interfaces. In: *Encyclopedia of Emulsion Technology*, Vol. I: *Basic Theory*. Marcel Dekker, New York.

Taiwo, K., Karbstein, H., Schubert, H. 1997. Influence of temperature and additives on the adsorption kinetics of food emulsifiers. *J. Food Process Eng*. 20:1.

Tharp, B. W. 1982. The effect of certain colloid/emulsifier blends and processing procedures on emulsion stability. *Prog. Food Nutr. Sci*. 6:209.

Tiberg, F. 1996. Physical characterization of non-ionic surfactant layers adsorbed at hydrophilic and hydrophobic solid surfaces by time-resolved ellipsometry. *J. Chem. Soc., Faraday Trans*. 92:531.

Tiberg, F., Jonsson, B., Tang, J., Lindman, B. 1994a. Ellipsometry studies of the self-assembly of nonionic surfactants at the silica-water interface: equilibrium aspects. *Langmuir* 10:2294.

Tiberg, F., Jonsson, B., Lindman, B. 1994b. Ellipsometry studies of the self-assembly of nonionic surfactants at the silica-water interface: kinetic aspects. *Langmuir* 10:3714.

Titus, T. C., Mickle, J. B. 1971. Stability of milk fat-water emulsions containing single and binary emulsifiers. *J. Food Sci*. 36:723.

Tornberg, E., Hermansson, A. M. 1977. Functional characterization of protein stabilized emulsions: effect of processing. *J. Food Sci*. 42:468.

Torrey, S. 1983. *Edible Oils and Fats: Developments since 1978*. Noyes Data Corp., Park Ridge, NJ.

Torrey, S. 1984. *Emulsions and Emulsifier Applications: Recent Developments*. Noyes Data Corp., Park Ridge, NJ.

Trumbetas, J., Fioriti, J. A., Sims, R. J. 1978. Use of pulsed nuclear magnetic resonance to predict emulsion stability. *J. Am. Oil Chem. Soc*. 55:248.

Tsen, C. C. 1975. Fatty acid derivatives and glycolipids in high-protein bakery products. *J. Am. Oil Chem. Soc*. 51:81.

Tsen, C. C., Peters, E. M., Schaffers, T., Hoover, W. 1973. High protein cookies. 1. Effect of soy fortification and surfactants. *Bakers Dig*. 47:34.

Tsen, C. C., Bauck, L. J., Hoover, W. J. 1975a. Using surfactants to improve the quality of cookies made from hard wheat flours. *Cereal Chem*. 52:629.

Tsen, C. C., Farrel, E. P., Hoover, W. J., Crowley, P. R. 1975b. Extruded soy products from whole and dehulled soybeans cooked at various temperatures for bread and cookie fortifications. *Cereal Foods World* 20:413.

Tsuda, T., Nakanishi, H. 1983. Gas–liquid chromatographic determination of sucrose fatty acid esters. *J. Assoc. Off. Chem*. 66:1050.

Tsutsumi, H., Nakayama, H., Shinoda, K. 1978. Some characteristics of polyoxyethylene sorbitol tetraoleate: oligomer type emulsifier. *J. Am. Oil Chem. Soc*. 55:363.

Tsuzuki, W., Ito, T., Hatakayema, E., Kobayashi, S. Suzuki, T. 1993. Synthesis and emulsion properties of a novel nonionic surfactant, dialkyl glucosylglutamate. *J. Agric. Food Chem*. 41: 2272.

Tung, M. A., Jones, L. J. 1981. Microstructure of mayonnaise and salad dressing. *Scanning Electron Microscopy* 3:523.

Van Lonkhuysen, H., Blankestijn, J. 1974. Interaction of mono-glycerides with starches. *Staerke* 26:337.

Van Nieuwenhuyzen, W. 1981. The industrial uses of special lecithins: a review. *J. Am. Oil Chem. Soc*. 58:886.

Van Velthuijsen, J. A. 1979. Food additives derived from lactose: lactitol and lactitol palmitate. *J. Agric. Food Chem*. 27:680.

Villota, R. V. 1985. Functional assessment of Cab-O-Sil in food systems. Annual Report to Cabot

Corporation, submitted by R.V. Villota, Department of Food Science, University of Illinois, Urbana, IL.

Von Pardun, H. 1982. Eine empirische methode zur bestimmung der der emulgierfahigkeit von pflanzenlecithinen in O/W-systemen. *Fette-Seifen-Anstrichmittel* 84(8):291.

Waginaire, L. 1997. Place of lipids as emulsifiers in the future. *Ocl-Oleagineux Corps Gras Lipides* 44:271.

Walker, N. A. 1983. EEC production and usage of emulsifiers. *Food Flavour Ingred. Process Packag.* 5:38.

Walstra, P. 1983. Formation of Emulsions. In: *The Encyclopedia of Emulsion Technology*, Becher, P. (Ed.). Marcel Dekker, New York.

Watson, K. S., Walker, C.E. 1986. The effect of sucrose esters on flour–water dough mixing characteristics. *Cereal Chem.* 63: 62.

Wheeler, E. L. 1979. Quantitative determination of sodium stearoyl-2-lactylate in soy-fortified wheat flour blends. *Cereal Chem.* 56:236.

Wei, J.-J. 1984. Synthesis and feeding studies of sucrose fatty acid polyesters utilized as simulated milk fat. Ph.D. Dissertation. Washington State University, Pullman, WA.

Williams, A., Janssen, J. J. M., Prins, A. 1997. Behaviour of droplets in simple shear flow in the presence of a protein emulsifier. *Colloids and Surfaces A—Physicochemical and Engineering Aspects* 125:189.

Wilton, I., Friberg, S. 1971. Influence of temperature induced phase transitions on fat emulsions. *J. Am. Oil Chem. Soc.* 48:771.

Wittman, J. S. 1984. Hydrated emulsifier for use in flour based baked goods. Batter-Lite Foods, Inc. U.S. patent 4,424,237.

Zlatanos, S. N., Sagredos, A. N., Papageorgiou, V. P. 1985. High yield monoglycerides from glycidol and carboxylic acids. *J. Am. Oil Chem. Soc.* 62:1575.

24

A Comprehensive Review of Commercial Starches and Their Potential in Foods

THOMAS E. LUALLEN

Cargill, Cedar Rapids, Iowa

I. INTRODUCTION

A. History

Starch has been a primary food source since ancient Egypt. Wheat, a staple for many foods created in early history of mankind, contained and still provides a substantial supply of starch to the world food market. It is well known that starch provides the principle form of carbohydrate for storage by plants. The structure of starch as *n* α-linked glucan contributes much of the food energy in all cereal grains, various tubers, fruits, and pulses. Starch has become a significant contributor to the commercial food industry. It probably has generated more research time and capital expense of all the biopolymers over several decades. Through research over 175 years ago by Kirchoff in 1811 on the hydrolysis of starch with acid, one of the most significant discoveries was made. Simple sugars such as glucose (dextrose) could be isolated. Later he announced the discovery of enzyme degradation using amylase and forming maltose, or malt sugar. These discoveries opened the way for academic and industrial research to flourish, especially in the brewing industry.

In the early 20th century research established the basic understanding of the starch polymer and described its large molecules with specific linkages of the structure glucose. Evidence was also noted at that time regarding the potential of branching within the structure matrix (1). It wasn't until midway into the century that another significant breakthrough came. In 1940 research noted that starch was a heterogeneous polymer. It contained two distinct structures: amylose, a linear polymer, and amylopectin, a branched polymer (2,3). Research continued in the areas of enzymes and their sources. The discovery of amylolytic, synthesizing, and debranching enzymes broadened our basis for the development of the fine structure of amylose and amylopectin (4). All of these terms will be referenced in this chapter.

As mentioned starch was and is considered of commercial importance. It is isolated commercially primarily from cereals and root crops. Starch, although processed and sold within and to the food industry, has also found economic and commercial value in nonfood industries (paper, textiles, cosmetics, detergents, and pharmaceuticals) (5–7). For the remaining portion of this chapter, we will concentrate on only the food sector for describing starch and its functionality. In foods we consider starch as a source for nutrition, an extender thickener, stabilizer, texturizer, and/or processing aid (8). These are basic descriptors for starch and do not always denote the overall functional attribute(s) generated within a food matrix. As a food ingredient, starch today is offered not in its pure form but as a hydrolyzed (converted) product to one of the largest consuming industries. It is sold as a sweetener.

For foods, starch is typically isolated or processed commercially from maize, wheat, rice, potato, and cassava (tapioca) (5,9,10). In this chapter I may refer to maize as corn and cassava as tapioca. Those starch sources identified as corn may be derived from common, waxy, or high amylose varieties. Common corn starch is also referred to by other terms such as dent corn, No. 2 yellow dent, native, and/or regular corn starch. As is evident one must know exactly what and how starch is identified. The waxy and high-amylose are hybrids of common corn and thus have distinctly differing functional attributes. These will be noted in detail later in Sections IV and VIII. The starch cassava will most likely be discussed and referred to as tapioca.

B. Starch Granules

As you will examine the sources of starch, you will note the diversity in shape and size of starch granules (see Fig. 1 and Table 1). Granules can exist as oval, sphere, polygonal (polyhedral), or disk-shaped particles (11–14). Their size ranges from submicron (<1

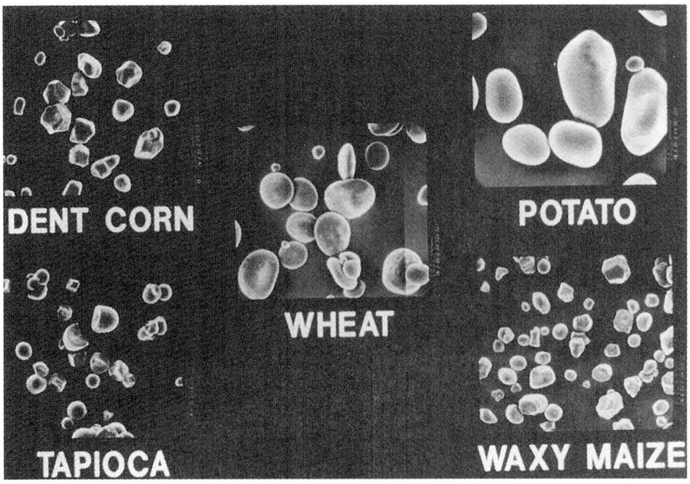

Figure 1 Starch granules.

Table 1 Granule Characteristics

Source	Type	Diameter range (μm)	Shape	Iodine stain
Corn	Cereal	5–26	Round, polygonal	Blue
Waxy corn	Cereal	5–26	Round, polygonal	Red-violet
Tapioca	Root	5–25	Truncated, round, oval	Blue
Potato	Root	15–100	Round, oval	Blue
Rice	Cereal	3–8	Round, polygonal	Blue

μm) to greater than 100 μm. Utilizing both light and scanning electron microscopy we have been enabled to study starch granule macrostructure in detail (13). Most expert microscopists and starch chemists have the ability to identify starches via light microscopes, utilizing only their experience and technique to differentiate the sources.

Granules are colorless and translucent in their native state. Surfaces appear smooth upon macroinspection. However, it was noted by Sterling that tiny pores (0.7–20 Å) were present on the granule surface (15). Deformations have occurred on the granule as angularities or concave indentations. These can result from collisions with other granules or other cellular components such as protein microbodies (13). Granules have been said to have a scaly appearance. This has been attributed to remnants of amyloplast membranes or thin incomplete outer shells. Most starch granules are thought to grow as a single entity inside an amyloplast or chloroplast of a cell. A granule's growth originates at the nucleus or hilum of the granule and continues by apposition. Considering that this growth produces a single type starch granule, we can conclude that a single nucleus yields one granule. However, in an imperfect world, a second type of granule can be found. This is referred to as a compound granule. Or many starch granules can develop in a single amyloplast. Rice contains only compound granules, while oat starch contains both single and compound granules (16). Maize contains spherical and/or polygonal granules. Wheat contains two distinct populations of granules (bimodal). These are usually easy to identify and isolate. The smaller is spherical and of 1–10 μm, and the other significantly larger of lenticular and 15–25 μm. Oats contain large elliposoidal compound granules (13).

C. Microscopic Assay of Starch Granules

When attempting to identify starch granules utilizing the light microscope we at times incorporate the use of polarized light. It is considered that starch granules are birefringent and produce a characteristic pattern when using polarized light. This pattern is referred to as the ''Maltese cross'' (Fig. 2). Starch granules are partially crystalline in structure and therefore produce this X-ray diffraction pattern. Nonhydrated starch granules have a mean refractive index of 1.50. However, hydrated granules can produce concentric shell ''growth rings,'' or layers of alternating high and low refractive index. Potato starch is an excellent example. This phenomenon also relates to density, crystallinity, and a starch's resistance to chemical or enzymatic degradation. The Maltese cross, or intersection of light pattern refraction, occurs at the hilum of the starch granule.

As mentioned earlier, the starch granule is composed of two molecules, amylose and amylopectin. These two polyglucan molecules are organized within the granule as radially anisotropic and semicrystalline structures. It should be noted that amylopectin has an average of 4 branches per 100 glucose residues. Whereas amylose could contain

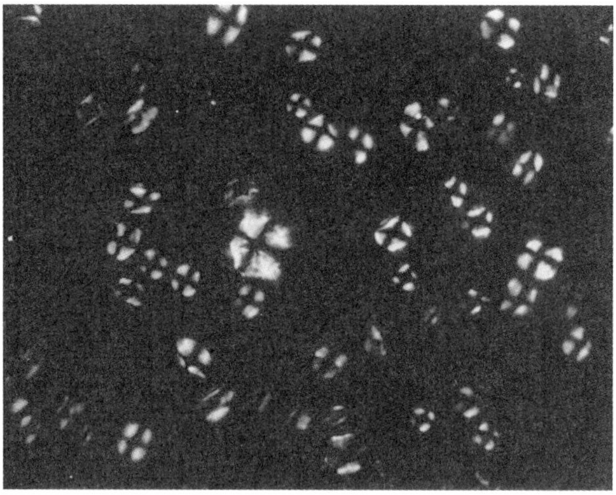

Figure 2 Maltese cross.

9 to 20 chains per molecule. These chains are perpendicular or parallel to the surface of the granule and therefore under the polarized light appear as dark regions (17). Granule surface is not flat, but rather has curvature, therefore some chains are not oriented perpendicular nor parallel to the light and thus the light is observed through the second prism. Studies utilizing X-ray diffraction has confirmed the presence of both ordered and amorphous regions of partially crystalline structure (18).

II. STARCH SOURCES, STRUCTURE, CHARACTERISTICS, AND PROPERTIES

A. Sources

As previously noted, there are several sources for starch. However, for the quantity necessary and the quality needed to make it industrially economical, we are somewhat restricted today to the following sources: maize, potato, rice, wheat, and tapioca. These dominate the world as primary raw material sources (Fig. 3). There are a few, such as sorghum, arrowroot, and pea starches, that have some commercial use. However, they represent very selective regions of the world or are in very limited supply. The aforementioned primary sources also contain hybrid derivatives. These hybrids are referred to as the waxy or high-amylose varieties. Maize contributes both. Rice and potato starches, commercially, supply only the waxy variety. Wheat, however has not evolved commercially with hybrid varieties as of this writing (Table 2).

B. Components of Starch

Let us now consider the amylose and amylopectin role in starch (see Fig. 4), the two major polymeric components, which contribute significantly to the structure, characteristics, and properties of the different starch sources. As shown in Table 3 amylose content and structure varies with the botanical source. Significant differences can be noted when reviewing

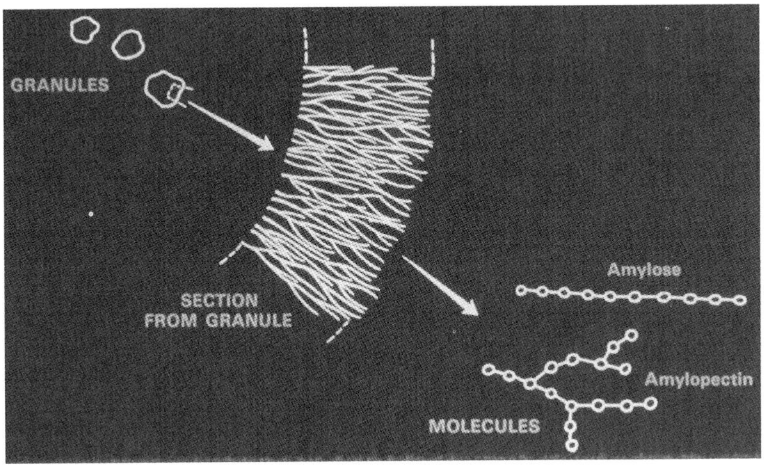

Figure 3 Progressive enlargements of a typical starch granule.

Table 2 Starch Sources

Commercial		
Common corn	Wheat	Pea
Waxy maize	Potato	Waxy rice
High-amylose corn	Rice	Tapioca
Intermediate-amylose corn		
New hybrids		
Waxy potato		
Waxy wheat		

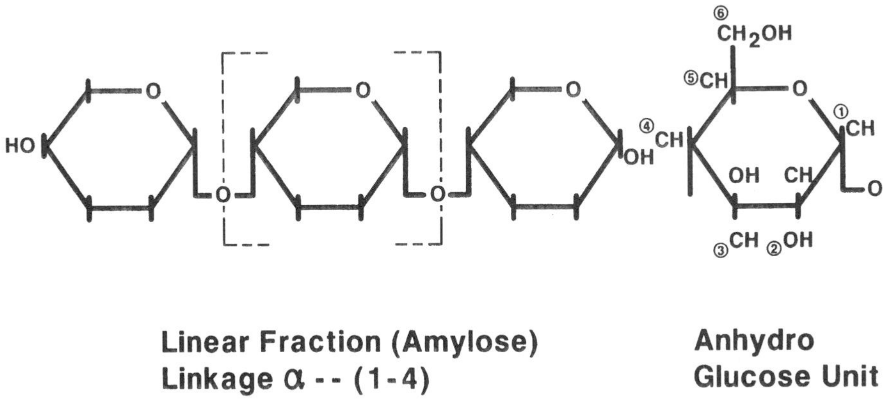

Figure 4 Amylose molecular structure.

Table 3 Starch Characteristics

	Percent amylose	Texture	Flavor
Common corn	28	Firm gel	Cereal
Waxy maize	<1	Paste	Slight cereal
Tapioca	22	Soft gel	Neutral
Potato	18	Sauve	Earthy
Wheat	30	Firm gel	Slight grain
High amylose corn	55–70	Rigid gel	Cereal
Rice	24	Soft gel	Slight grain
Waxy rice	<2	Paste	Neutral

the hybrids of maize. These can be less than 1% to greater than 70% present in varieties today.

Research by Colona and Mercier, revealed the presence of a third component (intermediate fraction). This varied in percentage, but was considered to be approximately 5–10% in most starches (19).

1. Amylose

Amylose (Fig. 5) is known as a linear polymer, but is not defined as just a straight chain molecule. It frequently forms a helix and is thought to inter-twin, even through the several layers of amylopectin (20). Not only does amylose have this unique shaping, but it has been shown that it also consists of limited but distinctly measurable branch points. Amylose consists predominately of α-1,4-D-glucose, bonds (21,22). Amylose also forms a very strong complex with iodine. This complex yields a very intense blue color producing a λ_{max} of 620 nm. Because of this characteristic one can use a light microscope to assay qualitatively for amylose in starch.

2. Amylopectin

Differing significantly from amylose, amylopectin (Fig. 6) is a highly branched polymer. These branch points are α-1,6-D-glucose bonds. Due to the vast number of chains in the

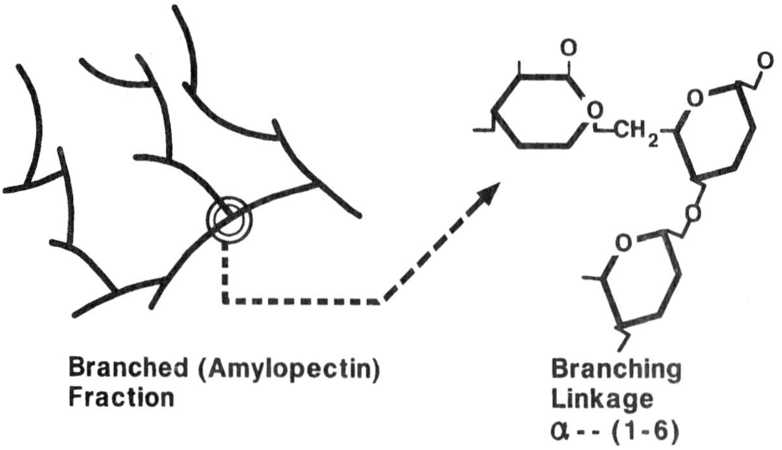

Figure 5 Amylopectin molecular structure.

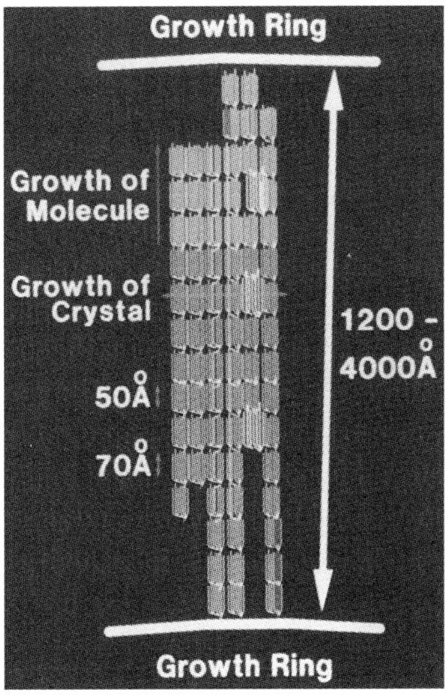

Figure 6 Growth ring of starch granule.

structure of amylopectin, we have an infinite number of potential branch points. Because of this it has been very difficult to determine the exact structure representation of amylopectin. Most carbohydrate scientists have accepted structural models depicting clusters, double helices, and irregular chain lengths (see Figs. 7–9) (23–27). It has yet to be determined as to how amylose and amylopectin are distributed within a starch granule. Amylose was first considered to exist in the amorphous region. Recent research has indicated that amylose is located in the granule as bundles between amylopectin clusters and is randomly dispersed. Thus they could be located among the amorphous and crystalline regions of the amylopectin clusters (27).

3. Minor Constituents

As with most ingredients supplied to industry and the food chain, starch does not exist as a pure entity. Due to the vast regions of the world where starch sources are grown,

Figure 7 Structural organization of a starch granule.

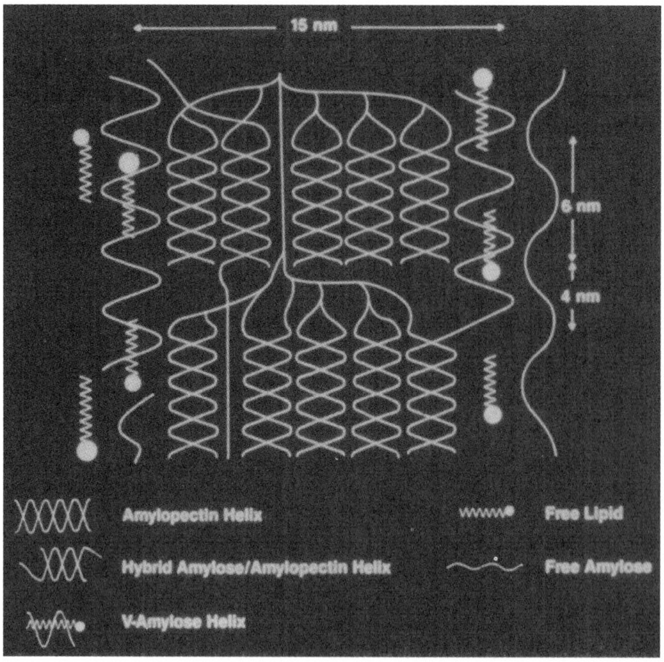

Figure 8 Model of a starch crystallite showing the possible positioning and interactions of various components.

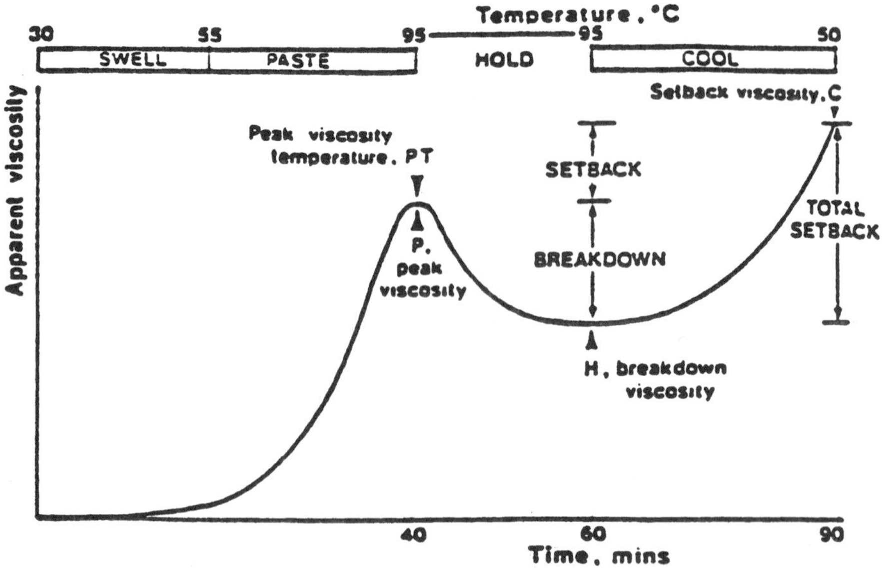

Figure 9 Brabenber model curve.

Table 4 Composition of Starch Granules

Starch Components	Potato	Maize	Wheat	Tapioca	Waxy maize
Amylose content (% on dry stubtance)	21	28	28	17	0
Amylopectin content	79	72	72	83	100
Lipids (% on dry substance)	0.05	0.7	0.8	0.1	0.15
Proteins (% on dry substance)	0.06	0.35	0.4	0.1	0.25
Ash (% on dry substance)	0.4	0.1	0.2	0.2	0.1
Phosphorus (% on dry substance)	0.08	0.02	0.06	0.01	0.01

minor constituents differ somewhat and vary in amount. Typically in starch the most common minor constituents are moisture (water), lipids (fats), nitrogen (protein), phosphorus, and trace elements (minerals). See Table 4.

a. Lipids. Commercial starches supplied to the food industry usually contain less than 1% fat. Levels greater than this are removed, typically via extraction or hydrolysis. A small percentage of lipids are thought to be bound within the matrix of the starch amylose/amylopectin configuration. This would account for the small percentage usually found through acid hydrolysis assay. It is also postulated that some lipids are in association with internal proteins (28).

b. Protein. Protein content varies based on the source of the starch. All protein analysis is reported as percent nitrogen. For those starches isolated from high protein containing flours (wheat, barley, etc.) the starchy phase is isolated from the protein. Other starches commercially sold within the food industry contain less than 1% protein. As with lipids it is also postualted that the protein present may be structurally bound within the matrix of the starch granule, thus making it unavailable for simple extraction (28).

c. Phosphorus. The root (tuber) and legume sources of starch contain esterified phosphorus (phosphate monoesters). It is present as phosphate linked to the C-6 and C-3 hydroxyl groups of the glucose units. Most cereal starches contain very small amounts of phosphorus. If present, it is typically analyzed as phospholipids (29).

d. Trace Elements. In addition to the forementioned compounds present in starch, starch can and does contain very small amounts of minerals and inorganic salts. During the isolation of starch products for commercial application, these compounds are assayed for and reported collectively as "ash." Ash content does vary on the native starches. This variance is primarily dependent upon source or origin and the regions of the world from which they have been produced. The ash content for most commercial starches is reported to be <0.5% based on a dry starch basis (dsb).

e. Moisture. We have considered starch an almost pure carbohydrate with typically very small amounts of trace materials (contaminants). However, we usually overlook one of the most common materials in nature, water. Water or moisture content of starch varies significantly in its native state. Therefore, as a refined product for the food industry, we expect and receive a more consistent moisture level than what is found in nature. Starch as prepared commercially contains on the average approximately 12.0% moisture. There are exceptions, ranging from as low as 3.0% to as high as 18% for some commercial starch products.

C. Characteristics and Properties

1. Gelatinization

First we need to be certain that we do not confuse one event or definition with another. When considering what happens to starch granules during heating in excess water, several events take place. As heat is introduced and the starch granules begin to hydrate (swell), the crystallites within the granule disappear over a range of temperatures. A loss of birefringence and X-ray diffraction occur. With this change in crystallites, referred to as "melting," is thought to occur the process known as gelatinization. This term has been attempted to be defined for all researchers, however to date only one definition has been somewhat collectively generated:

> Starch gelatinization is the collapse (disruption) of molecular orders within the starch granule manifested in irreversible changes in properties such as granular swelling, native crystalline melting, loss of birefringence and starch solubilization. The point of initial gelatinization and the range over which it occurs is governed by starch concentration, method of observation, granular type and heterogenities within the granule population under observation (30).

Today we also have available the differential scanning calorimeter (DSC) for interpreting gelatinization. Utilization of this technique has significantly improved the measurement of water-heat effect on starch. It not only confirmed and refined past data pertaining to the gelatinization temperatures of known starches, but it simplified the data generation and time involvement (31–36). However important this is for basic research, for the commercial use of starch it plays only a minor role in the formation of many food products. Some baked items (cookies, cakes, etc.) are formulated in limited water systems. This affects the hydration and functionality of the selected starch(es). Thus, knowledge of gelatinization can be a key indicator as related to functionality.

2. Swelling

It is now appropriate to discuss what happens when incorporating starches into a food system with excess water. This phenomenon is what generates swelling, pasting, and even-

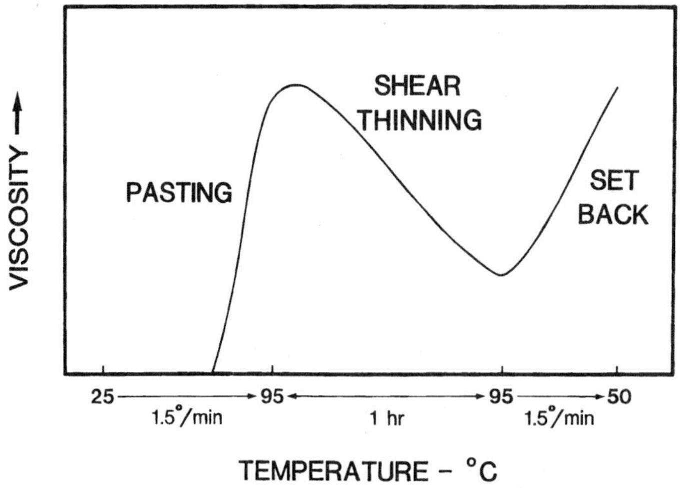

Figure 10 Pasting curve.

tually solubilization of some starch components and the generation of viscosity. This behavior of swelling is generally monitored via some mechanical instrument. Those commonly used in the food industry for measuring the continuous swelling are the Visco-amylo-graph (Brabender) (C. W. Brabender, Inc., South Hackensack, NJ) and Rapid-visco-analyzer (RVA) (Newport Scientific, Warriewood, Australia). These instruments yield data points illustrated as curves representing data markers referred to as the pasting temperature, peak viscosity, stability profile, and the change related to cooling for the product being analyzed (Fig. 10). If the product being analyzed generates an increased consistency during or after cooling, this characteristic is generally referred to as ''set-back.'' (see Fig. 11).

Studies have shown that tuber (root) starches have a tendency to swell more rapidly as compared to cereal grain starches. This faster swelling usually corresponds to a rapid increase in viscosity. This is particularly true of potato starch. There can be a drawback to producing more highly swollen granules. They can be more readily disintegrated by excessive shear or continue to introduce heat. Disintegration is typically measurable and depicted by a loss of viscosity (see Fig. 12) (37,38). Examples such as monoglycerides, sodium stearoyl 2-lactylate, salts, and sucrose esters have shown the ability to inhibit the swelling of starch and lower the viscosity of the final pastes.

3. Pasting

Pasting is more of a sequence of events rather than a fixed point or defined region. There has been a proposal made to differentiate pasting from gelatinization. ''Pasting is the

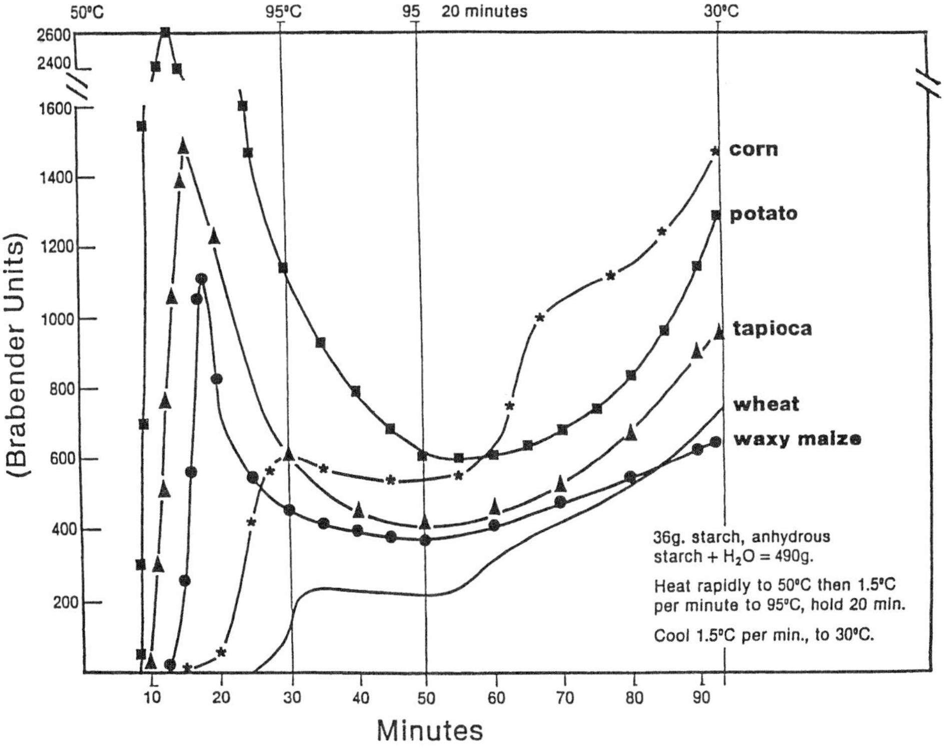

Figure 11 Brabender curves of native starches.

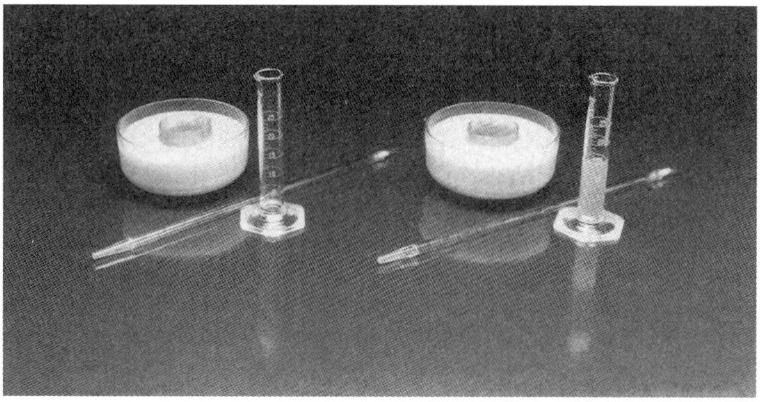

Figure 12 Syneresis example.

phenomenon following gelatinization in the dissolution of starch. It involves granular swelling, exudation of molecular components from the granule and evenutally, total disruption of the granules'' (30). It should be noted again that pasting is a sequence. This then creates the potential for gelatinization to occur during pasting and, therefore, pasting to occur prior to and after gelatinization. In turn, this explains why to best utilize starch as a source of viscosity and/or water-binding matrix, one needs to fully hydrate or complete pasting during heating of starch granules. Doing so then offers the greatest potential for maximum use of the starch for developing viscosity, clarity, and textural characteristics (see Figs. 13 and 14).

4. Retrogradation

It was mentioned earlier that as viscosity is increased during cooling we generate a phenomenon referred to as set-back. This sometimes is misinterpreted as retrogradation. As with pasting and gelatinization an attempt has been made to offer a definition for consensus agreement: ''Starch retrogradation is a process which occurs when starch chains begin to re-associate in an ordered structure. In its initial phases, two or more starch chains may form a single juncture point which then may develop into more extensively ordered regions. Ultimately, under favorable conditions, a crystalline order appears'' (30). This definition offers a general guideline for consideration. However, the event of retrogradation is more detailed and deserves some additional discussion. Starch in its cooked form is a viscous or semisolid starch paste, which upon cooling forms or sets to a gel. This gel system is generally considered to be a three-dimensional mass formed from the amylose containing starches. The mass is generated by a mechanism known as ''entanglement.'' The entirety of the structure is not exclusively due to the amylose fraction of starch. Because some short-chain branches can be sheared from the amylopectin structure, they too can contribute to the gel. As cooling continues these entangled molecules lose their translational motion, thus entrapping water within the matrix. As crystallites form, the gel slowly increases in rigidity. At a point in formation the entrapped water is slowly released or squeezed out of the gel. This process of freeing moisture is referred to as ''syneresis.'' Syneresis is very common with native or nonchemically derived starches (Fig. 15). The process of retrogradation is influenced by starch concentration, acidity, time and temperature of cooking, as well as the rate and temperature of cooling (39).

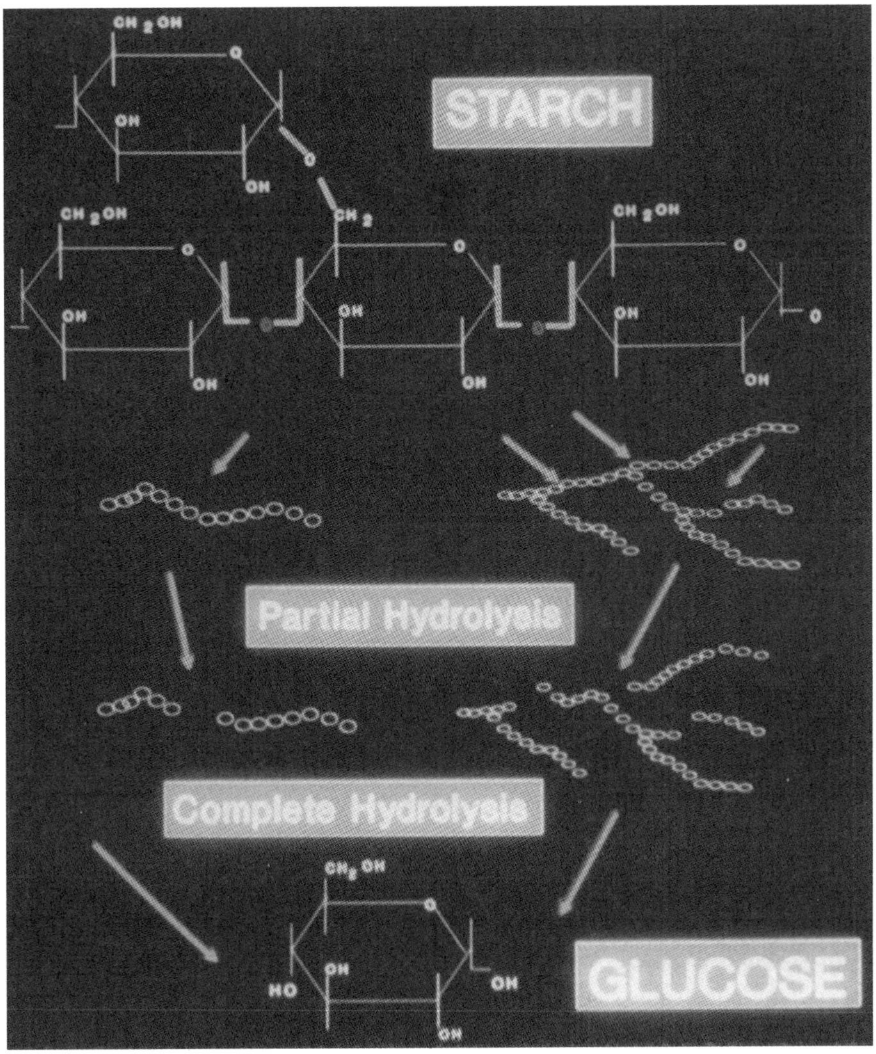

Figure 13 Hydrolysis

Color / Form	White Powder
Moisture, %	6.0
pH	5.0

Saccharide Distribution	1	5	10	15	20
Mono, %	.03	0.7	1.0	1.1	1.3
Di, %	.01	0.9	1.3	2.2	3.5
Tri, %	.02	1.2	2.3	3.6	5.4
Tetra & Higher, %	99.4	97.2	95.4	93.1	89.8

Figure 14 Typical analysis of maltodextrins.

D.E.	Average Chain Length
100.0	1
50.0	2
20.0	5
10.0	10
5.0	20
2.0	50
1.0	100
.5	200

Figure 15 Average chain length versus dextrose equivalents.

III. THE CHEMICAL AND PHYSICAL MODIFICATION OF STARCH

For decades starch has been and still is a primary ingredient for use in processed foods. It has been used to contribute functional characteristics such as viscosity, shelf-stability, clarity, opacity, flavor, as well as many others. In the native or nonderviatized form starch contributes only partial functionality and to a very limited degree. Through the use of chemical modification and/or physical processing the starch industry now can offer a vast product line of starch-based food ingredients. This section identifies those modifications commonly utilized, their functional attributes, and derivatives of their physical state that further expands their functional characteristics.

A. Chemical Modification

First, we note that chemical modification is an accepted process throughout the world. However, not all chemicals and/or derivatives are allowed either in certain foods or in given combinations around the globe. Differences will be noted as we discuss the various modifications. Refer elsewhere in this book for more details related to world use and labeling for food ingredients. Functional properties will be discussed in detail in Section IV. Modifications discussed can be performed on all native or common starches as previously outlined. Their ultimate functional properties are still influenced by their origin or source.

1. Hydrolysis

Hydrolysis (Fig. 16) of the bonds (acetyl) in starch creates the potential for a vast number of products. Acid hydrolysis is a very random reaction, thus yielding similar but inconsistent finished products. Functional characteristics may be very similar and difficult to measure analytically. Chemical distribution by the constituents are usually quite different. With the inclusion of enzymes for the hydrolysis of starch, we are now offered methodology that produces very specific products. During hydrolysis you can form simple modified starches, dextrins, and other differentiated sweeteners. These are also referred to as amylodextrins, maltodextrins, and pyrodextrins.

a. Amylodextrins. Amylodextrins are also referred to as acid-thinned or thin boiling starches. These have been traditionally produced via acid hydrolysis of starch on a commercial scale. Depending upon the degree of hydrolysis, yields vary for the finished prod-

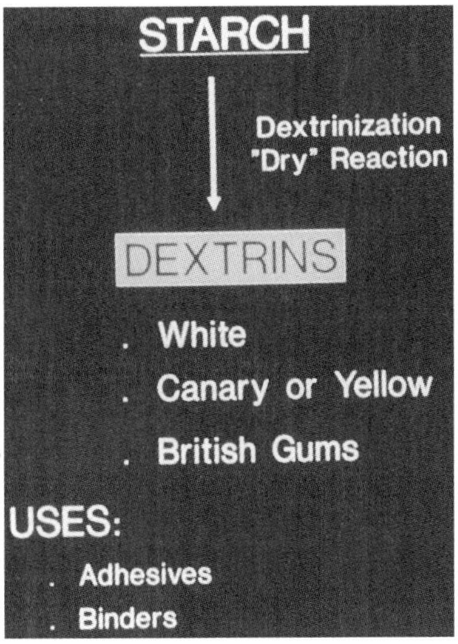

Figure 16 Starch to dextrin.

uct. Because the process causes the cleavage of molecular chains within the starch granule, potential viscosity development as compared to the native starch is less. Amylodextrins are still granular starches and therefore retain characteristic structure and birefringence. They are known for their solubility in water, rather than their swelling. Thin boiling starches produce enhanced gel properties and significantly lower viscosity sols. Molecular weights can be determined for analytical differences. This can usually be accomplished by size exclusion chromotography or chain length distribution by anion-exchange chromotography. However, for commercial production a fluidity number is associated with a specific product's functionality (40). A fluidity number is the volume of slurry (aqueous) that flows through a standard funnel within a fixed time. This time is usually relative to a fixed volume of water flowing through the funnel. In some cases of hydrolysis, an alkaline system may be used instead of pure water. The same funnel could be used for several products. This is accomplished by utilizing differing starch weights and solution pH as contributing factors. Amylodextrins are commonly used in the food industry for their high gelling characteristic.

b. Maltodextins. Maltodextrins (see Table 5) are known throughout the world. However, when attempting to utilize this type of sweetener solid, be aware that there are differences in definition based on country of use. In the United States, maltodextrins are defined by dextrose equivalence (DE). Saccahride polymers consisting of D-glucose units linked primarily by α-1,4 bonds with an average degree of polymerization of 5 or higher are generally recognized as the measurement of reducing sugars, calculated and reported as percent dextrose. The value must be <20 DE (Fig. 17). At this time there is no lower limit in the Unites States. However, a minimum is regulated for other countries around the world. Sweetener products exceeding the 20-DE level move into a category referred to as low DE sweetener solids (see Fig. 18). Today they must be identified by starch origin. In years past they were called "corn syrup solids," primarily because all products produced were derived from maize (common and waxy). Maltodextrins are primarily uti-

Table 5 Dry Corn Sweeteners/Maltodextrins

Feed stock	Starch from common corn, waxy corn, tapioca and potato
Process	Starch hydrolysis
	Acid
	Enzyme
	Acid/enzyme
Products	Maltodextrins
	Corn syrup solids
	Crystalline dextrose
	Crystalline fructose

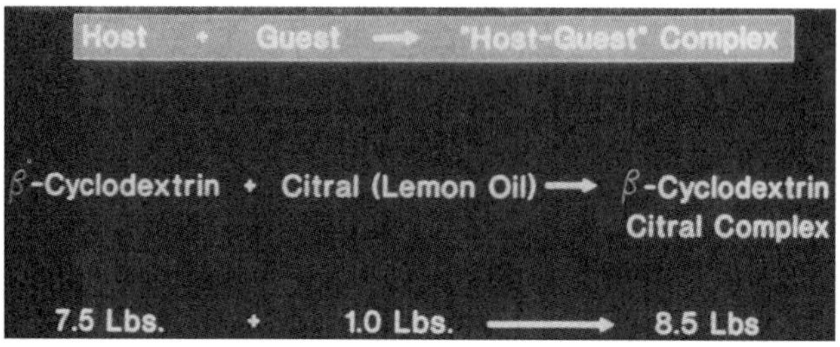

Figure 17 Cyclodextrin complex relative weights.

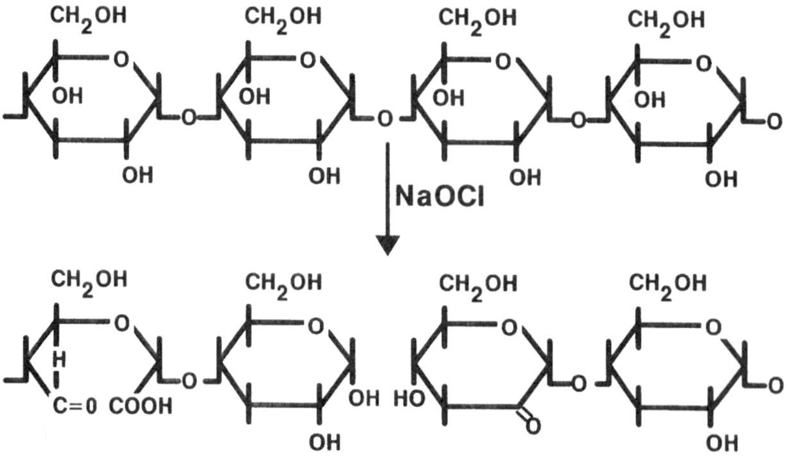

Figure 18 Hypochlorite oxidation. The nature of the reaction is highly pH dependent.

lized in the food industry for their low viscosity, low sweetness, clarity, and bland flavor. More will be said about their functionality in the section on applications.

c. Dextrins (Pyrodextrins). Dextrins fall under the guidelines for starch modification, however they differ in that they must be labeled as dextrins, not ''Food Starch Modified.'' Dextrins are produced by elevated temperatures, with or without added acid. This roasting of the starch at high temperatures can create yellow (canary) dextrins (Fig. 19). These are water soluble, low in viscosity, and contain a variety of glycosidic linkages. Again, based on the source of starch, commercial products can offer a wide array of functional properties for food application. Dextrins are usually limited in their food use due to the flavor profile generated during production. As this is a dry (nonaqueous) process, several compounds are present in the finished product. Dextrins typically contain high levels of salts (ash) and therefore can contribute to off-flavors.

Cyclodextrins will not be covered in this chapter. They possess properties unique for encapsulation of essential oils, but have received only limited approval in food applications around the world. The technology of forming a cyclic configuration has significant potential (fig. 20).

2. Oxidation Modification

There are two primary reactions used to produce commercial quantities of oxidized starches for food use. They involve the use of sodium and calcium hypochlorite compounds (fig. 21). Sodium hypochlorite is more commonly used for large volume production. Other compounds are permitted for the bleaching of starch, however at significantly

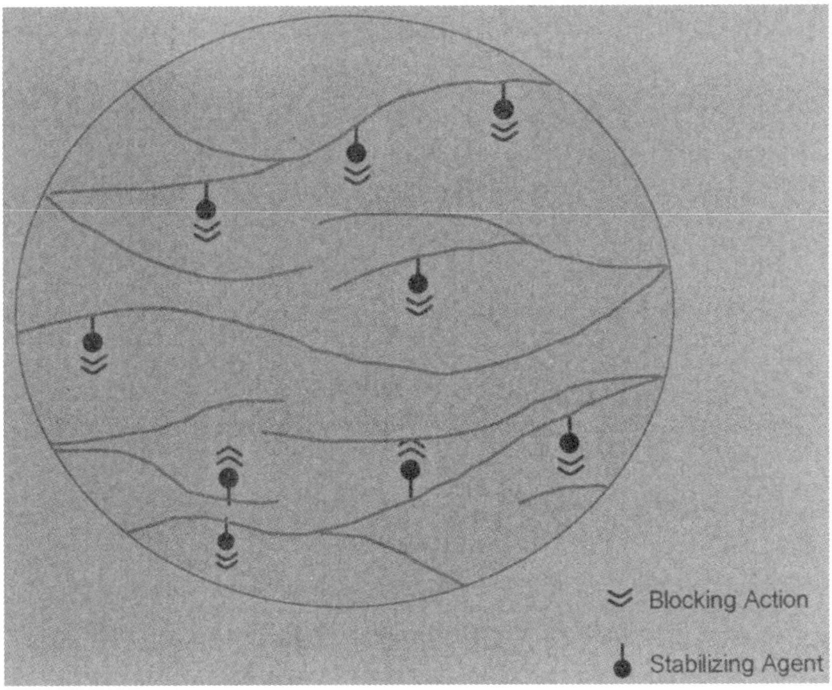

Figure 19 Monosubstitution.

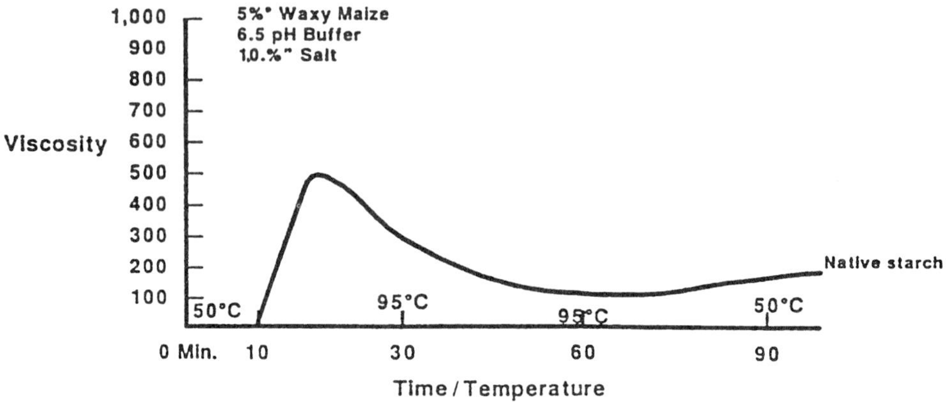

Figure 20 Native starch Visco-Amylo-Graph.

reduced levels as compared to those for oxidation (Table 6). Sodium hypochlorite is an aqueous reaction, while the calcium hypochlorite modification is done via dry blending at typical moisture levels for starch (8–12%). The oxidation reaction constitutes the cleavage of polymer chains resulting in the oxidation of alcohol groups into carbonyl and carboxyl groups. As in hydrolysis, the depolymerization of amylose and amylopectin significantly reduces the granules swelling and paste viscosity. The introduction of the carbonyl and carboxyl groups results in a reduction of the gelatinization temperature, increased solubility, and decreased gelling. With the introduction of some reducing ends to the molecular structure one could expect an effect causing browning. Highly oxidized starches exhibit exceptional dry flow properties as a powder. This can have functional advantages in some food applications.

3. Monosubstitution of Starch

Monosubstitution (Fig. 22) is in actuality the esterification or etherification of starch with monofunctional reactants. The common reagents utilized throughout the world for starch organic and inorganic monoesters for food use areacetic anhydride, succinic anhydride,

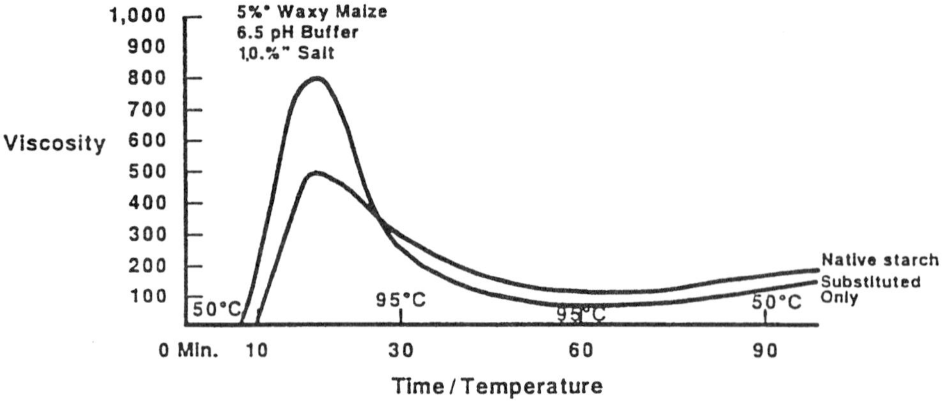

Figure 21 Effect of modification upon starch viscosity—substitution.

Table 6 Regulations for Modification of Food-Grade Starches in the United States

172.892 Food starch-modified.

Food starch-modified as described in this section may be safely used in food. The quantity of any substance employed to effect such modification shall not exceed the amount reasonably required to accomplish the intended physical or technical effect, nor exceed any limitation prescribed. To insure safe use of the food starch-modified, the label of the food additive container shall bear the name of the additive ''food starch-modified'' in addition to other information required by the Act. Food starch may be modified by treatment, prescribed as follows:

(a) Food starch may be acid-modified by treatment with hydrochloric acid or sulfuric acid or both.

(b) Food starch may be bleached by treatment with one or more of the following:

Use	Limitations
Active oxygen obtained from hydrogen peroxide and/or peracetic acid, not to exceed 0.45% of active oxygen	
Ammonium persulfate, not to exceed 0.075% and sulfur dioxide, not to exceed 0.05%.	
Chlorine, as calcium hypochlorite, not to exceed 0.036% of dry starch.	The finished food starch-modified is limited to use only as a component of batter for commercially processed foods.
Chlorine, as sodium hypochlorite, not to exceed 0.0082 pound of chlorine per pound of dry starch.	
Potassium permanganate, not to exceed 0.2%.	Residual manganese (calculated as Mn), not to exceed 50 ppm in food starch-modified.
Sodium chlorite, not to exceed 0.5%.	

Table 6 Continued

(c) Food starch may be oxidized by treatment with chlorine, as sodium hypochlorite, not to exceed 0.055 pound of chlorine per pound of dry starch.
(d) Food starch may be esterfied by treatment with one of the following:

Use	Limitations
Acetic anhydride	Acetyl groups in food starch-modified not to exceed 2.5%.
Adipic anhydride, not to exceed 0.12%, and acetic anhydride	Do.
Monosodium orthophosphate.	Residual phosphate in food starch-modified not to exceed 0.4% calculated as phosphorus.
1-Octenyl succinic anhydride, not to exceed 3%.	
1-Octenyl succinic anhydride, not to exceed 2%, and aluminum sulfate, not to exceed 2%.	
1-Octenyl succinic anhydride, not to exceed 3%, followed by treatment with a beta-amylase enzyme that is either an approved food additive or is generally recognized as safe.	Limited to use as stabilizer or emulsifier in beverages and beverage base as defined in §170.3(n)(3) of this chapter.
Phosphorus oxychloride, not to exceed 0.1%	
Phosphorus oxychloride, not to exceed 0.1%, followed by either acetic anhydride, not to exceed 5%, or vinyl acetate no to exceed 7.5%.	Acetyl groups in food starch-modified not to exceed 2.5%.
Sodium trimetaphosphate	Residual phosphate in food starch-modified not to exceed 0.04%, calculated as P.
Sodium tripolyphosphate and sodium trimetaphosphate.	Residual phosphate in food starch-modified not to exceed 0.4% calculated as ~P.
Succinic anhydride, not to exceed 4%.	
Vinyl acetate	Acetyl groups in food starch-modified not to exceed 2.5%.

(e) Food starch may be etherified by treatment with one of the following:

	Limitations
Acrolein, not to exceed 0.6%.	
Epichlorohydrin, not to exceed 0.3%.	
Epichlorohydrin, not to exceed 0.1%, and propylene oxide, not to exceed 10% added in combination or in any sequence.	Residual propylene chlorohydrin not more than 5 ppm in food starch-modified
Epichlorohydrin, not to exceed 0.1%, followed by propylene oxide, not to exceed 25%.	Do.
Propylene oxide, not to exceed 25%.	Do.

(f) Food starch may be esterified and etherified by treatment with one of the following:

	Limitations
Acrolein, not to exceed 0.6% and vinyl acetate, not exceed 7.5%	Acetyl groups in food starch-modified not to exceed 2.5%
Epichlorohydrin, not to exceed 0.3%, and acetic anhydride.	Acetyl groups in food starch-modified not to exceed 2.5%.
Epichlorohydrin, not to exceed 0.3%, and succinic anhydride, not to exceed 4%	
Phosphorus oxychloride, not to exceed 0.1%, and propylene oxide, not to exceed 10%.	Residual propylene chlorodrin not more than 5 ppm in food starch-modified.

(g) Food starch may be modified by treatment with one of the following:

	Limitations
Chlorine, as sodium hypochlorite, not to exceed 0.055 pound of chlorine per pound of dry starch; 0.45% of active oxygen obtained from hydrogen peroxide; and propylene oxide, not to exceed 25%.	Residual propylene chlorohydrin not more than 5 ppm in food starch-modified.
Sodium hydroxide, not to exceed 1%.	

(h) Food starch may be modified by a combination of the treatments prescribed by paragraphs (a), (b), and/or (i) of this section and any one of the treatments prescribed by paragraph (c), (d), (e), (f), or (g) of this section, subject to any limitations prescribed by the paragraphs named.

(i) Food starch may be modified by treatment with the following enzyme:

Enzyme	Limitations
Alpha-amylase (E.C. 3.2.1.1).	The enzyme must be generally recognized as safe or approved as a food additive for this purpose. The resulting nonsweet nutritive saccharide polymer has a dextrose equivalent of less than 20.

Source: Code of Federal Regulations, Title 9.

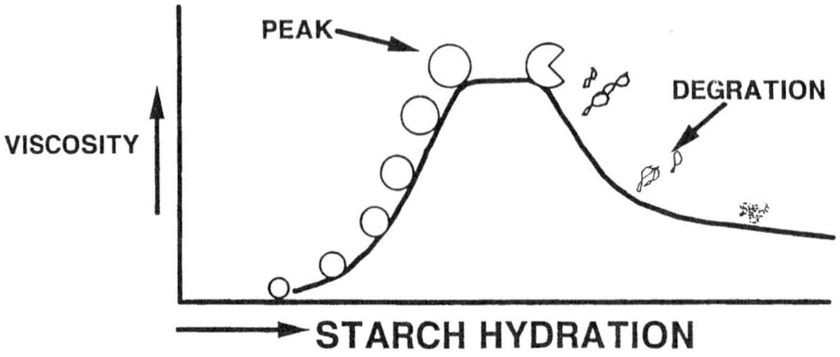

Figure 22 Granule hydration.

vinyl acetate, sodium tripolyphosphate, 1-octenyl succinic anhydride, and propylene oxide (Table 6). Although still approved, vinyl acetate is rarely used today. We should also note that the succinate esters are anionic polymers. Acetylation and propylation are the two most common forms of reactions utilized for the production of food starches. These modifications contribute significantly to the stabilization of food systems for refrigeration and freezing. Differences will be explained in more detail within the applications section. Determination of the level of substitution by hydroxypropylation is by proton NMR (41–43). Substitution position determination is also possible utilizing the same proton spectra (43). Commercial differentiation of starch products is generally characterized by change in viscosity. This change is usually measured using the Brabender or RVA (Figs. 23–25). Monosubstitution with propylene oxide only can be done at an addition level up to 25% by weight of starch (Table 6). These higher levels of substitution do yield water-soluble starch products.

All of the modifiers discussed thus far produce hydrophilic products. The introduction of these monoesters or ethers contributes significantly to the functional properties of the starch product. Regardless of starch origin, the effect is the same. Monosubstitution

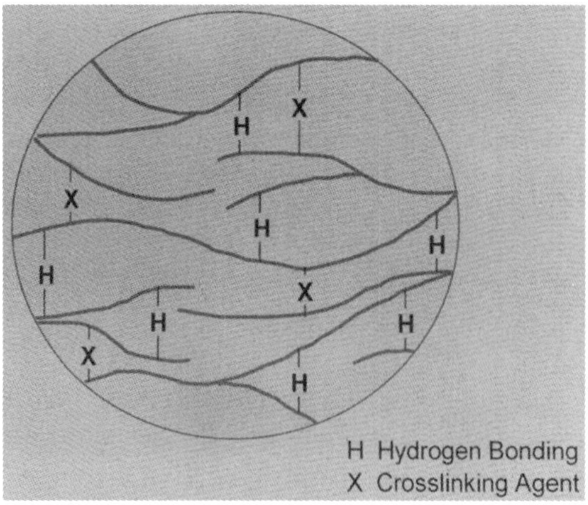

Figure 23 Crosslinked starch.

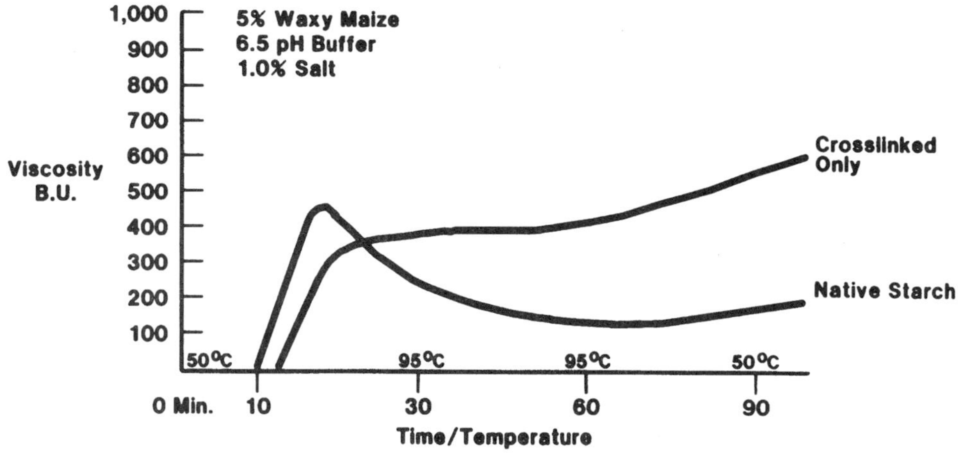

Figure 24 Effect of modification upon starch viscsoity—crosslinking.

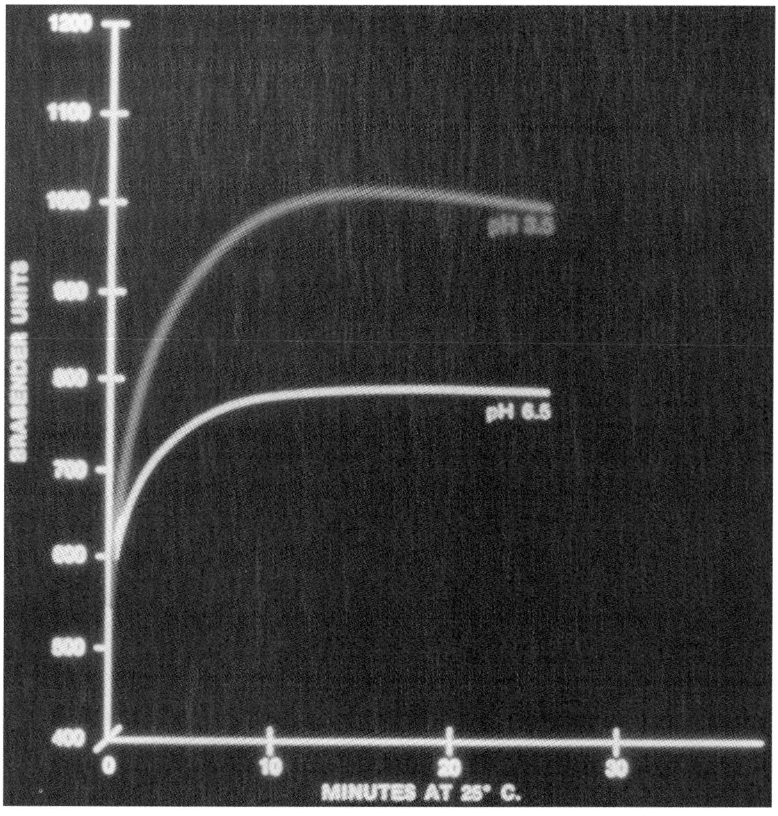

Figure 25 Visco-Amylo-Graph of pregelatinized waxy starch (5% dsb starch; 15% sucrose).

reduces the gelatinization temperature, increases water holding capacity, raises viscosity, and reduces shear tolerance.

The 1-octenyl succinate ester forms food grade starches hydrophobic in nature. They are considered more lipophilic in functional characteristics. This functional property lends itself to unique food application.

4. Crosslinking

The crosslinking (Fig. 26) of starch for commercial application is done today via the introduction of various multifunctional reagents. Those permitted today have been self-regulated by processors in the United States to exclude epichlorohydrin. Others that are still utilized both in the United States and other facilities around the world are phosphorus-oxychloride (POCl3), acetic anhydride, adipic anhydride, and sodium trimetaphosphate with or without sodium tripolyphosphate. (Table 6). Today the anhydrides and POCl3 are the most common reagents used. Phosphates are utilized, but for very specific products requiring given functional parameters. The effect of crosslinking on starch is dramatic. Altered functional properties include an increase in gelatinization temperature, reduction in viscosity, and increase in acid, heat, and shear stability (fig. 27). These later characteristics significantly improve the total functional contribution of starch to a food formulation.

B. Physical Modification

All that we have discussed thus far has been the chemical modification of starch. The alteration of the internal structure through the introduction of compounds to replace a hydroxyl group or the degradation of structure via molecular breakdown with acid or

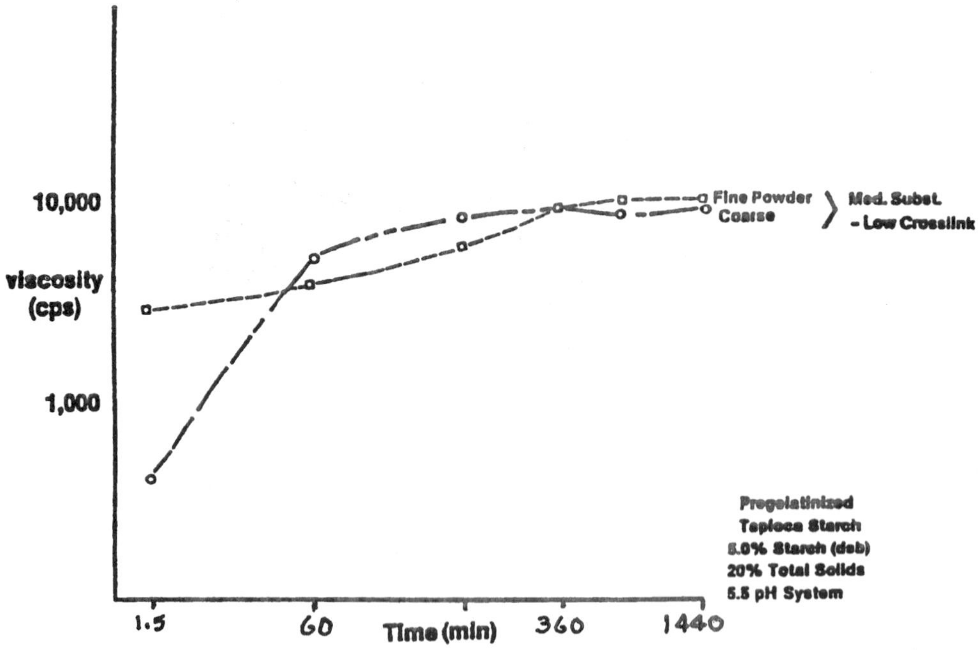

Figure 26 Brookfield viscosity versus particle size.

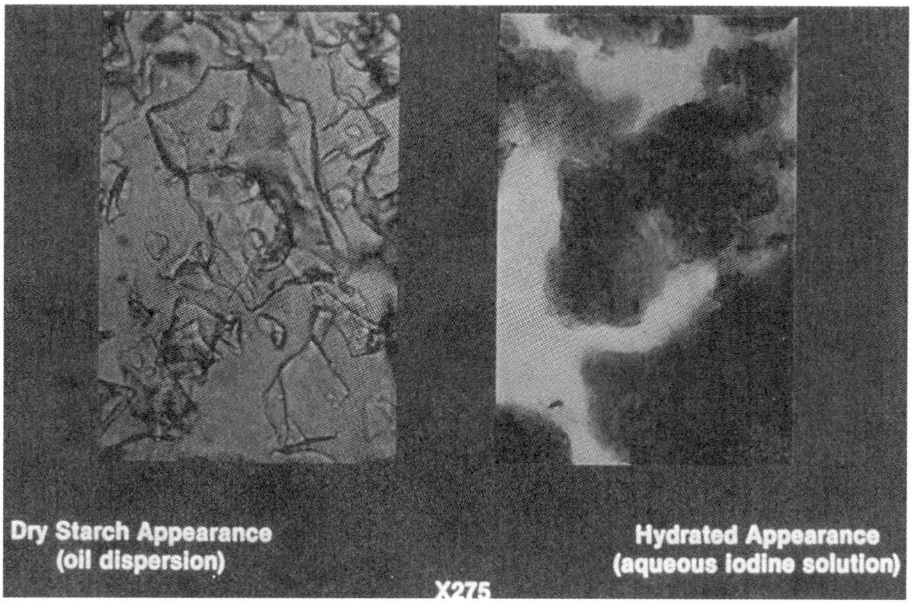

Figure 27 Pregelatinized waxy starch.

enzymes was the methodology used. For many years we have utilized some form of physical modification, however not to its fullest extent. Within the dry milling industry physical modification is and was common practice. The grinding and milling of grain yielded a variety of products. Use of air or screen classification also enhanced the functionality of specific milled products. In addition to particle size, moisture classification was incorporated. The use of drying techniques offered a wide variety of dry milled products containing starch for food use.

In the starch industry, those manufacturers that do not dry mill flour but isolate relatively pure starch via wet milling also found physical processing advantageous for producing unique and value added products. These processes went beyond those for physically processing flour; they offered new opportunities for native starches (Table 7).

1. Pregelatinization

The first of these is a process called pregelatinization. This process starch is cooked beyond the gelatinization point and dried, utlimately producing an instant hydrating, starch product

Table 7 Starch Types

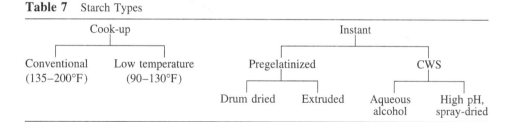

Cook-up		Instant			
Conventional (135–200°F)	Low temperature (90–130°F)	Pregelatinized		CWS	
		Drum dried	Extruded	Aqueous alcohol	High pH, spray-dried

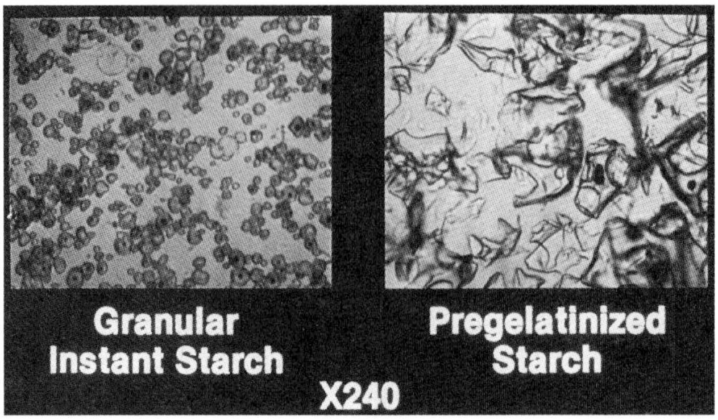

Figure 28 Granular starch versus pregelatinized starch.

(Fig. 28). Depending upon the native starch and whether it has been chemically modified, the functional properties it possesses are determined. Starches produced via pregelatinization are cold water soluble and hydrate very rapidly. The rate of hydration can be controlled either by modification, particle size, the use of dispersing aids, or a combination of all of these (Fig. 29). Today, most commercial pregelatinized starches are prepared using single drum dryers. A small percentage of food grade instant starches are still prepared on double drum drying systems. With most starches using either of these drying procedures the granules are greater than 80% fragmented (Fig. 30).

2. Instant Granular Preparation

A second method to produce instant starches differs in that the process forms an instant hydrating product that is not fragmented but retains the granules intact (Fig. 31). This is

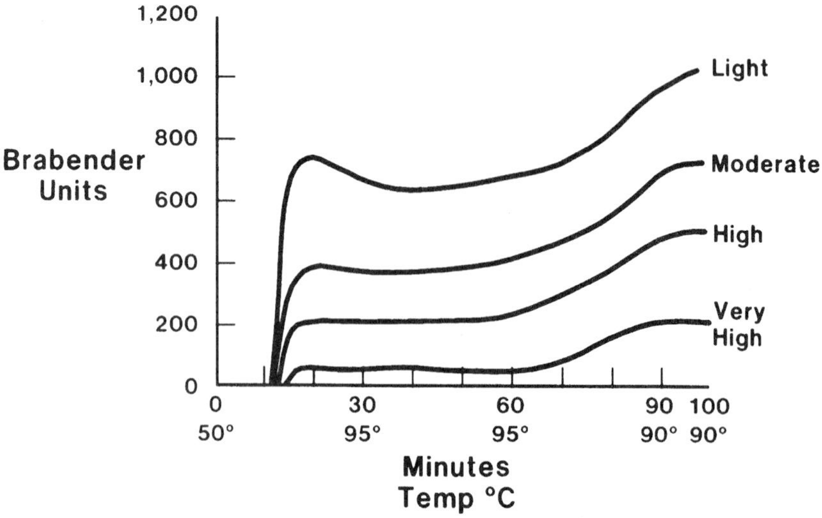

Figure 29 Crosslinked waxy starches (pH 6.5; 5% ds. starch; all highly substituted with various degrees of crosslinking).

Figure 30 Solution and solubility Demo

accomplished today via two different commercial processes. One is that of spray-drying starch specifically pretreated through a uniquely designed nozzle (44). The other process is that of dispersing starch in a mixture of alcohol and water. Subjecting this mixture to time, temperature, and pressure creates an instant starch. The alcohol retards the granules from becoming soluble, thus retaining birefringence and granule integrity.

Considering the number of native starches we have to select from for commercial food grade products along with the various methods of modification, both chemical and physical, we now have the potential for hundreds of unique and application-specific starch products.

3. Extrusion

Extrusion is another process where external heat is transmitted into a starch/water mixture. This is somewhat similar to drum drying, however a significant amount of mechanical shear is introduced with extrusion. Drum drying does not have this mechanical impact upon the starch matrix. Also, the starch matrix for extrusion is generally considerably lower in moisture content. Drum drying is done with a starch slurry, while extrusion presents more of a dough. Experience with these two systems has proven that similar

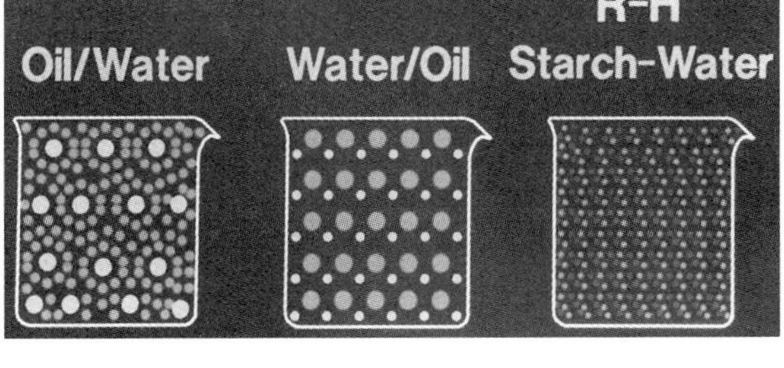

Figure 31 Emulsions.

products can be produced analytically. However, the functional properties are significantly different. To produce a functional match via extrusion usually involves changing one of two things. Either the analytical values are significantly varied from those of drum dried starch or the types/degrees of modification are altered to produce a functional match. In either case, an identical match does not occur. This is not to say that quality products cannot be produced via extrusion. On the contrary, they just cannot be marketed as one-for-one replacements for drum dried starches. They will need to be marketed for their individual unique characteristics.

Extrusion is usually accomplished using one of two types of extruders. A single and a twin-screw system are available. Each yields a different product, even with the same starting material. An important factor to remember when considering extrusion is the physical force or energy implied; this can be a benefit or a drawback. Shear is still an important factor. Granule degradation or fragmentation can and does occur. With it come characteristic and functional changes.

4. Heat Treatment of Starch

One last product type which has been introduced in just that past couple of years, is referred to as resistant starches (Table 8). These starches, have been uniquely processed to contain retrograded amylose resistant to α-amylose digestion. Resistant starch has generated considerable interest as a food additive. The smooth consistency, lower caloric contribution, and low water affinity are desirable characteristics. In addition to these properties it has also exhibited possible health benefits as related to cardiovascular disease, diabetes, and colon cancer (45–48). How do we obtain these unique starches? One process is the heating of an aqueous suspension of starch granules at a temperature just below the gelatinization point for an extended period of time. The lowering of the Tg of the amorphous phase allows the polymer chains to be mobile or in a rubbery state (49). Annealing results with an increase in gelatinization temperature. This can be done in excess salt solutions for heating starch. Essentially any medium that retains the starch just below the gelatinization temperature offers the potential to generate this type of physically modified starch. It has been shown that some heat and acid tolerance can be introduced into the starch granule via this process. Therefore producing a native starch with similar properties to lightly crosslinked starch via chemical modification is possible.

We have completed the discussion for both chemical and physical modifications as independent processes. Within the scope of starch production, regardless of what part of the world or the origin of the starch, the industry has been granted the clearance to incorpo-

Table 8 Resistant Starch Classifications

Category listing	Accepted definitions
Type I	Physically inaccessible starch which is locked in the plant material, e.g., milled grains, seeds, legumes
Type II	Native granular starch, found in food containing uncooked starch, e.g., bananas
Type III	Indigestible starch that forms after heat and moisture treatment; may be present in foods such as cooked potatos
Type IV (proposed)	Resistant starch that has been produced by chemical or thermal modifications

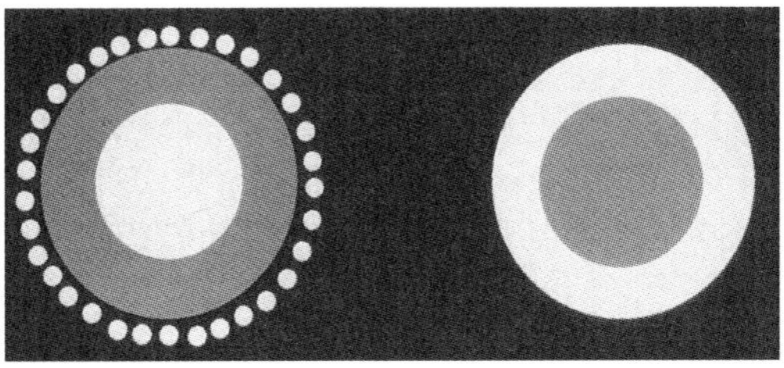

Figure 32 Emulsion versus encapsulation.

rate multiple modifications upon or within the starch granule. This agreed upon procedure creates the opportunity for the production of hundreds of modified starches, each possessing moderately to significant differing functional characteristics. Most governing bodies of the world that regulate food production and the ingredients utilized within have granted such approval to their respective starch industries. Many of the food grade starches discussed in the applications section offer varying functional characteristics primarily due to the use of two or three modifications within the starch granule (Fig. 32). This does not include the opportunity to bleach the granule for the purpose of providing a whiter dry starch product to the consumer.

IV. FOOD APPLICATIONS FOR NATIVE AND MODIFIED STARCHES

A. Native Starches

The dictionary defines carbohydrates as ''any of various neutral compounds of carbon, hydrogen and oxygen (as sugars, starches and cellulose), most of which are formed by green plants and which constitute a major class of animal foods.'' Around the world we find differing tastes and desires for what is considered palatable. The native starch for regional food application could be based on wheat, maize, potato, rice, arrowroot, or sorghum. Much of this dominance is due to climate, growing conditions, and history. What is preferred in the United States (primarily maize and wheat) would not be considered acceptable for general food use in Europe or Asia, and what is traditional to Europe is not always desirable in other sectors of the world (50). This continues all around the globe.

Native starches are usually used because of economics. In some instances native starch is added to foods for a specific functional property. It could include texture, flavor, and/or clarity—or lack there of (Table 2). They should not be added to food systems as a stabilizer for the water phase, at least not at this time. Possibly in years to come, with continued research and genetic engineering, we may see such functional properties contributed from nonchemically modified starches.

One of the first characteristics associated with starch is the flavor aspect. This can and does limit native starch use in neutral or sensitive flavored foods. Even for flour, where wheat starch is a primary component, flavor can be an issue. But at this point there is a fact that must be noted. Although I have not been able to locate documentation to verify or dispute it, however I have studied many food products within the food industry and I contend that starch masks flavors to a lesser degree than sucrose. This was highlighted during the introduction of high-fructose corn syrup in the 1970s for use in spoonable salad dressings. During the replacement of sucrose with the liquid sweetener, many companies discovered they could significantly reduce the added seasonings due to a significant increase in flavor intensity. The change took place without altering the starch usage. In many cases a native starch was one of the added components. An important factor to remember when utilizing native starches is to identify what flavor each starch may contribute, which is usually identifiable from its origin.

Consider now the texture aspect of native starches. In many instances it is directly proportional to the percent and the molecular size of the amylose molecule. As was discussed earlier, amylose becomes significant when attempting to utilize native starches in processed food products. Amylose tends to cause gelling, or aggregates of starch crystallites. These aggregates generate grainy to very rigid textures from native starches. In starches that are predominately amylopectin (waxy), the texture can be short and smooth to long or stringy. They are also be referred to as cohesive. How can we take advantage of the extreme variability in textures produced by native starches? I've already mentioned dressings, but also consider such foods as confections, sauces, gravies, and fillings. Depending on the strength of the gel or the contribution to flavor and/or texture of savory fillings these will most likely dictate your starch choice.

Native starches can also be used as a dry ingredient. If you were to reduce the moisture content of the native starch to less than 7% (redried starch), it could be added to mixes for moisture control as well as for providing some thickening and economics. Native starches can be and are used just as a source of carbohydrate, however as we see now there are many functional possibilities. One other very common use for native starch is that of producing powdered sugar. In this application it acts as a flow aid.

As the third characteristic of native starches, clarity (or opacity) can be advantageous (or in some cases cause a serious problem). Those products containing short chain amylose, such as the cereal starches generally contribute opacity to a finished food product. This goes along with the retrogradation process. For some foods this may not be a significant issue, however in many this change from clear is unacceptable. Gravies, sauces, dressings, puddings, and other savory foods usually are opaque. The foods requiring more clarity such as fruit fillings and jellies require the use of the waxy variety of starch. In some cases the longer chain tuber starches have been found to be functional.

For native starches that have not been modified, a food processor should only expect a few hours of shelf-life. Physically modified starches have shown in some tests longer stability, but do not compare directly with the chemically modified starches as yet.

B. Modified Starches

With the introduction of modified starches to the food industry, starch usage in food systems has increased dramatically over the past 50 years. One reason for this shift in use is the capability of modified starches to be utilized over a broad range of processing parameters, e.g., acidity, thermal conditions, and mechanical shear. Their stability throughout these parameters of processing and the other identified functional characteristics have lead to use increase. Compared to other ingredients within a food matrix (other than water), starch cost is proportionally minimal.

The modification of starch, as well as the residual components after modification are regulated worldwide. There are differing regulations depending on the chemical and/or process used, the country of preparation, and location of use. It is very important to be aware of all regulatory issues, restrictions, and definitions before utilizing a starch product in a food system for export. Surprises can be costly to the unaware.

Modified starches are typically used in foods around the world. They are added to thicken, stabilize, or texturize a food product. In actuality, they are used for many more functional characteristics. Prior knowledge of the product parameters, processing conditions, and other key factors will save time and expense.

V. A STARCH SELECTION

Too many times a food technologist or formulation scientist initiates a formulation profile without analyzing the ingredients. In addition the same person will rely on past recipes or formulations to establish their baseline. This can be a critical error. Not to say that the use of existing formulas is not a good idea but too many times we do not consider all the parameters before beginning our research. By simply asking a few questions pertaining to what you want and requesting input from suppliers about their respective ingredients, research could save you a lot of time and expense. Unfortunately, many suppliers are not aware of how their ingredients might interact with other ingredients under a variety of conditions. What they have studied may not be relevant to your system. The following questions should be asked when selecting a starch. For additional information see Table 9.

A. What Is the Desired Function of the Starch Being Added

Considering the food product, is the starch expected to provide only one function or act as a multiple function ingredient? The answer to this question could help identify whether one starch or a blend of starches is required. If it is a blend that is required, then the ratio between starches becomes critical as related to the physical properties ultimately desired. Answering just this one question directs many of the choices regarding starch selection.

B. What Is the Method of Processing Being Considered?

This question is designed to identify whether you are going to cook the starch-containing food or not. If it is a dry mix or a noncooking preparation, we have possibly identified the need to consider instant starches, perhaps not exclusively, but we should consider them. If cooking is required, from a cost and functional standpoint a cook-up starch would be first choice. For foods that are dry mixes, we will have to consider what happens when the consumer interacts with the product before we can make a selection.

C. What Is the Food System pH?

The acidity of the food is extremely important. Because foods can be hot filled at pH <4.6, knowing the finished food pH is important. Acids possess different pK values and therefore food technologists should consider the total acidity of the food system. Using titration for total acidity is more accurate. The pH considered neutral for starch is not 7. Those foods having a measurable pH of at least 4.6 or higher are considered neutral. Typically a pH value above 7.5 is self-regulating due to off-flavors being generated. Neutral foods expected to retain long shelf-life require sterilization. The sterilization process requires temperatures above atmospheric and thus dictates starch selection. Foods having extreme or high acidity, less than pH 3.5, possess the potential to hydrolyze starch. Therefore starch selection will be determined by degree of crosslinking and textural properties.

Table 9 Questions to Ask Before Selecting a Starch

1. What is the desired function for the added starch?
   - a. thickener
   - b. stabilizer
   - c. texturizer
   - d. emulsifier
   - e. binder
   - f. bulking agent
   - g. ___________
2. What is the processing method?
   - a. cooking (liquid)
   - b. noncook (liquid)
   - c. dry mix
   - d. ___________
3. What is the pH of the food system?
   - a. >4.5
   - b. <4.5
   - c. <3.5
   - d. ___________
4. What is the percent soluble solids (ingredients that hydrate water)?
   - a. >45%
   - b.<45%
   - c. >55%
   - d. <20%
   - e. ___________
5. What is the shelf-life expectancy?
   - a. <30 days
   - b. 1–6 months
   - c. >1 year
   - d. ___________
6. How will the finished product be stored?
   - a. ambient temperature
   - b. frozen
   - c. ___________
7. What is the desired product texture?
   - a. smooth
   - b. nonsmooth
   - c. grainy
   - d. ___________
8. What is the desired eating quality (mouth-feel)?
   - a. creamy
   - b. fudge
   - c. pasty
   - d. jelly
   - e. ___________
9. What is the desired surface appearance?
   - a. sheen
   - b. opaque
   - c. dull
   - d. translucent
   - e. ___________
10. How and when will you measure or determine viscosity?
    - a. Visco-Amylo-Graph
    - b. Brookfield
    - c. Bostwick
    - d. FMC brabender
    - e. Fluidity (funnel)
    - f. Rapid-Visco-Analyzer
    - g. ___________
11. Is high or excessive shear introduced?
    - a. homogenization
    - b. milling
    - c. vacuum
    - d. direct steam injection
    - e. pumping >25 ft hot
    - f. ___________

Table 9 Continued

12. Will the process involve one or more of the following?
    a. hot filling
    b. ambient filling
    c. refrigeration
    d. slow freezing
    e. blast or quick freezing
    f. reheating (reconstitution)
    g. steam tables
    h. ___________
13. If fat is used, what type?
    a. liquid
    b. solid
    c. shortening
    d. lard
    e. ___________
14. Are salts utilized?
    a. what type?
    b. percent?
    c. blends?
    d. ___________
15. Are other hydrocolloids used?
    a. what type?
    b. blends?
    c. ___________
16. Is the final product a dry mix?
17. If yes to No. 16, does the process involve the following?
    a. blending
    b. agglomerating
    c. extruding
    d. ___________
18. Is moisture content of the added starch critical?
19. If a dry mix, what is the anticipated packaging?
    a. paper
    b. wax coated
    c. poly-lined
    d. heat sealed
    e. ___________
20. If not a dry mix, what is the anticipated packaging?
    a. glass
    b. can
    c. pouch
    d. drums
    e. paper
    f. plastic
    g. tote
    h. ___________
21. Is the added starch to be used in more than one product? If yes, consider similar events for the other products and identify the critical functions contributed by the starch.
22. How important is ingredient economics?
    a. very
    b. not very
    c. quality dependent
    d. ___________

The food systems that range between pH 3.5 and 4.6 offer several starches for selection. The texture, process, and other handling parameters will determine the starch of choice.

D. Does the Process Contribute High Shear?

Knowing the answer to this question aids the technologist to determine how to maintain granule integrity. If this is not important, an instant starch may be acceptable. However, if water management is critical, granule stabilization is also. Shear is similar to heat and acidity; it can destroy starch granules. Simple milling can fracture granules; consider what high shear homogenization or vacuum could do to starch under stress conditions.

E. What Percent Soluble Material Will Be Present?

Ingredients that hydrate water limit the available water for granule swelling and generating viscosity. They do so not so much by taking on the water, but by reducing the water directly available. Starch requires minimum levels of water to be fully functional. When other soluble material is present and elevates the total soluble solids to greater than 45%, the starch has a tendency to limit hydration under atmospheric conditions. At soluble solids less than 20%, starch typically is incapable of adequately managing the water that it could if the solids were greater than 20% but less than 45%. Therefore, to do so the technologist must increase the percentage of added starch.

F. Is One of the Following Used: Fats, Salts or Gums?

As with soluble solids two of these ingredients hydrate water and can significantly alter the hydration and functional potential of any incorporated starch. Gums, like starches, contribute to viscosity and texture. They can also influence mouth-feel or eating quality. Because they compete for water, they can cause latent starch hydration, or not allow it to hydrate at all. This then could contribute to a starchy flavor (off-taste). Salts, too, hydrate water, but more importantly they can be thermally active. In other words they may retain or store heat, thus potentially causing starch over cook, as well as alter flavors. Lipids or fats obviously do not hydrate the water, but can alter the hydration rate of added starches. This can be accomplished by coating the starch with fat or using the fat to coat other ingredients. Therefore knowing the type of fat is important. If it is a hard fat, knowing the melting point is important. All of these are critical for preparing a dry mix. Incorrect ingredient addition can not only change hydration rates, but also contribute to final product characteristics.

G. Is the Final Product Subjected to Postprocessing?

Having the knowledge of what is going to happen to the finished food is critical. If the consumer is going to bake, microwave, or subject it to steam heating for prolonged periods, the technologist's choice of starch and processing will be very important. The technologist should ensure the product going to the consumer is capable of meeting the requirements demanded. Therefore, process specifications from the consumer are necessary prior to formulating.

H. How Will the Product Be Stored?

Storage of the food product is sometimes as important as the preparation. Knowledge of how it is to be stored as well as the conditions of storage are critical. Without this information, the technologist preparing the food product may not know the best process to prepare the food for storage. An example is a food that is to be frozen and kept in the freezer for longer than a year. This will aid with the decision as to slow or fast freeze. If that is not an option, then it will aid with starch selection. All starches that offer good to excellent freeze–thaw stability do not always guarantee extended shelf-life, e.g., up to a year or more. However, those starches that can give a year's freezer life, will always provide adequate freeze–thaw. Refrigeration is a simple shelf profile to meet. Extended freezer life at −20 to −40°C is very difficult to achieve. Starch alone is not the best answer. This is one of those times when mixed hydrocolloids are advantageous. A properly modified starch (crosslinked/monosubstituted) with a small percentage of a gum (xanthan) can provide the needed stability.

VI. THE EFFECT OF PROCESSING ON STARCH

There are many facets to a process that must be understood prior to formulating a food product. We have touched on some processes and their relationship to starch selection. We now need to relate what effects different processes may contribute, both positive and negative. The following issues have been identified as critical to the production of a successful food product.

A. Temperature Control

Temperature control does not mean just the temperature for gelatinization. The rate for heating the system and, just as importantly, the rate of cooling must be considered. Both of these are essential for providing a fixed set of conditions that can be monitored and repeated. Knowing the rates for both conditions will also help to identify the necessary equipment for scaling the system for commercialization.

B. Shear

Earlier shear was described for its degrading effect upon starch. Knowing about equipment and its shear potential is very helpful to a food technologist. Knowing when and where shear can and cannot be introduced is critical also. Many of the modified starches designed for food processing can withstand some shear. A few are specifically engineered to withstand very high shear systems. In all cases though, starch is much more shear stable at ambient or cool temperatures as compared to elevated temperatures. Studies have shown that below 135°F starch solutions can be processed with little shear damage. Starch in solution prior to gelatinization is very shear stable (Table 10).

C. Packaging

One mistake common to many products prepared with starch relates to packaging. Too many times the same product is packaged in a variety of containers. This in most cases is not a reliable practice. Many starch-stabilized food products cannot be packaged over a large volume range of filled quantities. An example would be hot fudge topping. Utilizing a single correctly modified starch to achieve viscosity, texture, and stability is possible for product packaged in containers from 1 pint (0.5 L) to 5 gallons (20 L). However, attempts to package this same product in 55 gallon drums (110 L) will most likely be unsuccessful. Experience has shown that to facilitate this practice required the use of two

Table 10 Shear Intensity Based on Equipment

Equipment	Intensity of shear
Dixie Mixer	Low
Lightnin' Mix	Moderate
Likwifier	Moderate to high
Colloid mill	High
Homogenizer	Very high
Piston pump	Low to moderate
Gear pump	Moderate
Moyno Pump	Moderate
Centrifuge	Moderate to high
Kettle cooker	Low
Steam infusion cooker	Low to moderate
Swept surface cooker	Moderate to high
Plate heat exchanger	High
Flash cooler	High
Jet cooker	Very high
Spray-drying	Moderate to high

modified starches or a starch/gum blend. In both cases starch levels had to be adjusted, and some characteristic difference was noted in the finished topping.

D. Storage

All too often the storage of product is overlooked when initiating product development. The storage conditions used for the is food product just as important as the container. In addition to conditions, the time of storage is critical. Some foods can be stored much longer than others under the same conditions. Changes usually occur in texture during storage. Some products thin; others thicken. Others could become lumpy, grainy, or eventually separate. Warehouse, cooler, or freezer conditions are essential to be specified. Also, the food product should be pretested in as many storage conditions as possible; one never knows what the consumer or transportation department will do.

VII. OTHER INGREDIENTS AND THEIR EFFECT ON STARCH

A. Water

Water may be the most common ingredient utilized with starch. It is also one of the most underestimated ingredients affecting starch negatively. Water as a food ingredient has been emphasized as essential to make starch, both native and modified multifunctional, in food products. This being true, then the variety of water must be considered, not only from home to home but also the differences around the globe. Water obtained from a well has certain measurable properties and constituents. The same can be said about water taken from a city water system. However, these two waters are significantly different, which could dramatically alter the functional properties of added starch.

Consider the treatment facility for any city and the number of cities around the world. We now have several approved treatments for the purification of water and few cities use exactly the same process, let alone similar levels of reagents within a process. Regulatory guidelines leave a broad range providing room for variability as to the effect on starches. Experience has proven that if a water supply smells like a swimming pool, it probably has enough chlorine present to oxidize starch. Also, well water systems can offer a vast variety in mineral content. Some of these can lead to starch complexes that inhibit the functionality of the added starch. Water softened with salts may be great for bathing or washing dishes, but it has the potential to negatively affect added starch in foods. Because of these differences, many development labs formulate with distilled or de-ionized water. Unfortunately, as soon as a product is taken to the plant for production, a complete reformulation is necessary. What is often thought to be the simplest ingredient for foods is seen to be a very complex supply of components.

B. Sweeteners

Sweeteners (see Fig. 33 and Table 11) offer a lot more to the food product than just sweetness. Sweeteners for this review include maltodextrins, sweetener solids greater than 20 DE, high-fructose syrup, syrups classified as medium and high conversion, as well as sucrose. The high-intensity sweeteners contribute sweetness in foods, but are predominately carried by maltodextrins. Therefore, I will forgo discussion about their effect, as it relates primarily to that of maltodextrins. When utilizing maltodextrins with starch, the food technologist is attempting to increase solids, control sweetness, or generate a desired color. In some instances a maltodextrin may be added to assist with stickiness (tackiness). This is best done with very low-DE (<5) maltodextrin products. These same maltodextrins have proven to provide fatlike properties. Because of the blandness and solubility of maltodextrins, they are very compatible with the incorporation of instant starches. They conveniently offer a method for premixing without affecting the functionality of the added starch. In fact, the low-DE maltodextrins are somewhat similar to modified starches processed with enzyme. Sweetener solids having a DE greater than 20 are not classified as maltodextrins. In fact, those derived from maize starches are called corn syrup solids. Similar products are commercially available today, derived from other starch sources, i.e., rice, tapioca, potato, etc. They now represent a small but reasonable share of the dry solids and liquid sweetener market. A primary functional characteristic is that of adding

Table 11 Conventional Corn Syrups

Common Name	D.E.	Dextrose	Solids	Confectionery Functionality
Low D.E. Syrup	25	5	77.5	High viscosity . . . low hygroscopicity
	35	15	80	Limited use in hard candy
Regular Corn Syrup "Glucose"	42	20	80	All purpose syrup . . . high cooked candies . . . soft candies, when blended with dextrose, fructose, or invert sugar
High Maltose Syrup	42	7	80	Low dextrose content . . . good color stability . . . low hygroscopicity . . . special hard candy uses
Intermediate Syrup	55	30	81	"Comparative syrup" used in both hard & soft candies
Sweetose (high conversion)	64	40	82 84	Best choice for soft candies . . . good moisture affinity . . . best economics

sweetness to food products differing from that of maltodextrins. However, one similar characteristic is the higher-DE sweetener solids are also excellent humectants. Water activity can be significantly reduced through the utilization of low-DE sweetener, solids.

C. Salts

Most technologists consider this ingredient predominately sodium chloride. For many food systems this may be correct. However, salts can complex with a number of compounds creating unique structures, thus resulting in very unique eating quality. An example of such a complex is the preparation of instant puddings. Many food scientists as well as consumers consider the starch as the building block for producing the set and eating quality for instant mix puddings. The starch however is added to control the water phase of the mixture, enhance mouth-feel, and provide stability during storage. The structure and eating quality is generated from the salt complex created with the protein from the dairy portion of the mixture and added salts. The protein can come from added liquid milk or dry milk solids. Quantity and blend ratio of the added salts influence the rate of set, strength of the gel, and the mouth-feel of the finished pudding. Calcium and potassium salts, either phosphate or chloride derivatives, significantly add to or detract from the quality of the puddings. Calcium usually increases gel strength and shortens set time, whereas potassium has the reverse effect. The chloride derivatives generally produce a distinct off-flavor and are considered unacceptable for commercial products, however the effect is worth noting.

D. Other Food Ingredients

Many food ingredients (spices, fruits, flavors, etc.) can contribute to soluble solids of the food matrix. This includes ingredients that compete for water. Ingredients competing for water typically retard the hydration of the added starch. However, many ingredients are added for flavor, particulate characteristics, or enhancement of other functional ingredients. Many times these unique additives contain α-amylase. α-Amylase is an enzyme utilized in degrading starches for the production of sweeteners. It is also the type of enzyme utilized by the body to break down carbohydrates. Starch is a carbohydrate and therefore will be digested by this enzyme. Peppers, some smoke flavors, fruits such as pineapple, and fresh blueberries are known to contain significant quantities of α-amylase. Dairy products such as cheeses (blue cheese) and low heat processed milk solids can be a source.

In the early 1990s the flour industry experienced a significant setback with flour. For a 2-year period the wheat crop experienced a very wet season and a second growth occurred within the crop prior to harvest. Not only were protein-related enzymes formed, but so was α-amylase. Many bakery products experienced breakdown of the starch phase of the flour as well as the digestion of added starches. In many products the effect was not noticed until the product (e.g., refrigerated and frozen dough) reached the consumer. There are test procedures available to assist with analysis. Many contract laboratories assay for enzyme presence. A simple nonquantitative test can be utilized using triplicate containers of a control starch paste and inoculating with suspect material. It only requires 24 h at room temperature to study the effect, or at elevated temperatures (140°F) about 4–6 h yields an indication of enzyme activity. As with all tests always use a control sample for comparison.

E. Proteins and Other Starches

Ingredients that contain proteins typically contain carbohydrates. Therefore, always consider the source of the protein and its anticipated contribution to to the food product. Proteins, as mentioned earlier, complex very easily with salts. These complexes can create gray to dark particulates. These particulates can create undesirable textures. Foods that will be fried, baked, or subjected to long-term processing at high temperatures can potentially produce dark spots on the surface (Mallard reaction). Protein can also contribute to texture. In breads it may be desirable, however the elastic effect of glutenlike dough in some foods is not acceptable. Agglomerated proteins generate small to large particulates that can create uneven surface texture as well as a gray color.

Starch added from other sources or as another ingredient, not intended to yield functional properties, are often overlooked. One source that can contribute a significant percentage of starch that has already been mentioned is that found in powdered sugar. The 3% starch allowed in powdered sugar could significantly alter the cook, flavor, clarity, texture, and shelf-life of a food product.

When a technologist has identified all sources of starch or other hydrophilic compounds prior to formulating the final product, a relative amount of correctly modified starch can be added allowing for the effect from the other ingredients. In many cases the technologist may need to seek advice from suppliers as to ingredient composition and potential functionality when mixing with other ingredients. Some ingredients utilized can be preblended compounds from many sources. A supplier may not quantify the amounts, and in some instances not disclose the detailed listing, of ingredients. An example of such a mixture could be a blend of common and waxy corn starches. The supplier of such a product may only identify the product as containing "corn starch". The supplier would be correct in doing so.

VIII. SPECIFIC FOOD APPLICATIONS AND PROCESSES

Most of the examples in Section V were generic in description. This section identifies types of starch, modifications, and possible levels of use for a few specific food processes.

A. Thermal Processing

Thermal processing can involve severe heat treatment. Typically a food technologist would recommend a modified starch possessing the treatment of monosubstitution in conjunction with moderate to high levels of crosslinking. These modifications will provide stability, texture, and consistency. Processes such as retorting or heat-sterilization require such starches. The correct starch will depend upon other functional attributes, such as flavor, length of shelf life, and appearance. Usually a modified common or waxy corn is a good choice to initiate studies. In the high acid or low pH foods, heat treatment is usually minimal. A crosslinked only starch could be the best choice. However, monosubstitution could be necessary, depending upon desired product characteristics and storage, e.g., refrigeration and/or freezing.

Starches that possess unique characteristics can offer process advantages in thermally processed foods. Native or specifically modified starches offering either thin-to-thick or thick-to-thin viscosity profiles aid with processing and particulate integrity. Starches that provide thin-to-thick properties may reduce sterilization time. This could increase productivity and improve product quality. Quality issues could relate to color, texture, and mouth-feel. These starches are usually very highly crosslinked and monosubstituted to various degrees depending upon final viscosity parameters. Waxy maize is one starch used widely for this process. The thick-to-thin concept is limited in starch choice and utilization. This concept is generally applied for food preparations that either require specific particulate addition to containers prior to processing or a very critical viscosity profile postprocessing. Native potato starch is a natural thick-to-thin starch and has been used for such. The final texture can be undesirable for some food formulations. Due to this, the starch industry offers modified starches, some just monosubstituted and others with the addition of specific acids blended with the starch to create a hydrolysis reaction during processing. A smoother consistency to the food results. These starches have also found unique functionality in microwave foods today.

Some foods are not designed to be clear, they are intended to be opaque (cloudy). Starches have been engineered to provide such a characteristic. Foods such as gravies, sauces, or beverage mixes are excellent applications for such a modified starch. These starches are stable in high-temperature processes. Modified common corn and/or wheat starches are used. Starches designed for opacity are not suggested for use as water binders or phase stabilizers.

Considering the variety of base starches utilized in thermally processed foods and the various levels of modifications, a technologist has an extensive spectrum of starch stabilizers and thickeners to choose from.

B. Freezing

The freezing of food is a method of preservation. It can offer greater than 1 year's stability. Studies have shown that freezing typically maintains a high quality profile for foods, rivaling that of fresh.

Many foods sold as frozen require thermal processing prior to storage. This has been done to enhance flavor, consistency, texture, and appearance. In reality, it is done to stabilize the water phase. Some foods can be prepared without heat and those usually utilize instant starches, with or without the incorporation of gums. In many cases this frozen food is then heated prior to consumption.

A technologist should be aware of the type of freezing used as well as how long the product is going to be subjected to freezing. The use of rapid or blast freezing reduces ice crystal size. The use of spiral freezers, which are much slower, is more common and generally creates larger ice crystals. The slower the freeze, the greater the chance of ice crystal buildup. This can be detrimental to the added starch. Ice crystals can cause the fragmentation of granules and/or food particulate cell walls. Depending upon the storage conditions, ice crystals can grow, thus increasing storage or product degradation. One positive attribute for ingredient stabilization, regardless of the method of freezing, is that freezing retards bacterial and enzyme activity.

If extended shelf-life is desired (greater than 1 year), industrial studies have shown that starch alone does not maintain stability as well as when blended with a small amount of gum (hydrocolloid). For foods that are to be frozen, it is suggested that a minimum of 1.5% of a properly modified starch and approximately 0.01% of gum (xanthan) be used (formula weight). Actual amounts required will vary based on the product characteristics desired. Any starch can be used with proper modification. The desired source is up to the technologist and product parameters. It should be noted that neither crosslinking nor monosubstitution, as an individual modification, yield a stable frozen product. These modifications alone do not support extended freezer or freeze–thaw for the purpose of maintaining freshlike characteristics. To achieve these functional properties, the proper levels for both crosslinking and monosubstitution are required. As with other foodstuffs, the starch of choice must be determined based on several parameters. However, experience has shown that tapioca, wheat, and waxy maize starches are extremely stable in frozen foods, especially those anticipated for extended freezer life. Freeze–thaw can be accomplished with just about any properly modified starch. Waxy maize starch can yield exceptional stability when modified with the proper degrees of crosslinking and monosubstitution. As described in Section VI, the ultimate starch selection must include all facets of preparation and handling.

C. Instant Products

Most instant foods sold commercially today are dry mixes. These mixes thus require the consumer to prepare them by reconstitution. Usually this involves water or other basic liquids commonly sold, e.g., juices, milk, broth, etc. The reconstitution may or may not require the use of heat. The commercial process for manufacturing these dry mixes usually involves the blending of several ingredients prior to packaging. Starches used in these foods are typically of low moisture content. Instant starches are usually less than 5% moisture, and if cook-up starches are incorporated, they would most likely possess a moisture of less than 10% (8–15% based on mix weight).

Instant foods are produced commercially via spray-drying preblended mixtures or utilization of freeze drying, drum drying, or extrusion. Starches used in foods of this type generally contribute to the product functionality and characteristics. Viscosity, texture, stability, appearance, and eating quality can be affected by the incorporation of starch. Instant foods processed via spay-drying, drum drying, or extrusion may not support the use of added starches. The process may adequately gelatinize any natural starches within the mix.

Instant mixes formulated with instant modified starch usually require the addition of ingredients for dispersion. Such products may already be included in the product formula, but if not ingredients such as sucrose, dextrose, low-DE sweetener solids, maltodextrins, and/or flours can be used as dispersing aids. These will eliminate or significantly reduce the potential of lumping or ''fish eyeing,'' as it is sometimes called (Fig. 34). Some operations have utilized agglomeration as a method to reduce lumping. Another method is to request not only a modified instant starch, but one of coarser particle size. The larger particle reduces the surface area available for hydration and thus results in slower water uptake. It can produce a somewhat grainy texture compared to the other accepted procedures. Also the reverse is possible, utilizing a finer particle size to increase surface area and hydration rate.

1. Functional Differences Based on Process of Production

Instant starches available commercially today offer a vast range of viscosity and functional parameters for the food technologist. It is very difficult to attempt to prepare similar functional properties with instant starches processed via different production systems. Therefore, when attempting to secure two starches having similar functional characteristics, they should be acquired from similar production processes. As mentioned earlier regarding drum drying and extrusion, it is similar for spray-drying, freeze drying, or other physical modifications being incorporated today. As to the potential for genetically derived instant starches, they have not been developed to date, therefore we must wait for property comparison.

D. Snack Foods

Both native and modified starches are commonly utilized in snack products. They have been used, for example, for expansion capability, relating to final product texture.

1. Puffed Extruded Products

Common maize, tapioca, potato, wheat, and other grains are commonly used. Modified starches however, usually require specific modifications and are used to produce specific textures. These textures are as much related to the starch origin as well as the degree and type of modification incorporated. Monosubstituted and/or hydrolyzed starch generates unique expanded products.

2. Fried Products

Native or unmodified starches are not predominately used in these products. Generally, these snacks consist of modified starches and flavorings. Forming may be used depending upon the snack product. Those products formed are usually done so prior to frying. They are sometimes frozen prior to frying, but can also be frozen after *par frying*. Par frying is the process of only partially cooking the food item. For many of the fried snack products, blends of ingredients are used. Starches, both modified and native, as well as flours, dextrins, and low-DE sweeteners are common ingredients. Level of usage is dictated by texture, storage, and frying conditions. Typically though for simple coats (batters) 20 to 80% starch is used. More complex coatings can involve several starches at significantly differing levels (1–10%). All ingredients are chosen based on product quality and functional attributes to be contributed.

3. Baked, Microwaved and Impinged Air Processed Products

These terms can also be referred to as nonfried. For many snack products, this is one of the fastest growing market areas. For starch use, this too has been a significant growth area. Starches have been able to create the unique products required for this growth. Such products are call ''half-products.'' A half-product represents a formulation that for commercial use requires the consumer or commercial operation to complete the texture or functional properties incorporated into the product. An example could be an extruded particulate that requires baking or microwaving to complete the expansion, viscosity development, or crisping of the ultimate product for consumption. Depending upon texture and other functional properties, starch choice could range from common corn to a highly modified starch (origin optional based on function).

E. Dressings, Sauces, Gravies, and Other Condiments

Products such as dressings, sauces, gravies, etc., possess very high water content. This excess water usually demands stabilizers that are also excellent water binders. When formulating dressings always consider that you will be working with high acid conditions, and typically some mechanical shear will be implied at some point. For sauces, gravies, and other condiments, formulations are typically that of neutral pH as considered for starch use, however mechanical shear still may be involved. Therefore, knowledge of the process is necessary prior to selecting the correct starch. Many gravies and sauces are prepared utilizing the dry mix procedure initially, followed by heating for serving. In some instances, such as food service preparations, it is necessary to use modified starches capable of providing long periods of stability at elevated temperatures. Water control is very important, as well as the ability to reconstitute some foods after storage by refrigeration and/ or freezing overnight. Again, the choice of starch depends on the textural, eating, and process characteristics desired. Always refer back to the starch origin for basic properties. Usage could range from 1.5–4.0%, wet basis.

F. Bakery Products

In this section we discuss bakery systems more typical to baked products recognized today. Those previously mentioned as half-products were unique to that discussion and are not included here. Products considered are totally processed utilizing a bakery operation. In most bakery products the starch source is wheat flour. However, to develop and retain the unique properties in today's baked foods, the incorporation of modified starches and gums has taken place. Balance of moisture in differing systems has become a very difficult formulation task. Water activity and the knowledge of ingredient interaction has become a dominate challenge for the food scientist. This becomes more significant as the product undergoes mixing, baking, and storage. Now not only do the ingredients have to interact correctly, but packaging must also compliment the final product to retain quality.

Those baked products that are formulated as a dough require special handling when attempting to incorporate starch. Some may consider this a half-product, although for this writing we are considering dough as a finished mix or a transition phase. Moisture hydration must be considered, primarily due to the dough characteristics: mixing, sheeting, handling, etc. It is not just the total hydration capacity of the added starch, but also the rate of hydration that could be significant. Also important are the rate at which the starch loses moisture and what physical properties are generated after processing. Wheat starch is a starch for consideration, both as a native or modified source. It derives from wheat and therefore offers synergy with the flour. Tapioca and waxy maize have also performed very well in many baked foods. Depending upon how the baked item is to be handled, cook-up starches can be used. However, if boil-out or blow-out is a problem, either high-soluble solids are needed or utilization of cook-up and instant starches. Gums may contribute to controlling viscosity problems, but texture can become an issue if use levels are too great. The same can be said for high levels of starch.

As with any food system, always consider the other ingredients when selecting a starch. The fat and sweetener matrix can become very critical as related to total functionality produced by the added starch. As discussed earlier in Section VII, the type of fat, when it is added, and what it is added with can significantly contribute to the hydration and functionality of the starch. The same can be said regarding the sweetener system. It is important to balance the sucrose and flour for certain baked goods, but remember the effect that sucrose has on starch. Will this cause a problem with your starch of choice? It may be necessary to evaluate more than one sweetener or fat system.

G. Pet Products

Do not be fooled by the word *pet* when making considerations for product choices. Pet foods undergo as many regulatory issues as do food products for human consumption. There are few ingredients utilized within pet foods that we do not use in human food products. The care of formulation and product safety as well as the standards are very high for pet foods. Formulation, processing, and consumer satisfaction are taken very seriously.

Starch utilization is very similar to foods formulated and processed for humans. The canned foods are sterilized with similar standards, therefore requiring starches with similar food approval and functional characteristics. Those pet foods extruded or formed for snacks also require starch and ingredient mixtures requiring similar property considerations as for human products. In some instances, because of the different eating characteristics between humans and animals, standards for functionality may need to be altered from that for humans. Hardness or bite may be an issue. What a human may consider hard, a pet would not consider hard enough. This then requires animal eating, tasting, and quality studies for product analysis, similar to human feeding studies. The same holds true for diet contribution, digestion, health benefit, fecal discharge, and other physical and physiological issues. Starch as utilized in pet foods will be dictated by the basic functional properties of the native starch and the acceptance of it by the consumer, in this case the pet.

H. Meat Products

Starch has recently gained approval for general use in many meat products. Most meat items commercially produced had very strict standards of identity. For the improvement of storage and process, starch and related products were approved for use. The use of starch (modified) in meat products is regulated by the United States Department of Agriculture's Food Safety and Inspection Service. Specifics are available in Title 9, Part 318.7 of the Code of Federal, Regulations.

Prior to the approval of starch and related products the limitation of water management in many meat products was controlled through the use of high levels of salt. In meats such as frankfurters, bologna, luncheon loafs, etc., starch contributes a method to reduce purge (free water/brine). Flour and other grain-based ingredients have been evaluated, none possessing the water absorption and control as compared to modified starch.

In products of seafood derivatives, such as surimi (seafood analogs), starch contributes more than water management. Texture, process improvement, and most significantly economic advantage were achieved with added starch. It was determined that not only could modified starch be used, but significant levels of native starch also. Unmodified tapioca, potato, and wheat are commonly used in conjunction with one or more modified starches. Modified waxy maize, tapioca, and potato are typical starches of choice. Again, the desired product and texture determines the starches utilized. As can be seen, this type of meat analog is a complex formulation of starches. It took several evaluations to determine the ideal blend for the production of a commercial analog.

I. Cereals and Related Products

Cereals today utilize predominately whole grains, with or without added starch. Many cereals depend on the expansion capability of the flour(s) used to develop and maintain the texture and eating quality of the finished product. Referring back to the sections pertaining to extrusion and instant starches, we can relate them to the contribution of starch to cereals. Many cereals utilize extrusion as the method of preparation. Partial to total gelatinization is developed during the process of cereal production. Therefore, the process effect as well as the functional properties will be generated during and after processing by the starch within the flour or grain used. Today, cereal producers utilize added starch to enhance texture, processing, and functional properties with the finished cereal. In addition, several of the cereal products commercially available today are coated with an individual ingredient or a complex blend, thus requiring the cereal to have added functional attributes.

A property that many cereal producers have researched for years is that of extended bowl life. A property that offers longer crisp texture to a cereal when mixed with a liquid, usually milk. The coating of cereal pieces has been an accepted method for years, however the desire for a cereal product requiring no coating and equally stable still exists. Improvement has been achieved, but none to date match what can be accomplished with a coating.

The coating process has now led to another line of cereallike products. It has also created an extension to the snack industry with cereallike products: the granola or whole grain coated snacks. We have bars, clusters, and bits for our enjoyment, all a step from the cereal itself.

Wheat, oat, barley, and rice flours are popular starters for cereals. Wheat, oat, rice, tapioca, and potato starches are commonly used in many cereals as well as the cereallike snacks.

J. Confections and Candy

When discussing confections and candy products (51) the food technologist should consider the different selections of confections there are. Confections consist of three categories: (1) soft, (2) gummy (chewable), and (3) hard (nonchewable) candies.

Soft candies are those products such as chocolate-based or flavor-coated products (e.g., circus peanuts) that may or may not be chewable. Gummy type candy include gum drops, jelly beans, etc. The hard candies would be those confections such as cough drops, lozenges, etc.

Starches used for these types of confections differ dramatically. The soft candies usually require starches tender in texture or nongelling. Some could require little heat stability and total enzyme degradation. This would be for a chocolate covered cherry type product. Others may be used in caramel or fruit centers.

The starches for the gummy and hard candies are similar but differ with degree of modification and are blended as necessary to accomplish the texture. Gummy candies are produced using hydrolyzed starches (refer to Section III.A). In some instances lightly modified or unmodified high-amylose starch may be utilized. For the production of hard candies this type of starch would be used because of the set generated by the greater percentage of amylose (refer to Sections II.A and II.B).

In all confections the food scientist will have to work with a sweetener. It could be sucrose or more likely it will be some type of liquid sweetener derived from a carbohydrate modification. It is probable that it would be a corn syrup, as liquid sweeteners derived from other starch bases are not as commercially available and do not offer quite the economics as yet. Solids for most confections are in excess of 80%, thus requiring elevated or greater than atmospheric conditions to process. To develop the textures desired for the gummy or hard candies starch is typically added at a 6–13% level. The added starch could be either a single unit or more likely a blend of starches to achieve the desired results. Flavors and colors are usually minimal in these formulations.

Only in confections being marketed today as low or no sugar is there much variance from these levels. However, in such low calorie products the food scientist has the challenge of balancing excess water with other carbohydrates and hydrocolloids that will, it is hoped, duplicate the original product. In many cases it is easier with the no sugar claim than with the low or no fat claim. The area of fat reduction will be discussed a little later in the chapter.

K. Dairy and Related Products

Dairy foods are those foods that contain all or a high percentage of dairy-derived ingredients and meet a given set of criteria for a standard of identity. Dairy-related products are those products usually sold in the dairy section of the commercial outlet and although they have been formulated to simulate a standard dairy product, they do not meet the identity standards for dairy products. Many of these products contain dairy products as a portion of their ingredient base. Starches are widely used in the dairy related products and only recently have some standards changed to allow their use.

Dairy products such as puddings, yogurts, toppings, cheese, sour cream, and ice cream can utilize starch. For many years starch-related products such as the sweeteners, maltodextrins, and dextrins have been accepted for use.

In puddings, yogurts, and sour cream, starches are added to stabilize the final product. Most of the time it is the management of the water phase that is critical for the stabilization. In the yogurt, starch could be in the dairy portion and/or in a fruit portion, depending on the type of yogurt product. The sour cream products require added starch for use in food service facilities. Many foods are served hot and natural sour cream does not possess sufficient heat stability to not cause a serving issue. In the cheese industry added starch is used to either improve slicing, shelf stability, or package release. Except for package release, a modified starch is required. Considering that most dairy products are neutral in flavor or rather sensitive flavors are at issue, starches of not strong overtones are desired. Those products requiring heat and or process stability should be crosslinked accordingly.

In dip, sauces, or dairy condiments it may be possible to utilize instant starches. Depending upon the texture and manufacturing process involved the finished food item will determine the starch of choice. Remember to consider appearance and dispersion when selecting—a granular instant may be preferred.

L. Starch as Fat Replacer, Substitute, or Mimetic

As we saw great interest in the 1980s for the inclusion of bran in foods, the 1990s brought the reformulation of foods to reduce the content of fat. Thus arose the need for specific ingredients to either totally replace the fat in a food or substitute or mimic the properties of part of the fat.

The starch and protein industries responded with an abundance of so-called fat replacers. Unfortunately, only the pourable salad dressing industry was lucky enough to quickly and successfully institute a complete line of fat free products. The rest of the food products being researched to totally replace fat have fallen well short of success. A few baked products have acceptable but not highly recognized fat free products in the marketplace. There may be a few, but as a whole the attempt for a single ingredient fat replacer is at this time unsuccessful. The industry soon realized that complex formulations were required and most likely a two or three ingredient mix was required to somewhat simulate the removed fat (Tables 12 and 13).

This holds true for fat free foods, however as we move to new definitions for low fat and reduced fat, the research community has been very successful with fulfilling new

Table 12 Liquid Sweeteners: Representative Chemical and Physical Data

Degree of conversion	Very low	Low	Regular	Regular	Intermediate	Intermediate	High	High
Type of conversion	Acid–enzyme	Acid	Acid	Acid–enzyme	Acid–enzyme	Acid	Acid–enzyme	Acid–enzyme
Dextrose equivalent (DE), %	26	35	43	42	50	54	63	63
Dextrose (monosaccharides), %	5	13	19	9	10	27	37	37
Maltose (disaccharides), %	8	10	14	34	42	22	29	29
Maltotriose (trisaccharides), %	11	11	13	24	22	15	9	9
Higher saccharides, %	76	66	54	33	26	36	25	25
Baume at 100°F, degrees	42	43	43	43	43	43	43	44
Total solids, %	77.5	80.0	80.3	80.4	80.7	81.0	81.6	83.6
Moisture, %	22.5	20.0	19.7	19.6	19.3	19.0	18.4	16.4
pH	4.7	4.7	4.7	4.7	4.7	4.7	4.7	4.7
Viscosity, Poises at 100°F	220	200	125	125	90	75	55	155
Boiling point, °F	222	226	227	227	228	229	233	234
Weight, lb/gal at 100°F	11.70	11.81	11.81	11.81	11.81	11.81	11.81	11.93

Note: All values are typical, they should not be construed as specifications.

Table 13 Commercial Starches Utilized as Fat Replacers

Product	Source of starch/modification	Company of origin
Amalean I	High amylose corn/modified	Cerestar USA
Amalean II	High amylose corn/modified/instant	Cerestar USA
Frigex	Tapioca/modified	National Starch
Gel N Creamy	Tapioca/modified	National Starch
Instant Pure Flo	Waxy corn/modified/instant	National Starch
Instant Stellar	Common corn/modified/instant	A.E. Staley
Instant W-11	Waxy corn/modified/instant	Cerestar USA
Leanbind	Corn/modified	National Starch
Mira Thick 468	Waxy corn/modified/instant	A.E. Staley
N-Lite CM	Tapioca/modified	National Starch
N-Lite L	Waxy corn/modified	National Starch
N-Lite LP	Waxy corn/modified/instant	National Starch
Paselli BC	Potato/modified	Avebe
Pure Gel B-990	Corn/modified	GPC
Remygel	Rice	Remy Industries
Remyline	Rice	Remy Industries
Slenderlean	Tapioca/modified	National Starch
Starch Plus SPR	Rice	California National Products
Starch Plus SPW	Waxy rice	California National Products
Sta-Slim 142	Potato/modified/instant	A.E. Staley
Sta-Slim 143	Potato/modified	A.E. Staley
Sta-Slim 150	Tapioca/modified/instant	A.E. Staley
Sta-Slim 151	Tapioca/modified	A.E. Staley
Sta-Slim 171	Waxy corn/modified	A.E. Staley
Stellar	Corn/modified	A.E. Staley

Table 14 Commercial Maltodextrins Used for Fat Replacers

Product	Native Starch Source	Company offering Product
Dairytrim	Oat flour	Quaker Oats/Rhone Poulenc
Instant N-Oil II	Tapioca	National Starch
Maltrin M040	Common corn	GPC
N-Lite B	Waxy corn	National Starch
N-Lite D	Common corn	National Starch
Novadex 120–01	Oat flour	National Oats
Novadex 120–10	Oat flour	National Oats
Oatrim 5	Oat flour	Quaker Oats/Rhone Poulenc
Oatrim 5Q	Oat flour	Quaker Oats/Rhone Poulenc
Paselli SA2	Potato	Avebe
Rice Trim	Rice flour	Zumbro
Star Dri-1	Common corn	A.E. Staley
Trimchoice 5	Oat flour	ConAgra/A.E. Staley
Trimchoice OC	Oat/corn flour	ConAgra/A.E. Staley

formulated foods. The technology for many of the starch products utilized as fat substitutes or mimetics was developed from that used for production of starches used for emulsification and encapsulation. Refer to the discussion in Section III. A. 3.

M. Emulsification and Encapsulation

Starches specifically modified for emulsification offer the food scientist the capability to utilize starch not as a water binder or stabilizer, but rather as an ingredient to form a noncontinuous phase within which is a matrix of water, lipid, and starch. Typically, if the food product is to be water phase stable an additional starch or gum will be required. Starches usually referred to as alkenylsuccinates are successful in these applications. Emulsifying starches also generally do not develop significant viscosity and are limited in their heat tolerance. Basically what these starches support are surface reactions and not an internal absorption (Fig. 35).

Whereas with products referred to as encapsulating starches, a different physical effect takes place. Encapsulation refers to the entrapment of lipids (oils) with starch and/or related products (Fig. 36). Significant success for encapsulation has been noted and commercialized through the utilization of octenylsuccinic acid anhydride reacted starch (OSA). This treatment is also FDA approved (Table VI). It was through the inclusion of these starches and gums that the pourable dressing industry was successful with the development of low and fat free products. Starches derived from oxidization and dextrinization are also utilized for this functional property. Typically though one must always consider the total characteristics produced from a reaction and the ultimate effect it will have on the finished food product. Flavor, color, etc., should always be evaluated.

One additional mechanism has been identified as significant for utilizing starch as an encapsulating agent. It was discovered that the inclusion of polyvalent metal ions (magnesium, calcium) produced a water repellent starch when reacted with an OSA-treated starch (52). It was then shown that this related to a slower release of flavor from encapsulated oils. Usage can range from 10–30% based on oil weight.

Starches used for encapsulation can vary. The food scientist must consider the volatility of the flavors and then determine what starch base would be correct. For those non-volatile compounds a maltodextrin may suffice as a carrier. However, this would be a surface coating and not an encapsulation. This is done quite commonly. For the volatile flavors, a high-amylose starch reacted with OSA may be of choice, taking advantage of the quick setting properties of high-amylose starch and the encapsulating effect of the reactant.

REFERENCES

1. Irvine, J. C., MacDonald, J. 1926. The constitution of polysaccharides. Part X. The molecular unit of starch. *J. Chem. Soc.* 1502.
2. Meyer, K. H., Bernfeld, P. 1940. Recherches sur l'amidon. V. L.' amylopectine. *Helv. Chim. Acta* 23:875–885.
3. Schoch, T. J. 1941. Physical aspects of starch behavior. *Cereal Chem.* 18(2):121–128.
4. Manners, D. J. 1979. The enzymic degradation of starches. In: *Polysaccharides in Food.* Blanshard, J. M. V., Mitchell, J. R. (Eds.). Butterworths, London, pp. 75–91.
5. Whistler, R. L., Paschall, E. F. 1967. *Starch: Chemistry and Technology*, Vol. II. Acamedic Press, New York; Whistler, R. L., BeMiller, J. N., Paschall, E. F. 1984. *Starch: Chemistry and Technology*, 2nd ed, Academic Press, New York.

6. Radley, J. A. 1976. *Industrial Uses of Starch and Its Derivatives*. Applied Science Publishers, London.
7. Schoch, T. J. 1967. Mechano-chemistry of starch. *J. Technol. Soc. Starch* (Japan) 14(2–3): 53.
8. Wurzburg, O. B. 1972. Starches in the food industry. In: *Handbook of Food Additives*, 2nd ed., Furia, E. (Ed.). CRC Press, Cleveland, OH, pp. 361–395.
9. Radley, J. A. 1976. *Starch Production Technology*. Applied Science Publishers, London.
10. Greenwood, C. T. 1976. Starch. In: *Advances in Cereal Science and Technology*, Vol. 1, Pomeranz, Y. (Ed.). American Association of Cereal Chemists, St. Paul, MN, pp. 119–157.
11. French, D. 1975. Chemistry and biochemistry of starch. In: *Biochemistry of Carbohydrates*, Whelan, W. J. (Ed.). Butterworths, London, pp. 267–335.
12. Greenwood, C. T. 1970. Starch and glycogen. In: *The Carbohydrates: Chemistry and Biocheemistry*, 2nd ed., Pigman, W., Horton, D. (Eds.). Academic Press, New York, pp. 471–513.
13. Evers, A. D. 1979. Cereal starches and porteins. In: *Food Microscopy*, Vaughan, J. G. (Ed.). Academic Press, New York, pp. 139–191.
14. French, D. 1983. Physical and chemical organization of starch granules. In: *Starch: Chemistry and Industry*, Whistler, R. L., Paschall, E. F., BeMiller, J. N. (Eds.). Academic Press, New York.
15. Sterling, C. 1964. Starch-primilin flourescence. *Protoplasma* 59:180–192.
16. Betchel, D. B., Pomeranz, Y. 1981. Ultrastructure and cytochemistry of mature oat endosperm: the aleurone layer and starchy endosperm. *Cereal Chem*. 58:61–69.
17. Hizukuri, S., Takeda, Y., Yasuda, M., Suzuki, A. 1981. Multibranched nature of amylose and the action of debranching enzymes. *Carbohydrate Res*. 96(2):143–159.
18. Blanshard, J. M. V. 1979. Physicochemical aspects of starch gelatinization. In: *Polysaccharides in Food*, Blansheard, J. M. B., Mitchell, J. R. (Eds.). Butterworths, London, pp. 139–152.
19. Colona, P., Mercier, C. 1984. Macromolecular structure of wrinkled and smooth-pea starch components. *Carbohydrate Res*. 131:117.
20. Hizukuri S., Takeda, Y., Yasuda, M., Suzuki, A. 1981. Multibranched nature of amylose and the action of debranching enzymes. *Carbohydrate Res*. 94:205
21. Takeda, Y., Hizukuri, S., Juliano, B. O. 1986. Purification and structure of amylose from rice starch. *Carbohydrate Res*. 148:299.
22. Takeda, Y., Shitaozono, T., Hizukuri, S. 1988. Molecular structure of corn starch. *Starke* 40: 51.
23. Nikuni, Z. 1978. Studies on starch granules. *Starke* 30:105.
24. French, D. 1984. Organization of starch granules. In: *Starch: Chemistry and Technology*, 2nd ed., Whistler, R. L., BeMiller, J. N., Paschall, E. F. (Eds.). Academic Press, New York, p. 184.
25. Lineback, D. R. 1984. The starch granule: organization and properties. *Bakers Digest* 58(2): 16.
26. Biliaderis, C. 1992. *Food Technol*. 46:98.
27. Robin, J. P., Mercier, C., Charbonniere, R., Guilbot, J. A. 1974. Lintnerized starches: gel filtration and enzymatic studies of insoluble residues from prolonged acid treatment of potato starch. *Cereal Chem*. 51:389.
28. Jane, J., Xu, A., Radosavljevic, M., Seib, P. A. 1992. Location of amylose in normal starch granules. I. Susceptibility of amylose and amylopectin to cross-linking reagents. *Cereal Chem*. 69:405.
29. Morrison, W. R. 1989. Uniqueness of wheat starch. In: *Wheat is Unique*, Pomeranz, Y. (Ed.). American Association of Cereal Chemists, St. Paul, MN, p. 193.
30. Lim, S. T., Kasemsuwan, T., Jane, J. 1994. Characterization of phosphorus in starch by ^{31}P-nuclear magnetic resonance spectroscopy. *Cereal Chem*. 71:488.
31. Atwell, W. A., Hood, L. F., Lineback D. R., Varriano-Marston, E., Zobel, H. F. 1988. The

terminology and methodology associated with basic starch phenomena. *Cereal Foods World* 33:306–311.
32. Stevens, D. J., Elton, G. A. 1971. Thermal properties of starch/water system. *Starke* 23:8–11.
33. Donovan, J. W. 1977. A study of the baking process by differential scanning calorimetry. *J. Sci. Food Agric.* 28:571–578.
34. Wooton, M., Bammarachchi, A. 1979. Application of differential scanning calorimetry to starch gelatinization. *Starke* 31:262.
35. Kugimiya, M., Donovan, J. W., Wong, R. Y. 1980. Phase transitions of amylose–lipid complexes in starches: a calorimetric study. *Starke* 32:265–270.
36. Eliasson, A. C., Carlson, T. L. G., Larsson, K., Miezis, Y. 1981. Some effects of starch lipids on the thermal and rheological properties of wheat starch. *Starke* 33:130–134.
37. Kugumiya, M., Donovan, J. W. 1981. Calorimetric determination of the amylose content of starches based on formation and melting of the amylose–lysolecithin complex. *J. Food Sci.* 46:765–770,777.
38. Schoch, T. J. 1965. Starch in bakery products. *Bakers Digest* 39(2):48–57.
39. Miller, B. S., Derby, R. I., Trimbo, H. B. 1973. A pictorial explanation for the increase in viscosity of a heated wheat starch suspension. *Cereal Chem.* 50:271.
40. Osman, E. M. 1967. Starch in the food industry. In: *Starch Chemistry and Technology*, Vol. II, Whistler, R. L. Paschall, E. F., (Eds.). Academic Press, New York, pp. 163–215.
41. Wurzburg, O. B. 1986. In: *Modified Starches: Properties and Uses*, Wurzburg, O. B. (Ed.). CRC Press, Boca Raton, FL, pp. 18–40.
42. de Graff, R. A., Lammers, G., Janssen, L. P. B. M., Beenackers, A. A. C. M. 1995. *Starke* 47:469–475.
43. Ostergard, K. Bjork, I., Gunnarsson, A. 1993. *Starke* 40:58–66.
44. Xu, A., Seib, P. A. 1997. *J. Cereal Science* 25:17–26.
45. Pitchon, E. P., O'Rourke, J. D., Joseph, T. H. 1986. Apparatus for cooking or gelatinizing materials. U. S. patent 4,600,472.
46. Bjork, I. 1996. In: *Carbohydrates in Food*, Eliasson, A. C. (Ed.). Marcel Dekker, New York, pp. 505–553.
47. Englyst, H. N., Kingman, S. M., Hudson, G. J., Cummings, J. H. 1996. *Br. J. Nutrition.* 75: 749–755.
48. Eerlingen, R. C., Delcour, J. A. 1995. *J. Cereal Science* 22:129–138.
49. Pomeranz, Y. 1992. *Eur. J. Clin. Nutr.* 46 (2):S63–S68.
50. Gough, B. M., Pybus, J. N. 1971. Effect on the gelatinization temperature of wheat starch granules of prolonged treatment with water at 50°C. *Starke* 23:(6) 210–212.
51. Luallen, T. E. 1988. Structure, characteristics, and uses of some typical carbohydrate food ingredients. *Cereal Foods World* 33(11):924–927.
52. Zallie, J. *The Role and Function of Speciality Starches in the Confection Industry*. A National Starch and Chemical Publication.
53. Wurzburg, O. B., Cole, H. M. U.S. patent 3,091,567.
54. Alexander, R. J. 1995. Fat replacers based on starch. *Cereal Foods World* 40(5):366–368.

25

Food Phosphates

LUCINA E. LAMPILA

Albright & Wilson Americas, Inc., Rhodia, Cranbury, New Jersey

JOHN P. GODBER

Albright & Wilson Canada, Ltd., Rhodia, Aubervilliers, France

The purpose of this review is to focus on the most recent reports in the literature and trends in food processing which involve the food phosphates. Excellent reviews of phosphorus chemistry have been published by Van Wazer (1958), Toy (1976), and Kirk-Othmer (1982). Ellinger (1972a,b) and Molins (1991) have reviewed the applications of the food grade phosphates in great detail. The reader is invited to enjoy the previous reviews as well as the references cited herein for further, more detailed information.

I. PHOSPHATE CHEMISTRY RELEVANT TO FOODS

The goal of this section is to offer an overview of the chemistry that is relevant to the use of phosphates in food systems. Those seeking to develop new uses for phosphates should find some useful information in the references. This discussion does not examine phosphorus chemistry in general since it has been covered in excellent works by Van Wazer (1958) and Corbridge (1986). Reviews covering food phosphate chemistry can also be found in Molins (1991), Ellinger (1972a,b) and DeMan and Mehnychyn (1971). Nomenclature is covered since it is often archaic with many names referring to the same chemical. Phosphates are almost exclusively used in aqueous media and for this reason, the rate of dissolution and the hydrolysis of phosphates are discussed. Sequestration of metal ions by condensed phosphates is described. Methods of analysis are also reviewed.

A. Nomenclature

Phosphates are defined and distinguished from other phosphorus-containing molecules as those in which the central phosphorus atom is surrounded by four oxygen atoms. The oxygen atoms spatially occupy a structure resembling a tetrahedron with the oxygen atoms at the corners. Sharing of oxygen atoms between two phosphorus atoms gives rise to a great number of different phosphate types. For mainly historical reasons, different names have often been given to the same chemicals. Attempts to improve the naming systems have not always met with universal acceptance but are still used. For these reasons, Table 1 is offered indicating the different names and which compounds they chemically refer to. Ideally, phosphate nomenclature provides information on (1) the structure of the phosphate and (2) the type and number of charge balancing cations.

The structural arrangement of the phosphate tetrahedron forms the basis of its nomenclature. Phosphates can form structures containing from one to hundreds or even thousands of phosphate tetrahedra. Structurally, the simplest is orthophosphate containing one tetrahedron. Joining two orthophosphates together produces a diphosphate. Diphosphates of this type are also commonly known as pyrophosphates. Continuing in the same vein leads to, sequentially, triphosphate, tetraphosphate, etc. Triphosphates, or tripolyphosphates as they are also known, are the longest ''pure'' phosphate oligomer commercially available. Higher polymers containing the phosphate tetrahedron are commercially sold as mixtures. For example, sodium hexametaphosphate is actually a mixture of polyphosphates containing from about 13 up to 35 phosphate tetrahedra linked in a linear fashion. If a polyphosphate is joined back on itself to form a ring, it is referred to as a true metaphosphate (in contrast to hexametaphosphate which retains this name today for historical reasons). Trimetaphosphates are available commercially. One other phosphate structural type is also known, the ultraphosphates. These are phosphate anions which contain a structural unit in which a phosphate tetrahedron is linked to three other phosphate tetrahedra. It is a branched phosphate structure.

There are then structurally only four phosphate types: (1) orthophosphates which are not attached to any other phosphate tetrahedron; (2) polyphosphates (including pyrophosphates) in which a phosphate tetrahedron is linked to at most two other phosphate tetrahedra; (3) true metaphosphates in which the polyphosphate joins back upon itself to form a ring; and (4) ultraphosphates which contain at least one phosphate tetrahedron joined to three others. Polyphosphates, metaphosphates, and ultraphosphates are all prepared from orthophosphates by heating to drive off water. For this reason, they are called, as a group, condensed phosphates.

Composed of (formally) pentavalent phosphorus (P^{+5}) and divalent oxygen (O^{-2}) the phosphates are negatively charged and require charge-balancing cations. These can be protons (H^+), metal ions, or molecular ions like ammonium. Phosphate nomenclature then includes the type of cation and a code describing the type of phosphate structure. For example, potassium tripolyphosphate is a polyphosphate containing three phosphate tetrahedra and the phosphate anion is charge balanced by potassium ions. Other common terms are shown in Table 1.

Calcium phosphates are a bit confusing since the calcium ion is doubly positive. The naming for this orthophosphate is a carryover from the past when phosphates were expressed as an oxide and related to its P_2O_5 content. Monocalcium phosphate contains one calcium oxide molecule for each P_2O_5 moiety; it is $CaO:P_2O_5:2H_2O$ or $Ca(H_2PO_4)_2$.

Dicalcium phosphate is $2CaO:P_2O_5:H_2O$ or $CaHPO_4$, containing two molecules of calcium oxide for each of P_2O_5.

B. Production Methods

The production of phosphates begins with phosphoric acid (Ullmann's, 1991; Kirk-Othmer, 1982). The acid itself is made by one of two routes. ''Thermal'' phosphoric acid is made by burning phosphorus (P_4), which is itself prepared by smelting phosphate ore (shown as $Ca_x[PO_4]_y$ in Reaction [1]) in an electric furnace using coke (C in Reaction [1]) as a reducing agent. Silica (SiO_2) is added as a flux to lower the melting point of the mixture.

$$Ca_x[PO_4]_y + C + SiO_2 \rightarrow CaSiO_3 + P_4 + CO \qquad [1]$$

The phosphorus is then burned in specially designed torches in an excess of air and quenched with water, yielding phosphoric acid:

$$P_4 + O_2 \rightarrow \text{``}P_2O_5\text{''} \qquad [2]$$

$$\text{``}P_2O_5\text{''} + H_2O \rightarrow H_3PO_4 \qquad [3]$$

Since phosphate rock naturally contains arsenic and the chemistry of arsenic compounds are similar in many respects to phosphorus the arsenic is carried over into the acid. To bring it up to food grade quality, the arsenic can be removed using various methods.

''Purified wet'' phosphoric acid starts from fertilizer grade phosphoric acid made by reacting phosphate rock with sulphuric acid. This is purified using solvent extraction methods in which essentially pure phosphoric acid is extracted into an organic solvent and then in a subsequent step released into an aqueous phase. The impurities remain in the fertilizer grade acid. Both the thermal process and purified wet process result in the production of food grade phosphoric acid.

To produce phosphates from phosphoric acid, reaction with an alkali is required. The alkali can be any of a large number of different reagents. The following list gives an idea of the variety of alkali sources that can be used: NaOH, Na_2CO_3, NaCl, KOH, K_2CO_3, KCl, CaO, $Ca(OH)_2$, $CaCO_3$, and NH_3. Depending on the source of the alkali raw material, impurities common to its origin can be carried through into the phosphate. Food grade alkali sources are used to make food grade products.

The product from the neutralization of phosphoric acid with alkali is then usually dried to give an orthophosphate. While the orthophosphates are themselves available commercially, they are also the raw materials for subsequent processing to give condensed phosphates. The mechanisms by which these reactions occur can be complex (e.g., in the case of STPP), but all involve the removal of the constituents of water from the orthophosphates; for example, the synthesis of tetrapotassium pyro-(or di-) phosphate from dipotassium phosphate is

$$2\ K_2HPO_4(s) + \text{Heat} \leftrightarrow H_2O(g) + K_4P_2O_7(s) \qquad [4]$$

The fact that the synthesis of condensed phosphates requires significant amounts of heat to effect the transformation immediately alerts one to the fact that they are thermodynamically unstable to decomposition by water. This will be discussed in the section on hydrolysis.

Table 1 Nomenclature, Acronyms, and Chemical Formulae of Phosphates Used in Food Applications

Common names	Acronyms	Formulas	pH	Solubility
Phosphoric acid	PA	H_3PO_4	Very low	High
Orthophosphoric acid				
Monoammonium phosphate	MAP	$NH_4H_2PO_4$	4.5	38 (20°)
Ammonium phosphate monobasic				120 (80°)
Monoammonium dihydrogen phosphate				
Primary ammonium phosphate				
Diammonium phosphate	DAP	$(NH_4)_2HPO_4$	8.0	69 (20°)
Ammonium phosphate dibasic				97 (60°)
Diammonium hydrogen phosphate				
Secondary ammonium phosphate				
Monocalcium phosphate monohydrate	MCP-1	$Ca(H_2PO_4)_2 \cdot H_2O$	3.4	Disproportionates
Calcium phosphate monobasic				
Acid calcium phosphate				
Monocalcium phosphate anhydrous	MCP-O	$Ca(H_2PO_4)_2$	3.4	Disproportionates
Coated monocalcium phosphate	cAMCP			
Dicalcium phosphate anhydrous	DCP-O	$CaHPO_4$	7.5	Insoluble
Calcium phosphate dibasic	DCPA			
Calcium hydrogen orthophosphate				
Dicalcium phosphate duohydrate	DCP-2	$Ca(HPO_4) \cdot 2H_2O$	7.5	Insoluble
Tricalcium phosphate	TCP	$Ca_5(PO_4)_3(OH)$[a]	7.2	Insoluble
Calcium phosphate tribasic				
Precipitated calcium phosphate				
Synthetic hydroxy apatite				
Ferric orthophosphate	—	$FePO_4$[a]	3.8–4.4	Insoluble
Monopotassium phosphate	MKP	KH_2PO_4	4.4	20 (20°)
Potassium phosphate monobasic				
Monopotassium dihydrogen phosphate				
Potassium acid phosphate				
Potassium biphosphate				
Dipotassium phosphate	DKP	K_2HPO_4	9.5	120 (10°)
Potassium phosphate dibasic				260 (50°)
Dipotassium hydrogen phosphate				
Dipotassium acid phosphate				
Secondary potassium phosphate				
Tripotassium phosphate	TKP	K_3PO_4	12	51 (20°)

Potassium phosphate tribasic				
Hemisodium phosphate	HSP	$NaH_2PO_4 \cdot H_3PO_4$	2.2	High
Monosodium phosphate	MSP	NaH_2PO_4	4.4	85 (20°)
Sodium phosphate monobasic				212 (80°)
Monosodium dihydrogen phosphate				
Disodium phosphate	DSP, DSP-2,	Na_2HPO_4	8.8	7.7 (20°)
Sodium phosphate dibasic	DSP-7			93 (80°)
Disodium hydrogen phosphate				
Trisodium phosphate dodecahydrate	TSP-12	$Na_3PO_4 \cdot 1/4NaOH \cdot 12H_2O$	12	13 (20°)
Sodium phosphate tribasic				
Trisodium phosphate anhydrous	TSP-0			
Sodium aluminum phosphate, acidic	SALP	$NaAl_3H_{14}(PO4)_8 \cdot 4H_2O$	2.4–2.7	Slight
		$Na_3Al_2H_{15}(PO_4)_8$		
Calcium acid pyrophosphate	CAPP	$CaH_2P_2O_7$	3	Slight
Calcium pyrophosphate	—	$Ca_2P_2O_7$	6.0	Insoluble
Calcium diphosphate				
Dimagnesium phosphate	DMP	$MgHPO_4 \cdot 3H_2O$	6.5	Slight
Tetrapotassium pyrophosphate	TKPP	$K_4P_2O_7$	10.4	180 (20°)
Potassium diphosphate				260 (50°)
Sodium acid pyrophosphate	SAPP	$Na_2H_2P_2O_7$	4.2	12 (20°)
Acid sodium pyrophosphate	ASPP			22 (60°)
Disodium dihydrogen diphosphate				
Disodium pryophosphate				
Tetrasodium pyrophosphate	TSPP	$Na_4P_2O_7$	10.2	6 (20°)
Sodium diphosphate				50 (80°)
Potassium tripolyphosphate	KTPP	$K_5P_3O_{10}$	9.6	178 (20°)
Pentapotassium triphosphate				216 (80°)
Pentapotassium tripolyphosphate				
Sodium tripolyphosphate	STPP	$Na_5P_3O_{10}$	9.8	15 (20°)
Pentasodium triphosphate				17 (80°)
Pentasodium tripolyphosphate				
Potassium metaphosphates (Kurrol's salt)	—	$(KPO_3)_n$	4–8	Insoluble
Sodium hexametaphosphate (Graham's salt)	SHMP	$(NaPO_3)_n$	7.0	High
Sodium trimetaphosphate	STMP	$Na_3(PO_3)_3$	6.7	30 (35°)

Note: The pH values are for a 1% solution or, for insoluble materials, a 10% slurry. The solubility data are in grams of phosphate per 100 g of water. MCP reacts with water leading to phosphoric acid and DCP.

[a] Approximate chemical composition.

C. Solubility in Water

The generally accepted values for the solubility of the different phosphates are collected in Table 1. Some condensed phosphates are difficult to obtain in solution at the maximum solubility level due to both slow solubility and hydrolysis, which can give the appearance of an insoluble phosphate due to precipitation of orthophosphates. Since phosphates are seldom used by themselves in pure water, the values shown in Table 1 are acceptable for most applications.

The rate at which a phosphate dissolves in water is important for practical, commercial reasons; food applications generally occur in aqueous solutions, while phosphates are almost always sold as solids. Two examples will serve to illustrate this concept. First, control over their rate of dissolution is a useful feature of the use of phosphate-based acids for chemically leavened doughs. This is accomplished either by barrier methods operating on quickly dissolving leavening acids (MCP, SAPP) or by producing poorly soluble phosphate salts (SALP, DMP). Second, the *method* of preparation of a phosphate solution is of great importance in preparing brines since the phosphate is usually the most difficult compound to dissolve. It is usually recommended that the phosphate be the first thing dissolved in water before any of the other ingredients. Due to solubilization difficulties, a number of phosphate manufacturers have focused on producing rapidly dissolving phosphates and phosphate blends for food uses. This is accomplished by either increasing the surface area of the phosphate that is available to dissolve in water or by increasing the thermodynamic driving force to dissolve, or both. Both these aspects are discussed in more detail in the next section.

1. STPP Dissolution

Improving the dissolution rate of STPP has been an important advance in the use of STPP. This is the most poorly soluble sodium phosphate and among the most important in food uses. STPP can be the slowest useful food phosphate to dissolve, but by increasing the thermodynamic driving force to dissolution, this situation can be improved. Two anhydrous crystalline phases of STPP exist with differing heats of solution (the amount of energy released when dissolved), a high temperature crystalline phase, denoted I, and a lower temperature phase, II (see Fig. 1). In commercially available STPP for food use, there is generally a mixture of the two different crystalline phases. Phase I releases a greater amount of heat during its reaction and dissolution in water than does phase II. It also dissolves more rapidly, which is thought to be a direct consequence of the environment of some of the sodium ions in the phase I crystal. By adjusting the production conditions, the proportion of the different phases can be controlled (see Fig. 2).

The proportion of phase I crystals in STPP can be determined by at least a few methods. The most accurate, but most resource intensive, is via X-ray diffractive methods. Rapid estimations of the phase I content can be achieved by either an infrared spectroscopic method (AWA, 1997) or by measurement of the amount of heat released by the phase I fraction (McGilvery, 1953).

2. Dissolution of Other Phosphates

While they have very high maximum solubility, the rate of dissolution of long chain polyphosphates is generally slow. High shear mixers can rapidly dissolve SHMP solutions with minimal heating of the solution, which leads to hydrolysis. The rate of dissolution

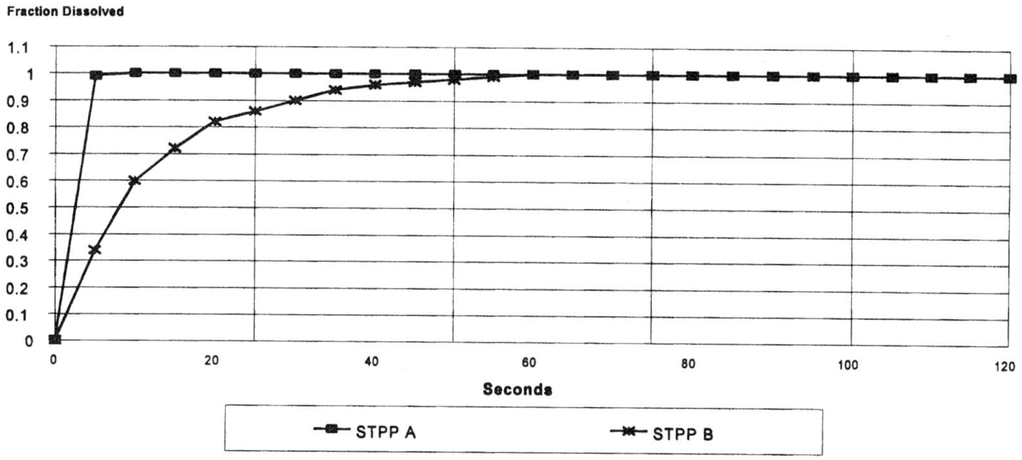

Figure 1 A comparison of the rates of dissolution of typical food grade STPP to that of a rapidly dissolving STPP at 5°C. The difference in the initial rate of dissolution can make a large difference in commercial applications.

of SAPP and MCP-1 are important in their application to chemically leavened doughs and the amount of ''bench action'' carbon dioxide that is generated. Doping of the orthophosphate in the preparation of SAPP can change the rate of dissolution either by introducing poorly soluble impurities or more rapidly dissolving ones. This is accomplished by doping with K^{+}, Ca^{+2}, or Al^{+3}. Coated MCP-0 has a polyphosphate coating which retards the rate of reaction of the MCP-0 with bicarbonate. The history behind the development of this coating makes interesting reading (Toy, 1987).

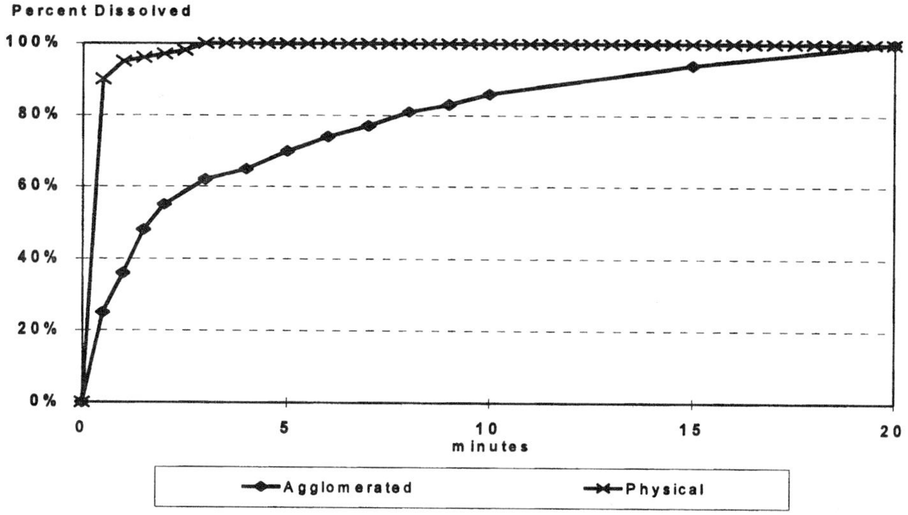

Figure 2 A comparison of the rate of dissolution of a physical blend (simple mixing of powders) of STPP and SHMP with that of an instantized blend with the same composition.

3. Measurement of the Rate of Dissolution

The rate at which phosphates dissolve can be readily measured indirectly using a conductivity probe (AWA, 1994). This is a useful tool that can be used to predict the solubility characteristics of various phosphates alone or in combination under a wide variety of conditions like water hardness, temperature, pH, previously dissolved additives, etc.

D. Phosphate Hydrolysis

The hydrolysis of complex phosphates is of interest in two areas: in solution and in use. The major difference between the two being that in food there are more avenues for hydrolysis and thus the rates are faster. The formation of orthophosphates from polyphosphates is of concern to meat workers since sodium orthophosphates effloresce on the surface of meat in a phenomenon known as ''snow formation,'' or ''whiskering.'' The reduction in favorable properties such as microbial inhibition is also of importance. To gain an understanding, it is useful to examine the thermodynamic and kinetic aspects of hydrolysis (Osterheld, 1972).

1. Thermodynamics of Hydrolysis

As noted, condensed phosphates are thermodynamically unstable with respect to hydrolysis over all pH and temperature ranges. An examination of the free energy changes (Table 2) shows that hydrolysis is spontaneous for all condensed phosphates. Since the mechanism of hydrolysis is molecularly similar for all polyphosphate chain lengths, it is not necessary to examine all the different polyphosphates separately, and only the energetics of pyrophosphate hydrolysis are shown in the table.

To penetrate further into the thermodynamics of polyphosphate hydrolysis, investigators have resorted to complex calculations of the energy changes and modeling of the reaction (Saint-Martin et al., 1991; Romero and DeMeis, 1989; Dupont and Pugeois, 1983). The interest in polyphosphate hydrolysis is driven by a desire to more fundamentally understand the driving force behind many biochemical reactions. This began in about 1940 with the discovery of ATP and its involvement in energy storage and release in living tissues. According to the current understanding (Saint-Martin et al., 1994) about half of the energy that is released in the hydrolysis comes from a rearrangement of chemical bonds, and the other half comes from changes in the hydration of the products and reactants. The interactions of the products and reactants with water are quite different. The contribution of hydration energy to the enthalpy of the pyrophosphate hydrolysis accounts for about half of the energy released during hydrolysis.

Table 2 The Equilibrium Constant (K), Enthalpy ($\Delta H°$), and Entropy ($\Delta S°$) for the Hydrolysis of Pyrophosphate

Reaction	K	$\Delta H°$ (kcal/mole)	$\Delta S°$ (eu)
$H_4P_2O_7 + H_2O \rightarrow 2\ H_3PO_4$	6×10^2	−5.3	−5
$H_2P_2O_7^{-2} + H_2O \rightarrow 2\ H2PO_4^-$	2×10^2	−6.7	−12
$P_2O_7^{-4} + H_2O \rightarrow 2HPO_4^{-2}$	8×10^3	−6.6	−5

Source: Wagman et al. (1968).

2. Kinetics

It is interesting to note that the mechanisms of energy transfer in living organisms are possible because at in vivo temperatures and pH values, the hydrolysis of condensed phosphates is slow. The rate is dependent on a number of factors, the main ones are described herein.

Hydrolysis of condensed phosphates can occur in two ways. The majority of the hydrolysis occurs via end group "clipping" in which a terminal phosphate tetrahedron is cleaved from the chain. Much less probable is "random cleavage" of the middle of the chain, but it has been established that this does occur with long chain phosphates (McCullough et al., 1956).

End group clipping:

$$\begin{array}{c}\text{O} \quad\quad \text{O} \quad\quad \text{O} \quad\quad \text{O} \quad\quad \text{O} \quad\quad \text{O} \\ \parallel \quad\quad \parallel \quad\quad \parallel \quad\quad \parallel \quad\quad \parallel \quad\quad \parallel \\ \text{O}-\text{P}-\text{O}-\text{P}-\text{O}-\text{P}-\text{O}-\text{P}-\text{O}-\text{P}-\text{O}-\text{P}-\text{O} \\ | \quad\quad | \quad\quad | \quad\quad | \quad\quad | \quad\quad | \\ \text{O} \quad\quad \text{O} \quad\quad \text{O} \quad\quad \text{O} \quad\quad \text{O} \quad\quad \text{O} \end{array}$$

$$\downarrow H_2O$$

$$\begin{array}{c}\text{O} \quad\quad\quad \text{O} \quad\quad \text{O} \quad\quad \text{O} \quad\quad \text{O} \quad\quad \text{O} \\ \parallel \quad\quad\quad \parallel \quad\quad \parallel \quad\quad \parallel \quad\quad \parallel \quad\quad \parallel \\ \text{O}-\text{P}-\text{O} \;+\; \text{O}-\text{P}-\text{O}-\text{P}-\text{O}-\text{P}-\text{O}-\text{P}-\text{O}-\text{P}-\text{O} \quad\quad [5] \\ | \quad\quad\quad\; | \quad\quad | \quad\quad | \quad\quad | \quad\quad | \\ \text{O} \quad\quad\quad \text{O} \quad\quad \text{O} \quad\quad \text{O} \quad\quad \text{O} \quad\quad \text{O} \end{array}$$

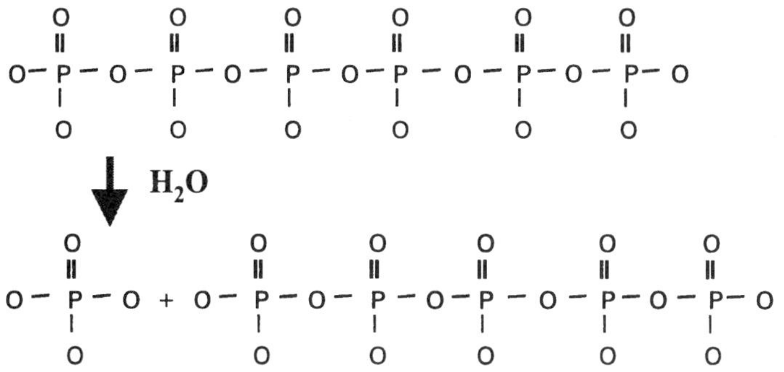

Random cleavage:

$$\begin{array}{c}\text{O} \quad\quad \text{O} \quad\quad \text{O} \quad\quad \text{O} \quad\quad \text{O} \quad\quad \text{O} \\ \parallel \quad\quad \parallel \quad\quad \parallel \quad\quad \parallel \quad\quad \parallel \quad\quad \parallel \\ \text{O}-\text{P}-\text{O}-\text{P}-\text{O}-\text{P}-\text{O}-\text{P}-\text{O}-\text{P}-\text{O}-\text{P}-\text{O} \\ | \quad\quad | \quad\quad | \quad\quad | \quad\quad | \quad\quad | \\ \text{O} \quad\quad \text{O} \quad\quad \text{O} \quad\quad \text{O} \quad\quad \text{O} \quad\quad \text{O} \end{array}$$

$$\downarrow H_2O$$

$$\begin{array}{c}\text{O} \quad\quad \text{O} \quad\quad \text{O} \quad\quad\quad \text{O} \quad\quad \text{O} \quad\quad \text{O} \\ \parallel \quad\quad \parallel \quad\quad \parallel \quad\quad\quad \parallel \quad\quad \parallel \quad\quad \parallel \\ \text{O}-\text{P}-\text{O}-\text{P}-\text{O}-\text{P}-\text{O} \;+\; \text{O}-\text{P}-\text{O}-\text{P}-\text{O}-\text{P}-\text{O} \quad\quad [6] \\ | \quad\quad | \quad\quad | \quad\quad\quad | \quad\quad | \quad\quad | \\ \text{O} \quad\quad \text{O} \quad\quad \text{O} \quad\quad\quad \text{O} \quad\quad \text{O} \quad\quad \text{O} \end{array}$$

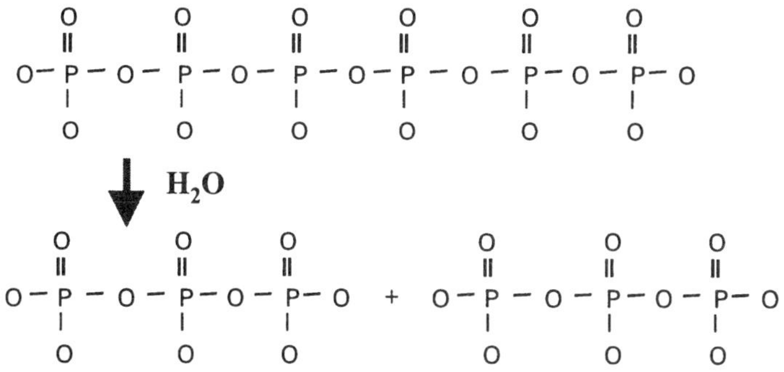

The rate of this type of hydrolysis is negligibly slow for polyphosphates up to about hexaphosphate. Whether the loss of orthophosphate from the end of a polyphosphate occurs via a concerted mechanism in which water is split apart at the same time the P–O–P bonds are broken or by a stepwise mechanism has been debated. Most of the work has been done on organic substituted phosphates. The precise details of the mechanism has been extensively studied (Jencks, 1992). It is believed that hydrolysis occurs because of an increase in the positive character of the central P atom in the (soon to be) orthophosphate leaving group. Either via a concerted or stepwise loss mechanism, the chain is shortened by one phosphate tetrahedron.

The rates of hydrolysis are strongly affected by temperature and pH. They are slower at lower temperatures and higher pH values. The exact rates are so dependent on the presence of other species that it is not possible to state much more in general. However, in pure solution and varying temperature a rule of thumb is that for every 5°C increase in temperature, a doubling of the rate is expected (Van Wazer, 1958). The effect on rate of other factors is shown in Table 3.

The variation in hydrolysis rates with pH is not actually a linear relationship. Plateaus occur at different pH values indicating that certain protonated or unprotonated species are more resistant to the type of polarization that is necessary to occur before the orthophosphate leaving group is cleaved from the chain. A useful nomograph relating the half-life of a phosphate under different conditions of pH and temperature has been published (Griffith, 1959). This is useful for estimating the amount of hydrolysis under different pH and temperature conditions.

Metal ions in solution also have an effect on the hydrolysis rates. The effect is not completely understood, but studies have found that hydrolysis is often accelerated by the presence of some metal ions like calcium, and inhibited slightly by the presence of magnesium. The influence the metal ion has on a complex phosphate is thought to involve polarization of the P–O bonds and thus affects the electropositive character of the central phosphorus atom (Osterheld, 1972).

3. Trimetaphosphate Formation

While not truly a hydrolysis, the formation of trimetaphosphates in solution from long chain phosphates is a chemically intriguing reaction. It is mentioned here since the sequestration ability of trimetaphosphate is essentially nonexistent. As well, the polyanionic character of trimetaphosphates is distinctly different from that of polyphosphates, and therefore a solution which starts out as polyphosphate can rapidly change its sequestering power even though condensed phosphates remain in solution. The reaction is simply

$$[NaPO_3]_n \rightarrow Na_3P_3O_9 + [NaPO_3]_{n-3} \tag{7}$$

Metaphosphate abstraction requires the presence of sodium ions. This was demonstrated in studies using tetramethyl ammonium salts of oligophosphates which showed essentially no metaphosphate abstraction products (Iler, 1952). It is thought that the metaphosphate abstraction reactions occurs via a coiling of the phosphate chain around a sodium ion, followed by an intramolecular scission to yield the trimetaphosphate ion and a polyphosphate chain shorter by three units (Thilo, 1962).

Table 3 The Relative Effects of Varying Certain Parameters on the Rate of Hydrolysis of Condensed Phosphates

Factor	Approximate effect on rate
Temperature	10^5–10^6 faster from freezing to boiling
pH	10^3–10^4 slower from strong acid to base
Enzymes	At most 10^5–10^6 faster
Cations	Manyfold increase in some cases
Concentration	Roughly proportional
Ionic environment	Severalfold change
Colloidal gels	As much as 10^4–10^5 faster

Source: Van Wazer (1958).

4. Hydrolysis of Complex Phosphates in Meats

The hydrolysis chemistry in meats is similar to that occurring in solution, but the rates are faster (Sutton, 1973). This is due to the presence of phosphatase enzymes, which continue to be active for a period of time after death. The phosphatase activity decreases with time. After rigor mortis, little change in the phosphatase activity is observed. For example, the hydrolysis of STPP in cod muscle proceeds even when stored at 0°C (Sutton, 1973). Large deviations in the rates of hydrolysis have been observed depending on the type of animal and on the animal themselves (Neraal and Hamm, 1977a,b). Interestingly, although the condensed phosphates are degraded in the tissue, their effects on the food often continue. This is best demonstrated by the fact that the increase in the water holding capacity of meat due to the presence of pyrophosphates is maintained after hydrolysis to orthophosphate (Hamm and Neraal, 1977c). This is suggested to be due to irreversible changes induced in the structures of the muscle (Sutton, 1973).

E. Sequestration

Sequestration of metal ions by condensed phosphates is an important function of complex phosphates in food applications. In its most general definition sequestration is the elimination of chemical effects of metal ions initially dissolved in aqueous media. Sequestration generally refers to the formation of a stable, water-soluble complex with a metal ion that prevents the ion from participating in reactions. Orthophosphates can be said to reduce the effect of hardness ions (e.g., Ca^{+2}, Mg^{+2}, $Fe^{+2/+3}$, etc.) by precipitation, but this is not strictly sequestration. Sequestration requires a complex phosphate since it needs at least two separate sites of attachment on the same molecule.

The ability of a particular polyphosphate to sequester a particular metal ion is a function of pH, temperature, and the other species that are present. The sensitivity of the sequestration capacity to pH value of a polyphosphate decreases with increasing chain length. Sequestration is best understood with respect to water softening and reducing scale. Reference to the original papers is recommended for those interested in the subject (Van Wazer and Campanella, 1950; Van Wazer and Callis, 1958; Thilo, 1955; Irani and Callis, 1962; Irani and Morgenthaler, 1963). The sequestration of cations in foods is strongly associated with the inhibition of certain reactions involving metal ions like catalysis of lipid oxidation or enzymatic browning. Because of the flexible polyphosphate backbone and the presence of many anionic ''sites,'' polyphosphates can accommodate the ''needs'' of many different cations in forming complexes with them (Table 4).

Table 4 Sequestering Power of Selected Sodium Phosphates

	STPP	TSPP	SHMP
Ca	13.4	4.7	19.5
Mg	6.4	8.3	2.9
Fe	0.2	0.3	.03

Note: The table illustrates the number of grams of the metal ion that can be sequestered by 100 g of the indicated phosphate.
Source: Albright & Wilson Americas, Inc., unpublished data.

F. Buffering of Solutions

Mixtures of orthophosphates are excellent buffers. Condensed phosphates are less useful at buffering. Diphosphates (pyrophosphates) are of some utility, but chains longer than 2 are not good buffers at all. The ranges of pH from 2 to 3.5, 5.5 to 7.5, and 10 to 12 can all be buffered by the ortho- and pyrophosphate anions. Long phosphates can perform some buffering in the range of 5.5 to 7.5, but it is not very cost effective compared to orthophosphates. The problem with long chain phosphates is that except for the end groups there are no weak acid titratable protons.

G. Analytical Methods

Analytical methods for phosphates can be broadly separated into two groups, those for total phosphate and those sensitive to the different phosphate species present (that is ortho-, pyro-, triphosphate, etc.). The difficulty with the analysis of phosphates in foods is the presence of compounds other than phosphate. Chemical methods of analyzing phosphate species usually suffer from interferences. It is necessary to somehow remove these, leaving behind the phosphate. There are many methods found in the literature. If the total amount of phosphate is all that is of concern, then exhaustive alkali or acid digestion techniques, either with an oxidizer or by charring, will eliminate the interferences caused by the food. Physical methods (like NMR spectroscopy) will in many cases succeed since it is sensitive only to the presence of a certain nuclei with specific chemical shift ranges specific to certain phosphates, and J–J coupling indicating the types and relative amounts of the phosphate species.

1. Analytical Methods for Total Phosphate

The best methods for total phosphate are colorimetric. Both vanado-molybdate and ammonium molybdate (Vogel, 1989) methods are excellent for determining the amount of pentavalent phosphate present as orthophosphate. The vanado-molybdate method is best suited for when phosphate is the major component in solution. The molybdenum blue method is more suited for solutions when phosphate is a minor component. All phosphate species present in a sample can be confidently converted to orthophosphates by refluxing with nitric or hydrochloric acids. The ashing of all organic components of the food is required to prevent the development of turbidity in the colored solutions created by addition of the molybdo reagents. This can be time consuming. Less sensitive to the presence of impurities are inductively coupled plasma–atomic emissions spectroscopy (ICP-AES) methods. While the equipment is more expensive than other analytical laboratory equipment, it does allow for the determination of many elemental components from one sample. Gravimetric methods of P determination may suffer from interferences from insoluble monophosphate sources.

2. Analytical Methods for Different Phosphate Species

Ion chromatography is the method of choice for speciation of phosphates. The method is very cost effective, accurate, and straightforward. This method relies on the fewest assumptions of any analytical method aimed at quantifying the different phosphate species present. Like all other chemical methods, it requires careful pretreatment of the sample to remove other components in the food. Once this is achieved, quantitative analysis of the phosphate species is readily achieved. Depending on the chromatography column and choice of eluent, isocratic methods are available. Gradient techniques have also been used.

The application of ^{31}P NMR spectroscopy to the quantitative analysis of condensed phosphates has been recently reviewed in which it is compared to other methods of analysis. It is a precise and accurate method for the quantification of phosphate species with the potential for automation allowing for the analysis of a large number of samples. Two-dimensional J resolved spectroscopy can be used to semiquantitatively determine the mixtures of phosphate chains that exist in phosphate melts (Gard et al., 1992). The use of ^{31}P NMR in the study of whole cells and in particular polyphosphates has been achieved. Addition of chelating agents, like EDTA, to reduce signal broadening by metal ions attached to the polyphosphates is required. The potential for this technique in terms of the reduced need for extensive sample preparation is great (Roberts, 1987).

II. USES AND APPLICATIONS OF FOOD GRADE PHOSPHATES

Food grade phosphates are widely used in many foods for a variety of reasons. The phosphates may be used to adjust pH (either acid or alkaline); buffer; sequester minerals; supplement minerals; either aid or inhibit coagulation; modify protein; disperse ingredients, and inhibit caking. Table 5 shows the functions of a variety of the food grade phosphates. The direct benefits would be to provide antioxidant activity; to thicken or gel dairy products; for emulsification (meats and cheeses); for color protection or cure color development; for water binding, and for chemical leavening. Foods in which phosphates are used include meats, poultry, seafood, dairy products, bakery products, fruits, vegetables, sugars, oils (refining), confections, beverages, pet foods, and personal care products. Applications of the phosphates by food product are shown in Table 6 and by food phosphate (Table 7).

Food phosphates must be manufactured according to good manufacturing practices (GMP) (21 CFR §110) and either meet or exceed those standards identified in Food Chemicals Codex IV (FCC IV), when used in the United States, Canada, and most European countries. Globally, there are nine major manufacturers (in aggregate supplying >75% of the products in commerce) of food grade phosphates, which include Albright & Wilson, Budenheim, FMC Corporation, Haifa Chemicals, Kemira, Prayon, Rhodia, BK Giulini Rotem, and Solutia. In 2000, Rhodia acquired Albright & Wilson, and FMC and Solutia formed the joint venture Astaris. It is important to use only those phosphates produced in accordance with FCC IV to assure safety (low arsenic and heavy metals). High purity products are also critical to avoid equipment failure in manufacturing operations and to prevent sensory defects. Other reasons for using food grade phosphates are shown in Table 8. In 2000, Rhodia acquired Albright & Wilson, and FMC and Solutia formed the joint venture Astaris.

The most commonly used food phosphates are generally recognized as safe (GRAS) by the U.S. Food and Drug Administration (FDA) under Title 21 of the Code of Federal Regulations, Parts 182 and 184. Uses of the food grade phosphates are also covered in 9 CFR for meats; 21 CFR for foods, boiler water additives, adhesives, and indirect contact, and 27 CFR for alcoholic beverages.

Canadian regulations currently differ from U.S. regulations. In 1996, a proposal was introduced to streamline the format of Canadian regulations with the intention to harmonize with U.S. standards. The proposed changes divided food ingredients into categories which required either minimal or rigorous oversight. Table 9 summarizes the differences and similarities between the two NAFTA countries.

Phosphates are almost always certified as kosher and there is an increasing desire (a requirement in Malaysia) for halal certification. Requirements will vary among customers.

Table 5 Functions of Phosphates as Food Ingredients

Function	Ingredient	
Acidulant	Monoammonium phosphate	Phosphoric acid
	Monocalcium phosphate	Sodium acid pyrophosphate
	Monopotassium phosphate	Sodium aluminum acid phosphate
	Monosodium phosphate	Calcium acid pyrophosphate
Adsorbent	Tricalcium phosphate	
Alkalinity	Disodium phosphate	Sodium tripolyphosphate
	Dipotassium phosphate	Tetrasodium pyrophosphate
	Potassium tripolyphosphate	Trisodium phosphate
Anticoagulant	Sodium hexametaphosphate	
Buffering agent	Diammonium phosphate	Monopotassium phosphate
	Disodium phosphate	Monosodium phosphate
	Monoammonium phosphate	Trisodium phosphate
	Monocalcium phosphate	
Coagulant	Phosphoric acid	Tetrasodium pyrophosphate
	Sodium acid pyrophosphate	
Dispersing agent	Disodium phosphate	Sodium acid pyrophosphate
	Sodium hexametaphosphate	Tetrasodium pyrophosphate
	Sodium tripolyphosphate	
Emulsifiers	Disodium phosphate	Tetrasodium phosphate
	Trisodium phosphate	Sodium hexametaphosphate
Flow conditioner	Tricalcium phosphate	
Leavening agent	Calcium acid pyrophosphate	Monocalcium phosphate
	Dicalcium phosphate	Sodium acid pyrophosphate
	Dimagnesium phosphate	Sodium aluminum phosphate
Mineral supplement	Dicalcium phosphate	Monopotassium phosphate
	Dimagnesium phosphate	Monosodium phosphate
	Dipotassium phosphate	Tricalcium phosphate
	Disodium phosphate	
Nutrient	Diammonium phosphate	Phosphoric acid
	Monoammonium phosphate	Tripotassium phosphate
	Monopotassium phosphate	
Protein modifier	Disodium phosphate	Sodium tripolyphosphate
	Monocalcium phosphate	Tetrasodium pyrophosphate
	Potassium tripolyphosphate	Trisodium phosphate
	Sodium acid pyrophosphate	
Sequestrant	Sodium hexametaphosphate	Sodium tripolyphosphate
	Sodium acid pyrophosphate	Tetrasodium pyrophosphate

A. Bakery Applications

In North America, the largest demand for the food grade phosphates is by the cereal and baking industries. These uses notably include the leavening of bakery products; pH adjustment and buffering; dough conditioning; enrichment; growth factors for yeasts; starch modification; and the manufacture of quick cooking cereals.

Chemical leavening to be discussed here is based upon the neutralization of common baking soda (typically sodium bicarbonate, but may be potassium or ammonium bicarbonate) by acidic phosphate salts. Expansion of the dough or batter is based upon the evolution

Table 6 Food Product Ingredients

Product	Ingredients
Baked goods	
Cakes, mixes	MCP, SALP, SAPP, DCP, MAP, DAP, SALP, DMP
Doughnuts	SAPP
Refrigerated dough	SAPP, SALP
Baking powder	MCP, SAPP, SALP
Beverages	
Noncarbonated	SHMP, STPP, MSP
Cola	H_3PO_4
Root beer	H_3PO_4
Milk-based	H_3PO_4, DSP, TCP, MCP, TSPP, SHMP
Dry mix	MSP, TCP, MCP
Candy and confections	
Chocolates	TSPP, TSP, SHMP
Icings and frostings	SAPP, TSPP, SHMP
Marshmallows	TSPP, SHMP
Cereals	
Dry cereal	DSP, TSP
Quick cooked	DSP
Cheese	
Cottage cheese	H_3PO_4, MCP
Imitation cheese	TSP, SHMP, STPP
Process cheese	DSP, TSP, SHMP, TSPP, STPP, SALP
Starter media	DSP, DKP
Chip dips	STPP, SHMP, DSP
Coffee	
Flavored, instant	DKP
Dairy Products	
Nondairy creamers	DKP, SHMP, TSPP
Frozen desserts	DSP, MCP, TSPP, SHMP
Sour cream	SHMP, STPP
Whipped toppings	TSPP, DSP, DKP
Eggs	
Whole	SHMP, MSP, MKP
Egg whites (dried)	SHMP, MSP, MKP
Fruit	
Canned fruits	MCP
Canned tomatoes	MCP
Juices	MCP, H_3PO_4
Gelatin desserts	DSP, MSP
Gums	
Alginate, agar	DKP, TSPP
Carrageenan, other gums	DSP, SHMP, STPP
Ice cream	
Hard, soft, and imitation	DSP, TSPP, SHMP
Jams and jellies	H_3PO_4
Lactose from whey	SHMP

Table 6 Continued

Product	Ingredients
Meat products	
Ham, corned beef	STPP, phosphate blends
Bacon	STPP, phosphate blends
Sausage, franks, bologna	STPP, phosphate blends
Roast beef	STPP, phosphate blends
Blood processing	SHMP
Milk products	
Beverages	DSP, TSPP
Buttermilk	TSPP, H_3PO_4
Cream	DSP, TSPP, SHMP
Evaporated and condensed	DSP
Nutrition products	
Commercial liquid diets	DCP, TCP, DKP, DSP
Enteral feedings	DCP, TCP, DKP, DSP
Infant foods	MCP, DCP, TCP, DKP, DSP
Isotonic beverages	DKP, MKP, DKP
''Instant'' breakfast preparations	DCP, TCP
Nutritional supplements	DCP, TCP, MCP, DSP, DKP, DMP
Vitamins	DCP, TCP, MCP, DSP, DKP
Oils	H_3PO_4
Pasta	DSP
Peanuts	STPP, SHMP
Pet foods	STPP, MCP, H_3PO_4, DCP
Processed potatoes	SAPP
Poultry	STPP, phosphate blends
Puddings	MCP, DSP, TSPP
Seafood	
Canned crab meat	SAPP
Fish and seafood	STPP, phosphate blends
Shrimp	STPP
Scallops	STPP
Surimi	Combination of TSPP, SAPP, STPP, and SHMP
Canned tuna	SAPP
Sugar	H_3PO_4
Vegetables	
Potatoes	SAPP
Peas and beans	MCP
Yeast	H_3PO_4

of carbon dioxide (CO_2), air, and steam. During the wet-mixing of doughs and batters, bubble formation is achieved by entrapment of air and/or CO_2 evolved from the neutralization by sodium bicarbonate of the leavening acid and/or other acidic components by fruit, flour, or fat. During bench time, additional CO_2 may evolve thus further expanding the dough or batter. Upon heating, final volume develops as a result of the CO_2 from any remaining active leavening agent; release of CO_2 dissolved in the aqueous portion; the generation of steam; and the thermal expansion of the gases.

(*text continues on p. 849*)

Table 7 Applications of Phosphates as Food Ingredients by Product

A&W product	Applications		
Acids			
Phosphoric Acid	Beer	Fat and oils	Pet foods
	Cola beverages	Fillings	Sugar
	Cottage cheese	Jams and jellies	Yeast
Orthophosphates			
Monoammonium phosphate	Breads and doughs	Cheese starter cultures	Yeast
Diammonium phosphate	Breads and doughs	Cookies	Yeast
	Cheese starter cultures	Crackers	
Monosodium phosphate	Cola beverages	Egg yolks	Instant pudding
	Dry powder beverages	Gelatin desserts	Isotonic beverages
	Egg whites	Instant cheesecake	Starch
Disodium phosphate	Breakfast cereal	Half-and-half	Nonfat dry milk
	Cheese powders	Ice cream	Pasta
	Chip dips	Imitation cheese	Pet food
	Condensed milk	Infant food	Process cheese
	Cream	Instant cheesecake	Starch
	Evaporated milk	Instant pudding	Vitamin capsules
	Flavored milk powders	Isotonic drinks	Whipped topping
	Gelatin desserts		
Trisodium phosphate	Cereals	Imitation cheese	Process cheese
	Cheese powders	Isotonic beverages	
Monopotassium phosphate	Breads and doughs	Isotonic beverages	Starter cultures
	Dry powder beverages	Mineral supplement	Yeast
	Eggs		
Dipotassium phosphate	Dairy creamers	Isotonic beverages	Starter cultures
	Dry powder beverages	Mineral supplement	

Table 7 Continued

A&W product	Applications		
Monocalcium phosphate	Bakery mixes	Flour	Pet food
	Baking powder	Fruit juices	Pudding
	Canned fruits	Infant food	Yogurt
	Dough conditioner	Microbial inhibitor	
	Dry powder beverages	Milk-based beverages	
Dicalcium phosphate	Bakery mixes	Food bars	Multivitamin tablets
	Cereals	Infant food	Pet food
	Dry powder beverages	Milk-based beverages	Yogurt
	Flour	Mineral supplementation	
Tricalcium phosphate	Cereal	Milk-based beverages	Salt
	Dry powders	Mineral supplementation	Spice blends
	Grated and powder cheese	Multivitamins	Sugar
	Infant food	Pet food	Yogurt
	Lard	Polymers	
Sodium Aluminum Phosphate, Acidic	Bakery mixes	Baking powder	Self-rising flour
	Pancake mixes	Waffle mixes	
Sodium Aluminum Phosphate, Basic	Process Cheese		
Dimagnesium Phosphate	Cakes, biscuits	Muffins	Pancakes
Pyrophosphates			
Sodium acid pyrophosphate	Bakery mixes	Imitation cheese	Refrigerated doughs
	Baking powder	Potatoes	Seafood
	Canned seafood	Poultry	Vegetables
	Cured meats	Process cheese	
	Icing and frostings	Processed meat	

Tetrasodium pyrophosphate	Buttermilk	Instant pudding	Seafood
	Cured meat	Marshmallows	Starch
	Flavored mild powders	Pet food	Whipped topping
	Ice cream	Poultry	
	Instant cheesecake	Processed meat	
Tetrapotassium pyrophosphate	Cured meat	Pet food	Seafood
	Flavored milk powders	Poultry	Starch
	Instant cheesecake	Processed meat	Whipped topping
	Instant pudding		
Calcium acid pyrophosphate	Baking powder	Crackers	Whole grain breads
	Dough conditioner	Frozen doughs	
Polyphosphates			
Sodium tripolyphosphate	Dips	Poultry	Vegetables
	Egg whites	Process cheese	Whey
	Eggs	Seafood	Whipped toppings
	Meat	Sour cream	Yogurt
	Noncarbonated beverages	Table syrup	
	Pet food	Vegetable protein	
Sodium hexametaphosphate	Carrageenans and other gums	Half-and-half	Process cheese
	Chip dip	Ice cream	Seafood
	Cream	Jams and Jellies	Table syrup
	Dairy beverage	Marshmallows	Vegetables (canned)
	Eggs	Meat	Whey
		Noncarbonated soft drink	Whipped toppings
		Poultry	

Table 8 Guide to the Use of Food Grade Phosphates

Factor	Food grade	Lesser grades
Export	Food ingredients meeting FCC IV are mandated for products exported to the United States, Canada, and Europe.	Grades not meeting FCC IV standards are not permitted by many importing countries.
Safety	Restrictions on heavy metals, toxins, and microbial pathogens for responsible use and limits liability.	May not be a safe substance for human consumption.
Purity	Inhibit equipment failures such as clogged injector needles due to formation of mineral phosphate precipitates.	May cause clogged injector needles.
	Mineral impurities may cause the formation of crystalline precipitates at product (e.g., scallop) surface.	Sensory defects which have caused customs to impound imported product.
	Mineral contaminants may catalyze oxidative reactions (rancidity of lipids and loss of flavor and color).	Rancidity of fatty acids, greening of meat color, and development of objectionable flavor.
	Assayed for minimum and consistent levels of desired phosphate species.	Assurance of product quality for consistent performance.
	Nitrite/nitrate impurities may cause undesirable cured pigment development.	Pink color may be a defect and is illegal in most seafood species.
Appearance and color	Phosphate should be a white powder/granular substance free of visible contaminants and discoloration.	Gray, brown, or black specks or tones may carry over into the finished product.
Solubility	Should be rapid and complete for most sodium and all potassium phosphates.	Slow and incomplete dissolution is not time and process efficient.
Good manufacturing practices and self-audits	Required for food grades.	Standards tend to be more for worker protection rather than product safety.
Uniformity of methods for product evaluation	Described in FCC IV	Few, if any, exist.

Table 9 Deviation of Proposed Canadian Regulations from Existing Canadian and U.S. Regulation

Product	Food group	Existing Canadian regulations	Proposed Canadian use level	United States use level	Exceptions (U.S.)
H_3PO_4					
	Chocolate, cocoa, cottage, cheese, cream cottage cheese, gelatin, light milk chocolate, mono- and diglycerides, sweet chocolate	GMP[X]		GMP	
	Cream and processed cheeses, cold-pack cheese food, whey cheese	GMP[X]		0.5%-cream cheese; GMP, pasteurized Neufchatel cheese	
	Fish protein	GMP[X]		GMP	
	Unstandardized foods	GMP[X]		GMP	
	Mono- and diglycerides	0.02%[XII]		GMP	
	Ale, beer, light beer, malt liquor, porter, stout	GMP	Deleted	GMP	
a	Mayonnaise, French dressing, salad dressing	GMP[XII]		GMP	
a	Ice cream mix	GMP[X,XII]			
a	Sherbet	0.75%[XII]		As in dry ingredients	
a	Ice milk mix	GMP[X,XII]			
a	Cream	GMP[X]			
a	Grape juice	GMP[X]		Not permitted	
a	Jams	GMP[X]		GMP	
a	Marmalade	GMP[X]		GMP	
a	Jelly	GMP[X]		GMP	
a	Liquid, dried, or frozen whole egg	GMP[X]		Pasteurization aid	
a	Liquid, dried, or frozen yolk	GMP[X]		do	
a	Liquid, dried, or frozen egg white	GMP[X]		do	

Table 9 Continued

Product	Food group	Existing Canadian regulations	Proposed Canadian use level	United States use level	Exceptions (U.S.)
DAP					
	Bread	2,500 ppm of flour	2,500 ppm	GMP	
	Ale, bacterial cultures, baking powder, beer, malt liquor, porter, stout	GMP[X,XIV]	Deleted	GMP	
	Unstandardized bakery foods	GMP[X]		GMP	
	Cottage cheese	GMP[X]			
a	Ice cream mix	GMP[X]			
a	Ice milk mix	GMP[X]			
a	Cream	GMP[X]			
a	Grape juice	GMP[X]			
a	Jams	GMP[X]			
a	Marmalade	GMP[X]			
a	Jelly	GMP[X]			
a	Liquid, dried, or frozen whole eggs	GMP[X]			
a	Liquid, dried, or frozen egg yolks	GMP[X]			
a	Liquid, dried, or frozen egg white	GMP[X]			

MAP					
	Bread	2500 ppm of flour	2500 ppm	GMP	
	Ale, bacterial cultures, baking, powder, beer, malt liquor, porter, stout	GMP[X]	Deleted	GMP	
	Unstandardized bakery foods	GMP[X,XIV]		GMP	
	Uncultured buttermilk	0.1%[X]		GMP	
	Cider, honey wine, wine	GMP[XIV]		GMP	
a	Cottage cheese	GMP[X]			
a	Ice cream mix	GMP[X]			
a	Ice milk mix	GMP[X]			
a	Cream	GMP[X]			
a	Grape juice	GMP[X]			
a	Jams	GMP[X]			
a	Marmalade	GMP[X]			
a	Jelly	GMP[X]			
a	Liquid, dried, or frozen whole eggs	GMP[X]			
a	Liquid, dried, or frozen egg yolks	GMP[X]			
a	Liquid, dried, or frozen egg white	GMP[X]			
DCP					
	Cream cheese and processed cheese products	3.5%[IV]	3.5%	0.5%	
	Bread/baked goods		2,500 ppm	GMP	
	Desserts, toppings, and fillings		0.25%	GMP	
	Gravy and sauces		2.0%	GMP	
	Dry mixes		2.5%	GMP	
	Chewing gum		5.0%	GMP	
	Dried whey products		1.5%	GMP	
	Fruit fillings and toppings		0.08%	GMP	Canned berries, NTE 0.035% as Ca^{2+}

Table 9 Continued

Product	Food group	Existing Canadian regulations	Proposed Canadian use level	United States use level	Exceptions (U.S.)
	Seasonings		0.15%	GMP	
	Unstandardized foods	$GMP^{IV,V,X}$		GMP	
	Flour	900 ppm^{VIII}		Flour—U.S. use level: flour bleaching, see 21 CFR 137.105; self-rising enriched flour, see 21 CFR 137.185	
	Bread	2,500 ppm of flour		600 mg Ca^{2+}/lb finished product	
	Unstandardized bakery foods	GMP		GMP	
a	Cottage cheese	GMP^{X}			
a	Ice cream mix	GMP^{X}			
a	Ice milk mix	$GMP^{X,IV}$ at 0.5%			
a	Cream	$GMP^{IV,X}$			
a	Grape juice	GMP^{X}			
a	Jams	GMP^{X}			
a	Marmalade	GMP^{X}			
a	Jelly	GMP^{X}			
a	Liquid, dried, or frozen whole eggs	GMP^{X}			
a	Liquid, dried, or frozen egg yolks	GMP^{X}			
a	Liquid, dried, or frozen egg white	GMP^{X}			
a	Cocoa products	GMP^{IV} at 0.5%			
a	Sherbet	GMP^{IV}			
a	Canned vegetables	GMP^{VI}		21 CFR 155.170; 155.190; 155.200	Firming agent
a	Mincemeat	GMP^{IV}			

MCP					
	Processed fruit and vegetables		GMP	145.170	Canned berries, 0.035% as Ca^{2+}
				150.110	Fruit butter, GMP
				150.140	Fruit jelly, GMP
				150.160	Fruit preserves and jams, GMP
				155.170	Canned peas, 350 ppm
				155.172	Canned dry peas, Do.
	Canned tomatoes	0.026% as Ca^{2+}		155.190	Canned tomatoes 0.046 to 0.08% Ca^{2+} of food
	Frozen dairy desserts	GMP^{XII}	GMP	not allowed	Ice cream
	Desserts, toppings, and fillings		0.35%	GMP	
	Confectionery		0.5%	GMP	See 21 CFR §163
	Beverages		0.13%	GMP	Juices—not permitted
	Breakfast cereals		100 ppm	GMP	
	Baking powder	GMP^{X}	1.0%	GMP	(Range 5 to 40%)
	Sauces		0.04%	GMP	
	Dairy-based flavoring preparations		3.0%	GMP	
	Soups, snack foods		0.5%	GMP	
	Unstandardized foods	$GMP^{V,X}$		GMP	
	Canned apples	0.026% as Ca^{2+}			
	Canned vegetables, frozen apples	0.026% as Ca^{2+}		21 CFR 155.170; 155.190; 155.200	Firming agent
	Ale, beer, malt liquor, porter, stout	GMP^{X}	Deleted		
	Unstandardized dairy products	GMP^{XII}		GMP	
	Bread	7500 ppm of $flour^{XII}$		600 mg CA^{++} finished product enrichment	
	Flour	7500 ppm^{XII}		960 mg Ca^{2+}/lb; enrichment, 0.25 to 0.75% by wt, phosphated	

Table 9 Continued

Product	Food group	Existing Canadian regulations	Proposed Canadian use level	United States use level	Exceptions (U.S.)
	Unstandardized bakery foods	GMP		GMP	
a	Cottage cheese	GMP[X]			
a	Cream	GMP[X]			
a	Grape juice	GMP[X]			
a	Jams	GMP[X]			
a	Marmalade	GMP[X]			
a	Jelly	GMP[X]			
a	Liquid, dried or frozen whole eggs	GMP[X]			
a	Liquid, dried, or frozen egg yolks	GMP[X]			
a	Liquid, dried, or frozen egg white	GMP[X]			
a	Mayonnaise, French dressing, salad dressing	GMP[XII]			
a	Sherbet	GMP[XII]			
a	Canned vegetables	GMP[VI]			
TCP					
	Processed cheese products	1.0%	1.0%	not permitted	
	Frozen dairy desserts	GMP[XII]	GMP	not permitted	Ice cream
	Baked goods		0.5%	GMP	
	Fruit fillings and toppings		0.2%	21 CFR §145.120	Only canned berries
	Breakfast cereals		0.1%	GMP	
	Condensed soups, dry soup mixes		1.0%	GMP	
	Batter and breading		50 ppm	GMP	
	Sauces		3.0%	GMP	
	Beverages		0.5%	GMP	Juices not permitted
	Desserts, toppings, and fillings		0.02%	GMP	
	Dressings		0.3%	GMP	

	Salt, seasoned salt	to 2.0%[I]	2.0%	GMP
	Prepared starch-based foods		0.2%	GMP
	Dry cure	GMP[I]		0.5% in finished product
	Unstandardized dry mixes	GMP[I]		0.5% in finished product
	Oil soluble annatto	GMP[I]		GMP
	Icing sugar	to 1.5%[I]		GMP
	Unstandardized foods	GMP[IV,X]		GMP
	Flour	900 ppm[VIII]		21 CFR 137.105 bleaching
	Liquid whey for drying (not infant formula)	0.04% dried whey product[VIII]		GMP
	Unstandardized bakery foods	GMP		GMP
	Cocoa products	GMP[IV]		
a	Mince meat	GMP[IV]		
a	Sherbet	GMP[IV,XII]		
a	Mayonnaise, French dressing, salad dressing	GMP[XII]		
a	Cottage cheese	GMP[X]		
a	Ice cream mix	GMP[X,XII]		
a	Ice milk mix	GMP[X,XII]; GMP[IV] at 0.5%		
a	Cream	GMP[IV,X]		
a	Grape juice	GMP[X]		
a	Jams	GMP[X]		
a	Marmalade	GMP[X]		
a	Jelly	GMP[X]		
a	Liquid, dried, or frozen whole eggs	GMP[I,X]		
a	Liquid, dried, or frozen egg yolks	GMP[I,X]		
a	Liquid, dried, or frozen egg white	GMP[X]		
a				
a				

Table 9 Continued

Product DKP	Product DSP	Food group	Existing Canadian regulations	Proposed Canadian use level	United States use level	Exceptions (U.S.)
√	√	Baked goods		0.2%	GMP	
	√	Processed cheese and cream cheese products	3.5%^{IV}	3.5%	3.0%	
√		Processed cheese and cream cheese products	3.5%^{IV}	3.5%	Not permitted	
√	√	Cottage cheese products	DSP, 0.5%^{IV}	0.5%	Not permitted	
√	√	Evaporated milk products	DSP, GMP	0.1%	GMP	
√	√	Sour cream	DSP, 0.05%^{IV}	0.05%	GMP	
√	√	Cured meat and poultry products	DSP, 0.5% as DSP^{XII}	0.3% PO_4	0.5% by wt	
√	√	Solid cut meat and poultry and byproducts	DSP, 0.5% as DSP^{XII}	Do.	Do.	
√	√	Other beverages		0.4%	GMP	Fruit juices—not permitted
√	√	Breakfast spreads		0.1%	GMP	
√	√	Cocoa mixes		0.5%	GMP	21 CFR 163
√	√	Desserts, toppings, and fillings		3.5%	GMP	
√	√	Breakfast cereals, rice products		0.4%	GMP	
√	√	Seasonings		0.3%	GMP	
√	√	Soups		1.0%	GMP	
√	√	Icings and table syrups		0.1%	GMP	
√	√	Batter and breading		0.04%	GMP	
√	√	Sauces		0.5%	GMP	
√	√	Snack foods		0.5%	GMP	
√	√	Vegetable oil creaming agents and their emulsions		2.5%	GMP	

√	√	Prepared starch based foods		0.5%	GMP	
√	√	Frozen entrees		0.5%	GMP	
	√	Milk and fluid milk products	GMP[IV]		GMP	Emulsifiers/stabilizers
	√	Mustard, pickles, relishes	GMP[IV]			
	√	Evaporated and concentrated milk	0.1%[IV]		GMP	Emulsifiers/stabilizers
	√	Frozen fish, glazed	GMP[VIII]		GMP	
	√	Frozen mushrooms	GMP[VIII]		GMP	
	√	Ale, bacterial cultures, beer, cream, light beer, malt liquor, porter, stout	GMP[X]	deleted	GMP	
	√	Unstandardized foods	GMP[X,XIII]		GMP	
		Ice cream mix, ice milk mix, sherbet	GMP[XII]		0.2%, ice cream as, contained in added ingredients	Sherbet
√		Ale, beer, cider, honey wine, light beer, malt liquor, porter, stout, wine	GMP[XIV]	deleted	GMP	
√	√	Cocoa products	GMP[IV]			
√	√	Mincemeat	GMP[IV]			
√	√	Mayonnaise, French dressing, salad dressing	GMP[XII]			
√	√	Sherbet	GMP[IV]; GMP[XII] at 0.5%			
√	√	Cottage cheese	GMP[X]			
√	√	Ice cream mix	GMP[X,XII]			
√	√	Ice milk mix	GMP[X,XII]; GMP[IV] at 0.5%			
√	√	Cream	GMP[IV,X]			
√	√	Grape juice	GMP[X]			
√	√	Jams	GMP[X]			
√	√	Marmalade	GMP[X]			

Table 9 Continued

Product		Food group	Existing Canadian regulations	Proposed Canadian use level	United States use level	Exceptions (U.S.)
√	√	Jelly	GMP[X]			
√	√	Liquid, dried or frozen whole eggs	GMP[X]			
√	√	Liquid, dried, or frozen egg yolks	GMP[X]			
√	√	Liquid, dried, or frozen egg white	GMP[X]			
MKP	MSP					
√	√	Frozen dairy desserts		GMP	Not permitted	Ice cream
√	√	Cured meat and poultry products	MSP, 0.5% as DSP[XII]	0.3% PO_4	0.5% by wt	
√	√	Solid cut meat and poultry and byproducts	0.5% as DSP[XII]	Do.	Do.	
	√	Processed cheese and cream cheese products	3.5%[IV]	3.5%	3.0%	
√		Processed cheese and cream cheese products	Not permitted	3.5%	Not permitted	
√		Desserts, toppings, and fillings		0.4%	GMP	
√		Flavors		3.0%	GMP	Vanilla—not permitted
√		Unstandardized foods	GMP[X,XII]			
√	√	Ice cream mix, ice milk mix, sherbet	GMP[XII]			
√	√	Ale, beer, cider, honey wine, light beer, malt liquor, porter, stout, wine	GMP[XIV]	deleted		
√	√	Seasonings		3.0%	GMP	
√	√	Sauces		0.1%	GMP	

√	√	Baked goods		0.35%	GMP	
√	√	Beverages		0.4%	GMP	Fruit juices—not permitted
√	√	Batter and breading		0.15%	GMP	
√	√	Frozen entrees, starch-based foods		0.7%	GMP	
√	√	Snack foods		0.2%	GMP	
	√	Unstandardized foods	GMP[IV,X,XII]		GMP	
	√	Ale, beer, light beer, malt liquor, porter, stout	GMP[X]			
	√	Solid cut, prepared and byproducts of meat and poultry	0.5% as DSP[XII]		0.5%	
√[a]	√[a]	Mayonnaise, French dressing, salad dressing	GMP[XII]			
√[a]	[a]	Ice cream mix	GMP[XII]			
√[a]	[a]	Sherbet	GMP[XII] at 0.5%			
√[a]	[a]	Ice milk mix	GMP[XII]			
[a]	√[a]	Cocoa products	GMP[IV]			
[a]	√[a]	Mincemeat	GMP[IV]			
[a]	√[a]	Sherbet	GMP[IV]; GMP[XII] at 0.5%			
[a]	√[a]	Cottage cheese	GMP[X]			
[a]	√[a]	Ice cream mix	GMP[X,XII]			
[a]	√[a]	Ice milk mix	GMP[X,XII]; GMP[IV] at 0.5%			
[a]	√[a]	Cream	GMP[IV,X]			
[a]	√[a]	Grape juice	GMP[X]			
[a]	√[a]	Jams	GMP[X]			
[a]	√[a]	Marmalade	GMP[X]			
[a]	√[a]	Jelly	GMP[X]			
[a]	√[a]	Liquid, dried, or frozen whole eggs	GMP[X]		MSP, MKP; 0.5%	Frozen only
[a]	√[a]	Liquid, dried, or frozen egg yolks	GMP[X]		GMP	Pasteurization aids
[a]	√[a]	Liquid, dried, or frozen egg white	GMP[X]		GMP	Whipping/pasteurization aids

Table 9 Continued

Product: TKP	Product: TSP	Food group	Existing Canadian regulations	Proposed Canadian use level	United States use level	Exceptions (U.S.)
	√	Cream cheese and processed cheese products	TSP, 3.5%[IV]	3.5%	3.0%	
√		Cream cheese and processed cheese products		3.5%	Not permitted	
√	√	Beverages		0.2%	GMP	Fruit juices—not permitted
√	√	Fruit snacks		0.05%	GMP	
√	√	Sauces		1.5%	GMP	
√	√	Baked goods		0.5%	GMP	
√	√	Confectionery		0.3%	GMP	21 CFR 163
√	√	Flavors		3.0%	GMP	Vanilla—not permitted
√	√	Seasonings		3.0%	GMP	
√	√	Dry cure	GMP[I]			
√	√	Unstandardized dry mixes	GMP[I]		GMP	
√	√	Oil soluble annatto	GMP[I]			
√	√	Icing sugar	To 1.5%[I]			
√	√	Unstandardized foods	GMP[IV,X]		GMP	
√	√	Flour	900 ppm[VII]			
√	√	Liquid whey for drying (not infant formula)	0.04% dried whey product[VIII]			
√	√	Ice cream mix, ice milk mix	GMP[XII]			
√	√	Unstandardized bakery foods	GMP			
	√	Ale, beer, light beer, malt liquor, porter, stout	GMP[X]	deleted		
	√	Unstandardized foods	GMP[IV,X]		GMP	

[a]	√[a]	Cocoa products	GMP[IV]			
[a]	√[a]	Mincemeat	GMP[IV]			
[a]	√[a]	Sherbet	GMP[IV]			
[a]	√[a]	Cottage cheese	GMP[X]			
[a]	√[a]	Ice cream mix	GMP[X]			
[a]	√[a]	Ice milk mix	GMP[X]; GMP[IV] at 0.5%			
[a]	√[a]	Cream	GMP[IV,X]			
[a]	√[a]	Grape juice	GMP[X]			
[a]	√[a]	Jams	GMP[X]			
[a]	√[a]	Marmalade	GMP[X]			
[a]	√[a]	Jelly	GMP[X]			
[a]	√[a]	Liquid, dried, or frozen whole eggs	GMP[X]			
[a]	√[a]	Liquid, dried, or frozen egg yolks	GMP[X]			
[a]	√[a]	Liquid, dried, or frozen egg white	GMP[X]			
KPMP	SHMP					
√	√	Frozen dairy desserts (not sherbet)	SHMP, 0.5%[XII,IV]	GMP	0.2% SHMP	Ice cream
√	√	Sherbet	SHMP, 0.75%[IV]	0.75%	No direct addition	
√	√	Infant formula	SHMP, 0.05%[IV]	0.05%	GMP	
	√	Cream cheese and processed cheese products	SHMP, 3.5%[IV]	3.5%	3.0%	
√		Cream cheese and processed cheese products	KHMP, 3.5%[IV]	3.5%		
	√	Mustard, pickles, relishes	GMP[IV]		GMP, SHMP	
	√	Unstandardized foods	GMP[IV,X,XII]		GMP, SHMP	
	√	Prepared fish and meat blends	0.1% SHMP[IV]	0.5%		
√		Cream cheese and processed cheese products		3.5%	Not permitted	

Table 9 Continued

Product		Food group	Existing Canadian regulations	Proposed Canadian use level	United States use level	Exceptions (U.S.)
√		Surimi-based products and canned seafood	0.1% SHMP, canned seafood[IV,XII]	0.1%	0.5% SHMP	Not canned seafood
√		Beef blood	SHMP 0.2%[VIII]	0.2%	GMP, SHMP	
√		Frozen fish, crustaceans and molluscs	SHMP, 0.5% as DSP[VIII]	0.3%	0.5% by wt SHMP	
√		Cured meat and poultry products	SHMP[XII], 0.5% as DSP	Do.	Do.	
√		Solid cut meat and poultry and their byproducts	Do.	Do.	Do.	
√		Gelatin for marshmallows	SHMP, 2.0%[VIII]	2.0%	GMP, SHMP	
√		Fruit spreads		2.5%	Varies, SHMP	
√		Baked goods		0.05%	GMP, SHMP	
√		Beverages		0.25%	GMP, SHMP	Not vanilla
√		Flavors		3.0%	GMP, SHMP	Not vanilla
√		Breakfast cereals		0.05%	GMP, SHMP	
√		Fruit fillings and toppings		0.06%	GMP,SHMP	Canned—not permitted
√		Snack foods		0.12%	GMP, SHMP	
√		Table syrups		500 ppm	GMP, SHMP	
√		Sweetened glazes		0.3%	GMP, SHMP	
√		Sauces		0.1%	GMP, SHMP	
[a]	√[a]	Cocoa products	GMP[IV]			
[a]	√[a]	Mincemeat	GMP[IV]			
[a]	√[a]	Sherbet	GMP[IV,XII]			
[a]	√[a]	Mayonnaise, French dressing, salad dressing	GMP[XII]		GMP	
[a]	√[a]	Cottage cheese	GMP[X]			
[a]	√[a]	Ice cream mix	GMP[X,XII]			

[a]	√[a]	Ice milk mix	GMP[X], GMP[IV]	0.5%		
[a]	√[a]	Cream	GMP[IV,X]		GMP	
[a]	√[a]	Grape juice	GMP[X]			
[a]	√[a]	Jams	GMP[X]			
[a]	√[a]	Marmalade	GMP[X]			
[a]	√[a]	Jelly	GMP[X]			
[a]	√[a]	Liquid, dried, or frozen whole eggs	GMP[X]	MSP, MKP; 0.5%	Frozen only	
[a]	√[a]	Liquid, dried, or frozen egg yolks	GMP[X]		GMP	Pasteurization aids
[a]	√[a]	Liquid, dried, or frozen egg white	GMP[VII,X]		GMP	Whipping/pasteurization aids
TKPP	TSPP					
√	√	Solid cut meat and poultry prepared or byproducts	0.5% as DSP[XII]	0.3% PO_4	0.5% by wt	
	√	Processed cheese and cream cheese prodcuts	TSPP, 3.5%[IV,X,XII]	3.5%	3.0%	
√		Processed cheese and cream cheese products		3.5%	Not permitted	
√		Surimi-based products		0.1% PO_4	0.5%	
√		Frozen fish, crustaceans, or molluscs	In blends w/ STPP, and SAPP, NTE 0.5% as DSP[VIII]	do	do	
	√	Frozen dairy desserts	GMP[XII]	GMP	0.2%	Ice cream
√		Frozen dairy desserts		GMP	Not permitted	
√		Cured meat and poultry products	STPP, 0.5% as DSP[XII]	0.3% PO_4	0.5% by wt	
√		Marshmallows		0.25%	GMP	
√		Desserts, toppings, fillings		1.0%	GMP	
√		Soups		0.4%	GMP	

Table 9 Continued

Product		Food group	Existing Canadian regulations	Proposed Canadian use level	United States use level	Exceptions (U.S.)
√		Sauces		1.0%	GMP	
√	√	Meat tenderizers	GMP[XII]		0.5% by wt	
√		Unstandardized foods	GMP[IV,X,XII]		GMP	
√		Prepared fish and meat blends	0.1%[IV]		0.5% by wt	
√	√	Cured meat and poultry	0.5% as DSP[XII]		0.5% by wt	
√[a]	[a]	Ice cream mix	GMP[XII]			
√[a]	[a]	Sherbet	GMP[XII] at 0.5%			
√[a]	[a]	Ice milk mix	GMP[XII]			
√[a]	[a]	Mayonnaise, French dressing, salad dressing	GMP[XII]			
[a]	√[a]	Cocoa products	GMP[IV]			
[a]	√[a]	Mincemeat	GMP[IV]			
[a]	√[a]	Sherbet	GMP[IV,XII]at 0.5%			
[a]	√[a]	Mayonnaise, French dressing, salad dressing	GMP[XII]			
[a]	√[a]	Cottage cheese	GMP[X]			
[a]	√[a]	Ice cream mix	GMP[X,XII]			
[a]	√[a]	Ice milk mix	GMP[X,XII]; GMP[IV] at 0.5%			
[a]	√[a]	Cream	GMP[IV,X]		GMP	
[a]	√[a]	Grape juice	GMP[X]			
[a]	√[a]	Jams	GMP[X]			
[a]	√[a]	Marmalade	GMP[X]			
[a]	√[a]	Jelly	GMP[X]			
[a]	√[a]	Liquid, dried, or frozen whole eggs	GMP[X]			
[a]	√[a]	Liquid, dried, or frozen egg yolks	GMP[X]			
[a]	√[a]	Liquid, dried, or frozen egg white	GMP[X]			

KTPP	STPP				
	√	Surimi-based products		0.1%	0.5%
	√	Frozen fish, crustaceans, or molluscs	0.5% as DSP	0.3% PO_4	0.5% by wt
	√	Cured meat and poultry products	0.5% as DSP	Do.	Do.
√		Cured meat and poultry products	0.5% as DSP	Do.	Do.
	√	Solid cut; prepared meat and poultry or their byproducts	0.5% as DSP[XII]	0.1%	0.5% by wt
	√	Sauces		1.5%	GMP
	√	Vegetable oil creaming agnts or their emulsions		25 ppm	GMP
√		Prepared fish and meat blends	0.1%[IV]		0.5% by wt
√		Unstandardized foods	GMP[X,XII]		GMP
√		Meat tenderizers	0.5% as DSP[XII]		0.5% by wt
√		Starch modification	0.4% phosphate calculated as P		STPP, GMP
[a]	√[a]	Cocoa products	GMP[IV]		
[a]	√[a]	Mincemeat	GMP[IV]		
[a]	√[a]	Sherbet	GMP[IV]; at 0.5%[XII]		
[a]	√[a]	Mayonnaise, French dressing, salad dressing	GMP[XII]		
[a]	√[a]	Cottage cheese	GMP[X]		
[a]	√[a]	Ice cream mix	GMP[X,XII]		
[a]	√[a]	Ice milk mix	GMP[X,XII]; GMP[IV] at 0.5%		
[a]	√[a]	Cream	GMP[IV,X]		
[a]	√[a]	Grape juice	GMP[X]		
[a]	√[a]	Jams	GMP[X]		
[a]	√[a]	Marmalade	GMP[X]		
[a]	√[a]	Jelly	GMP[X]		

Table 9 Continued

Product	Food group	Existing Canadian regulations	Proposed Canadian use level	United States use level	Exceptions (U.S.)
a √[a]	Liquid, dried, or frozen whole eggs	GMP[X]			
a √[a]	Liquid, dried, or frozen egg yolks	GMP[X]			
a √[a]	Liquid, dried, or frozen egg white	GMP[X]			
SAPP					
	Cream cheese or processed cheese products	3.5%[IV]	3.5%	3.0%	
	Frozen fish, crustaceans, or molluscs	In blends w/ STPP and TSPP, total P NTE 0.5% as DSP	0.3% PO_4	0.5% by wt	
	Canned seafoods	0.5% as DSP[XII]	0.3% PO_4	Specific species	
	Frozen dairy desserts	GMP[XII]	GMP	Not permitted	Ice cream
	Cured meat and poultry products	0.5% as DSP[XII]	0.3% as PO_4	0.5% by wt	
	Solid cut; prepared meat and poultry or their byproducts	0.5% as DSP[XII]	Do.	Do.	
	Baked goods		2.75%	GMP	
	Desserts, toppings, fillings		0.04%	GMP	
	Icings		0.2%	GMP	
	Dehydrated potato products, soup mixes		2.0% as consumed	GMP	
	Batter and breading		1.0%	GMP	
	Snack foods		0.1%	GMP	
	Prepared starch-based foods		0.1%	GMP	
	Baking powder	GMP[X]		GMP	

	Unstandardized foods	GMP[X,XII]		GMP	
	Cocoa products	GMP[IV]			
	Mincemeat	GMP[IV]			
	Sherbet	GMP[IV]; at 0.5%[XII]			
	Mayonnaise, French dressing, salad dressing	GMP[XII]			
	Cottage cheese	GMP[X]			
	Ice cream mix	GMP[X,XII]			
	Ice milk mix	GMP[X,XII]; GMP[IV] at 0.5%			
	Cream	GMP[IV,X]			
	Grape juice	GMP[X]			
	Jams	GMP[X]			
	Marmalade	GMP[X]			
	Jelly	GMP[X]			
	Liquid, dried, or frozen whole eggs	GMP[X]			
	Liquid, dried, or frozen egg yolks	GMP[X]			
	Liquid, dried, or frozen egg white	GMP[X]			
SALP					
	Cream cheese or process cheese products	3.5%[IV]	3.5%	3.0%	
	Baked goods	GMP	1.8%	GMP	
	Dry mixes		2.5%	GMP	
	Egg-based foods		0.2%	GMP	Not permitted in eggs alone
	Batter and breading		2.0%	GMP	
	Unstandardized foods	GMP[X]		GMP	
	Baking powder	GMP		GMP	
a	Cocoa products	GMP[IV]			
a	Mince meat	GMP[IV]			
a	Sherbet	GMP[IV]			

Table 9 Continued

Product	Food group	Existing Canadian regulations	Proposed Canadian use level	United States use level	Exceptions (U.S.)
a	Cottage cheese	GMP[X]			
a	Ice cream mix	GMP[X]			
a	Ice milk mix	GMP[X]; at 0.5%[IV]			
a	Cream	GMP[IV,X]			
a	Grape juice	GMP[X]			
a	Jams	GMP[X]			
a	Marmalade	GMP[X]			
a	Jelly	GMP[X]			
a	Liquid, dried, or frozen whole eggs	GMP[X]			
a	Liquid, dried, or frozen egg yolks	GMP[X]			
a	Liquid, dried, or frozen egg white	GMP[X]			
STMP					
	Starch modification	400 ppm calculated as P			

Notes: Assume cold breakfast cereals. Seafood definitions proposed to be generic. Phosphates added to proposed regulations: TKP, KTPP, KPMP.

[a] Citations listed in divisions other than Division 16.

[I] Anticaking agents: TCP.

[IV] Emulsifying, gelling, stabilizing, and thickening agents: DCP, TCP, DKP, SAPP, SALP, SHMP, DSP, MSP, TSP, TSPP, STPP.

[VI] Firming agents: DCP, MCP.

[VII] Miscellaneous food additives: DCP, TCP, SAPP, SHMP, DSP, TSPP, STPP.

[X] pH adjusting agents, acid reacting materials, and water correcting agents: DAP, MAP, DCP, MCP, TCP, H_3PO_4, DKP, SAPP, SALP, SHMP, DSP, MSP, TSP, TSPP, STPP.

[XII] Sequestering agents: MCP, TCP, H_3PO_4, MKP, TKPP, DKP, SAPP, SHMP, DSP, MSP, TSPP, STPP.

[XIII] Starch modifying agents: STMP, STPP.

[XIV] Yeast foods: MAP, DAP (see p. 67–55), DCP, MCP, TCP, H_3PO_4, DKP, MKP.

Phosphates currently used in leavening applications include monocalcium phosphate, [anhydrous and monohydrate (MCP-0 and MCP-1, respectively)], dicalcium phosphate dihydrate (DCP-2), sodium aluminum phosphate (SALP), sodium ammonium sulfate (SAS), sodium acid pyrophosphate (SAPP), diammonium phosphate (DAP), calcium acid pyrophosphate (CAPP), and dimagnesium phosphate (DMP). The SAPPs, MCPs, and SALPs are the most commonly used leavening acids (Table 10).

1. Rate of Reaction

In order to better understand the selection of leavening phosphate for specific product application, it is important to understand the dough rate of reaction (DROR). DROR measures the reactivity (CO_2 generated) of leavening acid with soda during mixing (2 to 3 min) and, subsequent bench time (5 to 6 min at 27°C) of a standard biscuit dough (Table 11). In the United States, the CO_2 measured at 8 min is fairly standard while the 2-min

Table 10 Food Phosphates for Use in Bakery and Cereals

Application	Phosphate functionality
Baking powder	MCP-0, MCP-1, SAPP-28, or SALP phosphates function as leavening agents.
Batter and breading	SAPP-40, SAPP-28, or SALP for stable bench time with rapid (leavening) action upon heating.
Biscuit mixes	SAPP-28 or SALP, slow acting at the bench, provide reactivity when heated and result in the desired coarse crumb texture.
Cake mixes (layered)	SALP, MCP, SAPP, or DCP-2 used for leavening.
Cake mixes (angel food)	MCP, SALP, or SAPP-40 used for leavening.
Cookie mixes	SAPP-28, SALP, or MCP-1 used for leavening.
Corn meal	Enrichment, MCP to 1.1 to 1.7 mg Ca/kg.
Dough conditioners	MCP or DCP used for leavening.
Doughnut mixes	SAPP-40 or SAPP-43 used for leavening.
Fat free and reduced fat snacks	MCP or SAPP used for leavening.
Flour products	Enrichment, MCP to 960 mg/lb or 2.1 g/kg; Phosphated, MCP to 0.25 to 0.75% by weight; Self-rising, SAPP, MCP, or SALP + $NaHCO_3$ to a standard weight; Self-rising enriched flour, DCP may be added.
Frozen doughs	DCP or SALP used for leavening.
Hot cereals	DSP to hasten cooking, 0.2 to 2.0% wt/wt.
Hush puppy mixes	SAPP-28, SAPP-40, or MCP-0 used for leavening.
Macaroni products	DSP to hasten cooking to 0.5 to 1.0% wt/wt.
Modified starch	MSP, DSP, STPP, or PA for low temperature gel formation, to inhibit freeze–thaw weep or high acid retorted foods.
Pancake mixes	MCP-1, SALP, SAPP-28, or SAPP-40 used for leavening.
Pizza mixes	MCP, SALP, SAPP, or DCP-2 used for added leavening (with yeast).
Refrigerated biscuits	Very slow ROR SAPP (22) or SALP for long-term chilled storage without gas production (leavening).
Snack crackers	SAPP-28 and MCP-1 for a double action leavening system at the bench (primary) and in the oven (secondary).
Tortillas	Wheat only (corn nonleavened).

Table 11 Properties of Commercially Available Phosphate Leavening Acids

Phosphate leavening acid	Neutralizing value	Percent leavening gas released: 2-min mix stage	8-min bench action	During baking
SAPP 22	70	22	11	67
SAPP 28	70	30	8	60
SAPP 40	70	40	8	52
MCP-1	80	60	8	52
MCP-0	83	15	35	50
DCP-2	33	0	0	100
CAPP	67	44	6	50
SALP-4	100	22	9	69
SALP-A	100	21	4	75

measure is often cited in Europe. A soda blank, which is responsible for about 20% of the gas evolved, includes all ingredients except the leavening acid.

DROR is skewed upward by high temperatures, high moisture levels, aged shortening, or flour (high acid value). It is skewed downward by sugar (which competes for water) and cations (such as calcium), which interact with acid salts thus hindering either their hydration or dissolution (Heidolph, 1996; Lajoie and Thomas, 1991).

Common baking powder is prepared with sodium bicarbonate, the leavening phosphate and a diluent (i.e., starch) to inhibit reactivity during storage. In the preparation of a dough or batter, wetting of the dry ingredients initiates the reaction between the bicarbonate and the acidulant to generate CO_2. (The exceptions would include DCP-2, which is insoluble, and those acids that have been coated to inhibit reactivity, e.g., coated MCP-0.) The rate of gas production and the degree to which the reaction progresses is dependent upon the speed with which the acidic compound dissolves.

2. Neutralizing Value

Neutralizing value (NV) is defined as the weight of sodium bicarbonate neutralized by 100 parts of the acidulant. It is a measure of the acid required within a specific bakery formulation. Not all bakery products benefit by a neutral pH. Ellinger (1972a) reported that the end pH should range between 6.9 to 7.2 for white cakes, 7.2 to 7.5 for yellow cakes, and 7.1 to 8.0 for chocolate and devil's food cakes. Color and flavor development is affected by end pH; a pH that is too low may result in a tartness and off-color, while excess bicarbonate (high pH) may result in a soapy flavor.

3. SAPP

The SAPPs range in dough rate of reaction (ROR) from 22 to 43. SAPP-22 is the slowest of its family and produces about 70% CO_2 at oven temperatures. This makes it most suitable for refrigerated biscuit doughs and cookies and frozen doughs and batters.

SAPP-28 and -33 release about 60% CO_2 at oven temperatures and are more applicable to prepared cake mixes. SAPP-28 is most frequently used in baking powder. SAPP-40 releases a theoretical 40% CO_2 in mixing and 50% at cook temperatures. SAPP-40 and -43 are the fastest reacting grades and are used in batters and breadings and cake doughnut preparations. Mixtures of the SAPPs or blends with other leavening acids may be used depending upon the end use. A limitation of the SAPPs is a ''pyro'' aftertaste in products with little sweetness (Reiman, 1977).

4. SALP

SALP is also used in leavening systems that require a delayed ROR or 20 to 30% CO_2 evolved in mixing and bench times and the balance upon heating. Two types of SALP (the molar ratios indicating Na:Al:P) are available, SALP 1:3:8 [$NaH_{14}Al_3(PO_4)_{14} \cdot 4H_2O$, or SALP4] and SALP 3:2:8 [$Na_3H_{15}Al_2(PO_4)_8$, or SALPA]. The SALPs may be used in combination with MCP and are included in mixes for waffles, pancakes, refrigerated biscuit doughs, frozen doughs, baking powder, and self-rising flours. SALP is resistant to cold temperatures and thus ideally suited to pancake preparations which may be prepared days in advance in institutional settings.

Concerns over the effect of aluminum on the development of Alzheimer's disease have been debated and resulted, primarily in Europe, in the reformulation of many baked goods in favor of other leavening acids. Concerns about aluminum have now been determined invalid.

5. DCP-2

DCP-2 is only sparingly soluble in liquids and reacts only at temperatures exceeding 135°F when it decomposes to MCP, free phosphoric acid, and hydroxyapatite (Toy, 1976). Its application is best suited for some cake mixes with a long bake time and high pH; cakes with high sugar content; frozen doughs; and, now, microwave cake mixes.

6. MCP-0

MCP-0 reacts to evolve CO_2 during mixing (15%), bench time (35%), and the balance during heating. Most MCP-0 is coated with potassium or aluminum phosphates to delay its reactivity by 3–5 min at the bench (Chung, 1992). It is commonly blended with other leavening phosphates in dry mixes but may be used singly in self-rising flours and in baking powders.

7. MCP-1

MCP-1 is a very fast acting leavening phosphate. Its primary function is to rapidly create a large number of gas cells during mixing which later serve as nuclei for expansion during heating (Chung, 1992). In mixing, up to 60% CO_2 is evolved. The balance of leavening is released after heating temperatures exceeding 140°F since the initial reaction generates DCP. It has applications in phosphated flours, pancake and cookie mixes, double acting baking powder, reduced fat snack foods, and inhibiting ropiness in bread.

Ropiness in bread is caused by the bacterium *Bacillus subtilis* (*B. mesentericus*). This microorganism is a spore former and it secretes a mucilagenous material which resembles stringiness in the center of a loaf of bread. Active or living bacilli will be destroyed by normal baking temperatures, however loaf internal temperatures may not reach a point sufficient to kill its spores. After the bread cools, surviving spores revert to a growth stage and begin to multiply rapidly in the favorable warm, moist and nutrient-rich environment provided by the fresh crumb. As the bacteria grow, enzymes are secreted which break down protein and starch in the bread. The bacteria also produce shiny (mucous) capsules which form the fine threads (rope) observed when the bread is pulled apart. An off-odor

Table 12 Bread Defects: Ropiness in Bread

Factors affecting development of ropy bread	Solution
1. Natural breads or those without preservative ingredients	Add monocalcium phosphate (MCP), 0.25%, to inhibit *B. subtilis* growth.
2. Recycling small amounts of baked product into the dough	Eliminate use of recycled product.
3. Failure to frequently and thoroughly sanitize equipment used for fluid ingredients	Frequently wash and sanitize equipment.
4. Use of improperly wet processed ingredients	Control and reduce the temperature and holding time for wet processed ingredients.
5. The presence of *Bacillus* spores in primarily flour and sometimes in yeast or malt	Control factors 2, 3, 4, 6, and 7. Use bacterial inhibitors such as acetic acid and MCP.
6. Cracks and crevices in dough handling equipment	Repair/seal cracks and crevices.
7. Holding dough too long at ambient or warmer temperatures	Shorten holding times.
8. Dough pH not sufficiently acidic	Add MCP to reduce dough pH.

similar to an overripe melon and a brown discoloration develop. Factors causing ropiness in bread are described in Table 12.

8. CAPP

CAPP is manufactured by only a limited number of phosphate producers and is utilized in a few specific applications. It has an NV of 67 and a DROR of 44 at 2 min and 51 at 8 min (Brose, 1993). It's recommended uses include rye flour doughs, crackers, and frozen (yeast) doughs and for dough strengthening.

9. DAP and MAP

The ammonium phosphates are used as leavening agents on a limited basis in the United States. Their application is restricted to very low moisture cookies and crackers since higher moisture leads to ammoniacal off-flavors. They are more widely used as yeast nutrients and dough strengtheners.

10. DMP

This is a recently patented leavening agent (Gard and Heidolph, 1995). It is heat activated (40.5 to 43.5°C) and intermediate between SALP (38 to 40.5°C) and DCP-2 (57 to 60°C). As a consequence, it may need to be used in combination with a faster leavening agent.

11. Sodium Aluminum Sulfate

While not a phosphate, sodium aluminum sulfate (SAS) is worthy of mention. It is not used alone as a leavening agent since it has too slow a DROR and may accelerate rancidity in flour-based mixes (Smith, 1991). Leavening occurs late in the baking cycle, which may make it desirable for products in which either a tunneling or blister effect is desirable, but not in cakes which require a uniform texture. SAS is used in commercial baking powders with other leavening acids.

12. Baking Powders

Baking powders are formulated to be either single or double acting. They consist of the leavening acid(s), sodium bicarbonate, and a diluent such as starch. Most commonly, MCP-1 and SAPP are the acidulants of double acting powders. Single acting powders generally contain MCP-1. In prepared cake mixes, the proportions of the leavening agents range from 10 to 20% of fast acting phosphates (MCP-0 or MCP-1) and 80 to 90% of delayed acting phosphates. Physical characteristics of either the bicarbonate, leavening acid, or added ingredients may cause defects in the finished product. A brief troubleshooting guide is shown in Table 13.

13. Dough Conditioning

Dough conditioning is another important function of the phosphates. The polyelectrolyte behavior of both gluten protein and of the phosphates assists in the strengthening of the dough and of the water holding capacity.

B. Cereal Applications

1. Quick Cooking Cereals

DSP is a commonly added ingredient to cereals in order to expedite cooking. This is due to both increasing the pH of the cooked cereal and partially gelatinizing the starch of those products soaked in phosphate solutions prior to cooking.

2. Pasta Products

DSP may be added to macaroni formulations at levels ranging from 0.5 to 1.0% to hasten cooking (21 CFR 139.110). Phosphates are used in the Asian noodle market in the form of *kansui*, which is a mixture of potassium and sodium carbonates and sodium and potassium phosphates. Kansui is permitted in Chinese but not Japanese noodles. The use level of kansui ranges from 1 to 2 g per kilogram of flour in dry noodles and from 0.5 to 1.7% in noodles for soup (Kubomura, 1998). Kansui interacts with the gluten to form the gum-like texture characteristic of the noodles and is associated with the characteristic yellow color and flavor development.

3. Starch Modification

Modified starches play an important role in the development of processed prepared foods. In all likelihood, their applications will be augmented with efforts to manufacture a greater array of low and reduced fat foods.

Starches are modified with any number of phosphates to form either starch phosphate monoesters or diesters. The resulting starch phosphate esters (SPE) are highly resistant to retrogradation, form gels at lower temperatures, and produce gels that are clearer and less viscous. The SPE monoesters are well suited to cold set puddings and pie fillings, while the SPE diesters are highly stable to acidic pH and to long and intense periods of heating. Kawana et al. (1990) reported on the use of a phosphoric crosslinked wheat flour, egg, and calcium phosphate coating to deep-fat fry shrimp that was subsequently vacuum packaged and retorted. There was no loss of the shrimp shape or texture.

Fortuna et al. (1990), increased the level of phosphorus substitution of starches. It was concluded that phosphorylated starches had improved solubility and water holding

Table 13 Troubleshooting Guide for Bakery Products

Problem	Solution
1. Dry mix reactivity	Monitor moisture of finished mix (≤3% moisture).
	Assure low moisture flour (<10% moisture).
	Increase soda granulation.
	Check storage temperature of mix; reduce if necessary.
	Check acid value of shortening.
	Check age of flour.
	Check packaging materials to assure transmission of excess moisture.
	Monitor relative humidity of storage; geographic constants and/or seasonal fluctuations may require its control.
2. Bench over-reactivity	Cool added fluids (water temperatures may fluctuate up to 60°F seasonally).
	Cool mix areas.
	Change to leavening agent that has reduced bench time (oven reactive and/or coated leaveners).
	Reduce holding time.
	Increase soda granulation.
	Check acid value of shortening.
	Check age of flour.
	Evaluate acidity of other added ingredients (e.g., fruits); coat if necessary.
3. Poor after-bake volume	Increase amount of leavening agents.
	Change leavening acid to be high temperature reactive.
	Avoid overmixing.
	Alter bicarbonate levels to allow for either reduced sodium plus increased ammonium bicarbonate to increase crown or use larger granulation sodium bicarbonate or add DCP-2 if a long bake time is possible.
4. Dry Texture	Alter bicarbonate source to a potassium salt which encourages humectancy.
	Evaluate different SAPP grades.
	Avoid overmixing.
5. Color	
Black streaks	Reduce granulation of acid leavening.
Brown spots	Reduce particle size of bicarbonate
Dark color	Reduce pH of mix.
	Reduce bake time and/or temperature.
	Increase fluids to slow evaporation losses.
Pale color	Increase pH.
	Increase bake time/temperature.
	Reduce added fluids.

capacity compared to untreated starches. The greater the degree of phosphorylation, the lower the pasting temperature but with slightly greater viscosity.

In an extension of their studies, Fortuna (1991) evaluated distarch phosphates of potato, maize, wheat, rye, and triticale origin for a variety of physicochemical factors. Distarch phosphates of potato origin showed lower reducing capacity, higher water binding capacity, and higher gel viscosity than those of cereal origin. This may be related to the naturally greater level of phosphorus indigenous to the potato.

In efforts to identify useful characteristics of other starches, Teo et al. (1993) evaluated the functionality of sago and tapioca roasted in the presence of urea and phosphate (3M) at pH 3.5 and 120°C. Chang and Lii (1992) reduced the amount of phosphate required to modify corn and cassava starches by employing extrusion. This process, however, may have led to damage of the starch molecule, which resulted in lower gelatinization temperatures, lower enthalpies, and reduced paste viscosities.

C. Meat, Poultry, and Seafood

Processed meat, poultry, and seafood benefit from the use of phosphates in many ways. These include buffering; pH control; solubility; water binding; emulsion development and stability; color development and stability; and inhibition of oxidation (Table 14) (Strack and Oetker, 1992).

Early studies focused on the use of phosphates in muscle foods to decrease cook–cool losses. Froning (1965) held whole poultry carcasses in 6% solutions of ''mixed'' polyphosphates for 15 h. It was determined that there were no significant differences in total water uptake by the carcasses prior to cooking. After heating, the polyphosphate soaked-meat had significantly ($p < 0.5$) less cook–cool loss, pH was elevated (from 6.1 to 6.6), and moisture of the meat was 6.2% greater than the control.

By and large, the most commonly used phosphate is STPP, followed by blends. Blends commonly consist of STPP and SHMP as a base with varying levels of SAPP and TSPP. There is increasing interest in the potassium phosphates (TKPP, KTPP) for end products with reduced sodium content. The alkaline phosphates (STPP, TSPP, KTPP, TKPP) increase pH and serve to increase water binding and emulsion development. Cure color development may be delayed by elevated pH, and therefore time may be required before heat processing. SAPP is an acid phosphate (pH 4.0 to 4.2) and is not normally used alone due to the adverse effect of low pH on water holding capacity. SHMP has

Table 14 Function of Phosphates in Meat Systems

Reduce requirement for NaCl—phosphate and NaCl act synergistically
Reduce development of ''warmed-over'' flavor (WOF)
Protect color
Assist color development in cured products
Reduce cook–cool loss
Reduce thaw-drip loss
Protect proteins in freezing and frozen storage
Inhibit lipid oxidation
Develop and stabilize emulsions
Allow myosin to form a ''sol'' and a ''gel'' upon heating
Enhance succulence of the cooked product

little effect on protein, but is included for its sequestration effects to reduce rancidity and discoloration and to soften water. The potassium phosphates are hygroscopic, but are the most soluble and tend to be more expensive. Orthophosphates contribute the least to meat, poultry, and seafood processing (Rust and Olson, 1987).

1. Phosphates, Sodium Chloride, and Their Interactions

Before the advent of phosphate use meat processing frequently required the use of high levels of sodium chloride in order to promote emulsion development. Maesso et al. (1970) studied the effects of mechanical action (beating for 3 min) and pH adjustment (with either NaOH or HCl to pH 5.0, 6.5, or 8.0) on poultry muscle in order to elucidate the mechanism of phosphate action. The investigators determined that tensile strength (Instron force to pull apart the meat) increased with pH, and therefore polyphosphate and NaCl in combination acted by simply altering surface pH.

Schults et al. (1972) evaluated the effects of (0.5%) TSPP, STPP, SHMP, and two blends of STPP and SHMP with and without NaCl on the physicochemical aspects of beef muscles (*longissimus, biceps femoris*, and *semimembranosus*). TSPP in combination with NaCl had the greatest effect on pH rise in all three muscle types. Naturally, the pH of *biceps femoris* was about 0.2 greater than the other two muscle types. Swelling of muscle fibers was markedly greater with pyrophosphates plus NaCl compared to the other phosphate treatments. Shrink was consistently reduced by the synergism with NaCl but less so with the SHMP treatment. It was theorized that under low (1 to 2%) NaCl concentration, there was an ion exchange of calcium for sodium on the meat proteins. At 3 to 4% NaCl swelling is reduced due to the exchange of potassium and magnesium by sodium. At levels of 5 to 10%, NaCl exhibits solely an ionic effect.

Schwartz and Mandigo (1976) evaluated the effect of 20 combinations of salt and STPP on restructured pork after 4 weeks storage at −23°C. There was a synergistic effect of STPP and NaCl upon TBA values, thaw-drip loss, improved cooked color, aroma, flavor, eating texture, cook–cool loss, raw color, and improved juiciness of the restructured chops. Optimal levels were determined to be 0.75% salt and 0.125% STPP.

Clarke et al. (1987) evaluated the concentration of NaCl (1.3, 2.0, 2.6, or 3.3%) needed to reduce cook–cool loss in beef comminutes. Use of 2.6% NaCl resulted in the least amount of cook–cool loss (20 to 35%); however; this was reduced to 1.0 to 1.4% upon the addition of 0.4% STPP to the batter. Strack and Oetker (1992) have described the polyelectrolyte behavior of the polyphosphates in a muscle food system. This behavior contributes to protein hydration and dispersion.

Offer and Trinick (1983) determined that pyrophosphate (10 mM) of beef myofibrils, in combination with reduced levels of sodium chloride, extracted the A-band completely beginning at both ends. This effect was confirmed by Voyle et al. (1984) with pork. In the absence of pyrophosphate, however, only the center of the A-band was extracted. Lewis et al. (1986) determined from 5-g pork, beef, chicken, and cod samples that an A/I overlap composed of denatured actomyosin and connection was formed, while unassociated myosin and actin were probably dispersed (sol) through the meat structure in the form of a water-holding gel (post–heat treatment). Trout and Schmidt (1987) concluded that at high ionic strengths (>0.25) pyrophosphate affected hydrophobic interactions which stabilize the protein structure and thus the thermal stability of the protein. Elevating pH (1 M NaOH), in combination with pyrophosphate, increased the temperature (from 70 to 87°C) for and the extent of protein aggregation. Yagi et al. (1985) confirmed that inorganic polyphosphate offered a high degree of protection to carp myofibrils from thermal denaturation.

Water retention is correlated with increased pH and is normally associated with the use of alkaline polyphosphates such as sodium tripolyphosphate. Orthophosphates have virtually no effect on water binding. Pyrophosphates are associated with improved protein solubility (myosin) and water binding. Consequently, water binding is dependent upon the type of phosphate used, and specific physicochemical reactions may require the use of blends.

SAPP, SHMP, or STPP (0.4%) were used in reduced salt (20 to 40%) turkey frankfurters (Barbut et al., 1988a). In formulations containing 40% less salt, phosphates improved emulsion stability and yields. While STPP improved firmness, flavor was not considered to be fresh. SAPP improved plumpness and enhanced salt flavor intensity. This was attributed to pH of the frankfurter (pH 6.1, SAPP and pH 6.5, STPP) and the sequestration of metal ions by the SAPP. If flavor had been altered by sequestration of metal ions alone, SAPP and SHMP should have shown a similar effect.

Studies by Kim and Han (1991) evaluated the gel strength of mixed pork myofibrillar and plasma protein. The gel strength of the mixture and myofibrillar protein solubility showed a significant increase when NaCl content was increased from 2 to 3%. By adding 0.3% polyphosphate in the presence of 2% NaCl, the gel strength increased fourfold.

Steinmann and Fischer (1993) investigated the level of NaCl and temperature on emulsion development and stability of a frankfurterlike sausage. It was determined that cooling lean meat with liquid nitrogen while mixing and the addition of SAPP resulted in greater extraction of salt soluble proteins and an increased stability of the emulsion.

Schantz and Bowers (1993) evaluated the effects of varying and mixed levels of NaCl (1.5 to 2.0%), STPP (0 to 0.5%), and SAPP (0 to 0.25%) on turkey sausages. A trained sensory panel showed increased saltiness with the presence of either polyphosphate. In the presence of 1.5% of NaCl, high SAPP, and low STPP, firmness decreased. It should, however, be noted that SAPP has a pH 4.2 and STPP a pH of 9.5 to 10.2. Therefore the pH effect was of some importance. Sensory evaluation also indicated that as the level of STPP increased and NaCl decreased, a slight soapy off-flavor was detected. The effects of pH and sensory evaluation clearly indicate another practical aspect of the use of polyphosphate blends.

Robe and Xiong (1993) evaluated the effect of phosphates on *longissimus dorsi* (LD, white), *serratus ventralis* (SV, red), and *vastus intermedius* (VI, red) salt soluble proteins (SSP). The SSP suspensions showed pseudoplastic flow. STPP (0.25%) decreased the shear stress of both LD and SV samples. Addition of STPP resulted in reduced dynamic shear storage modulus of all three muscle types. This effect was not duplicated in the presence of similar ionic strengths of NaCl.

Anjaneyulu et al. (1990a) studied the difference between phosphate action versus pH adjustment on end pH, water holding capacity (WHC), salt soluble protein content, emulsifying capacity (EC), cook–cool loss (CCL), emulsion stability (ES), patty yield, and shrinkage of finely minced buffalo meat (*biceps femoris*). The treatments consisted of 2% NaCl plus 0.5% polyphosphate (65% tetrasodium pyrophosphate, 17.5% sodium tripolyphosphate, and 17.5% sodium acid pyrophosphate), 2% NaCl with pH adjustment (with NaOH) to equal that of the phosphate treatment, and a control without either NaCl or added polyphosphate. The results indicated improved EC, ES, patty yield, and WHC and decreased CCL and shrinkage in the treatments in the following sequence: polyphosphate > NaOH pH adjustment > control. This negates the notion that the use of polyphosphate is responsible only for a pH effect.

In an extension of their studies, Anjaneyulu et al. (1990b) evaluated buffalo chunked and ground meat treated with 2% NaCl, 2% NaCl plus 0.5% phosphate, or no added ingredients (control) during holding at 2 to 3°C for up to 9 days. The results indicated that phosphate addition overcame the negative effects of NaCl by showing improved color and sensory quality. There was a synergistic effect on WHC and reduced CCL after holding for 5 days. Curiously, pH, EC, SSP, TBA, ES, and yield were thought to be superior for the control versus the NaCl plus phosphate treatment at 9 days. Incorporation of oxygen via equivalent mix times may have proven different for parameters associated with oxidation. In addition, short preblend holding prior to heating is frequently employed with commercial manufacture of restructured meats. Here, a long holding after preblending and overmixing may have adversely affected EC and ES. In native protein (not thermally denatured), STPP is broken down rapidly by muscle phosphatases. However, the pyrophosphatases are more slowly hydrolyzed (Hamm, 1986). This would explain the lost (positive) effect normally attributed to the polyphosphates.

Later studies by Anjaneyulu and Sharma (1991) indicated that EC and ES of precooked buffalo meat patties during refrigerated storage were favorably affected by the presence of a polyphosphate blend. Additionally, oxidative rancidity, as indicated by TBA values, was inhibited by the presence of polyphosphates.

2. Protein Functionality—Process Effects

Barbut (1988) evaluated the effect of serial levels of NaCl with and without added polyphosphate (STPP, SHMP, and SAPP), chop time, and speed on the ES of mechanically deboned chicken meat (MDCM). The results indicated that slow chop speed (40 versus 100 rpm), reduced time (66 versus 165 s) and low NaCl (1.5 versus 2.0 or 2.5%) plus polyphosphate enhanced ES. Ostensibly, pH was 6.71 versus 6.55 versus 6.23 with the use of STPP, SHMP, and SAPP, respectively. It is interesting to note that SHMP imparted greater stability than either STPP or SAPP in terms of cook–cool loss. SHMP is normally better associated with sequestration rather than meat protein functionality. Since calcium is associated with reduced STPP functionality (Regenstein and Rank Stamm, 1979a,b,c), it may be presumed that excess calcium naturally occurring in the MDCM may interfere with emulsion stability. Here, SHMP sequestered the calcium. STPP tends to form colloidal masses with calcium and thus is less available to react at the protein surface. SAPP would be highly active under these conditions and would act synergistically with NaCl to yield enhanced coating of the fat by myofibrillar proteins.

Gariepy et al. (1994) compared the effects of hot and cold boning (HB and CB, respectively) of beef chucks from electrically stimulated cattle carcasses at varying levels of NaCl (0 to 2%) with and without STPP. The results indicated that bologna prepared from HB mince containing 2% NaCl and CB mince containing STPP showed similar cooking yields and were greater than those of CB minces without STPP ($p > 0.05$). STPP provided an equalizing effect to compensate for process differences.

3. Protein Functionality—Chilled and Frozen Storage

Chen et al. (1991) evaluated the effectiveness of TSPP and STPP with and without sucrose and sorbitol (1 : 1) on chicken surimi during a 14-week period of frozen storage (−18°C) or at 4°C. The results indicated that a combination of sucrose (4%) and sorbitol (4%) and STPP (0.3) was the most effective cryoprotectant. At 4°C elasticity properties remained stable for greater than 4 weeks. However, pH, EC, ES, WHC, and gel strength decreased at 2 weeks. This is likely due to alkaline phosphatases and the hydrolysis of STPP to orthophosphates. Properties of the surimi were stabilized for 2 months at −18°C. Deterioration of protein functionality was attributed to oxidation of sulfhydryl groups.

These results are interesting but different from those observed with finfish-based surimi. Most phosphate-based cryoprotectants consist of a blend of STPP and TSPP and are used at a level of 0.4 to 0.5% and in combination with 8% sucrose and sorbitol (1:1). Similarly, functionality of sulfhydryl groups is associated with positive gel attributes of finfish-based surimi (Chang-Lee et al., 1989,1990; Pacheco-Aguilar et al., 1989).

Park et al. (1993) evaluated the effect of polydextrose (8%) with and without sodium tripolyphosphate and tetrasodium pyrophosphate (1:1) at the 0.55 level on the functional properties of pre- and postrigor beef. Although the authors did not determine a synergistic, cryoprotective effect on the treated prerigor beef, addition of the ingredients to an untreated control thawed after 5 months frozen storage did result in superior protein functionality. This effect was attributed to the presence of phosphate.

4. Lipid Oxidation

Akamittath et al. (1990) evaluated the impact of NaCl with and without polyphosphate on the stability of lipid and color in restructured beef, pork, and turkey steaks stored at −10°C for 16, 8, and 8 weeks, respectively. The polyphosphates were effective in delaying lipid oxidation in beef (4 weeks), turkey (6 weeks), and pork (8 weeks). Discoloration and lipid oxidation occurred simultaneously in pork and turkey, but discoloration preceded oxidation in beef. This work reaffirmed that lipid oxidation may be accelerated by pigment oxidation.

Mikkelsen et al. (1991) evaluated polyphosphates for lipid oxidation and color quality during retail display. As with the previous researchers, color stability, which was an initial ''blooming'' on the fresh ground beef patties, was reduced in frozen storage. The surface discoloration was linearly correlated with the metmyoglobin level of the total myoglobin extracts and, therefore, was used to track oxymyoglobin oxidation in frozen storage.

Oxidation of oxymyoglobin varied by phosphate type and in aqueous extracts or minced beef. Levels of sodium di-, or triphosphate or trimetaphosphates were added at levels ranging from 0.2 to 5.0%. Lipid oxidation was inhibited by no phosphate < trimetaphosphate < di- or triphosphate. The latter phosphates also counteracted the oxidative effect of added NaCl.

Craig et al. (1991) added STPP or sodium ascorbate monophosphate (SAMP) at levels ranging from 0.3 to 0.5% to ground turkey that was cooked, vacuum packaged, and frozen. Although a soapy flavor was present, the phosphate salts inhibited the development of rancid flavor, hexanal, and bathphenathroline-chelatable (nonheme) iron; and the salts decreased cook–cool losses, which is reflected by increased moisture content of the meat.

5. Thermal Effects

Trout and Schmidt (1987) evaluated the effects of ionic strength (0.12 to 0.52), pH (5.5 to 6.0), concentration of TSPP (0 to 0.31%), and cook temperature (52 to 87°C) on cook yield and tensile strength of beef homogenates. Generally, ionic strength between 0.32 to 0.42 prevented cook–cool loss at all temperatures with lower ionic strength required at higher pH and concentration of TSPP. TSPP had a positive effect on cook yield and tensile strength at ionic strength >0.25.

Since TSPP and NaCl are insoluble in nonpolar lipids, their concentration in the aqueous phase in fat free products is considerably lower than in fat-containing products, i.e., 0.5% in low fat systems and 0.3% in high fat systems. This work would be analogous to 20% fat products.

Trout and Schmidt (1987) also noted that at higher ionic strength, TSPP tended to negate the adverse impact of hydrophobic interactions with heat. NaCl increased the temperature of aggregation and the tendency to undergo syneresis; this impact was enhanced by the presence of TSPP. At pH 5.5 in the absence of TSPP, syneresis began at 70°C and at pH 6.0; in the presence of TSPP, syneresis was initiated at 87°C.

Robe and Xiong (1992) investigated the effect of phosphates on thermal transitions of pork salt soluble protein aggregation. The SSP transitions of the control (pH 6.0) were one or two for red muscle and three for white muscle. Addition of ortho-, pyro-, tripoly-, and hexametaphosphates up to 1% increased SSP transition temperature and altered transition patterns. This effect did not occur in the presence of NaCl at comparable ionic strengths. SSP transitions were most affected by STPP (0.15 to 0.25%) at pH 6.0 or lower. Red and white SSP responded differently to phosphate and showed different thermal properties. The authors concluded that red and white muscle types should undergo different processing treatments for optimal quality meat products.

Robe and Xiong (1994) extended their studies on the three aforementioned muscle types to the thermal aggregation of SSP in the presence of STPP. The SSP solutions were heated between 40 to 70°C with 0 to 0.2% STPP with protein aggregation measured by turbidity. SSP aggregation followed first order kinetics with rate and extent mediated by red fiber content. STPP increased the temperature for aggregation, but the rate varied by muscle type. STPP caused a 10 to 11.5% reduction in the activation energy for SSP aggregation. Muscles vary in thermal properties and response to treatment with STPP.

6. Nontraditional Phosphates

Zorba et al. (1993a) prepared emulsions with beef (fat level adjusted with sheep tail fat) using a model system of a Turkish style meat emulsion to evaluate the effects of oil, temperature, NaCl, and DKP. Emulsion stability (ES) and emulsion viscosity (EV) were enhanced in the presence of DKP (to 0.75%). ES was increased by an average of 3.8% in the presence of 0.5 to 0.75% DKP over a control (no phosphate). The EV increased by 22.3 to 27.0 by 0.5 and 0.75% DKP, respectively. It is interesting to note that NaCl (2.5%) had no statistical effect on EV.

In a continuation, Zorba et al. (1993b) determined that emulsion capacity (EC) increased an average of 9.5% in the presence of DKP over a control. Using electron microscopy, it was observed that DKP resulted in diminished protein aggregates and a homogenous emulsion.

7. Novel Applications of Phosphates in Muscle Foods

Farouk et al. (1992) investigated the effects of postexsanguination infusion on the composition, tenderness, and functional properties of beef. Between infused and control animals, there were no significant differences in WHC, and very low correlations were determined between tenderness, moisture, and ether-extractable fat.

8. Seafood Applications

Among the legitimate functional goals for the use of phosphates in seafoods are retention of natural moisture and flavor, inhibition of fluid losses during shipment and prior to sale, emulsification, inhibition of oxidation of flavors and lipids by chelation of heavy metals, and cryoprotection to extend shelf-life. Properly used, phosphates impart no flavor. Key applications are shown in Table 15.

a. Application of Phosphates. Phosphates are generally applied by dipping in, spraying with, or tumbling in a phosphate solution. Injector needle systems may also be used with and without added tumbling. Dry addition is used in comminuted meat systems, e.g., surimi and fish sausage formulations.

The most predictable way to apply phosphates is through vacuum tumbling, if done properly and the structure of the flesh can withstand mechanical action. Contrary to some practices, tumbling in an excess of solution results in protein extraction rather than absorption of solution. This uniform and rapid means of treating the muscle offsets the inefficiency of protracted holding in phosphate-based solutions (soaking).

It has been demonstrated that treating finfish prior to smoking requires different phosphate concentrations depending on the dimensions of the fillets and/or pieces. For example, with the same size pieces of flesh (within selected species), a 5% phosphate dip requires 24 h treatment time, while a 25% phosphate dip requires only 2 s (Wekell and Teeney, 1988) to reach equal processing effects, i.e., inhibition of surface curd formation and reduced cook–cool losses. This is especially valuable when delicate muscle structure eliminates tumbling as an option. Caution should be exercised when applying phosphates to fish of different muscle thickness, muscle types (e.g., interspecies variation), and initial moisture content (spawning).

Table 15 Food Phosphates for Use in Seafood

Application	Phosphate functionality
Canned salmon	STPP or STPP/SHMP combinations to inhibit curd formation
Canned tuna	SAPP to inhibit struvite formation.
Pasteurized crab	SAPP to inhibit blue discoloration of the meat
Canned abalone	SAPP and citric acid to inhibit blackening
Mechanical peeling of shrimp	STPP to assist cleavage of immature collagen and to firm the flesh
Kamaboko/surimi-based analogs	Mixtures of SAPP, TSPP, STPP, and SHMP
Frozen fish blocks	STPP or STPP/SHMP combinations for solubilizing surface proteins to prevent voids
Fresh scallops	STPP or phosphate blends to inhibit excessive exudate after harvest
Smoked fish	STPP or STPP/SHMP blends to retain flavor
Peeled shrimp	STPP treatment before freezing to decrease thaw-drip loss, or STPP treatment prior to cooking to decrease cook–cool loss
Fresh or frozen fillets	STPP or phosphate blends to retain natural moisture, inhibit color and lipid oxidation, reduce drip, and protect native protein

b. Methods to Determine Phosphate Application. Some methods to monitor phosphate use are based upon total moisture content of the muscle. One example would be the French HP (Loreal and Etienne, 1990) method, which is used to monitor the ratio of protein to water within muscle. In scallops, the ratio is considered to be between 4.0 and 4.9:1.0 (water:protein). The moisture content of commercially harvested seafood muscle is 80% or greater in species including, but not limited to, soft-shell blue crab, some molluscs, and postspawned finfish. Webb et al. (1969) determined that the moisture content of bay scallop meats was significantly different at the 5% level between harvest years, sounds, locations within the sounds, and among months and within locations. These researchers (Webb et al., 1969) also determined that the moisture content (monthly sampling) of land-shucked bay and Calico scallops ranged between 74.15 to 83.66 and 76.12 to 81.86%, respectively.

The HP ratio then would not be realistic for many species or at certain times of year. This value is based upon Kjeldahl protein to moisture (overnight drying at 100 to 105°C).

In theory, determination of total phosphorus in seafoods might be a useful marker of phosphate treatment; however, it is not necessarily accurate. For example, Crawford (1980) determined that the natural level of phosphorus in fresh shrimp (*Pandulus jordani*) ranged from 537 to 727 mg/100 g. Shrimp of the same history showed increases of 81 ± 39 and (base not given) ±110 mg of phosphorus, respectively, after treatment with either 1.5 or 6% phosphate solutions for 5 min. In shrimp (*Pandulus jordani*), the natural variation in phosphorus exceeded that added by responsible treatment.

Total natural phosphorus has also been reported to vary in lobster, blue mussels, squid, anchovies, carp, capelin, catfish, Atlantic cod, eel, hake, herring, yellow leatherjacket, European pilchard, and albacore tuna (Sidwell, 1981). Penetration of phosphate, and therefore phosphorus content, will also vary according to concentration of solution used, variations in muscle thickness, subsequent processing, etc.

Other methods to screen for added phosphates include high pressure liquid chromatography (HPLC), ion chromatography, and thin layer chromatography. Wood and Clark (1988) have reviewed the difficulties associated with these phosphate determinations.

Biochemical decomposition of condensed phosphates necessitates assaying immediately after treatment of the seafood species. Hydrolysis of condensed phosphates occurs due to muscle alkaline phosphatase activity during the posttreatment (lag) time prior to cooking. Sutton (1973) determined that sodium tripolyphosphate is rapidly hydrolyzed to pyrophosphate (phosphate dimer) and orthophosphate (phosphate monomer) in cod muscle at either 0 or 25°C.

It has also been determined that after 2 weeks of frozen storage (−26°C), only 12% of the total phosphorus in raw shrimp muscle corresponded to the originally added sodium tripolyphosphate. By 10 weeks, phosphorus levels corresponded to 45% orthophosphate (Tenhet et al., 1981). Clearly, in treated seafood muscle the condensed phosphates were unstable over time.

C. Mechanism of Action

PHOSPHATES AS PROCESSING AIDS. Crawford (1980) was instrumental in developing a protocol for the treatment of Pacific shrimp (*Pandulus jordani*) to be mechanically cooked and peeled. Use of the phosphates resulted in a firming of the flesh and more effective cleavage of immature collagen, which connects the body to the shell. These effects resulted in improved efficiency of shell removal by pinch rollers (nips). Recommendations for the use of phosphates were developed based on 3-day-old, fresh, iced shrimp. Conditions included exposure of smaller shrimp to a 3% iced phosphate solution and larger sizes to a 6% iced solution for 3 to 5 min prior to steam cooking for 90 to 110 s. With the responsible use of phosphates in treating Pacific shrimp to be mechanically cooked and peeled, meat yield increased an average of 12%. There was no significant uptake of moisture, and there was an added ex-plant income (in Oregon alone) of greater than $65 million in the first 8 years of use (Crawford, 1988). STP is now the phosphate of choice due to cost and the natural, relatively soft water supply of the Pacific Northwest.

PRESERVATION OF FRESHNESS. A process for using low concentrations (1 to 2%) of sodium tripolyphosphate in either flaked or crushed ice was patented by Stone (1981). Use of this ice increased the yield of shrimp and effectively reduced moisture and nutrient loss. Shrimp stored in phosphated ice could be overexposed to polyphosphates if treated again during further in-plant processing, which could cause either off-flavor, >0.5% residual phosphate, or both.

d. Specialty Blends. Among products for extending the shelf-life of fish fillets, Crawford (1984,1985) developed a patented blend consisting of sodium tripolyphosphate, sodium hexametaphosphate, citric acid, and potassium sorbate. Fish fillets were dipped into either distilled water or (ca.) 12% treatment solutions. The shelf-life (aerobic plate count $\leq 1 \times 10^6$ CFU/g) for treated samples was 12.4 days, and that of the control (water-dipped) was 6.8 days. Both control and treated fillets increased in weight by 4% after 60 s of immersion. Those dipped in the patented blend remained at their stated package weight throughout the 14 days of storage at 5°C, while the controls, dipped in water, dropped below the initial weight within 4 days of chill storage. Shelf-life extension would most likely be increased due to (1) the antimicrobial activity contributed by the sorbic acid and (2) the sequestration by phosphates of enzyme (metal) cofactors.

Indian researchers reported on the preservation of salted pink perch (*Nemipterus japonicus*) without preservatives or with a combination of sodium benzoate, potassium sorbate, SAPP, and butylated hydroxyanisole (Khuntia et al., 1993). The preserved fish showed decreased total volatile base nitrogen, amino acid nitrogen, TBA, free fatty acids, and aerobic plate counts over their controls when held in either chilled or ambient temperature. The preservative effect was somewhat enhanced at ambient temperature.

e. Frozen Seafood. Researchers at Texas A&M University reported that sodium tripolyphosphate dissolved slowly in seawater (Duxbury, 1986). In addition, fresh, shell-on, and peeled shrimp (Gulf of Mexico) became translucent and slippery to the touch after dipping in solutions of phosphate–sea water. This led to subsequent treatments which included 5 minute dips in water and 2, 4, or 5% condensed phosphates. Using a blend of sodium tripolyphosphate and hexametaphosphate resulted in rapid solubilization of the condensed phosphate, and more desirable sensory (touch) properties. The dipped shrimp were frozen and stored at −26°C for 2 weeks. Upon thawing and cooking (4 min), those shrimp dipped in the 4% blend for 5 min lost 0.8% weight after frozen storage (control, 2.0% loss) and 19.8% after cooking (control, 25.3% loss). It was concluded that addition of these phosphate blends imparted a cryoprotective effect.

Woyewoda and Bligh (1986) dipped Atlantic cod fillets into 12% solutions of sodium tripolyphosphate, sodium metaphosphate blends, or no solution and a control, respectively, for 45 s and stored each treatment at either −12°C or −30°C for up to 26 weeks. Phosphate-treated cod showed decreased thaw and cooked drip loss and resulted in higher moisture content in both raw and cooked product. After 26 weeks (at −30°C), all phosphate-treated fillets were judged the most tender and highly acceptable by sensory evaluation. The use of tripolyphosphate significantly reduced expressible water after holding at −30°C up to 26 weeks and up to 24 days at −12°C.

Ohta and Yamada (1990) evaluated the effect of K, P, Na, and Cl on the denaturation of fish (chub mackerel, black porgy, yellowfin tuna, and rock bream) species held at −6°C. It is agreed that the ion concentration of the unfrozen portion of muscle juices would be higher in K and P than in Na and Cl. Ohta and Yamada (1990) indicated that phosphates and chlorides may be related to denaturation of protein in frozen stored fish. It is more likely that high phosphorus, and especially potassium phosphates, lead to freezing point depression at a far lower molarity than their chloride counterparts. Furthermore, the lower the frozen storage temperature, the greater the ATPase activity and soluble myofibrillar protein over time.

f. Surimi. Surimi is the minced, washed, and refined flesh of finfish, which is now used in the manufacture of imitation shrimp, crab, and lobster. Originally, surimi was prepared and steam cooked as a means of preservation for later consumption. Typically, lean, white flesh species with a low flavor profile are preferred in the manufacture of surimi. The fish are mechanically skinned and deboned; the flesh is washed in cold, potable water to remove pigment and some lipid and sarcoplasmic proteins (low levels of NaCl may be added to the wash water), refined to remove residual pieces of skin and bone, and dewatered (since the flesh tends to absorb moisture during the washing). Depending upon the species, refinement may follow dewatering in order to minimize losses due to fines and to maximize yield. The mince is mixed with 8% sucrose and sorbitol and up to 0.5% sodium phosphates for cryoprotection of the concentrated myofibrillar protein; the product is packaged in 40-lb blocks and blast frozen. Alaska pollock has been processed in great quantity and Pacific whiting is now caught for this purpose.

Surimi quality is based upon gel (or jelly) strength of the kamaboko. Surimi is blended with water, salt, and phosphate to prepare a sol. After heating the sol becomes a protein crosslinked gel (kamaboko). Methods to determine gel strength may be based upon the fold test (no breaks after a circle is folded twice into a quarter) or mechanical methods (Instron testing for hardness, gumminess, cohesiveness, and chewiness; torsion testing and like procedures). The highest quality (SA grade) is positively correlated with myofibrillar protein solubility ($r = 0.9849$) and Ca^{2+}-ATPase activity ($r = 0.9584$) as reported by Kim and Cho (1992). It has been well documented that the myofibrillar proteins of seafood protein are sensitive to the effects of chilled and frozen storage and that their solubility and the activity of the ATPase decrease with the length and temperature of frozen storage; hence the need for added cryoprotectants.

Sucrose and sorbitol are commonly used cryoprotectants, but sweetness and expense limit their addition to 4% each. Matsumoto and Noguchi (1992) reported that some Japanese processors now use a mixture of sorbitol, polyphosphates, and a glycerine ester of fatty acids for cryoprotection. Sucrose may also lead to discoloration of the surimi over time (Yu et al., 1994). These researchers reported that sucrose acts by increasing the surface tension of water and by preventing withdrawal of water from proteins. Yu et al. (1994) studied the cryoprotective effect of SAPP and STP and concluded that STP was superior. This could be explained by the greater anionic nature of STP than SAPP. The polyphosphates are sorbed onto the surface of the protein and cause the dispersion shown in the electron micrographs of Sato et al. (1990). Dziezak (1990) explained that the polyphosphates attach themselves to positively charged sites of large molecules, as proteins, to increase water binding and gel formation. Commercially, blends of polyphosphates (STPP, TSPP, SAPP, SHMP) are used for cryoprotection of surimi. Sample formulations for surimi-based analogs have been published (Lee et al., 1992).

g. Thermally Processed Seafoods. Struvite, or magnesium ammonium phosphate, may be formed in thermally processed seafoods (e.g., canned tuna and crab). Sodium acid pyrophosphate can be used to sequester magnesium ions and thus inhibit struvite crystals, which resemble broken glass.

Salmon may develop a surface curd (denatured protein) if either held on ice for a protracted length of time or frozen prior to canning. The curd may constitute up to 4% of the pack by weight and may be considered questionable by many consumers.

Curd was significantly ($p \leq 0.05$) reduced by dipping sockeye salmon steaks for 2 to 120 s in 15 to 20% solutions of a condensed phosphate blend (sodium tripolyphosphate and sodium hexametaphosphate) and by dipping for 30 to 120 s in 5 to 10% solutions (Wekell and Teeney, 1988). To avoid dipping, Wekell and Teeny (1988) verified that there was a 68% reduction in curd formation by dry addition of the phosphate blend prior to sealing the can. Although it was estimated that 1.0% polyphosphate would be needed to completely inhibit curd formation, this would exceed the legal limits for phosphate in canned salmon.

Domestically, phosphate is not uniformly allowed in canned salmon except for a temporary allowance granted to several processors. Its use in canned salmon has, however, been given provisional approval by the Canadian government.

Phosphates provide significant benefit to the seafood industry when there is a large harvest within close proximity, and, conversely, there are limited quotes (i.e., freezing fillets to extend wholesale/retail availability). Spawning salmonids may represent one of the most important applications since the muscle has been physicochemically altered. Such finfish contain reduced levels of myofibrillar proteins which lead to impaired muscle water holding capacity. This is parallel with elevated levels of sarcoplasmic proteins and total moisture, a combination conducive to curd development.

Crapo and Crawford (1991) evaluated the effect of holding backed (removal of the back and viscera) Dungeness crab (*Cancer magister*) in 10% solutions of a blend of STP and SHMP at 2 to 4°C for 0 to 240 min. Treated crabs were either boiled or steam cooked for 4 to 20 min. The results indicated that moisture was not significantly increased in treated crabs versus the controls; phosphorus content was increased by 78 mg/100 g wet weight basis; yield was improved by approximately 8.5%, and sensory acceptance was improved. Furthermore, the authors determined that cooking time over 8 min (in steam) resulted in reduced yields. It is important to note that this protocol would not be applicable to all crab species since most mid-Atlantic and Gulf states mandate that live crabs be retort cooked prior to picking.

h. Troubleshooting. Often when phosphates are added in excess, a glassine look develops. This is particularly noticeable on shrimp. There are regulatory constraints to the use of polyphosphates along with organoleptic problems (a soapy taste) if the phosphates are used in excess. The glassine appearance probably occurs more in error than through intentional overuse of phosphates since there are no standard or defined procedures for their application. Most industrial protocols have been developed by trial and error and/ or have been based upon far more resilient terrestrial muscle.

Combining sodium tripolyphosphate with sea water will frequently promote the formation of a ''floc'' on the surface of certain species. Mineral content and pH of the muscle will exacerbate the formation of this crystalline precipitate.

Polyphosphate insolubility is related to water quality and to the individual type of condensed phosphate. Minerals in hard water will compete with some types of polyphosphate for solubility. In addition, not all forms of polyphosphate are readily soluble in water.

Erratic functionality of phosphates also may be caused by either heating phosphates to promote solubility or using old solutions. Many of the polyphosphates are prone to hydrolysis, and the monomeric forms will not perform the same as the polymers.

i. Economic Fraud and Legal Limits. Recently, the use of phosphates in some segments of the seafood industry has been subject to government scrutiny. When improperly used, excessive absorption of moisture may lead to charges of economic fraud by the U.S. Food and Drug Administration. It is important to note, however, that seafood myofibrillar proteins readily denature at refrigeration temperatures (5°C) and may lose up to 80% of their water-binding capacity within 5 days (Morey et al., 1982), while similar changes to beef muscle take in excess of 45 days at >20°C (Lampila, 1991). Responsible use of food grade phosphates prevents economic fraud by seafood processors by aiding the processed muscle to retain natural juices and thus avoid excessive drip loss. Excessive drip loss can lead to fluid weeping to the extent that net stated weight is not met and the product becomes violative.

Research conducted by Crawford (1984,1985) verified that phosphate treated and controlled (water dipped) rockfish fillets increased 4% in weight after a 60-s immersion. Those dipped in the phosphate remained at their stated package weight throughout the 14 days of storage at 5°C, while the controls, dipped in water, dropped below the initial weight within 4 days of chill storage.

Fresh, shell-on and peeled shrimp (Gulf of Mexico) were treated for 5 min in water and 2, 4, or 5% polyphosphate solutions. The shrimp were frozen and stored at −26°C for 2 weeks. Upon thawing and cooking (4 min), those shrimp dipped in the 4% blend for 5 min lost 0.8% weight after frozen storage (control, 2.0% loss) and 19.8% after cooking (control, 25.3% loss). It was concluded that addition of these phosphate blends imparted a cryoprotective effect (Duxbury, 1986).

Food grade phosphates and their products should not be cited only as ingredients to promote economic fraud. They impart numerous positive attributes within the seafood and food industry. These negative connotations related to economic adulteration reflect adversely on an ingredient which provides many advantages to the industry.

Phosphates are used as processing aids in the mechanical peeling of Pacific Northwest shrimp (Crawford, 1980,1988); to prevent struvite formation in canned tuna (CFR, 1997); to inhibit curd formation in canned salmon (Wekell and Teeny, 1988); as cryoprotectants to seafood quality during freezing and frozen storage (Woyewoda and Bligh, 1988); and in specialty blends to better control bacterial outgrowth and inhibit drip loss in finish (Crawford, 1984,1985). Recent researches by Applewhite et al. (1992,1993) have demonstrated that trained sensory panels prefer phosphate treated scallops and shrimp. Juiciness and tenderness were cited as some of the preferred characteristics over the untreated controls.

Proposals have been submitted to limiting moisture content (to 80% or less) of selected seafood species as a means of predicting economic fraud promulgated by the use of food grade phosphates. Basing decisions related to adulteration and economic fraud solely upon muscle moisture content (≥84%) is unrealistic, may limit a fishery, and results in the disposal of edible food for unjust causes. As evidenced by the following references it is well documented that moisture content of a given species will vary widely.

The moisture content of commercially harvested seafood muscle is 80% or greater in species, including, but not limited to, soft-shell blue crab, some molluscs, and postspawned finfish (Lampila, 1992,1994; Sidwell, 1981).

Webb et al. (1969) determined that the moisture content of bay scallop meats was significantly different at the 5% level between harvest years, sounds, locations within the sounds and among months, and within locations. These researchers (Webb et al., 1969) also determined that the moisture content (monthly sampling) of land-shucked bay and Calico scallops ranged between 74.15 to 83.66% and 76.12 to 81.86%, respectively.

In New Zealand (Hughes et al., 1980), scallops were determined to contain between 82.3 to 82.4% moisture. It should be noted that these meats were shucked into plastic bags (presumably to prevent hydration via melting ice), iced, and transported directly to the laboratory for analysis. This does not relate to current industry practice but was done to preclude moisture addition due to melting ice.

In April 1994, the FDA Office of Seafood conceded that the use of sodium tripolyphosphate in the treatment of mechanically cooked and peeled Pacific shrimp did not result in a humectant effect and was indeed a processing aid (Billy, 1994). Treatment of shrimp with phosphate solutions prior to cooking and mechanical peeling served to better cleave immature collagen and therefore increased meat recovery and reduce added water.

The maximum permitted legal level in processed meat and poultry is 0.5% by weight of the final product and serves as the current guideline where their use is permitted. Polyphosphates are now allowed in certain seafood species (CFR, 1997). They are, however, self-limiting. If much more than 0.5% of the high pH phosphates, such as sodium tripolyphosphate, is used, flavor and appearance will be adversely affected. A guide to troubleshooting the most common problems associated with meat, poultry, and seafood is shown in Table 16.

D. Dairy Applications

1. Cheese

The primary use of phosphates in dairy products is in the production of pasteurized process cheeses and sauces. Wines, due to tartrates, were the original emulsifier for fondue. Sodium potassium tartrate (Rochelle salt) was used early in the 20th century for emulsification. It was abandoned due to a tendency to cause sandiness and resulted in cheese that was brittle, mealy, and had poor slice characteristics.

In 1912, J. L. Kraft developed a process to heat, melt, and stir pieces of cheddar cheese to produce a homogenous, preserved product. Nine years later, Elmer Eldredge at the Phenix Cheese Company was the first to use DSP-2 in process cheese to make a product with a uniform melt, smooth texture, and good slice characteristics (Zehren and Nusbaum, 1992).

The principles for manufacturing process cheese are essentially unchanged. Cheese emulsification is a process that undergoes the physical transition of gel–sol–gel. It takes place due to chemical (emulsifying salts), thermal (heating), and mechanical (stirring) action.

Process cheese is manufactured by blending young and mature ground cheeses with water and emulsifying salts using heat to an end temperature up to 98°C and mechanical

Table 16 Troubleshooting Guide for Meat, Poultry, and Seafood

Problem	Solution
Phosphate dissolves slowly	Dissolve phosphate before NaCl and other added ingredients. If H_2O hardness is a factor, either softening or use of a phosphate blend may be required.
Clumping of phosphate	Phosphate should be introduced into water agitated with sufficient vortex to prevent settling. A slow rate of addition best assures wetting and dissolution.
''Whiskering''	Crystalline precipitates are orthophosphate, which appear due to surface desiccation, excess solids competing for available H_2O, or uneven distribution of phosphate. Prevent drying by monitoring cooking end point temperature and introducing well-circulated humidity, if possible. Reduce added solids, usually NaCl. Introduce physical measures to assure even distribution of brines.
Discoloration	Biochemical reactions involving muscle pigments, lipids, minerals, oxygen, or combinations may result in discoloration. Using a polyphosphate blend (containing SHMP) can be useful to sequester copper, iron, and magnesium to inhibit discoloration. Pockets of alkalinity may cause browning. Brines must be evenly distributed through the muscle.
Flavor defects	Uneven distribution of solutions can result in soapy flavors. An alkaline ''bite'' can result from either inadvertent overuse or treatment of products with extremely delicate flavor profiles.
Glassy surface appearance	The absorption of phosphates during dip applications will vary significantly among species, and particularly with seafood. For most seafoods, only seconds (at high brine concentrations) to a few minutes (at lower brine concentrations) may be required to assure adequate treatment.
Freezer burn	Polyphosphates are used as cryoprotectants for isolated and native protein. Temperature fluctuations and package headspace can result in significant migration of the moisture phase from the muscle surface. Shrink and vacuum packaging will minimize this occurrence.
Reduced yields	Insufficient treatment of the product; loose packaging, slow freezing, and fluctuating temperatures in frozen storage; acidic pH; naturally hard or highly alkaline water; overcooking; and /or inadequate smokehouse humidity or air circulation may lead to reductions in yield.

agitation. Cheese sauces contain more added water and, frequently, modified starch. The salts, added in order to peptize the protein, undergo ion exchange with calcium to form the soluble sodium paracaseinate and bind water through electrostatic bonds of the peptide chains (Zehren and Nusbaum, 1992; Berger et al., 1989). The net effect is to form a stable emulsion with even slice, melt, or spread/flow characteristics. Most notably, process cheeses are desirable to prevent protein agglomeration and fat separation in subsequent heating.

Preferably, melting salts will have polyvalent anions, form alkaline solutions, complex calcium, have ion exchange capacity, disperse proteins, be water soluble, and have a low flavor profile. Termed "melting salts," DSP, SALP, basic, and DSC (disodium citrate) are most frequently used. TSP (to raise pH), IMP (for freeze–thaw stability in sauces), and SHMP (enhanced thermostability for cheese-stuffed pizza crusts) are used in lesser quantities.

The phosphates permitted in cheeses according to 21 CFR 133 include MSP, DSP, DKP, TSP, SHMP, SAPP, TSPP, and SALP. Use levels are permitted to 3.0% in the United States. Canada is more liberal with the permitted phosphates and allows use to 3.5%. Care should be taken when using the maximum allowable level of either DSP or TSP, since it may cause crystal formation (Table 17).

Recently, a patent was granted for a "superphosphate" which is described to contain 73 to 77% P (as P_2O_5) with less than 2% metaphosphate (Merkenich et al., 1993). It is an acidic, long chain (8 to 20) polyphosphate which is possibly crosslinked. It has been reported to have some preservative effects in process cheese presumably due to sequestration of metal ions and its resistance to hydrolysis by heat and enzymes.

Zboralski (1986) patented a formulation for calcium supplemented process cheese containing TCP (0.4 to 1.6%). Normally, this would be considered an anomaly since too high a level of calcium phosphate causes a sandy mouth-feel. The formulation was described to impart increased creaminess and gloss to the finished product. It is important to note that TCP sold as a calcium supplement for orange juice or as an anticaking agent would not be suitable. TCP for the described formulation would need to be 50 to 100% beta-TCP, be micronized (particle size 1 to 20 μm), have a large surface area, and consist of a $CaO:P_2O_5$ ratio between 2.5:1 to 3.1:1. By current manufacturing methods, this would be an expensive form of TCP.

Table 17 Troubleshooting Guide to Crystal Formation in Process Cheese

Event	Cause
High pH	Too much TSP-12.
	Too much DSP-2.
	Raw ingredient pH is too high (very mature cheese tends to have a high pH).
	In combination with low (<50°F) storage temperature precipitates crystallization.
Presence of DKP	DKP (75%) in combination with DSP (25%) at a level of 2.3% emulsifying salts.
Citrates	May result in the formation of tricalcium citrate tetrahydrate under conditions of high pH and low (37 to 45°F) temperature.
Other added ingredients	Cheeses containing high levels of tyrosine which act as "seed" for precipitation.
	Excess addition of either whey or skim milk powder, which results in the formation of lactose crystals (i.e., the concentration of lactose in the water phase is 16.9% at 60°F)
Tartrates	Although not commonly used, the presence of tartrates can lead to the formation of calcium tartrate crystals.
Physical factors	Uneven distribution of DSP within the cheese mixture.
	Inadequate time/temperature for melting of the crystals.
Mechanical	Conditions conducive to the formation of crystals as deposits on the cooling belt, uneven surfaces on the deflector rollers, etc.

As previously indicated, process cheese is typically manufactured with a blend of young and mature cheeses. Kalab et al. (1991) described a study in which white or unripened cheeses were incorporated at 8, 16, or 33% in the manufacture of processed cheese. Characteristically, white cheeses which do not melt when heated alone would lend intact casein and whey proteins as well as the advantage of blending without the need for aging the cheese. At all three levels of addition, the meltability of the cheese increased in the presence of TSP. Firmness of the processed cheese was unaffected at the 8 and 16% levels of white cheese, but did increase at the 33% level. This lends a broader application of cheeses for the production of cheese dips, spreads, and sauces.

Cheese powders were prepared from cheddar cheeses ranging from 2 to 8 months. Slurries of the cheeses were prepared with total solids (TS) ranging from 21 to 35%. Either trisodium citrate (TSC) in equal combination with DSP or DSP alone (2.5, 3.0, or 3.5%) were added to the slurries. The results indicated that sensory and physical properties of the reconstituted cheese spread were maximally achieved by using 3.0% DSP (Vipan and Tewari, 1992).

Trecker and Monckton (1990) have evaluated the use of DSP to parmesan cheese either before or after drying (to 19–24% moisture) to prevent subsequent oiling off. DSP added to the cheese dissolved into the matrix to inhibit agglomeration and oil-off at ambient and greater storage temperatures.

Agusti (1993) used MCP in the production of mancheo cheese (a natural cheese). Its presence resulted in a reduction of coagulation time by 5 to 10 min, a firmer curd, lower moisture loss during ripening, and more intense flavor and aroma in the finished product.

Less than 9% of calcium in skim milk is retained in the curd of cottage cheese. This led Demott (1990) to evaluate the impact of $CaCl_2$, MSP, DSP, SAPP, or STPP to skim milk to determine if the calcium content of cottage cheese curd could be increased. Levels of phosphate used were approximately 0.5 g/L. SAPP and STPP resulted in curd containing about nine times the calcium of the control. Curd containing STPP was firmer than either the control or SAPP treatments but was mealy.

2. Milk Products

Phosphates are also used in the manufacture of milk products such as chocolate milk (TSPP aids in the suspension of cocoa), in drying milks to aid in rehydration, and in concentrated milks (Table 18). Either SHMP or TSPP may be added to ice cream to prevent churning (a sandiness or grittiness) of the fat. Allowable uses of phosphates to standardized dairy products are cited in 21 CFR parts 131, 133, and 135.

Pouliot and Boulet (1991) surveyed the variation in heat stability of concentrated milk over a 12-month period. The buffer capacity of the milk showed a seasonal variation being greatest in the summer and lowest in the winter. This effect was overcome by adding DSP to increase the pH by one unit to between pH 6 to 7 to enhance thermostabilization.

The seasonal variation is related to diet. Feeding silage tends to cause milk which is naturally low in phosphates and citrates (Hegenbart, 1990). After homogenization, the emulsion may break down with separation of the milk components. Added phosphates and citrates will overcome this problem.

Mil and van de Koning (1992) confirmed the seasonal variation in heat stability of milk described by Pouliot and Boulet (1991). By adding DSP to milk to be spray-dried, the pH was favorably adjusted to cause a shift in heat stability and enhanced rehydration capabilities. MSP was ineffective.

Table 18 Food Phosphates for Use in Dairy Foods and Desserts

Application	Function
Buttermilk	H_3PO_4 used as an acidulant or TSPP used as a stabilizer.
Chocolate/malted milk	TSPP is added for dispersion of the chocolate and to maintain suspension of the solids.
Cottage cheese	Phosphoric acid or MCP may be added to reduce the pH to 4.5 to 4.7.
Evaporated milk Low and whole fat dry milk	DSP is often used (0.1%) to protect protein (casein) from coagulating during canning/drying.
Fermented dairy products	Sour cream may contain those phosphates (STPP, SHMP) that improve texture, prevent syneresis, or extend shelf-life. Yogurt may contain phosphates that act as stabilizers.
Fluid dairy products (includes low fat milk, cream, half-and-half, and eggnog)	Phosphates may be added as stabilizers and/or emulsifiers.
Ice cream	DSP, TSPP, or SHMP (0.1 to 0.2%) may be added to prevent churning of the fat.
Imitation coffee creamers	DKP is most often used to adjust pH and prevent ''feathering.'' For neutral flavor characteristics, DKP is preferred over DSP (1.0 to 2.0%, dry weight). SHMP or STPP may also be used.
Instant puddings Cheesecake filling	TSPP alone (to set the gel) or in combination with either DSP-O or MCP-1 (to hasten setting) are used.
Processed cheese	''Emulsifying salt'' (primarily orthophosphates and sometimes pyrophosphates) and SHMP are added (to 3%) in order to alter the melt, flow, and slicing properties of cheese. Sodium phosphates are used for ion exchange with calcium in the cheese base.
Starter cultures	Nutrients: DSP, MSP, MAP, DAP, MKP, DKP.
Whipped toppings	DSP (0.1 to 1.0%) is added as a foam stabilizer; TSPP or SHMP.

Khan et al. (1992) confirmed the stabilizing effect of phosphates in sterilized buffalo milk. The effect was greatest at pH > 6.5 using MSP, DSP, and SHMP at levels ranging between 25 to 100 mg/m. Although MSP has a detrimental effect due to pH reduction in cow's milk, Loter (1983) reported that MSP could be used to make acidified milk heat stable to protein coagulation.

Montilla and Calvo (1997) used a commercial phosphate blend consisting of MSP, DSP, and TSP (Turrixin ST, now BK Giulini Rotem) to enhance the heat stability of goat milk. Goat milk is less resistant to thermal denaturation than cow milk. Goat milk has low (5% total) alpha$_{s1}$-casein relative to cow milk (38% of total) (Mora-Gutierrez et al., 1993). Levels (0.09%) of Turrixin ST were determined to be effective against UHT denaturation without altering pH. The effect was attributed to changes in ionic calcium and equilibrium of micellar calcium phosphate.

DSP has been repeatedly reported to stabilize the proteins of evaporated, condensed, and dried milk products. Harwalkar and Vreeman (1978a) demonstrated that DSP did cause accelerated age gelation in UHT-treated milk and that SHMP (1.5 g/kg) in fact resulted in enhanced shelf-life stability. Age gelation was described (Harwalkar and Vreeman, 1978b) as a rise in viscosity of the milk concurrent with the development of thread-like tails on the perimeters of the casein micelles. As viscosity increased, micelle pairs or triplets developed. As storage time increased, micelles aggregated with longer chains to form a continuous three-dimensional network (gelation). SHMP-treated micelles did not change noticeably during storage.

Kocak and Zadow (1986) extended this work and added either sodium citrate (3 g/kg) or SHMP (1 g/kg) to UHT milk. It was determined that apparent viscosity of the SHMP-treated sample was unchanged at 500 days storage and that this resulted in a sixfold extension in shelf-life. This effect was dependent upon a low initial psychrotrophic count in the raw milk (2.7×10^3 CFU/mL).

Earlier work (Kocak and Zadow, 1985) indicated that the effect of SHMP to inhibit age gelation in UHT milk could vary tremendously depending upon the phosphate manufacturer. The previous study described the use of thin layer chromatography to evaluate the SHMP from different suppliers. Indeed, SHMP from one supplier contained large amounts of unpolymerized material (orthophosphates, pyrophosphates, STPP, and TSPP). Caution should be exercised when ordering samples of SHMP since it is manufactured in short, medium, and long chain forms. Each will have some different physicochemical properties.

3. Protein Gelation

Xiong (1992) evaluated the effect of ions and ionic strength on the thermal aggregation of whey proteins. Whey protein isolate was dialyzed against distilled, deionized water to remove calcium and small molecules. Sodium phosphate (5 mM) inhibited protein aggregation (1.2 mg/mL) when heated to 100°C, while 20 to 50 mM concentrations caused aggregation at 77°C. The addition of 0.6M NaCl plus 50 mM sodium phosphate (pH 7.0 to 7.5) resulted in a dramatic increase in protein aggregation.

Matsudomi et al. (1992) evaluated the effect of gelation and gel properties by varying the ratios of alpha-lactalbumin (La) and beta-lactoglobulin (Lg) in a 100 mM potassium phosphate buffer (pH 6.8). Four percent solutions of Lg gelled, while 8% solutions of La did not gel upon heating at 80°C for 30 min. The addition of 6% La to 2% Lg formed gels at lower heating temperatures due to sulfhydryl–disulfide interchange. A similar effect was noted using bovine serum albumin (BSA) and La in a 100 mM sodium phosphate buffer (Matsudomi et al., 1993).

Calcium caseinate (25 to 45% solids), SHMP (long chain, 0.5% concentration), and carrageenan (1 to 4%) or starches (wheat or potato, 2.5 to 5%) were used to prepare gels with properties similar to kamaboko (Konstance, 1993). At 35 to 45% levels of calcium caseinate, 0.5% SHMP, and carrageenan (1 to 2%) gels, properties of hardness, cohesiveness, and water holding capacity (not elasticity) were similar to kamaboko. It was determined that cohesiveness was dependent upon the added phosphate. The long chain SHMP was responsible for calcium binding and linkages that improved the gel matrix.

4. Reduced Fat Formulations

Reduced fat formulations have gained popularity in the United States. This is related to recommendations to reduce dietary fat to 30% or less of total daily caloric intake.

Keogh (1993) described a method for the production of a water-in-oil emulsion (75:25 w/w) for low fat spreads. Sodium caseinate, starch, and sodium and hydrogen ions contributed to water binding, increased aqueous phase viscosity, and emulsion stability. Citrates and DSP bound calcium ions to enhance emulsion stability.

Melachouris et al. (1992), described a technique for reforming casein micelles to be suitable in coffee whiteners or in reduced fat formulations (frozen and refrigerated desserts, puddings, whipped toppings, sauces and dips). High shear was used to process a casein material (0.1 to 0.4 μm) without a micellar structure; soluble calcium salts, sodium hydroxide (to adjust pH to 6.0), and phosphates (DSP, DKP, TSP, TKP, or TSPP) were added to stabilized the reformed ''micelle.'' The resultant product can be spray-dried and will have a fat content of about 0.5%.

E. Vegetable and Fruit Processing

The two most important functions of phosphates in vegetable and fruit processing are chelating metal ions and acidification. Ellinger (1972b) covered this topic extensively, and Molins (1991) updated this topic noting that little work had been published in the intervening two decades (Table 19). Greatest attention will therefore be provided to the primary uses of phosphates in vegetable and fruit processing.

Potato processing is the main use of phosphates in the processing of vegetables. It is used to inhibit the after-cook blackening of the cut potato surface. Talley et al. (1969) published an extensive bibliography related to the after-cook discoloration of potatoes. The discoloration is due to oxidation of o-diphenolic compounds in the presence of metal ions. Typically, a dilute (ca. 1%) SAPP solution heated to 140–160°F is sprayed onto the cut potato surface in order to complex ferrous ions in a colorless form. Iron content of the potato will vary by region and from year to year, and the concentration of the diphenols will increase according to the size of the potato (Siciliano et al., 1969; Heisler et al., 1969) and with its length of post-harvest storage. Sweet potatoes have grown in popularity, and they are also treated with SAPP to inhibit discoloration.

F. Beverages

Carbonated (cola and root beer) beverages contain phosphoric acid, which functions as an acidulant and contributes specific, tart flavor properties. Typically about 0.05 and 0.1%

Table 19 Use of Food Phosphates in Fruits and Vegetables

Application	Function
Canned berries	MCP (0.035% calculated as Ca) in combination with pectic acid (polygalacturonic acid) results in a firming of texture.
Canned peas, tomatoes, bean sprouts, lima beans, carrots, green sweet peppers, red sweet peppers, and potatoes	Calcium salts may be added as firming agents (peas, 0.035%; tomatoes, 0.045 to 0.08%, calculated as calcium).
French fries	A dilute (1 to 2%) SAPP spray is applied to the cut French fries to sequester iron. This inhibits blackening of the potato after the par-fry.
Fruit butter	Phosphoric acid may be added to acidify or to preserve (amount varies).
Fruit jelly, jam, and preserves	Phosphates may be added as acidifiers or buffers.
Peas and beans	SHMP is applied to the peas and beans to inhibit toughening of the skin (by sequestering calcium).

phosphoric acid is used in colas and root beers, respectively. Other phosphates and beverage applications are described in Table 20.

Orange juice has been supplemented with calcium (TCP) in the United States for the past 12 years, and approval is now pending by the Canadian government (Solomon, 1997). The levels added range from 0.1 to 0.2% by weight, which is in the range for that occurring in milk (0.12%) (Burkes et al., 1995). Calcium supplementation of fruit juice has gained in popularity due to the association of inadequate dietary calcium and the development of osteoporosis. In the United States, milk consumption generally decreases with age due to an increased consumption of soft drinks, coffee and tea, and alcoholic beverages.

Canton (1994) has patented a dry formulation for use with coffee to form a foamed head typical of that of steam generator prepared coffee (espresso or Cuban coffee). This novel preparation consists, in approximate amounts, of sugar (90%), gelatinized starch (5%), sodium bicarbonate (3%), MCP (1%), and SALP (1%). The leavening phosphates are employed to react to generate gas bubbles in the range of 1–2 mm in diameter (foam) and the gelatinized starch stabilizes the foam and aids to form the skin.

In the United States, there is a large market for isotonic beverages intended for consumption after vigorous exercise and during physical work in hot and humid climates. An example of an isotonic beverage to replace electrolytes, nutrients, and energy lost during exercise has been patented (Hastings, 1992). This formulation consisted (in percentages) of fructose (72.2), maltodextrin (16.7), citric acid (5.2), lemon flavor (1.6), TCP (1.4), potassium citrate (0.9), salt (0.7), hydrogenated soy oil (0.6), vitamin C (0.3), MgO (0.2), zinc monomethionate (0.1), chromemate (0.1), and a trace of beta carotene. This beverage would supply approximately 50% of the adult daily requirement of calcium. Phosphorus loading is of importance as well, since Cade et al. (1984) demonstrated that

Table 20 Phosphoric Acid and Phosphates in Beverages

Application	Phosphoric acid or phosphate	Function
Colas and root beer	H_3PO_4	Acidulant and flavor
Dry beverage mixes	MCP	Nonhygroscopic partial (50%) replacement for citric acid
	TCP	Anticaking agent; clouding agent; nutritional supplement
Fruit juices/cider	MCP; H_3PO_4	Acidulants
	SHMP	Stabilizer; shelf-life extension
	TCP	Calcium supplement
Isotonic formulations	MSP; MKP	Sodium or potassium supplement
Noncarbonated beverages	SHMP; STPP	Sequestrant; shelf-life extension
	TCP	Calcium supplement
Nutritional formulations	MCP; DCP; TCP; MKP; DKP; TKPP; MSP	Mineral supplement
UHT milk	SHMP	Inhibit age gelation
Evaporated, condensed, or dried milk	DSP	Inhibit protein coagulation
Nondairy creamer	DKP; MKP	Inhibit feathering
Beer and wine	DAP	Yeast nutrient
	SHMP	Prevent clouding

increased concentration of phosphate led to elevating 2,3-diphosphoglycerate and oxygen loading of the red blood cells. This would be of use for short-term oxygenation of tissues for aerobic exercise or when oxygenation may be reduced in disease states including vascular obstruction.

The effect of DSP and DKP on emulsion stability of liquids prepared with soy protein isolate (SPI) has been reported by Hwang et al. (1992). Emulsion stability was greatest when either DSP or DKP was added prior to emulsification and under unfavorable conditions: pH 4.5, the isoelectric point of the SPI.

Barnes et al. (1992) investigated the effect of various stabilizers in a model milk beverage system which was sweetened, acidified, and carbonated in order to prevent a two layer separation from occurring while simulating the viscosity of 2% milk. DSP, DAP, or sodium bicarbonate at levels of 0.3% prevented separation of the beverage for up to 21 days.

G. Oils and Dressings

The refinement of oils has been well documented (Ellinger, 1972b). The phosphates are used in both alkalai and acid refining protocol as well as during bleaching. Recently, Taylor et al. (1992) described the manufacture of zirconium silicate (an intercalate), bonded using phosphate, phosphonate, or phosphites, to remove carotenoids (yellow to red pigments) of palm, cottonseed, and grapeseed oils and chlorophyll (green pigment) soya oil.

Attempts to reduce dietary fat have led to the development of nonstick spray coatings for cookware. Clapp and Campbell (1992) described a formulation that consisted of phosphate salt derivatives of mono- and diglycerides, vegetable oils, a blocking agent (one of which may be baking powder), a suspending agent, and a propellant for spraying from an aerosol container. This formulation is claimed to be stable to 400 to 500°F (204 to 260°C), which is superior to lecithin-based sprays, which tend to char at temperatures of 350°F (176.6°C).

Reduced fat, low fat, and fat free products have shown explosive growth in recent years. The two major limitations include texture and flavor defects due to the use of hydrocolloids, which have distinctly different texture and flavor carrier profiles. Coutant and Wong (1994) patented a formulation to include 1.5 to 3.0% TCP in reduced fat liquid products, focusing on salad dressings. The presence of TCP enhanced opacity and smoothness, while concomitantly reducing ''gloppiness'' or the tendency to pour unevenly or have a '' gloppy flow.'' Although it is not stated, it is presumed that a TCP with a small particle size would be preferred for greatest smoothness. Other uses of phosphates in oils and dressings are shown in Table 21.

H. Eggs

Phosphates are used in many egg products for a variety of technical and functional effects. Shell-on eggs may be washed in weak phosphate solutions to sequester iron, an accelerator of egg spoilage. Preservation of natural color, improved whipping, foam height, and sensitization of *Salmonella* are shown in Table 22.

Reduction of cholesterol in egg yolks using acid alone or a combination of salt and acid has also been patented (Lombardo and Kijowski, 1994). This process involves the use of high shear and centrifugal force to separate the mixture into a stream of egg yolk

Table 21 Food Phosphates for Use in Salad Dressings, Fats, and Oils

Application	Function
Diet salad dressings	H_3PO_4 acts as an acidulant to impart flavor; acidity may act to control pH and inhibit microbial growth.
French dressing	SHMP sequesters calcium to aid added thickeners.
Margarine	STPP acts as a stabilizer.
Mayonnaise	SHMP acts as a sequestrant.
Oil refining	H_3PO_4 is used in acid and alkalai refining.
Salad dressing	Phosphates act as sequestrants, stabilizers, and thickening aids.

and water and a second stream of oil and cholesterol. Although the preferred acid is acetic, phosphoric, ascorbic, and the like are also suitable.

I. Oral Care Products

Common oral care products include gels, toothpastes, mouthwashes, lozenges, and chewing gum. Sodium monofluorophosphate (SMFP) has been used in toothpaste for decades to supply additional fluoride. DCP-2 is included for its abrasive properties, which include polishing and whitening the teeth (Nathoo et al., 1991, 1992; Chan et al., 1990). More recently, reports have been issued relating to restoration of the tooth surface or a repair of scratches and early enamel caries and remineralization of exposed roots (Schumann et al., 1992; Tung, 1992; Cheng et al., 1991) by DCP which is strengthened by fluoride. Another patent (Winston and Usen, 1996) has included any one or a combination of DKP, DCP-2, DSP, SHMP, SMFP, and MKP to remineralize teeth.

TSPP is often added to inhibit demineralization of the enamel (Featherstone and Mazzanobile, 1993). Saito (1991) determined that a combination of TSPP (20 ppm) and casein phosphopeptide (40 ppm) synergistically inhibited hydroxyapatite formation (dental

Table 22 Food Phosphates for Use in Eggs

Application	Function
Egg yolks—dried, frozen, or refrigerated	STPP and SHMP protect protein from coagulation and MSP or MKP aid to preserve natural color.
Egg whites—dried	H_3PO_4 to adjust pH for reduction of glucose. STPP (1.3 to 2.0%) and SHMP (2.3 to 3.0%) improve subsequent whipping ability and stability to frozen storage. SHMP (2.5% dry wt) aids to decrease whip time, increase foam height, and stability and tolerance to overbeating.
Egg whites—frozen	MSP or MKP preserve color. SHMP and STPP improve whipping ability.
Egg whites—refrigerated	MSP lowers the pH (from 9.3 to ca. 8.0) for better tolerance to overbeating of pasteurized egg whites. SHMP (2%) overcomes effects of yolk contamination on foam development.
Liquid eggs	MSP or MKP (0.5%) preserves color. SHMP (0.5%) plus $CaCl_2$ during pasteurization (130 to 135°F; 54 to 57°C) enhance protein stability and reduction of *Salmonella*.

calculus) and patented their inclusion in a toothpaste formulation. Winston et al. (1991) patented a dentifrice (anticalculus) formulation which contained sodium bicarbonate to displace TKPP (which may impart a bitter flavor at high levels) and reduced TSPP. Drake et al. (1994) investigated the effect of pyrophosphate on oral bacteria (*Streptococcus sanguis, S. mutans* (serotype C), *Actinomyces viscosus, A. naeslundii*) commonly associated with supragingival plaque. All species were susceptible to TSPP at minimal inhibitory concentrations (0.67% wt/vol; 25mM). Pronounced growth inhibition (over 24 h) of *S. mutans* was observed after two 5-min exposures to TSPP in combinations with sodium dodecyl sulfate (Drake et al., 1994).

Toothpaste and mouthwash formulations to prevent dental calculus containing other phosphates have also been patented. White et al. (1990) included the use of STPP and either MSP or TSP; Gaffar et al. (1990) included combinations of TSPP, SMFP, and polyvinyl phosphonate (to inhibit enzymic hydrolysis of TSPP) and Gershon et al. (1991) described an oral rinse for removing dental plaque containing DSP-0 and MSP.

J. Pet Foods

Phosphates are included in many pet food applications, which include wet and dry food and snacks and treats. STPP and TSPP are frequently included in canned foods to maintain the integrity of the pack or to bind the ingredients into a cohesive mass or chunk. MCP, DCP, and TCP are used as calcium sources. Phosphoric acid is used as an acidulant to treat meat byproducts during the production of either a high or low viscosity hydrolysate ingredient used in, for example, extruded pellets and flavor sprays, respectively.

Awareness of human dental health and the implications that bacterial gum infections can lead to an infected pericardium has lead to research into the prevention of calculus in dogs and cats. Scaglione et al. (1989) determined that application of a solution containing TSPP (5.42%) and trisodium pyrophosphate (1.85%) showed a significant reduction in tartar accumulation in dogs. These researchers patented a dry biscuit formulation containing flour, soy meal, meat and bone meal, TSPP, wheat meal, tallow, natural flavors, salt, trisodium pyrophosphate, DCP, bone meal, dough conditioners, CaCO, and a vitamin premix. The dry biscuit, when chewed, cleans tooth surfaces, removes tartar, and exercises and massages the gums. Scaglione and Staples (1989a) later patented a dry biscuit formulation without DCP but containing the two aforementioned sodium phosphates in lesser amounts. This was followed by a patent for a biscuit coating containing only trisodium pyrophosphate (Scaglione and Staples, 1989b).

Spanier et al. (1989) applied reduced levels of TSPP and trisodium phosphate (1.5% and 0.5%, respectively) to dogs' teeth and noted reduction of tartar over a 30-day period. These observations resulted in a patent for processing rawhide treats containing sodium phosphates to reduce tartar when chewed/eaten. Spanier and Ekpo (1989) subsequently patented a process to coat the rawhide treat with a flavored, tartar control solution.

K. Miscellaneous Uses of Food Grade Phosphates

1. Candy and Confections

DCP agglomerated with lecithin and sugar is used to produce a nonchalky chewing gum base (Carroll et al., 1985). It also acts to control hydration of the gum and control the amorphous, cohesive mass.

2. Pigment Integrity

Washino and Moriwaki (1989) patented a process to protect rutin from UV irradiation. This process involved the conversion of rutin to quercetin 3-O-monoglucoside and subsequently a water-soluble flavonol glycoside. Sugar syrup was mixed with purple corn pigment (5 ppm), flavonol glycoside (0.5 ppm), sodium erythorbate (1 ppm), and SHMP (0.1 ppm) and after 16 h of UV treatment resulted in 83% remaining pigment.

3. Flavor

Ugawa et al. (1992) investigated the effects of either NaCl or sodium phosphate on the sweetness of the amino acids glycine, alanine, and serine. It was determined that 30 mM NaCl added to 100 mM of amino acid was equivalent to 500 to 600 mM amino acid while sodium phosphate affected sweetness only slightly. Aoki and Hata described the addition of apatite heated to 1100°C (beta-TCP) to improve the flavor of rice.

4. Insect Control

TCP (2%) was added to whole and dehulled sorghum flour (Rao and Vimala, 1993). It was determined to control insect infestation during storage. This has previously been reported to be effective in wheat flour and works by abrasive action under the insect wing.

L. Microbial Inhibition

A number of reports have been published related to the antimicrobial effects of the food phosphates. While the food phosphates are not considered direct preservatives, they can impart some desirable properties when used as acidulants or in combination with other food ingredients.

Rajkowski et al. (1994) evaluated the antimicrobial effect of SHMP [13 (medium chain) at 0.5 and 1.0%] in either the presence or absence of NaCl in UHT sterilized milk. Either *Listeria monocytogenes* or *Staphylococcus aureus* were inoculated into the substrate at a level of 10^3 CFU/mL. Cultures were incubated at 12, 19, 28, or 37°C for up to 48 h. No significant inhibition of growth was determined due to the presence of SHMP alone or in combination with NaCl. This was attributed to the presence of calcium and magnesium in the milk. SHMP is known to form soluble calcium complexes with milk casein. This phosphate (1% solution) typically has a pH of 6.9; it is also a very weak buffer and therefore would not have exerted a strong pH effect.

Lee et al. (1994) confirmed that *S. aureus* would not be inhibited by sodium ultraphosphate (UP, sodium acid hexametaphosphate) and SHMP (chain length, ca. 20) in the presence of either 10 mM calcium or magnesium or by TSPP in the presence of 10 mM iron. These effects were noted if the minerals were added prior to inoculation or 1 h after inoculation. Chang and Lee (1990) confirmed that bacterial inhibition mediated by SHMP was decreased by $CaCl_2$, KCl, and $MgSO_4$. Jen and Shelef (1986) noted that inhibitory effects of STPP were overcome by calcium and iron but not zinc or magnesium.

A number of studies have described the autoclave sterilization of the phosphates within the culture media. This will result in hydrolysis of pyro-, tripoly-, and hexa- metaphosphates to orthophosphate. Therefore, the results will not reflect anything related to any phosphate longer than an orthophosphate. Phosphate solutions should be filter sterilized and added to cooled, sterile media.

The antimicrobial effect (initial load 10 CFU/mL) use of SHMP (chain length 3 to 100) at levels of 900 to 3000 ppm with a preservative (sorbic acid, benzoic acid, or combinations thereof) in noncarbonated beverages (pH 2.5 to 4.5) has been patented (Calderas et al., 1995). It is important to note that this patent refers specifically to noncarbonated juice and tea preparations which would typically provide minimal nutrient value to microflora. The authors noted specifically that fortification of such beverages with calcium, magnesium, and iron or the use of hard water would nullify the beneficial antimicrobial effect of the SHMP. A second patent (Smith et al., 1996) extends the shelf-life of similar juice- and tea-based noncarbonated beverages and cites inhibition of *Zygosaccharomyces bailii* (a preservative-resistant yeast) and an acid tolerant, preservative-resistant bacteria.

These patents are significantly narrower relative to antibacterial and antifungal claims cited by Kohl and Ellinger (1972). These authors evaluated the impact of polyphosphates (chain length 16 to 37, 0.5 to 1.0% wt/wt) on products ranging from apple juice and egg whites to fish fillets. In some products the synergistic effect of sorbates, benzoates, and proprionates to inhibit fungi was noted.

Nisin in combination with SHMP was investigated for inhibitory effects on the growth of *Escherichia coli* 0157:H7 and *Salmonella typhimurium*. Although SHMP enhanced the effect of nisin, EDTA was more effective in reducing bacterial populations in vitro (Cutter and Siragusa, 1995a). These effects were not reproducible in situ using lean beef tissue as the substrate (Cutter and Siragusa, 1995b).

Yen et al. (1991) evaluated, individually and in combination, the effect of cure ingredients [NaCl, dextrose, nitrite, erythorbate, and phosphates (0.4% STPP plus SHMP)] on the inhibition of *L. monocytogenes* in ground pork. Fresh pork was mixed with the cure ingredients, inoculated with 10^7 to 10^8 CFU/g *L. monocytogenes*, and cooked in a water bath to 63°C. It was not determined if the phosphates enhanced heat lethality or inhibited recovery of injured cells (0.8 log reduction). Most disturbing was the fact that the greatest protective effect to the organism was noted in the presence of all cure ingredients.

Li et al. (1993) determined no STPP mediated microbial inhibition in either cooked or uncooked ground turkey meat. In ground, raw turkey meat, the hydrolysis of STPP to orthophosphate was complete after 1 day storage at 5°C. Frederick et al. (1994) studied the effect of microbial growth of 95% lean beef, German (cooked) sausage containing either 0.5 or 1.0% phosphate (unspecified). Microbial growth was somewhat reduced in products containing phosphate plus 0, 10, or 20% added water, but was enhanced in the high fat control and the sample containing 35% added water. Hydrolysis of phosphate during the cook cycle or during frozen storage to orthophosphate may actually have been a nutrient microbial growth.

Phosphates have been patented as carcass wash treatments to either reduce or eliminate microbial contamination. Pathogen reduction has been of greatest interest due to recent outbreaks of foodborne illness.

Bender and Brotsky (1991) and Bender and Brotsky (1993a,b) patented processes for treating poultry carcasses and red meat (to control *Salmonella* and bacterial contaminants, respectively) with phosphates and, particularly, 8 to 12% TSP solutions. This is a very alkaline treatment (pH ca. 11–12). A second treatment includes a chealating agent (SAPP, acidic SHMP, EDTA, etc.), a fatty acid monoester, and a food grade surfactant (Andrews and Munson, 1995) and was described as being as effective or moreso than an 8% TSP wash. Various researchers have evaluated the efficacy of a variety of carcass washes and arrived at differing conclusions (Bautista et al., 1997; Lillard, 1994; Somers and Schoeni, 1994). Among factors in dispute are the carryover of an alkaline wash into either diluent peptone or plating media, poor adherence of bacteria when carcasses are artificially inoculated, and sublethal injury. Pathogen reduction does not equate to elimination.

Kim and Slavik (1994) determined that a 10% TSP wash to beef pieces resulted in effective removal of *Escherichia coli* and *Salmonella typhimurium* from the fat, but less so from the fascia. *E. coli* was determined to be more efficiently removed than *S. typhimurium*.

A disinfectant containing ethanol (8.7%), lactic acid (0.87%), and phosphoric acid (0.43%) for eliminating coliforms from blocks of kamaboko was described by Ueno et al. (1981). This work was expanded to include a group of organic acids (acetic, citric, tartaric gluconic, malic ascorbic, or phytic) to substitute for lactic when used in combination with ethanol and phosphoric acid (Ueno et al., 1987). This formulation was described to be suitable for treating kamaboko, onion, cucumber, chicken eggs, chicken portions, Vienna sausage, ham, raw vegetables, noodles, and pasta. It was described as disinfecting a kamabokolike crab cake after a contact time of 30 s (Ueno et al., 1984).

III. NUTRITIONAL EFFECTS

From the numerous uses and applications described within this chapter, the reader could question how much dietary phosphorus is consumed due to the use of food grade phosphates and phosphoric acid and whether there is any toxicological effect attributable to the phosphates. In addition, the specter of revitalizing substandard foods and adding excessive water have been issues raised by the FDA and other international regulatory organizations.

A. Phosphate Consumption

The total consumption of food grade phosphates has been estimated by many investigators. Such estimates have been based upon dietary recall disappearance data on food, and physically analyzing subjects' diets and industry production estimates. All are subject to error. Dietary recall is typically clouded by selective memory of the subject not wanting to divulge true amounts of food and alcohol consumed, inability to accurately estimate quantities consumed, and poor documentation habits. Disappearance data of food assume that all food was consumed, that a given ingredient was indeed used and at a maximum permitted level, and there must be corrections for imports and exports and ingredients used as processing aids (those ingredients used but removed in further processing, are present in insignificant amounts, or are present in levels naturally occurring in the food). Chemically analyzing a given diet only provides a total concentration for a given substance; it fails to provide for natural variation and is prone to laboratory error and efficacy of the method of determination. Industry estimates can provide valuable insight into production, but must be corrected for processing aids, nonfood uses, and oral care products which are expectorated (Table 23).

Greger and Krystofiak (1982) estimated, based upon industry production in 1973, that 376 mg phosphorus were consumed per person per day as food additives. The authors admitted that this was a crude estimate but that it was close to the 50th percentile for intake of the top 17 additives based upon the National Academy of Sciences Survey of the Industry on the Use of Food Additives (Committee on the GRAS List Survey, 1979). It was therefore estimated that phosphates contributed between 20 to 30% of phosphorus in the average adult diet (Greger and Krystofiak, 1982). Feldheim and Kratzat (1978) estimated that the level was closer to 10% phosphorus attributable to food phosphates in the German diet, and that these numbers had been stable on a consumption pattern between

Table 23 Nonfood and Processing Aid Uses of Food Grade Phosphates

Phosphate	Nonfood use	Processing aids
Sodium orthophosphate	Cosmetics (all categories apply), animal feed, enema formulations	Antimicrobial, poultry denuding/scald, rendering aid, starch modification
Sodium polyphosphates	Mouthwash, anticoagulants, water treatment, cosmetics	Poultry scald, whey processing
Tetrasodium pyrophosphates	Pet food, dentifrice, mouthwash	Poultry scald, packaging material, plastic wrap extrusion
Sodium acid pyrophosphate	Metal finishing, coatings, dental applications	Poultry scald, french fries processing, packaging material
Sodium tripolyphosphate	Pet food, dental applications, cosmetics, animal feed	Poutry scald, mechanical peeling of Pacific shrimp, egg cleaning, starch modification
Phosphoric acid	Pet food, metal finishing, water treatment (binds lead), animal feed	Sugar refining, oil degumming, lubricating/can manufacturing, coreactant (polydextrose), starch modification, cleaning equipment, yeast/bacterial nutrient
Others		
Ca^{2+}	Pet food supplement, chewing gum, photocopy toners, dentifrice, pharmaceutical excipients, animal feed	Rendering aid, stabilizer in packaging material, styrofoam polymerization, mechanical poultry deboning
NH_4^+/K^+	Tobacco, wine/spirits, microbial (yeast/bacterial) nutrient, animal feed, dental applications	

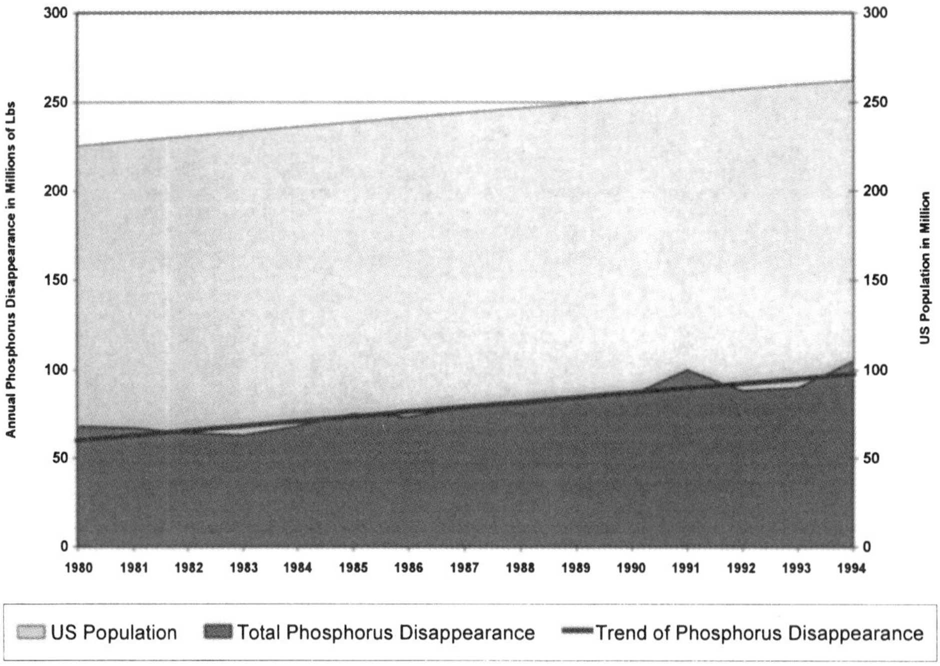

Figure 3 Disappearance of phosphorus (P_4).

1909 to 1978. More recent data (IFAC, 1997) have shown, based on industry production estimates that there has been no increase in the consumption pattern of phosphates between 1980 and 1996, corrected for population (Fig. 3).

B. Safety of Food Phosphates

The U.S. Food and Drug Administration has classified the food phosphates as GRAS (generally recognized as safe) when used at levels not exceeding that required for the intended, functional effect. Standardized foods have maximum levels for use and phosphates, and other added ingredients are described in 21 CFR Parts 131 to 169. Phosphates supply phosphorus and, depending upon the type, contribute calcium, sodium, or potassium—all elements essential to human life. There are recommended daily intakes (RDIs) for each of these minerals depending upon age (Williams, 1993). The FAO/WHO (1964,1982,1986) has established levels for the consumption of phosphorus at <30 mg/kg unconditionally and 30 to 70 mg/kg conditionally. The phosphates are used medicinally and as nutritional supplements. Each topic will be covered separately.

C. Medicinal Uses of Food Phosphates

Pharmaceutical grade products require the use of U.S.P. grade ingredients, while over-the-counter (OTC) medications may include either food grade or U.S.P. quality. Oral care products and mouthwash formula products have been covered in a previous section.

Unmilled DCP-2 is a commonly used excipient used in tabletting. Both DCP and TCP are used in calcium supplements and in vitamin preparations.

MSP and DSP are used in combination in commercial preparations to induce laxation either for constipation or in preparing patients for lower gastrointestinal procedures (*Physicians Desk Reference*, 1996). The concentration used is very high (2.4 g MSP and 0.9 g DSP) and would not normally be consumed in any food product due to an objectionable flavor and mouth-feel. Extremely high levels of DSP and MSP (7 and 19 g, respectively) are used as enemas (*Physicians Desk Reference*, 1996).

D. Calcium and Phosphorus Status

There has been a concern that excess dietary phosphorus will cause calcium depletion and exacerbate the onset and severity of osteoporosis. This theory has been based on animal data, and the hypothesis that it is assumed to carry over to humans. Greger (1988) noted the vast difference in bone metabolism between rats and humans, thus showing unreliability of the rodent model. Anderson et al. (1977), saw no evidence of bone disease in monkeys fed a low calcium, high phosphorus diet for as long as 7 years.

Zemel and Linkswiler (1981) evaluated the effect of high and low orthophosphate (MKP or MSP) and polyphosphate (SHMP) and calcium supplemented diets on men to determine the effects of absorption and retention of calcium and phosphorus and to measure parathydroid hormone (PTH) secretion and measures of bone resorption. The results indicated the following: that phosphate has little if any effect on calcium absorption when calcium intake is low; calcium loss was decreased by the orthophosphate supplement due to a decrease in urinary calcium and an increase in calcium absorption; calcium and phosphorus supplements acted synergistically to improve calcium balance; phosphorus retention is not improved by high phosphate intake; and markers (cyclic AMP and hydroxyproline) suggested a decrease in PTH-mediated bone resorption. SHMP did not completely hydrolyze (20% recovered from the feces) and was associated with increased calcium [during a low calcium (399 mg/day) regimen] in the feces attributable to sequestration by the intact phosphate. It is important to note that SHMP is typically used in meats to a (typical) level of 0 to 250 mg/100-g serving and beverages to a level of 1000 ppm or less. The level consumed in this dietary regimen was 1000 mg phosphorus per day, which translates to 3.29 g of SHMP (30.4% phosphorus). Since SHMP is either enzymatically cleaved or hydrolyzes to short chain and orthophosphate in a food system, it is unlikely one's intake would approach that level.

Silverberg et al. (1986) evaluated the effect of administering 2 g elemental phosphorus daily to 13 healthy subjects aged 19 to 36 for 5 days to determine the impact on activating bone remodeling. The results indicate that there were three factors to support bone remodeling mediated by short-term phosphate feedings. First, PTH levels (a physiologic stimulus to new bone formation) became significantly elevated by day 3 concurrent with a reduction in hydroxyproline (a marker indicative of bone resorption). Second, elevated PTH leads to phosphaturia and decreased urinary excretion of calcium concomitant with their serum reduction, which may be indicative of their deposition into the bone. Third, gamma-carboxyglutamic acid protein (BGP)—a marker of osteoblast activity, the metabolite thought responsible for bone formation—was elevated by day 2. There is a negative impact of long-term phosphorus supplementation, but it was suggested that short-term cyclic administration would yield positive effects.

Smith et al. (1989) evaluated the effect of consumption of nonalcoholic carbonated beverages (1.4 L/day) containing phosphoric acid on calcium metabolism in healthy women. Over an 8-week period, mean serum levels of calcium, ionized calcium, phosphorus, alkaline phosphatase, parathyroid hormone, 1,25-dihydroxyvitamin D_3, and osteocalcin levels were similar to those consuming the control diet. Kohls (1991) supplemented cola with calcium and noted calcium balance as well.

Barger-Lux and Heaney (1993) evaluated the impact of either calcium restriction (5 mM/day) or supplementation (70 mM/day) on markers of bone modeling in 28 healthy premenopausal women. By 8 weeks, the low-Ca women showed elevated levels of PTH and urine hydroxyproline, markers of bone resorption. Both groups were infused with calcium and the low-Ca group evidenced greater serum levels and lower urinary levels of Ca than the high-Ca group. Low urinary excretion of Ca tended to last for 2 days after inducing the hypercalcemic state. These researchers concluded that there were detectable differences in the Ca regulatory system between high- and low-Ca subjects. Calcium starving showed higher levels of PTH and bone remodeling. Ability to spare Ca was also noted and has been confirmed by Fairweather-Tait et al. (1992). These results are important since none of the subjects had consumed diets high in phosphorus and therefore that Ca restriction alone can evoke a PTH response in the absence of high P.

Other dietary factors must be considered relative to calcium depletion to include protein, electrolytes, fiber and phytate, oxalate, fat, lactose, vitamin D, and calcium supplements. Excellent reviews on calcium bioavailability have been published by Miller (1989) and Greger (1988).

Dietary protein causes increases in urinary calcium excretion (Miller, 1989). Fortuitously, most high protein foods contain substantial levels of (naturally occurring) phosphorus, which tends to offset the hypercalciuric effect of protein (Zemel, 1985). The hypercalciuric effect of high dietary protein (53 g), but not of moderate protein (23 g), supplemented with phosphorus (7.5 mM) was confirmed by Whiting et al. (1997).

It has been well documented that phytate (inositol hexakisphosphate) binds calcium, thus reducing its bioavailability. Miller (1989) noted that little was actually published on the effects of processing on hydrolysis of phytate and the reduced sequestration of phytate metabolites. Morris and Hill (1995) evaluated the concentrations of inositol tris- (IP3), tetra- (IP4), penta- (IP5), and hexakisphosphate (IP6) in 43 different breakfast cereals. The results indicated that the levels of IP6 and IP5 were greatest in bran or bran added cereals (17.9 μmol/g) > oat (10.9) > wheat (8.0) > mixed grain cereals (4.9) > rice (1.5) > corn-based cereals. These results indicate the relative quantities of mineral binding phytate in the different cereal types and the importance of determining type as opposed to total phytate.

Calcium binding ($CaCl_2$) of 18 different fiber sources was evaluated by Weber et al. (1993). The results indicated that calcium binding capacity ranged from 480 μg/g to 20,137 μg/g for cellulose and orange fiber, respectively. This indicates that fiber source, in addition to phytate, of high fiber diets is an important consideration to calcium bioavailability.

Bioavailability of calcium sources is important for selecting the appropriate salt for supplementation or fortification. A consumer test was reported which involved placing calcium tablets in vinegar to determine their disintegration as a measure of efficacy. Mason et al. (1992) tested this method against standard methods (Simulate Gastric Fluid Test) to determine its accuracy. Although their results indicated agreement; it was concluded that disintegration is not a valid measure of calcium bioavailability. Heaney et al. (1990)

Table 24 Nutrient Content of Food Grade Phosphates

Trade names	Product code	P (g/100 g)	K (g/100 g)	Ca (g/100 g)	Na (g/100 g)	Moisture (%)	Ash (%)
Diammonium phosphate	DAP	23.5	0	0	0	0	80.3
Monoammonium phosphate	MAP	26.9	0	0	0	0	69.6
Dicalcium phosphate, anhydrous	DCP-0 or DCP-A	22.8	0	29.4	0	0	93.4
Dicalcium phosphate, duohydrate	DCP-2	18.0	0	23.3	0	20.9	73.8
Monocalcium phosphate, monohydrate	MCP-1	24.6	0	15.9	0	7.1	78.6
Tricalcium phosphate or calcium phosphate, tribasic	TCP	18.5	0	39.9	0	0	100
Calcium acid pyrophosphate	CAPP	24.4	0	31.0	0	0	90.3
Sodium aluminum phosphate, basic	SALP	19.0	0	0	28.2	0	97.2
Sodium aluminum phosphate, anhydrous	SALP-A	26.9	0	0	7.4	0	84.9
Sodium aluminum phosphate, tetrahydrate	SALP-4	26.1	0	0	2.4	7.6	78.8
Dimagnesium phosphate, trihydrate	DMP	17.8	0	0	0	31.0	62.6
Dipotassium phosphate, anhydrous	DKP	17.8	44.9	0	0	0	94.8
Monopotassium phosphate	MKP	22.8	28.7	0	0	0	86.7
Tetrapotassium pyrophosphate	TKPP	18.8	47.3	0	0	0	100
Disodium phosphate, anhydrous	DSP-0, DSP-A	21.8	0	0	32.4	0	93.7
Disodium phosphate, duohydrate	DSP-2	17.4	0	0	25.8	20.2	74.7
Monosodium phosphate	MSP	25.8	0	0	19.2	< 1	85
Sodium acid pyrophosphate	SAPP	27.9	0	0	20.7	0	91.9
Sodium hexametaphosphate or sodium polyphosphate, glassy	SHMP or SPG	30.4	0	0	22.5	0	100
Tetrasodium pyrophosphate	TSPP	23.3	0	0	34.6	0	100
Trisodium phosphate, dodecahydrate	TSP-12	8.2	0	0	18.1	55.4	44.6

determined that even under controlled, chemically defined conditions, solubility of a calcium source has little influence on its absorbability and that absorption is determined by other food components. Miller (1989) best summarized the bioavailability in tablets as being dependent upon the solubility in the gastrointestinal tract, pH and ionic strength, binders used in tableting, and the formulation and compression used.

E. Behavioral Effects

In the 1970s, much attention was given to the correlation between diet and hyperkinetic behavior in children. For the most part, food additives were cited as the instigator to hyperactivity, and exceptional media attention was then given to the elimination diets. The most notable example was probably the Feingold diet, which promoted the avoidance of foods containing food additives.

Steinhausen (1980) evaluated 13 peer-reviewed studies published between 1976 and 1978 related to the impact of food additives on hyperkinetic behavior. It was concluded that there is ''no scientific evidence to support dietary change in therapeutic intervention with hyperkinetic children.''

Stolley et al. (1979) evaluated the so-called diet poor in phosphate for the dietary treatment of hyperkinesis in children. It was determined that the diet was not really poor in phosphate, that is, the control and phosphate poor diets contributed 1370 and 1230 mg phosphorus per day, respectively. The intake of phosphate as an ingredient to processed foods accounted for only 3% of total dietary phosphorus in six different groups of children ranging in age from 2 to 14.9 years. Furthermore, the diet poor in phosphate was low in carbohydrate, dietary fiber, and ascorbic acid, while notably rich in calories from animal protein, total fat, and cholesterol.

Walther et al. (1980) conducted a study with 35 hyperactive children, of which 14 of the parents observed that behavior had improved after following a low phosphate diet publicized by the media. Trials consisted of one of three treatments that included a control, a low phosphate diet, and an additive free diet. Standardized observation protocol and testing did not show any behavioral differences attributable to diet. In a double-blind trial, acute behavioral deterioration followed both phosphate and placebo consumption. The investigators indicated that behavioral disorders in children could be neither generated nor maintained by dietary phosphates.

In almost two decades, research into the mechanisms of hyperkinesis has shifted from food additives. Much work has focused on the effects of allergy, toxin, nutrient, and environmental deficiency. In fact, it has been determined that administration of pyroidoxine hydrochloride (Vitamin B_6) significantly increased serum levels of serotonin and pyridoxal phosphate, compounds which were routinely depressed in hyperactive children.

F. Nutrition Content of Food Phosphates

The nutrition content of the food grade phosphates are shown in Table 24. This information is representative of those products described throughout this chapter and is based on information provided by the industry. It is important to contact the manufacturer for actual specifications for each ingredient.

REFERENCES

Agusti, C. 1993. New applications of monocalcium phosphate. *Alimentacion Wquipos y Tecnolgia* 12:77.

Akamittah, J. G., Brekke, C. J., Schauss, E. G. 1990. Lipid oxidation and color stability in restructured meat systems during frozen storage. *J. Food Sci.* 55:1513.

Anderson, M. P., Hunt, R. D., Griffiths, J. H., McIntyre, K. W., Zimmerman, R. E. 1977. Long-term effect of low dietary calcium: phosphorus ratio on the skeleton of *Cebus albifronis* monkeys. *J. Nutr.* 107:834.

Andrews, J. F., Munson, J. F. 1995. Disinfectant composition. International patent WO 95/07616.

Anjaneyulu, A. S. R., Sharma, N., Kondaiah, N. 1990a. Specific effect of phosphate on the functional properties and yield of buffalo meat patties. *Food Chem.* 36:149.

Anjaneyulu, A. S. R., Sharma, N., Kondaiah, N. 1990b. The effect of salt and phosphate preblending of buffalo meat on the physicochemical properties during refrigerated storage. *Fleischwirtschaft* 70:177.

Anjaneyulu, A. S. R., Sharma, N. 1991. Effect of fat and phosphate on the quality of raw and precooked buffalo meat patties. *J. Food Sci. Technol.* (India) 28:157.

Aoki, H., Hata, M. 1989. Food taste–improving materials containing calcium phosphate ceramics for cooking. Japanese patent 2,215,352.

Applewhite, L., Otwell, W. S., Garrido, L. R. 1992. Initial studies to measure consumer perception of water added to shrimp with phosphate treatments. Proceedings of the Seventeenth Annual Tropical and Subtropical Fisheries Technological Conference of the Americas. Merida, Mexico. November 4–6, 1992, pp. 86–89.

Applewhite, L., Otwell, W. S., Garrido, L. R. 1993. Consumer evaluations of phosphated shrimp and scallops. Proceedings of the Joint Meeting of the Atlantic Fisheries Technologists and Tropical Subtropical Fisheries Technologists. Williamsburg, VA. August/September, 1993. p. 101.

AWA. 1997. *Procedure for estimating the Phase I content of STPP using FT-IR spectroscopy.* Albright & Wilson Americas Laboratory Method AP-04-17.

AWA. 1994. *Rate of solution test.* Albright & Wilson Americas Laboratory Method E-099.

Barbut, S. 1988. Microstructure of reduced salt meat batters as affected by polyphosphates and chopping time. *J. Food Sci.* 53:1300.

Barbut, S., Maurer, A. J., Lindsay, R. C. 1988. Effects of reduced sodium chloride and added phosphates on physical and sensory properties of turkey frankfurters. *J. Food Sci.* 53:62.

Barger-Lux, M. J., Heaney, R. P. 1993. Effects of calcium restriction on metabolite characteristics of premenopausal women. *J. Clin. Endocrinol. Metab.* 76:103.

Barnes, D. L., McGuire, R. P., Bodyfelt, F. W., McDaniel, M. R. 1992. Effect of selected stabilizing and buffering agents on a sweetened acidified, carbonated milk beverage. *Cultured Dairy Products Journal* 27:21.

Bautista, D. A., Sylvester, N., Barbut, S., Griffiths, M. W. 1997. The determination of efficacy of antimicrobial rinses on turkey carcasses using response surface designs. *Int. J. Food Microbiol.* 34:279.

Bender, F. G., Brotsky, E. 1991. Process for treating poultry carcasses to control salmonellae growth. U.S. patent 5,069,922.

Bender, F. G., Brotsky, E. 1993a. Process for treating red meat to control bacterial contamination and/or growth. U.S. patent 5,192,570.

Bender, F. G., Brotsky, E. 1993b. Process for treating red meat to control bacterial contamination and/or growth. U.S. patent 5,268,185.

Berger, W., Klostermeyer, H., Merkenich, K., Uhlmann, G. 1989. *Processed Cheese Manufacture: A. Joha Guide.* BK Laudenburg, Ladenburg, Germany.

Billy, T. 1994. FDA Office of Seafood. Letter to Dr. Michael Morrisey, Oregon State University.

Brose, E. 1993. ‘‘Leavening Agents.’’ Chemische Fabrik Budenheim Rudolf A. Oetker. Budenheim bei, Mainz, Germany.

Burkes, A. L., Butterbaugh, J. L., Fieler, G. M., Gore, W. J., Zuniga, M. E. 1995. Storage stable calcium-supplemented beverage concentrates. U.S. patent 5,422,128.

Cade, R., Conte, M., Zauner, C., Mars, D., Peterson, J., Lunne, D., Hommen, N., Packer, D. 1984.

Effects of phosphate loading on 2,3-diphosphoglycerate and maximal oxygen uptake. *Med. Sci. Sports Exercise.* 16:263.
Calderas, J. J., Graumlich, T. R., Jenkins, L., Sabin, R. P. 1995. Preparation of noncarbonated beverage products with improved microbial stability. U.S. patent 5,431,940.
Canton, S. T. 1994. Additive for foaming coffee. U.S. patent 5,350,591.
Carroll, T. J., Rubin, M., Piccolo, D. J., Glass, M. 1985. Process for preparing a non-chalky, organoleptically pleasing chewing gum composition. U.S. patent 4,493,849.
Chan, A. S., Wiedemann, J. R., Hull, S. L. 1990. Zinc tripolyphosphate compounds as anticalculus and antiplaque agents. Canadian patent 90-2,028,249.
Chang, Y. H., Lee, C. 1992. Preparation of starch phosphates by extrusion. *J. Food Sci.* 571:203.
Chang, D., Lee, T. 1990. Effects of treatment method and environmental factors on the bacteriostatic activity condensed phosphates. *Bull. Korean Fisheries Soc.* 23:394.
Chang-Lee, M. V., Pacheco-Aguilar, R., Crawford, D. L., Lampila, L. E. 1989. Proteolytic activity of surimi from Pacific Whiting (*Merluccius productos*) and heat-set gel texture. *J. Food Sci.* 54:1116.
Chang-Lee, M. V., Lampila, L. E., Crawford, D. L. 1990. Yield and composition of surimi from Pacific Whiting (*Merluccius productos*) and the effect of various protein additives on gel strength. *J. Food Sci.* 55:83.
Chen, M. T., Pan, S. W., Liu, D. C. 1991. Preparation and characteristics of chicken surimi. IV. Effects of cryoprotectants and cold storage on functionality and physical properties of chicken surimi. *J. Chinese Soc. Animal Sci.* 20:503.
Cheng, C., Gao, J., Wang, Q., Du, B., Qi, H. 1991. The effect of calcium ion and phosphate on remineralization of early enamel caries in vitro. *Beijing Yikee Daxue Xuebao.* 23:305.
Chung, F. H. Y. 1992. Chemical leavening agents. In: *Kirk-Othmer Encyclopedia of Chemical Technology*, 4th ed., Vol. 3. John Wiley & Sons, New York.
Clapp, C. P., Campbell, J. 1992. Parting composition for cooking foodstuffs. U.S. patent 5,156, 876.
Clarke, A. D., Means, W. J., Schmidt, G. R. 1987. Effects of storage time, sodium chloride and sodium tripolyphosphate on yield and microstructure of comminuted beef. *J. Food Sci.* 52:854.
Code of Federal Regulations. 1997. Title 21. Food and drugs. Parts 100–199. Office of the Federal Register, National Archives, and Records Administration, Washington, D.C.
Committee on the GRAS List Survey. 1979. Phase III. National Academy of Sciences. Washington, D.C.
Corbridge, D. E. C. 1986. Phosphorus, an outline of its chemistry, biochemistry and technology. In: *Studies in Inorganic Chemistry 6*, 3rd ed. Elsevier, Amsterdam.
Coutant, A. F., Wong, P. 1994. Tricalcium phosphate to generate smoothness and opaqueness in reduced fat liquid food products and method. U.S. patent 5,292,544.
Craig, J. B. Bowers, J. A., Seib, P. 1991. Sodium tripolyphosphate and sodium ascorbate monophosphate as inhibitors of off-flavor development in cooked, vacuum-packaged, frozen turkey. *J. Food Sci.* 56:1529.
Crapo, C. A., Crawford, D. L. 1991. Influence of polyphosphate soak and cooking procedures on yield and quality of dungeness crab meat. *J. Food Sci.* 56:657.
Crawford, D. L. 1980. Meat yield and shell removal functions of shrimp processing. Oregon State University Extension Marine Advisory Program. Special Report 597.
Crawford, D. L. 1984. Composition for treating fish fillet to increase yield and shelf-life. U.S. patent 4,431,679.
Crawford, D. L. 1985. Composition for treating fish to increase fish yield. U.S. patent 4,517,208.
Crawford, D. L. 1988. Personal communication.
Cutter, C. N., Siragusa, G. R. 1995a. Population reductions of gram-negative pathogens following treatments with nisin and chelators under various conditions. *J. Food Protection* 58:977.
Cutter, C. N., Siragusa, G. R. 1995b. Treatments with nicin and chelators to reduce *Salmonella* and *Escherichia coli* on beef. *J. Food Protection* 58:1028.

DeMan, J. M., Mehnychyn, P. 1971. *Symposium: Phosphates in Food Processing*. AVI Publishing, Westport, CT.

Demott, B. J. 1990. Influence of addition of calcium chloride or sodium phosphates upon calcium retention and body and texture characteristics of cottage cheese curd. *Cultured Dairy Products Journal* 25:24.

Drake, D., Grigsby, B., Krotz-Dieleman, D. 1994. Growth-inhibitory effect of pyrophosphate on oral bacteria. *Oral Microbiol. Immunol.* 9:25.

Dupont, Y., Pugeois, R. 1983. Evaluation of H_2O activity in the free or phosphorylated catalytic site of Ca^{++}-ATPase. CA2 *FEBS Lett.* 156:93.

Duxbury, D. 1986. Phosphate blends in shrimp, fish reduce thaw shrink, cook loss . . . research evaluates cryoprotection. *Food Processing* 47:18.

Dziezak, J. D. 1990. Phosphates improve many foods. *Food Technol.* 44:79.

Ellinger, R. H. 1972a. *Phosphates as Food Ingredients*. CRC Press, Cleveland, OH.

Ellinger, R. H. 1972b. Phosphates in food processing. In: *Handbook of Food Additives*. CRC Press, Cleveland, OH.

Fairweather-Tait, S., Prentice, A., Heumann, K. G., Jarjou, L. M. A., Stirling, D. M., Wharf, S. G., Turnlund, J. R. 1995. Effect of calcium supplements and stage of lactation on the calcium absorption efficiency of lactating women accustomed to low calcium intakes. *Am. J. Clin. Nutr.* 62:1188.

FAO/WHO Expert Committee on Food Additives. 1986. *Evaluation of Certain Food Additives and Contaminants*, 29th ed. WHO Tech. Rep. Ser. No. 733. World Health Organization, Geneva.

FAO/WHO Expert Committee on Food Additives. 1982. *Evaluation of Certain Food Additives and Contaminants*, 26th ed. Tech. Rep. Ser. No. 683. World Health Organization, Geneva.

FAO/WHO Expert Committee on Food Additives. 1964. *Specification for the Identity and Purity of Food Additives and their Toxicological Evaluation: Emulsifiers, Stabilizers, Bleaching and Maturing Agents*. WHO Tech. Rep. Ser. No. 281. World Health Organization, Geneva.

Farouk, M. M., Price, J. F., Salih, A. M., Burnett, R. J. 1992. The effect of postexsanguination infusion of beef on composition, tenderness, and functional properties. *J. Animal Sci.* 70: 2773.

Featherstone, J. D. B., Mazzanobile, S. 1993. Dental formulations for preventing tooth enamel demineralization. International patent WO 9319728.

Feldheim, W., Kratzat, J. 1978. Untersuchungen zum calcium- und phosphor- stoffwechsel beim menschen. I. Mitteilung: Bergleich der aufnahme mit der kost in Deutschlend/BRD und in den USA in den Jahren 1909–1976. *Akt. Erhaehrung*. 2:58.

Fortuna, T. 1991. Comparative studies on the properties of distarch phosphates of various origin. *Acta Aliment. Pol.* 17:275.

Fortuna, T., Gibinski, M., Palasinski, M. 1990. Properties of monostarch phosphates depending on the degree of phosphorus substitution. *Acta Aliment. Pol.* 16:27.

Frederick, T. L., Miller, M. F., Tinney, K. S., Bye, L. R., Ramsey, C. B. 1994. Characteristics of 95% lean beef German sausages varying in phosphate and added water. *J. Food Sci.* 59:453.

Froning, G. W. 1965. Effect of polyphosphates on binding properties of chicken muscle. *Poultry Sci.* 44:1104.

Gaffar, A., Afflitto, J., Smith, S. F. 1990. Anticalculus oral composition having alculus-inhibiting polyphosphate salt and polymeric vinyl phosphate as enzymic hydrolysis inhibitor. U.S. patent 5,094,844.

Gard, D. R., Heidolph, B. B. 1995. Leavening composition. U.S. patent 5,405,636.

Gard, J. K., Gard, D. R., Callis, D. F. 1992. In: *Phosphorus Chemistry Developments in American Science*. Walsh, E. N., Griffith, E. J., Parry, R. W., Quin, L. D. (Eds.). ACS Symposium Series No. 486. American Chemical Society, Washington, D.C.

Gariepy, C., Delaquis, P. J., Aalhus, J. L., McGinnis, D. S., Robertson, M., Leblanc, C., Rodrigue, N. 1994. A modified hot processing strategy for beef: functionality of electrically stimulated and hot-boned meat preblended with different NaCl concentrations. *Food Res. Int.* 25:519.

Gershon, S., Fox, C., Garfinkle, N. 1990. Oral rinse and method for plaque removal. International patent WO 9,104008

Greger, J. L. 1988. Calcium bioavailability. *Cereal Foods World* 33:796.

Greger, J. L., Krystofiak, M. 1982. Phosphorus intake of Americans. *Food Technol.* 36:78.

Griffith, E. J. 1959. A nomograph for the hydrolytic degradation of pyro and tripolyphosphates. *Ind. Eng. Chem.* 51:240.

Hamm, R. 1986. Functional properties of the myofibrillar system and their measurements. In: *Muscle as Food.* Bechtel, P. J. (Ed.) Academic Press, Orlando, FL.

Harwalkar, V. R., Vreeman, H. J. 1978a. Effect of added phosphates and storage on changes in ultra-high temperature short-time sterilized concentrated skim-milk. 1. Viscosity, gelation, alcohol stability, chemical and electrophoretic analysis of proteins. *Netherlands Milk Dairy J.* 32:94.

Harwalkar, V. R., Vreeman, H. J. 1978b. Effect of added phosphates and storage on changes in ultra-high temperature short-time sterilized concentrated skim-milk. 2. Micelle structure. *Netherlands Milk Dairy J.* 32:204.

Hastings, C. 1992. Isotonic energy composition for use as a beverage. U.S. patent 5,294,606.

Heaney, R. P., Recker, R. R., Weaver, C. M. 1990. Absorbability of calcium sources: the limited role of solubility. *Calcif. Tissue Intl.* 46:300.

Hegenbart, S. 1990. Processing aids: the hidden helpers. *Prepared Foods* 159:83.

Heidolph, B. B. 1996. Designing chemical leavening systems. *Cereal Foods World* 41:118.

Heisler, E. G., Siciliano, J., Porter, W. L. 1969. Relation of potato composition to potato size and blackening tendency. *Am. Potato J.* 46:107.

Hughes, J. T., Czochanska, Z., Pickston, L., Hove, E. L. 1980. The nutritional composition of some New Zealand marine fish and shellfish. *New Zealand J. Science* 23:43.

Hwang, J., Kim, Y., Pyun, Y. 1992. Effect of phosphate salts on the emulsion stability of soy protein isolate. *J. Korean Agric. Chem. Soc.* 35:152.

IFAC (Ebert, A. G.) 1997. Letter to Panel on Calcium and Related Nutrients, Standing Committee/ Scientific Evaluation of Dietary Reference Intakes, Food and Nutrition Board, National Academy of Sciences, July 8, 1997.

Iler, R. K. 1952. *J. Phys. Chem.* 56:1086.

Irani, R. R., Callis, C. F. 1962. *J. Am. Chem. Soc.* 84:156.

Irani, R. R., Morgenthaler, W. W. 1963. *J. Am. Chem. Soc.* 84:283.

Jen, C. M. C., Shelef, L. A. 1986. Factors affecting sensitivity of *Stapylococcus aureus* 196E to polyphosphates. *Appl. Environ. Microbiol.* 52:842.

Jencks, W. P. 1992. Phosphorus chemistry developments in American science. ACS Symposium Series No. 486. Walsh, E. N., Griffith, E. J., Parry, R. W., Quin, L. D. (Eds.). American Chemical Society, Washington, D.C.

Kalab, M., Modler, H. W., Caric, M., Milanovic, S. 1991. Structure, meltability and firmness of process cheese containing white cheese. *Food Structure* 10:193.

Kawana, T., Yamane, Y., Inaguma, T., Ishiguro, Y. 1990. Manufacture of deep fried foods packaged in retort pouch starch phosphate. Japanese Patent 4,210,563.

Keogh, M. K. 1993. Contribution of ingredients and their interactions to the stability of a water-in-oil emulsion. *Intl. Dairy Fed.* 1992 special issue 9303:352.

Khan, S., Ghatak, P. K., Bandyopadhyay, A. K. 1992. Heat stability of buffalo milk as affected by phosphate salts addition. *Indian J. Dairy Sci.* 45:461.

Khuntia, B. K., Srikar, L. N., Reddy, G. V. S., Srinivasa, B. R. 1993. Effect of food additives on quality of salted pink perch (*Nemipterus japonicus*). *J. Food Sci. Technol.* 30:261.

Kim, Y., Cho, Y. 1992. Relationship between quality of frozen surimi and jelly strength of kamaboko. *Bull. Korean Fish Soc.* 25:73.

Kim, C., Han, E. S. 1991. Effect of sodium chloride, phosphate and pH on the functional properties of a mixed system of pork myofibrillar and plasma proteins. *Han'guk Sikp'um Kwahakhoechi* 23:428.

Kim, J., Slavik, M. F. 1994. Trisodium phosphate (TSP) treatment of beef surfaces to reduce *Escherichia coli* O157:H7 and *Salmonella typhimurium*. *J. Food Sci.* 59:20.

Kirk-Othmer. 1982. *Encyclopedia of Chemical Technology*. Mark, H. F., Othmer, D. F., Overberger, C. G., Seaborg, G. J. (Eds.). John Wiley & Sons, New York.

Kocak, H. R., Zadow, J. G. 1985. Polyphosphate variability in the control of age gelation in UHT milk. *Australian J. Dairy Technol.* 40:65.

Kocak, H. R., Zadow, J. G. 1986. Storage problems in UHT milk: age gelation. *Food Technol. Australia* 38:148.

Kohl, W. F., Ellinger, R. H. 1972. Method of preserving food materials, food product resulting therefrom, and preservative composition. U.S. patent 3,681,091.

Kohls, K. 1991. Calcium bioavailability from calcium fortified food products. *J. Nutr. Sci. Vitaminol.* 37:319.

Konstance, R. P. 1993. Axial compression properties of calcium caseinate gels. *J. Dairy Science* 76:3317.

Kubomura, K. 1998. Instant noodles in Japan. *Cereal Foods World* 43:194.

Lajoie, M. S., Thomas, M. C. 1991. Versatility of bicarbonate leavening bases. *Cereal Foods World* 36:420.

Lampila, L. E. 1991. Comparative microstructure of red meat, poultry and fish muscle. *J. Muscle Foods*. 1:247.

Lampila, L. E. 1992. Functions and uses of phosphates in the seafood industry. *J. Aquatic Food Product Technol.* 1 (3/4):29.

Lampila, L. E. 1994. Polyphosphates: rationale for their use and functionality in seafood and seafood products. Proceedings of the Joint Meeting of the Atlantic Fisheries Technologists and Tropical Subtropical Fisheries Technologists. Williamsburg, VA. August/September, 1993.

Lee, C. M., Wu, M., Okada, M. 1992. Ingredient and formulation technology for surimi-based products. *Food Sci. Technol.* 50:273.

Lee, R. M., Hartman, P. A., Olson, D. G., Williams, F. D. 1994. Metal ions reverse the inhibitory effects of selected food-grade phosphates in *Stapylococcus aureus*. *J. Food Protection* 57:284.

Lewis, D. F., Groves, K. H. M., Hodgate, J. H. 1986. Action of polyphosphates in meat products. *Food Microstructure* 5:53.

Li, W., Bowers, J. A., Craig, J. A., Perng, S. K. 1993. Sodium tripolyphosphate stability and effect in ground turkey meat. *J. Food Sci.* 58:501.

Lillard, H. S. 1994. Effect of trisodium phosphate on *Salmonellae* attached to chicken skin. *J. Food Protection*. 57:465.

Lombardo, S. P., Kijowski, M. 1994. Reduction of cholesterol in egg yolk by the addition of either acid or both salt and acid. U.S. patent 5,312,640.

Loreal, H., Etienne, M. 1990. Added water in frozen scallop muscles: French specifications and methodology. 20th WEFTA Meeting. Reykjavik, Iceland.

Loter, I. 1983. Acid cheese curd production—using salt of acid to reduce pH without causing casein precipitation. U.S. patent 4,374,152.

Maesso, E. R., Baker, R. C., Vadehra, D. V. 1970. The effect of vacuum, pressure, pH and different meat types on the binding ability of poultry meat. *Poultry Sci.* 49:697.

Mason, N. A., Patel, J. D., Dressman, J. B., Shimp, L. A. 1992. Consumer vinegar test for determining calcium disintegration. *Am. J. Hosp. Pharm.* 49:2218.

Matsudomi, N., Oshita, T., Sasaki, E., Kobayashi, K. 1992. Enhanced heat-induced gelation of beta-lactoglobulin by alpha-lactalbumin. *Biosci. Biotech. Biochem.* 56:1697.

Matsudomi, N., Oshita, T., Kobayashi, K. Kinsella, J. E. 1993. Alpha-lactabalbumin enhances the gelation properties of bovine serum albumin. *J. Agric. Food Chem.* 41:1053.

Matsumoto, J. J., Noguchi, S. F. 1992. Cryostabilization of protein in surimi. *Food Sci. Technol.* 50:357.

McCullough, J. F., Van Wazer, J. R., Griffith, E. J. 1956. *J. Amer. Chem. Soc.* 78:4528.

McGilvery, J. D. ASTM Bulletin No. 191. July 1953.

McMahon, D., Savello, P. A., Brown, R. J. Kalab, M. 1991. Effects of phosphate and citrate on the gelation properties of casein micelles in renneted ultra-high temperature (UHT) sterilized concentrated milk. *Food Structure* 10:27.

Melachouris, N., Moffitt, K. R., Rasilewicz, C. E., Tonner, G. F. 1992. Reformed casein micelles. U.S. patent 5,173,322.

Merkenich, K., Maurer-Rothmann, A., Scheurer, G., Tanzler, R. 1993. Acid sodium phosphates, their manufacture and their use in cheesemaking. U.S. patent 4,128,124.

Mikkelsen, A., Bertelsen, G., Skibsted, L. H. 1991. Polyphosphates as antioxidants in frozen beef patties. Lipid oxidation and color quality during retail display. *Z. Lebensm. Unters. Forsch.* 192:309.

Mil, P. J. J. M., van de Koning, J. 1992. Effect of heat treatment, stabilizing salts and seasonal variation on heat stability of reconstituted concentrated skim milk. *Netherlands Milk Dairy J.* 46:169.

Miller, D. D. 1989. Calcium in the diet: food sources, recommended intakes, and nutritional bioavailability. *Advances Food Nutr. Res.* 33:103.

Molins, R. A. 1991. *Phosphates in Food.* CRC Press, Boca Raton, FL.

Montilla, A., Calvo, M. M. 1997. Goat's milk stability during heat treatment: effect of pH and phosphates. *J. Agriculture Food Chem.* 45:931.

Mora-Gutierrez, A., Farrell, H. M., Jr., Basch, J. J., Kumosinski, T. F. 1993. Modeling calcium-induced solubility in caprine milk caseins using a thermodynamic linkage approach. *J. Dairy Sci.* 76:3698.

Morey, K. S., Satterlee, L. D., Brown, W. D. 1982. Protein quality of fish in modified atmospheres as predicted by the C-PER assay. *J. Food Sci.* 47:1399.

Morris, E. R., Hill, A. D. 1995. Inositol phosphate, calcium, magnesium, and zinc contents of selected breakfast cereals. *J. Food Comp. Anal.* 8:3.

Nathoo, S. A., Chmielewski, M. B., Fakhry-Smith, S. 1991. Aqueous tooth whitening dentifrice. U.S. patent 5,171,564.

Nathoo, S. A., Prencipe, M., Chmielewski, M. B., Drago, V., Smith, S. F. 1992. Abrasive tooth-whitening dentifrices containing dicalcium phosphate and peroxide. European patent 92-308294.

Neraal, R., Hamm, R. 1977a. Enzymatic breakdown of tripolyphosphate and diphosphate in minced meat. II. Occurrence of tripolyphosphatase in muscular tissue. *Z. Lebensm. Unters. Forsch.* 163:18.

Neraal, R., Hamm, R. 1977b. Enzymatic breakdown of tripolyphosphate and diphosphate in minced meat. V. Change in diphosphatase activity of muscle post-mortem. *Z. Lebensm. Unters. Forsch.* 163:208.

Neraal, R., Hamm, R. 1977c. Enzymatic breakdown of tripolyphosphate and diphosphate in comminuted meat. XI. *Z. Lebensm. Unters. Forsch.* 164:101.

Offer, G., Trinick, J. 1983. A unifying hypothesis for the mechanism of changes in the water-holding capacity of meat. *J. Sci. Food Agriculture* 34:1018.

Ohta, F., Yamada, T. 1990. On the correlation between the concentrations of phosphates and chlorides in the unfrozen portions of the frozen fish muscle juices and the denaturation rate of fish muscle protein during frozen-storage. *JAR* (trans.) 7:85.

Osterheld, R. K. 1972. Non-enzymic hydrolysis at phosphate tetrahedra. In: *Topics in Phosphorous Chemistry*, Vol. 7. Griffith, E. J., Grayson, M. (Eds.). John Wiley & Sons, New York.

Pacheco-Aguilar, R., Crawford, D. L., Lampila, L. E. 1989. Procedures for the efficient washing of minced whiting (*Merluccius productos*) flesh for surimi production. *J. Food Sci.* 54:248.

Park, J. W., Lanier, T. C., Pilkington, D. H. 1993. Cryostabilization of functional properties of pre-rigor and post-rigor beef by dextrose polymer and/or phosphates. *J. Food Sci.* 58:467.

Physicians Desk Reference. 1996. R. Arky (Ed.). Medical Economics Company, Montvale, NJ.

Pouliot, Y., Boulet, M. 1991. Seasonal variations in the heat stability of concentrated milk: Effect of added phosphates and pH adjustment. *J. Dairy Sci.* 74:1157.

Rajkowski, K. T., Calderone, S. M., Jones, E. 1994. Effect of polyphosphate and sodium chloride on the growth of *Listeria monocytogenes and Stapylococcus aureus* in ultra-high temperature milk. *J. Dairy Sci.* 77:1503.

Rao, A., Vimala, V. 1993. Efficacy of tricalcium phosphate on the storage quality of sorghum flour. *J. Food Sci. Technol.* (India). 30:58.

Regenstein, J. M. 1984. Protein–water interactions in muscle foods. *Reciprocal Meat Conference Proceedings* 37:44.

Regenstein, J. M., Rank Stamm, J. 1979a. The effect of sodium polyphosphates and of divalent cations on the water holding capacity of pre- and post-rigor chicken breast muscle. *J. Food Biochem.* 3:213.

Regenstein, J. M., Rank Stamm, J. 1979b. Factors affecting the sodium chloride extractability of muscle proteins from chicken breast, trout white, and lobster tail muscles. *J. Food Biochem.* 3:191.

Regenstein, J. M., Rank Stamm, J. 1979c. A comparison of the water holding capacity of pre- and post-rigor chicken, trout and lobster muscle in the presence of polyphosphates and divalent cations. *J. Food Biochem.* 3:223.

Reiman, H. M. 1977. Chemical leavening systems. *Bakers Digest* 51:33.

Robe, G. H., Xiong, Y. L. 1992. Phosphates and muscle fiber type influence thermal transitions in porcine salt-soluble aggregation. *J. Food Sci.* 57:1304.

Robe, G. H., Xiong, Y. L. 1993. Dynamic rheological studies on salt-soluble proteins from three procine muscles. *Food Hydrocolloids* 7:137.

Robe, G. H., Xiong, Y. L. 1994. Kinetic studies of the effect of muscle fiber type and tripolyphosphate on the aggregation of porcine salt-soluble proteins. *Meat Sci.* 37:55.

Roberts, M. F. 1987. *Phosphorus NMR in Biology*. Burt, C. T. (Ed.). CRC Press, Boca Raton, FL.

Romero, P. J., De Meis, L. 1989. Role of water in the energy of hydrolysis of phosphoanhydride and phosphoester bonds. *J. Biol. Chem.* 264:7869.

Rust, R., Olson, D. 1987. Phosphates in processing workshop. *Meat & Poultry*.

Saint-Martin, H., Ortega-Blake, I., Les, A., Adamowicz, L. 1991. Abinitio calculations of the pyrophoshate hydrolysis reaction. *Biochim. Biophys. Acta* 1080:205.

Saint-Martin, H., Ortega-Blake, I., Les, A., Adamowicz, L. 1994. The role of hydration in the hydrolysis of pyrophosphate—a Monte-Carlo simulation with polarizable-type interaction potentials. *Biochim. Biophys. Acta* 1207:12.

Saito, T. 1991. Dentifrices containing polypeptides and polyphosphates. Japanese patent 4,244,012.

Sato, S., Nakagawa, N., Terui, S., Tsuchiya, T. 1990. Effects of polyphosphates on dispersion state of protein in Alaska pollock frozen surimi. *Nippon Suisan Gakkaishi* 56:1299.

Scaglione, F., Staples, L. C. 1989a. Nutritionally balanced biscuits containing an inorganic pyrophosphate for tartar control. U.S. patent 5,000,973.

Scaglione, F., Staples, L. C. 1989b. Antitartar dog biscuits having a coating containing an inorganic pyrophosphate. U.S. patent 5,015,485.

Scaglione, F., Staples, L. C., Ympa, J. W. 1989. Tartar-control dog biscuits containing alkalai metal pyrophosphate for cats and dogs. U.S. patent 5,000,943.

Schantz, R., Bowers, J. 1993. Response surfaces of sensory characteristics for reduced sodium chloride and phosphate salts in emulsified turkey sausages. *J. Sensory Studies* 8:283.

Schults, G. W., Russell, D. R., Wierbicki, E. 1972. Effect of condensed phosphates on pH, swelling and water-holding capacity of beef. *J. Food Sci.* 37:860.

Schumann, K., Foerg, F., Laska, H. 1992. Remineralizing dentifrice. German patent 4,237,500.

Schwartz, W. C., Mandigo, R. W. 1976. Effect of salt, sodium tripolyphosphate and storage on restructured pork. *J. Food Sci.* 41:1266.

Siciliano, J., Heisler, E. G., Porter, W. L. 1969. Relation of potato size to after-cooking tendency. *Am. Potato J.* 46:91.

Sidwell, V. D. 1981. Chemical and nutritional composition of finfishes, whales, crustaceans, mollusks, and their products. NOAA Technical Memorandum NMFS F/SEC-11. U.S. Department of Commerce, National Oceanic and Atmospheric Administration, National Marine Fisheries Service.

Silverberg, S. J., Shane, E. Clemens, T. L., Dempster, D. W., Segre, G. V., Lindsay, R., Bilezikian, J. P. 1986. The effect of oral phosphate administration on major indices of skeletal metabolism in normal subjects. *J. Bone Mineral Res.* 1:383.

Smith, J. 1991. *Food Additive Users Handbook.* Blackie Academic & Professional, London.

Smith, J. A., Graumlich, T. R., Sabin, R. P., Vigar, J. W. 1996. Preparation of noncarbonated beverage products having superior microbial stability. International patent WO 96/26648.

Smith, S., Swain, J., Brown, E. M., Wyshak, G., Albright, T., Ravnikar, V. A., Schiff, I. 1989. A preliminary report of the short-term effect of carbonated metabolism in normal women. *Arch. Intl. Med.* 149:2517.

Solomon, C. K. 1997. Canada takes tentative steps toward food fortification. *Food in Canada* 57: 13.

Somers, E. B., Schoeni, J. L. 1994. *Int. J. Food Microbiol.* 22:269.

Spanier, H. C., Ekpo, B. O. 1989. Raw hide having a coating containing an inorganic pyrophosphate. U.S. patent 5,011,679.

Spanier, H. C., Staples, L. C., Scaglione, F. 1989. Raw hide containing an inorganic pyrophosphate for prevention or reduction of tartar accumulation in dogs. U.S. patent 5,047,231.

Steinhausen, H. C. 1980. Hyperkinetisches slyndrom und diat—eine therapeutische verbindung? *Klin. Padiat.* 192:179.

Steinmann, R., Fischer, A. 1993. Frankfurter-type sausage manufacture. Protein solubility and stability in frankfurter-type sausage mixture as a function of salt concentration, nitrogen cooling and diphosphate. *Fleischwirtschaft* 73:435.

Stolley, H., Kesting, M., Droese, W., Reinken, L. 1979. Bemerkungen zu einer sogenannten phosphatarmen diat fur kinder mit hyperkinetischem sydrom. *Monatsschr. Kinderheilkd.* 127: 450.

Stone, E. W. 1981. Method of treating fresh shrimp to reduce moisture and nutrient loss. U.S. patent 4,293,578.

Storey, R. Albright & Wilson Americas, unpublished results.

Strack, H. J., Oetker, R. A. 1992. Phosphate key ingredient in meat products. *Intl. Food Ingreds.* 5:45.

Sutton, A. H. 1973. The hydrolysis of sodium triphosphate in cod and beef muscle. *J. Food Technol.* 8:185.

Talley, E. A., Zaehringer, M. V., Reeve, R. M., Hyde, R. B., Dinkel, D. H., Heisler, E. G., Pressey, R. 1969. Bibliography: The after-cooking discoloration of potatoes. *Am. Potato J.* 46:302.

Taylor, R. S., Davies, M. E., Whittle, M., Johnstone, R. 1992. Zirconium phosphate compounds and their use to treat edible oils. International patent WO 92/06782.

Tenhet, V., Finne, G. Nickelson, R., Toloday, D. 1981. Phosphorous levels in peeled and deveined shrimp treated with sodium tripolyphosphate. *J. Food Sci.* 46:350.

Teo, S. K. S., Oates, C. G., Wong, H. A. 1993. A response surface methodology approach to the optimization of the functional properties of urea modified alkali metal starch phosphates. *J. Food Process. Preserv.* 16:381.

Thilo, E. 1955. *Agnew. Chem. Intl. Ed. Engl.* 67:141.

Thilo, E. 1962. *Adv. Inorg. Chem. Radiochem.* 4:1.

Toy, A. D. F. 1976. *Phosphorus Chemistry in Everyday Living*, American Chemical Society, Washington, D.C.

Trecker, G. W., Monckton, S. P. 1990. Grated hard parmesan cheese and method for making same. U.S. patent 4960605.

Trout, G. R., Schmidt, G. R. 1987. The effect of cooking temperature on the functional properties of beef proteins: The role of ionic strength, pH, and pyrophosphate. *Meat Sci.* 20:129.

Tung, M. S. 1992. Dentifrice compositions for mineralizing and fluorinating teeth containing amorphous calcium compounds. International patent WO 9,404,460.
Ueno, R., Kanayama, T., Fujita, Y., Yamamoto, M. 1981. Liquid bactericide for foods and food supplies and food supply handling machines or devices. German Patent 3,138,277.
Ueno, R., Kanayama, T., Fujita, Y., Yamamoto, M. 1984. Liquid bactericide for foods and food processing machines or utensils, employing a synergistic mixture of ethyl alcohol, an organic acid and phosphoric acid. U.S. patent 4,647,458A.
Ueno, R., Kanayama, T., Fujita, Y., Yamamoto, M. 1987. Liquid bactericide for foods and food processing machines or utensils, employing a synergistic mixture of ethyl alcohol, an organic acid and phosphoric acid. U.S. patent 4,647,458.
Ugawa, T., Shoji, K., Kenzo, K. 1992. Enhancing effects of sodium chloride and sodium phosphate on human gustatory responses to amino acids. *Chem. Senses* 17:811.
Ullmann's Encyclopedia of Industrial Chemistry. 1991. John Wiley & Sons, New York.
Van Wazer, J. R. 1958. *Phosphorus and Its Compounds*. Vols. 1 and 2. Wiley Interscience, New York.
Van Wazer, J. R., Callis, D. A. 1958. *Chem. Rev*. 58:1011.
Van Wazer, J. R., Campanella, D. A. 1950. *J. Am. Chem. Soc*. 72:655.
Vipan, K., Tewari, B. D. 1992. Effect of processing variables on physico-chemical properties of cheese powder. *Japanese J. Dairy Food Sci*. 41:A23–A28.
Vogel, A. 1989. *Textbook of Quantitative Chemical Analysis*, 5th ed. (revised by Jeffery, G. H., et al.). Longman Scientific with John Wiley & Sons, New York.
Voyle, C. A., Jolley P. D., Offer. G. W. 1984. The effect of salt and pyrophosphate on the structure of meat. *Food Microstructure* 3:113.
Wagman, D. D., Evans, W. H., Parker, V. B., Halow, I., Bailey, S. M., Schumm, R. H. 1968. Selected values of chemical thermodynamic properties. N.B.S. Technical Note. TN 270.3.
Walther, B., Dieterich, E., Spranger, J. 1980. Verandert nahrungsphosphat neuropsychologische functionen und verhaltensmerkmale hperkinetischer und impulsiver kinder? *Monatsschr. Kinderheilkd*. 128:382.
Washino, K., Moriwaki, M. 1989. Stabilization of anthocyanins. Japanese patent 2,214,780.
Webb, N. B., Thomas, F. B. Busta F. F., Monroe. R. J. 1969. Variations in proximate composition of North Carolina scallop meats. *J. Food Sci*. 34:471.
Weber, C. W., Kohlhepp, E. A., Idouraine, A., Ochoa, L. J. 1993. Binding capacity of 18 fiber sources for calcium. *J. Agric. Food Chem*. 41:1931.
Wekell, J. C., Teeney, F. M. 1988. Canned salmon curd reduced by use of polyphosphates. *J. Food Sci*. 53:1009.
White, D. J., Jr., Cox, E. R., Hunter, M. A. 1990. Phosphate-containing anticalculus dentifrice and mouthwash. U.S. patent 5,096,701.
Whiting, S. J., Anderson, D. J., Weeks, S. J. 1997. Calciuric effects of protein and potassium bicarbonate but not of sodium chloride or phosphate can be detected acutely in adult women and men. *Am. J. Clin. Nutr*. 65:1465.
Williams, S. R. 1993. *Nutrition and Diet Therapy*. Mosby, St. Louis, MO.
Winston, A. E., Usen, N. 1996. Stable single-part compositions and the use thereof for remineralization of lesions in teeth. U.S. patent 5,571,502.
Winston, A. E., Miskewitz, R. M., Walley, D. R., Berschied, J. R. 1991. Anticalculus dentifrices comprising sodium bicarbonate. U. S. patent 5,180,576.
Wood, H. G., Clark, J. E. 1988. Biological aspects of inorganic polyphosphates. *Ann. Rev. Biochem*. 57:235.
Woyewoda, A. D., Bligh, E. G. 1986. Effect of phosphate blends on stability of cod fillets in frozen storage. *J. Food Sci*. 51:932.
Xiong, Y. L. 1992. Influence of pH and ionic environment on thermal aggregation of whey proteins. *J. Agriculture Food Chem*. 40:380.
Yagi, H., Sakamoto, M., Wakameda, A., Arai, K. 1985. Effect of inorganic polyphosphate on ther-

mal denaturation of carp myofibrillar protein at low ionic strength. *Bull. Japanese Soc. Scientific Fisheries* 51:667.

Yen, L. C., Sofos, J. N., Schmidt, G. R. 1991. Effect of meat curing ingredients on thermal destruction of *Listeria monocytogenes* in ground pork. *J. Food Protection* 54:408.

Yu, S. Y., Ngu, T. I., Abu-Akbar, A. 1994. Cryoprotective effects of sucrose, sorbitol and phosphates on *Nemipterus tolu* surimi. *ASEAN Food J.* 9:107.

Zboralski, U. 1986. Processed cheese preparation with addition of insoluble calcium phosphate—having special lime to phosphorus pentoxide ratio and beta-tricalcium phosphate content to fusion salts. U.S. patent 4,609,553.

Zehren, V. L., Nusbaum, D. D. 1992. *Process Cheese.* Cheese Reporter Publishing Company, Madison, WI.

Zemel, M. B. 1985. Phosphates and calcium utilization in humans. In: *Nutritional Bioavailability of Calcium*, Kies, C. (Ed.). American Chemical Society, Washington, D.C.

Zemel, M. B., Linkswiler, H. M. 1981. Calcium metabolism in the young adult male as affected by level and form of phosphorus intake and level of calcium intake. *J. Nutr.* 111:315.

Zorba, O., Gokalp, H. Y., Yetim, H., Ockerman, H. W. 1993a. Model system evaluations of the effects of different levels of dibasic potassium phosphate, sodium chloride, and oil temperature on emulsion stability and viscosity of fresh and frozen Turkish style meat emulsions. *Meat Sci.* 34:145.

Zorba, O., Gokalp, H. Y., Yetim, H., Ockerman, H. W. 1993b. Salt, phosphate and oil temperature effects on emulsion capacity of fresh or frozen meat and sheep tail fat. *J. Food Sci.* 58:492.

APPENDIX 1

Functional Classes, Definitions, and Technological Functions

Functional Classes, Definitions, and Technological Functions

Functional classes (for labeling purposes)	Definition	Subclasses (technological functions)
1. Acid	Increases the acidity and/or imparts a sour taste to a food.	Acidifier
2. Acidity regulator	Alters or controls the acidity or alkalinity of a food	Acid, alkali, base, buffer, buffering agent, pH adjusting agent
3. Anticaking agent	Reduces the tendency of particles of food to adhere to one another.	Anticaking agent, antistick agent, drying agent, dusting powder, release agent
4. Antifoaming agent	Prevents or reduces foaming.	Antifoaming agent
5. Antioxidant	Prolongs the shelf-life of foods by protecting against deterioration caused by oxidation, such as fat rancidity and color changes.	Antioxidant, synergist, sequestrant
6. Bulking agents	A substance, other than air or water, which contributes to the bulk of a food without contributing significantly to its available energy value.	Bulking agent, filler
7. Color	Adds or restores color in a food.	Color
8. Color retention agent	Stabilizes, retains, or intensifies the color of a food.	Color fixative, stabilizer
9. Emulsifier	Forms or maintains a uniform mixture of two or more immiscible phases such as oil and water in a food	Emulsifier, plasticizer, dispersing agent, surface active agent, surfactant, wetting agent
10. Emulsifying salt	Rearranges cheese proteins in the manufacture of processsed cheese in order to prevent fat separation.	Melding salt, sequestrant

11. Firming agent	Makes or keeps tissues of fruit or vegetables firm and crisp or interacts with gelling agents to produce or strengthen a gel.	Firming agent
12. Flavor enhancer	Enhances the existing taste and/or odor of a food.	Flavor enhancer, flavor modifier, tenderizer
13. Flour treatment agent	A substance added to flour to improve its baking quality or color.	Bleaching agent, dough improver, flour improver
14. Foaming agent	Makes it possible to form or maintain a uniform dispersion of a gaseous phase in a liquid or solid food.	Whipping agent, aerating agent
15. Gelling agent	Gives a food texture through formation of a gelling agent.	Gelling agent
16. Glazing agent	A substance which, when applied to the external surface of a food, imparts a shiny appearance or provides a protective coating.	Coating, sealing agent, polish
17. Humectant	Prevents food from drying out by counteracting the effect of a wetting agent atmosphere having a low degree of humidity.	Moisture/water retention agent, wetting agent
18. Preservative	Prolongs the shelf-life of a food by protecting against deterioration caused by microorganisms.	Antimicrobial preservative, antimycotic agent, bacteriophage control agent, chemosterilant/wine maturing agent, disinfection agent
19. Propellant	A gas, other than air, which expels a food from a container.	Propellant
20. Raising agent	A substance or combination of substances which liberates gas and thereby increases the volume of a dough	Leavening agent, raising agent
21. Stabilizer	Makes it possible to maintain a uniform dispersion of two or more immiscible substances in a food.	Binder, firming agent, moisture/water retention agent, foam stabilizer
22. Sweetener	A nonsugar substance which imparts a sweet taste to a food.	Sweetener, artificial sweetener, nutritive sweetener
23. Thickener	Increases the viscosity of a food.	Thickening agent, texturizer, bodying agent

Source: Adapted from Codex Alimentarius Commission, January, 2001. Class names and the international numbering system for food additives. Codex Alimentarius, Vol. 1A, *General Requirements*. www.fao.org/es/esn/codex/standard/volume 1a/vol 1a_e.htm.

APPENDIX 2

Numbering System for Food Additives

Numbering System for Food Additives (listed in alphabetical order)

No.	Name of food additive	Technical function(s)
370	1,4-heptonolactone	Acidity regulator, sequestrant
950	Acesulfame potassium	Sweetener
260	Acetic acid, glacial	Preservative, acidity regulator
472a	Acetic and fatty acid esters of glycerol	Emulsifier, stabilizer, sequestrant
929	Acetone peroxide	Flour treatment agent
355	Adipic acid	Acidity regulator
406	Agar	Thickener, gelling agent, stabilizer
400	Alginic acid	Thickener, stabilizer
956	Alitame	Sweetener
103	Alkanet	Color
129	Allura red AC	Color
307	Alpha-tocopherol	Antioxidant
173	Aluminum	Color
523	Aluminum ammonium sulfate	Stabilizer, firming agent
522	Aluminum potassium sulfate	Acidity regulator, stabilizer
559	Aluminum silicate	Anticaking agent
521	Aluminum sodium sulfate	Firming agent
520	Aluminum sulfate	Firming agent
123	Amaranth	Color
264	Ammonium acetate	Acidity regulator
359	Ammonium adipates	Acidity regulator
403	Ammonium alginate	Thickener, stabilizer
503(i)	Ammonium carbonate	Acidity regulator, raising agent
503	Ammonium carbonates	Acidity regulator, raising agent
510	Ammonium chloride	Flour treatment agent
380	Ammonium citrates	Acidity regulator
368	Ammonium fumarate	Acidity regulator
503(ii)	Ammonium hydrogen carbonate	Acidity regulator, raising agent
527	Ammonium hydroxide	Acidity regulator
328	Ammonium lactate	Acidity regulator, flour treatment agent
349	Ammonium malate	Acidity regulator
923	Ammonium persulfate	Flour treatment agent

342	Ammonium phosphates	Acidity regulator, flour treatment agent
452(v)	Ammonium polyphosphates	Emulsifier, stabilizer, acidity, regulator, raising agent, sequestrant, water retention agent
442	Ammonium salts of phosphatidic acid	Emulsifier
517	Ammonium sulfate	Flour treatment agent, stabilizer
1100	Amylases	Flour treatment agent
160b	Annatto extracts	Color
323	Anoxomer	Antioxidant
163	Anthocyanins	Color
163(i)	Anthocyanins	Color
409	Arabinogalactan	Thickener, gelling agent, stabilizer
938	Argon	Packing gas
300	Ascorbic acid (L-)	Antioxidant
304	Ascorbyl palmitate	Antioxidant
305	Ascorbyl stearate	Antioxidant
951	Aspartame	Sweetener, flavor enhancer
927a	Azodicarbonarnide	Flour treatment agent
122	Azorubine	Color
408	Bakers yeast glycan	Thickener, gelling agent, stabilizer
901	Beeswax, white and yellow	Glazing agent, release agent
162	Beet red	Color
558	Bentonite	Anticaking agent
210	Benzoic acid	Preservative
906	Benzoin gum	Glazing agent
928	Benzoyl peroxide	Flour treatment agent, preservative
160f	Beta-apo-8′-carotenic acid, methyl or ethyl ester	Color
160e	Beta-apo-carotenal	Color
160a(i)	Beta-carotene (synthetic)	Color
459	Beta-cyclodextrin	Stabilizer, binder
163(iii)	Black currant extract	Color
542	Bone phosphate (essentially calcium phosphate, tribasic)	Emulsifier, anticaking agent, water retention agent
151	Brilliant black PN	Color
133	Brilliant blue FCF	Color
1101(iii)	Bromelain	Flour treatment agent, stabilizer, tenderizer, flavor enhancer
443	Brominated vegetable oil	Emulsifier, 154 Brown FK Color
155	Brown HT	Color
943a	Butane	Propellant
320	Butylated hydroxyanisole	Antioxidant
321	Butylated hydroxytoluene	Antioxidant
629	Calcium 5′-guanylate	Flavor enhancer
633	Calcium 5′-inosinate	Flavor enhancer
634	Calcium 5′-ribonucleotides	Flavor enhancer
263	Calcium acetate	Preservative, stabilizer, acidity regulator
404	Calcium alginate	Thickener, stabilizer, gelling agent, antifoaming agent
556	Calcium aluminum silicate	Anticaking agent
302	Calcium ascorbate	Antioxidant
213	Calcium benzoate	Preservative

924b	Calcium bromate	Flour treatment agent
170(i)	Calcium carbonate	Surface colorant, anticaking agent, stabilizer
170	Calcium carbonates	Surface colorant, anticaking agent, stabilizer
509	Calcium chloride	Firming agent
333	Calcium citrates	Acidity regulator, firming agent, sequestrant
450(vii)	Calcium dihydrogen diphosphate	Emulsifier, stabilizer, acidity regulator, raising agent, sequestrant, water retention agent
385	Calcium disodium ethylene-diamine-tetra-acetate	Antioxidant, preservative, sequestrant
538	Calcium ferrocyanide	Anticaking agent
238	Calcium fonnate	Preservative
367	Calcium fumarates	Acidity regulator
578	Calcium gluconate	Acidity regulator, firming agent
623	Calcium glutamate	Flavor enhancer
383	Calcium glycerophosphate	Thickener, gelling agent, stabilizer
170(ii)	Calcium hydrogen carbonate	Surface colorant, anticaking agent, stabilizer
352(i)	Calcium hydrogen malate	Acidity regulator
227	Calcium hydrogen sulfite	Preservative, antioxidant
526	Calcium hydroxide	Acidity regulator, firming agent
916	Calcium iodate	Flour treatment agent
318	Calcium isoascorbate	Antioxidant
327	Calcium lactate	Acidity regulator, flour treatment agent
399	Calcium lactobionate	Stabilizer
482	Calcium lactylates	Emulsifier, stabilizer
352(ii)	Calcium malate	Acidity regulator
352	Calcium malates	Acidity regulator
482(ii)	Calcium oleyllactylate	Emulsifier, stabilizer
529	Calcium oxide	Acidity regulator, flour treatment agent
930	Calcium peroxide	Flour treatment agent
341	Calcium phosphates	Acidity regulator, flour treatment agent, firming agent, texturizer, raising agent, anticaking agent, water retention agent
452(iv)	Calcium polyphosphates	Emulsifier, stabilizer, acidity, regulator, raising agent, sequestrant, water retention agent
282	Calcium propionate	Preservative
552	Calcium silicate	Anticaking agent
203	Calcium sorbate	Preservative
486	Calcium stearoyl fumarate	Emulsifier
482(i)	Calcium stearoyllactylate	Emulsifier, stabilizer
516	Calcium sulfate	Flour treatment agent, sequestrant, firming agent
226	Calcium sulfite	Preservative, antioxidant
354	Calcium tartrate	Acidity regulator
902	Candelilla wax	Glazing agent
161g	Canthaxanthin	Color

150a	Caramel I, plain	Color
150b	Caramel II, caustic sulfite process	Color
150c	Caramel III, ammonia process	Color
150d	Caramel IV, ammonia sulfite process	Color
927b	Carbamide (urea)	Flour treatment agent
152	Carbon black (hydrocarbon)	Color
290	Carbon dioxide	Carbonating agent, packing gas
469	Carboxymethyl cellulose	Thickener, stabilizer
120	Carmines	Color
952	Cyclarnic acid (and Na, K, Ca salts)	Sweetener
265	Dehydroacetic acid	Preservative
472e	Diacetyltartaric and fatty acid esters of glycerol	Emulsifier, stabilizer, sequestrant
342(ii)	Diammonium orthophosphate	Acidity regulator, flour treatment agent
450(vi)	Dicalcium diphosphate	Emulsifier, stabilizer, acidity regulator, raising agent, sequestrant, water retention agent
341(ii)	Dicalcium orthophosphate	Acidity regulator, flour treatment agent, firming agent, texturizer, raising agent, anticaking agent, water retention agent
940	Dichloroditluoromethane	Propellant, liquid freezant
389	Dilauryl thiodipropionate	Antioxidant
450(viii)	Dimagnesium diphosphate	Emulsifier, stabilizer, acidity regulator, raising agent, sequestrant, water retention agent
343(ii)	Dimagnesium orthophosphate	Acidity regulator, anticaking agent
242	Dimethyl dicarbonate	Preservative
480	Dioctyl sodium sulfosuccinate	Emulsifier, wetting agent
230	Diphenyl	Preservative
450	Diphosphates	Emulsifier, stabilizer, acidity regulator, raising agent, sequestrant, water retention agent
628	Dipotassium 5′-guanylate	Flavor enhancer
450(iv)	Dipotassium diphosphate	Emulsifier, stabilizer, acidity, regulator, raising agent, sequestrant, water retention agent
340(ii)	Dipotassium orthophosphate	Acidity regulator, sequestrant, emulsifier, texturizer, stabilizer, water retention agent
336(ii)	Dipotassium tartrate	Stabilizer, sequestrant
627	Disodium 5′-guanylate	Flavor enhancer
631	Disodium 5′-inosinate	Flavor enhancer
635	Disodium 5′-ribonucleotides	Flavor enhancer
903	Carnauba wax	Glazing agent
410	Carob bean gum	Thickener, stabilizer
160a	Carotenes	Color
407	Carrageenan and its Na, K, NH_4 salts (includes furcellaran)	Thickener, gelling agent, stabilizer
1503	Castor oil	Release agent
460	Cellulose	Emulsifier, anticaking agent, texturizer, dispersing agent

925	Chlorine	Flour treatment agent
926	Chlorine dioxide	Flour treatment agent
945	Chloropentafluoroethane	Propellant
140	Chlorophyll	Color
141(i)	Chlorophyll copper complex	Color
141(ii)	Chlorophyllin copper complex, sodium and potassium salts	Color
1000	Cholic acid	Emulsifier
1001(i)	Choline acetate	Emulsifier
1001(ii)	Choline carbonate	Emulsifier
1001(iii)	Choline chloride	Emulsifier
1001(iv)	Choline citrate	Emulsifier
1001(vi)	Choline lactate	Emulsifier
1001	Choline salts and esters	Emulsifier
1001(v)	Choline tartrate	Emulsifier
330	Citric acid	Acidity regulator, antioxidant, sequestrant
472c	Citric and fatty acid esters of glycerol	Emulsifier, stabilizer, sequestrant
121	Citrus red 2	Color
141	Copper chlorophylls	Color
468	Croscaramellose	Stabilizer, binder
519	Cupric sulfate	Color fixative, preservative
100(i)	Curcmnin	Color
100	Curcumins	Color
450(i)	Disodium diphosphate	Emulsifier, stabilizer, acidity regulator, raising agent, sequestrant, water retention agent
386	Disodium ethylene-diamine-tetra–acetate	Antioxidant, preservative synergist, sequestrant
331(ii)	Disodium monohydrogen citrate	Acidity regulator, sequestrant, emulsifier, stabilizer
339(ii)	Disodium orthophosphate	Acidity regulator, sequestrant, emulsifier, texturizer, stabilizer, water retention agent
335(ii)	Disodium tartrate	Stabilizer, sequestrant
390	Distearyl thiodipropionate	Antioxidant
312	Dodecyl gallate	Antioxidant
127	Erythrosine	Color
488	Ethoxylated mono- and diglycerides	Emulsifier
324	Ethoxyquin	Antioxidant
462	Ethyl cellulose	Binder, filler
313	Ethyl gallate	Antioxidant
467	Ethyl hydroxyethyl cellulose	Thickener, stabilizer, emulsifier
637	Ethyl maltol	Flavor enhancer
214	Ethyl p-hydroxybenzoate	Preservative
143	Fast green FCF	Color
570	Fatty acids	Foam stabilizer, glazing agent, antifoaming agent
381	Ferric ammonium citrate	Anticaking agent
505	Ferrous carbonate	Acidity regulator
579	Ferrous gluconate	Color retention agent
537	Ferrous hexacyanomanganate	Anticaking agent

585	Ferrous lactate	Color retention agent
1101(iv)	Ficin	Flour treatment agent, stabilizer, tenderizer, flavor enhancer
161a	Flavoxanthin	Color
240	Formaldehyde	Preservative
236	Formic acid	Preservative
297	Fumaric acid	Acidity regulator
418	Gellan gum	Thickener, stabilizer, gelling agent
574	Gluconic acid (D-)	Acidity regulator, raising agent
575	Glucono delta-lactone	Acidity regulator, raising agent
1102	Glucose oxidase	Antioxidant
620	Glutamic acid (L(+)−)	Flavor enhancer
422	Glycerol	Humectant, bodying agent
445	Glycerol esters of wood resin	Emulsifier, stabilizer
915	Glycerol-, methyl-, or penta-erithrytol esters of colophane	Glazing agent
640	Glycine	Flavor modifier
958	Glycynhizin	Sweetener, flavor enhancer
175	Gold	Color
163(ii)	Grape skin extract	Color
142	Green SC	Color
314	Guaiac resin	Antioxidant
626	Guanylic acid	Flavor enhancer
412	Guar gum	Thickener, stabilizer
414	Gum arabic (acacia gum)	Thickener, stabilizer
419	Gum ghatti	Thickener, stabilizer, emulsifier
241	Gum guaicum	Preservative
939	Helium	Packing gas
209	Heptyl p-hydroxybenzoate	Preservative
239	Hexamethylene tetrarnine	Preservative
586	4-Hexylresorcinol	Color retention agent, antioxidant
507	Hydrochloric acid	Acidity regulator
907	Hydrogenated poly-1-decene	Glazing agent
463	Hydroxypropyl cellulose	Thickener, emulsifier, stabilizer
464	Hydroxypropyl methyl cellulose	Thickener, emulsifier, stabilizer
132	Indigotine	Color
630	Inosinic acid	Flavor enhancer
1103	Invertases	Stabilizer
172(i)	Iron oxide, black	Color
172(ii)	Iron oxide, red	Color
172(iii)	Iron oxide, yellow	Color
172	Iron oxides	Color
315	Isoascorbic acid (erythorbic acid)	Antioxidant
943b	Isobutane	Propellant
953	Isomalt (isomaltitol)	Sweetener, anticaking agent, bulking agent, glazing agent
384	Isopropyl citrates	Antioxidant, preservative, sequestrant
416	Karaya gum	Thickener, stabilizer
425	Konjac flour	Thickener
161c	Kryptoxanthin	Color
920	L-Cysteine and its hydrochlorides, sodium and potassium salts	Flour treatment agent

921	L-Cystine and its hydrochlorides, sodium and potassium salts	Flour treatment agent
641	L-Leucine	Flavor modifier
270	Lactic acid (L-, D-, and DL-)	Acidity regulator
472b	Lactic and fatty acid esters of glycerol	Emulsifier, stabilizer, sequestrant
966	Lactitol	Sweetener, texturizer
478	Lactylated fatty acid esters of glycerol and propylene glycol	Emulsifier
913	Lanolin	Glazing agent
344	Lecithin citrate	Preservative
322	Lecithins	Antioxidant, emulsifier
1104	Lipases	Flavor enhancer
180	Litholrubine BK	Color
161b	Lutein	Color
160d	Lycopene	Color
642	Lysin hydrochloride	Flavor enhancer
1105	Lysozyme	Preservative
504(i)	Magnesium carbonate	Acidity regulator, anticaking agent, color retention agent
504	Magnesium carbonates	Acidity regulator, anticaking agent, color retention agent
511	Magnesium chloride	Firming agent
345	Magnesium citrate	Acidity regulator
580	Magnesium gluconate	Acidity regulator, tinning agent
625	Magnesium glutamate	Flavor enhancer
504(ii)	Magnesium hydrogen carbonate	Acidity regulator, anticaking agent, color retention agent
528	Magnesium hydroxide	Acidity regulator, color retention agent
329	Magnesium lactate (DL-)	Acidity regulator, flour treatment agent
530	Magnesium oxide	Anticaking agent
343	Magnesium phosphates	Acidity regulator, anticaking agent
553(i)	Magnesium silicate	Anticaking agent, dusting powder
553	Magnesium silicates	Anticaking agent, dusting powder
518	Magnesium sulfate	Firming agent
553(ii)	Magnesium trisilicate	Anticaking agent, dusting powder
296	Malic acid (DL-)	Acidity regulator
965	Maltitol and maltitol syrup	Sweetener, stabilizer, emulsifier
636	Maltol	Flavor enhancer
421	Mannitol	Sweetener, anticaking agent
353	Metatartaric acid	Acidity regulator
461	Methyl cellulose	Thickener, emulsifier, stabilizer
911	Methyl esters of fatty acids	Glazing agent
465	Methyl ethyl cellulose	Thickener, emulsifier, stabilizer, foaming agent
489	Methyl glucoside—coconut oil ester	Emulsifier
218	Methyl-p-hydroxybenzoate	Preservative
900b	Methylphenylpolysiloxane	Antifoaming agent
460(i)	Microcrystalline cellulose	Emulsifier, anticaking agent, texturizer, dispersing agent
905c(i)	Microcrystalline wax	Glazing agent
905a	Mineral oil, food grade	Glazing agent, release agent, sealing agent

472f	Mixed tartaric, acetic, and fatty acid esters of glycerol	Emulsifier, stabilizer, sequestrant
306	Mixed tocopherols concentrate	Antioxidant
471	Mono- and diglycerides of fatty acids	Emulsifier, stabilizer
624	Monoammonium glutamate	Flavor enhancer
342(i)	Monoammonium orthophosphate	Acidity regulator, flour treatment agent
341(i)	Monocalcium orthophosphate	Acidity regulator, flour treatment agent, firming agent, texturizer, raising agent, anticaking agent, water retention agent
343(i)	Monomagnesium orthophosphate	Acidity regulator, anticaking agent
622	Monopotassium glutamate	Flavor enhancer
340(i)	Monopotassium orthophosphate	Acidity regulator, sequestrant, emulsifier, texturizer, stabilizer, water retention agent
336(i)	Monopotassium tartrate	Stabilizer, sequestrant
621	Monosodium glutamate	Flavor enhancer
339(i)	Monosodium orthophosphate	Acidity regulator, sequestrant, emulsifier, texturizer, stabilizer, water retention agent
335(i)	Monosodium tartrate	Stabilizer, sequestrant
160a(ii)	Natural extracts	Color
959	Neohesperidine dihydrochalcone	Sweetener
375	Nicotinic acid	Color retention agent
234	Nisin	Preservative
941	Nitrogen	Packing gas, freezant
918	Nitrogen oxides	Flour treatment agent
919	Nitrosyl chloride	Flour treatment agent
942	Nitrous oxide	Propellant
411	Oat gum	Thickener, stabilizer
946	Octafluorocyclobutane	Propellant
311	Octyl gallate	Antioxidant
182	Orchil	Color
231	Orthophenylphenol	Preservative
338	Orthophosphoric acid	Acidity regulator, antioxidant synergist
948	Oxygen	Packing gas
387	Oxystearin	Antioxidant, sequestrant
1101(ii)	Papain	Flour treatment agent, stabilizer, tenderizer, flavour enhancer
160c	Paprika oleoresins	Color
905c(ii)	Paraffin wax	Glazing agent
131	Patent blue V	Color
440	Pectins	Thickener, stabilizer, gelling agent, emulsifier
451(ii)	Pentapotassium triphosphate	Sequestrant, acidity regulator, texturizer
451(i)	Pentasodium triphosphate	Sequestrant, acidity regulator, texturizer
429	Peptones	Emulsifier
905b	Petrolatum (petroleum jelly)	Glazing agent, release agent, sealing agent 905c Petroleum wax Glazing agent, release agent, sealing agent
391	Phytic acid	Antioxidant
235	Pimaricin (natamycin)	Preservative

1200	Polydextroses A and N	Bulking agent, stabilizer, thickener, humectant, texturizer
900a	Polydimethylsiloxane	Antifoaming agent, anticaking agent, emulsifier
1521	Polyethylene glycol	Antifoaming agent
475	Polyglycerol esters of fatty acids	Emulsifier
476	Polyglycerol esters of interesterified ricinoleic acid	Emulsifier
432	Polyoxyethylene (20) sorbitan monooleate	Emulsifier, dispersing agent
433	Polyoxyethylene (20) sorbitan mono-o-oleate	Emulsifier, dispersing agent
434	Polyoxyethylene (20) sorbitan monopalmitate	Emulsifier, dispersing agent
435	Polyoxyethylene (20) sorbitan monostearate	Emulsifier, dispersing agent
436	Polyoxyethylene (20) sorbitan tristearate	Emulsifier, dispersing agent
431	Polyoxyethylene (40) stearate	Emulsifier
430	Polyoxyethylene (8) stearate	Emulsifier
452	Polyphosphates	Emulsifier, stabilizer, acidity regulator, raising agent, sequestrant, water retention agent
1202	Polyvinyl polypyrrolidone	Color stabilizer, colloidal, stabilizer
1201	Polyvinyl pyrrolidone	Bodying agent, stabilizer, clarifying agent, dispersing agent
124	Ponceau 4R	Color
125	Ponceau SX	Color
261(i)	Potassium acetate	Preservative, acidity regulator
261	Potassium acetates	Preservative, acidity regulator
357	Potassium adipates	Acidity regulator
402	Potassium alginate	Thickener, stabilizer
555	Potassium aluminium silicate	Anticaking agent
303	Potassium ascorbate	Antioxidant
212	Potassium benzoate	Preservative
228	Potassium bisulfite	Preservative, antioxidant
924a	Potassium bromate	Flour treatment agent
501(i)	Potassium carbonate	Acidity regulator, stabilizer
501	Potassium carbonates	Acidity regulator, stabilizer
508	Potassium chloride	Gelling agent
332	Potassium citrates	Acidity regulator, sequestrant, stabilizer
261(ii)	Potassium diacetate	Preservative, acidity regulator
332(i)	Potassium dihydrogen citrate	Acidity regulator, sequestrant, stabilizer
536	Potassium ferrocyanide	Anticaking agent
366	Potassium fumarates	Acidity regulator
577	Potassium gluconate	Sequestrant
501(ii)	Potassium hydrogen carbonate	Acidity regulator, stabilizer
351(i)	Potassium hydrogen malate	Acidity regulator
525	Potassium hydroxide	Acidity regulator
632	Potassium inosinate	Flavor enhancer
917	Potassium iodate	Flour treatment agent
317	Potassium isoascorbate	Antioxidant

326	Potassium lactate	Antioxidant synergist, acidity regulator
351(ii)	Potassium malate	Acidity regulator
351	Potassium malates	Acidity regulator
224	Potassium metabisulfite	Preservative, antioxidant
252	Potassium nitrate	Preservative, color fixative
249	Potassium nitrite	Preservative, color fixative
922	Potassium persulfite	Flour treatment agent
340	Potassium phosphates	Acidity regulator, sequestrant, emulsifier, texturizer, stabilizer, water retention agent
452(ii)	Potassium polyphosphate	Emulsifier, stabilizer, acidity regulator, raising agent, sequestrant, water retention agent
283	Potassium propionate	Preservative
560	Potassium silicate	Anticaking agent
337	Potassium sodium tartrate	Stabilizer, sequestrant
202	Potassium sorbate	Preservative
515	Potassium sulfates	Acidity regulator
225	Potassium sulfite	Preservative, antioxidant
336	Potassium tartrates	Stabilizer, sequestrant
460(ii)	Powdered cellulose	Emulsifier, anticaking agent, texturizer, dispersing agent
407a	Processed euchema seaweed (PES)	Thickener, stabilizer
944	Propane	Propellant
280	Propionic acid	Preservative
310	Propyl gallate	Antioxidant
216	Propyl p-hydroxybenzoate	Preservative
1520	Propylene glycol	Humectant, wetting agent, dispersing agent
405	Propylene glycol alginate	Thickener, emulsifier
477	Propylene glycol esters of fatty acids	Emulsifier
1101(i)	Protease	Flour treatment agent, stabilizer, tenderizer, flavor enhancer
1101	Proteases	Flour treatment agent, stabilizer, tenderizer, flavor enhancer
999	Quillaia extracts	Foaming agent
104	Quinoline yellow	Color
128	Red 2G	Color
161f	Rhodoxanthin	Color
101(i)	Riboflavin	Color
101(ii)	Riboflavin 5′-phosphate, sodium	Color
101	Riboflavins	Color
908	Rice bran wax	Glazing agent
161d	Rubixanthin	Color
954	Saccharin (and Na, K, Ca salts)	Sweetener
470	Salts of fatty acids (with base Al, Ca, Na, Mg, K, and NH_4)	Emulsifier, stabilizer, anticaking agent
166	Sandalwood	Color
904	Shellac	Glazing agent
551	Silicon dioxide, amorphous	Anticaking agent
174	Silver	Color

262(i)	Sodium acetate	Preservative, acidity regulator, sequestrant
262	Sodium acetates	Preservative, acidity regulator, sequestrant
356	Sodium adipates	Acidity regulator
401	Sodium alginate	Thickener, stabilizer, gelling agent
541	Sodium aluminium phosphate	Acidity regulator, emulsifier
541(i)	Sodium aluminium phosphate—acidic	Acidity regulator, emulsifier
541(ii)	Sodium aluminium phosphate—basic	Acidity regulator, emulsifier
554	Sodium aluminosilicate	Anticaking agent
301	Sodium ascorbate	Antioxidant
211	Sodium benzoate	Preservative
452(iii)	Sodium calcium polyphosphate	Emulsifier, stabilizer, acidity, regulator, raising agent, sequestrant, water retention agent
500(i)	Sodium carbonate	Acidity regulator, raising agent, anticaking agent
500	Sodium carbonates	Acidity regulator, raising agent, anticaking agent
466	Sodium carboxymethyl cellulose	Thickener, stabilizer, emulsifier
331	Sodium citrates	Acidity regulator, sequestrant, emulsifier, stabilizer
266	Sodium dehydroacetate	Preservative
262(ii)	Sodium diacetate	Preservative, acidity regulator, sequestrant
331(i)	Sodium dihydrogen citrate	Acidity regulator, sequestrant, emulsifier, stabilizer
215	Sodium ethyl p-hydroxybenzoate	Preservative
535	Sodium ferrocyanide	Anticaking agent
237	Sodium formate	Preservative
365	Sodium fumarates	Acidity regulator
576	Sodium gluconate	Sequestrant
500(ii)	Sodium hydrogen carbonate	Acidity regulator, raising agent, anticaking agent
350(i)	Sodium hydrogen malate	Acidity regulator, humectant
222	Sodium hydrogen sulfite	Preservative, antioxidant
524	Sodium hydroxide	Acidity regulator
316	Sodium isoascorbate	Antioxidant
325	Sodium lactate	Antioxidant synergist, humectant, bulking agent
481	Sodium lactylates	Emulsifier, stabilizer
487	Sodium lamylsulfate	Emulsifier
350(ii)	Sodium malate	Acidity regulator, humectant
350	Sodium malates	Acidity regulator, humectant
223	Sodium metabisulfite	Preservative, bleaching agent, antioxidant
550(ii)	Sodium metasilicate	Anticaking agent
219	Sodium methyl p-hydroxybenzoate	Preservative
251	Sodium nitrate	Preservative, color fixative
250	Sodium nitrite	Preservative, color fixative
232	Sodium o-phenylphenol	Preservative
481(ii)	Sodium oleyllactylate	Emulsifier, stabilizer

339	Sodium phosphates	Acidity regulator, sequestrant, emulsifier, texturizer, stabilizer, water retention agent
452(i)	Sodium polyphosphate	Emulsifier, stabilizer, acidity, regulator, raising agent, sequestrant, water retention agent
281	Sodium propionate	Preservative
217	Sodium propyl p-hydroxybenzoate	Preservative
500(iii)	Sodium sesquicarbonate	Acidity regulator, raising agent, anticaking agent
550(i)	Sodium silicate	Anticaking agent
550	Sodium silicates	Anticaking agent
201	Sodium sorbate	Preservative
485	Sodium stearoyl fumarate	Emulsifier
481(i)	Sodium stearoyllactylate	Emulsifier, stabilizer
514	Sodium sulfates	Acidity regulator
221	Sodium sulfite	Preservative, antioxidant
335	Sodium tartrates	Stabilizer, sequestrant
539	Sodium thiosulfate	Antioxidant, sequestrant
200	Sorbic acid	Preservative
493	Sorbitan monolaurate	Emulsifier
494	Sorbitan monooleate	Emulsifier
495	Sorbitan monopalmitate	Emulsifier
491	Sorbitan monostearate	Emulsifier
496	Sorbitan trioleate	Stabilizer, emulsifier
492	Sorbitan tristearate	Emulsifier
420	Sorbitol and sorbitol syrup	Sweetener, humectant, sequestrant, texturizer, emulsifier
909	Spermaceti wax	Glazing agent
512	Stannous chloride	Antioxidant, color retention agent
484	Stearyl citrate	Emulsifier, sequestrant
483	Stearyl tartrate	Flour treatment agent
960	Stevioside	Sweetener
363	Succinic acid	Acidity regulator
472g	Succinylated monoglycerides	Emulsifier, stabilizer, sequestrant
446	Succistearin	Emulsifier
955	Sucralose (trichlorogalactosucrose)	Sweetener
474	Sucroglycerides	Emulsifier
444	Sucrose acetate isobutyrate	Emulsifier, stabilizer
473	Sucrose esters of fatty acids	Emulsifier
220	Sulfur dioxide	Preservative, antioxidant
513	Sulfuric acid	Acidity regulator
110	Sunset yellow FCF	Color
441	Superglycerinated hydrogenated rapeseed oil	Emulsifier
309	Synthetic delta-tocopherol	Antioxidant
308	Synthetic gamma-tocopherol	Antioxidant
553(iii)	Talc	Anticaking agent, dusting powder
181	Tannins, food grade	Color, emulsifier, stabilizer, thickener
417	Tara gum	Thickener, stabilizer
334	Tartaric acid (L(+)−)	Acidity regulator, sequestrant, antioxidant synergist

472d	Tartaric acid esters of mono- and diglycerides of fatty acids	Emulsifier, stabilizer, sequestrant
102	Tartrazine	Color
319	Tertiary butylhydroquinone	Antioxidant
450(v)	Tetrapotassium diphosphate	Emulsifier, stabilizer, acidity regulator, raising agent, sequestrant, water retention agent
450(iii)	Tetrasodium diphosphate	Emulsifier, stabilizer, acidity, regulator, raising agent, sequestrant, water retention agent
957	Thaumatin	Sweetener, flavor enhancer
479	Thermally oxidized soya bean oil with mono- and diglycerides of fatty acids	Emulsifier
233	Thiabendazole	Preservative
388	Thiodipropionic acid	Antioxidant
171	Titanium dioxide	Color
413	Tragacanth gum	Thickener, stabilizer, emulsifier
1518	Triacetin	Humectant
341(iii)	Tricalcium orthophosphate	Acidity regulator, flour treatment agent, tinning agent, texturizer, raising agent, anticaking agent, water retention agent
1505	Triethyl citrate	Foam stabilizer
343(iii)	Trimagnesium orthophosphate	Acidity regulator, anticaking agent
451	Triphosphates	Sequestrant, acidity regulator, texturizer
332(ii)	Tripotassium citrate	Acidity regulator, sequestrant, stabilizer
340(iii)	Tripotassium orthophosphate	Acidity regulator, sequestrant, emulsifier, texturizer, stabilizer, water retention agent
331(iii)	Trisodium citrate	Acidity regulator, sequestrant, emulsifier, stabilizer
450(ii)	Trisodium diphosphate	Emulsifier, stabilizer, acidity, regulator, raising agent, sequestrant, water retention agent
339(iii)	Trisodium orthophosphate	Acidity regulator, sequestrant, emulsifier, texturizer, stabilizer, water retention agent
100(ii)	Turmeric	Color
927b	Urea (carbamide)	Texturizer
153	Vegetable carbon	Color
161e	Violoxanthin	Color
910	Wax esters	Glazing agent
415	Xanthan gum	Thickener, stabilizer
967	Xylitol	Sweetener, humectant, stabilizer, emulsifier, thickener
107	Yellow 2G	Color
557	Zinc silicate	Anticaking agent

Source: Adapted from Codex Alimentarius Commission, January, 2001. Class names and the international numbering system for food additives. Codex Alimentarius, Vol. 1A, *General Requirements.* www.fao.org/es/esn/codex/standard/volume 1a/vol 1a_e.htm.

APPENDIX 3

List of Modified Starches

List of Modified Starches

No.	Name of food additive	Technical function(s)
1422	Acetylated distarch adipate	Stabilizer, thickener, binder
1423	Acetylated distarch glycerol	Stabilizer, thickener
1414	Acetylated distarch phosphate	Emulsifier, thickener
1401	Acid-treated starch	Stabilizer, thickener, binder
1402	Alkaline-treated starch	Stabilizer, thickener, binder
1403	Bleached starch	Stabilizer, thickener, binder
1400	Dextrins, roasted starch, white and yellow	Stabilizer, thickener, binder
1411	Distarch glycerol	Stabilizer, thickener, binder
1412	Distarch phosphate esterified with sodium trimetaphosphate and esterified with phosphorus oxychloride	Stabilizer, thickener, binder
1443	Hydroxypropyl distarch glycerol	Stabilizer, thickener
1442	Hydroxypropyl distarch phosphate	Stabilizer, thickener
1440	Hydroxypropyl starch	Emulsifier, thickener, binder
1410	Monostarch phosphate	Stabilizer, thickener, binder
1404	Oxidized starch	Emulsifier, thickener, binder
1413	Phosphated distarch phosphate	Stabilizer, thickener, binder
1420	Starch acetate esterified with acetic anhydride	Stabilizer, thickener
1421	Starch acetate esterified with vinyl acetate	Stabilizer, thickener
1450	Starch sodium octenyl succinate	Stabilizer, thickener, binder, emulsifier
1405	Starches, enzyme treated	Thickener

Note: The Codex General Standard for the Labelling of Prepackaged Foods (CODEX STAN 1, 1985) specifies that modified starches may be declared as such in the list of ingredients. However, as some countries presently require the specific identification of modified starches the following numbers are provided as a guide and as a means of facilitating uniformity. Where these starches are specifically identified in the list of ingredients, then it would be appropriate to include them under the relevant class name, e.g., thickener.

Source: Adapted from Codex Alimentarius Commission, January, 2001. Class names and the international numbering system for food additives. Codex Alimentarius, Vol. 1A, *General Requirements*. www.fao.org/es/esn/codex/standard/volume 1a/vol 1a_e.htm.

Index